S0-DTD-897

Semiconductor Memories

Second Edition

Semiconductor Memories

A Handbook of Design, Manufacture and Application

Second Edition

Betty Prince
Texas Instruments, USA

JOHN WILEY & SONS
Chichester • New York • Brisbane • Toronto • Singapore

Second edition of the book *Semiconductor Memories* by B. Prince and G. Due-Gundersen

Reprinted with corrections February 1995
Reprinted January 1996

Other Wiley Editorial Offices

John Wiley & Sons, Inc., 605 Third Avenue,
New York, NY 10158-0012, USA

Jacaranda Wiley Ltd, G.P.O. Box 859, Brisbane,
Queensland 4001, Australia

John Wiley & Sons (Canada) Ltd, 22 Worcester Road,
Rexdale, Ontario M9W 1L1, Canada

John Wiley & Sons (SEA) Pte Ltd, 37 Jalan Pemimpin #05-04,
Block B, Union Industrial Building, Singapore 2057

British Library Cataloguing-in-Publication Data

A catalogue record for the book is
available from the British Library

ISBN 0 471 92465 2 ; 0 471 94295 2 (pbk)

Typeset by Techset Composition Limited, Salisbury, Wiltshire
Printed in Great Britain by Bookcraft (Bath) Ltd.

CONTENTS

INTRODUCTION

The goal of this book is to provide a basic handbook of the various aspects of semiconductor memories including history, market, applications, technology, design, various product types, manufacturing, testing, and a glimpse at the future.

Chapter 1 describes basic memory market trends and indicates why memories are of strategic importance to both industry and governments worldwide, while Chapter 2 gives an overview of the book for the casual reader.

The applications requirements which determine the direction of memory development are covered in Chapter 3.

Chapters 4 and 5 cover the technical aspects of the memories themselves—process technology, device characteristics, and basic design architectures.

Chapters 6 to 12 then give descriptions and usage of the various types of memories—SRAMs, DRAMS, VDRAMs, EPROMS, EEPROMs, ROMs, etc. Chapter 13 deals with memory packaging and large scale integration.

Chapters 14 and 15 consider the manufacturing aspects of the memories. Chapter 14 deals with memory reliability and test, and Chapter 15 with yield and other productivity considerations ending with a description of a modern submicron memory factory.

Future trends in memories are considered in Chapter 16.

There is no attempt here to deal exhaustively with these topics, rather, it is hoped that the reader will be given some basic familiarity with the subject and provided with references to use in researching more detailed questions.

ACKNOWLEDGEMENTS

I would like to thank all those who contributed information and the many experts in various memory fields who proof read chapters and offered suggestions. In particular Roelof Salters, who is technical advisor to the Philips Advanced Memory Design Center and Design Manager for the JESSI Design Team at Philips, for advice and suggestions throughout the book.

Fred Jones of Dataquest, Con Gordon and Pieter te Booij of Philips, and Jim Benson of TI for reading and offering helpful suggestions on Chapters 1 and 2.

Hein van Bruck and Rien Galema of the Philips Central Applications Laboratory, and Bill Vogley of TI for reviewing and for helpful suggestions on Chapter 3.

Ad Bermans, of the European JESSI Office, formerly Advanced Technology Manager of the Philips I.C. Technology center, and Maartin Vertregt of the Philips Advanced Memory Design Center for reviewing and for suggestions on Chapter 4.

Cormack O'Connell, Design Team Leader at Mosaid, and Howard Sussman, Design Manager at NEC, for reviewing and for many suggestions on Chapter 5.

Howard Sussman, and Ad van Zanten, Design Manager of the Philips Advanced Memory Design Centre, for suggestions on Chapters 6 and 7 and Joe Hartigan of TI for comments on Chapter 7.

Frans List, SRAM Design Managers at Inmos, and Ad van Zanten for reviewing and for many helpful suggestions on Chapters 8 and 9.

Roger Cuppens, Non-volatile Design Manager of the Philips Advanced Memory Design Center, for reviewing and offering suggestions on Chapters 10, 11 and 12. Sebastiano D'Arrigo, flash EPROM Design Manager, of T.I. for reviewing and offering helpful suggestions on Chapter 11, Don Knowlton of Waferscale Integration for reviewing Chapters 11 and 12 and Howard Sussman of NEC for reviewing Chapter 10.

Henk Kiela of the Philips Package Development Group and Daniel Baudouin of TI for suggestions on Chapter 13.

Reese Brown, who is now a private consultant and formerly Components Manager of Unisys, for helpful suggestions on the reliability aspects of Chapter 14, and J. Weidenhofer of the DRAM Test Department at Siemens and John Salter, Production Engineering Manager at the Philips Submicron Technology Facility, for help on the test sections of this chapter.

Albert Maringer of Siemens for contributing material and for helpful suggestions on Chapter 15. Frank Stein and Sharon Phelan for helpful suggestions on editing.

Pallab Chatterjee of TI whose inspiring opening talk of the 1990 VLSI Technology

Symposium contributed to the writing of Chapter 16, and Earnest Powell of TI for suggestions on Chapter 16.

Finally I would like to thank Dr Theo Claasen, Director of Design at Philips, for his inspiration and encouragement throughout this book.

While the author gratefully acknowledges the above noted comments and suggestions, all errors in the final version remain strictly the responsibility of the author, who would greatly appreciate being notified of such errors so they can be corrected in subsequent printings.

1 THE STRATEGIC NATURE OF SEMICONDUCTOR MEMORIES

Semiconductor memories were, by the end of the 1980s, a fifteen billion dollar market, 20% of the total semiconductor market and maintaining a 70% bit growth annually. During the decade a trade war had been fought over them and the worldwide memory industry had been totally restructured. In the 1990s the future growth of the electronics industry will still depend on memories and the technologies developed with them.

1.1 TWENTY YEARS OF MEMORIES

The 1970s saw the dawn of the electronic revolution. During this time, electronic devices pervaded our lives. The first simple electronic semiconductor calculators grew into complex microcomputer systems on tiny silicon chips rivaling the room size mainframe computers of the 1960s. Worldwide communications networks spanned the globe, and factories with industrial robots first became a reality. Looking forward into the 1980s, we appeared to stand on the threshold of a data communications and robotics revolution which would totally transform factory, home, and office.

All of these technological advances depended on the ability to store and retrieve massive amounts of data quickly and inexpensively. They all depended on the development of the semiconductor memory.

During the 1970s, semiconductor memories were the fastest growing market in the semiconductor industry. Revenues grew from a mere $15 million in world sales in 1970 to $2.4 billion in 1980—a spectacular 65% compound growth rate [9].

During the early 1980s, as manufacturing techniques improved and the cost of semiconductor memories continued to fall, memories found their way into such diverse applications as data storage banks for mainframe computers, and large telecommunication central switching systems. Memories in electronic dashboards remembered the mileage when the ignition was turned off, and memories in cable television systems metered how many programs had been watched. Memories in credit cards retained the owner's credit balance, and memories in gaming machines remembered the actions

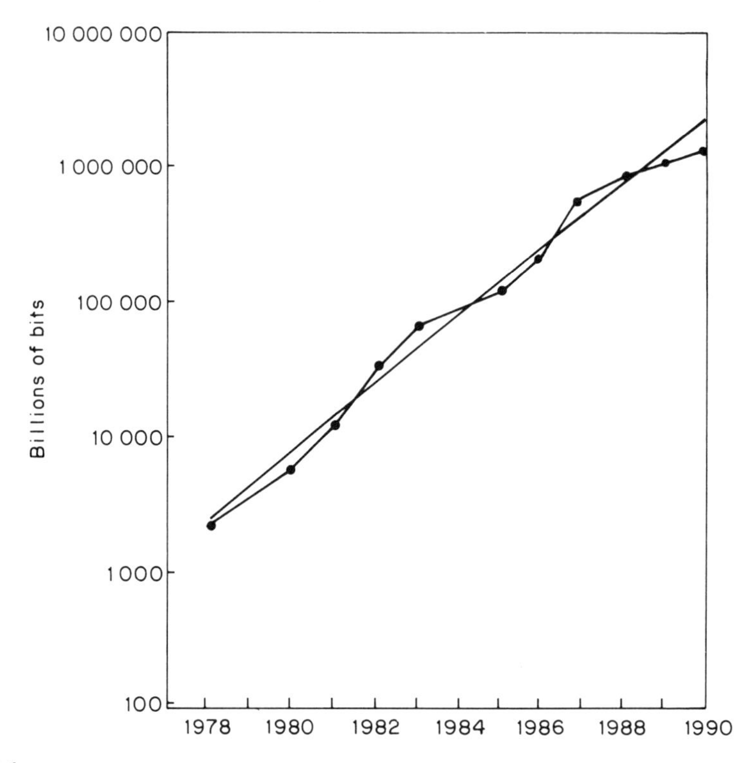

Figure 1.1
Shipments worldwide of memory bits from 1978 to 1990 in billions of bits, showing 70% compound annual growth rate. (Source Dataquest, Inc.)

of the particular game being played. The number of new applications grew as supplementary intelligence and low cost memory became more available. Throughout the 1980s the number of bits of memories shipped grew at a 70% compound annual growth rate as shown in Figure 1.1.

The mid 1980s, however, were a time of consolidation for the electronics industry. Worldwide communications networks were hampered by geopolitical standardization problems. Personal computers proliferated but could not communicate with each other. Software was slow to develop, and being complex to master, tended to frighten people away from using office and home computers. Early attempts at automated factories, lacking standards, became nightmares of equipment which did not quite work together.

Owing to these structural problems which resulted from rapid expansion, annual growth in the giant computer industry, which uses more than 70% of all memories, slowed, from its long period of growth between 1976 and 1984, and fell in 1985 and 1986.

The semiconductor memory industry suffered financially during this time as the slowdown in electronic equipment sales, trade issues, economic realities, and the management inexperience of the young industry combined to plunge the companies involved in this glamor growth product into huge losses.

In 1985, after three years of high growth following a disastrous industry-wide price drop in 1981, memory prices again collapsed, plunging revenues from a growth rate of over 60% in 1984 to minus 40% in 1985, as shown in Figure 1.2. They were not to recover to the 1984 levels until 1988.

Many Western manufacturers, after two severe declines within five years, withdrew from the memory market after 1985. Some re-entered the market in 1988 when it rose again to a new high of $12 billion, followed by another significant increase in 1989 to over $15 billion. Memories had an overall 23% compound annual revenue growth rate during the 1980s. This was lower than the 65% compound growth rate of their first decade, but still one of the fastest growing segments of the semiconductor industry [4].

The significant impact of the computer segment on memory consumption can be seen in Figure 1.2, showing where slight declines in computer growth rates in 1980, 1985, and 1988 were followed by severe declines in memory consumption. Likewise, small increases in growth rate of the computer production in 1981 and 1987 were followed by huge upswings in memory growth rates the following year. Some of the causes of this volatility in memories are discussed in the following section.

The late 1980s saw a resurgence of the earlier promise of the electronics revolution. World standards began to be adopted in such areas as communications with ISDN (Integrated Systems Digital Network), factory automation with MAP (Management Automated Protocol), and software such as computer operating systems like UNIX (developed by AT&T) and MS DOS (developed by Microsoft Inc. for the IBM personal computer).

Figure 1.2
Percentage annual growth rate for MOS memory consumption (x—x) and for computer production (——) worldwide in current US dollars (Source Dataquest, Inc.).

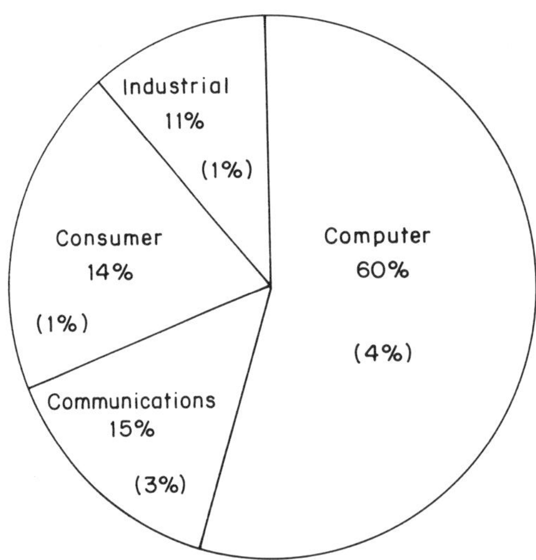

Figure 1.3
Electronic end equipment consumption worldwide in 1989 by market segment showing memory value in the average system. Total is $460 billion. Figures in parentheses show percentage memory content in the average system. (From Electronics Annual Consumption Survey, January 1990 and N.V. Philips.)

With these new standards the computer industry, and with it the memory industry, showed renewed growth. Figure 1.3 shows the $460 billion world electronic equipment industry in 1989 split by equipment segment [1] and also indicates an estimate for memory content in an average system for each segment. The reason for the significant impact of computers on memory sales is clear both from the size of the computer industry and from the relatively high percentage of memory content in the average computer system.

There are other factors in the renewed prospects of the electronics revolution as we enter the 1990s. For example, government telecommunications monopolies are being ended worldwide, thus freeing this high memory content market segment for growth.

A new factor which only emerged in the late 1980s was video applications in both the computer automated design (CAD) and the workstation area. Together with video applications in the consumer television segment, these also promise to increase significantly the memory value in the average consumer system in the future.

Government cooperation with the worldwide electronics industry has increased, not only in Japan where it has always been a factor, but in Europe and the United States as well. All these effects combine to renew the promise of office, home, and factory totally transformed by electronics.

Global restructuring of the electronics industry by the end of the 1980s had concentrated high technology integrated circuits production into the large companies which could both afford the high capital investment in semiconductor process equipment and provide the management experience and vision to succeed with this demanding but rewarding product. As a result, most of the major manufacturers of semiconductor memories today are also major manufacturers in the electronics equipment industry.

At the start of the 1990s, the vision of a new era in electronics is matched by a new excitement in the area of semiconductor memories. New worlds of electronics appear to be just around the corner, and the ubiquitous memory continues to be an indispensable factor in the new developments.

1.2 THE LARGEST AND MOST VOLATILE INTEGRATED CIRCUIT MARKET

Throughout the eighties memories were one of the largest of the semiconductor markets as shown in Figure 1.4 which compares the split of the total semiconductor market by product in 1980 and 1988. Their growth since 1980 has been approximately the same as that of the electronics systems industry. They have remained on average about 20% of the total value of semiconductors shipped due to the relatively constant level of memory usage in the average electronic system. This varies in any one year, however, due to the price fluctuations common in the memory industry.

Although some memory has been incorporated in the various processor chips, it has not reduced the percentage of standalone memory being used. This perhaps indicates a tendency for the amount of memory in the average electron system, including systems with embedded memory, to increase slowly over time.

In the 1980s, memories were also among the most volatile of the semiconductor markets. This was due mainly to severe short term price fluctuations even though the growth in units shipped remained fairly stable, reflecting the relative stability of demand for memories in systems in the short term. Figure 1.5, which shows memory revenue, average selling price, and units shipped from 1980 to 1990, indicates that this market is subject to multibillion dollar year-to-year price fluctuations while shipments of units remain fairly stable over time [1].

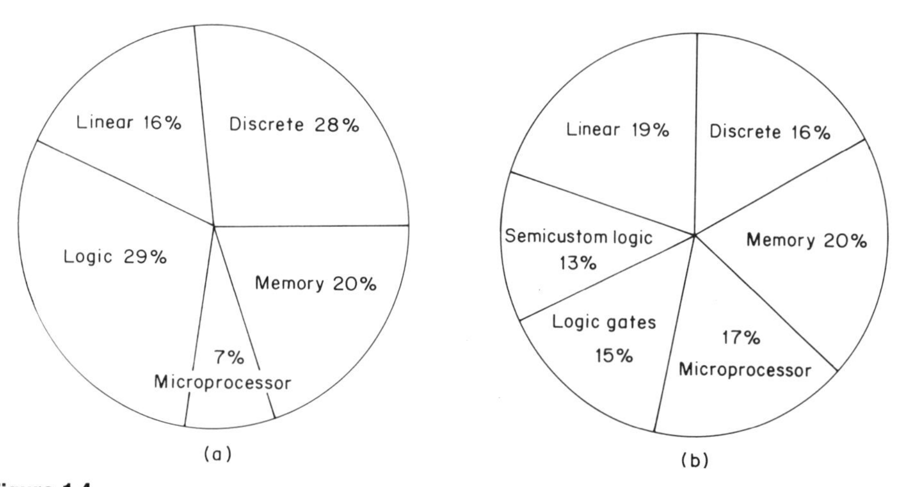

Figure 1.4
Total semiconductor consumption split by product in (a) 1980 ($14.1 billion) and (b) 1988 ($46 billion). (Source Electronics Magazine Annual Consumption Survey 1982 and 1989)[1].

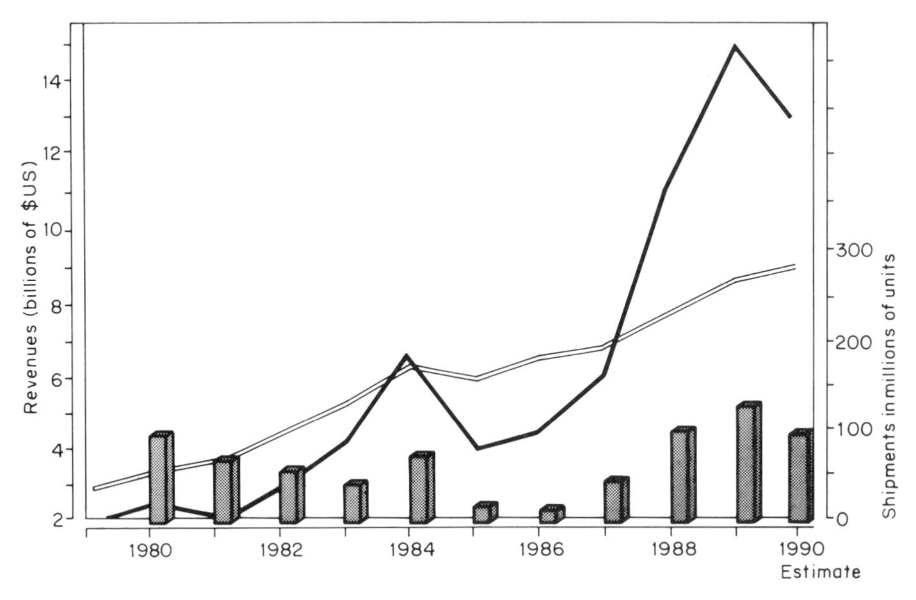

Figure 1.5
MOS memory revenues (——), units shipped (═══) and average selling price (■) 1980 to 1990. (Source Dataquest, Inc.)

This volatility is a result of fairly well understood industry structural factors and of an interplay of short term elastic economic factors such as unit price and short term inelastic (long term elastic) economic factors such as unit demand which affect the unit price of the memory, although they do not influence appreciably in the short term the actual number of units shipped.

1.3 STRUCTURAL AND ECONOMIC FACTORS AFFECTING VOLATILITY

Structural factors affecting volatility include the large size of the memory market, the small number of participants, the high capital entry barriers, and the high levels of government intervention in a market which is considered strategic.

1.3.1 Structural supply factors

The huge multibillion dollar size of the world memory market, coupled with the small number of suppliers, requires a single manufacturer to ship large volumes in order to establish a market share large enough to gain manufacturing experience and so be able to reduce costs fast enough to remain competitive.

In 1987, according to Dataquest, the top five suppliers accounted for 61% of the worldwide memory market and the top 20 accounted for 91%. This means that a single supplier must hold about a 5% market share to be competitive. A 5% market share is 500 million US dollars annually at the 1988 level of shipments which was equivalent to the output of two full size standard production facilities at that time.

The cost of manufacturing facilities is a significant capital cost entry barrier which contributes to keeping the number of participants low. This is illustrated in Figure 1.6 using the Dynamic RAM (DRAM) memory product line which represented more than half of the total memory market. This figure shows the approximate cost of a DRAM wafer fabrication facility by year, total industry DRAM revenues by year, and DRAM revenues by individual product generation.

An average advanced DRAM volume manufacturing facility in 1980 cost $50 million, in 1985 cost $200 million, and in 1990 cost $500 million. By 1995 the cost of a full size memory production area will be over one billion dollars.

The revenues, which are also shown, indicate that the large capital expenditure is financed by the increase in revenues over time, as shown by the dotted line on the figure. It appears that, on the average over the long term, the rate of increase in cost of a manufacturing area is balanced by the rate of increase of industry revenues for the product.

It is clear from the figure that for a supplier who participates in this market over a long period of time each product generation on average pays for the manufacturing facilities of the next generation.

The memory supply market is, therefore, a bit like the legendary California housing market: once a supplier is in, the future can be funded from the past. For those trying to enter or re-enter, however, the entry barrier becomes increasingly higher and the future revenues must pay back the present investment while at the same time providing funding for the next generation.

It is also interesting to note that the delay in revenues for the 256k DRAMs in 1985 to 1986 at a time when the 1Mb wafer fabs were being built meant that the 256k generation did not fund the 1Mb facilities as would have been expected. The result was that many vendors left the market at that time.

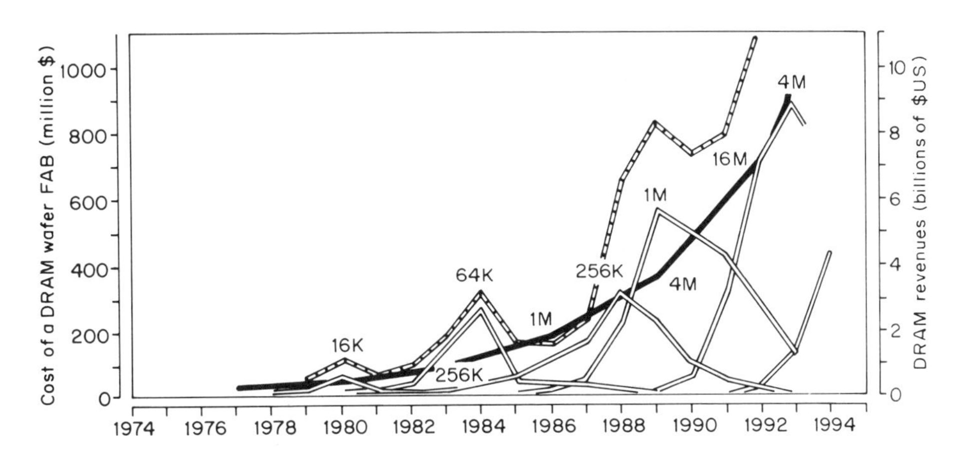

Figure 1.6
DRAM revenues (⌶⌶⌶), wafer facility cost by generation (——) and DRAM product life cycles (═══). (From Prince [8], Philips 1989.)

Figure 1.7
Submicron MOS Factory of N.V. Philips at the Eindhoven, Research Labs in The Netherlands. (Source N.V. Philips Research.)

The high cost of fabrication areas for memories leads to a high level of commercial competition as manufacturers strive to cover the overheads of these expensive facilities. A photograph of a modern memory manufacturing facility is shown in Figure 1.7.

There is also significant supply risk inherent in this industry due to the leading edge technology required to produce the memory chips cost effectively and the competitive pressure to bring these technologies into production quickly.

The high cost of the memory manufacturing facility is reflected in the unit cost of the basic element of production, the silicon wafer. Memory chips are manufactured on the surface of such a wafer. A photograph of a memory silicon wafer with memory chips embedded in its surface is shown in Figure 1.8.

Costs can be most effectively reduced by reducing the area of the memory chips on a wafer and so increasing the number of memories produced per unit of manufacturing cost. Since memory chips in any given technology are made as small as possible, reducing the size of the memory chip involves continually moving to increasingly advanced technologies.

The technology factor increases the supply volume risk of the total market since

the temporary loss of the production of even one of these giant manufactu
can affect as much as 5% of the total volume of the market. The large si
manufacturing facilities also affects the volatility of the market as they com
production.

1.3.2 General economic factors

To some extent the volatility of the memory market can be attributed to general economic factors which tend to upset the market forces balancing this capital intensive, high volume worldwide competitive product. For example, the high US interest rates in the early 1980s affected severely the more capital intensive US industries such as the memory manufacturers allowing the Japanese companies, prepared by the government sponsored VLSI project, to dominate the DRAM segment of the memory market during most of the 1980s [6].

The volatility can also be attributed, again to some extent, to attempts by various governments to intervene in this market. For example, following the US–Japanese trade accord on memory products in 1986, the Japanese government attempted to stabilize the DRAM market by matching the DRAM market supply with demand. Unfortunately an unexpectedly strong increase in the growth rate of the data processing market at that time produced widespread shortages of DRAMs and took prices in 1989 to their highest levels ever [4].

Figure 1.8
Photograph of a silicon wafer with memory chips embedded on the surface. (Source N.V. Philips Research.)

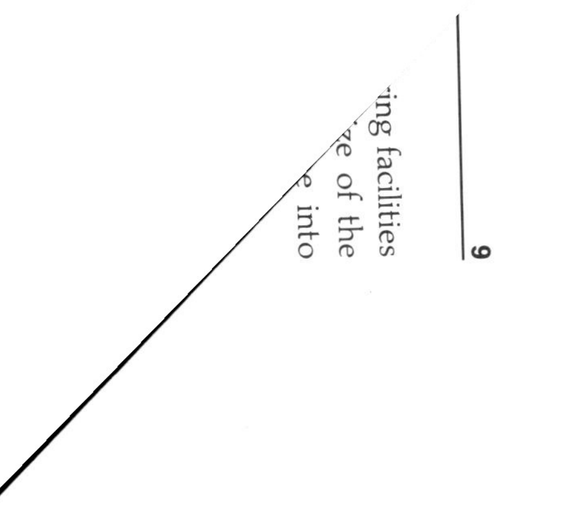

stic short term)
e rate of increase in demand for bits
e rate of technological development

y' rate of increase of production
substitution possible between product types

elasticity
aphical flexibility and fire sales
ory fluctuations

Underlying economic mechanisms which affect memory product life cycles. (From Prince [8], Philips 1989.)

A corresponding effort by the Japanese suppliers to reduce production levels when DRAM prices again began to fall too rapidly in 1989, had less of an impact due to market re-entry in 1988 of American suppliers and to the entry of European and Korean suppliers into this market at that time.

1.3.3 Interactions of short term and long term factors

The short term volatility of the memory industry can be explained by the interaction of long term and short term economic effects, as indicated in Figure 1.9. The long term effects are stable, that is relatively inelastic, in the short term. The short term effects are elastic and act as perturbing influences on the system.

On the demand, that is the user, side the long term effect is the stable rate of increase in demand for memory bits over time. On the supply side it is the stable rate of technological development perturbed occasionally by technological risk factors.

The price volatility is caused by the interaction of the short term demand inelasticity of the electronic equipment market coupled with the risk factor of the high technology wafer production facilities on the down side, and the large increments of increased supply represented by one of these high overhead facilities coming on stream on the high side. If the forecast for memory usage is inaccurate, or if the computer industry (which represents some 77% of total memory usage worldwide) slows or grows faster than expected, or if a number of such production facilities come on stream in the same time period, then in the short term the only elastic variable is the price.

Even a slight mismatch between the perceived growth in demand of the systems market and the capital intensive capacity of the memory manufacturers has led to severe price volatility as manufacturers drop prices far below cost to keep expensive manufacturing facilities filled or raise prices in times of shortages.

There are also medium term effects on the demand side possible from substitutions between product types. Geographical flexibility and inventory fluctuations also contribute to the short term volatility.

1.3.4 Stabilizing factors

As an attempt to minimize the technological risk factor, the manufacturers and users of memories have joined forces to create a high level of standardization in memories. This gives the systems manufacturers second sources for most memory components, thus improving the supply possibilities, and it keeps the memory manufacturers from making costly mistakes in product definitions which might be unacceptable to the user community.

While the standardization level is high, the number of such standardized products is small. For example, more than 95% of the $6.7 billion DRAM market in 1988 (97% according to Dataquest) is revenue from just two products, the 256k DRAM and the 1Mb DRAM. Since the DRAM market represents more than half of the total MOS memory market (54% in 1988), this means that these two standardized products represent about 54% of total $12.1 billion 1988 MOS memory revenues [4].

In a further attempt to reduce the risk in memory production and usage, an information industry has become a major by-product of this market. A high level of information exchange normally exists in the technical and scientific community. This is represented in the memory industry by the IEDM (International Electron Devices Meeting) of the US IEEE, where technology advances are routinely announced, and by the ISSCC (International Solid State Circuits Conference) of the IEEE, where historically circuit design advances have been announced. In recent years the VLSI Technology and Circuits Conferences in Japan have also become a significant place to watch for new memory product announcements.

In addition, a large number of consultants have appeared, including Dataquest Inc., Integrated Circuits Engineering Corporation (ICE), etc. The oldest is Dataquest, whose tracking of the details of memory production and of the memory price curve over the years has made memories the only semiconductor product with a widely recognized industry cost forecast against which the various companies which use memories as a production technology driver can measure the average level of cost competitiveness of their MOS technology.

More recently, the WSTS (World Semiconductor Trade Statistics), in which most of the companies in the industry participate, has assumed a major role in maintaining the overall statistics of the memory industry.

Owing to the high volume shipments of a few standardized products, the high level of information exchange, and the intense commercial competition, the memory market tends to be a close approximation of an economic 'perfect competition' where marginal prices approach marginal costs and only the best (highest volume) manufacturers make money.

1.4 A STRATEGIC COMPONENT IN ADVANCED ELECTRONIC SYSTEMS

Since the level of capability of intelligent systems is related to the memory capacity, the continued fall in price of semiconductor memory with time has put increased memory content within the range of an increasing number of systems applications.

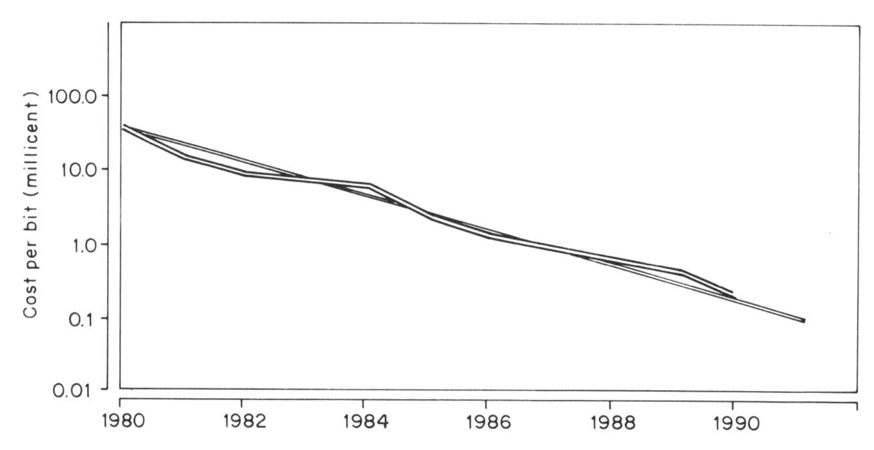

Figure 1.10
MOS memory price per bit from 1980 to 1990. (Source Dataquest, Inc.)

Figure 1.10 shows the historical price curve of the memories in millicents-per-bit. The long term stable trend of falling price of the memories has provided the fuel to drive the development of the electronics industry by continually expanding the range of possible applications.

Since memories are truly ubiquitous, being present in almost every electronic system, but can be produced profitably only by a few large suppliers the maintenance of adequate memory supply has become a priority both with major systems manufacturers and with governments intent on maintaining a strategic presence in the world computer industry. This was evident historically in the maintenance of captive internal memory production capacity by several of the larger electronics equipment suppliers. Not even these giant companies, however, attempt to supply the majority of their memory usage internally in competition with the multibillion dollar external memory suppliers.

There has also been increased intercompany cooperation between the giant multinational companies that participate in this market. This was evident in the technology cooperatives of the early 1980s—the American Microelectronics and Computer Consortium (MCC) and the European megaproject. These were joined in the late 1980s by the American Sematech Cooperative and the European JESSI (Joint Electronic Semiconductor Silicon Initiative) Project, both of which are expected to show results in the mid 1990s.

1.5 A MAJOR TRADE ISSUE

Since memories tend to be the development vehicle for advanced MOS semiconductor technologies and also strategic components in electronic systems, government participation in advanced memory development is commonplace. This intervention includes support for technological development such as the Japanese VLSI project in the late 1970s, the US VHSIC military project of the early 1980s, and the European Esprit programs, the Megaproject of the mid-1980s and the JESSI cooperative of the early 1990s.

Intervention also includes tariffs and other trade restrictions, such as the EEC tariff on import of memory products which compete with similar products produced by a manufacturer in the Common Market, and the American–Japanese trade agreement which sets base values, called Fair Market Values (FMVs), for the various memory products imported from Japan. These FMVs are based on actual costs of the various Japanese semiconductor suppliers submitted bi-annually in voluntary undertakings by these suppliers to the US government. These undertakings were a result of dumping charges lodged against the Japanese companies during the severe memory price decline of 1985 when many US companies left the memory market.

Similar floor prices stemming from dumping charges in this same time frame in the DRAM and EPROM product lines have been enacted by the European Commission of the EEC.

The effect of these various government interventions can be seen, for example, in the dip in prices in the 1981 time frame, which was due in part to large scale entry of Japanese companies into a market previously dominated by the Americans, following the completion of the Japanese VLSI project in the late 1970s.

The price increase of 1988 was in part due to monitoring of the various memory markets by Japan as a result of U.S. protests over third world sales activity following the US–Japan trade agreement of 1986. The effects of the European Megaproject will not be seen until the early 1990s when the volume sales from this multi-billion dollar investment begin.

A major reason for government interest in the memory market is the historical supply and demand imbalances among the three major regions of the world in what

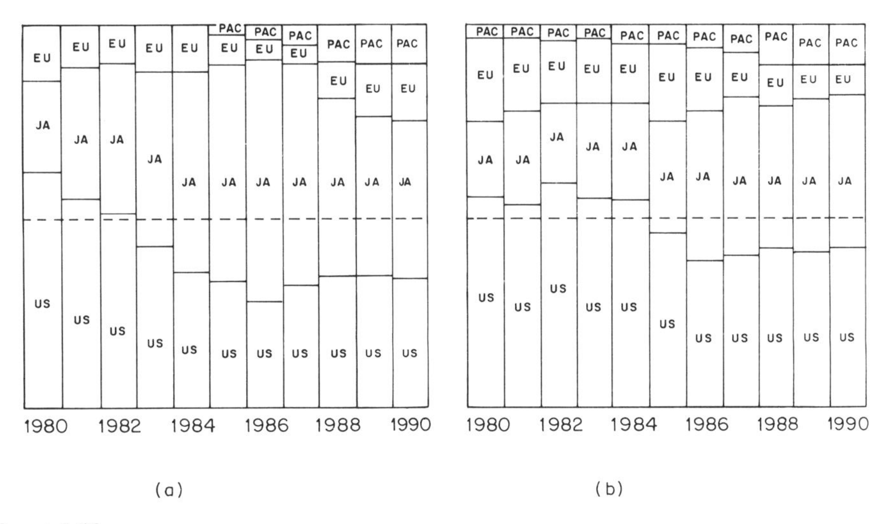

Figure 1.11
Normalized percentage of (a) supply and (b) demand by regions (US, USA; JA, Japan; EU, Europe; PAC, Pacific) from 1980 to 1990. (Sources: (a) Dataquest/WSTS, (b) Electronics Magazine Consumption Surveys, 1980–1990.)

has come to be considered a strategic component. Figure 1.11 shows the normalized percentage of supply and demand by region of the world over the 1980 to 1990 time frame.

1.5.1 Supply shifts in memory trade

The United States, with its historical leadership position in the world computer market, was in the early 1980s the leader in the development, supply, and use of MOS memories. By 1986 there was a significant shift in supply toward Japan as US suppliers left the market.

Toward the end of the 1980s this balance began to shift back somewhat. US and European suppliers, who had left the memory market in the mid-1980s began to return, and new suppliers, most noticeably Korean companies entered the market. All were encouraged by the higher than normal prices in this time frame. The better regional balance in trade was also due to Japanese companies establishing regional supply bases in the second part of the 1980s.

By 1989 the Korean manufacturers were becoming a significant factor in memory supply. They had the advantage of lower labor costs. They also did not suffer the impact of the decline in value of the dollar during the late 1980s as much as the Japanese manufacturers did since Korea's currency was tied more tightly to the dollar.

The economic integration of the European Economic Community expected in 1992 along with the EEC decision to determine European content in memories by the site of wafer fabrication may become a significant factor in memories in the 1990s. Many memory manufacturers have announced new wafer fabrication sites in Europe. This could result, after a period of consolidation, in Europe becoming a net supplier region of memories in the late 1990s.

1.5.2 Demand shifts in memory trade

In Japan, an emphasis on computers and computer technology throughout the last decade coupled with an emphasis on worldwide computer sales generated a correspondingly rapid growth in memory consumption. Japan's memory demand grew from 1983 to 1987 at a much faster rate than that of either the United States or Europe. The Japanese supply of memories, however, grew even faster, so that Japan became a net exporter of memories.

In Europe, each country attempted to have its own major computer manufacturer, most of which did not have a large enough customer base to support the research, development, and production levels necessary to be at the forefront of computer technology. This fragmented country-based computer effort resulted in a low rate of MOS memory demand over the entire period.

In the United States, the multinational computer firms rationalized their memory demands world-wide to match regional manufacturing trends.

By the end of the 1980s there appeared to be a more reasonable balance between supply and demand by region of the world. Companies based in the Far East, however, continued to dominate supply with 75% of the $15.6 billion memory sales in 1989, according to ICE, coming from companies based in Korea and Japan [9].

1.6 AN ESSENTIAL FACTOR IN MOS TECHNOLOGICAL DEVELOPMENT

Memories are the ideal technology driver for advanced semiconductor processes. This is because, as a result of the price pressure on this product and the high cost of a square area of silicon, they are designed to be as small as possible. This in effect means that they are designed to be as tight on the technology as is permitted by the physical limits of the technology and so provide excellent process monitors for the MOS technology which will later be used for other large scale integrated circuits such as microprocessors and ASICs.

Since the market price of the memories drops continually, tighter geometry technologies must continually be developed with resulting smaller area memories to lower the overall cost of the product which is proportional to the area of silicon. This trend of designing memories tightly on the technology continues over time.

Another factor in making memories the ideal test devices for the development of MOS integrated circuit technology is their simple repetitive design. This makes the memories easy to analyze and test. A picture of a memory silicon chip is shown in Figure 1.12. The simple repetitive array can readily be seen to be some 60% of the area of the chip.

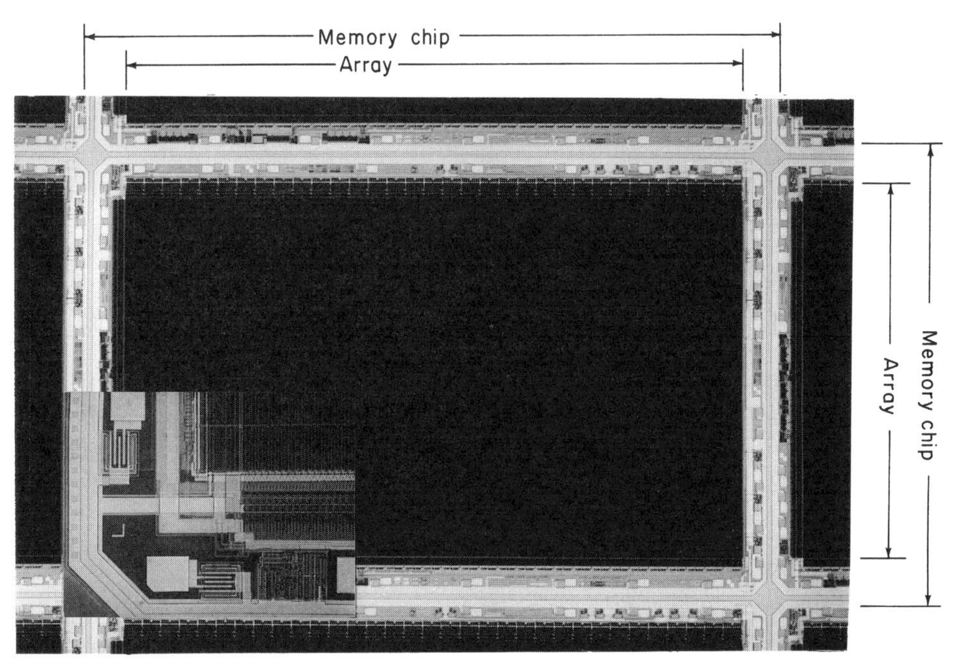

Figure 1.12
Photograph of 1Mb SRAM chip. (Reproduced with permission from N.V. Philips Research.)

1.7 A KEY TO LOW COST MOS MANUFACTURING TECHNOLOGY

Memories are not merely the process development tools of the technology, but have a number of advantages which make them ideal as a volume manufacturing process driver.

They are the only semiconductor product where manufacturing costs are publicly known and forecast. The continual drop in prices forces a continual lowering of cost. Since the competition forces the marginal pricing to be close to the marginal cost it means that the published price history and forecast must be close to the required cost to be competitive. Since memories are the development vehicle of the new technologies, their cost is indicative of general manufacturing line cost for all MOS integrated circuit products. That is, if the memories are competitive, it can be assumed that the other products which follow them on the same technology are also cost competitive.

The market demand is high enough to justify the volume required to detect statistically low process and manufacturing problems which can be corrected and thereby lower the defect level and hence the cost of the manufacturing facility.

Memories are tightest on the technology of any product and so will show up manufacturing problems earlier than any other product.

1.8 THE NEW MEMORY REVOLUTION

The memory market is undergoing continual rapid changes due both to multinational trade factors and to the general rapid development of the electronics systems market. Both technical and market related trends are outlined in this book.

BIBLIOGRAPHY

[1] 1980–1990 World Market Forecasts *Electronics Magazine*
[2] Newman, P. (1982) 1982 Seminex Paper. As reported in *Microcomputer News International,* **6**(2), April 1982.
[3] Dataquest forecast and market share estimates, May 1991.
[4] Memory Market Forecast, *Dataquest Research Newsletter,* 1985–1988.
[5] U.S. chipmakers are back in the race. *Fortune,* June 28, 1982, 79.
[6] World Markets. *Electronics,* January 13, 1983, 125–156.
[7] Equipment Market Production. *Dataquest Research Newsletter.*
[8] Prince, B. (1989) Life cycles of high technology products, VLSI Plenary Session invited paper, ICCD 1989.
[9] ICE Memory Market Report, January 1990.
[10] WSTS MOS Memory Forecasts, September 1989.

2 THE BASICS OF MEMORIES: MARKET, TECHNOLOGY AND PRODUCT

While memories are today a leading semiconductor component with multibillion dollar sales, they came into being only in the late 1960s. In the meantime they have become the major driving force of many aspects of semiconductor technology from production, to process development, to reliability, through package and test development and on into the era of wafer level integration. A short review of semiconductor memories will give some indication of why they continue to have such an impact on the electronics world.

This chapter is intended as both an overview of the memory market and an introduction to the highlights of the technical chapters to follow.

2.1 A BRIEF HISTORY OF THE BULK MEMORY STORAGE MARKET

The original bulk memory storage device was the core memory, which stored bits of digital data, '1's and '0's, in magnetic wire-wound coils configured in large matrices in mainframe computers. Core memory was reliable, but slow and bulky without much potential for significant cost reduction. These factors placed significant limitations on the realistic size, speed, and cost of a computer built with core memory.

In the 1970s the major contest was between the established core memory and two new technologies—magnetic disk technology and the relatively unproven semiconductor memories. Gradually, by the end of the 1970s, magnetic disk and semiconductor data storage had reduced the older core technology to less than 1% of the total data storage market.

During this time there were other challengers for the data storage market—charge coupled devices, magnetic bubble memories, GaAs, etc., all of which had relatively short lives due to the unrelenting feature enhancement, cost reduction, and quality improvement of the semiconductor memories.

Semiconductor integrated circuits with their microscopic dimensions offered the potential for greatly increasing memory performance, while at the same time

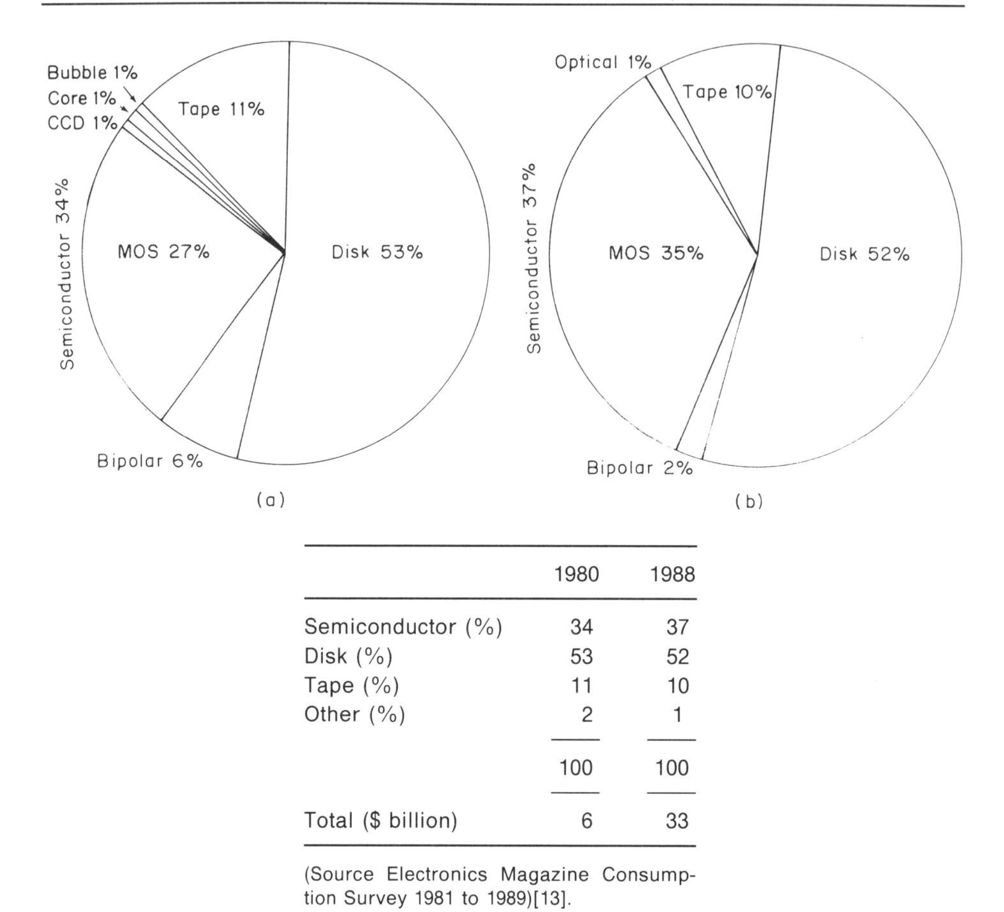

	1980	1988
Semiconductor (%)	34	37
Disk (%)	53	52
Tape (%)	11	10
Other (%)	2	1
	100	100
Total ($ billion)	6	33

(Source Electronics Magazine Consumption Survey 1981 to 1989)[13].

Figure 2.1
Data storage market by memory type (a) 1980 and (b) 1988.

dramatically reducing cost. The demand for high density database storage devices in computer systems provided the volume needed to drive the cost of production down.

The simplicity of the basic semiconductor memory storage element, the cell, which is duplicated once for every bit in an easily manufacturable repetitive matrix, has made it the obvious test device to extend the limits of semiconductor technology. Together these factors made memories the devices to drive semiconductor production to higher performance and greater cost effectiveness.

Microprocessors and other more complex logic devices have extended the frontiers of innovative logic design, but have followed a generation behind memories in use of process technology.

During the 1980s the fast but still expensive semiconductor memories and the slow but low cost disk technology dominated the data storage market. Figure 2.1 shows the market size of various types of memory storage in 1980 and 1988[15]. During this decade the market split between the high performance semiconductor memories and the low cost magnetic disk media has remained relatively constant.

The main shift during this time period was between the MOS and the bipolar semiconductor memories. Even by the early 1980s the lower cost MOS memories dominated the older but faster bipolar memories. By the end of the 1980s MOS had increased in speed by an order of magnitude and decreased in cost by three orders of magnitude, effectively reducing the bipolar memory segment to a small niche of the market.

In the 1980s the optical disk appeared as a new contender for the slow, but high density, market dominated by magnetic disk technology. Both offered high density in the multi-megabyte levels and cost effective data storage. Both technologies have low performance compared to semiconductor memories.

As we enter the 1990s, disk technology, which has not decreased in cost as fast as the MOS memories, is being challenged by compactly packaged systems of MOS memories, by large scale integration of these memories which offer higher performance, and by the even lower cost optical technology[9].

Since MOS offers other advantages in performance and reliability over the various competing technologies, there is significant encouragement for the semiconductor industry to pursue the relentless drive to lower the cost of semiconductor memories.

There has been a shift in technologies also within the MOS memories between NMOS and CMOS, as shown in Figure 2.2 which indicates the dramatic switch the memory industry made to CMOS technology in the late 1980s. This battle has been one of cost and performance. The technology with the best overall performance has been the one where the manufacturing and development emphasis has been placed, gradually bringing that technology down in cost faster than competing technologies. The increase in slope in 1986 is due to the ramp into production of the 1 megabit

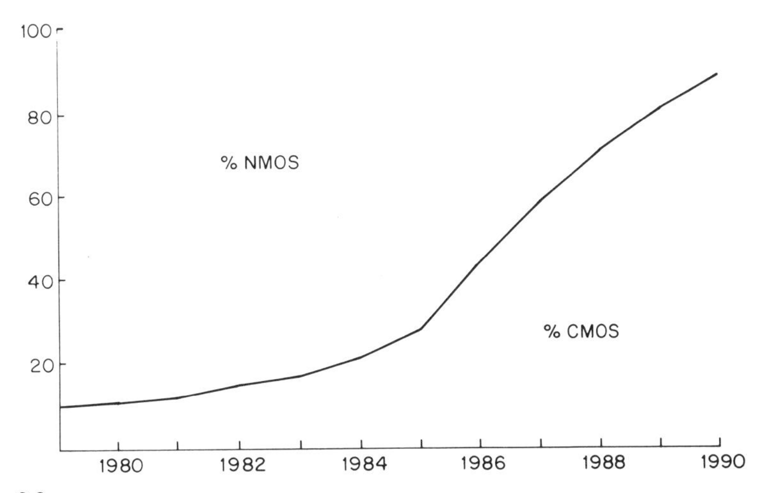

Figure 2.2
Percent CMOS and NMOS of world MOS memory market. (Source Dataquest, Inc.)

DRAM which was the first of the DRAM products to be designed predominantly in CMOS.

In the 1970s NMOS was the clear leader in cost and speed and dominated the MOS memories in spite of its reliability problems and higher power consumption. CMOS, with superior stability characteristics but with much higher cost, lagged far behind, as did bipolar which had superior speed characteristics but much higher power and less potential for size reduction.

Then in the late 1970s a mixture of CMOS and NMOS, which we will call 'Mix-MOS' appeared from the manufacturers of NMOS who were the mainstream memory manufacturers. This combined some of the power and noise margin advantages of CMOS with the smaller chip size and hence lower cost of NMOS.

In the mid 1980s another challenger appeared, BiCMOS, this time from the ranks of the bipolar manufacturers who saw their market share dwindling significantly from the onslaught of the Mix-MOS. BiCMOS is made by combining bipolar and CMOS circuitry in one memory chip. It combines the speed advantages of bipolar with the size and hence cost reduction advantages of CMOS. BiCMOS, in its ECL compatible variation, has advantages in many high speed system applications over CMOS. It does not, however, have the low power characteristics of CMOS, nor does it scale to smaller geometries or lower voltages as well.

As the 1990s begin, CMOS is being made ECL compatible without the use of bipolar. Lower voltage and lower power implementations of BiCMOS are being explored. Chip sizes are being taken into the deep submicron geometry regions and new materials and technologies are being explored.

2.2 A MARKETING VIEW OF MEMORIES AND THE LEARNING CURVE

The learning curve, or experience curve, has been used widely in the semiconductor industry as a guide for producing the cost effectiveness in high volume which has made MOS memories the multibillion dollar industry they are today and as a result introducing electronics into every aspect of modern life. This theory was originally developed by the Boston Consulting Group[3]. An early participant in semiconductor memories, Texas Instruments, popularized it by using it to drive technological advances in MOS memories.

The learning curve concept states simply that as cumulative volume of a product increases, the cost decreases due to the experience gained in the course of production. MOS memories have traditionally followed a 68% learning curve, which means that each time the cumulative industry production volume (in bits) doubles, the cost of production drops to 68% of its former value. Figure 2.3 illustrates the price learning curve of MOS memories from 1975 to 1990.

This learning curve concept was used to drive the volume up by projecting ahead what the cost would be. Pricing for future delivery was put at correspondingly low prices which were not yet supported by actual manufacturing costs. The lower prices encouraged the systems manufacturers to use more memory, increasing the demand volume to match the projected supply volume. The lower pricing on the commercial

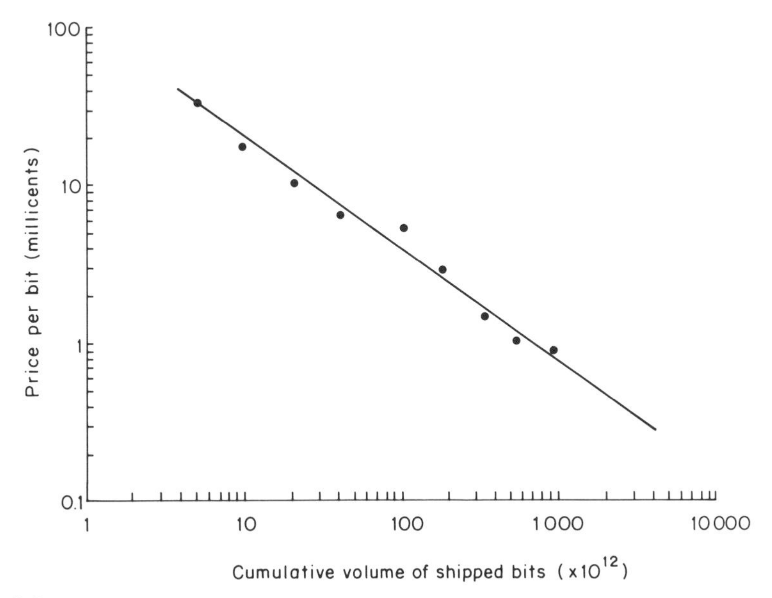

Figure 2.3
MOS memory learning curve showing historical 68% slope (price (millicents per bit) plotted against cumulative volume of shipped bits). (Source Dataquest, Inc.)

backlogs then forced the manufacturers to increase the volume to drive the cost down the actual curve to meet the theoretical curve to which they had priced the backlog.

The result has been that even in an age of widespread inflation, the cost and the selling price of these memory storage devices have dropped continually. Not only has the price of each individual memory product decreased with time, but the cost per bit for all MOS memories has also decreased steadily over time.

This continuous reduction in cost per bit of memories has made electronics more easily accessible in widespread uses that have permeated all aspects of modern life.

A new look at the learning curve concept is being taken today as the continuous drive to smaller and lower cost memories has brought the manufacturers into the submicron geometry technology range. The expense of the equipment and facilities needed to manufacture such devices is increasing non-linearly with decrease in geometry, as shown in Figure 2.4.

This increased capital expense has resulted in discussion among memory manufacturers of a re-evaluation of the basic learning curve concept. It has also resulted in driving many of the earlier manufacturers of MOS memories out of the market and centralizing supply in the hands of the giant companies who can afford the capital investment. It has also brought governments into the arena as whole country industries have been affected, as discussed in Chapter 1.

A glance at Figure 2.3 reveals that the learning curve is still being followed by the intensely competitive electronics industry. It is, therefore, a matter of speculation whether making the learning curve shallower would serve the industry well.

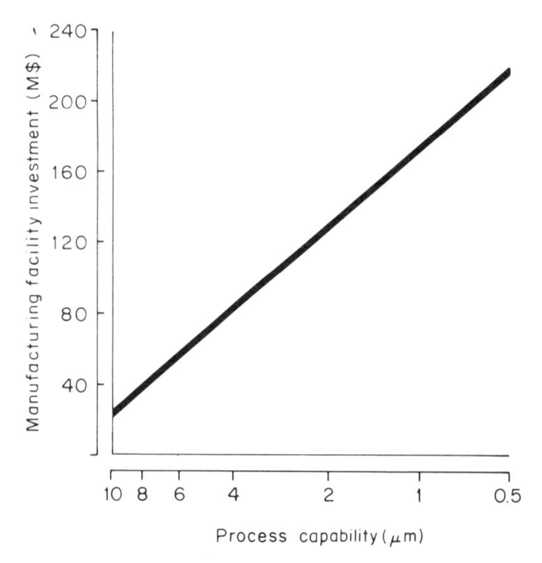

Figure 2.4
Trend in manufacturing facility investment plotted against minimum geometry process capability. (Source N.V. Philips Research.)

While the cost reduction of semiconductor technology has certainly dramatically affected manufacturers, it has been equally dramatic in bringing electronics into every aspect of modern life. A slowdown in the decrease in cost per bit of memory could spill over into a decrease in cost reduction for all of MOS technology and a slowing of the rate of development of new electronic systems such as artificial intelligence with the high memory consumption of speech recognition and other 'closer to human' functions. It could also result in a slowing of the introduction of memories and other semiconductors into older systems where it has previously not been feasible from a cost standpoint to introduce them. A good example is the digital television industry which today stands on the brink of bringing significantly improved video features within the price range of the average consumer through the use of low priced semiconductor memory.

2.3 A MARKETING VIEW OF MEMORY PRODUCT LIFE CYCLE THEORY

The product life cycles, which result from the application of learning curve theory, have been of prominent concern in MOS memories because the life cycle is short for a single generation of memories. This has had a significant impact on the planning and capital investment levels of both memory and system manufacturers[4].

Figure 2.5 illustrates the sales volume of a product over the four phases of the standard product life cycle. During the introduction phase, product and technology development costs are high and suppliers are few. Profit margins are low even though selling prices are high due to the amortization of development costs and the high initial manufacturing costs.

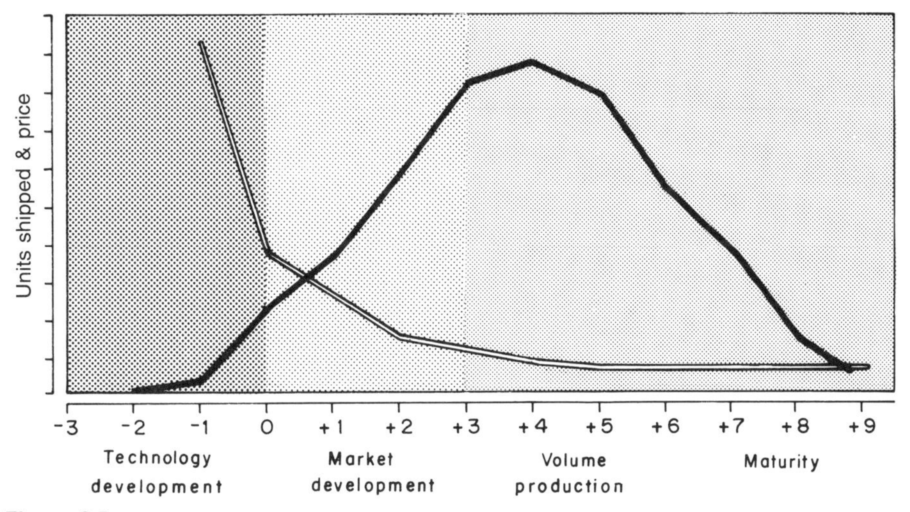

Figure 2.5
Typical product life cycle (——) and price learning curve (═══) of a high technology component. (From Prince [12], Philips 1989.)

During the second phase the number of suppliers remains small and non-linear growth in volume occurs as the market develops for the product. Prices decrease somewhat but remain high during this phase. As the production costs fall development costs begin to be amortized. The high profits mean that competition begins to enter the field.

During the maturity phase, the increased competition drives the prices and profits down and some of the suppliers begin to leave the market. At this point in standard economic theory the manufacturers with the largest market share and hence the lowest costs on the experience curve become the manufacturers who survive in the final end-of-life phase where some profit margin can be maintained due to the relative efficiency of the remaining supplier(s) and the lack of significant competition.

A small variation on this standard economic picture has developed in MOS memories, where the companies who have focused on innovation and introduction of new products are not always the ones with the lowest cost production potential to benefit from it. This has resulted to some extent in one set of companies participating in the development phase and another set picking up the volume production. This has been a source of some turbulence within the memory industry.

Today, this division of competency continues to some extent, but in a more orderly manner, as smaller, more flexible and innovative companies specializing in product design cooperate with and are financed by the larger companies who can afford the huge capital technology development and production investment but whose bureaucracies tend to suppress fast design innovation.

A further variation on the simple product life cycle theory for some memory products has been a tendency for the large companies based in the higher cost geographical manufacturing areas, which are also the areas which are more attractive to highly trained development personnel, to develop the technology and to man-

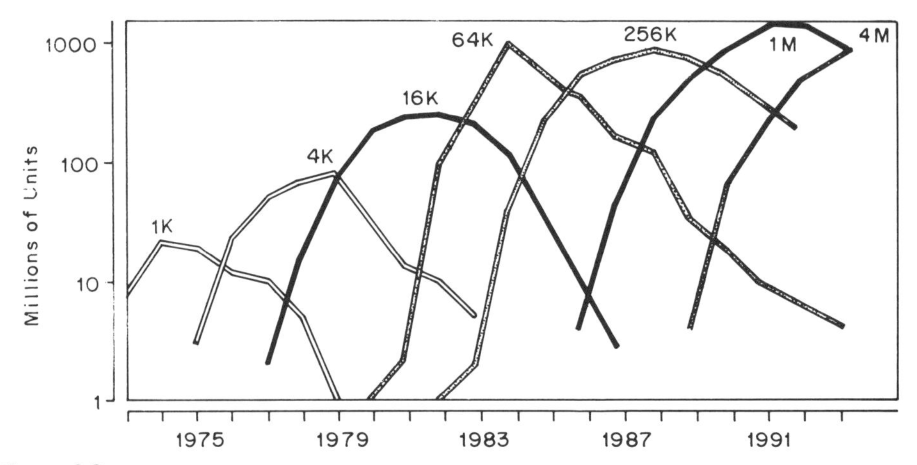

Figure 2.6
Illustration of product life cycles of seven generations of dynamic RAMS in millions of units shipped [12]. (Source Modified ICE.)

ufacture in volume through the early maturity phase and be replaced by second-tier suppliers from the lower cost labor regions of the world during the end-of-life phase of the products and the accompanying technology. This occurs instead of the most cost effective manufacturer maintaining his competitive manufacturing advantage during the end of the life cycle.

The entire cycle between the introduction of one generation of a product and the next is in the range of three to four years, as shown in Figure 2.6 which plots successive generations of units shipped of DRAMs from the 1k to the 16Mb.

Not only is a new generation part introduced every three to four years, but the sales of each succeeding part have been larger than those of the previous part due to increased usage generated by the decreasing prices. In addition, each new product generation is also a new technology generation.

The short life cycle results in an intense design and production effort with at least three technology generations in the development stage and three in the production phase at any one time. New generations of memories are introduced at a rate at which the density doubles in bits every year and a half.

In summary, the MOS memory life cycles have been dominated by bit level cost reduction. This has taken the form of reducing the size by introducing new technologies and increasing the density by introducing new designs of the memory silicon chip because the limiting factor and highest cost have been the silicon wafer and its processing. The less silicon area a bit of memory takes, the less it costs to produce.

2.4 A TECHNOLOGY VIEW OF LIFE CYCLES AND LEARNING CURVES

This brings us to the technology view of life cycles and learning curves. In this section product life cycles and learning curves are viewed from the vantage point of technology and production optimization[12].

For a new generation of MOS memory products, the development part of the life cycle begins about three years before the products are actually introduced into production in the market. Once in production, engineering effort shifts to cost reduction and yield enhancement to minimize the cost and maximize the production volume on a given manufacturing line. Volume increases and costs fall until higher cost suppliers begin to leave the market about four years into the cycle. Volume then declines as suppliers leave the market. High volume memory products tend to have limited eight to nine year life cycles, after which the technology in which they are manufactured is no longer maintained for the production of memories.

The unit price tends to fall, in the case of a 'normal' life cycle by about the fourth year of the cycle, to a value which is approximately constant over time for all memory products in terms of price per square millimeter of silicon. This 'constant' reflects the basic unit of manufacturing in the memory fabrication process—the silicon wafer. Over time more bits of memory are fit on a single silicon wafer while at the same time the cost per bit falls at about the same rate. The result is a fairly constant price of a unit area of memory silicon. This is an approximation and not a constant of nature.

It is this increase in bit density of the silicon wafer which leads to a series of product life cycles in these high technology products. This is because memory bits are not set in a 'sea of cells' on a wafer but are divided into elements of manufacturable size—the individual memory chip.

It is the memory chip which contains an increasing number of bits as it moves up from one generation to the next, generally in factors of four. This occurs as technology improvements permit a fourfold increase in bits to be contained in a manufacturable chip area.

This process is shown for the DRAMs as an illustration in Figure 2.7. The size of each of the generations of DRAMs is shown relative to the early 16k DRAM. Also shown is the equivalent chip size that the next generation part would have been if it were manufactured in the technology of the previous generation. For example, the size of a 256k is shown as if composed of four 64ks and compared to the actual

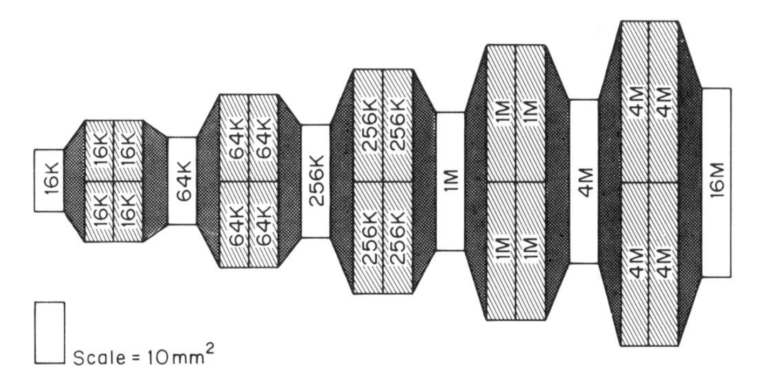

Figure 2.7
Size reduction of successive generations of DRAMs taking the 16k DRAM as base. (□ actual chip size, ▨ equivalent chip size is made in previous generation technology). (From Prince [12], Philips 1989.)

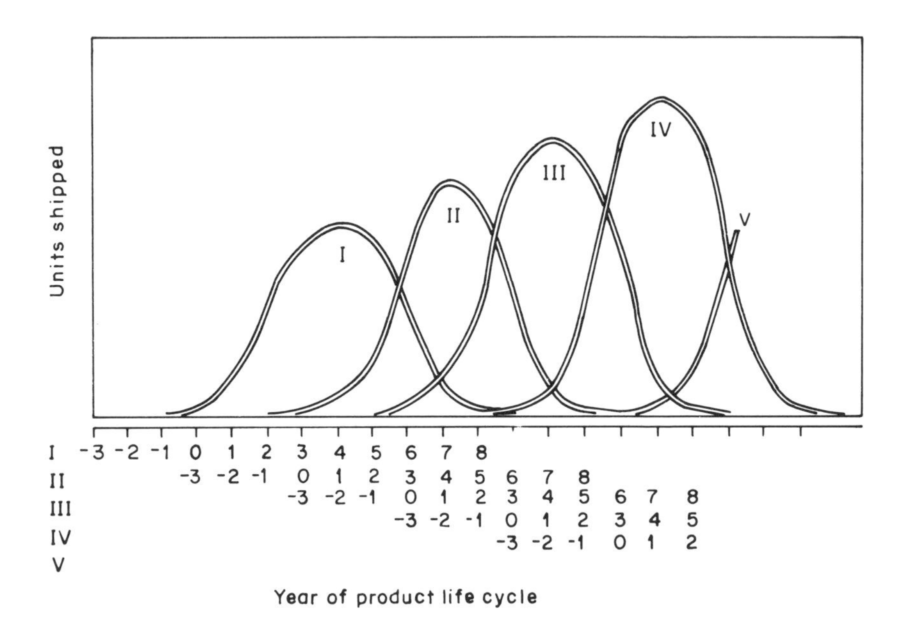

Figure 2.8
The high technology life cycle as a series of individual life cycles with product development starting three years before the part is introduced. (From Prince [12], Philips 1989.)

relative size of the 256k, which is also shown. While the chip size does tend to increase over time, the size of the wafers has also been increased, keeping the average yield at about the same level.

The increase in bits over time per unit area of silicon is implemented in a series of product generations, as shown in Figure 2.8. Each generation reflects a higher bit density than the last. In the case of technology driver products like the DRAMs or commodity SRAMs or EPROMs, each product generation is also a technology generation. In the steady state, a company can afford to support about four of these technology generations at one time. This factor helps explain the eight year product life cycles of the memory component. For example, in the figure, the life cycle of product I is phased out as product IV moves into production and a new generation of technology development is begun for product generation V.

Returning to the stable factors in the memory market discussed in Chapter 1. On the demand, that is the user side, the long term effect is the stable rate of increase in demand for bits. On the supply side it is the stable rate of technology development.

2.5 STANDARDIZATION OF MEMORIES

Production implementation of the technology has not always been as stable in the short term so that insuring stability of supply was a major problem for memory users particularly in the early years of MOS memories.

The short product life cycle also meant that devices became obsolete, forcing systems manufacturers into expensive systems redesigns unless the memory devices were upgradable in the system. This high cost of designing systems and the need for systems manufacturers to ensure stable supply meant that a demand for standardization of memory components developed. This was discouraged by both the highly competitive nature of the industry, which led to almost paranoid attempts at secrecy, and by antitrust legislation, which kept the US semiconductor suppliers, who were the early participants in this industry, from sharing information which could be considered as lessening competition except under very controlled circumstances.

To meet this need the Joint Electronics Device Engineering Council (JEDEC) was developed as an outgrowth of the American Electronic Industries Association. The memory committee of this Council is composed today of representatives from about 40 companies including most semiconductor memory vendor companies in the United States, the large international MOS memory companies, and most of the multinational systems companies. This committee meets regularly to agree on memory standards. Gradually, through the efforts of this group, a certain amount of standardization has been achieved in memories[5].

While other standardization bodies do exist, they have for the most part accepted the lead of the JEDEC committee in memories. The other major component standardization bodies are the Japanese Electronic Industries Association of Japan (EIAJ) and the International Electrotechnical Commission (IEC).

A broad range of standards have been developed for memories through the ongoing efforts of the memory JEDEC committee. These standards will be indicated where appropriate in the technical sections of this book. The reader should be aware that these standards are revised from time to time and that the latest versions should be obtained from the JEDEC Administrative offices to ensure having the most up-to-date versions. It is also wise to note that on occasion standards are passed that are not used. Non-standard devices are also sold.

2.6 AN OVERVIEW OF VARIOUS DATA STORAGE DEVICES

The various data storage devices can be divided by technology into the semiconductor types and the moving media types which require mechanical equipment for operation. The five basic semiconductor types are bipolar, N-channel and P-channel MOS, complementary MOS, and charge coupled devices (CCD). The moving media types include magnetic disk, optical disk, and holographic storage. While magnetic bubbles are not mechanical, they require equipment for supplying a magnetic field for operation and will be considered with the mechanical types. Applications for these various data storage devices will be considered in Chapter 3.

Figure 2.9 compares these data storage devices in terms of cost and performance. Since the semiconductor memories are decreasing in cost per bit faster than the other types of data storage, various attempts to configure them for the disk application are occurring. These developments will be documented throughout this book.

There also exist devices using various combinations of the basic semiconductor

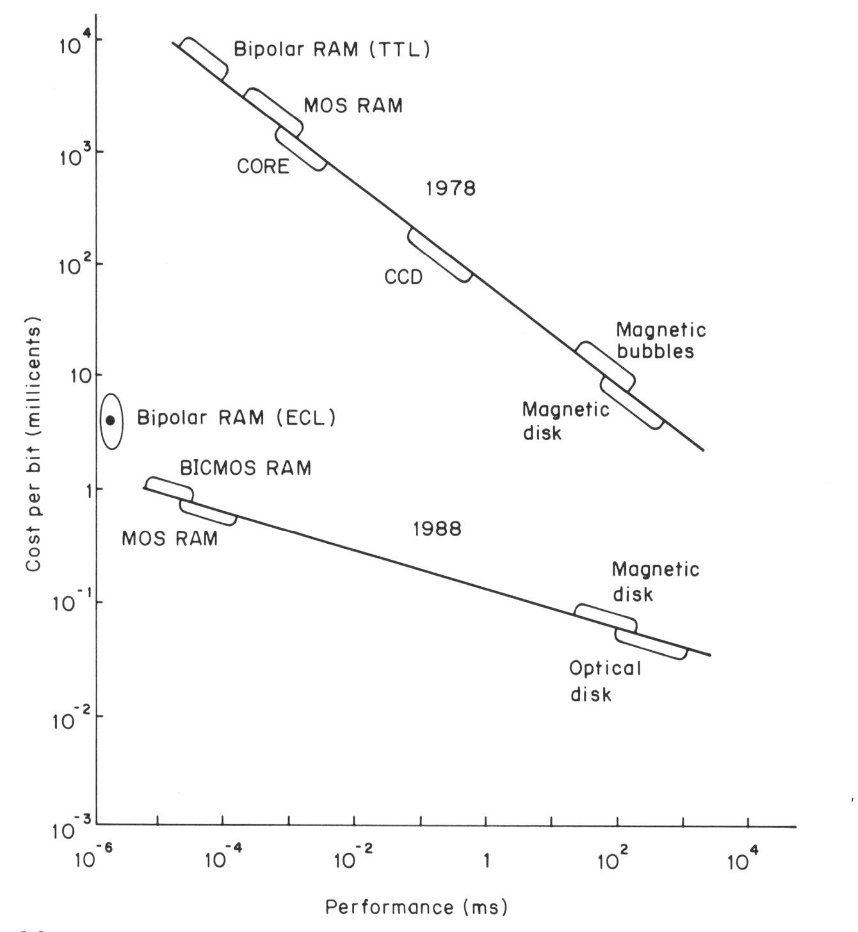

Figure 2.9
Comparison of cost and performance of various types of memories 1978 to 1988. (Source Dataquest, Inc.)

technologies such as CMOS–NMOS (Mix-MOS) and bipolar–CMOS (BiCMOS), and devices made in exotic technologies such as gallium arsenide (GaAs).

A brief review of the various data storage devices follows.

2.7 CHARGE COUPLED DEVICES

In the mid 1970s when 16k NMOS dynamic RAMs were just being introduced, CCDs were already available in 64k densities. CCDs work by pulsing charge stored under depletion-biased electrodes along the surface of the chip, as shown in Figure 2.10(a) and (b). They are slower than MOS memories with speed comparable to core memories.

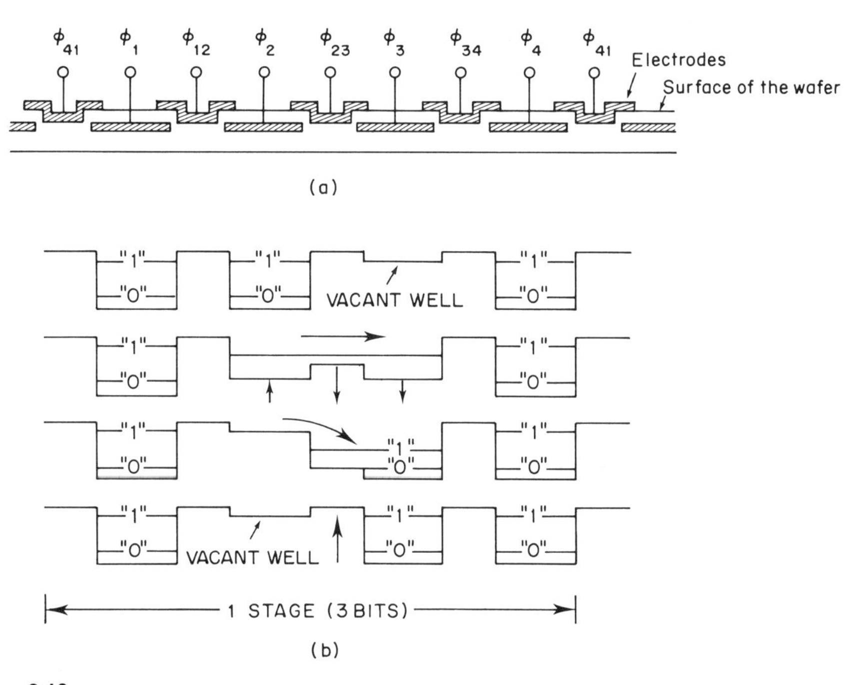

Figure 2.10
(a) Schematic cross-section of a CCD device showing the charge transfer scheme along the surface. (b) Schematic of the surface electrical potential associated with each of the gates as the charge is transferred along the surface of the device. (Reproduced by permission of the IEEE.)

At one time it appeared that they might fill the requirement for high density mass memories in very large storage systems; however, the 64k DRAM became a reality earlier than expected in 1978–1979 with introductions by Fujitsu, TI, and Motorola. The disadvantages of the CCD relative to random access memories caused the CCDs to be retired out of the mainstream market into the semi-obscurity of specialized memory uses just as they were about to be formally introduced by several major semiconductor companies.

CCDs were not only slower than NMOS memories but were limited by being serial access devices. RAMs (Random Access Memories) are distinguished by the ability to access any address at random in the memory. This gives them a flexibility which is a significant advantage in a complex system. Simple solid state imaging systems, such as video cameras which demand neither high performance nor the flexibility of random access memory, can be made more cost effectively with CCDs. CCDs also have analog storage capabilities which make them useful in some telecommunications transmission systems. For such speciality markets, CCDs continue.

As a final tribute to the CCDs we should note that the original planar one transistor DRAM cell was developed from the older CCD technology and was charge coupled as will be described in Chapter 6.

2.8 BIPOLAR MEMORIES

Although bipolar memories have represented a shrinking part of the total semiconductor memory market, this technology has continued to be important in very high speed applications and has recently returned to the spotlight through integration with CMOS in various BiCMOS memory technologies which give nearly the speed of bipolar and nearly the density of MOS.

The basic bipolar technology memory range can be split in two product types—static RAMs and user-programmable ROMs (PROMs). They are used predominantly in high speed applications. This includes computer systems using cache, buffer, and scratch pad memories where short cycle times and high access times are important. Other areas are in small mainframe systems with auxiliary, add-on memory boards, military computers, and high speed data processing systems utilizing bit slice designs.

Historical memory technologies include ECL (Emitter Coupled Logic), which has very high speed but has the disadvantage of high power consumption, simple bipolar TTL (Transistor–Transistor Logic), and I^2L (Integrated Injection Logic), which had a more moderate speed–power performance, but with the possibility of size reduction.

Interface voltage levels in many memory systems continue to be either TTL compatible or ECL compatible. These interface levels can be attained with either bipolar or CMOS technology and should be distinguished from the initial bipolar technologies where they originated.

Bipolar memories in general lag behind MOS parts with respect to size, density, and organization. The current state of the art is 64k density in bipolar static RAMs with access times in the 3 ns access time range[8]. A simple bipolar static RAM memory cell is shown in Figure 2.11(a).

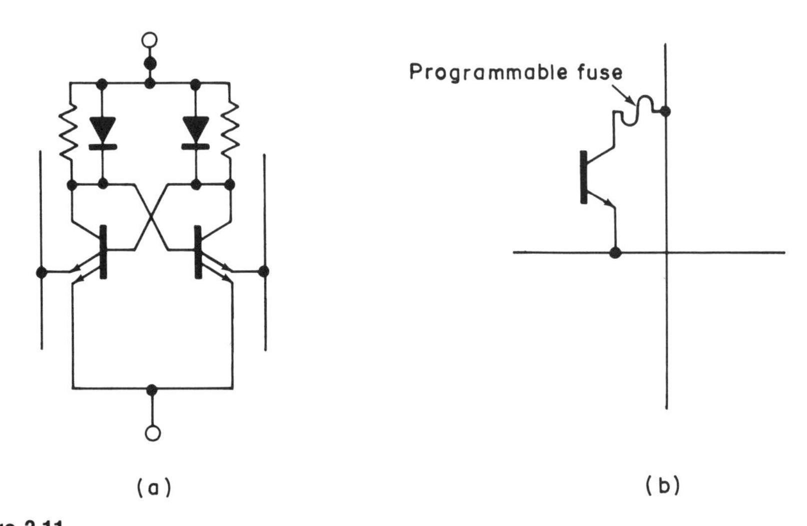

Figure 2.11
Basic bipolar memory storage cells. (a) Bipolar memory (RAM), (b) Bipolar memory (PROM). (Source Hitachi Databook 1988/1989 [14].)

Bipolar ECL technology for very fast static RAMs has been in increasing demand for first level cache in mainframe computer systems. The state of the art in speed is 4k to 5k bit devices with access times in the 1 ns range and below.

BiCMOS static RAMs have recently appeared which combine many of the speed advantages of the bipolar RAMs with the size and lower power characteristics of the CMOS SRAMs.

Also available in bipolar technology are high speed programmable ROMs (PROMs) that are user programmable. They are available as fuse-link types in which a fuse is blown electrically to program them and as EPROM Programmable Devices (EPLD) in which an EPROM cell is programmed electrically. A circuit schematic of a fuse link bipolar PROM memory cell is shown in Figure 2.11(b).

Bipolar memories continue to suffer from being more difficult to scale to smaller (lower cost) sizes than MOS and from being higher than MOS in power dissipation. These disadvantages have pushed bipolar memories also into those speciality markets where the higher speed is essential and the high power dissipation acceptable.

The bipolar market is small and is covered by only a few memory manufacturers.

2.9 MOS MEMORIES

The ideal memory would be low cost, high performance, high density, with low power dissipation, random access, non-volatile, easy to test, highly reliable, and standardized throughout the industry.

Those memory technologies which did not offer these advantages to some extent were one by one successfully challenged by the MOS memories. Unfortunately a single memory having all these characteristics has not yet been developed, although each of the characteristics are held by one or another of the MOS memories.

The MOS memories fall into two broad categories.

(a) Read–Write memories, dynamic RAMs and static RAMs, allow the user both to read information from the memory and to write new information into the memory while it is still in the system,

(b) Read Only memories, ROMs, EPROMs, EEPROMs, are used primarily to store data: however, the EEPROMs can also be written into a limited number of times while in the system. Read-Only-Memories (ROMs) are non-volatile, that is, they retain their memory if the power is turned off whereas Read–Write memories do not.

A comparison of the relative usage of these various MOS memories is shown in Figure 2.12. Clearly the dynamic RAM has gradually become the dominant MOS memory device. Brief descriptions of the various types of MOS memories and memory cells follow. All MOS memory cells are constructed of the basic MOS transistors

A simplified cross-section of the two types of MOS transistors, NMOS and PMOS, are shown in Figure 2.13. Both consist of a gate electrode made of conducting material which is separated from the semiconductor by an electrically insulating material. A

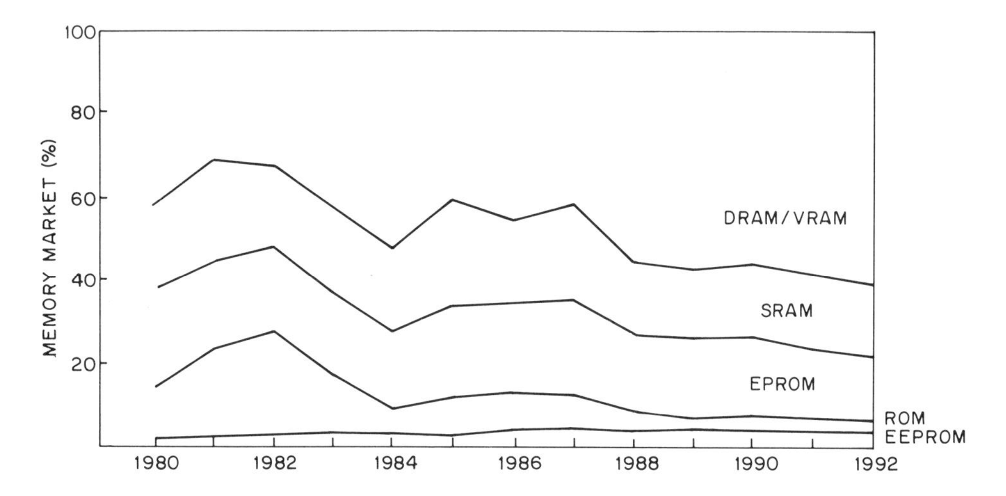

Figure 2.12
Percentage of various MOS memory types out of total annual MOS memory market. (Source Dataquest, Inc.)

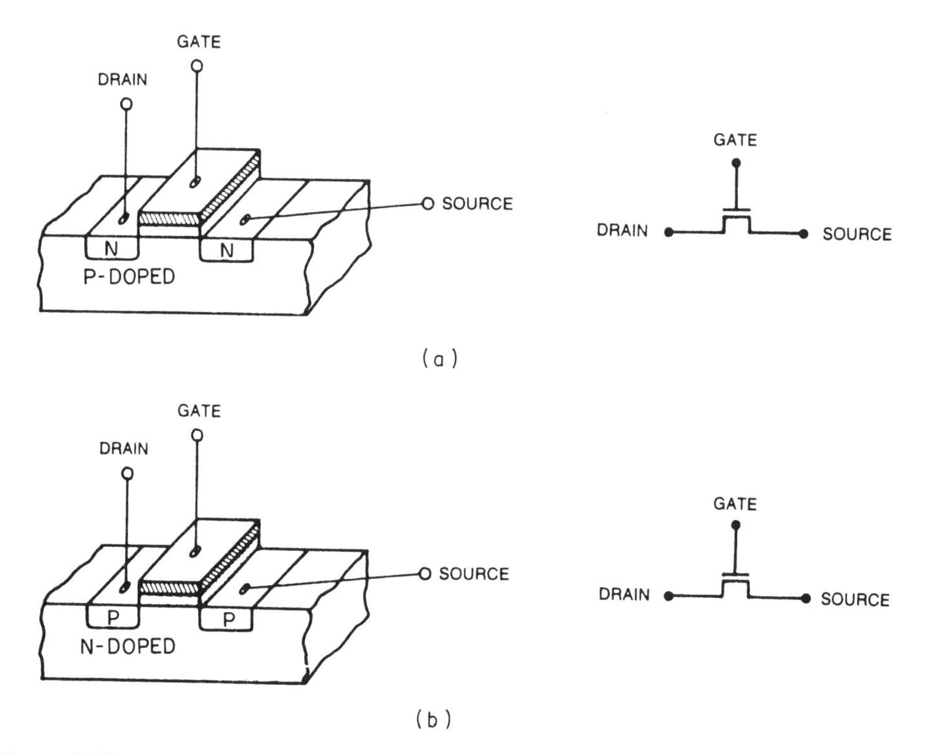

Figure 2.13
Cross-section and circuit symbol of (a) NMOS and (b) PMOS transistors.

voltage applied on this gate electrode can control the flow of current through the semiconductor below the gate. This current flows between the drain and source regions on either side of the gate. The drain and source are formed by positive ('p' doped) or negative ('n' doped) ionized regions in the silicon.

2.9.1 Read–write memories

2.9.1.1 Dynamic RAMs

A dynamic RAM is a MOS memory which stores a bit of information as a charge on a capacitor. Since this charge decays away in a finite length of time (milliseconds) a periodic refresh is needed to restore the charge so that the dynamic RAM retains its 'memory'.

There are many advantages with dynamic RAMs. The basic memory cell, which consists of a single transistor and a capacitor, is small and a very dense array can be made using these cells (see Figure 2.14(a)). The major cost of a semiconductor memory is usually in the cost of the silicon wafer, thus the more chips on a wafer, the lower the cost of a single chip. Dynamic RAMs therefore have a lower cost per bit than memories with less compact arrays. Dynamic RAMs are also fast for a system to access, giving them a high performance rating.

The disadvantage of a dynamic RAM is that it is volatile. The memory cells do need to be refreshed. They normally need additional external circuitry to refresh the memory cells. Some memories with dynamic cells, however, called 'pseudostatics', have refresh on the chip to overcome this particular disadvantage.

Dynamic RAMs also need external address counters to keep track of the memory cells that are being refreshed. A dynamic RAM, therefore, requires some engineering sophistication by the user which many small system manufacturers historically did not have. For this reason, dynamic RAMs have tended in the past to be used primarily by the larger systems manufacturers who can afford the initial cost of the engineering effort to produce systems which utilize the performance characteristics and cost advantages of dynamic RAMs.

Since large systems can normally use a deeper memory than smaller systems, dynamic RAMs have been available primarily in the 1 bit and 4 bit wide organizations which have the advantage of being packaged in smaller packages than wider devices, giving a cost advantage to the user at the board level.

During the 1980s the market for small computers grew significantly while the mainframe computer market growth was significantly less. At the same time the sophistication of systems designers of small computer systems tended to increase so that the low cost of the DRAMs brought them into the memory of all but the smallest of the computer systems.

Another factor in the move to small computers has been the percentage of the average small computer system memory that is main storage and the percentage that is display memory. With the increase in importance of the high resolution graphics display, a new memory type has branched off the original DRAM technology—the

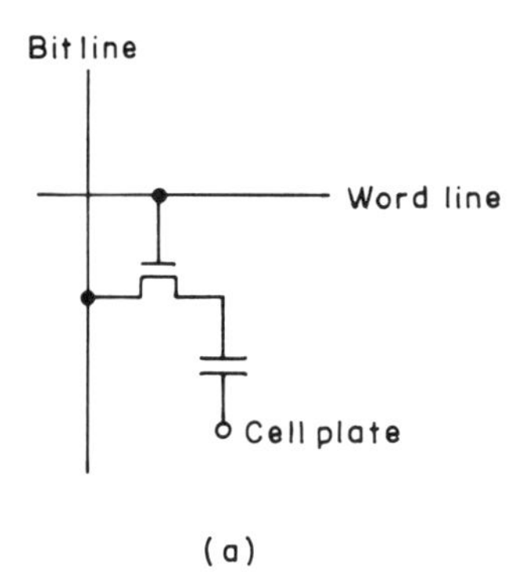

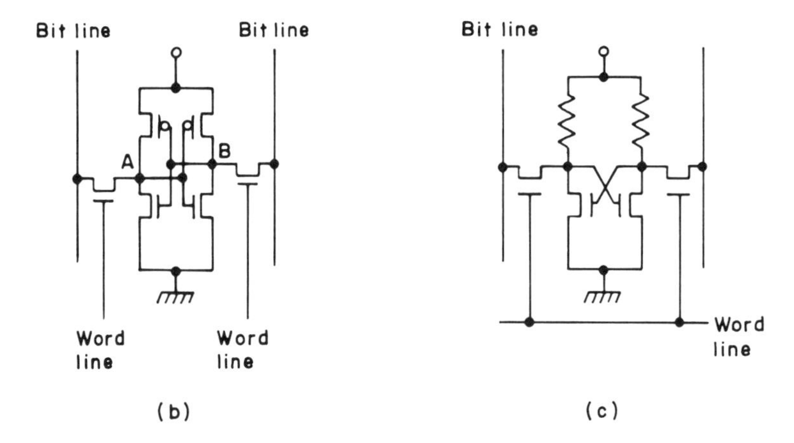

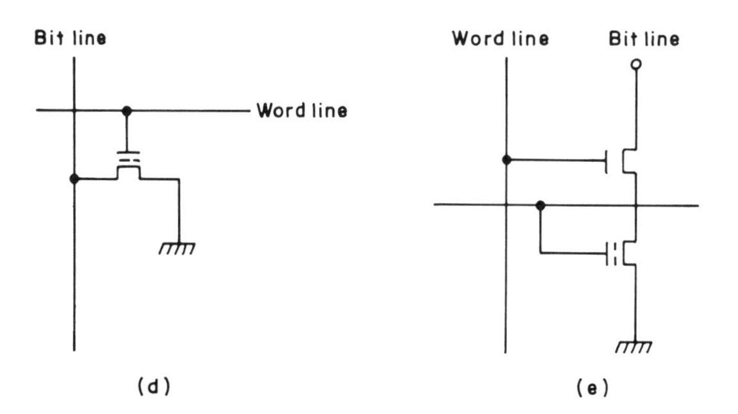

Figure 2.14
Basic MOS memory storage cells. (a) Dynamic RAM, (b) CMOS static RAM, (c) NMOS static RAM, (d) EPROM, (e) EEPROM.

video display memory. This is a DRAM with both a serial port that feeds data to the display and a random port that interfaces with the processor.

While it is not a systems requirement that video memories be made from DRAMs, the significant cost advantage of the DRAM combined with the competitive nature of the small computer market seems certain to keep the low cost, high density DRAM technology used in the video RAM area.

Further discussion of standard DRAMs can be found in Chapters 5 and 6, and of applications specific DRAMs in Chapter 7.

2.9.1.2 Static RAMs

A static RAM cell consists of a basic bistable flipflop circuit which needs only a dc current applied to retain its memory. It contains four transistors plus either two transistors or two polysilicon load resistors as pull-up devices, as shown in Figure 2.14(b) and 2.14(c). From Figure 2.14(b), it can be seen that data are stored with either a high potential at point A and a low at point B or vice versa. This means that two stable states are available which are defined as a logic '1' or '0'. No periodic refresh is required. This eliminates the need for external address counters and refresh circuitry as used in dynamic RAMs. This lack of need for external support circuitry and consequent ease of use is a major advantage.

There are several historical disadvantages of the static RAM.

The basic cell is larger due to the use of six transistors rather than one transistors as in DRAMs. This means they are more expensive to build than DRAMs and have as a result followed a generation behind the DRAMs in density.

Historically SRAMs had higher power consumption than DRAMs due to the need to have dc power continuously applied. This disadvantage has been overcome by the general development of low power CMOS SRAMs. Today one of the main advantages of CMOS SRAMs is their very low power standby characteristics which are used in battery back-up of large memory systems.

Also, again historically, SRAMs had slower speed due to being manufactured in older, lower performance technologies than DRAMs. Today one of the main advantages of SRAMs is their high speed capability.

Today the only disadvantage remaining compared to DRAMs is that of size and, as a result, of cost.

Historically the slow static RAMs have tended to be used by smaller systems manufacturers who needed to invest less in engineering sophistication to use them. Since these were usually manufacturers of microprocessor-based systems, the static RAM has traditionally been organized in 4 or 8 bit wide configuration. This configuration requires a larger, more expensive package than the dynamic RAM and consumes more space in the system, which leads to increased system cost.

A tradeoff between speed and power dissipation in static RAMs has developed resulting in the development of two variations. High density, and low power but slow devices are used for battery back up applications. Very fast static RAMs are used in the cache type of applications where traditionally bipolar SRAMs have been used and also in the newer supercomputer type of applications.

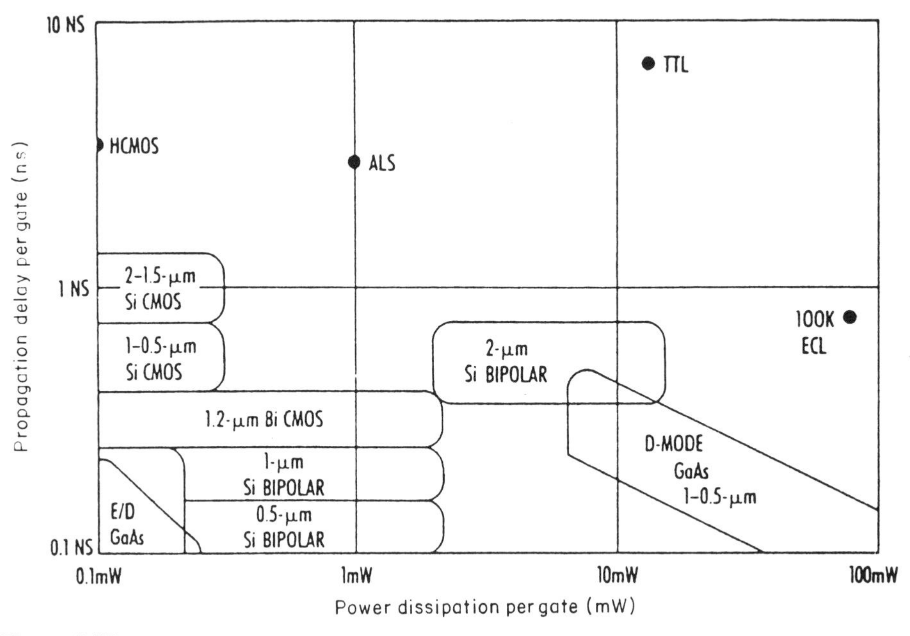

Figure 2.15
Comparison of power dissipation and propagation delay per gate for various types of MOS memory technologies. (With permission of Electronics[16].)

The requirement for high density but very fast SRAMs has led to the combination of CMOS and bipolar technologies which is referred to as BiCMOS. BiCMOS is usually a CMOS static RAM with bipolar gates added on the periphery. The CMOS reduces the power consumption and permits larger and higher density matrices. The bipolar devices improve the speed of the peripheral circuits and in the case of ECL circuits allow compatible interface into the system, which also improves the effective speed of the chip in the system.

A comparison of the power dissipation and propagation delay (speed) per gate for the various memory types is shown in Figure 2.15 for CMOS, bipolar, ECL, BiCMOS, and GaAs memory technologies.

Further discussion of static RAMs can be found in Chapters 5, 8 and 9.

2.9.2 Non-volatile memories

The major disadvantage of RAMs, both dynamic and static, is that they are volatile, that is, they lose their memory if the power is turned off. It is more important in many memory system uses that the memory be retained with the power off than that the memory be alterable.

For these non-volatile applications another class of memory has developed called the ROM or Read Only Memory. ROMs come in a number of versions. The mask programmable ROM is preprogrammed in the silicon by the memory manufacturer.

The EPROM (Erasable Programmable ROM) can be reprogrammed by the user after removing it from the system. The EEPROM (Electrically Erasable Programmable ROM) can be reprogrammed electrically by the user in the system a limited number of times.

2.9.2.1 Mask programmable ROMs

A ROM cell is a transistor which is effectively either present or absent at a given location in the memory. The main advantage of a mask programmable ROM is the very small cell size possible compared to that of the EPROM or EEPROM which, in high enough volume, makes it a low cost option and allows very high density chips to be manufactured compared to other less dense memory technologies.

The disadvantages of ROMs, aside from the obvious fact that their contents cannot be altered by the user, is in the high initial, one time, cost of producing a custom product. The volume to be produced must be sufficiently high to be able to amortize the initial customizing (or masking) cost over the volume and still take advantage of the low per-unit manufacturing cost. The system design must be stable since any change will entail a remasking charge and soon negate the cost advantage of using a ROM.

ROMs are the best memory solution for any system where a standard program is being stored and the system will be manufactured in sufficient volume to justify the initial cost. Examples of such uses are program storage in gaming machines, language character generators, and operating system storage in computer systems. One application which has required very high density ROMs is Japanese Kanji character generators for word processing systems.

While ROMs are useful in large volume applications where the programming is fixed, many systems manufacturers do not have this ideal situation. The need for a non-volatile memory often occurs in small volume applications or in applications where the software is undergoing constant updating in order to remain current in a competitive market situation. In these cases, where the requirement is for a non-volatile memory in which the programming can be changed occasionally, the initial programming cost and minimum volume requirements can make ROMs prohibitively expensive.

Further discussion of mask ROMs can be found in Chapters 5 and 10.

2.9.2.2 EPROMs

EPROMs provide a solution for the low volume application. The EPROM has a cell containing only a storage transistor which holds excess charge when programmed, as shown in Figure 2.14(d). EPROMs can be erased and reprogrammed by the user, eliminating the need for remaking the silicon chip. They are then non-volatile like a ROM and retain their memory when the power is off. EPROMs are more expensive than ROMs in per-unit cost for volume quantities but are a solution to the need for a reprogrammable non-volatile memory.

There are several disadvantages to EPROMs. They are historically not as fast in performance as RAMs or ROMs. They cannot be reprogrammed in the system but must normally be removed from the system manually and reprogrammed using special ultraviolet light equipment to erase the old programming followed by another piece of equipment to program the new.

EPROMs are in expensive packages with quartz windows for entry of the ultraviolet light. Once in the system EPROMs tend to be light sensitive to some degree through the quartz window, even to standard fluorescent lights.

Finally, even though the manufacturing problems that plagued EPROM manufacturers in the early years have mostly been solved, long term reliability results are in general still not in the same range as comparable results for standard RAMs and ROMs.

Further discussion of EPROMs can be found in Chapters 5 and 11.

2.9.2.3 EEPROMs

Some of the disadvantages of the EPROM have been overcome by the Electrically Erasable Programmable Read Only Memory. EEPROMs are electrically erasable in the system, so the need to remove the EEPROM from the system and to buy expensive ultraviolet erase equipment is eliminated. EEPROMs are available in standard plastic packages without the quartz window which eliminates light sensitivity problems and reduces packaging cost. The EEPROM also erases considerably faster than the EPROM—in milliseconds compared with many minutes for high density EPROMS. A cell for an EEPROM contains a select transistor and a storage transistor, as shown in Figure 2.14(e).

A major disadvantage of the EEPROM is in the large size of the two-transistor memory cell compared to the EPROMs one-transistor cell. This has kept the cost high enough that a significant market has not developed among EPROM users.

The EEPROM is also at a disadvantage compared to static RAMs except in those applications where the non-volatility is critical and the use of a battery backed up SRAM system is not feasible, that is, it is not a non-volatile static RAM. The technology is more difficult to control than SRAM technology. It is a 'read-mostly' memory since the number of times that the EEPROM can be erased and reprogrammed is limited by technology, as opposed to the RAM technology which effectively has an infinite number of write–read cycles. The access time for erase and reprogramming is much longer than that normally accepted for the write cycle in a RAM.

Since applications for a high cost, non-volatile, read-mostly memory device are limited, the EEPROM has remained a niche market part with only a few competitors specialized in this type of memory. It developed during the late 1970s and by the mid-1980s had branched into two application-specific product types:

(1) small, less than 1k, EAROMs (Electrically Alterable ROMs) used in volume in the consumer and industrial control markets, and

(2) medium density to 256k EEPROMs which were targeted at various market niches where the reprogrammability requirement offset the high cost.

Further discussion of EEPROMs and EAROMs can be found in Chapters 5 and 12.

2.9.2.4 *Flash memories*

The flash memory is a relatively new type of memory device which uses a technology similar to that of the EPROM and also has a one transistor cell so that it has the low cost of an EPROM. It is, however, electrically reprogrammable a limited number of times in the system. This makes it ideal for those applications where only a few changes in the programming of the system are expected over the lifetime of the memory device. Electrical erasure in the system is for either the entire memory array or for blocks of it. This property gave rise to the name 'flash'. See Chapter 11 for more on flash memories.

2.10 ALTERNATIVES TO SEMICONDUCTOR MEMORIES

Three main alternatives exist to semiconductor memories for high density data storage—magnetic bubbles, magnetic disk, and the newer optical disk.

2.10.1 Magnetic bubbles

Magnetic bubbles were another early candidate for the high volume mainstream memory market due to their high density and large storage capacity. Since they are non-volatile, serial access, and slow compared with RAMs, they were considered mainly for magnetic disk replacements in computer systems. Bubble memories had exceeded the 256k data storage level by 1978 and the Mb level by 1980, and reached the 4Mb level by 1984.

Many of the largest MOS semiconductor suppliers started into the magnetic bubble market including: TI, Rockwell, National Semiconductor, Motorola, Siemens, Fujitsu, and Intel. The cost of the bubble technology did not, however, decrease rapidly enough in production. The magnetic disk manufacturers themselves also responded to the potential threat by producing disks which were more cost effective than bubbles and faster, and managed to retain their market position.

Bubbles also suffered from the need for interface and control circuits which add significantly to the cost at the system level. By 1982 a number of the original suppliers had dropped the bubble product line, and by the mid-1980s bubble memories had been relegated also to the realm of a specialty memory.

In the early 1990s it is the MOS dynamic RAMs that are challenging disk technology with the combined advantage of large storage capacity at the 16Mb level and new high density packaging which makes them nearly comparable in size to disk technology. In addition to the increasing density and decreasing price, DRAMs have the significant advantage of being random access, although this particular feature is not currently being used in disk compatible applications. They do still have the disadvantage in the disk application of being volatile. At least one manufacturer, Anamartic of the UK, has even developed a wafer scale integration of DRAMs which has the high density needed to compete with disks although it is slow and serial access.

2.10.2 Magnetic disk

The magnetic disk technology has been driven by the demands of the computer industry for higher density and more compact data storage, and by the competition from MOS memories to significant density increases and size and cost reductions.

The major disadvantage of the magnetic disk is that it has mechanical moving parts which are inherently not as reliable as semiconductors. The significant cost advantage which the disk technology has managed to maintain has given it a competitive advantage over the semiconductor competition. The disk is slow and useful mainly for archival storage for mainframe computer systems.

2.10.3 Optical disk

Optical disk technology is relatively new. To date, large data storage archives with gigabytes of write-once optical memory have been introduced. Multiwrite optical is now entering the scene and promises to be significant competition for both high density semiconductors and magnetic disk applications.

The basic technology involves etching data bits in glass with a laser and reading the resulting bits also with a laser using a mechanical disk drive. The optical disk technology has the same disadvantage as the magnetic disk. It has moving parts which are inherently less reliable than solid state memory. Its density is sufficient to keep it useful some distance ahead of the semiconductor memory for the time being.

The optical disk shares with the magnetic disk the problem of being slow so that it is also useful mainly in archival storage for mainframe computer systems. The use of slow archival storage requires a level of faster memory between the archive and the main memory to increase the speed of the data to match that of the main memory. Optical systems with MOS RAM caches have been introduced. Another possibility for a faster, higher density interface between the archive and the main memory is a holographic cache.

2.10.4 Holographic storage

Photorefractive volume holographic storage (PVHS) as a technology was first explored in the 1970s and is being rediscovered as a possible new high density storage medium for the 1990s. PVHS stores data images as holograms in photorefractive crystals. The storage mechanism is a redistribution of photoelectrons within the crystals. The resulting pattern of spots is read with a reference laser beam which accesses the data as an interference pattern from the set variations in the local refractive index. Data are written by variations in light intensity from a laser beam which produces changes in the local refractive index of the crystal. Each two-dimensional array of spots is a page. Arrays of stacks of pages can be overlaid in the same region by changing the angle of recording. This could lead to very high density storage which is also potentially faster than optical or magnetic disk and could be used in mainframe computer systems between the archival stacks and the first level main memory.

2.11 HIGH DENSITY MEMORY PACKAGING

From the beginning packaging has been an integral part of the product development of a memory integrated circuit.

The memory silicon chip itself is only a few millimeters on a side. This near microscopic size has made handling the chip with conventional circuit wiring tools impossible. It needs to be in a package of normal circuit element dimensions. The electrical leads of the package must make good contact with the rest of the circuit and the package also must help dissipate some of the heat which is generated by the many transistors on this tiny piece of silicon.

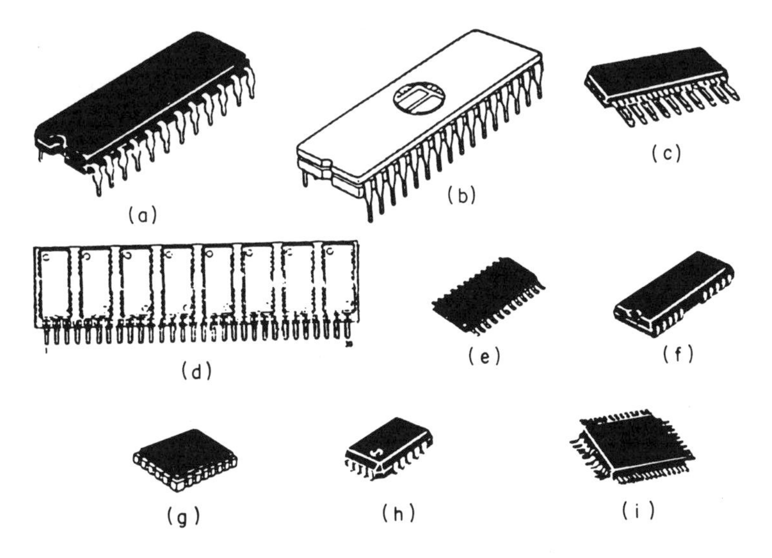

Figure 2.16
Various packages used for MOS memories. (a) Dual-in-line (DIP), (b) ceramic dip (CERDIP) with quartz window, (c) Zigzag-in-line (ZIP), (d) single-in-line module (SIM), (e) small outline package (SOP), (f) SOJ-leaded (SOJ), (g) ceramic leadless chip carrier (CLCC), (h) plastic leaded chip carrier (PLCC), (i) Flatpack.

The traditional memory package is the 'dual-in-line', so-called because of the two lines of electrical leads on either side of the package. The dual-in-line package (or 'DIP' for short) is shown in Figure 2.16(a), which shows the various types of memory packages available. The DIP is available in either ceramic or plastic.

There are two variations of the DIP package: the sidebraze package is a single-unit ceramic package with a cavity for the chip covered by a small gold lid. This package has good power dissipation and its gold-plated leads make good electrical contact, but it is too expensive for most systems applications. The other is a plastic dual-in-line package which is very cheap and was the standard DIP package throughout the 1980s. Its disadvantages are that it dissipates less heat than the ceramic package and it is not hermetic, that is, moisture can be absorbed through the plastic. Since the surface passivation on most MOS memory chips today is almost hermetic in its own right, the moisture problems with the standard plastic DIP package are mostly historical.

An older package, the ceramic DIP (CERDIP), was formed by two pieces of ceramic sandwiched together with a molten glass seal. This package tended to be fragile and subject to cracking of the glass seal. It has mostly vanished from standard applications except for EPROMs, where it has been retained with a transparent quartz window in the top through which ultraviolet light can pass to reprogam the EPROM, see Figure 2.16(b).

In the last few years the trend to system miniaturization and to high system packing density has led to a requirement for smaller packages with tighter packing capability in the system. These fall into two categories, those with leads intended to pass through holes in the printed circuit board, and those intended to be welded to the surface of the board.

The two high density packages used on through-hole boards are the zigzag-in-line package (ZIP) which is shown in Figure 2.16(c) and the single-in-line memory module (SIMM) package shown in Figure 2.16(d). The SIMM package is itself a tiny printed circuit board with memories mounted on it and with a single-in-line through-hole connector.

Two-sided miniature packages for surface mount assembly on printed circuit boards include the small outline package (SOP) which is a thin plastic package with leads bent out on two sides as shown in Figure 2.16(e). There is a similar package with leads bent in (like a 'J') on both sides (SOJ) as shown in Figure 2.16(f). The SOJ has been standardized for surface mount applications by the JEDEC memory committee. It is compatible in pinout in each standard with the corresponding DIP pinout. These two-sided packages are slightly larger in area than the corresponding four-sided packages but have the advantage that lines can be run under them on the PC board.

There are also various four-sided miniature packages such as the ceramic leadless chip carrier (LCC) which is shown in Figure 2.16(g). This package is expensive but highly reliable and tends to be used in high density, multilayer PC board systems with severe environment restrictions. The plastic leaded chip carrier (PLCC) shown in Figure 2.16(h) is an inexpensive package used commonly in high density systems and for mounting on SIMMs. The flatpack, as shown in Figure 2.16(i), is used uncommonly for memories in high density systems. An even smaller package is being developed called the thin small outline package, which is a thinner version of the SOP.

Figure 2.17
1 Mb static RAM made of four 256k SRAMs flipped and thermal compression bonded onto a gold bumped silicon carrier chip. (Source N.V. Philips Research.)

The continued drive toward higher density and smaller systems is leading now to packaging concepts that eliminate the package, such as tape automated bonding (TAB), and chip-on-substrate. See Figure 2.17 for a picture of a 1Mb static RAM made up of four smaller 256k SRAMs flipped upside down and bonded to a lower silicon substrate. These base substrates must then also be packaged.

Much work has been done in the packaging area in the last few years to overcome both the historical limitations and to develop new packaging for high density systems. Further discussion of packaging can be found in Chapter 13.

2.12 TEST AND TEST STRATEGIES

Memory test has gone from being, in the early days of memories, a screening exercise to sift out an acceptable level of quality to being a highly sophisticated debugging tool used by the vendor and assumed by the user. Just in time (JIT) deliveries have put the responsibility for test in the 1980s and beyond onto the manufacturer. The level of manufacturing quality is approaching the point where extensive production screening is unnecessary. Defects in both quality and reliability are measured in a few parts per million (ppm). While the price for memory test equipment approaches half a million dollars or more, the capability of that equipment has increased significantly. Scanning electron microscopes (SEMs) enable defects to be seen visually that would be difficult to trace electrically. Sophisticated voltage and current contrast systems work with the SEMs to pinpoint defects. The increased computer capability has augmented this trend. Finally, self test on the chip itself is becoming more prevalent. Further discussion of memory test can be found in Chapter 14.

2.13 RELIABILITY CONCERNS

The fabrication processes and equipment for building MOS devices in the 1970s and early 1980s were experimental. The newness of the technology meant that the various failure mechanisms to which these devices were susceptible were not well understood. In the last 20 years, MOS reliability engineering has changed from being an art which was regarded with some skepticism to a highly respected science.

In the early days process 'recipes' were lost with unfortunate regularity. Contamination problems, the bane of MOS technologists, were barely controlled. Packaging technology was in its infancy and not as susceptible to control due the use of offshore assembly in low technology areas of the world for cost control. Manufacturers had not developed the test techniques needed to screen their devices adequately, hence, the development of adequate quality screen control fell upon the large systems users of memories. These users contributed heavily to forcing manufacturers to the high level of quality found today in MOS memories. The early recognition by the Japanese semiconductor suppliers of the need for high quality, probably stemming from the close links with their own internal systems users, also had a significant impact on the world semiconductor quality.

A revolution of sorts occurred in the reliability area in the 1980s. Quality control in the manufacturing area to a large extent replaced the extensive screening of former years. Memory quality and reliability today is so high that most systems manufacturers do not even perform reliability monitors on MOS memories incoming to their factories. Even in the qualification stage they increasingly depend on the manufacturer's information. Modern efforts in reliability are traced in Chapter 14.

2.14 THE SUPER-FACTORY AND CLUSTERED CLEAN EQUIPMENT

The high quality and reliability of the MOS memories has historically depended on the high technology factories in which they are manufactured. These huge expensive facilities are vibration free, dust free, partially automated, and capable today of producing microscopic circuits with over 16 million transistors etched on a surface the size of a postage stamp. By the year 2000 such a factory will cost close to one billion dollars. A photograph of a 'clean room' in a submicron MOS factory is shown in Figure 2.18.

The high cost of such a factory and the increasingly stringent clean room requirements mean that attempts are being made to move to smaller factories or to move the clean environment into the manufacturing equipment. Clusters of equipment which automatically perform a series of process steps have been developed.

At the same time efforts are being made to make the wafers less sensitive to the environment by implementing redundant circuits on the chip, and better understanding of the dependance of yield on the clean room environment is being gained. These trends are traced in Chapter 15.

2.15 MEMORY APPLICATIONS

Two of the biggest factors affecting semiconductor memory technology in recent years have been the migration of memory devices into semi-specialized applications areas, and the coming of age of CMOS.

High speed system applications have increased in importance with the advent of the supercomputer, and the high speed 32 bit microprocessors have driven MOS memory speeds down into the range previously dominated by bipolar while also demanding high densities.

Figure 2.18
Clean room in a submicron MOS wafer fabrication facility. (Source N.V. Philips Research.)

Fast static RAMs have appeared as cache and TAG RAMs for control store, which is designed to match the processor speed to the main memory speed and also improve the apparent speed of the main memory. The fastest low density MOS static RAMs have migrated into this area.

DRAMs have moved to wider organizations, and to specialized configurations to enhance serial data rate such as nibble mode, page mode and static column. DRAMs with on-chip refresh to simulate static RAMs at a fraction of the cost have appeared, as have SRAMs with batteries in the package to compete with the EEPROMs in non-volatile applications.

As long as the memory enhances the systems' overall performance, the increased cost of the applications specific design is saleable.

Graphics applications have dominated the scene in requiring applications specific memory. Personal computers, mainframes, and above all the workstations, required advanced graphics systems with specialized memories. The video memories are also in demand for high volume consumer applications such as high definition, or special feature television. Dual-port DRAMs with on-chip serial shift registers accessible from a second port have been developed and standardized for graphics and television applications. It is estimated that graphics applications alone will consume from one quarter to one half of all DRAM production in the 1990s.

These applications have been enhanced by the trend toward CMOS technology. While CMOS is slightly more complex to manufacture than NMOS, requiring more masks, the logic circuits implemented in CMOS are simpler, and use fewer transistors than their NMOS equivalents. Also, memory circuits in CMOS require significantly fewer and simpler peripheral circuits than in NMOS.

CMOS also permitted the development of high density chips that consume small amounts of power at low operating temperatures. CMOS speeds now rival those of bipolar circuits, while still maintaining a power consumption and temperature advantage. CMOS circuits also are relatively immune to soft errors and preserve noise immunity as geometries are scaled to submicron dimensions.

From the applications standpoint, CMOS memory technology opens the world of portable and battery powered systems. For further discussion of memory applications see Chapter 3.

2.16 EMBEDDED MEMORIES

Semiconductor memory storage exists not only as stand-alone memory devices, but also embedded in processor chips. The performance of an embedded memory can be better since bandwidth problems are reduced and interface circuitry and package leads are eliminated. It can also have characteristics tailored to the specific application rather than being a standardized compromise between many factors such as power consumption and speed.

The limiting factor for an embedded memory is cost. Since the manufacturing yield of a single silicon chip decreases with area, the cost of a chip increases with area. A very large processor chip with embedded memory is likely to have higher cost than the combined costs of a smaller processor chip and a smaller memory device made separately. This is particularly true if the memory device is bought from a volume memory supplier.

A separate memory also gives the systems designer more flexibility to upgrade or change both the memory and the logic.

For these reasons, the memory embedded in processor chips has tended to remain relatively small and tightly tied to the function of the processor. The percentage of memory embedded on processors, however, has increased with time.

2.17 LARGE SCALE INTEGRATION OF MEMORIES

The history of large scale integration has been one of difficulties and disappointments. It is only recently that some of the problems have begun to be overcome for special applications. Considerable development remains to be done.

Large scale integration of memories has been attempted for several reasons.

One reason is identical to a reason for embedding memories—performance. Very fast systems require the interconnects between memories and between memories and processors to be as short as possible. Very large silicon integrated circuits, however, tend to have defects which must be wired around. For this reason high density packaging of pretested chips currently provides the shortest interconnects.

Another reason for attempting large scale integration of memories is to reduce the size of the memory system. Again because of the presence of defective areas in the silicon, the highest level of integration currently can be provided by dense packaging techniques.

An exception is serial wafer integration, which is being used in one case to integrate sufficient memory in a small enough area to be competitive with magnetic disk technology in standard disk format package.

Further discussion of high level integration in memories can be found in Chapter 13.

2.18 WHERE TO FROM HERE

All of the trends traced here are explored further in the following chapters of this book.

The future is just hinted at in the last chapter and as always remains to surprise and challenge us.

BIBLIOGRAPHY

[1] Altman, l. (1978) *Memory Design: Microcomputers to mainframes,* Electronic Book Series, McGraw-Hill, New York.

[2] *Scientific American,* Special Issue on Microelectronics, September 1977.

[3] *Perspectives on Experience.* The Boston Consulting Group, Inc.

[4] Clifford, Jr, D. K. (1980) Managing the product life cycle. In: P. Kotler and K. Cox (Eds) *Marketing Management and Strategy, A Reader,* Prentice-Hall, Englewood Cliffs, N.J.

[5] JEDEC Standard 21C (1990) Electronic Industries Association, Washington, D.C.

[6] Chan, Y.H. *et al.* (1988) A 3ns 32k Bipolar RAM, *ISSCC86 Proceedings,* p210.

[7] Chuang, Ching-te *et al.* (1988) A subnanosecond 5k bit Bipolar ECL RAM, *IEEE Journal of Solid State Circuits,* **SC-23**, No. 5, October 1988, p1265.

[8] Ogiue, Katsumi *et al.* (1986) A 13ns 500mW 64k bit ECL RAM, *ISSCC86 Proceedings,* p212.

[9] Murphy, M. and Morgenthaler, L. (1987) Why chips will soon kill off disk drives, *Electronic Business,* November 1, 49.

[10] Gamou, Y. *et al.,* (1980) All TTL Compatible CCD Memory with CCD Clock Generator, *IEEE Journal of Solid State Circuits,* **SC-15**, No. 5, Oct. 1980, 881.

[11] Weste, N.H.E. and Eshraghian, K. (1985) *Principles of CMOS VLSI Design,* Addison-Wesley, Reading, MA, p 6.

[12] Prince, B. (1989) Life Cycles of High Technology Products, Invited paper in Plenary Session of ICCD, November 1989.

[13] Prince, B. and Due-Gundersen, G. (1983) *Semiconductor Memories,* John Wiley, Chichester.

[14] *MOS Memory Databook, Hitachi 1988/1989.*

[15] World Market Forecasts, *Electronics Magazine,* 1981–1989.

[16] *Electronics,* June 1988, p. 67.

3 TRENDS IN MEMORY APPLICATIONS

3.1 VARIETIES OF DATA STORAGE DEVICES BY MEDIA

Over the years many different media have been tried for the basic data storage function as shown in Figure 3.1. Logical bits have been stored in mechanical devices such as paper tape and cards, in magnetic moving media such as bubbles, hard disks, floppy disks and core, in optical moving media such as magneto–optical disk and holographic devices, and in solid state media which include semiconductors and ferroelectrics. Semiconductors divide by technology into MOS, bipolar and charge coupled devices, and by function into RAM, ROM, and SAM. MOS and bipolar are available in all three functions, while charge coupled devices are available only as SAM.

3.2 FUNCTIONAL CHARACTERISTICS OF VARIOUS SEMICONDUCTOR MEMORIES

The basic differences in function in semiconductor memories have come about as a result of specific requirements of the differing systems applications in which they are used. No single set of memory characteristics is optimal for all systems, nor has any single memory yet been made which has all of the optimum characteristics.

Functional characteristics that are significant for the systems environment include performance (speed), power dissipation (heat), memory density (number of storage bits per chip), chip size (memory cost), size of package (system cost), external organization of the memory, reprogrammability (endurance of the memory to repeated write–erase cycles), long term reliability characteristics, ability to retain data when the power is off (volatility), length of time the data is retained when dc power is on without an active refresh of the data (data retention), interface voltage levels into the system (TTL, ECL, CMOS), optimal power supply voltage levels, moisture resistance (hermeticity) of the package, and the amount of logic integrated on the memory rather than used separately in the system.

There are several major product types of semiconductor memories differing by various sets of these characteristics. This was determined historically by the basic

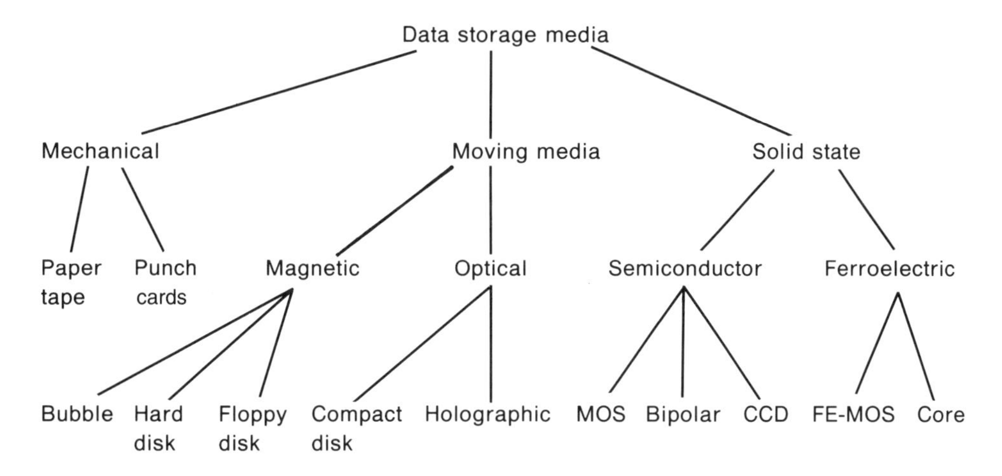

Figure 3.1
Various catagories of data storage devices.

Characteristic	DRAM	SRAM	EPROM	ROM	EEPROM	NVRAM
Number of devices in cell	1.5	4–6	1.5	1.0	2.5	8–9
Relative cell size	1.5	4–6	1.5	1.0	3–4	9–10
Density (1990)	4Mb	1Mb	4Mb	4Mb	1Mb	16k
Overhead cost	yes refresh logic	no	yes UV erase programmer	yes mask charges	no	no
Volatile (power off)	yes	yes	no	no	no	no
Data retention (d.c. power on)	4 ms	∞	10 years	∞	10 years	10 years
In system reprogrammable	yes	yes	no	no	yes	yes
Number of reprogram times (endurance)	∞	∞	100 PROM(1)	0	10 000 –1000 000	10 000 ∞ (write)
Typical write (Reprogram) speed	100 ns	25 ns	30 min	—	2.5 s	—
Typical read speed (ns)	100	25	100	100	200	200
Number of read cycles	∞	∞	∞	∞	∞	∞

Figure 3.2
Characteristics of MOS memory product types.

memory cell. Semiconductor memories are split by major cell type into DRAMs, SRAMs, SAMs, EPROMs, EEPROMs, and ROMs. General characteristics of the basic MOS memory types are shown in Figure 3.2. Minor variations include the NVRAM and the CAM.

The structure of the memory array and periphery determines the input and output organization. The logic in the periphery also determines functional subcategories of memories such as video and multi-port DRAMs, synchronous and asynchronous SRAMs.

A further subset is defined by the Input and Output (I/O) levels and configuration of the memory. I/O levels can be TTL, ECL, or CMOS. Inputs, outputs, and addresses can be multiplexed or non-multiplexed, serial or parallel.

3.2.1 MOS memory selection by system requirement

Different systems require different MOS memory characteristics. For example, a memory system requiring an infinitely reprogrammable memory would use an SRAM or a DRAM. If it is a large memory system with commercial cost pressure then it will use the highest density, lowest cost option—the DRAM. If it is a large very fast system requiring a random access memory, then it will either use SRAM for the main memory or have a small SRAM cache which can make a lower cost but slower DRAM memory appear faster in the system. If it is important that the main memory not lose its data when the power fails, there are again several options. Either SRAM can be used for the main memory due to its low standby power dissipation and backed up by a battery, or the main memory can be backed up by an EEPROM.

As another example a system which needs a small quantity of fast infinitely writable non-volatile memory will use an NVRAM, bearing in mind that its non-volatile reprogrammability is the same as the EEPROM—about 10 000 reprogram cycles.

In general DRAMs and EPROMs tend to be optimized for high density and low cost and used in large systems where many parts need to be assembled in a small space. The large quantity of parts used mean that the per-unit overhead costs are low.

DRAMs have the lowest cost of any RAM and the highest density because they have the smallest memory cell, consisting of one transistor and a capacitor. The capacitor stores the charge when the memory is in the '1' state. The capacitor occupies a large percentage of the room in the memory cell. In high density DRAMs it is usually either stacked on top of the transistor or lowered into a narrow trench in the silicon. A DRAM, therefore, effectively has a one plus transistor cell area.

The penalty that is paid for this small cell size and consequent high density is the need for the memory system to provide a periodic refresh to the cell to restore the charge to the capacitor as it leaks away. During the time that the memory cell is being 'refreshed' no other operation can take place in that area of the array. This 'dead time' slows down the effective speed of the DRAM in the system. While the percentage of the operating time that is occupied with refresh is decreasing as higher quality capacitors are constructed, refresh is still a significant factor in the timing of the system.

All of the capacitors in the memory array must be periodically refreshed. This is done by accessing each row in the array, one row at a time. When a row is accessed, it is turned on and voltage is applied to the row. This recharges each capacitor on the row to its initial value. For example, on a 1Mb DRAM refresh can be performed as either a single burst of 512 consecutive refresh cycles (one cycle per row) every 8 ms, or it can be distributed over time, that is, one 200 ns refresh cycle every 15.6 μs (8 ms per 512 rows = 15.6 μs per row). A combination of these two methods is also permitted as long as each row is refreshed at least every 8 ms. Details of refreshing the DRAM will be considered in Chapter 6. The requirement for refresh does, however, add to the complexity of the system and hence to the overhead cost of a system using DRAMs.

EPROMs also have an overhead cost. They are erasable, but must be removed from the system and placed in an ultraviolet light which erases the memory through a transparent window in the EPROM package. They must then be reprogrammed in a separate programmer before being returned to the system. This makes EPROMs cumbersome to use.

SRAMs and EEPROMs, on the other hand, are easy to use from the system sophistication viewpoint and do not have the overhead costs, but are more expensive per unit. In a small system where the cost of system overhead is a relatively higher part of the system cost, the total cost of the SRAMs and EEPROMs moves closer to that of the DRAMs and EPROMs, and the determining factor becomes the ease of use.

Where in-system reprogrammability is required, neither EPROMs, nor ROMs can be used: however systems requiring data to be retained when the power goes off must use one of the non-volatile memories—ROMs, EPROMs, EEPROMs, non-volatile RAMs, or CMOS SRAMs with battery back-up. If the high performance of the semiconductor memory is not required these applications can also use one of the slower non-volatile moving media or mechanical memory types.

ROMs and EPROMs are used primarily to hold software which will not be changed. In lower densities and speeds, they were used mainly for holding setup and configuration data for peripheral or bootstrap programs. As higher speeds and densities appeared they were used to hold entire operating systems, real-time kernels and canned applications programs, and in consumer toys and games.

Battery backed up SRAMs (BRAMs) are used for applications in peripherals requiring high speed random access combined with non-volatility. The parts, developed first by Mostek, contain lithium batteries in a compartment of the SRAM package. They also contain various types of voltage sensing circuits that switch over to the batteries when external power is removed.

With relatively small numbers of setup parameters needing to be sorted and changed only occasionally, non-volatile SRAMs, which are also called shadow RAMs, can be used. These combine RAM and EEPROM on one chip. BRAMs can also be used. Shadow RAMs (developed by Xicor) combine a normal static RAM array with an EEPROM array. When the chip is powered up, the contents of the EEPROM are written into the RAM, providing default configuration parameters. These parameters can be changed while the system is on, and the EEPROM can be reprogrammed to change the default setting.

Other characteristics are common choices to all the memory types but differ in various system applications, such as organization of the memory, common or separate inputs and outputs, and whether the memory is synchronous or asynchronous. Depth of memory required in the system affects the required organization of the memory and also the width of memory word, or number of inputs and outputs. Considerations of bus width and error correction are often important determiners of desired memory width. Whether a memory is required to be synchronous or can be asynchronous can be determined by the presence or absence of a system clock, the desired speed of the system, and the required retention of the last accessed data while new data are incoming.

3.3 MEMORY USAGE IN COMPUTER APPLICATIONS

To see how differing systems applications require these differing characteristics, a few typical systems using memories from various end system market segments are considered now.

Figure 3.3 divides the total memory usage for DRAMs and SRAMs in major applications segments for 1988 and forecast for 1992.

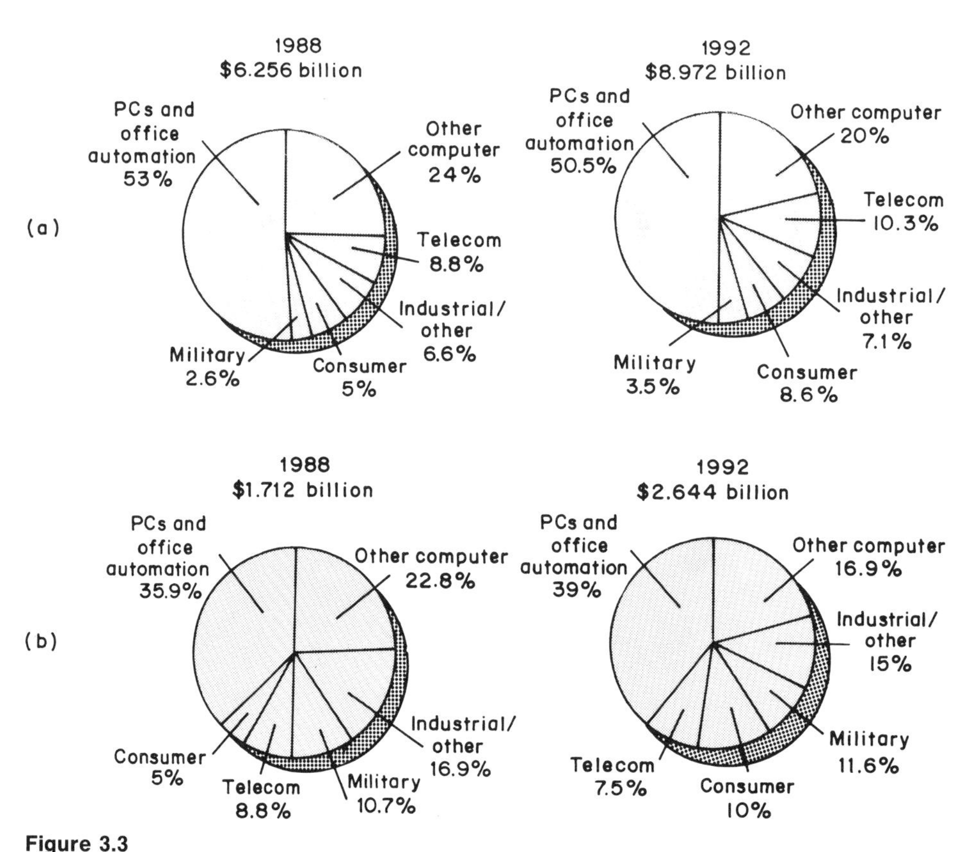

Figure 3.3
(a) DRAM and (b) SRAM usage by user market segment for 1988 and 1992. (From HTE Research Inc.)

Computer application is both the largest of the electronic systems market and also uses the highest average value of memory per system. As a result 73% of all memories go into some type of computer system. These include mainframes, supercomputers, parallel processing systems and graphics workstations.

3.4 COMPUTER STORAGE HIERARCHIES

Throughout the 1980s there was an increasing mismatch between the speed of the DRAMs, which are used for main memory, and the speed of the microprocessors, used as computing elements in all but the largest computers. Figure 3.4 illustrates this for the RISC (reduced instruction set computers), CISC (complex instruction set computer), MPU (microprocessor units), and for four generations of DRAM. Fast SRAM speeds increased at the same rate as that of the processors. A hierarchy of storage layers divided by speed developed.

A typical computer storage hierarchy is shown in simplified form in Figure 3.5. Indications are given of memory types frequently used at each level. Memory is organized in several layers of decreasing speed going from fast cache to slow archival memory. Progress in mass storage technology continues to add layers to the traditional system storage hierarchy.

If infinite amounts of low cost, random access, non-volatile memory bits were available at the speed required by the processors with no wait states, there would be little reason to use differing types of memory. Since this is not the case, memory hierarchies are used to provide a reasonable facsimile of the ideal memory in the system.

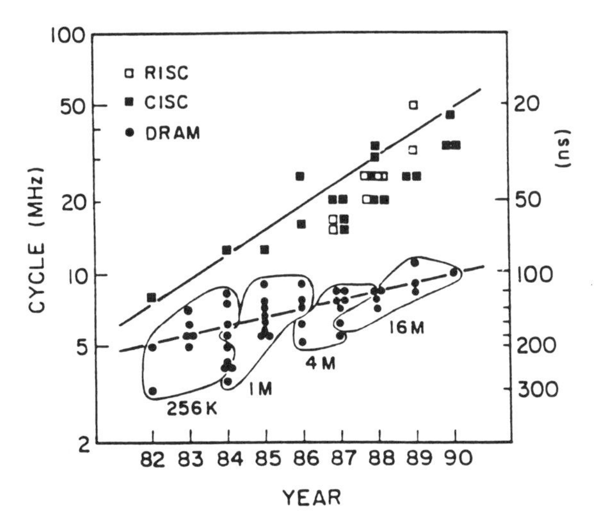

Figure 3.4
Speed trends of MPUs and DRAMs. (From Kushiyama [29], Toshiba 1991, with permission of IEEE.)

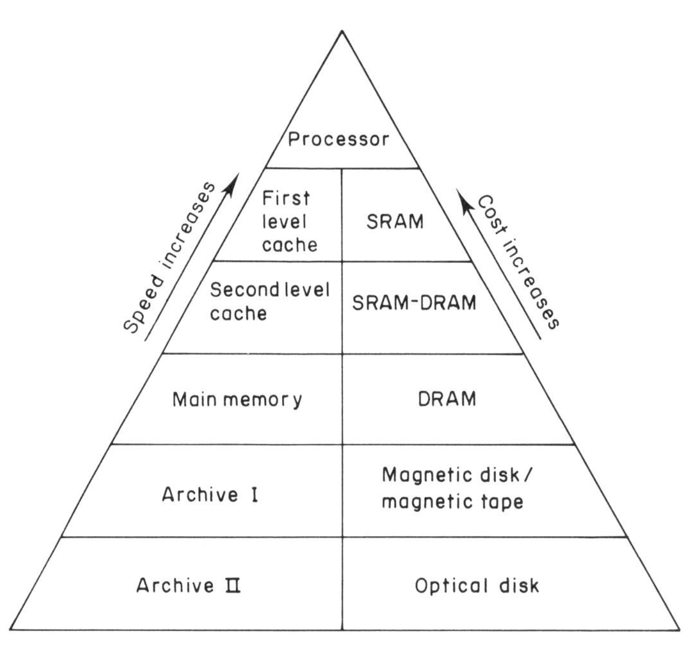

Figure 3.5
Typical computer system data storage hierarchy.

3.4.1 Main memory

The storage hierarchy exists to support the main memory. The storage device used most frequently as main memory in large computers is the dynamic RAM. This is because the DRAM is optimized for low cost and high density, which are the major needs when fitting very large quantities of memories into a system. What is traded off is a higher cost in system overhead needed to refresh these dynamic RAMs, volatility, and a speed penalty.

When it is necessary to interface a fast processor with a slower memory, the processor is slowed down since its cycle time is faster than the memory address access time. One of the ways to handle this interface situation is to use a cache to decrease the effective time to access the main memory.

3.4.2 Overview of cache

Cache memories are buffers that hold selected information from larger and usually slower main memory so that processors can reach this information more rapidly. The cache is based on the principle that certain information has a higher probability at any given time of being selected next by the processor than other information. If these higher probability data are stored in a cache with a faster access time then the average speed of access of the data will be increased.

There is a trade off between memory performance, size, and cost, which encourages system designers to find ways to use memory more efficiently. System designers have historically had two choices available in RAM memory devices: fast and expensive SRAMs or slow and cheap DRAMs. While SRAMs with 25–35 ns access times can keep pace with a 10 MHz processor, a DRAM with even 80 ns access time will force this same processor to wait. Given the large size requirement in current memory arrays, the importance of finding an optimum trade-off between size and performance is, however, of increased importance.

Caching schemes are common which mix a large amount of slower less expensive DRAM with a small amount of faster more expensive SRAM. A variety of such schemes can be constructed to allow the processor to operate at maximum speed, and are a way to ensure that the data a processor needs will be available when the processor needs it without having to wait for the data to be read from their normal location in main DRAM memory.

In general most caches are designed on the assumption that a processor is likely in the near future to need the information it is currently working on, plus other data which are stored near it in the main memory. Cache performance depends on two factors: access time and hit ratio.

The hit ratio is defined as the probability of finding the needed information in the cache, which is a hit, divided by the probability of having to go to the main memory, which is a miss. This hit ratio, which is the ratio of the number of hits divided by the number of tries, is dependent on the size of the cache and on the number of bytes that are brought in on any given fetch from main memory.

The access time for the cache is the mean time it takes to access the cache on a hit. The access time for the main memory must include the initial attempt to find the data in the cache. The overall performance of the memory system will then be a function of the cache access time, the cache hit ratio and the access time of the main memory for references that are not found in the cache.

There are two general types of caches: a direct mapped cache and an associative or tag cache.

3.4.3 Direct mapped cache

The contents of a direct mapped cache correspond exactly to the data that were obtained from main memory and are addressed by the same addressing scheme as the main memory. The latest data obtained are normally retained by the cache along with adjacent data from the main memory. The least used data in cache at the time are replaced by the latest data.

Direct mapped caches are used primarily with smaller main memory having direct address mapping schemes. Most direct mapped caches consist of small amounts of SRAMs embedded in microprocessors.

3.4.4 Associative cache

An associative cache, also called a cache tag comparator, is more useful with large memory systems which include large main memory and perhaps even lower levels of archival storage.

Frequently these large memory systems are organized as demand paged virtual memory in which memory is organized in pages of a fixed size that have a given virtual address. Pages beyond the range of existing physical RAM can be stored in large, slower archival memory, usually disk, and swapped into physical RAM when needed and where space is freed by another page being swapped back to disk. (In this case the main memory also acts as a cache for the disk.) When pages reside in physical memory, they also have a physical address which is different from their virtual address.

There is a trade-off in page size between the frequency with which one must swap pages from the disk and the time it takes to make such a swap. A memory management unit (MMU) is used to translate between virtual and physical addresses. Addresses are organized in terms of a page address, called a 'TAG', and an 'offset' which locates the data within that page.

An associative cache, upon receiving an address from the processor, associates or compares the TAG with the page address to see if that page address is in cache. If so, the offset address is checked and if there is a 'hit', that is, if the data are in the cache, they are immediately transferred from the cache to the processor.

An example of a single set associative cache is shown in Figure 3.6. This cache includes a TAG indicating the page address of data stored in the cache, the offset address which locates the data within the page, and the actual data. When a request is received from the processor, the TAG field in the address requested is compared to the TAG field of data held in the cache. If the TAG field is valid and the page offset is present then the data in the cache are returned to the processor.

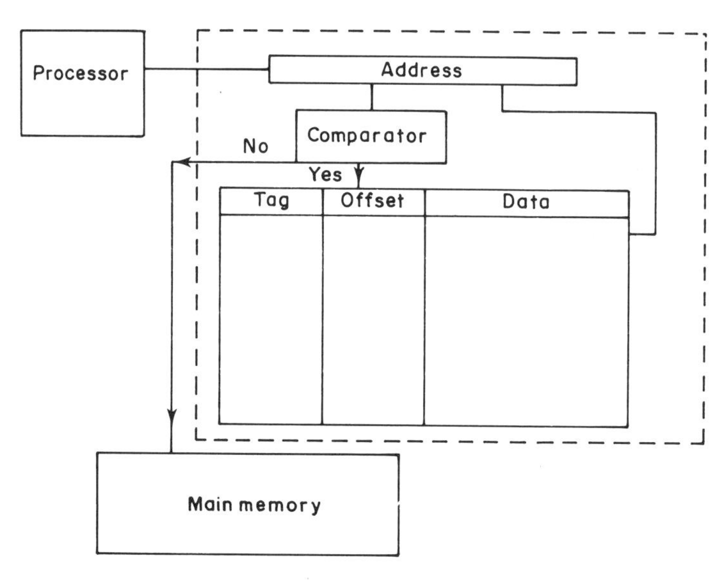

Figure 3.6
Single set associative cache.

A two-way set associate cache would allow the same offset from two different pages to be cached. A fully associative cache would allow any number of pages to be cached. A fully associative cache is, however, more complex and hence more expensive than a simpler cache.

Cache hierarchies historically included a processor, main memory, and a cache to buffer the speed of the main memory for input to the processor. Now cache hierarchies can be multilevel with fast processor, cache 1, virtual main memory, cache 2, and archive memory. Speeds range down from a few tens of nanoseconds for the first-level cache to a few milliseconds for the archive.

3.4.5 Application-specific cache chips

A cache can be constructed of logic and RAM chips or it can be bought as an SRAM with the logic integrated on the chip. A number of applications specific SRAMs are available as caches for memory systems, although there is always a trade-off for the system designer between using a more expensive cache chip, or buying less expensive standard SRAM and implementing the cache logic off-chip.

Many levels of sophistication are possible in cache design, usually determined by the speed requirements and cost constraints of the system.

3.4.6 Dual-port SRAMs as cache buffers

An ideal cache could be accessed simultaneously by both a processor and another processor or a peripheral. This could result in bus contention unless the cache had two ports and the system had two independent buses.

Even with two ports, an ideal cache buffer would need to be able to handle both external processors accessing the same address simultaneously without resulting in indeterminate data in the cache if one processor was attempting to read when another was writing to the same location. In reality, this situation is frequently handled by having a wait state for one of the processors or by allowing the possibility of having information incorrect. Contention arbitration logic can be provided to eliminate overlapping operations to the same memory location. It can either allow contention to be ignored and both operations to proceed, or it can arbitrate and delay one port until the other port's operation is completed. When this occurs, a busy flag is sent to the side delayed. A priority port can be specified. Another problem that arises is having different width buses on the processor and the peripherals side of the dual port.

3.4.7 Interleaving memory to gain system speed

An additional option for the fast processor and slow memory combination is interleaving the memory to increase the data rate. The most basic type of interleaving scheme is the two-way interleave in which odd addresses are stored in one memory array and even addresses are stored in another. The first access will be at the speed of the memory array, but consecutive accesses from separate blocks will be at about twice the speed of the memory arrays.

This works best in the case of consecutive addressing from the processor where there is a high level of interleaving. This scheme works since consecutive fetches are directed at alternate memory banks so there is a time penalty only on the first access. For example, two 100 ns memory arrays could be interleaved to give an average access time close to 50 ns, rather than using a more expensive 50 ns memory.

The simple two-way interleaving scheme totally fails if, for example, complex numbers are being used where the real and imaginary parts are accessed consecutively but stored separately. Since each address access involves both memory arrays, the time advantage of alternating slower arrays disappears, leading to a failure of the interleaving scheme. A four-way interleave solves this problem. The interleaving scheme helps to speed up both instruction and data fetches from memory.

Caches and interleaving can also be combined, for example, to speed up instruction fetches in time to be executed when needed. Ideally, the fetching would occur on machine cycles where no other task is using main memory. The cache memory acts as a buffer between the main memory and the control unit, greatly reducing the memory bandwidth consumed by instruction fetches.

3.4.8 Archive or mass storage

In addition to main memory and cache, system requirements for more memory and for higher capacity peripherals have resulted in the use of a variety of storage devices whose main requirements are very high density, very low cost, and non-volatility. These include magnetic disk, floppy disk, optical disk, and magnetic tape. Optical disks are available in three forms which include optical read-only-memory (ROM), optical write-once-read-many (WORM), which is a PROM, and most recently optical multiwrite, which is a form of EEPROM.

Reel-to-reel has mostly been replaced by tape cartridges. Hard disks are not being displaced because their capacities continue to increase and their size to shrink. For example, the $5\frac{1}{4}$ inch hard disks have increased from the 10 to 20Mbyte capacities common in the mid 1980s to 200Mbyte and higher in 1989, while 14 inch hard disks now have capacities in the gigabyte range.

Interface standards for optimizing the speed and transparency of links between computers and peripherals have developed, such as SCSI (Small Computer System Interface) which is used primarily for small systems and IPI which was developed for larger systems. These interface standards allow designers to optimize the storage capacities of their devices.

To accommodate these additional options the system I/O interface must become more flexible and supply faster data transfers.

Solid state memory is still too expensive to be a major contender for most mass storage applications. It is, however, used in special applications whose reliability requirements preclude the use of mechanical storage systems. Since the cost per bit of MOS memory continues to drop faster than that of the older media, an increasing number of applications could potentially be filled with MOS memory in the future. The low cost DRAM and flash memories are the main contenders among the MOS memories for a share of the archive memory market.

3.5 SUPERCOMPUTERS AND PARALLEL PROCESSORS

Performance in today's supercomputers is determined in part by the high performance communications networks connecting the various processors in the total computer system. Supercomputers run in parallel processing type clusters linked by these high speed networks.

Applications where parallel processing makes sense include those where a massive amount of data must be processed, for example, image processing. The quantity of information represented by a typical digital image means that the total processing time is likely to be minutes or hours for a single image when a single micro- or miniprocessor is used. The use of parallel processing can significantly reduce the time for such computations.

Parallel processors also require parallel input–output capabilities and mass storage arrays that bring enough I/O and disk performance to match the required levels of computational performance.

Since speed is critical in these super fast systems, circuit elements need to be as close as possible to shorten the length of interconnects. Some supercomputer systems use high density cassettes of SRAM memory chips without packages to create the high density memory arrays needed. Other systems embed the data storage and data acquisition subsystems modules in the array of processors. In either case significantly enhanced computational power results so that complex processing of analog data becomes possible at very high speeds. Image and signal processing are among the possible applications.

3.6 MEMORIES FOR VIDEO PROCESSING AND GRAPHICS WORKSTATIONS

The advent of the microprocessor in the 1980s increased the demand for memories in small systems such as personal computers, word processors, computer terminals, work stations and CAD–CAM systems. The applications of RAMs in small systems can largely be divided into main memory and display memory.

The ratio of number of RAMs used in display memories for storing character data and processing graphic data to the number used in the main memory is much larger in small systems than in mainframe systems, and in fact in small systems the number of display memories is on a par with the number of memories used in the main memory of the small system.

The size of display memories per system ranges from several tens of kilobytes to as many as several megabytes. It is expected that during the 1990s a significant percentage of DRAMs will be used for display purposes. Display memories use standard DRAMs integrated into the system with stand-alone logic chips and also include several applications specific DRAMs such as frame buffers and video DRAMs for bit-mapped displays.

The basic function of a frame buffer is to send display data to the video display monitor at a designated speed. The performance level of a system is determined by the efficiency with which the processor can refresh the display data while continuing this basic function. A frame buffer works best with two access ports. One port intakes new data from the CPU and the other sends display data to the CRT.

There are two ways to use a standard single port DRAM in this application. One is to permit the CPU access to the DRAM only during the blanking phase of the display as illustrated in Figure 3.7(a). CPU access during the data display phase is restricted to sending display data to the CRT. This restriction, however, reduces the access efficiency of the processor considerably.

The other method is the cycle steal method as illustrated in Figure 3.7(b) where the timing intervals are minutely divided to allow the processor access to the memory during the display phase. This requires complex high speed peripheral circuits although it improves the operational efficiency of the processor.

DRAMs with special modes giving fast or serial access can be configured in such an application. Operating modes for high serial access speed include static column, nibble mode and page mode access. These modes are methods for reading and writing successive blocks of data at a faster rate than the normal per bit access time. They are useful in applications that require high speed serial access such as video bit map graphics monitors or RAM disks.

A comparison of typical operating characteristics for these modes is shown in Figure 3.8. It is clear that the nibble mode is the fastest, at less than half the access time of random access mode. Nibble mode works by allowing serial access of up to four bits of data by cycling one of the DRAM clocks (CAS) while holding the other (RAS) active. The first access is at the normal random access time, but the three sequential reads are at the faster nibble access time. This mode can be used to give a sustained higher serial data rate by interleaving DRAMs so that effectively only the higher data rate accesses appear on the system bus. Additional complexity of the system then is the cost of higher speed.

Page mode allows access to any of the column locations on a specific row and gives an access time which is typically $\frac{1}{3}$ of the random access time. Static column is similar to page mode and offers a speed typically $\frac{1}{2}$ of the random access time.

Although page mode is faster than static column mode, it is more difficult to interface in a system due to extra column address strobe pulses that are required in page mode.

Static column generates less noise than page mode, because output buffers and chip select are always active in this mode. Noise transients, generated every time the column address strobe is cycled, are thus eliminated in static column mode so trade-offs for the higher speed of page mode are additional complexity and a higher level of system noise.

These special DRAM modes were originally developed with the graphics needs of workstations and computers with the ability for extended graphics processing

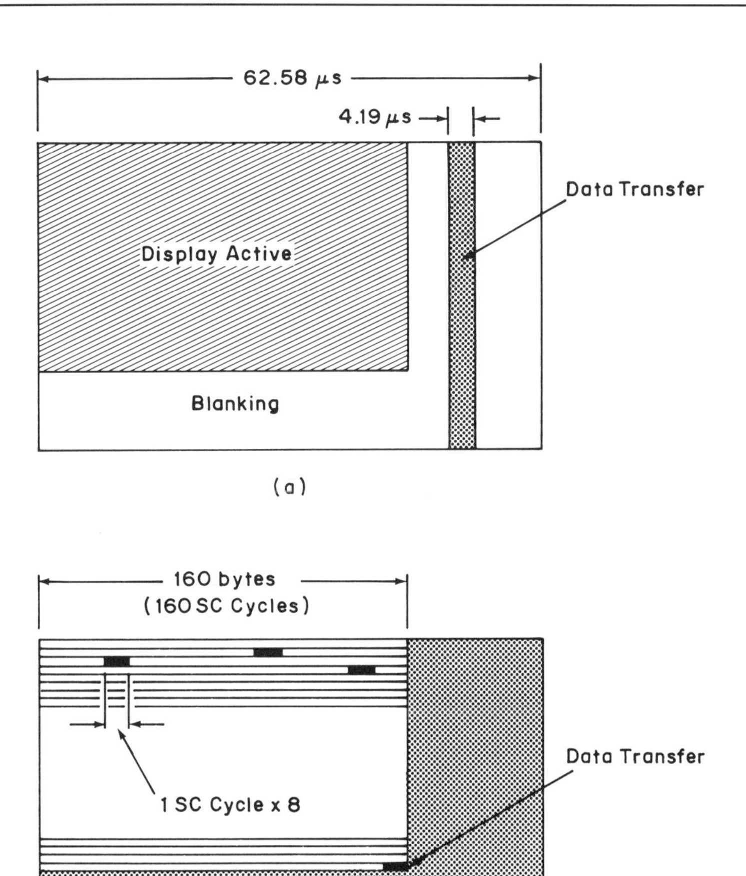

Figure 3.7
Methods for using a single-port DRAM as a display memory. (a) CPU access to DRAM during the blanking period. Number of data transfer = 260, CPU working efficiency = 100 × [1 − 4.19/62.58] = 93.03%. (b) Cycle steal method for CPU access during display phase. Number of data transfer = 125, CPU working efficiency = 100 × [1 − (279.3 × 8 × 125)/(62.58 × 1000 × 260) = 98.2%. (From NEC Databook 1986 [19].)

capabilities in mind and are not really optimized for simple frame buffer applications.

A further solution developed for those graphics applications requiring significant amounts of image processing capability is multi-port DRAMs, also called video DRAMs, with on-chip shift registers where one port is for random access and the other port outputs data at a fast serial access rate which is several times the random access port data rate. Data can be written into the RAM via the random access port at the same time they are being shifted out of the serial port. The array is unavailable for updating only during the loading of the shift registers.

Parameter		Page	Nibble	Static Column	Random
Access time (ns)*	t_{CAC}	25	—	—	—
	t_{NCAC}	—	20	—	—
	t_{AA}	—	—	45	—
	t_{RAC}	—	—	—	85
Cycle time (ns)*	t_{PC}	50	—	—	—
	t_{NC}	—	40	—	—
	t_{SC}	—	—	50	—
	t_{RC}	—	—	—	165
Accessible bits		1024	4	1024	All
Order of accessible bits		Random	Fixed	Random	Random
Conditions	$\overline{RAC}$	Active	Active	Active	Cycle
	$\overline{CAS}$ or $\overline{CS}$**	Cycle	Cycle	Active	Cycle
	Addresses	Cycle	N/A	Cycle	Cycle
	Outputs	Cycle	Cycle	Active	Cycle
Time to read 4 bit (ns)*		235	205	235	660
Time to read 1024 unique bits (ns)*		51 235	70 400	51 235	168 960

* Values for a 1M × 1 85-ns device.
** CS on Static Column.

Figure 3.8
Operating characteristic comparison for 1 Mb DRAM operated in various modes. (From Motorola Memory Databook 1988 [23].)

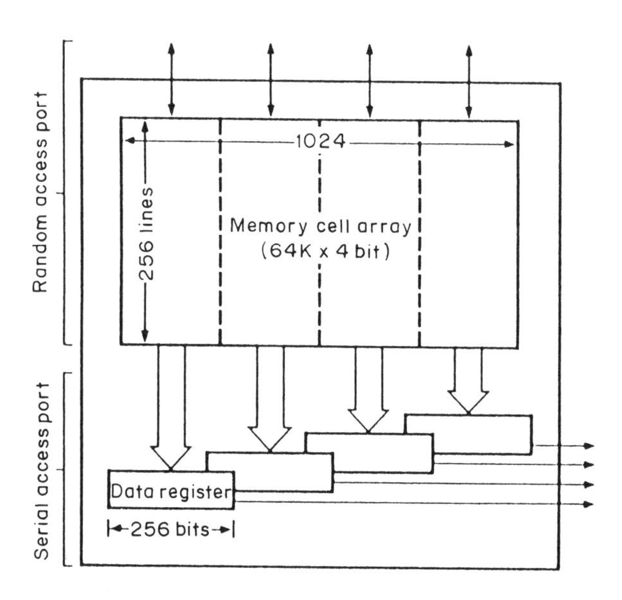

Figure 3.9
Basic configuration of a dual-port DRAM. (From NEC Memory Databook 1986 [19].)

A VDRAM solves many of the frame buffer speed problems. The serial port can allow serial output at a high rate, and can be used as a dedicated port for sending display data to the CRT. Several such DRAMs can be multiplexed to increase the data rate manyfold. For example, if one VDRAM has a serial output rate of 25 megabits per second, then four of them with a parallel serial converter attached can send display data at 100 megabits per second. Meanwhile the processor can access the VDRAM through the random port with nearly 100% efficiency.

A basic configuration for a VDRAM is shown in Figure 3.9. This standard 1 Mb multi-port Video DRAM is a high speed 256k × 4 DRAM with on-chip registers. The random port sees the RAM organized as 256k words of 4 bits each. The serial port is interfaced to four internal 256 bit data registers which make the memory appear organized as 256 four-bit words that are accessed serially.

A consideration in using a VDRAM in place of a standard single port RAM is the increase in chip size, and hence cost, due to the serial registers. This in turn translates into increased device and system cost. Clearly, high performance graphics systems can afford the higher cost better than low performance systems or consumer systems such as high performance television.

The low cost consumer oriented television applications tend to use another applications specific DRAM called a frame buffer which adds only the logic for making the input and output serial but does not offer the full random access port capability of the VDRAM.

The charge-coupled memory device (CCD) which has a high speed serial access mode could also be used in this application. The historical belief in the slow access of the CCD related to comparisons with random access capability. In serial access mode the CCD is in theory competitive with the DRAMs in both speed and cost.

Systems using high performance graphics memories such as the VDRAM include the various advanced graphics workstations.

The highest performance graphics workstations, called super graphics workstations, are implemented with multiple processors combined in a network so that the effective speed is much higher than the basic speed of the memory or processor. These systems range from the advanced graphics workstations, in which 5% to 10% of the content is memory, to the supergraphics workstations with features like animation (the Hewlett Packard Animation super workstation for example) and run from 15% to 20% of the cost in memory content.

A supergraphics workstation development project, backed by the Japanese government, performs interactive three-dimensional graphics using a hypercube design in which multiple processors are linked in multidimensional cube arrangements with direct connections between neighboring nodes. An n-dimensional cube has 2^n nodes, so for example, a three-dimensional cube has $2^3 = 8$ nodes. Each processor board node contains many megabytes of memory and can run serial transfer rates over 100 megabits per second.

Along with high performance graphics, another memory intensive computer system development is high performance character recognition systems, at least one of which is based on neural network technology that simulates the organization of neurons in the human brain and has the potential of improved accuracy over conventional techniques. The technique used is a repeated character learning method called back-propagation learning algorithm where the neural network checks characters and corrects them if necessary before the character is output by the system. Such a system would be expected to use RAM to store the characters during comparison.

3.7 EXAMPLE OF TYPICAL COMPUTER MEMORY STORAGE CONFIGURATION

Many of the various ways of using memory in a system can be illustrated by the Sony NEWS network station which is shown in Figure 3.10. This workstation uses two 32 bit microprocessors, a CPU and an I/O processor in a master–master arrangement. It has a version of Unix, integrates into an Ethernet local area network, is multiprocessor, has a multiwindow user environment, and supports graphics. It has a SCSI bus between the peripherals and the main memory, a VME interface bus for VME peripherals, and NFS to provide distributed processing among networked stations.

Memory devices or functions in this system include the following:

(a) An 8k byte instruction cache using a 35 ns 8k × 8 static RAM which allows the I/O processor to operate at 16.67 MHz (60 ns average cycle time) with no wait states. This is achieved by averaging the time for instructions obtained from the fast SRAM cache at 35 ns access and from the slower dynamic RAM in the main memory at 120 ns (one wait state). Even with a relatively low hit rate in the cache, the 60 ns average cycle time is achievable. Clearly the 16 bit and 32 bit system buses require the wide × 8 memory.

(b) The I/O processor provides memory management for its program memory which is a 128k × 8 SRAM on the 32 bit processor bus. This can be a slower than 100 ns memory since it is used in communicating mainly with slower peripheral devices. The isolation of the I/O processor and its program memory from the main memory bus means that operation of the CPU is slowed down only during data transfer and communication between the two processors.

(c) The 32 bit I/O processor and its program memory are isolated from the 16 bit I/O bus by a 16k byte bidirectional transmission and reception buffer which has a capacity of 10 data packets. This buffer could be implemented with a serial memory.

(d) The main memory is shared by the two processors. Part of the 8Mb of main memory is reserved for communications between the two processors. Commands from either processor pass through this section of the memory. Either processor

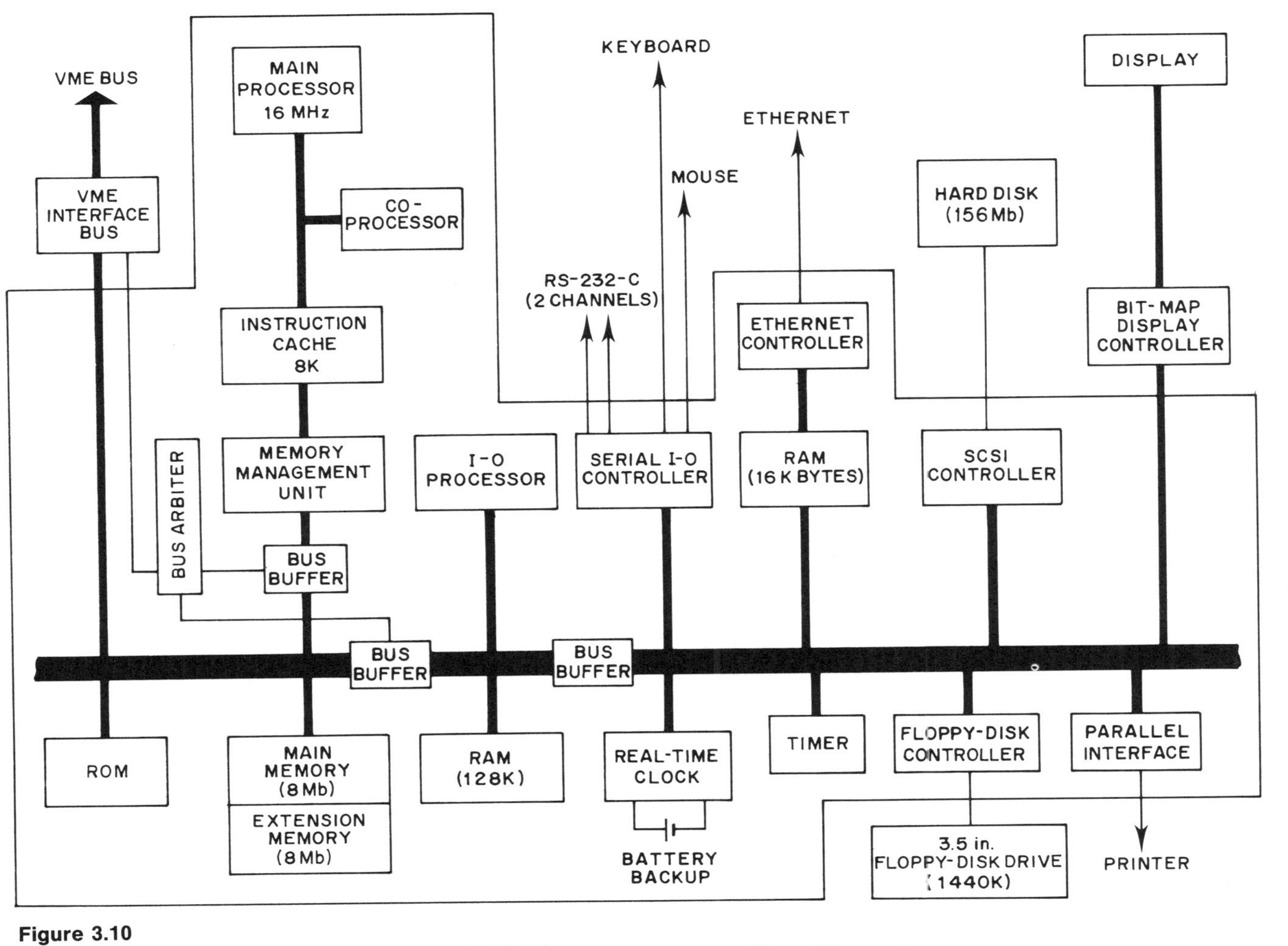

Figure 3.10
Typical bus oriented system configuration highlighting memory usage. (From Electronics [6] 1987, Sony Corporation.)

can transmit an interrupt signal. Interrupt signals are used when two processors can access the same memory location at the same time to avoid an indeterminate state occurring in that memory location if one tries to read while the other is writing. Another feature which can be used in a memory system, where contention between two processors accessing the same bit of memory is possible, is a 'busy' signal which indicates that the data location is being accessed.

(e) The memory management unit gives the system single level paging, virtual memory architecture. It supports demand-paging virtual memory. The MMU acts as a 35 ns cache for faster access of the pages in most recent use. One level paging with a page size of 4k bytes is used. The originals of the MMU page-table entries are stored in the main memory and the MMU serves as the cache for faster access of these pages.

(f) The high speed multiwindow display system requires hardware with the 'bit boundary block' (bitblt) transfer function for the bit-map displays. This could be implemented either in an applications specific memory, or, as in this case, with a 2000 gate array. The color display uses one gate array to hold the current data for each color plane all of which operate concurrently so that multiwindow processing in color can be done as fast as for monochrome.

(g) There is an option for connecting peripherals on a VME bus. In a master–slave arrangement, the CPU operates as master for controlling access to the peripherals. Both addresses and data words are 32 bits wide.

(h) Archive files are provided with 156Mb of hard disk and a 1440k byte floppy disk drive, both on the 8 bit SCSI control bus. This bus is capable of parallel data transfer at 1.5 Mb s^{-1} (or 83 ns/bit data rate) for asynchronous transmission and 4 Mb s^{-1} for synchronous. This very slow data rate and large size of memory means that a high density, low cost alternative must be chosen for the archive memory. Clearly either optical or magnetic memory could be used for the hard disk.

(i) The system also supports distributed processing, or networking. The directories of the memory files from a remote machine can be entered in the users machine and accessed just like local files through a virtual file system in the Unix.

(j) The system also supports Ethernet using 16k bytes of SRAM in this interface onto the 16 bit peripherals bus.

(k) Finally, the system contains a real-time clock which can store the time either in static RAM with battery back-up or in an EEPROM. In this particular system SRAM with battery back-up was used.

3.8 SOME OTHER SYSTEM APPLICATIONS OF MEMORIES

3.8.1 Systems for non-volatile memories

The advent of distributed processing and the availability of inexpensive computers has spread electronics into every aspect of modern life, forming vast information networks throughout the society.

Memories which can be updated remotely are required throughout these systems. The general interface functions of society will be linked and require remote updating (see Figure 3.11(a), (b) and (c)).

The home will be the center of such a network. Remote updating of memory will be required in the home for telephone repertory dialers to remember phone numbers; in utility meters to take remote readings; in television sets as synthesizers, channel selectors, and reprogrammable charge meters; in automobile engine tuning systems, which can modify themselves to compensate for engine wear; in coded garage door openers; in reprogrammable trip computers; in portable calculators; in personal computers; for in-system software updates, and in other such diverse home uses as electronic keys.

The automated office and factory is already becoming a reality and will depend heavily on remote updating for such things as industrial controller personality, terminal identifiers, printer characterizers, photocopier configurations, equipment calibrator constants, remote field services, complex industrial controller parameters, system software libraries, and, of course the ultimate usage, the computer which can update itself to adjust to its environment, which is a major step in robotics toward true artificial intelligence.

Society at large will also need remote reprogramming capability for weapon guidance parameters, traffic light controllers, automated communication telephone networks, automated train network controllers, EDP access codes, postage meter price updating, electronic credit cards, in medicine for remote diagnostics, medical electronics, and electronic credit cards. Some of these applications already exist and are filled currently with an EPROM, DIP switch, or CMOS static RAM in battery back-up mode. Others are areas of potential new development where the EEPROMs offer a unique advantage.

3.8.2 Computer communications

Distributed computer networks, such as point of sale terminals can be updated remotely using reprogrammable non-volatile memory. Either low power SRAMs with battery back-up or EEPROMs can be used. An example of such an application would be a system of point-of-sale terminals for a market chain. Pricing for items could be modified throughout a system by remotely modifying the price look-up table in each terminal. Similarly, gasoline prices at computer controlled service station pumps may be remotely updated. Airline reservations systems are another example where remote updates would be useful.

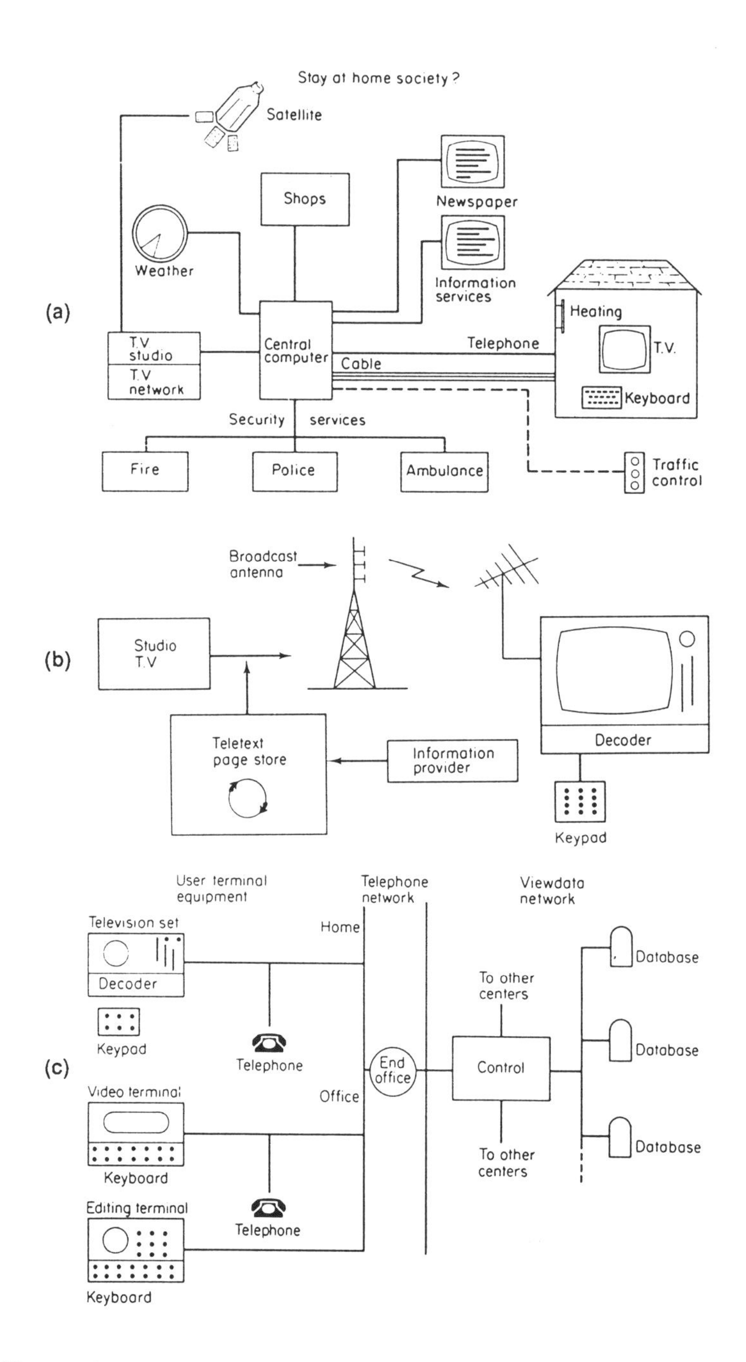

Figure 3.11
(a) Stay at home society; (b) home information system. Information available may include news headlines, weather, newsflashes, financial news, sports headlines and travel news. (c) telephone network. Information available may include news, weather, job advertisements, want ads, real estate, travel timetables, and professional services. (From Prince and Due-Gundersen [26] 1983.)

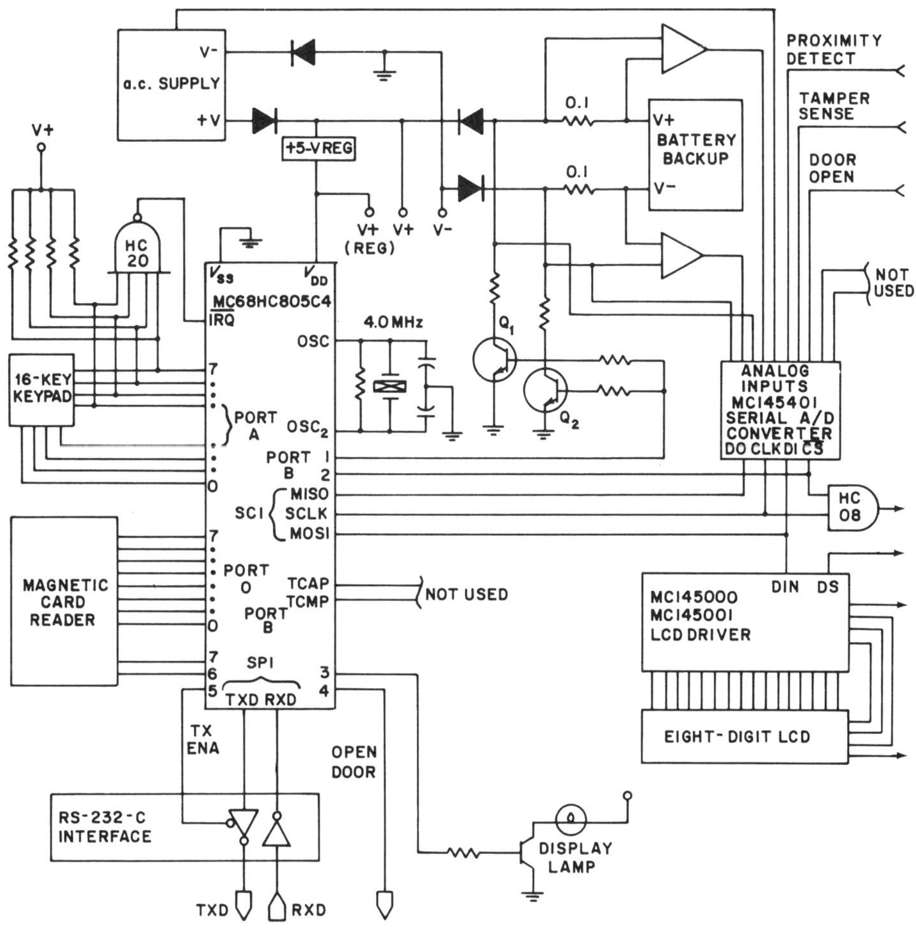

Figure 3.12
A battery-backed security system using an 805C4 microprocessor with embedded EEPROM to monitor entrance to a door. (Source Motorola.)

3.8.3 Distributed processing and process control

Microcomputers with on-chip embedded memory are used in many new control and distributed processing types of applications. Embedded SRAMs, ROMs, EPROMs and EEPROMs are now included. An example of such an application is a battery backed security system to control the entrance through a door which is one point in a distributed control security network.

This particular system used a Motorola 6805 microcontroller with 32k of EEPROM embedded as shown in Figure 3.12. Entrance is gained by inserting a security card into a magnetic card reader and then entering a private access code on a keypad. The access codes are stored in the on-chip EEPROM along with switch debouncing and key decoding functions, thereby eliminating the need for a separate keypad interface chip. Port A of the 805C4 is dedicated to the keypad from which data entry generates an interrupt.

3.8.4 High and extended definition television

True HDTV produces a picture with 1000 lines of resolution (compared to today's 525) and has a wide 5:3 width-to-height ratio screen. It is forecast that HDTV will account for about 32 billion dollars in sales by the year 2000. Standards so far include the European HD–MAC and the Japanese MUSE.

Meanwhile, improved picture quality using the older technology is being pursued by such improvements as the European 100 Hz systems, where visible screen flicker due to the 50 Hz scan rate is reduced by cycling each line in a serial frame buffer and playing it back onto the screen so that every line is effectively doubled. Such a buffer does not need to have the additional random access port found in the VDRAMs intended for graphics uses on computers since it is used as a shift register only.

In the United States the visible spacing between the lines can be lessened by a method called line progressive scan, which consists of storing the odd (even) lines and recycling them onto the screen at double the line rate between the even (odd) lines. Other possible features include multiple picture-in-picture, zoom, and image freeze. Other applications for VRAMs include transmission in electronic mail for automated office environments and facsimile transmission.

3.8.5 Teletext

Interactive television systems, such as the French Minitel or the British Teletext, permit user selection of information and permit updates on various topics of current interest such as news, weather or shopping.

A typical teletext system uses internal buffer RAM to eliminate speed problems inherent in the use of external RAMs, and a character generator ROM which stores programmed characters from as many as eight languages.

3.8.6 A programmable compact disk video processing system

An example of multiple memory applications in a consumer system was presented in an experimental compact disk video system developed by the Philips Advanced Products Laboratories, as shown in Figure 3.13.

In this system an erasable compact disk, which can hold more than 600M of data, is used to store color television pictures on the disk on one or more tracks. When a picture is to be displayed it is transferred over the SCSI bus into a solid state picture memory CD-RAM with high speed read-out capability before being transferred through a FIFO buffer memory to the monitor. The system controller consists of a 68000 microprocessor with RAM and PROM where tables of all pictures stored in

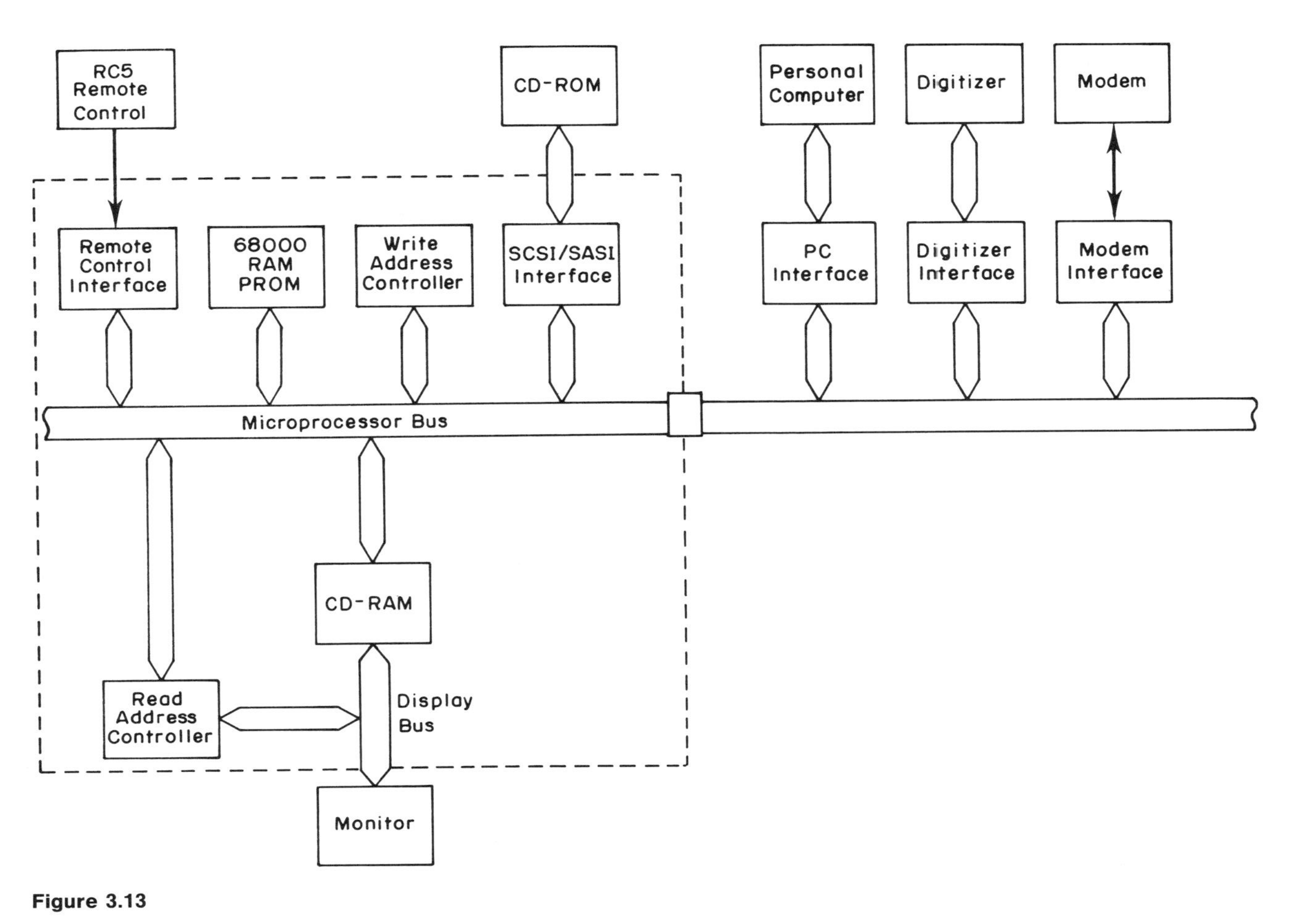

Figure 3.13
Programmable compact disk picture memory and video processing system. (Source N.V. Philips Research.)

the compact disk are kept by number. If a PC is added to this system then it will allow image processing to be carried out on the data, such as data compression, convolution and rotation. The slow access time of the DRAMs in the CD-RAM compared to the data transfer rate of the digitizer, thus 32 bits are stored in parallel to enable a fast enough data transfer.

3.8.7 Portable systems

Higher density memory has given rise to an array of special purpose hand held computers. These include thesauruses, agendas, memo pads, language translators and telephone dialers. This application requires the memory to be both packaged densely and either to be non-volatile or to drain very low current from the battery when not in use. Usually the speed of the memory can be very slow, in the millisecond range. EEPROMs and SRAMs with special write protect features either on the chip or implemented in the system are needed.

If standard static RAMs are used in such a back-up application, the additional support system can be as shown in Figure 3.14(a). This system works using the following sequence. The detect circuit signals failure of the system power supply and provides signals to the RAM which cause it to go into standby mode. The circuit then separates the RAM from the system power supply. It also switches from the system power supply to the back-up power supply. When the system power is again available, the circuit confirms stability of the power supply, then signals the memory to return to active status.

An example of an application-specific SRAM with the support circuitry implemented on chip is the low power standby SRAM with a write protect function which turns off the write enable function of the memory when the applied voltage drops below a specified value. Other features available include a battery detect scheme which notifies the circuit when the battery power is low. A block diagram of the memory circuit for such a battery backup RAM (BRAM) is shown in Figure 3.14(b).

Another memory useful in portable systems is a SRAM which runs at lower voltage, such as the JEDEC 2.8 V standard for battery powered applications, or the JEDEC 3.3 V standard for low voltage TTL compatible applications. Such an SRAM allows a smaller, lighter weight battery to be used to reduce the weight of portable or hand held systems.

3.8.8 Large memory smart cards

Large memory smart cards are used as disk cards in computers and as program cartridges in instruments and other industrial equipment. These cards can contain many megabytes of static memory either SRAM with battery back up to keep it non-volatile, or ROM if the program is fixed, or the new electrically reprogrammable, non-volatile flash memories if the data is changed occasionally.

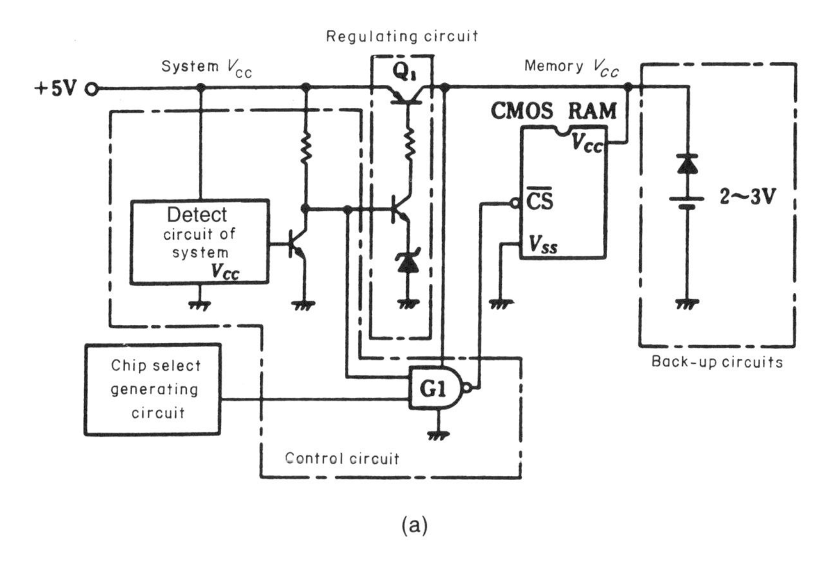

(a)

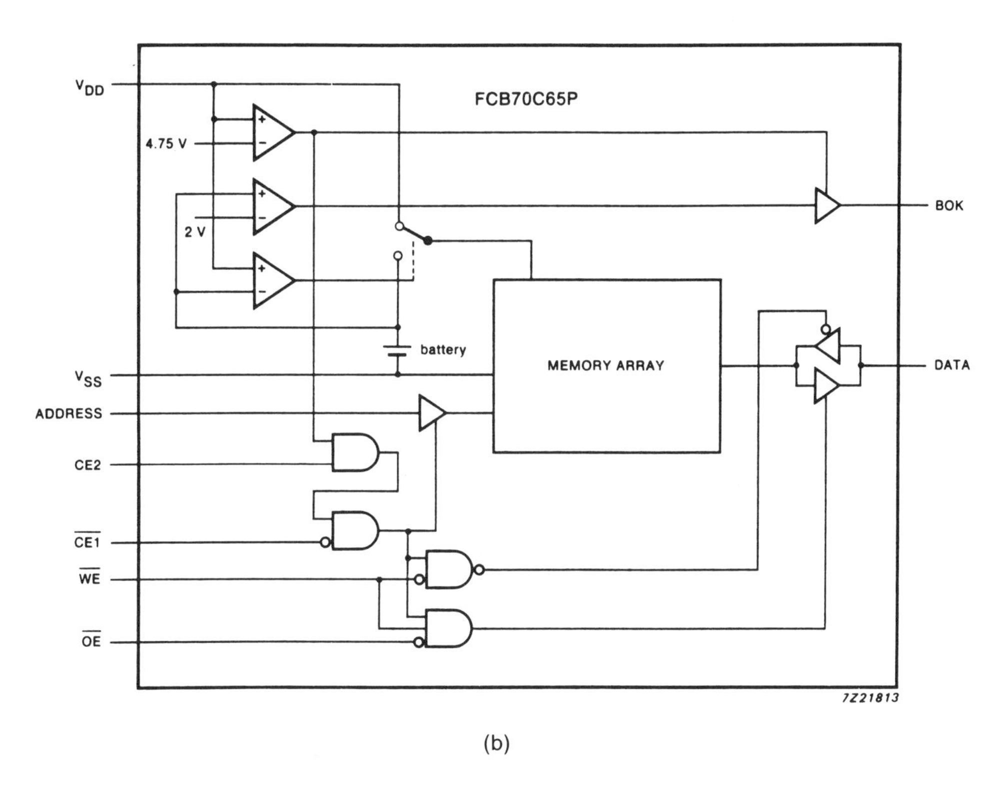

(b)

Figure 3.14
SRAMs in battery back-up applications. (a) Standard SRAM. (Source [24] Hitachi Memory Databook 1988/89.) (b) Application-specific SRAM. (Source N.V. Philips Research.)

3.8.9 Consumer games

While the high period of consumer games peaked before the personal computer became common in the home, game systems which use cartridge read-only memories to hold particular games are still sold. These cartridges can be tape or they can be solid state ROM.

3.8.10 Memories in automotive electronics

Automotive electronics accounted for $800 of the price of the average car in 1987. This market is expected by the year 1993 to reach $30 billion and by the year 2000 as high as $60 billion. Applications will be divided between navigation, instruments, engine controls, safety and security, and comfort and entertainment.

The key requirements are safety, performance, and luxury. For the electronic component, this translates into ruggedness, temperature stability, and reliability. Safety features include intelligent windshield wiper controls, anti-lock braking systems, electronic locks, controls for seat, and mirror adjustments. Performance features include anti-lock braking systems, anti-knock devices for engines and automatic load-level controls. Luxury features include navigational computers, cellular telephones and anti-theft systems.

Engine management is also a significant potential market for electronics in the future. Memories will be needed in adaptive engine management systems such as fuel controls, ignition timing systems, power train control and intelligent suspension.

For the component manufacturer the automotive industry is taking an integrated systems approach where the systems design and component design are integrated from the earliest development stages. As part of this new approach, automakers are focusing more on functionality and integration of electronics into the complete vehicle. This requires much closer user–vendor relationships in the design phase than in the past which extend into the production phase. This is particularly important for a product like memories which has very high development costs and a history of extreme market volatility.

These closer ties between supplier and user extend not just through the development stages, but throughout the production cycle. The competitive nature of this market means that it is important for the automotive manufacturer to minimize supply inventories. The result has been an increasing reliance on Just In Time (JIT) delivery arrangements with components suppliers. This in turn helps the component supplier by tieing his production rates to the real system production usage of his component relating to user inventory management.

Because of the ruggedness and extended temperature requirements of the automotive environment, the memories which have gone into this application have met near military market specifications. A typical automotive market temperature requirement is -40 to $+125$°C for under the hood applications, and -40 to $+85$°C for dashboard and interior applications.

Both the temperature and the ruggedness requirement have tended in MOS devices to move this application toward CMOS technology with its wider noise margins. There has also been a trend in the past toward static rather than dynamic memories for noise margin and high temperature leakage reasons. EPROMs continue to be preferred to EEPROMs in the automotive environment due to the potential for EEPROMs being accidentally reprogrammed by the high voltages present under the hood.

One of the future trends is toward higher module integration in automotive electronics with perhaps in the future only three master control modules. One would handle engine and transmission functions. Another would manage body-embedded functions such as regulating information and comfort features such as instrumentation and navigation systems, door locks, automatic climate control and seat adjustment. The third would handle chassis requirements such as anti-lock braking systems and sophisticated active adaptive suspension systems.

This integration will require from the component manufacturers both higher density packaging techniques and for memories perhaps the ability to embed more memory or smarter memory onto the controller chips themselves.

3.8.11 The automated factory and robotic advances

Most robots function as part of an integrated manufacturing system. Many can be simple mechanical fixtures. The more complex ones, however, require sophisticated vision systems, and a higher level of local computing power and memory.

There are six features that are particularly common to intelligent robots. These are off-line programming, high level languages, system control by the robot, configurability for a specific task through plug-in hardware, freedom from the constraints of the workstation, and integrated sensing devices.

Off-line programming offers the advantage that the robot can continue in use while the central computer reprograms the robot. The robots current program is stored in local memory which can be 'plug-in' reprogrammable.

Calibration data for the robot can be stored in electrically reprogrammable memory in the robot. This memory should be non-volatile since it is expected to retain the calibration information while the robot is turned off, such as overnight.

Increasingly a CPU is being used as the controller in the robot. The local CPU and the plug-in hardware programming, lets the robot be free from the workstation and hence has led to adaptable robots.

Integrated senses, such as vision, enhance both the flexibility and affordability of robots since custom user designed versions are eliminated.

3.8.12 Computer integrated manufacturing

Computer integrated manufacturing is becoming a reality. The advanced robots in such a factory environment need sophisticated controller and calibrating. Embedded

EEPROMs in microprocessors can provide non-volatile storage of the calibration data. Larger EPROMs can store the operating program for these robots. In the future both non-volatile memory functions may be able to be processed on one chip. For a flexible operating environment, the program may need to be changed from a central location, this would involved fast, writable, high density non-volatile RAM. The various battery RAMs may be useful in this type of application.

3.8.13 Industrial applications

An increasing number of medical systems use memory intensively. Medical systems need to store X-ray pictures for instant recall during surgery. Since several frames need to be stored at any one time a significant amount of memory is required. A disk could be used; however, the reliability requirements of the medical system preclude moving parts. In such a system the best option might be solid state memory.

Another example is high performance logic analyzers which run at over 300 MHz data acquisition rates. The memory speeds in these very fast systems can be obtained by using slower memories in an interleaved structure.

3.9 BASICS OF READING TIMING DIAGRAMS FROM A MEMORY DATASHEET

In order to design a system using memories, it is important to understand the characteristics of the memories available. The best indicator of these characteristics is the data sheet specification for the particular memory. A few comments on reading such a specification follow.

In a memory system, there are signals going from the processor via the bus into the inputs of the memory and signals coming from the outputs of the memory onto the bus and to the processor.

Inputs from the system processor to the memory include:

(1) addresses, which indicate the memory locations selected.

(2) write enable, which chooses between read and write mode and also controls writing of new information into the memory.

(3) chip select(s), which select one memory out of several in a system. If a chip select is off, the memory is deselected.

(4) output enable, which can be used to control the output buffer.

(5) data input(s), to be written into the memory

Outputs from the memory include

(1) data output being read from the memory
Some memories such as SRAMs with wider datapath can have a common pin for input and output.

3.9.1 Timing diagrams

Timing diagrams specify the minimum required and maximum expected timing requirements for system actions. Two sets of timing symbols are shown in most cases. The first is the JEDEC standard timing symbol which is self explanatory, the second given in parenthesis is an older set of symbols less self explanatory but still in widespread usage.

A timing diagram for a basic read cycle during which the system reads out information that is stored in a static RAM is shown in Figure 3.15 for a wide bus, common I/O SRAM with chip enable, and output enable functions. It consists of

(1) System selects the RAM by turning the chip enable on ($\bar{E}$ low).

(2) System sets the correct addresses (A set).

(3) System turns the output enable on ($\bar{G}$ low).

(4) System must make sure that the time that old data from other sources is still on the common I/O bus is less than the minimum of $t_{GLQX}(t_{OLZ})$ or $t_{ELQX}(t_{CLZ})$.

(5) The system must wait a minimum time of t_{AA}, t_{AC}, or t_{OE} in order to be sure of correct data.

After the maximum required wait time (which is the minimum time the system must wait) the system may read the information stored in the memory.

A simple write cycle for changing data in an SRAM (writing into it) is shown in Figure 3.16. It consists of

(1) System sets the correct addresses (A set).

(2) System selects the RAM by turning the chip enable on ($\bar{E}$ low).

(3) System waits a minimum required amount of time after changing the addresses for the RAM to do internal 'set up' of the addresses $t_{AVWL}(t_{AS})$, and then turns the write enable on ($\bar{W}$ low).

(4) System waits a minimum required amount of time after turning on the write enable $t_{WLQZ}(t_{WZ})$ for the memory to disable the data output driver 'Q' in preparation for using these lines for data input.

(5) System inputs the new data and waits a minimum required amount of time for the memory to write the data before turning off the write enable $t_{DVWH}(t_{DW})$.

(6) System waits a minimum required amount of time after turning the write enable on before turning it off $t_{WLWH}(t_{WP})$. This is to be sure the write enable pulse width is wide enough for correctly writing the data into the RAM.

READ CYCLE

Parameter	Symbol								
	Standard	Alternate	Min	Max	Min	Max	Min	Max	Unit
Read Cycle Time	t_{AVAV}	t_{RC}	25	–	30	–	35	–	ns
Address Access Time	t_{AVQV}	t_{AA}	–	25	–	30	–	35	ns
Chip Enable Access Time	t_{ELQV}	t_{ACS}	–	25	–	30	–	35	ns
Output Enable Access Time	t_{GLQV}	t_{OE}	–	12	–	15	–	15	ns
Output Hold from Address Change	t_{AXQX}	t_{OH}	5	–	5	–	5	–	ns
Chip Enable Low to Output Active	t_{ELQX}	t_{LZ}	5	–	7	–	10	–	ns
Chip Enable High to Output High-Z	t_{EHQZ}	t_{HZ}	0	10	0	12	0	15	ns
Output Enable Low to Output Active	t_{GLQX}	t_{LZ}	5	–	8	–	10	–	ns
Output Enable High to Output High-Z	t_{GHQZ}	t_{HZ}	0	10	0	12	0	15	ns

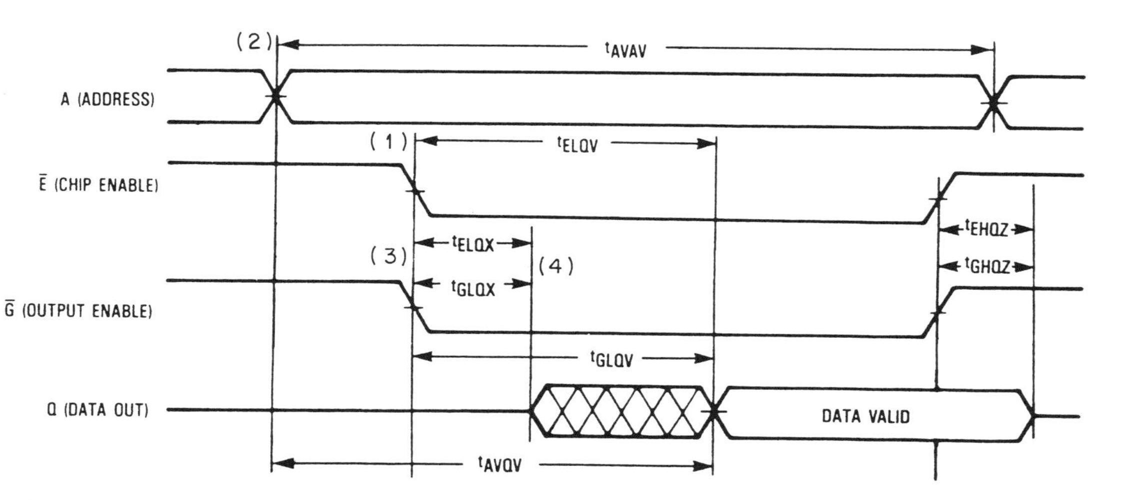

Figure 3.15
Typical read cycle timing diagram for a static RAM with chip enable and write enable control pins. (From Motorola Memory Databook 1988 [28].)

WRITE CYCLE

Parameter	Symbol								Unit
	Standard	Alternate	Min	Max	Min	Max	Min	Max	
Write Cycle Time	t_{AVAV}	t_{WC}	25	—	30	—	35	—	ns
Address Setup Time	t_{AVWL}	t_{AS}	0	—	0	—	0	—	ns
Address Valid to End of Write	t_{AVWH}	t_{AW}	20	—	25	—	30	—	ns
Write Pulse Width	t_{WLWH}	t_{WP}	20	—	25	—	30	—	ns
Data Valid to End of Write	t_{DVWH}	t_{DW}	10	—	12	—	15	—	ns
Data Hold Time	t_{WHDX}	t_{DH}	0	—	0	—	0	—	ns
Write Low to Output High-Z	t_{WLOZ}	t_{WZ}	0	10	0	12	0	15	ns
Write High to Output Active	t_{WHOX}	t_{OW}	5	—	8	—	10	—	ns
Write Recovery Time	t_{WHAX}	t_{WR}	0	—	0	—	0	—	ns

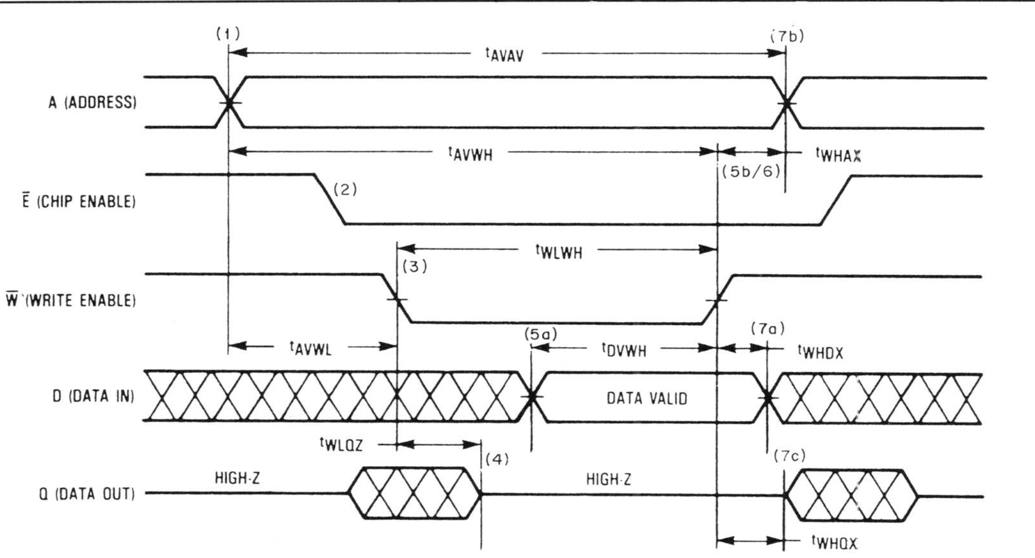

Figure 3.16
Typical write cycle timing diagram for a static RAM with chip enable and output enable control pins. (From Motorola Memory Databook 1988 [28].)

(7) System waits a minimum required amount of time after turning the write enable off

(a) before changing the data $t_{WHDX}(t_{DH})$. This is called the 'data hold time', and it ensures that the data is stable during the entire write cycle.

(b) before changing addresses to start the next cycle $t_{WHAX}(t_{WR})$. This is called the 'write recovery time' and ensures that the addresses are stable during the entire write cycle.

(c) makes sure the data have disappeared before the RAM turns the data output drivers back on $t_{WHQX}(t_{OW})$

The RAM is now ready to begin the next cycle. It should be clear that the minimum timings are periods when the system must wait for the RAM to do something, and the maximum timings are guaranteed limits within which the system will act.

3.10 FACTORS SLOWING DOWN A MEMORY SYSTEM

3.10.1 Bus contention

Since most memory applications in computer systems are bus oriented, a factor to consider is bus contention. Bus contention occurs when two or more devices try to output opposite logic levels on the same common bus line. This can happen when one device has not completely turned off before another device is turned on. Bus contention can result in large transient current spikes which not only generate system noise, but can also affect the long term reliability of the devices on the bus.

Very fast RAMs with common input and output (I/O) pins are the most likely to suffer from bus contention due to the tight timing requirements that are needed to achieve high speed operation. The most common form of bus contention for memories occurs when switching from a read mode to a write mode or vice versa.

In the example given in the last section of a write cycle on a static RAM, if the system does not wait for the output 'Q' to turn off before turning on the input driver 'D' (violating the t_{WLQZ} minimum time), the common input output bus will be in an indeterminate state. The RAM will be attempting to pull the output low at the same time that the input driver is trying to pull it high. The result will be contention on the bus line.

This example of bus contention due to the input driver being enabled prior to disabling the RAM output is shown in the timing diagram in Figure 3.17(a). The proper operation, which is to disable the input driver prior to enabling the RAM output, is shown in the timing diagram in Figure 3.17(b).

Another form of bus contention can occur when switching from a write to a read mode. It is necessary to wait to change the addresses or the data-in

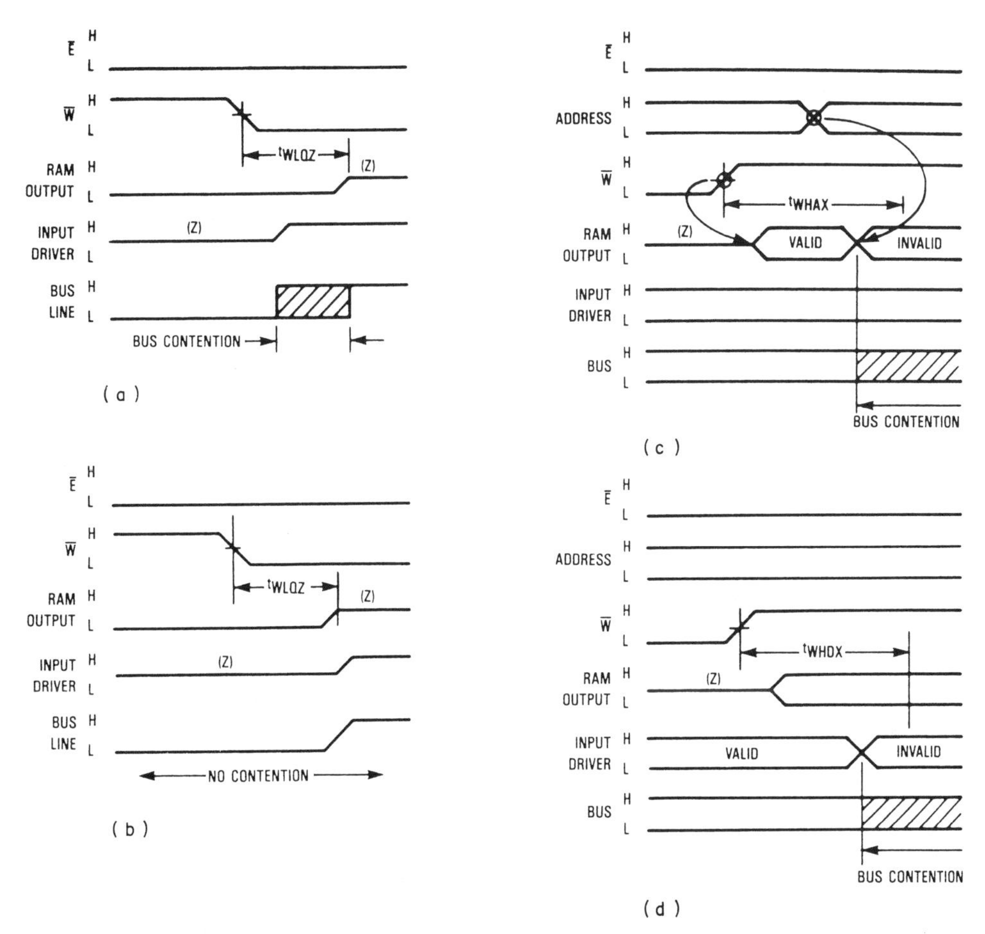

Figure 3.17
Two examples of bus contention with two methods in each case of avoiding the bus contention. (a) Input driver enabled prior to disabling RAM output, (b) Input driver disabled prior to enabling RAM output, (c) Data setup time violation, (d) Data hold time violation. (From Motorola Memory Databook 1988 [23].)

until a minimum amount of time has elapsed after the system has turned off the write enable to allow the memory to terminate the write mode. This form of bus contention is shown in the timing diagrams in Figure 3.17(c) which shows a data setup time violation where the addresses have switched too early, and Figure 3.17(d) which shows a data-hold time violation in which the input driver has tried to switch too early.

While many of the minimum timing hold requirements that have resulted in bus contention in the past have been shortened to zero on modern fast SRAMs, it is still good system design practice to make allowances for minimum hold times. Bus contention should always be eliminated by careful system design and close attention to the data sheet specifications of the memory devices used.

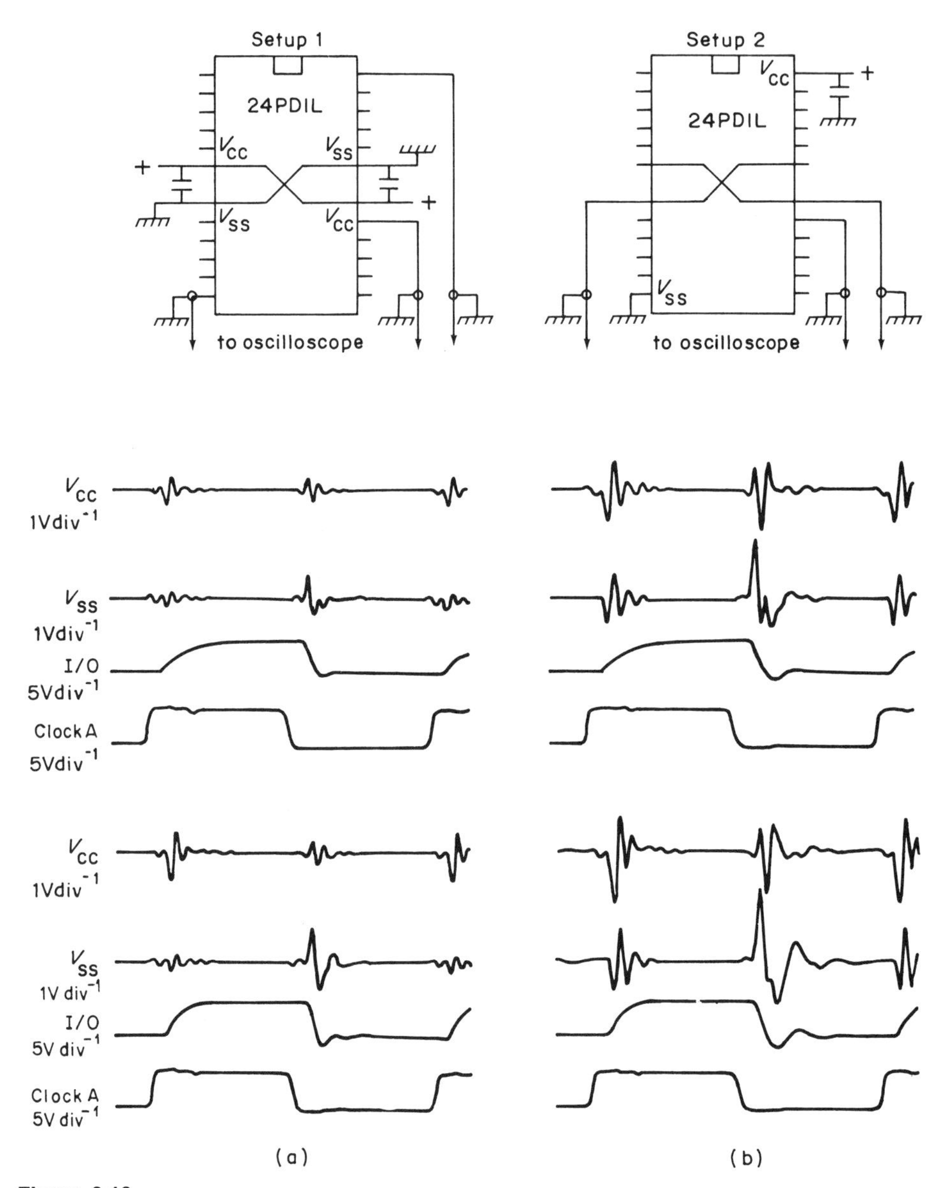

Figure 3.18
Example of ground bounce reduction in V_{CC} and V_{SS} lines for (a) center multiple power and (b) ground package compared to end single power and ground package. (From Salters [27], Philips 1988.)

3.10.2 Ground bounce

Another similar system problem encountered in high speed, wide bus systems is ground and supply bounce. This occurs when transient voltage spikes on the power supply lines resulting from the simultaneous switching of several memory outputs.

When these outputs switch simultaneously they cause transient currents to flow through the self-inductance of the ground and supply connections giving rise to the voltage spikes.

One effect of ground bounce can be access time push-out resulting from the need to wait until the ringing has stopped before reading the data out from the device.

While a decoupling capacitor or the use of ground planes on the printed circuit board can reduce the ringing effect on the system, the ground bounce can be significantly reduced on fast memory devices by changing from the older pinouts for slow parts which have power and ground pins at the ends of the package to the newer faster memories with pinouts which have power and ground pins doubled and adjacent at the center of the package.

The two test setups shown in Figure 3.18 represent two configurations of a test chip which had both conventional end power and ground bonding pads and double center power and ground bonding pads. In test setup 1 the center power and ground were connected and in setup 2 the end power and ground were connected.

The photographs in Figure 3.18 show the reduction in ground bounce on the power (V_{CC}) and ground (V_{SS}) lines for a similar RAM device having (a) center power and ground connections as opposed to one having (b) end power and ground connections.

This topic is discussed further in Chapters 6, 8 and 14.

3.10.3 System bandwidth

Another factor which slows down a memory system is the limitations imposed by system bus bandwidth which is the number of bytes that can be transferred per second on the system bus. This is a factor driving the move toward higher speed and wider bus in RAMs. Since the size of systems keeps increasing the amount of data that needs to be moved on the system bus also keeps increasing. System bus bandwidth problems therefore need to be considered.

One approach is to bypass the system bus with a private memory bus. Another help is the pipelining features of the current microprocessors that allow them to fetch a word of data at the same time they are generating the address for the next memory reference. This gives the potential for reducing some of the delays inherent in using RAMs.

The system bus band width can be represented as

$$BW = \frac{(\text{bus utilization}) \times (\text{bytes transferred})}{(\text{cycle time in ns}) \times (\text{clock cycles/transfer})}$$

or

$$BW = \frac{\text{bus utilization} \times \text{bytes transferred}}{\text{system access time}}$$

If the bus utilization or the number of bytes being transferred simultaneously increases then the system access time must increase accordingly to maintain the same bandwidth.

Wider bus memories allow more bytes to be transferred in a given amount of time thereby potentially allowing an increase in system bandwidth. Faster memories decrease the system access time also resulting in improved bandwidth.

BIBLIOGRAPHY

[1] Special report on Semiconductor memories, *Computer Design*, August 8, 1984, pp105, 114.
[2] U.S. Orders Collision Alarms on Planes, *International Herald Tribune*, January 9, 1989.
[3] Iversen, W. R. (1988) Auto Electronics Grows Even Faster, But It Gets Harder To Win Business, *Electronics*, July, p.53.
[4] Here's the Net for Superfast Computers, *Electronics*, June 1988, p.30.
[5] Cohen, C. L. and Manuel, T. (1988) Japan's Big Splash in Graphics Stations, *Electronics*, October p.28.
[6] Sony Packs Power into Low-cost Work Stations, *Electronics*, March 19, 1987, p.79.
[7] Gustafson, J. and Heinrich, M. (1985) Memory-Mapped VLSI and Dynamic Interleave Improve Performance, *Computer Design*, November 1, p.93.
[8] Sachs, H. (1985) Improved Cache Scheme Boosts System Performance, *Computer Design*, November 1, p.83.
[9] Deubert, R. (1983) Digital Prowess Increases in Color Television Sets, *Electronics*, September 8.
[10] Killmon, P. (1986) For Computer Systems and Peripherals, Smarter is Better, *Computer Design*, January 15, p.57.
[11] Kaminsky, J. (1988) The Great Global Consumer Electronics Race, *Management Review*, July, p.35.
[12] Woodcock, R. and Carone, R. (1983) Intelligent Robots Lend a Hand in Automating High Technology, *Electronics*, June 16, p.120.
[13] Analyzer Doubles as PC, *Electronics*, May 26, 1988, p.104.
[14] Manuel, T. (1988) As the World Turns Parallel, Transputer Applications Explode, *Electronics*, December, p.110.
[15] *Application Specific Memory Products Data Book 1988*, VLSI Technology, Inc.
[16] 5 Volt Only EEPROM Mimics Static RAM Timing. Reprint from *Electronics*, in the Xicor DATA Book 1988.
[17] Willions, T. (1988) Mainframe memory schemes address needs of new micro's, *Computer Design*, November 1, p.69.
[18] Smith, A. J. (1987) Cache Memory Design: an Evolving art, *IEEE Spectrum*, December, p.40.
[19] *Memory Products Databook*, NEC Electronics Inc, USA, April 1986.
[20] *Memory Products Databook*, Texas Instruments, September 1986.
[21] *Philips Static RAM Specifications*. Philips Research, June 5, 1988.
[22] Page, Nibble, and Static Column Modes: High speed Serial access Option on 1 M-bit DRAMs AN986, *Motorola Memory Data Book*, DLE 113R4/D, 1988.
[23] Avoiding Bus Contention in Fast Access RAM Designs AN 971 Rev.1, *Motorola Memory Databook*, DLE 113R4/D, 1988.
[24] *Hitachi Memory Databook* 1988/1989.

[25] Silvey, J. (1985) With EEPROM on a Single-Chip uC Changing Programs is a Snap, *Electronics Products,* October 15.

[26] Prince, B. and Due-Gundersen, G. (1983) *Semiconductor Memories,* John Wiley, Chichester.

[27] Salters, R. (1988) Revolutionary Center Power and Ground Pinout for Fast SRAMs, Presentation to the EIA JEDEC 42.3 Committee meeting, December 1988.

[28] *Motorola Memory Databook 1988.*

[29] Kushiyama (1991) A 12 MHz Data cycle 4 Mb DRAM with Pipeline Operation, *IEEE Journal of Solid State Circuits,* **26**, No. 4, April 1991, p. 479.

4 MEMORY DEVICE AND PROCESS TECHNOLOGY

4.1 REQUIRED DEVICE CHARACTERISTICS

Desirable characteristics in memories include a potential for significant cost reduction during the life of the circuit, high speed, low power, and good reliability.

Cost improvement must be effected both by initially having an adequate design and by circuit density and process yield improvements during the life of the product.

Speed must be continually increased to provide system makers with state-of-the-art components which will keep them competitive with the latest high speed systems. This speed is determined primarily by technology both for memory gates and for microprocessors. In 1990 this was in the 100 MHz range up from 10 MHz in 1980 and 50 MHz in 1985.

The power consumed by each memory must not exceed certain limits as higher densities of circuits are integrated on the chip. Power dissipation has tended to remain fairly constant over time and is a function of design, technology, the materials characteristics of the memory, and the operating power supply voltage.

Finally, the technology must meet reliability standards which are becoming increasingly stringent at the device level as larger numbers of devices go into single systems.

All of the required memory characteristics are a result of process technology and good design. In this chapter process technology is considered, and in Chapter 5, the basics of design.

4.2 TRENDS IN MOS DEVICE CHARACTERISTICS

4.2.1 Increased density

Since a major cost in semiconductor manufacturing is the high initial capital investment in the process manufacturing area, the ability to make more circuits on a given manufacturing line is the major factor in cost improvement. MOS integrated circuits are fabricated on the surface of disks of silicon called wafers which range in size currently from 100 to 200 mm^2. Figure 4.1 shows 150 mm^2 memory wafers in a production type wafer carrier.

A given manufacturing line is designed to output a maximum specified number of wafers of a specified size per period of time.

An increase in the unit output of the wafer fab is, therefore, dependent on increasing the number of good devices made on a wafer. This increase in good circuits or die per wafer can be accomplished by increasing the ratio of good to total die on the wafer (the wafer yield), and also by decreasing the size of the individual circuit so that more of them can fit on the surface of the wafer.

While per unit manufacturing cost can be reduced by shrinking the die size, sales price can be increased by fitting more memory capacity in a single circuit. This increase in capacity increases the size of the die which again needs to be reduced to the smallest possible size to be competitive.

The result is that circuit density has been increasing in MOS memories at the rate of a doubling of memory density approximately every 1.5 years since the 1k DRAM was introduced in the early 1970s up to the 16Mb DRAM in 1990 as shown in Figure 4.2(a).

Since SRAMs have six devices per memory cell and DRAMs have effectively less than two, the SRAM cell area is about four times that of the DRAM cell in a similar technology. As a result SRAMs tend to be at a quarter the density level of the corresponding DRAM generations so that the chips are of comparable size to improve the efficiency of manufacturing. Since silicon chip cost is related to the area of the silicon wafer, comparable size DRAMs and SRAMs in the same technology would tend to have similar manufacturing costs and be priced to reflect this cost.

4.2.2 High speed

As the size and scale of the silicon chip is reduced, the channel length and the gate oxide thickness of the transistors are also reduced. The gate delay switching times decrease proportionately. The increased switching speed of the transistors in the periphery of the memory and the resulting performance improvement of the entire circuit is therefore a direct result of the decrease in geometry. This is illustrated in Figure 4.2(b) which shows speed (average access time) and minimum geometry (L_{eff}) of the process technology for static RAMs by year.

As more advanced technologies are developed, current generations of memories make use of them, so, for example, the 256k SRAM was presented at the ISSCC over a 5 year period with speed increasing from 50 to 8 ns and minimum geometry decreasing from 1.5 to 0.7 μm. Memory speed can be optimized at any given technology level by innovative design techniques.

4.2.3 Power consumption

Power dissipated when the chip is running tends to increase in proportion to the speed of operation. Increased power consumption means increased heat generation in

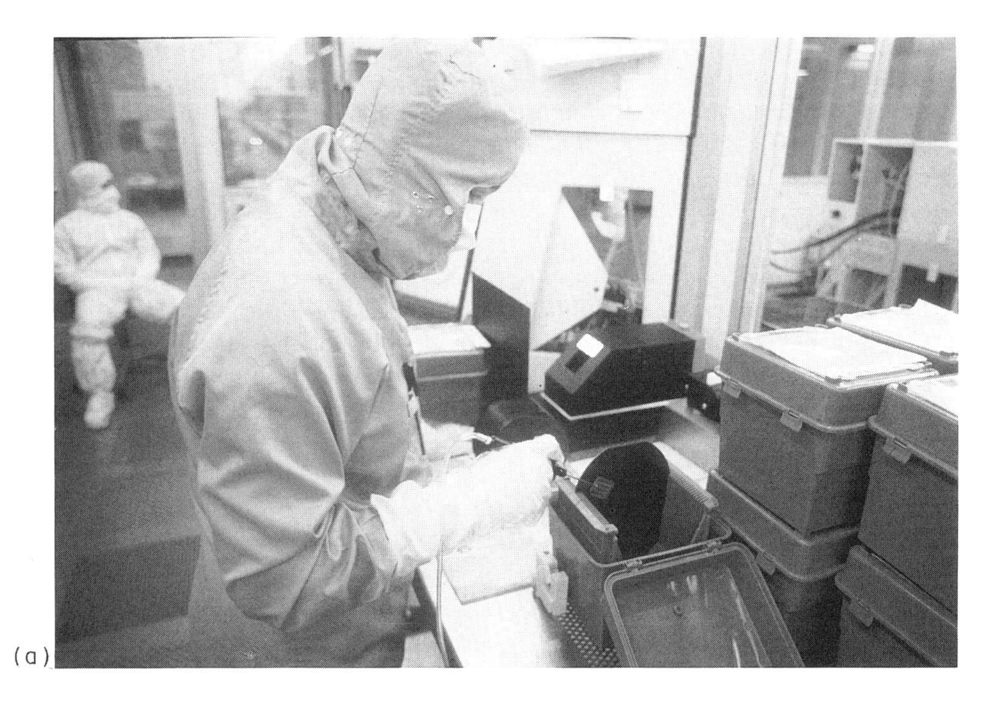

(a)

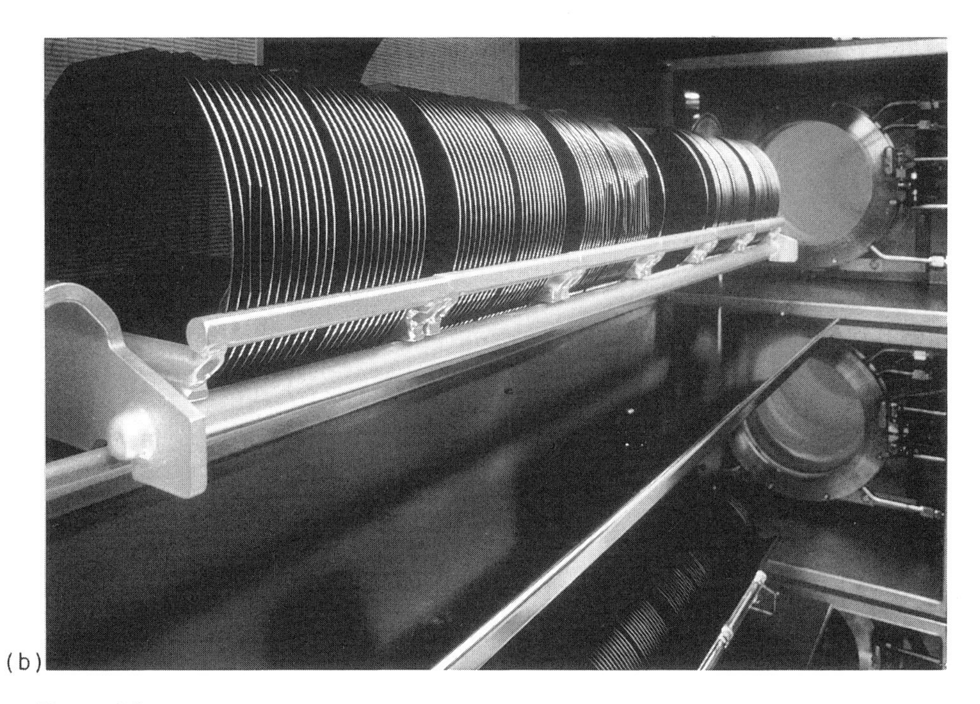

(b)

Figure 4.1
(a) 150 mm^2 (6 inch) silicon wafers being loaded into a production 'clean room' carrier. (b) A quartz track of wafers being moved into a high temperature diffusion furnace by an automated handler. (Source N.V. Philips Research.)

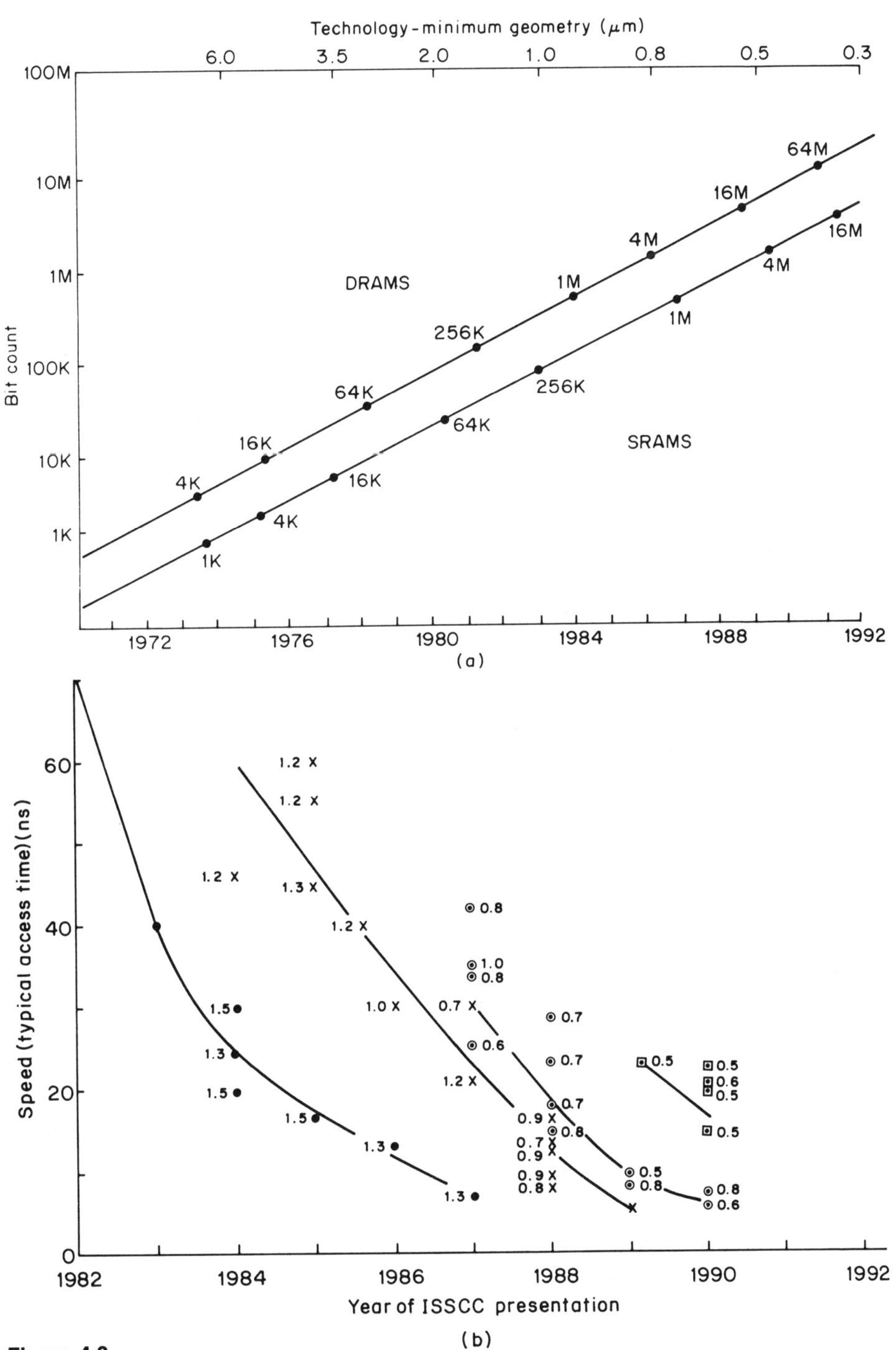

Figure 4.2
(a) Bit count and technology plotted against time for DRAMs and SRAMs. (b) Typical access time and minimum geometry and process technology (L_{eff} of various densities of SRAMs by year (● 64 k, × 256 k, ⊙ 1Mb, ⊡ 4M), numbers shown are L_{eff} (μ^2)). (Source ISSCC Proceedings 1982-1990.)

the system containing the memory. Since higher temperature shortens the expected life of the memory circuit and also requires expensive cooling apparatus in the system, it is desirable to minimize the power consumption of the memories.

In CMOS the majority of the active power dissipation is in charging the capacitors; therefore the dissipation is not only a function of the design but also of the actual speed at which the memory is being run, the power supply voltage and the capacitance (C). In practice the internal voltage swing for CMOS memories is equal to the supply voltage (V_{CC}) so that the major part of the power consumption is $\mathrm{PD} = V_{CC}^2 Cf/2$, where f is the frequency of operation. In addition there may be some dc current, which contributes to the power dissipation, from TTL level input buffers and sense amplifiers.

There are also design and operational tradeoffs possible. Since power is proportional to speed, the circuit will consume less power if it is run more slowly. High speed applications are, therefore, constrained by the ability of the system to dissipate the heat generated. Other applications require low power and can relax speed requirements somewhat.

From the design standpoint there is a narrow window in which speed–power tradeoffs are possible. Design related power reduction primarily tends to focus on the peripheral circuitry of the memory and on clever partitioning of the parts of the memory array that are active at a given time to reduce the capacitance and resistance of long interconnect lines.

Beyond the point where it is possible by design and process techniques to reduce the power consumption of the memory, there remain the materials characteristics. Transition to materials other than silicon, silicon dioxide and aluminum may prove in the future a potential area for further power reduction.

4.3 BASIC MOS MEMORY PROCESS

The basic MOS integrated circuit process begins with the growth of large cylindrical single crystals of pure silicon which are sliced into 'wafers'. These wafers are thinned to about 400 μm after processing and are between 4 and 8 inches in diameter. The surface of these wafers must be perfectly flat and with a minimum of surface defects.

Silicon is used as the basic material since as a semiconductor it has a resistivity between that of an insulator and a conductor. This resistivity can be changed by the introduction of ionic impurities.

The actual MOS memory circuit is constructed on the surface of the wafer using a series of successive photolithographic steps to form patterns alternated with diffusion or ion implantation of various chemical dopants into the openings of the pattern. A simplified description of a single step in patterning the silicon wafer in this process follows.

Step 1. Thermal silicon dioxide is grown on the surface of the silicon wafer or a layer of various materials is deposited on the wafer (silicon dioxide, metal, silicon nitride, etc.)

Silicon dioxide (SiO_2)
Silicon wafer (Si)

Step 2. A photolithographic (photoresist) material is applied to the surface of the wafer.

Photoresist
SiO_2
Si

Step 3. A patterned mask is set over the wafer through which ultraviolet light exposes and hardens the photoresist in the pattern. The rest of the photoresist is then washed away. This defines the desired geometries. The exposed surface, in this case silicon dioxide, is then removed using an electronic plasma or chemical etch. Following this, the rest of the photoresist is removed. An etched pattern in the surface results.

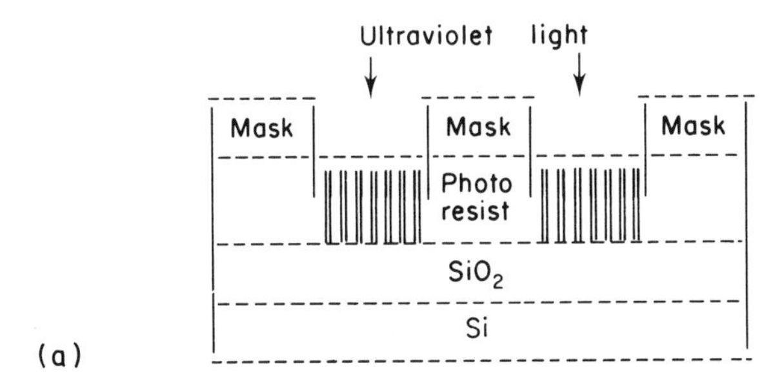

(a)

Step 4. Dopant atoms are then introduced into the silicon. This is done either in a high temperature gas environment or by an ion beam accelerator which implants ions in the pattern defined on the silicon surface of the wafer. The atoms cause the silicon to act either N-type or P-type doped.

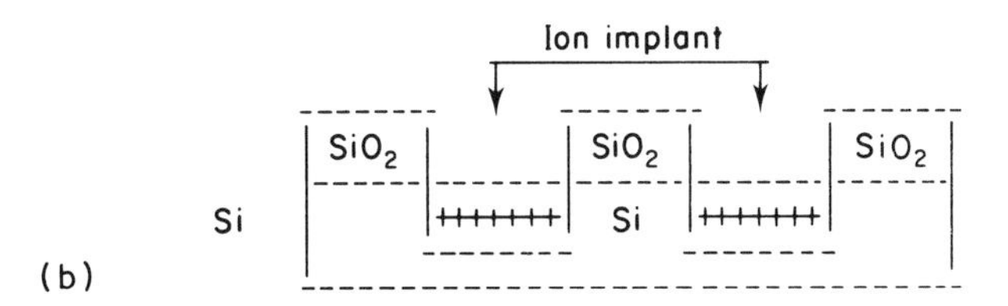

(b)

Another step, which is not shown, would involve depositing a doped amorphous polysilicon or metal layer on the surface. This is then patterned, using the technique described above, to form interconnects between the elements on the wafer surface.

This series of steps will be repeated from six to twenty times to define a series of N-type and P-type doped patterns in the surface of the silicon. These form MOS transistors and also pattern various materials which act as interconnects between these transistors.

Modern lithography and etch methods integrate as many of the steps of this process as possible to improve the throughput and reduce the defect density on the surface of the wafer due to particle contamination. Figure 4.3 shows photographs of systems for integrated lithography and etch processes. Figure 4.3(a) shows a system which coats the wafer with photoresist, exposes the wafer to light to pattern the resist and develops or rinses the soft areas of the resist from the hardened pattern on the wafer.

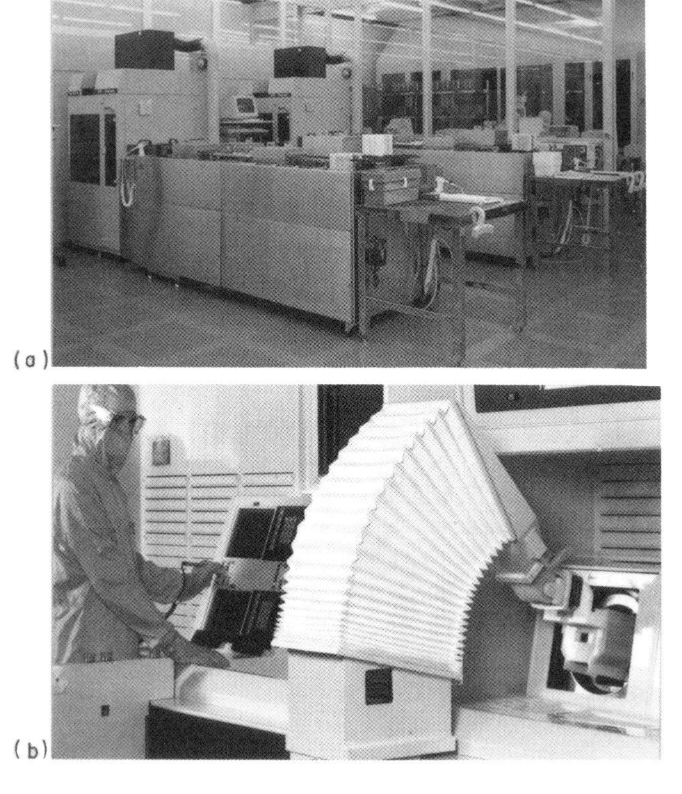

Figure 4.3
Modern integrated lithography and etch processes. (a) Direct interfaced coat–expose–develop units to improve process uniformity and increase throughput. (b) Automatic wet bench for etching which allows multiple processes to run in parallel. (Source N.V. Philips Research 1989.)

Figure 4.3(b) shows an automatic chemical 'wet bench' which allows multiple chemical steps to run in parallel such as etching the oxide surface exposed through the hardened photoresist pattern and then removing the resist from the wafer surface.

4.4 VARIOUS MOS FIELD EFFECT TRANSISTOR GATES

4.4.1 NMOS and PMOS field effect transistors

A completed NMOS Field Effect Transistor (MOS FET) cross-section is shown in Figure 4.4. The gates of an NMOS integrated circuit are formed of FET devices with a lightly P-type doped substrate and N-type doped drain and source regions. It is also possible to manufacture PMOS field effect transistors which have N-type doped substrate and P-type doped drain and source regions.

As speed is an important factor in MOS memories and because the NMOS gates are intrinsically faster than PMOS gates, NMOS was historically the preferred configuration for memories throughout the late 1970s and early 1980s. NMOS is faster than PMOS due to the higher intrinsic mobility of the negative electrons compared with the positively charged holes.

In spite of the speed advantage of NMOS, the earliest FETs were made using PMOS due to the greater controllability of the PMOS process. Contamination of NMOS devices posed a serious problem to early NMOS manufacturers and has led to expensive clean room procedures in modern wafer manufacturing facilities to minimize this problem.

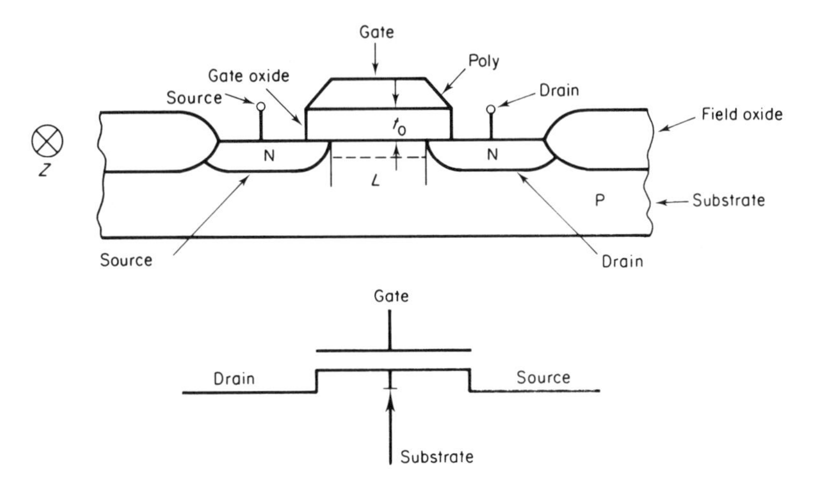

Figure 4.4
Cross-section of an NMOS silicon gate transistor. (From Prince and Due-Gundersen [66] 1983.)

4.4.2 MOS transistor characteristics

A brief review of MOS transistor characteristics follows. A MOS transistor is a device in which a current in a conducting channel between the source and drain is modulated by a voltage applied to the gate. In an n-channel transistor the majority carriers are electrons so a positive voltage applied on the gate with respect to the substrate enhances the number of electrons in the channel which is the region immediately under the gate and hence increases the conductivity of the channel. When the gate voltage reaches a value called the threshold voltage the channel region begins to conduct. If there is a voltage differential between the drain and source regions a current will flow.

Those devices that are normally non-conducting with no gate bias (gate-to-source voltage) are called enhancement mode transistors and those which conduct with zero gate bias are called depletion mode devices. Typical graphs of drain current plotted against gate-to-source voltage for NMOS enhancement and depletion mode transistors are shown in Figure 4.5(a) and (b).

MOS transistors have three normal modes or regions of operation controlled by the gate–source voltage.

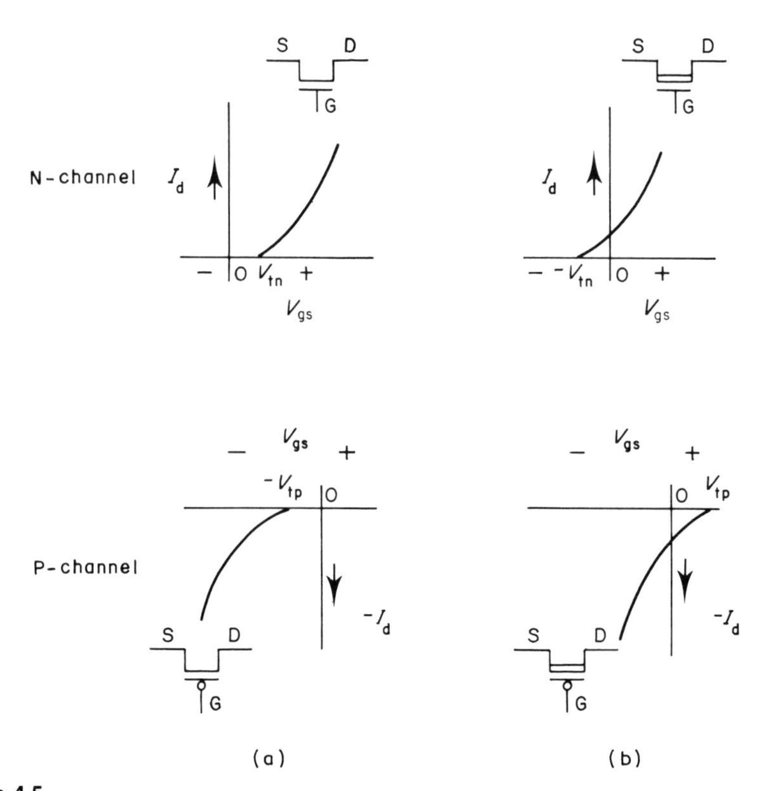

Figure 4.5
Typical transfer characteristics of (a) enhancement mode and (b) depletion mode MOS Transistors. (V_{gs} gate-to-source voltage, I_d drain current). (Adapted from Hodges and Jacksen [69] 1983.)

(a) The 'cut-off' where the gate voltage is low enough that the only current flow is due to source drain leakage current.

(b) The 'linear' region where the gate voltage is high enough to turn the transistor on but is still low and the resulting channel inversion is weak. In this region the drain current increases linearly with gate voltage.

(c) 'Saturation' where the gate voltage is high and the channel region is strongly inverted. In this region the drain current is independent of the drain voltage.

There is also an abnormal mode of operation called avalanche breakdown or punch-through which occurs when the drain–source voltage is very high. Under this condition the gate no longer has control over the drain current.

The ideal first-order equations describing the behavior of an NMOS device in the three regions are

$$I_{ds}\begin{cases} \text{(a)}\ 0 & -V_{gs}-V_t\leq 0 \quad \text{cut-off} \\ \text{(b)}\ \left[B(V_{gs}-V_t)V_{ds}-\dfrac{V_{ds}^2}{2}\right] & 0<V_{ds}<V_{gs}-V_t \quad \text{linear} \\ \text{(c)}\ \left[\dfrac{B}{2}(V_{gs}-V_t)^2\right] & 0<V_{gs}-V_t<V_{ds} \quad \text{saturation} \end{cases}$$

where I_{ds} is the source–drain current, V_{gs} the gate–source voltage, V_t the threshold voltage, and B the MOS transistor gain. B is dependent on the process parameters, materials characteristics and thickness of the gate insulator, and the geometry of the MOS transistor.

Schematic cross-sections of the basic enhancement and depletion mode transistors are shown in Figure 4.6 along with characteristic output I–V curves for N-channel and P-channel enhancement and depletion mode devices.

The enhancement mode transistor is normally off and needs a minimum applied threshold voltage on the gate to turn it on. With no gate–source voltage difference applied, it does not have a channel between the source and drain.

A depletion mode transistor is normally on and a voltage must be applied to the gate to turn it off. Figure 4.6 also shows the output characteristics of the NMOS and PMOS enhancement and depletion mode transistors.

These characteristics are such that the depletion mode is superior for using as a load device in a CMOS invertor since it increases the speed and lowers the supply voltage. In the four transistor CMOS cell, its similarity to resistor characteristics is used to substitute a resistor for the depletion load devices.

The threshold voltage of the transistor can be affected by ion doping of the silicon under the gate, or by using different gate materials. It can also be affected by different substrate voltages encountered, for example, when transistors are arranged in series so that the source to substrate voltages differ as you move along the chain from ground. This difference in substrate voltage is called the body effect and results in an effective threshold voltage given by $V_t = V_t(0) + g(V_{sb})$ where g is a constant

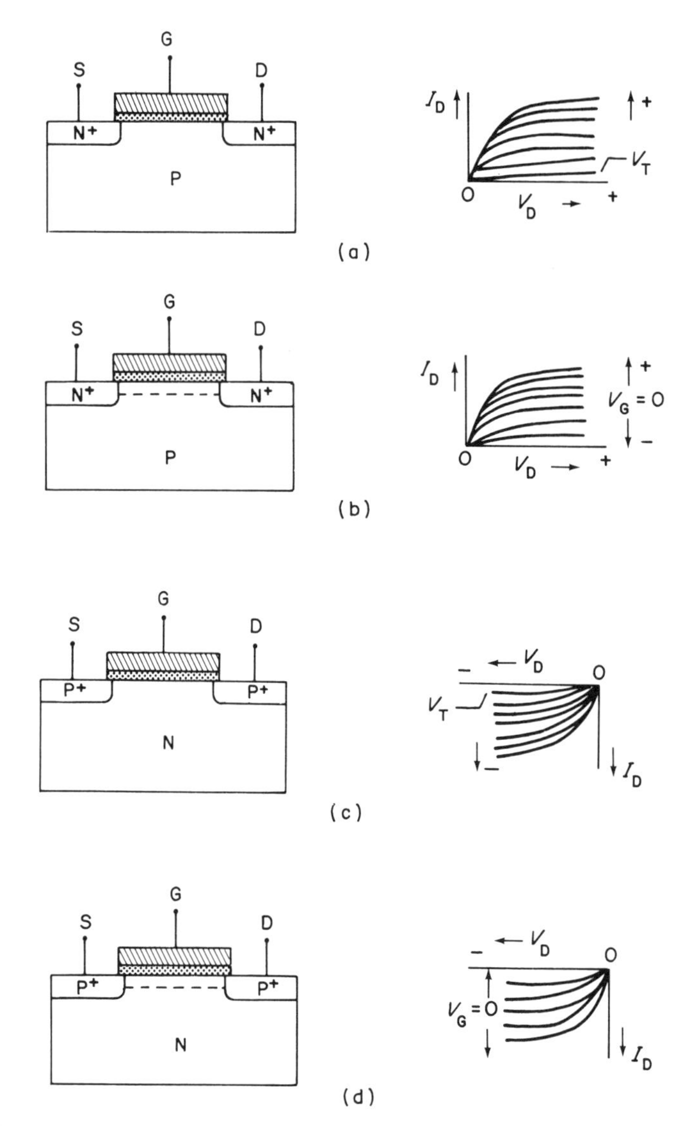

Figure 4.6
Schematic cross-sections and output characteristics of N-channel and P-channel depletion and enhancement mode transistors. (a) N-channel enhancement (normally off), (b) N-channel depletion (normally on), (c) P-channel enhancement (normally off), (d) P-channel depletion (normally on).

which depends on substrate doping, and the thickness and materials properties of the silicon dioxide. The sign is negative for a PMOS transistor.

As a transistor is scaled to smaller sizes 'short channel' effects occur which are neglected in the basic transistor equations. For a short channel transistor the channel length is replaced by the 'effective' channel length.

4.4.3 CMOS field effect transistors

The advantages of a combination of NMOS and PMOS transistors was investigated in the late 1960s and began to be used widely in building integrated circuits in the early seventies. Much of this early development work was done by the RCA Laboratories. This combination called complementary MOS (CMOS) consists of a series combination of the basic NMOS and PMOS gates as shown in Figure 4.7 which indicates both the circuit symbols and a cross section of a basic CMOS invertor structure in silicon.

The historical CMOS process used a ion dopant which was diffused into the surface of the silicon to form both the wells and the drain and source regions of the transistors. This historical process also used aluminum gates on the transistors, and was as a result called 'metal gate' CMOS.

The more modern CMOS memory process uses ion implantation to form positive and negatively doped regions of the silicon and uses ion doped polysilicon gates for the transistors. In addition to adding the flexibility of a second interconnect layer,

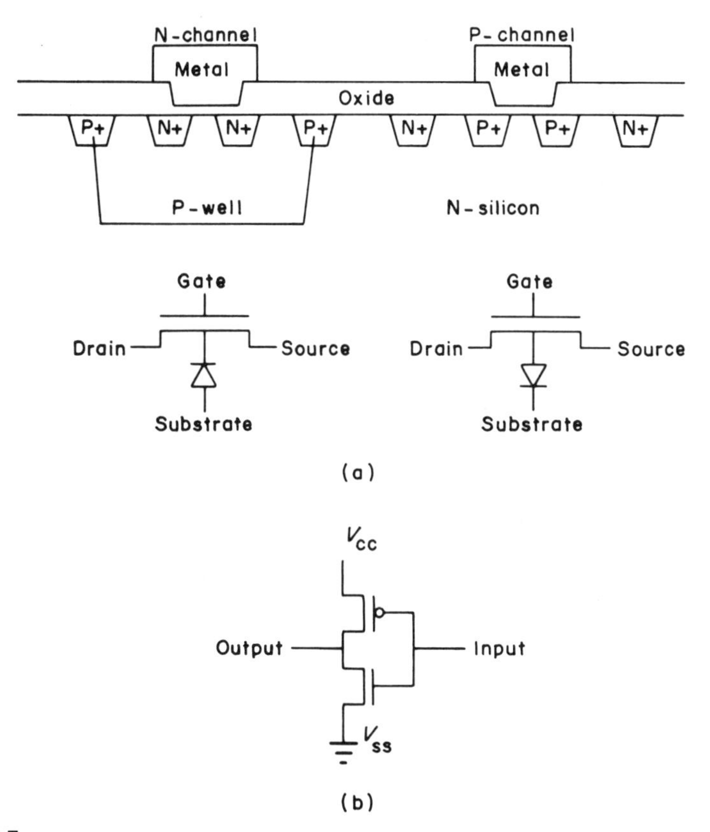

Figure 4.7
(a) Schematic cross-section of a metal gate CMOS process. (b) schematic circuit diagram of a basic CMOS invertor. (From Prince and Due-Gundersen [66] 1983.)

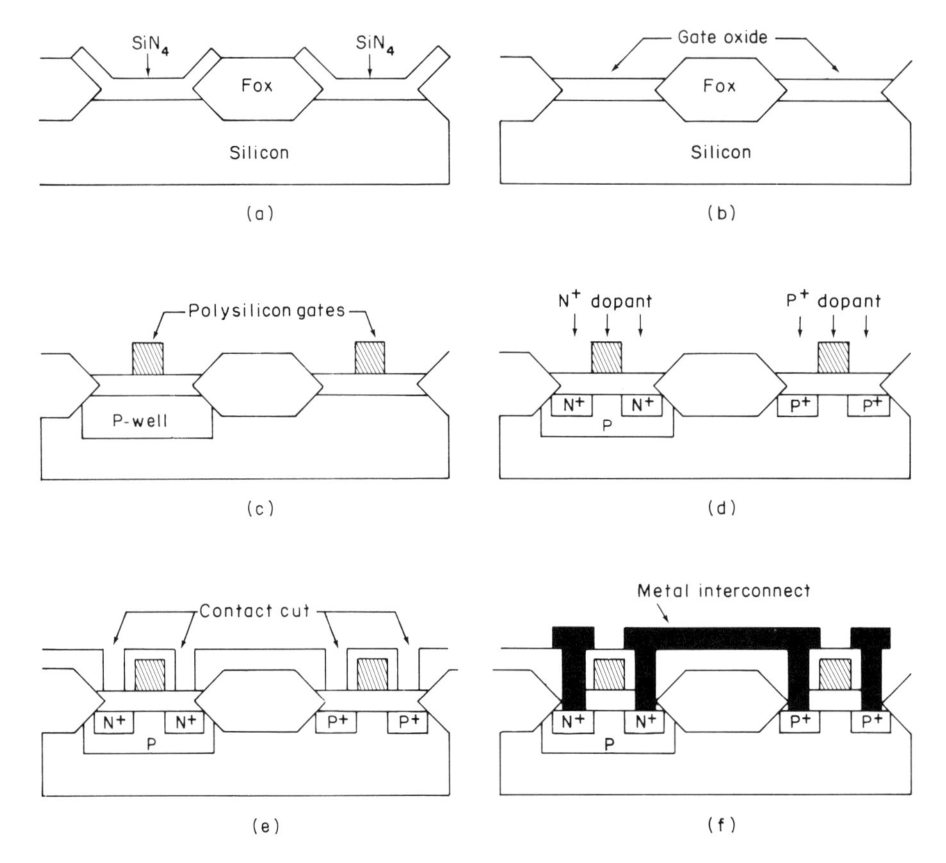

Figure 4.8
Silicon gate CMOS process using ion implantation. (a) SiN_4 is deposited on the surface of the wafer, patterned, and a thick field oxide (FOX) is grown in the opened regions of silicon (LOCOS process). (b) The SiN_4 is removed and the gate oxide grown. (c) The P-wells are implanted through the gate oxide and amorphous polysilicon is deposited and patterned to form the transistor gates. (d) The P^+-source and drain regions of the P-channel transistor are implanted through a mask and are self-aligned by the silicon gate. The reverse mask is then used to implant the N^+ transistor source and drain regions which are also self aligned by the silicon gates. (e) A SiO_2 layer is deposited or spun onto the surface, and contact cuts are made in this oxide layer to the source and drain regions of the transistors. (f) Aluminum, or other metal, is deposited and patterned to form interconnects between the transistors and the surface is passivated for protection with a layer of oxide or other material. (Source N.V. Philips Research.)

the polysilicon gates allow reduced geometries by precisely defining the position of the source and drain regions. Undoped polysilicon has a high resistivity but, when doped with either negative (N) or positive (P) ions, the resistivity is lowered. An even lower resistivity can be obtained in modern processes by metal–polysilicon combinations, called silicides, which will be discussed in a later section.

A representation of the silicon gate CMOS process is shown in Figure 4.8. While P-well technology is shown here, N-well is also used for CMOS. The simplified process shown uses six masking steps. A modern CMOS process has between 15 and 20 masking steps.

Notice that in addition to the normal transistor structures formed under the gate oxides, transistor structures are also formed under the sections of field oxides. These are called parasitic field oxide transistors and must be controlled to avoid interference with the intended function of the circuit.

CMOS memories have a considerable advantage in being much lower in power than comparable NMOS circuits. This lower power consumption is due to the fact that either the NMOS or the PMOS transistor in a single CMOS gate is always off so that there is ideally no direct power drain in standby mode. The momentary power drain during switching when both transistors are half on plus the charging of the capacitance of the next gate accounts for the characteristic tendency of CMOS to drain more power during operation at higher speeds. CMOS memories also share the traditional CMOS advantage of operating over a wide voltage and temperature range and having wider noise margins. They have the additional advantage of reducing susceptibility to soft errors.

Historically the CMOS process was more expensive than the NMOS process, not only because of the larger size of the chip, but because of the increased complexity of the CMOS process. Additional processing steps are required for the 'P' channel devices. Over time, however, the complexity of the optimized NMOS process increased to the point where the incremental cost of CMOS was relatively small compared to the circuit advantages gained.

Several basic types of CMOS process technologies are available. CMOS can be formed of N-channel devices in P-type wells, P-channel devices in N-type wells, or both devices in 'twin-wells' of the opposite polarity. The three options are illustrated in Figure 4.9. Any one of these three techniques has all the advantages of standard CMOS.

Early logic compatible CMOS memory processes utilized P-wells because boron was the only impurity available that could form lightly doped wells in a silicon substrate of the opposite type. P-well CMOS was also compatible with the existing metal gate PMOS logic technology of the time.

N-well CMOS was used initially by memory manufacturers since it combined the low power of CMOS with the NMOS speed in the memory array and the then familiar NMOS manufacturing methods. Its processing complexity was similar to the optimized scaled NMOS process. The difficulty with forming N-wells was largely overcome by the advent of ion implantation which allowed 'N-type' impurity doping with a high degree of control.

There are a number of advantages historically cited for using N-well CMOS.

The speed of N-well CMOS came from putting the faster NMOS devices into the high resistivity P-type substrate. Since the standard NMOS process was also made with the same substrate, processing techniques developed to optimize the NMOS process could be directly transferred to N-well CMOS. The high resistivity minimized drain and source to substrate parasitic capacitances. This minimized the *RC* constant of the circuit and increased the speed.

The twin-well process was first developed by Bell Laboratories, and is now

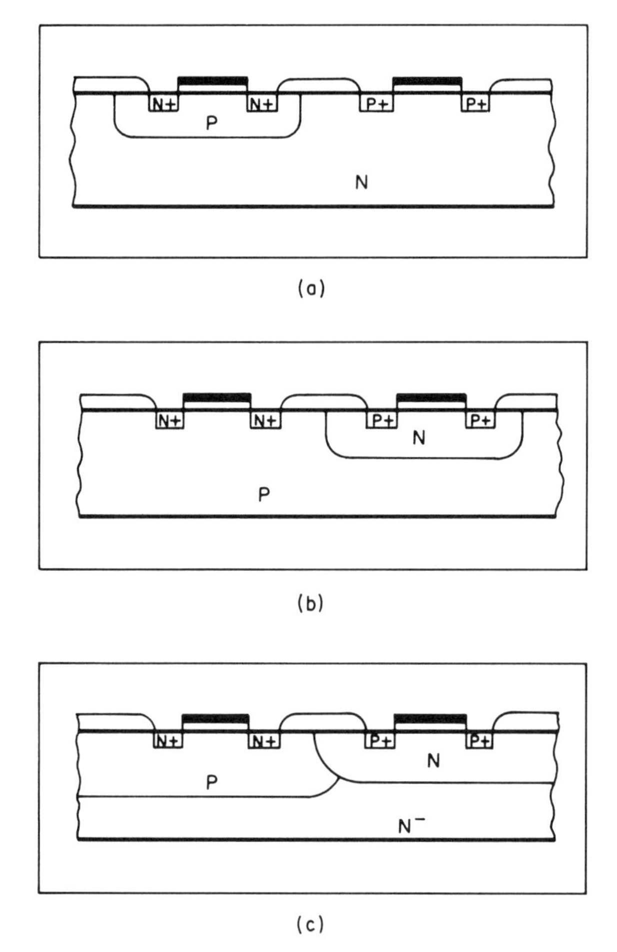

Figure 4.9
Various CMOS well structures. (a) P-well CMOS, (b) N-well CMOS, (c) Twin-well CMOS. (From Prince and Due-Gundersen [66] 1983.)

used as the predominant memory process by most volume CMOS SRAM manufacturers for the production of high performance CMOS static memories.

The use of twin-well CMOS results in several advantages over single-well technology. The diffusion of separate N- and P-type wells simultaneously optimizes the performance of both P- and N-channel transistors. Each well can be implanted with the optimum dose for the required device characteristics. Punch-through, where the drain depletion region extends to the source, is suppressed. One mask can be used for both wells. Most important, however, the ability to dope the wells separately provides flexibility as existing designs are scaled to smaller geometries.

Speed can be increased by use of twin-well technology. This is because of an inherent problem of single-well CMOS that, for example, in P-well technology, the P-type diffusion must overcompensate for the N-type substrate (in N-well the reverse must happen). In order to achieve a reasonable amount of control this diffusion has to be much higher in concentration than the N-type substrate. Heavy doping, however, leads to higher drain–source capacitance. This in turn increases the RC time factor of the circuit and results in slower speeds. The use of twin wells with intrinsic or undoped substrate reduces this problem and improves speed. Another factor which improves with the better dopant control of twin-well technology is the susceptibility of the device to latch-up.

Full CMOS SRAMs can be made with any of these three technologies. Mix-MOS SRAMs, which are an SRAM which is a combination of a CMOS periphery and an NMOS array, and can also be built with either P-well, N-well, or twin-well technology. In the case of N-well, the central NMOS array is implanted directly into the P-type substrate. This has the disadvantage of not affording the protection the shallow well gives to alpha particle induced errors, but permits the full utilization of scaling techniques developed for NMOS technologies.

In the case of P-well and twin well CMOS, the central NMOS array is implanted in a shallow P-well which protects against alpha particle sensitivity but was less familiar historically to the NMOS DRAM manufacturers who initially manufactured the high performance NMOS and Mix-MOS SRAMs.

A triple well technology has also been developed in which the periphery of the memory and array are each in their own wells which can be separately optimized.

4.5 CMOS INVERTOR CHARACTERISTICS

The CMOS invertor, consisting of an NMOS and a PMOS transistor in series as shown in Figure 4.7(b), is the basic gate in the CMOS circuit. It is used in the peripheral logic of the memory and in the input and output buffers as well as in the flipflop in a full CMOS SRAM cell.

4.5.1 An analytical approach

Basic CMOS invertor characteristics can be discussed using an analysis of an n-stage cascades network of invertors from a paper by Ahrons [26] 1970. An analytical expression for the delay time of a pair of CMOS transistors is

$$T_d \simeq \frac{0.9C_0}{V_0}\left[\frac{1}{K_n(1-\alpha_n)^2}+\frac{1}{K_p(1-\alpha_p)^2}\right]$$

with limits $\alpha_n + \alpha_p < 1.0$ and $K_p/K_n \leq 0.2$ where

$$\alpha_n = \frac{V_{tn}}{V_0}; \quad \alpha_p = \frac{V_{tp}}{V_0}; \quad K_n = \frac{\mu_n W_n \xi_1}{2l_n \tau_{0x}}, \quad K_p = \frac{\mu_p W_p \xi_2}{2l_p \tau_{0x}}$$

where C_0 is the total capacitance represented by a fanout of unity, V_0 the supply voltage,

K_n and K_p constants representing the conductances of N-type and P-type devices, respectively,

α_n and α_p the ratio of threshold to supply voltage of the n- and p-transistors, respectively,
W_n and W_p the channel width of an NMOS and a PMOS transistor, respectively and
l_n and l_p are channel lengths of an NMOS and a PMOS transistor, respectively.

The node capacitance 'C_0' is the sum of the various inherent capacitances in the invertor stages

$$C_0 = 2C_{gd} + C_w + C_g + 2C'_{gd}$$

where C_{gd} is the gate-to-drain capacitance of the driving stage,
C_d the total drain diffusion capacitance,
C_w is the wiring capacitance and
C_g represents a single gate of the total gate–drain capacitance of the next stage.

The total power dissipation P is given by

$$P = P_s + P_a$$

where P_s is the standby power, and P_a the active, that is, switching or transient power.

The active power is given by

$$P_a = C_0 V_0^2 f$$

where f is the operating frequency.

The standby power is given by

$$P_s = I_L V_0$$

where I_L is the total leakage current of the devices. In CMOS circuits this is very small.

The parameters which must be optimized in high performance CMOS circuits are the conductance, the node capacitance and the threshold voltages of the transistors. The parameters which can be controlled are the geometrical factors of channel length, channel width, the threshold voltages and gate oxide thickness, and the process dependent parameters such as effective carrier mobility.

The substrate also has an effect on the conductance of the MOS transistor as the source to substrate voltage is varied. The effective conductance of a transistor is

$$g_m = g_0 - g_b = \frac{g_0}{\left(1 + \dfrac{g_b}{g_m}\right)} = \frac{g_0}{\left(1 + \dfrac{V_g}{V_b}\right)}$$

where g_0 is the intrinsic device conductance, g_b the conductance due to the substrate gate, V_g the gate voltage and V_b the source to substrate bias.

4.6 EPROM MEMORY CELL: FLOATING GATE AVALANCHE INJECTION MOS

The storage transistor used in the EPROM is a variation of the NMOS transistor. This cell stores charge on a section of polysilicon which is floating in a sea of silicon dioxide. The silicon dioxide is about 50 to 100 nm thick and insulates the polysilicon

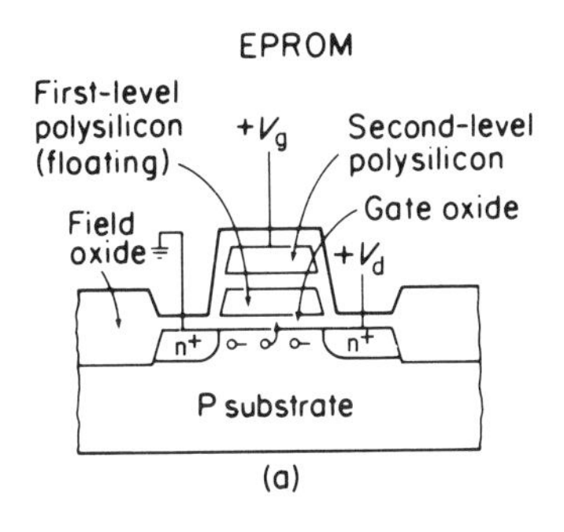

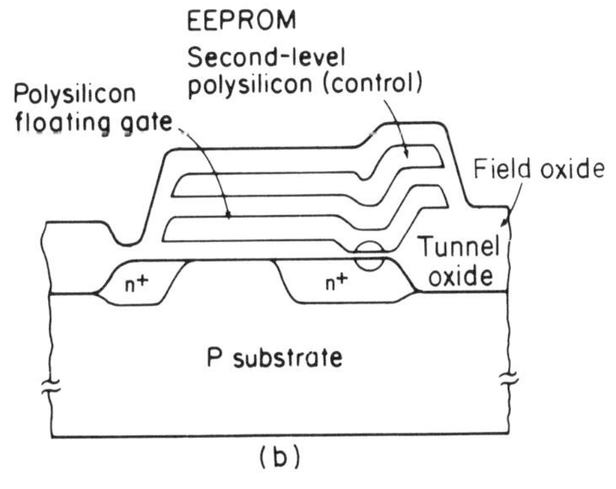

Figure 4.10
Two basic electrically programmable ROM cross-sections. (a) EPROM using floating gate avalanche injection MOS (b) EEPROM using a thin oxide region over the drain for Fowler-Nordheim tunneling of electrons during programming and deprogramming. (From Prince and Due-Gundersen [66] 1983.)

from the control gate above and the N-channel device below as shown in Figure 4.10(a). An electrically induced avalanche injection mechanism is used to charge the floating gate with electrons from the substrate. This charge is then retained until it is discharged by either ultraviolet light or X-rays. In practice, ultraviolet light is used. X-ray equipment has never been commercially developed for this purpose. The charge tends to leak off over time, see also Chapters 5 and 11.

4.7 EEPROM MEMORY STORAGE TRANSISTOR

An electrically erasable PROM (EEPROM) transistor cross-section is shown in Figure 4.10(b). While the construction of the cell is similar to that of the EPROM, the region of very thin 'tunnel' oxide, from 5 to 10 nm thick, over the drain region permits an electrically induced Fowler–Nordheim tunneling mechanism to be used to charge (discharge) the gate by electrons being attracted through the thin oxide to (from) the floating gate. For further discussion, see Chapter 5.

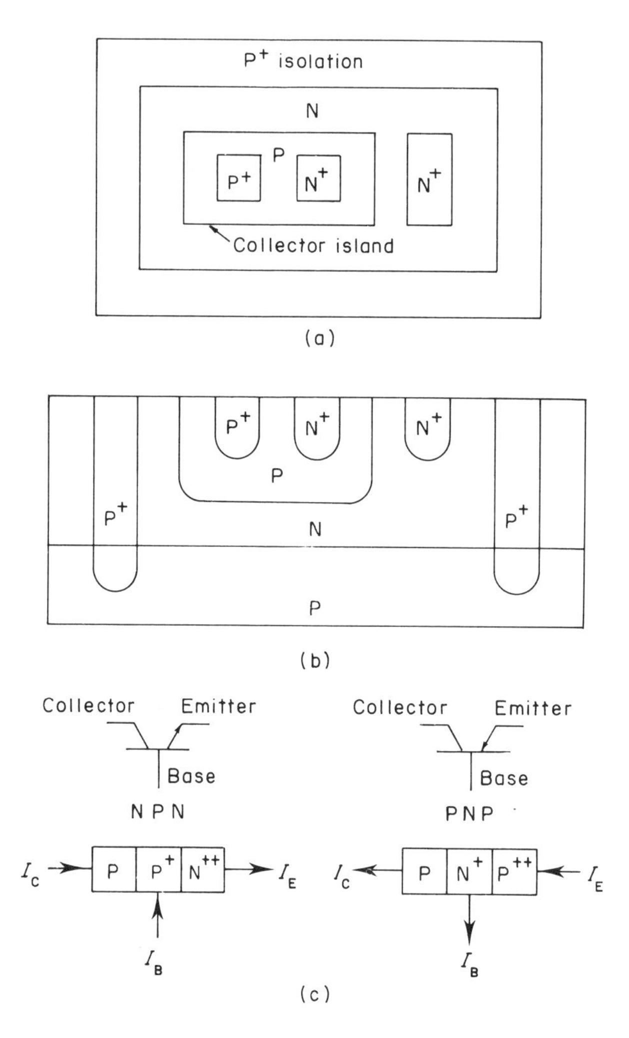

Figure 4.11
(a) Top view of NPN transistor. (b) Schematic cross section. (c) Schematic circuit diagram and bulk structure of NPN and PNP transistors.

4.8 BIPOLAR CIRCUITS

Other possible memory storage cells include the various bipolar gate configurations. Bipolar differs from MOS in including the bulk of the silicon wafer in the memory cell where MOS uses only the surface. Consequently refinements in both the vertical and the horizontal dimension on the wafer are critical to bipolar, whereas reductions in the lateral dimension are most critical to MOS. The critical dimension in MOS for speed and area efficiency is the channel length and in bipolar is the width of the base region for speed and the horizontal dimensions for speed and area. The top view and a schematic cross section of a bipolar transistor together with its isolation region is shown in Figure 4.11(a) and (b). Circuit schematics of an NPN and PNP bipolar transistor are shown in Figure 4.11(c).

A bipolar memory built with Emitter Coupled Logic (ECL), which is the most common bipolar circuitry, has a very high speed and drive capability but low cell density on the wafer due to the size of the bipolar transistor with its isolation region. Bipolar TTL circuits have the potential for higher density than ECL but are not as fast.

4.9 BiCMOS GATES

Combinations of CMOS and bipolar transistors have yielded good results both for improving the drive of the MOS circuits as they are shrunk to smaller and smaller geometries and for allowing critical areas in a CMOS circuit to be speeded up by the inclusion of a bipolar transistor with larger current drive capability for heavy loads. Another reason for using a combination CMOS and bipolar technology is to interface a high density CMOS circuit into a bipolar ECL system. BiCMOS can be made from a basic CMOS process or a basic bipolar process or it can be optimized for both transistors at the expense of a much more complex process.

An example of a BiCMOS transistor cross-section which is optimized for CMOS is shown in Figure 4.12(a). A comparison of gate delay as a function of load capacitance for a CMOS and BiCMOS ring oscillator is shown in Figure 4.12(b).

Combinations of bipolar and CMOS transistors on the same chip have been used to advantage in memories in the micron geometry range. The major performance advantage of bipolar is in the case of high capacitance loading.

Bipolar devices have suitable characteristics for the analog parts of the circuit, such as the differential amplifiers, as a result of their smaller threshold voltage variations and their inherent higher sensitivity for small signals. They can also provide a high speed sense amplifier with low offset voltage. The input signal swings for the same output level swing are less for bipolar than for CMOS and the maximum gains of bipolar are greater. In the output buffers of a circuit bipolar offers high current driver capability and high speed.

CMOS technology, on the other hand, does better in the digital parts of the circuit because the delay, area of chip occupied, and the power dissipation capacity of CMOS are superior.

The problem encountered in BiCMOS combinations is the higher process complexity and hence higher cost of both optimized CMOS and bipolar transistor structures. A compromise is normally sought of either making CMOS in a bipolar process, or making bipolar in a CMOS process. Various levels of complexity can then be added depending on which approach is chosen to optimize those features which are needed in the circuit. There have been indications that the use of ion implantation techniques can aid in optimizing both transistor types.

4.10 LATCH-UP MECHANISM

Latch-up is an effect of a parasitic bipolar circuit which exists intrinsically in CMOS. It can result in the shorting of the power and ground lines resulting in either

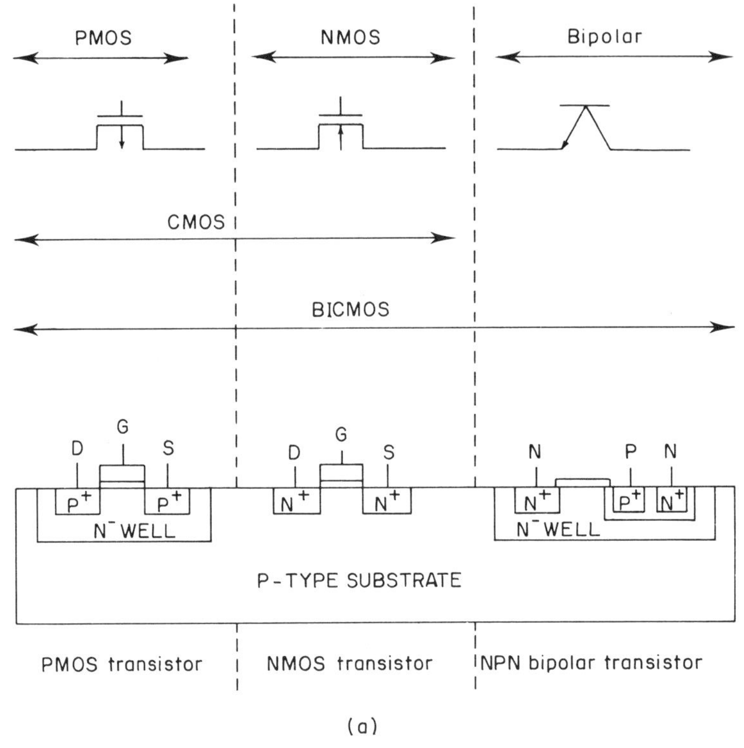

Figure 4.12
(a) Schematic cross-section of a simplified N-well BiCMOS process. (b) Comparison of CMOS and BiCMOS delay time plotted against load capacitance for a 15-stage ring oscillator. (Reproduced with permission of Toshiba.)

destruction of the chip or system failure. The parasitic bipolar effect is shown in Figure 4.13. The source of the latch-up effect is the switching on of the two parasitic bipolar transistors T_1 and T_2. This draws current from the power supply which can be large enough to result in power failure.

While latch-up was considered a serious problem in the early years of CMOS, it is well enough understood now to be controlled in modern design sufficiently that it is no longer a major problem in CMOS although still a consideration in any CMOS circuit design.

There are trade-offs between the input protection circuitry and latch-up protection which must be carefully thought about.

The advent of combining bipolar and CMOS circuitry on one memory chip has brought a new concern about latch-up since latch-up can be triggered by the active bipolar transistors switching into saturation. The base–collector junction becomes in this case forward biased and current is injected into the substrate. CMOS diodes in the neighborhood of the active bipolar transistor can be forward biased causing the parasitic bipolar transistors to exhibit the latch-up phenomena. One solution has been

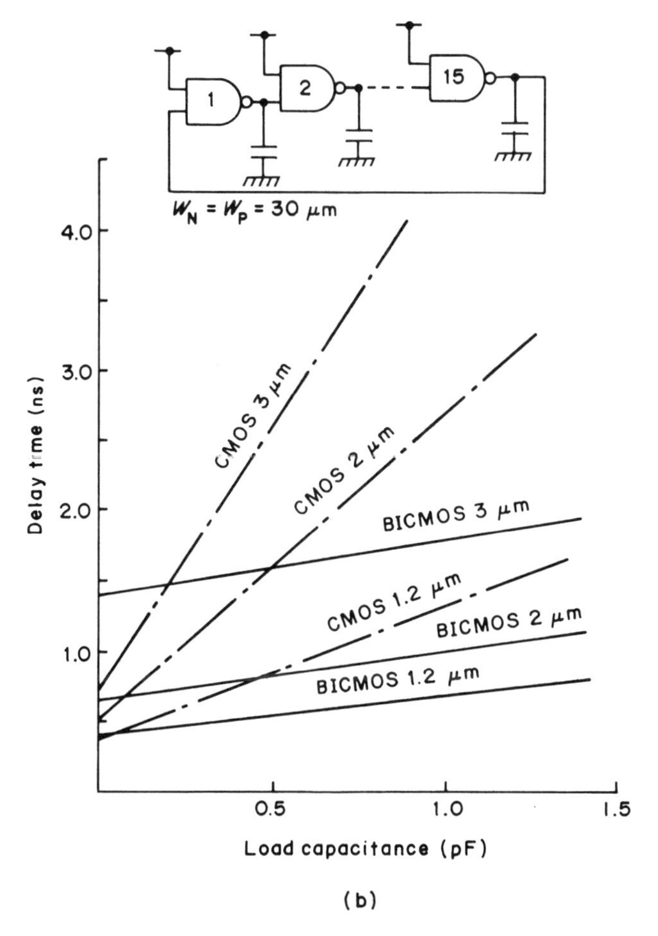

Figure 4.12 *(continued)*

shown to be the use of a buried layer in the well, which is also the highly doped part of the collector of the active transistor.

4.11 MOS SCALING

Significant performance improvements were obtained in MOS memories in going from the 6 μm to the 1.5 μm minimum geometry range. This derived from simple scaling of the chips to smaller dimensions, since the speed of the memories is enhanced by a decrease in effective channel length. Figure 4.14 shows L_{eff} plotted against density for first and for scaled second generations of SRAMs.

Speed enhancements in a single product generation can be obtained from these chip size reductions. Also the greater number of potential chips on a wafer results in higher manufacturing output for the same yield and hence lower costs.

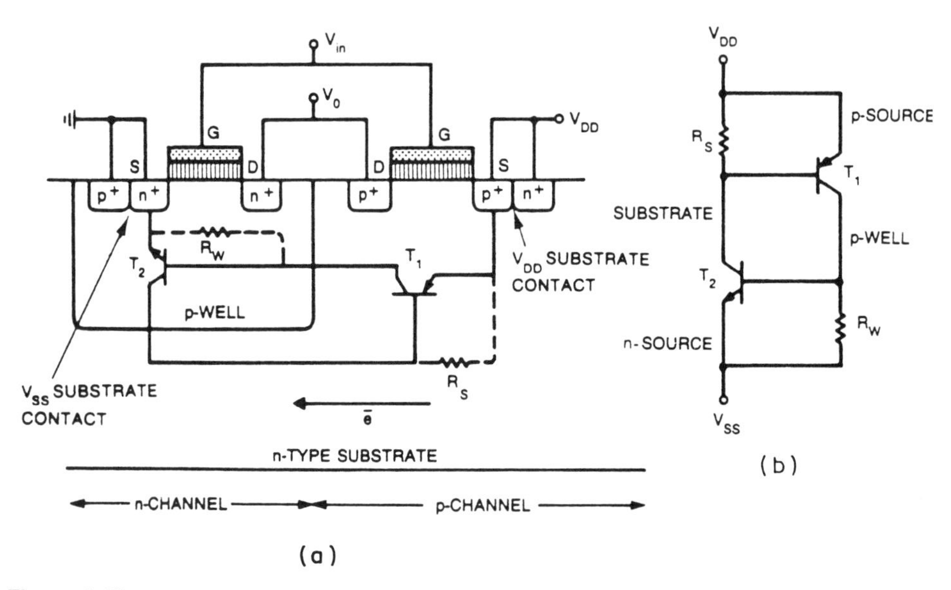

Figure 4.13
Source of latch-up in CMOS circuitry. (a) Schematic cross-section showing parasitic bipolar structure in CMOS. (b) Equivalent circuit for parasitic transistor. (From Jackson and Hodges [69] 1983.)

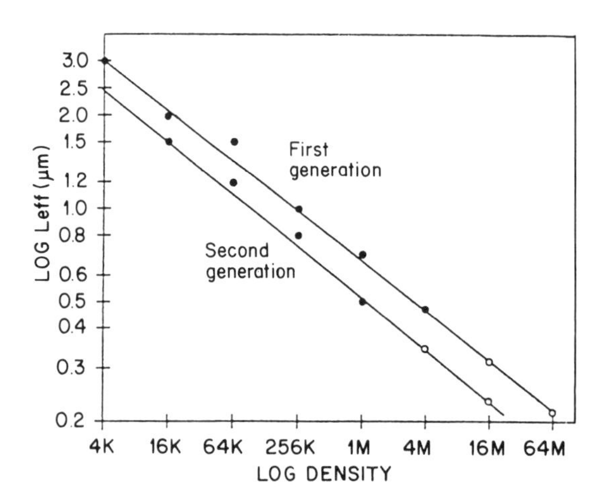

Figure 4.14
Density plotted against L_{eff} for first and second generation SRAMs. (From Flannagan [15], Motorola 1988, with permission of IEEE.)

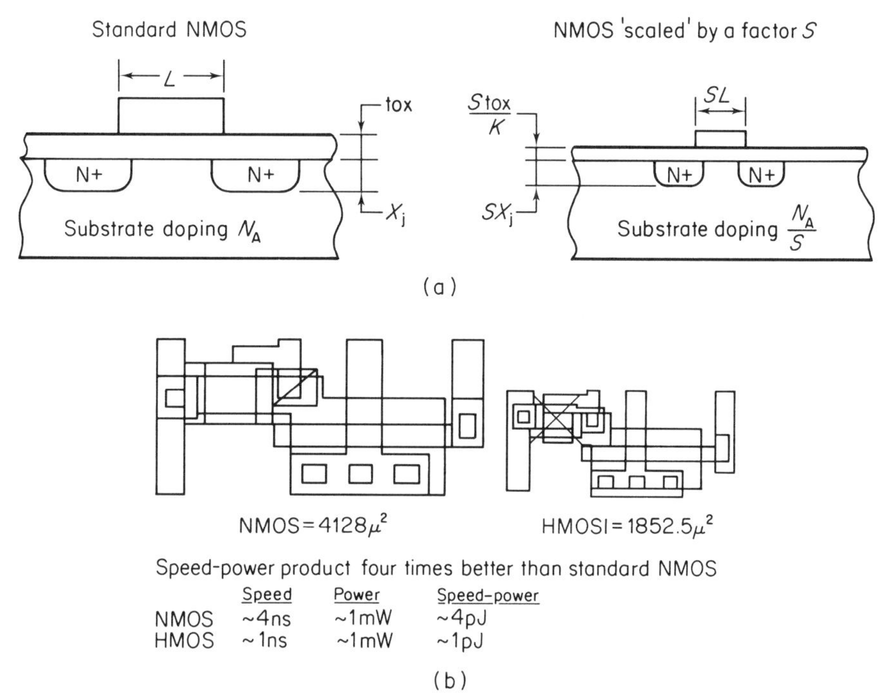

Figure 4.15
Illustration of scaling factors for >2.0 μm NMOS. (a) Cross-section, (b) top view. (From Prince and Due-Gundersen [66] 1983.)

This potential for improvement in MOS technology is a result of the ability of MOS to be scaled to smaller sizes while retaining and even improving its performance characteristics. Scaling is not, however, simply a matter of shrinking the lateral dimensions of the pattern on the surface of the chip. All dimensions both parallel and perpendicular to the surface of the wafer must be scaled as shown in Figure 4.15 where S is the scaling factor. All deposited or grown layers such as silicon dioxide and polysilicon must be thinned. In addition the dopant charge concentrations and the power supply voltage must also be reduced to gain the full benefits from scaling.

If all device dimensions and voltages are scaled properly then a reduction in channel length by one half results in approximately a four-fold reduction in both the speed–power product, and a reduction in the surface area of the MOS device to approximately 25%.

Scaling the technology reduces chip size and permits higher densities of MOS memories to be made more cost effectively. It also gives a significant improvement in performance. These scaling techniques have been transferred directly from NMOS to CMOS technologies and provide similar chip size and speed reduction advantages in both.

An example of the effects of scaling on the performance of a memory is

	Intel 2147* 4k × 1 static RAM	Intel 2147H* 4k × 1 static RAM	Intel HMOS-III** development vehicle 1k × 4-bit static RAM
Technology	HMOS	HMOS II	HMOS III
Year of introduction	1977	1979	n/a
Address access time (ns)	55	35	14
Chip select address time (ns)	55	35	21
Organization	4k × 1	4k × 1	1k × 4
Automatic power-down	yes	yes	yes
Common I/O	no	no	yes
TTL compatible	yes	yes	yes
Active power (mA)	150	180	100
Standby power (mA)	30	30	20
Power supply	+5 V only. ± 10%	+ 5 V only. ± 10%	+ 5 V only. ± 10%

* Data sheet
** Typical

(a)

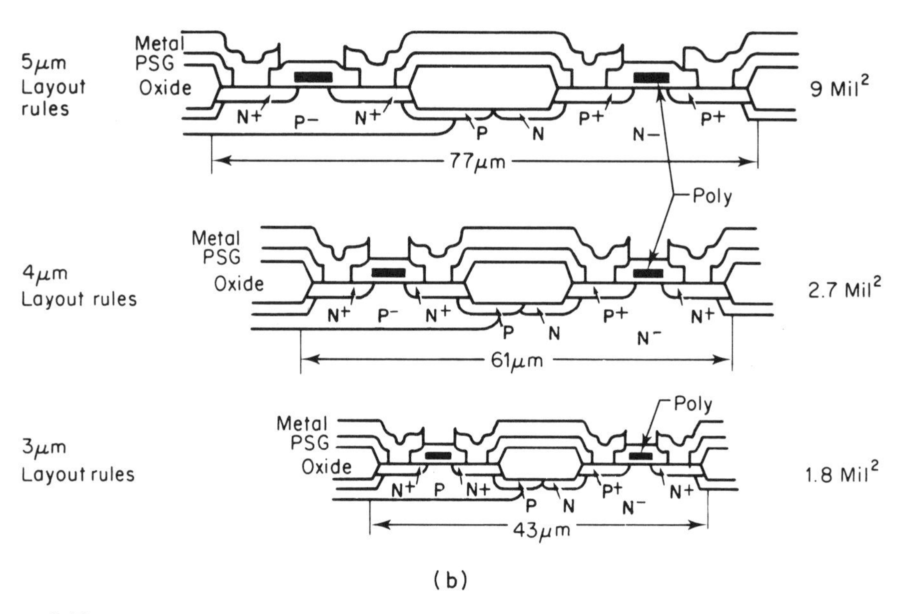

(b)

Figure 4.16
(a) Comparison of performance of a 4k SRAM for three generations of HMOS process technology. (b) Cross-section of a CMOS SRAM cell comparing scaling dimensions for three generations of HCMOS technology. (From Prince and Due-Gundersen [66] 1983.)

shown on an early NMOS SRAM, the Intel 2147, which is illustrated in Figure 4.16(a) for three generations of process technology.

Figure 4.16(b) shows schematic cross-sections of the scaling process for a CMOS memory gate for similar CMOS processes in 5, 4 and 3 μm design rules. It is clear that each reduction in process technology gives a significant improvement in the timing parameters as well as reduction in cell size.

For geometries below 2.0 to 3.0 μm the earlier simple scaling did not work due to the lack of possibility to scale the power supply voltage also. New scaling laws were investigated below the 2.0 μm level. Figure 4.17(a) indicates a set of scaling laws which can apply to VLSI from 3 to 0.8 μm minimum geometries. S is the scaling factor.

Column 3 indicates the simple classical scaling of circuit and device parameters, as proposed by Dennard. As indicated simple scaling could be applied down to the 2 μm minimum geometry range for all circuit and device characteristics except supply voltage which is constrained to a standard 5 V.

From 2 μm to about 0.8 μm the same scaling is effective except that the resistance and line delay are increased by the increased resistance of the narrower interconnects and the resulting higher RC time constant. Power consumption increases below about 2 μm due to the increase in packing density. At 0.5 μm and below it becomes desirable to scale the supply voltage also to avoid drain–source breakdown and hot electron effects. One example of device characteristics for NMOS for design rule scaling are shown in Figure 4.17(b).

Since memory applications have historically required a 5 V external power supply, the full benefits of scaling have not been completely available. Various design methods are used to compensate for the 5 V external power supply requirement which will be explored in the following chapter. Significant benefits in device characteristics would result from a shift to a lower voltage standard such as reductions in power–delay product, punchthrough, gate–dielectric breakdown, and hot electron effects.

Many other factors than simple transistor geometry must be considered in scaling memories in order to get higher densities of arrays on similar size chips. For example, while scaling down device dimensions has the effect of increasing the speed of the device, scaling the interconnects between devices while at the same time increasing the density four fold can have the opposite effect of decreasing the speed of the circuit. This is because scaling a resistor decreases its cross-sectional area which increases the resistance, while going to the next density chip keeps the interconnect lines approximately the same length. The result is an overall increase in resistance. This in turn increases the RC time constant of the circuit which tends to slow the chip down.

One solution to this problem is to use lower resistivity interconnect materials such as refractory metals for the gate and silicides for the polysilicon interconnection lines. Such materials tend, however, to add to the cost and complexity of the process.

It is necessary to optimize the product and the technology. For example, in static RAMs with resistor–load cells the high value resistors required for low dc load current can result in slower speeds. This is because the current through the load resistor must be balanced by the leakage current of the storage transistor so that low leakage, high

Circuit and device parameter	Symbol	Simple scaling	3–2 μm V_D const.	1.3 μm V_D const. L const.	0.8 μm V_D scaling L const.
Supply voltage	V_D	$1/S$	$1/S$	$1/S$	$1/S$
Drain current	I_D	$1/S$	$1/S$	$1/S$	$1/S$
Capacitance (τ_{ox})	C_L	$1/S$	$1/S$	1	1
Capacitance (x_j)	C_d	$-1/S$	$1/S$	1	1
Conductance	G_m	1	S	$<S$	1
Tail constant	α	1	1	1	1
Resistance	R	S	S	S^2	S^2
Line delay	τ_L	1	1	S^2	$>S^2$
C/G_m	τ	$1/S$	$1/S^2$	$>1/S$	1
Power	P_w	$1/S^2$	S	S	$-1/S^2$
Power and delay prod.	$P_w\tau$	$1/S^2$, $1/S^2$	S, $1/S$	S^3, 1	1, $1/S^2$

L means an interconnection length.

(a)

Device characteristic	Design rules 3 μm	2 μm	1.3 μm	0.8 μm
V_{th} lowering				
ΔV_{th} (V)	0.3	0.3	0.3	0.45
Tail constant				
α (mV/decade)	70	70	70–80	80
Back bias constant				
K ($V^{1/2}$)	0.2	0.1	0.1	0.1
Triode				
G_m (μS μm)	1.8	3.5	8.3	20
Saturation				
G_m (μS μm)	18	47	82	140
BV_{DS} (V)	9.6	8.7	7.7	5.5
BV_{HC} (V)	6.5	5.3	~4.0	~3.0

(b)

Figure 4.17
(a) Examples of scaling laws for VLSI MOS memory circuits. (b) Corresponding device characteristics. (From Takeda *et al.* [20], Hitachi 1985, with permission of IEEE.)

threshold storage transistors must be used. These in turn impair speed, and hence there is a trade-off between high resistivity of the load resistor and speed.

This particular problem can be solved in modern memories by using more layers of polysilicon interconnect allowing different layers to be optimized for different circuit functions.

Another factor to consider in scaling is hot electron trapping which is a reliability consideration resulting from scaling both the gate oxide thickness and channel length without proportionately scaling down the power supply voltage. This results in a substantially higher electric field being created in the MOS device which will increase the energy of the electron carrier in the channel region. With enough energy, these 'hot electrons' can inject through the interface energy barrier and become trapped in the gate oxide, thereby causing a shift in the transistor threshold voltage, as well as a degradation in the performance. Careful optimizing of the processing can reduce the impact of this ionization current and minimize the hot electron effect in greater than 1 μm geometries.

While reducing the power supply voltage on a process which is optimized for a 5 V supply can result in speed degradation, for any technology an optimal voltage that maximizes the speed (stage delay) can be found. See Figure 4.18 for a plot of the stage delay plotted against power supply curve for an 0.25 μm process. The optimal power supply voltage indicated to maximize speed is 2.5 V.

The upper limit of operating voltages is determined by the hot carrier allowable voltage. Hot carrier minimizing structures such as double-diffused drains (DDD) and lightly doped drains (LDD), which will be described in a later section, can be used but result in a trade-off between performance and device breakdown voltage and design margins. These hot-carrier resistance device structures have a drawback of causing a decrease of the transconductance since channel doping must be increased which decreases the electron mobility.

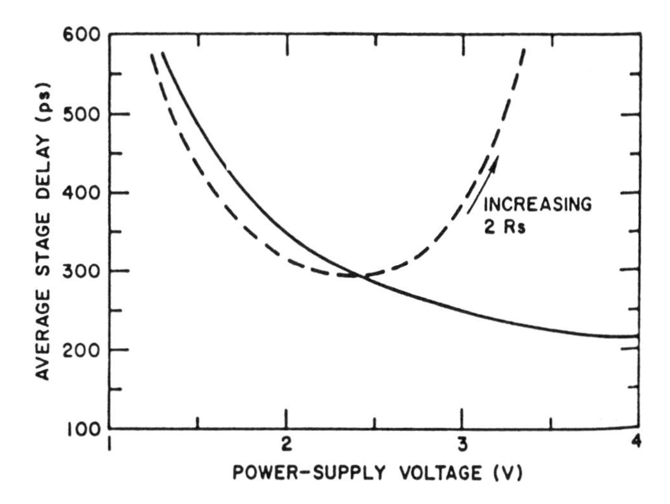

Figure 4.18
Simulated power supply voltage for maximum speed for an 0.25 μm MOS technology. The solid line represents a fixed technology (fixed S–D junctions). The broken line represents fixed hot carrier margin and varying S–D junction series resistance with supply voltage (FI = FO = 3 NAND, $W/L = 15/0.25$, Cwire = 200 fF). (From Davari *et al.* [28], IBM 1988, with permission of IEEE.)

Problems in scaling include not only the hot carrier effects due to the degraded channel region becoming comparable with the channel length, but also the parasitic resistances of the interconnects and of the contact become more important. In order to keep the sheet resistance low there is a need to scale the vertical dimensions of the interconnects by less than the horizontal ones. This causes problems with planarization and yield particularly in higher density devices since the more advanced processes tend to have more layers of interconnects.

Moving into the submicron realm of process technologies, simple scaling techniques are no longer sufficient. Special techniques have been developed for submicron MOS processing. Further discussion of deep submicron device theory can be found in a 1991 paper by Hitachi [86].

4.12 DEVELOPMENT OF MOS PROCESS TECHNOLOGY

As MOS technologies are scaled to submicron feature sizes, increasingly sophisticated techniques are required. The chart in Figure 4.19 traces some of the past advances in lithography, oxide isolation and interconnect structures and makes some forecasts for trends in the future. The following section provides a brief description of these processes.

DRAM capacity	4Mb	16Mb	64Mb	256Mb	$\geqslant$1Gb
Feature size	0.8 μm	0.5 μm	0.3 μm	0.2 μm	0.15 μm
Lithography in production	g-Line	i-Line	i-Line (+Phase shift) Excimer	Excimer +Phase shift EB, X-ray?	EB, X-ray?
Capacitor oxide	9–7 nm	6–4 nm	~3 nm	2 nm	1.5–1 nm
Insulator	Si_3N_4/SiO_2	Si_3N_4/SiO_2	Ta_2O_5	Ta_2O_5?	?
Junction leakage current			$<10^{-16}$ A/μm^2,	$<10^{-18}$ A/μm^2	
Isolation	1.0 μm	0.8–0.7 μm	0.5–0.4 μm	~0.3 μm	0.2–0.15 μm
	LOCOS		Improved LOCOS		Trench
Interconnection	Al	Al W	Al W	Al W, Cu	Al W, Cu
Voltage (array/external)	3.3/5.0	3.3/5.0	2.3/3.3	1.5/3.3	1.0/1.5

Figure 4.19
Development forecast for various process technologies through the 1Gb DRAM. (Adapted from presentation by K. Itoh at Panel Discussion VLSI Technology Symposium 1990. With permission of Hitachi).

4.12.1 Silicon dioxide as the basis of MOS processing

The ability to grow layers of SiO_2 on the surface of a silicon wafer is the basis of MOS process technology. The SiO_2 is a good electrical insulator and is used both between vertical layers of conducting material deposited on the wafer surface, and is used vertically as an electrically insulating barrier between transistors on the surface of the wafer. Layers of SiO_2 on the surface of the wafer can be grown, chemically deposited, or spun on as liquid glass.

Grown SiO_2, called thermal oxide, is possible because silicon, if heated in the presence of oxygen, will combine chemically with the oxygen to form SiO_2 which eats into the silicon surface as it grows. Grown oxide is sufficiently defect free to be used for the thin gate oxide of a high quality MOS transistor.

Deposited oxide, called chemical vapor deposition (CVD) oxide, normally is used to provide an electrically insulating layer so the defect density is less important as long as there are no holes in the oxide. Since oxide is normally deposited after high temperature diffusion of ionized dopants into various profiles in the surface of the wafer, a criteria is that it be deposited at relatively low temperature to avoid disturbing the diffusion profiles. Spin-on-glass (SOG) has similar requirements and properties.

4.12.2 Local oxidation of silicon (LOCOS)

Early MOS fabrication processing began with a thick field oxide grown over the entire surface of the wafer. Areas were then opened in this thick layer for the transistor formation. This process had the problem of steep steps on the oxide which resulted in the breaking of metal lines deposited over these steps (step coverage problem).

Modern MOS process techniques usually use the 'LOCOS' process developed by Philips which is a selective field oxidation of areas on the surface of the silicon which will not be used for active devices. This was shown in Figure 4.8(a). Selective oxidation has the advantage of a slope in the oxide step on the wafer surface, called a bird's beak, so that breaks are less likely to occur in the layers of interconnect material which run over these steps.

The LOCOS process involves depositing SiN_4 on the wafer in patterns which cover areas of the silicon in which active transistor devices will be formed in later steps. Silicon dioxide will not grow under SiN_4 deposited over a silicon layer. A thick field oxide is then grown in the areas which are not covered with the SiN_4 then the SiN_4 is removed leaving the bare silicon exposed in the active areas.

4.12.3 'Bird's beak' effect

The bird's beak oxide slope in the LOCOS process occurs because thermal silicon dioxide grows fairly uniformly in all directions so there is some lateral growth of

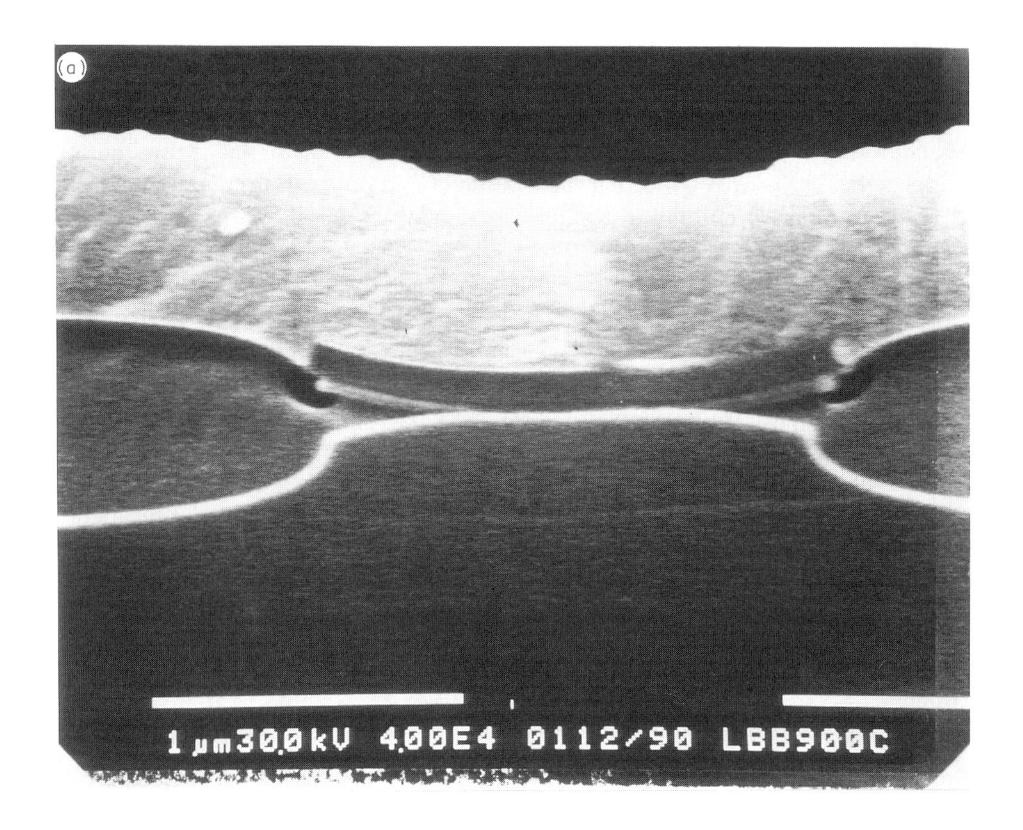
(a)
1 μm 30.0 kV 4.00E4 0112/90 LBB900C

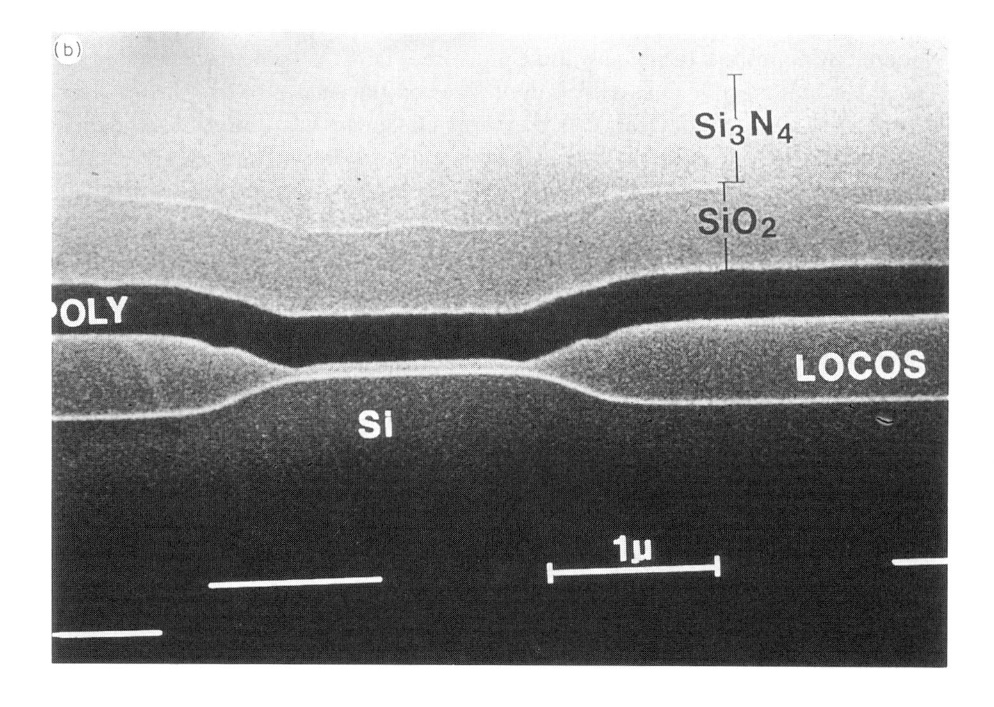
(b)
Si3N4
SiO2
POLY
LOCOS
Si
1μ

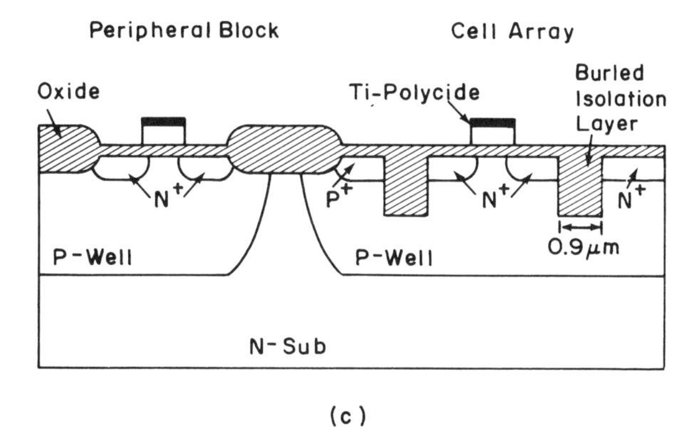

Figure 4.20
Different types of improved isolation techniques: (a) traditional LOCOS isolation, (b) improved LOCOS isolation from 0.7 μm 1Mb SRAM process (From Bastiaens and Gubbels [58] 1988, with permission of Philips), (c) trench isolation of n-channel MOSFET in the cell area combined with LOCOS isolation in the periphery (from Kobayashi *et al.* [61], NEC 1985, with permission of IEEE.)

oxide under the SiN_4 protective layer in the form of a bird's beak as shown earlier in Figure 4.8(a). While the 'bird's beak' improves step coverage, it also has an undesirable effect in modern small geometry scaled processes that it both occupies needed area and has electrical field effects that permit current leakage.

The basic LOCOS technology needed modification to reduce the extent of the 'bird's beak' effect in the submicron channel lengths.

Philips [64] developed a modified LOCOS giving a 0.6 μm P^+–P^+ spacing by both substituting an oxinitride stress relieving layer for the silicon dioxide stress relieving layer in the original LOCOS process and by doing a sacrificial oxidation step prior to growth of the channel oxide layer. Sony [73] also used an improved LOCOS technology which reduced the 'bird's beak' length to less than 0.2 μm as shown.

A comparison of the original LOCOS process and the advanced LOCOS as modified by Philips is shown in Figures 4.20(a) and 4.20(b) respectively. The new LOCOS processing is expected to be used though at least the 0.3 μm feature size level.

4.12.4 Trench isolation

Isolation of the individual devices that eliminates the bird's beak can be done by etching shallow vertical 'trenches' in the silicon between neighboring devices as shown in the cross-section in Figure 4.20(c). Trench isolation allows the transistors to be moved much closer together and minimizes the problem of effective channel width control since it eliminates the need for a field implant. If a device is surrounded by trenches, the width is well defined without additional process steps. It also results in a more planar surface by avoiding the formation of the bird's beak. The trade-off is in the more complex process required than for the LOCOS field oxide isolation.

In the illustration in Figure 4.20(c), which is from a 256k SRAM by NEC [61], the width of the buried isolation layer is 0.9 μm. This is about 60% of the 1.4 μm width of an isolation layer fabricated in a similar geometry process in the LOCOS technique. The depth of the buried isolation layer is 1.5 μm. The buried isolation layer of amorphous silicon dioxide is fabricated by using both reactive ion etching and chemical vapor deposition. In this illustration trench isolation is used for the NMOS transistor in the array and LOCOS is used for an NMOS transistor in the periphery where density is not so critical as in the array.

Mitsubishi [57, 60] also used a trench isolation technique in the array with a silicon dioxide filled 0.8 μm trench for both of their first generation 1Mb SRAMs. A 0.7 μm deep trench was used on the faster part. The subthreshold current characteristics of the NMOS cell transistors were shown to be similar to the LOCOS process. The conventional LOCOS isolation process was used in the periphery.

Trench isolation may be used in deep submicron processes, below 0.3 μm, as was indicated in Figure 4.19, where it will permit the tighter line definition, and the greater planarity required.

4.13 PHOTOLITHOGRAPHY

A significant limiting factor of MOS technology development has been the photolithographic techniques used to pattern the tiny gates on the silicon chip. Increasingly sophisticated pattern definition technologies have been needed as the critical geometries have shrunk from the 6 μm of the late 1970s to the submicron technologies of the late 1980s to the deep submicron regions of the 1990s.

The earliest mask patterns were hand drafted and cut by hand from a gelatin-like surface called a 'ruby'. This patterned ruby was then photographically reduced in size and reproduced in a repeating matrix on a mask for a single wafer. The mask, which was made from photographic emulsion, was placed in a contact printer which held it in direct contact with the wafer while an ultraviolet light was shone through the mask exposing the photolithographic resist on the surface of the wafer as illustrated in Figure 4.21(a).

Since perfect alignment of the mask with the pattern defined on the wafer in previous steps was not always achieved, the mask and wafer frequently had to be separated and realigned. This tended to damage the mask and resulted in a high surface defect density on the wafer. The original NMOS memory process used this technique.

An improvement in alignment technique called proximity alignment was made by separating the mask slightly from the wafer and projecting a shadow of the mask on the wafer. This reduced the production of surface defects but caused focusing problems. It was never widely adopted.

The next true progress was made both in mask production and in alignment techniques. Computer aided design and computer pattern generators provided the means to design masks digitally.

Alignment techniques were greatly advanced by the introduction of the projection

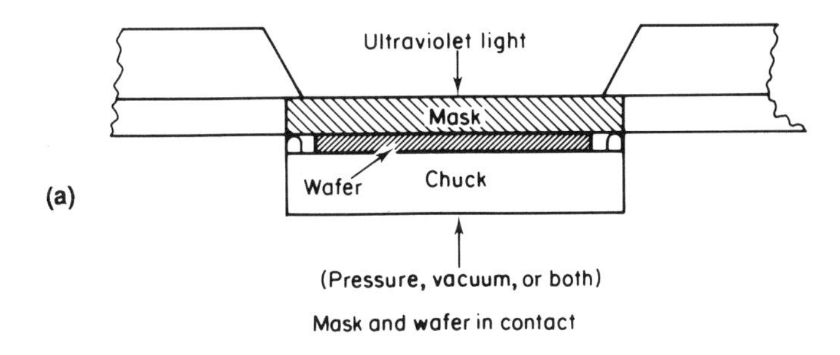

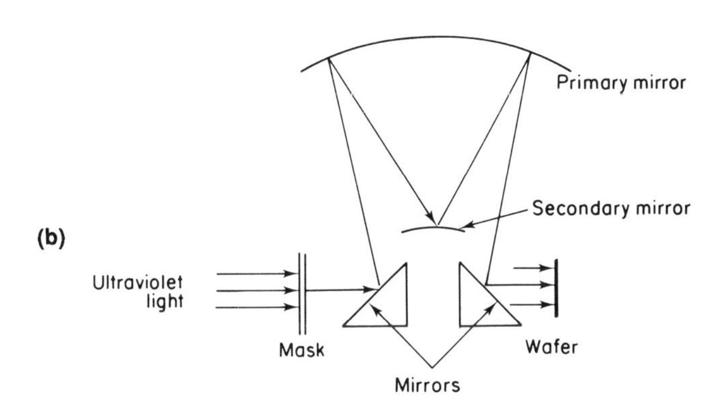

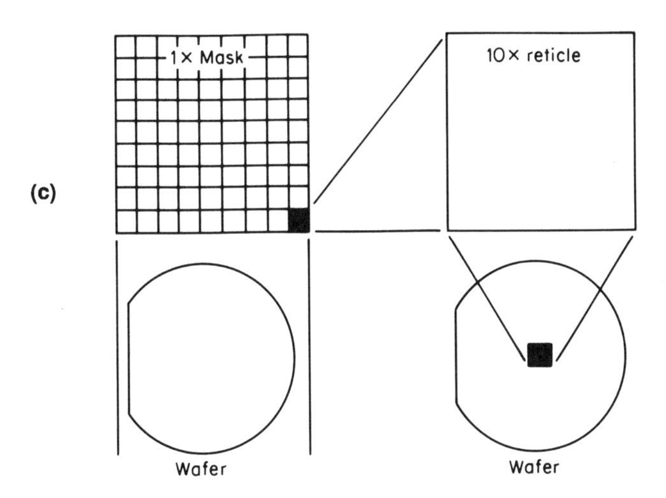

Figure 4.21
Four historical lithograph methods. (a) contact printing. (b) projection printing, (c) comparison of 1:1 and 10:1 projection systems for optical wafer steppers. (From Prince and Due-Gundersen [66] 1983.)

aligners as shown in Figure 4.21(b). Alignment with a projection aligner is easier because both the mask and wafer are in proper focus at all times. The other major cause of mask and wafer defects, the contact between the mask and wafer, is also eliminated. Since the mask is not abraded by wafer contact, the photomask lifetime is extended from less than 100 exposures per mask to tens of thousand of exposures per mask.

Projection alignment made high quality masks economically feasible. An added benefit was the higher consistency achieved by using a single mask for volume production which allows tighter control over other process parameters. Projection printers were used widely in the industry from 4 μm to 2 μm geometries.

For less than 2 μm geometries, a new photolithographic process was needed to improve resolution, depth of field, and focus parameters while maintaining alignment accuracy and decreasing random defects. The direct-step-on-wafer technique offered those advantages.

With a direct wafer stepper, the entire wafer with its multiple chip masks was no longer exposed at one time; rather, a single chip mask or cluster of chips is aligned and exposed, and then a step is made to the next position successively exposing the entire surface of the wafer. Figure 4.21(c) compares a single mask projection system with a 10 × single chip stepping system. The mask or reticle is larger and hence permits more accurate detail. Wafer flatness

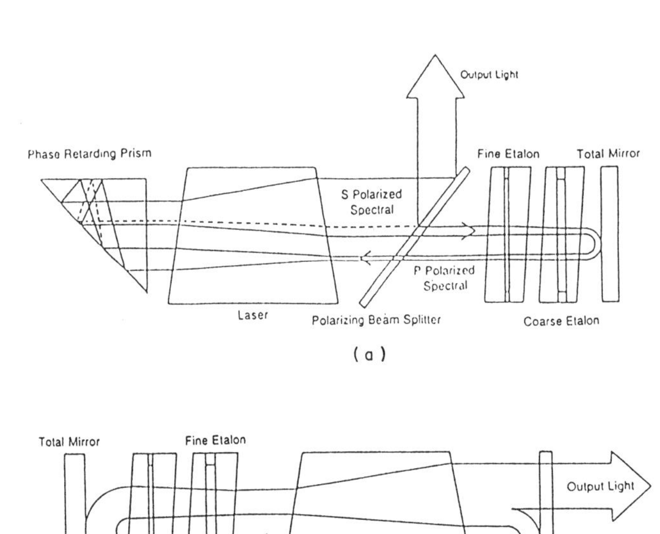

Figure 4.22
KrF excimer laser stepper lithography. (a) Conventional method, (b) new high power laser with polarization coupled resonator. (From Tani *et al.* [78], Matsushita 1990, with permission of Japan Society of Applied Physics.)

has less effect since automatic focusing is included on the wafer stepper and a smaller surface area of the wafer is exposed.

Rather than aligning with microscopes as in previous techniques, operator control is maintained by a high resolution closed circuit television which isolates the operator from the temperature and contamination controlled chamber containing the wafer.

The initial disadvantage of the direct wafer stepper was that it took longer to expose a single wafer which slowed the production process down and increased the cost.

The increased potential for accuracy of the alignment system was matched by increasingly accurate mask production by using electron beam rather than optical mask-making techniques. E-beam maskmaking systems offer higher resolution than optical systems since electrons have shorter wavelengths than light.

The improved masks that can be made with E-beam systems, together with continued improvements in the resolution of lenses used with steppers and improved throughput productivity have worked together to keep the wafer stepper the optical alignment system of choice down to at least the 0.5 μm level. Steppers can print features about twice the wavelength of the light being used. One possible wavelength is 250 nm, which can print features at roughly 0.3 μm. Another limitation than the wavelength of the light is the small depth of focus of high resolution optical lenses which for submicron geometry features can require multiple level resist and considerable planarization of the chip topology.

Referring to Figure 4.22, optical g-line and i-line steppers are expected to be widely used for the 0.3 μm features of the 64Mb DRAM and 16Mb SRAMs with the excimer laser stepper beginning commercial production use at the 64 Mb and phasing perhaps into E-beam or X-ray wavelength lithography at the 0.2 μm minimum geometry level. Figure 4.22(a) shows a conventional excimer laser stepper and (b) shows a new high power laser stepper from Matsushita [78] with polarization coupled resonator.

For deep submicron geometries beyond the range of conventional i-line lithography, phase shift lithography techniques have been used [85]. The principle of this technology is the interference between the field diffracted by adjacent apertures with an opposite optical phase. High resolution can, as a result, be obtained with the use of a conventional stepper. The amplitude of transmitted light at the shifter edge is exactly zero while adjacent transmitted light interferes with pattern resolution so that a narrow space pattern is obtained.

The dominant production lithography below 0.2 μm is not yet clear. Some of the contenders for the position include electron beam lithography, X-ray lithography and ion beam lithography all of which have been demonstrated to be effective below 0.1 μm geometries.

Direct-write-on-wafer using an E-beam system for the lithography is also possible for the deep submicron geometries but current production equipment is extremely slow.

Electron beam direct-write-on-wafer is available commercially, although very expensive and with slow throughput. It is installed in a very few production areas around the world, but is used more extensively in prototyping as it can be quickly altered for experimentation. Its advantages are programmability, lower tooling cost and short turnaround times. The disadvantage is throughput speed and cost. It is used, however, extensively in mask making.

X-ray lithography is still in the development phase but appears to be the preferred technique for the 0.1 to 0.2 μm geometries. The advantage is high image resolution, down to the 0.1 μm level which it can reach because of the short 1 nm wavelength of X-rays. Line widths as small as 0.25 μm have been printed with X-rays. Problems include: developing a low cost and compact radiation source, developing resists that are opaque to X-rays and developing masks that are transparent to X-rays.

Beyond X-ray may be ion beam lithography which can get below 0.1 μm line definition. The advantage is that ions have a high energy intensity and do not scatter in resists. Disadvantages are mask heating which can lead to distortion of the projected pattern, and slow throughput resulting from low beam brightness.

4.14 ETCHING TECHNIQUES

Etching techniques of fine patterns on a silicon surface are another important part of MOS manufacturing technology. Etching of the very fine lined patterns defined by the mask must be equally as well controlled as the lithography. With early MOS technology etching was performed with chemicals. Micron and submicron dimensions require the control of an ionized plasma etch.

The ionized plasma works by being absorbed chemically onto the wafer surface. There it forms volatile compounds from the surface material which are then desorbed into the gas again, leaving the desired pattern etched in the surface material of the chip.

Plasma etching moved from the laboratory into the production line around 1973–1976 in response to the large scale production requirements of silicon gate NMOS memory technology. Development has been rapid since then.

Reactive Ion Etching (RIE), which is a low pressure plasma etch technique, has been used since 1980 to form deep trenches in a silicon surface for vertical capacitors in dynamic RAMs and more recently as an alternative isolation technique to the LOCOS process. Modern RIE equipment can form vertical trenches in silicon with depth-to-width ratios of 20:1.

4.15 DOPANT INTRODUCTION

Introduction of doped ions into the silicon to form the source and drain regions of the MOS FETs has been accomplished by several methods. Two basic techniques are diffusion from doped oxides and ion implantation.

Diffusion from doped oxides is the older technique. The use of this technique permitted the narrow channel lengths necessary for speed in MOS transistors. The process consists of depositing N-doped and/or P-doped oxides on the surface of the silicon. These oxides are then patterned over the required drain and source regions of the transistors so that the channel silicon is exposed between the etched layers of doped oxide. The wafer is then placed in an oxidation furnace and reoxidized to both drive in the dopants and grow the thin gate oxide.

Ion implantation is the other common technique used for ion introduction into the silicon. In this case the gate material is used as a mask with the silicon in the source and drain regions exposed during ion implantation. This technique is increasingly used for finer lines and for other doping processes such as graded junctions where the level of doping is varied with depth in the silicon.

4.15.1 Structured doping techniques

Structured doping techniques can optimize the electrical characteristics of transistors and improve their reliability.

These include special tailoring of: enhancement and depletion mode transistors to optimize transistor electrical characteristics, graded ionization such as 'lightly doped drains' and 'double doped drains' to reduce such reliability hazards as the 'hot electron' effect, retrograde wells to reduce the threat of 'punch-through' and twin-well structures to improve latch-up performance.

4.15.2 Doping for enhancement and depletion mode transistors

A depletion mode transistor has an additional ion implantation in the gate area which gives it a current channel between the source and drain without an external applied gate voltage. This channel has the effect of lowering the threshold voltage required to turn the device on. If the threshold voltage is negative then an external voltage must be applied to turn it off. Typical electrical characteristics of depletion mode transistors are shown in Figures 4.5 and 4.6.

4.15.3 Lightly doped drains (LDD)

A 'lightly doped' grading of the diffusion of the MOS transistor near the channel provides resistance to the hot electron carrier effect. See Figure 4.23(a) for an example of a lightly doped drain (and source) structure profile for an NMOS and a PMOS transistor.

4.15.4 Double diffused MOSTs (DMOS)

The double diffused MOS transistor structure consists of a second diffusion of the source and drain side of the transistor which reduces the apparent L_{eff} of the transistor as shown in Figure 4.23(b). The DMOS transistor can improve the

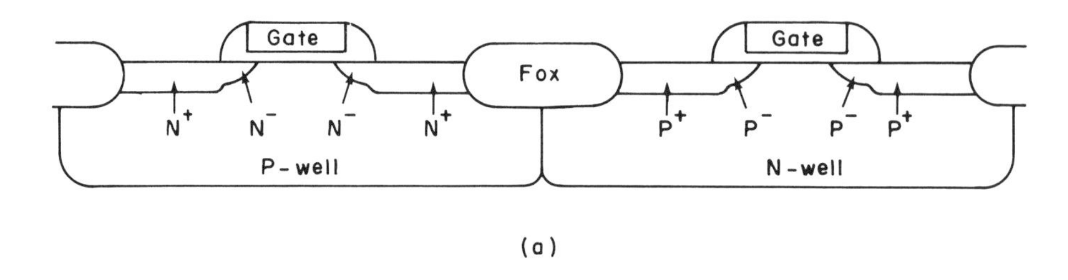

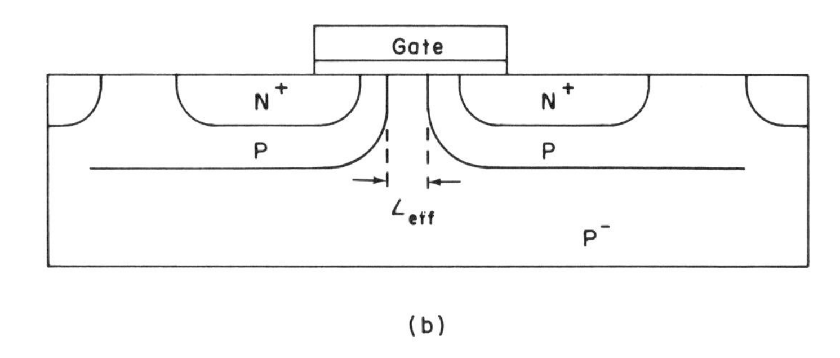

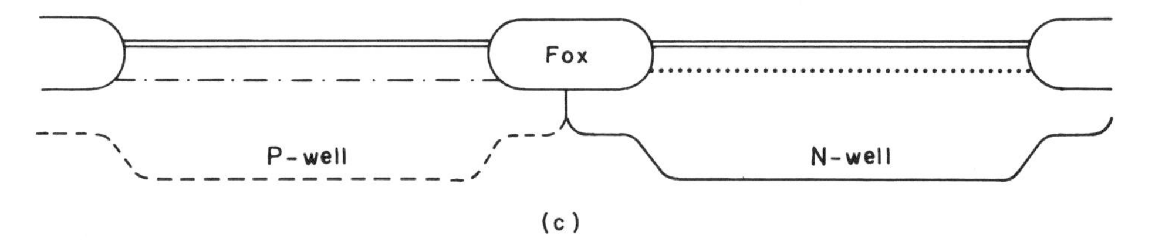

Figure 4.23
Examples of structured doping showing: (a) lightly doped P and N drain profiles, (b) a double diffused MOS transistor with reduced effective channel length, (c) retrograde P and N well profiles. (Source N.V. Philips Research.)

high frequency response of the MOS structure by reducing the effective channel length.

4.15.5 Retrograde wells

The retrograde P-well process developed by Intersil involved implanting the P-well through part of the field oxide at the edges of the active regions so that the well underlies the field oxide. Examples of retrograde P- and N-well profiles are shown in Figure 4.23(c). This not only acts as a channel stop to improve leakage between transistors. It also permits precise definition of the dimensions of the P-well for smaller geometry processes.

Shaping of the ionization profile of the transistor well in CMOS also provides some protection against the phenomena known as latch-up which was described earlier. This can result in the device being destroyed. An example of the structure of a retrograde well is also shown in Figure 4.8(c).

4.16 INTERCONNECT TECHNOLOGY—METALLIZATION, POLYSILICON

Another important factor which limits the reduction in size of the memory chip is the area which is occupied by interconnecting lines between the devices. The original MOS devices had metal interconnections only. A significant advance was achieved by the introduction in the early 1970s of doped polysilicon for the gates of MOS FETs. It is doped to reduce the resistivity. By using doped polysilicon as gates and as interconnects between gates, in addition to using metal interconnects, a two-layer structure was obtained which greatly reduced the size of the chip.

4.16.1 Polysilicon interconnects

Double polysilicon was introduced for dynamic RAMs in about 1976. This added a third possible layer to the interconnect structure and permitted further reductions in the size of the chip. In a double poly process there are two main alternatives. Both levels can be used for gates; alternatively one level can be used for gates and the other for interconnection and resistors.

Triple polysilicon can also be used and reduces the area of the chip still further, although the complexity of the process is much higher. With triple poly the storage capacitor of a dynamic RAM can be stacked on top of the access transistor, resulting in significant space saving.

Polysilicon straps are also used. These consist of short runs of polysilicon between different levels in the structure. They can, for example, be used to connect buried contacts to overlaying interconnect structures.

4.16.2 Metal interconnects

Multiple layers of metalization can also be used to increase chip density. These metal layers can be evaporated onto the surface of the chip or deposited using a sputtering machine, patterned with a protective photoresist and etched using either chemical or plasma etching techniques. Chemical etching is the historical method for larger than 1 μm geometries. It is, however, difficult to use it to define accurate lines for submicron geometries since the chemical etch tends to undercut the protective photoresist as shown in Figure 4.24(a).

Very fine line metal definition can be obtained by using a metal lift-off technique

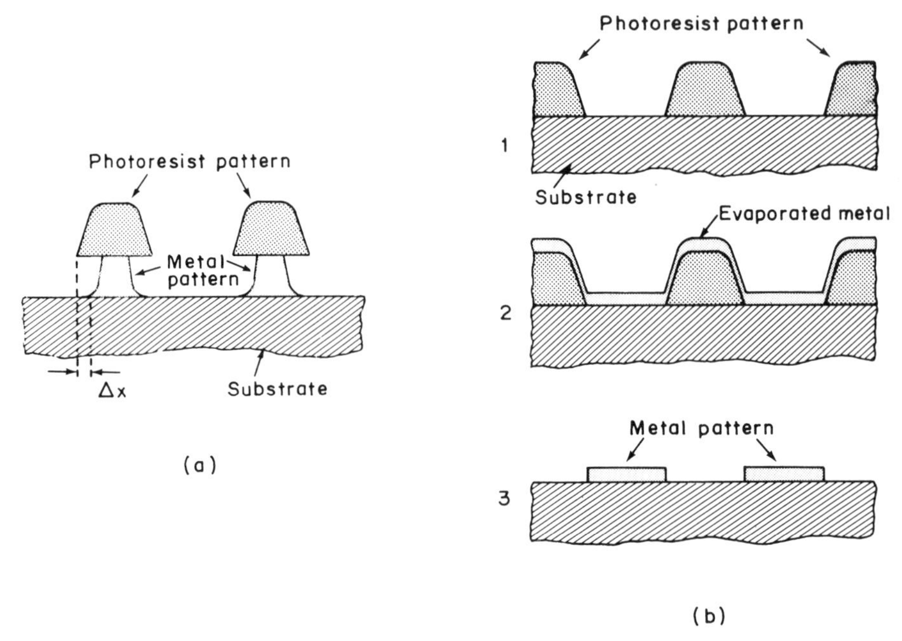

Figure 4.24
Illustration of various methods of patterning metal interconnects where; (a) is patterned by chemical etching and (b) is patterned by a resist lift-off technique. (From Widmann [83], Siemens 1973, with permission of IEEE.)

to define the metal lines. This consists of depositing the metal over the patterned photoresist then washing away the resist as shown in Figure 4.24(b). In practice the lift-off process has particle contamination problems and is seldom used in production. Fine line metal definition can also be obtained for submicron geometries by modern methods of ion etching which are more common in production today.

Progress in interconnect technology has been instrumental not only in scaling down the size of the chip, it has also contributed to increasing the speed. Since the speed of a circuit is limited by the *RC* time delay of the polysilicon interconnects, much development has also been done on various methods of reducing the resistivity of these interconnects.

4.16.3 Metal and polysilicon combination interconnects

Metal silicides are used to reduce the resistivity of the polysilicon interconnects and reduce the propagation delay. These refractory metal interconnects can be formed by either depositing a layer of metal on top of polysilicon, or alternatively, the metal ions can be implanted into the polysilicon and annealed. The silicide offers the advantage of lower resistivity while retaining silicon gate characteristics. The most suitable silicides to date appear to be WSi_2, $TiSi_2$, and $MoSi_2$. Silicides were first widely used in the memory industry in the 256k dynamic RAMs.

When silicides are applied using a self-aligned process, they are frequently referred to as salicides (self-aligned silicides). Since the salicides are self-aligned by the transistor gate, they also result in low resistance silicided drain and source region contacts.

4.17 PLANARIZATION TECHNIQUES

Since MOS technology is a surface process, planarization of the surface of the wafer both originally and during each step of the manufacturing process is important. At submicron geometries it becomes critical.

Planarization is required for several reasons. One is that the manufacturing process entails fine definition optical lithography with short focal length lenses. Focus is critical and must be maintained over a wide area of the wafer surface compared to the depth of field.

A second reason is that several layers are patterned and etched out successively one on top of the other. This requires that each subsequent layer run over the steps etched in the previous layer. If the surface is not planarized following each operation, these steps become relatively large and breaks can occur which affect both the functionality and the reliability of the circuit.

Two general methods are available to planarize the wafer surface. The first is to fill in any gaps, and the second is to etch off any rises.

Filling may take the form of depositing layers of passivation such as SiO_2 over large gaps, or of depositing small plugs, of metal for example, in tiny contact holes. Both of these filling methods entail a separate step to remove the fill material from the raised areas leaving it only in the valleys.

Etching may take the form of depositing a protective layer on a low region so that the high regions can be etched back.

4.18 PROCESS CONSIDERATIONS FOR REDUNDANCY TECHNIQUES

Redundant cells can be added to a memory design to replace defective cells which are identified at electrical test after wafer processing.

Redundancy has been achieved by three methods: current blown fuses, laser blown fuses, and laser annealed resistor connections.

Fuses incorporated on the wafer can be blow electrically at wafer electrical test to swap defective cells. The advantage is the elimination of the laser which opens the surface passivation on the chip. The disadvantage is the necessity of using silicon area to incorporate extra high current transistors on the chip to blow the fuses.

Laser blown fuses and laser annealed resistor connections do not add to the area of the chip but since the surface passivation of the chip is broken they require the incorporation of the laser and wafer probe into the clean room environment and the addition of an extra passivation step to reseal the surface of the chip. Alternately the area containing the fuse on the surface of the chip can have space consuming guard-

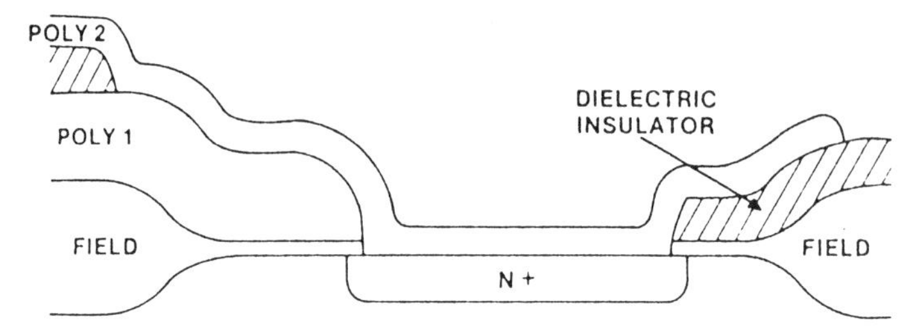

Figure 4.25
Schematic cross-section of shared contact used to reduce silicon area in SRAMs. (From Sood *et al.* [67], Motorola, 1983 with permission of IEEE.)

rings around it which protect the rest of the chip from the effects of the laser; this, however, consumes space on the chip.

4.19 CONTACT TECHNOLOGY

Contacts between transistors and the various interconnect layers and between different interconnect layers occupy area in the memory array. A shared contact can be used in a double poly process to reduce cell area. In the example shown in the schematic diagram in Figure 4.25 a contact is made from the first polysilicon layer to the N^+ layer through an intermediate contact to the second polysilicon layer.

Barrier materials such as titanium–tungsten or titanium nitride are frequently used in advanced processes to underlay aluminum alloy interconnects. These prevent interactions between the metal and the substrate. They also lower contact resistance and improve adhesion. Contacts are also occasionally silicided, or have a refractory interconnect material such as molybdenum or tungsten silicide.

4.20 AN OPTIMIZED CMOS SUBMICRON SRAM MEMORY PROCESS

An example of a submicron production process used for a 1Mb SRAM by Philips [27] is shown in the following. Some of the features used in this process to achieve the small geometries are as follows.

- 0.7 μm n-channel gate lengths with 0.6 μm L_{eff}.
- Tight 1.0 μm field oxide isolation using a $SiNiO_2$ masking technique.
- Small 0.25 μm N^+ to P^+ spacing.
- Thin epi material to suppress latch-up.

- Retrograde twin wells to suppress parasitic channel formations and punch-through currents for the field transistors.

- LDD transistor structures for both n and p channel devices to reduce the hot electron effect.

- Strap technology as local interconnects to connect buried contacts to overlying interconnects and gates in minimal spacing.

- Metal salicide technology to reduce the resistivity of the polysilicon interconnects.

- Double layers of metal interconnects to reduce chip size.

- Planarization following all major process steps to permit extremely small line definition and reduce step coverage related reliability and yield problems. This includes both types of fills, the small area metal via contact fills and large area SiO_2 deposition fills. It also includes etches of raised areas such as the silicon nitride protection of the surface during etch-back of the LOCOS oxide.

The process steps are described below and illustrated in Figure 4.26.

(a) The field oxide is formed using a modified LOCOS process. A sandwich layer of silicon–oxynitride and silicon nitride is used to suppress the 'bird's beak' formation which is restricted to 0.25 μm, and the lateral encroachment at 0.1 μm. This works since oxygen cannot pass through silicon nitride to grow SiO_2 on the silicon surface during the high temperature LOCOS oxidation.

Planarization of this first layer is then accomplished by a plasma etchback of the LOCOS oxide while the active region is still protected by the silicon nitride. The silicon nitride is then removed leaving a relatively planar surface area for the subsequent processing steps.

(b) High energy graded ion implants, using a 1 MeV ion implanter, are then used in preparation for forming retrograde twin well structures as shown. The implant energies are chosen such that the maxima of the implant profiles in the field oxide regions are located just beyond the Si–SiO_2 interface. In the active area the profiles are somewhat countersunk. The gate oxide is grown during the high temperature implant anneal step. However, the conventional high temperature well drive is eliminated so that the very fine geometries can be attained.

(c) Polysilicon is then deposited and patterned to form the gate region of the transistors and interconnects which run over the field oxide. The high dose n^+ and p^+ drain implants form the MOS transistors with the lightly doped drain area appearing under the edge of the gate structure as a result of the earlier implants. Metal is deposited and patterned to form the contacts with the MOSTs and the $TiSi_2$ straps are deposited and patterned to form local connections for the source and drain.

(d) A layer of deposited silicon dioxide (TEOS) is then put down and planarized. Contact holes (vias) to the $TiSi_2$ gate contacts are cut and filled with tungsten plugs

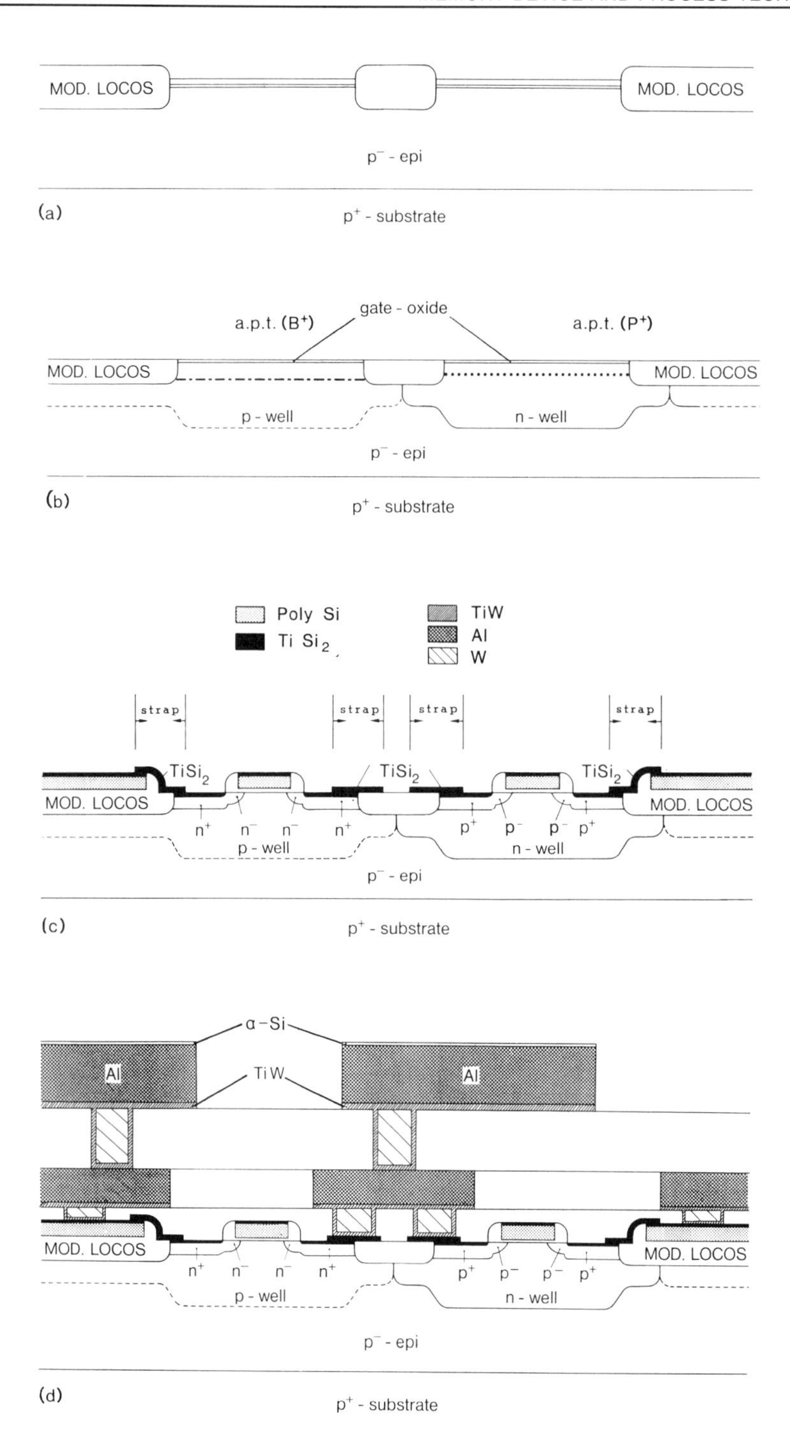

Figure 4.26
Process steps of an 0.7 μm CMOS process. (From de Werdt [27], Philips 1987, with permission of IEEE.)

using TiW as the adhesive. This planarizes the surface so that the first layer aluminum interconnects do not have to step down into the 0.9 micron contact holes. The first layer aluminum interconnect layer is then deposited, patterned, and the openings in the pattern filled with another deposited oxide (spin-on-glass) This surface is then covered with another layer of oxide (TEOS) which is planarized in an etching step and the contact holes to the first layer aluminum are cut and filled again with tungsten plugs. The second layer of aluminum is then deposited and patterned. Aluminum with 2% copper is used for the interconnect layers to reduce electromigration and suppress hillock formation.

This process was used to make a six transistor cell CMOS 1Mb SRAM.

4.21 A SUBMICRON BiCMOS PROCESS

An example of a BiCMOS process which combines the features of the advanced CMOS process techniques with the isolation advantages of the buried layers present in the bipolar process is shown in Figure 4.27. This process has self-aligned gates, double-layer polysilicon and double layer metal with the first layer metal interconnect being tungsten. Titanium silicide can be seen on the second polysilicon interconnect structure. It uses twin P- and N-wells for the NMOS and PMOS transistors. A lightly doped drain structure can be seen on the NMOST. The planarization of successive steps required by advanced lithography techniques can be seen. This process was used to make a BiCMOS 1Mb SRAM.

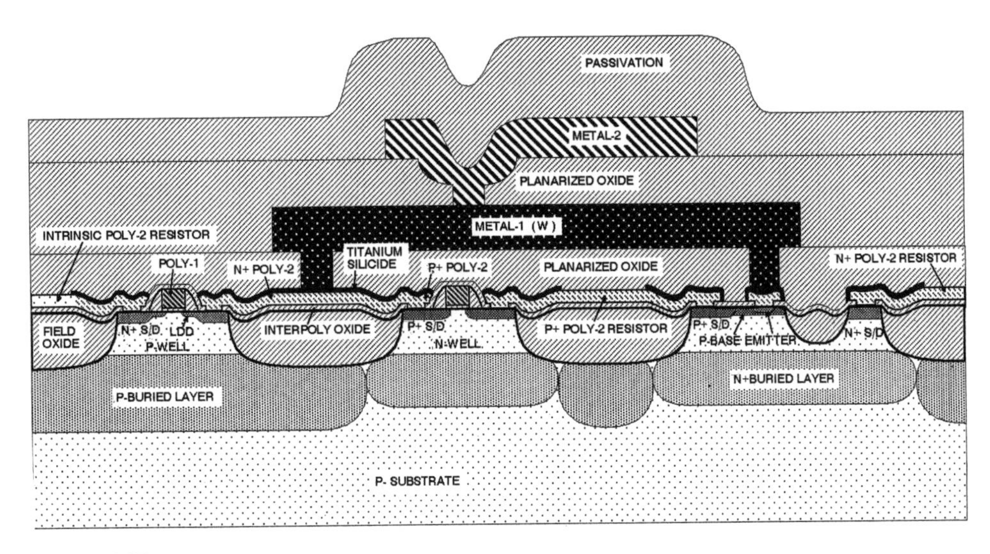

Figure 4.27
Schematic cross-section of submicron BiCMOS process. (Reproduced with permission from National Semiconductor.)

4.22 PROCESS DEVELOPMENT FOR THE ONE-TRANSISTOR DRAM CELL

The ability to scale the size of the memory cell for successive generations of products underlies the progress of memory technology. Scaling the size of the DRAM memory cell is primarily a problem of reducing the area without decreasing the storage capacitance of the cell.

4.22.1 Planar DRAM cell theory

The cell which has been used in the one transistor DRAMs from the 16k to the 1Mb is called the planar DRAM cell. The capacitor used in DRAM cells has historically been placed horizontally beside the single transistor where it has occupied about 30% of the area of a typical cell. A cross-section of this cell structure is shown in Figure 4.28(a), along with an equivalent circuit representation. The charge storage in the cell takes place on both the oxide and the depletion capacitances.

The amount of charge in the storage region of a linear capacitor corresponding to a full charge well $Q(0)$ is given by

$$Q(0) = C_{01}(V_p - P_1) - K_1\sqrt{P_1}$$

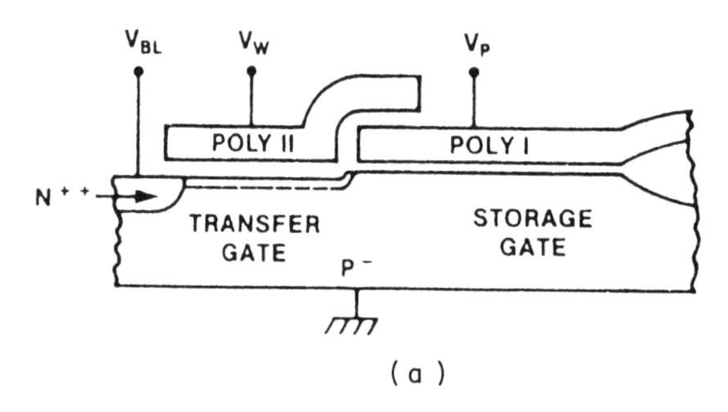

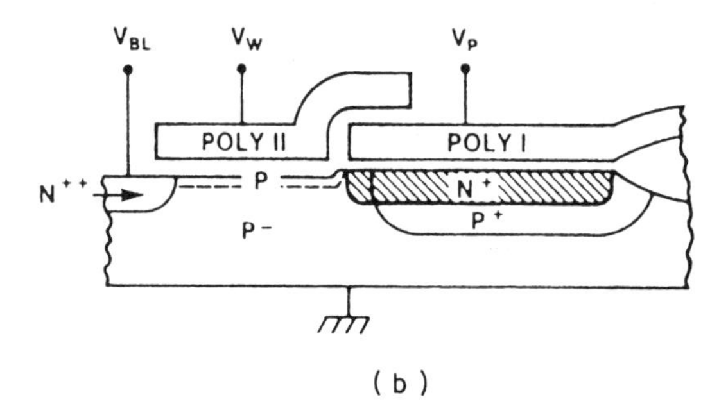

Figure 4.28
Schematic cross-section of planar DRAM cells with equivalent circuit representation. (a) conventional double poly planar cell, (b) Hi–C double poly planar cell. (From El-Mansy and Burghard [49], Intel 1982, with permission of IEEE.)

where C_{01} is the oxide capacitance per unit area for the storage region, K_1 the bulk doping concentration and P_1 is the surface potential corresponding to a full charge well.

The amount of charge in the storage region corresponding to an empty well is given by

$$Q(1) = C_{01}(V_p - P_3) - K_1\sqrt{P_3}$$

where P_3 is the surface potential in the transfer gate region for a gate voltage of V_w.

The signal charge is defined as the difference between $Q(1)$ and $Q(0)$ and is given by $\Delta Q = Q(0) - Q(1)$ or

$$\Delta Q = \int_{Q(0)}^{Q(1)} dQ = \int_{P_1}^{P_3} C_{01}\left(\frac{1}{V_{t1}}, \frac{1}{P_1}\right) dP + \int_{P_1}^{P_3} C_d(K_1, P_3)\, dP$$

Therefore to increase the charge in the storage region options are to

(1) increase the oxide capacitance (C_{01}) by thinning the oxide in the storage gate region,

(2) reduce V_t (threshold voltage in storage area) and P_1 by reducing the doping concentration and oxide thickness in the storage area,

(3) increase P_3 by thinning the oxide in the transfer area,

(4) increase the depletion capacitance (C_d) by increasing K_1, the doping concentration in the storage area.

The first three methods have been used as a part of good technology scaling procedures on the standard planar DRAM cell. The fourth method is hard to implement in conjunction with the second method since it is difficult to increase the depletion capacitance while keeping the threshold voltage near zero. A high capacitance implanted cell was developed in 1982 [49] in an attempt to resolve these conflicting requirements.

4.22.2 High capacitance implanted DRAM cell theory

The Hi–C cell, which was developed by Intel [49], was an early method of maintaining capacitance while reducing the area of the capacitor. It involved implanting a deep substrate doping under the capacitor thereby increasing the effective capacitance. A schematic cross-section of a Hi–C DRAM cell is shown in Figure 4.28(b). The charge density, and hence effective capacitance, is determined by a sum of the charge stored between the N^+ layer and the grounded storage plate and the charge stored between the N^+ layer and the deep substrate implant.

The increase in substrate doping is achieved by implanting a P^+ (boron) layer in the storage area. This results in an increased doping density and hence increased depletion capacitance. It also results, however, in an unwanted increase in threshold voltage. The increase in threshold voltage is then offset by implanting a layer of N-type dopant at the surface. This implant can be tailored to give any threshold voltage desired.

The result is that the equivalent capacitance is the sum of the oxide capacitance (C_{01}), the capacitance of the surface depletion region (C_s), and the depletion capacitance under the storage gate (C_d).

A careful balance must be maintained between the surface doping and the substrate doping to be sure of an increase in capacitance. Care must also be taken since the implanted cell needs to be designed with enough voltage margin to avoid loss of information from low level avalanche currents preceding junction breakdown. This limits the maximum operating voltage allowed across the junction to below the actual breakdown voltage of the junction.

The planar implanted capacitor DRAM cell was generally used throughout the 1980s. The limits of this method for retaining sufficient capacitance were reached in the mid 1980s with the 1Mb DRAM.

4.23 NEW STRUCTURES FOR SUBMICRON GEOMETRY DRAMs

The submicron geometry MOS processes extended the basic processing described above to accommodate increased layers of deposited material, severe planarization requirements, and new vertical device methods which allowed megabit chip densities to be made in sizes not significantly larger than historic chip sizes. They also permitted new innovative types of structures such as trenches which burrow down into the silicon and stacked layers of devices rising up over the surface. The main requirement is that the area of the chip cannot increase for cost reasons even as the number of transistors in the chip increases.

Smaller horizontal geometries on the surface of the wafer can be achieved by making increased use of the vertical dimension both above the surface of the silicon and below it.

4.23.1 Trench capacitors

For the DRAM by the 1Mb density it was clear that scaling the cell to smaller sizes meant ways had to be found to scale the area occupied by the capacitor on the surface of the wafer without reducing the effective area of the capacitor.

One method of reducing the size of a horizontal DRAM capacitor cell, such as shown in Figure 4.29(a), was to fold it vertically into the surface of the silicon in the form of a trench as shown in Figure 4.29(b).

It was the development of reactive ion etching that enabled vertical trenches to be formed in the silicon [80]. A buried type of structure attempted previously, called VMOS, was formed by crystal orientation dependent etching but was not successful due to breakdown voltage degradation at the corner. The trench, however, retained suitable electrical characteristics for the formation of a buried capacitor.

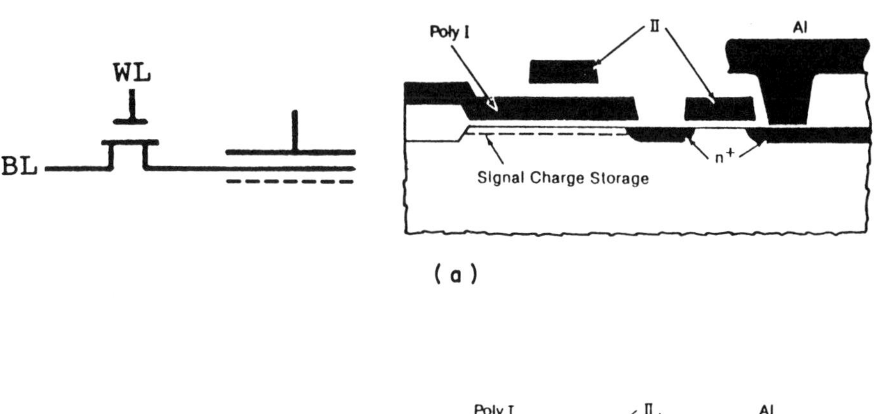

(a)

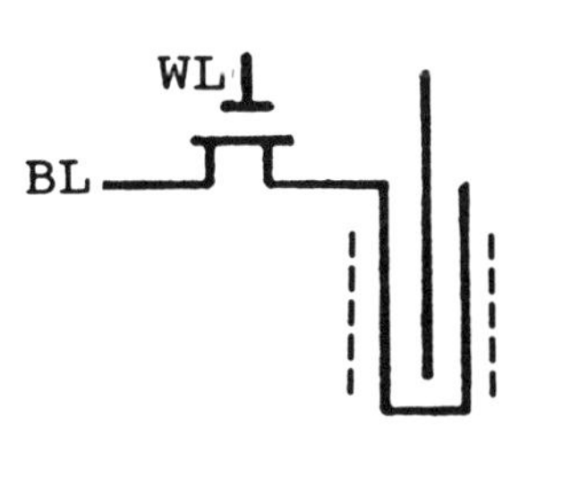

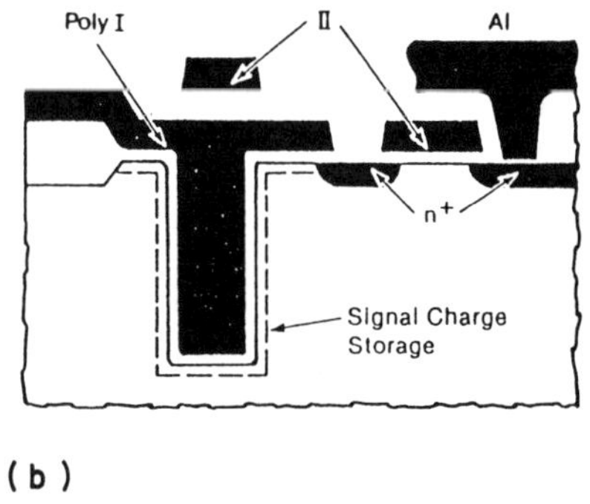

(b)

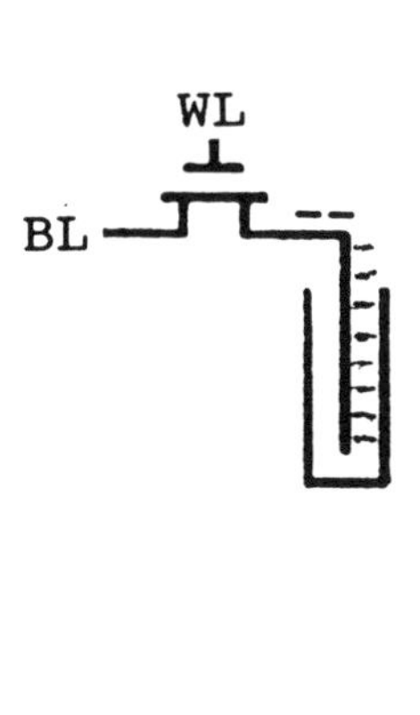

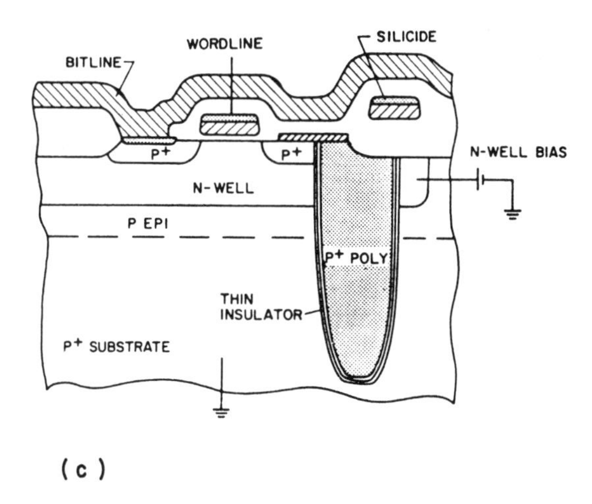

(c)

Figure 4.29
Four generations of DRAM trench cells. (a) Planar cell up to 1 Mb, (b) simple trench cell 1 Mb–4 Mb, (c) grounded substrate trench cell 4 Mb–16 Mb, (d) buried trench with horizontal transistor 16 Mb–64 Mb, (e) buried trench with vertical transistor 16 Mb–64 Mb. (a,c,d: from Lu [68,81], with permission of IBM; b: from Sunami *et al.* [79], Hitachi 1982, with permission of IEEE; e: from Richardson *et al.* [39], TI 1985, with permission of IEEE.)

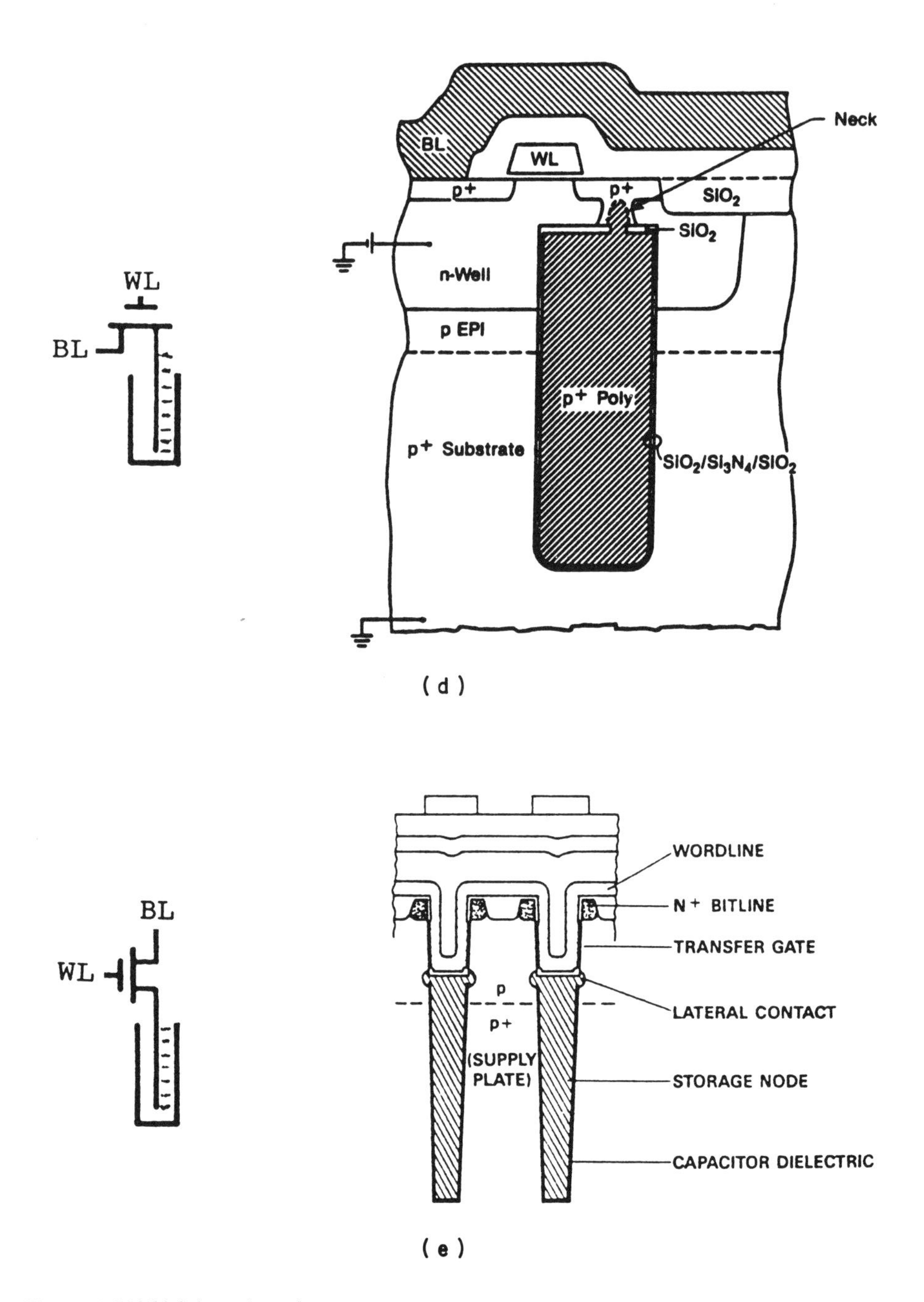

Figure 4.29(d)(e) *(continued)*

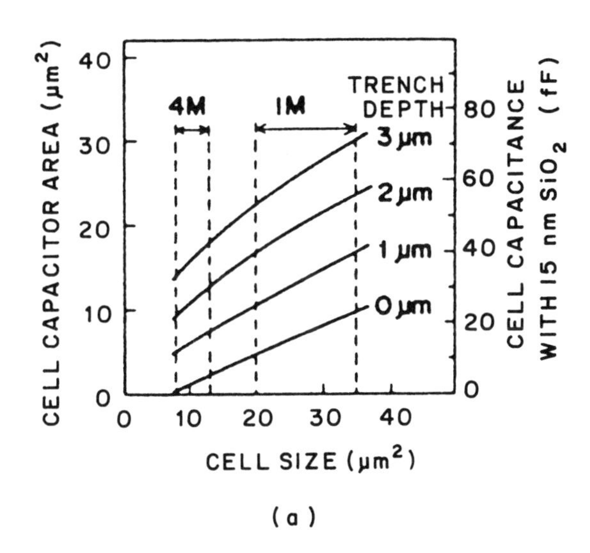

(a)

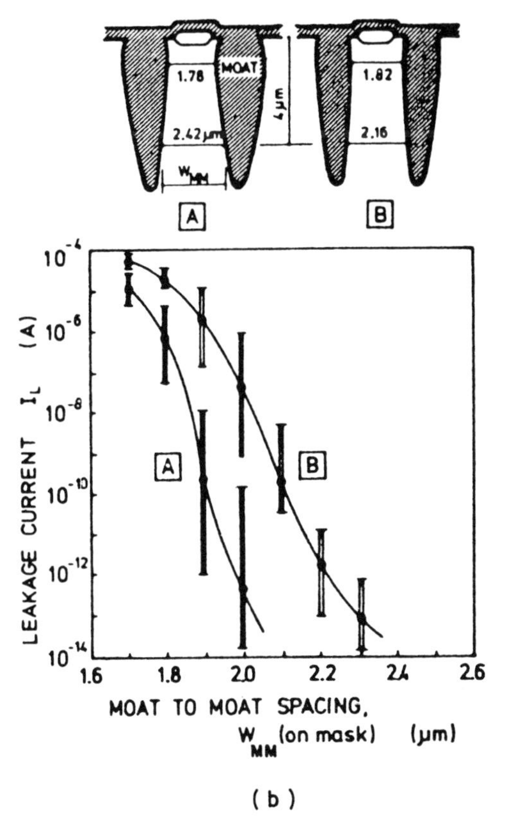

(b)

Figure 4.30
Parameters of typical trench cells for 1 Mb and 4 Mb DRAMs. (a) Cell size as a function of capacitor area, cell capacitance, and trench depth, (b) Leakage current as a function of trench-to-trench spacing. (a) with permission of IEEE; b:from Sunami *et al.* [79], Hitachi 1982, with permission of IEEE.)

The first widely known trench capacitor, developed by Hitachi in 1982 [79] is also shown in Figure 4.29(b). It stored charge in the substrate along the outside walls of the trench. A drawback was that current leakage between adjacent trenches resulted in punch-through and required a minimum trench separation. This constraint was not compatible with the smaller geometry cells required by higher density DRAMs.

A simple solution was to store the charge on the inside wall of the trench and use the outer wall as the capacitor plate with a grounded substrate. NEC [38] and IBM [81] reported this type of trench cell in 1985. The IBM cell is shown in Figure 4.29(c).

In 1988 IBM [82] reported a further development in reducing the size of this 'inner storage node' type of cell. This new cell had the trench buried in the silicon with the transistor placed horizontally above it as shown in Figure 4.29(d).

In 1985 TI [3] showed a similar buried trench cell for high densities with the transistor placed vertically in the trench as shown in Figure 4.29(e). The process complexity of these higher density trench cells led many companies to explore another route to high density DRAM cells which was to stack a capacitor formed of oxide sandwiched between layers of polysilicon above the surface of the silicon. The stacked capacitor will be described further in a later section.

Some of the characteristics of the trench capacitors are shown in Figure 4.30. Figure 4.30(a) illustrates the cell capacitor area and cell capacitance as a function of cell size for the 1Mb and 4Mb DRAM trench capacitors. Figure 4.30(b) indicates leakage current as a function of spacing between adjacent trenches in one of the early substrate storage node trench cells.

4.23.2 A basic trench DRAM capacitor process

A simplified description of the method of forming such a trench in the surface of the wafer is illustrated below [52].

(a) Locus and N-well formation

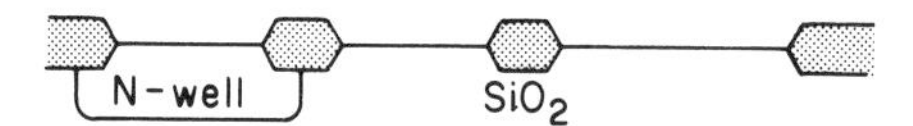

(b) Reactive Ion Etching (RIE) is used with SiO_2–Si_3N_4 as a mask to etch a trench a few microns deep in the silicon surface. The etch is optimized for a rectangular shape. The sidewalls of the trench can also be doped or oxide grown on them.

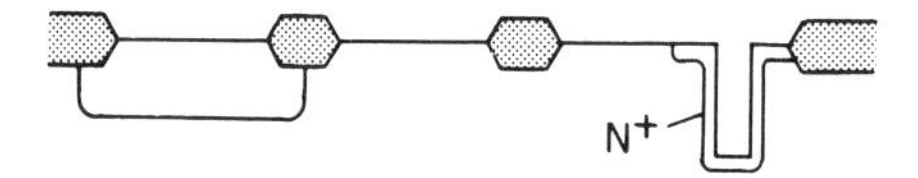

(c) Polysilicon is deposited and patterned for gates and in the trench.

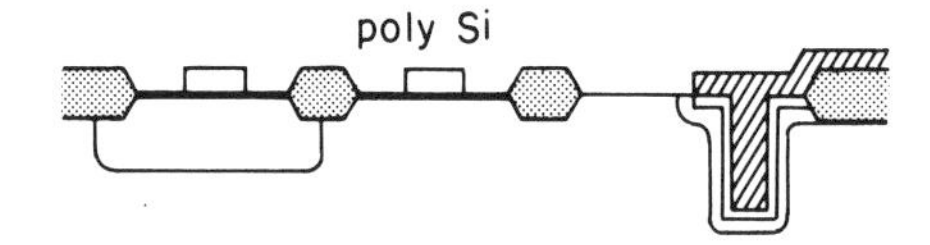

(d) Doping of the source and drain regions of the transistors.

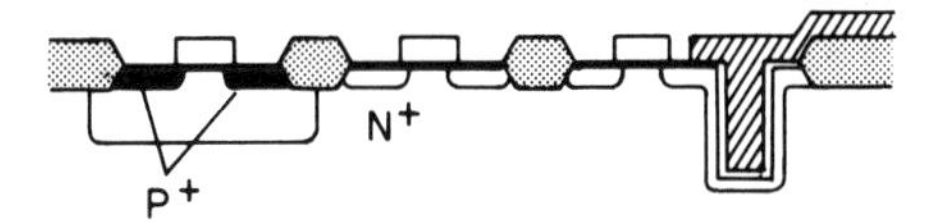

(e) Deposition and patterning of the metal interconnects.

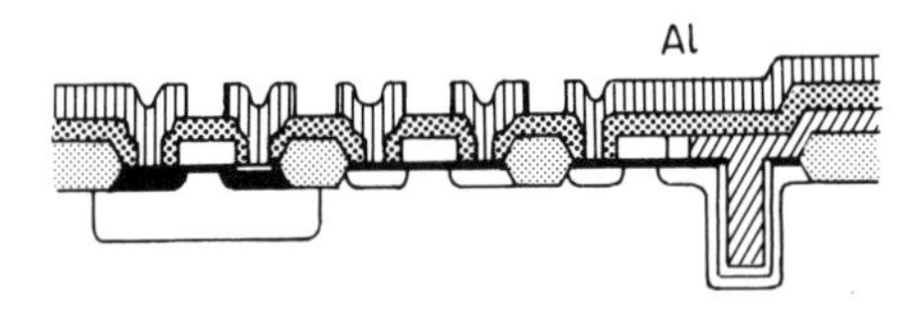

Major problems encountered with the trenches include: complexity and control of process, leakage between trenches, and parasitic transistors formed between the trenches and the various doped regions in the substrate.

4.23.3 Stacked capacitors for DRAMs

Since the industry had some difficulty mastering the trench technology, an alternative method for reducing horizontal capacitor size was developed. This method stacked the capacitor over the bit line above the surface of the silicon rather than trenching it down into the silicon.

The stacked capacitor is formed of a layer of deposited dielectric such as silicon dioxide sandwiched between two layers of deposited polysilicon. A schematic comparison of the double polysilicon planar cell, the simple trench capacitor cell and the simple stacked capacitor cell is shown in Figure 4.31(a).

The effective capacitance of a stacked cell is increased over that of a planar cell not only because of the increased surface area, but also by the curvature and sidewall field effects as shown in Figure 4.31(b).

The processing techniques used in the formation of the stacked capacitors, while complex, are still variations of historical MOS processing techniques. Two examples of the fabrication process of stacked capacitor cells are shown in Figure 4.32. Figure 4.32(a) shows a single-layer storage node. Following formation of the word lines and bit lines, the capacitor oxide followed by polysilicon for the storage node is deposited over the bit lines. This is followed by the electrode plate. The storage node of the capacitor is between the oxide and the polysilicon layer. The result is a single-layer capacitor which is stacked over the bit lines rather than occupying area on the chip.

A similar process which involves stacking more layers to increase the capacitance is shown the multi-fin structure in Figure 4.32(b). The fabrication process is as follows:

(a) the transistor is formed,

(b) SiN is deposited,

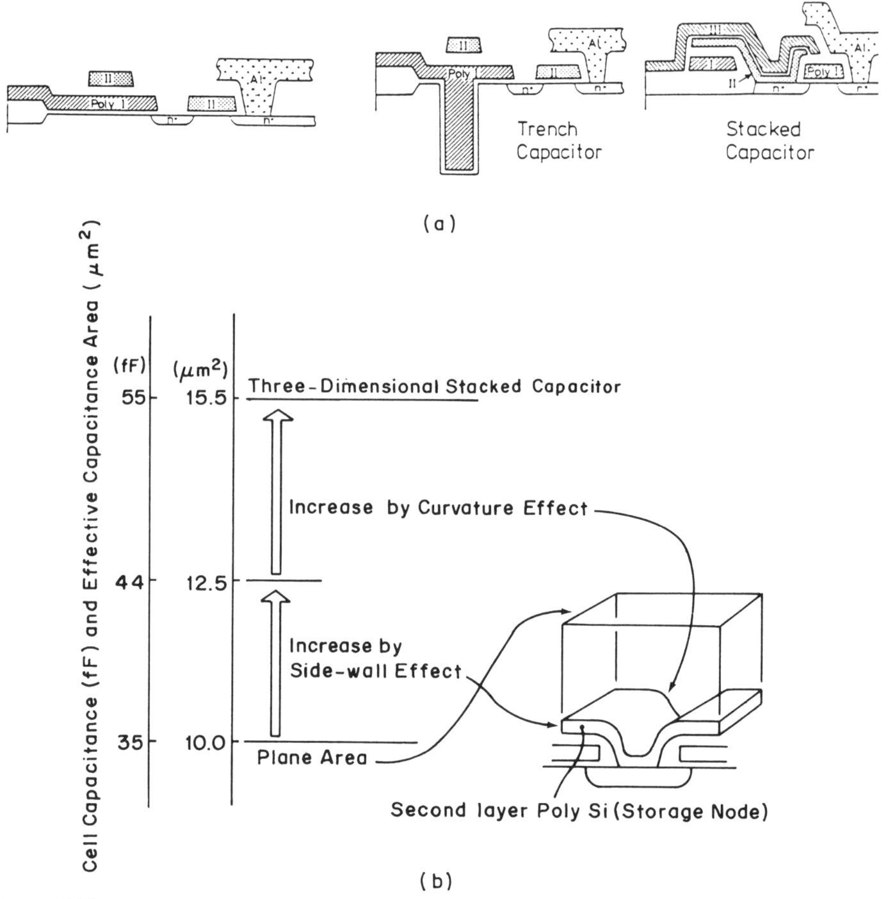

Figure 4.31
The stacked capacitor cell. (a) Comparison with the planar capacitor and the trench capacitor showing cell size reduction, (b) schematic of stacked capacitor cell showing components of the effective capacitance. (From Takemi *et al.* [75], Fujitsu 1985, with permission of IEEE.)

(c) oxide, polysilicon, then oxide are deposited and the contact holes are cut,

(d) polysilicon is deposited,

(e) the storage electrode is patterned,

(f) the oxide is removed and the capacitor film is grown.

(g) The polysilicon is deposited followed by the cell plate. The bit lines are formed.

Theoretically any number of such layers could be stacked.

A slightly more sophisticated process technology involves 'texturing' the surface of the polysilicon to form a surface layer composed of many small hemispheres side

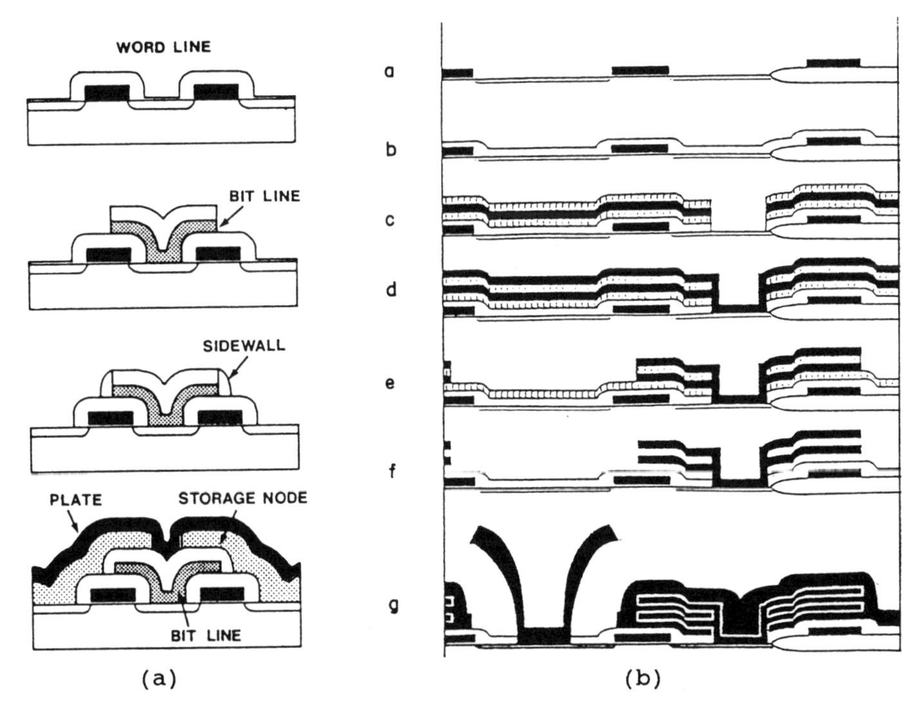

Figure 4.32
Two examples of process techniques for stacked capacitor cells for DRAMs. (a) From Kimera *et al.* [54], Hitachi 1988, with permission of IEEE, (b) From Ema *et al.* [63], Fujitsu 1988, with permission of IEEE.

by side. This increases the effective storage area of the capacitor as shown in the cross-section in Figure 4.33(a).

Capacitance as a function of texturing conditions, such as temperature and polysilicon film thickness, are shown in Figure 4.33(b). The current and voltage characteristics of a planar capacitor compared to a textured capacitor are shown in Figure 4.33(c).

4.24 SILICON ON INSULATOR STRUCTURES

Over the years many attempts have been made to isolate the MOS transistors from the bulk silicon to reduce parasitic interactions between transistors and resulting reliability problems such as latch-up.

The most ubiquitous method is the use of silicon epitaxial layers grown over a differently doped silicon substrate.

Early attempts at more complete isolation involved such exotic technologies as Silicon on Sapphire (SOS) which never was generally accepted in spite of considerable effort due to the high cost of the sapphire substrates.

Another such technique called SIMOX (Silicon IMplanted with OXide) involves

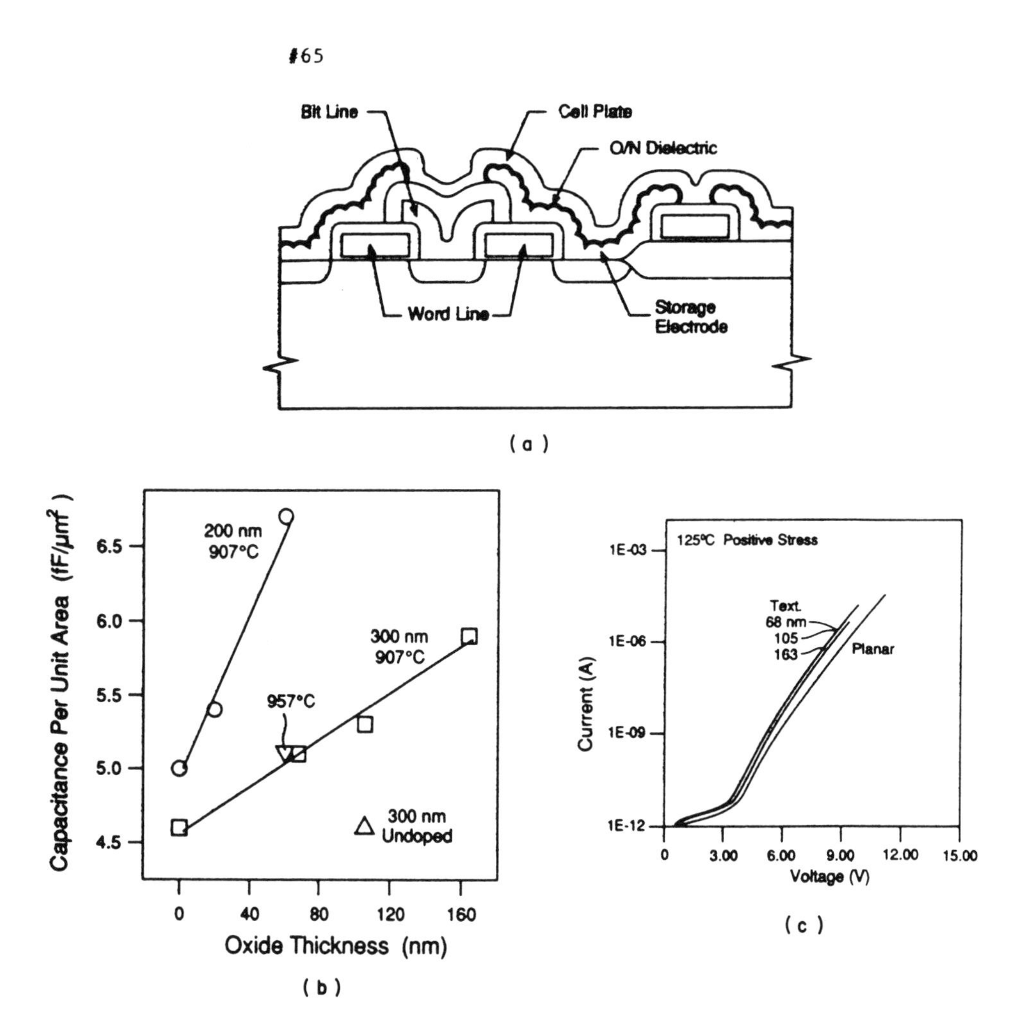

Figure 4.33
Illustration of DRAM cell using a textured capacitor. (a) Schematic cross-section of DRAM cell, (b) capacitance plotted against texturing conditions, (c) current and voltage characteristics of textured capacitor. (From Fazan and Ditali [84], Micron Technology 1990, with permission of IEEE.)

implanting a buried layer of silicon nitride or silicon dioxide into the silicon substrate. This forms an insulating layer with residual silicon remaining on top onto which an epitaxial layer of silicon may be grown. This would permit CMOS circuits without using isolation wells. It also could potentially reduce the invertor delay times significantly due to the absence of parasitic capacitance, and also eliminate latch-up problems. Process difficulties in implementing this technique have not yet been sufficiently overcome for it to receive widespread acceptance. See Chapter 9 for further discussion of SIMOX.

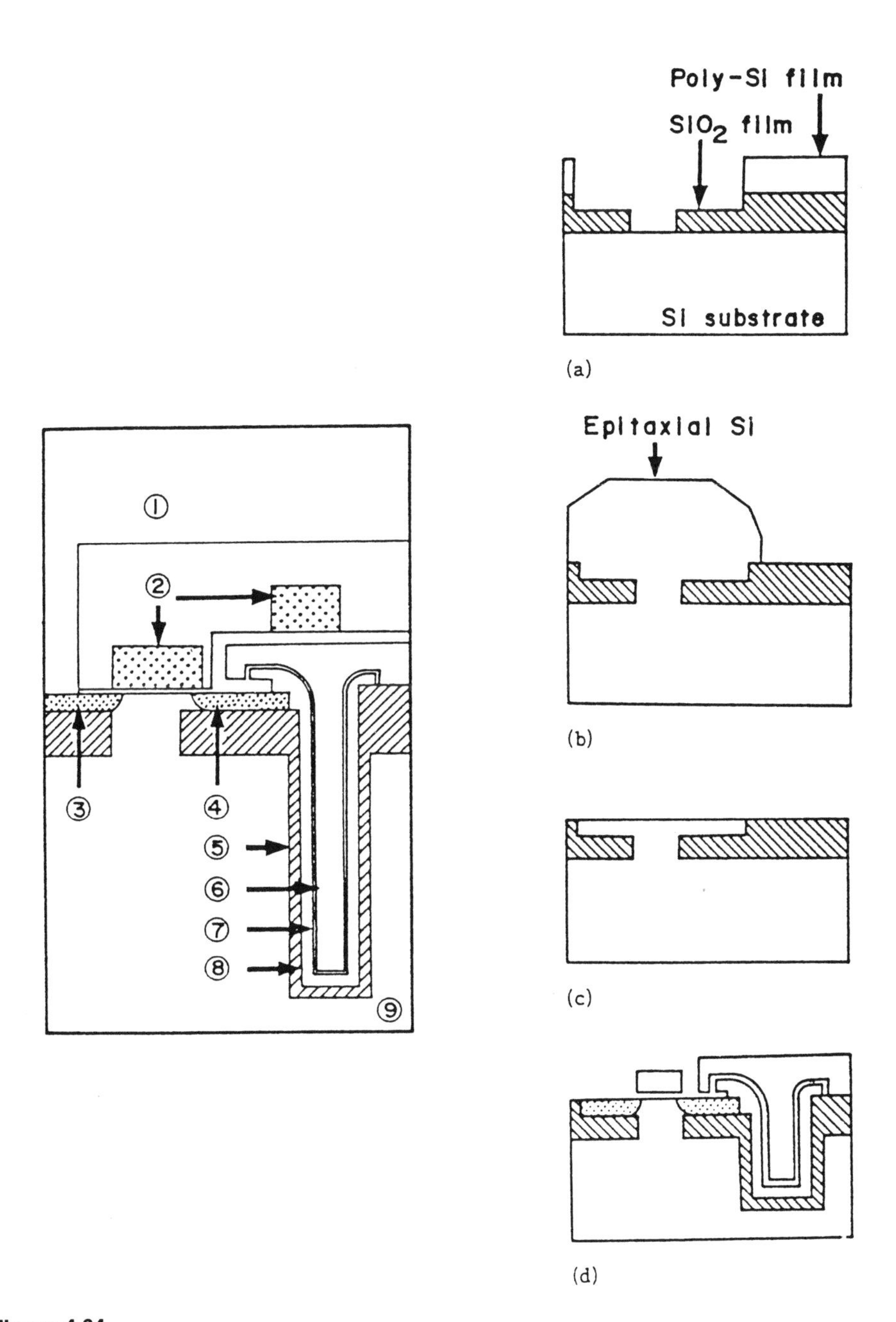

Figure 4.34
Process for a DRAM with transistor in SOI above a trench. TOLE cell structure: 1 bit-line, 2 word-line, 3 drain area, 4 source area, 5 SiO_2 film, 6 capacitor phase, 7 capacitor insulator film, 8 charge storage electrode, 9 Si substrate. (a) Device and seed area formation, (b) epitaxial lateral overgrowth, (c) preferential polishing, (d) capacitor formation. (From Kubota *et al.* [53], NEC 1987, with permission of IEEE.)

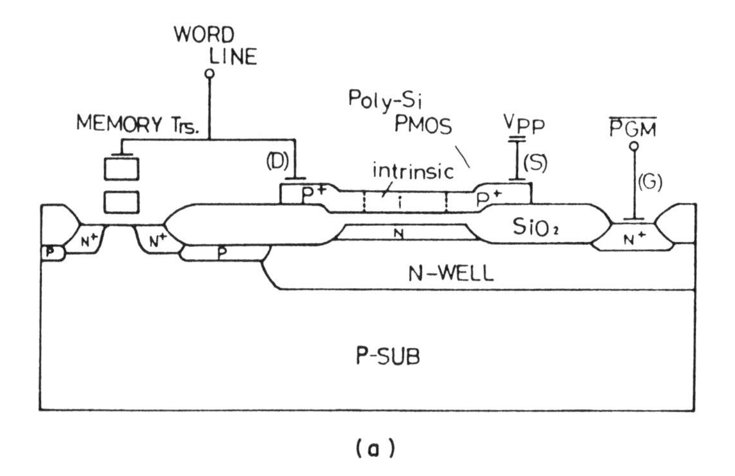

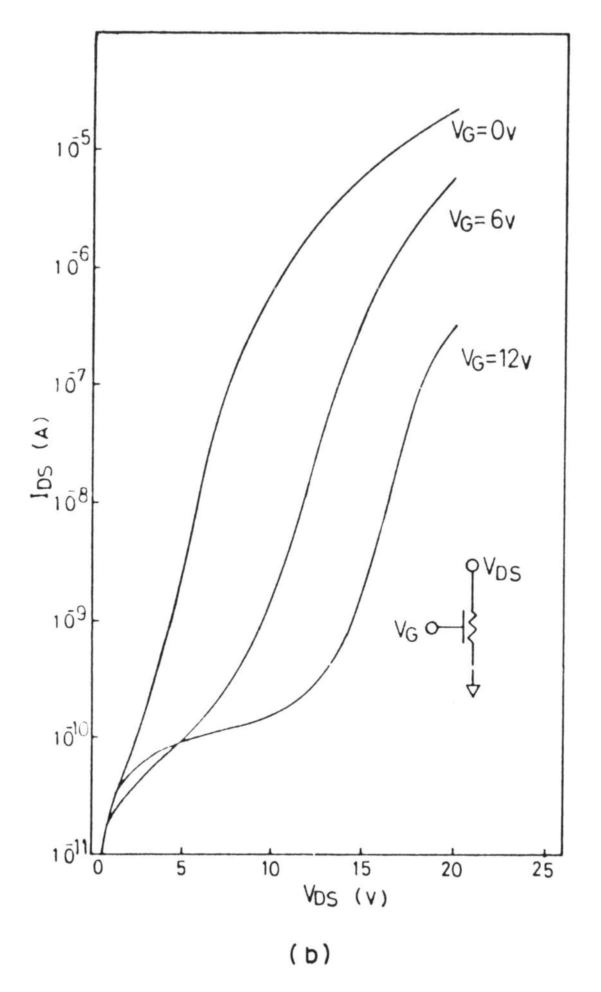

Figure 4.35
Stacked polysilicon PMOS transistor. (a) Schematic cross-section, (b) electrical *I–V* characteristics. (From Yoshitaki *et al.* [77], Hitachi 1985, with permission of IEEE.)

Another such method, developed by Sony, uses silicon seeds deposited in a regular pattern on the surface of a silicon dioxide layer. These are used to grow a selective epitaxial layer with boundaries. This layer can be polished to planarize it and transistors diffused into it. If the boundaries are observed RAM cells could be constructed on the epi layer.

An example of growing an SOI epitaxy from a seed, developed by NEC [53], is shown in Figure 4.34. Shown is the growth of the epitaxial layer in which a transistor in a DRAM cell is stacked horizontally over the trench.

Various silicon-on-insulator (SOI) methods have been investigated including: using laser beams, electron beams, and carbon heaters to recrystallize deposited polysilicon into single crystal silicon that MOS transistors can be diffused into.

4.25 NEW TRANSISTOR STRUCTURES FOR SRAMs

As minimum geometries moved into the sub 0.5 μm range, new transistor structures began to appear. One was a new technique of stacking the P-channel transistors in the polysilicon interconnect layer. An example from an early Hitachi [77] EPROM paper is shown in the schematic cross-section in Figure 4.35. This technique was widely used in the 4Mb CMOS SRAMs.

Also shown are the transistor *I–V* characteristics. It can be seen from these characteristics that the quality of the early amorphous polysilicon transistors did not equal that of single crystal silicon transistors. For further discussion of the polysilicon PMOS cell see Chapter 9.

4.26 POSSIBILITIES FOR DEEP SUBMICRON CELLS

New cell investigations in the sub 0.1 μm range include the heterogeneous devices which use combinations of exotic materials rather than the standard metal–oxide–silicon of the traditional MOS transistor. These devices fall into two basic types, those that use some form of heterojunction or superlattice structures consisting of GaAs and aluminum GaAs in alternating layers, and those that use silicon germanium heterojunctions.

BIBLIOGRAPHY

[1] *Business Week*, July 6, 1981, p.31.

[2] *Scientific American*, Special issue on Microelectronics, September 1977.

[3] Resor, G. L. and Tobey, A. C. (1981) The role of direct step-on-the-wafer in microlithography, strategy for the 80's. *Solid State Technology*, May 1981, p.65.

[4] Douglas, E. C. (1981) Advanced process technology for VLSI circuits, *Solid State Technology*, May 1981, p.65.

[5] Liaw, H. Ming (1982) Trends in semiconductor material technologies for VLSI and VHSIC applications, *Solid State Technology*, July 1982, p.66.

[6] Eidson, J. C. (1981) Fast electron-beam lithography, *IEEE Spectrum*, July 1981.

[7] Coe, M. E. and Rogers, S. H. (1982) Low frequency planer plasma etching of polycide structures in an SF6 glow discharge, *Solid State Technology*, August 1982.
[8] Borgeron, S. F. and Duncan, B. F. (1982) Controlled anisotrope etching of polysilicon, *Solid State Technology*, August 1982.
[9] Mohammadi, F. (1981) Silicides for interconnection technology, *Solid State Technology*, January 1981.
[10] Benzing, C. (1982) Shrinking VLSI dimensions demand new interconnection materials, *Electronics*, August 25, 1982.
[11] Tolliver, D. L. (1980) Plasma processing in microelectronics—past, present and future, *Solid State Technology*, November 1980.
[12] Asai, K. *et al.* (1983) 1 Kb static RAM using self-aligned FET technology, 1983 IEEE International Solid State Conference.
[13] Yokoyama, N. *et al.* (1983) A GaAs 1K static RAM using tungsten-silicide gate self-alignment technology, 1983 IEEE International Solid State Circuit Conference.
[14] Liu, Sheau-ming, S. *et al.* (1982) HMOS III Technology, *IEEE Journal of Solid State Circuits*, **SC-17**, No. 5, October 1982, p.810.
[15] Flannagan, S. (1988) Future Technology Trends for Static RAMS, *IEDM Proceedings* p.40.
[16] Gosch, J. (1983) Buried-nitride CMOS may double device density, *Electronics*, January 27, p.70.
[17] Waller, L. (1987) Fine tuning optical steppers, *Electronics*, October 15, p.134.
[18] Cole, B. C. (1987) Here comes the Billion-Transistor I.C., *Electronics*, April 2, p.81.
[19] Cohen, C. (1983) 3-D IC may augur denser VLSI circuitry; multiple layers are a possibility, *Electronics*, September 22, p.92.
[20] Takeda, E. *et al.* (1985) Constraints on the Application of 0.5 micron MOSFET's to ULSI Systems, *IEEE Journal of Solid State Circuits*, **SC-20**, No. 1, February 1985, p.242.
[21] Dennard, R. H. *et al.* (1974) Design of ion-implanted MOSFET's with very small physical dimensions, *IEEE Journal of Solid State Circuits*, **SC-9**, Oct. 1974, p.256.
[22] Appls, J. A. *et al.* (1970) Local Oxidation of Silicon and its Application in Semiconductor Device Technology, *Philips Res. Rep.*, **25**, p.118.
[23] Gosch, J. (1987) West Germany Grabs the Lead in X-ray Lithography, *Electronics*, February 5, p.78.
[24] Sakui, K. *et al.* (1988) A New Static Memory Cell Based on Reverse Base Current (RBC) Effect of Bipolar Transistor, *IEDM Proceedings* p.44.
[25] McLeod, J. (1987) Keys to the Future: CAD and Fab, *Electronics*, April 2, p.86.
[26] Ahrons, R. W. and Gardner, P. D. (1970) Interaction of Technology and Performance in Complementary Symmetry MOS Integrated Circuits, *IEEE Journal of Solid State Circuits*, **SC-5**, No.1, February 1970, p.24.
[27] de Werdt, R. *et al.* (1987) A 1M SRAM with Full CMOS Cells Fabricated in a 0.7 μm Technology, *IEDM Proceedings*, p.532.
[28] Davari, B. *et al.* (1988) A High Performance 0.25 μm CMOS Technology, *IEDM Proceedings*, p.56.
[29] Momose, H. *et al.* (1985) 1.0-μm n-Well CMOS/Bipolar Technology, *IEEE Journal of Solid State Circuits*, **SC-20**, No.1, February 1985.
[30] Maekawa, T. *et al.* (1987) A 60 μm^2 SOI CMOS SRAM Cell, *IEDM Proceedings* p.536.
[31] Lin, H. C. *et al.* (1976) Optimum Load Device for DMOS Integrated Circuits, *IEEE Journal of Solid State Circuits*, **SC-11**, No.4, August 1976.
[32] Wada, T. *et al.* (1978) A 15ns 1024-Bit Fully Static MOS RAM, *IEEE Journal of Solid State Circuits*, **SC-13**, No.5, October 1978.

[33] Arai, E. and Ieda, N. (1978) A 64k bit Dynamic MOS RAM, *IEEE Journal of Solid State Circuits*, **SC-13**, No.3, June 1978.
[34] Yamanaka, T. *et al.* (1988) *A 25 μm* New Poly-Si PMOS Load (PPL) SRAM Cell Having Excellent Soft Error Immunity, *IEDM Proceedings*, p.48.
[35] Iwai, H. *et al.* (1987) 0.8 micron BI-CMOS Technology with high ft Ion-Implanted Emitter Bipolar Transistor, *IEDM Proceedings*.
[36] Lyman, J. (1988) This Etcher Opens the Way to Superchips of the '90's *Electronics*, June 1988.
[37] Lu, N. (1986) *IEEE Journal of Solid State Circuits* **SC-1** Oct. 1986 p.627.
[38] Sakamoto, M. *et al.* (1985) Buried Storage Electrode Cell for Metabit DRAMs, *IEEE IEDM Tech. Digest*, p.710.
[39] Richardson, W. F. *et al.* (1985) A Trench Transistor Cross-point DRAM Cell, *IEEE IEDM Tech. Digest*, December, p.714.
[40] Nakamura, *et al.* (1984) Buried Isolation Capacitor Cell for Megabit MOS Dynamic RAM *IEDM Proceedings*, p.236.
[41] Nakajima *et al.* (1984) An Isolation-Merged Vertical Capacitor Cell for Large Capacity DRAM, *IEDM Proceedings*, December 1989, p.240.
[42] Wada (1984) A Folded Capacitor Cell for Future Megabit DRAMs *IEDM Proceedings*, December 1984, p.244.
[43] Mashiko, K. *et al.* (1987) A 90ns 4Mb DRAM in a 300mil DIP, *IEEE Journal of Solid State Circuits* ISSCC *Proceedings* February 1987, p.12.
[44] Taguchi, M. *et al.* (1986) Dielectrically Encapsulated Trench Capacitor Cell, *IEDM Proceedings*, p.136.
[45] Horiguchi, F. *et al.* (1987) Process Technologies for High Density High Speed 16Megabit Dynamic RAM *IEDM Proceedings*, p.324.
[46] Tsukamoto, K. *et al.* (1987) Double Stacked Capacitor with Self-Aligned Poly Source/Drain Transistor (DSP) Cell for Megabit DRAM, *IEDM Proceedings*, December 1987, p.328.
[48] Kaga, T. *et al.* (1987) A 4.2 μm^2 Half V_{CC} Sheath-Plate Capacitor DRAM Cell with Self-Aligned Buried Plate-Wiring, *IEDM Proceedings*, December 1987, p.332.
[49] El-Mansy, Y. A. and Burghard, R. A. (1982) Design Parameters of the Hi-C SRAM cell, *IEEE Journal of Solid State Circuits* **SC-17**, No.5 October 1982.
[51] Oguie *et al.* (1986) *IEEE Journal of Solid State Circuits*, **SC-21**, p.681.
[52] Minigishi, K. *et al.* (1983) A Submicron CMOS Megabit Level Dynamic RAM technology using doped Face Trench Capacitor Cell. *IEDM Proceedings*, p.321.
[53] Kubota, T. *et al.* (1987) A New Soft-Error immune DRAM Cell with a Transistor on a Lateral Epitaxial Silicon Layer, *IEDM Proceedings* p.344.
[54] Kimura, S. *et al.* (1988) A New Stacked Capacitor DRAM Cell Characterized by a Storage Capacitor on a Bit-Line Structure, *IEDM Proceedings* p.596.
[55] Lewyn, L. L. and Meindl, J. D. (1985) Physical Limits of VLSI DRAM's *IEEE Journal of Solid State Circuits*, **SC-20**, No.1, February 1985, p.231.
[56] Taguchi, M. *et al.* (1985) A Capacitance-Coupled Bit Line Cell, *IEEE Journal of Solid State Circuits*, **SC-20**, No.1, February 1985, p.210.
[57] Wada, T. *et al.* (1987) A 34-ns 1-Mbit CMOs SRAM Using Triple Polysilicon, *IEEE Journal of Solid State Circuits*, **SC-22**, No.5, October 1987, p.727.
[58] Bastiaens, J. J. J. and Gubbels, W. C. H. (1988) The 256-kbit SRAM: an important step on the way to submicron IC technology, *Philips Technical Review*, Vol. 44, No.2, April, p.33.
[60] Kohno, Y. *et al.* (1988) A 14-ns 1-Mbit CMOS SRAM with Variable Bit Organization, *IEEE Journal of Solid State Circuits*, **SC-23**, No.5, October 1988, p.1060.

[61] Kobayasi, Y. *et al.* (1985) A 10 μW Standby Power 256k CMOS SRAM, *IEEE Journal of Solid State Circuits* **SC-20**, p.935.
[62] Sunouchi, K. *et al.* (1989) A Surrounding Gate Transistor (SGT) Cell for 64/256 Mbit DRAMs, *IEDM Proceedings* p.23.
[63] Ema, T. *et al.* (1988) 3-Dimensional Stacked Capacitor Cell, *IEDM Proceedings*, p.592.
[64] Takemae, Y. *et al.* (1985) A 1Mb DRAM with 3-Dimensional Stacked Capacitor Cells, *ISSCC Proceedings*, p.250.
[65] van de Plas, P. (1985) Geometry Dependent Bird's Beak Formation for submicron LOCOS Isolation.
[66] Prince, B. and Due-Gundersen, G. (1983) *Semiconductor Memories* John Wiley, Chichester.
[67] Sood, L. C. *et al.* (1983) A 35ns 2k × 8 HMOS Static RAM, *IEEE Journal of Solid State Circuits*, **SC-18**, October 1983, p.498.
[68] Lu, N. (1988) DRAM Design, Memory Design Workshop, *IEDM Proceedings*, 1988.
[69] Hodges, D. A. and Jackson, H. G. (1983) *Analysis and Design of Digital Integrated Circuits*, McGraw-Hill, New York.
[70] Deferm, L. *et al.* (1988) Latch-up in a BiCMOS Technology *IEDM Proceedings*, p.130.
[71] Kohno, Y. *et al.* (1988) A 14-ns 1-Mbit CMOS SRAM with Variable Bit Organization, *IEEE Journal of Solid State Circuits*, **SC-23**, No.5, October 1988, p.1060.
[72] Kobayasi, Y. *et al.* (1985) A 10μW Standby Power 256k CMOS SRAM, *IEEE Journal of Solid State Circuits*, **SC-20**, No. 5, October 1985, p.935.
[73] Miyaji, F. *et al.* (1989) A 25-ns 4-Mbit CMOS SRAM with Dynamic Bit-line Loads, *IEEE Journal of Solid State Circuits*, **SC-24**, No.5, October 1989, p.1213.
[74] Kertis, R. *et al.* (1990) A 6.8ns 1Mb ECL I/O BiCMOS Configurable SRAM, VLSI Circuits Symposium, June, p.39.
[75] Takemae, Y. *et al.* (1985) A 1Mb DRAM with 3-Dimensional Stacked Capacitor Cells, *ISSCC Proceedings* February, p.250.
[76] Weber, S. (1988) Will Quantum-effect Technology Represent a Quantum Jump in IC's?, *Electronics*, October, p.143.
[77] Yoshizaki, K. *et al.* (1985) A 95ns 256k CMOS EPROM. *ISSCC Proceedings*, February, p.166.
[78] Tani, Y. *et al.* (1990) KrF Excimer Laser Lithography with High Sensitivity Positive Resist and High Power Laser. *Symposium on VLSI Technology Proceedings*, p.7.
[79] Sunami, H. *et al.* (1982) A Corrugated Capacitor Cell (CCC) For Megabit Dynamic MOS Memories, *IEDM Tech. Digest*, p.806.
[80] Chatterjee, P. *et al.* (1986) Trench and Compact Structures For DRAMs, *IEDM Tech. Digest*, p.128.
[81] Lu, N. *et al.* (1985) The SPT Cell—A New Substrate-Plate Trench Cell For DRAMs, *IEEE IEDM Tech. Digest*, Dec. 1985, p.771.
[82] Lu, N. *et al.* (1988) *IEEE IEDM Tech. Digest*, Dec. 1988.
[83] Widmann, D. W. (1976) Metallization for Integrated Circuits Using a Lift-off Technique, *IEEE Journal of Solid State Circuits*, **SC-11**, No.4. August 1976.
[84] Fazan, P. C. and Ditali, A. (1990) Electrical Characterization of Textured Interpoly Capacitors for Advanced Stacked DRAMs *IEEE IEDM Tech. Digest*, December 1990, p.663.
[85] Yamanaka, T. *et al.* (1990) A 5.9 μm^2 Super Low Power SRAM cell using a new Phase-shift Lithography *IEEE IEDM Tech. Digest*, December 1990, p.477.

5 BASIC MEMORY ARCHITECTURE AND CELL STRUCTURE

5.1 INTRODUCTION

Another possible set of classifications of semiconductor data storage devices is related to the basic architecture and cell structure of the memory circuit as shown in Table 5.1.

Static memory gates store their data in latches, while dynamic memory gates store their data on capacitors. The static memory cell has a non-destructive readout and retains its memory under the application of dc power. The one-transistor dynamic cell has a destructive readout. All DRAM cells lose their data without the application of a periodic refresh.

Both dynamic and static RAMs are called volatile since they lose their data when the power is turned off. Another class of memory cells is called non-volatile since the data requires no refresh and is retained even when all power is off. This class of non-volatile memories, which includes the ROMs, EPROMs, and EEPROMs, stores data in preprogrammed storage cells.

All of the devices mentioned so far have random access read capability and those that are writable have random access write. Memory devices that have only serial read and write capability are called Serial Access Memories (SAMs) and may be dynamic or static, volatile or non-volatile.

5.2 BASIC MEMORY ARCHITECTURE

The basic memory architecture has the configuration shown in Figure 5.1(a). This potentially includes: inputs, outputs, addresses, read control, write control and data storage. The three major divisions of memories are RAMs, ROMs, and SAMs. These subsets of this basic architecture are shown in 5.1(b), (c), and (d).

The basic RAM (random access memory), shown in Figure 5.1(b), is a memory in which any storage location can be randomly accessed for Read and Write by inputting

Table 5.1

	SRAM	DRAM	(E)EPROM	SAM
Data storage	latch	capacitor	programmed	XX
Volatile	yes	yes	no	yes
Destructive read	no	yes/no	no	XX
Access	Random	Random	Random	Serial

the coordinates of that bit on the address pins. It usually combines the Read (R) and Write (W) control pins into a single Write Enable (WE) control pin whose two states determine Read or Write.

The basic ROM (read-only memory), shown in 5.1(c), is a hardware preprogrammed memory in which any storage location can be randomly accessed for Read by inputting the coordinates of that bit on the address pins. The Data inputs and Write control pins are missing and the Read control pin usually becomes a Chip Select or Chip Enable (CS/CE) control pin.

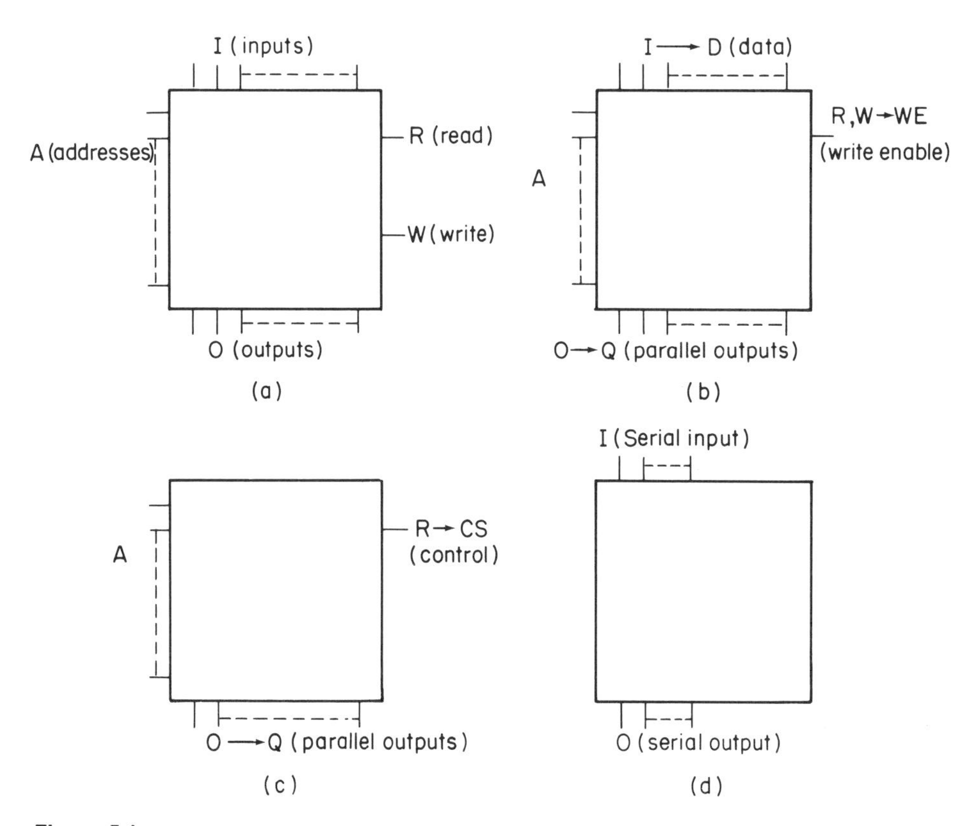

Figure 5.1
Schematics of memory configurations. (a) Basic memory, (b) RAM, (c) ROM, (d) SAM. (From Salters [50] 1989, with permission of Philips Research.)

Table 5.2 Basic semiconductor memory architecture.

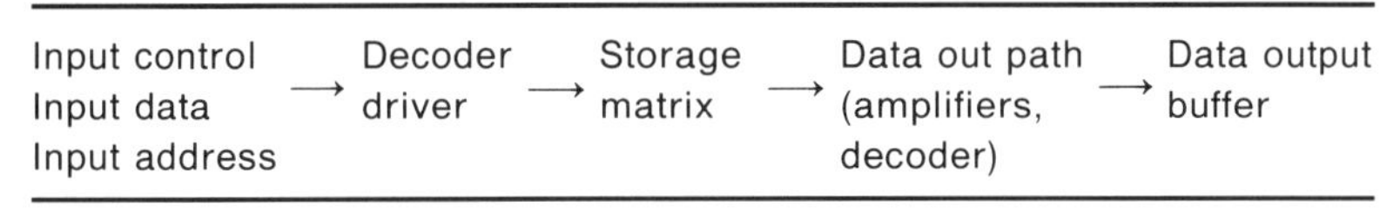

Input control Input data Input address	→	Decoder driver	→	Storage matrix	→	Data out path (amplifiers, decoder)	→	Data output buffer

The basic SAM (serial-access memory), shown in Figure 5.1(d), is a memory with serial input(s) and serial output(s) in which any address location can be serially accessed by counting the number of input cycles to that location.

The WOM is a mythical data storage device which can be only written but not read. It occurs most frequently when the Write Enable pin is inadvertently left off a RAM pinout.

The internal architecture of the memory is usually configured as shown in Table 5.2.

A typical random access memory (RAM) architecture is shown in Figure 5.2 and consists of a matrix of storage bits with bit capacity $2^N \times 2^M$ bits arranged in an array with 2^M columns (bit lines) and 2^N rows (word lines).

To read the data stored in the array, a row address is input and is decoded to select one of the rows or word lines. All of the cells along this word line are activated. The column (bit) decoder then addresses one bit out of the 2^M bits that have been activated and routes the data that is stored in that bit to a sense amplifier and then out of the array. Data in and Data out are controlled by the Read–Write control circuit shown on the lower left-hand side of the figure.

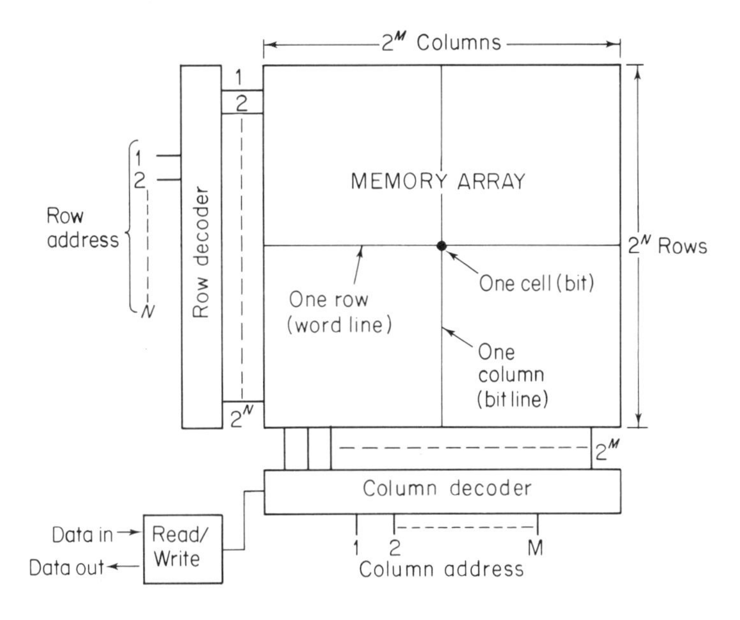

Figure 5.2
Basic random access memory architecture.

5.3 DATA STORAGE ELEMENTS

The most basic part of a memory is the data storage cell which can be anything that can store two well defined states in a specified and repeatedly accessible location. In semiconductor memories the most common storage elements are either a latch or a capacitor.

When the memory bit is stored in a latch, or bistable flipflop, we say that the cell is static since it does not need to have its data refreshed as long as dc power is applied. When the memory bit is stored in a capacitor, we say the cell is dynamic since the charge in the capacitor tends to leak away over time and needs to be periodically refreshed.

5.3.1 Static RAM (latched) cells—configuration and types

Latched semiconductor storage cells are bistable transistor flipflops in various configurations. An example is shown in Figure 5.3. The flipflop consists of two load elements (L) and two storage transistors (C). There are two access transistors in addition (D).

Semiconductor flipflops can be made of either MOS or Bipolar transistors. Figure 5.3 shows a MOS static cell with load devices which may be either enhancement or depletion mode transistors (in an NMOS cell), PMOS transistors (in a CMOS cell) or load resistors (in a Mix-MOS or R-load cell). The access and storage transistors are enhancement type NMOS.

Data are stored as voltage levels with the two sides of the flipflop in opposite voltage configurations, that is, node A is high and node B is low in one state and node A is low and node B is high in the other resulting in two stable states. The cell operation is described in more detail in a later section.

The purpose of the load device is to offset the charge leakage at the drains of the

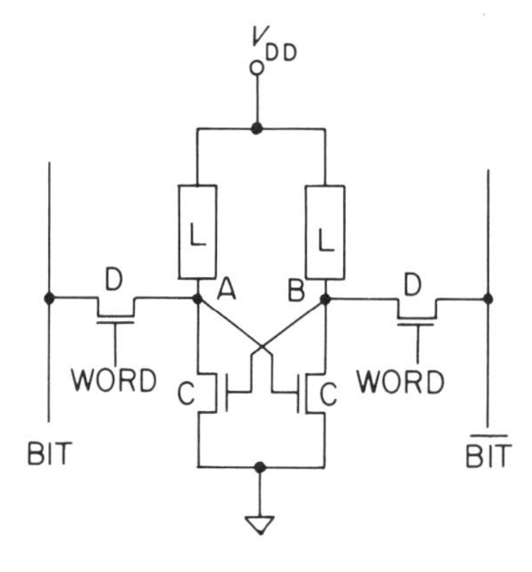

Figure 5.3
Schematic of generalized static RAM memory cell. (From Bastiaens and Gubbels [49]1988, with permission of Philips Research.)

storage and select transistors. When the load transistor is PMOS then the resulting CMOS cell has essentially no current drain through the cell except when it is switching. Either the NMOS or PMOS transistor is always off. The other two cells always have a low level of current flowing through the cell so that the standby power dissipation is higher than that of the CMOS cell. The select transistors are usually NMOS as are the storage transistors.

A depletion load transistor is normally used in the NMOS flipflop instead of a second enhancement transistor since it has better switching performance, high impedance, and is relatively insensitive to power supply variations. It is similar to a CMOS P-channel load in that internal voltage swings are essentially equal to the power supply voltage.

The polysilicon load resistor when used in place of the PMOS load transistor in the CMOS cell permits up to a 30% reduction in cell size if double polysilicon technology is used. This is true since the resistors can be stacked on top of the cell, therefore, reducing the cell size by two transistors. The resulting cell has higher leakage in standby when not being accessed than a cell with PMOS load since a small amount of current always flows through the resistor. Also the noise margin between the high and low state of the cell are reduced. The tradeoff is in performance against cost.

Design techniques can reduce the drawbacks of the R-load cell to some extent, for example power consumption of the total memory in standby is reduced if all peripheral circuits outside the memory array can be powered down without disturbing the stored data. It also helps if subthreshold leakages of 'off' transistors in the memory are at an adequately low level to maintain the stored data during standby.

Figure 5.4 shows various types of relatively common CMOS static RAM latched cells. Figure 5.4(a) shows the basic six transistor cell described above with PMOS load devices. This is the full CMOS static RAM cell. Figure 5.4(b) shows a four transistor static cell with resistor load pull-up devices which is the R-load static RAM cell. Figure 5.4(c) shows an eight transistor double-ended dual port static cell. Figure 5.4(d) shows a nine transistor static Content Addressable Memory (CAM) cell. All of these cells are static since the data is latched in the flipflop and periodic refresh of the data is not needed.

The dual ported static cell is useful in cache architectures, particularly embedded in a microprocessor chip, since this cell can be accessed simultaneously through both ports. Simultaneous access eliminates wait states for the microprocessor unless one port wants to read while the other is writing, or both want to write simultaneously. This could cause an undetermined state to appear in the cell unless this potential contention is resolved with support control circuits.

The content addressable, or associative, memory cell is used in applications where knowledge of the contents of the cell, as well as the location of the cell, is required. In normal operation it reads and writes like a normal static cell. A comparison operation is, however, also possible. In this operation in the cell shown, DATA is placed on $\overline{\text{BIT}}$ and $\overline{\text{DATA}}$ is placed on the BIT lines. This is the opposite of the normal read configuration. If the data match those in the cell, then the match transistor will stay turned off. If any cell has data that do not match, the match transistor pulls a previously precharged 'match' line low.

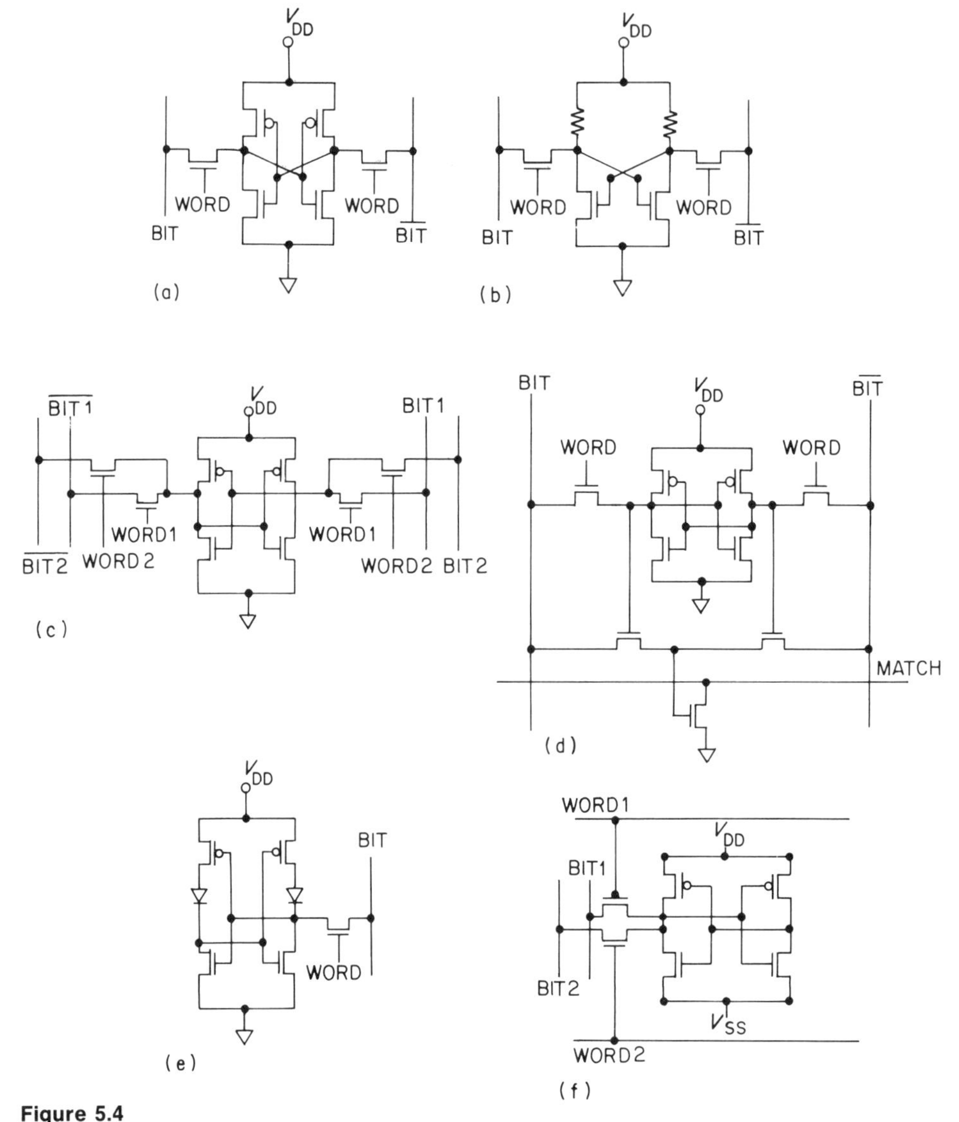

Figure 5.4
Various configurations of CMOS static RAM cells. (a) Six-transistor full CMOS, (b) four transistor with resistor load NMOS, (c) dual port with double-ended access, (d) content addressable, (e) five-transistor full CMOS having a single access transistor, (f) dual port with single-ended access.

While the content addressable cell shown has nine transistors, other applications require associative cells with even more transistors. Clearly the size of such a cell and the resulting memory size is much larger than for standard RAM cells. The application must require the particular associative features of this cell to justify the additional cost.

Two relatively unusual cells which are variations of two of the above cells are shown in Figure 5.4(e) and 5.4(f).

Figure 5.4(e) shows a single-ended five transistor static CMOS cell which can replace the basic six transistor full CMOS cell where a smaller array size is needed. The five transistor CMOS cell contains one less component and one less bit line per cell than the more common six transistor cell. It has not been widely used because of lower operating margins and difficulty in performing the Write operation reliably with standard power supply voltages. While there is no problem with writing a '0', writing a '1' is difficult since the NMOS transistor between the cell and the bit line operates in a source follower mode limiting the voltage actually transmitted to the cell. This makes the WRITE operation unreliable unless the word line voltage is increased above the V_{DD} power supply. In cases where it has been used for the density advantages it gives, careful control of the peripheral control architecture has been required to ensure reliable operation of the cell. The additional circuitry in the periphery can add to the chip area thereby reducing the only advantage this cell has.

The single-ended dual ported cell shown in Figure 5.4(f) can be used in place of the more common double-ended dual port cell shown in Figure 5.4(c) with a corresponding saving in silicon area resulting from eliminating two transistors. The comments made on the single-ended cell above also apply to this cell.

5.3.2 Dynamic RAM (capacitor storage) cell evolution

Dynamic RAM cells that store charge on a capacitor are shown in Figure 5.5. Since the charge stored on the capacitor tends to leak away with time, the data need to be refreshed periodically. Refreshing data consists of restoring it to the original '0' or '1' level stored in that cell. In the one transistor cell the storage capacitor is explicit. The storage capacitor in the three and four transistor cells are storage transistors.

The evolution of these cells began with the four transistor dynamic cell shown in Figure 5.5(a). This cell is derived from the six transistor static cell by removing the load devices. Since current is no longer supplied to the storage nodes to replace that lost to leakage, the cell must be periodically refreshed. The cell is activated by a clock pulse to the two access transistors. Since the charge eventually leaks off, the cell must be reactivated to refresh the data.

This four transistor cell is the simplest and one of the earliest dynamic cells. It is compatible with MOS logic circuitry since it contains only transistors and only simple control circuitry is required. It can be easily used as an embedded memory cell in logic if care is taken to provide refresh at sufficiently frequent intervals. A clock running in the system for other purposes can be used to provide the refresh timing.

Figures 5.5(b), (c), (d) and (e) show four possibilities for three transistor dynamic cells. These cells differ in the number of Read and Write Control lines and Read and Write Input–Output lines. All of them are 'inverting' cells in the sense that the information read out must be inverted before it is placed back into the cell upon refresh.

The cell shown in 5.5(b) has two control and two I/O lines. Its separate read and write select lines make it relatively fast, but the four lines with their additional contacts to every cell occupy area making it larger. This cell was used for the 1k DRAM [66] and also has been used in such devices as a dual-port FIFO buffer by NEC.

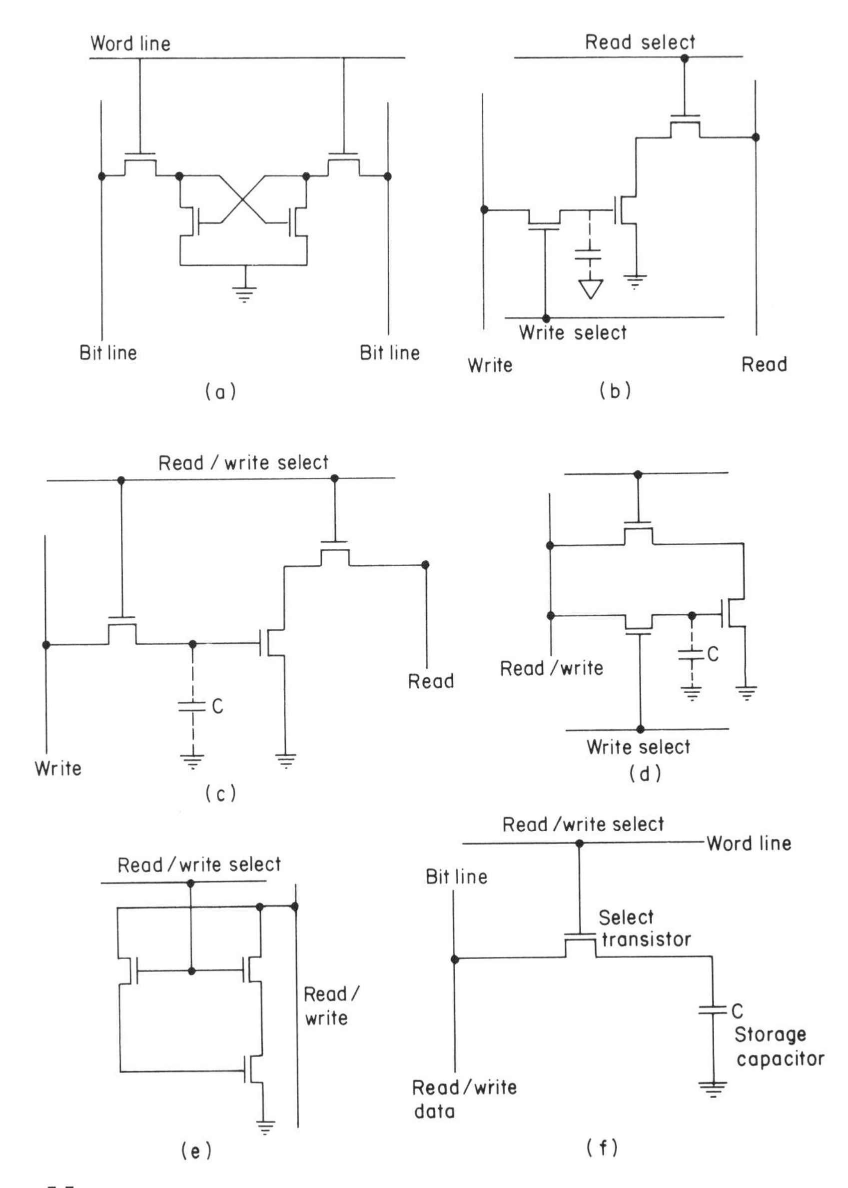

Figure 5.5
Historical evolution of the DRAM memory cell. (a) Basic bi-stable flipflop without load devices, (b) $2X - 2Y$, (c) $1X - 2Y$, (d) $2X - 1Y$, (e) $1X - 1Y$, (f) $1X - 1Y$ where X is the number of control lines and Y the number of I/O lines.

The cell shown in Figure 5.5(c) has one address select that serves both as a Read and Write control. It has, in addition, two I/O lines giving it three lines per cell. It unfortunately requires very accurate pulse shaping and voltage levels to operate in a stable manner and has not been used in practical applications.

The cell in Figure 5.5(d) has two controls and one I/O line giving it three lines per cell. It has also not been used in practical applications for control reasons.

The cell shown in Figure 5.5(e) with one Read–Write Control and one I/O line occupies the least area of any of the three transistor cells. It has relatively simple refresh but requires the overhead of a data control resistor to keep track of refresh status. This cell requires a three-state input row select voltage with refresh, read and write being each selected by one of the three voltage states. It has been used in a 4k DRAM.

The one transistor cell shown in 5.5(f) has become the industry standard dynamic RAM cell. With only one transistor and a capacitor, it has the lowest component count and hence smallest chip size of all the dynamic cells. It has one Read–Write control line (Word Line), and one I/O (Bit Line). It is operated by the Row decoder–driver turning on the word line which controls the gate of the transistor. This causes the charge stored on the capacitor to feed out onto the bit line. Since the capacitance of the bit line is normally about 10 times that of the cell capacitor, the signal voltage is reduced by a factor of 10, and the charge stored on the cell capacitor is lost and must be restored (refreshed) during the same access cycle.

5.4 STATIC RAM ARCHITECTURE

5.4.1 Basic architectural overview

A schematic block diagram of a total static RAM memory in a 4k deep by 1 bit wide (4k × 4) static RAM configuration is shown in Figure 5.6. Addresses are not multiplexed. Row addresses A_0 to A_5 select the word line, and column addresses A_6 to A_{11} select the bitline pairs to access the correct bit in the memory array. In a read operation this bit is then read out onto the bit lines, onto the databus, through the sense amplifier and out through the three-state output buffer.

For a write operation, the row and column address buffers and decoders again select the proper word line and bit line pairs. The data on the Data-in pin passes through the three-state input buffers onto the data bus and is written into the correct location in the memory array over the selected bit lines. The Read–Write control buffer controls whether read or write is selected on the Data-in and Data-out pins.

5.4.2 Basic SRAM cell operation

Since all MOS static RAMs have in common a basic cell consisting of two transistors and two load elements in a flipflop configuration with two select transistors, for this example we will use the illustration of a six transistor full CMOS cell as shown in Figure 5.7. For reference the layout of this cell showing the location of the various transistors is also shown.

Information is stored in the form of voltage levels in the flipflop which is formed by the two cross-coupled invertors. The flipflop has two stable states which we can designate '1' and '0'. If in logic state '1' point C_5 is high and point C_6 is low

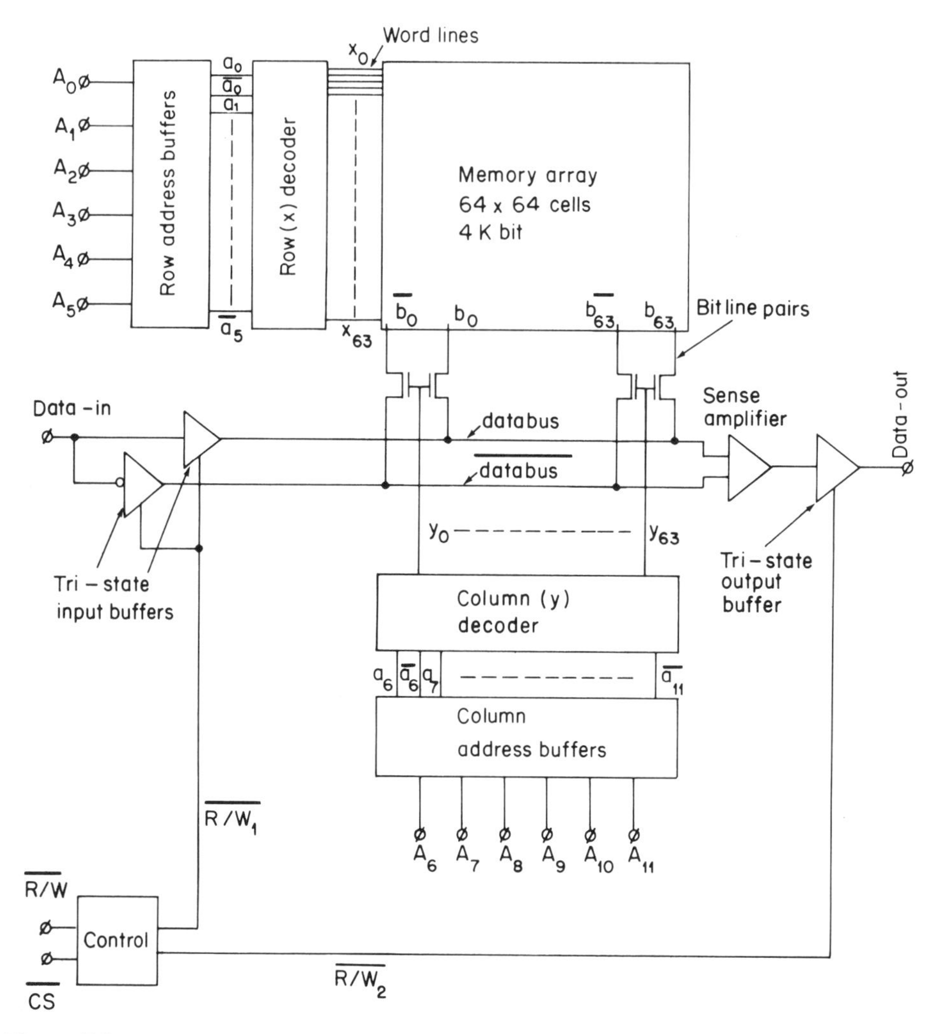

Figure 5.6
Block diagram of basic internal organization for a 4k static RAM. (From Salters [50] 1989, with permission of Philips Research.)

then T_1 is off and T_2 is on and T_3 is on and T_4 is off. The logic '0' state would be the opposite with point C_5 low and point C_6 high. Both states would be stable with neither branch of the flipflop conducting as long as the dc voltage is applied. No refresh is needed to retain data as in the case of the DRAMs.

The memory cell is embedded in an array of similar cells which are accessed by a row decoder which selects the word line and a column decoder circuit which selects the appropriate bit and $\overline{\text{bit}}$ lines as described in the last section.

The set up for reading the state that is stored in the cell or writing a new state to the cell is accomplished as follows. The row address of the desired cell is routed to the row address decoder which translates it and makes the correct word line of the addressed row in the matrix high. The NMOS transistors T_5

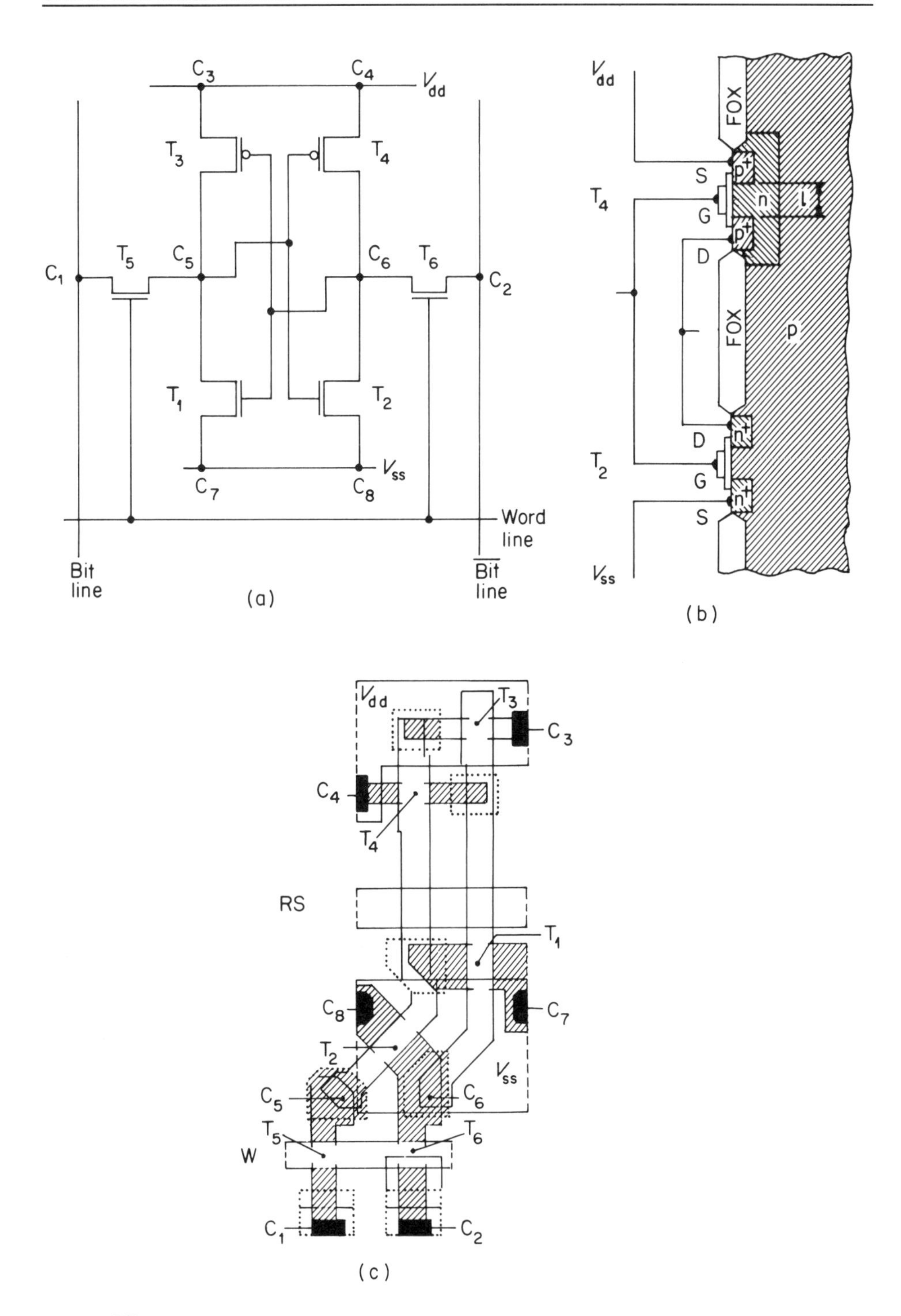

Figure 5.7
Full CMOS latched SRAM cell. (a) Schematic circuit diagram, (b) schematic cross-section, (c) layout. (From Bastaeins and Gubbels [49] 1988, with permission of Philips Research.)

and T_6 in all the cells of the row then switch high or 'on'. The column address decoder translates the column address and makes a connection to the bit line B and the inverse bit line $\bar{B}$ of all the cells in the column addressed.

In general a read is performed by starting with both the bit and $\overline{\text{bit}}$ lines high and selecting the desired word line. At this time the data in the cell will pull one of the bit lines low. This differential signal is detected on the bit and $\overline{\text{bit}}$ lines, amplified, and read out through the output buffer.

Referring to Figure 5.7, reading from the cell would occur if B and $\bar{B}$ of the appropriate bit lines are made high. If the cell is in state '1', then T_1 is off and T_2 is on. When the word line of the addressed cell becomes high, a current starts to flow from $\bar{B}$ through T_6 and T_2 to V_{SS} or ground. The level of $\bar{B}$ as a result becomes lower than B and this differential signal is detected by a differential amplifier connected to the bit and $\overline{\text{bit}}$ lines, amplified, and fed to the output buffer. Reading a '0' stored in the cell would be the opposite with the current flowing through T_5 and T_1 to ground and the bit line B becoming lower than $\bar{B}$.

Reading the cell of a static RAM is a non-destructive process and after the read operation the logic state of the cell remains the same.

In general to write into the cell, data are placed on the bit line and $\overline{\text{data}}$ are placed on the $\overline{\text{bit}}$ line. Then the word line is activated. This will force the cell to flip into the state represented on the bit lines so the new data are stored in the flipflop.

Again using the illustration in Figure 5.7, if information is to be written into a cell then B becomes high and $\bar{B}$ becomes low for the logic state '1'. For the logic state '0', B becomes low and $\bar{B}$ high. The word line is then raised causing the cell to flip into the configuration of the desired state.

No clocks or internal referencing are needed in pure static design and no external counters are required. This means that a multiple read operation can occur during a single chip enable cycle.

5.4.3 Basic peripheral circuit operation of the SRAM

Figure 5.8 shows a circuit schematic of an early SRAM which is similar to the block diagram shown earlier in Figure 5.6. This circuit is fully internally static. To select a location in the array, the row addresses pass in through the row address buffers and are decoded in the row decoders to select the word line for the correct bit in the array and thence into the word line drivers which turn on the word line of the selected cell. The column addresses pass through the column decoder which selects the proper set of bit line pairs then turns on the column switch for this pair. This bit of data feeds out of the selected cell onto the bit lines to the sense amplifiers where it is amplified and from there to the output buffer.

We will refer back to this diagram as we discuss the SRAM circuitry in more detail in the following sections. Clearly the particular circuitry used in this illustration is only one example of the types of circuitry possible to use in an SRAM.

In a static RAM, due to the simple peripheral circuitry, more than 60% of the chip

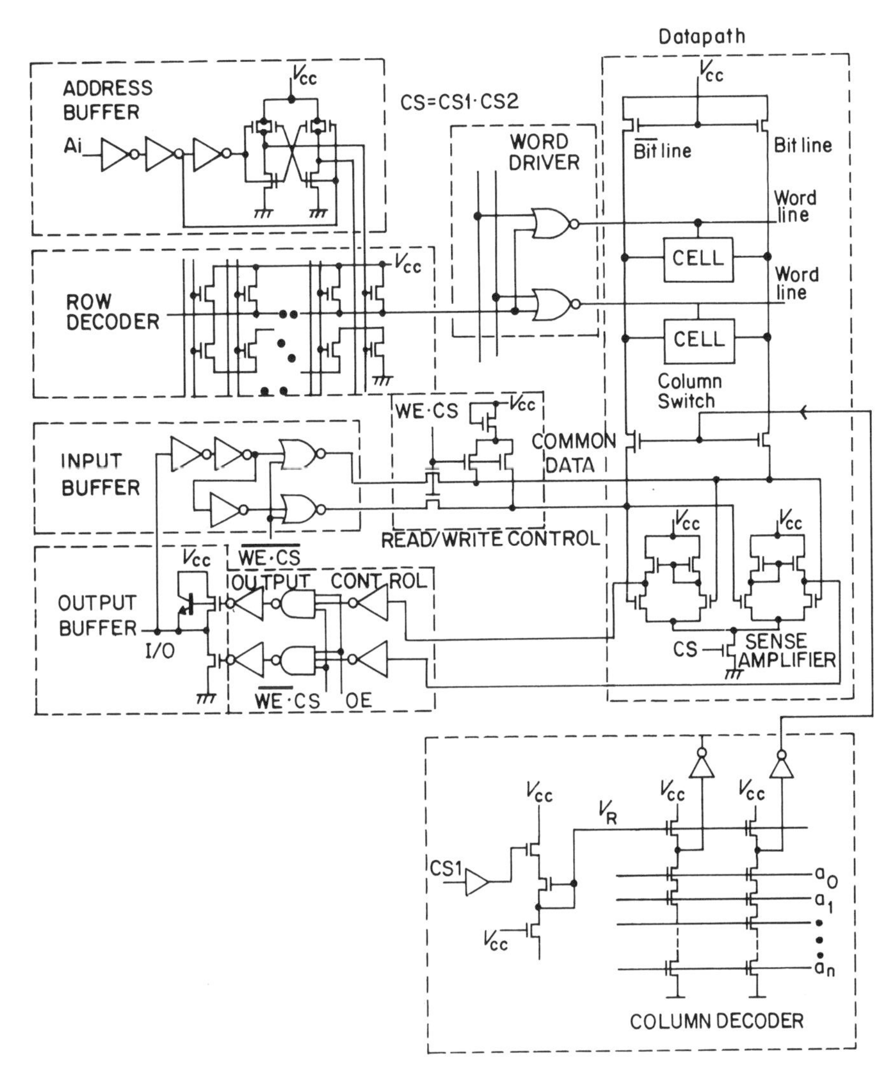

Figure 5.8
Schematic circuit diagram (——) overlayed with the block diagram (----) of a 64k SRAM. (Adapted from Minato [52] Hitachi 1982, with permission of IEEE.)

area is in the memory array. Statics have little or no pattern sensitivity and have the ability to tolerate a noisy system environment. Address inputs may be skewed and rise times different.

5.4.4 Read and write circuitry

Once a particular pair of bit lines have been selected, circuitry is required to write and read the cell. Write circuitry is relatively straightforward. Typical write circuitry is shown in Figure 5.8, consisting of invertors on the input buffer and a pass gate with a write control input signal to the bit and $\overline{\text{bit}}$ lines.

Read circuitry is more complex than write circuitry since it involves the use of a sense amplifier to read the low level of signal from the cell. A simple invertor can be used as a sense amplifier, however, a pair of single-ended differential sense amplifiers are used in the case shown in Figure 5.8 since they can sense and amplify a very small difference between level on the bit and $\overline{\text{bit}}$ lines. Two types of read data path circuitry, static and dynamic are discussed in the following section.

5.4.5 Data path circuitry

The data path is an important consideration with static RAMs since the power delay product is determined in large part by the load impedance along the data path. The data path typically consists of the bit line pull-ups, a memory cell, the column select pass device or latch, the sense amplifier, and output driver as shown in Figure 5.8.

In recent years SRAMs that are internally dynamic have been designed with clocks that turn the various parts of the data path on and off as needed to reduce the operating power of the device.

A typical data path of an internally clocked CMOS SRAM is shown in Figure 5.9(a). Shown from top to bottom are the static bit-line loads, a memory cell from the array, a bit-line equalizing device driven by the '$\overline{\text{REQ}}$' clock signal, a P-channel column selection device (column latch), a sense amplifier with a 'DEQ' clock for shorting out the load devices, and a three-state driver for the global read bus.

The bit-line load is the impedance of a bit line to V_{DD}. In a read operation with full static bit line loads there is dc current through the bit line pull-up and through the accessed cell to ground during data transfer from the cell to the bit lines. This current flow during operation increases the operating power of the memory. Dynamic bit-line loads can be used to reduce this power drain although not shown in this particular figure.

Clocked bit-line equalizing devices, driven in this case by the $\overline{\text{REQ}}$ (Read EQualize) clock, are used to short the bit lines bringing them to a common voltage level in preparation for the Read operation. The reduced swing of the equalized bit lines increases the speed of the read operation.

The speed of the static sense amplifier can also be increased by shorting the load devices during the early part of the read operation with the DEQ (Data EQualize) clock and the power consumption decreased by turning it on at the appropriate moment with the SEL (select) clock. This makes it a 'dynamic' sense amplifier.

The internal signals in an outwardly 'static' RAM are usually generated by a technique called 'Address Transition Detection' (ATD) in which the transition of an address line is detected and used to generate the various clock signals required.

The timing diagram of the dynamic data path is shown in Figure 5.9(b). An initial clock signal PEQ triggers the $\overline{\text{REQ}}$ and $\overline{\text{DEQ}}$ and SEL signals equalizing the bit lines and the load devices on the sense amplifier and read bus driver and turning on the sense amplifier. The equalizing effect of shorting the bit lines together is shown

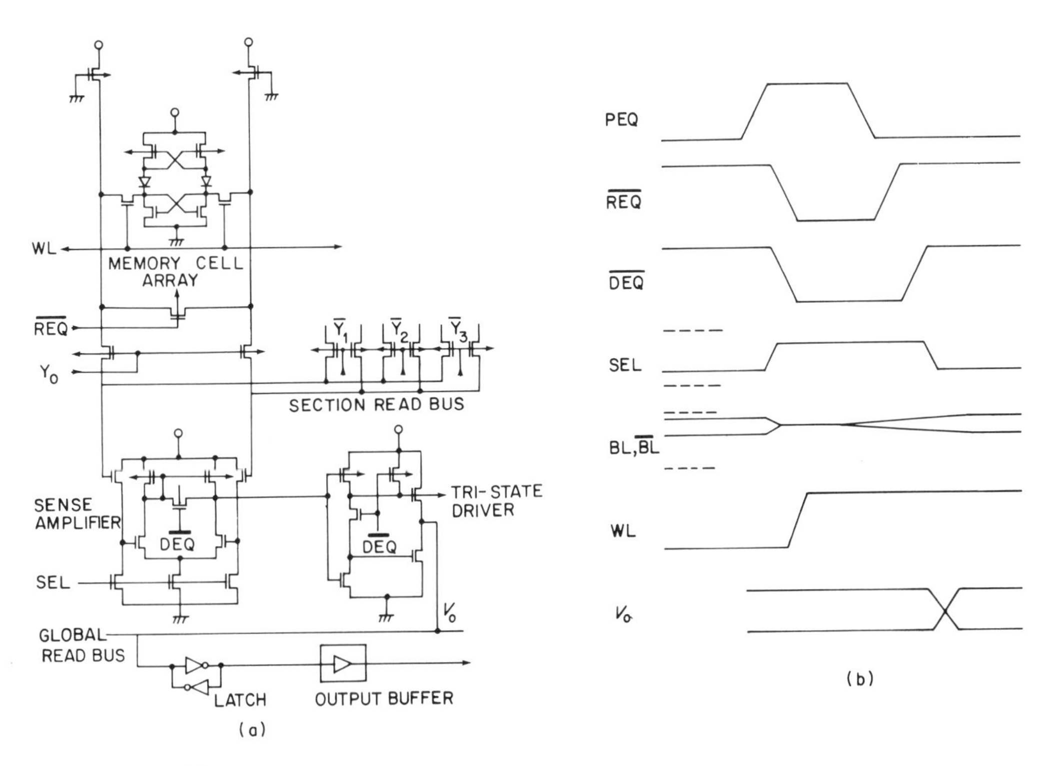

Figure 5.9
Typical 256k CMOS SRAM with clocked internal circuitry. (a) Data path with clocked sense amplifier and bit line equalization. (b) Timing diagram of the datapath. (From Gubbels [46], Philips 1987, with permission of IEEE.)

just prior to the word line going high and the differential signal from the cell appearing on the bit lines. The equalizing signals are then turned off and the amplified bit line signal appears on the global read bus line.

The internal design of modern internally dynamic SRAMs has become significantly more complex and will be treated in greater detail in Chapter 8.

5.4.6 Input circuitry for SRAMs

Input circuitry consists of address decoders, word line drivers and decoder controls as shown in Figure 5.8. The requirements for input circuits are: speed, high packing density, high driving ability into heavy loads, especially in the word line driver, and wide timing margins.

One of the tradeoffs involved in selecting optimum decoder and driver design is selecting the smallest W/L (channel width to channel length) ratio to achieve maximum packing density for transistors in the decoder and driver. This is determined considering the tradeoffs of switching speed and the noise immunity.

Row decoders are needed to select one row of word lines out of a set of rows in the array. A basic CMOS static row decoder is shown in Figure 5.8. When the row decoder selects a word line all the cells along that word line are activated and put their data on the bit lines. Clocked, or dynamic row decoders, are also used and an example of a simple static row decoder and a simple dynamic row decoder are shown in Figure 5.10(a) and 5.10(b). The clock signal on the dynamic circuit simply acts to turn what is otherwise a static circuit on or off.

Column decoders select the desired bit pairs out of the sets of bit pairs in the selected row. A typical static column decoder is shown in Figure 5.8.

5.4.7 Read and write circuitry

Once a particular pair of bit lines have been selected, circuitry is required to write and read the cell. Write circuitry is relatively straightforward. Typical write circuitry is shown in Figure 5.10(c) consisting of a pair of invertors and a pass gate with a write control input signal to the bit and $\overline{\text{bit}}$ lines.

Read circuitry is more complex than write circuitry since it involves the use of a sense amplifier to read the low level of signal from the cell.

5.4.8 Sense amplifier for SRAMs

Desirable qualities for the sense amplifier are: high speed, high packing density, ability to fit in the pitch of the bit lines, wide timing margin, high stability, and data holding ease.

A simple invertor can be used as a sense amplifier as shown in Figure 5.10(d). Differential sense amplifiers are used for speed since they can sense and amplify a very small difference between levels on the bit and $\overline{\text{bit}}$ lines. A simple single ended PMOS differential sense amplifier is shown in Figure 5.10(e). A double-ended PMOS cross-coupled amplifier is shown in Figure 5.10(f), and a current mirror amplifier composed of two cross-coupled PMOS single ended amplifiers is shown in Figure 5.10(g). The current mirror amplifier was also used in the circuit in Figure 5.8. The clocked sense amplifier used in Figure 5.9 was a single-ended differential sense amplifier with clock lines added.

5.4.9 Output circuitry for SRAMs

Output circuitry can include an amplifier to enhance the signal amplitude, a level shifter to interface to differing voltage levels and an output driver.

Desirable qualities for the static memory output circuit are: high speed, high packing density, wide timing margin, high stability, data holding ease, and compatibility with the voltage level standards of external circuits (TTL, ECL, etc.). The output driver in Figure 5.8 is a bipolar TTL level driver.

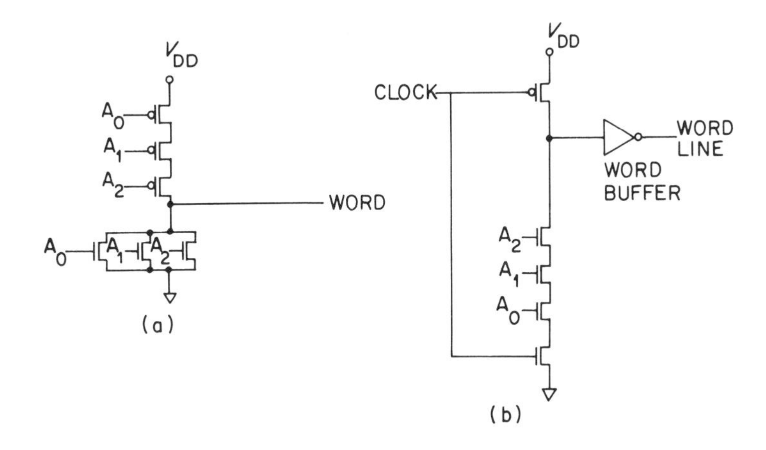

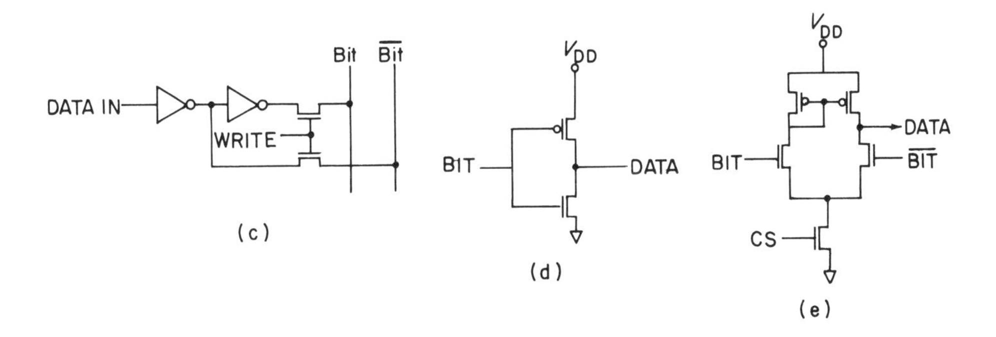

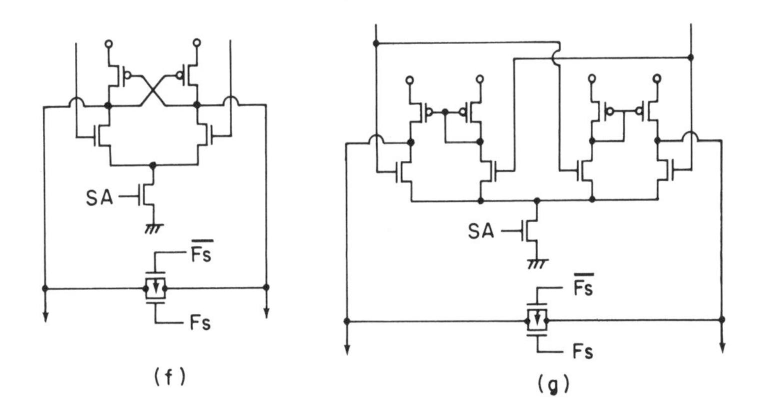

Figure 5.10
Various static RAM circuit elements. (a) Static row decoder, (b) dynamic row decoder, (c) simple write circuitry, (d) invertor amplifer, (e) differential sense amplifier, (f) PMOS cross coupled amplifer, (g) current mirror amplifer.

Further discussion on output interface circuitry, noise related output problems, and noise margins is included in a later section of this chapter.

5.5 SYNCHRONOUS (CLOCKED) AND ASYNCHRONOUS (NON-CLOCKED) SRAMs

Asynchronous SRAMs are most common. They require no external clocks and so have a very simple system interface. Clocked (synchronous) SRAMs do require system clocks. They are, however, faster in the system since all inputs are clocked into the memory on the edge of the system clock, whereas the asynchronous devices have internal timing delays while the various inputs are established in the memory. Clocked SRAMs also can require a larger package to accommodate the external clock pin.

Synchronous SRAMs tend to be used in systems where speed is critical, and Asynchronous SRAMs are usually used in systems where simplicity of system interface not speed is the criteria. RAMs are considered 'clocked' or not based on the external circuitry.

5.6 TERMINOLOGY FOR RAMs

This is probably the best place to discuss RAM terminology since there is some potential confusion arising from early terminology which was not specific enough to designate current memory distinctions.

In memory terminology 'static' refers to the cell that needs no refresh, and 'dynamic' refers to the cell that needs refresh.

Synchronous and asynchronous, in the name, always refer to whether a part is externally clocked or not. Most static RAMs today are internally clocked but are not distinguished in terminology from those that are not. For example, in Table 5.3, see A (internally asynchronous SRAM) and C (internally synchronous SRAM).

For historical reasons the word SRAM refers to an asynchronous SRAM. The alternative is a synchronous SRAM (SSRAM). Prefixes carry the added distinction of whether the part needs external refresh or not.

As an example, in Table 5.3 the synchronous SRAM and the pseudostatic DRAM (PSDRAM) look the same externally to the user (for the distinguishing features we are considering). We know from the name that the synchronous SRAM has a cell that needs no refresh (SRAM) and needs to be clocked (synchronous). We also know that the pseudostatic DRAM has a cell that needs refresh (DRAM) and externally needs no refresh (static). 'Pseudo' in this case means the same as 'synchronous'— externally it needs to be clocked.

The current terminology is unfortunately not yet well thought out for minor variants. For example, a variant of the PSDRAM called the virtually pseudostatic DRAM (VPSDRAM) is not clocked (asynchronous variant of synchronous pseudo-static DRAM).

The tendency to create words also results from the variety of minor distinctions covered under the current terminology. An example of the later exists with items A

Table 5.3 Common terminology for static-like RAMS.

	Cell		Internal circuitry		Exernal circuitry			
	REFRESH	NO REFRESH	CLOCK	NO CLOCK	REFRESH	NO REFRESH	CLOCK	NO CLOCK
A SRAM		X		X		X		X
B DRAM	X				X		X	
C SRAM		X	X			X		X
D SSRAM		X				X	X	
E PSDRAM	X		X			X	X	
F VPSDRAM	X		X			X		X

and C in Table 5.3, the internally clocked and non-clocked SRAMs. While the timing specifications for these parts appear similar, there are minor timing variations which can cause a system to malfunction if not taken into consideration in the system design.

Prefixes are also used to indicate features on the part. For example a Dual-Port SRAM (DPSRAM) is an asynchronous SRAM which externally has two essentially similar I/O ports, whereas a Multi-Port DRAM (MPDRAM) is a DRAM with two different types of I/O ports. The multi-port DRAM in which one of the ports is a SAM and the other port is a RAM is called the Video DRAM (VDRAM) which should not be confused with the VPSDRAM indicated in Table 5.3. For further reference, the JEDEC Standard 21 has definitions of the more common variations of RAM.

Many of these variations on the basic RAM will be considered further in the product specific chapters later in this book. Chapter 7 discusses application-specific DRAMs and Chapter 9 discusses application-specific SRAMs.

5.7 SERIALLY ACCESSED MEMORY (SAM) ARCHITECTURE AND CELLS

5.7.1 SAM architecture

Shift registers are memory architectures through which data are transferred in and out serially. They can be designed from MOS transistors or from CCD devices.

In MOS the major advantage of the shift register is smaller silicon area usage than SRAMs, simpler processing than the DRAM, and logic compatible processing due to the absence of the DRAM capacitor. It also uses lower pincount and hence less costly packaging due to eliminating the requirement for address pins. The penalty paid is loss of random access capability.

There are two basic types of MOS shift registers, static and dynamic. Examples of NMOS static and dynamic shift register cells are shown in Figure 5.11(a) and 5.11(b). The major differences between the two is the absence of transistor T_7 in the dynamic shift register. In the static shift register this transistor forms the feedback loop of a static flipflop enabling data to be retained under dc operating conditions.

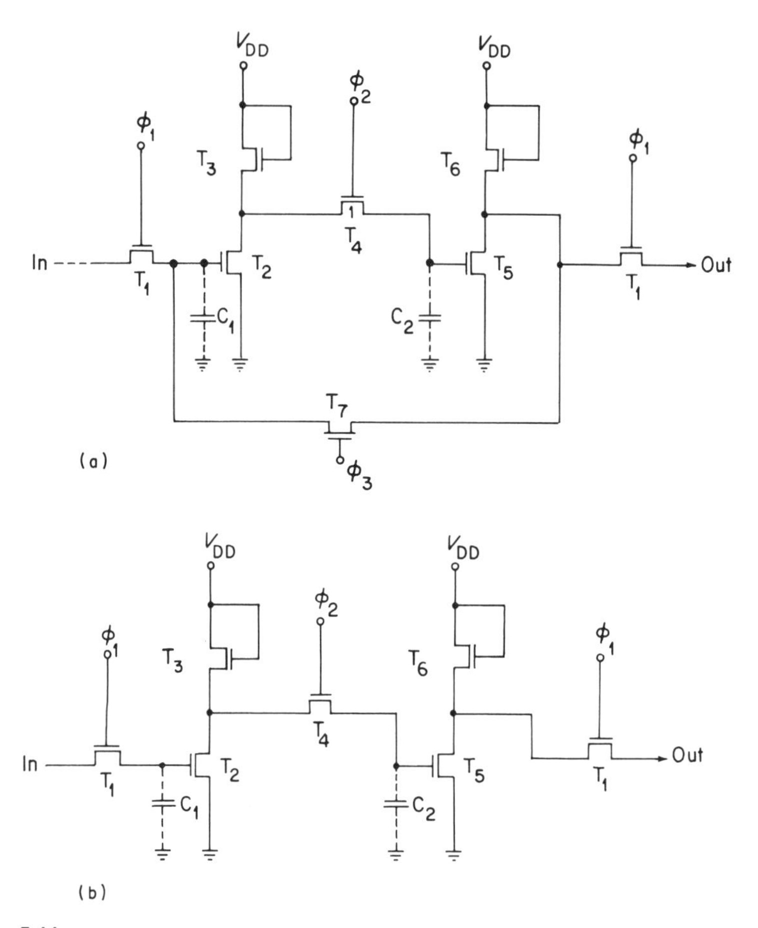

Figure 5.11
Examples of serially accessed memory architecture (SAM). (a) Static shift register, (b) dynamic shift register. (From Lueke [4] 1973.)

The third clock and associated clock lines are also absent in the dynamic shift register. The result is that if it is unnecessary to retain data in standby operation, then the dynamic shift register cell is smaller and more cost effective.

Data are transferred from cell to cell by the sequential operation of the clocks. In the dynamic shift register, the data are stored temporarily on the parasitic capacitance of the gate of the transistor.

5.7.2 SAM cells

Any RAM cell could theoretically be used in a shift register. As a practical matter small shift registers, like small FIFO buffers, are usually made with either standard SRAM cells or multiple transistor DRAM cells. Large shift registers, such as Video

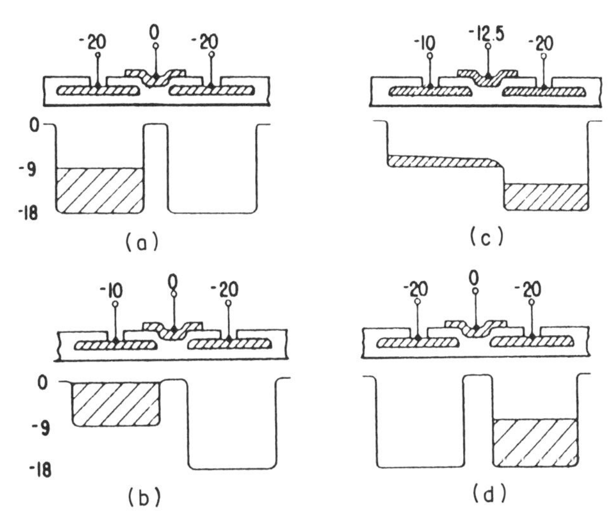

Figure 5.12
Charge coupled device (CCD) surface charge transfer cycle. Shown are electrodes and charge transfer through potential wells. (From Engeler [27] 1971, with permission of IEEE.)

field memories, are usually made with one transistor DRAM cells with the I/O ports being serial rather than random.

It is also possible to use charge coupled device (CCD) technology for shift registers. In CCDs a surface depletion region or 'well' is formed in the surface of the semiconductor by the application of an electric field from an electrode. While thermally generated minority carriers eventually fill the well to its equilibrium value, in the time interval before this occurs, the charge in the well may be used to store information. In order to use this phenomena for shift registers, it is necessary to transfer the charge from the one surface well to a neighboring one. This can be done by applying a greater potential to the adjacent electrode so that a deeper surface well forms and the charge flows into it.

The basic CCD cell is shown in Figure 5.12(a). There are two electrodes that produce surface depletion regions and a third electrode that overlaps them both called the transfer gate. A simple illustration of the CCD charge transfer cycle is shown in Figure 5.12(a) to (d).

While the CCD cell itself is much smaller than the MOS transistor shift register cell, the amount of overhead control circuitry of the CCD is somewhat greater.

5.8 DRAM SPECIFIC ARCHITECTURES

5.8.1 One-transistor cell operation

The basic one-transistor dynamic memory cell consists of a select transistor and a storage capacitor as was shown in Figure 5.5(f). The gate of the select transistor is controlled by a word line. When the word line is selected the select transistor turns

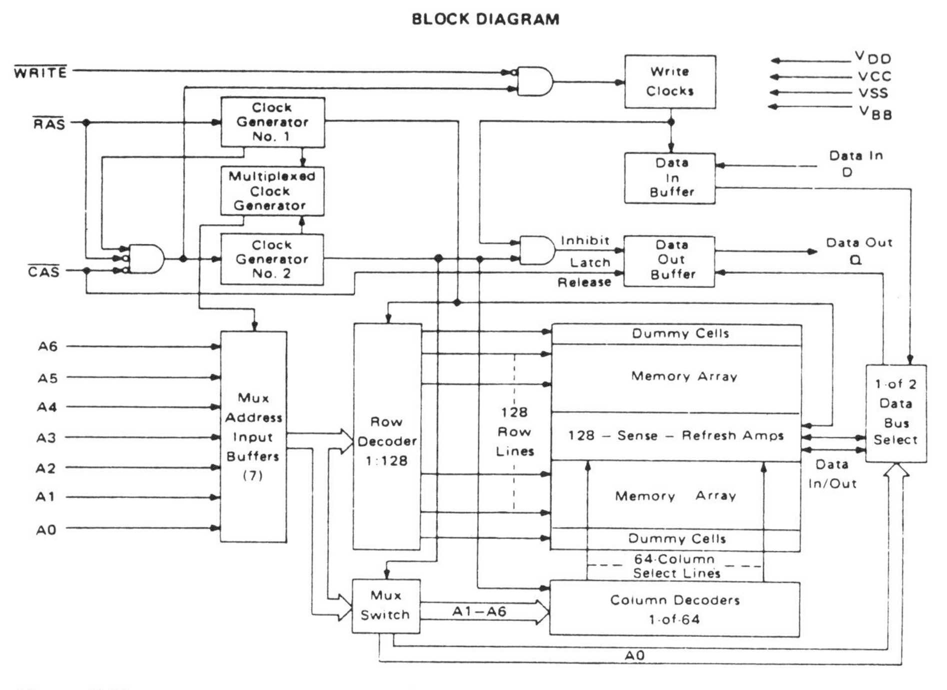

Figure 5.13
Block diagram of a typical multiplexed DRAM architecture—the 16k DRAM. (Reproduced with permission of Motorola.)

on and the charge stored on the capacitor is fed out onto a bit line and to a sense amplifier. The sense amplifier compares it with a reference cell and determines if the DRAM cell stored a logic '1' or a '0'. The read out from the cell discharges the capacitor so that the data are no longer stored in the cell. Before the read operation is complete the sense amplifier circuit must restore the original charge to the cell.

5.8.2 Basic architecture

A basic architectural configuration of a dynamic RAM is shown in Figure 5.13, using an early 16k DRAM as a simple example. The addresses are time multiplexed with two separate address fields on the A_0 to A_6 pins. These are clocked into the RAM by two external clock signals, the Row Address Strobe ($\overline{RAS}$) and the Column Address Strobe ($\overline{CAS}$). The first field of addresses goes to the row decoder to select one of 128 word-lines and the second field goes to the column decoders to select one of 64 bit-lines. The addresses are multiplexed to reduce the pin count and, as a result, the size of the package.

In a read operation, the data from the selected cell are taken out through the select transistor onto the bit line to the sense amplifier where they are amplified and read out. Amplification is required since the signal from the small cell capacitance is reduced still more by the ratio of the larger capacitance of the bit-line.

This also results in destruction of the data in the cell which must also then be restored to the cell in a 'refresh' operation. The operation and control of the DRAM cell are described in more detail in the following sections.

The two major problems of the one transistor cell DRAMs have remained that of ensuring sufficient charge storage in the cell and of finding sensitive enough sense amplifiers that are also small enough to fit into the area left between the bit lines.

5.8.3 Storage to bit-line capacitance ratios and signal levels

The storage capacitance to bit-line capacitance ratio is one of the important characteristics of the one transistor Dynamic RAM since sufficient charge must be stored on the capacitor to provide a readable signal on the bit lines to the sense amplifier.

The parasitic bit-line capacitance (C_b), due to the large number of cells on a single bit-line, is typically larger than the capacitance of the storage cell (C_s). When the cell is selected and the signal stored in the cell capacitor (V_s) is read out onto the bit-line, it is reduced by the ratio of the storage capacitance to bit-line parasitic capacitance. The signal magnitude on the bit-lines (V_b) is, therefore, frequently described by this ratio, i.e. $C_s/C_b = V_b/V_s$, or $V_b = V_s\ (C_s/C_b)$.

The specific bit-line capacitance C_b/C_s is determined by the lithography and technological parameters. The total value of C_b is strongly influenced by the number of storage cells on the bit-line. This reduction from the stored cell signal is the reason the signal on the bit sense line has to be amplified by a sensitive amplifier which is also able to refresh the information of the selected cell. The more sensitive the amplifier is, the smaller the storage capacitance can be. If the bit-line capacitance is large relative to the capacitance of the memory cell then the signal which appears across the bit-lines and is available to the sense amplifier may be too small to be read. For example, if the ratio of C_b to C_s is 10, and the voltage stored on the storage capacitor has decayed to 2 V at the time the cell is read, the signal which is read out onto the bit-line will be reduced by 1/10 to 200 mV. The sense amplifier must be sensitive enough to read this signal.

5.8.4 Bit-line architectures

The bit-line architecture consists of the configuration and number of cells on the bit line, the length and resistivity of the bit line, and the pitch or spacing of the bit-lines.

The bit-line pitch on a DRAM must be large enough to allow sufficient sense amplifiers to be placed in the same pitch as the bit-lines.

The number of cells per bit-line is determined by such considerations as

- required signal size which is dependent on the size of the capacitor in the cell,
- number of sense amplifiers and the power dissipation resulting from the number on at any one time,

- the number of cycles of the refresh sequence since this determines how much the signal has decayed before it is refreshed,
- the propagation delay of the signals through the bit lines which affects the access time of the chip.

5.8.5 Basic dynamic sense amplifier concept

Since the readout of stored data from the storage capacitance of the cell onto the larger capacitance of the bit line results in loss of the cell data, techniques are required to both sense the low level of data and to restore the original signal levels in the cell so data are retained.

A basic one-transistor storage cell structure with reference cell and a cross-coupled latch sense amplifier is shown in Figure 5.14(a). This was the refresh structure used in the first 16k DRAM introduced in 1976 by Intel [36]. The timing required for operation of this cell is also shown in Figure 5.14(b). Three clock signals are required for the operation of this cell of which one generates the other two on the chip.

The operation of the circuit with reference to the clock inputs is as follows. Clock $\overline{CE}$ is derived from the Row Address Strobe ($\overline{RAS}$) clock while clocks ϕ_0 (the word-line), ϕ_1, and ϕ_2 are internally generated on chip. Initially $\overline{CE}$ is held high causing the reference cell to be precharged to ground and ϕ_2 is held high causing both bit lines to be precharged. The $\overline{RAS}$ transition initiates row decoding and initiates the clock sequence shown. First the $\overline{CE}$ falls shutting off the shorting device and isolating the reference cell, and ϕ_2 falls shutting off the load devices. After row decoding is complete, ϕ_0 is turned on connecting the selected storage cell to one side of the sense amplifier, and a reference cell to the other. A small differential signal from the cell appears on the bit-lines $B_{1,2}$. ϕ_1 then rises causing this signal to be amplified. If the rise of ϕ_1 is controlled, the discharge of the bit line precharge is confined to the low side; however, a slight dip occurs on the high side as well so that ϕ_2 is used to turn on the load devices briefly to restore a maximum signal in the storage cell for a stored '1'.

In order to read and restore the data stored in the cell, the bit-lines are normally divided on either side of centrally placed balanced differential sense amplifiers.

5.8.6 Dynamic sense amplifiers with dummy cell structures

A dummy cell structure is normally associated with the sense amplifier to aid in sensing the low data signal from the cell by providing a more sensitive reference level. For example, in the sense amplifier arrangement shown in Figure 5.15, the bit and $\overline{\text{bit}}$ lines are either shorted together or at a common reference level before

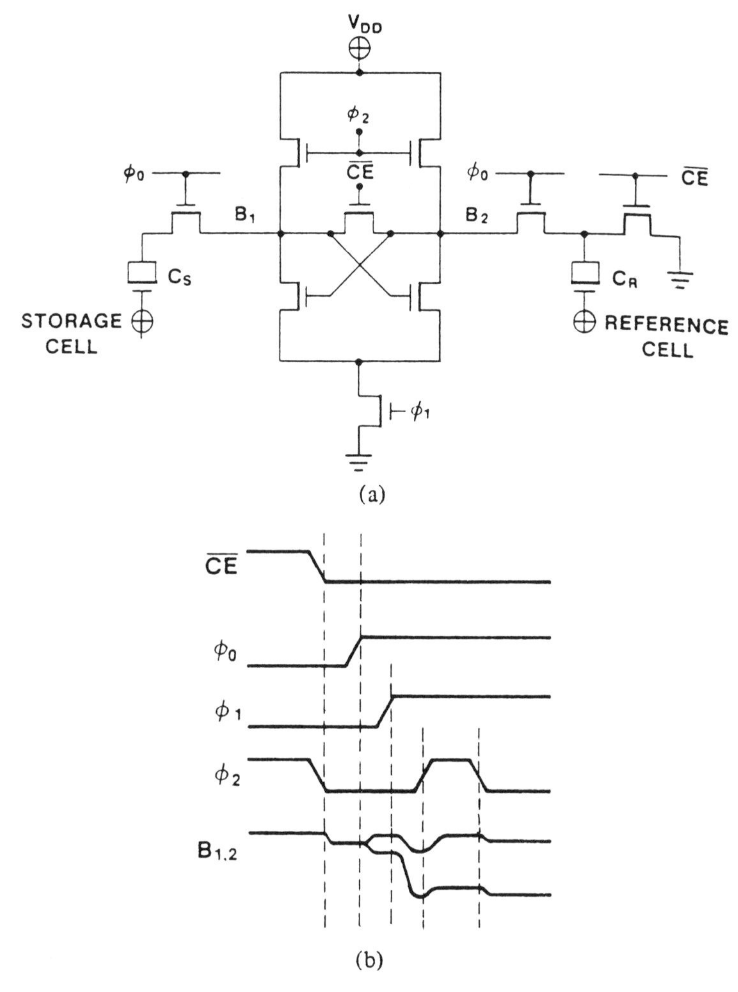

Figure 5.14
(a) Basic one-transistor storage cell together with cross-coupled latch sense amplifer. (b) Associated timing diagram. (From Ahlquist *et al.* [36], Intel 1976, with permission of IEEE.)

attempting to read the cell. During this time the reference voltage level is written into the dummy cell structures on either side of the invertor. When the memory is selected by the select clock the dummy cells are unshorted. As the word line is selected, the data from a storage cell are dumped onto one side of the sense amplifier at the same time as the reference level from the dummy cell is dumped on the other side. That causes a voltage differential to be set up across the two nodes of the sense amplifier. The load clock now turns on the two load transistors. Both nodes rise together until the higher side starts to turn on the opposite transistor. Regenerative action then results in amplification of the signal to full logic levels and the restoration of the zero or one level back into the storage cell.

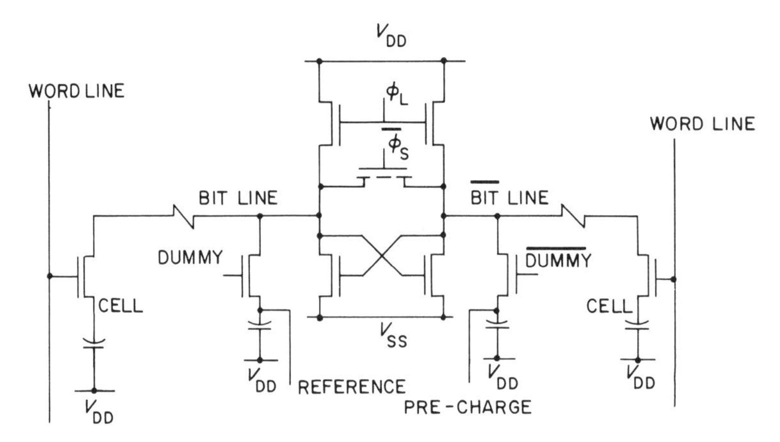

Figure 5.15
A DRAM differential sense amplifer showing dummy cell structure. (From Foss and Harland [8], Mosaid 1975, with permission of IEEE.)

Dynamic sense amplifiers are used in fast memory circuitry. During the resetting period both bit and $\overline{\text{bit}}$ lines are precharged to the applied power supply voltage level by the clock. The common source node 'n' is also precharged to the applied power supply voltage minus a transistor threshold.

Such a sense amplifier can provide very fast signal detection since it is sensitive to low voltage differential signals. It, however, cannot detect a signal smaller than the difference between the thresholds of the two active transistors in the flipflop.

5.8.7 Timing and sequencing

Clearly some degree of sophistication is required in the timing and sequencing of the various clocks associated with the operation of the dynamic RAM memory cell. Since several clocks are required in the internal operation of a dynamic RAM, it is advantageous to be able to generate secondary clock pulses on chip from the basic input clock pulses.

An example of a simple clock generator formed from the input clock and a bootstrapped output stage is shown in Figure 5.16(a). This particular illustration is in the early PMOS technology.

The bootstrap principle utilizes an output transistor with an enlarged gate-to-drain capacitance. When the gate voltage is more negative than the threshold voltage there will be an inversion layer under the gate resulting in a large capacitance between the gate and drain. When a clock pulse is now supplied, this voltage will be almost completely added to the gate voltage due to the strong capacitive coupling between gate and drain. The result is to keep the transistor in the unsaturated state resulting in fast charging of the load capacitance to the amplitude of the supplied clock signal.

Clock generation can be performed with an on-chip shift register formed from the basic clock generator described above as shown in Figure 5.16(b). This shift register timing generator was used in the early 4k DRAM described by Philips in 1973.

5.8.8 Advanced DRAM architectures

Further discussion on more advanced architectural concepts for both conventional DRAMs and application-specific DRAMs can be found in Chapters 6 and 7. A discussion of noise sources, redundancy, reliability, and test for DRAMs can be found in Chapter 14, and DRAM packaging including modules can be found in Chapter 13. DRAM cell technology was covered in Chapter 4.

5.9 OVERVIEW OF NON-VOLATILE MOS MEMORIES

The ROM family includes an entire class of MOS memories which have cells that are non-volatile, that is, they maintain their data when the power is removed from the chip. These include: the mask ROM, the ultraviolet erasable electrically programmable ROM (UV-EPROM), the electrically erasable electrically programmable ROM (EEPROM), the electrically alterable ROM (EAROM), the one-time-programmable EPROM (OTP-EPROM), and various modern variations such as the NV-SRAM and NV-DRAM.

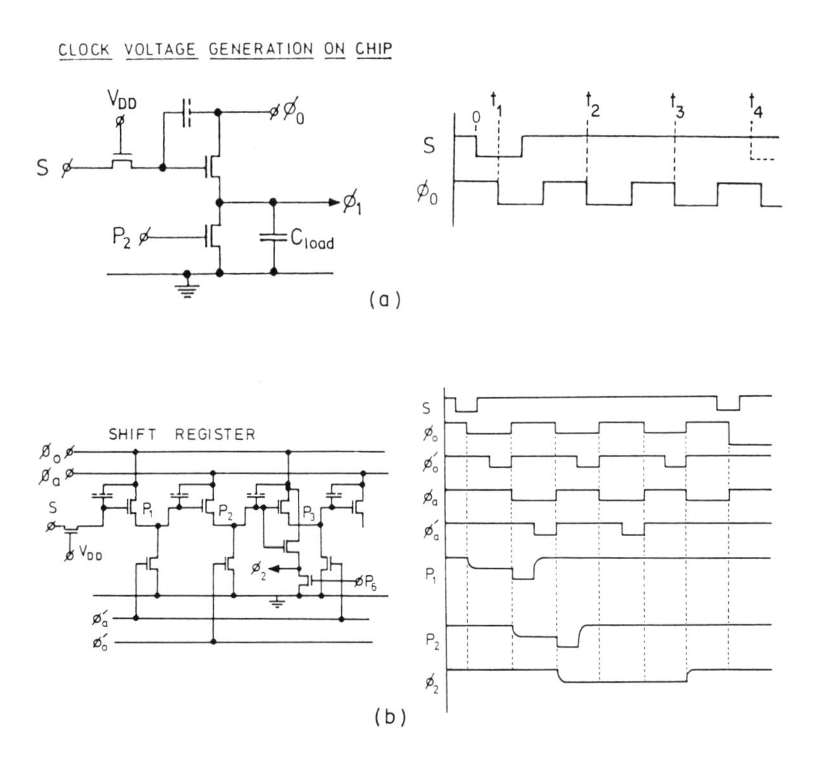

Figure 5.16
(a) Early PMOS bootstrapped clock generator. (b) Timing shift register used on an early 4k DRAM. (From Boonstra *et al.* [34], Philips 1973, with permission of IEEE.)

5.10 MASK ROM ARCHITECTURE AND CELLS (SINGLE TRANSISTOR)

Read-only memory (ROM) cells are different from either latched SRAM or capacitor DRAM cells. They contain a single transistor per bit whose state is hardware preprogrammed during the wafer fabrication process and maintained indefinitely without the application of voltage.

Two types of ROM arrays are possible, the parallel or NOR array as shown in Figure 5.17(a), and the series or NAND array as shown in Figure 5.17(b).

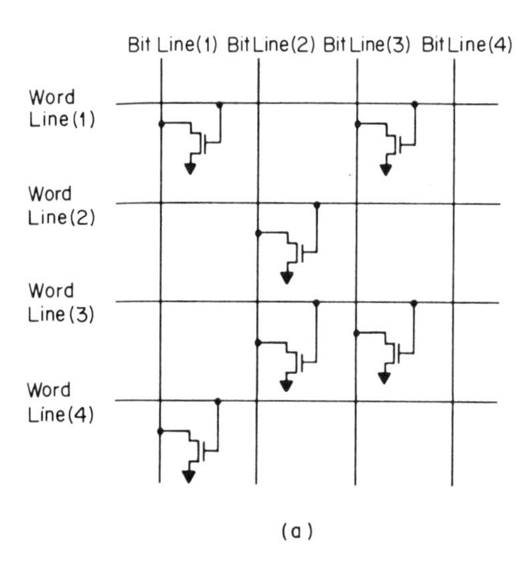

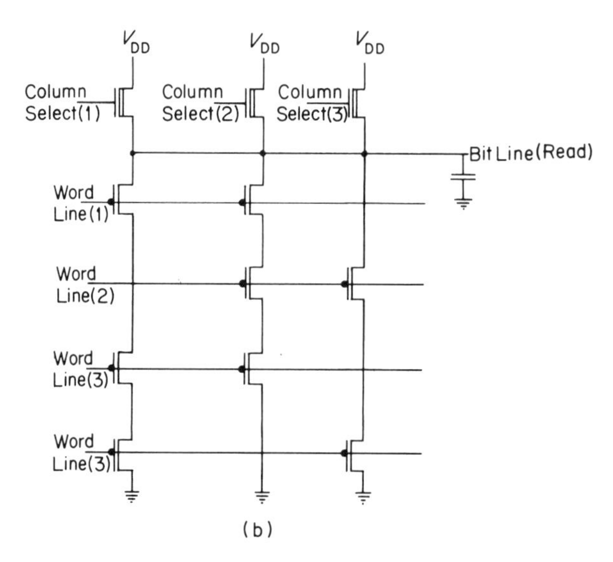

Figure 5.17
Two types of ROM cell arrays. (a) NOR array, (b) NAND array.

The NOR array is larger since there is a bit line contact to every cell; hence it is less cost effective than the NAND array which has only one bit line contact for a series of cells. The NOR array has the advantage that it can be programmed by either opening or omitting contact holes very late in the fabrication process. Historically the NOR cell was used to offer fast turn around time for ROM prototyping.

The NAND array can be much smaller in area since a contact is not needed to each series cell so it is more cost effective. The drawback historically was that this cell was programmed very early in the manufacturing process. The turn-around time of the NAND array is hence longer than that of the NOR array.

Mask ROMs are discussed in more detail in Chapter 10.

5.11 UV-EPROM CELL TECHNOLOGY AND ARCHITECTURE

EPROMs have memory arrays that are similar to the ROM with one transistor per cell, but rather than being hardware programmed they are programmed electrically and erased by ultraviolet light. The cells are static since they do not need to be refreshed so the peripheral architecture is similar to that of the static RAM.

5.11.1 UV-EPROM cell technology

The conventional field programmable UV-EPROM cell has two polysilicon gates as shown in the cross-section in Figure 5.18(a). The upper gate is used for selection and is connected to the row decoders by means of the word line. The bottom floating gate between the select gate and the substrate is isolated in the surrounding silicon dioxide. It is capacitively coupled to the select gate and the substrate rather than being connected.

Programming of a cell is achieved by applying a high positive voltage to both the drain and the control gate and grounding the source. Avalanche injection, also called hot electron injection, of high energy electrons then occurs from the substrate through the isolating oxide under the influence of the positive applied drain voltage. The electrons must gain sufficient energy to jump the 3.2 eV energy barrier at the interface between the silicon substrate and the silicon dioxide. The positive voltage on the select gate then pulls the electrons through the oxide towards the floating gate causing charge to be collected on the floating gate. As the gate becomes more charged the electrons in the oxide field begin to be repelled from the floating gate back to the substrate. For this reason the programming process is self limiting. It is difficult to overprogram an EPROM. Energy loss due to photon emission increases at higher temperatures and thus programming is easier at lower temperatures.

The cell is erased through internal photoemission of electrons from the floating gate to the select gate and the substrate. Incoming ultraviolet light increases the energy of the floating gate electrons to a level where they jump the 3.2 eV energy barrier between the floating gate and the silicon dioxide.

The charge on a memory cell's floating gate alters the threshold voltage of the

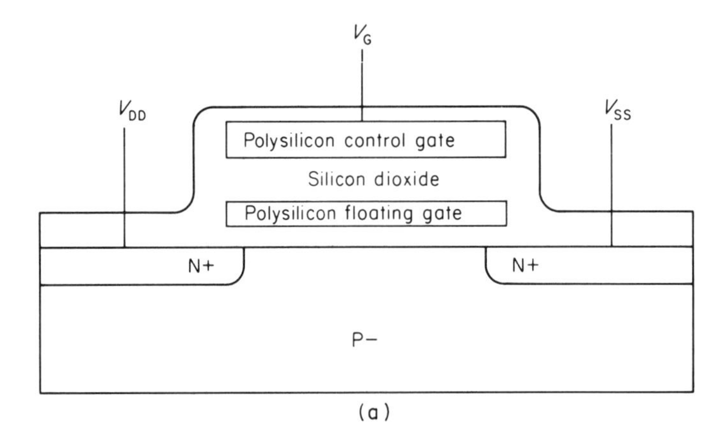

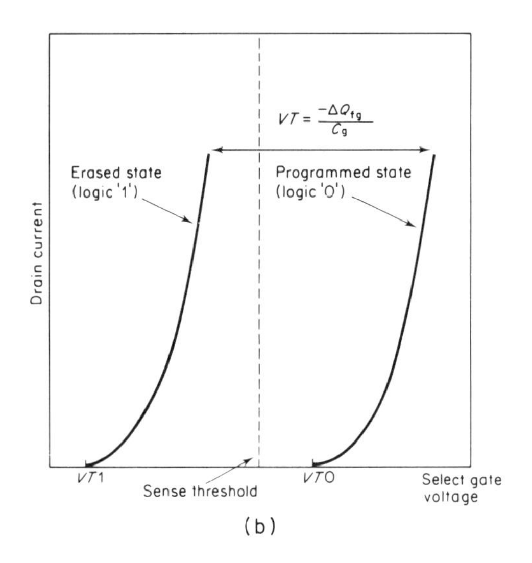

Figure 5.18
(a) EPROM storage transistor cross-section showing polysilicon control gate and polysilicon floating gate. (b) Threshold characteristics of an EPROM storage cell in the programmed and unprogrammed state. (Reproduced by permission of Elektronikcentralen.) (From Prince and Due-Gundersen [54] 1973.)

device as seen from the select gate. In the unprogrammed state with no charge on the floating gate, the transistor has a low threshold and when selected the transistor will turn on. If, on the other hand, the cell is in the programmed state with charge on the floating gate, the memory cell has a high threshold and even with the application of a positive voltage the transistor will not turn on.

The drain current and threshold voltage characteristics of the EPROM storage cell in the programmed (logic '0') state and unprogrammed (logic '1') state are shown in

Figure 5.18(b). A shift of voltage threshold characteristics occurs where Q_{fg} is the change of charge on the floating gate, and C_g is the capacitance between the floating gate and the select gate. The charge of the memory cell's floating storage gate changes the threshold voltage of the select gate by this amount. When the cell is programmed, the negative charge of the floating gate causes the floating gate source voltage to be negative. This turns the cell off, even with a positive reading voltage applied to the select gate.

Before an EPROM can be programmed, it must be completely erased. This is accomplished by exposing the entire array to high intensity ultraviolet light, with a wavelength of 253.7 nm. Typical sources for erasure are quartz-jacketed mercury arc lamps, mercury vapor lamps, or any of a range of commercially available erasure units. Normal erase time under an ultraviolet light intensity of 15 W cm^2 is about 35 minutes. Erasure can also be caused by ordinary sunlight or fluorescent lighting, although over a much longer time period. To prevent this accidental erasure, opaque covers are sometimes applied over the quartz window package after programming.

5.11.2 UV-EPROM cell reliability

The EPROM cell structure permits very high density of design since the floating storage gate and the select gate are directly above the transistor's channel. Even higher density can be obtained by modern self-alignment techniques. Self-alignment, however, leaves the edge of the floating gate visible to oncoming light and makes the device more easily erased. This is a benefit if the erasure is intentional and a liability if not.

Early EPROMS suffered from a number of reliability problems which technology advances have, for the most part, overcome. Wafer processing has a significant effect on the reliability of EPROMS. A major concern is electron trapping in defects in the silicon dioxide after several programming cycles. Trapped electrons decrease the rate of programming because the electrons that flow through the oxide now encounter locally expulsive fields. Their small optical cross-sections mean that electrons once trapped in the oxide are difficult to dislodge by UV erasure of the floating gate. Eventually, trapped electrons can raise the threshold to the point that the cell is sensed incorrectly. The solution is the growth of high quality defect-free silicon dioxide. This is also discussed in Chapter 14.

Another wafer processing effect is caused by pointed crystal outcroppings extending from the surface of the polysilicon layer into the oxide which are called 'asperities'. While the average electric field in the oxide between the floating gate and select gate is only about 1 mV cm^{-1}, the presence of asperities can increase this field to the point where partial erasure of the floating gate can occur.

We should note that asperities are only a reliability problem when unwanted. One manufacturer has used deliberate controlled production of asperities to produce a modern electrically erasable PROM which will be discussed in Chapter 12 under EEPROMs.

Methods of screening out devices with defective oxide during production testing have been developed. These screens, coupled with advances in wafer fabrication technology, have greatly increased the reliability of EPROMS.

Normal reading voltages of 5 V make the modern EPROM cell like that of an ordinary ROM. No programming, even at a low level, occurs during read since the high electric fields (greater than 10^5 Vcm^{-1}) needed to generate the hot electrons for programming are not obtained at 5 V but need a 20 V or greater applied voltage to generate. The 3.2 eV energy barrier for injection into the oxide also makes accidental programming virtually impossible at normal reading voltages.

5.11.3 UV-EPROM architecture

Since the cells of the EPROM are so small, the peripheral circuitry in an EPROM must be very compact to fit in the pitch of the bit-lines and word-lines. The EPROM is a static device and could use peripheral circuitry similar to that of the SRAM. Complexity is reduced, however, wherever possible. Circuitry developed for the one transistor cell dynamic RAM, which has similar space constraints, is frequently used.

Data sensing in EPROMs can be done with a simple single-ended sensing scheme, similar to that used in DRAMs, which senses the potential difference between a bit sense line and a reference line with a dummy reference cell as shown in Figure 5.19. The bit sense line potential depends on the state of a selected EPROM cell on that bit-line—either erased or programmed. If it is erased then the selected storage transistor turns on (logic '1') and the bit-line potential after a transition period is below the potential of the reference line. If it is programmed then the selected storage transistor remains off (logic '0') and after a transition period the bit sense line potential is above the potential of the reference line.

An example of a compact static row decoder and a clocked row decoder developed for a 64k EPROM are shown in Figures 5.20(a) and (b). The static row decoder circuit draws static current when selected. The clocked row decoder circuit has zero static current consumption. Power is dissipated in the selected clocked decoder only during the brief time after an address transition when the row clock is low.

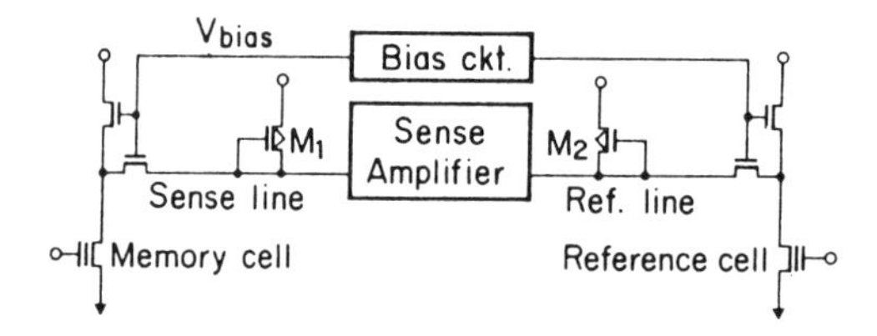

Figure 5.19
Single-ended EPROM data sensing for 4Mb showing memory cell, reference cell and sense amplifier. (From Imamiya *et al.* [42], Toshiba 1990, with permission of IEEE.)

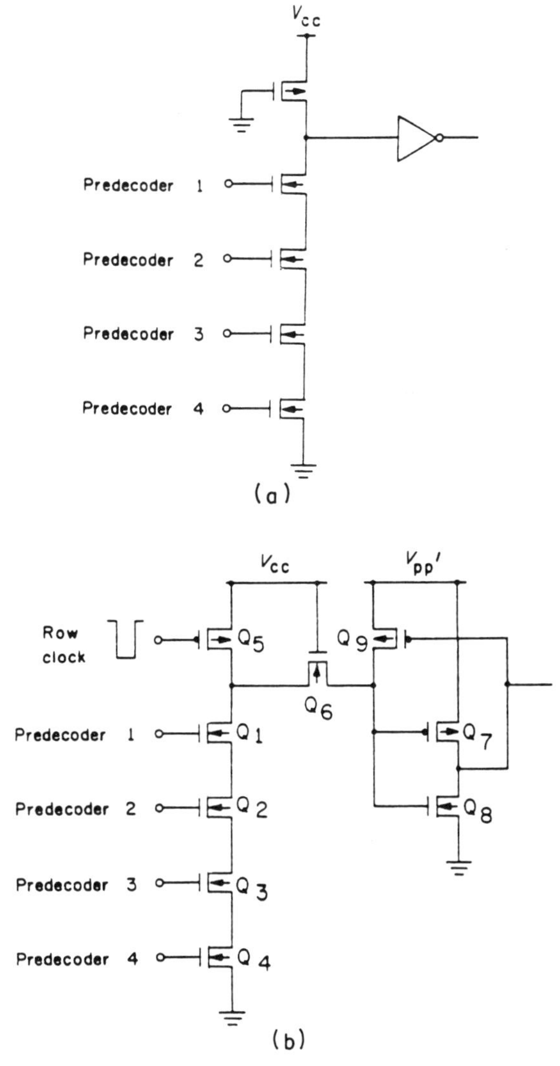

Figure 5.20
Schematic row decoder circuitry for a typical 64k CMOS EPROM. (a) Static row decoder circuit, (b) clocked row decoder circuit. (From Knecht [58], Philips 1983, with permission of IEEE.)

5.12 ONE-TIME-PROGRAMMABLE EPROM (OTP)

The UV-EPROM can be used for ROM prototyping before the system data configuration is final. EPROMs are, however, much more expensive than ROMs since they need to be contained in a special package with a window for the UV light to enter for erase.

For such applications as prototyping, the need to erase may not arise. In this case the EPROM can be packaged in a standard plastic package and programmed one time only. These plastic packaged EPROMs are called OTPs.

An OTP cannot be tested for programmability by normal test methods after it is packaged since it can only be programmed once. This has limited its market. For more discussion on the OTP see Chapters 11 and 14.

5.13 ELECTRICALLY ERASABLE PROM TECHNOLOGY AND ARCHITECTURE

5.13.1 Overview

The need to remove the EPROM from the system to erase it under UV light, combined with its high cost of packaging and test has led the industry to continue looking for the low cost, system reprogrammable non-volatile memory.

Various candidates have been the EEPROM, the non-volatile DRAM and the flash memory. The first two are covered in detail in Chapter 12 and the latter in Chapter 11.

5.13.2 Fowler–Nordheim tunneling

For electrical reprogrammability in the system, the EEPROM, NV-DRAM and some types of flash memory depend on a technology mechanism called variously 'Fowler–Nordheim' tunneling, 'cold electron' tunneling, or just 'tunneling'. Cold electron tunneling is a quantum-mechanical effect which allows the electrons to pass through the energy barrier at the silicon–silicon dioxide interface at a lower energy than the 3.2 eV required to pass over this energy barrier. This means it can occur at lower current density than hot electron injection.

Figure 5.21(a) shows an energy level diagram illustrating hot electron injection over the energy barrier at the Si-SiO_2 interface and Figure 5.21(b) shows an energy level diagram of cold electron tunneling through the energy barrier at the Si–SiO_2 interface.

5.13.3 EEPROMs and EAROMs

The EEPROM was the first attempt to use Fowler–Nordheim tunneling to make an electrically reprogrammable ROM. The EEPROM suffered from higher cost than the EPROM due to a two transistor cell and a difficult thin oxide process technology.

The basic storage cell consists of an access, or select, transistor and a double polysilicon storage transistor with a floating polysilicon gate isolated in silicon dioxide capacitively coupled to a second polysilicon control gate which is stacked above it. The process was used initially with NMOS periphery and later with CMOS.

The floating gate tunneling oxide cell, which was developed by Intel and has been one of the most common EEPROM cells, is shown in Figure 5.22(a). The storage

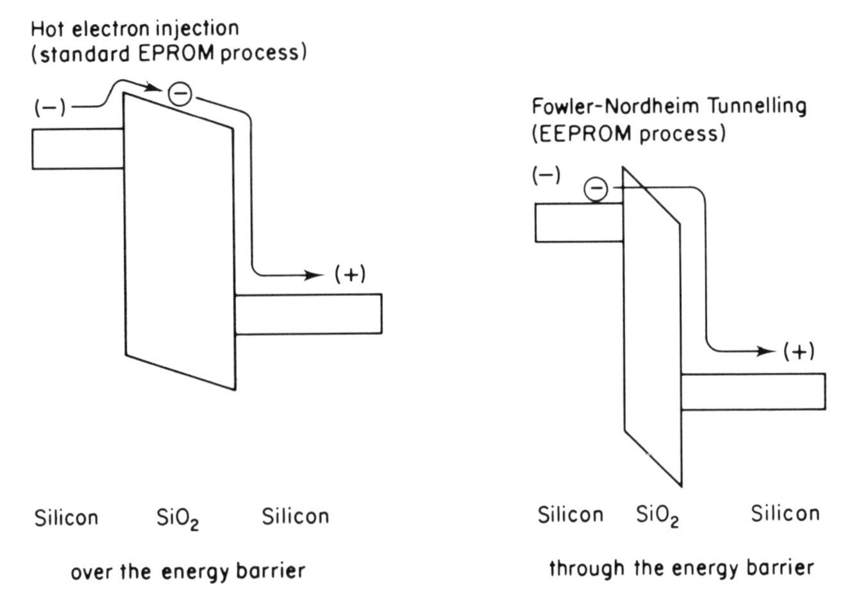

Figure 5.21
Energy diagrams for (a) hot electron (avalanche) injection over the energy barrier at the Si–SiO_2 interface, (b) cold electron (Fowler–Nordheim) tunneling through the energy barrier at the Si–SiO_2 interface. (From Prince and Due-Gundersen [54] 1983.)

transistor has a region between the floating storage gate and the drain where the oxide is thin enough to permit Fowler–Nordheim tunneling to occur.

The storage transistor is programmed by Fowler–Nordheim tunneling of electrons through this thin oxide layer between the gate and the drain of the transistor as shown in Figure 5.22(b). In the write, or program, mode the control gate is grounded and the drain is connected to a high voltage through the access transistor. Electrons tunnel from the floating gate to the drain leaving the floating gate relatively more positively charged. This shifts the threshold voltage in the negative direction so that in the read mode the transistor will be 'on'. The programmed state corresponds to a logical '0' state in the cell.

In the erase mode the control gate is at the high voltage while the drain is grounded as shown in Figure 5.22(c). Electrons tunnel to the floating gate and the threshold voltage shifts in the positive direction so that in the read mode the transistor will be 'off'. The erased state corresponds to a logical '1' stored in the cell.

To read the device, the state of the cell is determined by current sensing via the select transistor.

5.13.4 EEPROM reliability

The theory behind the small injection region over the drain is that the tunneling current has an exponentially increasing dependence on the electric field across the

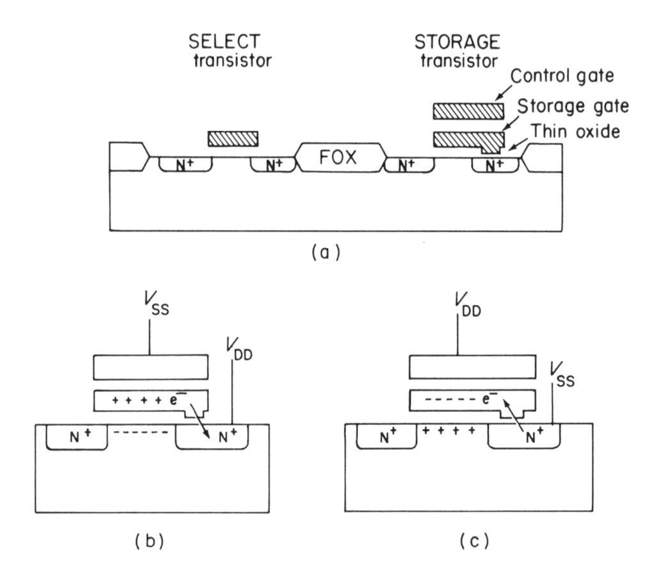

Figure 5.22
Basic EEPROM. (a) Cell showing select transistor and storage transistor with thin oxide over the drain region. (b) Programmed configuration is logical '0'. Threshold is lower. In read mode transistor is on. Gate is discharged. (c) Erased configuration is logical '1'. Threshold is higher. In read mode transistor is off. Gate is charged.

thin oxide area which is linearly dependent on the differential between the drain voltage and the floating gate voltage; thus, for a small cell size a method of coupling higher voltages to the floating gate is reducing the thin oxide capacitance by reducing the area of the thin oxide.

The program–erase operation of the cell depends on being able to apply a large reversible electric field to the thin oxide separating the floating gate from the substrate. This field must be sufficient to result in an appreciable current through the thin oxide by indirect tunneling.

At the same time, the field in the interpoly oxide layer must be maintained at a relatively low value to prevent unwanted transport between the floating gate and control gate. During program and erase, the thin oxide field is controlled by the relative voltage applied to the control gate and the drain overlap region resulting in much lower programming power than EPROMs.

The storage transistor normally exhibits an endurance of greater than 10^5 program–erase cycles with extrapolated data retention in excess of ten years.

Continued program–erase cycling of the cell results in variation of the threshold window due to the interaction of emission currents with the thin insulator as shown in Figure 5.23. Initially, the window width increases as trapped positive charge accumulates at the injecting interfaces and increases the emission current. After the first few cycles, the window width remains relatively constant until after approximately 10^4 cycles when the effects of electron trapping in the thin oxide become evident.

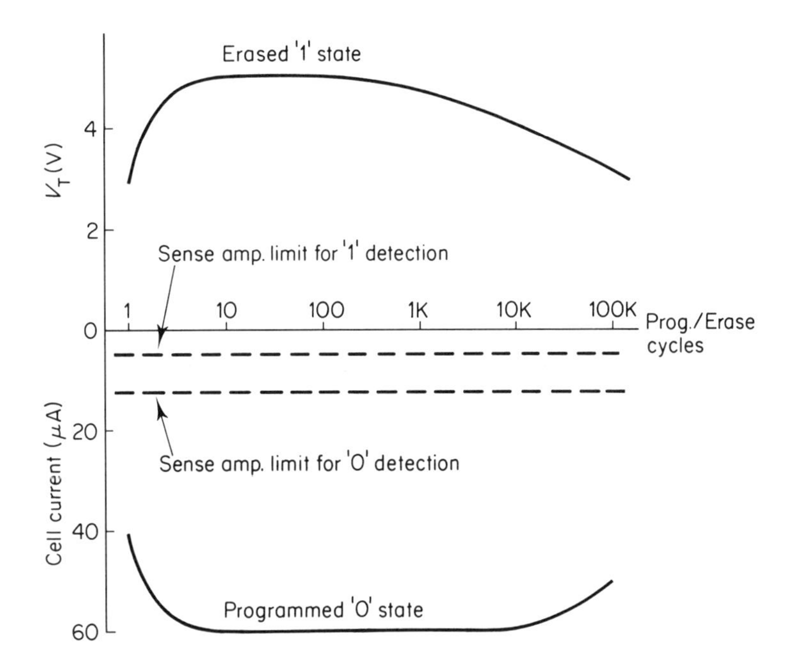

Figure 5.23
Program–erase cycle endurance of an EEPROM cell plotted against threshold voltage.

This trapping of electrons reduces the Fowler–Nordheim emission currents leading to a decrease in window width for fixed program–erase times. After 10^5 cycles the window width is still greater than 10 V allowing ample margin for data retention. Window closure occurs at approximately 10^6 cycles for the example shown.

Data retention in the read mode and during storage is affected by oxide fields in the range of 1 to 2 mV cm^{-1} that can exist in the structure. These fields result in the discharge of the floating gate over long periods of time, leading to eventual loss of data. Decay of the threshold window has been monitored on test cells stored at 250°C. The decay rate of the positive and negative states were found to be approximately the same.

5.13.5 Flash memories

Flash memories were a direct derivative of the one-transistor cell EPROM. They resulted from innovative cell designs and improved technology that allowed the one transistor cell EPROM to be reprogrammed electrically in the system.

Figure 5.24 shows schematic cross-sections of a series of cells that have been developed for the single cell flash memories showing the range of mechanisms from hot electron programming to low voltage cold electron erase and programming.

Figure 5.24(a) shows a one-transistor flash memory cell, used by Intel, with floating gate structure which overlaps the drain and source similar to an EPROM but with

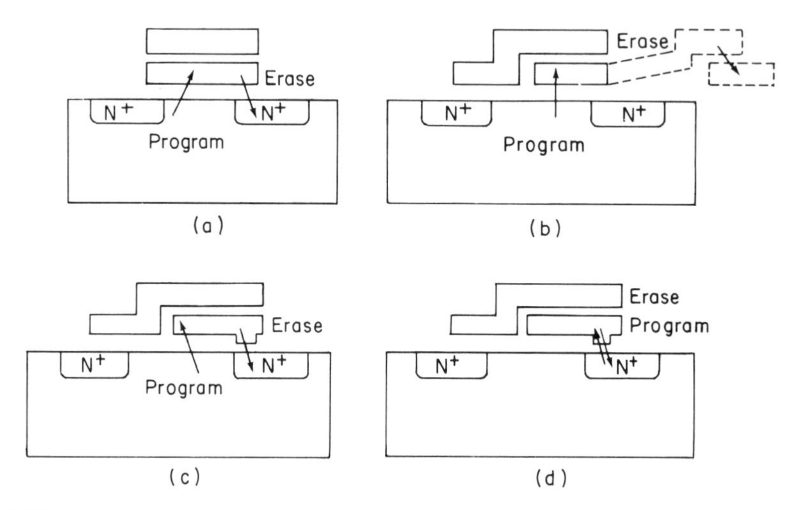

Figure 5.24
Schematic cross-sections showing the range of early flash EPROM cell development. (a) Single-transistor hot channel electron programming and cold electron erase. (b) Split transistor hot channel electron programming and erase to separate erase gate. (c) Split transistor hot channel electron programming and cold electron erase to source. (d) Split transistor cold electron program and erase to source.

thinner oxide under the floating gate. Programming is by hot electron injection from the channel near the drain like an EPROM and erase is by cold electron tunneling through the thin oxide between the floating gate and the source. An advantage of this cell is the small cell size. A disadvantage is the high programming current, as for the EPROM.

Many of the flash memory cells that have been developed use the split gate concept in which the separate select and storage transistor gates of the EEPROM are merged into a single device with the channel region shared by the two gates. This was done since the Fowler–Nordheim erase is not self limiting and the floating gate tends to become positively charged. The extra control gate over the channel of the storage transistor keeps it off when it is in the low threshold erased state unless it is selected to be read. The disadvantage is the slightly larger cell size.

Figure 5.24(b) shows a triple poly cell with a split gate used by Toshiba. Programming is by hot electron injection from substrate to the floating gate and erase is by hot electron injection from the floating gate to the separate erase gate. The disadvantage of this cell is the higher voltage required for the poly-to-poly erase.

Figure 5.24(c) shows another cell with a split gate used by Seeq. This one has a thin oxide region over the drain. Programming is by hot electron injection from the substrate to the floating gate and erase is by cold electron tunneling from the floating gate through the thin oxide to the drain. Using the same junction for both operations causes more stress on the gate oxide.

Figure 5.24(d) shows another split gate cell with thin oxide over the source region. In this case both program and erase are by cold electron tunneling through the thin oxide. The low current Fowler–Nordheim programming permits 5 V only operation of the cell with on-chip voltage generators. However, the longer programming time for the cold electron process may have long term stress effects on the thin oxide.

The flash memories appear to be on the way to replacing the UV-EPROM, the OTP and perhaps even a segment of the ROM markets. They are slower to write than SRAMs but could also replace some SRAMs in those low speed applications that require the low power of SRAMs in battery back-up, or in applications that predominantly read and seldom write.

The NV-DRAM is a new non-volatile technology development which also promises in system reprogrammability in a small cell size coupled with high speed.

5.14 ON-CHIP VOLTAGE GENERATION

All MOS memory architectures on occasion use on-chip voltage generation. While any voltage between ground and the power supply is available on a memory chip, it is also possible to generate on the chip voltages that are below ground and above the power supply value.

The most common on-chip voltage generation in MOS memories is back-bias voltage generation. Early MOS RAMs used a power supply which provided a negative voltage bias to the substrate. On-chip back bias generation has replaced one of these power supplies in many cases, although in other cases the substrate can be grounded with similar effect.

Other examples of on-chip voltage generation include the high programming voltages required for EPROM programming, and the grounded P-wells used on static RAMs.

An on-chip back bias generator works by pumping charge out of the ground and into the substrate which then acts as a capacitor in retaining the charge.

Another example of on-chip voltage generation is the voltage converter circuitry used in submicron technologies to protect the MOS array transistors from the standard external 5 V power supplies. A submicron MOS transistor with less than 0.5 μm effective channel length cannot operate without degradation under an applied voltage of 5 V or more due mainly to hot carrier injection.

The solution used in many cases is to reserve the short channel length devices for the memory array and use long channel MOS transistors which can sustain 5 V power supplies in the peripheral circuitry. Between the array and the periphery a voltage regulator circuit supplies converted 3 V power to the array.

5.15 BIPOLAR MEMORY BASIC GATES AND CHARACTERISTICS

Architectures for bipolar SRAMs are basically similar to those of the MOS SRAMs. The main difference is in the technology. A review of bipolar memory cells follows.

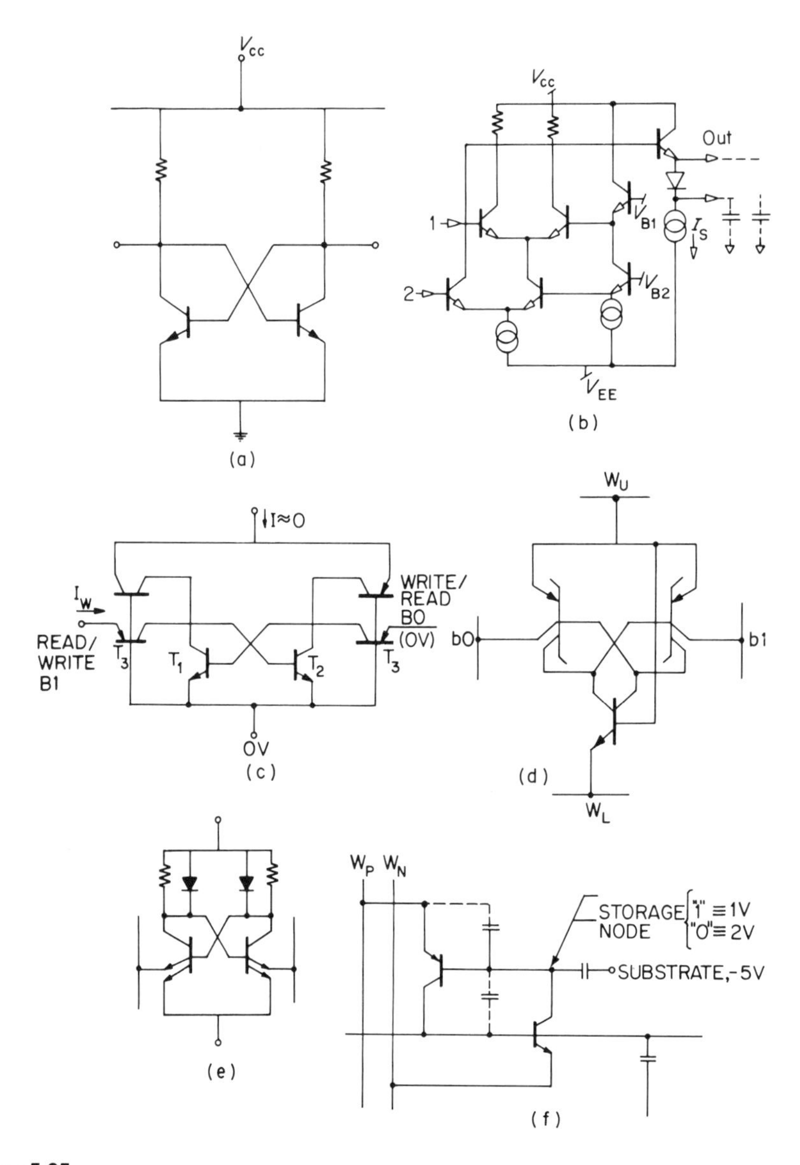

Figure 5.25
Various bipolar memory cells. (a) Basic bipolar flipflop, (b) ECL [61], (c) I^2L [55], (d) CCL [32], (e) multiple emitter [60], (f) SCR (IBM) (with permission of IEEE).

5.15.1 Basic bipolar flipflop

Transistor–transistor Logic (TTL) has been the most common form of bipolar circuitry historically. It replaced an earlier form called Diode Transistor Logic (DTL). A simple bipolar RAM memory cell is shown in Figure 5.25(a).

5.15.2 ECL memory cell

Emitter-coupled logic (ECL) memories provide the shortest access time of all semiconductor memories with typical propagation delays of less than 1 ns and clock rates approaching 1 GHz. This performance is achieved by preventing the transistors in the cells from ever entering into the saturation region. This improves the switching speed of the cells since it is a time consuming process to bring a transistor out of saturation. With ECL circuitry transistors are switched between two well defined levels in such a way that they never saturate. The high speed performance of ECL results from both its non-saturating mode of operation and from its relatively small voltage swing.

One of the drawbacks of ECL is the very low value of the load resistors required which results in the storage cell drawing current at all times. This constant current drain results in relatively high power dissipation compared to bipolar TTL or MOS circuitry. The well defined current levels are regulated by a voltage reference generator in the peripheral circuitry. A simple ECL RAM memory cell is shown in Figure 5.25(b).

5.15.3 Integrated injection logic memory cells (I^2L)

Integrated injection logic incorporates lateral PNP transistors as current sources and multicollector NPN transistors as invertors. It was explored as a possible competitor with MOS memories in the 1970s. An equivalent circuit of the cell is shown in Figure 5.25(c). While the circuit schematic appears complex, the layout on the silicon chip is very dense since all elements are in a common N buried layer. Since only active devices are used, the cell size is determined by the contact hole and metal line dimensions.

For operation the PNP bit line transistors provide a simple coupling means. They are operated in normal or inverse mode depending on whether a write or a read operation is being performed. The storage is in T_1 and T_2, which are emitter coupled, with T_3 and T_4 acting as load devices.

Integrated injection logic bipolar cells have been used in static RAM memories such as the Texas Instruments 4k clocked SRAM.

5.15.4 Collector coupled static RAM cell (CCL)

Another bipolar cell developed in the late 1970s in an attempt to compete with the density of MOS was the collector coupled logic SRAM. This SRAM uses two Schottky collector transistors as a switch with merged NPN load as shown in Figure 5.25(d). This is not a significant technology today.

5.15.5 Multiple emitter cell

The multiple emitter cell is a basic bipolar flipflop used in either TTL or ECL compatible circuitry as shown in Figure 5.25(e). It is in common usage today.

5.15.6 Bipolar SCR dynamic cell structure

This cell consists of a PNP and an NPN transistor connected in a Silicon Controlled Rectifier (SCR) configuration as shown in Figure 5.25(f). This cell was developed by IBM for use in bipolar dynamic RAM circuitry containing a bipolar sense amplifier and bipolar transistor word line driver. This cell is operated in the normal mode unlike the I^2L cell.

5.15.7 Combination MOS and bipolar cells

Possible combinations of bipolar and MOS memory cells have also been explored. One example is the JCMOS Dynamic RAM cell. This cell includes an N-channel MOST, an N-channel JFET, and a PNP bipolar transistor.

The only of these cells that have survived to compete with MOS memories are the ECL compatible types of bipolar cells.

5.16 BiCMOS ARCHITECTURES AND CONSIDERATIONS

BiCMOS is a combination of bipolar and CMOS technology transistors in the same circuit [43]. While the technology is more complex BiCMOS has had some advantages. Bipolar ECL input and output buffers have been used on CMOS memories both to interface the memory into a bipolar level logic system and to increase the performance. The CMOS memory cells have lower power consumption, superior stability, better alpha particle resistance and a smaller area than an equivalent bipolar cell.

There is increased flexibility to optimize a BiCMOS design since other circuits in the memory can be selectively chosen either bipolar or CMOS depending on the desired characteristics. For example since bipolar NPN transistors provide large output drive, BiCMOS gates can be effective at high capacitive nodes such as occur in decoders, wordline drivers, and output buffers. Control logic functions with small fanout can still use CMOS. Sense amplifiers, which require high input sensitivity, can use pure bipolar to provide the high gain and input sensitivity required for fast sensing of small differential bit line swings. Figure 5.26(a) shows the parts of the circuit which are frequently made in the two technologies. An example of BiCMOS circuit elements from a 1Mb BiCMOS DRAM [44, 45] are shown in Figure 5.26(b). The band gap reference and differential sense amplifier are bipolar, and the comparator is BiCMOS. An example of input decoder and driver circuits in both CMOS and in BiCMOS is shown in Figure 5.27(a). A comparison of a CMOS and a BiCMOS sense amplifier with output latches is shown in Figure 5.27(b).

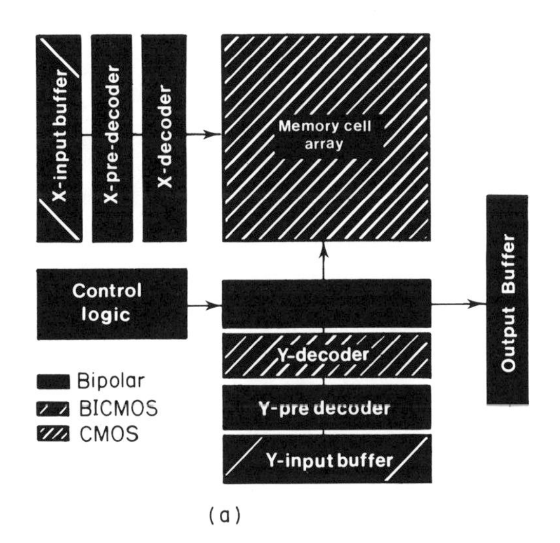

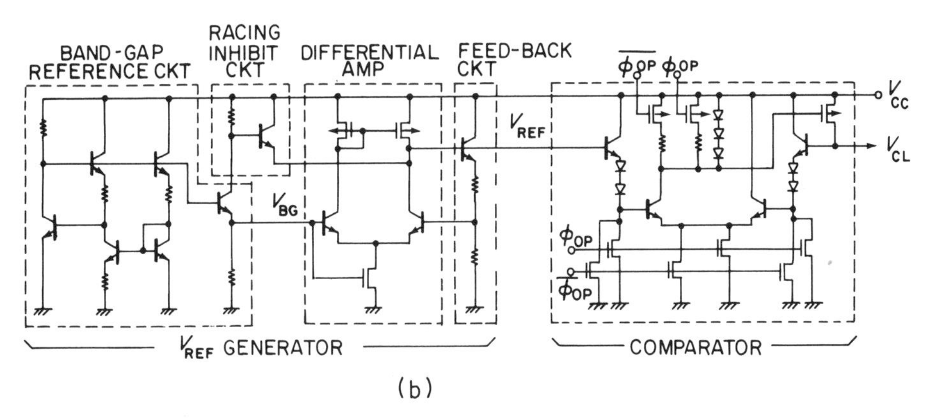

Figure 5.26
BiCMOS circuit elements. (a) Schematics of possible circuit blocks for a TTL compatible BiCMOS SRAM [43]. (b) Circuit schematic of various functions in a BiCMOS voltage generator. (From Kitsukawa *et al.* [44], Hitachi 1989, with permission of IEEE.)

5.17 I/O INTERFACE CHARACTERISTICS AND CIRCUITS

5.17.1 Interface characteristics (I/O) of memory circuits

As important to the memory user as high memory performance, density, and low cost is the ability to interface the chip easily into the system. This is a function of having the correct input and output voltage levels, designing the chip to use the correct power supply voltage, having the required output drive, and having the necessary noise margins in the memory to operate reliably in the system.

The differences in characteristics of invertors in different technologies has historically led to several interface standards with the most common being: 5 V TTL, 3.3 V

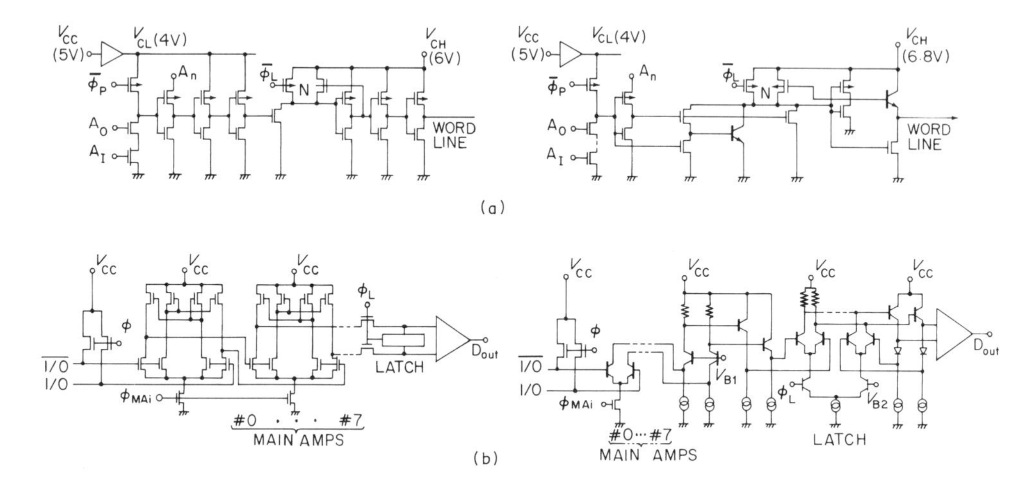

Figure 5.27
Comparison of CMOS and bipolar circuits. (a) Input x-decoder and driver circuits. (b) Sense amplifier output latch circuits. (From Watanabe *et al.* [45], Hitachi 1989, with permission of IEEE.)

TTL, CMOS, 10K ECL and 100K ECL. Today all the common memory technologies are available in these different interface standards.

5.17.2 Transfer characteristics of CMOS invertors

The transfer characteristics of the invertors in the input and output buffers determine the noise margin of the memory chip. The dc characteristics of a CMOS invertor are a combination of the characteristics of the NMOS and PMOS devices. These can be summarized in a characteristic curve for the invertor called 'a dc transfer characteristic' curve as shown in Figure 5.28(a). This consists of five operating regions as follows.

(a) $0 < V_{\text{in}} < V_{\text{tn}}$ so the NMOS device is cut-off and the PMOS is in the linear region. The output voltage is $V_0 = V_{\text{DD}}$.

(b) $V_{\text{tn}} < V_{\text{in}} < V_{\text{DD}}/2$ in which the PMOS is still in its linear region and the NMOS device is in saturation.

(c) $V_{\text{in}} = V_{\text{DD}}/2$ This region is actually a point where both devices are in saturation and a very steep transition occurs between the two states.

(d) $V_{\text{DD}}/2 < V_{\text{in}} < V_{\text{DD}} - V_{\text{tp}}$ where the PMOS device is in saturation and the NMOS in its linear region.

(e) $V_{\text{tp}} < V_{\text{in}} < V_{\text{DD}}$ The PMOS device is cut off and the NMOS is in its linear region. $V_0 = 0$.

The invertor transfer function is used by defining logic levels within which the transfer must occur. These are defined as voltage input and output high levels and voltage input and output low levels. A MOS invertor transfer function falling within a set of defined logic levels is shown in Figure 5.28(b).

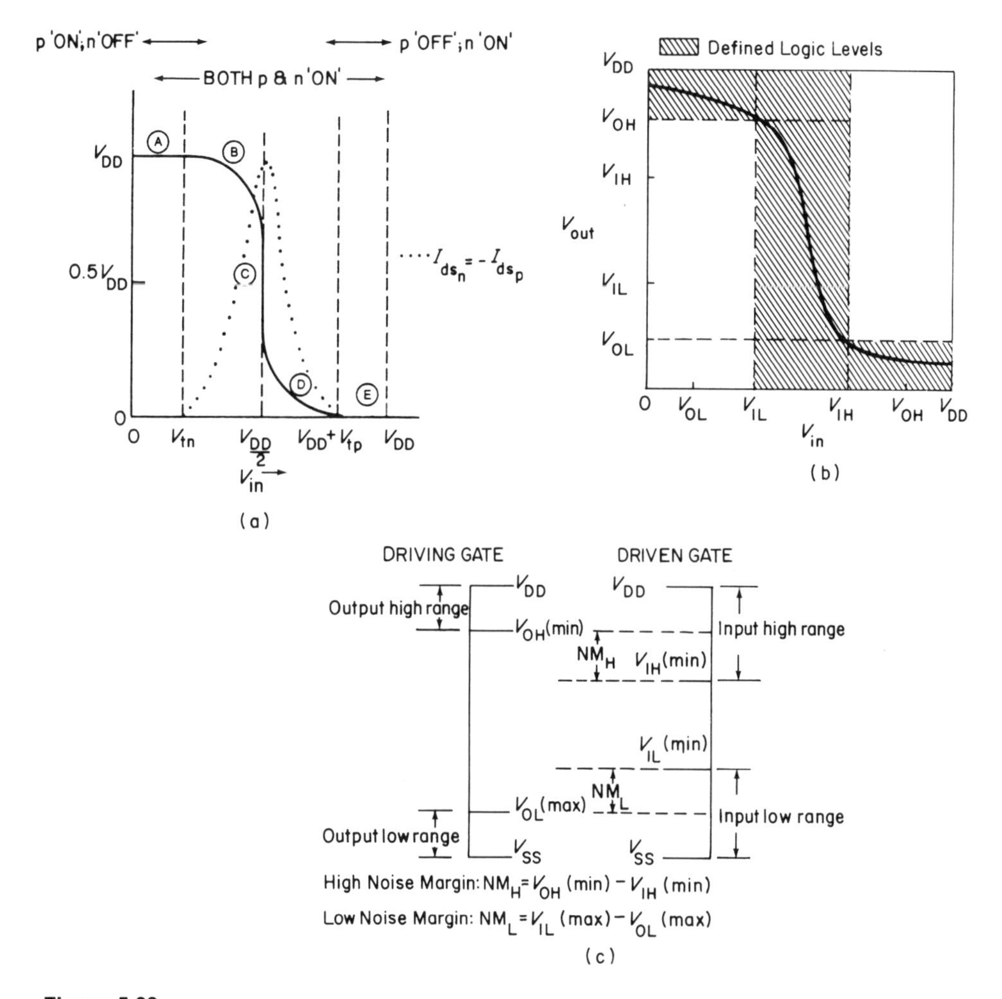

Figure 5.28
DC transfer characteristics of a CMOS invertor. (a) Transfer regions (—— output voltage, I_{DD} supply current). (b) Output and input levels ---- determining noise margin [93]. (c) Noise margin input and output definitions.

5.17.3 Transfer characteristics of ECL gates

A comparison of the transfer characteristics for conventional and fully compensated ECL is shown in Figure 5.29 with varying temperature and with varying supply voltage (V_{EE}).

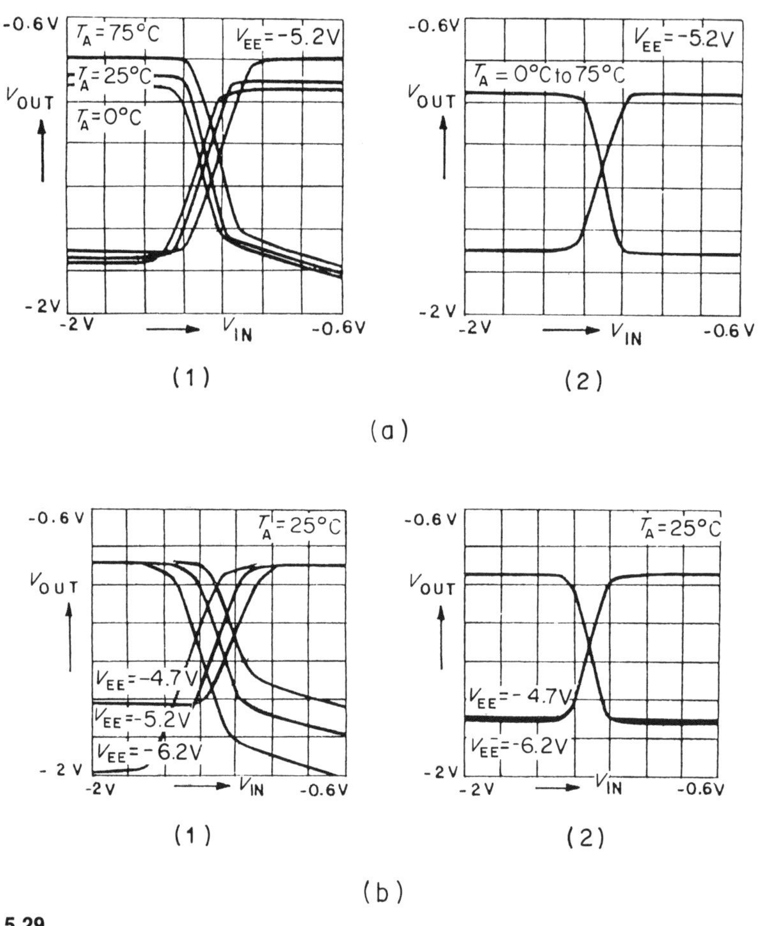

Figure 5.29
(a) Comparison of the transfer characteristics for conventional (1) and fully compensated (2) ECL with varying temperature. (b) Comparison of the transfer characteristics for conventional (1) and fully compensated (2) ECL with varying supply voltage V_{EE}. (From Muller *et al.* [31], IBM 1973, with permission of IEEE.)

The transfer switching characteristic of an ECL gate has much smaller voltage swings than a CMOS gate. This is one source of its speed advantage over other forms of bipolar circuitry and over CMOS.

5.18 INPUT AND OUTPUT LEVELS

5.18.1 Commonly used input and output voltage levels

There are several sets of commonly used interface logic levels corresponding to the various types of invertor logic and technologies available. These include TTL, CMOS and ECL as defined below.

Table 5.4

	Nominal supply	V_{IL} Max.	V_{OL} Max.	V_{IH} Min.	V_{OH} Min.
TTL	5.0	0.8	0.4	2.0	2.4
ECL(10K)*	−5.2	−1.4	−1.7	−1.2	−0.9
CMOS†	5.0	$0.3V_{CC}$	0	$0.7V_{CC}$	5.0

* The values given for 10K ECL are approximate since they differ at different temperatures.
† Different CMOS logic families exist and power supply voltages differ among families. Input levels also differ in different CMOS families. In full CMOS circuits they are generally a percentage of the power supply voltage.

These voltage levels are commonly specified as in Table 5.4. The input logic swing of ECL (200 mV) is much less than that of TTL (1200 mV), accounting in part for its higher speed.

5.18.2 Noise margin

Noise can be present in a digital circuit from a number of different sources. Voltage noise in the ground and power supply lines can be caused by ringing, voltage spikes, or voltage drops due to series resistances. There is also series voltage noise in the interconnection lines between gates and parallel current noise to inputs or outputs of gates.

Careful memory circuit design coupled with careful system design can reduce noise, but it is not possible to completely eliminate it. The concept of noise margin therefore exists for specifying limits to the amount of acceptable noise that can be present on the inputs of an invertor before the data in the invertor becomes indeterminate.

The noise margin of a CMOS invertor depends on its transfer characteristic curve, that is, it is related to its input and output voltage characteristics. The noise margin helps specify the noise allowed on the input voltage of a gate or invertor so that the output will not be affected.

Two parameters are used to specify noise margin, the low noise margin and the high noise margin. Referring to Figure 5.28, the low noise margin is the difference between the maximum low output voltage of the driving gate and the maximum low input voltage recognized by the driven gate. The high noise margin is the difference between the minimum high output voltage of the driving gate and the minimum input high voltage recognized by the receiving gate. If the noise margin is reduced then the gate may be more susceptible to switching noise.

The definition of the noise margins as shown in Figure 5.28(c) is

$$\text{Noise margin (high)} = V_{OH}(\text{min}) - V_{IH}(\text{min})$$

$$\text{Noise margin (low)} = V_{IL}(\text{max}) - V_{OL}(\text{max})$$

ECL has relatively low values of noise immunity resulting from the small voltage swing. Fully compensated ECL renders the output voltage high, the output voltage

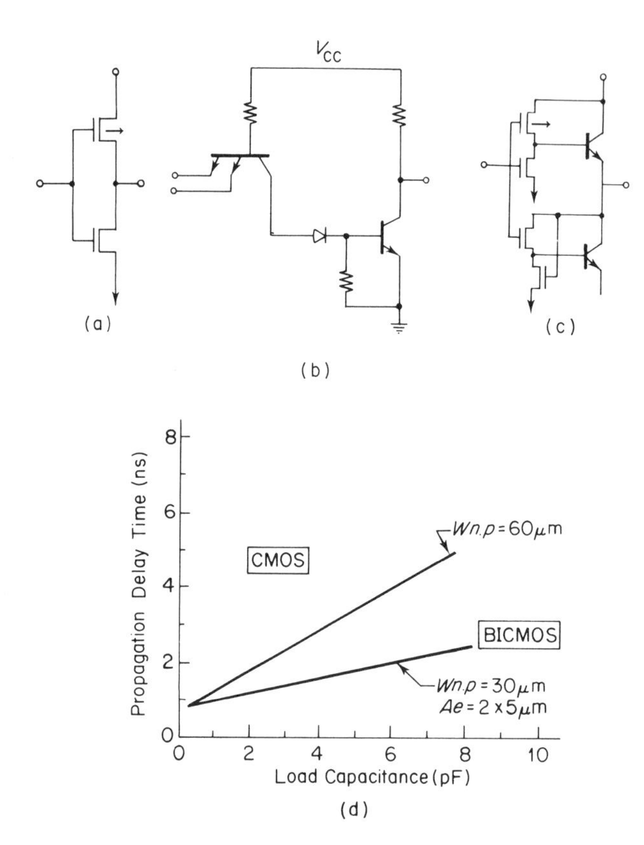

Figure 5.30
TTL compatible output drivers in various technologies: (a) CMOS, (b) Bipolar, (c) BiCMOS, (d) Propagation delay of CMOS and BiCMOS output buffers ([62], with permission of IEEE.)

low levels and the threshold invariant with changes in temperature and supply voltage. This ensures constant noise immunity for most system conditions.

5.18.3 Voltage level interfaces and level shifters

It is necessary to interface a memory device into a system. This means that the memory and the surrounding system need to have matched logic high and low levels. Since the common interface levels are TTL, ECL and CMOS, the memory must be designed with inputs and outputs that shift from the levels of the internal circuitry to one of these external interfaces.

Various configurations of invertors are used for such level shifting. Typical TTL level output buffers are shown in Figure 5.30. Figure 5.30(a) shows a TTL level output buffer made in CMOS technology for a CMOS circuit and Figure 5.30(b) shows a bipolar TTL level output buffer made for a bipolar circuit. Figure 5.30(c) shows a BiCMOS TTL output buffer which can be used in a memory with a CMOS cell array.

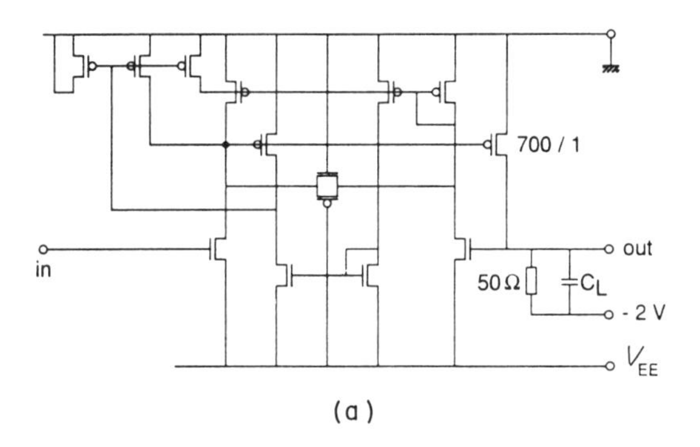

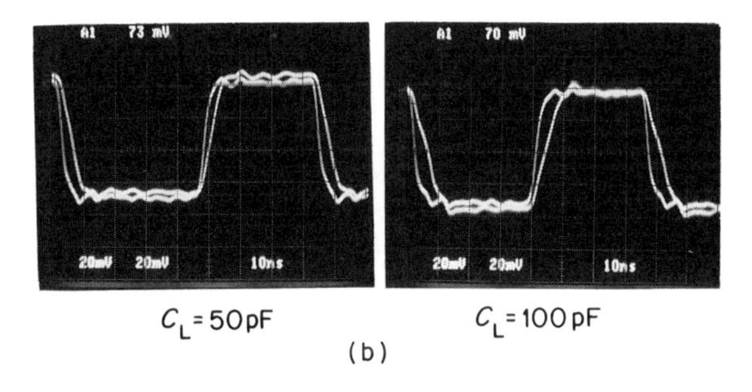

Figure 5.31
ECL-level compatible output buffer in CMOS technology. (a) Complete ECL-compatible output buffer circuit. (b) Switching waveforms of ECL compatible CMOS circuitry. (From Seevinck [35], Philips 1989, permission of Japanese Society of Applied Physics.)

Figure 5.30(d) illustrates the propagation delay for a CMOS and a BiCMOS output buffer for different capacitance loads.

It is also possible to make ECL-compatible input and output buffers in CMOS technology. An output buffer is shown in Figure 5.31(a). The sharply defined transfer characteristics of this buffer are shown in Figure 5.31(b). Further discussion of this output buffer can be found in Chapter 8.

A simple ECL input buffer in bipolar technology is shown in Figure 5.32. The basic emitter coupled current switch that gives ECL its name is shown.

ECL interfaces are available as 10K and 100K ECL. 100K ECL differs from 10K ECL in having reference voltages that are invariant to supply and temperature changes. This reduces the shifts in the voltage transfer characteristic with temperature found in 10K ECL. The negative supply voltage is also reduced from -5.2 to -4.5 V to reduce the power dissipation of the circuit.

Technologies are also mixed to accomplish input level shifting. For example, ECL and CMOS technologies are sometimes combined in a 'BiCMOS' circuit configuration where, for example, the input buffer may be made in bipolar ECL technology, the translator made in CMOS technology, and the driver in BiCMOS.

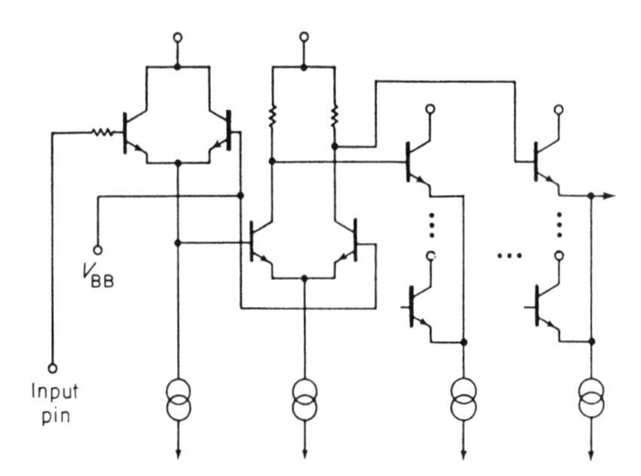

Figure 5.32
ECL level compatible input buffer in bipolar ECL technology. (Source Electronic Design, 1988 [63].)

5.19 SELF AND MUTUAL INDUCTANCE EFFECTS ON OUTPUT TIMING

A factor which can slow down the external interface of a memory is ringing due to the self inductance effects of the bonding wires from the chip to the package. An equivalent circuit for the effect at the chip level is shown in Figure 5.33(a).

At the individual output buffer level, the power supply bounce or ringing effect is due to both the output capacitance to ground and the inductance to ground of the switching transistor and the output capacitance to the ground and the inductance to the power supply through the load transistor. A schematic of a TTL output buffer with the inductive and capacitive loops to ground indicated is shown in Figure 5.33(b).

The switching of the output buffer from one state to the other results in an *LC* ringing effect which can be decreased at the chip level by the addition of additional power and ground lines so that each supply line handles a reduced capacitive load. Ringing can also be reduced by the mutual inductive effect between parallel power and ground lines which is opposite in effect to the self inductance. This mutual inductance effect is stronger when the power and ground lines are closer together. Ringing is also reduced by making the bond wires as short as possible to reduce their intrinsic inductance. Normally the center pins on a packaged chip have the shortest bond wires.

Significant design effort has gone into reducing this effect at the chip level. The JEDEC standard pinouts for fast SRAMs which double the power and ground pins and move them to the center of the package also help reduce this effect. A comparison of the older standard with the power supply pins at the end of the package and the new standard for fast SRAMs is illustrated in Figure 5.34.

Ringing can also be controlled in the circuit's output buffer. For example, a circuit technique for suppressing switching noise in an output driver was described by Digital Equipment [34] in 1991. It consisted of a series transistor in a predriver with a modulating voltage applied to its gate to adjust its conductance. The result is a more controlled change of current over time as the output changes.

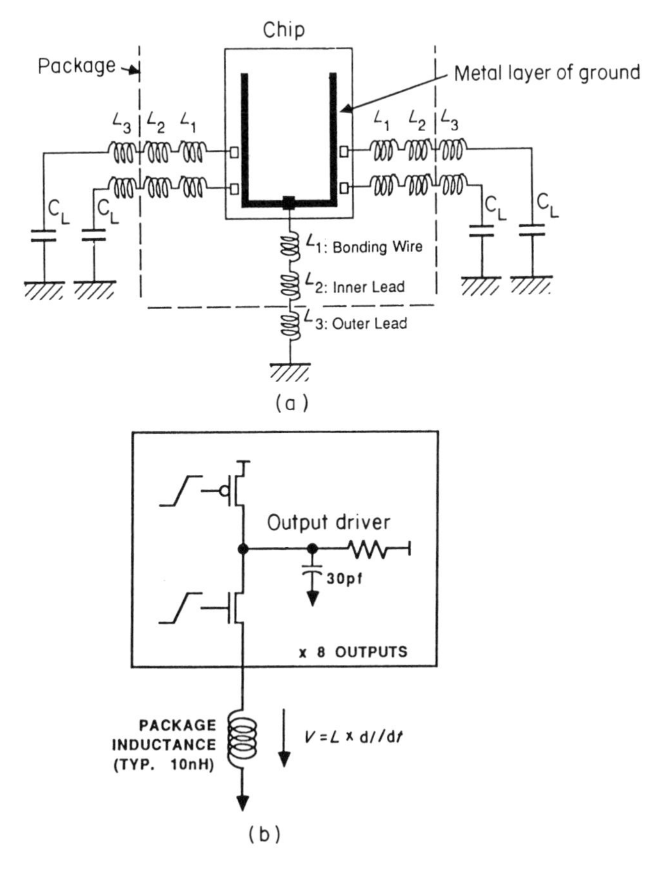

Figure 5.33
(a) Model of ringing effect at the package level Vol bounce $= (L_1 + L_2 + L_3)di/dt$ (with permission of NEC). (b) Model of ringing effect at the chip level (from Memory Design Workshop 1988 IEDM, with permission of IEEE.)

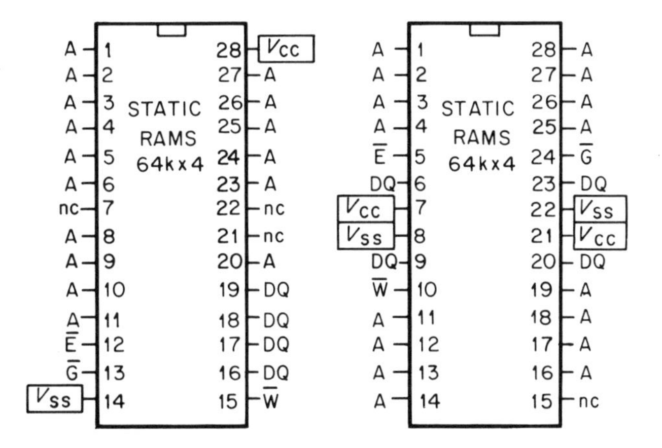

Figure 5.34
Comparison of JEDEC standard end power and ground pinouts for slow SRAMS and JEDEC standard center multiple power and ground pinouts for fast SRAMS. (From Salters [25] 1988, with permission of N.V. Philips Research.)

5.20 MULTIPLEXING OF INPUT SIGNALS TO THE MEMORY

Multiplexing of input signals is used in memory circuits to reduce the pincount and permit smaller more cost effective packaging to be used. The possibilities for multiplexing are shown conceptually in Figure 5.35.

The best known example of multiplexed input signals is the multiplexed addressing of the standard dynamic RAM memory. The number of address pins is reduced by half, but the timing requirements become more complex. Input and output multiplexing, and address and data line multiplexing has also been used.

5.20.1 Address multiplexing in DRAMs

A block diagram of a multiplexed address DRAM was shown in Figure 5.13. To address the RAM, the row addresses are applied to the address pins then clocked in with the row address strobe ($\overline{RAS}$) clock. The column address pins are then applied and clocked in with the column address strobe ($\overline{CAS}$) clock.

It is clear that the access time of the memory is penalized by this arrangement, and for very fast memories separate address input pins are occasionally used. This is referred to as 'broadside' addressing.

5.20.2 Common as opposed to separate I/O

The same kinds of trade-offs exist in multiplexing the inputs and outputs as in multiplexing the addresses. If they are multiplexed then the package is smaller and more cost effective, but the speed of the chip in the system is slower since the same

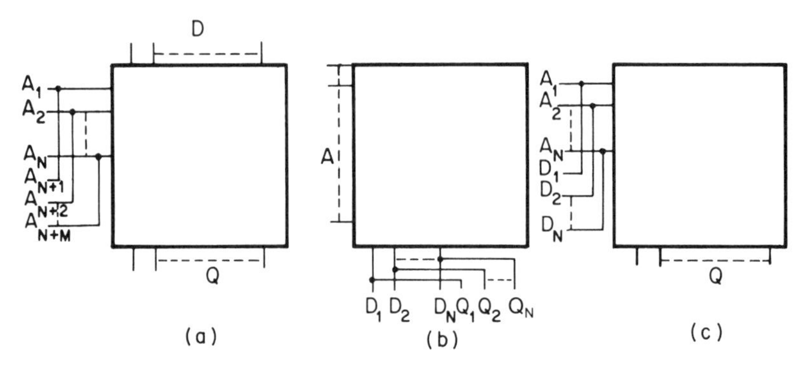

Figure 5.35
Examples of multiplexing in RAMs. (a) Multiplexed addressing (used in DRAMs). (b) Common I/O (used in SRAMs). (c) Multiplexed address/data (standardized by JEDEC but infrequently used). (Adapted from Salters [50] 1989, with permission of Philips Research.)

lines are used for input as for output. If the inputs and outputs are separate, then the chip is faster in the system but it requires a larger package. Only systems which must have the speed will pay the price for the more expensive part.

5.20.3 Other types of multiplexing

Consideration has also been given to other types of multiplexing such as address–data multiplexing for wide bus memories. The JEDEC JC42.3 memory standardization committee has standardized families of such parts although few actual representative circuits using this multiplexing have appeared on the market to date.

5.21 REDUNDANT CIRCUITS FOR ENHANCED YIELD

A few redundant rows or columns can significantly enhance yield of a memory circuit since many devices are rejected for single bit failure or failures in a single row or column. Figure 5.36 indicates the selection of redundant rows using the x-row decoder or redundant columns using the y-column decoder. Normally the redundant rows or columns are selected at wafer electrical test and implemented by either having interconnections blown by a laser, or by blowing fuses already designed into the circuit. Further discussion of redundancy can be found in Chapter 15.

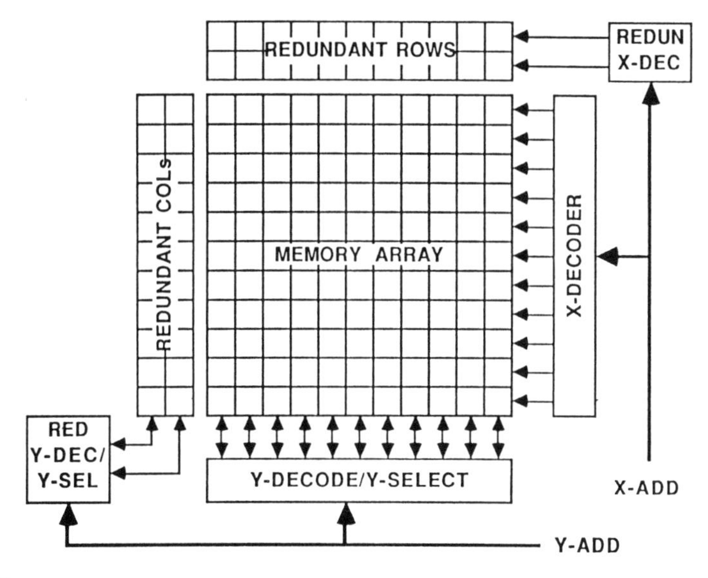

Figure 5.36
Row and column redundancy. (From Bateman 1988 [65], with permission of IEEE.)

5.22 ERROR CORRECTION ON THE MEMORY CHIP

On-chip error correction is used on some memory chips with a number of benefits. When implemented on EEPROMs it can be used to replace final electrical test and hence reduce the cost of the device. Alternatively, the device can be tested and the on-chip error correction can be used to improve the life time of the device. In RAMs error correction has been suggested as one possibility for reducing alpha induced soft errors which are more common in submicron parts.

The simplest form of on-chip error detection–correction is through the use of Hamming codes. A single additional bit in each partial row and column can identify the x and y coordinates of a bit requiring correction. A quick check can detect if the parity of the output word is still the same, if not then there is evidence that an error in the stored information has occurred. This data can then be checked and corrected. Multiple errors in a single word cannot, however, be corrected.

A slight variation due to Edwards [21] implements an error detection–correction algorithm on a block basis rather than for the entire memory array. In this case a horizontal parity bit is added to each word and a vertical parity is generated for each block. More sophisticated techniques have also been explored such as the Osman [22] technique for clocked memories. Further discussion on error correction can be found in Chapter 15.

5.23 COMPUTER AUTOMATED DESIGN AND SIMULATION

Computer aided design (CAD) has evolved over the last 20 years from the device modeling, circuit simulation, and interactive layout of the 1970s to the more sophisticated system design tools of today. These include macromodeling, automated layout, and design rule checking. The photograph in Figure 5.37(a) shows a design engineer at work on a computer automated layout tool—a familiar sight in any integrated circuit design area today. Figure 5.37(b) shows a team of designers checking the layout done on the computer.

Gradually the focus of advanced memory development is turning from the almost total concentration on process technology to a greater emphasis on the development of design tools.

In the early days of MOS memory design, available computer time was sufficiently expensive that much effort went into both simplifying mathematical simulation models, and into partitioning the simulation and design effort so that small blocks of computer time were utilized.

Early models for the MOS transistor such as the Shichman–Hodges model and the Frohman–Bentchkowsky and Vadasz models were used extensively and remain the basis today of much of the computer simulation work done. More recently the increasing power of computer workstations has enabled increasingly complex models and circuits to be designed, laid out, and the circuitry simulated.

Libraries of components, including compilers for memory cells which create memory arrays to be embedded in logic circuits, are being created. These libraries, however,

(a)

(b)

Figure 5.37
Integrated circuit designers at work. (a) I.C. designer using a computer automated design layout tool. (b) A team of memory designers visually checking the layout of a memory done with computer aided design tools. (With permission of N.V. Philips Research.)

serve primarily the logic world. Pure memories are still design labor intensive in order to minimize silicon area. Circuit and transistor simulations are, however, advancing at a rapid rate as required by the exacting requirements of submicron technologies for memories.

Expert systems which can replace the design creativity are still a long way off.

BIBLIOGRAPHY

[1] Inoue, M. *et al.* (1988) A 16-Mbit DRAM with a Relaxed Sense Amplifier Pitch Open-Bit-Line Architecture, *IEEE Journal of Solid State Circuits*, **SC-23**, No. 5, October 1988, p.1104.

[2] Ahrons, R. W. and Gardner, P. D. (1970) Interaction of Technology and Performance in Complementary Symmetry MOS Integrated Circuits, *IEEE Journal of Solid State Circuits*, **SC-5**, No.1, Feb. 1970 p.26.

[3] Weste, N., *et al.* (1985) *Principles of CMOS VLSI Design, A System Perspective*, Addison-Wesley, Reading MA.

[4] Lueke, G. *et al.* (1973) *Semiconductor Memory Design and Application*, McGraw-Hill, New York.

[5] Glasser, L. A. and Dobberpuhl, D. W. (1985) *The Design and Analysis of VLSI Circuits*, Addison-Wesley, Reading, MA.

[6] Taub, H. and Schilling, D. (1977) *Digital Integrated Electronics*, McGraw-Hill, Kogahusha, Tokyo.

[7] Seevinck, E. *et al.* (1987) Static Noise Margin Analysis of MOS SRAM Cells, *IEEE Journal of Solid State Circuits*, **SC-22**, No.5, October 1987.

[8] Foss, R. C. and Harland, R. (1975) Peripheral Circuits for One Transistor Cell MOS RAMs, *IEEE Journal of Solid State Circuits*, **SC-10**. No. 5, October 1975.

[9] Hodges, D. A. and Jackson, H. G. (1983) *Analysis and Design of Digital Integrated Circuits*, McGraw-Hill, New York.

[10] Deferm, L. *et al.* (1988) Latch-up in a BICMOS Technology, *IEDM Tech. Dig.*, Dec 1988.

[11] Mano, T. *et al.* (1983) Circuit Techniques for a VLSI Memory, *IEEE Journal of Solid State Circuits*, **SC-18**, No. 5, October 1983.

[12] Rountree, R. N. (1988) ESD Protection for Submicron CMOS Circuits, Issues and Solutions, *IEDM Tech. Dig. Dec.* 1988, p.580.

[13] Goser, K. and Pomper, M. (1973) Five-Transistor Memory Cells in ESFI MOS technology, *IEEE Journal of Solid State Circuits*, **SC-8**, No.5, October 1973, p.324.

[14] Haraszti, T. P. (1982) A Novel Associative Approach for Fault-Tolerant MOS RAMs, *IEEE Journal of Solid State Circuits*, **SC-17**, No.3, June 1982, p.539.

[15] Abbott, R. A. *et al.* (1973) A 4K MOS Dynamic Random-Access Memory, *IEEE Journal of Solid State Circuits* **SC-8**, No.5, October 1973, p.292.

[16] McLeod, J. (1987) Keys to the Future: CAD and Fab, *Electronics*, April.

[17] O'Connor, K. J. (1987) The Twin-Port Memory Cell, *IEEE Journal of Solid State Circuits*, **SC-22**, No.5, October 1987, p.712.

[18] Miyanaga, H. *et al.* (1986) A 0.85ns 1k bit ECL RAM, *IEEE Journal of Solid State Circuits*, **SC-21**, No.4, August 1986, p.501.

[19] Eldin, A. G. and Elmasry, M. I. (1985) A Novel JCMOS Dynamic RAM Cell for VLSI Memories, *IEEE Journal of Solid State Circuits*, **SC-20**, No.3, June 1985.

[20] Lewyn, L. L. and Meindl, J. D. (1983) Physical Limits of VLSI DRAMs, *IEEE Journal of Solid State Circuits*, **SC-20**, No.1, February 1983.

[21] Edwards, L. (1981) Low Cost Alternative to Hamming codes corrects memory errors, *Computer Design*, July 1981, p.132.

[22] Osman, F. I. (1982) Error-Correction Technique for Random-Access Memories, *IEEE Journal of Solid State Circuits*, **SC-17**, No.5, October 1982, p.877.

[23] Schrader, L. (1978) A New Circuit Configuration for a Static Memory Cell with an Area of 880 microns squared, *IEEE Journal of Solid State Circuits*, **SC-13**, No.3, June 1978.

[24] O'Connell, T. R. *et al.* (1977) Two Static 4K Clocked and Nonclocked RAM Designs, *IEEE Journal of Solid State Circuits*, **SC-12**, No.5, October 1977.

[25] Salters, R., Presentation at SRAM Panel Discussion ISSCC, February 1989.

[26] Dingwall, A. G. F. and Stewart, R. G. (1979) 16k CMOS/SOS Asynchronous Static RAM, *IEEE Journal of Solid State Circuits*, **SC-14**, No.5, October 1979, p.867.

[27] Engeler, W. E. *et al.* (1971) A Memory System Based on Surface-Charge Transport, *IEEE Journal of Solid State Circuits*, **SC-6**, No.5, October 1971, p.306.

[28] Lohstroh, J. (1979) Static and Dynamic Noise Margins of Logic Circuits, *IEEE Journal of Solid Statc Circuits*, **SC-14**, No.3, June 1979, p.591.

[29] Wiedmann, S. K. (1973) Injection-Coupled Memory: A High Density Static Bipolar Memory, *IEEE Journal of Solid State Circuits*, **SC-8**, No.5, October 1973, p.332.

[30] de Troye, N. C. (1974) Integrated Injection Logic—Present and Future, *IEEE Journal of Solid State Circuits*, **SC-9**, No.5, October, 1974, p.206.

[31] Muller, H. H. *et al.* (1973) Fully Compensated Emitter-Coupled Logic: Eliminating the Drawbacks of Conventional ECL, *IEEE Journal of Solid State Circuits*, **SC-8**, No.5, October 1973.

[32] Hewlett, F. W. Jr. (1979) The Collector-Coupled Static RAM Cell, *IEEE Journal of Solid State Circuitry* **SC-14**, No.5, October 1979, p.865.

[33] Cilingiroglu, U., (1979) A Charge Pumping Loop Concept for Static MOS RAM Cells, *IEEE Journal of Solid State Circuits*, **SC-14**, No.3, June 1979, p.599.

[34] Boonstra, L. *et al.* (1973) A 4096b One Transistor per Bit Random Access Memory with Internal Timing and Low Dissipation, *IEEE Journal of Solid State Circuits*, **SC-8**, No.5, October 1973, p.73.

[35] Seevinck, E. *et al.* (1989) CMOS Subnanosecond True ECL Output Buffer, VLSI Seminar Digest, May 1989.

[36] Ahlquist, C. N. *et al.* (1976) A 16k 384-Bit Dynamic RAM, *IEEE Journal of Solid State Circuits*, **SC-11**, No.3, October 1976.

[37] Masuda, H. *et al.* (1980) A 5V—Only 64k Dynamic RAM Based on High S/N Design, *IEEE Journal of Solid State Circuits*, **SC-15**, October 1980, p.846.

[38] Penoyer, *et al.* (1980) An 18K Bipolar Dynamic Random Access Memory, *IEEE Journal of Solid State Circuits* **SC-15**, October 1980, p.861.

[39] El-Mansy, Y. A. and Burghard, R. A. (1982) Design Parameters of the Hi-C DRAM Cell, *IEEE Journal of Solid State Circuits*, **SC-17**, No.5, October 1982, p.951.

[40] Masahiro, A. and Ohara, M. (1979) New Input/Output Designs for High Speed Static CMOS RAM, *IEEE Journal of Solid State Circuits*, **SC-14**, No.5, October 1979, p.823.

[41] Dumbri, A. C. and Rosenzweig, W. (1980) Static RAMs with Microwatt Data Retention Capability, *IEEE Journal of Solid State Circuits* **SC-15**, No.5, October 1980, p.826.

[42] Imamiya, K, *et al.* (1990) A 68ns 4-Mbit CMOS EPROM with High Noise Immunity Design, *IEEE Journal of Solid State Circuits*, **SC-25**, No.1, February 1990, p.72.

[43] Alvarez, A. R. *et al.* (1989) "Tweaking BICMOS Circuits by Optimizing Device Design", *Semiconductor International*, May, p.226.

[44] Kitsukawa, G. *et al.* (1989) A 1-Mbit BICMOS DRAM Using Temperature-Compensation Circuit Techniques, *IEEE Journal of Solid State Circuits*, **SC-24**, No.3, June 1989, p.597.

[45] Watanabe, T. *et al.* (1989) Comparison of CMOS and BiCMOS 1-Mbit DRAM Performance, *IEEE Journal of Solid State Circuits* **SC-24**, No.3, June 1989, p.771.
[46] Gubbels, W. *et al.* (1987) A 40ns/100pF Low-Power Full-CMOS 256k (32k × 8) SRAM, *IEEE Journal of Solid State Circuits,* **SC-22**, No.5, October 1987, p.741.
[47] Akiya, M. and Ohara, M. (1979) Input/Output Designs for High Speed Static CMOS RAM, *IEEE Journal of Solid State Circuits,* **SC-14**, No.5, October 1979, p.823.
[48] Ohzone, K. l (1980) 8k × 8 bit Static MOS RAM, *IEEE Journal of Solid State Circuits,* **SC-15**, No.5, October 1980.
[49] Bastiaens, J. J. J. and Gubbels, W. C. H. (1988) The 256k bit SRAM: An Important Step on the Way to Submicron IC Technology, *Philips Technical Review,* **44**, No.2, April, p.33.
[50] Salters, R. (1989) Basics of Memories, from *Philips Memory Training Manual,* (B. Prince, R. Salters, and A. van Zanten, eds.) 1989.
[51] Sze, S. M. (1981) *Physics of Semiconductor Devices,* 2nd Edn., John Wiley, Chichester.
[52] Minato, O. *et al.* (1982) A Hi-CMOS II 8k × 8 bit Static RAM. *IEEE Journal of Solid State Circuits,* **SC-17**, No.5, October 1982, p.773.
[53] Salters, R. (1988) Advantage of Center Power—Ground Pinout, Presentation to EIA JC42.3 JEDEC Committee, February.
[54] Prince, B. and Due-Gundersen, G. (1973) *Semiconductor Memories,* John Wiley, Chichester.
[55] Wiedmann, S. K. (1973) Injection Coupled Memory—A High Density Static Bipolar Memory, *IEEE Journal of Solid State Circuits,* **SC-8**, No.5, October 1973, p.332.
[56] Hoffman, W. K. and Kalter, H. L. (1973) An 8kb Random Access Memory Chip Using the One-Device FET Cell, *IEEE Journal of Solid State Circuits,* **SC-8**, No.5, October 1973.
[57] Kuo, C. *et al.* (1973) Sense Amplifier Design in Key to One Transistor Cell in 4096 bit RAM, *Electronics,* September, 1973.
[58] Knecht, M. W. *et al.* (1983) A High-Speed Ultra-Low Power 64K CMOS EPROM with On-Chip Test Functions, *IEEE Journal of Solid State Circuits,* **SC-18**, No.5, October 1983, p.554.
[59] BiCMOS Memories, *Semiconductor International,* May 1989, p.231.
[60] *MOS Memory Databook,* Hitachi 1988/1989.
[61] Wilson, G. (1990) Invited paper at VLSI Circuits symposium.
[62] Ogive, *et al.* (1986) Invited paper at VLSI Circuit Symposium.
[63] Bursky, D. (1988) Memory IC's, *Electronic Design,* February, p.76.
[64] Partovi, H., (1991) Noise suppression techniques for logic and memory circuits, Symposium on VLSI Circuits, June 6–3, p.51.
[65] Bateman, D. (1988) Memory Design Seminar, IEEE IEDM, February, 1988.
[66] Regitz, W.M. and Karp, J. (1970) A Three-Transistor Cell, 1024-bit, 500 ns MOS RAM, *Proceedings of the IEEE ISSCC,* p.42.

6 DYNAMIC RANDOM ACCESS MEMORY TRENDS— 4k TO 256Mb

6.1 AN OVERVIEW OF DYNAMIC RAMs

The dynamic RAM is the flagship product of the semiconductor industry. It is the memory most closely tied to the technology, and also the one with the highest volume production. DRAMs account for more than half of the total MOS memory market.

A single 4Mb dynamic RAM chip about the size of a thumbnail contains more bits of memory than a room size computer did 30 years ago. The cost, however, of a megabyte of DRAM memory in 1975 was around $8000, in 1985 it was $500, and in 1995 it is expected to be around $8.

The cost of a DRAM manufacturing area doubles every three years with factories for 4Mb DRAMs costing as much as $350 million and factories for 16Mb DRAMs costing around $700 million as shown previously in Figure 1.6. This makes the minimum entry barrier for a potential DRAM supplier $500 million in 1990 and $1 billion by 1994 if technology development costs are also included. Since the industry revenues double about every three years, a company whose DRAM sales grow at the average rate of the DRAM industry will have a nearly constant ratio of revenues to capital investment cost.

The bit density of DRAMs, however, quadruples about every three years so that the revenue and capital cost per bit of the product fall continuously with time. See Figure 6.1 for the DRAM industry learning curves in terms of price per bit plotted against cumulative volume of units shipped.

The limiting factor of production in a DRAM manufacturing area is the silicon wafer so that the more DRAM bits that can be made on a single wafer, the lower the average cost per bit. The history of DRAM manufacturing has been about the struggle to increase the number of bits on a single wafer. This has, as a practical matter, taken the form of increasing the number of bits in an individual DRAM chip. While the chip size has increased slightly, by less than a factor of ten over six generations, the cell size has decreased two orders of magnitude and the density of bits on a chip has increased by three as shown in Figure 6.2.

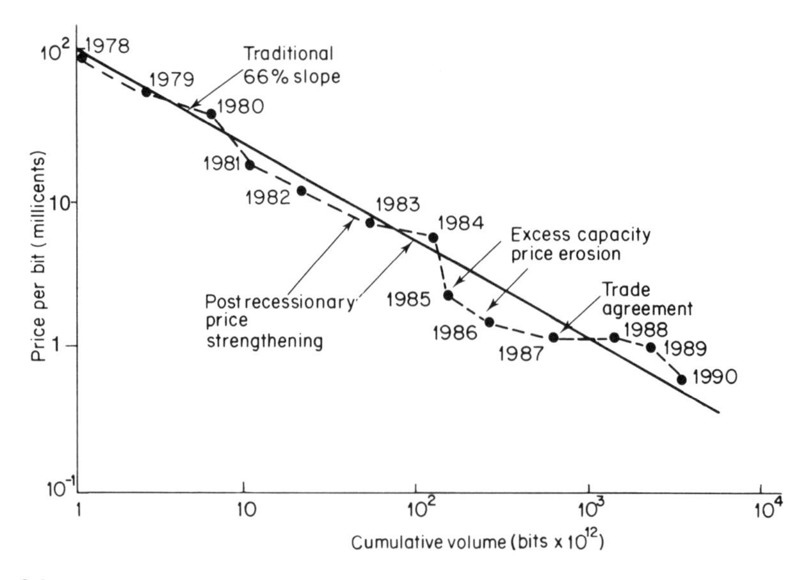

Figure 6.1
Price learning curve for MOS DRAMs. (Source ICE.)

As a result of this rapid decrease in price per bit over the last 20 years, DRAMs have replaced many other data storage media including core memory during the early 1970s, as the dominant mainframe data storage technology. They have replaced static RAMs in the 1980s in many small system and microprocessor applications. New operating modes and on-chip serial registers have allowed them to compete with serial memories such as FIFO buffers and charge coupled devices in graphics applications. They are now beginning to move into the large magnetic disk storage market as their cost per bit becomes competitive with high end magnetic media and as design innovations reduce their power consumption to the level of being able to run reliably on battery backup.

Advances in packaging technology have also improved the density of DRAMs in the system. Dual-in-line packages with 100 mil (1 mil = 25.4 μm) lead spacings are being replaced by smaller surface mount packages with 50 mil or smaller lead spacings or by zigzag-in-line packages which stand on edge side by side and reduce the area occupied on the PC board even more. Single-in-line modules carry this further by standing cards with multiple DRAMs on the PC mother board.

6.2 EARLY DYNAMIC RAM DEVELOPMENTS

During the early years of the DRAM era, from about 1970 to 1973 there was a period of intensive innovative design effort in memory cell structures with differing numbers of transistors in an attempt to both reduce the size of the memory array and increase its performance. During this period of time cells ranging from six

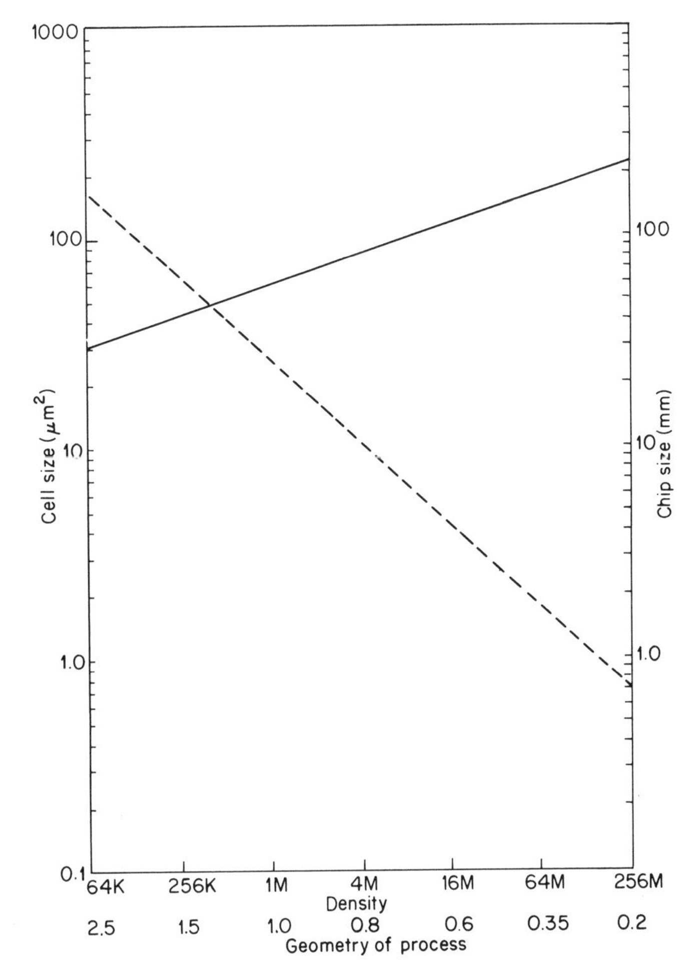

Figure 6.2
Chip size (——) and cell size (----) for six generations of DRAMs. (Source ISSCC Proceedings 1980–1990, with permission of IEEE.)

transistors down to one transistor were proposed. These early dynamic cells are discussed in Chapter 5.

After the one transistor cell became the standard in the 4k DRAMs about 1973, efforts at innovation changed to perfecting the operation of this cell. This effort consisted of development of sense amplifiers sufficiently sensitive to detect the small differential voltage stored on the cell capacitor and accessed through a capacitive bit line structure. Development also took place in the technology and structure of this cell and its accompanying bit lines.

The introduction of the double polysilicon transistor cell structure by Intel in 1976 ushered in the next era. The widespread adoption of this optimized cell led the DRAM industry to a period during the early 1980s of innovative process scaling which lasted to the 1Mb DRAM. A period of intensive innovation in DRAM cells followed throughout the late 1980s, during which three-dimensional cells were developed.

6.3 THE FIRST MODERN DRAM—THE ONE-TRANSISTOR CELL 4k

The history of the modern DRAM really begins with the advent of the one transistor memory cell. The first DRAM devices with this cell, early 4k dynamic RAMs, were shown in 1973. The one-transistor cell emerged the victor because of its small size in spite of having complex signal timing requirements and the need to be refreshed to retain its memory.

6.3.1 An example of an early 4k DRAM

These first of the modern DRAMs were made in the older P-channel MOS technology. A typical 4k DRAM, which was described by Philips [1] in 1973, regulated the timing needed for operation with an on-chip shift register. The 13.4 mm^2 chip was in single polysilicon, single aluminum technology and featured a 'small' cell size of 864 μm^2 with 10 μm minimum feature sizes. The array to chip area ratio was 26%. The sense amplifier was a simple structure with two static invertors and two switches distinguished mainly by a bootstrapped output stage which resulted in fast charging of the load capacitances. A 50 mV signal could be sensed.

It had non-multiplexed addressing, a chip select control pin and a bonding option choice of common or separate input and output in an 18 or 22 lead package, respectively.

The shift register generated the three regulated timing signals needed to operate the sense amplifier from a single external clock. A schematic of the sense amplifier with bootstrapped output stage, and the three signal timing diagram for refresh are shown in Figure 6.3(a) and 6.3(b). This is one of the earliest examples of clock pulse generation and regulation on a memory chip.

6.3.2 Address multiplexing and the 16 pin package

The first address–address multiplexed DRAM was introduced by Mostek [2], now a part of SGS-Thomson, also at the 4k bit density level. It was subsequently widely sourced by other companies in the industry.

With time multiplexed addressing, the 4k used six address pins per device to select one of 64 rows followed by one of 64 columns to select a particular bit instead of using twelve address pins to access one specific location. The pinout of this 4k DRAM is shown in Figure 6.4. The six multiplexed addresses were clocked into the row and column decoders by the row address strobe ($\overline{RAS}$) and the column address strobes ($\overline{CAS}$). Input and output data latches were also used. Data to be written into a selected storage cell were first stored in an on-chip latch. This simplified memory system design reduced the total package count.

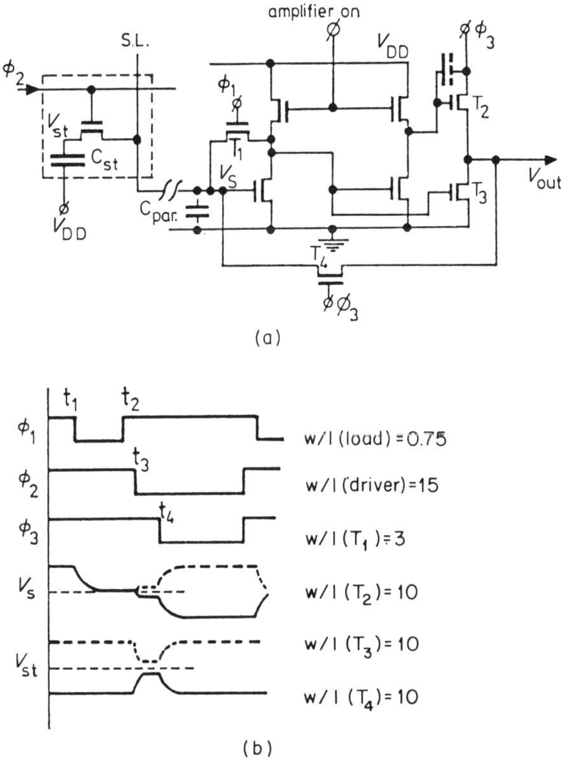

Figure 6.3
One-transistor PMOS cell on an early 4k DRAM showing sense amplifer and three signal sense and refresh operation. (From Boonstra *et al.* [1], Philips 1973, with permission of IEEE.)

Due to the multiplexed addressing, this part fit in a small 16 pin package rather than the 22 pin package of the non-multiplexed 4k DRAMs. This 16 pin package became the standard DRAM package through four generations of DRAMs from the 4k to the 256k shown also in Figure 6.4(b). The tradeoffs made at each generation to remain in this standard package can be seen in this figure. The 16k dropped the chip select pin to accommodate the additional address pin. Next the technology permitted dropping two of the power supplies enabling the 64k and 256k DRAMs to fit also.

The address multiplexing required more complexity in use but, by reducing the size of the package, significantly increased the potential board density of a DRAM system. This was the first widely alternately sourced standard DRAM.

6.4 THE 16k DRAM—BEGINNING OF THE ERA OF TECHNOLOGY INNOVATION

The 16k DRAM was the first in a series of four generations of DRAMs up to the 1Mb DRAM which were essentially similar in basic cell construction and design. During this time period the focus of the industry fell on scaling the production

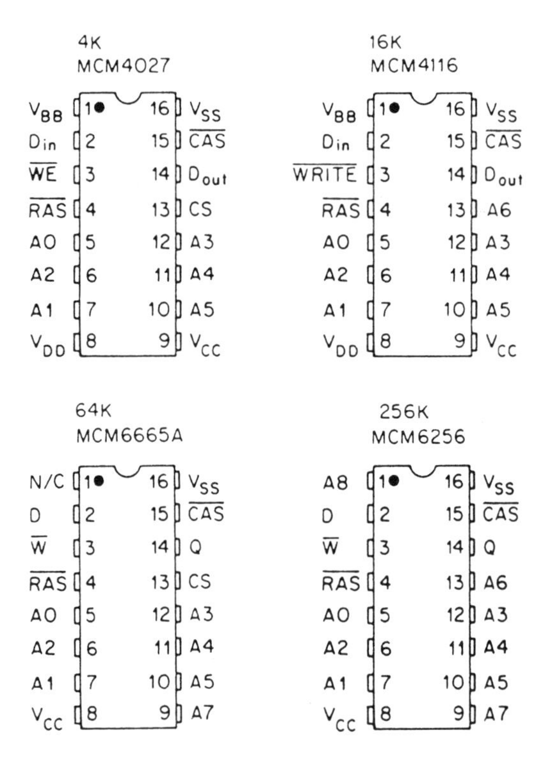

Figure 6.4
Four generations of dynamic RAMs in the standard 16 lead 300 mil width dual-in-line package. (Reproduced with permission of Motorola.)

technology to achieve the higher densities and smaller chip sizes needed to remain competitive.

In 1976 the first 16k DRAM was introduced by Intel and rapidly became the industry standard at this density level [3]. It followed the standard set by Mostek at the 4k density level of multiplexed addressing. The continued use of the 16 pin package permitted systems to be upgraded without changing the board size. This first 16k DRAM had 150 ns typical address access time and 350 ns cycle time. I_{DD} average was 40 mA typical at 500 ns cycle while I_{DD} standby was typically 1mA. It used three power supplies including +12 V for V_{DD}, +5 V for the output driver, and −5 V for the substrate bias.

6.4.1 Array efficiency

The chip size of the Intel chip was 25.8 mm^2 and cell size was 455 μm^2 giving a 34% array-to-chip area ratio or array efficiency.

Array efficiency is important since the user pays for the total chip area but wants only the data storage capacity. There is generally indifference to the internal operational overhead of the memory which makes up the periphery of the chip unless it adds value as it does in some high performance and applications specific memories. This means that the area of cell matrix to periphery needs to be as high as possible.

The array efficiency of the DRAMs increased from that of the 4k at 26% to the 16k at 34% to the later 64k and 256k DRAMs at 50%. 50% appeared to be the optimum level for NMOS DRAMs. The array efficiency improved again to about 60% only at the 1Mb DRAM generation when the industry converted from NMOS to CMOS due to the simpler structure of CMOS logic compared to NMOS logic.

Even with the change to CMOS in the late 1980s, it is, however, clear that array efficiency has not improved as much over time as various other DRAM benchmarks partially due to the inefficiency of the logic used in the peripheral control circuitry.

6.4.2 $\overline{\text{RAS}}$ and $\overline{\text{CAS}}$ clock access

The addresses on the standard 16k DRAM were multiplexed in two groups of seven using external Row Address Strobe ($\overline{\text{RAS}}$) and Column Address Strobe ($\overline{\text{CAS}}$) clocks. To initiate the cycle $\overline{\text{RAS}}$ sampled the address and started a timing chain. Column addresses are not required until later in the cycle. As long as the user provided the column addresses within the t_{RCD} window, there was no access time penalty resulting from the multiplexing.

The read, write, and read–modify–write timing for this part are shown in Figure 6.5. To read or to write first the row or word addresses were strobed in with a downward transition of the $\overline{\text{RAS}}$ clock to select one word-line. The column or bit addresses were then strobed in with the $\overline{\text{CAS}}$ clock to select one bit from this word-line. This external timing sequence became the standard basic addressing sequence for all subsequent generations of standard dynamic RAMs and remains essentially unchanged through today.

6.4.3 The double poly one-transistor cell structure

The cell used on the Intel 16k DRAM was in 5 μm (13 μm metal pitch) N-channel technology rather the P-channel used on the older 4k devices. It also had the double polysilicon one transistor cell which became the standard cell structure for DRAMs up to the 1Mb density. See Figure 6.6 for a layout and cross-section of this cell which became known as the 'planar' DRAM cell. The three generations of production technology scaling that followed resulted from the ability to scale this cell while retaining sufficient charge to be detected.

In the double polysilicon DRAM cell, an N^+ diffused bit line is used with a second-level polysilicon access gate. The cell is charge coupled. The signal consists of electrons stored in the inversion layer under the first level of polysilicon. Contacts are shared between two cells so there is $\frac{1}{2}$ contact per cell, a structure commonly used to reduce array size. The technology of this cell was derived from the double

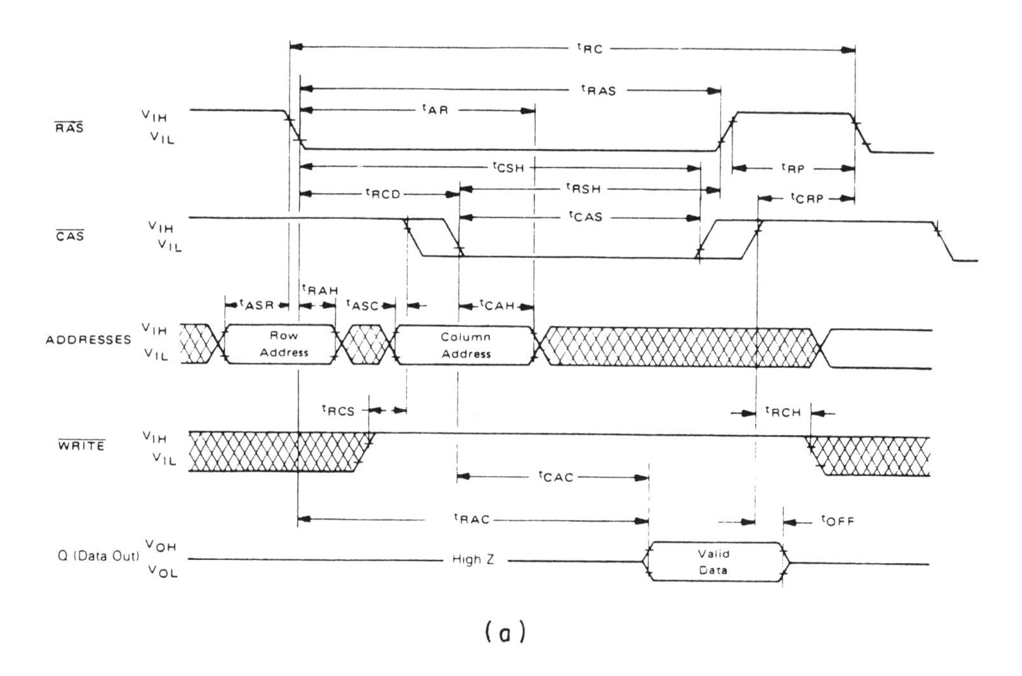

(a)

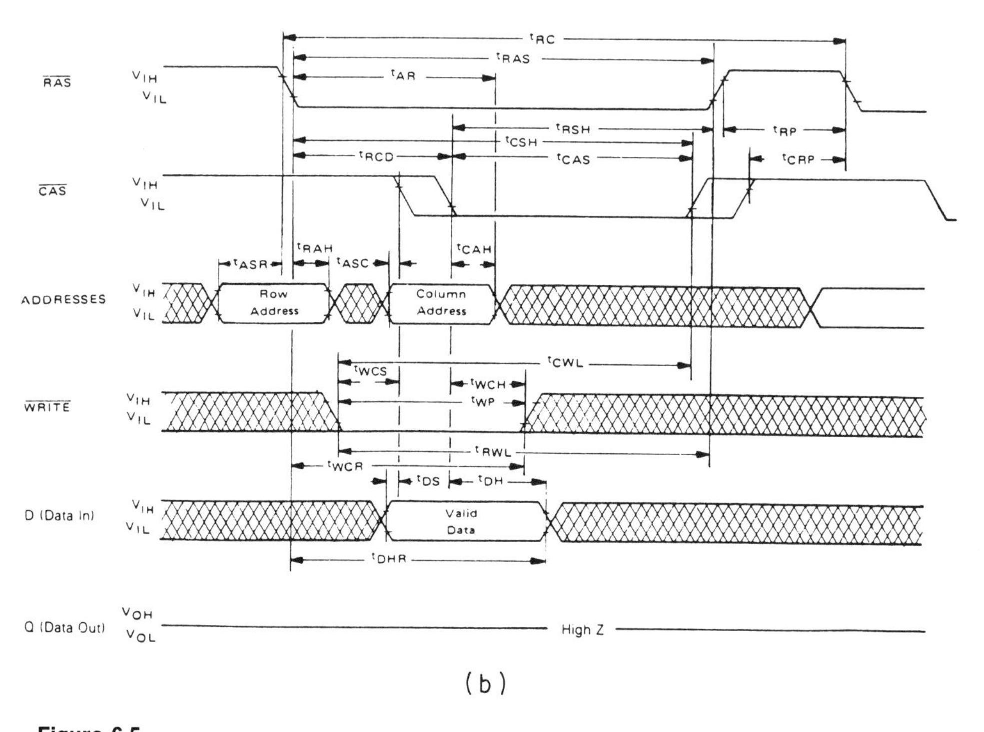

(b)

Figure 6.5
16k address multiplexed DRAM basic timing: (a) read cycle timing, (b) write cycle timing, (c) read–modify–write cycle timing. (Reproduced with permission of Motorola.)

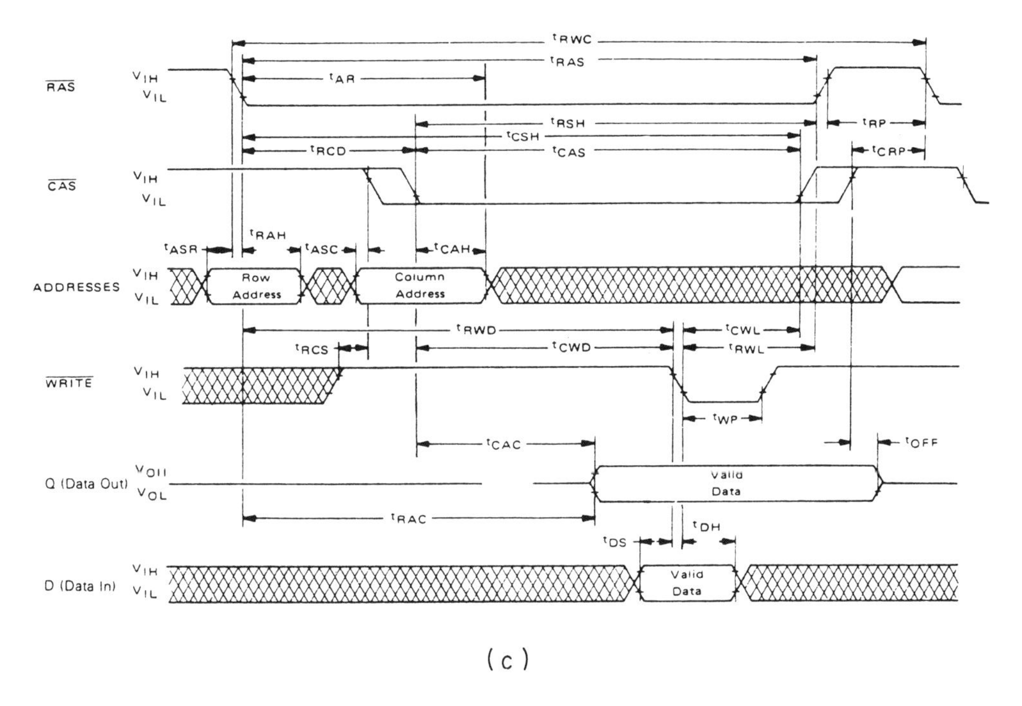

(c)

Figure 6.5 (*continued*)

polysilicon charge coupled device (CCD) technology of the time. The storage to bit line capacitance ratio C_s/C_b was between 3 and 6.

6.4.4 16k DRAM sense amplifier

The sense amplifier in the Intel 16k DRAM was the simple cross-coupled latch type whose operation was described in Chapter 5. The sense amplifier circuit included dummy cell generation of reference potential, equalized precharge of the potential on the bit lines, clocked pull down of the sense nodes and delayed clocking of the load devices. A worst case differential signal of about 100 mV could be read. The bit-line capacitance to storage capacitance ratio (C_b/C_s) was between 3 and 6. Since the signal on the storage capacitor decayed to 1 V on the average by the time the cell was refreshed, this resulted in a signal on the bit-lines of about 200 mV, well within the 100 mV sensitivity of the sense amplifier.

6.4.5 16k bit-line architecture and refresh cycles

The architectural floor plan of the 16k consisted of two lines of 128 sense amplifiers with bit-lines consisting of 32 cells on either side of the sense amplifiers as shown in

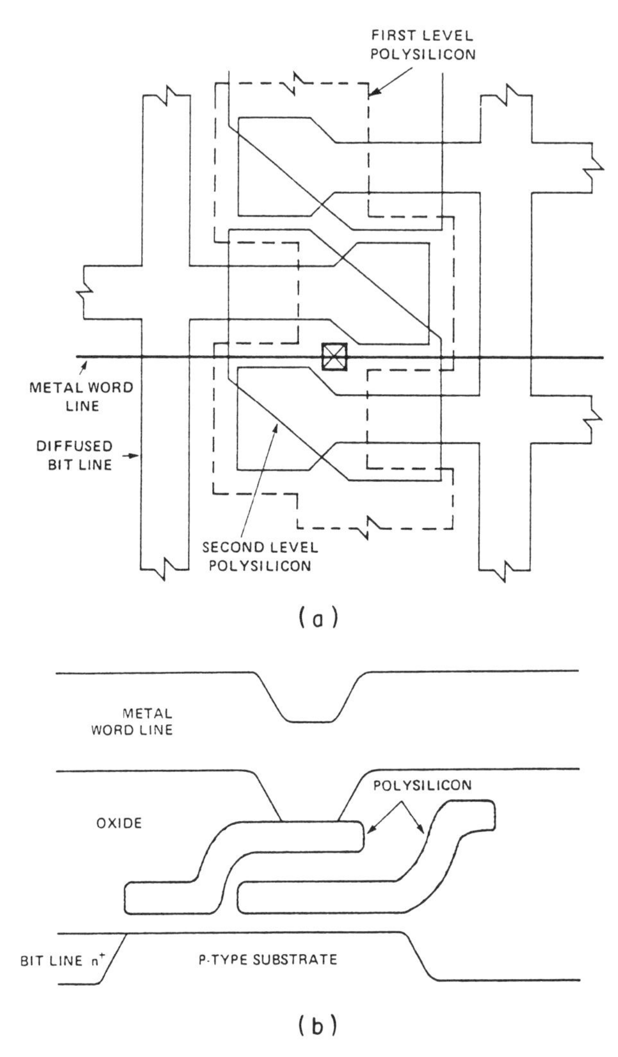

Figure 6.6
Layout (a) and storage cell (b) of double layer polysilicon one-transistor planar DRAM cell. (From Ahlquist *et al.* [3], Intel 1976, with permission of IEEE.)

Figure 6.7(a). The 16k bits of memory is hence divided into two blocks of 64 rows by 128 columns. Since all the bits on the word-line are refreshed whenever a row is selected, this permitted a 64 cycle refresh sequence if both sets of sense amplifiers are activated at once. At the 350 ns cycle time, the entire memory could be refreshed in 22.4 microseconds or 1.1% of the 2 ms refresh period.

A possible alternative architecture would have been a 128 × 128 array such as that shown in Figure 6.7(b) with 128 bits per sense amplifier and a single set of 128 sense amplifiers, that is, the number of cells on the bit-line is doubled and the number of sense amplifiers is cut in half. This means that the bit-line capacitance doubled, so

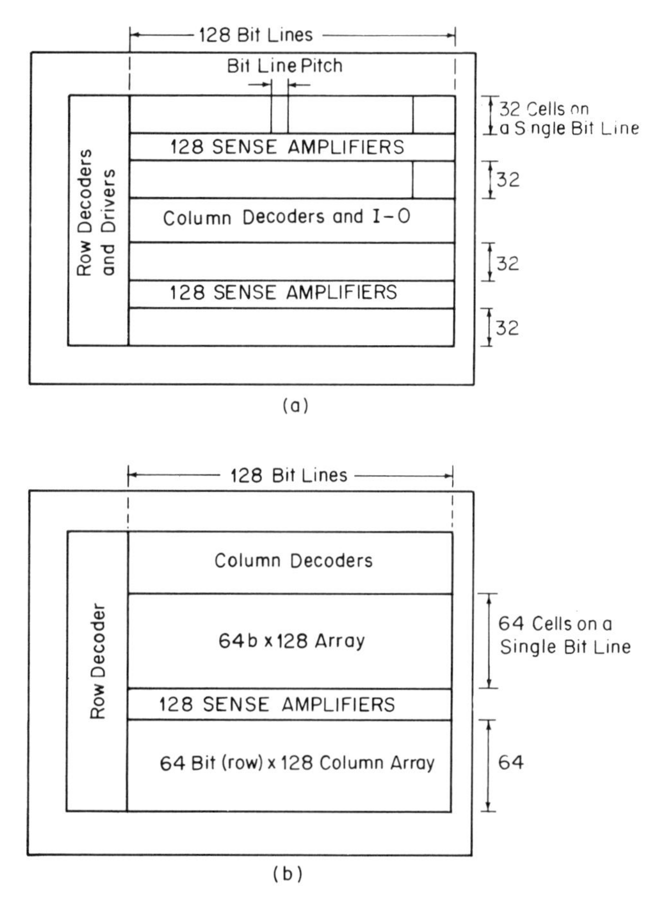

Figure 6.7
16k DRAM floor plans showing design tradeoffs between numbers of sense amplifiers and bit-line architecture for (a) Intel 16k DRAM and (b) alternative 16k DRAM architecture with twice the number of cells on a bit-line and half the number of sense amplifiers.

that the storage cell capacitance would need to be increased to maintain a constant C_b/C_s ratio. The estimate was that the cell size would have to be increased by about 35%.

The increase in length of the bit-lines would also mean increased resistance for the signal to propagate through so that the propagation delay time would be increased resulting in increased access time and write times. A 64 cycle refresh in 2 ms would only refresh half the array at the 350 ns cycle time since each of 128 rows must be accessed, so a time of 4 ms was used to refresh all the bits in the memory. This option was, therefore, not upward compatible. The benefit of such a choice would have been that the number of sense amplifiers could be reduced. With only half as many sense amplifiers on at any one time the dc power consumption would be halved.

This is the first example of the refresh cycle–power–speed–upgradability trade-off which was subsequently revisited for each new DRAM generation.

6.4.6 $\overline{\text{RAS}}$ only refresh

The 16k DRAM retained the read operation refresh common to all single transistor cell memories. That is, a read operation always refreshes the data in the cell being read. It also had a $\overline{\text{RAS}}$ only refresh which became standard on address multiplexed DRAMs. This operation permitted refresh using only the $\overline{\text{RAS}}$ clock without activating the $\overline{\text{CAS}}$ clock. As a result, any sequence in which all row addresses were selected was sufficient to refresh the entire memory.

6.4.7 TTL compatible MOS level shifters

While the 4k DRAMs had some of the input levels TTL compatible, by the 16k DRAMs TTL compatibility on all inputs and outputs was standard. The simple invertor initially used as a level shifter at the 4k DRAM level was replaced by a lower power and faster alternative at the 16k level.

The Intel version of a fast TTL compatible MOS input buffer for the 16k DRAM is shown in Figure 6.8 along with a simulation of the circuit operation. A large input transistor Q_1 is in series with a much smaller transistor Q_2 which in this case is about 10% of the size of Q_1 and is biased on by a 5 V reference voltage. When $\overline{\text{RAS}}$ is high, Q_2 has little effect. When $\overline{\text{RAS}}$ goes low, Q_2 cuts off when the output rises to V_R minus a threshold. At this point the output rises more rapidly to V_{DD} than would have been the case without Q_2 in the circuit.

6.5 THE 64k DRAM

An era of production technology development followed. During this time the focus of the DRAM industry was on scaling of the basic cell developed in the 16k to smaller geometries and developing the operational control needed for these smaller cells. There was also work on sense amplifier architectures sensitive enough to support the smaller amount of signal charge stored in the capacitor. Development from the 16k to the 1Mb occurred during this phase over an eight year time period from 1976 to 1984.

The first factory production lines produced 4k DRAMs in 1974 and 16k DRAMs in 1978 within a year after the parts were announced. Production of succeeding generations lagged the initial announcement by an increasing amount as shown in Figure 6.9 due to the increasing sophistication of the production technology. The 1Mb lagged in introduction even more due to initial uncertainty whether the 1Mb could be obtained by scaling the existing cell or whether a new cell would be required. It did ultimately follow the scaling path of the previous generations. A second period of basic cell innovation then followed for the 4Mb and greater densities.

The first 64k DRAMs were introduced in 1977 and 1978 still in the classical 16 pin package. The technology had advanced to the point that the 12 V power supply could be reduced to 5 V and V_{DD} eliminated. An on-chip substrate bias generator

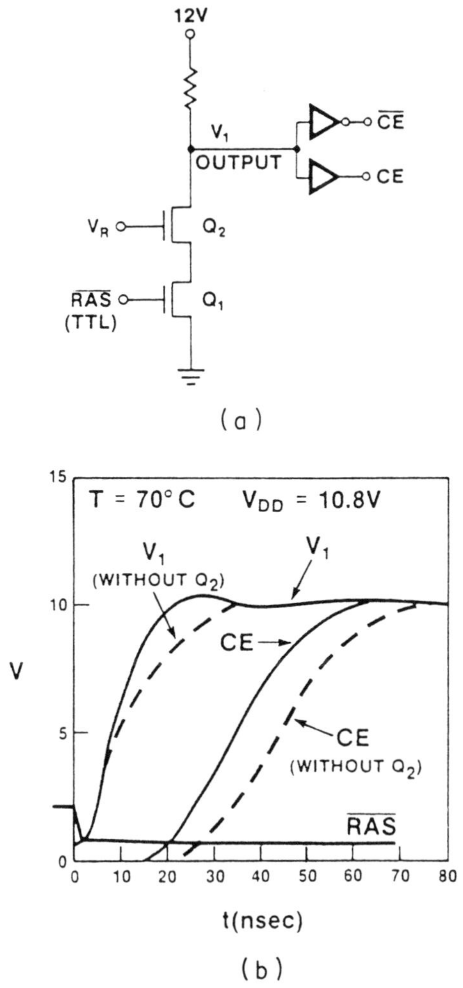

Figure 6.8
Early static TTL-MOS level inverter used on $\overline{RAS}$ input. (a) Circuit, (b) simulation of circuit operation. $T = 70°C$, $V_{DD} = 10.8V$. (From Ahlquist *et al.* [3], Intel 1976, with permission of IEEE.)

permitted the -5 V substrate bias to be generated on chip. The two pins previously devoted to these power supplies were then free for the 64k and 256k address pins.

The 64k DRAM generation saw further refinement of the one transistor cell and its operation. Smaller and more sensitive sense amplifiers were investigated including dynamic load and capacitive coupled sense amplifiers. The first redundancy was developed as yields diminished due to increasing chip size and investigations into alpha particle related soft errors were begun. Wide output parts also appeared at the 64k level. The sensitivity of the smaller cells to noise led to voltage boosted word-lines, precharged bit-lines and investigations of various voltage levels for the

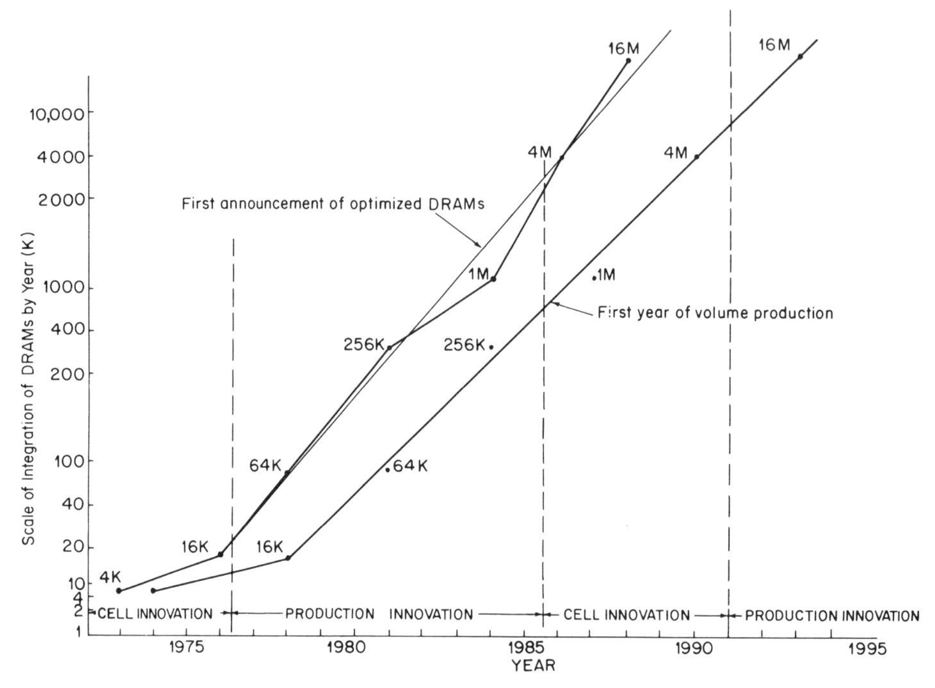

Figure 6.9
First announcement of each generation of optimized DRAMs compared to the first year of volume production. Periods of emphasis on cell innovation and on-production innovation are also indicated.

storage capacitor cell plate. Various bit line architectures were investigated including open and folded bit lines to reduce the effects of common mode noise.

6.5.1 Architecture and refresh on the 64k

Two options in architecture at the 64k level paralleled the choices made on the 16k DRAM. The one [7] used an architecture with two blocks of 128 rows by 256 columns. Within a block there were 64 cells on a bit-line on each side of a row of 256 sense amplifiers. Both blocks of 128 rows could be refreshed simultaneously in 2 ms thereby maintaining compatibility with the 16k DRAM 2 ms refresh time. The other [6] used an architecture with one block of 256 rows by 256 columns having 128 cells on a bit-line on each side of 256 sense amplifiers. Since the cycle time was the same on both parts, 4 ms was used to refresh the 256 rows. The first architecture had a more complex array structure, but maintained the upward compatibility in refresh time with the 16k DRAM. The second part had half the number of sense amplifiers on and so lower power.

From the point of view of the systems users of these parts the upward compatibility in the system was desirable and encouraged the use of the more complex array structure required. The 128 cycle refresh also had the systems advantage of less time spent in refresh activities than with a 256 cycle refresh so that a larger percentage of time was available for random access operations in a 2 ms refresh period.

NTT [4] described an alternative architecture which divided the chip into four blocks of 16k bits which were in turn subdivided into four blocks of 4k bits each. This division shortened the bit-lines thereby reducing the bit-line capacitance. This had the effect of increasing the speed at the cost, as before, of doubling the number of sense amplifiers required and so increasing the power dissipation.

6.5.2 64k DRAM sense amplifiers and bit-line structures

Sense amplifiers for the 64k needed to be more sensitive to detect the smaller signal from the larger bit-line capacitance. They also needed to be lower power to compensate for the increased number of sense amplifiers, and finally smaller to fit in the reduced bit-line pitch of the tighter geometry technology. One of the disadvantages of maintaining the 128 cycle refresh was having 512 sense amplifiers rather than 256 sense amplifiers in the circuit. More sense amplifiers both take extra silicon area and consume more power.

In addition the increased number of cells on the bit lines increased the capacitive bit-line load, leading to investigations of no-load and capacitively coupled sense amplifiers.

NEC for the 64k [7] described a typical no-load cross-coupled latched sense amplifier with two balancing precharge transistors connected in parallel with two sensing transistors as shown in Figure 6.10(a) with its simulated waveforms shown in Figure 6.10(b). This resulted in lower power consumption and better precharge balancing at the expense of greatly increased complexity due to an increase in the number of clock signals. The operation is described below.

(1) Precharge clock ϕ_{po}, which is derived from $\overline{RAS}$ going high precharges the bit-lines to 10 V.

(2) Clock ϕ_t conducts the high level to the sense nodes.

(3) Clock ϕ_{p1} charges the dummy cell to ground.

(4) The $\overline{RAS}$ clock transition to low resets clocks ϕ_{po}, ϕ_{p1}, ϕ_{p2}.

(5) After decoding, word and dummy word-line clocks ϕ_w and ϕ_{dw} connect the selected storage cell to one half bit-line and a dummy cell to the other half.

(6) The stored voltage differential appears on the bit-line pair.

(7) Clock ϕ_{s1} for slow discharge and clock ϕ_{s2} for fast current sink rise and amplify the sense node signal.

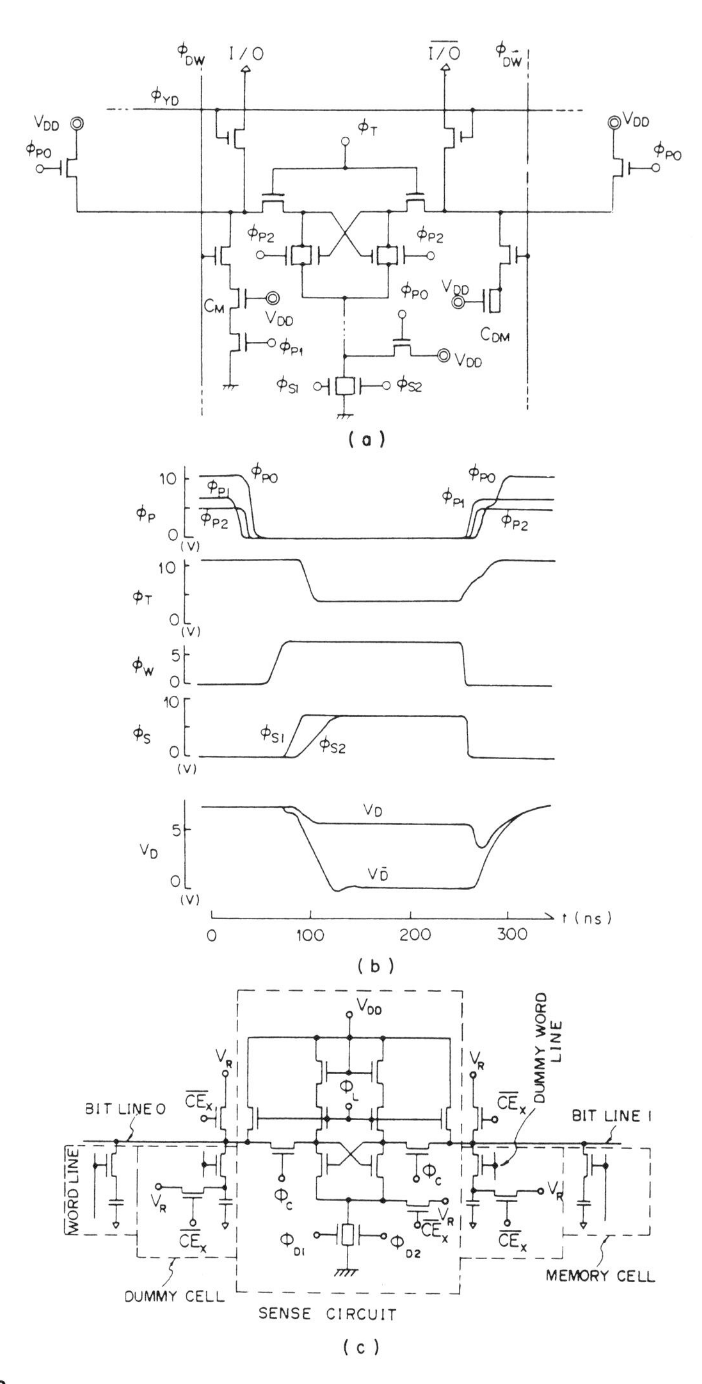

Figure 6.10
64k DRAM dynamic direct load sense amplifiers: (a) sense amplifier circuit, (b) timing signals for the sense operation of (a), (c) sense amplifier showing split load transistors designed to decrease capacitive load on the sense amplifer during clocking. (a,b: From Wada *et al.* [6], NEC 1978, with permission of IEEE; c: From Arai and Ieda [4], NTT 1978, with permission of IEEE.)

(8) Clock ϕ_t goes to the intermediate level of 5 V to prevent the discharge of the higher bit-line voltage level through the transfer transistor. It also isolates the bit-line capacitance from the internal sense nodes of the amplifier permitting a much faster amplification.

(9) Clock ϕ_{yd} connects the complementary bit-line pair with the complementary I/O bus pair.

(10) ϕ_{s1} and ϕ_{s2} fall to reset clocks ϕ_w, ϕ_{dw}, and ϕ_{yd}. Clocks ϕ_{p1}, ϕ_{p2}, ϕ_t and ϕ_{po} rise sequentially. ϕ_{po} rises to the full 10 V precharge level to restore a full high storage level of 7 V to the refreshed cell.

The bit-line capacitance to storage cell capacitance ratio C_b/C_s in this case is between 8 and 10 and the sensitivity of this sense amplifier is adequate to sense a worst case differential signal input of 80 mV in this case.

A drawback of the direct load dynamic sense amplifier is the complexity and speed penalty resulting from the number of clocks which must be used.

NTT [4] developed a dynamic sense amplifier for the 64k DRAM with dummy cell structure distinguished by a separation of the load transistor in the flipflop into two transistors to decrease the capacitive load of these transistors when clocked. It is shown in Figure 6.10(c).

The NTT sense amplifier circuitry could sense a differential signal as small as 30 mV. This circuitry used an intermediate precharge of the bit-line at about 2 V which was utilized as a reference level. After a memory cell was selected and the memory cell signal was transmitted to the sense amplifier, the switching transistors were turned off. The signal was then amplified dynamically in the flipflop.

This type of sense circuit is very sensitive because of the disconnection of the bit-line capacitor during the sensing operation. It dissipates low power because of the dynamic amplification.

There were also investigations of smaller less complex sense amplifiers since the tighter geometries of the 64k made it harder to fit the sense amplifiers in the pitch of the bit-lines.

Investigations into the bit-line pitch of various cell structures with the traditional N^+ diffused bit-line, compared to polysilicon and metal bit-lines, were done by NEC for their 16k DRAM compatible 64k chip with results as shown in Figure 6.11. The small bit-line pitch of the metal bit line was actually used, but required a very simple sense amplifier to be small enough to fit in the space between the bit-lines.

The trade-off here is between sufficient sense amplifier sensitivity and the small chip size permitted with the smaller bit-line pitch. The differential signal between the pair of bit-lines was on the average about 100 mV with C_b/C_s equal to 10.

A capacitive coupled sense amplifier which occupied a small amount of silicon area was proposed by NTT [9] in 1978. This sense amplifier, which required only one clock, is shown with its waveforms in Figure 6.12(a). Each bit line is connected to the clock ϕ_{sad} through the capacitor. When the signal from a cell appears as the potential difference between two bit-lines, the amplifier is driven by the clock. During the first time period the clock boosts the bit-line voltage and amplifies the signal slightly. In

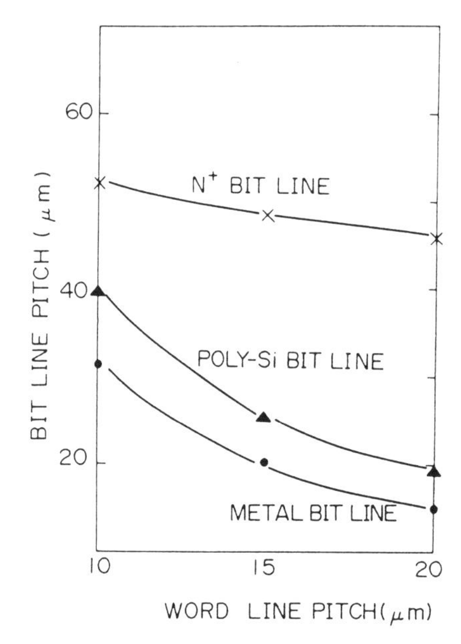

Figure 6.11
Relation between address and bit line pitch for 64k DRAM configurations (× N^+ bit line, ▲ polysilicon bit line, ● metal bit line). (From Wada *et al.* [6], NEC 1978, with permission of IEEE.)

the second period the clock reaches the threshold voltage of Q_3. In the third period the amplifier operates as a conventional dynamic flipflop and the high signal becomes adequate for refreshing. The refresh level depends on the ratio of coupling capacitance to effective bit-line capacitance.

The advantage of the capacitive coupled sense amplifier is not only its speed and simplicity of operation, but its lack of complexity of design which allows it to fit in the small bit-line pitch in the array. Its drawback is its lower sensitivity and lower noise margins.

An improved capacitive coupled sense amplifier for the 256k DRAM was announced by NTT in 1980 [18] and is shown in Figure 6.12(b) with its operational wave forms. The new sense amplifier took the best features of the previous generation dynamic amplifiers which generally fell into the two categories described above: those that were very sensitive, but also quite complex with intricate clocking systems, and those that were very fast, but not as sensitive such as the one clock capacitive coupled sense amplifier.

This improved sense amplifier with three clocks was a variation of the capacitive coupled amplifier. Clock ϕ_{D1} is used for both detection and for refreshing the cell. ϕ_{D2} is the accelerating clock, and ϕ_B is the refresh control clock. Sensing is completed with two clocks so high performance is achieved and the full refresh level is restored to the cell upon the ϕ_B clock. This sense amplifier has the additional benefit that its relative simplicity permitted it to fit between the smaller bit line pitches on the 256k DRAM. It could sense a signal of ± 50 mV with a 35 ns sensing period.

6.5.3 5 V only power supplies appear

Various early experimental 64k DRAMs were discussed with requirements for multiple external power supplies such as used in the 16k DRAM. One early production 64k from Fujitsu freed up the address pin needed to remain in the 16 pin package by using two +7 V and −2 V power supplies.

There were other early attempts to free the address pin needed by eliminating the +5 V V_{CC} pin and retaining the +12 V V_{DD}. An example of circuitry designed for this purpose was an early 64k DRAM with +12 V V_{DD} from Toshiba [13] which used a feedback circuit to restrict the high level voltage applied to the gate of the load transistor in the data output invertor thereby eliminating the necessity for the external 5 V V_{CC} pin.

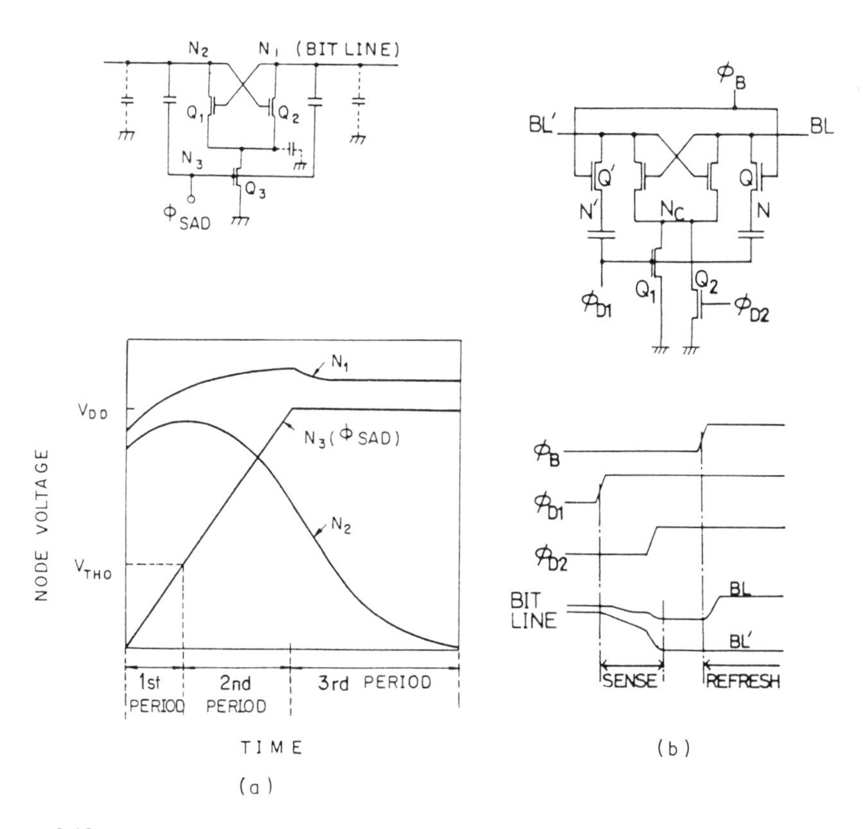

Figure 6.12
Capacitive coupled dynamic sense amplifier.(a) Basic capacitive coupled sense amplifier developed for the 64k DRAM with fundamental waveforms also shown. (b) Improved capacitive coupled sense amplifier with improved sensitivity due to addition of switching transistors to decouple the bit-line load developed for the 256k DRAM. (a: From Kondo *et al.* [9], NTT 1978, with permission of IEEE; b: From Mano *et al.* [18], NTT 1980, with permission of IEEE.)

During 1979 and 1980 several 5 V 64k DRAMs were announced by companies including Texas Instruments, Motorola, Fairchild and Hitachi.

The introduction of the 5 V only 64k freed two pins in the standard 16 pin package. Referring to the pinouts of 4k to 64k DRAMs shown in Figure 6.5, one sees the reduction in number of power supplies from four at the 4k level to two at the 64k DRAM level.

A number of methods were used to generate the higher voltage on chip which had previously been supplied externally on the + 12 V V_{DD} pin. Motorola used bootstrapping to achieve the higher voltage. NEC [6] used an on-chip shift register clock generator, similar to that described earlier for the Philips 4k DRAM, to generate and regulate the various timing signals required by the one transistor cell read and refresh sequence. Enhancement–depletion mode technology was used for the flipflops in sense amplifiers and in the peripheral circuitry.

The combination of lower supply voltages and the lower threshold voltages of the NMOS transistors on the scaled 64k DRAM caused mismatch problems with the TTL I/O circuitry. The transistor threshold voltage of around 0.8 V was very close to the 0.4 to 0.8 V TTL input low level which allowed very little margin for noise. For this reason, NTT [4] developed a new level detecting input circuit in which a logic threshold voltage could be established independently of the MOS transistor threshold voltage.

A conventional invertor circuit and the new level detecting circuit, which was a Schmitt trigger, are compared in Figure 6.13. In the conventional dynamic invertor, shown in Figure 6.13(a), an output signal began to drop when an input signal reached the threshold voltage V_{th}. In the level detecting circuit, shown in Figure 6.13(b), the output signal was not reversed even if an input signal exceeded the threshold voltage. The transistor Q_2 turned on when the input signal V_{in} exceeded the threshold voltage plus the voltage on node A. The voltage on node A was determined by the transconductance ratio of Q_4 to Q_3. Since the transconductance ratio could be changed, different logic threshold voltages could be obtained independently of the MOST threshold voltage.

6.5.4 Elimination of external back bias voltage

Elimination of the external −5 V power supply was accomplished by eliminating the external back bias voltage and, either grounding the substrate, or replacing it with an on-chip back bias generator.

The option of grounding the substrate was simpler and did not add to the standby power dissipation as a back bias generator would have. It had the advantage of reduced leakages associated with the decay of a stored 'one'. This gave the potential for longer refresh intervals. It also had the advantage of a solid reference voltage to reduce substrate noise. A grounded substrate, however, suffered from injection of electrons into the substrate which can act to reduce the refresh time unless compensated for by more complex design techniques. It also could lead to local forward biasing of junctions unless compensated for by careful design.

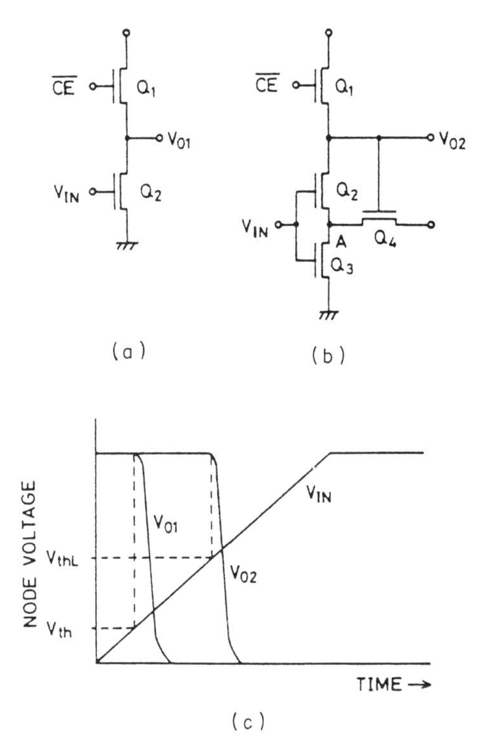

Figure 6.13
Dynamic output invertor circuits. (a) Conventional invertor used in 16k DRAM. (b) Level detecting invertor for improving noise margins in 64k DRAMs. (c) Comparative outputs of invertors (a) and (b). (From Arai and Ieda [4], NTT 1978, with permission of IEEE.)

The use of a back bias generator achieved the goal of a single 5 V power supply part while at the same time preserving the advantages of substrate bias in NMOS DRAM technology.

The most important advantage of the on-chip substrate bias generator was the reduced probability of localized forward biasing of junctions. Such forward biasing injects electrons into the substrate leading to dynamic circuit problems and reduced refresh times due to the collection of electrons beneath the capacitor plate for stored 'ones'. Speed and power characteristics also improve both because transistors operate in a flatter part of the body-effect curve and so lose less drive capability as their sources rise and because the lower junction capacitance reduces the load. A final advantage is that the C_b/C_s ratio improves because junction capacitance contributions more to the bit line capacitance than to the total storage capacitance so that the differential signal to the sense amplifiers will increase.

The Motorola 64k DRAM [14] was one of the early DRAMs with back bias generation on chip. The generator circuit developed for this 64k DRAM avoided several problems of previous V_{BB} on-chip generation by pumping to a known regulated voltage. This avoided substrate drift with variations in substrate current which result from changes in cycle time. This drift changes device characteristics and degrades storage level. This part used two level reference schemes to avoid changes in substrate bias voltage that otherwise could result from the shift in V_{BB} between precharged and active memory states when memory duty cycle changes.

6.5.5 64k pinout and package issues

The elimination of the 12 V V_{DD} and the external back bias voltage freed up two pins, only one of which was needed for addressing at the 64k level. This led to various attempts to use the other pin to advantage. It could be used for special test purposes which was mainly an advantage for the manufacturer. It could be used for a new control mode such as a second chip select or a refresh control but this would mean a risk for the manufacturer of having a non-standard part. Such a risk is only justified if the new function is clearly an advantage in some known system application. One supplier, Motorola, did in fact offer the part with a refresh control function on this pin. This function is described in more detail in Chapter 9 in the section on pseudostatic DRAMs.

The second and more common option for the spare pin was to leave it 'no connect' so that the system manufacturer could design memory boards with the address line for the 256k DRAM already connected to that pin. This allowed easy drop-in upgrading to the next generation of DRAM density without board redesign. It turned out that this upgradability option was the more popular in larger systems. Whereas the pin one refresh control was used to a small extent in very small systems which benefited from the reduction in refresh control overhead.

The 128 cycle 2 ms as opposed to 256 cycle 4 ms refresh issue also needed to be resolved. While the 16k DRAM eventually standardized with 128 refresh cycles every 2 ms, the 64k DRAM vendors divided on this issue. A 128 cycle refresh permitted upward compatibility with the 16k DRAM, whereas a 256 cycle refresh was compatible with the next generation 256k DRAM where the trade-offs in sense amplifier design and power dissipation led to a 256 cycle refresh being the standard.

Clearly those vendors who left pin one unconnected for upward compatibility with the 256k DRAM were motivated to use the 256 cycle refresh. It meant, however, that they had to convince users to redesign boards at the 64k level when the users did not have to since 128 cycle refresh 64k DRAMs were also commonly available.

This problem of when to upgrade boards has reoccurred many times in the history of memories whenever a clear change in the next generation part is expected. It creates risk for the users since they do not know for sure which selection of memory will be generally available, and it creates a risk for the vendor who is not sure that the part being designed will be the standard part.

The need for a lower cost package to meet the expectation of continuing cost reduction in DRAMs led first to the ceramic dual-in-line package and then to the plastic package. The DRAM industry began to convert to the plastic dual-in-line package at the 64k generation. Hermiticity of this new package was a major barrier to user acceptance initially. This will be described further in Chapter 10 on memory packaging.

6.5.6 Noise related problems—alpha radiation

The soft error problem due to alpha radiation emitted from the packaging material, although discovered on 16k RAMs, had a more serious, negative effect on 64k RAM designs due to the smaller charge capacity of the scaled capacitor.

Alpha particles are emitted from trace contaminants in the integrated circuit packaging materials and can generate millions of electron–hole pairs in the silicon substrate. This discharges the storage cell capacitors, the bit-lines, particularly if they are diffused, and the sensing nodes in the sense amplifiers. Errors induced by alpha particles are related almost exclusively to the amount of stored charge. Pattern related errors also occur but are normally due to noise caused by leakage current between adjacent bits or by capacitive coupling between adjacent bit-lines or peripheral circuit signals.

Initially it is was believed that alpha particles also resulted from cosmic radiation, but experimental tests in caves eliminated this as a source of the radiation.

This problem was initially reduced by various techniques such as covering the chip surface with polyamide coating to impede the passage of the alpha particles. It was also addressed directly by efforts to increase cell capacitance. Cell capacitance at about the 40 to 50 fF level was found acceptable to avoid alpha radiation related soft error problems. This level of capacitance remained fairly constant as a target for succeeding generations of DRAMs in spite of scaling of the area occupied on the silicon surface by the DRAM capacitor.

On the 256k DRAMs metal bit-lines tended to be used to reduce the alpha hit problem associated with diffused bit ones. The folded layout, which connects adjacent bit-lines to the differential sense amplifier's inputs, was also used to ensure that noise coupled locally to the bit-lines formed a common mode input to the amplifier and was rejected.

A further discussion of the alpha radiation problem is given in Chapter 14.

6.5.7 Other noise considerations in 5 V only DRAMs

The early DRAMs, which had external power supply voltages ranging up to 17 V or more, had relatively loose cell geometries plus thicker oxides which were able to withstand these high voltages. In turn these high voltages permitted sensing of the data signal in spite of high levels of noise on the chip. Even the 16k DRAM had a 12 V V_{DD} signal.

With the advent of the low voltage 5 V only 64k DRAM with its small differential signals the various noise sources became much more of a problem. An excellent analytical study of small signal noise sources in dynamic RAMs which explored this problem was done by Hitachi [16]. This study found that common sources of noise in NMOS dynamic RAMs included the following.

(1) Imbalance of the dynamic sense amplifiers caused both by fabrication variance and by the $C_s/2$ sized dummy cells.

(2) Array noise caused by capacitive effects between the bit-line and word-lines during row access when many bit lines are swung between V_{CC} and ground simultaneously.

(3) Noise from peripheral circuitry such as clock pulse generators.

(4) Noise from bumps on incoming power supply lines, which is called V_{DD}-bump.

6.5.8 Grounded word-lines

Word-line noise was a common cause of pattern sensitivity bit failures in DRAMs. One solution commonly used was a grounded word-line technique in which an additional dc low conductance path to ground was applied to all word-lines. This had the disadvantage, however, that the grounding was a load on the selected word line, and thus could not be applied with capacitively bootstrapped word-line drivers. Various techniques were used to minimize this effect. Active pull down transistors were provided for the non-selected word-lines as on the Fairchild 64k DRAM [17]. Motorola [20] used grounded word-lines except during sensing with much the same effect. These techniques gave the device a high immunity to V_{CC} changes and minimal noise in the memory array.

6.5.9 Bit-line architectures to reduce noise

The margin between the signal which appears on the bit-lines and the sensitivity of the sense amplifier must be maintained. An approach was developed at the 64k DRAM level of folding the bit-lines to improve the balance of the lines and to reduce the transmission of substrate and junction noise which was coupled onto the bit-lines.

Figure 6.14(a) shows both conventional open bit-lines and the new folded bit-line scheme as used by Motorola [20]. With the folded bit-lines the column decoders are at one end of the array and sense amplifiers on the other. A single pair of bit-lines is folded back on itself between the two so that any fluctuations on the column decoders will be common mode noise on the lines and will hence not be transferred to the differential signal on the bit-lines [146].

Open bit-lines use a row of sense amplifiers in the middle with memory arrays on each side. The noise injected in this layout will not be balanced and an error could be introduced in the cell due to wrong sensing. It should be noted that the bit-lines

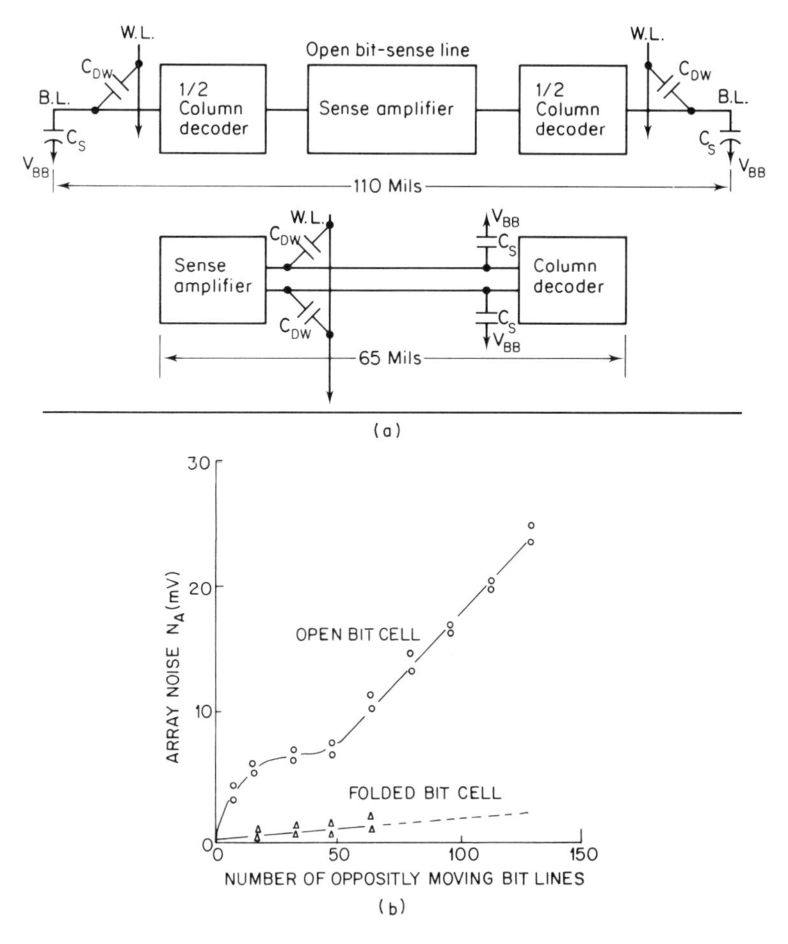

Figure 6.14
(a) Schematic comparison of open and folded bit sense architectures. (b) Measured array noise in open (○) and folded (△) bit-lines. (a: From Prince and Due-Gundersen [20] 1983; b: Masuda *et al.* [16], Hitachi 1980, with permission of IEEE.)

go through the column decoder area where a lot of coupling will normally take place. The advantage of the open-bit-line approach is the smaller area occupied by the bit-lines.

Figure 6.14(b) shows measured results of array noise plotted against number of oppositely moving bit-lines for an open bit and a folded bit-line configuration. In addition, the chip area with folded bit-line and 128 cycle refresh was larger than the chip area for open bit-lines with 256k cycle refresh [16].

A 64k by National (Fairchild) [17] combined the two approaches using an architecture with open bit-lines during a read operation and folded or 'twisted' bit-lines during a write operation. In this scheme an ultrasensitive sense amplifier was placed in the center of two pairs of cross connected bit-lines, as shown in Figure

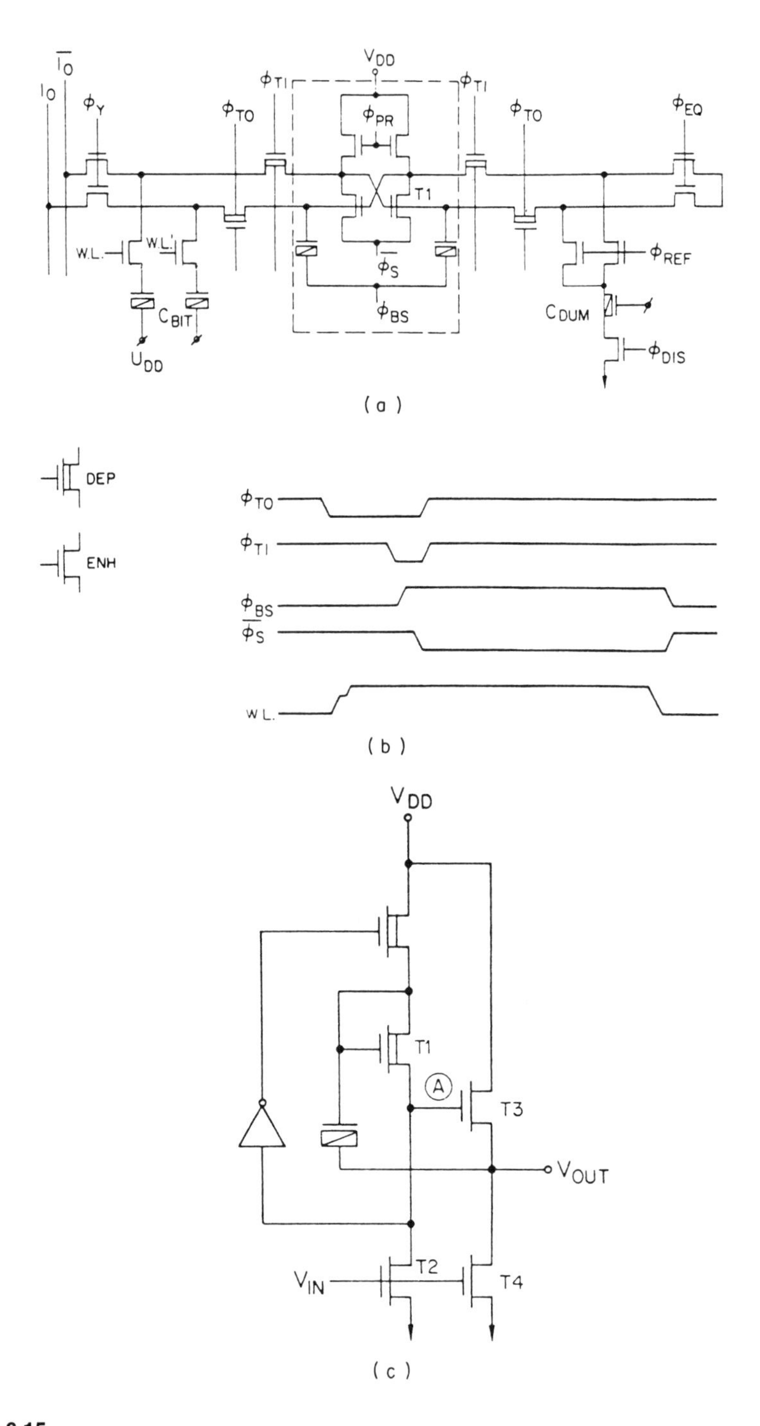

Figure 6.15
64k DRAM. (a) Sense amplifier circuit showing combination open and folded (twisted) bit-lines and (b) associated timing diagrams. (c) Depletion mode invertor used to generate 7.5 V on-chip V_{DD}. (From Barnes and Chan [17], National Semiconductor 1980, with permission of IEEE.)

6.15(a). Each bit-line section was connected to 64 cells instead of 128 cells as in the open bit-line approach. During a read, write or refresh operation, only one of the two bit-line pairs was initially selected during sensing. Subsequently the two bit-line pairs were reconnected on a single-folded metal bit-line.

The sense amplifier was capable of detecting a 30 mV signal under worst case timing conditions. It contained preamplification sense capacitors to restore a full V_{DD} level in the cell. Depletion mode devices were used to isolate non-selected bit-line halves during sensing. Since the dummy reference cell was shared between two bit-line halves, a full sized reference cell could be used instead of the half size reference capacitors frequently used. This also contributed to a reduction in noise from sense amplifier imbalance.

The timing diagram shown in Figure 6.15(b) shows the operating sequence. During the precharge period, the bit-lines are precharged to V_{DD} by ϕ_{PR}. During memory access ϕ_{PR} goes low and either ϕ_{t0} or ϕ_{t1} goes low to deselect the unselected bit-line halves. The selected word-line WL goes high and the signal is read out onto the bit-lines. Meanwhile the reference word-line clock ϕ_{ref} goes high establishing a reference level onto the reference bit-lines. ϕ_{ref} then goes low before preamplification and sensing. As the signal to the sense amplifiers is established, ϕ_{t1} goes low to isolate the sense amplifiers from their bit-lines. Sense operation begins by bringing $\bar{\phi}_{bs}$ high. Capacitance sharing between the low side sensing capacitor and the latch node $\bar{\phi}_s$ through ϕ_{t1} then establishes a much larger differential signal across the sense flipflop. $\bar{\phi}_s$ then goes low and ϕ_{t0} and ϕ_{t1} return to the high state to discharge low side bit-lines thus restoring a full V_{DD} level to the cells.

It frequently happens that a technique that is useful for one density of memory develops problems at the next density level. Open and folded bit-line schemes were later compared in a paper by Fujitsu [33] for the 256k DRAM. Technologies assumed were triple polysilicon for the open bit-line, and polycide gate technology for the folded bit-line. Each bit-line had 128 cells in both schemes. Cell array configurations and cell layouts are shown in Figures 6.16(a) and (b) for the folded and open bit-line schemes.

Fujitsu found that the pattern sensitivity components remained smaller for the folded bit-line schemes; however, as the bit-line spacing narrowed, the pattern sensitivity increased due to the capacitive coupling of the adjacent bit-lines which were now closer. Also silicide or polycide was required to decrease the word-line propagation delay with the longer word-lines required in the folded bit-line approach at the 256k density, thereby increasing the complexity of the process. Some array characteristics found for these two approaches are shown in Figure 6.16(c).

A further consideration with the larger 256k die was that the length of the short side of the chip was longer with the folded than with the open bit-line configuration which made it difficult to fit in the standard 300 mil dual-in-line package.

An alternative method to the folded bit-lines was needed to reduce common mode noise.

The bit-line shield for an open bit-line structure was described by Mitsubishi [35] in 1984 as a further alternative to the simple open and folded bit-line structures for 256k DRAMs. The open bit-line architecture was preferred because its aluminum word-lines and short bit-lines enhance fast array access. The bit-line shield consisted

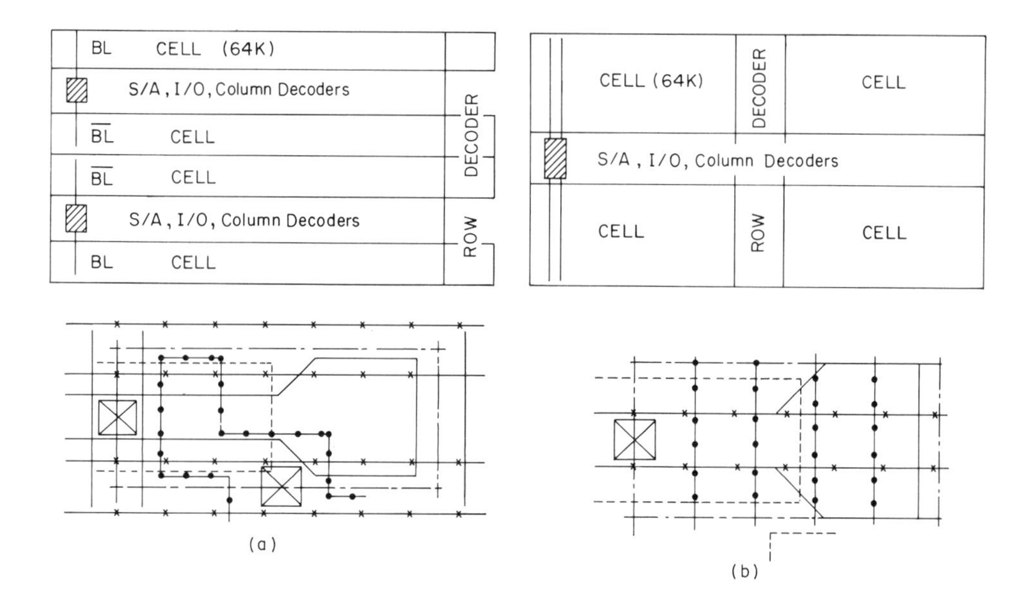

	Open BL	Folded BL
Storage cell capacitance (fF)	36.5	36.5
Bit-line capacitance (fF)	500	640
Capacitance ratio (C_{BL}/C_s)	13.7	17.5
Signal voltage at S/A (mV)	154	120
Sense amplifier imbalance (mV)	15	15
Pattern sensitivity (mV)	10	5
Effective signal voltage (mV)	129	100
Cell size (μm)	71	74
Die size (mm)	8.7 × 3.9	8.3 × 4.2
Word-line time-cont (ns)	0	21
Array access time (ns)	24	67

(c)

Figure 6.16
Bit-line architectures for the 256k DRAMs. (a) Open bit-line scheme cell configuration and layout. (b) Folded bit-line scheme cell configuration and layout. (c) Array characteristics for the two schemes. (From Nakano *et al.* [33], Fujitsu 1983, with permission of IEEE.)

of a third-level polysilicon layer placed between bit-lines and signal lines to shield the bit-lines from the non-common mode noise which is inherent in open bit-line architectures.

6.5.10 Boosted bit-lines

One of the drawbacks of the return to the open bit-line architectures for the higher density DRAMs was the need to fit increasingly complex circuitry into the narrower cell pitch of the higher density 256k DRAMs.

To solve this problem, NEC developed a simplified sense and restore circuit with boost capacitors connected to each bit-line. These boost capacitors were placed below a power supply line so that no extra area was used. It was shown that the boosted bit-line technique resulted in this case in a reduction in soft error rate at $V_{CC} = 5$ V of 50%.

6.5.11 Boosted word-lines

Another problem encountered when going down from 12 V V_{DD} to 5 V only power supply (V_{CC}) on the 64k was the loss of cell signal due to noise in the circuit. This is a more serious problem with a 5 V V_{CC} since on the cell capacitor a signal of word-line voltage minus a threshold is the maximum. That means, for example, for 12 V V_{CC} a reduction to 10 V in the cell which was not a problem; however, for 5 V V_{CC} it means only 3.5 V. This can be improved by boosting the word-line voltage over V_{CC}.

The boosted word-line technique was initially developed for the 64k DRAM. An example of boosted word-line operation was shown in the timing diagram in Figure 6.15 for the Fairchild 64k DRAM.

This technique was refined for the higher density 256k DRAM. An example was given by Fujitsu [33] for a 256k DRAM which used a sense amplifier circuit with a fast word-line driving technique including a new type of dummy cell called the capacitance coupled dummy cell and a high level boosted clock generator for the word-line driver.

The sense amplifier circuit used along with its operational waveforms is shown in Figure 6.17(a) and (b). Operation begins with the bit-line and $\overline{\text{bit-line}}$ shorted together and charged to V_{CC}. After reset the selected word-line is driven higher than $V_{CC} + V_{th}$ to connect cell and bit-line. By discharging the dummy word-line connected to the opposite side of the bit-line from V_{CC} to V_{SS}, the reference voltage is generated for BL by capacitive coupling of the dummy cell. When adequate signal voltage is obtained at the input of the sense amplifier clocks A and B are activated to amplify the voltage difference between the bit-line and $\overline{\text{bit-line}}$. Then the active restore operation is done to recharge the bit-line on the high side to the full V_{CC} level.

To increase noise margin with respect to soft errors, the entire potential difference between V_{CC} and V_{SS} in the cell must be used as an input to the sense amplifier. For

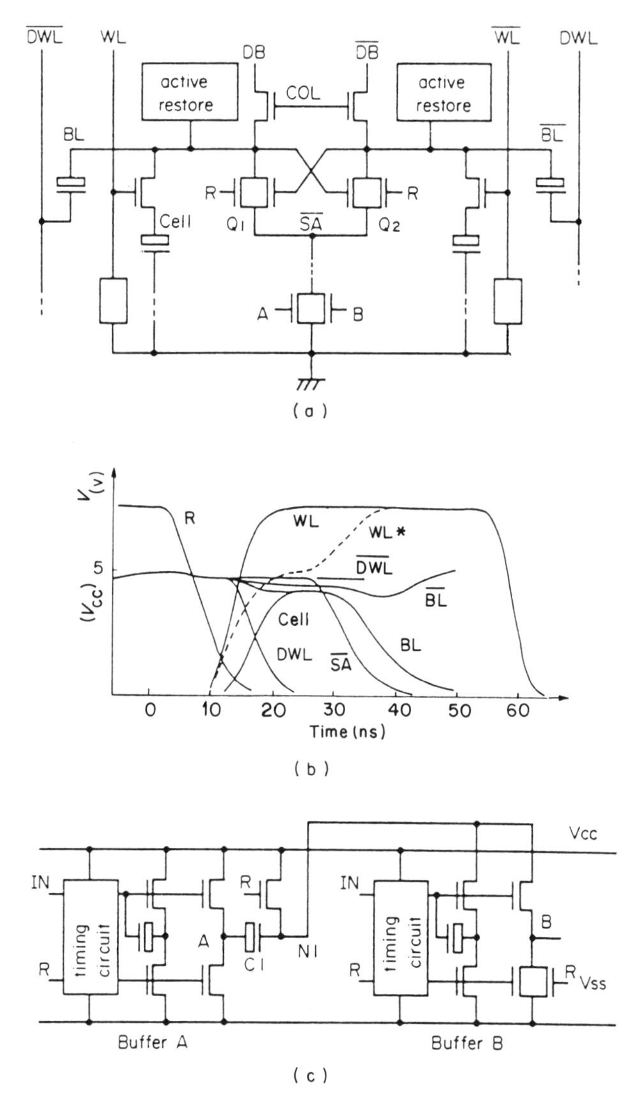

Figure 6.17
Sense amplifier technique using boosted word-lines. (a) Sense amplifier circuit, (b) associated timing diagram, (c) boosted high level clock generator. (From Nakano *et al.* [33], Fujitsu 1983, with permission of IEEE.)

this reason, the word-line must be boosted above $V_{CC} + V_{th}$ to eliminate the threshold loss of the access transistor before the sense amplifier begins its operation. The signal from a conventional boosted high level clock generator is shown in the figure by the dotted line. Since the capacitance coupled dummy word-line is no longer connected to the word-line drive, the capacitive loading on the word-line driver is halved.

For the higher speed operation required by the 256k DRAM a new type of boosted high level clock generator for driving the word-lines was also developed as shown in Figure 6.17(c). Preceding operation, node N_1 is charged to V_{CC} by clock R and a full charge is stored in capacitor C_1. The input signal of IN simultaneously triggers buffers A and B. The output of buffer A pushes node N_1 to twice V_{CC}. Buffer B, powered from node N_1, operates as if it were supplied twice the V_{CC} supply voltage and generates the boosted high level clock. The advantage is that boost capacitor C_1 no longer needs to be connected in parallel with the loading capacitance.

In another innovation for the 256k DRAM, Mitsubishi [24] in 1982 described a method of boosting the voltage stored under the plate without using the boosted word-line technique developed at the 64k DRAM level. This eliminated the additional time needed to boost the word-lines. A schematic equivalent circuit is shown in Figure 6.18(a). This technique uses a three-transistor cell plate controller, T_1–T_3. The cell plate and the word-line are paired so that the selected word-line controls the cell plate potential. When a word-line is accessed, the word-line signal turns on transistor T_3 and discharges the cell plate voltage V_g. After the sense amplifier refreshes the cell signal the voltage of the bit-line becomes V_{DD} for high level and V_{SS} for low. ϕ_g then goes high and V_g is charged up to the V_{DD} level through transistor T_3. If the cell previously stored the high level of $V_{DD}-V_{th}$, this rising motion of the cell plate boosts the storage node over the V_{DD} level. The voltage potential diagram and timing diagram are shown in Figure 6.18(b) and 6.18(c).

6.5.12 Implanted capacitor storage cell and grounded cell plate

A reduced area was required for the storage capacitor in the higher density DRAMs, but the capacitance had to remain around the 40 to 50 fF established at the 64k level to prevent alpha particle related soft errors. Various techniques were tried to increase the capacitance including ion implants beneath the capacitor and different voltages on the cell plate.

The Hi–C storage capacitor DRAM cell was first described by Intel [28] in 1976, and introduced by Intel in 1978 [27]. It is a one-transistor cell where the storage capacitance is enhanced by using two cell implants to increase the depletion capacitance. The Hi–C DRAM cell technology was illustrated in Chapter 4.

The doping of the DRAM capacitor in this cell consisted of a positive ion implant layer which increased the depletion capacitance in the storage area. Unfortunately this implant also increased the threshold voltage in the storage region. This increase in threshold voltage is then offset by an additional negative ion implant at the surface in the capacitor region as shown in Figure 4.28.

Since the dosage of the N^+ layer can be tailored to obtain any desired threshold voltage value, it permitted the cell to be operated with a grounded plate. A grounded plate gives improved storage area isolation characteristics as well as better tolerance to interpoly defects. It also improves immunity to power supply bumps since the cell plate is no longer connected to V_{CC}.

An implanted capacitor cell was used also on the Motorola and the AT&T parts. It reduced soft error sensitivity due to alpha particle hits on the cell by increasing the storage capacitance of the cell.

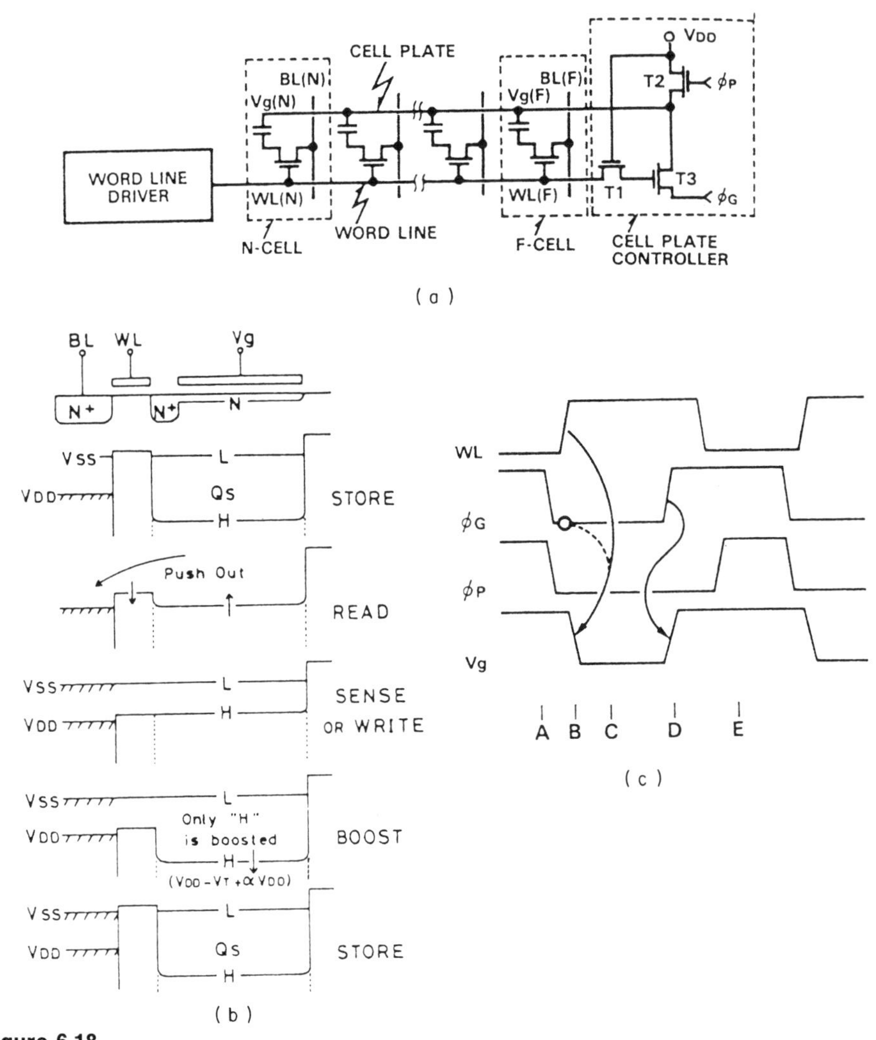

Figure 6.18
Boosted storage node(SNB) technique of boosting the capacitor plate voltage on a DRAM. (a) Schematic equivalent circuit of SNB DRAM, (b) potential diagrams of SNB DRAM action, (c) associated timing diagram. (From Fujishima *et al.* [24], Mitsubishi 1982, with permission of IEEE.)

The advantage of the implanted capacitor cell was the ability to ground the capacitor cell plate to reduce problems related to noise injection from the V_{CC} power supply. The drawback of the cell was the added complexity of the extra ion implantation steps in the process. This led to the search for a further solution.

6.5.13 Cell plate biased at V_{CC}

The Fairchild 64k used a biased cell plate to address the issue of charge loss in the cell due to noise injection from the V_{CC} supply. This part used an internally generated constant bias high voltage supply to the memory cell plate. This 7.5 V reference voltage was generated by using the oscillator of the substrate bias generator and an enhancement depletion mode invertor buffer with bootstrapped capacitor as shown in Figure 6.15(c). When V_{in} is in the high level state a voltage close to V_{CC} is stored across the capacitor. As V_{in} goes to a low level state, the charge stored in the capacitor is shared with node A through t. The positive feedback from V_{out} enhances the node A voltage until V_{out} reaches a maximum supply level of V_{CC}. The invertor then is used to isolate the capacitor-to-V_{CC} path thus allowing the capacitor plate voltage to be bootstrapped to greater than V_{CC}.

In a 256k DRAM paper, Mitsubishi [35] found that the storage node capacitance of a V_{CC} cell plate is many times larger than that of a grounded (V_{SS}) cell plate. This is because in the case of the V_{CC} cell plate the distance between the memory cell region and the transfer gate can be minimized because the inversion layer is formed under the cell plate. In the case of a V_{SS} cell plate, the memory cell capacitor region must be away from the transfer gate region to allow the alignment tolerance for ion implantation of the depletion layer. It is also indicated that the power supply bounce problem arising from the V_{CC} cell plate scheme can be compensated by increasing the storage node capacitance. This problem is explored further in the following section.

6.5.14 Noise due to voltage bump on a V_{CC} biased cell plate

At the 256k density level a problem was found with the use of the cell plate boosted to V_{CC}. Mitsubishi [32] in 1983 described the effects of voltage bumping on the boosted cell plate described above. Voltage bumping refers to fluctuations of the power supply voltage V_{CC} which can cause erroneous data to be read out. It particularly affects those DRAMs in which V_{CC} is directly connected to the cell plate.

In the conventional DRAM, voltage fluctuation problems can occur when V_{CC} during the write operation is lower than V_{CC} during the read operation. This is called a positive *V*-bump. The fluctuation of V_{CC} lowers the read-out voltage from the memory cell and may cause a read error in which a high level is read as a low level.

In the case of a boosted cell plate, a positive voltage-bump raises the stored level in the cell by the coupling of the cell capacitance. Since the dummy reference cell level is kept at ground and is not affected by the positive supply voltage fluctuation, a low level in the cell may be 'read' as a high level when it is compared with the dummy cell. A half V_{CC} supply level of the cell plate is easier to isolate from V_{CC} and was found to be an optimal condition for reducing positive *V*-bump sensitivity. The readout voltage in this case remains constant and the DRAM becomes tolerant of the positive voltage bumping.

6.5.15 New operational modes—nibble mode and $\overline{CAS}$ before $\overline{RAS}$ refresh

An early DRAM with a fast serial access mode, called nibble mode, was presented by Inmos in 1981 [23]. This second generation 64k DRAM operated in standard RAS access mode with 100 ns access time, and in nibble mode with 20 ns serial access. Nibble mode allows up to four bits of data to read serially in a continuous stream before another full access must be made. This was accomplished by an internal 16k × 4 memory organization with a 1-in-4 decoder designed to operate as a four-bit shift register during nibble cycles. Thus four bits stored in four output sense amplifiers defined by a common column address could be serially transferred to the data output pin at high speed before another column address was needed. The shift register wrapped around so that if another address is not selected the same four bits are accessed again.

The real advantage of this new mode was that nibble mode was designed to enable several parts to be interleaved to give a continuous stream of data at the fast nibble rate. A nibble mode timing diagram is shown in a later section (Figure 6.23(c)).

Another new operational mode presented on this same Inmos DRAM was the '$\overline{\text{CAS}}$-Before-$\overline{\text{RAS}}$ refresh mode'. In this refresh mode the refresh address is generated internally. It is triggered by activating the column address strobe ($\overline{\text{CAS}}$) prior to the row address strobe ($\overline{\text{RAS}}$).

Timing diagrams for the various DRAM refresh modes are shown in Figure 6.19. These include $\overline{\text{RAS}}$-only refresh, hidden refresh, and $\overline{\text{CAS}}$-before-$\overline{\text{RAS}}$ refresh. These modes all appeared by the 64k DRAM generation, were generally implemented by the 256k, and continued unchanged in subsequent DRAM generations.

6.6 AN OVERVIEW OF THE 256k DRAMs

When the 256k DRAMs were introduced in 1983 only about 30% of DRAM sales were currently going into the mainframe computers which had traditionally driven the industry to higher density devices. Small computers and terminals of various types used the rest. These were applications that were more cost sensitive than they were demanding of higher densities at that time. This led to a slower rate of acceptance of the higher density memory.

There was also a reluctance on the part of the DRAM suppliers to undermine the 64k DRAM market having just emerged from the severe price recession of 1981–1982 on the 16k DRAM. For these reasons, the introduction of the 256k into volume production took longer than expected. Capacity was only shifted to the 256k significantly following the severe price decline of the 64k DRAM in 1985.

The small computer application could benefit from more application-specific treatment of the standard DRAMs. Features which appeared in response to this demand at the 256k level included: combinations of the various operational modes developed developed at the 64k level, wider outputs, and various features such a CAS-before-RAS refreshing intended to make the DRAMs more convenient to use in systems with small numbers of memories. The trend to applications specific DRAMs will be investigated more fully in Chapter 7.

Figure 6.19

Various refresh modes in dynamic RAMs. (a) $\overline{\text{RAS}}$-only refresh cycle timing, (b) hidden refresh cycle timing, (c) automatic ($\overline{\text{CAS}}$-before-$\overline{\text{RAS}}$) refresh cycle timing. (Source Texas Instruments MOS Memory Databook.)

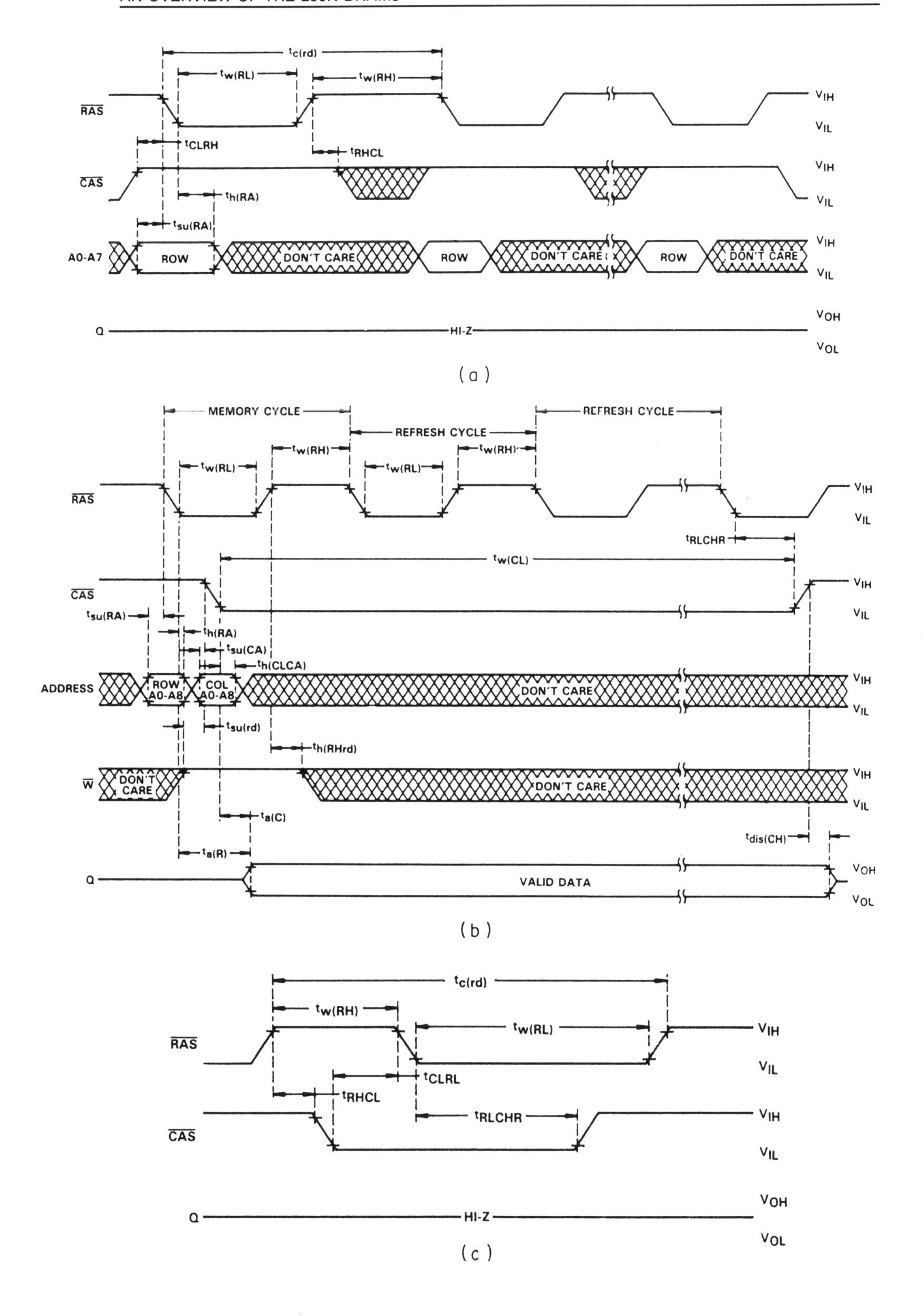

$t_{c(rd)}$
$t_{w(RL)}$
$t_{w(RH)}$
RAS
V_{IH}
V_{IL}
t_{CLRH}
t_{RHCL}
CAS
$t_{h(RA)}$
$t_{su(RA)}$
A0-A7
ROW
DON'T CARE
Q
HI-Z
V_{OH}
V_{OL}
(a)
MEMORY CYCLE
REFRESH CYCLE
t_{RLCHR}
$t_{w(CL)}$
$t_{su(CA)}$
$t_{h(CLCA)}$
ADDRESS
ROW A0-A8
COL A0-A8
$t_{su(rd)}$
$t_{h(RHrd)}$
W
DON'T CARE
$t_{a(C)}$
$t_{a(R)}$
$t_{dis(CH)}$
VALID DATA
(b)
t_{CLRL}
(c)

A COMPARISON OF 256 K DYNAMIC RANDOM ACCESS MEMORIES

Company	NEC	Motorola	Fujitsu	Toshiba	Mitsubishi
Technology (word line/bit line)	Double polysilicon (metal-poly/metal)	Single or double polysilicon (titanium silicide/metal)	Triple polysilicon (poly/poly)	Double polysilicon (molybdenum silicide/metal)	Double polysilicon (molybdenum silicide-poly/metal)
Design rule (μm)	1.3	2.0	2.5	2.0	2.0
Cell area (μm^2)	67	84	68	77	98
Chip area (mil^2)	52 700	71 600	52 900	71 300	73 700
Typical access times (ns)					
row	90	90	<80	94	100
column	(not available)	50	35	34	50
nibble	—	15	15	—	25
cycle	(not available)	250	200	260	210
Power consumption (mW)					
active	250	225	300	170	250
standby	10	15	15	15	12
CAS before RAS refresh	no	yes	yes	no	yes
Redundancy	none	laser; 8 rows, 4 columns	2 rows, 1 column	laser; 4 rows, 4 columns	laser; 4 rows, 4 columns

Figure 6.20
Comparison of 256k DRAM device characteristics. (Source various ISSCC Proceedings.)

First generation 256k technology ranged from NTT's experimental 1.0 μm technology to Fujitsu's relaxed, production oriented, 2.5 μm parts which were built on the same lines that ran their 64k DRAMs.

A chart which includes various of the 256k DRAMs introduced is shown in Figure 6.20. All of these first generation devices were NMOS. Three out of the five included the nibble mode operation, $\overline{\text{CAS}}$-before-$\overline{\text{RAS}}$ refreshing mode, and silicide technology. Four had redundancy provisions.

The technology used in the 256k DRAMs was approximately that of the second generation 64k DRAMs. The slightly larger chip encouraged the development and use of redundancy techniques, either fuses or laser blown, to maintain the yields at the same levels as on the 64k DRAMs. Yields and redundancy techniques will be discussed further in Chapter 15.

Lithography became more difficult since the full wafer optical projection techniques used in the 64k DRAMs did not provide the resolution needed for the 1.0–2.0 μm geometries of the 256k DRAMs. The choice fell primarily on the 5:1 or 10:1 wafer steppers.

6.6.1 256k DRAM devices

The first two 256k DRAMs appeared in 1980 from NTT [18] and NEC [19]. The NTT part was a technology oriented experimental chip in an aggressive 1 μm geometry process attained by using e-beam direct-write-on-wafer, positive and negative resists, dry etching techniques and, polysilicon bit-lines and molybdenum word-lines. It featured redundancy using on-chip fuse programmable cells which was a first in DRAMs and an improved capacitive coupled sense amplifier. A 7 V V_{DD} was applied to the higher threshold voltage molybdenum gate transistors in the memory array. These were used to reduce the size of the memory cell. Transistors in the periphery used the polysilicon gate. NTT used the new sense amplifier which was shown previously in Figure 6.12. The fuse programmable circuitry on this chip is examined further in the chapter on test and yield.

While the 256k described by NTT appeared to be primarily a research device, the one announced by NEC was clearly intended for production. It was TTL compatible with single 5 V power supply, standard 300 mil 16 pin DIP and was intended for easy upgrade from the 64k DRAM with 256 cycle refresh. It used poly word-lines and metal bit-lines organized in two 128k blocks each organized as two balanced 128 by 512 bit memory arrays sharing a common row of 512 sense amplifiers. An open bit-line configuration was used. The additional address was on pin 1.

The part included folded bit-lines which reduce common mode noise which can result from soft error hits on bit-lines as well as from other sources. The folded bit-line structure also permitted a sense amplifier to be connected to a bit-line pair giving more room for the sense amplifier in the smaller geometry of the 256k device. The bit-line pitch was 9 μm so that the allowed sense amplifier pitch was 18 μm.

The NEC [31] 256k used double polysilicon and double layer aluminum. The double poly was used in the cell and the double aluminum was used both for the bit and word-lines and also for logic circuit design in the periphery which contributed to a 15% reduction in die size. It was recognized that the reduction in size of the DRAM chips needed to be accomplished by efforts both in the memory array and also in the logic of the periphery which occupies 50% of the chip array in an NMOS DRAM.

In addition, the longer and narrower word-lines in the higher density DRAMs required a technology which could reduce the increasing word-line resistance, resulting in a realization of faster access times. While the basic word-lines were formed out of polysilicon, a second set of aluminum word-lines ran parallel to the poly word-lines and were connected at intervals of 32 bits. This resulted in reducing the word-line resistance to less than 3% of that with only polysilicon.

While the size of the DRAM chip increases as the cell size and hence array size increases, the use of double layer aluminum both helped to reduce the chip size for a given cell size by reducing the percentage of the chip used in the periphery resulting in increased array efficiency.

Another effect which became more pronounced at the 256k DRAM level was transient effects of supply current variations. This occurred due to the fast cycle rates, approaching 10 MHz, and due to the large matrix capacitance which needed to be precharged. With long interconnect wires the propagation delay from one end of a wire to the other can be significant giving rise to classical transmission line effects. With higher resistance interconnect systems, this delay could exceed the gate delays.

A paper from Mostek (now SGS-Thomson) described the use of double metal interconnect lines to shorten these delays and the use of a controlled precharge ramp rate to confine the Ldi/dt effects to the maximum acceptable value [37]. Sequential precharge of the matrix was used as a controllable means of approximating the ideal current waveform. The matrix was divided into quadrants with the occurrence of bit-line precharge for each quadrant being controlled by timed clock phases.

6.7 HISTORY OF CMOS DRAMs

The first CMOS dynamic RAM was a 64k density device described by Intel [29] in 1983. In the memory array the transistors were P-channel and the array was in an N-well. CMOS was used for the peripheral circuits. The Intel 64k CMOS DRAM implemented a new fast page mode permitting rapid random access of column bits.

Another 64k CMOS DRAM was described in 1983 by Fujitsu [30]. This part used N-channel transistors in the array with implanted capacitors and CMOS peripheral circuits. It reduced timing skews by replacing the $\overline{\text{CAS}}$ clock with a chip select.

The Fujitsu part also featured a true static column operation in which the DRAM operated like a static memory for a time limited only by the maximum $\overline{\text{RAS}}$ low time or refresh requirement.

The Fujitsu device also featured an 'edge-triggered' write technique used to minimize speed deterioration caused by timing skew. When write enable ($\overline{\text{WE}}$) became low, address, data in and data out were internally latched to start the write operation. Once latched, address and data in were freed and $\overline{\text{WE}}$ could go high after a fixed time. The write operation period was automatically determined in the chip. If the time required for write elapsed, a write completion signal was executed and the write operation was automatically terminated. Address and data in for the new cycle could already be set during the previous cycle since these signals were latched and freed at the beginning of the write operation. Critical timing between a write cycle and the following read or write cycle is therefore removed resulting in shorter times in the system.

In a 1984 CMOS 256k DRAM paper Intel [38] outlined the rational for using CMOS technology in higher density DRAM devices. This 256k device, like the Intel 64k DRAM, used PMOS transistors in an N-well. Low standby power and wide noise margins resulted from the use of CMOS in the peripheral circuitry.

The static CMOS circuits reduced the clock delays of NMOS and eliminated the necessity of bootstrapping used in NMOS dynamic circuit design. The use of static CMOS circuitry reduced power consumption of clock drivers, address buffers, column decoders, and data input and output paths. Dynamic CMOS circuits were also used where needed to provide better area utilization and to match the memory cell word line pitch where needed.

A low resistivity epitaxial layer on the silicon substrate was used to increase the immunity to latch-up, and also to minimize the voltage drops resulting from substrate currents generated by the short channel effects of the NMOS transistors.

The cell was embedded in an n-well which was biased at the V_{DD} potential. This resulted in improved refresh characteristics over NMOS and reduced soft errors from alpha particles.

While it appeared that there were many advantages of using CMOS for DRAMs, it was not until the 1 Mb DRAM that the industry generally converted to the CMOS technology although many NMOS 1Mb papers were presented.

6.7.1 Advantages of CMOS for DRAMs

Advantages from using CMOS were found to include: simplified logic circuit designs, shorter design and debug times, microwatt standby power, wide noise margins, extended refresh intervals, high data bandwidth output modes, and a usable high speed access with wide T_{RCD} window. T_{RCD} is the maximum usable low time for $\overline{RAS}$.

An advantage of using CMOS was the reduced complexity of the periphery circuit design. In this case, the total number of transistors in the periphery was reduced by 30% over a comparable design in NMOS. This resulted in an improvement in array efficiency from 40% with the NMOS DRAMs to 50% or higher for the CMOS DRAMs. A DRAM address buffer circuit, for example, used 16 transistors in CMOS compared to 32 in NMOS with the number of clocks reduced from 5 to 2.

Another advantage of CMOS was in reduction in noise sensitivity, including soft error sensitivity, due to the effect of having the P-channel memory cell imbedded in the N-well. The N-well tends to isolate the storage cell from substrate current noises such as alpha particle hits and also from impact ionization current or minority carrier injection due to external signal undershoot at the input pins. One result of the reduced sensitivity of the cell transistor to noise is that a shorter effective channel length can be used in the transfer gate. The limiting factor of effective channel length then becomes the high temperature subthreshold leakage required for charge holding stability.

The use of CMOS periphery in DRAMs also gives the possibility of utilizing various static column modes of operation without adding dc power consumption as happens if a static column function is designed in NMOS.

CMOS can improve refresh characteristics over those of NMOS DRAMs which tend to be limited by thermally generated carriers from defect sites in the substrate within a diffusion length of the memory cells. These minority carriers collected by the memory cells degrade the refresh times. Embedding the memory array in a reverse biased N-well dramatically reduces this thermal leakage current.

The N-well to substrate reverse biased junction also acts as an effective minority carrier barrier to reduce the soft error rates induced by incident alpha particles. The barrier also isolates the memory cells from spurious injections into the substrate by the high speed peripheral circuits. In addition to the memory cells, the bit-lines and sense amplifiers were all embedded in the N-well to reduce the alpha particle induced soft error rate.

6.7.2 Drawbacks of CMOS with DRAMs

A disadvantage was a greater number of mask steps used for the early CMOS DRAMs.

Another potential problem was latch-up which can occur as the result of both internal and external noise sources. Examples of internal noise sources include: impact ionization currents, transient voltage drops on major bus lines, and capacitively induced voltage transitions from the substrate and N-well bounces. External noise sources arise from spurious signals, overshoots on input and output pins, or supply transients and ringing. In order to provide high latch-up immunity and low standby current, no substrate, or well, charge pump circuits were used. The P-substrate and N-wells were connected to V_{SS} and V_{DD}, respectively.

A substrate bias generator can help prevent latch-up during operation as well as improve the isolation characteristics of parasitic transistors and reduce the possibility of minority carrier injection. The substrate bias also allows the chip to tolerate input level undershoot. The back bias generation itself could, however, cause latchup during power-up by capacitive coupling to V_{CC} since it is a relatively high output impedance. AT&T on their 1Mb DRAM limited this potential source of positive voltage excursions by including a ballasting capacitance of the grounded cell plate and devices to clamp the substrate to V_{SS} until the back bias generator was active.

6.8 THE 1Mb DRAMs

The initial 1Mb DRAMs included both NMOS and CMOS devices as shown in Figure 6.21. Technology advances included both those required by the smaller geometries,

Manufacturer	Process	Access time (ns)	Active power (mW)	Standby power (mW)	Chip size (mm^2)	Cell size (μm^2)	Cell structure (fF)	Transister design rule
Toshiba	CMOS	100–120	330 (190 ns)	11	4.40 × 12.32 (54.2)	3.24 × 9.0 (29.16)	Planar (40)	LDD Tr 1.2 μm
Hitachi	CMOS	100–150	385 (190 ns)	11	4.66 × 13.74 (64.0)	3.4 × 10.5 (35.7)	Planar (36)	LDD Tr 1.3 μm
Fujitsu	NMOS	120–150	550 (230 ns)	24.8	4.44 × 12.32 (54.7)	3.15 × 8.4 (26.46)	Stacked (55)	Conv. Tr 1.4 μm
	NMOS	100–150	550 (210 ns)	24.8	4.70 × 12.82 (60.3)	3.4 × 8.6 (29.24)	Planar (40)	Conv. Tr 1.0 μm
Mitsubishi	CMOS	100–150	412 (190 ns)	22	4.70 × 13.80 (64.9)	3.4 × 8.8 (29.92)	Planar (40)	Conv. Tr 1.0 μm
NEC	NMOS	100–150	550 (200 ns)	27.5	4.70 × 10.33 (48.6)	3.4 × 6.0 (20.4)	Trench (60)	Conv. Tr 1.0 μm
OKI	NMOS	100–120	412 (200 ns)	27.5	4.92 × 11.89 (58.5)	3.8 × 7.4 (28.12)	Stacked (50)	LDD Tr 1.2 μm
Matsushita	NMOS	100–120	522 (190 ns)	35.8	4.98 × 11.14 (55.5)	4.5 × 7.8 (35.1)	Planar (53)	LDD Tr 1.2 μm
TI	CMOS	120–150	357 (230 ns)	11	4.30 × 11.70 (50.3)	2.5 × 8.5 (21.25)	Trench (—)	LDD Tr 1.0 μm

Figure 6.21
Comparison of 1Mb production dynamic RAMs. (Source Toshiba Memory Presentation 1988, with permission of Toshiba.)

larger chip sizes and higher densities, and also those due specifically to the change in technology from NMOS to CMOS.

An early experimental 1Mb CMOS DRAM was described by NTT [42] in 1984. It had a distributed sensing circuit and used an 0.8 μm double polysilicon process with molybdenum silicide word-lines and a 1.5 μm deep trench capacitor with 30 fF capacitance giving a small 20 μm^2 cell size. The small capacitance meant that an error detection and correction circuit (ECC) was needed for protection against alpha induced soft errors. This circuitry used the bidirectional parity which is discussed in Chapter 15.

In 1984 NEC [46] also presented one of the first 1Mb DRAMs. This NMOS part used non-multiplexed addressing and a planar 44 μm^2 cell with 50 fF capacitance. This gave a 75.9 mm^2 chip size in a 1.0 μm process. Double poly and double aluminum interconnects were used. The focus was on suppressing bit line noise using a technique which permitted low noise $V_{CC}/2$ bit-line precharge by sandwiching the bit-line between two dummy bit-lines so the interference from adjacent bit-lines is reduced.

Mostek [40] showed a 1Mb CMOS DRAM in 1985. The part was fabricated of NMOS cells directly on the grounded P-type substrate. A divided bit-line architecture, which combined the earlier shared sense amplifier and folded bit-line concepts, was used. The Mostek DRAM also used $V_{CC}/2$ bit-line precharge and word-lines boosted above V_{CC} to restore a full V_{CC} level into the memory cell. Small charge pumps, driven by an asynchronous oscillator, were used to keep boosted nodes above V_{CC}. This permitted extended RAS active cycles up to 8 ms long. A planar cell was used with a 36 μm^2 cell size in an 0.9 μm technology with static column operating mode.

In addition, the static I/O and column circuits permitted the $\overline{\text{CAS}}$ function to be replaced by an output enable function for controlling the common I/O interface.

NEC in 1985 showed a second 1Mb chip also in NMOS with trench capacitors giving a 20.4 μm^2 cell size and 60 fF capacitance in a 43.2 mm^2 chip. This part featured a test mode which was entered by applying a voltage higher than V_{CC} to the NC pin. The test mode reconfigured the 1Mb $\times$ 1 part as $\times$ 4 so that testing in parallel at four times the normal test speed could take place.

Mitsubishi [44] in 1985 showed a 1Mb NMOS DRAM with a four-bit test mode similar to that of the NEC device, and a new shared sense amplifier. A planar cell with 45 fF capacitance was used giving a 35.7 μm^2 cell and 65 mm^2 chip in a 1.2 μm minimum geometry. The 512 cycle 8 ms refresh which became standard for the 1Mb DRAMs was used. The part used a $V_{CC}/2$ cell plate to reduce the electric field across the thinner oxide of the 1Mb DRAM memory capacitor. A grounded epitaxial substrate was used to reduce the voltage bounce of the cell plate.

Toshiba presented two 1Mb DRAMs in 1985, one in NMOS [53] with a capacitor that was planar with a mini-trench extension. The other was CMOS [41] with a planar cell and offered the fast page mode, developed earlier by Intel, as a mask option with the static column mode. The part used a $V_{CC}/2$ bit-line precharge and the selected word-line was bootstrapped above V_{CC} to restore full V_{CC} to the cell. At the end of the active cycle both bit-line pairs and dummy word-lines were equalized to $V_{CC}/2$. Partial activation of one of four blocks of the memory array was used to reduce the active power dissipation to 1/4. The capacitor plate was biased to $V_{CC}/2$ to protect the thin oxide of the capacitor.

IBM [55] showed an NMOS 1Mb DRAM in 1.0 μm minimum feature size with a planar implanted 36 μm^2 cell in a 57.75 mm^2 chip. The part had a feature to improve the data rate of the fast page mode. It used boosted word-lines.

Hitachi [57] showed a 1Mb CMOS DRAM with static column mode also in 1985 which used a 24.12 μm^2 trench cell to give a 47.3 mm^2 chip size. The part used a folded bit-line structure. It used ATD circuitry to reduce power dissipation during long cycles and for additional power reduction this chip also used a dual back bias generator with the low power generator supplying the substrate leakage current and the other supplying the charge pumping capability.

AT&T [58] showed a CMOS 1Mb with planar cell in 1.3 μm design rules giving a 36.7 μm^2 cell and a 69.6 mm^2 chip. The chip used a back bias generator to improve reliability and a grounded cell plate to clamp the substrate to ground until the generator was active. The part included a fast column timing mode clocked by chip enable which is described in a following section. The part had boosted word-lines, folded bit-lines, and precharge of the bit-lines to $V_{CC}/2$.

In 1986 there were another three standard 1Mb DRAMs introduced from Intel, TI and a third part from Toshiba, along with an application-specific 1Mb which will be considered in the next chapter. All three were CMOS.

The Intel [59] part used a compact version of their planar cell. They returned to the simple differential sense amplifier. The memory array was in an N-well that was biased at 2.5 V above V_{CC} to eliminate injection from the substrate and forward biased diodes and hence improve the soft error rate.

The TI part included a trench capacitor in a contactless field plate cell that used depletion mode trenches and buried N^+ bit-lines. The top plate of the capacitor was tied to ground for improved skew margin. As a result of the trench the cell size was reduced to 21.25 μm^2 and the chip size to 50.32 mm^2 in a 1.0 μm technology. Bit-lines precharged to $V_{CC}/2$ and boosted word-lines were used, along with segmented bit-lines intended to reduce parasitic capacitance resulting in an 8:1 C_b/C_s ratio.

Toshiba [47] showed a version of the earlier 1Mb which was optimized for alpha induced soft error reduction by using an improved N^+ diffused bit-line.

6.8.1 A typical production 1Mb DRAM

The Toshiba 1Mb DRAM is described as typical of the production generation 1Mb DRAMs. It was CMOS, organized 1Mb $\times$ 1 and $\times$4, used 5V power supply, had 512 cycles/8ms refresh and had as refresh functions, RAS only, hidden refresh, and CAS-before-RAS. Operating modes included the new fast page mode and static column mode. While the trench and stacked capacitor cells were experimented with at the 1Mb level, the simpler planar cell was used by most companies in production. Three to four levels of interconnects were standard implemented in a 1.0 μm technology. Precharged bit-lines and boosted word-lines were common. Cell plates were frequently tied to $V_{CC}/2$ to reduce the field across the thin oxide or were tied to ground.

A comparison of the Toshiba 256k and 1Mb DRAM process parameters and memory cell structures are shown in Figure 6.22 as typical of the two generations.

	256k DRAM	1 Mb CMOS DRAM
Feature Size	2.0	1.2
contact hole (μm^2)	2.0 × 2.0	1.2 × 1.4
metal pitch (μm)	4.0	2.6
Process	NMOS Double level poly Si	N-well CMOS Triple level poly Si LDD structure
Gate material (2nd/1st)	$MoSi_2$-polycide/poly Si	poly Si/poly Si
Gate oxide (2nd/1st)	350A/250A	250A/120A
Gate length (μm)	2.0	1.2 (NMOS) 1.6 (PMOS)
Bit-line	Al	$MoSi_2$-polycide
Word-line	$MoSi_2$-polycide	Al/poly Si
Cell size (μm^2)	6.4 × 12.0	3.8 × 9.9

Figure 6.22
A comparison of the characteristics of a typical 256k and 1Mb DRAM. (Reproduced with permission from Toshiba.)

6.8.2 New timing modes

The CMOS DRAMs featured several new rapid access modes in addition to the older nibble mode and page modes. Figure 6.23 gives examples of the various access modes which were common by the 1Mb DRAM generation.

The new fast page mode was introduced first on the Intel 64k CMOS DRAM. In this mode after the row address information has been latched into the row decoder, the input port of the address buffer will be opened to the external signal. The column address signal is latched into the chip on the falling edge of $\overline{CAS}$ as in a conventional DRAM. Then, along a selected row, the 256 individual column bits can be rapidly accessed and read out randomly. As in a standard DRAM, the output enable signal is controlled by the external $\overline{WE}$ and $\overline{CAS}$ inputs. A metal mask option was also provided to allow the address buffer port to remain open during the read operation. In this case the external column address is held stable and the $\overline{CAS}$ signal acts as an output enable clock. The timing wave forms for the standard page mode and the fast page mode are shown in Figures 6.23(a) and (b) along with timing waveforms for the nibble mode described earlier in Figure 6.23(c).

The IBM 1Mb DRAM increased the data rate of the fast page mode by reducing the column precharge time. An internally generated early restore phase is used to reset the column access phases and bit decoders as soon as their functions are complete.

The static column mode eliminated the precharge time normally required by conventional DRAMs during a page mode operation and permitted access times to be achieved that were nearly equal to cycle times as in a static RAM. A timing diagram of this operation is shown in Figure 6.23(d). The static column mode for CMOS DRAMs was also generally adopted for the 1Mb density and above.

● **PAGE MODE READ CYCLE**

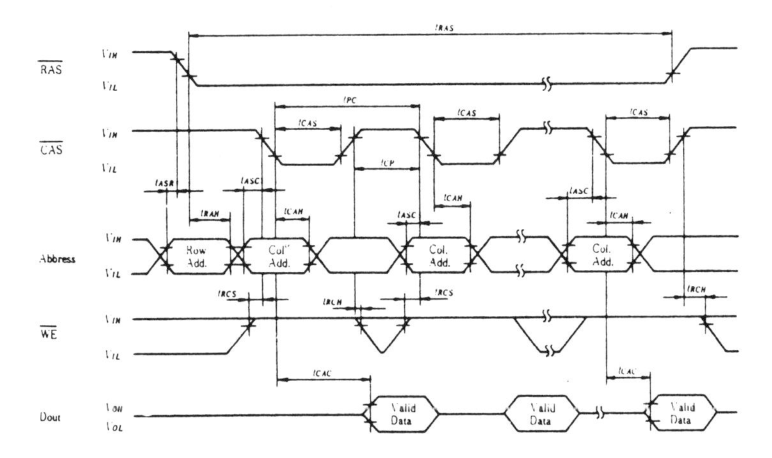

● **PAGE MODE WRITE CYCLE**

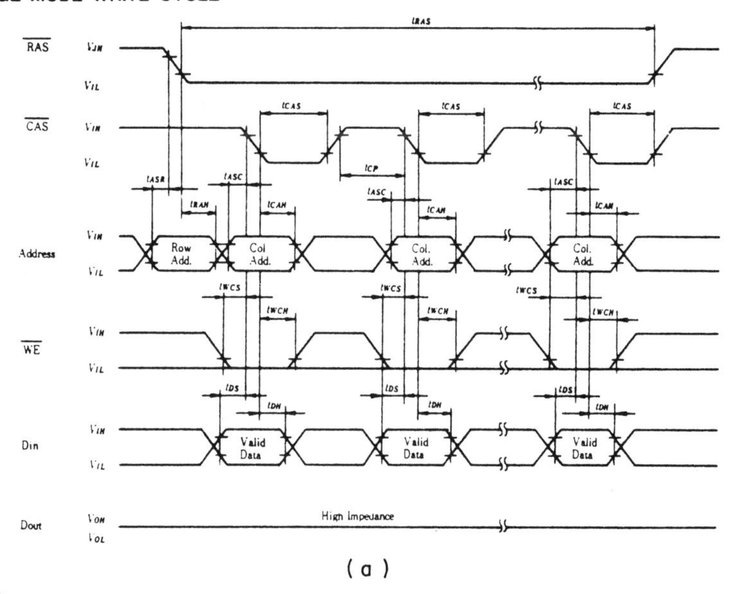

Figure 6.23
Timing waveforms for various dynamic RAM operational modes. (a) Page mode, (b) fast page mode, (c) nibble mode, (d) static column mode. (Source Hitachi MOS Memory Databook 1988.)

In the static column operation a row address is latched, like in conventional DRAMs, on the leading edge of the $\overline{\text{RAS}}$ clock and the column address is then applied like in a static memory or address transition detection circuits may be used so that no $\overline{\text{CAS}}$ clock is required. After the address has been changed to the selected address and held for a fixed interval, called the $\overline{\text{CAS}}$ access time, the output data are obtained. This allows true static operation with a DRAM during this mode of operation within

● High Speed Page Mode Read Cycle

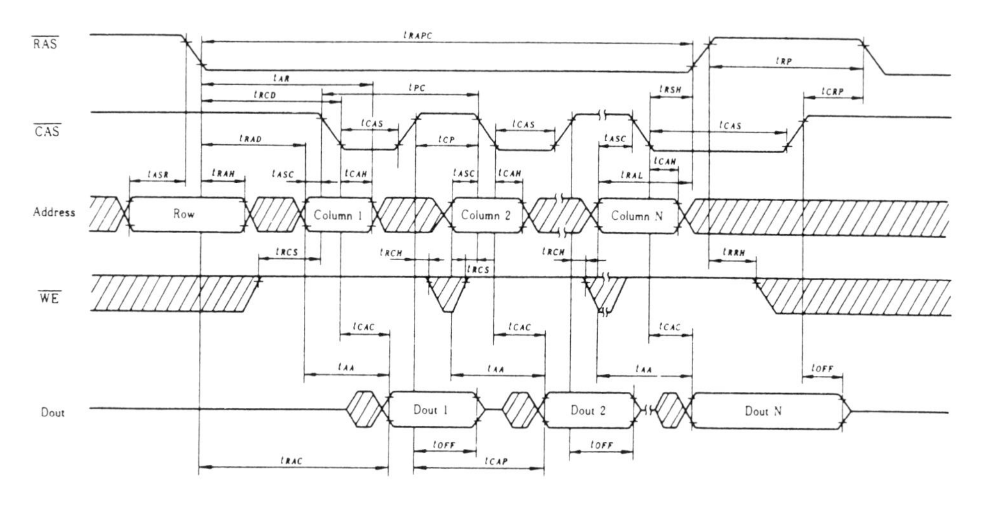

● High Speed Page Mode Write Cycle

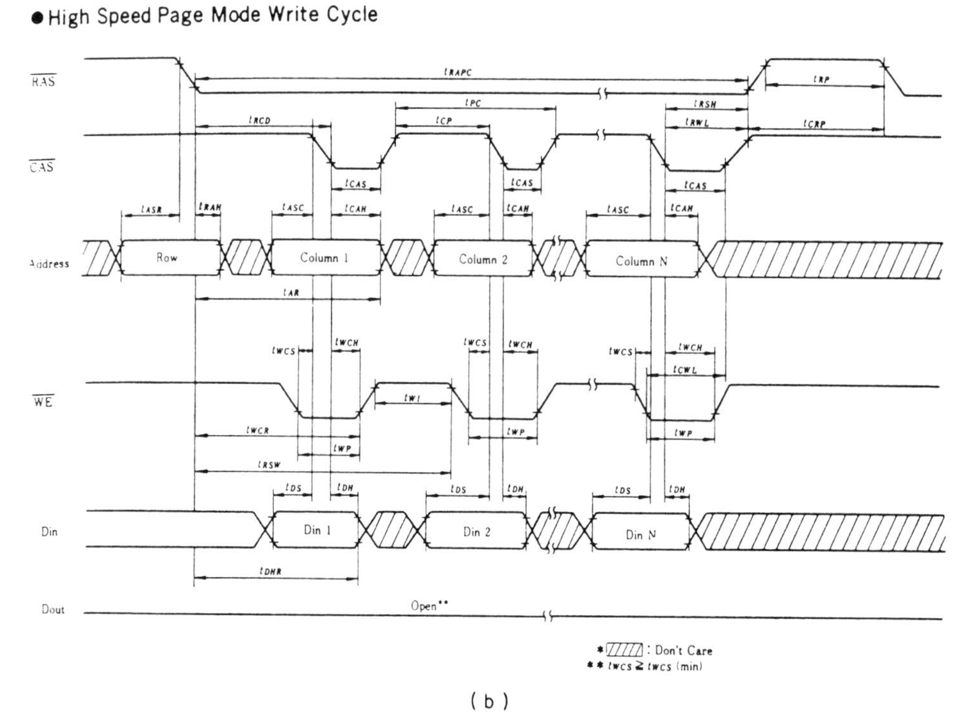

(b)

Figure 6.23(b) *(continued)*

- **Nibble Mode Read Cycle**

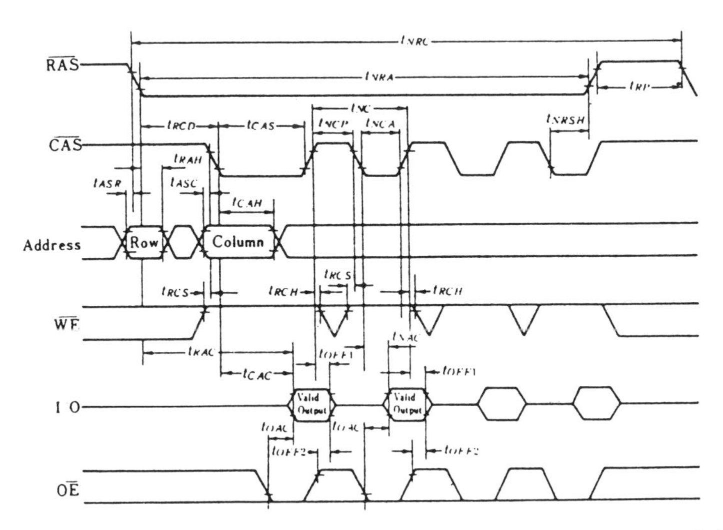

* ▭ : Don't care

- **Nibble Mode Write Cycle**

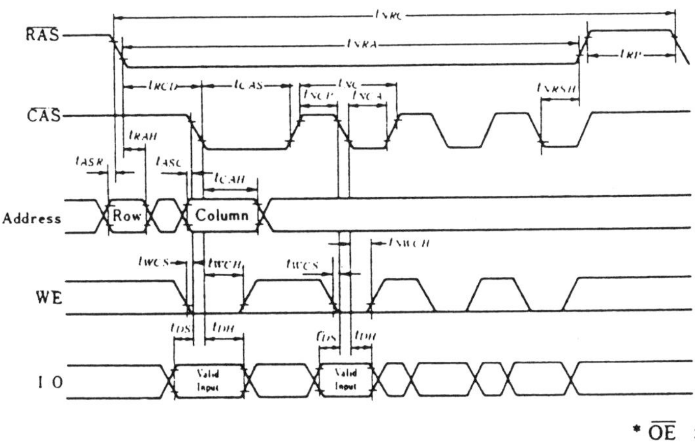

* $\overline{OE}$: Don't care

* ▭ : Don't care

(c)

Figure 6.23(c) *(continued)*

- **Static Column Mode Read Cycle**

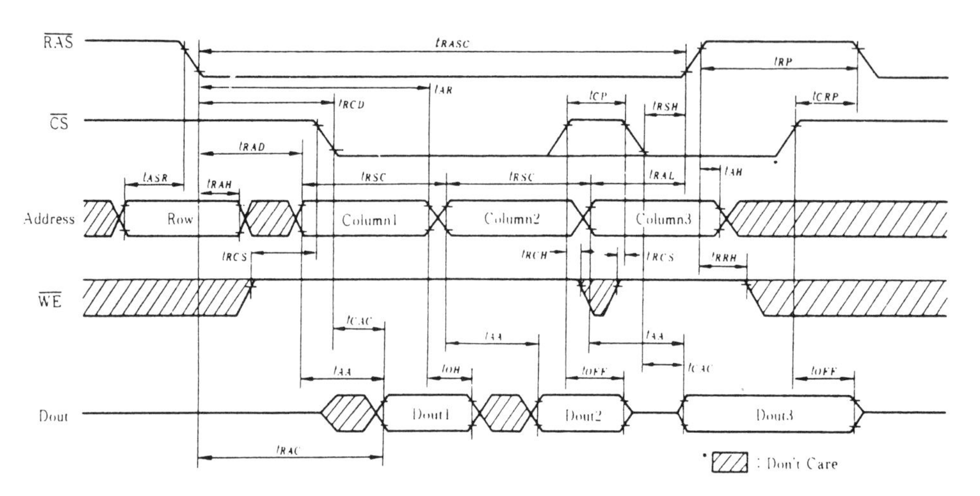

- **Static Column Mode Write Cycle**

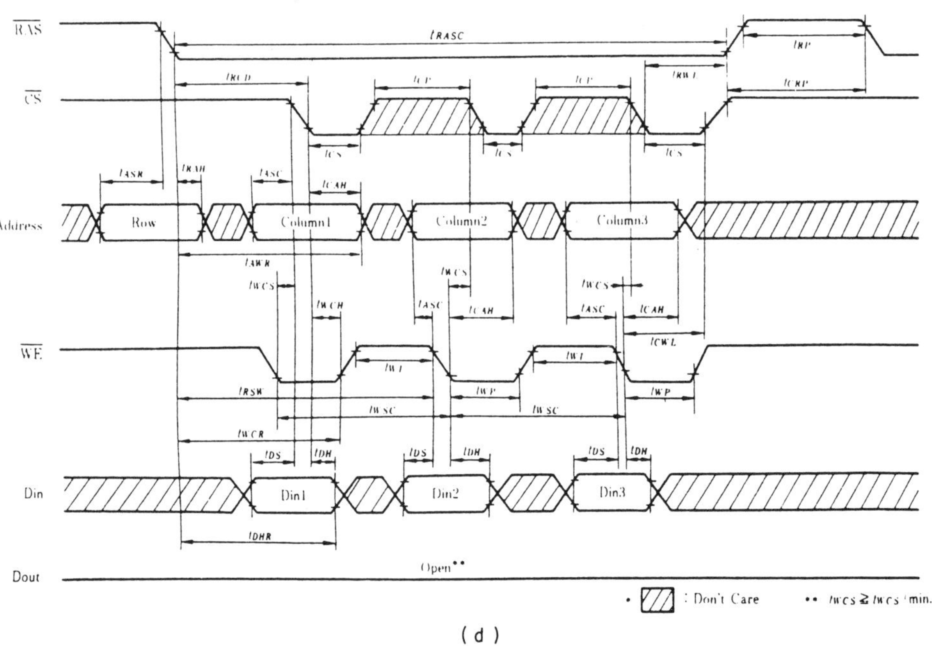

(d)

Figure 6.23(d) *(continued)*

the maximum $\overline{RAS}$ low time. Static column operation has been subsequently offered on most versions of CMOS DRAMs.

In conventional DRAMs, the CAS signal has three functions: it activates the column circuitry, it is the strobe for column address inputs, and it is the strobe for the output enable signal.

The static column and I/O circuits of the CMOS DRAMs permit the $\overline{CAS}$ function in triggering the column circuitry to be eliminated by an internal read operation of the column circuitry which is signalled by a transition of the address inputs rather than by a falling edge of the $\overline{CAS}$ signal. This excludes the circuit delays in a $\overline{CAS}$ buffer. The role of CAS as strobe for column address inputs is retained during use of the fast page mode however it is eliminated in the static column mode since this mode requires no toggling of the $\overline{CAS}$ clock. The role in controlling the output enable is retained.

It can then be replaced by a chip select ($\overline{CS}$) terminal which enables or disables the output allowing wired-ORs at the output. $\overline{CS}$ is not a clock and has no timing restriction for $\overline{RAS}$. It can be applied at any time. If it is not required, it can be grounded to simplify operation. It also controls the write function and must be low along with the $\overline{WE}$ pin for write to start. Figure 6.23(d) shows the static column mode timing for a part with the $\overline{CS}$ function rather than the $\overline{CAS}$ function. The replacement of $\overline{CAS}$ by the $\overline{CS}$ function was generally adopted on higher density CMOS DRAMs. Both the Mostek and Toshiba 1Mb DRAMs were among those to replace $\overline{CAS}$ with $\overline{CS}$.

A fast column mode clocked by chip enable was suggested by AT&T [58] for their 1Mb DRAM. This mode was similar to page mode but the output was kept active between successive $\overline{CE}$ read cycles. Column addresses were strobed in at the falling edge of $\overline{CE}$ so that address skew did not affect the access time. In this mode, the rising edge of $\overline{CE}$ did not three-state the output. Also a column cycle once initiated was allowed to finish even if $\overline{CE}$ went high again before the access was complete. In this case internal timing was provided for the recovery of the column circuitry

6.8.3 Improved sense amplifiers for the 1Mb DRAMs

The increasing length of the bit-lines in the higher density DRAMs which resulted in higher capacitance ratios and increasing RC delay times required new sense architectures.

In the conventional shared sense amplifier [50], shown in Figure 6.24(a), the switching devices placed at the right-hand side of the sense amplifier selectively disconnect the right-hand segment bit-line pair from the sense amplifier when the left-hand side is selected. The voltage in the left-hand segment bit-line pair is sensed and restored by the center NMOS latch through the left switching devices and the left PMOS latch. After the above operations, the right switching devices reconnect the right segment bit line pair to the sense amplifier which introduces the voltage difference to the right segment bit-lines. The voltage difference is then amplified by the center NMOS latch through the right switching devices and the right PMOS latch. The switching devices act as both cutoff and reconnect transistors in each cycle which brings an increase in the access time penalty.

Mitsubishi in 1987 on a 1Mb DRAM [50] introduced an alternative to the conventional shared sense amplifier, which they called the distributed sense and restore amplifier (shown in Figure 6.24(b)). This was designed to minimize the access time penalty of the shared sense amplifiers. In this case the bit-line is also divided into

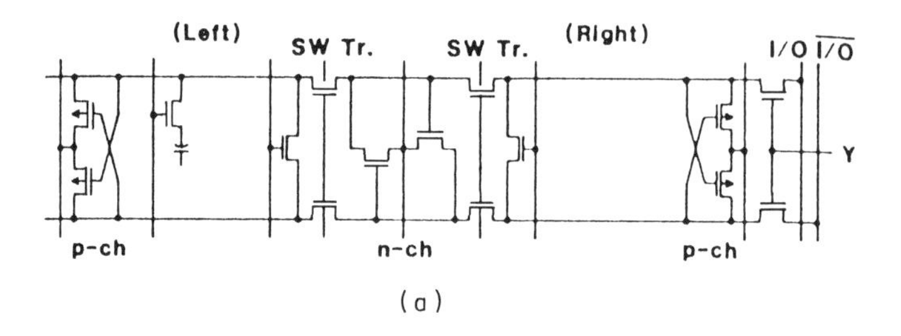

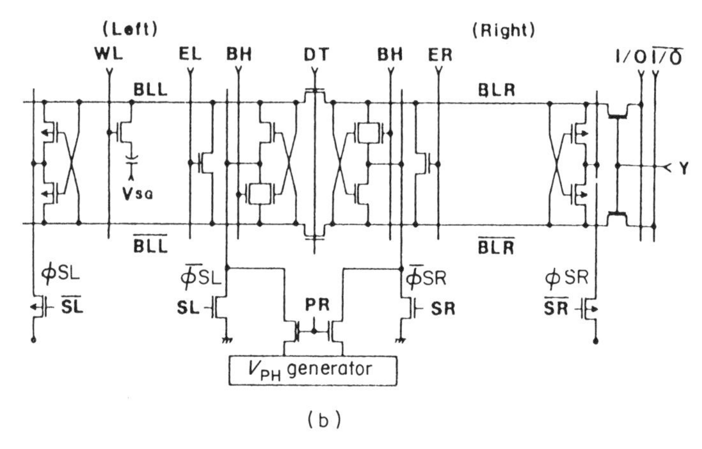

Figure 6.24
Schematics illustrating shared sense amplifier concept. (a) Conventional shared sense amplifier, (b) distributed sense and restore amplifier (DSR). (From Miyamoto *et al.* [50], Mitsubishi 1987, with permission of IEEE.)

two segments and the NMOS sense amplifier and the PMOS restore circuit are connected to each segment. The column decoder and I/O bus lines are shared by eight 64k memory arrays.

This sense circuit halves the bit-line length thus improving the C_b/C_s ratio. The segment bit-line pair is sensed and restored by the latches in each side in the amplifier structure, which allows earlier reconnection of the left segment bit-line pair to the right one and as a result the access time penalty in the new structure is much smaller than in the conventional shared sense amplifier scheme.

Mostek [40] dealt with the *RC* delay and capacitive coupling effects by using an innovative divided bit-line architecture which was evolved from a combination of the folded-bit line techniques and the shared sense amplifier concept as shown in Figure 6.25. This bit-line architecture divided the long columns into 16 polysilicon bit-line segments of 64 cells each with eight segments arranged end to end in a line on either side of a central column decoder. Adjacent segments were grouped into pairs of open bit-lines to form eight memory blocks of 128k bits each. Blocks were then formed of bit-lines running parallel to the segments in folded bit-line manner. This disconnected $\frac{3}{4}$ of the matrix capacitance in any one RAS cycle so that both a high signal level and low power consumption were obtained.

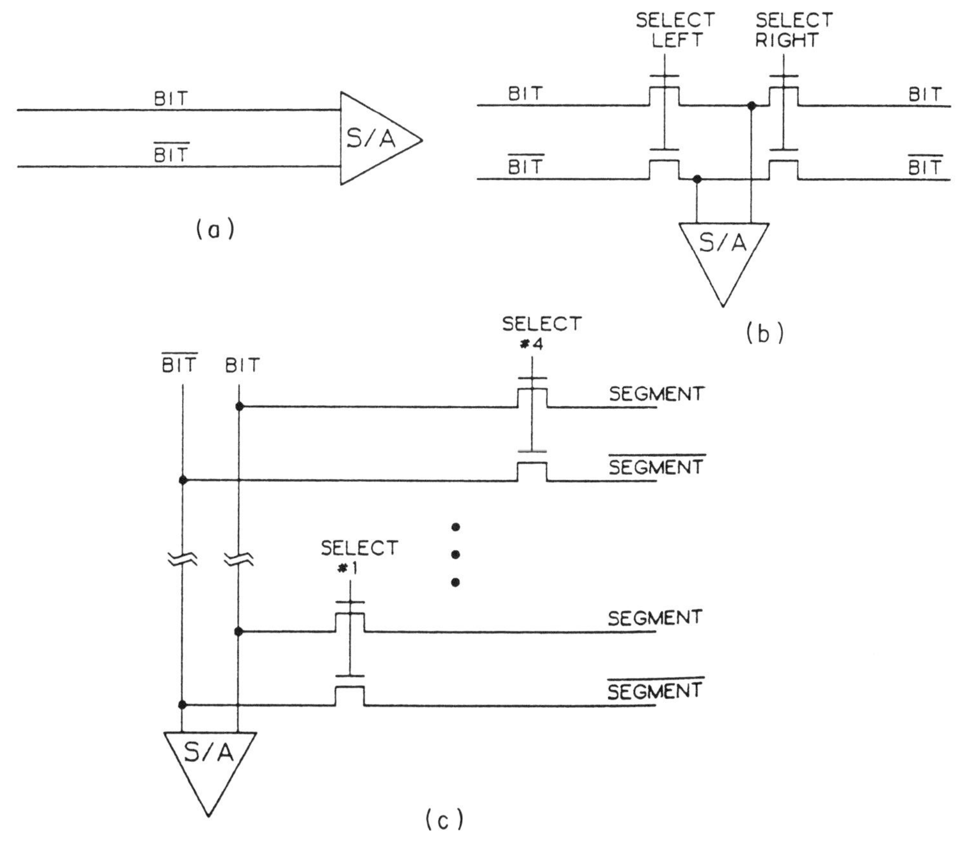

Figure 6.25
Evolution of multiplexed DRAM sense amps. (a) Folded bit-line. (b) Shared sense amplifier. (c) Divided bit-line. (From Taylor and Johnson [40], SGS-Thomson (Mostek) 1985, with permission of IEEE.)

A high speed sense amplifier was introduced by Toshiba [47] in 1986 on a 1Mb DRAM as shown in Figure 6.26 along with simulated waveform comparisons of the conventional and the new sense amplifier. This new amplifier includes a pair of barrier transistors in the flipflop itself. The gates of the barrier transistors are in the ON state during sensing and hence act as resistors to improve the response of the N-channel flipflop nodes in the latch operation. This amplifier affords faster latching for small signals than conventional sense amplifiers. It was also used by Toshiba in their 4Mb DRAM introduced in 1986 [49].

6.8.4 Half V_{CC} bit-line sensing schemes on DRAMs

On the Intel part the sensing circuitry used bootstrapped word-lines to allow a full supply voltage level to be written into the memory cell. The bit-lines were actively

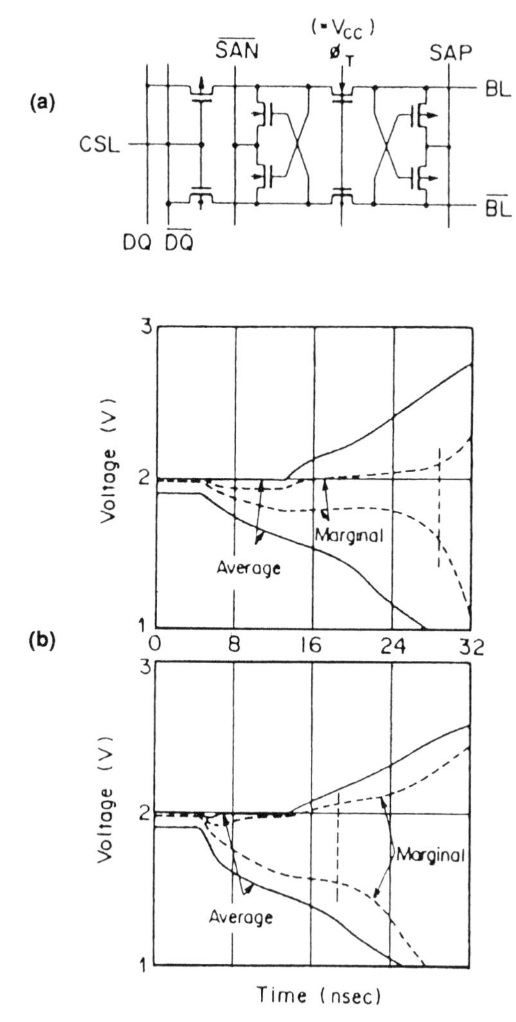

Figure 6.26
Sense amplifier for the IMb DRAM with barrier transistors in the flipflop for improved sensitivity. (a) Sense amplifier circuit schematic, (b) simulated waveforms. (From Fujii *et al.* [47], Toshiba 1986, with permission of IEEE.)

restored by N-channel latches. They were actively held at V_{DD} or V_{SS} during the normal precharge time. The signal sensing was done with half V_{CC} level on the bit-lines to eliminate the need for a dummy cell. Half the bit-lines were sensed from half V_{CC} to V_{CC} while the other half were restored from half V_{CC} to V_{SS} during the active cycle. The half V_{CC} sensing scheme speeded up the charge transfer from the memory cell to the bit-line and enhanced access time.

Also in 1984, IBM discussed a half-V_{CC} bit-line sensing scheme in N-well CMOS DRAMs [39]. The study showed that half-V_{CC} bit-line sensing has several advantages for high performance high density CMOS DRAMs with N-well CMOS arrays. One is the reduction of peak currents at both sensing and bit-line precharge by almost a factor of two due to the half V_{CC} swing. This reduces the electromigration problem and the IR drop so chip reliability can be increased. Also the resulting narrower metal lines decrease the parasitic wiring capacitances giving better speed.

Half V_{CC} bit-line sensing also reduces the dI/dt by a factor of two during bit-line precharge and discharge which decreases the voltage bounce noise due to wiring inductance and means that the precharge and discharge time can be shortened by a factor of two. It also reduces the ac power for charging and discharging the bit-lines because the precharge voltage is obtained by charge sharing between the two bit-line halves instead of charging bit-lines to V_{CC}.

In addition, since the pullup and pulldown of bit-lines are balanced and have only half V_{CC} swing, by using folded bit-lines, coupled noises due to bit-line swing to the memory cell plate, the array substrate and word-lines can be reduced. This also relaxes the requirements for using a low-impedance cell plate during voltage bumping.

The only drawback seen was a longer period during which the bit-lines float making them susceptible to radiation induced soft errors. It was suggested that in addition to other measures that this might be a reason to use a dummy cell to generate reference potential so that the absolute bit-line precharge level and the word-line to bit-line coupling noise are not of first-order magnitude. Dummy cells are also frequently included when a $V_{CC}/2$ bit-line precharge is used to compensate for capacitive coupling of the word-line pulse [71].

6.8.5 $V_{CC}/2$ biased cell plate

A $V_{CC}/2$ bias on the cell plate has a number of advantages. It reduces the stress of the electric field across the thinner oxide of the megabit generation storage capacitor, and reduces the influence of supply voltage variation. V_{CC} bump becomes symmetrical for stored '1' and '0' so that disturbs are minimal. It also reduces power dissipation which is a function of the square of the voltage.

While the experimental work on the cell plate biased at $V_{CC}/2$ was done earlier, the 1Mb DRAMs were the first commercial devices to use this technique extensively. Both the Mitsubishi [44] and the Toshiba [41] 1Mb DRAMs presented in 1985 had $V_{CC}/2$ biased cell plates.

6.8.6 Voltage generators for the 1Mb

By 1985 when the majority of the 1Mb DRAMs appeared the $V_{CC}/2$ bit-line precharge was a familiar feature, as was the cell plate biased to $V_{CC}/2$ to limit the stress on the thin oxide. Since this voltage was generated on the chip, low power and high drive current voltage generators were needed.

An example of a $V_{CC}/2$ voltage generator from the 1986 Toshiba 1Mb DRAM paper [47] is shown in Figure 6.27. This generator had milliampere level current

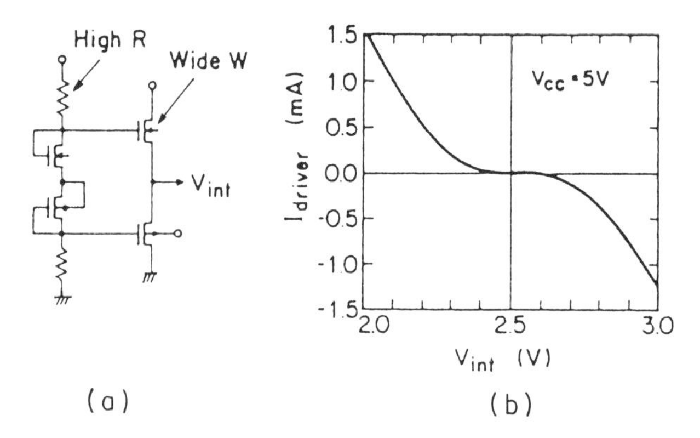

Figure 6.27
Half V_{CC} voltage generator. (a) Circuit schematic of generator, (b) example of *I–V* characteristics. (From Fujii *et al.* [47], Toshiba 1986, with permission of IEEE.)

drivability with microampere level current dissipation. It consisted of a bias stage and a driver stage. An example of the *I–V* characteristics of this generator is shown in the figure.

Another example of a 1Mb DRAM voltage generator was a backbias generator used by Toshiba to bias the P-channel substrate at a negative potential. The use of back bias provided reduced junction capacitance, protection of data against external signal undershoot and electrical isolation between n-type junctions.

6.8.7 Address transition detection for power reduction

The Toshiba 1Mb DRAM used clocked circuitry with address transition detection for high speed operation. This permitted partial activation of one block out of four of the memory array to reduce power consumption.

The Hitachi CMOS 1Mb also used ATD techniques to reduce power dissipation during long cycles. The ATD clock activated a column driver clock, a column select signal and a main amplifier for a limited period of time.

6.8.8 Test modes for the 1Mb DRAM

Since the 1Mb DRAM had four times the number of bits as the 256k, it took about four times as long to test the memory. Several types of parallel test modes were developed which allowed multiple bits to be tested in the time that one normally is tested. Entry into these test modes was typically by applying a voltage higher than V_{CC} to an extra pin which was usually the no-connect (NC) in the standard pinout. Mitsubishi [44] and NEC [54] both used a 4 bit test mode which could test four bits in parallel.

TI [60] used an 8 bit parallel test mode which compared the output data against the expected data on the tester. If the data differed then the complement of the original data appeared on the outputs. In addition the TI part included various other tests such as refresh disturb, leakage, margin and V-bump. For more on DRAM test modes see Chapter 14.

6.8.9 Reducing the resistance of long poly word-lines

The resistance of the long polysilicon word-lines was reduced by using a layer of aluminum wiring as a jumper which connected to the long poly word-line at regular intervals.

For example, NEC [46] used aluminum wiring to connect to a polysilicon word-line at periodic intervals to reduce the resistance of the poly word-line on a 128k × 8 DRAM introduced in 1984. Toshiba also on their 1Mb DRAM [49] formed the word lines in the second polysilicon and connected it to the aluminum wiring running parallel and at 64-bit intervals. The third polysilicon–molybdenum polycide had an order of magnitude lower sheet resistance than conventional polysilicon and was used for the bit-line.

6.9 THE 4Mb DRAM

6.9.1 Overview

The multi-megabit DRAMs brought in a new age of innovation.

Vertical cell techniques were developed for the submicron line widths of the technology. Special fast access modes were introduced to avoid wait states with the new high performance microprocessors. Low voltage internal circuitry was developed and low voltage interfaces appeared.

The first 4Mb DRAM papers began to appear in 1984 with three 4Mb DRAMs discussed at the 1986 ISSCC from TI, NEC and Toshiba. Five more followed in 1987 from Mitsubishi, IBM, Hitachi, Matsushita and Fujitsu, and one in 1988 from Siemens. Eight of these early 4Mb DRAMs used the trench capacitor cell. Two used the stacked capacitor initially and for production chips several later converted to the more easily manufactured stacked capacitor cell.

The 4Mb DRAMs began to ramp into volume production in 1989. This was also a generation of exploration of applications specific variations of DRAMs which will be covered in more detail in Chapter 7. A chart of various early 4Mb DRAMs is shown in Figure 6.28.

The 4Mb generation saw continued experimentation with the vertical cell capacitors. The parallel test modes were adopted by many suppliers to shorten test time. Refresh was standard at 1024 cycles in 16 ms. Most devices used 0.8 to 0.9 μm minimum geometries and four to five layers of interconnects. Chip size remained, after some experimentation, on the historical trend line.

Year	Company	L_{ef} (μm)	Technology	Chip size (mm^2)	Cell size (μm^2)	Capacitor type	Array efficiency (%)
1986	TI	1.0	2P2M	100	8.9	Trench	36
1986	Toshiba	1.0	3P1M	137	17.4	Trench	52
1986	NEC	0.8	2P1M	99.2	10.6	Trench	44
1987	IBM	0.8	1P2M	78	10.6	Trench	62
1987	Mitsubishi	0.8	3P1M	72.3	10.9	Trench	62
1987	Hitachi	0.8	2P2M	110.8	14.7	Stack	54.3
1987	Matsushita	0.8	2P2M	67.1	8.0	Trench	48.8
1987	Fujitsu	0.7	4P1M	63.7	7.5	Stack	48.2
1987	Toshiba	0.9	2P2M	111.2	13.7	Trench	50.5
1988	Siemens	0.75	2P2M	91.3	10.6	Trench	49

Figure 6.28
Comparison of some early 4Mb DRAMs. (Source ISSCC Proceedings 1986–1987.)

6.9.2 Vertical cell structures—trench and stacked capacitors

The size requirement of the 1Mb DRAM to fit into the 300 mil standard package determined the need to investigate other cell concepts than the planar capacitor structure which had been used since the 16k DRAM. This ended the era of simple scaling of DRAMs and brought in another era of cell innovation.

These innovations included the trench capacitor cell which was developed by Hitachi [115] and the stacked capacitor cell developed by Fujitsu [56], both of which are briefly described in Chapter 4 for the technology.

The main benefit of the trench concept was that the capacitance of the cell could be increased by increasing the depth of the trench without increasing the surface area of silicon occupied by the cell. The major disadvantage was the difficulty of manufacture.

The benefit of the stacked capacitor was the ability to use standard planar silicon processing with the capacitor being contained in the multiple interconnect layers above the silicon. The disadvantage initially was smaller capacitance than the trench, requiring special high dielectric constant materials for the insulator. A later disadvantage at higher densities was the additional height of the array above the level of the circuit periphery. The reduced depth of focus of the lithography equipment used for deep submicron geometries required separate processing steps for the array and the periphery in some cases, thereby increasing the cost of the stacked capacitor process.

Most of the early 4Mb DRAMs used the trench capacitors. The vertical evolution of the simple trench capacitor DRAM cell was shown in Chapter 4. The evolution of the DRAM capacitor to more complex structural types is shown in Figure 6.29.

The earliest trench cell, as developed by Hitachi [115] in 1982, is shown in Figure 6.29(a) and consists of a planar capacitor cell with the capacitor folded down into the trench. The charge is stored in the substrate and oxide isolation was grown on the wall of the trench and a polysilicon field plate was deposited in the trench.

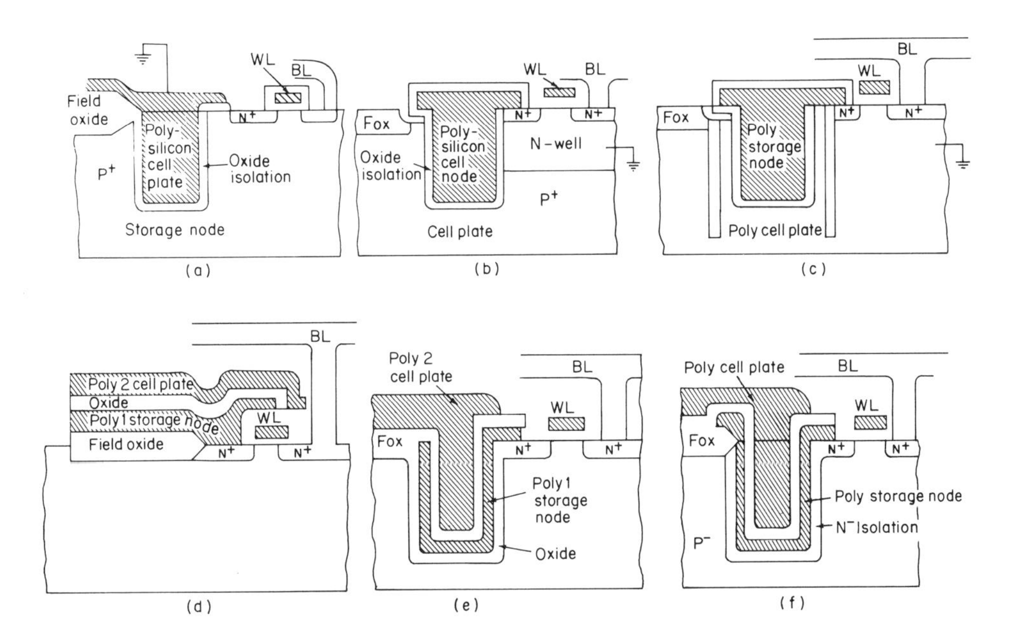

Figure 6.29
An evolution of vertical DRAM cells. (a) Early trench capacitor cell with substrate storage node. (corregated capacitor). (From Sunami *et al.* [115], Hitachi 1982, with permission of IEEE). (b) Trench cell with isolated storage node (From Sakamoto *et al.* [118], NEC 1985, with permission of IEEE). (c) Trench cell with isolated cell plate and storage node (From Taguchi *et al.* [122], Fujitsu 1986, with permission of IEEE). (d) Stacked capacitor cell (From Takemae *et al.* [56], Fujitsu 1985). (e) Buried stacked capacitor cell (From Yoshioka *et al.* [116], Oki 1987, with permission of IEEE). (f) Stacked trench capacitor cell (From Furuyama *et al.* [49], Toshiba 1986, with permission of IEEE.)

Cell to cell spacing was limited due to the leakage current which could result in punch through between adjacent trenches. This leakage phenomena increased significantly at trench-to-trench spacings below 2.0 μm.

An additional problem was the large surface storage area exposed to alpha radiation generated by minority carriers in the substrate. This was difficult to compensate for with shallow CMOS wells due to the deep trenches.

In 1985 TI [117], IBM [119] and NEC [118, 123] proposed a variation of the trench cell with the storage node in the trench isolated from the substrate and the cell plate in the substrate. NEC's cell is shown in Figure 6.29(b). The IBM and TI cells were shown previously in Chapter 4. In this cell the punchthrough problem is eliminated except in the region where the charge is introduced into the cell. NEC in 1986 showed the cell concept with the addition of a self aligned contact for use at a higher densities [123].

Even with the cell with the buried storage electrode there was still a limitation to the trench spacing since the cell plates needed to be separated by a minimum distance. The ideal cell would have both the storage node and the cell plate isolated from the substrate. The difficulties in manufacture of the trench cell were an additional problem.

One attempt to deal with this problem by increasing the complexity of the trench was demonstrated by Fujitsu [122] in 1986 and is illustrated in Figure 6.29(c). This cell encapsulated the buried polysilicon cell plate in a cylinder of oxide to isolate it from the substrate. This cell, however, increased the difficulties of manufacture of the trench and investigations turned in other directions.

Meanwhile, many manufacturers for their production 4Mb DRAMs turned to the stacked capacitor which was described in Chapter 4 and earlier in this chapter. The stacked capacitor cell, which was first shown in 1980 [135], provided an initial solution to the problems of the trench capacitor. A stacked cell presented by Fujitsu [56] in

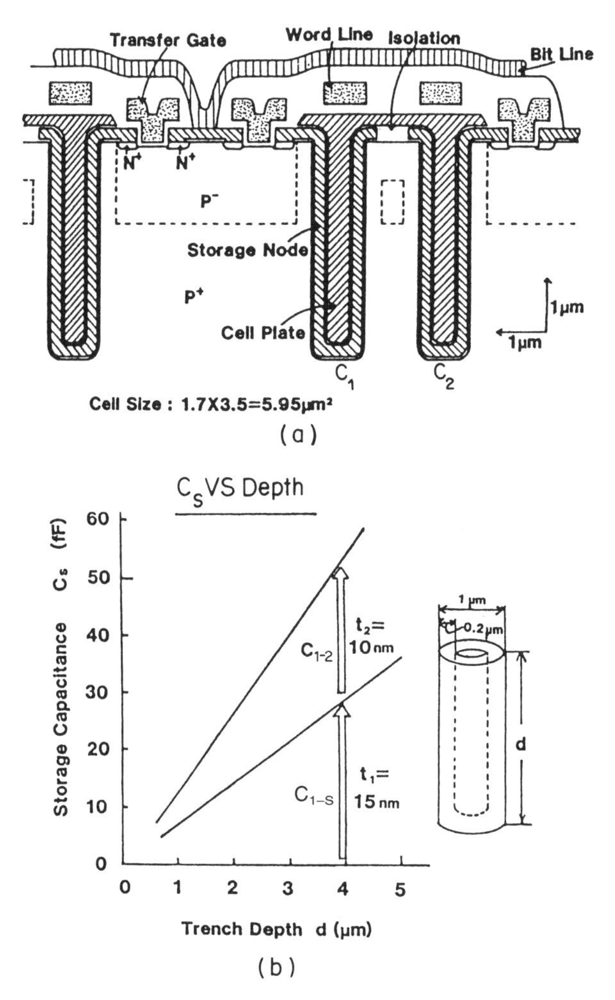

Figure 6.30

(a)'Stacked' trench capacitor (DSP cell). Cell size 1.7 × 3.5 = 5.95 μm^2. (b) Storage capacitance for concentric capacitors C_{1S} and C_{12} showing approximate doubling of capacitance. (From Tsukamoto *et al.* [83], Mitsubishi 1987, with permission of IEEE.)

1985 is shown in Figure 6.29(d). It forms the capacitor between two dielectrically separated sheets of polysilicon which are isolated from the substrate by the field oxide. The capacitor is stacked up and over the word-line providing additional surface area for the storage of charge. Unfortunately, the additional area was not able to provide sufficient storage with a simple stacked capacitor unless exotic dielectric materials are used again increasing the cost of the process.

The stacked capacitor concept, however, led the industry in two directions. The first was to more architecturally complex stacks as described in Chapter 4. The other was to more sophisticated trench architectures which combined the advantages of the stacked and the trench capacitor.

One of these was a buried stacked capacitor cell demonstrated by Oki in 1987 [116]. This cell combined the two concepts by folding the stacked capacitor into the trench to provide additional storage area as shown in Figure 6.29(e). This cell has both the storage node and the cell plate in the trench isolated from the substrate and from each other by layers of oxide. This cell solves the isolation problem of the trench and the capacitance problem of the stacked capacitor. However, it is also complex and difficult to manufacture.

In 1986 Toshiba [49] also presented their stacked trench capacitor which kept the buried stacked capacitor concept but simplified the manufacturing by replacing the field oxide layer with an outdoping of 'N' type dopant from the first polysilicon deposition in the trench thereby eliminating the additional oxide layer again, see Figure 6.29(f). It retained the effective isolation of the cell plate from the substrate thereby allowing the use of a $V_{CC}/2$ cell plate to reduce the voltage stress at smaller geometries.

Further innovations in both stacked and trench capacitor technology and architecture followed. Mitsubishi [83] in 1987 also showed a stacked trench capacitor which was effectively composed of concentric capacitors stacked in a trench as shown in Figure 6.30. The storage capacitance is almost doubled over that of a simple trench capacitor as shown in the graph in Figure 6.30(b). The cell area is 5.95 μm^2 using 0.7 μm geometries. The Toshiba cell, which has similar construction, is 6.12 μm^2 and uses 0.7 μm design rule (0.5 μm minimum geometry).

6.9.3 DRAM cell architectures

One of the first 4Mb DRAM papers from IBM [36] in 1984 was on a shared word line DRAM cell which yielded a low performance device but with twice the density of conventional DRAM architecture.

In the shared word-line DRAM cell, a pair of cells is connected to the same bit sense line and word-line. Unique read and write operations are accomplished by controlling the plate of the storage capacitor. A schematic of the shared word-line cells and operational pulse diagram are shown in Figure 6.31. Shared word-line cells can be placed in mirror image position about the bit-line, sharing the same word-line, thus the sense amplifier pitch is inherently two cell pitches. This reduces the impact of bit-line pitch on sense amplifier design and permits an efficient sense amplifier layout.

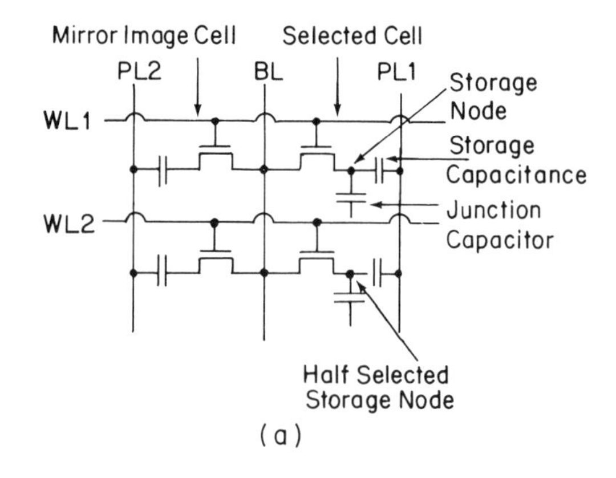

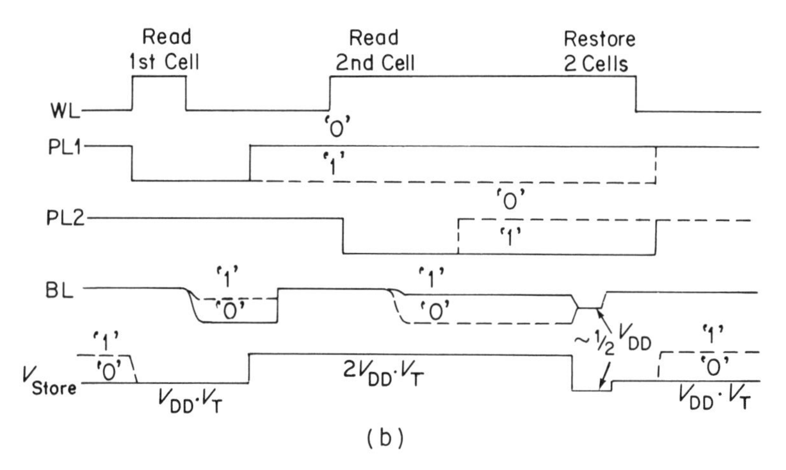

Figure 6.31
(a) Schematic of shared word-line DRAM cell. (b) operational pulse diagram. (From Scheuerlein *et al.* [36], IBM 1984, with permission of IEEE.)

6.9.4 On-chip voltage convertors

The use of on-chip voltage limiters or converters was due to the requirement for shorter channel length MOS transistors in the memory array than in the peripheral circuitry to obtain the needed chip size with a 1Mb density. The shorter channel length devices in the array could not withstand the 5 V operating voltage without hot electron injection effects into the thin oxide gate film which affected reliability. A voltage of less than 4 V came to be commonly used in the arrays with voltage limiters between the array and the peripheral circuitry. The latter was formed of transistors with longer channel lengths which could withstand the 5 V external voltage supplied to the chips.

One of the problems inherent in the use of such convertors was that it was difficult to bypass the convertors to have access to the transistors in the matrix for reliability stress testing.

One solution to this problem was offered in a paper from Toshiba [49] which offers the means to bypass such a convertor by means of a metal mask option for reliability testing. The voltage convertor used by Toshiba is shown in Figure 6.32 along with its output characteristics.

This convertor consists of a reference voltage generator and a current mirror amplifier with an output control transistor. The reference voltage generator produces a constant reference level of 3.5 V which is relatively independent of the external V_{CC} level. Since fast response is important, the current mirror has two load transistors.

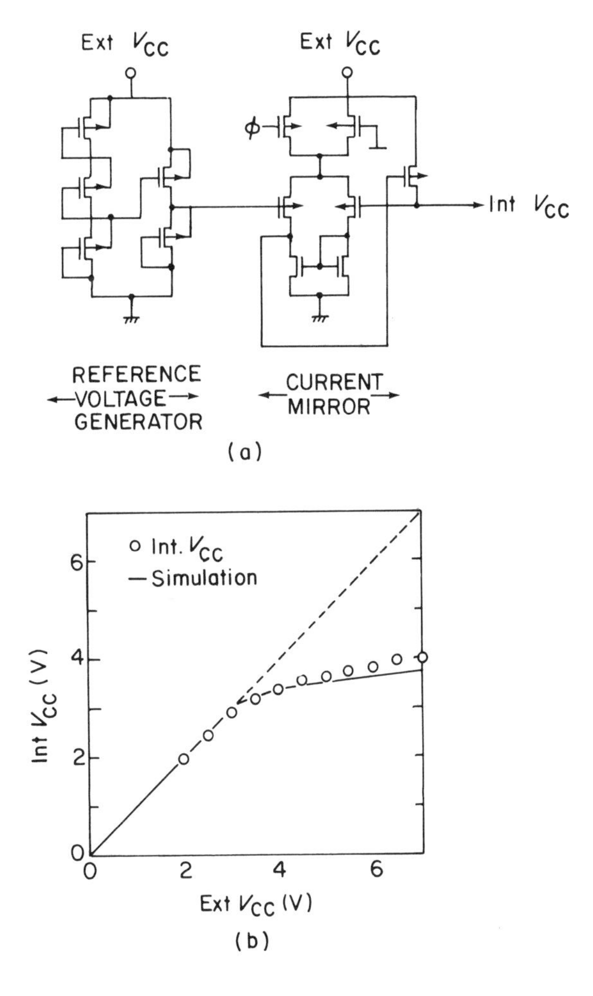

Figure 6.32
On-chip voltage limiter–converter and output characteristics. (a) On-chip voltage converter circuitry, (b) output characteristics as a function of external V_{CC} level. (From Furuyama [49], Toshiba 1986, with permission of IEEE.)

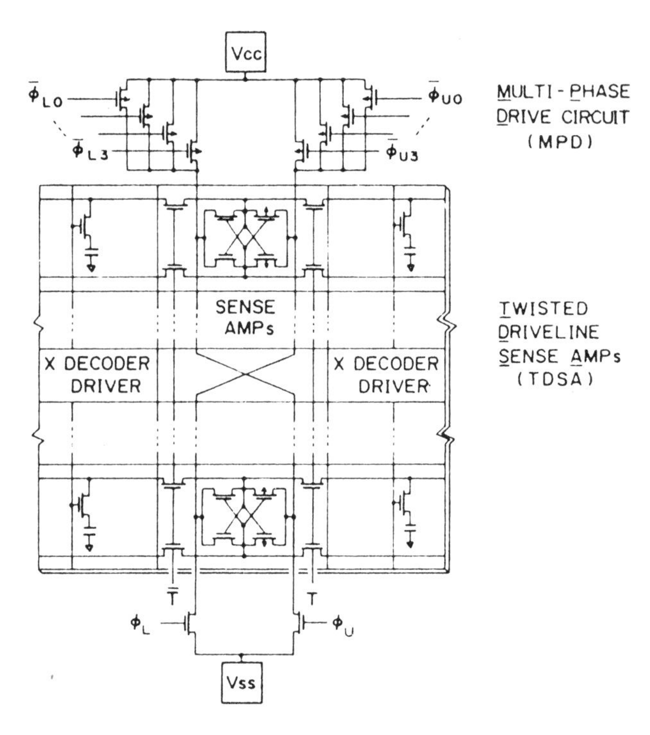

Figure 6.33
Array architecture for the 4Mb DRAM showing the twisted drive-line sense amplifier. (From Kimura *et al.* [51], Hitachi 1987, with permission of IEEE.)

A smaller conductance transistor is always on and the other transistor with larger conductance is controlled by a signal ϕ. As the current through the amplifier increases, so does the response speed. The larger conductance transistor is turned on in order to minimize the internal V_{CC} bounce due to the large transient current during bit-line charging and is off during the rest of the cycle to save power.

6.9.5 Circuits for noise reduction in 4Mb DRAMs

A further innovation at the 4Mb level was from Hitachi [51] who introduced a twisted driveline sense amplifier configuration and a multiphase drive circuit.

One of the problems inherent in the higher density of the multimegabit DRAMs is the increased parasitic capacitance and resistance of the large number of word and bit-lines. The two major problems arising from this are slow sensing speed and high peak power supply current. The increase in number of bit-lines and resistance increased the RC time delay and limits the sensing speed. The noise margin is impaired by the peak power supply current.

The parasitic capacitance problem can be reduced by the twisted driveline sense amplifier scheme from Hitachi [51] as shown in Figure 6.33. In the conventional scheme, n pairs of bit-lines are charged and discharged simultaneously. In the new scheme the two drivelines of the CMOS sense amplifiers crisscross in the middle of the array so that when the sense amplifiers in the upper half of the drivelines are activated, the sense amplifiers in the lower half stay inactivated. In this way the capacitance associated with each driveline is cut in half resulting in a faster sense amplifier operation.

The reduction of noise margins due to the increase in peak power supply current is handled with the introduction of a multiphase drive circuit, also shown as part of the diagram in Figure 6.33, to replace the usual two-phase drive circuit. This results in the usual sharp triangular waveform of the supply current being averaged into a trapezoidal-like shape.

6.9.6 Voltage bump at the 4Mb level

Matsushita [67] found that the $V_{CC}/2$ biased cell plate was not sufficient to solve the problem of high speed voltage bumping on 4Mb devices. A reduced cell plate bias of 2.3 V was used along with a 2.3 V bit-line precharge. This weaker dependence of V_{CC} reduced voltage loss due to V_{CC} bumping.

6.9.7 Transistor reliability concerns

There were several options explored for improving the hot electron immunity of the transistor at the 4Mb DRAM level. One was using the lightly doped drain which graded the electrical field and reduced the possibility of formation of hot electrons. Most of the 4Mb DRAMs used the lightly doped drain type structure.

In some cases an on-chip voltage convertor was used to maintain the external 5 V while providing the lower voltage across the cells required to reduce the hot electron effect as discussed in the previous chapter. This scheme was used by Hitachi [94] on their 16Mb DRAM.

The Hitachi internal voltage generator consisted of a voltage regulator which generated a stable voltage free from fluctuations of V_{CC} or temperature. The transistors in the memory array were scaled to 0.6 μm minimum channel length with 3.3 V applied voltage. In the periphery, with the larger 5 V applied voltage, the channel length is scaled only to 0.9 μm to avoid breakdown and hot electron effects.

Another solution, for reducing the hot electron degradation effect of the N-channel transistors, involved using a dual gate invertor structure, also called a 'cascode' structure elsewhere in the literature. This structure was described earlier by Terletzki and Risch [81] and used by Siemens [85] in their 4Mb DRAM. It is shown in Figure 6.34(a).

In this structure an additional N-channel transistor is connected in series to the N-channel switching transistor in the CMOS invertor structure. This distributes the high drain source voltage (V_{DS}) over two transistors. The gate of the upper transistor is connected to a reference voltage which is generated on chip. The

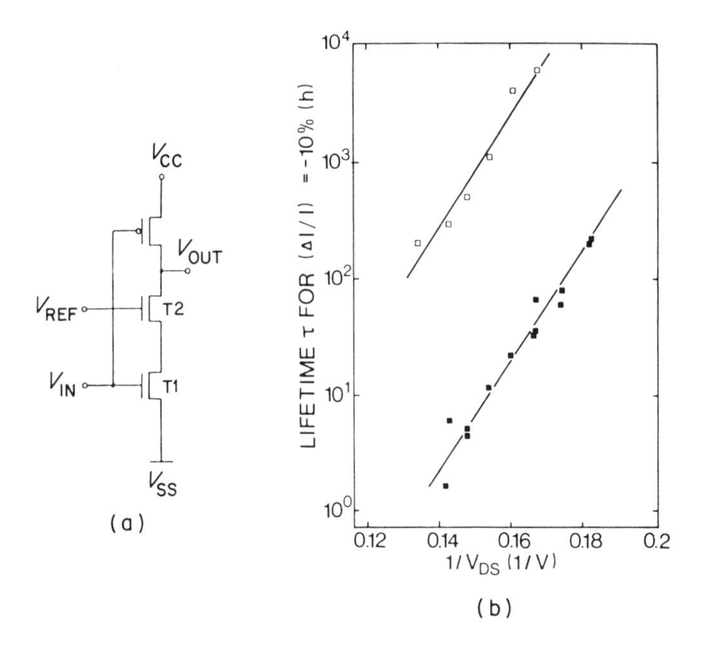

Figure 6.34
(a) Dual gate invertor (cascode) structure. (b) Life data for dual (□) as opposed to single (■) gate invertor structure. (From Harter [85], Siemens 1988, with permission of IEEE.)

reference tracks with V_{CC} to optimize speed and limit the maximum drain source voltage applied to each transistor. As shown in Figure 6.34(b), the lifetime of the dual gate transistors is two orders of magnitude higher than that of a single gate transistor.

6.9.8 New access and test modes

Other innovations occurred at the 4Mb density level to increase the data rate and reduce the test time of these high density parts. Parallel test modes similar to those developed for the 1Mb DRAM were also used at the 4Mb level.

Fujitsu [68] introduced a serial access mode on the 4Mb DRAM which used on-chip address counters that generate column addresses that could be mask optioned to give either serial access or nibble access mode.

Toshiba [69] implemented a 16 bit parallel test mode on their 4Mb DRAM, similar to the 4 bit parallel test modes that were introduced on the 1Mb DRAM. They also discussed a self test generator which will be described more fully in Chapter 14.

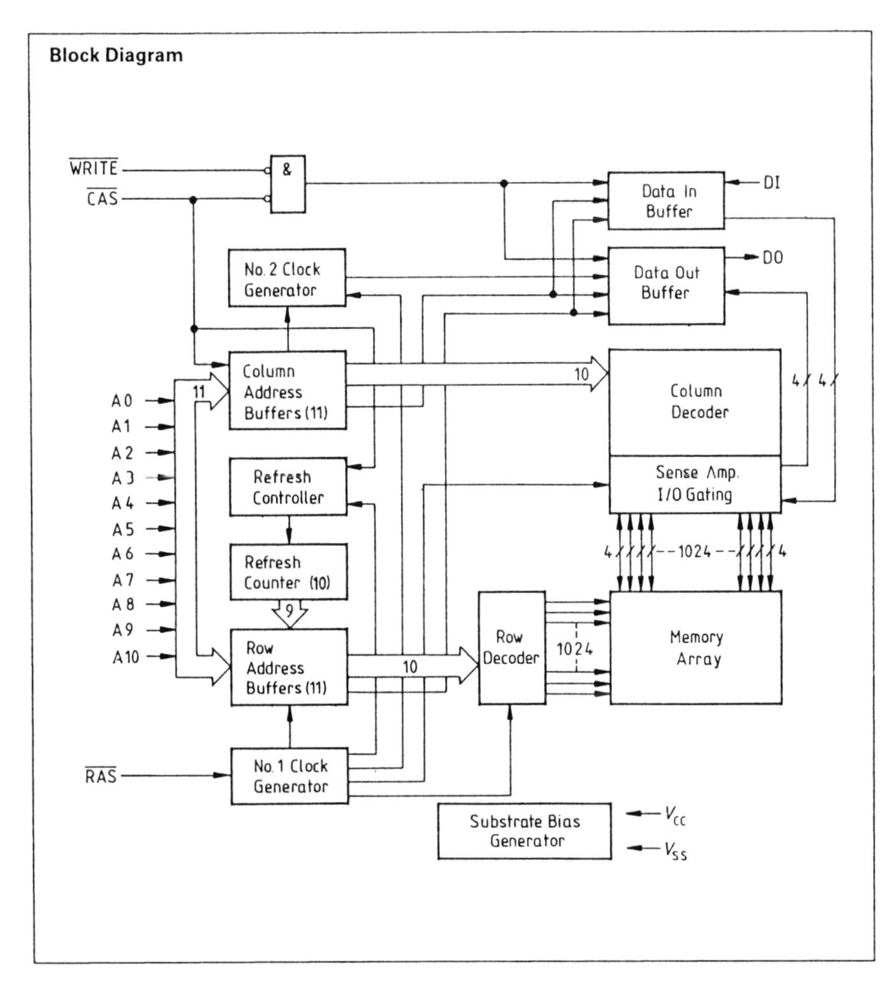

Figure 6.35
Block diagram of a typical 4Mb DRAM [52] (Reproduced with permission of Siemens 1989.)

6.9.9 A typical 4Mb production DRAM

The Siemens 4Mb CMOS DRAM is an example of a typical production part. It is 4Mb × 1 with 80 ns access time and 160 ns cycle time with a fast page mode cycle time of 45 ns. Maximum power dissipation is 495 mW active and 5.5 mW in standby.

It has the JEDEC standard pin configuration for the 4Mb DRAM. It includes an on-chip substrate bias generator as shown in the block diagram in Figure 6.35. This diagram also shows the on-chip refresh controller and refresh counter for the various self-refresh modes.

The part has the standard DRAM operational timing modes including read, write, read–modify–write and fast page mode. Refresh modes include RAS-only refresh, CAS-before-RAS refresh and hidden refresh during read.

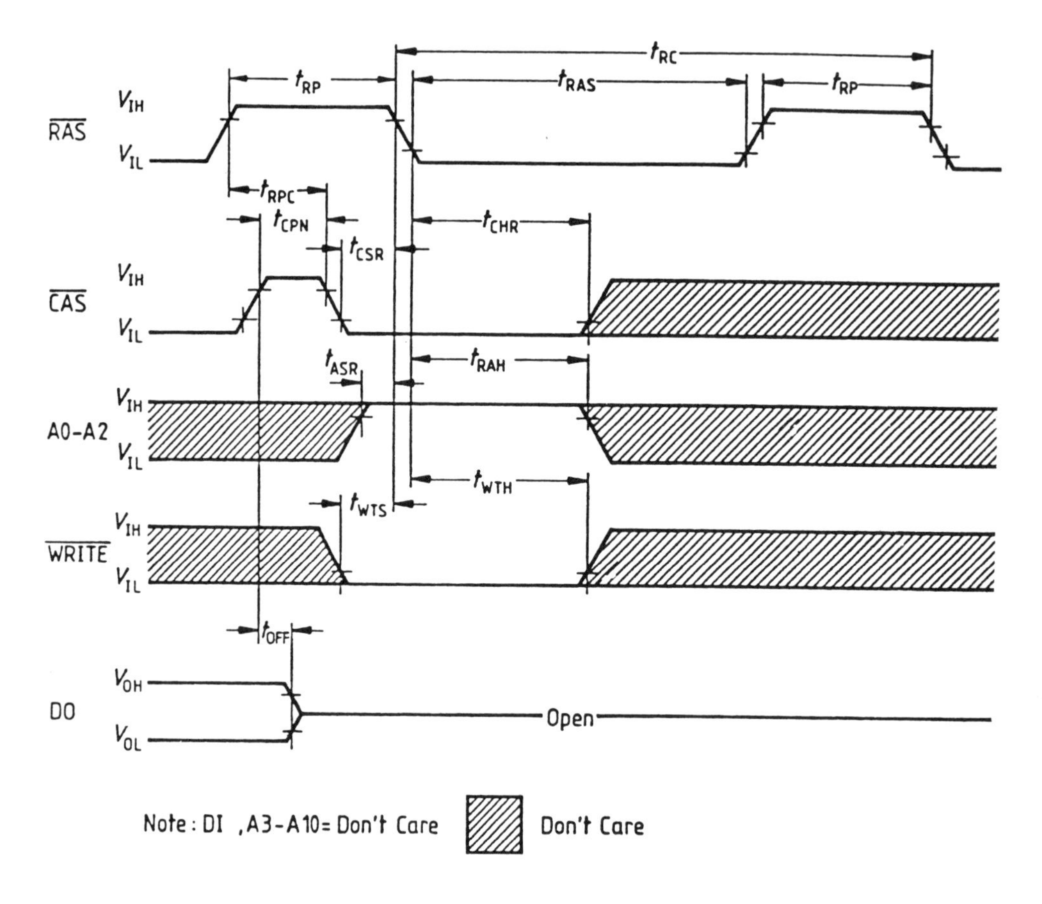

Figure 6.36
Schematic timing diagram of test mode entry cycle for 4Mb DRAM [52]. (Permission of Siemens 1989.)

In addition a test mode is included for enabling rapid test which can reduce the test time by a factor of $\frac{1}{8}$. Even though the part is organized externally as 4Mb × 1, it can be internally configured as 512k × 8. In test mode data are written into eight sections in parallel and retrieved the same way. If upon reading all bits are equal, that is all are '1' or all are '0', the data output pin indicates a '1'. If any bits differ, the data output pin indicates a '0'.

In test mode the 4Mb DRAM can be tested in 512k segments. The test mode entry cycle is shown in Figure 6.36 and is defined by the state of CAS, write, and the trailing edge of RAS. It consists of a write, and a CAS-before-RAS cycle. To exit test mode, a CAS-before-RAS refresh cycle or a RAS only refresh cycle will put it back into normal operating mode.

This test mode is consistent with the JEDEC standard basic test mode for 4Mb DRAMs as outlined in the JEDEC Standard 21 bulletin issued periodically by the American Electronic Industries Association.

6.10 16Mb DRAMs AND BEYOND

6.10.1 Overview of 16Mb DRAMs

By 1989, the 4Mb DRAM volume market was well underway, and by early 1990 at least eight companies had described 16Mb DRAMs at technical conferences, see Figure 6.37 for a chart of these early 16Mb DRAMs. In 1990 early samples of the 16Mb DRAM began to appear in the market from Samsung, TI, and others although early production did not begin until 1991.

Among the technical challenges at the 16Mb level were: to reduce the cell size without decreasing the sense amplifier input signal level and without decreasing the soft error immunity, to improve speed by increasing the transistor performance without increasing hot electron instability caused by the smaller geometries, and to reduce the gate material resistivity in order to lower word-line delay time and improve the speed. The first efforts to provide DRAMs with lower than 5 V power supply levels also were made with the 16Mb DRAM.

Sense amplifier considerations included concerns such as large peak currents when the sense amplifiers turn on, and asymmetries in the sense amplifiers due to the normal process variations. At the bit-line architecture level concerns were to layout the circuits to fit the sense amplifiers in the pitch of the bit lines which with the vertical capacitors were now close to the minimum design pitch. Innovative architectures were needed to reduce common mode noise in the bit sense lines. Finally it was needed to find ways of handling the reduced noise margin resulting from the impedance effects of long interconnect lines interacting with each other.

The functional block diagram and switching characteristics of a typical first generation 16Mb (4Mb × 4) DRAM are shown in Figure 6.38.

6.10.2 Vertical capacitors for the 16Mb DRAM

One of the problems faced by the multi-megabit DRAMs was that the cell capacitance was significantly reduced by making cells small enough to keep the DRAM chips in the size range required. Advanced vertical capacitors in the form of stacked capacitors which rose above the surface of the chip and of trench type capacitors which were buried in the surface of the chip were generally used in production in the 4Mb DRAM and beyond.

Year	Comapny	L_{eff} (μm)	Technology	Capacitor	Cell size (μm^2)	Chip size (mm^2)	TRAC (ns)	Voltage	P. D. (Active) (mW)
1987	NTT	0.7	2P/2M	N/A	4.9	147.7	80	3.3	500
1988	Toshiba	0.5	3P/3M	Stacked Trench	4.8	136.9	70	4/5	600
1988	Matsushita	0.5	2P/2M	Trench	3.3	93.8	65	3/4/5	450
1988	Hitachi	0.6	/2M	Stacked Trench	4.2	141.9	60	3.3/5	420
1989	Mitsubishi	0.5		Stacked Capacitor	4.8	134	60	3.3	450
1989	Samsung	0.55	3P/3M	Stacked Capacitor	4.4	134.7	60	4/5	500
1990	TI	0.6	3P2M	Trench	4.8	144	60	3/4/5	450
1990	IBM	0.5	2P/2M	Trench	4.13	140.9	50	3.3/5	

Figure 6.37
Examples of early 16Mb DRAMs. (Source ISSCC Proceedings 1987–1990.)

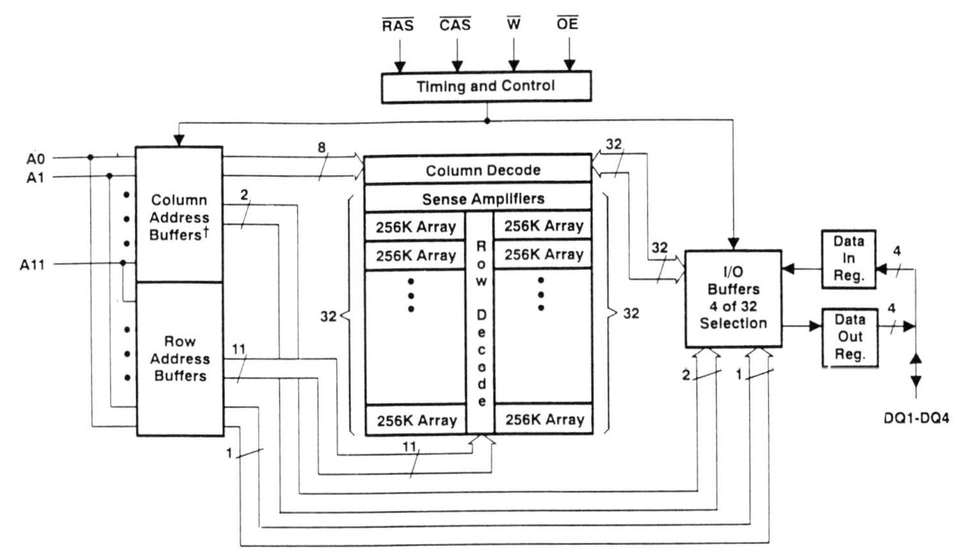

† Column Address 10 and Column Address 11 are not used.

(a)

switching characteristics over recommended ranges of supply voltage and operating free-air temperature

PARAMETER		TMS416400-60		TMS416400-70		TMS416400-80		TMS416400-10		UNIT
		MIN	MAX	MIN	MAX	MIN	MAX	MIN	MAX	
t_{AA}	Access time from column-address		30		35		40		45	ns
t_{CAC}	Access time from $\overline{CAS}$ low		15		18		20		25	ns
t_{CPA}	Access time from column precharge		35		40		45		50	ns
t_{RAC}	Access time from $\overline{RAS}$ low		60		70		80		100	ns
t_{OEA}	Access time from $\overline{OE}$ low		15		18		20		25	ns
t_{CLZ}	$\overline{CAS}$ to output in low Z	0		0		0		0		ns
t_{OH}	Output disable start of $\overline{CAS}$ high	3		3		3		3		ns
t_{OHO}	Output disable time start of $\overline{OE}$ high	3		3		3		3		ns
t_{OFF}	Output disable time after $\overline{CAS}$ high (see Note 6)	0	15	0	18	0	20	0	25	ns
t_{OEZ}	Output disable time after $\overline{OE}$ high (see Note 6)	0	15	0	18	0	20	0	25	ns

NOTE 6: t_{OFF} is specified when the output is no longer driven.

(b)

Figure 6.38
First generation 16Mb DRAM (a, functional block diagram of a 4Mb × 4 part; b, typical switching characteristics). (With permission of Texas Instruments.)

With the added vertical dimension in the cell the chip size continued to grow in the 4Mb and 16Mb densities on the trend line of the previous generations as shown in Figure 6.39 which shows DRAM chip size plotted against cell size for seven generations of DRAMs.

The vertical capacitors increased the capacitance relative to the cell size, but the absolute capacitance of the cell decreased. An example of this is shown in Figure 6.40 which shows the available cell capacitance for two types of stacked capacitor cells for a 1Mb, 4Mb and 16Mb DRAM with cell size, respectively, 22.6, 8.8 and 3.1 μm^2 [79]. Capacitance drops from 50fF for the 1Mb DRAM with conventional stacked capacitor to less than 30fF for the 16Mb DRAM even using the new stacked capacitor.

Further developments in the theory of vertical cells have been pursued for the 16Mb DRAM level and beyond. While it is not yet clear which of these will become the dominant production technology a few of the interesting theories and the problems they address will be described here.

Relatively simple trench and stacked capacitor structures were sufficient for the 4Mb DRAM. For the 16Mb and 64Mb with cell size requirements of 3 to 6 μm^2 and 1.5 to 3 μm^2 various innovations were attempted.

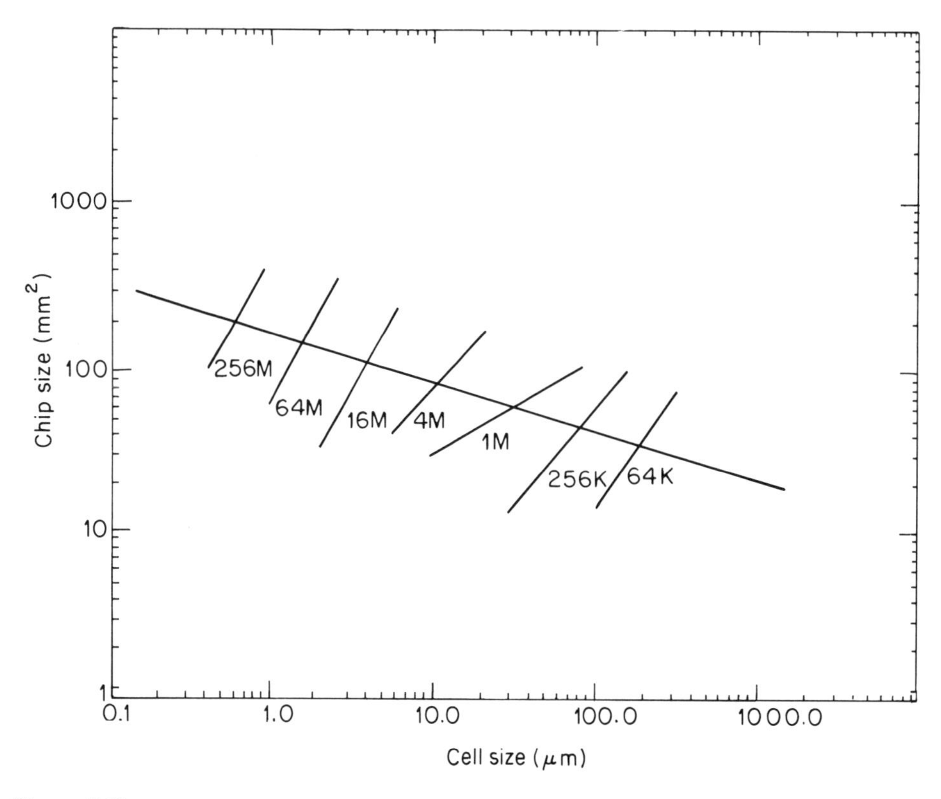

Figure 6.39
DRAM chip size plotted against cell size. (Source various ISSCC Proceedings, reproduced with permission of IEEE.)

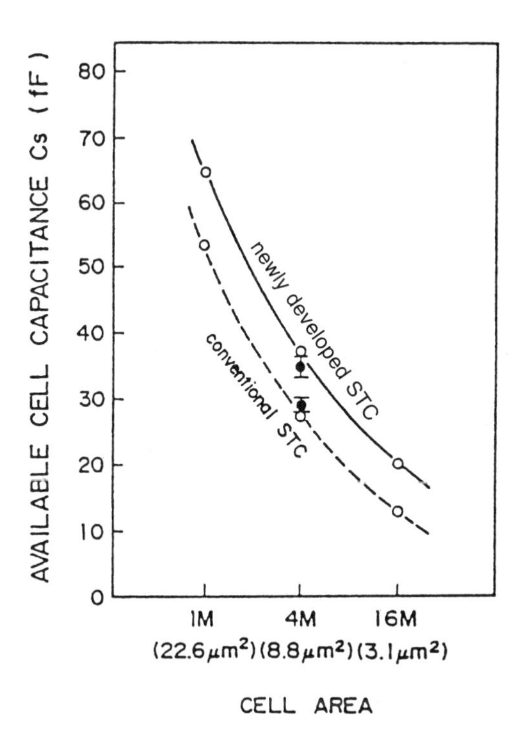

Figure 6.40
Cell capacitance by generation for two types of stacked capacitor DRAM cells (—— newly developed stacked capacitor, --- conventional stacked capacitor). (From Watanabe [79], Toshiba 1988, with permission of IEEE.)

Two problems discovered with the early trench capacitors were trench-to-trench current leakage and reduction of alpha induced soft error immunity.

Toshiba proposed a triple well structure [96] to suppress trench-to-trench punch-through current. This structure consisted of placing an N-well lying between two P-wells in its own P-well. A feature of the triple well structure is that independent bias voltage can be supplied for each well and substrate. This improves the body effect so that high performance transistors can be obtained.

To address these problems Hitachi [86] proposed for the 16Mb DRAM a cell which has an inner storage node to isolate the storage node from the substrate and an isolated storage plate similar to the buried storage plate trench capacitors described earlier for the 4Mb DRAM. This cell also permitted half V_{CC} plate operation through the use of buried plate wiring as shown in Figure 6.41(a). A small cell area of 4.2 μm^2 is attained using a self aligned process and 0.6 μm geometry.

The improvement in leakage current characteristics between this cell and the conventional substrate storage node cell is shown in Figure 6.41(b).

The simple stacked capacitor, as originally developed for the 4Mb DRAM, was pushed one step further by Hitachi [86] who stacked it on top of the bit line to gain

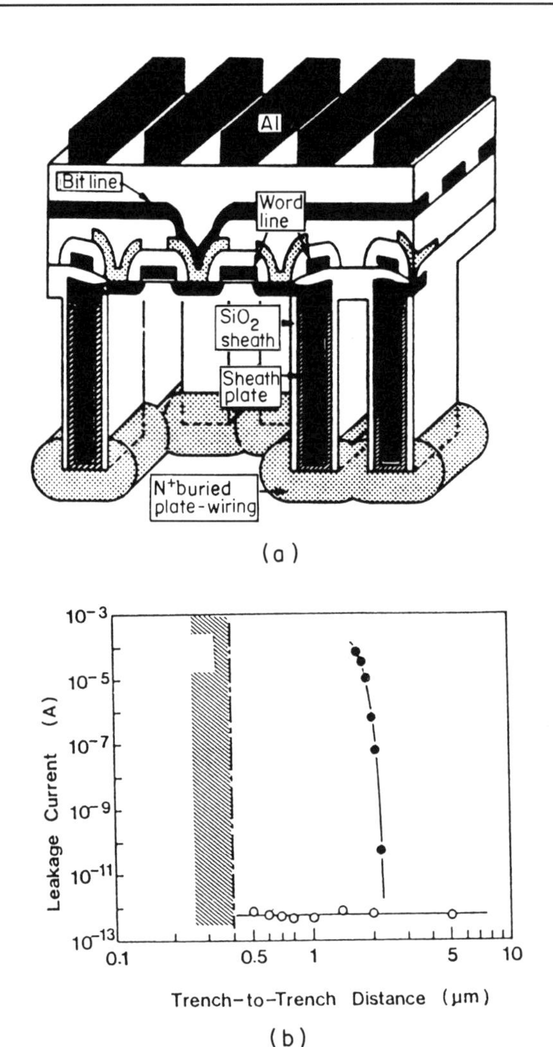

Figure 6.41
A trench capacitor cell for the 16Mb DRAM. (a) Schematic cross-section of the half V_{CC} sheath plate trench capacitor cell with isolated storage node and isolated cell plate with the cell plate wiring buried in the trench. (b) Trench-to-trench current leakage characteristics of the sheath plate capacitor cell with buried wiring (○) compared to the conventional trench cell (●) with unshielded storage node. (From Kaga *et al.* [86], Hitachi 1987, with permission of IEEE.)

extra space as shown in Figure 6.42(a). Toshiba [19] showed an advanced stacked capacitor they called their 'Type I STC' by forming a vertical area in the stacked space above the transfer transistor which is shown in Figure 6.42(b). The process sequence for forming a stack cell with a trench in the stack is shown in Figure 6.42(c) for a 4Mb DRAM from Samsung [91, 110].

The drawback of the extreme vertically stacked capacitor cells is the large steps required in the technology. Toshiba indicated that this was one of the reasons that

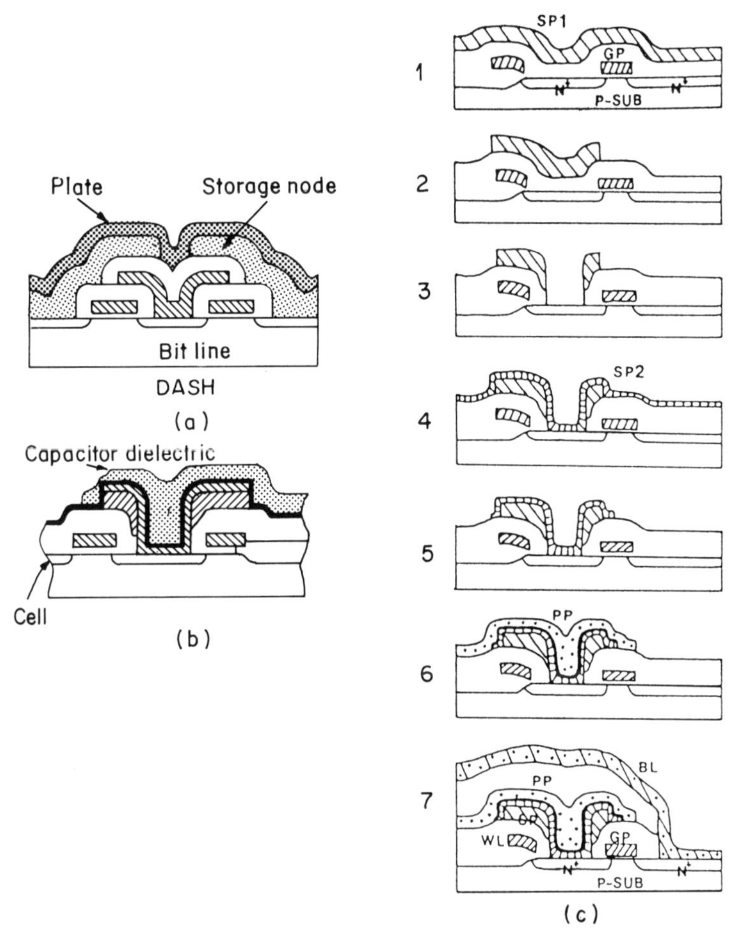

Figure 6.42
Stacked capacitor cells for the 16Mb DRAM. (a) Capacitor stacked over bit-line. (From Kimura [88], Hitachi 1988, with permission of IEEE.) (b) Stacked capacitor with trench formation in the stack (STC cell). (From Watanabe *et al.* [79], Toshiba 1988, with permission of IEEE.) (c) Process steps for a stacked capacitor DRAM cell for a 16Mb DRAM with the capacitor formed in the shape of a trench. (1) First storage polysilicon deposition, (2) saddle patterning, (3) contact etch, (4) second storage polysilicon deposition, (5) storage node definition, (6) deposition of ONO dielectric and plate, polysilicon patterning, (7) cell formation. (From Chin *et al.* [110], Samsung 1989, with permission of IEEE.)

their actual 4Mb and 16Mb designs used the stacked trench described in the previous paragraph.

6.10.3 Refresh related issues with 16Mb DRAMs

The reduced capacitance of the smaller 16Mb cells, combined with the larger size of the array which had to be refreshed, led to serious considerations of the refresh related issues that are present in each new DRAM generation.

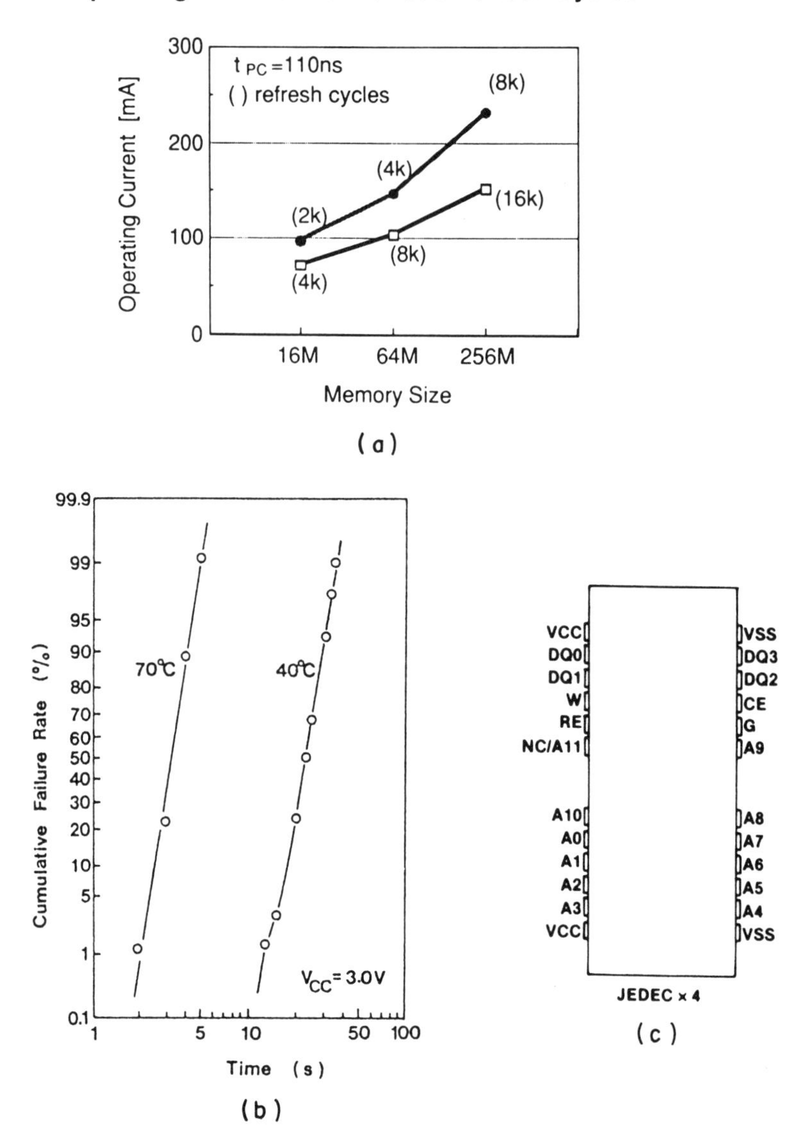

Figure 6.43

Refresh related issues on the 16 Mb DRAM. (a) Comparison of the two choices for refresh cycle for various generations of DRAM as a function of average current. (Source Texas Instruments.) (b) Memory retention characteristics of a typical stacked trench cell showing reduced retention time at high temperatures. (From Kaga *et al.* [86], Hitachi 1987, with permission of IEEE.) (c) JEDEC standard pinout for 16Mb DRAM showing the extra optional address (A_{11}) for 12/10 RAS/CAS addressing with 4k cycle 64 ms refresh. (From Kalter *et al.* [124], IBM 1990, with permission of IEEE.)

At the 16Mb density it was a straightforward option to refresh at the rate of four thousand cycles every 64 ms. The other option was to turn on twice the number of sense amplifiers at one time and refresh at two thousand cycles every 32 ms. The 4k cycle refresh has the advantage of lower power dissipation, while the 2k cycle refresh has the advantage of being upward compatible with the 4Mb DRAM which had a 2k cycle 32 ms refresh. A comparison of the operating current for these two options for several generations of DRAMs is shown in Figure 6.43(a). The lower power dissipation of the refresh architecture using the fewer activated sense amplifiers in each generation is clear.

The retention time characteristics of a typical advanced DRAM cell, the Hitachi sheath plate capacitor cell [86], is shown in Figure 6.43(b). The retention time is defined as the time at which 50% of the bits fail unless refreshed. The retention time decreases sharply with increasing temperature ranging in this case from 23 s at 40°C to 3.5 s at 70°C. Because memory suppliers were unable to agree on either a 2k or 4k refresh as a single standard for the 16Mb DRAM, a dual refresh standard developed at the 16Mb DRAM level. A pinout showing the option for using the 12th address pin (A11) as either an address pin or a 'no connect' is shown in Figure 6.43(c). A 4k cycle 64 ms part is refreshed with 12/10 RAS/CAS addressing while a 2k cycle 64 ms part is refreshed with 11/11 RAS/CAS addressing. Because of the different advantages of the two options, most companies in production offered both devices with some systems makers preferring to upgrade 4Mb DRAM boards using the compatible 2k cycle refresh and others preferring the lower power dissipation of the 4k cycle refresh DRAMs.

6.10.4 Sense amplifier considerations

In high density DRAMs a large peak current can occur when a large number of sense amplifiers start latching. Samsung [21] attempted to minimize this problem in a 16Mb DRAM with a distributed pull-down driver scheme in which the pull-down drivers turned on consecutively tracking the word-line delay.

Asymmetries due to the normal process variations affected sense amplifiers at the 16Mb density level. Matsushita [92] proposed a new sense amplifier structure for their 16Mb DRAM, shown in Figure 6.44, which minimizes these asymmetries. A schematic of a simple conventional sense amplifier is shown in Figure 6.44(a) and of the new sense amplifier in Figure 6.44(b). In this latter sense amplifier each of the paired transistors is divided into two transistors which have reverse source and drain arrangements geometrically on the wafer. The current through transistors 1 and 4 flow in the same direction and opposite to that in 2 and 3. Differences are cancelled and the entire current voltage characteristics of the pair transistors can be symmetrical. This arrangement also has symmetrical gate–source capacitances. This balanced circuit makes it possible to enhance the sensitivity of the sense amplifier.

Also at the 16Mb DRAM level, to avoid the reliability problems associated with applying 5 V across the shorter gates of the transistors in the cell, many parts used a 3 V supply on the memory matrix converting to a 5 V supply in the periphery and

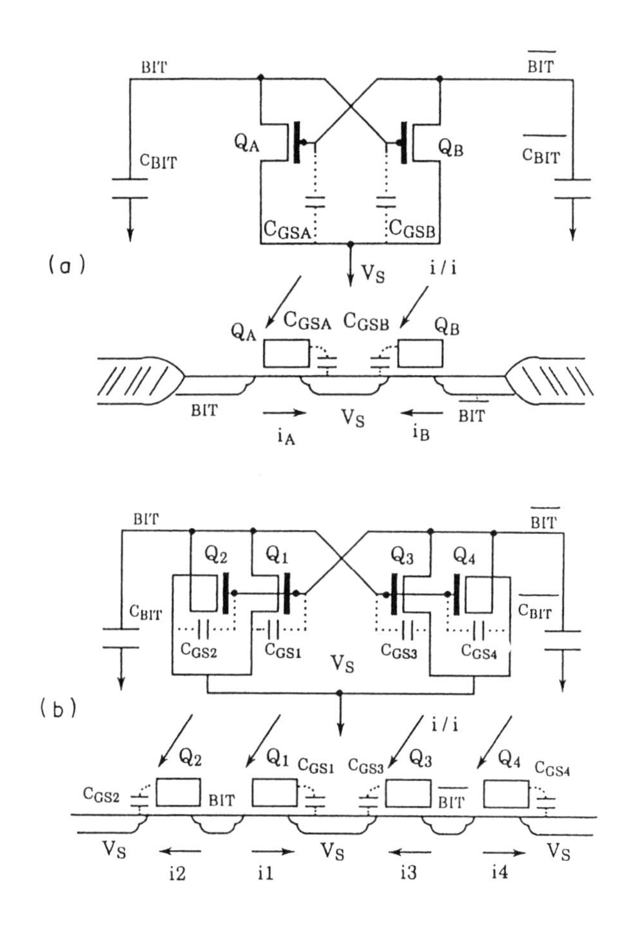

Figure 6.44
Equivalent circuit and schematic cross-sections of (a) a conventional sense amplifier circuit, (b) new sense amplifier circuit for 16Mb DRAM. (From Yamauchi *et al.* [92], Matsushita 1989, with permission of Japan Society of Applied Physics.)

I/O circuits. This lower operating voltage had the effect of reducing the drive capability of the sense amplifier which has to drive not only bit lines but also longer common I/O lines to the main sense amplifier. A large associated transmission line delay of the common I/O lines slows down the speed of the circuit.

To reduce this problem, Hitachi [94] proposed a new common I/O scheme as shown in Figure 6.45(a) and a new current sense amplifier consisting of a differential amplifier and a feedback transistor MF ($\overline{\text{MF}}$) as shown in Figure 6.45(b). This feedback transistor causes a negative feedback effect which reduces the voltage increase in the I/O lines. Therefore the sense amplifier has to charge the capacitance C(I/O) only up to this reduced voltage.

This enhances the speed of the data transmission even with a low level signal current. The conventional common I/O scheme is shown for comparison in Figure 6.45(c).

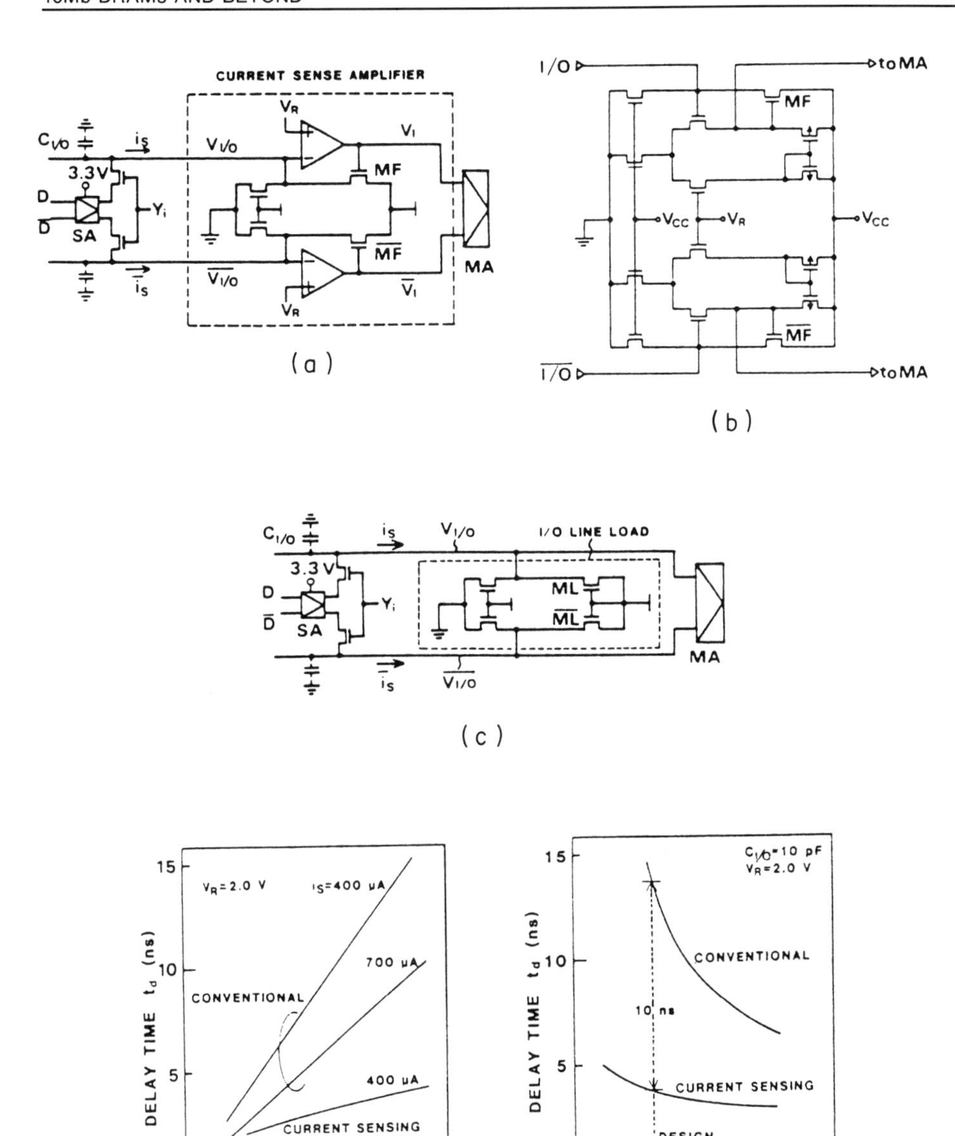

Figure 6.45
New common I/O for 16Mb DRAM. (a) Block diagram of new common I/O scheme including current sense amplifier. (b) Circuit schematic of current sense amplifier. (c) Conventional common I/O scheme. (d) Comparison of delay time plotted against capacitance for conventional I/O and new current sensing I/O. (e) Comparison of delay time plotted against current for conventional I/O and new current sensing I/O. (From Aoki *et al.* [94], Hitachi 1988, with permission of IEEE.)

Figure 6.46
Photographs of typical 16Mb DRAMs. (a) 16Mb chip photograph (with permission of IBM), (b) 16Mb in SOJ package against a background of 16 Mb DRAM chips (with permission of Texas Instruments.)

The two graphs shown in Figure 6.45(d) and 6.45(e) indicate the improvement in delay time for this new current sense amplifier compared to conventional sense schemes.

IBM [100] also investigated sensing schemes to increase speed and reduce bit-line coupling noise. They proposed a $\frac{2}{3}V_{DD}$ sensing scheme.

Two cases were investigated, with and without boosted word-lines. With a delay boosted word-line, the signal development time with the $\frac{2}{3}V_{DD}$ precharge improved over that of the conventional $\frac{1}{2}V_{DD}$ sensing. The increased precharge, however, resulted also in increases in power consumption, peak current, dI/dt, bit-line coupling noise, and bit-line capacitance.

IBM also investigated the case where the word-line is not boosted and the bit-line pull-down is clamped at a voltage below the P-channel threshold of the cell transfer device. This, used in conjunction with a precharge voltage of $\frac{2}{3}V_{DD}$, balances the bit-line pull-up and pull-down with a reduced voltage swing of $\frac{1}{3}V_{DD}$. This scheme gives a faster DRAM access time, reduced power consumption, smaller peak current, less bit-line coupling noise and a decrease in dI/dt. The trade-off is that limiting the bit-line swing increases the write time. A photograph of an IBM 16Mb DRAM chip is shown in Figure 6.46(a) and one of a TI 16Mb DRAM wafer and packaged part is shown in Figure 6.46(b). The IBM photograph shows the divided word and bit-line structure. In addition, the bond pads are located in the center of the chip rather than along the perimeter as was conventionally done.

6.10.5 Bit-line architectures for 16Mb DRAMs

The earliest DRAMs used the open bit-line architecture for the array. This architecture provides a dense cell array and is simple to implement if the memory cells are large enough to allow the decoders and sense amplifiers to fit in the layout pitches. However, the signal-to-noise (S/N) ratio of the open bit-line architecture degrades for the long lines in large arrays and the layout pitch becomes too small for implementation of complex sense amplifiers and decoders.

For the 64k to 1Mb DRAMs, the folded bit-line approach was widely adopted to improve the S/N ratio, ease the layout pitch problem and reduce the level of soft errors caused by common mode noise on the bit-lines. As a result the array area was not as dense as it would have been with the open bit-line approach.

At the 4Mb density level there was an overriding requirement to reduce the size of the chip by something like an open bit-line approach. The potential minimum cell size of the trench architecture using the open bit-line was only as large as one word-line pitch and one bit-line pitch. The drawback was the S/N ratio effects of the long transmission lines was even more extreme in the high density 4 Mb DRAM. It was, therefore, attractive to attempt to some architectural variation which would combine the density of the open bit-line architecture with the S/N benefits of the folded approach.

TI [84] projected using a 'cross-point' cell array in place of the conventional cell array as shown in Figure 6.47(a). This segmented bit-line architecture has the noise immunity and larger pitch advantages of the folded bit-lines combined with the density benefits of the open bit-lines. A $V_{CC}/2$ sensing scheme combined with a conventional CMOS sense amplifier with a full size reference cell was used for the sense circuit. An on-chip reference voltage generator provided the 2.2 V for the reference cell. Various design techniques were implemented to balance the folded bit-lines.

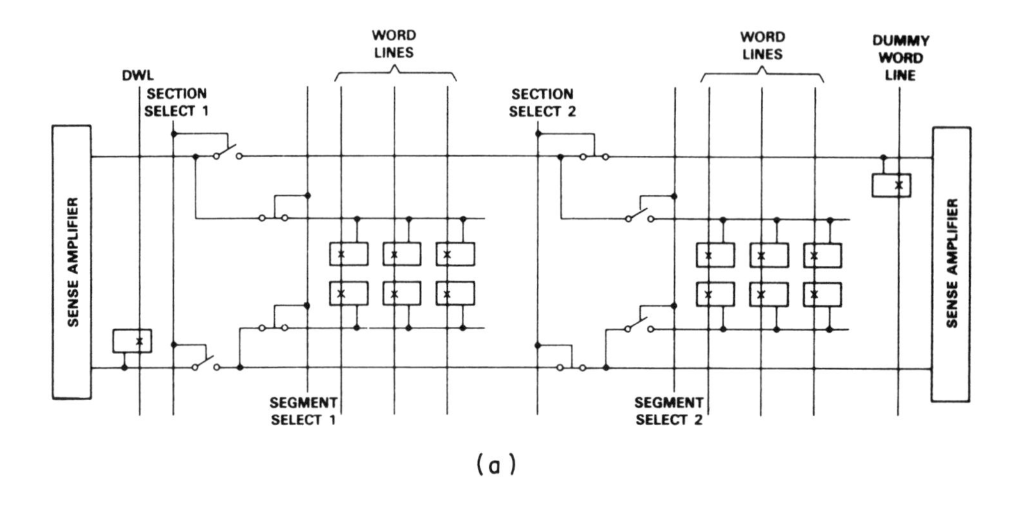

(a)

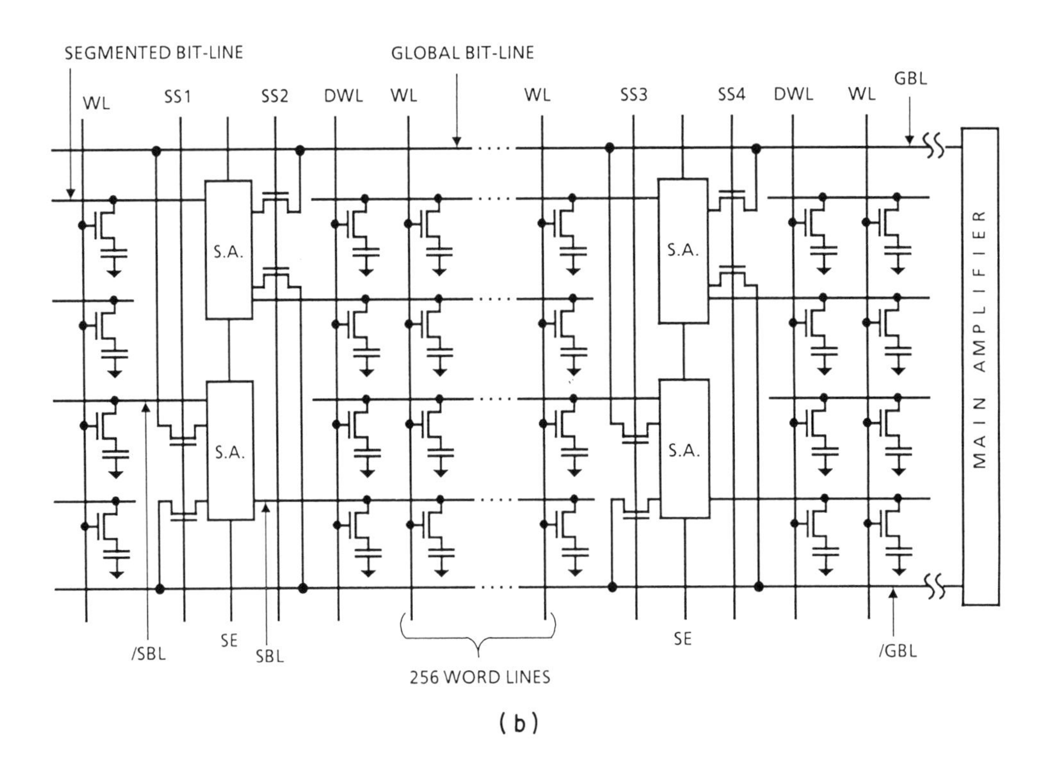

(b)

Figure 6.47

New bit-line concepts for the multi megabit DRAMs. (a) 'Cross-point' cell array for the 4Mb DRAM. (b) Transposed data-line structure for the 16Mb DRAM. (a: From Shah *et al.* [84], Texas Instruments 1986, with permission of IEEE; b: From Inoue *et al.* [93], Matsushita 1988, with permission of IEEE.)

The cross-point technique allows a cell to be placed at every intersection of the word-line and segmented bit-lines, something a conventional segmented folded bit-line does not permit. A word-line selects two cells for every pair of global bit-lines. When a word-line is selected, the four select transistors switch, breaking each of the global bit-line pairs into two shorter global bit-line pairs. Each of these pairs is in turn connected to each of the sense amplifiers at the opposite ends.

A similar 'pseudo-open' bit-line layout was proposed by Toshiba [87] in an experimental 16Mb DRAM to increase the packing density. A block of 32 columns was laid out as an open bit-line array structure and each neighboring block was connected to a sense amplifier in a folded bit-line manner. Furthermore, Toshiba [35] in their 16Mb used a refinement of this concept in a cell with 'quarter pitch' layout in which 4 word-lines and 4 bit-lines make the unit pitch of the cell array.

Matsushita [93] proposed a new bit-line configuration to allow sense amplifier layout in the very small bit-line pitch of the 16Mb DRAM. This was a staggered bit-line arrangement in which open bit-lines are divided into several segments and alternately connected to either side of the sense amplifiers as shown in Figure 6.47(b). Sense amplifiers are connected to two staggered bit-lines of adjacent segments. The staggered arrangement means an additional memory segmentation is needed, but the width of the individual sense amplifier is reduced so that a smaller memory array can be attained.

Matsushita attempted to deal with the lower noise immunity of the open bit-lines compared to the folded bit-lines by reversing one dummy word-line in each array segment.

Hitachi [94, 95] found that at the 16Mb density level the bit-line interference noise due to the increased inter-bit-line capacitance became more than 25% of the signal causing a serious noise problem. Their 16Mb used a transposed bit-line structure to attempt to deal with the problem.

In further investigations Hitachi [70] found that another source of noise is that generated by interference between wires laid over the sense amplifiers. It was indicated that this noise is reduced by the transposition of the amplifier in addition to the transposed bit-lines.

A capacitive coupling model of this structure using two adjacent data-line pairs is shown in Figure 6.48(a). A differential noise is generated in the data-line $\overline{DO}$ when a word-line is activated and the signal voltage appears in the adjacent data-line D_1. A second component of noise, shown in Figure 6.48(b), occurs due to different amplification rates of the data-line D_0 and D_1 due to the first differential noise component. This second component contributed less than 3% noise to the signal in the 4Mb DRAMs. In the 16Mb, however, this was shown to increase to over 25%.

One possible solution would be a single transposed data-line scheme as shown in Figure 6.48(c). This, however, cancels only half of the capacitive interference. As a solution Hitachi proposed the double transposed data-line scheme as shown in Figure 6.48(d). This scheme converts the differential noise into common mode noise, suppressing it to less than 5% of the signal. The trade-off is in chip area which was increased by 7% by the transposed data-line structure.

Mitsubishi [74] also investigated a twisted bit-line scheme which is similar

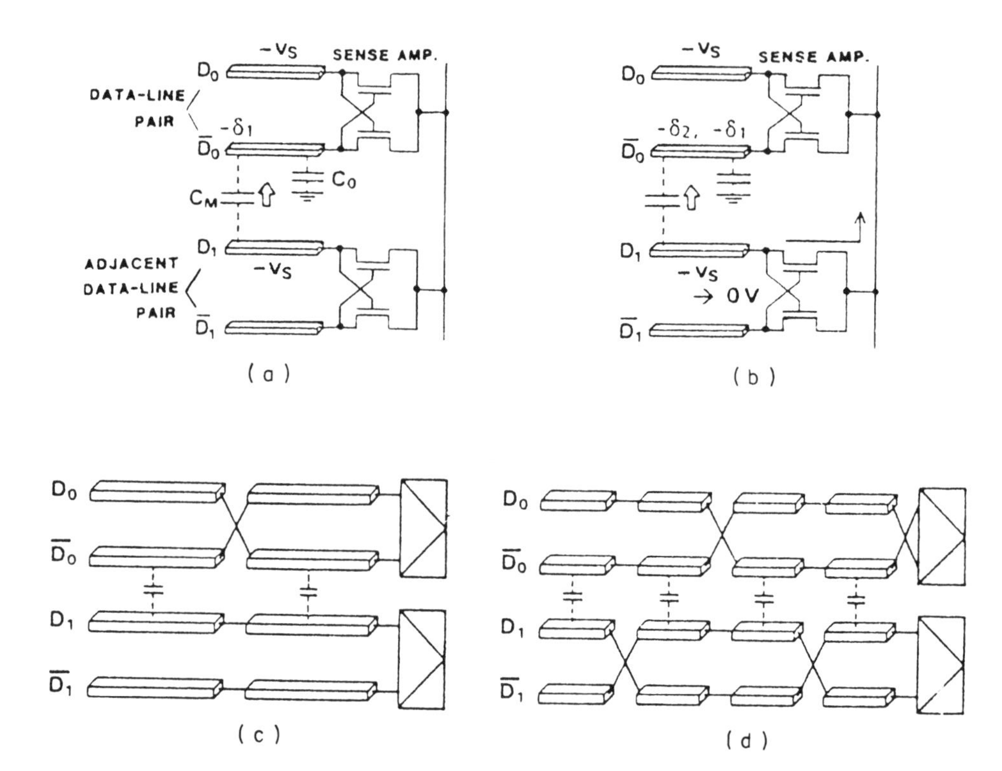

Figure 6.48
Capacitive coupling model of transmission line effects in 16Mb DRAMs in long data-lines. (a) Two adjacent data-line pairs showing previously known first order effects. (b) Second-order component of noise which becomes significant at the 16Mb DRAM density level. (c) Single transposed data-line structure to suppress first-order effect. (d) Double transposed structure to suppress second-order effect. (From Aoki *et al.* [94], Hitachi 1988, with permission of IEEE.)

to the Hitachi transposed bit-lines. They estimated that the signal loss due to inter-bit-line coupling noise was about 40% of the signal amplitude in the bit-lines of the 16Mb DRAM. This could lead to significant levels of soft errors due to alpha radiation. Mitsubishi found, from examining soft error test results, that the effective critical charge was improved by 35% with the twisted bit-lines.

In a separate investigation Mitsubishi [75] also analyzed the effects of different bit-line structures, materials, widths, spacings and passivation materials on the coupling noise between adjacent bit-lines in Mb DRAMs.

6.10.6 Low voltage power supplies and I/O interfaces

The first DRAMs using power supplies lower than 5 V appeared at the 16Mb density. These parts followed the JEDEC standards 8.0 and 8.1 which specified 3.3 V as the

		<4 Mb	16 Mb	64 Mb	256 Mb	1 Gb
Devices (L_g/t_{ox}) (μm/nm)		>1/20	0.6/15	0.4/10	0.3/7	0.2/5
Maximum operating voltage for devices (V)		>5.0	3.3	2.3	1.5	1.0
Power supply V_{cc} (V)	I Transition at every generation with minimizing power dissipation	—	3.3	2.3	1.5	1.0
	P(I)/P(II)	—	0.66	0.7	0.45	0.3/0.66
	II Maintenance of the same V_{cc} user's benefit	5.0	4.0 converted to 3.3	3.3 converted to 2.3	3.3 converted to 1.5	3.3 or 1.5 converted to 1.0

Figure 6.49
Operating power supply trends forecast for megabit DRAMs. (From Itoh [136], Hitachi 1990, with permission of IEEE.)

next voltage step after 5 V for systems using regulated power supplies. It was not until the 64Mb DRAM, however, that 3.3 V supplies replaced 5 V supplies for DRAMs. Further voltage reduction steps were expected with the possibility existing of going to a lower standard by the 1 Gb density as shown in Figure 6.49.

6.11 64Mb DRAMs

The first 64Mb DRAM was shown by Hitachi [73] in June 1990. The 197.5 mm^2 chip was fabricated in an 0.3 μm triple well CMOS technology with a 1.28 μm^2 cell. A 1.5 V internal array voltage was used [145]. Row address access time was shown to be 50 ns at 1.5 V with 29 mA power consumption. The storage node was a stacked cylindrical capacitor. This chip was followed by more 64Mb DRAMs in early 1991 from Matsushita [130], Fujitsu [131], Mitsubishi [132], Toshiba [133] and TI [138]. Characteristics of these early 64Mb DRAMs are shown in Table 6.1.

6.11.1 Cell structures for the 64Mb DRAM and beyond

For 64Mb DRAM cells there are two main problems. One is the need to have suitable work functions for the P-channel and N-channel devices. The other is the hot electron effect which requires the scaling of the voltage across the transistors. At the 64Mb DRAM density, the external system voltage will be reduced to the 3 V range

Table 6.1 Early 64 Mb DRAMs.

Year	Company	Power Supply (V)	Geometry (μ)	Cell size (μm^2)	Chip size (mm^2)	O_{RG}	T_{RAS} (ns)	Capacitance (fF)	Capacitor Type
1990	Hitachi	1.5/3.3	0.3	1.28	199.5	$\times 4$	50	44	Stack
1991	Matsushita	3.3	0.4	2.0	234.4	$\times 16$	50	35	Stack
1991	Fujitsu	3.3	0.4	1.8	224.7	$\times 1 \times 4$ $\times 8$	40		Stack
1991	Mitsubishi	3.3	0.4	1.7	233.8	—	45	30	Stack
1991	Toshiba	3.3	0.4	1.53	176.4	$\times 1 \times 4$	33	34	Trench
1991	TI	3.3	0.4	2.0	268.1	$\times 8$	40	44	Trench

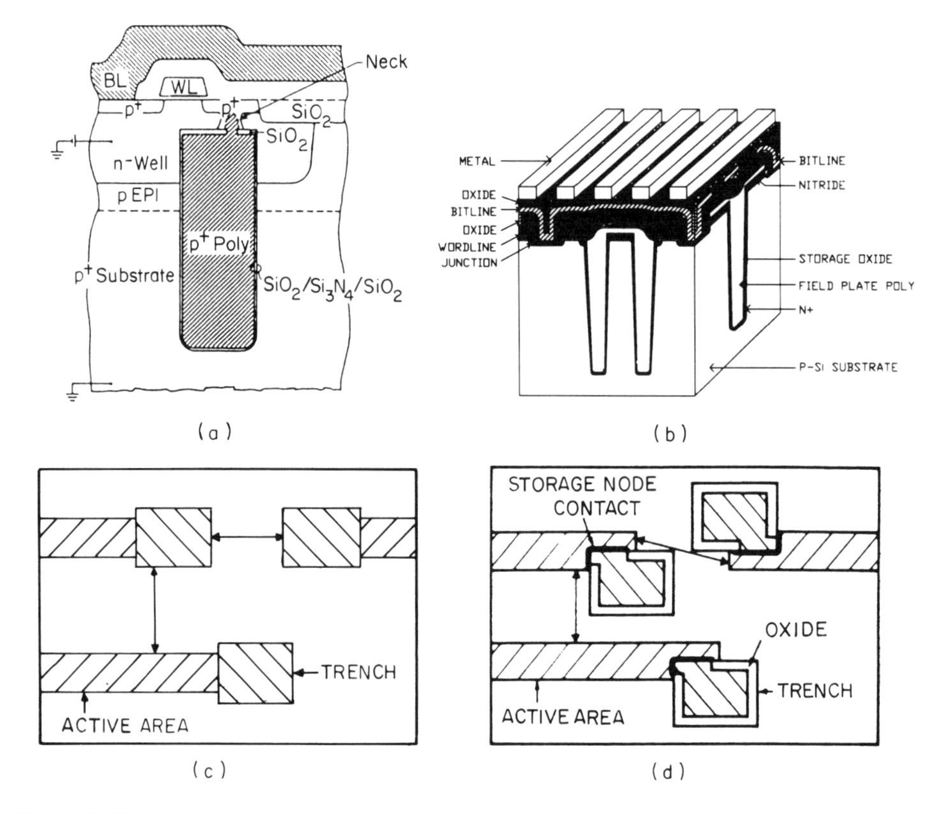

Figure 6.50
Trench cells for the 64Mb DRAM. (a) Buried trench cell with a transistor stacked above the cell in an epitaxial layer grown over the trench. (From Lu *et al.* [107], IBM 1988, with permission of IEEE.) (b) Field plate isolated trench capacitor cell using n^+ diffused field plate isolation. (From Shen *et al.* [127], Texas Instruments 1989, with permission of IEEE.) (c) Conventional trench cell layout. (d) New asymetrical trench cell layout (From Sunouchi [128], Toshiba 1990, with permission of IEEE.)

permitting reduction below the 5 V external voltage currently demanded by the market for the DRAMs.

A DRAM cell investigated by the IBM Research Labs [107] for 64Mb DRAMs and beyond is a buried trench DRAM cell using an epitaxial technique to grow silicon over the trench, see Figure 6.50(a). This technique allows placement of the transistor above the capacitor.

A second trench cell used for 64Mb DRAMs was the field plate isolated trench capacitor cell developed by Texas Instruments in 1989 [127] which is shown in Figure 6.50(b).

The advantages of the TI cell were large storage capacitance, planar topography and processing simplicity. The current leakages historically associated with the trench cell were reduced by the use of field plate isolation over an oxide–nitride dielectric stack. A pass gate transistor with 0.3 μm effective channel length was used.

A third trench cell used for the 64Mb DRAM was the asymmetrical stacked trench capacitor cell shown by Toshiba in 1990 [128]. Figure 6.50(c) shows the layout of a

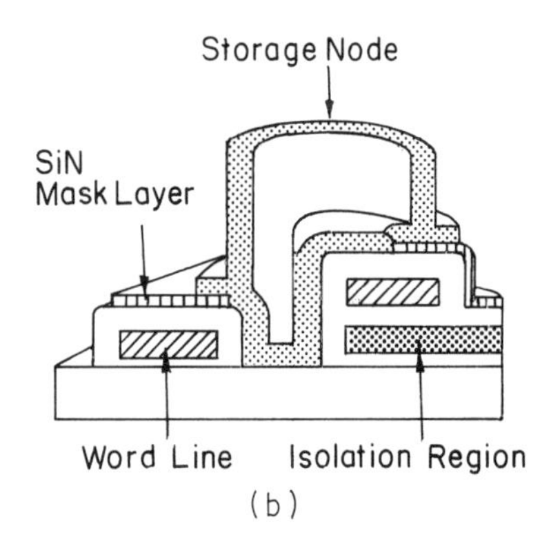

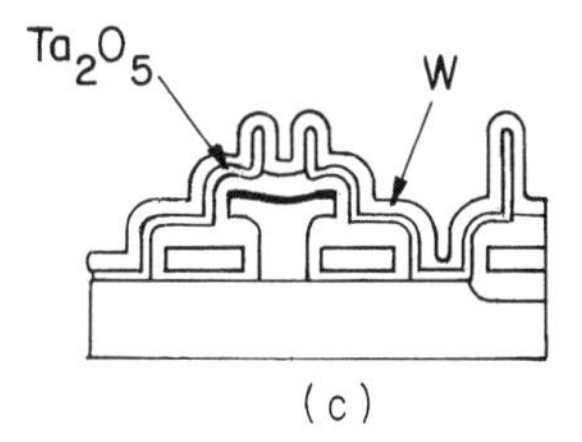

Figure 6.51
64Mb DRAM cylindrical stacked cells. (a) SEM of cylindrical storage node capacitor cell for an 0.25 μm 64Mb DRAM. Both sides of the cylindrical structure can be used as a capacitor and the storage capacitance increases with height. (b) Cross-section of Mitsubishi cylindrical capacitor. (c) Cross-section of Hitachi cylindrical capacitor. (a,b: From Wakamia [108], Mitsubishi 1989, with permission of IEEE; c: From Aoki, *et al.* [94], Hitachi 1988, with permission of IEEE.)

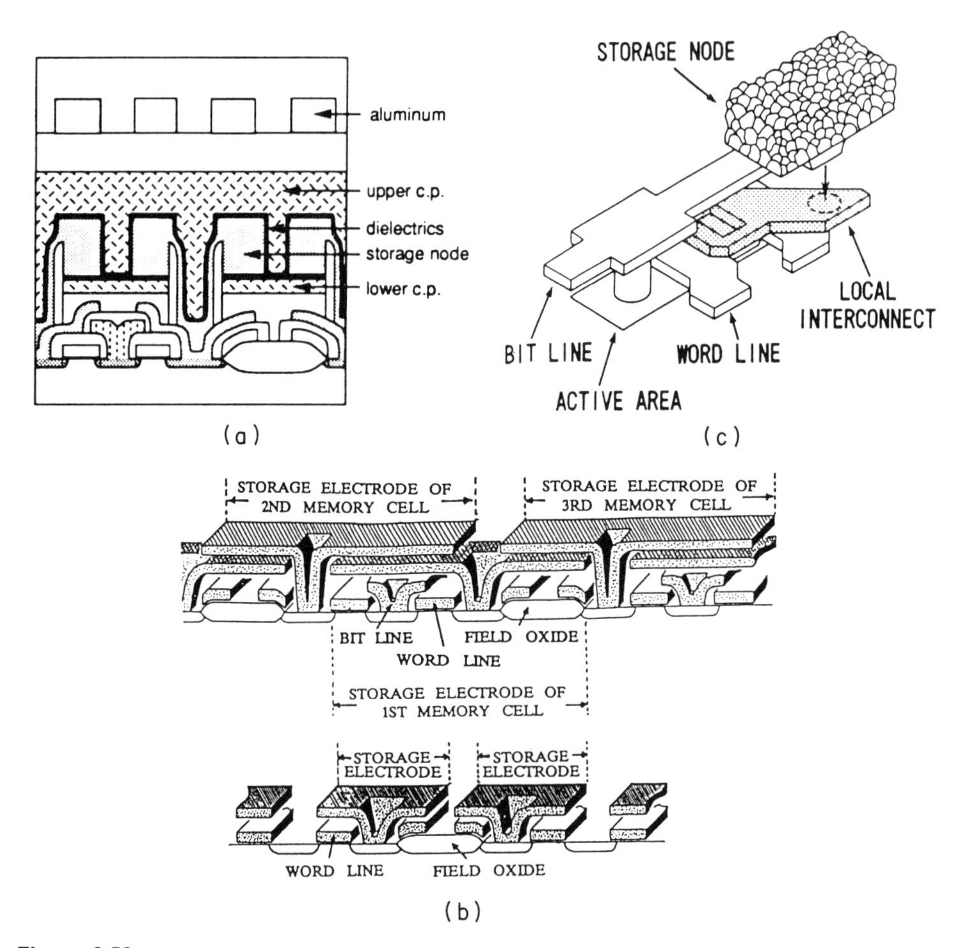

Figure 6.52
Stacked capacitor cells for the 64Mb DRAM. (a) Stacked capacitor with dual cell plates above and below the stacked storage node. (From Arima *et al.* [125], Mitsubishi 1990, with permission of IEEE.) (b) Stacked capacitor with overlapping horizontal planar storage nodes shown compared to more conventional stacked capacitors. (From Inoue *et al.* [111], Toshiba 1989, with permission of IEEE.) (c) Stacked capacitor over bit-line cell using textured polysilicon capacitors to increase storage area. (From Sakao *et al.* [126], NEC 1990, with permission of IEEE.)

conventional trench cell and Figure 6.50(d) shows that of the asymmetrical stacked cell. The spacing between the storage node contacts and the active areas was increased over that of the older trench cell layout providing greater protection against leakage.

Two novel stacked cylindrical capacitors for the 64Mb type of DRAM density were discussed by Mitsubishi [108] and Hitachi [113]. These use capacitors which rise as vertical cylinders above the surface of the wafer with varying heights possible to vary the capacitance. A scanning electron microscope (SEM) photograph of the Mitsubishi 64Mb DRAM cylindrical storage capacitors is shown in Figure 6.51(a).

A cross-section of the Mitsubishi cylindrical cell is shown in Figure 6.51(b) in which the capacitor can be seen to be stacked above the bit line. A cross-section of the Hitachi 'crown' cylindrical cell is shown in Figure 6.51(c). This cell uses Ta_2O_5 which has a higher dielectric constant than oxide–nitride combinations, to increase the effective capacitance of the cell. This cell has the capacitor stacked over the bit-line which is in turn stacked over the word-line.

Three other innovative stacked capacitor cells investigated for the 64Mb DRAM are shown in Figure 6.52. These include a cell shown by Mitsubishi [125] in 1990 in Figure 6.52(a). This cell has dual cell plates stacked above and below the storage node which is stacked above the bit-line and the word-line. One advantage of this cell is that the cell plates completely surround the surface of the storage node which significantly increase the capacitance. Another advantage is the relatively planar layered topology which increases the ease of manufacturing.

The stacked cell proposed by Toshiba [111] in 1989, shown in Figure 6.52(b), has interleaved horizontal storage nodes which are also planarized for ease of manufacturing. Since each node extends over the area of two cells the overlapping plates effectively double the capacitance of the more conventional stacked capacitor cell also shown.

A new innovation for the 64Mb DRAM was shown in a cell from NEC [126] in 1990 which is illustrated in Figure 6.52(c). The textured polysilicon surface exhibits hemispherical grains which increase the effective storage area of the capacitor. The formation of the textured surface is dependant on several process parameters which were discussed in the section on textured capacitors in Chapter 4.

Finally a 256Mb DRAM cell proposal from Toshiba is a pillar construction using trench process technology and is called the surrounding gate transistor. A schematic view of this cell is shown in Figure 6.53(a). The process steps for fabrication are shown in Figure 6.53(b). A similar transistor developed for high density SRAMs is described in Chapter 9.

Beyond that, perhaps new materials with very high dielectric constants, such as Ta_2O_5 which is already in use at the 64Mb DRAM density and ferroelectrics, may be used for a whole new generation of DRAM cells.

A DRAM which used a ferroelectric capacitor was shown in 1990 by National Semiconductor in conjunction with the University of California, Berkeley [141]. The fatigue problems of the ferroelectric material upon repeated cycling were reduced in this case by not cycling between opposite polarization states during the DRAM operation.

6.11.2 Architectural considerations for the 64Mb DRAM

Major concerns architecturally for the 64Mb DRAM are shielding, power consumption and distribution, the need to fit the circuitry into the 0.1 μm cell pitch, access time in a chip with 65 million transistors, and noise considerations with wide I/O organization.

Toshiba [105] in 1989 proposed a new technique for reducing the capacitive

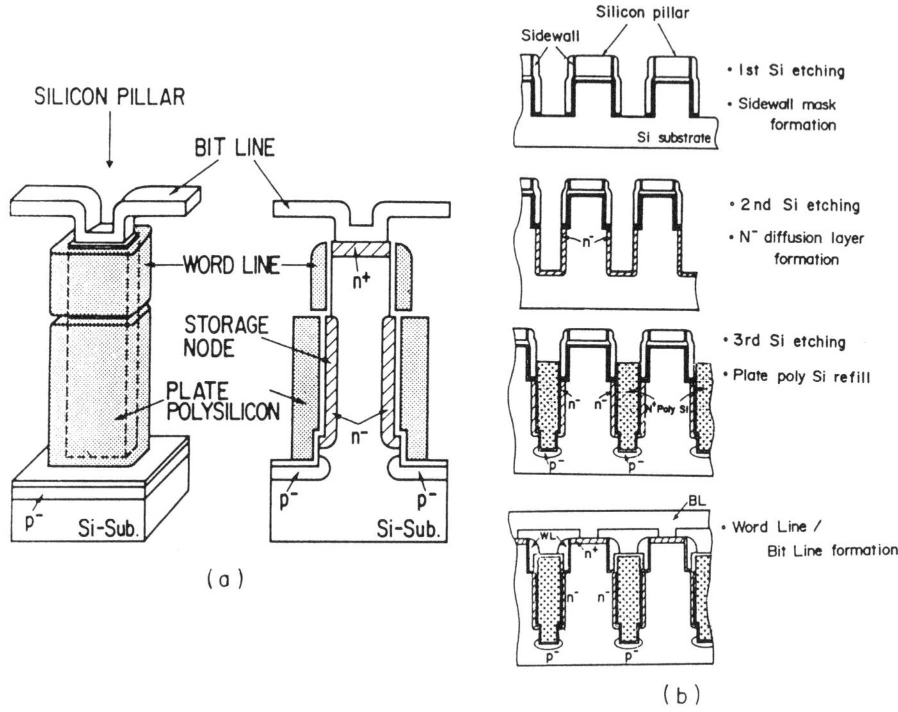

Figure 6.53
Potential cell for the 256Mb DRAM. (a) 'Pillar' capacitor DRAM cell. (b) Process sequence for formation of Pillar DRAM cell. (From Sunouchi [112], Toshiba 1989, with permission of IEEE.)

coupling effects between bit lines for 64Mb DRAM and called it the 'stabilized reference line' technique. In this technique, the bit-lines and the reference lines are placed alternately. Only the reference lines are connected to the reference voltage source, which is at $V_{CC}/2$, by a switch. Interference noise is therefore absorbed into the $V_{CC}/2$ generator. Figures 6.54(a,b,c) compare the folded bit-line structure, the twisted bit-line structure and the stabilized reference line structure.

Mitsubishi [114] also proposed for the 64Mb DRAM a new divided–folded–shared bit-line scheme which suppressed the inter-line coupling noise by combining folded bit-lines with activation of alternative bit-lines. It then utilized the alternate ion-active bit-lines for shielding the coupling noise throughout the sensing period. This was also indicated to be compatible with a twisted bit-line architecture. The divided–shared bit-line architecture is shown in Figure 6.54(d) and the combination with the twisted bit-line architecture is shown in Figure 6.54(e).

Hitachi [73] on their 1990 64Mb DRAM used an alternated sharing arrangement for sense amplifiers and I/O circuit to enable them to fit the required circuitry in the bit-line pitch. This arrangement also allowed shielded electrodes to be incorporated between neighboring bit-lines to minimize interference noise. Figure 6.55(a) shows the conventional sense amplifier and I/O sense line configuration and Figure 6.55(b) shows the alternate shared arrangement.

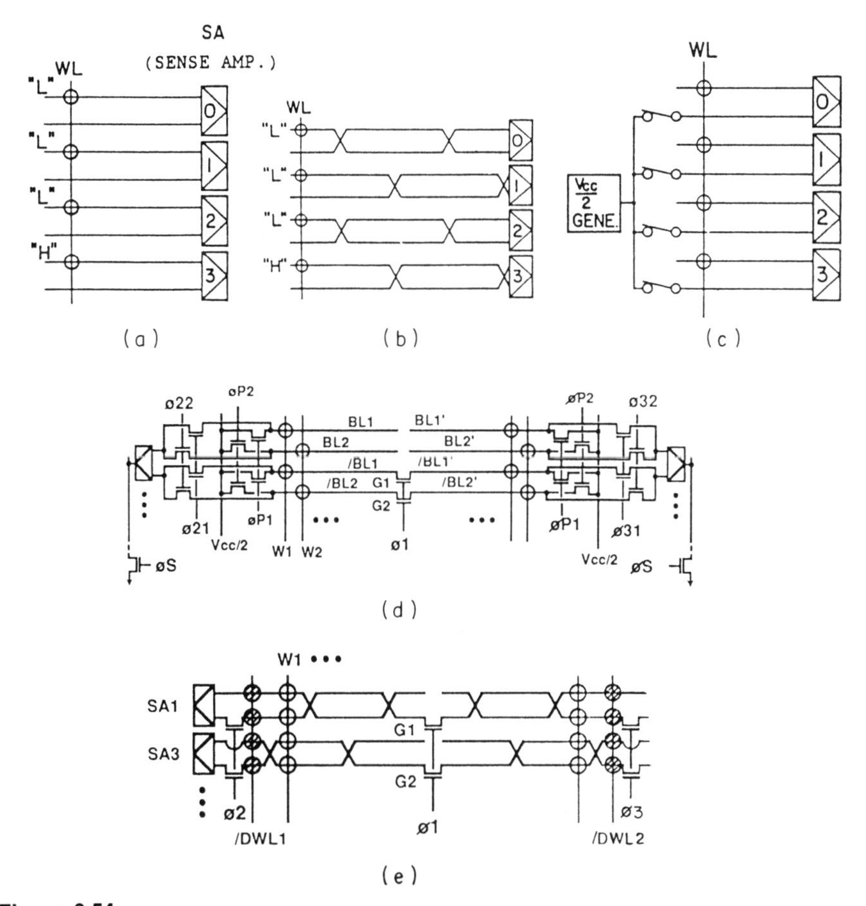

Figure 6.54
Comparison of different bit-line structures. (a) Conventional folded bit-line. (b) Conventional twisted bit-line. (c) 'Stabilized reference line' techniques for the 64Mb DRAM showing alternating bit-lines and reference lines. (d) Divided–shared folded bit-line technique for the 64Mb DRAM. (e) Divided–shared bit-line technique combined with twisted bit-lines for 64Mb DRAM. (a,b,c: From Tsuchida [105], Toshiba 1989, with permission of Japan Society of Applied Physics; d,e: From Hidaka [114], Mitsubishi 1990, with permission of Japan Society of Applied Physics.)

A new $V_{CC}/2$ voltage generator was also proposed. Drawbacks of the conventional $V_{CC}/2$ generator, shown in Figure 6.55(c), are that threshold voltage variation of the NMOS or the PMOS transistors affect the voltage level and the output transient response is very slow because of the source–follower configuration. The new generator, shown in Figure 6.55(d), achieved better accuracy by separating the voltage divider from the bias stage and improved speed by using a push–pull current mirror amplifier which converts small voltage errors at the output to a current which drives the push–pull output stage.

To compensate for oxide fluctuations and temperature variations with the on-chip voltage generators, Toshiba [139] in 1991 proposed a word-line control circuit for

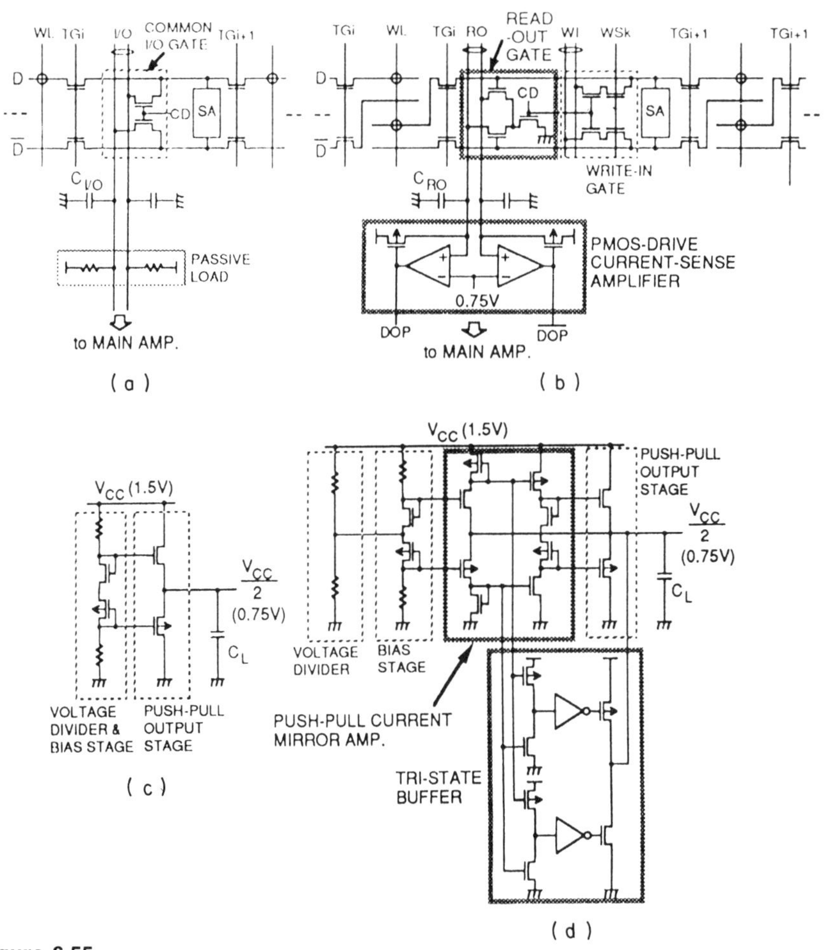

Figure 6.55
64Mb DRAM circuit technology. (a) Conventional sense circuit. (b) Low voltage complementary current sense circuit. (c) Conventional $V_{CC}/2$ voltage generator. (d) Fast low deviation $V_{CC}/2$ voltage generator. (From Nakagome [73], Hitachi 1990, with permission of IEEE.)

the 64Mb DRAM. This circuit consisted of two reference circuits, two linear amplifiers, and a constant gain word-line booster.

Matsushita [130] in 1991 addressed both the shielding problem and the power consumption of the wide 16 bit I/O using a meshed power line scheme. This consisted of a V_{CC} and V_{SS} mesh spread over the entire memory array. This mesh supplies all operating sensing current. Each memory block has 16 power and ground lines and distributed sense amplifier drivers distributed throughout the memory array at the corners of each unit memory block. To reduce the interaction between metal layers a V_{SS} shield plate scheme is used in the periphery.

Fujitsu [131] in 1991 described a 64Mb DRAM with 40 ns access time and a chip architecture designed for wide parallel data I/O and for the 64 bit internal test function used widely to reduce test time for the 64Mb DRAM. This permits test of the total chip in 150 μs.

Mitsubishi [132] also in 1991 described a 64Mb DRAM which focused on parallel test architecture with 150 μs test time. They used the dual cell plate stack capacitor cell described in an earlier section. The DRAM was divided into four 16Mb segments which operated separately with output data merged by a four-bit AND circuit resulting in 64k bits of data at the output. Using this operation all 64Mb are tested in 1024 cycles.

Toshiba [133] in 1991 focused their attention on speed to produce a 33 ns part, which was the fastest of the five 64Mb DRAMs shown to that time. They also had the smallest cell size at 1.53 μm^2 resulting in the smallest chip at 176.4 mm^2. The small cell was partly due to the use of the asymmetrical trench cell discussed earlier which has a stacked capacitor structure buried in a trench. Noise immunity was improved with the use of an interdigitated twisted bit-line in which four bit-lines in a unit pitch form two interdigitated columns. The bit-lines are twisted alternately in the center of the array. Inter-bit-line noise is shielded by having the bit-lines inserted in this structure.

6.12 IMPROVED COMPUTER DESIGN TOOLS FOR MULTIMEGABIT DRAMs

The three-dimensional structures of the multi-megabit DRAMs would not have been possible without the aid of increasingly sophisticated computer design tools.

Such tools as a two-dimensional device simulator (PISCES) [14] from Stanford University was used by Texas Instruments to model the trench transistor and adjust implants. The measured device characteristics were within 10% of the results of the simulations in both the linear and saturation regions.

BIBLIOGRAPHY

[1] Boonstra, L., Lambrechtse, C. W. and Salters, R. H. W. (1973) A 4096-b One-Transistor Per Bit Random-Access Memory with Internal Timing and Low Dissipation, *IEEE Journal of Solid State Circuits*, **SC-8**, No.5. October 1973, p.305.

[2] Green, R. and Proebsting, R. (1973) A TTL compatible 4096 bit n-channel RAM, *ISSCC Dig. Tech. Papers*, p.26–27, 570.

[3] Ahlquist, C. N. *et al.* (1976) A 16k 384-bit Dynamic RAM, *IEEE Journal of Solid State Circuits*, **SC-11**, No.3, October 1976.

[4] Arai, E. and Ieda, N. (1978) A 64k bit Dynamic MOS RAM, *IEEE Journal of Solid State Circuits*, **SC-13**, No.3, June 1978, p.333.

[5] Meusburger, G. *et al.* (1978) An 8mm^2, 5V 16k Dynamic RAM Using A New Memory Cell, *IEEE Journal of Solid State Circuits*, **SC-13**, No.5, October 1978, p.708.

[6] Wada, T. *et al.* (1978) A 64k × 1 Bit Dynamic ED-MOS RAM, *IEEE Journal of Solid State Circuits*, **SC-13**, No.5, October 1978, p.600.

[7] Wada, T. *et al.* (1978) A 150ns, 150mW, 64k Dynamic MOS RAM, *IEEE Journal of Solid State Circuits*, **SC-13**, No.5, October 1978, p.607.

[8] Schuster, S. E. (1978) Multiple Word/Bit Line Redundancy for Semiconductor Memories, *IEEE Journal of Solid State Circuits*, **SC-13**, No.5, October 1978, p.698.
[9] Kondo, M. *et al.* (1978) A High Speed Molybdenum Gate MOS RAM, *IEEE Journal of Solid State Circuits*, **SC-13**, No.5, October 1978, p.611.
[10] Hosticka, B. J. (1980) Dynamic CMOS Amplifiers, *IEEE Journal of Solid State Circuits*, **SC-15**, No.5, October 1980, p.887.
[11] Yoshimura, H. *et al.* (1978) A 64kbit MOS RAM, *ISSCC Dig. Tech. Papers*, February 1978.
[12] Schroeder, P. R. and Proebsting, R. J. (1977) A 16k × 1 bit Dynamic RAM, *ISSCC Dig. Tech. Papers*, February.
[13] Natori, K. *et al.* (1979) A 64k bit MOS Dynamic Random Access Memory, *IEEE Journal of Solid State Circuits*, **SC-14**, No. 2, April 1979, p.482.
[14] Martino, W. L. *et al.* (1980) An On-Chip Back-Bias Generator for MOS Dynamic Memory, *IEEE Journal of Solid State Circuits*, **SC-15**, No.5, October 1980, p.820.
[15] Chan, J. Y. *et al.* (1980) A 100ns 5V Only 64k × 1 MOS Dynamic RAM, *IEEE Journal of Solid State Circuits*, **SC-15**, No.5, October 1980, p.839.
[16] Masuda, H. *et al.* (1980) A 5V-Only 64k Dynamic RAM Based on High S/N Design, *IEEE Journal of Solid State Circuits*, **SC-15**, No.5, October 1980, p.846.
[17] Barnes, J. J. and Chan, J. Y. (1980) A High Performance Sense Amplifier for a 5V Dynamic RAM, *IEEE Journal of Solid State Circuits*, **SC-15**, No.5, October 1980, p.831.
[18] Mano, T. *et al.* (1980) A Fault-Tolerant 256k RAM Fabricated with Molybdenum-Polysilicon Technology, *IEEE Journal of Solid State Circuits*, **SC-15**, No.5, October 1980, p.865.
[19] Matsue, S. *et al.* (1980) A 256k bit Dynamic RAM, *IEEE Journal of Solid State Circuits*, **SC-15**, No.5, October 1980, p.872.
[20] Prince, B. and Due-Gundersen, G. (1983) *Semiconductor Memories*, John Wiley, Chichester.
[21] Ieda, N. *et al.* (1977) A 64k MOS RAM design, 9th Conf. Solid State Devices, *Dig. Tech. Papers*, August, p.17.
[22] May, T. C. and Woods, M. H. (1978) A New Physical Mechanism for Soft Error in Dynamic Memories, Proc, 1978 Int. Reliability Phys. Symp., San Diego, CA, April 1978, p.33.
[23] Eaton, S. S. *et al.* (1981) A 100ns 64k Dynamic RAM using Redundancy Techniques, *IEEE ISSCC81 Digest of Tech. Papers*, p.84.
[24] Fujishima, K. *et al.* (1982) A Storage-Node-Boosted RAM with WordLine Delay Compensation, *IEEE Journal of Solid State Circuits*, **SC-17**, No.5, October 1982, p.872.
[25] Benevit, C. A. *et al.* (1982) A 256k Dynamic Random Access Memory, *IEEE Journal of Solid State Circuits* **SC-17**, No.5, October 1982, p.857.
[26] El-Mansy, Y. A. and Burghard, R. A. (1982) Design Parameters of the Hi-C DRAM Cell, *IEEE Journal of Solid State Circuits*, **SC-17**, No.5, October 1982, p.951.
[27] Tasch, A. *et al.* (1978) The Hi-C RAM Cell Concept, *IEEE Trans. Electron Devices*, **ED-25**, p.33, Jan. 1978.
[28] Sodini, C. G. and Kamins, T. I. (1976) Enhanced Capacitor for one Transistor Memory Cell, *IEEE Trans. Electron Devices*, Vol. **ED-23**, p.1187.
[29] Chwang, R. J. C. *et al.* (1983) A 70ns High Density 64k CMOS Dynamic RAM, *IEEE Journal of Solid State Circuits*, **SC-18**, No.5, October 1983, p.457.
[30] Baba, F. *et al.* (1983) A 64k DRAM with 35ns Static Column Operation, *IEEE Journal of Solid State Circuits* **SC-18**, No.5, October 1983, p.447.
[31] Fujii, T. *et al.* (1983) A 90ns 256k × 1 Bit DRAM with Double-Level A1 Technology, *IEEE Journal of Solid State Circuits*, **SC-18**, No.5, October 1983, p.437.
[32] Fujishima, K. *et al.* (1983) A 256k Dynamic RAM with Page-Nibble Mode, *IEEE Journal of Solid State Circuits*, **SC-18**, No.5, October 1983, p.470.

[33] Nakano, T. *et al.* (1983) A sub-100ns 256k DRAM with Open Bit Line Scheme, *IEEE Journal of Solid State Circuits*, **SC-18**, No.5, October 1983, p.452.

[34] Beresford, R. and Lineback, J. R. (1983) Opening Salvos Launched in 256k Battle, *Electronics*, June 30, p.107.

[35] Mashiko, K. *et al.* (1984) A 70ns 256k DRAM with Bit-Line Shield, *IEEE Journal of Solid State Circuits*, **SC-19**. No. 5, October 1984.

[36] Scheuerlein, R. E. *et al.* (1984) Shared Word Line DRAM Cell, *IEEE Journal of Solid State Circuits*, **SC-19**, No. 5, October 1984. p.640.

[37] Kertis, R. A. *et al.* (1984) A 60ns 256k × 1 Bit DRAM Using LD3 Technology and Double-Level Metal Interconnection, *IEEE Journal of Solid State Circuits*, **SC-19**, No.5, October 1984, p.585.

[38] Mohsen, A. *et al.* (1984) The Design and Performance of CMOS 256k Bit DRAM Devices, *IEEE Journal of Solid State Circuits*, **SC-19**, No.5, October 1984, p.610.

[39] Lu, N. C. and Chao, H. H. (1984) Half-VDD Bit-Line Sensing Scheme in CMOS DRAMs *IEEE Journal of Solid State Circuits*, **SC-19**, No.4, August 1984, p.451.

[40] Taylor, R. T. and Johnson, M. G. (1985) A 1-Mbit CMOS Dynamic RAM with a Divided Bitline Matrix Architecture, *IEEE Journal of Solid State Circuits*, **SC-20**, No.5, October 1985, p.894.

[41] Saito, S. *et al.* (1985) A 1Mbit CMOS DRAM with Fast Page Mode and Static Column Mode, *IEEE Journal of Solid State Circuits*, **SC-20**, No.5, October 1985, p.903.

[42] Yamada, J *et al.* (1984) A Submicron 1 Mbit Dynamic RAM with a 4-bit-at-a-time Built-In ECC Circuit, *IEEE Journal of Solid State Circuits*, **SC-19**, No.5, October 1984, p. 627.

[43] You, Y. and Hayes, J. P. (1985) A self-Testing Dynamic RAM Chip, *IEEE Journal of Solid State Circuits*, **SC-20**, No.1, February 1985, p.428.

[44] Kumanoya, M. *et al.* (1985) A Reliable 1-Mbit DRAM with a Multi-Bit-Test Mode, *IEEE Journal of Solid State Circuits*, **SC-20**, No.5, October 1985, p.909.

[45] Hori, R. *et al.* (1984) An Experimental 1 Mbit DRAM Based on High S/N Design, *IEEE Journal of Solid State Circuits*, **SC-19**, No.5, October 1984.

[46] Suzuki, S. *et al.* (1984) A 128k × 8 Word × 8 Bit Dynamic RAM, *IEEE Journal of Solid State Circuits*, Vol. **SC-19**, No.5, October 1984, p.624.

[47] Fujii, S. *et al.* (1986) A 50-microAmp Standby 1M × 1/256k × 4 CMOS DRAM with High-Speed Sense Amplifier, *IEEE Journal of Solid State Circuits*, Vol. **SC-21**, No.5, October 1986.

[48] McAdams, H. *et al.* (1986) A 1-Mbit CMOS Dynamic RAM with Design for Test Functions, *IEEE Journal of Solid State Circuits*, Vol. **SC-21**, No.5, October 1986, p.635.

[49] Furuyama, T. *et al.* (1986) An Experimental 4-Mbit CMOS DRAM, *IEEE Journal of Solid State Circuits*, Vol. **SC-21**, No.5, October 1986, p.605.

[50] Miyamoto, H. *et al.* (1987) A Fast 256k × 4 CMOS DRAM with a Distributed Sense and Unique Restore Circuit, *IEEE Journal of Solid State Circuits*, Vol. **SC-22**, No.5, October 1987.

[51] Kimura, K. *et al.* (1987) A 65ns 4-Mbit CMOS DRAM with a Twisted Driveline Sense Amplifier, *IEEE Journal of Solid State Circuits*, **SC-22**, No.5, October 1987, p.651.

[52] *Siemens 4M DRAM Preliminary Datasheet*, 1989.

[53] Horiguchi, F. *et al.* (1985) A 1Mb DRAM with a Folded Capacitor Cell Structure, *ISSCC Proceedings*, February, p.244.

[54] Inoue, Y. *et al.* (1985) An 85ns 1Mb DRAM in a Plastic DIP, *ISSCC Proceedings* February, p.238.

[55] Kalter, H. L. *et al.* (1985) An Experimental 80ns 1Mb DRAM with Fast Page Operation, *ISSCC Proceedings*, February, p.248.

[56] Takemae, Y. *et al.* (1985) A 1Mb DRAM with 3-Dimensional Stacked Capacitor Cells, *ISSCC Proceedings,* February, p.250.
[57] Sato, K. *et al.* (1985) A 20ns Static Column 1Mb DRAM in CMOS Technology, *ISSCC Proceedings,* February, p.254.
[58] Kirch, H. C. *et al.* (1985) A 1Mb CMOS DRAM, *ISSCC Proceedings,* February, p.256.
[59] Webb, C. (1986) A 65ns SMOS 1Mb DRAM, *ISSCC Proceedings,* February, p.262.
[60] Neal, J. *et al.* (1986) A 1Mb CMOS DRAM with Design-for-Test Functions, *ISSCC Proceedings,* February, p.265.
[61] Takada, M. *et al.* (1986) A 4Mb DRAM with Half Internal-Voltage Bitline Precharge, *ISSCC Proceedings,* February, p.270.
[62] Furuyama, T. *et al.* (1986) An Experimental 4Mb CMOS DRAM, *ISSCC Proceedings,* February, p.272.
[63] Shah, A. H. *et al.* (1986) A 4Mb DRAM with Cross-Point Trench Transistor Cell, *ISSCC Proceedings,* February, p.269.
[64] Mashiko, K. *et al.* (1987) A 90ns 4Mb DRAM in a 300mil DIP, *ISSCC Proceedings,* February, p.12.
[65] Parent, R. M. *et al.* (1987) A 4Mb DRAM with Double-Buffer Static Column Architecture, *ISSCC Proceedings,* February, p.14.
[66] Shimohigashi, K. *et al.* (1987) A 65ns CMOS DRAM with a Twisted Driveline Sense Amplifier, *ISSCC Proceedings,* February, p.18.
[67] Sumi, T. *et al.* (1987) A 60ns 4Mb DRAM in a 300mil DIP, *ISSCC Proceedings,* February, p.282.
[68] Mochizuki, H. *et al.* (1987) A 70ns 4Mb DRAM in a 300mil DIP using 4-layer Poly, *ISSCC Proceedings,* February, p.284.
[69] Ohsawa, T. *et al. A 60ns 4Mb CMOS DRAM with Built-in Self-Test.*
[70] Aoki, M. *et al.* (1989) New DRAM Noise Generation Under Half-VCC Precharge and its Reduction Using a Transposed Amplifier, *IEEE Journal of Solid State Circuits,* Vol. **24**, August 1989, p.889.
[71] Kraus, R. and Hoffman, K. (1989) Optimised Sensing Scheme of DRAMs, *IEEE Journal of Solid State Circuits,* Vol. **24**, August 1989, p.895.
[72] Arimoto, K. *et al.* (1989) A 60ns 3.3V Only 16Mbit DRAM with Multi purpose register, *IEEE Journal of Solid State Circuits,* Vol. **24**, No.5, October 1989, p.1184.
[73] Nakagome, Y. *et al.* (1990) A 1.5V Circuit Technology for 64Mb DRAMs, Symposium on VLSI Circuits, 1990, p.17.
[74] Hidaka, H. *et al.* (1989) Twisted Bit-Line Architectures for Multi-Megabit DRAMs, *IEEE Journal of Solid State Circuits,* Vol. **24**, No.1, February 1989, p.21.
[75] Konishi, Y. *et al.* (1989) Analysis of Coupling Noise Between Adjacent Bit Lines in Megabit DRAMs, *IEEE Journal of Solid State Circuits,* Vol. **24**, No.1, February 1989.
[76] *Motorola Memory Data Manual,* 1984, DL113R2.
[77] *Hitachi IC Memory Products Databook,* 1987/88, 10-2S.
[78] *Philips Static RAM Product Guide,* October, 1988.
[79] Watanabe, H. *et al.* (1988) Stacked Capacitor Cells for High density Dynamic RAMs, *IEDM Digest of Tech. Papers,* p.600.
[80] Nakajima, S. *et al.* (1985) A Submicrometer Megabit DRAM Process Technology Using Trench Capacitors. *IEEE Journal of Solid State Circuits,* Vol. **SC-20**, No.1, February 1985, p.130.
[81] Terletzki, H. and Risch, L. (1986) Operating Conditions of Dual Gate Inverters for Hot Carrier Reduction, *Proceedings of the ESSDERC,* Sept., 1986, p.191.
[82] Pinto, M. R. *et al. PISCES-II User's Manual.* Stanford CA, Stanford University, 1984.

[83] Tsukamoto, K. *et al.* (1987) Double Stacked Capacitor with Self-Aligned Poly Source/Drain Transistor (DSP) Cell for Megabit DRAM *IEDM Proceedings*, p.328.

[84] Shah, A. *et al.* (1986) A 4-Mbit DRAM with Trench-Transistor Cell, *IEEE Journal of Solid State Circuits*, Vol. **SC-21**, No.5, October 1986, p.618.

[85] Harter, J. *et al.* (1988) A 60ns Hot Electron Resistant 4M DRAM with Trench Cell, *IEEE ISSCC Proceedings*, p.244.

[86] Kaga, T. *et al.* (1987) A 4.2 km^2 half-V_{CC} Sheath-Plate Capacitor DRAM Cell with Self-Aligned Buried Plate-Wiring, *IEDM Proceedings*, p.332.

[87] Horiguchi, F. *et al.* (1987) Process Technologies for High Density High Speed 16 Megabit Dynamic RAM, *IEDM Proceedings*, p.324.

[88] Kimura, S. *et al.* (1988) A New Stacked Capacitor DRAM Cell Characterized by a Storage Capacitor on a Bit-Line Structure, *IEDM Proceedings*, p.596.

[89] Watanabe, H. *et al.* (1988) Stacked Capacitor Cells for High Density Dynamic RAMs, *IEDM Proceedings*, p.600.

[90] Yoshikawa, S. *et al.* (1988) Process Technologies for a High Speed 16M DRAM with Trench Type Cell, *VLSI Technology Proceedings*.

[91] Chin, D. *et al.* (1989) An Experimental 16M DRAM with Reduced Peak-Current Noise, *VLSI Circuits Proceedings*, 1989, p.113.

[92] Yamauchi, H. *et al.* (1989) A Circuit Design to Suppress Asymmetrical Characteristics in 16Mbit DRAM Sense Amplifier, *VLSI Seminar* 1989, p.109.

[93] Inoue, M. *et al.* (1988) A 16-Mbit DRAM with a Relaxed Sense-Amplifier-Pitch Open-Bit-Line Architecture, *IEEE Journal of Solid State Circuits*, Vol. **23**, No.5, October 1988, p.1104.

[94] Aoki, M. *et al.* (1988) A 60ns 16-Mbit CMOS DRAM with a Transposed Data-Line Structure, *IEEE Journal of Solid State Circuits*, Vol. **23**, No.5, October 1988, p.1113.

[95] Nakagame, Y. *et al.* (1988) The Impact of Data-Line Interference Noise on DRAM Scaling, *IEEE Journal of Solid State Circuits*, **Vol. 23**, No.5. October 1988, p.1120.

[96] Fujii, S. *et al.* (1989) A 45ns 16Mb DRAM with Triple-Well Structure, *IEEE ISSCC Proceedings* p.248.

[97] Watanabe, S. *et al.* (1988) An Experimental 16Mb CMOS DRAM Chip with a 100MHz Serial Read/Write Mode, *IEEE ISSCC Proceedings*, p.248.

[98] Arimoto, K. *et al.* (1989) A 60ns 3.3V 16Mb DRAM, *IEEE ISSCC Proceedings*, p.244.

[99] Takeshima, T. *et al.* (1989) A 55ns 16Mb DRAM, *IEEE ISSCC Proceedings*, p.246.

[100] Shong, S. H. *et al.* (1988) High-Speed Sensing Scheme for CMOS DRAMs, *IEEE Journal of Solid State Circuits*, Vol. **23**, No.1, February 1988 p.34.

[101] Horiguchi, M. *et al.* (1988) Dual-Operating-Voltage Scheme for a Single 5-V 16-Mbit DRAM, *IEEE Journal of Solid State Circuits*, Vol. **23**, No.5, October 1988, p.1128.

[102] Mano, T. *et al.* (1987) Circuit Technologies for 16Mb DRAMs, *IEEE ISSCC Proceedings*, February 25, p.22.

[103] Arimoto, K. *et al.* (1989) A 60ns 3.3V 16Mb DRAM, *IEEE ISSCC Proceedings*, p.244.

[104] Scheuerlein, R. E. and Meindl, J. D. (1988) Offset Word-Line Architecture for Scaling DRAM's to the Gigabit Level, *IEEE Journal of Solid State Circuits*, Vol. **23**, No.1, February 1988, p.41.

[105] Tsuchida, K. *et al.* (1989) The Stabilized Reference-Line (SEL) Technique for Scaled DRAMs, *VLSI Circuits Proceedings* 1989, p.99.

[106] Kasai, N. *et al.* (1987) 0.25 μm CMOS Technology Using P+ Polysilicon Gate PMOSFET, *IEDM Proceedings*, p.367.

[107] Lu, N. C. C. *et al.* (1988) A Buried-Trench DRAM Cell Using A Self-aligned Epitaxy Over Trench Technology, *IEDM Proceedings* p.588.

[108] Wakamiya, W. *et al.* (1989) Novel Stacked Capacitor Cell for 64mb DRAM, *VLSI Technology Proceedings.*

[109] Fujii, S. *et al.* (1989) A 45ns 16Mbit DRAM With Triple Well Structure, *IEEE Journal of Solid State Circuits,* Vol. **24**, No.5, Oct. 1989, p.1170.

[110] Chin, D. *et al.* (1989) An Experimental 16-M bit DRAM with Reduced Peak-Current Noise, *IEEE Journal of Solid State Circuits,* Vol. **24**, No.5, October 1989, p.1191.

[111] Inoue, S. (1989), A Spread Stacked Capacitor (SSC) Cell for 64M bit DRAMS, *IEDM Proceedings*, December, 1989, p.31.

[112] Sunouchi, K. (1989), A surrounding gate transistor (SGT) cell for 64/256M bit DRAMS, *IEDM Proceedings*, December, 1989, p.23.

[113] Kawamoto, Y. *et al.* (1990) A 1.28um2 Bit-line shielded Memory Cell technology for 64Mb DRAMs, *Symposium on VLSI Technology*, p.13.

[114] Hidaka, H. *et al.* (1990) A Divided/Shared Bitline Sensing Scheme for 64Mb DRAM Core, *Symposium on VLSI Circuits*, p.15.

[115] Sunami, H. *et al.* (1982) A Corrugated Capacitor Cell (CCC) for Megabit Dynamic MOS Memories, *IEEE IEDM Tech. Digest,* Dec. 1982, p.806.

[116] Yoshioka, S. *et al.* (1987) 4Mb Pseudo/Virtually SRAM, *IEEE ISSCC Proceedings*, February 1987, p.20.

[117] Richardson, W. F. *et al.* (1985) A Trench Transistor Cross-point DRAM Cell, *IEEE IEDM Tech. Digest* Dec. 1985, p.714.

[118] Sakamoto, M. *et al.* (1985) Buried storage Electrode Cell for Megabit DRAMs, *IEEE IEDM Tech. Digest,* Dec. 1985 p.

[119] Lu, N. *et al.* (1985) The SPT Cell—A New Substrate Plate Trench Cell for DRAMs, *IEEE IEDM Tech. Digest,* Dec. 1985, p.771.

[120] Rao, K. V. *et al.* (1986) Trench Capacitor Design issues in VLSI DRAM Cells, *IEEE IEDM Tech. Digest,* Dec. 1986, p.140.

[121] Cottrell, P. (1988) N-Well Design For Trench DRAM Arrays, *IEEE IEDM Tech. Digest,* Dec. 1988, p.584.

[122] Taguchi, M. *et al.* (1986) Dielectrically Encapsulated Trench Capacitor Cell, *IEEE IEDM Tech. Digest,* Dec. 1986, p.136.

[123] Yanagisawa, M. *et al.* (1986) Trench Transistor Cell with Self-Aligned Contact For Megabit MOS DRAM, *IEEE IEDM Tech. Digest,* Dec. 1986, p.132.

[124] Kalter, H. *et al.* (1990) A 50ns 16Mb DRAM with a 10ns Data Rate, *IEEE ISSCC, Proceedings,* February 1990, p.232.

[125] Arima, H. *et al.* (1990) A Novel Stacked Capacitor Cell with Dual Cell Plate For 64Mb DRAMs, *IEEE IEDM Tech. Digest,* Dec. 1990, p.651.

[126] Sakao, M. *et al.* (1990) A Capacitor-over-Bit-Line (COB) Cell with a Hemispherical-Grain Storage Node For 64Mb DRAMs, *IEEE IEDM Tech. Digest,* Dec. 1990, p.655.

[127] Shen, B. W. *et al.* (1989) Scalability of a Trench Capacitor Cell for 64 Mbit DRAM, *IEEE IEDM Tech. Digest,* Dec. 1989, p.27.

[128] Sunouchi, K. *et al.* (1990) Process Integration for 64M DRAM using An Asymmetrical Stacked Trench Capacitor (AST) Cell, *IEEE IEDM Tech. Digest,* Dec. 1990, p.649.

[129] Kaga, T. *et al.* (1988) Half-V_{CC} Sheath-Plate Capacitor DRAM Cell with Self-Aligned Buried Plate Wiring, *IEEE Trans. on Electron Devices,* **ED-35**, No.8.

[130] Yamada, T. *et al.* (1991) A 64-Mbit DRAM with MPL (Meshed Power Line) and distributed sense-amplifier driver. *ISSCC Proceedings*, February 1991, p.108.

[131] Taguchi, M. *et al.* (1991) A 40-ns 64Mb DRAM with Current-sensing Data-bus Amplifier, *ISSCC Proceedings*, February 1991, p.112.

[132] Mori, S. *et al.* (1991) A 45-ns 64Mbit DRAM with a Merged Match-line Test Architecture, *ISSCC Proceedings*, February 1991, p.110.

[133] Oowaki, Y. *et al.* (1991) A 33ns 64Mb DRAM, *ISSCC Proceedings*, February 1991, p.114.

[134] Fazan, P. C. and Ditali, A. (1990) Electrical Characterization of Textured Interpoly Capacitors For Advanced Stacked DRAMs, *IEEE IEDM Tech. Digest*, Dec. 1990, p.663.

[135] Koyanagi, M. *et al.* (1980) A 5V only 16Kb stacked Capacitor MOS RAM, *IEEE Journal of Solid State Circuits* **SC-15**, No.4, August 1980, p.661.

[136] Itoh, K. (1990) Trends in Megabit DRAM Circuit Design, *Journal of Solid State Circuits*, **25**, No.3, June 1990, p.786.

[137] Kalter, H. L. *et al.* (1990) A 50ns 16Mb DRAM with a 10ns Data rate and on-Chip ECC, *IEEE Journal of Solid State Circuits*, **25**, No.5, October 1990, p.1118.

[138] Komatsuzaki, K. *et al.* (1991) Circuit Techniques for a Wide Word I/O Path 64 Meg DRAM, *1991 Symposium on VLSI Circuits*, June, 14–5, p.133.

[139] Takashima (1991) Word-line Architecture for Constant Reliability 64Mb DRAM, *1991 Symposium on VLSI Circuits*, June, 14–5, p.57.

[140] Chin, D. *et al.* (1989) An Experimental 16M-bit DRAM with Reduced Peak-current Noise, *IEEE Journal of Solid State Circuits*, **SC-24**, No. 5, October 1989, p.1191.

[141] Moazzami, R. *et al.* (1990) A Ferroelectric DRAM Cell for High Density NVRAMs, *Symposium on VLSI Technology*, June 1990, p.15.

[142] Evans, J. J. and Womack, R. (1988) An Experimental 512-bit Nonvolatile Memory with Ferroelectric Storage Cell, *IEEE Journal of Solid State Circuits*, **SC-23**, No. 5, October 1988, p.1171.

[193] Kumanoya, M. *et al.* (1985). A 90ns 1Mb CMOS DRAM with Multi-bit Test Mode, *IEEE ISSCC 85*, p.240.

[144] Taylor, R. and Johnson, M. (1985) A 1Mb CMOS DRAM with Divided Bit-line Matrix Architecture, *IEEE ISSCC 1985*, February, p.242.

[145] Nakagome, Y. *et al.* (1991) An Experimental 1.5 V 64Mb DRAM, *IEEE Journal of Solid State Circuits*, **SC-26**, No. 4, April 1991, p.465.

[146] Karp, J.A. and Reed, J.A. (1977) Interlaced Memory Matrix Array Having Single Transistor Cells, *U.S. Patent*, 4,025,907, May 24 1977.

7 APPLICATION-SPECIFIC DRAMs

7.1 OVERVIEW

The early 1990s were a time of excitement and innovation in dynamic RAM technology due both to the demand for new cell technology for the multi-megabit densities and for new circuit technologies to support lower voltages, higher speeds and battery operation. The systems manufacturers increasingly expect DRAMs more closely tailored to the system application.

As the memory world has become more knowledgeable about the single transistor cell, there has been an increasing sense of comfort at expanding the DRAM out of its traditional role as main memory for large computers and into many diverse applications areas.

DRAM suppliers have followed the market trend away from large systems to smaller systems, to networks of smaller systems, to multi-processor systems and to systems with advanced graphics displays. See Figure 7.1 for the percentage of DRAM sales in each of these traditional DRAM applications areas.

The small consumer systems have traditionally used SRAMs. These systems generally require ease of operation, wide buses, miniature packaging, and low power for battery operation and back-up. Small computer systems require low cost standard DRAM components or SIMMs with wide buses, high speed and for graphics applications they require serial interface options. The technological developments for the multi-megabit DRAMs have followed these systems demand trends.

As a result of the new applications for DRAMs, there has been an increase in numbers of market specific product types and a movement away from the old trend of one standard volume product per density generation.

The rate of development of the technology has been significant. The extent of innovation of new cell technologies was rivaled only in the early 1970s. The period of the late 1980s saw the development of very high density cells permitting product densities to at least the 256Mb level as shown in Figure 7.2 which illustrates the

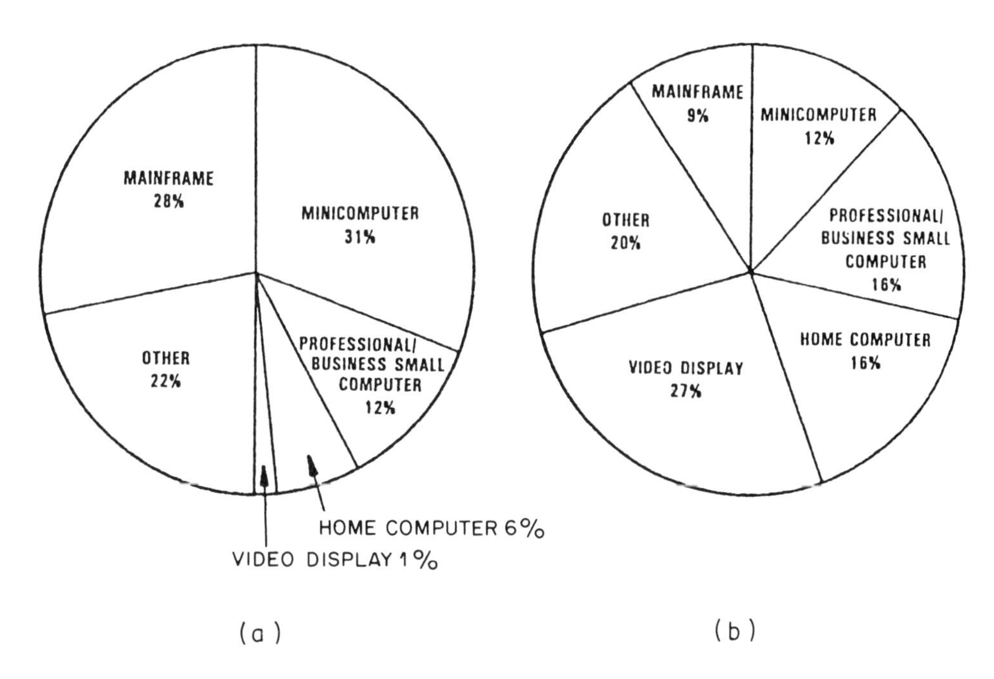

Figure 7.1
Usage of DRAM bits by market segment showing growth of usage in small systems and video display operations from (a) 1983 (25 trillion bits) to (b) 1988 (461 trillion bits). (Source Texas Instruments.)

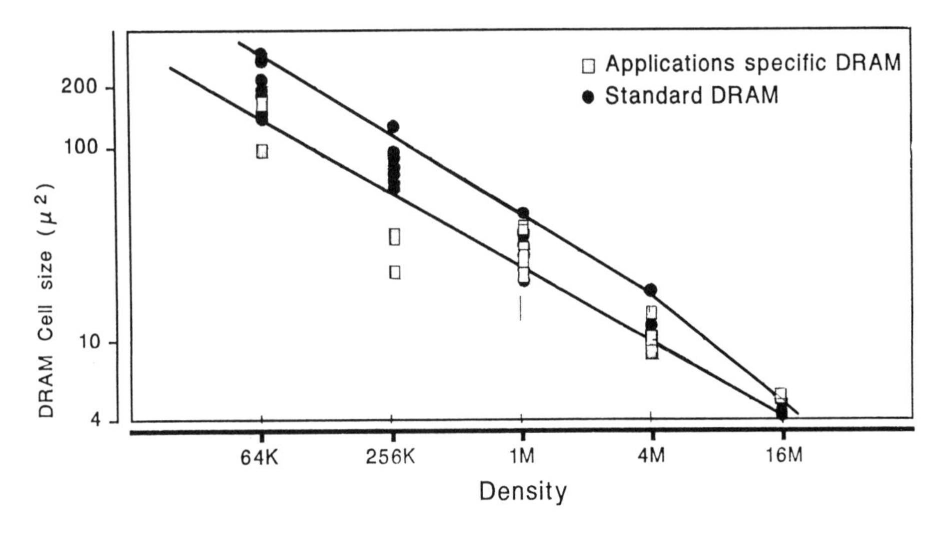

Figure 7.2
DRAM cell size plotted against density (□ applications specific DRAM, ● standard DRAM). (From Prince [77], Philips 1989.)

trend in cell sizes for each DRAM density. It also indicates the trend toward application-specific variations of each generation.

Feature size by the 256Mb is expected to be in the 0.2 μm region. Lithographic systems will phase to shorter wavelengths to X-ray and possibly e-beam systems. Higher dielectric constant materials will need to be developed for the submicron sized capacitors with materials such as Ta_2O_5 beginning to be used at the 16Mb densities to improve capacitance.

Test time reduction innovation required to test multi-megabit memories in a realistic time have produced both innovative test modes and on-chip test generators.

The extremely difficult yield and reliability questions faced by the multi-megabit DRAMs have been answered by such techniques as error correction on chip and improved redundancy schemes.

Further developments took place in simplifying the complex control of the one transistor cell including the advent of the pseudostatic DRAM with its potential for enabling DRAM technology to be embedded in large VLSI chips. There was also the lure of supplying applications, which use slow SRAMs for their simplified control, with the pseudostatic DRAMs which are lower in cost, available almost three years earlier and almost as easy to control in many applications.

New generations of microprocessors and other wide bus systems also made new demands on the DRAMs encouraging the development of DRAMs with wide input and output. The new graphics applications also required the wider bus parts.

The ever increasing demand for higher performance systems has driven the DRAMs both to further new operational modes and to new interface innovations. The rapidly mushrooming video graphics market has encouraged the development of high speed serial interfaces. This has resulted in both the dual port video DRAM with one serial and one random port, and the frame buffer with two serial ports. It has also given rise to new high speed pipelined interface modes.

The requirement for low power memories for smart battery operated systems and for high density fast applications where power dissipation is a concern has led to requirements from the systems side for low voltage DRAMs. This has coincided with the need of the DRAMs to move to lower voltages for reliability reasons in the submicron geometries. While some 16Mb DRAMs are being prepared to be offered with the 3.3 V TTL compatible standard interface as a mask option, it is not until the 64Mb generation that the DRAMs are expected to convert to the low voltage interface.

The move in the high performance systems area to higher speed bipolar logic has resulted in the development of BiCMOS technology for DRAMs.

Finally, as the price per bit of DRAMs approaches that of disk memory, the lure of the huge disk memory market has led to significant attempts to reduce the power consumption of DRAMs and couple this with the new ease-of-control functions to produce solid state archive memories.

Developments in miniature packaging have occurred to meet the demand of high performance systems for shortened interconnects between chips and for high density applications.

The challenge for the DRAM industry with the multi-megabit DRAMs was to pack multiple millions of memory cells in chips of a size to be manufacturable and to fit into the small packages demanded by the customers. Even so the chip size increased steadily.

Increased speed was also demanded by the faster processors appearing on the market. High density systems and portable systems demanded lower power.

7.2 DRAMs FOR THE SMALL SYSTEM ENVIRONMENT

The success of the personal computer during the 1980s led the DRAM designers in search of ways to make the DRAMs more system compatible with this fast growing small system applications. The traditional large computer main store memory needed to be deep and inexpensive. This meant $\times 1$ organization and as simple an organization as possible to reduce cost. Most of the control overhead was off-loaded onto the system which amortized the expense over the large number of DRAM memories used.

This began to change with the advent of the small system. The small systems needed fewer memories and since they were controlled by the wide bus microprocessors, they needed wider word memories. The smaller system also made it less attractive to try to amortize the memory control over the system. This, coupled with the early reliance of the systems on SRAMs, made the small system designer expect to find his memories easy to control and to work with.

The DRAM designers responded with such innovations as wider DRAMs, easier to control DRAMs, DRAMs with the refresh function and/or the address function handled on chip, DRAMs with de-multiplexed addressing, and finally with DRAMs which had the lower standby power characteristics which were previously the exclusive territory of the static RAMs.

7.2.1 Non-multiplexed addressing for DRAMs

An early example of a non-multiplexed, wide word DRAM designed with the small system one chip memory in mind was the 128k $\times$ 8 DRAM from NEC [45] which was described in 1984. The non-multiplexed addresses made the part similar to the static RAMs. The pinout used was an attempt to be static RAM compatible, although the package used had 30 pins rather than the 32 pins later used for the 128k $\times$ 8 SRAMs and pseudostatic DRAMs.

7.2.2 Simplifying the external control requirements of the DRAM

One of the continuing efforts of DRAM designers has been to simplify the complex external control requirements on the system of the one-transistor cell DRAM. Since this is one of the major differences between using a cost effective DRAM and the easier to control SRAM in a system, and because the SRAM market already exists, efforts to simplify the control of the DRAM have tended to move in the direction of making it more SRAM compatible. The DRAM needed to appear to the system more like a static RAM.

This development was significant because it eliminated the cost penalty for DRAMs of the refresh control overhead in the system and allowed the DRAMs to compete with the SRAMs in small systems applications where the main criteria for choosing the more expensive SRAMs was the cost of the DRAM overhead control circuitry. It also made the DRAMs easier to use by relaxing the tight timing requirements inherent in the external control function of a DRAM.

7.2.3 The first 64k DRAMs with simplified refresh control

One of the early attempts to make a DRAM operate more like an SRAM was one version of the Motorola [1] 64k DRAM which had a self-refresh pin which added an internal address counter on chip. Introduced in 1980, this chip used the spare location on pin 1 in the standard 64k DRAM pinout for a refresh control which generated refresh addresses on-chip so that the external system design could be simplified. Pin 1 still needed to be clocked externally to control this new operating mode so the ac parameters were not completely compatible with the ac parameters of a standard asynchronous SRAM which needs no external system clock.

$\overline{\text{RAS}}$ and $\overline{\text{CAS}}$ clocks were still needed as in a conventional DRAM and the part still used the address multiplexed DRAM pinout rather than the non-address multiplexed SRAM pinout. See Figure 7.3 for the various refresh modes which were available on this DRAM.

7.3 THE PSEUDOSTATIC AND VIRTUAL STATIC DRAMs (PSRAM–VSRAM)

Subsequent efforts at putting refresh control on the DRAM were implemented in pinouts which were more compatible with the small systems environment and hence more similar to the SRAMs which were designed for this environment. These various attempts to modify the lower cost DRAMs for use in SRAM sockets were called either pseudostatic RAMs (PSRAMs), or Virtual Static RAMs (VSRAMs) depending on the function implemented.

The JEDEC standardized pinouts of the 256k through 8Mb PSRAMs in the small outline 'J'-leaded (SOJ) package are shown in Figure 7.4(a). These are identical to the comparable JEDEC standard for the SRAMs except that where pin 1 is marked # for the PSRAMs, it is a no connect (NC) for the comparable SRAM.

A pseudostatic DRAM (PSRAM) is a 'one transistor cell' dynamic RAM with non-multiplexed addresses, an on-chip refresh address counter, and an external 'refresh' pin to control the refresh of the cell to maintain data in the memory. The exception on the 'refresh' pin was the 256k PSRAM which lacked the extra pin in the standard

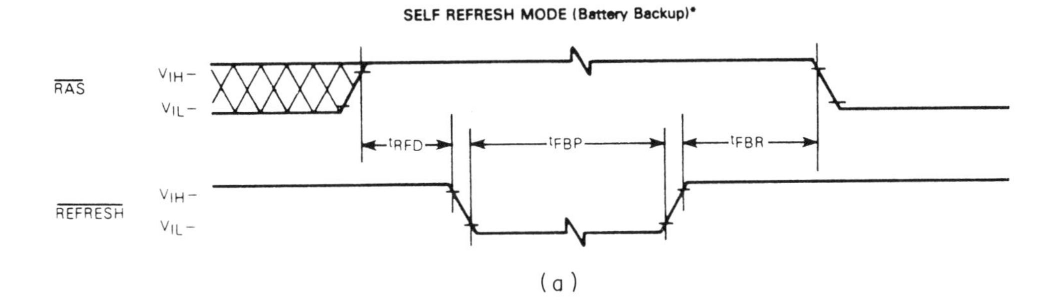

(a)

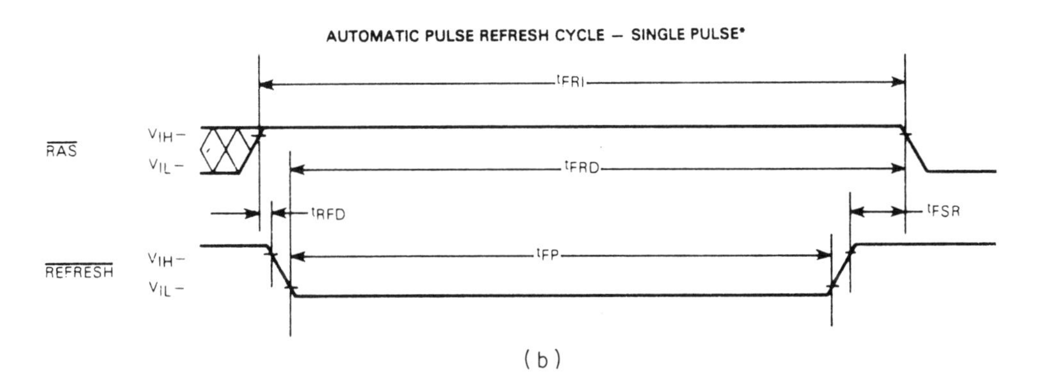

(b)

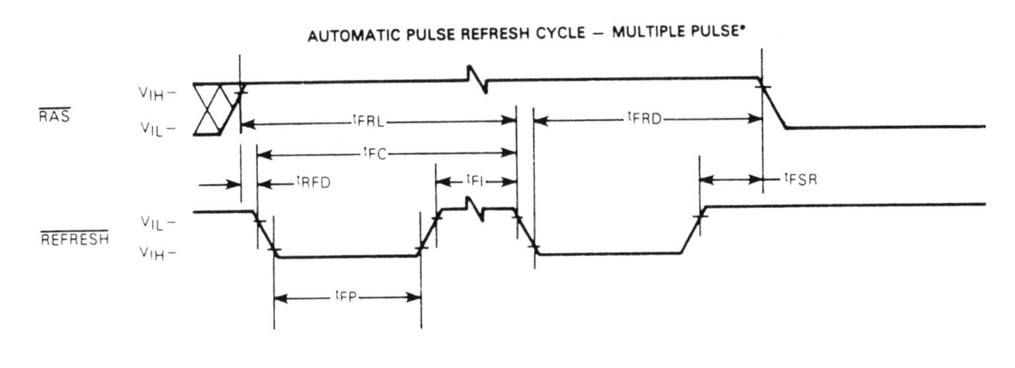

*Addresses, data-in and WRITE are don't care, CAS is high

(c)

Figure 7.3
Various refresh timing diagrams of the 64k DRAM with pin 1 refresh control. (a) Self-refresh mode (battery back up). (b) automatic pulse refresh cycle—single pulse. (c) automatic pulse refresh cycle—multiple pulse. (Source Motorola Memory Databook, reproduced with permission of Motorola.)

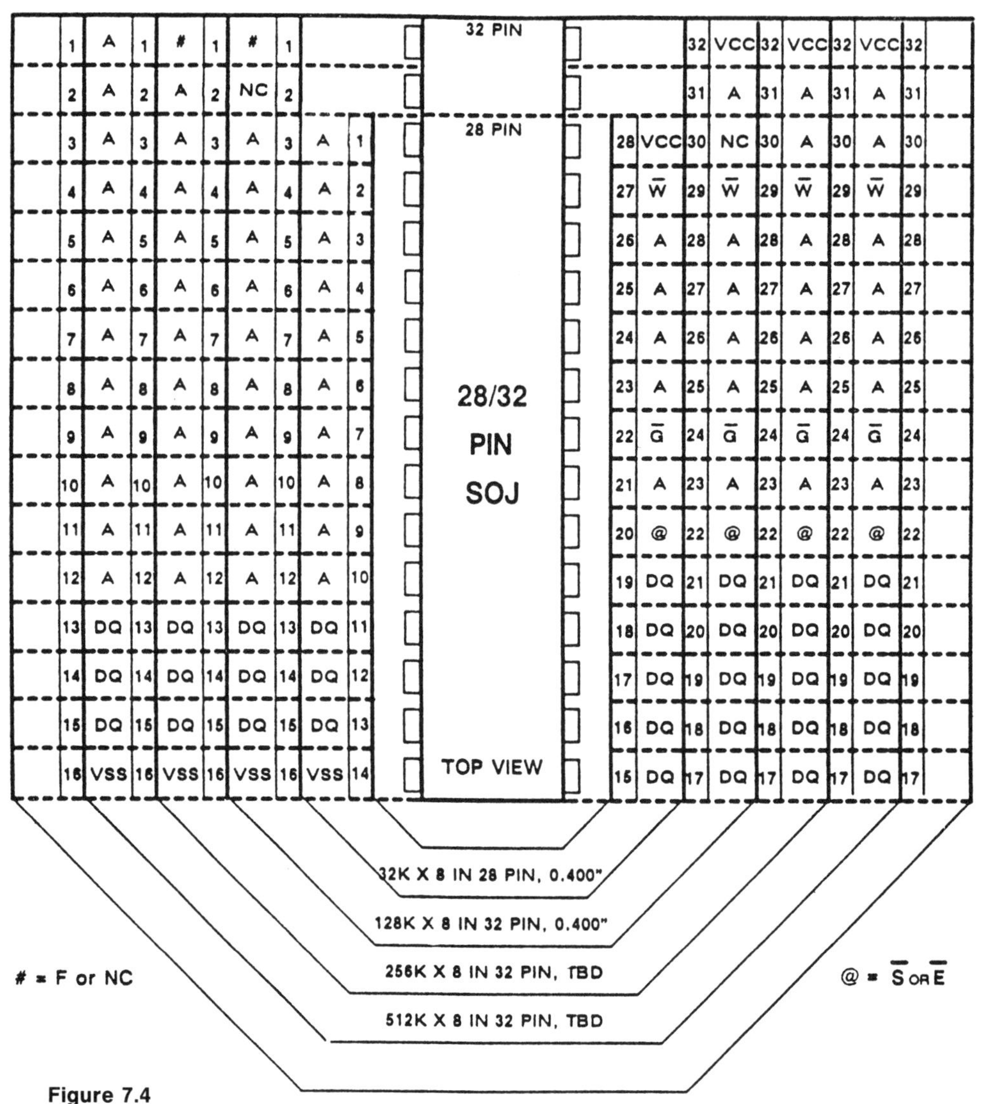

Figure 7.4
JEDEC standard pinout family of PSRAMs in SOJ package. (From JEDEC Standard 21-C, with permission of EIA.)

package and compensated for it by a more complex timing structure. The parts were byte wide (had 8 bit wide inputs and outputs) like the commodity static RAMs.

A PSRAM is synchronous since it operates from a system clock applied to the chip enable pin. Its access time and cycle time are identical to the underlying DRAM memory with address access time normally $\frac{1}{3}$ to $\frac{1}{2}$ of the cycle time.

The refresh of the memory cells occurs during the part of the cycle time when the memory is not being accessed. While there is the potential for the processor having to wait for an access if the previous cycle is not yet complete, the fact that the refresh is under the external control of the system minimizes this risk in the PSRAM.

A PSRAM appears to the system like a slow synchronous SRAM. If a pseudostatic is run in a system needing a system clock anyway and needing a speed no faster than the cycle time of the pseudostatic cycle time, then the pseudostatic appears to the system like a standard SRAM.

Systems which can use PSRAMs interchangeably with SRAMs must have the following

(1) be designed initially to use PSRAMs,

(2) have a system clock running,

(3a) have memory access time requirements slower than the cycle time of the PSRAM, or

(3b) faster memory access times but with intervals between accesses longer than the PSRAM cycle time.

This means that an SRAM can be dropped into a system designed for a PSRAM, but not the reverse.

7.3.1 Early PSRAMs

One of the earliest production PSRAMs was a 64k PSRAM from NEC [6] which had the standard 8k × 8 SRAM pinout with the addition of a refresh control pin in place of the 'no connect' on pin 1. This external refresh control pin was also found on the 1Mb and 4Mb PSRAMs, although not on the 256k PSRAMs. The part was synchronous and clocked from chip enable. There was an on-chip refresh address counter as shown in the block diagram in Figure 7.5(a). It had a mixture of dynamic storage cells in the array and had static input and output circuitry in an attempt to attain high speed and low power in the same device. The timing diagram for a 'pulse refresh cycle after read cycle is complete' is shown in Figure 7.5(b).

The 256k PSRAMs from Hitachi [2] also were early volume production parts. Typical of PSRAMs the chip was clocked externally from the chip enable pin on the standard byte wide SRAM pinout like a synchronous SRAM. They did not have the separate refresh control pin typical of the PSRAMs since the 256k SRAM pinout had no extra pin available. Lacking a separate refresh control pin the refresh was controlled from the output enable. The pinout was in accordance with the JEDEC standard and exactly the same as the comparable SRAM so that the two parts were socket compatible.

7.3.2 A comparison of a 256k SRAM with a 256k pseudostatic RAM

The block diagram of the 256k 'pseudostatic' RAM from Hitachi is compared with that of the 256k SRAM from the same manufacturer in Figures 7.6(a) and (b). The

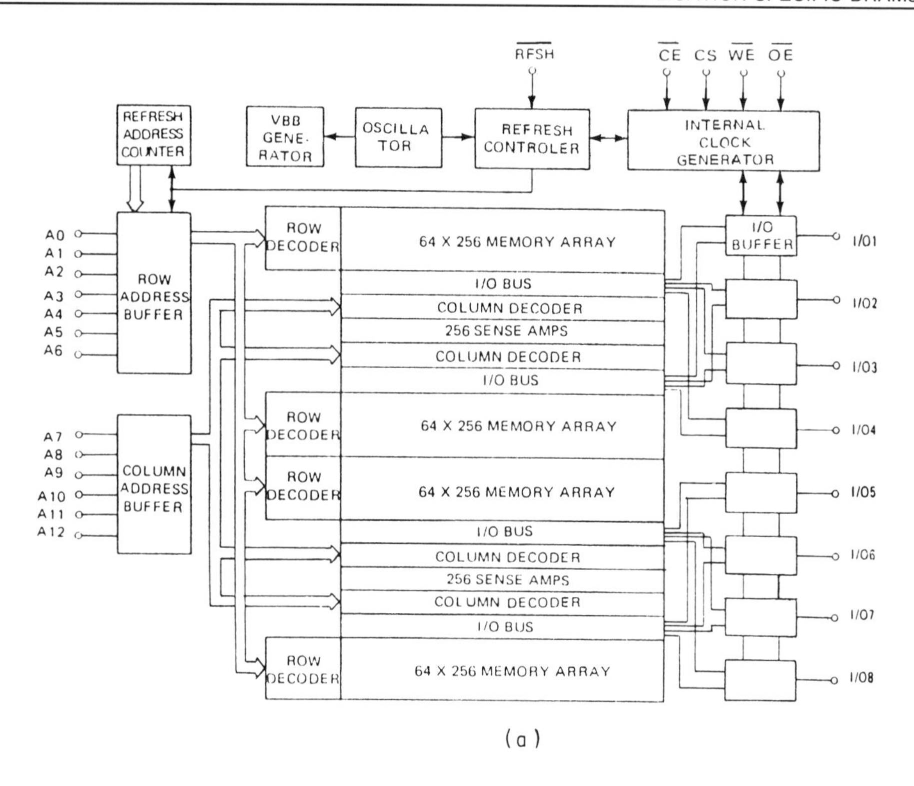

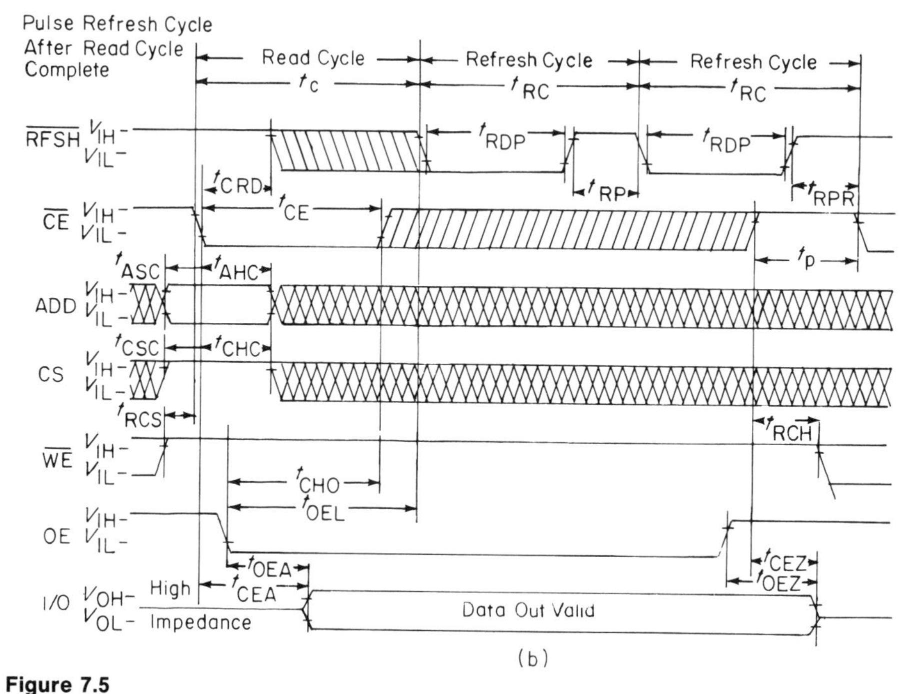

Figure 7.5
64k PSRAM. (a) Block diagram, (b) timing diagram. (Source NEC Memory Products Databook 1987 [76].)

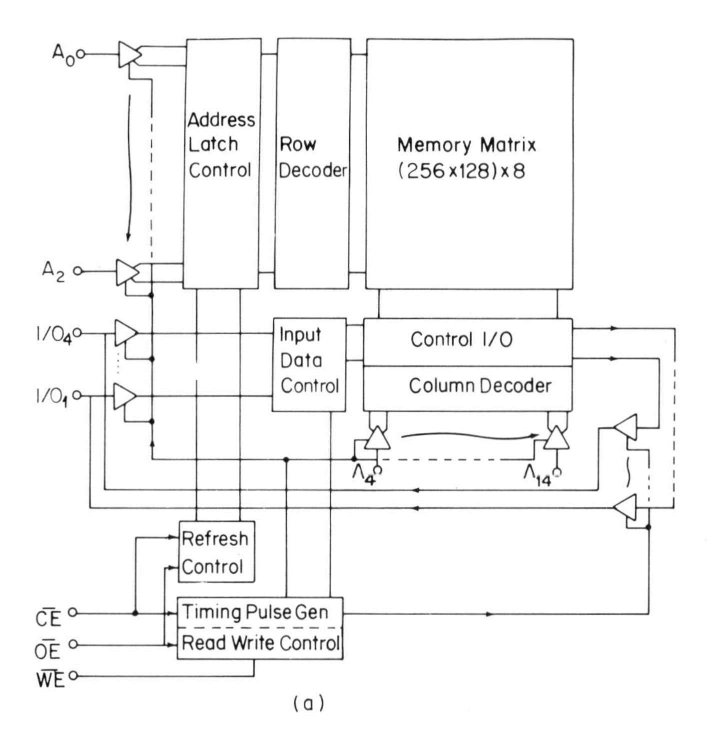

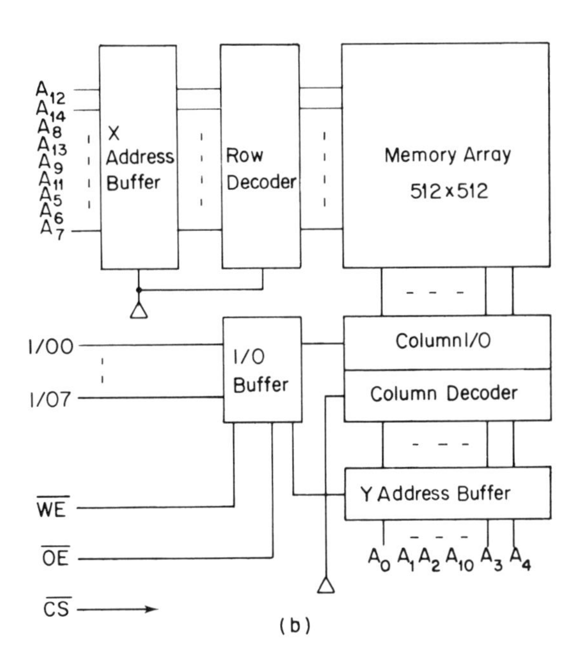

Figure 7.6
Block diagrams for comparison of (a) 256k PSRAM, (b) 256k SRAM. (From Hitachi IC Memory Products Databook 1987/88 [2].)

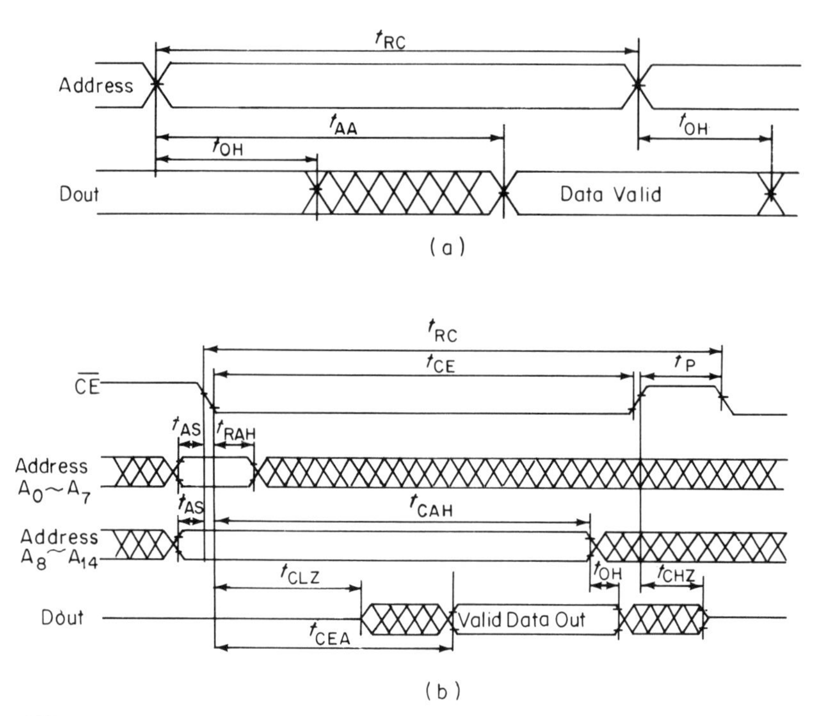

Figure 7.7
Simple read cycle for (a) 256k SRAM, (b) 256k PSRAM. (From Hitachi IC Memory Products Databook 1987/88.)

added internal complexity of the PSRAM is clearly shown in the address latch control, and in the refresh control and the timing pulse generator which are controlled by the chip enable and output enable pins.

There is an additional requirement for the PSRAM during read and write cycles that the output enable pin be high to disable the refresh at the time of chip enable going low. The PSRAM is synchronous and must be clocked, while the SRAM is able to be operated asynchronously. There are also some timing complexities associated with the internal refresh control requirements of the one transistor cell.

Two simple read cycles are compared for the two parts in Figure 7.7. A pure asynchronous static RAM as shown in Figure 7.7(a) needs only to have the addresses changed to have the new data appear after a delay at the output. The addresses must be held for the full read cycle time. The minimum read cycle time is equal to the maximum read access time. Although not shown on this timing diagram, when chip enable is used, the minimum read cycle time is also equal to the maximum chip enable access time.

The PSRAM, being synchronous, clocks in the addresses from chip enable. The required address hold times are different for the row and column addresses as in any DRAM. A chip enable precharge is required following each chip enable pulse which extends the length of the cycle time.

The cycle time of the SRAM is specified to be the same as the access time from

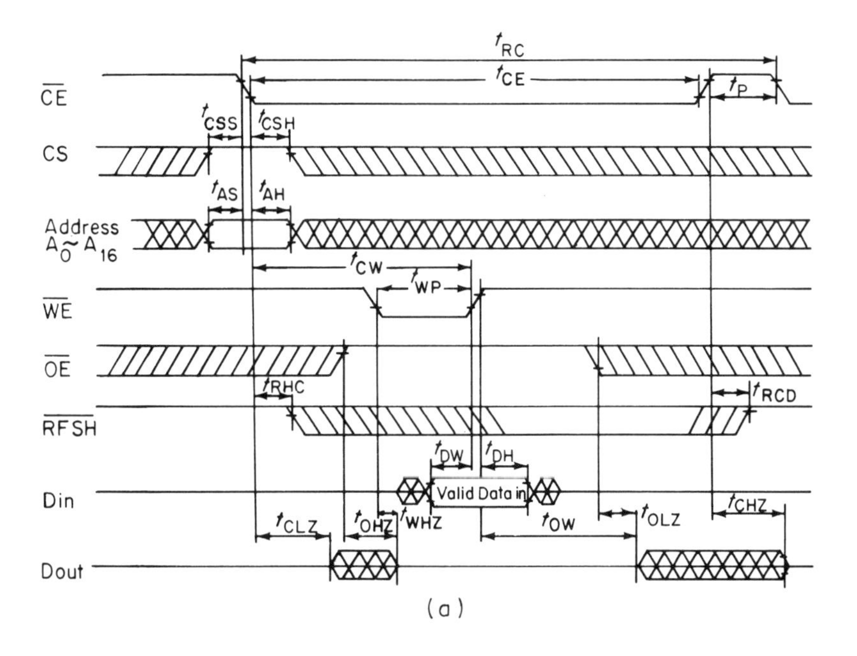

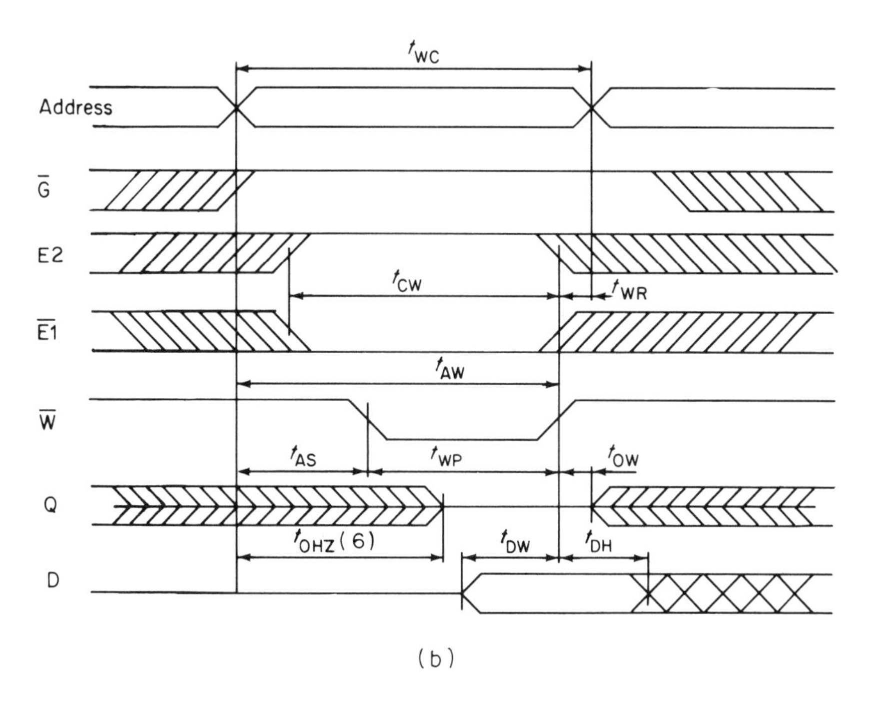

Figure 7.8
Write cycles using chip enable and output enable for: (a) 1Mb PSRAM (from Hitachi IC Memory Products Databook 1987/88), (b) 1Mb SRAM (from Philips Static RAM Product Guide).

address transition to valid data out, while the cycle time of the PSRAM is significantly longer than the access time from chip enable to valid data out due to the precharge pulse requirement. This means that the system has to wait a full DRAM cycle time for another memory access of the PSRAM, so that in a standard SRAM socket the PSRAM is as slow as a DRAM. However, if immediate sequential access of the same PSRAM is not required in the particular application, then the address access time is what counts and the PSRAM will appear to have similar timing characteristics to the SRAM.

The requirement that the chip enable ($\overline{CE}$) pin be clocked is also different from the standard asynchronous SRAMs' lack of dependance on a system clock. Many systems do have a system clock running anyway, so again there are some systems to which this PSRAM can be made to appear like a standard SRAM.

The dc characteristics for TTL input levels are comparable between the two parts. The CMOS typical standby power is, however, more than an order of magnitude higher for the PSRAM. It should be noted that the SRAM being used for comparison here has a Mix-MOS cell with four transistors plus two resistors rather than a CMOS cell with six transistors which normally has even another order of magnitude lower CMOS standby power dissipation (for further discussion of this point see Chapters 5 and 8).

An example is the Philips 256k full CMOS SRAM [3] which had 0.02 μA typical standby power. Typical standby power characteristics of these three parts are shown in Table 7.1.

Toshiba also had a 256k PSRAM which is similar in function and pinout to the Hitachi part. It also suffered from the lack of an extra pin to use for the $\overline{RFSH}$ function and combined it with the $\overline{OE}$ function.

7.3.3 Two 1Mb PS-DRAMs

Hitachi [2] and Toshiba [55] both had 1Mb PS-DRAMs in production by 1989 which returned to the typical PS-DRAM option of the separate refresh control pin as shown previously in the pinout in Figure 7.4. The Toshiba part had a 'virtual static' mode which eliminated the need for the refresh control pin. This mode is described later in this chapter. Both parts were synchronous like all PSRAMs.

Table 7.1 Standby power comparison of PSDRAMs and SRAMS.

	ISB $\overline{CS} = V_{IH}$ (mA)		ISB1 $\overline{CS} > V_{CC} - 0.2$ V (μA)	
1. 256k PSRAM	1	2	50	100
2. 256k Mix-MOS SRAM	0.5	3	2	50
3. 256k FULL CMOS SRAM	0.5	2	0.02	2

A write timing diagram for the 1Mb PS-DRAM is shown in Figure 7.8(a). The PS-DRAM was, as before, synchronous and clocked from chip enable. Also shown for comparison, in Figure 7.8(b), is the chip enable write timing diagram of the Philips 1Mb SRAM. Although not shown, it was also possible to read and write the SRAM which was asynchronous without using chip enable.

Clearly in a system where both chip enable and chip select are used the timing diagrams of the two parts are quite similar. The PSRAM, however, had the additional need to clock the $\overline{\text{RFSH}}$ pin. It also needed a chip enable precharge pulse at the end of the cycle which the SRAM did not need. The chip enable access times was nearly equal to the cycle time for SRAM, while it was about half of the cycle time for the PSRAM. Again the cycle time for the PSRAM was a full DRAM cycle time.

The standby power characteristics of the PSRAM, although similar to those of comparable 1Mb Mix-MOS SRAMs, were significantly higher than those of the CMOS SRAM as was the case with the 256k SRAM and PSRAM.

7.3.4 The 1Mb PS-DRAMs compete for market with the 1Mb SRAMs

The 1Mb PS-DRAM appeared in production before the 1Mb SRAM. The two 4Mb PS-DRAMs were described during 1987 and 1988, two years before the first 4Mb SRAM was described. This points up one of the significant advantages of the PS-DRAM, that being a DRAM, its product life cycle is simultaneous with that of the DRAMs. This gives the PS-DRAM its main advantage over the SRAM which is that it can be available for production three to four years earlier than the comparable density SRAM. This of course is true since SRAM densities tend to run a life cycle behind that of DRAMs.

The life cycles of several generations of PS-DRAMs and SRAMs are shown in Figure 7.9. It is clear that the PS-DRAM gains its main advantage over the SRAM in being available two to three years earlier in any given density than the comparable density SRAM.

This difference in timing of introduction means that the PSRAM can go down the cost learning curve about three years ahead of the comparable density SRAM, although ending with only a small price difference if price is based strictly on chip size.

If we consider for example the Toshiba 1Mb DRAM in 1.2 μm geometry, the chip size was 54 mm^2, and the Toshiba VSRAM, which was similar to the PSRAM, in the same 1.2 μm technology was 82.7 mm^2. The extra control circuitry on the VSRAM increased its chip size by 53% over that of the standard DRAM in the same technology. The Toshiba 1Mb PSRAM was made in the smaller 1.0 μm technology and is 73.2 mm^2. This is still a 35% increase in chip size over the standard DRAM in the older 1.2 μm technology.

Now comparing this 1Mb PSRAM with 1Mb SRAMs, the average size of a 1Mb SRAM was about 95 mm^2 and was typically in the next generation 0.8 μm technology, so that while the 1Mb SRAM was 75% larger than the 1Mb DRAM, it was only about 15% larger than the 1Mb PS-DRAM.

If we assume that the price per square millimeter of silicon is approximately constant

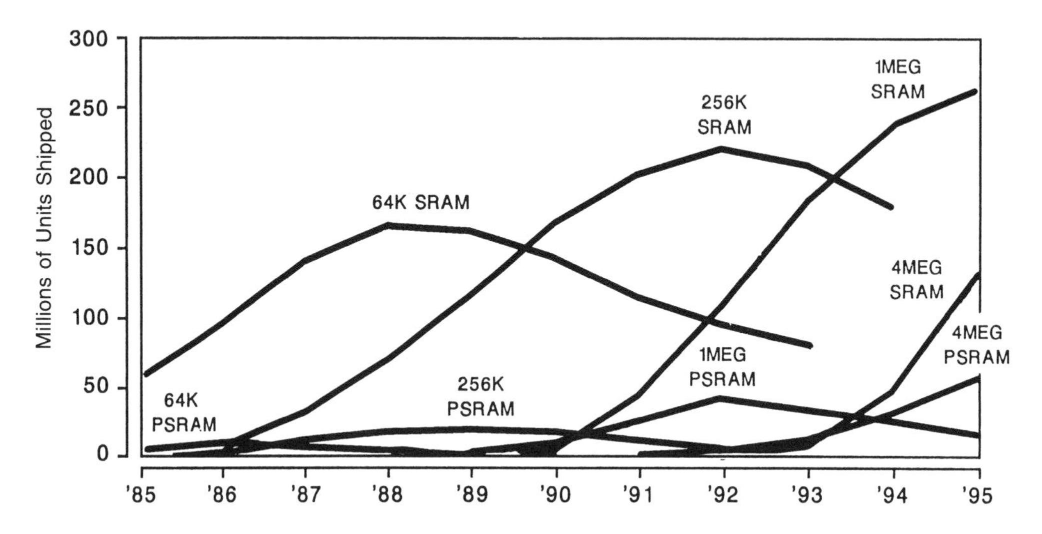

Figure 7.9
Shipments of SRAMs and PSRAMs. (Source Dataquest May 1988/July 1989.)

for volume memory products, then the 1Mb PSRAM would be expected to be about 15% less expensive than the 1Mb SRAM at end of life cycle prices for both products.

Figure 7.10 illustrates this also for the 4Mb DRAM and the 4Mb PSRAM which were in the 0.8 to 1.0 μm technology. 4Mb SRAMs were in 0.5 μm technology by comparison.

Historically this price differential has not been sufficient to compensate for the added complexity in the system of the PSRAM, in particular the requirement that the system be synchronous and the need to clock the refresh control pin. Another

Year	Company	Density	Technology (μm)	Chip size (mm^2)	Cell size (μm^2)	Array efficiency	Type
Typical standalone DRAMS							
1982	Hitachi	256k	2.0	46.8	98.0	55.0	DRAM
1985	Toshiba	1Mb	1.2	54.0	29.2	57.0	DRAM
1986	Toshiba	4Mb	1.0	137.0	17.4	52.0	DRAM
1986	NEC	4Mb	0.8	99.0	10.58	44.0	DRAM
Typical DRAMS with refresh control							
1984	Hitachi	288k	2.0	55.0	92.5	48.4	PSRAM
1988	IBM	500k	1.0	78.0	75.2	49.0	PSRAM
1986	Toshiba	1Mb	1.2	82.7	29.4	36.0	VSRAM
1988	Toshiba	1Mb	1.0	73.2	29.5	41.9	P/VSRAM
1987	Oki	4Mb	1.0	145.8	16.8	47.1	P/VSRAM
1987	Hitachi	256k	1.0	24.5	39.2	41.6	SCL/SRA

Figure 7.10
(Source ISSCC Proceedings 1982 to 1987.)

concern among users was the lack of significant numbers of second sources for the PSRAMs.

What was needed was a refresh controlled DRAM that was identical in the system to an SRAM. The VDRAM attempted to provide this option.

7.3.5 The 'virtually' static transparent refresh RAMs

Virtually static RAMs are DRAMs that have the refresh totally transparent to the user. They are basically pseudostatics that have a refresh timer on chip that generates the refresh request signal for the timing generator. See Figure 7.11 for a block diagram of the architecture and basic operation of a 1Mb virtually static RAM [4]. To the refresh address counter of the PSRAM has been added a refresh timer and timing generator.

Since the refresh is now totally transparent to the user, the cycle time and address access time must be equal as in a standard asynchronous SRAM. This means that what the processor sees as the address access time is really the access time plus the refresh time. The virtual SRAM, therefore, has a longer apparent access time than the pseudo SRAM, but the same cycle time which, in both cases, is just the DRAM cycle time.

A disadvantage of the virtually static RAM is that the on-chip timers require power, and the standby current of a virtually static RAM tends to be high. In the case of the Toshiba 1Mb VSRAM whose block diagram is shown in Figure 7.11, the typical standby current was 400 μA compared to the comparable Toshiba PSRAM at 30 μA.

An example of an early virtually static RAM was a 64k RAM described by Intel [5] in 1982 and called the 'iRAM', or 'internal refresh RAM'. This device was organized 8k $\times$ 8 bit and had non-multiplexed addresses, internal refreshing without the necessity for an external clock pulse, circuitry for arbitration between external memory access requests and internal memory refresh requests and the standard byte wide 64k SRAM pinout with the exception that a READY function was added on pin 1.

The READY pin acted as an arbitrator between access requests of the processor and the internal refresh circuitry of the memory. If the processor requested an access during a refresh cycle a wait state was generated. Similarly, if the memory refresh cycle time occurred when a processor access was in process then the refresh waited.

The arbitration logic theory of the iRAM was also interesting and will be covered more thoroughly under the section on metastability in dual port RAMs in Chapter 9.

The iRAM was operated from the chip enable with three control inputs required for operation. The device latched addresses into the memory on the leading edge of the chip enable ($\overline{CE}$) clock going low. A read cycle occurred if the output enable ($\overline{OE}$) clock went low, or a write cycle if the write enable clock ($\overline{WE}$) went low.

While the refresh was timed and controlled internally and the address counter for the refresh was also internal, the disadvantage of the wait state during refresh was serious on this part. If a processor access request occurred just after a refresh cycle was initiated then the memory cycle would be deferred as much as 325 ns.

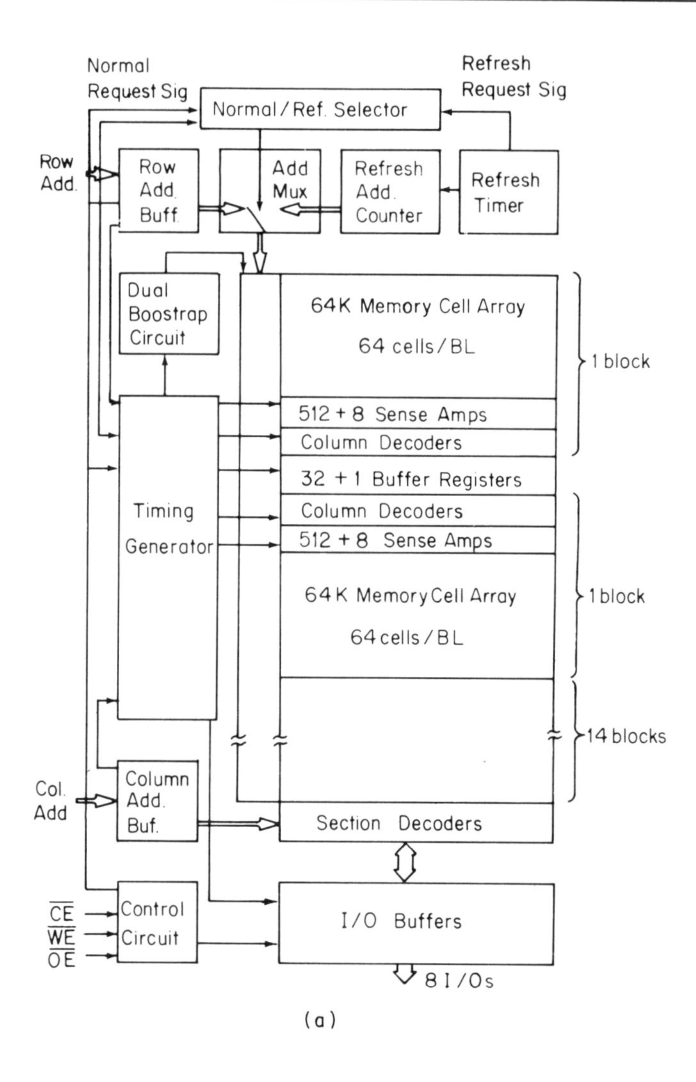

(a)

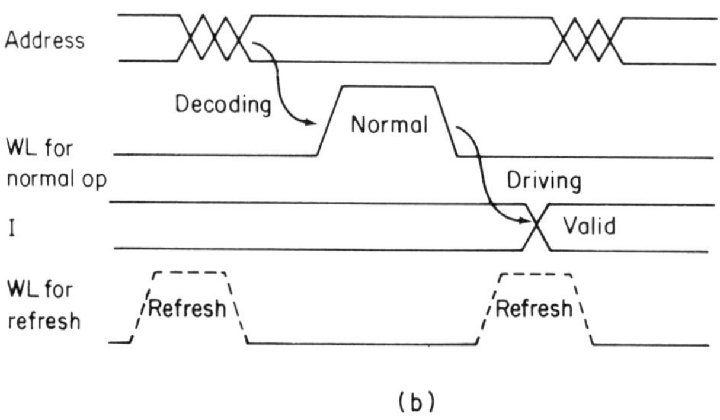

(b)

Figure 7.11
Virtual SRAM. (a) Block diagram showing basic structure. (b) Basic operation of VSRAM. (From Sakurai [4], Toshiba 1986, with permission of IEEE.)

A later example of a virtually static RAM which was much more SRAM compatible was that described by Toshiba [4] in 1986. In this VSRAM, unlike the PSRAM, the refresh is totally transparent to the user. It can therefore be used as a normal asynchronous SRAM without special considerations on the part of the system designer.

The transparent refresh was accomplished by an on-chip refresh timer which indicated when a refresh operation was needed and generated a refresh request signal intermittently. If the memory cell array of the RAM was busy with a normal access then the refresh operation waited until the cell data were handed to a buffer register which drove the output circuits. If the refresh operation had already started then a normal access waited until the refresh operation ended. Clearly long wait periods for a system processor were unattractive to system designers, so the core of this memory was made faster by being divided into 16 blocks and using double bit-line architecture to reduce capacitance. A double metal interconnect system was used for speed.

The effective access time of the memory was slowed down by the wait time for the refresh. The access time without refresh was 48 ns and with refresh was 62 ns giving an access time overhead of the background refresh of 29%.

Clearly, if a system has long enough cycle time between memory accesses then it is desirable not to have a wait state when accessing the memory, that is to have a PSRAM. A processor would, however, have added flexibility if able to act in both modes.

7.3.6 The virtually pseudostatic DRAM

To meet this demand for flexibility the 1Mb PSRAM from Toshiba [55], which was mentioned earlier in the chapter, had a virtually static optional mode which could be electrically selected in the system by grounding the refresh control pin.

On this part the VSRAM mode access time at 66 ns was slower than the PSRAM mode access time at 36 ns, but it was completely refresh control free as shown in the timing diagrams in Figure 7.12 which compare the PSRAM and VSRAM modes of this chip. The part was synchronous in both modes.

The PSRAM mode had control of refresh operation determined by the $\overline{\text{RFSH}}$ input and could be operated in either auto refresh or self refresh mode. The VSRAM mode controlled all refresh operations internally when the $\overline{\text{RFSH}}$ pin was fixed low. It was therefore usable in the system as a slow synchronous SRAM.

The VSRAM mode included an arbiter circuit which judged which of the refresh and the normal operations would be active when contention occurred between a normal operation request and an internal refresh request, relieving the system of this responsibility. Hence in the VSRAM mode the RAM could be directly connected to the CPU without any refresh controller in between. The VSRAM mode was slower because the normal operation might have to wait until an internal background refresh ends. There was hence a tradeoff between convenience and speed.

The data-retention current of this part was only 30 μA which was a significant improvement over previous generations of refresh control DRAMs and comparable

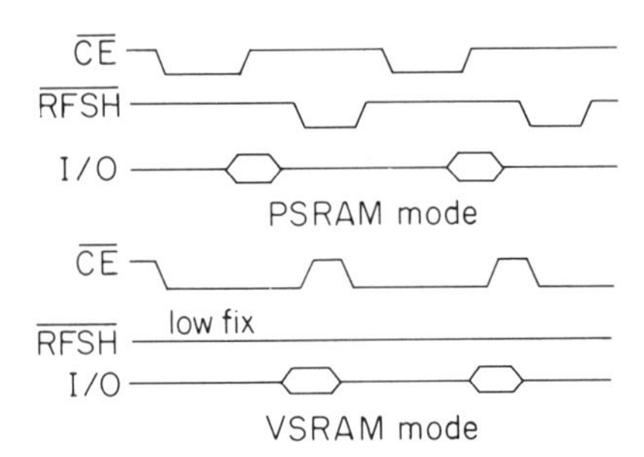

Figure 7.12
Comparison of refresh timing for PSRAM mode and VSRAM mode in the Toshiba PSRAM with VSRAM mode. (From Sawada [55], Toshiba 1988, with permission of IEEE.)

to Mix-MOS SRAMs. This was accomplished both by eliminating substrate bias by placing the transistors in a p-well and by a low power dissipation current mirror ring-oscillator timing generator.

7.3.7 4Mb P/VSRAMs

Two 4Mb PSRAMs were discussed between 1987 and 1989, a PSRAM by Hitachi [7] and a PSRAM with virtually static bond option by Oki [8].

The Oki 4Mb was the first PSRAM to use the new trench type DRAM cell. It used the JEDEC Standard 512k × 8 pinout, as shown previously in Figure 7.4, with the refresh control pin optioned with the output enable since the 4Mb has the same problem that the 256k PSRAM had of lacking an extra pin in the standard SRAM compatible pinout.

Grounding the output enable let the part be used in the virtual static mode where the device operated at the optimum refresh interval timing without external refresh control. In this case the worst case access time was slightly longer than in the pseudostatic mode due to the inherent arbitration delay of an access request occurring during a cell-row refresh operation.

In the pseudostatic mode the device operated at the specified refresh interval under the control of an external refresh controller on the output enable similar to other PSRAMs.

To reduce the capacitive cross-talk of the long signal lines in a chip as large as a 4Mb DRAM, Oki used a macro cell design technique in which the chip area of this 145 mm^2 chip was divided into four independent 1Mb RAM areas which included both the cell array matrix and some of the normal DRAM overhead circuitry.

The outer peripheral segment served these macro cells with low impedance supply lines, address buffers, and refresh circuitry. The interference between the lines in the activated cells was thus reduced. Typical access times were 60 ns in the PSRAM mode and 95 ns in the VSRAM mode. Average standby power was about 1 mA. Figure 7.13(a) shows the schematic block diagram with the macro cell concept of this part. The refresh timing counter, internal address counter and oscillator are shown.

The select option for the PSRAM mode is shown in Figure 7.13(b). A block diagram of one of these macro cells is shown in Figure 7.13(c) with the internal timing generators for $\overline{RAS}$ and $\overline{CAS}$ indicated.

The Hitachi 4Mb PSRAM focused on the problem of reducing the standby power dissipation of the PSRAM to approximate that of the SRAM when four million one-transistor cells must be periodically refreshed. It attained a data retention current of 36 μA, which is comparable to that of the Mix-MOS SRAM but still above the typical data retention current of the full CMOS SRAM.

This RAM was divided into 8 memory array blocks. In the self-refresh mode, if all 8 arrays are activated at the same time a large peak current was generated from the 8000 sense amplifiers, significantly reducing the noise margin of the part. Two methods were taken to reduce this peak current: one was reducing the drive of the sense amplifiers, and the other was shifting the phase of the sense amplifier activation timing in every array.

In 1991 Hitachi [81] showed a 4Mb pseudostatic DRAM which operated at 2.6 V with a 3 μA data retention current. The part used a cell leakage monitoring timer and special refresh controls. The peripheral circuits automatically disable the boosted word lines at high V_{CC} to improve the device reliability. This level of data retention current would enable this part, when used in a 20Mb RAM, to retain data for two months with a single lithium battery.

7.3.8 Other transparent refresh SRAMs with DRAM cells

In a 1987 paper, Hitachi [56] presented a 256k asynchronous SRAM with dynamic cells needing refresh but with this refresh supplied from an internal oscillator as in the 64k iRAM from Intel. The internal oscillator supplied a word line pulse for refresh. Internal arbitration logic decided conflicting requests between the refresh and the processor in favour of the processor with a 40 ns refresh operation added onto the processor operation.

The dynamic cells used on this part were not the usual dynamic one-transistor cells with capacitor, but were a four-transistor cell without the load devices needed to make the part static. The disadvantage of this part from the timing point of view was that the cycle time (86 ns typical) was twice the address access time (43 ns typical) as in all virtual SRAMs. Otherwise, for a slow commodity SRAM application this part looks exactly like a low power, asynchronous SRAM in the socket.

The disadvantage from the cost point of view is that the chip size of the part at 40 mm^2 is the same size as a standard six-transistor cell 256k CMOS SRAM in the same 0.7 μm geometry technology and it is also significantly slower than the comparable full CMOS SRAM. For example, the Philips 256k SRAM in 0.7 μm geometry technology was also 40 mm^2 and had a typical access and cycle time of 17 ns. The standby power dissipation of the two parts was comparable.

This is an illustration of the tradeoffs involved in attempting to use a smaller cell than the standard SRAM cell at the expense of more overhead control circuitry.

Another transparent refresh DRAM was described by LSI Logic [58]. This synchronous RAM employed a two-transistor and one-capacitor basic cell. It used a 50%

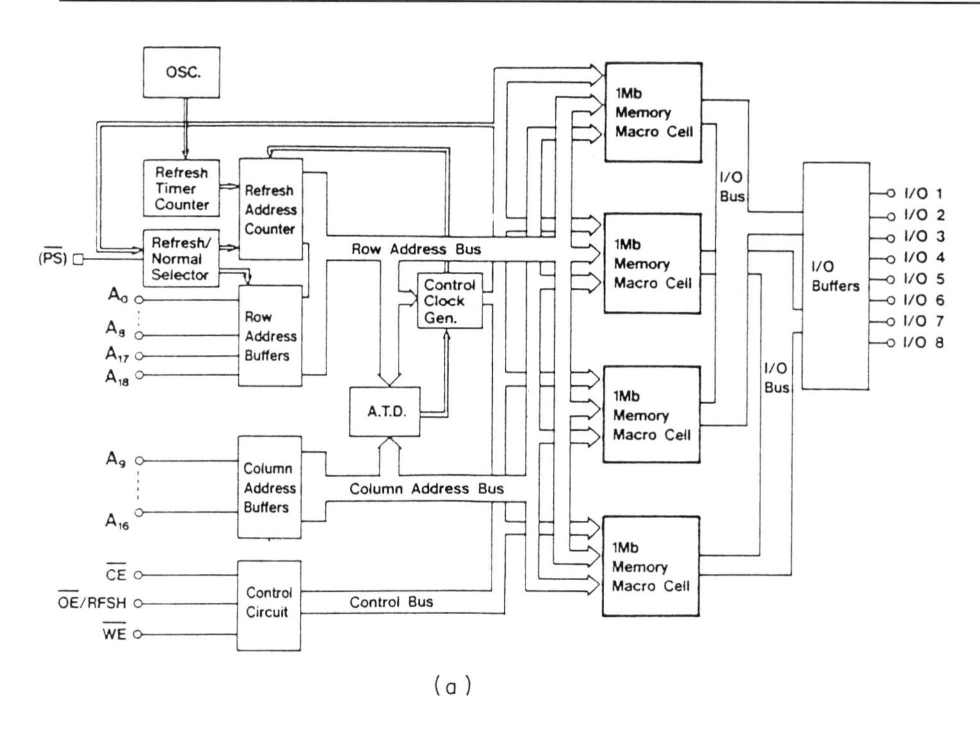
OSC.
Refresh Timer Counter
Refresh Address Counter
Refresh/ Normal Selector
(PS)
Row Address Bus
Control Clock Gen.
Row Address Buffers
A.T.D.
Column Address Buffers
Column Address Bus
Control Circuit
Control Bus
CE
OE/RFSH
WE
1Mb Memory Macro Cell
I/O Bus
I/O Buffers
I/O 1
I/O 2
I/O 3
I/O 4
I/O 5
I/O 6
I/O 7
I/O 8

(a)

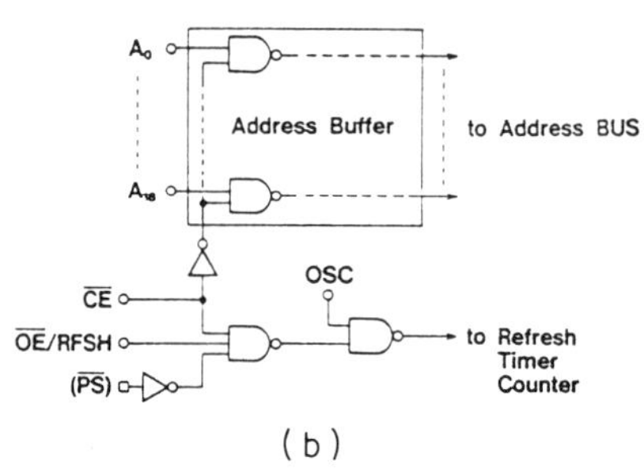
Address Buffer
to Address BUS
OSC
CE
OE/RFSH
(PS)
to Refresh Timer Counter

(b)

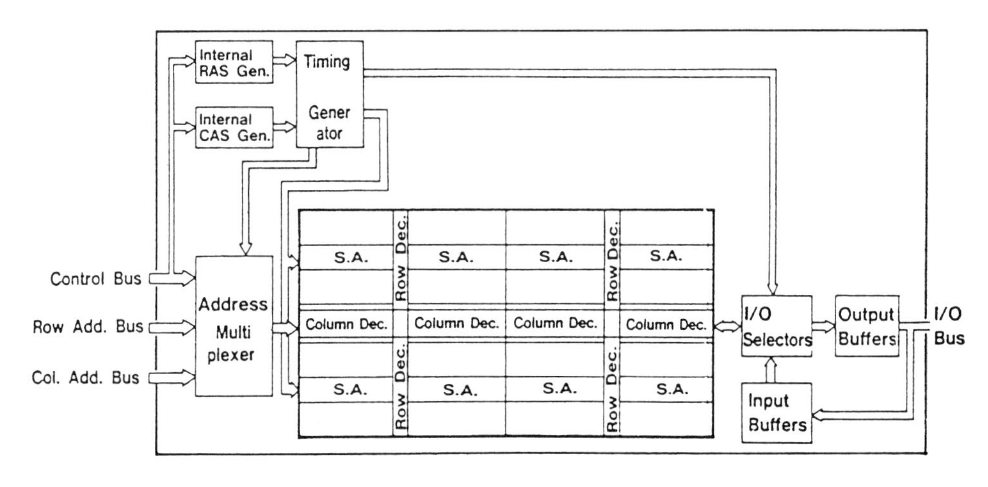
Internal RAS Gen.
Internal CAS Gen.
Timing Generator
Control Bus
Row Add. Bus
Col. Add. Bus
Address Multi plexer
S.A.
Row Dec.
Column Dec.
I/O Selectors
Output Buffers
Input Buffers
I/O Bus

(c)

duty cycle clock in which half the cycle was either used for read or for refresh and the other half for write operations. Since the refresh occurred periodically and could be interrupted by the read operation at any time, the part did not have a problem with simultaneous occurrence of a read and refresh, or a write and refresh operation. The read sense amplifier was designed to read as well as refresh in the same half cycle so no loss of data could occur.

A transparent refresh DRAM with a two-transistor and one-capacitor cell was shown by Mitsubishi in 1991 [82]. The cell configuration was dual port providing a background refresh 16Mb DRAM which was externally refresh free. This part was intended for use in a logic macrocell.

7.3.9 DRAM macrocells in logic I.C.s

One of the advantages over the DRAM that the CMOS SRAM has had historically has been its potential for being embedded as a RAM macrocell in large scale logic integrated circuits. The refresh and timing requirements of the DRAM kept it from being used in this application. The development of adequate refresh control on memories with dynamic cells gives SRAM-like ease of control and has potentially opened up this range of applications to devices with the smaller DRAM cells.

The 32k RAM chip from LSI Logic Corporation mentioned in the preceding section is a good example of a DRAM which was designed to be embedded. This RAM was designed to be used as an ASIC macrocell and hence can be reconfigured into various array organizations. Its immediate use was with a 100k gate logic array. Chip size was 11.5 mm^2 in a 1.5 μm CMOS technology. This was about half the size of the chip that an SRAM in 1.5 μm technology would have produced. The three-element cell consisting of two transistors and a capacitor and sensing circuitry for this chip are shown in Figure 7.14.

The Toshiba virtual PSRAM was also designed with the potential for embedding in logic in mind. For example, the Toshiba V/PSRAM used a double-layer polysilicon and double-layer aluminum interconnect structure rather than the triple-layer polysilicon and single-layer metal interconnect structure used on their standard standalone DRAM. Multiple polysilicon layers reduce the size of the memory cell but are not normally as useful in logic designs whereas at least double-layer metal is considered essential in advanced logic designs.

The double-metal, double-polysilicon structure of the Toshiba V/PSRAM meant that only one polysilicon layer needed to be added to the standard logic single polysilicon–double metal interconnect system to be able to embed this V/PSRAM with optimized DRAM cell. The double-layer metal also aided in reducing the size of the extra peripheral logic on the memory.

Figure 7.13
4Mb V/PSRAM. (a) Block diagram of overall architecture showing refresh control circuitry and macro cell structure. (b) PSRAM–VSRAM mode option select. (c) Block diagram of one macrocell block showing internal RAS–CAS timing generator. (From Yoshioka [8], Oki 1987, with permission of IEEE.)

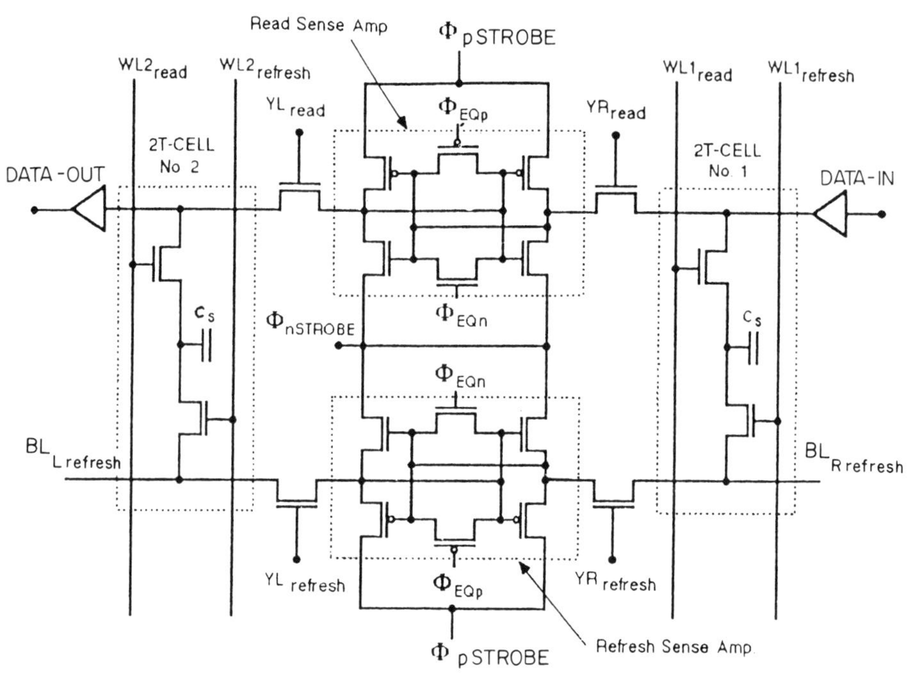

Figure 7.14
Schematic diagram of 2T cell and sensing circuitry for 32k ASIC synchronous RAM. (From Yuen [58], LSI Tech. 1989, with permission of IEEE.)

While some of the 1Mb application-specific DRAMs continued to use the traditional memory single-metal structure, at the 4Mb and 16Mb levels most of the application-specific DRAMs presented used the double-metal technology to reduce the silicon area required for the added logic.

Figure 7.15 illustrates the rapid rate of decrease in array efficiency of an application-specific DRAM with each increment of added logic.

An illustration of a PSRAM embedded in a significant amount of logic was given by Toshiba who embedded the V/PSRAM described above in a 72k gate array [57] producing a 223 mm^2 chip with 14% array efficiency. The cross-sectional view of the boundary region on the chip between the DRAM and the gate array is shown in Figure 7.16(a).

There are also power constraints inherent in the added logic considerations as can be seen in the graph from Toshiba shown in Figure 7.16(b). These are, however, beyond the scope of this book.

Historically the distinction between an embedded memory in logic and an application-specific memory has been legitimate and related to the following factors.

In a logic process the lack of the extra 'memory specific' technology features, such as double polysilicon, straps, and buried contacts, etc, have dictated that embedded memory cells were not optimized and were far larger than cells used in similar geometries in optimized memory processes. The time-consuming logic layout is increasingly being done by automated layout tools and use of cell libraries.

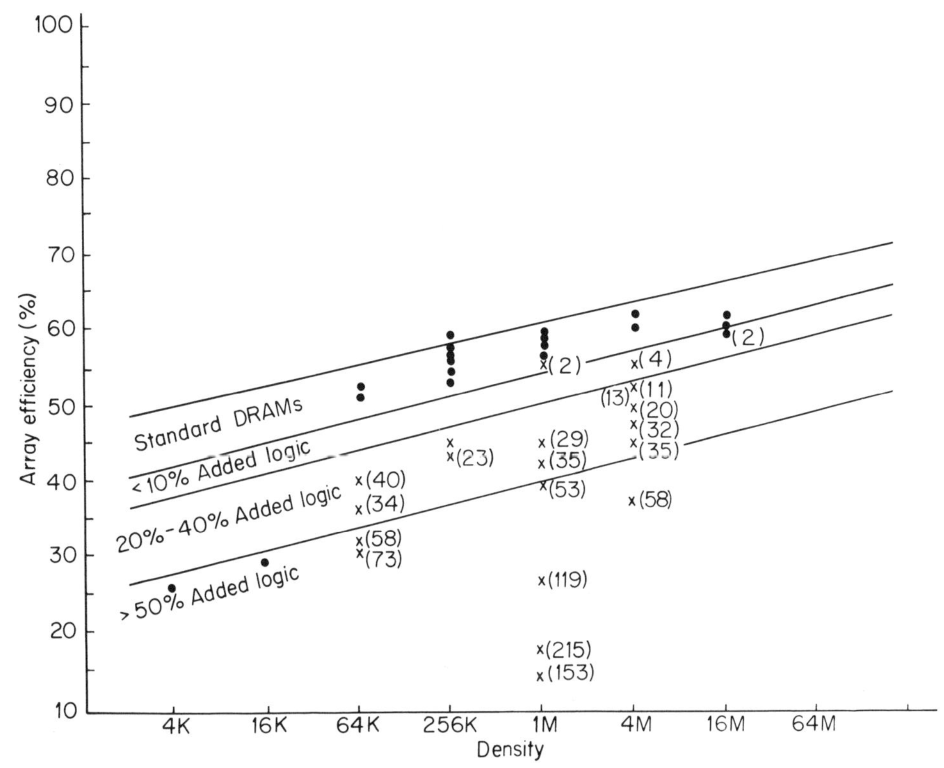

Figure 7.15
Array efficiency plotted against density for dynamic RAMs. (● Standard DRAM, X Application specific DRAM, (×) Percent logic area added to embedded DRAM. (Source ISSCC Proceedings 1973 to 1989.)

On the other hand, in an optimized memory process, not only is the memory cell optimized by the use of a more complex technology, but the logic in the periphery is hand crafted which is time consuming. There have been attempts recently to combine the two such as a 4Mb DRAM shown by Texas Instruments [11] in 1986 which used automated generation tools for designing the periphery. Undoubtedly other such attempts will follow.

A logic array with embedded SRAM which was shown by NEC at the 1989 IEDM stacked the logic array in two layers of polysilicon transistors over the memory cells which were in the substrate and first poly layer. This technique helped eliminate some of the space consuming aspect of the non-optimized logic processes but substituted instead the additional complexity of four layers of polysilicon. This is perhaps a manageable trade-off for memory technologists.

7.3.10 Self-refresh DRAMs in battery back-up applications

With the trend to battery operated systems in the 1990s came a demand for low cost and low power memory for battery back-up.

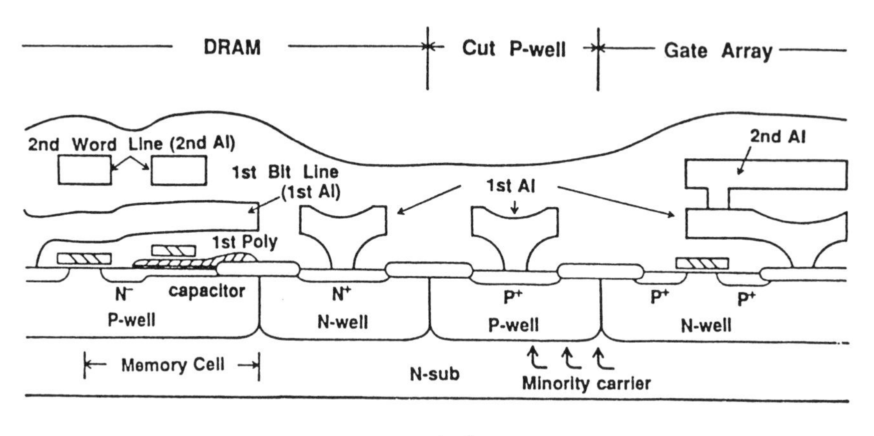

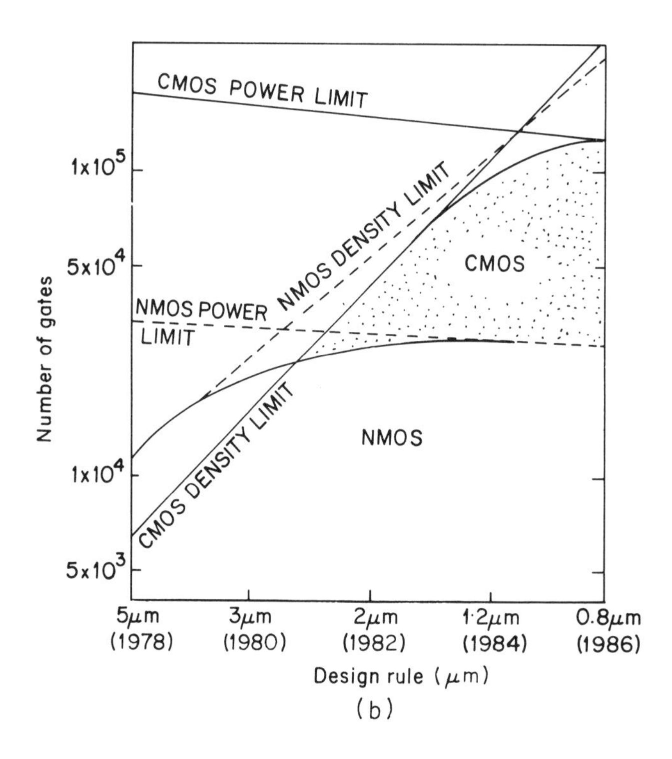

Figure 7.16

Technology aspects of memory embedded in Logic. (a) Cross-sectional view of the boundary region between DRAM and gate array for Toshiba 1Mb PSRAM embedded in a 72k gate logic array. (b) Power dissipation limits of random logic as a function of number of gates and technology. (Reproduced with permission of Toshiba.)

In 1990 Mitsubishi [78] described a 4Mb DRAM with a battery backup mode (BBU) which enabled automatic data retention with reduced power consumption. This automatic refresh scheme utilized a $\overline{CAS}$ before $\overline{RAS}$ refresh sequence. As shown in the timing diagram in Figure 7.17(a) if $\overline{CAS}$ is held low for over 16 ms without a $\overline{RAS}$ refresh cycle, the automatic refresh mode is triggered. The BBU mode continues as long as the $\overline{CAS}$ is held low regardless of the level of $\overline{RAS}$. The DRAM is reset to normal mode when $\overline{CAS}$ again goes high. As long as the timing cycle does not exceed 16 ms this part operates like a conventional 4Mb DRAM.

In operation this part is not compatible with the standard CBR refresh operation but can be refreshed by any of the other conventional refresh operations. An on-chip ring oscillator was used with binary counters to clock the 16 ms before activating the BBU mode as shown in the block diagram of the DRAM in Figure 7.17(b).

A comparison of the standard $\overline{RAS}$ only refresh with the BBU refresh is shown in Figure 7.17(c). The BBU used 4k cycle refresh with 64 μs cycle time and only activated $\frac{1}{32}$ of the array at one time compared to $\frac{1}{8}$ activated for the conventional 4Mb DRAM which utilized 1k refresh having 16 μs cycle time. The refresh period is therefore 256 ms rather than 16 ms. The peak current is reduced to $\frac{1}{4}$ and the array current to $\frac{1}{16}$ their usual values during operation.

7.3.11 Low voltage DRAMs for battery back-up applications

A low voltage DRAM suitable for battery back-up applications was made by Hitachi [53] in 1989. This 1.5 V DRAM has a reduced bit-line voltage swing combined with a pulsed cell plate for use in a 16Mb DRAM. The operating current for the 16Mb DRAM was estimated to be reduced from 15 mA for a conventional 5 V DRAM to 5 mA for a 1.5 V DRAM and to 3 to 4 mA for the reduced voltage DRAM with the pulsed cell plate.

Another low voltage 64Mb DRAM, designed to maintain a high signal-to-noise ratio, was discussed by Mitsubishi in 1991 [83]. This part had $V_{CC}/2$ cell plate lines which were connected to pairs of bit-lines by clocked transfer gates. This arrangement limited the voltage swing of the cell plate so that data stored in unselected cells was not disturbed during access.

7.4 DRAMs TARGET THE DISK MARKET

The magnetic disk market is an attractive target for MOS memories. Its requirements are high density, low cost and non-volatility. The DRAMs already provide high density and low cost, therefore, the advent of the low power and refresh control features of the PSRAMs which can be used in battery-backup mode for non-volatility open new possibilities for DRAM manufacturers.

7.4.1 DRAMs as silicon files

The disk market requires non-volatility and low operating power, but the access is serial and is orders of magnitude slower than that of standard DRAMs. An example

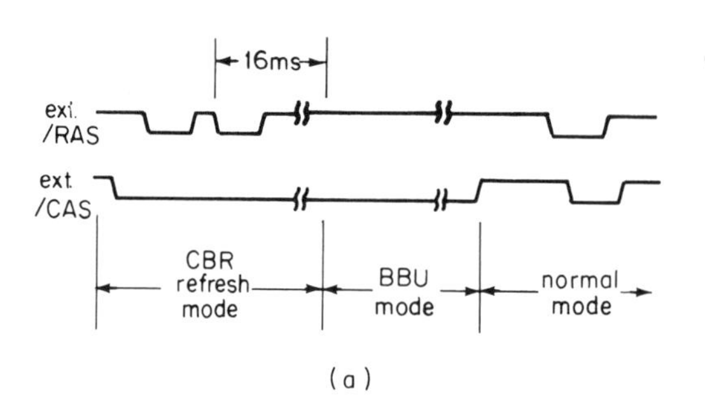

(a)

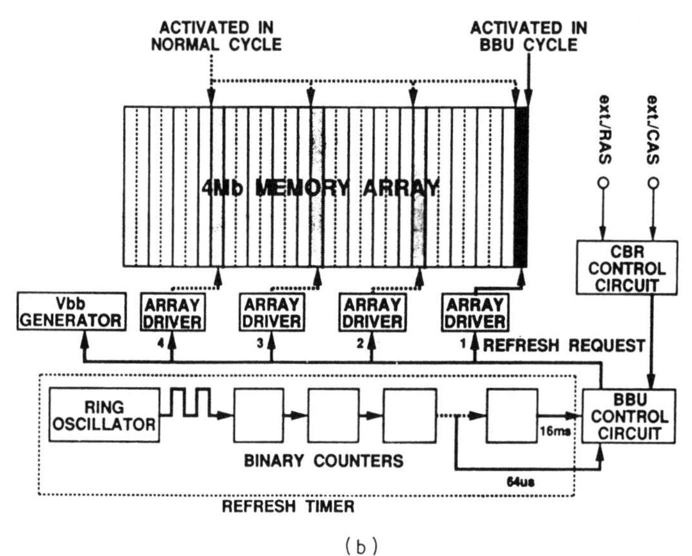

(b)

	Standard refresh	BBU refresh
Activated array	1/8	1/32
Turn around cycles	1024	4096
Refresh cycle time	16 μs	64 μs
Refresh	~16 ms	~256 ms
Peak current	1	~1/4
Array current	1	~1/16

(c)

Figure 7.17
A DRAM battery-back-up (BBU) mode. (a) Timing diagram for BBU mode. (b) Block diagram of 4Mb DRAM with BBU mode. (c) Comparison of BBU mode with conventional $\overline{RAS}$ only refresh mode. (From Konishi *et al.* [78], Mitsubishi 1990, with permission of IEEE.)

of a DRAM targeted at the disk market was the NEC [6] 'Silicon File' memory which was a 1Mb × 1 asynchronous DRAM with self-refresh on chip. A block diagram of the part is shown in Figure 7.18(a) and reveals the refresh address counters and timing generators typical of the PSRAMs. The pinout shown in Figure 7.18(b) indicates the small package possible with multiplexed addresses and one data-in and one data-out pin. It also reveals the $\overline{\text{RFSH}}$ pin expected on the PSRAM.

The NEC 'Silicon File', while much slower than standard DRAMs, is still faster at

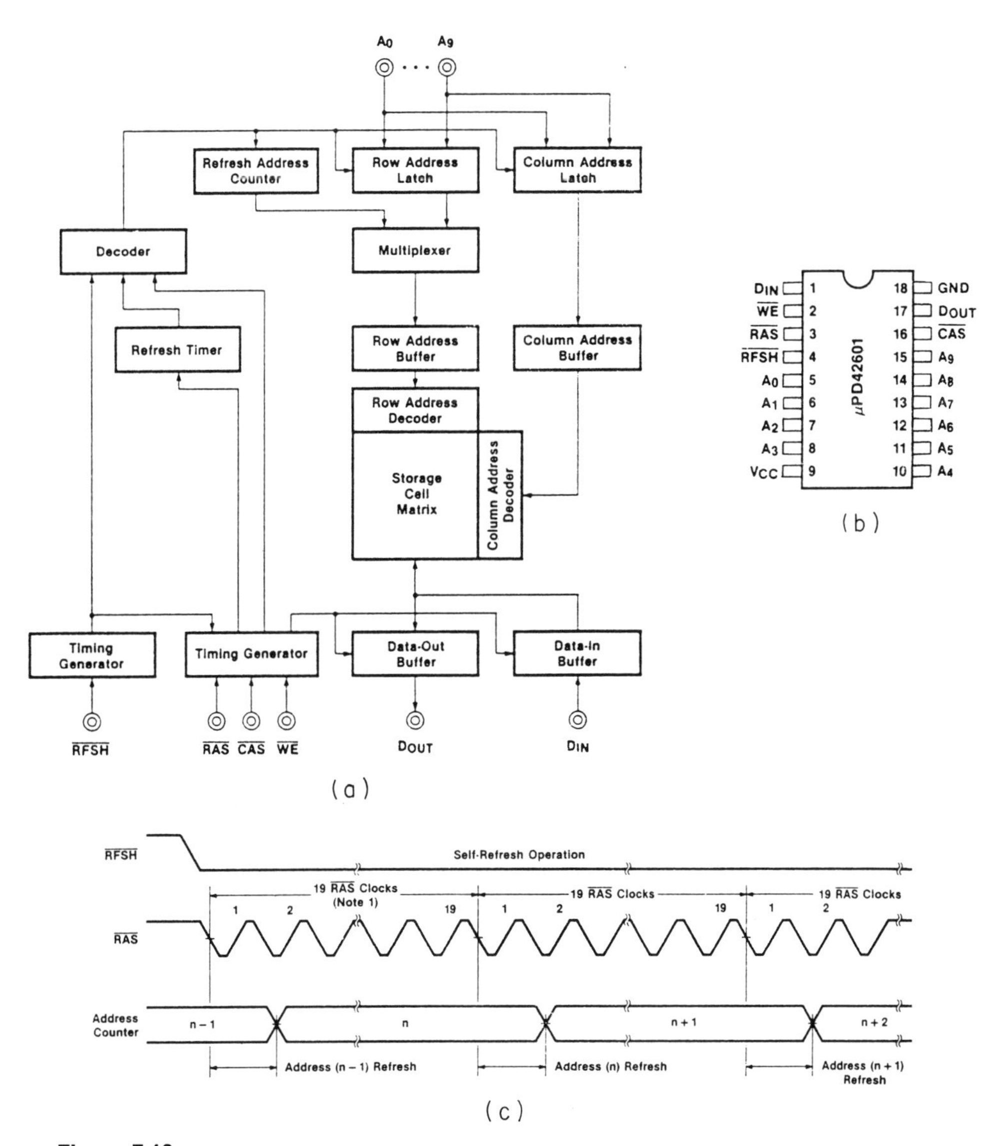

Figure 7.18
NEC 'Silicon File' DRAM: (a) block diagram, (b) pinout, (c) internal address generation in self-refresh operation. (From NEC Memory Products Databook, 1989.)

Table 7.2 Self-refresh conditions.

T_A (°C)	t_{RCF}(max) (μs)	Self-refresh current(max) (μA)	Clock frequency(min) (kHz)
50	20	30	50
60	10	60	100
70	5	120	200

600 ns random access time than a magnetic disk memory. It also can operate in page mode with 200 ns minimum cycle time, to approximate the serial format of the hard disk. In page mode, the first word is accessed in the same manner as in standard write and read operation with both the $\overline{RAS}$ and $\overline{CAS}$ clocks needed. Subsequent column addresses are, however, accessed by the $\overline{CAS}$ cycle which can be repeated during a period up to the maximum $\overline{RAS}$ pulse width of 100 μs.

A standard $\overline{CAS}$ before $\overline{RAS}$ refresh cycle is included with a built in address counter which makes external refresh addressing unnecessary. During the self refresh cycle refreshing is accomplished by maintaining the $\overline{RFSH}$ pin active low and cycling the $\overline{RAS}$ input as shown in Figure 7.18(c). Since the minimum required $\overline{RAS}$ cycle frequency depends on ambient temperature, the power consumption varies with temperature. The power required to maintain data in self refresh mode at various temperatures is shown in Table 7.2 which is taken from the 1989 NEC Memory Products Databook. Clearly at low temperatures this part is amenable to a battery back-up situation.

The emphasis in this development was on the reduced leakage of the DRAM capacitor at low temperatures which reduces the required frequency of refresh. This enhanced storage retention time is due to diminished leakages from thermal generation of carriers so that the retention time of the DRAM becomes long enough to be considered static.

This combination can be used in the disk application in mainframe computer environments which tend to be ambient temperature controlled. Clearly if the temperature is controlled at an even lower level even greater improvement in standby power consumption is obtained.

7.4.2 Multilevel storage techniques for high density

Another specialized DRAM file memory which used the on-chip refresh concept was proposed in 1988 by Hitachi [49] to provide still higher density and hence low cost per bit as required by the hard disk market. This was an experimental 1Mb DRAM which used multilevel storage techniques to provide a 4Mb serial file memory.

The device used one-transistor DRAM memory cells which store 16 levels (4 bits) of data each. These cells were divided into 4k bit sequential access blocks. Embedded peripheral circuits included a staircase-pulse generator for multilevel storage operation, a voltage regulator to protect against a power supply voltage bump, an error

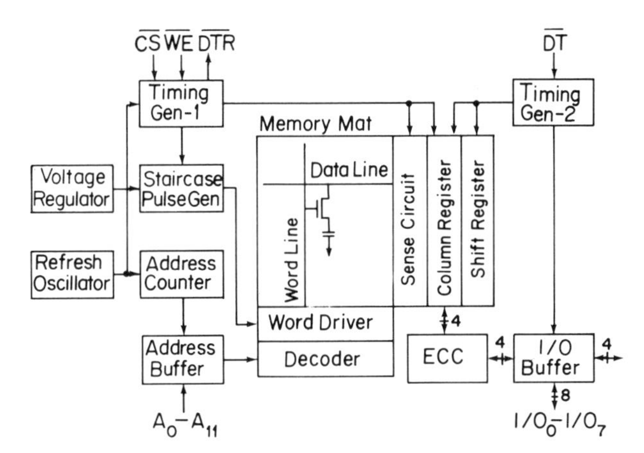

Figure 7.19
Block diagram of the Hitachi multilevel storage 'file' DRAM. (From Horiguchi [49], Hitachi 1988, with permission of IEEE.)

correction circuit to protect the data from alpha particle induced soft errors and timing generators to simplify test. A block diagram and the staircase pulse generator are shown in Figures 7.19 and 7.20(a). The word-lines were randomly accessed by the decode, while the data lines were sequentially accessed by the shift register.

The read operation is as shown in the timing diagram in Figure 7.20(b). When $\overline{CS}$ becomes low, the staircase pulse starts to ascend. It is applied to the selected word line. The dummy word line meanwhile is driven by a half step staircase pulse so that a half reference charge is read out from the dummy cell. The sense amplifier detects the signal and sends it to the column register where the data are stored in the form of 4-bit binary code. A data-transfer request signal $\overline{DTR}$ is generated when the four bits in each memory cell have been transferred. The data stored in the column register are then sequentially transferred to the data I/O terminals through the error correction circuit.

The bandgap reference voltage regulator protected against power supply voltage bump and temperature fluctuation so that the small signal voltage was accurately maintained.

A refresh oscillator and an address counter were implemented on chip to perform self-refresh operation. If a read–write request came during a refresh cycle, the request was deferred until the end of the refresh cycle.

The most serious drawback of this technique was the potential for soft error due to the small amount of charge that was stored in each of the four levels in the individual cell and also due to the fact that each incident alpha particle destroyed all four bits of data in a cell. A single particle could also destroy the data in more than one cell.

Another drawback was the slow speed. The random block selection time was 147 μs and serial data rate out was 210 ns. A block oriented architecture was adopted in an attempt to increase the efficiency of the data transfer process. The conversion between multi-level data and binary data was inherently slow due to slow iterative operation

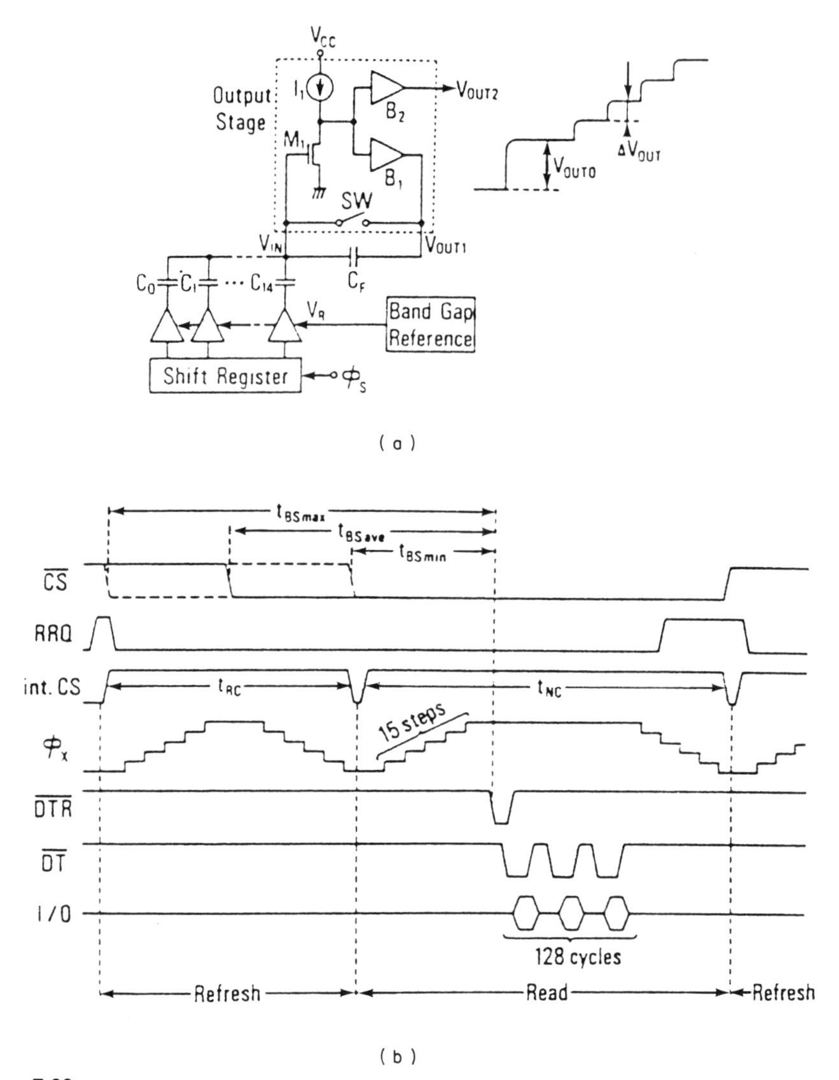

Figure 7.20
Hitachi multilevel storage file DRAM. (a) Staircase pulse generator. (b) Read operation timing chart showing operation of staircase pulse generator. (From Horiguchi [49], Hitachi 1988, with permission of IEEE.)

of the sense circuit. The standby power at 5 mA was also not as low as is usually wanted for battery back-up.

Another attempt at increasing density with multilevel storage was made in 1989 by Toshiba [50] with a 2 bit-per-cell DRAM in which four levels were stored in every cell. This part was targeted at the same embedded memory in logic application that Toshiba's earlier pseudostatic DRAMs targeted. The access on this part was completely random and typical access time was 170 ns.

This chip used 512k cells to give a 1Mb DRAM capacity of 37.8 mm^2 chip size compared to the typical 1Mb DRAM chip size of about 55 mm^2. It used, however, a relaxed 4Mb technology with a cell size of 17.1 μm^2. If this same cell had been used to make a standard 1Mb DRAM with 55% array efficiency the chip size would have been about 32 mm^2. The real comparison in density then is between 32 mm^2 for a standard 1Mb DRAM with this cell size and 37.8 mm^2 for the multilevel 1Mb DRAM. Clearly at this density level there is no chip size advantage to the multilevel cell.

The array efficiency of the 512k cell DRAM, at about 23.1%, is low due to the need to triple the number of sense amplifiers for the operation of the multibit cell. This technique could be interesting, however, if the number of cells per bit-line could be increased amortizing the extra sense amplifiers over more cells.

In the 2-bit/cell storage, one memory cell stored four levels and three reference levels were used to detect the four levels. Reference levels were positioned at the mid-levels between the four storage levels and were generated by reference voltage generators. A schematic of these levels is shown in Figure 7.21(a).

Each column of bit-line pairs was divided into three blocks with each block having the same number of memory cells, dummy cells, and a sense amplifier. There were, therefore, three sense amplifiers per column rather than the usual one.

The schematic of the read and write operations and a data conversion table are shown in Figures 7.21(b) and (c).

7.5 HIGH SPEED DRAMs

The trend to small systems has increased the incentive for DRAMs to be fast like the SRAMs that have traditionally been used in these systems so that the microprocessors running these systems can operate without wait states or without using additional SRAM cache memory. This requirement is for fast random access.

Another major application motivating the move to very fast DRAMs has been the serial graphics applications in these small systems where the requirement is for fast serial access.

The natural progression of the DRAMs into higher MOS technologies and more sophisticated design architectures has produced enhanced random access speed. The serial access speed has been produced first with special serial DRAM modes such as nibble mode, page mode, and static column as discussed in Chapter 6, and more recently with burst modes and a move to synchronous interfaces. The emphasis, however, has remained on low cost.

7.5.1 BiCMOS DRAMs

Efforts with BiCMOS technology for DRAMs primarily by Hitachi have also produced some speed enhancement. Hitachi [51, 52] compared the performance of a BiCMOS driver with a CMOS driver in a 1Mb CMOS DRAM with TTL interface

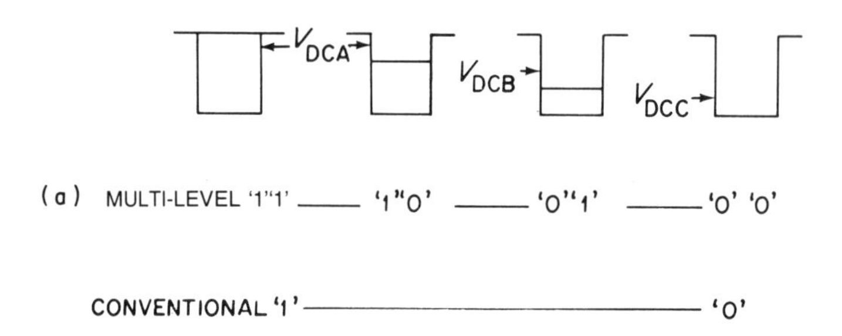

(b)

Cell	Dq_A	DQ_B	DQ_C	I/O_1	I/O_2
V_{cc}	1	1	1	1	1
$\frac{2}{3}$	0	1	1	0	1
$\frac{1}{3}$	0	0	1	1	0
V_{ss}	0	0	0	0	0

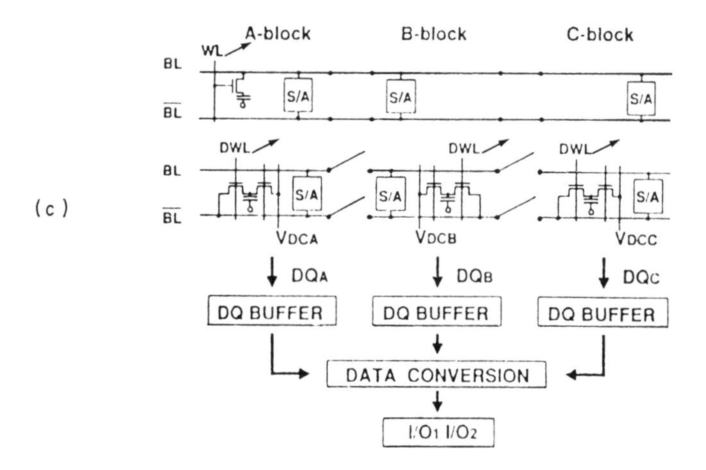

Figure 7.21
(a) Comparison of memory cell storage levels of multiple-level storage scheme and conventional DRAM. (b) Data conversion table for different storage levels. (c) Read operation showing schematic column circuit diagrams at two different timings and the data flow obtained. (From Furuyama [50], Toshiba 1989, with permission of IEEE.)

in 1.3 μm technology. The simulations showed that the access times of the CMOS and BiCMOS DRAMs were 35.3 and 27.2 ns respectively. The chip size of the BiCMOS chip was about 16% larger than a comparable CMOS chip using the same memory cell and the same design rules. This increase in chip size was attributed to having widened the A_1 wires to achieve low wiring resistance and high speed.

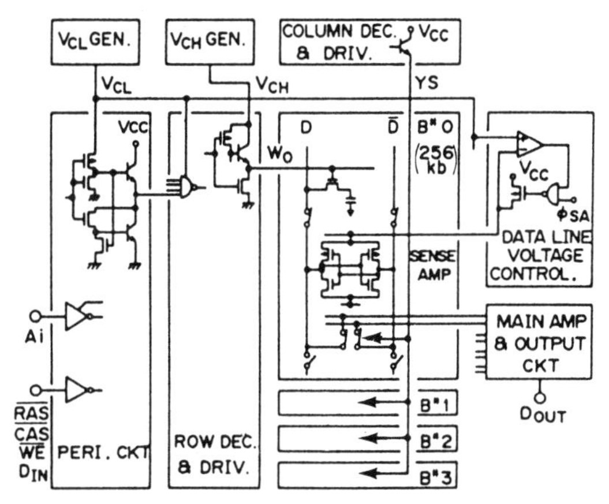

Figure 7.22
Block diagram of 1Mb BiCMOS DRAM. (From Kitsukawa [73], Hitachi 1989, with permission of IEEE.)

A block diagram is shown in Figure 7.22 of a 1Mb BiCMOS DRAM from Hitachi [73] showing the regulated internal 4.0 V V_{CL} generator which controlled the various BiCMOS circuits on the chip, as well as the data-line voltage controller V_{ch} and the CMOS datapath. It included a bipolar band gap reference generator which permitted the internal voltage to be immune from temperature and power supply variations. In 1991 Hitachi [84] described a 4Mb BiCMOS DRAM with 17 ns access time which was down from 23 ns on their 1Mb BiCMOS DRAM.

7.5.2 Special high speed DRAM modes

In another attempt at a high speed DRAM, IBM [47] described in 1984 a two-chip system of DRAMs which operated in nibble mode while interleaving the $\overline{CAS}$ cycles of the two chips so that a 20 ns serial output rate was obtained as shown in the schematic and timing diagram of Figures 7.23(a) and (b).

In 1988 IBM [70, 71] showed a 512k single-chip DRAM with 20 ns random access time and a high speed 12 ns page mode resulting in a 330Mb s^{-1} data rate. The high speed on this chip was obtained by straightforward design techniques targeted at optimizing for speed such as segmenting lines with excessively large *RC* constants, using wide power buses and multiple V_{DD} and V_{SS} pads to reduce ground bounce, and using separate sets of row and column address buffers and buses to reduce the delay from address multiplexing. There were some trade-offs made, the power dissipation was relatively high at 416 mW, 5 V typical at 50 ns cycle time and the array efficiency at 49% was somewhat lower than average.

A comparison chart of speed of various DRAMs by year of publication is shown in Figure 7.24(a). It is clear that this IBM chip is not on the trend line for the standard DRAMs which are optimized primarily for low cost. A comparison of the internal access time components of this 20 ns DRAM to a typical 1Mb CMOS

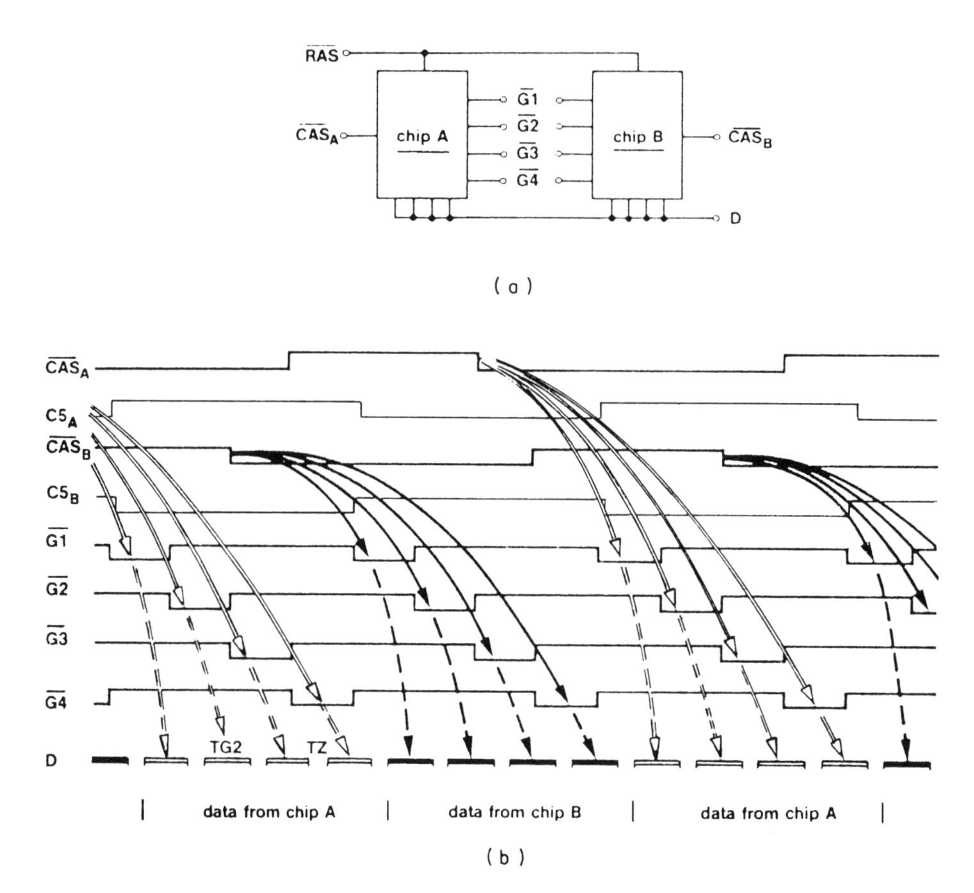

Figure 7.23
Combined page and serial buffer mode with CAS cycles of two chips interleaved showing (a) schematic of circuit with two chips, (b) timing diagram. (From Baier [47], IBM 1984, with permission of IEEE.)

DRAM are shown in Figure 7.24(b), showing that all components have been speeded up in the course of optimizing this part for speed.

In 1989 IBM described a 1Mb single chip CMOS DRAM [48] with a $\overline{RAS}$ access time of 27 ns. Page mode operation of this memory was also 12 ns for the $\overline{CAS}$ cycle. It was indicated that an improved version with 22 ns access time had been seen.

This later version was used by IBM in an interesting liquid nitrogen temperature experiment. It is known that an effect that low temperature has on CMOS transistors is to increase conductance which in turn increases the switching speed. IBM presented a paper describing the placing of this 1Mb DRAM [46], which had a 20 ns access time at room temperature, in a liquid nitrogen environment where it had a measured 12 ns access time. The soft error rate was also significantly reduced.

The required data rates of high performance microprocessors have been increasing faster than the access times of DRAMs have been increasing. Also in the 16Mb DRAM the number of memory cells per row became large so that higher speed

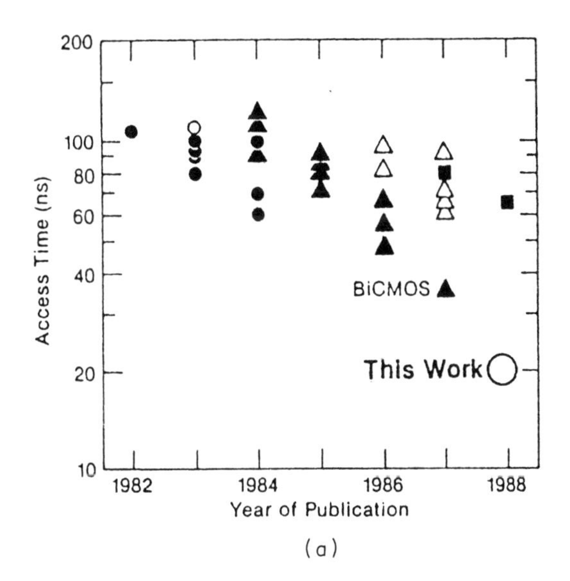

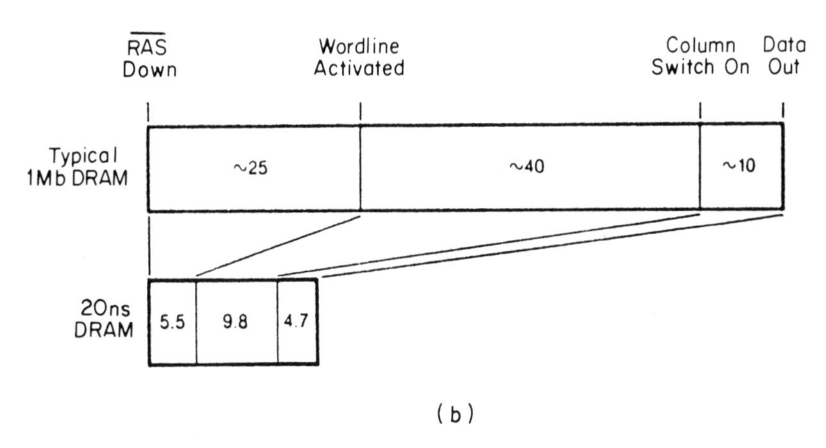

Figure 7.24
Comparison of high speed DRAM with typical cost optimized DRAM. (a) DRAM access time evolution plotted against year of publication. (b) Internal components of access time for a 1Mb DRAM also in 1.0 μm CMOS technology. (● 256k, ○ 512k, ▲ 1Mb, △ 4Mb, ■ 16Mb.) (From Lu [70], IBM 1988, with permission of IEEE.)

operations were required. Standard fast access modes did not have the speed required in these applications. DRAMs with fast access modes such as nibble mode can be used in high data rate systems by interleaving the parts in the system. Recently, however, a number of DRAMs with fast access modes suitable for high data rate systems have been described.

Page mode and static column mode can not be used in these applications since they are not able to operate at the needed 100 Mb s^{-1} data rate. Even though parts with these modes may specify 10 ns, these modes need extra time, such as setup and hold times, for the column address and $\overline{WE}$ signal before each column address is accessed. The cycle time is, therefore, slower than 100 MHz.

A 4Mb DRAM with a pipelined read access mode to obtain a high data transfer rate between the CPU and the memory was described by Toshiba [74] in 1990. The circuit used a latch in the output path to latch the output data from one read cycle during the precharge period of the next. In the second cycle this latched data was output while addressing took place, that is, a conventional read operation was divided into two parts with the first part of the second cycle read operation and the second part of the first cycle read operation being processed in parallel. A timing diagram of the pipelined read and non-pipelined write is shown in Figure 7.25(a). At the end of the string read operation a dummy cycle is needed to clear the latch as shown before a write cycle can take place.

The timing diagram for a typical application for such a pipelined DRAM is shown in Figure 7.25(b). The data transfer between the DRAM and a microprocessor is shown for a conventional DRAM and for the pipelined DRAM. In this application the pipelined DRAM has a 50% higher effective data rate than the conventional DRAM.

A 1988 16Mb part with 100 MHz serial read–write mode also from Toshiba [27] could serially access up to 2k bits on a selected row by clocking $\overline{CAS}$ at 100 Mb s^{-1}. It did not have the random access within the selected column needed in many microprocessor oriented applications and is therefore more like the parts described later under video field memories.

A fast access mode which does, however, have random access capability and is not limited in speed by the address set up and hold times is the nibbled-page architecture implemented by Toshiba [75] in 1989 also on a 16Mb DRAM. This mode can access all column addresses on a selected row randomly in units of eight bits at 100 Mb s^{-1} by interleaving on chip the data from 8 bit sections of different columns. The set-up and hold times are covered in the interleave so they do not delay the output data rate. A timing diagram for the nibbled page architecture is shown in Figure 7.26. Figure 7.26(a) shows the on-chip interleaving and Figure 7.26(b) shows the external data rate.

Another fast access pipelined mode called Toggle Mode was introduced by IBM [79] on their 16Mb DRAM. Toggle mode is a modification of the conventional static column mode. The address input transitions start the cycle by selecting two bits. The positive transition of the toggle pin drives the first bit and the negative transition drives the second bit. During the latter transition the address inputs can be changed to another random column address selecting two new bits for the next positive transition of toggle. This mode has 10 ns access time and 10 ns cycle time.

Another method of achieving effective fast access for a DRAM is to add an SRAM cache onto the DRAM chip. A 1Mb cache DRAM was discussed by Mitsubishi [80] in 1990. This part utilized an 8k SRAM with 12 ns access time embedded in a DRAM chip having a 80 ns access time chip. Another fast DRAM with on chip cache was the 4M "Enhanced" DRAM (EDRAM) which was introduced in the early 1990's by Romtron.

The mainstream of speed enhancement efforts on DRAMs, however, has fallen in the area of serial speed for high resolution graphics systems and is discussed in the following section after brief introduction to the requirements of the graphics application on the memories.

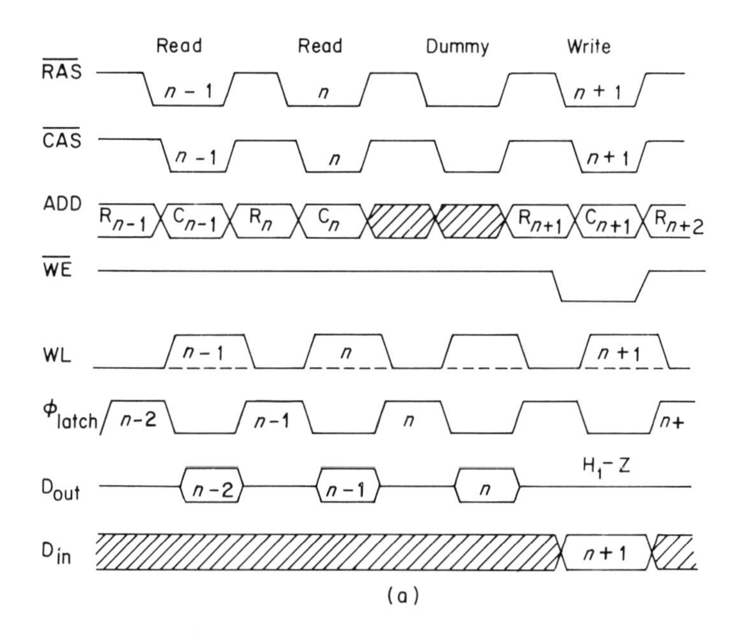

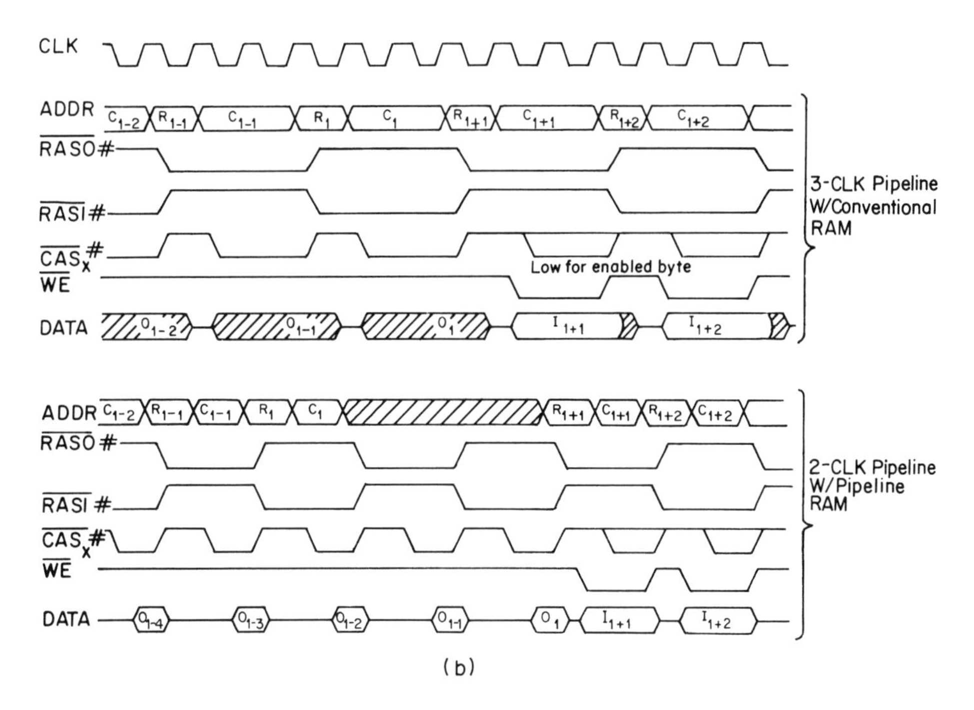

Figure 7.25
4Mb DRAM with fast pipelined read access mode. (a) Timing diagram of pipelined read operation and non-pipelined write operation. (b) Application example of timing diagram and data transfer between DRAMs and a microprocessor for conventional DRAM and for pipelined DRAM. (From Furuyama [74], Toshiba 1990, with permission of Japan Society of Applied Physics.)

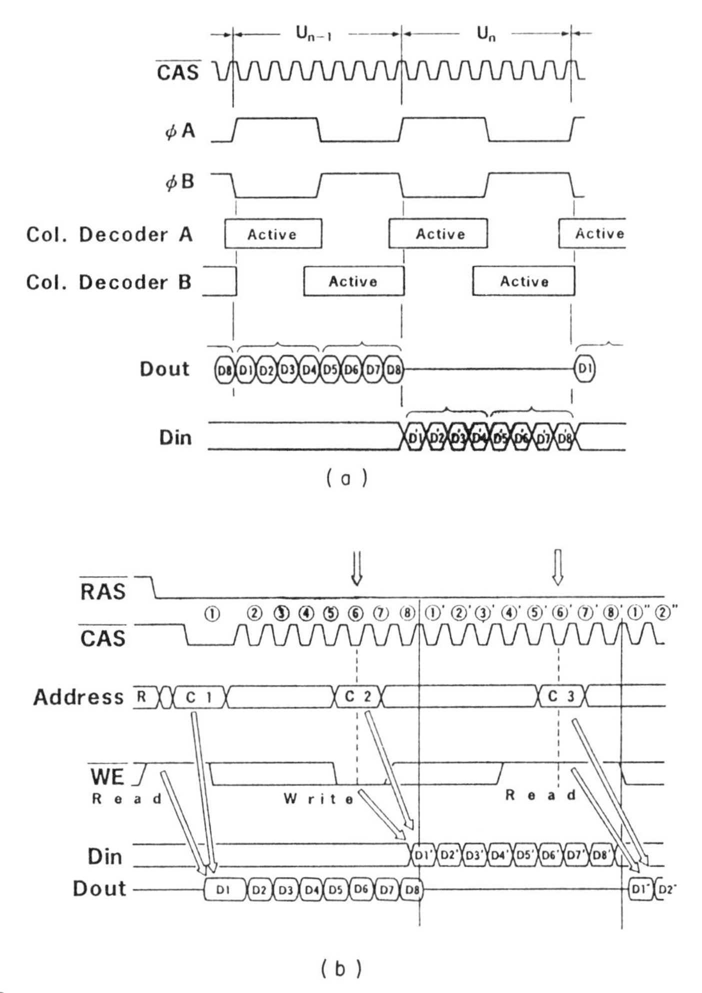

Figure 7.26
Page-nibbled mode on 16Mb DRAM example of internal interleaving of two standard modes to achieve data rate. (From Numata [75], Toshiba 1989, with permission of IEEE.)

7.6 A BRIEF HISTORY OF GRAPHICS APPLICATIONS FOR DRAMs

High resolution graphics applications in engineering work stations for computer aided design and computer aided manufacturing (CAD–CAM), and in high definition television for the consumer market, have provided a new and high volume demand for high density, low cost memory. Between 1983 and 1988 the video display market is thought to have grown from 1% to 27% of the total worldwide DRAM consumption as was shown in Figure 7.1. For the future applications of high resolution graphics are envisioned as expanding over the entire range of consumer television, computer and telecommunications markets as shown in Figure 7.27.

High resolution color graphics systems require multiple memory planes to achieve

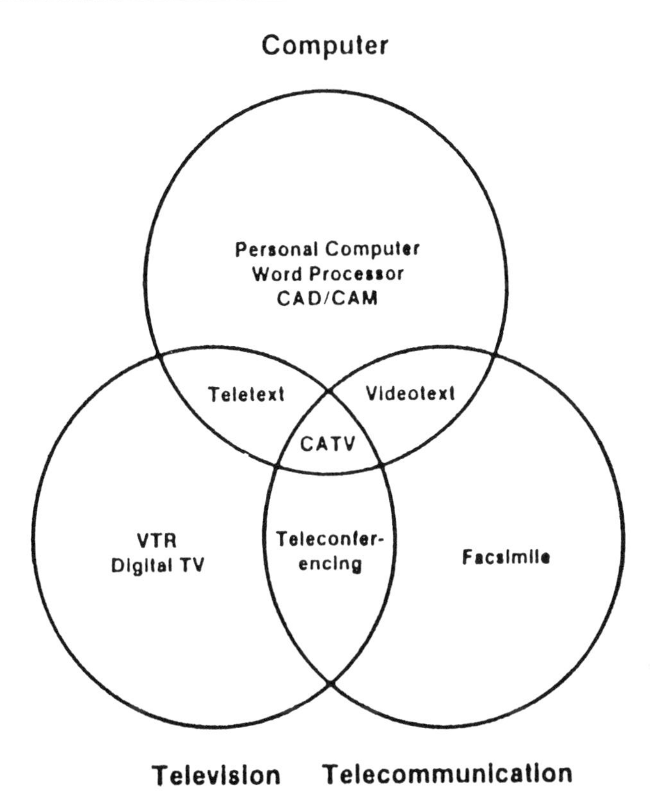

Figure 7.27
Applications of memories in high resolution graphics displays. (Source NEC Databook 1987 [76].)

the color capability necessary for a good user interface. In such a system features must be traded off against the size and cost. Some features are common to many designs such as efficient integration of text and graphics, time to redraw the screen image, the time to move objects on-screen, the amount of memory to map the display and the support logic to use the memory effectively.

A typical early video system, which evolved from the earliest text-only terminals, contained separate text and graphics controllers. To achieve more flexibility in image control, a bit mapped memory was added so the system could directly store images to be displayed. This mixed text and graphics system was a first attempt to add graphics capability to table driven systems.

As higher resolution and multiple color planes were added, the screen refresh required higher data rates from the memory, giving less time to the system processor for data management in the frame buffer. The video subsystem typically consumed 70% to 80% of the available memory accesses from the DRAM bit-mapped memory just to transfer pixel data to the display to keep the screen refreshed and avoid flicker. The graphics processor then tried to write the data to the bit-map memory during the remaining 25% of the time which comprised the horizontal and vertical refresh periods.

Bus contention in this mixed system also became a major problem. Graphics system controllers were added to the system to isolate the large bandwidth display bus from the system bus. Without this isolation, the databus would have been clogged with data passing from the frame buffer to the display.

The memory required for the frame buffer RAM in a color system was typically 40 (4 bits per pixel) times larger than the display list RAM in the mixed text system. Display data, transferred to the screen, load the data bus so that there was considerably less time available to update the frame buffer memory than the display list RAM. Pixel data, however, required more manipulating than display characters, that is, more memory had to be accessed more often and in less time.

A high resolution (1024 × 1024) graphics display, as used in many CAD systems, requires data from the refresh buffer at between 75 and 125 MHz from each plane depending upon the actual display device specification. This is independent of the graphics controller's need to access the refresh buffer to update the image stored in memory.

Timings involved in a high resolution video display are shown in Figure 7.28. The total frame time is the sum of active display time, and horizontal and vertical blanking intervals.

The first attempt at a specialized memory intended to serve the needs of this application was the wide-word 16k × 4 DRAM, initially introduced by TI. The 16k × 4 provided the large video bandwidth required in medium to high resolution video systems. A wide-word architecture provided more data-lines per depth of memory using the standard DRAM access timing, that is, addressing four times as many bits per device simplified the hardware needed to create the display frame buffer.

Many system design approaches were originated to relieve the bus contention problems. One method of improvement of performance was to use a double buffer technique to avoid contention problems between the graphics controller and the display refresh. In this type of system two display frame buffers were used. One provided information for the display and the other was available to the graphics controller for updates. When a new drawing was complete, the system switched the function of the two buffers. The disadvantages were first that even though this allowed more interaction time with the memory, it was at the expense of doubling the memory requirement, and secondly that when the buffers were switched, the graphics controller did not have a copy of the most recently available data image.

New systems in this period of time were also becoming more sophisticated and putting new demands on the performance capability of the memory. Systems were being required to support multiple windows and to allow data manipulation within one window without affecting the contents of another window, and to mix text and graphics information within a window.

The earliest attempt by the DRAM suppliers to increase performance in bit-mapped graphics systems by providing fast access modes was the 'nibble mode' developed by Inmos and introduced in 1981. This mode performed a burst transfer which reduced the memory cycle time when four contiguous memory locations were addressed from the DRAM array. The drawback of the nibble mode was that system timing and control tend to become more complicated while the problem of bus contention

Display Parameters		
Pixel clock frequency	88.00	MHz
Pixels per scan line	1380	MHz
Lines per frame	1063	MHz
Displayed pixels per scan line	1024	MHz
Displayed lines	1024	MHz
Horizontal blanking interval	4.05	μs
Vertical blanking interval	611.60	μs
Pixel time	11.36	ns

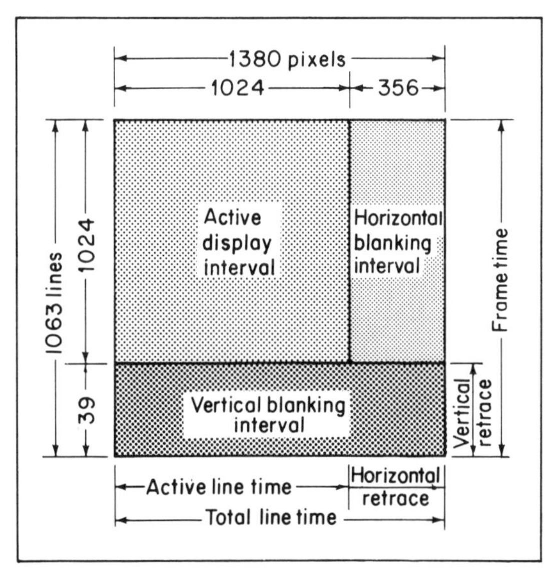

Figure 7.28
Frame time in a high resolution video display show as sum of active display time and horizontal and vertical blanking intervals [39]. (Reproduced with permission of Computer Design.)

between the display and the graphics processor was still not resolved. The CPU was still locked out of the memory most of the time.

In 1983, the first DRAM with static column architecture was introduced. Both static column and page mode have been used in graphics applications. Page mode has tended to be more widely used. The speed of page mode was improved in the early 1990's by the addition of an "extended data out" (EDO) feature which reduces the speed lost to CAS hold time. EDO is also called "hyperpage".

Although system performance was improved by using these fast modes, the basic problem of bus contention remained and required extra arbitration logic.

7.6.1 Early dual-port VDRAMs

In response to these applications requirements, multiport video DRAM development was begun. The basic configuration of a dual-port video DRAM includes a random access (RAM) port and a serial access (SAM) port. This is implemented by a chip architecture which includes a random access memory array coupled to serial data registers as shown in the outline of a 64k × 4 VDRAM in Figure 7.29. The video RAM organization makes possible unified text and graphics design. It virtually eliminates bus contention by decoupling the video data bus from the processor data bus.

By allowing simultaneous, asynchronous access to the two ports, the video DRAM allows the system processor and the display refresh to work independently. The need for double buffering is therefore removed. This gives maximum time for the system processor to access the memory. The random access port conforms to the signal and timing requirements of a standard DRAM. An on-chip shift register supports high resolution data rates and reduces video data shift logic and timing generation circuitry complexity.

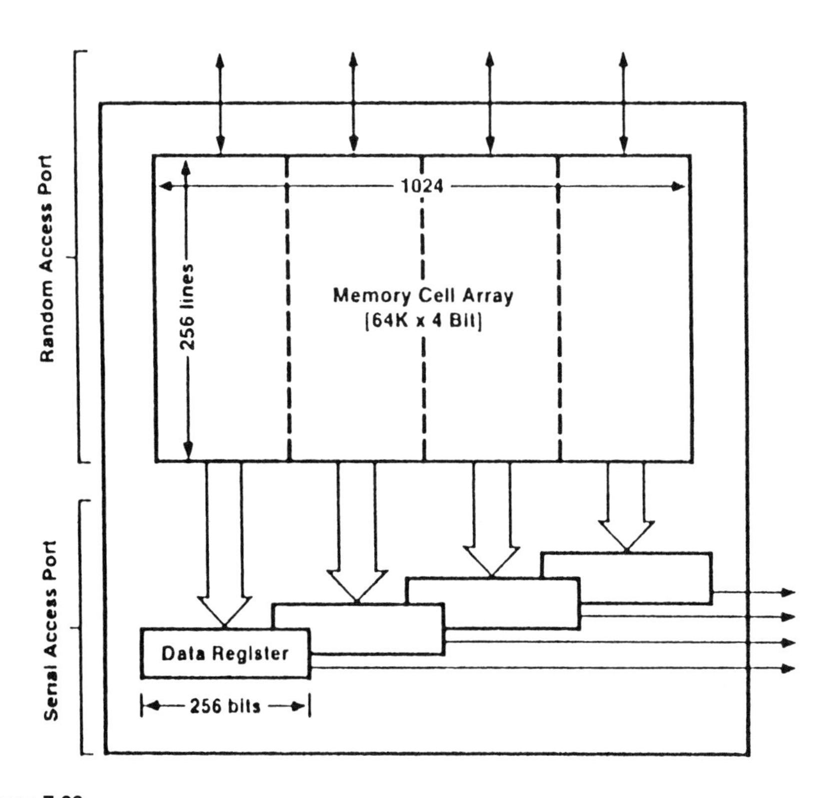

Figure 7.29
Basic configuration of a dual-port dynamic memory. (Source NEC Memory Products Databook 1987 [76].)

These VDRAMs are not independent dual-port memories since a cycle must be triggered on one port to allow a transfer on the other port. They fit well, however, with graphics display applications and show promise for applications in other fields.

The earliest VDRAM was a 64k × 1 DRAM with a 256 bit shift register on board that was introduced by TI [59] in 1983. The shift register on this early part was configured as 4 linked 64 bit shift registers able to provide shift lengths of 64, 128, 192 or 256 bits. The shift register could be loaded with 256 bits which corresponded to one row in the DRAM array in a single cycle.

Once the data had been loaded, the shift register was internally decoupled from the DRAM array and could be accessed serially while the DRAM was accessed simultaneously and asynchronously via the random port by the graphics processor.

To implement a 1280 × 1024 display using the 64k video DRAM, 20 memories were required. If, however, a 16 bit processor was used with 20 memories there was a mismatch between the 16 and the 20. Either the processor would access part of the

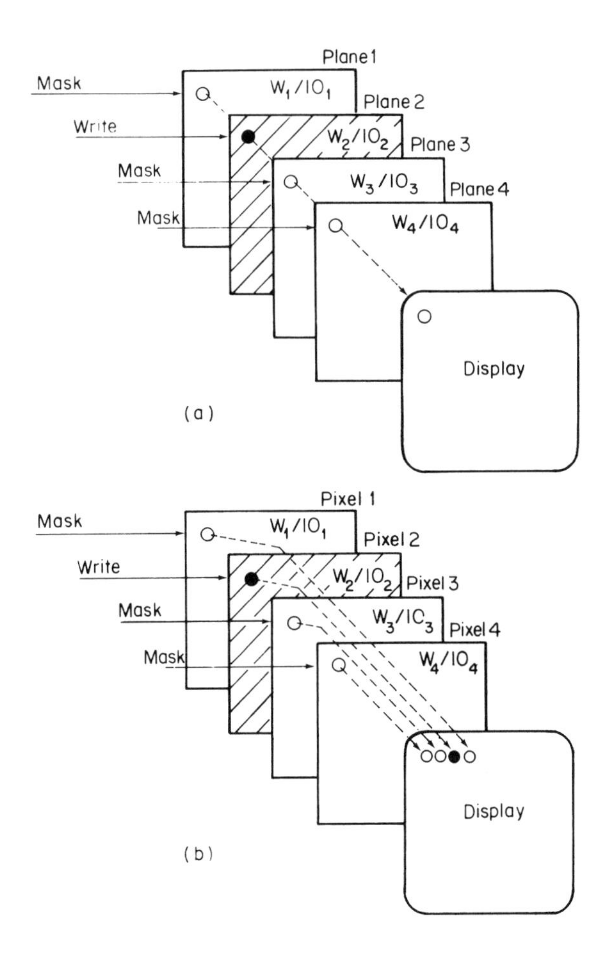

Figure 7.30
Applications of write-per-bit function. (a) Write of plane unit. (b) write of pixel unit [64].

memory as partial words or extra memory had to be used for video access. The added complexity of dealing with partial words made it simpler to use more memory. The extra memory could be used since graphics systems typically require large regions of scratchpad memory to be used by the processors for placing text fonts, display, lists, and for use in the calculations of drawing the displayed images.

It also became clear that in high density DRAMs the wider bus parts were desirable due to the granularity. For example, the minimum granularity of a 256k × 1 memory on a 16 bit CPU databus is 256k × 16, or 4Mb of memory. This means the processor must perform four million pixel operations to rewrite the entire screen which is not an efficient configuration. The result would be that much of the memory would be unused. In contrast with a 64k × 4 memory on a 16 bit databus, the minimum granularity is 64k × 16 or 1Mb of memory. Clearly the wider bus devices were needed.

One problem with the wider bus parts, however, was that, with conventional ×4 and ×8 DRAMs, it was not possible to write to separate random ports individually. That is, if the processor wanted to update one of the four or eight pixels from one memory chip, it had to execute a read–modify–write cycle on the four bits to change the one bit that was wanted. This increased the pixel update time.

The solution used was the maskable write feature which enabled or disabled the individual ports by loading a write mask onto the common I/O pins and latching it into an internal write mask register. Two applications of this 'write-per-bit' feature are illustrated in Figure 7.30. Figure 7.30(a) shows the write of a single memory plane out of four and Figure 7.30(b) shows the write of a single pixel out of four [64].

The first VDRAMs had serial output, but not serial input. Serial input is useful for such features as scrolling where the rows of memory can be reordered in the shift register without going back through the memory.

7.6.2 The 256k VDRAMs establish the market

The true VDRAMs, although not yet standardized, appeared at the 64k × 4 level from such companies as TI, Mitsubishi, Mostek [63], Hitachi [62], NEC [60] and Fujitsu among others.

A block diagram indicating the internal architecture of the NEC 64k × 4 VDRAM and the serial data registers is shown in Figure 7.31(a). A timing diagram is shown in Figure 7.31(b) indicating the CPU write to RAM and the data transfer to screen from the SAM in step with the video timing furnished by the system clock.

The 64k × 4 VDRAMs were all dual-port dynamic memories that include a shift register for providing a high speed secondary channel. The signals fell into two groups. One group corresponded to the usual signals for a DRAM, that is $\overline{RAS}$, $\overline{CAS}$, Data Transfer, Write enable, multiplexed address lines, and the four bi-directional data lines DQ.

In addition the serial access part of the memory, the SAM, consisted of four 256-bit registers which required six additional lines. These included a serial clock (SC) and a Serial Output Enable ($\overline{SE}$) signal which controled the shift register's I/O lines $S_3 \ldots S_0$. The $\overline{DT}$ signal was used in conjunction with the $\overline{RAS}$, $\overline{CAS}$, and $\overline{WR}$ signals

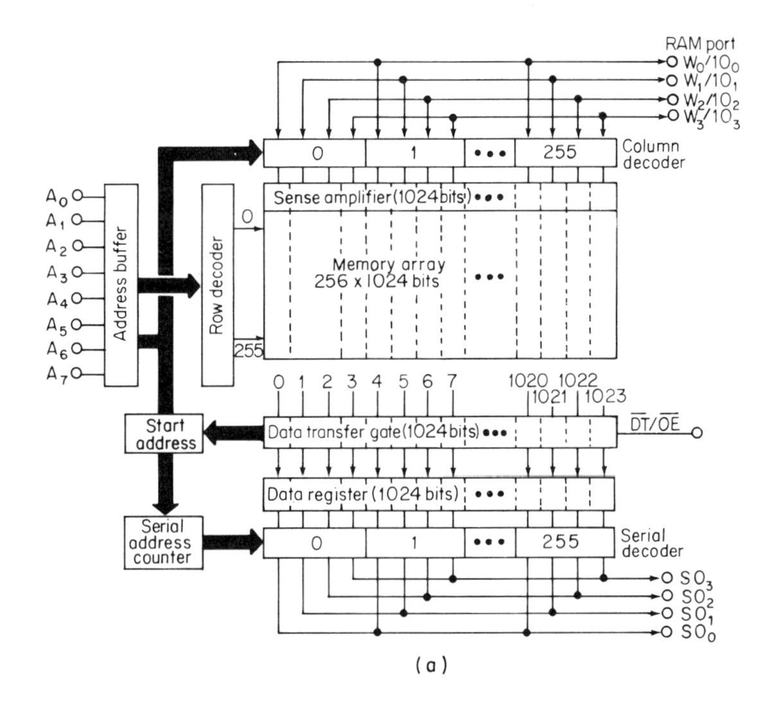

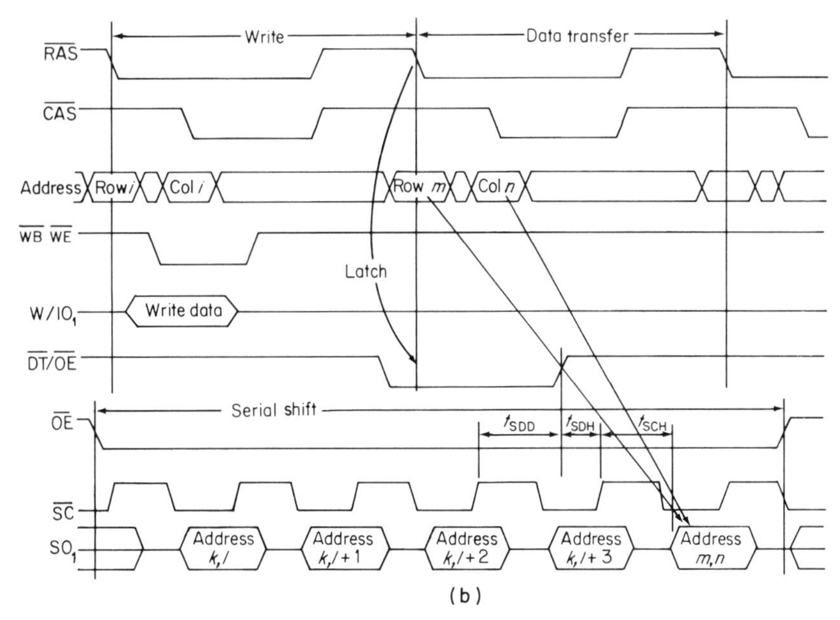

Figure 7.31
Block diagram of VDRAM showing (a) RAM array and port and SAM register and port. (b) Timing diagram showing the CPU writing to the RAM with the $\overline{DT}/\overline{OE}$ line high, then transfering data to the screen through the serial port with the $\overline{DT}/\overline{OE}$ line being pulsed by the system clock in step with the video timing [60]. (Reproduced with permission of Electronic Design, Penton Publishing Company.)

to transfer data between one selected memory row of 256 locations and the shift register.

For graphics display schemes several VDRAMs could be placed in series and in parallel to reach the required bandwidth and bitmap size. For example, a black and white screen with fair resolution, 70 MHz and 1024 × 800 interfaced to a 16-bit bus required only four 64k × 4 VDRAM circuits. The VRAM interface required only a small percentage of the bus bandwidth of a standard DRAM and the synchronization with the processor was easier since the transfer of data occurred during the horizontal retrace.

The SAM consisted of four serial access registers. Special internal data transfer cycles transfered a memory row to the SAM register. The address of the first bit of the SAM register to be transferred could be specified since the SAM register was not a usual shift register but rather a parallel latch with addressable read and write logic. A presettable counter, which scanned the latch in a circular manner, pointed to the transferred bit.

The SAM behaved like a simple shift register. The positive edge of the clock shifted the data out or stored incoming data. If the clock edge occurred when $\overline{SE}$ was not active, the counter was incremented, but the addressed SAM data were not read or written.

The $\overline{DT}$ controlled transfers between the DRAM and the SAM. Depending on whether $\overline{DR}$ was activated after or before $\overline{RAS}$ a normal memory access or a SAM data transfer occurred.

In normal write cycles a temporary mask register was loaded at the RAS active edge if the WR signal was already active. The masking function was highly useful for color displays.

Both static and dynamic technologies were used in the SAM cells. In most applications refresh was unnecessary because no delay occurred between the preparation of the SAM content and its use in the display.

Since the size of the SAM was large enough in most video display applications to correspond to more than a complete displayed line, the transfer from the RAM to the SAM could occur during the horizontal retrace so the problem of alternating transfer cycles and shift cycles did not occur. A continuous stream of shifted information was therefore possible with careful timing of the $\overline{RAS}$ and $\overline{DT}$ signals in relation to the SC signal. This was called SAM on-the-fly transfer.

When a write transfer followed a read transfer a special cycle, called a pseudowrite transfer cycle, had to occur to obtain the SAM start address. This was because the SAM start address was given in the transfer cycle and preceded a serial-in transfer in which writing to the RAM occurred, but the effective serial-in transfer was performed after the data were shifted. Hence a pseudowrite transfer had to be executed to obtain the SAM start address. This cycle, however, did not transfer the SAM data toward the memory. This was controlled by the $\overline{SE}$ line. If $\overline{SE}$ is active low at the RAS active edge a normal write cycle occurred. If $\overline{SE}$ was not active, a pseudowrite cycle occurred as shown in Figure 7.32.

VDRAMs could be used for fast communication between systems. By using the SAM port, transfer time between systems could be decreased from a 250 ns cycle time to 40 ns corresponding to a 25 MHz shift frequency.

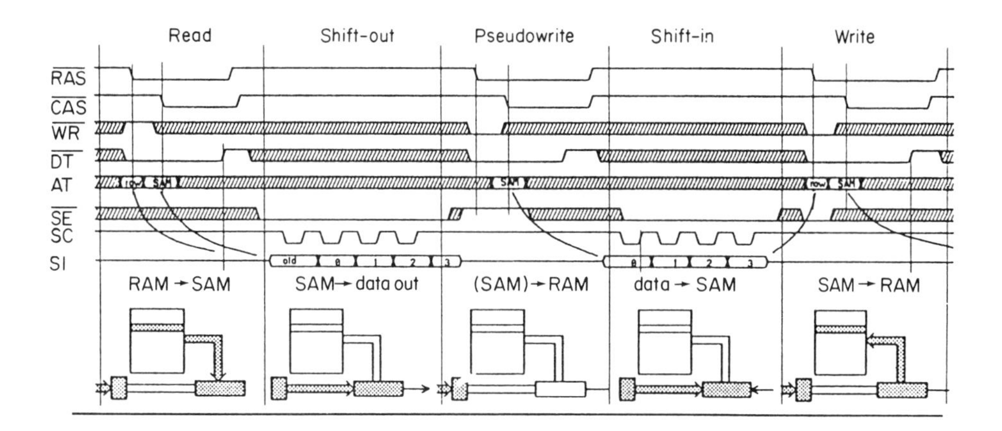

Figure 7.32
Example of a SAM pseudowrite transfer in a VDRAM. (From Nicoud [41] 1988, with permission of IEEE.)

7.6.3 1Mb multiport video DRAMs

A number of companies introduced 1Mb multiport video DRAMs. These parts were all in the JEDEC standard format with 256k × 4 memory organization for the DRAM and 512 words by 4 bits for the SAM. A typical example was the Texas Instruments part which had simultaneous and asychronous access from the DRAM and SAM ports, bi-directional data transfer between the DRAM and serial data register.

Features included: a 4 × 4 block write for fast area fill operations so that four memory address locations could be written per cycle from the on-chip color register. It included also a write-per-bit feature as discussed earlier. Page mode operation was provided along with $\overline{CAS}$ before $\overline{RAS}$ and hidden refresh modes. Multiplexing of I/O and address lines was permitted using the RAM output enable. A functional block diagram is shown in Figure 7.33.

The serial data stream from the SAM ran at up to 33 MHz in normal operation. Starting location in the SAM was any of the 512 words. There was also a split register operation possible with the register divided into a high and a low half. While one half was being read out of the SAM port, the other half could be loaded from the memory array. This real time register reload allowed continuous serial data without the necessity to interleave register externally. The timing diagram of the split register operating sequence is shown in Figure 7.34. In split register mode, data could be transferred from different rows to the low and high halves of the data register.

The color register was loaded with four bits of data using the DSF pin during a 'load–color–register' cycle as shown in Figure 7.35. Subsequently the contents of this register could be written to any combination of four adjacent column memory

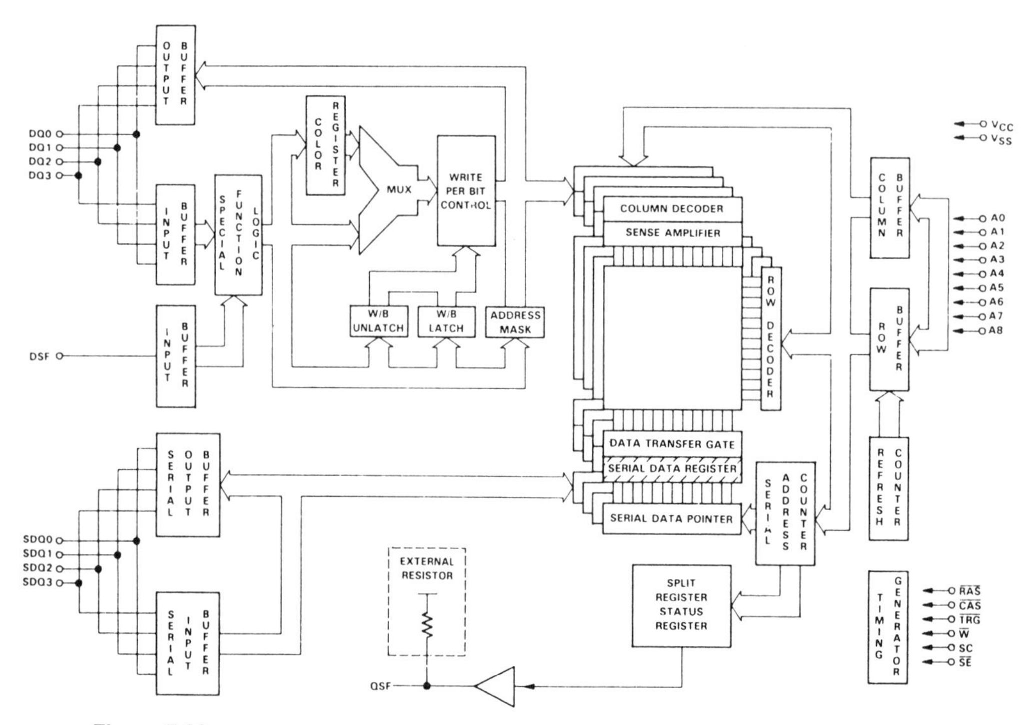

Figure 7.33
Functional block diagram of a 256k × 4 multiport video DRAM. (From Texas Instruments, MOS Memory Databook 1989.)

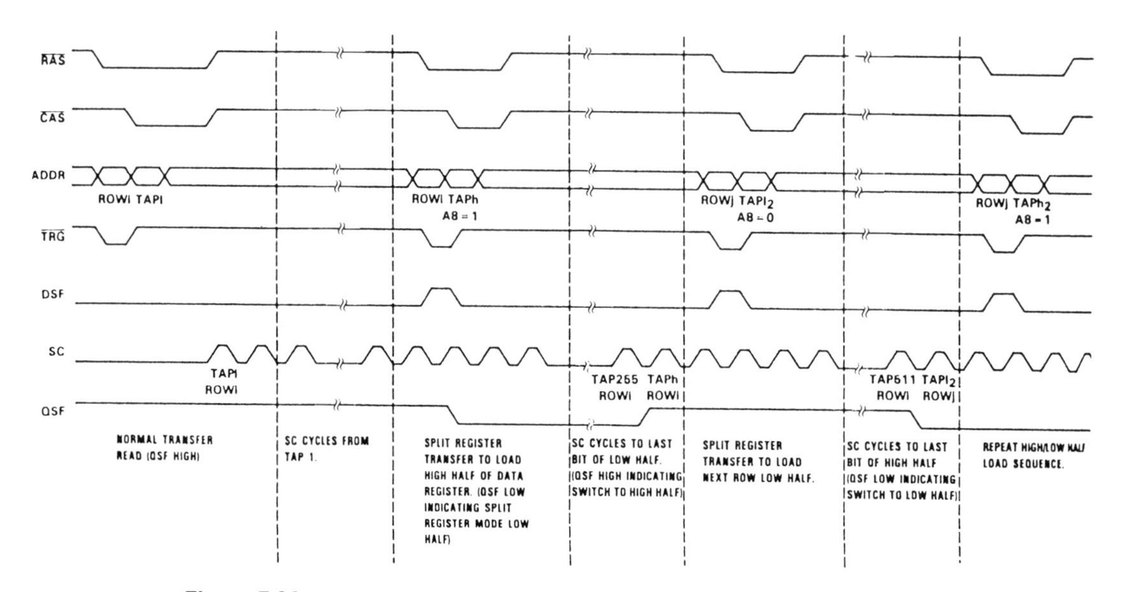

Figure 7.34
Split register operating sequence of 256k × 4 multiport VDRAM. (From Texas Instruments, MOS Memory Databook 1989.)

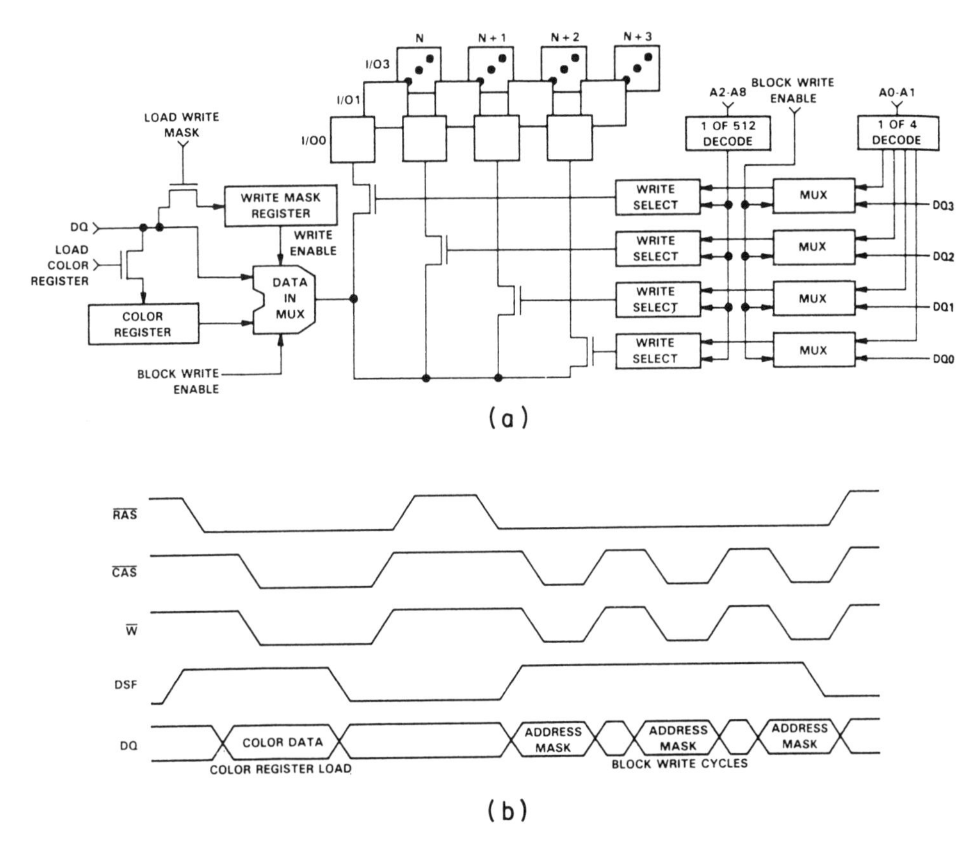

Figure 7.35
256k × 4 multiport VDRAM (a) color register block diagram, (b) load color register cycle timing. (From Texas Instruments, MOS Memory Databook 1989.)

locations using a 4 × 4 block write feature. During the block write cycle only the seven most significant column addresses were latched on the falling edge of $\overline{CAS}$. The two least significant addresses were replaced by the four DQ bits. These four bits were used as an address mask indicating which of the four column address locations were written with the contents of the color register during the write cycle.

A typical functional truth table for the T.I. 256k × 4 multiport video RAM is shown in Figure 7.36. Various timing diagrams are shown in Figure 7.37 including the serial data-in timing (7.37(a)), serial data register to RAM timing (7.37(b)), RAM to serial data register transfer timing (7.37(c)), serial data-out timing (7.37(d)), and write mode control pseudowrite transfer timing (7.37(e)).

Texas Instruments also discussed a 1Mb VDRAM in 1988 [68, 69] which could be organized as 128k × 8 or 256k × 4 thus potentially providing higher bandwidth in the system. Uninterrupted serial data streams of 70 MHz were achieved by combining pipelining and interleaving techniques on the chip with an internally triggered automatic memory-to-register transfer mechanism.

TYPE[†]	$\overline{\text{RAS}}$ FALL					$\overline{\text{CAS}}$ FALL	ADDRESS		DQ0-3		FUNCTION
	$\overline{\text{CAS}}$	$\overline{\text{TRG}}$	$\overline{\text{W}}$	DSF	$\overline{\text{SE}}$	DSF	$\overline{\text{RAS}}$	$\overline{\text{CAS}}$	$\overline{\text{RAS}}$	$\overline{\text{CAS}}$[‡] $\overline{\text{W}}$	
R	0	X	1	X	X	X	X	X	X	X	$\overline{\text{CAS}}$-BEFORE-$\overline{\text{RAS}}$ REFRESH
T	1	0	0	X	0	X	ROW ADDR	TAP POINT	X	X	REGISTER TO MEMORY TRANSFER (TRANSFER WRITE)
T	1	0	0	1	X	X	ROW ADDR	TAP POINT	X	X	ALTERNATE TRANSFER WRITE (INDEPENDENT OF $\overline{\text{SE}}$)
T	1	0	0	0	1	X	REFRESH ADDR	TAP POINT	X	X	SERIAL WRITE-MODE ENABLE (PSEUDO-TRANSFER WRITE)
T	1	0	1	0	X	X	ROW ADDR	TAP POINT	X	X	MEMORY TO REGISTER TRANSFER (TRANSFER READ)
T	1	0	1	1	X	X	ROW ADDR	TAP POINT	X	X	SPLIT REGISTER TRANSFER READ (MUST RELOAD TAP)
R	1	1	0	0	X	0	ROW ADDR	COL ADDR	WRITE MASK	VALID DATA	LOAD AND USE WRITE MASK, WRITE DATA TO DRAM
R	1	1	0	0	X	1	ROW ADDR	COL A2-A8	WRITE MASK	ADDR MASK	LOAD AND USE WRITE MASK, BLOCK WRITE TO DRAM
R	1	1	0	1	X	0	ROW ADDR	COL ADDR	X	VALID DATA	PERSISTENT WRITE PER BIT, WRITE DATA TO DRAM
R	1	1	0	1	X	1	ROW ADDR	COL A2-A8	X	ADDR MASK	PERSISTENT WRITE PER BIT, BLOCK WRITE TO DRAM
R	1	1	1	0	X	0	ROW ADDR	COL ADDR	X	VALID DATA	NORMAL DRAM READ/WRITE (NON MASKED)
R	1	1	1	0	X	1	ROW ADDR	COL A2-A8	X	ADDR MASK	BLOCK WRITE TO DRAM (NON MASKED)
R	1	1	1	1	X	0	REFRESH ADDR	X	X	WRITE MASK	LOAD WRITE MASK
R	1	1	1	1	X	1	REFRESH ADDR	X	X	COLOR DATA	LOAD COLOR REGISTER

[†]R = RANDOM ACCESS OPERATION; T = TRANFER OPERATION
[‡]DQ0-3 ARE LATCHED ON THE LATER OF $\overline{\text{W}}$ OR $\overline{\text{CAS}}$ FALLING EDGE.
ADDR MASK = 1 WRITE TO ADDRESS LOCATION ENABLED
WRITE MASK = 1 WRITE TO I/O ENABLED

Figure 7.36
Functional truth table of 256k × 4 multiport VDRAM. (From Texas Instruments, MOS Memory Databook 1989.)

7.6.4 Enhanced video DRAMs

Going even a step further in complexity than the standard video DRAM is the logic operation mode 64k × 4 video DRAM introduced by Hitachi. This part included an option for a logic operation between the memory data and the input data that could be done in one cycle. Various logic operations were controlled in accordance with a logic code truth table.

Another interesting VDRAM which was enhanced for high serial speed of 100 MHz was described by Mitsubishi [72] in 1988.

This part used a pipelined parallel bi-directional shift (PPB) architecture in which serial data streams were divided into even and odd bits which were shifted in parallel and pipelined manner. In the read mode the even bits of read data were shifted and synchronized with the phase of the odd bits. In the write mode only the odd bits of write data were shifted and were synchronized with the phase of the even bits. Both even and odd were then written into the SAM array in parallel in both cases. The result is a serial data I/O rate that is fast but with an array which operated at normal timing.

7.6.5 Triple-port video DRAMs

The next step in off-loading some of the logic calculations involved in advanced graphics processing from the processor was the triple-port video DRAM [42]. For sophisticated graphics manipulation as fast color filling, object rotation, shading etc., the intelligent VDRAM could restore the displays to real time.

An example is an NEC VDRAM which had 256k bits of RAM, an on-chip logic unit with registers for display manipulation, eight serial data channels, an 8-bit random access port and a 1 bit pixel access port. It could transfer data to the display in a matrix frame buffer architecture by using both the 8-bit common I/O port and the pixel access port. The serial 128 word by 8 bit I/O buffer could also be operated as two 64 × 8 serial data registers which could be loaded separately from the memory to allow continuous serial data flow to the display.

This part was well suited for pattern and color graphics manipulation such as painting, tiling, etc. Boundary detection can be implemented with compare registers which also ease the creation of windows and other image areas.

7.7 SIMPLIFIED DISPLAY MEMORIES FOR CONSUMER APPLICATIONS

Special graphics and display devices range from simple line buffers to complex triple port graphics buffers.

The video DRAMs are functionally set up well for enhanced graphics applications in a high performance computer environment; however, they increase the size of the DRAM silicon area by 40% to 50% and significantly increase the size of the package. Testing and additional power dissipation of the SAM register also become significant considerations.

The reduced amount of memory and extreme price pressure on the video features needed in consumer applications such as television and video tape recorder applications, mean that simpler, lower cost graphics DRAMs have also been pursued.

7.7.1 Line buffers

Line buffers basically are shift registers which enable up to 8 lines of a display to be held temporarily. These differ in line length depending on whether they are intended to be used in the US and Japanese NTSC system (910 bits) or the European (excluding France) PAL system (1135 bits). The NTSC line buffer typically is organized as 910 word × 8 bits and the PAL as 1135 words × 8 bits.

These devices have many applications in video processing. They can, for example, be used for image compression by enabling alternate lines to be skipped, or for expansion by storing a line and then recycling it to provide duplicate lines on the screen.

Double speed scan conversion Line buffers can also be used to convert interlaced scan to non-interlaced scan in both NTSC and PAL television systems. Interlaced scan

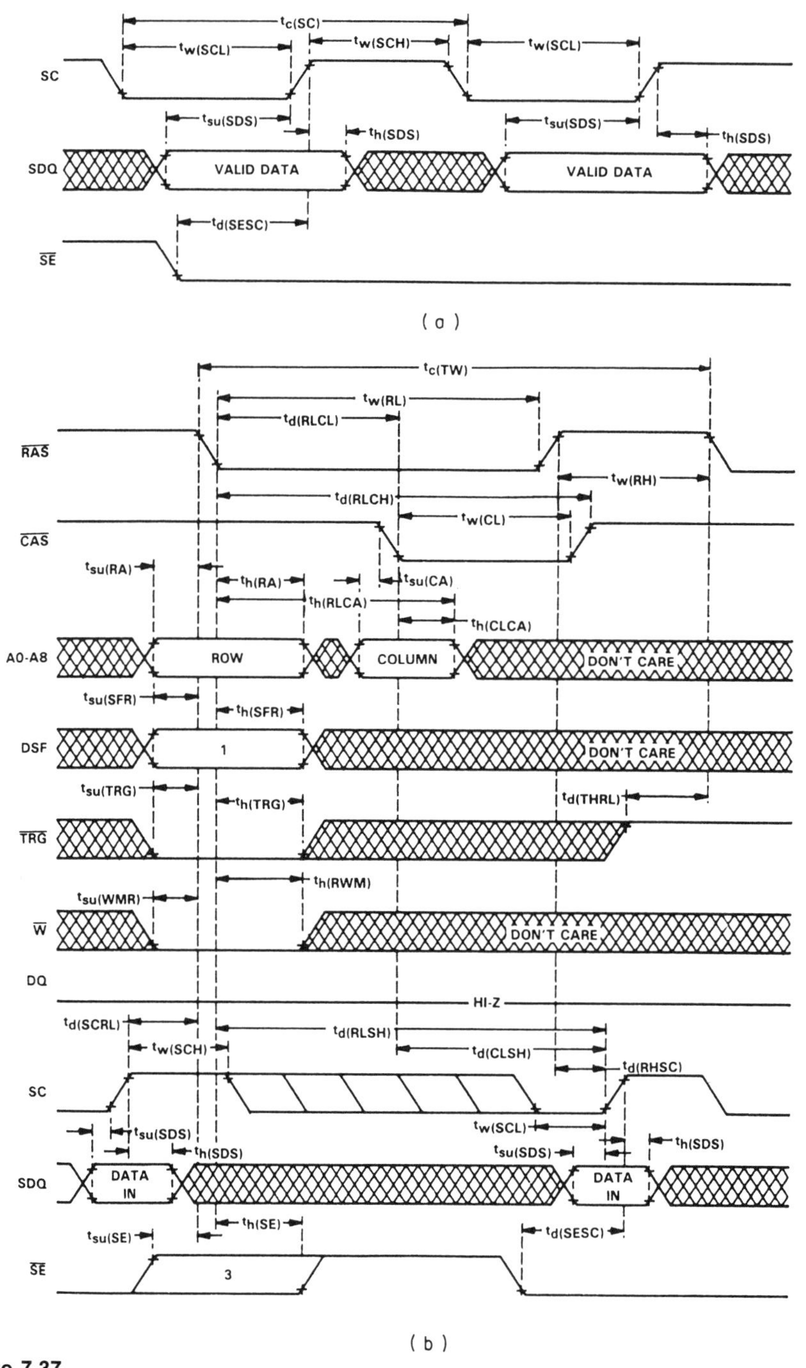

Figure 7.37
256k × 4 multiport VDRAM: (a) Serial data-in timing, (b) data register to memory timing, with serial input enabled, (c) memory to data register transfer timing, (d) serial data-out timing, (e) write mode control pseudowrite transfer timing. (From Texas Instruments, MOS Memory Databook 1989.)

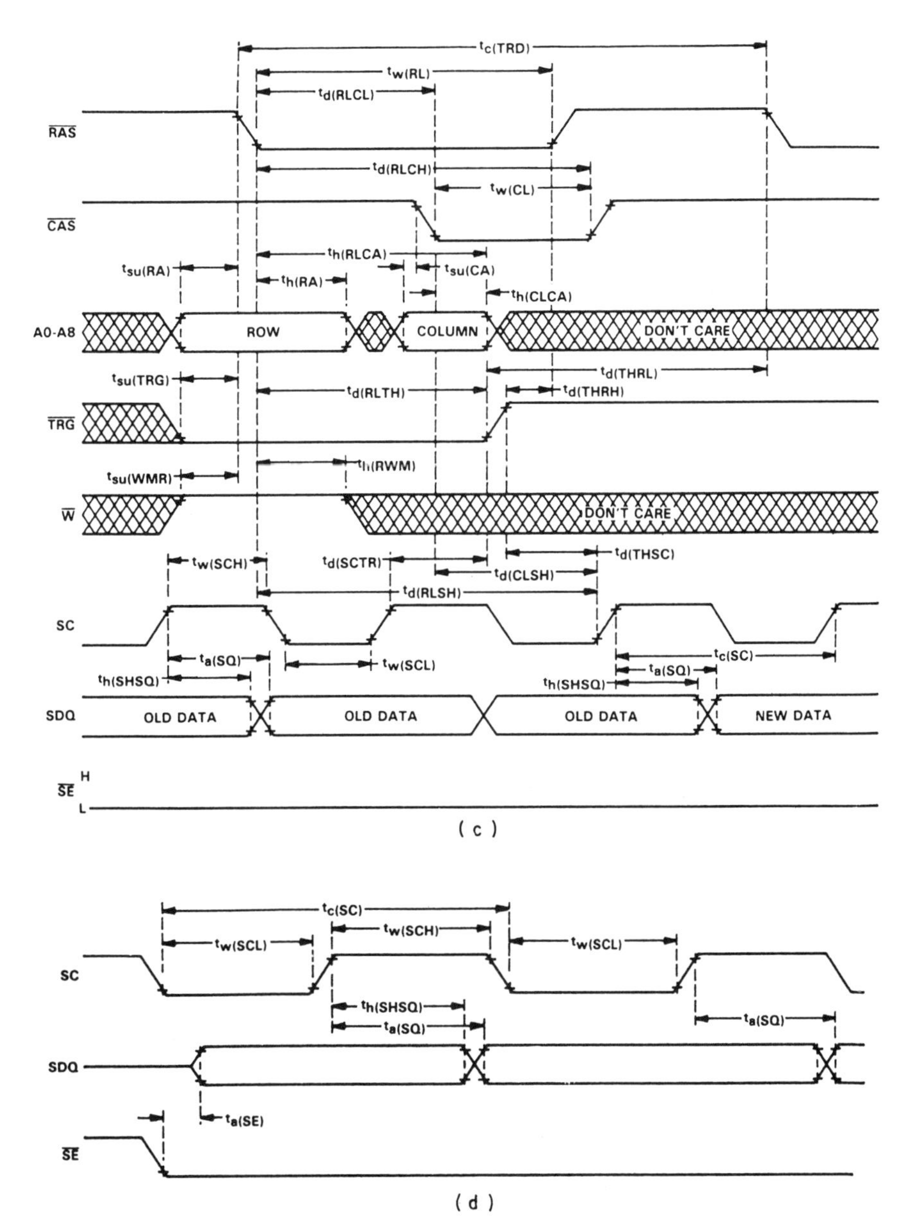

Figure 7.37 (*continued*)

is commonly used to reduce the flickering in motion scenes caused by field transition. In interlaced scanning a complete frame consists of two fields of scanning lines each with scanning being performed every two lines.

In the NTSC system a complete frame consists of two fields of 262.5 scanning lines at a field frequency of 60 Hz. In the PAL system a complete frame is made of two fields with a field frequency of 50 Hz which is the sum of 25 first fields per second and 25 second fields per second.

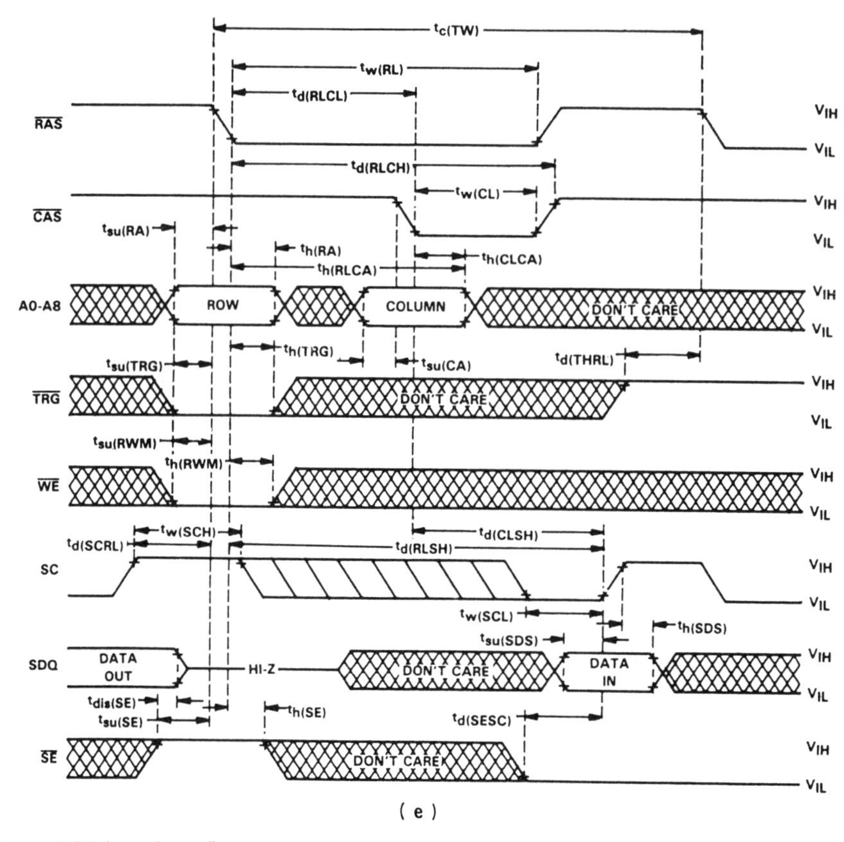

Figure 7.37 (*continued*)

A line buffer can be used to convert interlaced scanning between two fields to the simpler non-interlaced scanning of one field. This is done by doubling the number of scanning lines in a single field. Most simply this is done by creating the data of the skipped line, storing it in a line buffer memory and reading it out after the original line in an AABBCCDD format. It can also be done by using the data of that line in the last field, or by averaging the value of the lines before and after the skipped line.

Line buffers can also be used to cancel noise in a video tape recorder picture. If a line contains noise the portion of the previous line in the same position as the noise is reproduced instead. This is done by storing the previous line in a line buffer so it is available to be read out.

7.7.2 Field memories

A simple field memory stores a whole field of lines to be read out serially rather than storing 8 lines. The simplest field memory is made from a DRAM with serial input and output buffers. Typically these input buffers add only a few percent to the silicon

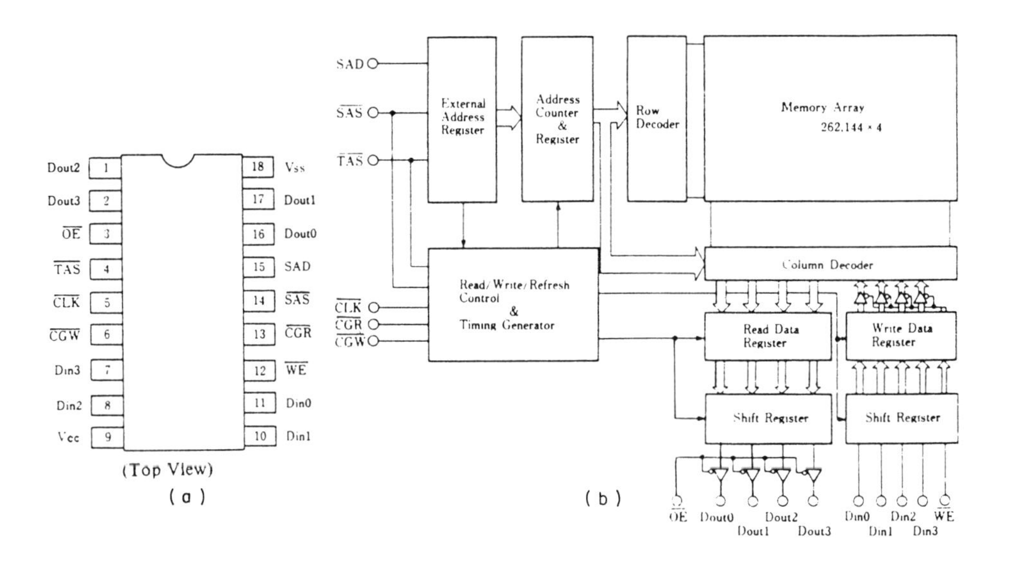

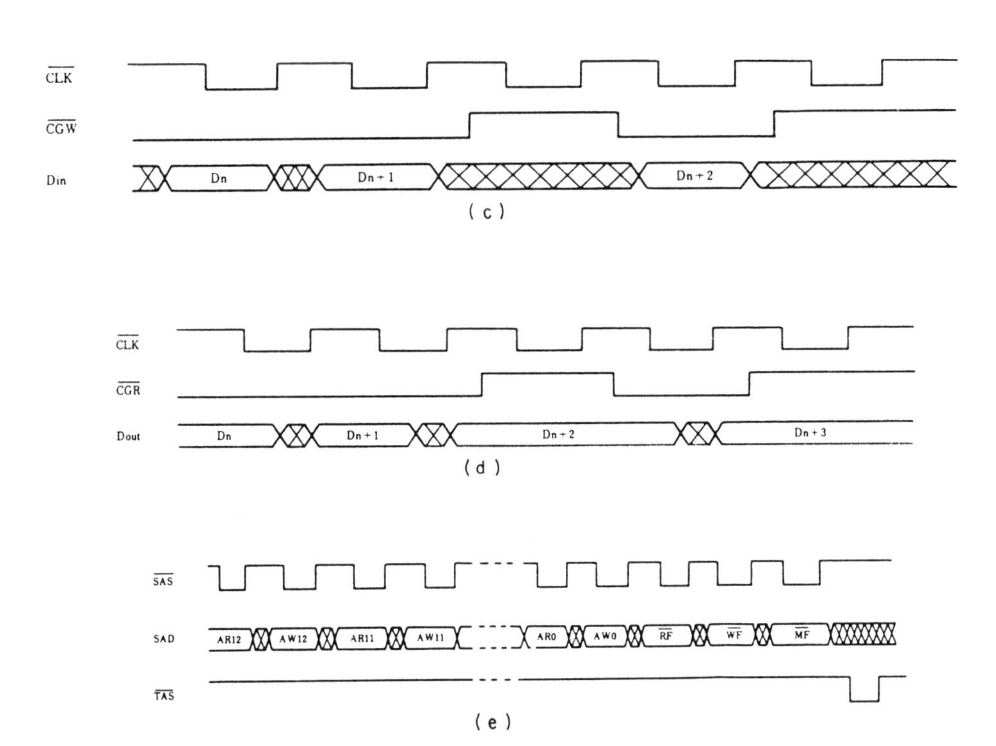

Figure 7.38
A 256k × 4 frame memory with block random accessing: (a) pin diagram, (b) block diagram, (c) write cycle, (d) read cycle, (e) 32 × 4 random access by serial input. (From Hitachi IC, Memory Products Databook 1987/1988.)

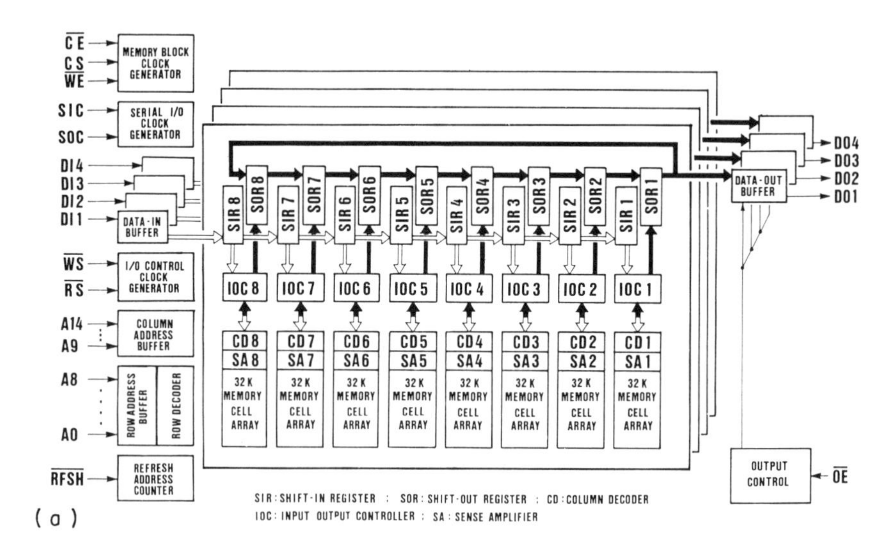
CE
CS
WE
MEMORY BLOCK CLOCK GENERATOR
SIC
SOC
SERIAL I/O CLOCK GENERATOR
DI4
DI3
DI2
DI1
DATA-IN BUFFER
WS
RS
I/O CONTROL CLOCK GENERATOR
A14
A9
COLUMN ADDRESS BUFFER
A8
A0
ROW ADDRESS BUFFER
ROW DECODER
RFSH
REFRESH ADDRESS COUNTER
DATA-OUT BUFFER
DO4
DO3
DO2
DO1
OUTPUT CONTROL
OE
SIR : SHIFT-IN REGISTER ; SOR : SHIFT-OUT REGISTER ; CD : COLUMN DECODER
IOC : INPUT OUTPUT CONTROLLER ; SA : SENSE AMPLIFIER
(a)

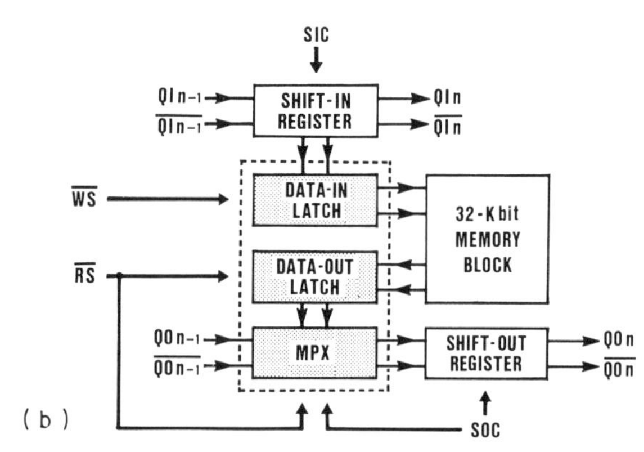
SIC
QIn-1
SHIFT-IN REGISTER
QIn
WS
DATA-IN LATCH
32-Kbit MEMORY BLOCK
RS
DATA-OUT LATCH
QOn-1
MPX
SHIFT-OUT REGISTER
QOn
SOC
(b)

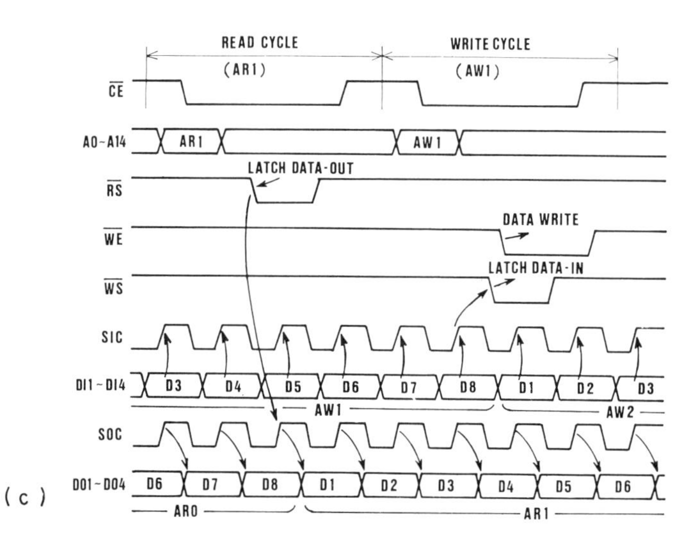
READ CYCLE
(AR1)
WRITE CYCLE
(AW1)
CE
A0~A14
RS
LATCH DATA-OUT
WE
DATA WRITE
WS
LATCH DATA-IN
SIC
DI1~DI4
SOC
DO1~DO4
AR0
AR1
AW1
AW2
(c)

area and can be packaged in small packages like the standard DRAM due to the reduction in pincount from having only serial input and output.

A simple field memory can be used for such features as 'picture freeze' by storing a frame of memory and reproducing it on the screen.

The next level of complexity incudes such features as random access by data block which allows the frame to be used for 'picture-in-picture' features and 'zoom' where some low level of random processing is required.

An example is a Hitachi [2] 256k × 4 frame memory which was packaged in a small 18 lead package as shown in Figure 7.38(a). This device performed serial access by internal address generator. It had 60 ns minimum serial cycle time and 40 ns maximum serial access time. The block diagram shown in Figure 7.38(b) shows the serial-in ports and serial-out ports. It also shows the on-chip address counter and register and refresh timing generator and control needed so the system did not have to handle the required DRAM refresh. The refresh timing typically worked off the pixel clock which is always available in such a system. The serial write cycle and serial read cycle are shown in Figures 7.38(c) and 7.38(d). The serial input and output were sufficient to handle such functions as picture 'freeze'.

Synchronization of the system clock permitted time compression or expansion for picture-in-picture, for example, which was implemented with block random accessing.

Block random accessing was performed by the serial address input, SAD, as shown in Figure 7.38(e). The unit of processing is 32 word by 4 bit. In order to output data continuously the address specified by the SAD pin incremented automatically.

Another 256k × 4 frame memory with 33 MHz (30 ns) serial cycle and 10 ns serial access time was described by Matshushita [66] in 1986. The block diagram of the memory is shown in Figure 7.39(a) and a schematic of the I/O structure in Figure 7.39(b), both illustrating the relatively simple serial input and output buffers of the frame memories compared to the VDRAM SAM registers.

The memory cell array was divided into four sets of 256k bit blocks which had an organization of 32k × 8 bit. Each 256k block had an 8 bit serial–parallel conversion circuit.

A timing diagram of the read–write operation is shown in Figure 7.39(c). At the falling edge of $\overline{WS}$, input data were transferred to the data-in latch and were written into the memory cell in the following write cycle when $\overline{WE}$ was low. Readout data from the memory cell were transferred to the data-out latch at the falling edge of $\overline{RS}$.

A larger 8Mb frame memory in 0.7 μm technology with 50 MHz serial data access was described by Matshushita [67] in 1989. This 214.7 mm^2 chip was organized

Figure 7.39
256k × 4 frame memory showing (a) block diagram of the memory with memory cell arrays divided into four sets of 32k × 8 bit blocks. Each 256k block has an eight bit serial–parallel conversion circuit. (b) I/O controller block showing serial data in and out. (c) Timing diagram for the same clock rate read–write operation. (From Ohta [66], Matsushita 1986, with permission of IEEE.)

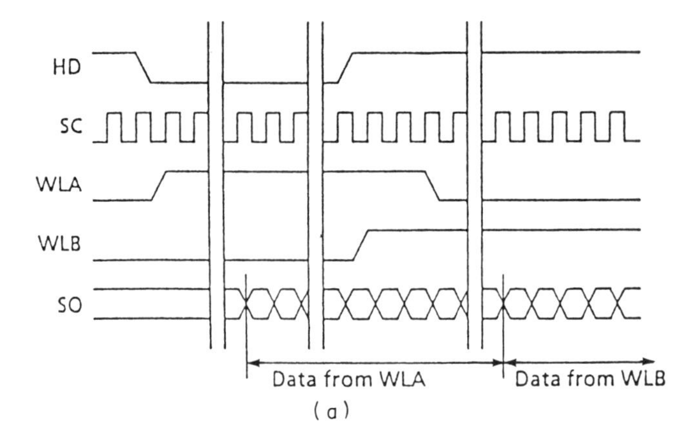

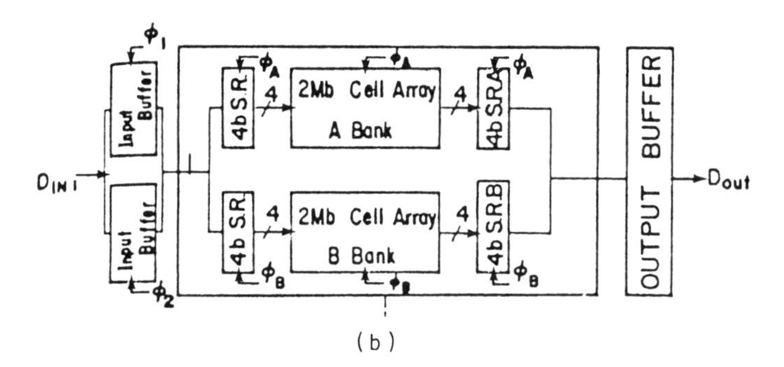

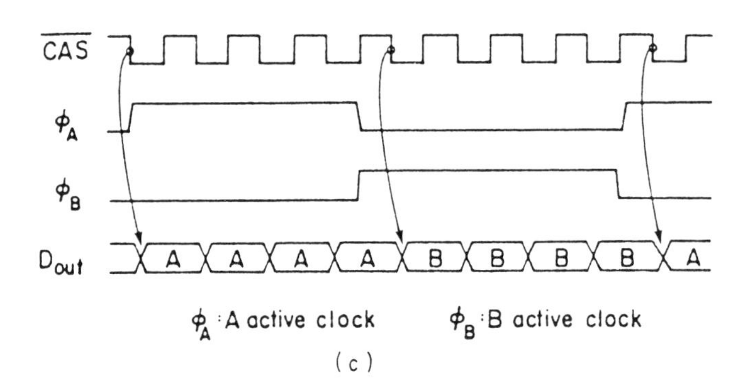

Figure 7.40
(a) 8Mb frame memory timing diagram showing interleaved serial readout of data from word-lines A and B. (b) 16Mb frame memory block diagram showing interleaved cell arrays for serial I/O. (c) Read cycle timing diagram for 16Mb frame memory. (a,b: from Kotani [67], Matsushita 1989, with permission of Japan Society of Applied Physics; c: From Watanabe [27], Toshiba 1988, with permission of IEEE.)

1Mb × 8 with 8 separate 1Mb arrays. It processed an internal 64k data word. The RAM cycle time was 160 ns and the serial cycle time was 20 ns.

Since 8 word-lines, one in each of the arrays, were driven at one time and 16 sense amplifiers were activated per bit-line, power was a primary consideration in this part. This was handled by a column drive scheme in which only 8 bits of data on one word-line rather than all 16 were transferred to the sense amplifier at the same time. The word-lines of the RAM were divided into two groups, A and B, and activated alternately as shown in the timing diagram in Figure 7.40(a).

Refresh meanwhile was done by separate circuitry using small load transistors to reduce power consumption so that it was completed with no penalty in timing during the read–write operation.

Since a frame of memory can be stored in an 8Mb DRAM array, it is a natural progression for the field memories to migrate to an 8Mb density initially rather than the 16Mb density which is the normal progression for the standard DRAMs.

7.7.3 Synchronous DRAMS

DRAMs with synchronous interfaces were announced in the early 1990's. These included the Cache DRAM from Mitsubishi mentioned previously, a synchronous DRAM module from Ramtron with a new control protocol and interface, and a synchronous DRAM standard developed by the EIA JEDEC committee which is called the SDRAM.

The JEDEC SDRAM was a direct evolution of the standard commodity DRAM. It was expected to run at speeds up to 100 MHz with a 3.3V TTL interface. It used a mode register to program high speed burst modes and had independent banks on the chip which could be interleaved to improve system speed. Most DRAM suppliers announced plans to produce it.

BIBLIOGRAPHY

[1] *Motorola Memory Data Manual*, 1984, DL113R2.

[2] *Hitachi IC Memory Products Databook*, 1987/88, 10-2S.

[3] *Philips Static RAM Product Guide*, October, 1988.

[4] Sakurai, T. *et al.* (1986) A 1Mb Virtually Static RAM, *ISSCC Dig. Tech. Papers*, Feb. p.252-253.

[5] Kung, R. I. *et al.* (1982) An 8k × 8 Dynamic RAM with Self-Refresh, *IEEE Journal of Solid State Circuits*, Vol. **SC-17**, No.5, October 1982, p.863.

[6] *NEC Electronics Inc., Memory Products Data Book* 1989.

[7] Kenmizaki, K. *et al.* (1989) A 36μA 4Mb PSRAM with Quadruple Array Operation, Symposium on VLSI Circuits, p.79.

[8] Yoshioka, S. *et al.* (1987) 4Mb Pseudo/Virtually SRAM, *IEEE ISSCC Digest*, p.20.

[9] Watanabe, H. *et al.* (1988) Stacked Capacitor Cells for High Density Dynamic RAMs, *IEDM Digest of Tech. Papers,* p.600.
[10] Nakajima, S. *et al.* (1985) A Submicrometer Megabit DRAM Process Technology Using Trench Capacitors, *IEEE Journal of Solid State Circuits,* Vol. **SC-20**, No.1, February 1985, p.130.
[11] Shah, A. *et al.* (1986) A 4 Mbit DRAM with Trench Transistor Cell *IEEE Journal of Solid State Circuits,* Vol. **SC-21**, No.5, October 1986, p.618.
[12] Terletzki, H., and Risch, L. (1986) Operating Conditions of Dual Gate Inverters for Hot Carrier Reduction, *Proceedings of the ESSDERC,* Sept. 1986, p.191.
[13] Harter, J. *et al.* (1988) A 60ns Hot Electron Resistant 4M DRAM with Trench Cell, *IEEE ISSCC Proceedings,* p.244.
[14] Pinto, M. R. *et al.* (1984) *PISCES-II User's Manual.* Stanford, CA, Stanford University, 1984.
[15] Tsukamoto, K. *et al.* (1987) Double Stacked Capacitor with Self-Aligned Poly Source/Drain Transistor (DSP) Cell Megabit DRAM. *IEDM Proceedings,* p.328.
[16] Kaga, T. *et al.* (1987) A 4.2μm^2 Half-VCC Sheath-Plate Capacitor DRAM Cell with Self-Aligned Buried Plate-Wiring, *IEDM Proceedings,* p.332.
[17] Horiguchi, F. *et al.* (1987) Process Technologies for High Density High Speed 16 Megabit Dynamic RAM, *IEDM Proceedings,* p.324.
[18] Kimura, S. *et al.* (1988) A New Stacked Capacitor DRAM Cell Characterized by a Storage Capacitor on a Bit-line Structure, *IEDM Proceedings,* p.596.
[19] Watanabe, H. *et al.* (1988) Stacked Capacitor Cells for High Density Dynamic RAMs, *IEDM Proceedings,* p.600.
[20] Yoshikawa, S. *et al.* (1988) Process Technologies for a High Speed 16M DRAM with Trench Type Cell, *VSI Technology Proceedings.*
[21] Chin, D. *et al.* (1989) An Experimental 16M DRAM with Reduced Peak-Current Noise, VLSI Seminar, p.113.
[22] Yamauchi, H. *et al.* (1989) A Circuit Design to Suppress Asymmetrical Characteristics in 16Mbit DRAM Sense Amplifier, *VLSI Circuits Proceedings* p.109.
[23] Inoue, M. *et al.* (1988) A 16-Mbit DRAM with a Relaxed Sense-Amplifier-Pitch Open-Bit-Line Architecture, *IEEE Journal of Solid State Circuits,* Vol. **23**, No.5, October 1988, p.1104.
[24] Aoki, M. *et al.* (1988) A 60ns 16-Mbit CMOS DRAM with a Transposed Data-Line Structure, *IEEE Journal of Solid State Circuits,* Vol. **23**, No.5, October 1988, p.1113.
[25] Nakagame, Y. *et al.* (1988) The Impact of Data-Line Interference Noise on DRAM Scaling, *IEEE Journal of Solid State Circuits,* Vol. **23**, No.5, October 1988, p.1120.
[26] Fujii, S. *et al.* (1989) A 45ns 16Mb DRAM with Triple-Well Structure, *IEEE ISSCC Proceedings,* p.248.
[27] Watanabe, S. *et al.* (1988) An Experimental 16Mb CMOS DRAM Chip with a 100 MHz Serial Read/Write Mode, *IEEE ISSCC Proceedings,* p.248.
[28] Arimoto, K. *et al.* (1989) A 60ns 3.3 V 16Mb DRAM, *IEEE ISSCC Proceedings* p.244.
[29] Takeshima, T. *et al.* (1989) A 55ns 16Mb DRAM, *IEEE ISSCC Proceedings,* p.246.
[30] Shong, S. H. *et al.* (1988) High-Speed Sensing Scheme for CMOS DRAMs, *IEEE Journal of Solid State Circuits,* Vol. **23**, No.1, February 1988, p.34.
[31] Horiguchi, M. *et al.* (1988) Dual-Operating-Voltage Scheme for a Single 5-V 16-Mbit DRAM, *IEEE Journal of Solid State Circuits,* Vol. **23**, No.5, October 1988, p.1128.
[32] Mano, T. *et al.* (1987) Circuit Technologies for 16Mb DRAMs, *IEEE ISSCC Proceedings,* February 25, p.22.
[33] Arimoto, K. *et al.* (1989) A 60 ns 3.3 V 16 Mb DRAM, *IEEE ISSCC Proceedings,* p.244.
[34] Scheuerlein, R. E. and Meindl, J. D. (1988) Offset Word-Line Architecture for Scaling

DRAMs to the Gigabit Level, *IEEE Journal of Solid State Circuits*, Vol. **23**, No.1, February 1988, p.41.

[35] Tsuchida, K. *et al.* (1989) The Stabilized Reference-Line (SEL) Technique for Scaled DRAMs, VLSI Seminar, p.99.

[36] Kasai, N. *et al.* (1987) 0.25 μm CMOS Technology Using P+ Polysilicon Gate PMOSFET, *IEDM Proceedings*, p.367.

[37] Lu, N. C. C. *et al.* (1988) A Buried-Trench DRAM Cell Using A Self-aligned Epitaxy Over Trench Technology, *IEDM Proceedings*, p.588.

[38] Wakamiya, W. *et al.* (1989) Novel Stacked Capacitor Cell for 64Mb DRAM, VLSI Seminar.

[39] Gulley, D. W. (1984) Joining Text and Graphics Enhances Video Performance, *Computer Design*, August, p.123.

[40] Pinkham, R. (1984) 'Video Memory Technology and Applications'. Texas Instruments, Inc., p.29/5.

[41] Nicoud, J. D. (1988) Video RAMs, Structure and Applications, *IEEE Micro*, February, p.8.

[42] Bursky, D. (1987) Triple-Port DRAM fuels Graphics Displays, *Electronic Design*, April, p.53.

[43] *Texas Instruments MOS Memory Data Book* 1989, Commercial and Military Specifications.

[44] Dynamic RAM looks Fully Static, *Electronics*, January 13, 1983, p.250.

[45] Suzuki, S. *et al.* (1984) A 128k Word × 8 Bit Dynamic RAM, *IEEE Journal of Solid State Circuits*, Vol. **SC-19**, No.5, October 1984, p.624.

[46] Henkels, W. H. *et al.* (1989) A 12ns Low Temperature DRAM, *IEEE ISSCC Proceedings*.

[47] Baier, E. K. *et al.* (1984) A Fast 256k DRAM Designed for a Wide Range of Applications, *IEEE Journal of Solid State Circuits*, Vol. **SC-19**, No.5, October 1984.

[48] Dhong, S. *et al.* (1989) An Experimental 27ns 1Mb CMOS High Speed DRAM, *VLSI Seminar Proceedings*, p.107.

[49] Horiguchi, M. *et al.* (1988) An Experimental Large-Capacity Semiconductor File Memory Using 16-Levels/Cell Storage, *IEEE Journal of Solid State Circuits*, Vol. **23**, No.1, February 1988, p.27.

[50] Furuyama, T. *et al.* (1989) An Experimental 2-bit/Cell Storage DRAM for Macrocell or Memory-on-Logic Applications. *IEEE Journal of Solid State Circuits*, Vol. **24**, No.2, April 1989.

[51] Watanabe, T. *et al.* (1989) Comparison of CMOS and BiCMOS 1 Mbit DRAM Performance, *IEEE Journal of Solid State Circuits*, Vol. **24**, No.3, June 1989, p.771.

[52] Kitsukawa, G. *et al.* (1987) An Experimental 1-Mbit BiCMOS DRAM, *IEEE Journal of Solid State Circuits*, Vol. **SC-22**, No.5, October 1987, p.657.

[53] Aoki, M. *et al.* (1989) A 1.5V DRAM for Battery-Based Applications, *ISSCC Proceedings*, February 17, p.238.

[54] Ohta, Y. *et al.* (1989) A Novel Memory Cell Architecture for High-Density DRAMs, *VLSI Circuits Proceedings*, p.101.

[55] Sawada, K. *et al.* (1988) A 30-μA Data-Retention Pseudostatic RAM with Virtually Static RAM Mode, *IEEE Journal of Solid State Circuits*, Vol. **23**, No.1, February 1988.

[56] Hanamura, S. *et al.* (1987) A 256k CMOS SRAM with Internal Refresh, *ISSCC Proceedings*, February 27, p.250.

[57] Sawada, K. *et al.* (1988) A 72k CMOS Channelless Gate Array with Embedded 1Mbit Dynamic RAM, *IEEE Custom Integrated Circuits Conference Proceeding*, p. 116.

[58] Yuen, A. *et al.* (1989) A 32k ASIC Synchronous RAM Using a Two-Transistor Basic Cell, *IEEE Journal of Solid State Circuits*, Vol. **24**, No.1, February 1989, p.57.

[59] Pinkham, R. *et al.* (1983) Video RAM Excels at Fast Graphics, *Electronic Design*, August 18, p.161.

[60] Forman, S. (1985) Dynamic Video RAM Snaps the Bond between Memory and Screen Refresh, *Electronic Design*, May 30, p.117.

[61] Ishimoto, S. *et al.* (1985) A 256k Dual Port Memory, *IEEE ISSCC Proceedings*, p.38.
[62] Sato, K. *et al.* (1986) The HM3462:64kbit × 4 Dual-Port Video RAM with Logic Functions, *Hitachi Review* Vol. **35**, No.5, p.259.
[63] Whiteside, F. *et al.* (1986) A Dual-Port 65ns 64k × 4 DRAM with a 50MHz Serial Output, *IEEE ISSCC Proceedings*, p.48.
[64] Nakane, M. and Tokushige, K. (1987) Video RAM Chips Designed for Computer Graphics, *JEE*, November, p.28.
[65] Price, S. M. (1985) CMOS 256-kbit video RAM, with wide two-way bus, picks up speed, drops power, *Electronic Design*, September, p.171.
[66] Ohta, K. *et al.* (1986) A 1-Mbit DRAM with 33MHz Serial I–O Ports, *IEEE Journal of Solid State Circuits*, Vol. **SC-21**, No.5. October 1986, p.649.
[67] Kotani, H. *et al.* (1989) A 50MHz 8Mb Video RAM with a Column Direction Drive Sense Amplifier, *Japanese VLSI Symposium Proceedings*, p.105.
[68] Pinkham, R. *et al.* (1988) A 128k × 8 70-MHz Multiport Video RAM with Auto Register Reload and 8 × 4 Block Write Feature, *IEEE Journal of Solid State Circuits*, Vol. **23**, No.5, October 1988, p.1133.
[69] Pinkham, R. *et al.* (1988) A 128k × 8 70MHz Video RAM with Auto Register Reload, *ISSCC Proceedings*, p.236.
[70] Lu, N. C. C. *et al.* (1988) A 20-ns 128kbit × 4 High-Speed DRAM with 330-Mbit/s Data Rate, *IEEE Journal of Solid State Circuits*, Vol. **23**, No.5, October 1988, p.1140.
[71] Lu, N. C. C. *et al.* (1988) A 20ns 512b DRAM with 83MHz Page Operation, *ISSCC Proceedings*, p.240.
[72] Morooka, Y. *et al.* (1988) Pipelined Parallel Bidirectional Shift (PPB) Architecture for High Speed Serial Access, *VLSI Circuits Proceedings*, p.89.
[73] Kitsukawa, G. *et al.* (1989) A 1-Mbit BiCMOS DRAM Using Temperature-Compensation Circuit Techniques, *IEEE Journal of Solid State Circuits*, Vol. **24**, No.3, June 1989, p.597.
[74] Furuyama, T. *et al.* (1990) A High Random-Access-Data-Rate 4Mb DRAM with Pipeline Operation, Symposium on VLSI Circuits, June, p.9.
[75] Numata, K. *et al.* (1989) New Nibbled-Page Architecture for High-Density DRAMs, *IEEE Journal of Solid State Circuits*, Vol. **24**, No.4, August 1989, p.900.
[76] *NEC Memory Products Databook* 1987.
[77] Prince, B. (1989) Life Cycles of High Technology Products, *ICCD Symposium* Invited Paper 1989.
[78] Konishi, Y. *et al.* (1990) A 38ns 4Mb DRAM with a Battery-Backup (BBU) Mode, *IEEE Journal of Solid State Circuits*, **25**, No.5, October 1990, p.1112.
[79] Kalter, H. L. *et al.* (1990) A 50-ns 16-Mb DRAM with a 10ns Data Rate and On-Chip ECC, *IEEE Journal of Solid State Circuits*, **25**, No.5, October 1990, p.118.
[80] Asakura, M. *et al.* (1990) An Experimental 1-Mbit Cache DRAM with ECC, *IEEE Journal of Solid State Circuits*, **25**, No.1, February 1990, p.5.
[81] Sato, K. *et al.* (1991) A 4Mb Pseudo-SRAM Operating at 2.6 ± 1 V with 3 μA Data Retention Current, *ISSCC Proceedings*, February 1991, p.268.
[82] Hidaka, H. *et al.* (1991) A High-density Dual-port Memory Cell Operation for VLSI DRAMs, *Symposium on VLSI Circuits*, June 1991, 7–5, p.65.
[83] Asakura, M. *et al.* (1991) Cell-plate Line Connecting Complementary Bit-line (C^3) Architecture for Battery Operating DRAMs, *Symposium on VLSI Circuits*, June 1991, 7–2, p.39.
[84] Miwa, H. *et al.* (1991) A 17 ns 4Mb BiCMOS DRAM, *IEEE ISSCC Proceedings*, February 1991, p.56.

8 TRENDS IN STATIC RAMs

.1 OVERVIEW OF SRAMs

Static RAMs today are fast, and low power. They are easy to use in the system since they require no overhead circuitry for refresh as do dynamic RAMs. Dynamic RAMs are still more common in large memory systems where their lower cost is attractive in spite of requiring more complex circuits for clocks and refresh cycles and although they frequently require a SRAM cache to reduce the number of wait states with the system processor.

Although dynamic RAMs have been moving into the small systems area as well, static RAMs are still often preferred as main memory in high-performance small systems where their higher cost is offset by simpler and less expensive systems design, and where their inherently higher speed permits operation without a cache. SRAMs, because of their low standby power, are used extensively for battery back-up and in battery operated systems.

Static RAMs are available in many varieties ranging from the super-fast bipolar and GaAs SRAMs to the slow commodity CMOS variety. A breakout of the RAM market for 1989 including the DRAMs is shown in Figure 8.1. The total static RAM market is about 28% of the semiconductor RAM market, with the rest being DRAMs.

Standard SRAM output width ranges from one bit wide to 32 bits wide. Fast bipolar SRAMs are usually ×1, while the fast CMOS and BiCMOS SRAMs tend to be ×1 and ×4. The commodity SRAMs generally tend toward the slow end of the speed range of the technology and have 8 or 16 bit wide outputs since they are used primarily in 8 and 16 bit microprocessor based designs. They sometimes appear as ×9 or ×18 if parity bits are required in the application.

Standard input and output interfaces include CMOS, TTL and ECL. Power supply standards include both the historical 5 V, and a new low voltage 3.3 V standards. The application determines the trade-offs between speed, power, cost and package size.

Static RAM share of the total MOS memory market has averaged about 20% over the last 5 years, and is expected to continue at about this level. The market is tending away from large centralized systems, such as mainframe computers.

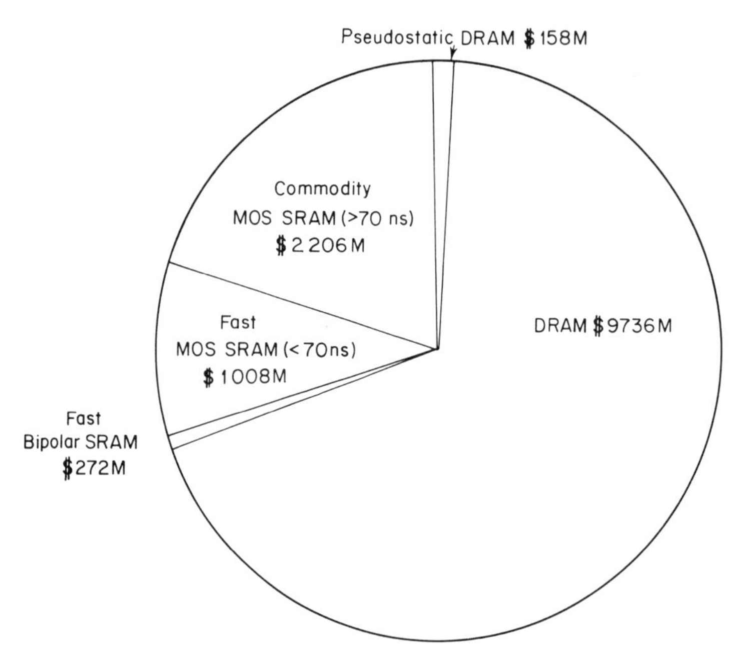

Figure 8.1
Semiconductor RAM market 1989 (total $12.03 billion). (Source Dataquest.)

An engine for growth should also be provided by the rapidly expanding market for battery operated and battery backup systems and for large memory cards which require the low standby power and ease of use of SRAM. Certainly the requirement for very fast SRAMs in caches and in high performance systems will continue. This should be balanced by the faster DRAMs eliminating the need for second level SRAM cache memory in some systems, the pseudostatic DRAMs taking a share of the slow commodity SRAM market, and the more cost competitive flash EPROM technology competing in the slow end of the battery back-up market.

Nearly all static RAMs today are made in CMOS technology since there is no longer a speed difference with the older NMOS technology, the complexity of manufacturing and hence the cost of production is similar, and the low power feature of CMOS is attractive for reducing heat in high density systems. CMOS is also attractive for providing simple battery back-up in systems with critical data requirements and for reduction of battery size and increase in battery life in transportable systems.

Statics have historically been a generation behind dynamics in functional density, due to the greater number of transistors in a static RAM cell. For example, a 256k static RAM has about the same number of transistors as a 1Mb DRAM giving it approximately the same chip size in similar technology generations. This means that the yields in the production process are roughly equivalent. Therefore, one would expect to see a particular density SRAM and the four times density DRAM following similar product life cycles and price learning curves.

As shown in the price learning curves in Figure 8.2, this was approximately true for the 16k SRAM and 64k DRAM, and the 64k SRAM and the 256k DRAM. The 256k SRAM and 1Mb DRAM followed slightly different price curves possibly due to the various international trade agreements in the late 1980s which affected the DRAM price and production levels but not those of the SRAM directly. The price of any particular part falls over time as the production volume increases and it moves down the experience curve. Also, the price per bit decreases significantly from generation to generation. However, the end-of-life price per chip increases slightly with each generation due to the increase in chip size combined with the fairly constant price per square millimeter of memory silicon.

There has also been a tendency for the SRAMs to slip more than a generation behind the DRAMs in recent years. For example, the first 16Mb DRAM papers were presented in 1987, but the first 4Mb SRAM product paper did not appear until 1989.

The commodity SRAMs tend to be used in microprocessor based applications primarily for their ease of use compared to the DRAMs and also in portable equipment where the lower standby power consumption of the SRAMs is useful in battery backup and battery operated applications.

Both of these applications use byte-wide memories with speeds matched to those of the mainstream microprocessors in use at the time. The speed of these processors and the associated memories increases over time. At any given period of time, however, this application tends to fall in the slower range of SRAM speeds available from the current generation of MOS processing technology. The speed distinguishing the commodity and the fast RAMs hence changes over time.

Fast SRAMs are used for first-level cache in medium size computer systems, and second-level cache in large computer systems and have tended to be organized $\times 1$ or $\times 4$.

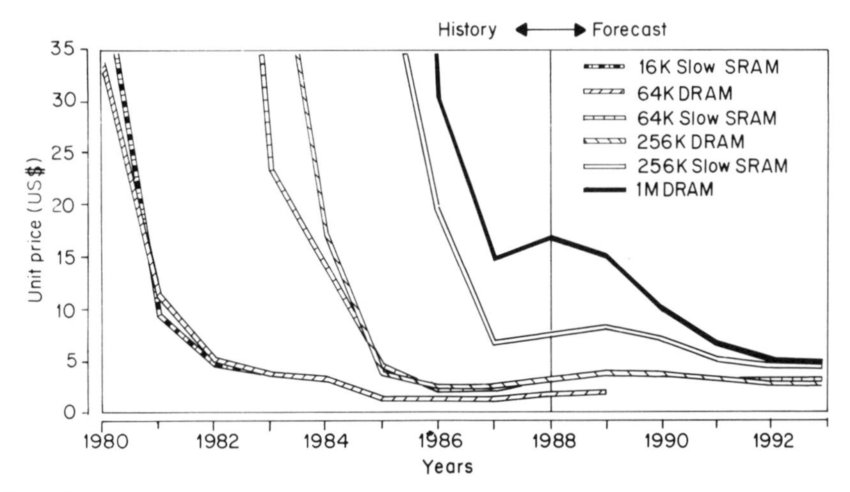

Figure 8.2
DRAM and commodity SRAM prices by year. (Source Dataquest 1989 [72].)

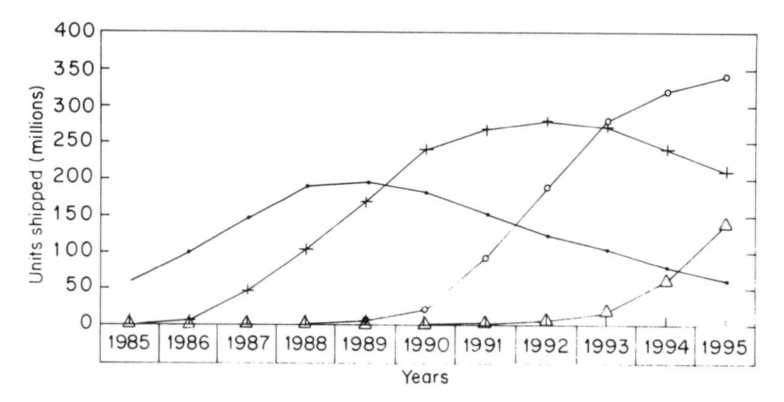

Figure 8.3
Life cycles of slow SRAMs by density (64 k ———, 256 k –+–+–,, 1 Mb —o—, 4 Mb –△–△–). (Source Dataquest 1990.)

A third category which is frequently referred to as 'very fast' represents the state-of-the-art in the current process technology and is used for first level cache in large computer systems and as main memory in supercomputers. These very fast SRAMs are usually organized $\times 1$ to reduce ground bounce effects. They frequently feature separate input–output pins and have various other options intended to speed up the part in the system, such as additional control pins.

Figure 8.3 illustrates product life cycles of three generations of byte-wide commodity SRAMs. These life cycles resemble those of the commodity DRAMs with one basic product type per generation.

Figure 8.4 illustrates the product life cycle of a typical fast SRAM, the fast 64k, both as a composite and broken out by speed. These fast SRAM life cycles are complex

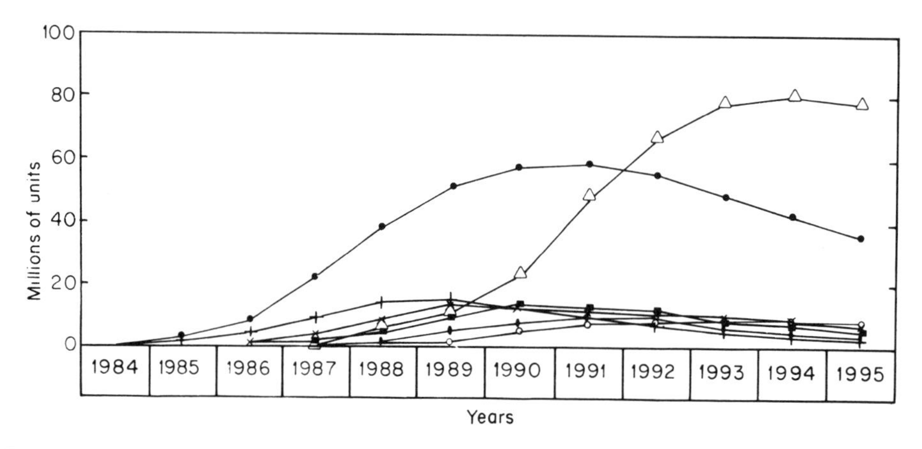

Figure 8.4
Fast 64 k SRAM lifecycle split by speed. (64 k total —•—, 45ns —+—, 35ns —*—, 25ns —■—, 20ns —◆—, 15ns —●—; 256 k Total —△—). (Source Dataquest 1990.)

since the definition of fast is upgraded throughout the life cycle as the state-of-the-art in technology permits. They consist, therefore, of composite life cycles of various speeds at a given density made in several different technologies and often designed into different systems.

In the case of the 'fast' 64k SRAM, shown in the figure, the 45 to 55 ns speeds were made in the 1.0 to 1.2 μm technology developed originally for the commodity 256k SRAM. The 15 ns to 25 ns speeds were introduced a year or so later in the 0.8 μm technology developed for the 1Mb SRAM and for the fast 35 to 45 ns 256k SRAM. The sub 10 ns speeds appeared several years later in the 0.6 μm technology developed for the 4M commodity SRAM.

While the memory industry has tried to pinpoint what is meant by 'fast' for the sake of historical traceability and of rational data collection, the result has not been totally accurate since the one category is continually passing into the other over time. If the progression of speeds over time are taken into account, then the fast SRAMs should probably be represented as a percentage of the total SRAM market, say 30%, all of which is increasing in speed over time.

8.2 A BRIEF HISTORY OF STATIC RAMs BY TECHNOLOGY

Early static RAMs developed along three separate technology paths: bipolar, NMOS and CMOS. These product lines had very different characteristics, were targeted at different markets, and, in many cases, were manufactured by different companies or different divisions of the same company without any particular overlap of design or production developments.

Bipolar SRAMs, although less than 1% of the total SRAM market today, still exist in lower densities for very fast applications. They will be discussed in more detail, along with GaAs SRAMs and other very fast RAMs, in Chapter 9.

Commodity SRAMs are represented by the 'Mix-MOS' technology which is a combination of CMOS and NMOS for high density commodity markets and by the full CMOS technology for the high density, very low power markets.

Fast high density SRAMs are represented by both CMOS and Mix-MOS technologies and by combinations of bipolar and CMOS, called BiCMOS, which provide fast bipolar compatible RAMs at densities not previously possible with only bipolar technology.

8.2.1 NMOS SRAMs convert to Mix-MOS (R-load CMOS) SRAMs

The RAM which was predominantly used in the late 1970s and early 1980s in microprocessor-based systems was the 4k NMOS static RAM, the 2114. This old industry standard part was originally developed by Intel. It was slow, low cost, easy to use and had a 4 bit wide architecture. A block diagram and bit map of this part was shown in Chapter 5.

This early NMOS workhorse of the industry was fully static with neither external nor internal clocks. It had non-multiplexed addresses and simple data access without the requirement of address set-up times. It was fully TTL compatible with common inputs and outputs. It had a chip select which connected and disconnected it from the system, but it could not be powered down. Address access times ranged from 150 to 450 ns and it used a 5 V power supply. Package power dissipation was normally specified at one watt.

The early 4 bit microprocessor-based systems, still used in the consumer and industrial markets, required very low cost from the memory but not much performance. Ease of use was a major criterion. Other major users of the 4k NMOS statics were less developed regions of the world and small companies where the engineering sophistication needed to use dynamics was lacking or where the volume was not sufficient to give the benefit of the lower per bit cost of dynamic RAM when the extra system costs were taken into account.

The early NMOS static RAM cells consisted of six transistors—four enhancement mode transistors and two depletion mode pull-up transistors. A significant improvement in both power dissipation and cost was achieved by substituting ion-implanted polysilicon load resistors for the two pull-up transistors. The power consumption was reduced over the depletion mode NMOS by the high resistivity of the polysilicon loads and by the increased accuracy of doping which was permitted by the newly developed ion-implantation process.

The successor of the low cost NMOS static RAMs in the commodity market in the early 1980s was the 'Mix-MOS' or 'resistor-load CMOS' SRAMs as they came to be called. These SRAMs consisted of an NMOS transistor matrix with high ohmic resistor loads and a CMOS periphery which gave the benefit of lower standby power consumption while retaining the smaller die size of the NMOS SRAM.

The transition from the R-load NMOS to the R-load CMOS SRAMs occurred also at the 4k density level, although the NMOS SRAMs persisted through the 16k density for a few very fast devices. A chart comparing the dc power characteristics of an NMOS 4k × 1 and an R-load CMOS 4k × 1 is shown in Figure 8.5. At the same 55 ns operating speed the power supply current of the NMOS was typically 120 mA and the CMOS was typically 15 mA. The power advantages of the Mix-MOS SRAM were significant.

The concept of a power down mode controlled by the chip enable pin ($\overline{E}$) appeared at this time and is shown in this illustration as the I_{SB} ($I_{StandBy}$) characteristic. I_{SB} is defined both with the inputs at TTL levels ($\overline{E} = V_{IH}$) and, for the CMOS part, with the inputs at the CMOS power supply rails ($\overline{E} = V_{CC} - 0.2$ V).

The next standard SRAM on the market after the NMOS 2114 was the Mix-MOS 16k SRAM. This 2k × 8 SRAM, which appeared in the early 1980s, was universally sourced by SRAM suppliers and lasted throughout the decade.

A block diagram of a standard 16k SRAM is shown in Figure 8.6(a). It retained the ease of use characteristics of the 4k with the only change visible to the user being the change of the chip select pin ($\overline{S}$) to a chip enable pin ($\overline{E}$), and the addition of an output control pin ($\overline{G}$). A typical read cycle timing diagram for the 16k SRAM indicating the use of these two control functions is shown in Figure 8.6(b).

The old chip select pin on the NMOS parts was used only as a switch to disconnect

Parameter	Symbol	MCM2147-55 Min	Typ*	Max	MCM2147-70 Min	Typ*	Max
put load current (all input pins, $V_{in} = 0$ to 5.5 V) (μA)	I_{IL}	—	0.01	10	—	0.01	10
ıtput leakage current (μA) $\bar{E} = 2.0$ V. $V_{out} = 0$ to 5.5 V)	I_{OL}	—	0.1	50	—	0.1	50
›wer supply current (mA) ($\bar{E} = V_{IL}$, outputs open, $T_A = 25°C$)	I_{CC1}	—	120	170	—	100	150
›wer supply current (mA) ($\bar{E} = V_{IL}$, outputs open, $T_A = 0°C$)	I_{CC2}	—	—	180	—	—	160
andby current (mA) ($\bar{E} = V_{IH}$)	I_{SB}	—	15	30	—	10	20
put low voltage (V)	V_{IL}	−0.3	—	0.8	−0.3	—	0.8
put high voltage (V)	V_{IH}	2.0	—	6.0	2.0	—	6.0
ıtput low voltage (V) ($I_{OL} = 8.8$ mA)	V_{OL}	—	—	0.4	—	—	0.4
ıtput high voltage (V) ($I_{OH} = -4.0$ mA)	V_{OH}	2.4	—	—	2.4	—	—

ypical values are for $T_A = 25°C$ and $V_{CC} = +5.0$ V.

(b)

Parameter	Symbol	MCM65L147-55 Min	Typ*	Max	MCM65L147-70 Min	Typ*	Max
put load current (μA) (all input pins, $V_{in} = 0$ to 5.5 V)	I_{IL}	—	0.01	1.0	—	0.01	1.0
utput leakage current (μA) ($\bar{E} = 2.0$ V, $V_{out} = 0$ to 5.5 V)	I_{OL}	—	0.1	1.0	—	0.1	1.0
ower supply current (mA) ($\bar{E} = V_{IL}$, output open)	I_{CC1}	—	15	35	—	15	35
tandby current ($\bar{E} = V_{IH}$) (mA)	I_{SB1}	—	5	12	—	5	12
tandby current ($\bar{E} = V_{CC} - 0.2$ V) (μA) ($0.2V \geqslant V_{in} \geqslant V_{CC} - 0.2$ V)	I_{SB2}	—	25	100	—	25	100
put low voltage (V)	V_{IL}	−0.3	—	0.8	−0.3	—	0.8
put high voltage (V)	V_{IH}	2.0	—	6.0	2.0	—	6.0
utput low voltage (V) ($I_{OL} = 12.0$ mA)	V_{OL}	—	—	0.4	—	—	0.4
utput high voltage** (V) ($I_{OH} = -8.0$ mA)	V_{OH}	2.4	—	—	2.4	—	—

Typical values for $T_A = 25°C$ and $V_{CC} = +5.0$ V

(a)

Figure 8.5
Dc power characteristics of an NMOS and a CMOS 4k × 1 SRAM in the same speed range. (a) CMOS 4k × 1 SRAM, (b) NMOS 4k × 1 SRAM. (Reproduced with permission from Motorola Memory Databook 1980 [68].)

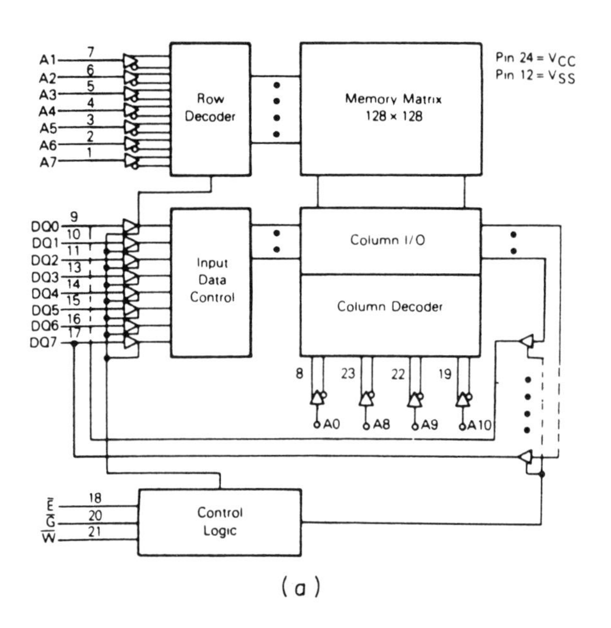

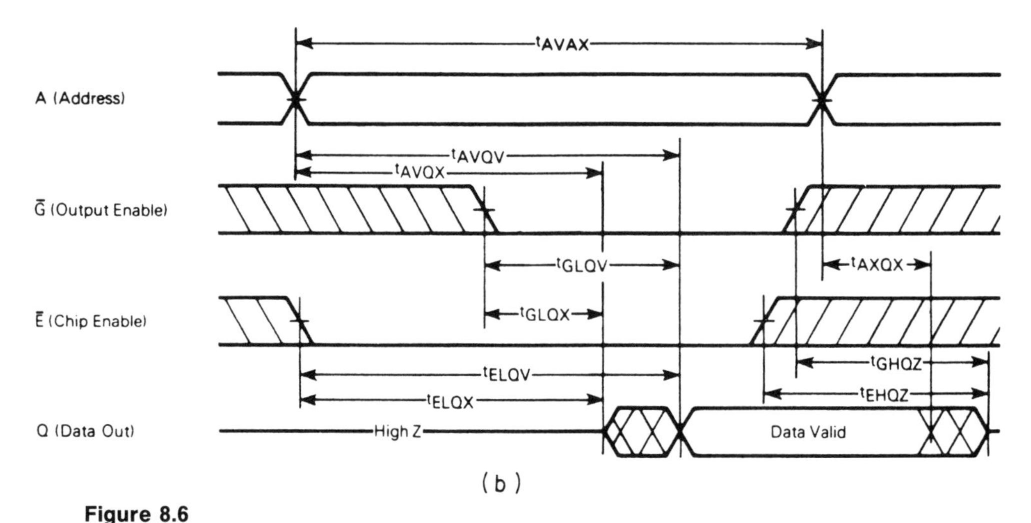

Figure 8.6
(a) Block diagram of standard 16k CMOS/Mix-MOS SRAM showing addition of chip enable ($\overline{E}$) and output enable ($\overline{G}$) control pins. (b) Typical read cycle timing waveforms using the new control features. (Reproduced with permission from Motorola Memory Databook 1982 [9].)

the RAM from the circuit. The chip enable pin, on the new CMOS SRAMs, functioned as a power-down pin permitting the use for the first time of a low voltage standby mode.

This concept of a low voltage standby mode controlled by the chip enable pin appeared first widely on the 16k Mix-MOS SRAM. It is illustrated in the timing diagram in Figure 8.7. In the standby mode, which is also called a retention mode,

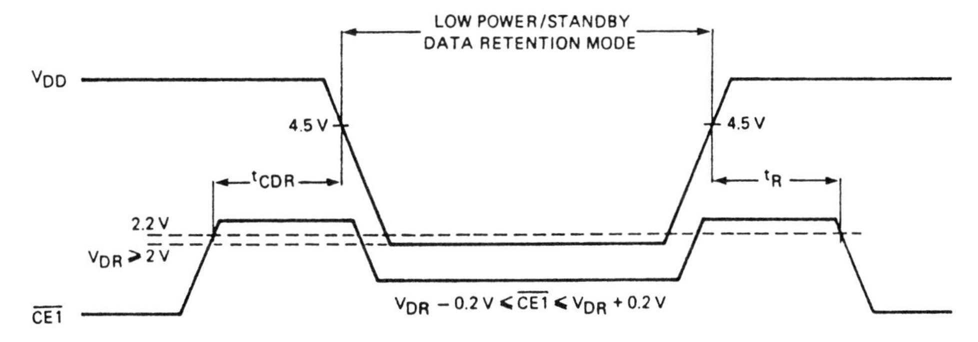

Figure 8.7
Typical timing waveform for low voltage data retention function on CMOS and Mix-MOS SRAMs. (Reproduced with permission from Philips Memory Databook 1989 [69].)

the power supply voltage can drop to a minimum of 2 V and the memory will still retain the data. Power consumption in this mode for Mix-MOS parts is typically in the microamp range and for full CMOS SRAMs in the nanoamp range.

The ability to drop to the microamp level of power in standby opened the potential for high density battery back-up applications using static SRAMs. Also, since the average operating power for a system is a sum of the power consumption of the memories that are operating plus the ones that are in standby at any given time, lowering the power consumption of the SRAMs in standby has the effect of lowering the average operating power of the system.

The lower standby power dissipation of the Mix-MOS compared to NMOS occurs since the peripheral parts of the circuit which consume the most power are in full CMOS. Dc operating conditions for a typical 64k bytewide Mix-MOS SRAM are given in Figure 8.8(a) along with some characteristic derating curves in Figure 8.8(b). As for any MOS RAM, the electrical characteristics vary across the voltage and temperature range. The worst case conditions may be either the high or low end of the range depending on the characteristic measured.

The Mix-MOS SRAMs were a logical spin-off from the CMOS technology since the polysilicon resistors merely replaced the PMOS transistors in the CMOS cell. There is, however, a power dissipation penalty for Mix-MOS over CMOS since the resistors always drain some current.

The Mix-MOS technique was able to combine the scaling advantages of NMOS and the lower power advantage of CMOS over NMOS. Thus many manufacturers used it in preference to either NMOS or CMOS to produce SRAMs which were fast, relatively low power, and with a small cost-effective chip size.

The commodity Mix-MOS SRAMs from the 16k to the 1Mb have been predominantly bytewide reflecting the transition of the small systems market from the 4 bit to the 8 bit processors. The width of the memories did not in general follow the move of the processors to the 16 and 32 bit bus widths due to the high power consumption and power line 'bounce' associated with switching a large number of outputs simultaneously. The JEDEC standard pinout for the bytewide family of SRAMs in the dual-in-line package (DIL or DIP) is shown in Figure 8.9.

Item	Symbol	Test condition	Min.	Typ.*	Max.
Input leakage current (μA)	I_{LI}	V_{IN} = GND to V_{CC}	−1	—	1
Output leakage current (μA)	I_{LO}	$V_{I/O}$ = GND to V_{CC} $\overline{CE}1 = V_{IH}$ or $CE2 = V_{IL}$ or $\overline{OE} = V_{IH}$ or $\overline{WE} = V_{IL}$	−1	—	1
Operating power supply current (mA)	I_{CC1}	$\overline{CE}1 = V_{IL}$ $CE2 = V_{IH}$ $V_{IN} = V_{IH}$ or V_{IL} $I_{OUT} = 0$ mA	—	30	60
Average operating current (mA)	I_{CC2}	Cycle = Min. Duty = 100% $I_{OUT} = 0$ mA	—	60	90
Standby current (μA)	I_{SB1}	$\overline{CE}1 \geqslant V_{CC} - 0.2$ V or $CE2 \leqslant 0.2$ V. $V_{IN} \geqslant V_{CC} - 0.2$ V or $V_{IN} \leqslant 0.2$ V	—	1	100
Standby current (mA)	I_{SB2}	$\overline{CE} = V_{IH}$ or $CE2 = V_{IL}$ $V_{IN} = V_{IL}$ or V_{IH}	—	10	25
Output high voltage (V)	V_{OH}	$I_{OH} = -4.0$ mA	2.4	—	—
Output low voltage (V)	V_{OL}	$I_{OL} = 8.0$ mA	—	—	0.4

(a)

* $V_{CC} = 5$ V, Ta = 25°C. $V_{CC} = 5$ V ± 10%. GND = 0 V. Ta = 0 to +70°C

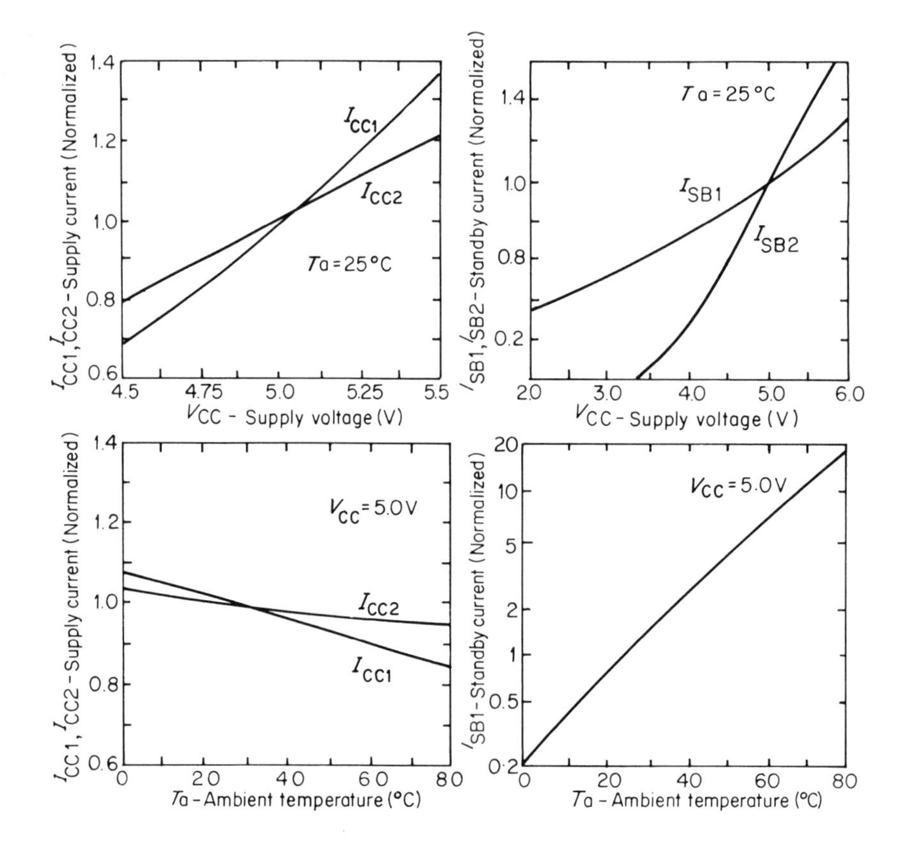

(b)

The fast MOS SRAMs developed from first the NMOS and then the Mix-MOS SRAMs in the early part of the 1980s. Both technologies had the speed and cost advantage of the optimized NMOS memory array technology derived from the DRAMs. Also the ability to scale NMOS to smaller geometries was well developed by DRAM manufacturers.

In the later half of the 1980s, the fast Mix-MOS SRAMs began to move into the speed range of the bipolar SRAMs. Since the speed of MOS circuits depends more directly on lateral dimensions than the speed of bipolar circuits, MOS technology benefited more from dimensional scaling than did the bipolar circuits. MOS was therefore able to reduce the speed gap with bipolar.

Eventually BiCMOS SRAMs formed from a combination of the Mix-MOS or CMOS technologies with bipolar transistors appeared in competition with the bipolar SRAMs in the submicron technologies and sub 10 ns speeds.

8.2.2 CMOS SRAMs

The third historic static RAM product was the low power, but slow six-transistor cell 'full CMOS' line. Small metal gate CMOS static RAMs have been in existence since early in the 1970s as part of standard CMOS logic portfolios. They were, however, slow with speeds in the 500 ns to 1 μs range, and small with sizes in the 64 to 256 bit range. Chips were large in area.

The early CMOS RAMs were used predominantly in military and other high reliability markets where their wide noise margins, wide voltage and temperature ranges, and low power consumption in battery back-up mode more than compensated for the expensive chips and complex manufacturing process.

They were developed mainly by CMOS logic manufacturers who designed them in their standard logic processes. These non-optimized processes produced very slow memories giving rise to the early reputation of full CMOS as a slow memory technology.

Modern mainstream CMOS static RAMs began with the development of the silicon gate technology whose two levels of interconnection (polysilicon and aluminum) permitted a larger, denser memory array. The 1024 bit '5101' was one of the first of the high volume silicon gate CMOS static RAMs. It became an industry standard. Many devices followed as CMOS overcame most of its traditional disadvantages without generally compromising its advantages.

These commodity CMOS SRAMs, however, were predominantly of the Mix-MOS (R-load) type since throughout the 1980s the full CMOS SRAMs were not preferred in higher density parts due to the higher cost of processing a larger cell. With the trend toward submicron geometries, the wide noise margins of full CMOS have again made it an attractive technology and a renewed interest in the technology has emerged.

Figure 8.8
(a) Dc operating conditions for a typical byte-wide 64k Mix-MOS SRAM, (b) characteristic derating curves. (Source Sony Memory Databook 1989 [70].)

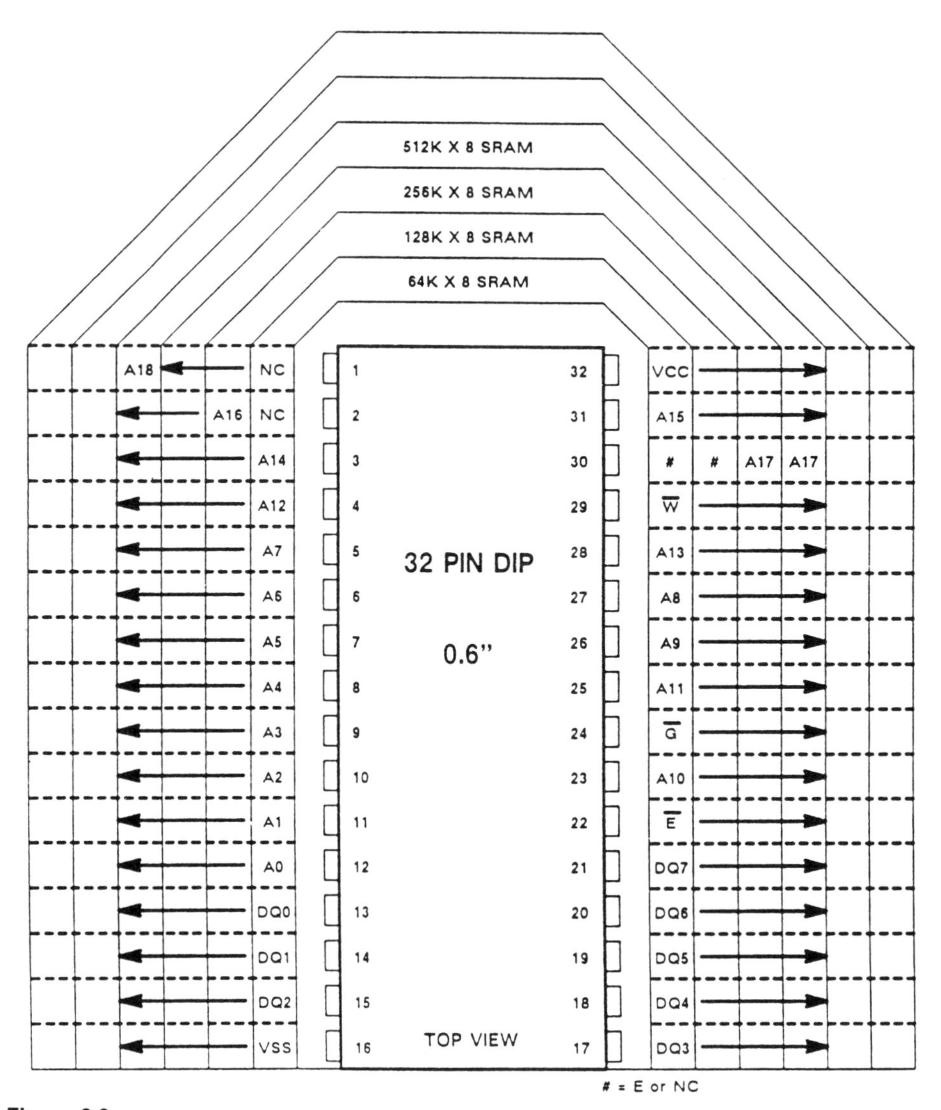

Figure 8.9

JEDEC standard pinout for bytewide family of CMOS SRAMs in dual-in-line package. (From JEDEC Standard 21C [71], with permission of EIA (1990).)

The dc power characteristics of a 64k full CMOS SRAM are shown in Figure 8.10(a), and the data retention characteristics are shown in Figure 8.10(b). While the operating power characteristics are similar to those of the 64k Mix-MOS part shown in the previous section, the standby and data retention characteristics of the full CMOS SRAM is at least an order of magnitude less than is typical for Mix-MOS SRAMs.

Today full CMOS SRAMs are as fast as NMOS or Mix-MOS SRAMs, have lower power in standby, and have the wide noise margins and insensitivity to noise

Parameter	Conditions	Symbol	Min.	Typ.	Max.
Input leakage current (μA)	$V_I = V_{SS}$ to V_{DD}	I_{LI}	−1.0	—	1.0
Output leakage current (μA)	$\overline{CE}$ or $\overline{OE} = V_{IH}$ or CE2 $= V_{IL}$; $V_{I/O} = V_{SS}$ to V_{DD}	I_{LO}	−1.0	—	1.0
Standby current (mA)	CE2 $\leqslant V_{IL}$ or $\overline{CE1} \geqslant V_{IH}$	I_{SB}	—	1.5	3.0
FCB61C65L only (μA)	all V_I = CMOS*	I_{SBL}	—	2	100
FCB61C65LL only (μA)	all V_I = CMOS*	I_{SBLL}	—	0.05	1
DC read current (mA)	$\overline{WE} = V_{IH}$; I/O = 0 mA	I_{DD1}	—	3	6
FCB61C65L only (μA)	all V_I = CMOS*	I_{DDL}	—	2	100
FCB61C65LL only (μA)	all V_I = CMOS*	I_{DDLL}	—	0.05	1
Average operating current (mA)	minimum cycle time; $I_{I/O} = 0$ mA	I_{DD}	—	40	70
Output voltage LOW (V)	$I_{OL} - 4$ mA	V_{OL}	—	—	0.4
Output voltage LOW (V)	$I_{OL} = 20$ μA	V_{OL}	—	—	0.2
Output voltage HIGH (V)	$I_{OH} = -1$ mA	V_{OH}	2.4	—	—
Output voltage HIGH (V)	$I_{OH} = -20$ μA	V_{OH}	$V_{DD} - 0.2$	—	—

* CMOS = CMOSH: $V_{DD} - 0.2 \leqslant$ level $\leqslant V_{DD} + 0.2$ or CMOSL: $-0.2 \leqslant$ level $\leqslant 0.2$ V.
$V_{DD} = 5$ V $\pm$ 10%; $T_{amb} = 0$ to 70°C. (Typical readings taken at $V_{DD} = 5$ V; $T_{amb} = 25$°C). All voltages are with references to V_{SS} (0 V) unless otherwise specified; standby current measurements are valid after thermal equilibrium has been established.

(a)

Parameter	Conditions	Symbol	Min.	Typ.	Max.
Supply					
Supply voltage for data retention (1) (V)	CE2 $\leqslant 0.2$ V or $\overline{CE1} \geqslant V_{DR} - 0.2$ V	V_{DR}	2.0	—	5.5
Supply current (1) during data retention	$V_{DR} = 3$ V; $\overline{CE1} \geqslant V_{DR} - 0.2$ V and CE2 $\geqslant V_{DR} - 0.2$ V or CE2 $\leqslant 0.2$ V				
FCB61C65L only (μA)		I_{DRL}	—	2	50
FCB61C65LL only (μA)		I_{DRLL}	—	0.05	1
Chip select to data retention time (ns)		t_{CDR}	0	—	—
Recovery time to fully active (ns)		t_R	t_{RC*}	—	—

* Read cycle time.
(FCB61c65L/LL only). $T_{amb} = 0$ to 70°C; $I_{DR/LL}$ measurements are valid after thermal equilibrium has been established.

(b)

Figure 8.10
Typical 64k byte-wide full CMOS SRAM. (a) Dc characteristics. (b) Data retention characteristics for low voltage standby mode. (Reproduced with permission from Philips Memory Databook [61] 1989.)

characteristic of CMOS technology. They remained, however, until the end of the 1980s, larger and more expensive due to the space needed for the six transistors in the cell.

8.3 ADVANTAGES OF CMOS SRAMs

By the middle of the 1980s the vast majority of all SRAMs were made from CMOS technology. Several basic types of CMOS process technologies were available. Either CMOS or Mix-MOS can be formed of N-channel devices in P-type wells, P-channel devices in N-type wells, or both devices in 'twin-wells' of the opposite polarity. The three options are described more fully in Chapter 4.

The traditional advantages of CMOS technology include low static power dissipation, superior noise immunity to either bipolar or NMOS, wide operating temperature range, sharp transfer characteristics, wide voltage supply tolerance, improved reliability without cooling fans, ease of computer modelling, versatility in combining CMOS with bipolar and analog logic on the same chip, and low susceptibility to alpha particle induced soft errors for transistors located in the wells.

In addition, the complementary CMOS transistor pairs permit circuit techniques, like using the current mirror referred to the positive supply to improve speed, which are missing in N-channel technology.

8.3.1 Low static power dissipation

CMOS is inherently lower power than NMOS as dc conducting paths between power and ground do not arise. In standard CMOS, the P- and N-channel devices in the periphery are in series and only on at the same time during switching. Current is, therefore, only drawn during switching. This makes SRAMs with CMOS peripheries low power in standby, when there are only surface, junction, and channel leakage currents.

CMOS SRAMs with R-load cells in the array tend to have higher standby power dissipation than a six-transistor CMOS memory array since there is always some current flowing in the resistor.

The power consumption in the active state is proportional to the clock rate of the circuit, since operating current flows mainly when the inputs are being switched. The low impedance path to power supply current occurs as the input voltage crosses through the switching threshold. The variation of active power consumption with frequency means CMOS has an automatic-power-down feature available. As frequency decreases, power dissipation also declines. Power is essentially off in a static state, or when the device is deselected. Also power down capability can be easily programmed, for example, when the pulsed word-line technique, which will be discussed in a following section, is used.

Since the trend is to pack an increasing number of faster circuits onto smaller size chips, the heat generated by the transistors becomes a serious problem. The low standby power consumption of CMOS is one of the answers to this problem and has certainly been a factor in turning the market toward CMOS usage.

The move to low cost plastic packaging during the mid 1980s also accelerated the trend to CMOS. Plastic packages do not dissipate heat as rapidly as the more expensive ceramic packages and therefore tend to heat up faster impairing reliability.

The low power consumption in standby, means that CMOS statics can retain their data for long periods of time in a battery back-up or data retention mode as discussed in the previous section. Even at a high temperature of 70°C the standby current is limited to 50 to 100 μA for R-load CMOS and about 1 μA for full CMOS.

8.3.2 High noise immunity

CMOS devices have a high noise margin (typically 45% of V_{CC}) because logic states are very close to the supply voltage rails, whereas Mix-MOS tends to have smaller noise margins. The good noise immunity combined with the low power consumption serves to minimize interaction or cross-talk between the various circuit elements when they are connected to a common supply bus.

A comparison of the static noise margin of full CMOS and R-load CMOS cells was done in a paper by Philips [5]. The static noise margin of the SRAM cells is affected the most strongly when cell nodes, which are written high, are read. During the read operation the voltage of the word-line is taken to V_{DD}, switching on the access transistors, which drops the high node voltage of the cell to V_{DD} minus a threshold. Since the passive resistors of the R-load cell have to maintain very high values in the 10 to 100 GΩ region [6] in order to keep the standby power dissipation as low as possible, their recovery is slower so they are less able to maintain stable high levels. The p-channel active loads of the full CMOS cell have low impedance when switched on, so they tend to maintain proper high levels in the cell, and the latch function of the flipflop is maintained even at low V_{DD}.

Figure 8.11(a) shows the static noise margin as a function of bit-line voltage V_{BL} of both types of cells with $V_{BL} = V_{WL} = V_{DD} = 5$ V. The 'write' region of the figure is where the static noise margin is zero and the 'read' region is where it is greater than zero. The cells are written in this case when V_{BL} is less than 2 V.

The R-load cell has smaller static noise margin values at high V_{BL} even though it has a larger cell ratio in this case. The cell ratio is the width of the driver transistor divided by the width of the access transistor.

8.3.3 Low voltage operation

Figure 8.11(b) shows the results of the Philips study on static noise margin plotted against supply voltage. The six-transistor cell is stable down to about 2 V whereas the R-load cell loses noise margin fairly rapidly as the power supply voltage decreases.

Since the high level of the cell will degrade to V_{DD} minus the threshold voltage of the access transistor if the cell is read, the six-transistor cell is much better for low

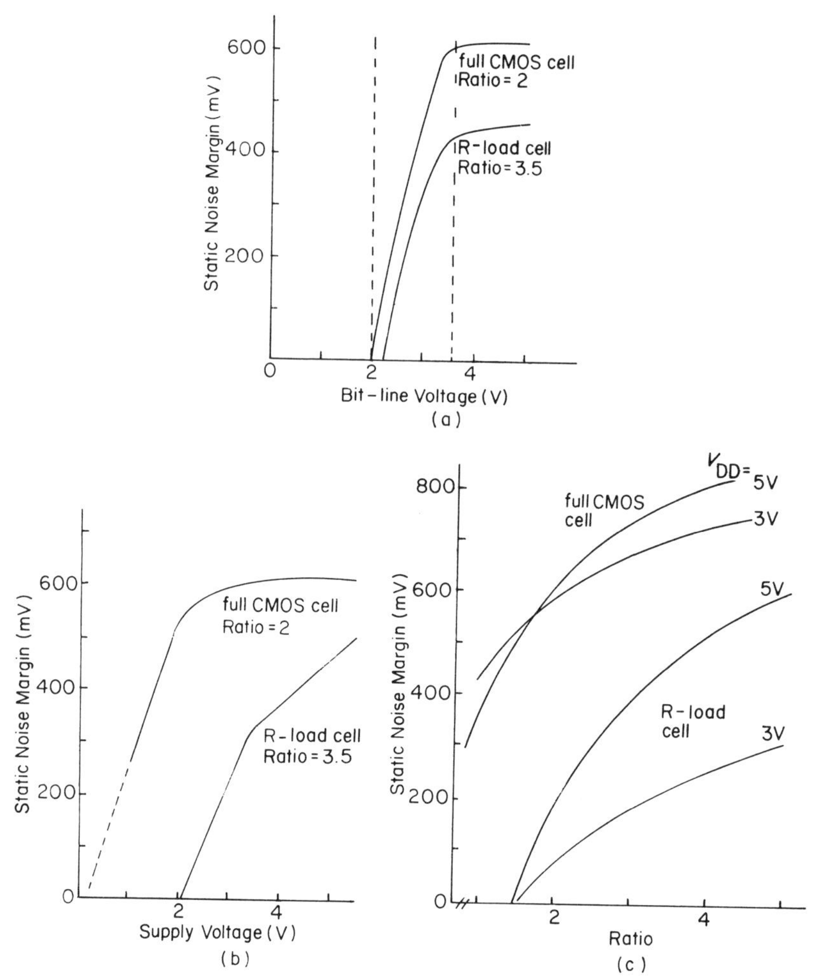

Figure 8.11
Static RAM noise margin for full CMOS and R-load CMOS SRAM cells; (a) as a function of bit-line voltage, (b) as a function of power supply voltage, (c) as a function of cell ratio for V_{DD} at 3 and 5 V. (From Seevinck *et al.* [5], Philips 1987, with permission of IEEE.)

voltage applications, where a threshold is a larger percentage of the power supply voltage, and the recovery is faster.

8.3.4 Cell size in low voltage applications

The size of a full CMOS cell as opposed to an R-load cell is another factor influenced by noise margin and power supply voltage. Figure 8.11(c) shows the static noise margin plotted against the cell ratio for V_{DD} at 3 V and 5 V where cell ratio is the ratio between the NMOS storage transistors in the cell and the NMOS access transistors.

A smaller cell ratio minimizes the cell area. It can be seen that to obtain the same static noise margin, the ratio of the full CMOS cell can be less than half that of the R-load cell at 5 V and less than 20% the size at 3 V. This tends to reduce the cell area advantage of R-load cells compared to full CMOS cells in low voltage applications.

8.3.5 Superior temperature characteristics and reliability

CMOS devices are able to operate at ambient temperatures which are higher than the NMOS or Mix-MOS operating range. This is because CMOS generally dissipates less heat than MNOS and, therefore, junction temperatures tend to be closer to ambient temperatures than equivalent NMOS parts. Historically standard operating temperature range for CMOS logic has been specified from −40 to +85°C. Since NMOS memories were, however, normally specified only from 0 to 70°C as a standard operating temperature range, this range has also been adopted as the industry has moved to CMOS SRAMs.

Since the long term reliability of the MOS memories decreases as the temperature of operation increases, CMOS failure rates may be ten to one hundred times lower than NMOS failure rates.

8.3.6 Resistance to alpha particles

CMOS is naturally immune to alpha particle sensitivity due to the shallow well. Any electron-hole pairs formed by alpha particles striking the chip will be primarily formed in the substrate below the well. The well forms an electrical barrier to the electrons (or holes in an N-well technology) and prevents them reaching the gate. Any carriers generated in the shallow well recombine quickly or are swept away by the built-in field of the well.

8.4 DISADVANTAGES OF CMOS

The major disadvantages of CMOS historically were high cost, slow speeds, large chip size, the added complexity of the CMOS manufacturing process, and the tendency to exhibit the phenomena of 'latch-up'.

The added complexity of the CMOS manufacturing process compared to NMOS historically resulted in lower yield and higher cost due to the fact that the CMOS silicon gate memory process was a 10–12 mask process compared with the 6 to 7 mask NMOS memory process. By the early 1980s the fully tweaked NMOS process used approximately the same number of masking steps as a CMOS silicon gate process eliminating the only major benefit of the older NMOS processing.

The high cost of full CMOS compared to the Mix-MOS technology was a result of the larger six-transistor cell. However, the relative cell size disadvantage of the full CMOS cell compared to the R-load cell has been reduced in the submicron geometries by the noise margin problems involved in scaling the R-load cell to these tight geometries.

In addition, as the industry moves to the 3.3 V power supply the size advantages of the six-transistor cell as described in the previous section also minimize the disadvantage. Finally the trend to the transistor load cell stacked in the poly over the remaining four transistors should basically eliminate any reason to maintain the R-load cell in high density or fast SRAM devices.

The higher cost has also been due to the larger size needed for the PMOS and NMOS gate combinations and the extra well spacings and guardbands to prevent latch-up. Latch-up results from a parasitic bipolar action which causes the device to draw large currents and destroy itself. It occurs in the thyristor type PNPN layers as described in Chapter 5. When this thyristor latches, it becomes self-maintaining requiring that the power supply be disconnected to stop it.

There are various process modifications which can decrease the tendency to latch-up conditions. The impurity doping in the well can be increased in order to decrease minority carrier lifetime in the parasitic base regions. This leads, however, to higher capacitance and slower speed. Guard rings can be used on the wells but they increase the size of the chip as they are space consuming. Another method of eliminating latch-up is to tie the wells and the substrate to the power supplies at every power or ground connection of a transistor. More recently trench isolation and retrograde wells, which have higher doping deeper in the well, have been used.

Historically all of these methods have been used to reduce susceptibility to latch-up.

8.5 EVOLUTION OF HIGH DENSITY COMMODITY STATIC RAMs

The commodity SRAMs are characterized by their low price, high density and ease of use. They are neither very fast nor very low power. They are, however, adequate for a vast range of small microprocessor controlled systems. They have migrated from the NMOS to the Mix-MOS technology and, to some small extent, to the pseudostatic DRAM technology in applications where low standby power is not a criteria. They have migrated to the full CMOS SRAMs in those applications where low standby power is a criterion.

8.5.1 Typical 256k byte-wide SRAM

A block diagram of a typical commodity 256k static RAM [69] is shown in Figure 8.12. The part was organized 32k × 8, and has common input and output pins. The addresses were not multiplexed. The cycle time was the same as the access time. A single five volt power supply was needed. Pinout was the JEDEC standard for slow byte-wide SRAMs which was shown in Figure 8.9. The second chip enable pin which appeared at the 64k density was gone. The number of pins available in the standard

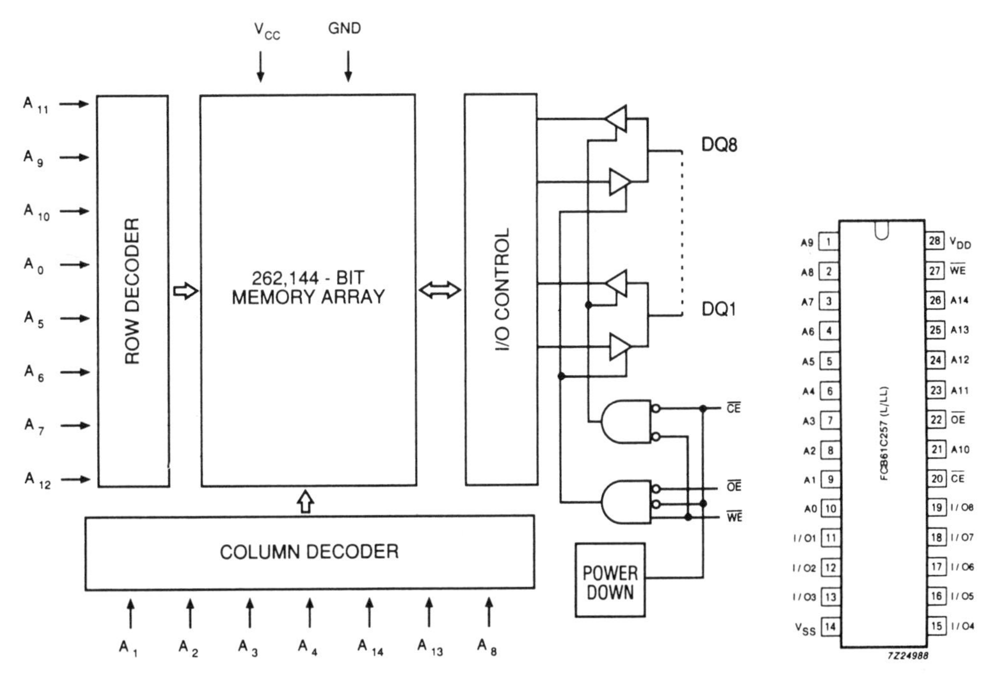

Figure 8.12
Block diagram and pinout of a typical commodity 256k SRAM. (Source Philips MOS Memory Databook. Reproduced with permission from Philips [69].)

package were limited and the benefit of the extra control was not enough to justify the larger package.

The width of 8 bits means that rather than having 2^{18} addresses with each address corresponding to a single cell, there are $2^{18}/8$, each corresponding to eight cells in which 8 bits can be read or written at the same time 'in parallel'.

This SRAM was fully TTL compatible with logic '1' input voltage specified between 2.0 V to 6.0 V and logic '0' input voltage specified between -0.5 to $+0.8$ V.

The particular part shown was made in a 1.2 μm minimum geometry full CMOS process and had address access times ranging from 45 to 70 ns. Both the geometry and the speed were typical for the commodity generation of 256k SRAMs, although the majority were in the Mix-MOS process. The difference affected primarily the dc power and data retention specifications which are similar to those shown earlier for the 64k full CMOS SRAM.

The read and write cycle timing diagrams and tables of ac characteristics are shown in Figure 8.13 where $\overline{CS}$ is the chip select, $\overline{WE}$ is the write enable and $\overline{G}$ is the output enable.

The background of some of the technological and architectural developments in the evolution of the modern commodity SRAM are described in the following sections.

A.C. PARAMETERS (Times in nano seconds)	Symb name	− 70		−100		−120	
		Min	Max	Min	Max	Min	Max
READ CYCLE: * SEE NOTE 6							
Read Cycle Time	t RC	70	–	100	–	120	–
Address Access Time	t AA	–	70	–	100	–	120
Chip Enable Access Time	tACE	–	70	–	100	–	120
Output Enable Access Time	t OE	–	35	–	45	–	55
Output Enable to Output Lo−Z	tOLZ	5	–	5	–	5	–
Chip Enable to Output Lo−Z	tCLZ	5	–	10	–	10	–
Output Disable to Output Hi−Z	tOHZ	–	35	–	40	–	45
Chip Disable to Output Hi−Z	tCHZ	–	35	–	40	–	45
Output Hold Time	t OH	5	–	10	–	10	–
WRITE CYCLE:							
Write Cycle Time	t WC	70	–	100	–	120	–
Chip Enable to end of Write	t CW	55	–	70	–	85	–
Address Valid to end of Write	t AW	60	–	80	–	100	–
Address Set−Up Time	t AS	15	–	15	–	15	–
Write Pulse Width	t WP	40	–	60	–	80	–
Write Recovery Time	t WR	0	–	0	–	0	–
Write Enable to Output Hi−Z	tWHZ	–	35	–	40	–	45
Data−to−Write Time Overlap	t DW	30	–	35	–	40	–
Data Hold from Write Time	t DH	5	–	5	–	5	–
DQ Low−Z from end of Write	t OW	0	–	5	–	10	–

Figure 8.13
256k SRAM timing diagrams and ac characteristics for typical read and write cycles. (From Motorola MOS Memory Databook [68] 1989.)

8.6 TECHNOLOGY EVOLUTION OF HIGH DENSITY SRAMs

8.6.1 Scaling reduces chip size and improves performance

Scaling the technology reduces chip size and permits higher density SRAMs to be made more cost effectively. It also gives a significant improvement in performance. These scaling techniques, as described in Chapter 4, have been transferred directly from NMOS to CMOS technologies and provide similar chip size and speed reduction advantages in both.

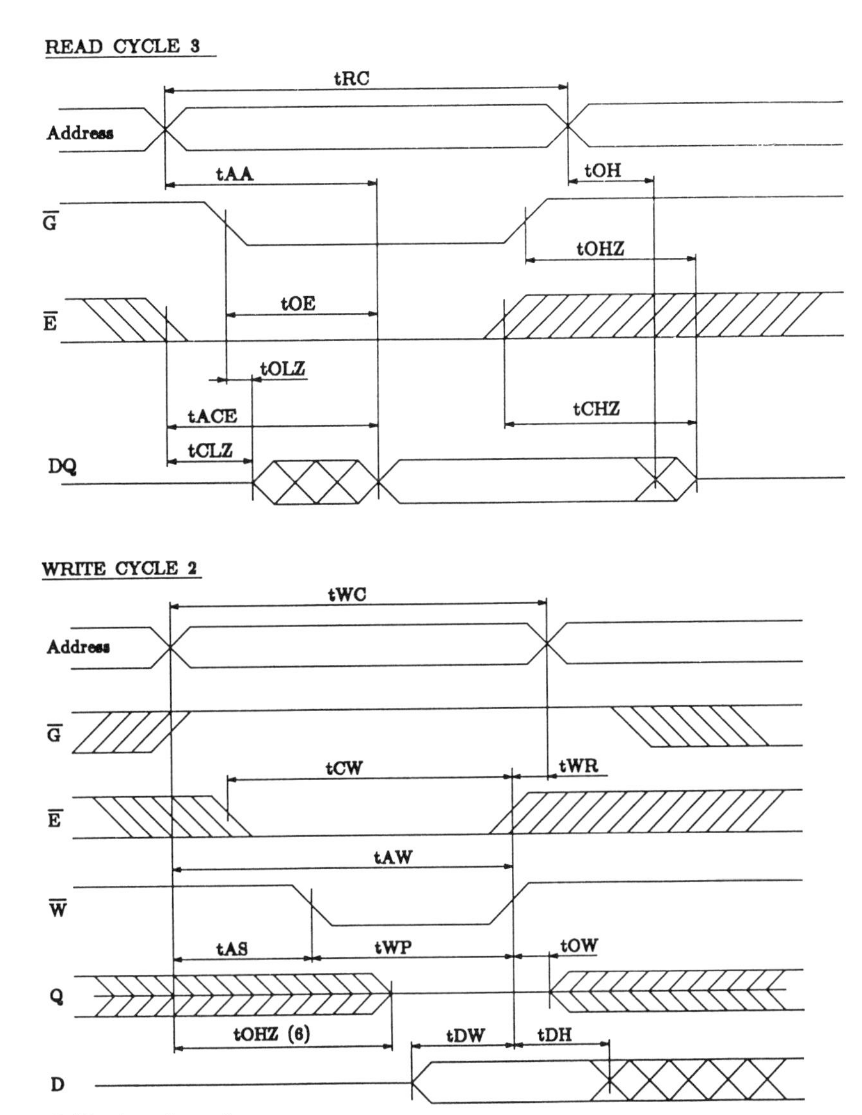

Figure 8.13 (*continued*)

8.6.2 Multiple interconnect layers

Historically, the advancement in the late 1970s from metal gate to polysilicon gate technology permitted cost reduction by providing an extra level of interconnection, with an associated reduction in cell size without going to smaller geometries on the chip.

In a double poly process, two alternatives present themselves. Both levels can be used for gates, or one level can be used for gates and the other for interconnection and load resistors. This provides the opportunity to optimize the ion doping for the particular function.

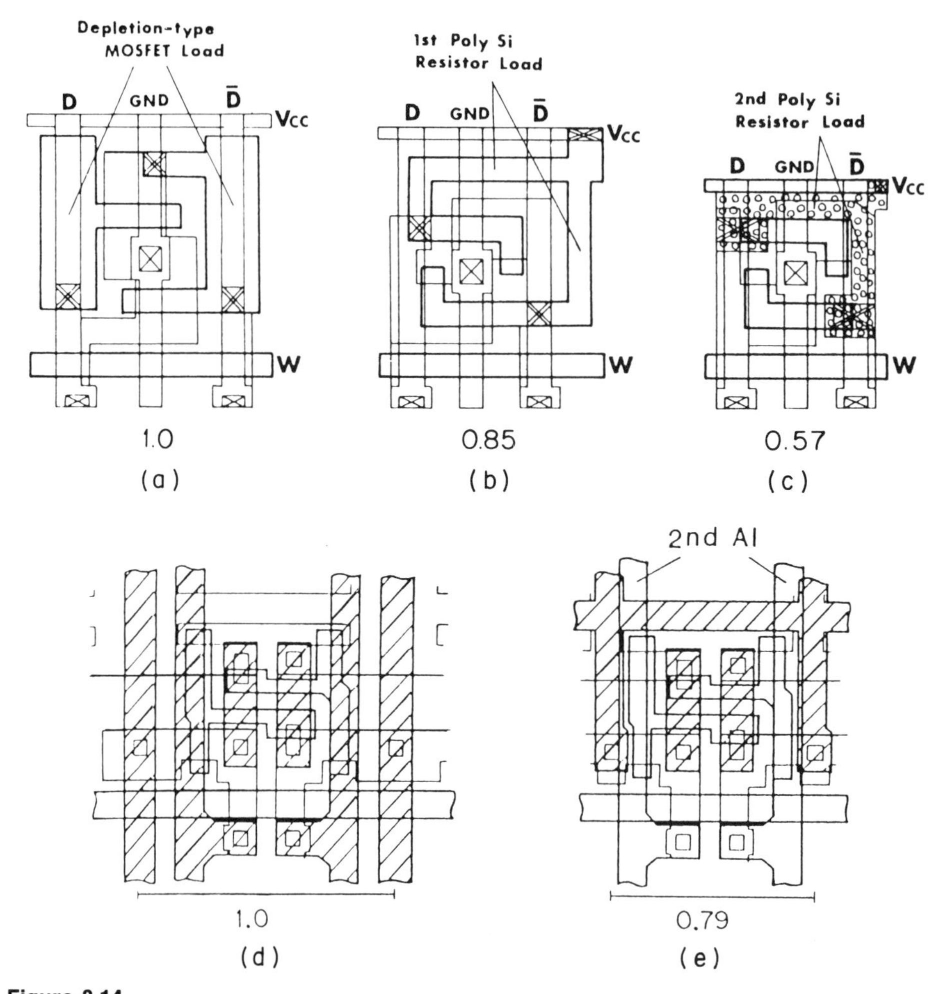

Figure 8.14
Comparison of layout of load devices in SRAM memory cells: (a) depletion mode MOSFET load. (b) single polysilicon memory cell. (c) double polysilicon memory cell with stacked load resistors. (From Ohzone *et al.* [6], Matsushita 1980, with permission of IEEE.) Comparison of two types of full CMOS memory cells showing 20% difference in cell size: (d) Single aluminum interconnect. (e) Double aluminum interconnect. (From Watanabe [25], NEC 1983, with permission of IEEE.)

Initially the load resistors were fabricated in single-level polysilicon and laid adjacent to the four transistors. The addition of a second level of polysilicon, however, permitted these load resistors to be stacked on top of the other four transistors in the cell, thereby reducing the cell size, and hence the chip size, and thus leading to a lower cost device. Some companies have even gone to three layers of poly in order to have a layer unique to the resistor for 1Mb and larger SRAMs.

Figures 8.14(a) to (c) shows a comparison of the cell size reduction for single-level polysilicon in scaling from 3 μm CMOS to 2 μm CMOS and then the additional size reduction in adding the second level of polysilicon for the stacked load resistor. This example is from a Mix-MOS SRAM described by Matshushita [6] in 1980 which combined P-well CMOS peripheral circuits and single-level polysilicon load resistor memory cells.

Double metal processing was more difficult for the manufacturers to master than double poly, but was used in higher density SRAMs both for chip size reduction, and from the 1Mb SRAM on was also used widely as global word and bit-lines to provide the required speed in high density SRAMs by shorting together blocks of memory. It is also compatible with standard logic processing which uses multiple metal interconnect layers. By the second generation 1Mb SRAM, all manufacturers had gone to double-layer metal and some were considering three layers.

A reduction in cell size for full CMOS SRAMs of about 20% was obtained by NEC [25] by going from single-level metal to double-level metal as shown in Figure 8.14(d) to (e) which compares the relative sizes of the layouts of the two cells.

More recently new types of cell structures have appeared using six transistor cells with stacked thin-film polysilicon load transistors. These polysilicon transistor cells have permitted six-transistor cell CMOS SRAMs with chip size almost as small as the polysilicon R-load cell SRAMs. This cell technology and others will be explored more thoroughly in Chapter 9.

The effect of technology scaling on higher density parts is not so easy to envision as it was on the 4k SRAM scaling, described in Chapter 4, because of changes in the number of polysilicon and aluminum layers as well as the geometry reductions.

Figure 8.15 indicates the typical progression in scaling the technology from the 16k CMOS SRAM to the 4Mb CMOS SRAM. This includes both scaling of the minimum geometry from 3 to 0.6 μm, and going from single polysilicon and single aluminum in the 16k SRAM to double polysilicon for the 64k and 256k, then up to five or six interconnect layers at the 4Mb density.

The table in Figure 8.16 shows the technology used in the first generation of 1Mb SRAMs presented in 1987 and 1988.

All of the 1Mb SRAMs used at least four layers of interconnects (if the polycide strap used by Philips is included as a separate interconnect layer). Mitsubishi and Fujitsu even used five layers in their faster 1988 parts. All added the second level of metal in the course of working out the faster first generation parts. Some faster parts also thinned the gate oxide and reduced the gate length, but otherwise kept the same basic technology to achieve the required speed. In spite of the addition of a second level of metal in the faster parts, the array efficiency stayed basically the same or decreased indicating the design trade-offs which were made to obtain the speed.

All of the Mix-MOS 1Mb SRAMs used the multiple polysilicon layers required for the load resistor with many going to three poly layers to improve the quality of the resistor by devoting a special interconnect layer to it. Mitsubishi [46, 53], Hitachi [50] and Fujitsu [54] used the third polysilicon for a high resistive polysilicon load resistor to increase the cell node capacitance. For Mitsubishi the use of a third layer avoided effects of side diffusion that might occur with only two poly layers since the second poly layer also received a diffusion of impurities for other functions.

Hitachi [50] made both layers out of polycide and used the first for the gate and

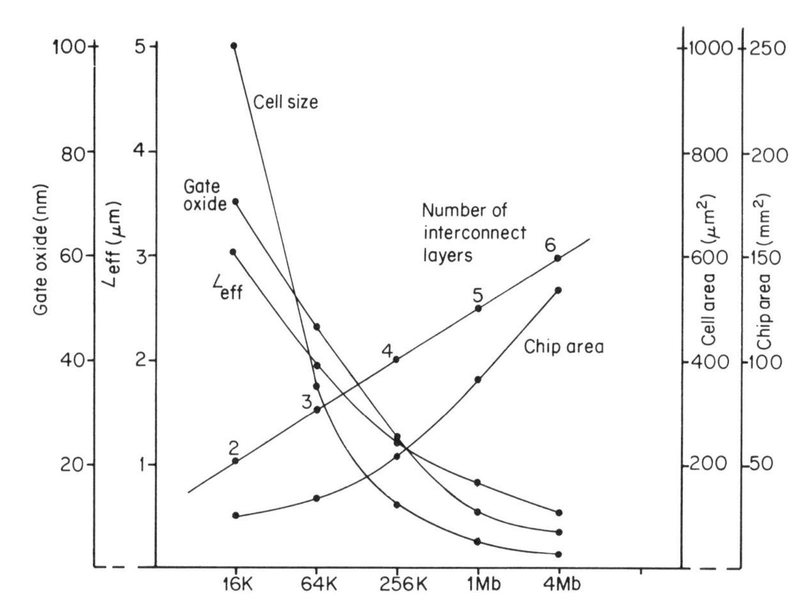

Figure 8.15
Typical CMOS SRAM technology progression for scaling of internal dimensions (L_{eff}) and vertical dimensions (gate oxide) plus increase in number of interconnect layers. (Source various ISSCC papers.)

	TACS+ (ns)	Levels		N-gate length (μm)	Oxide depth (nm)	Cell size (μm²)	Array efficiency (%)
Sony 1987	35	2P	2M	1.0	20	74	70
Toshiba 1987	25	2P	2M	0.6	16	53	50
Philips 1987	25	1P*	2M	0.7	17.5	60	65
Mitsubishi 1987	34	3P	1M	0.8	18	44	55
Mitsubishi 1989	14	3P	2M	0.7	18	42	49
Hitachi 1987	42	3P	1M	0.8		45	58
Hitachi 1988	15	2P	2M	0.8	17.5	44	48
Fujitsu 1987	41	3P	1M	0.8	20	41	43
Fujitsu 1988	18	3P	2M	0.7	17	41	46

* plus a polycide strap

Figure 8.16
First generation 1 Mb SRAM technology. (Source ISSCC Proceedings 1987, 1988.)

used the second in a double word-line technique to reduce current through the cell. Toshiba [48] used the second poly for the resistors and used selectively deposited epitaxially grown silicon for contact filling.

Philips [10] used a single poly technology since their six-transistor full CMOS cell did not have the poly load resistor which required multiple poly layers in Mix-MOS devices. All of the other first generation 1Mb SRAMs used the Mix-MOS cell with the poly load resistor.

In their second faster 1Mb, Fujitsu thinned the gate oxide and reduced the gate length of the transistors, both steps expected to increase the speed of the memory. In addition they added the second level of metal that is found in all the fast first generation 1Mb SRAMs.

Mitsubishi [53] on their fast part added the second aluminum layer for use in the array for the row group select lines which cross the first-level aluminum bit-lines every four memory cells and for use in the periphery as signal lines, power supply and ground.

There was some indication that the use of double-metal technology was also intended to improve implementation of logic on the SRAMs. For example, Philips indicated that the double-level metal technology was compatible with the requirements of their submicron logic process and they indicated an intent to use the same process for embedding SRAM in large VLSI chips. Sony [47] also indicated that they used double-level aluminum technology to get denser logic circuitry in the periphery increasing the array ratio to over 70%.

.7 DESIGN AND ARCHITECTURE OF HIGH DENSITY SRAMs

.7.1 Stability requirements of the R-load cell

A requirement for dc stability of the unaccessed R-load cell is that the load resistor connected to the high node must be the most conductive element connected to that node, that is, it must supply current to maintain the high node near V_{CC} in spite of leakages toward ground or toward the substrate.

These leakages include the subthreshold conduction of the 'off' storage transistor, the subthreshold conduction of the 'off' access transistor when the data line is low but the cell is unaccessed, and the leakage to substrate of the P–N diode created by the common drain diffusion region between these two transistors.

There is a trade-off, however, between the high resistance required for low power and the requirement to provide sufficient current to balance the leakages of the storage and access transistors to maintain the high node in the cell. It is necessary, therefore, to optimize the technology. Low standby power in Mix-MOS means high value resistors with correspondingly low static current.

Wide noise margins mean either a lower value resistor to maintain the high node or a low leakage storage transistor. Low leakage transistors, however, require high thresholds which impair speed. The trade-off, therefore, with the R-load cell is between low power and high speed.

8.7.2 Protection from hot carrier effects

Several design measures can be taken to prevent hot carrier degradation due to the submicron geometry channel length.

One measure used almost universally by the 1Mb SRAM level was the lightly doped drain structure described in Chapter 4.

Another was the use of cascode transistors as described in Chapter 6. For example, cascode transistors were used in the Philips 128k × 8 SRAM to protect the NMOS branches in all peripheral circuits from the full supply voltage.

They were also used on the bit-lines to protect the access transistors from hot carrier effects by preventing them from switching on when a high drain–source voltage was present. This allowed precharging the bit-lines to a threshold below V_{DD} in read mode thereby preventing the n-channel access transistors from being stressed when the low level in the cell and the high level on the bit-line were switched on.

In write mode, the cells are first selected before one of the bit-lines is pulled down so that a high level in the cell will be discharged without a large drain–source voltage difference across the access transistor. The driver transistors of the cell are not stressed because of the low trip point level of the cross-coupled cell invertors.

8.7.3 Substrate back-bias and grounded P-wells

In early NMOS devices a back-bias was applied to the substrate to adjust the threshold voltages and also to reduce the junction capacitance thereby increasing speed. This back-bias voltage was provided externally in the older RAMS requiring an extra external power supply.

Starting at about the 4k level a small charge pump was included on most NMOS statics to provide dc stability when the device was deselected or powered down without the necessity of a separate external power supply. Later, in R-load SRAMs, on-chip back-bias generators were also used to prevent current injection from peripheral circuits into the high-ohmic cells. This current resulted from negative undershoots at the inputs injecting electrons into the substrate. This increased the effective leakage current of the memory cells so that the data in the cells were lost.

One problem with back-bias generators is that they consume power, thereby increasing the power consumption of the SRAM in standby. This problem is handled in some designs by powering down the peripheral circuitry in standby and leaving the matrix on a low power consumption back-bias generator, a technique originated in a Bell Laboratories [19] paper of 1980.

With the advent of the CMOS SRAMs use of the back-bias generator was reduced. Since the Mix-MOS technology is usually an NMOS in a shallow P-well, injection from the peripheral circuitry can be reduced by using different wells for the array and the other circuits.

One common method was to use a grounded P-well on a N-type substrate. This provided a low soft error rate and in many cases eliminated the need for the power consuming back-bias generators. The grounded P-well technology was used in the Mix-MOS SRAMs by many manufacturers [33, 37, 38, 58]. A schematic cross-section of the technology is shown in Figure 8.17(a).

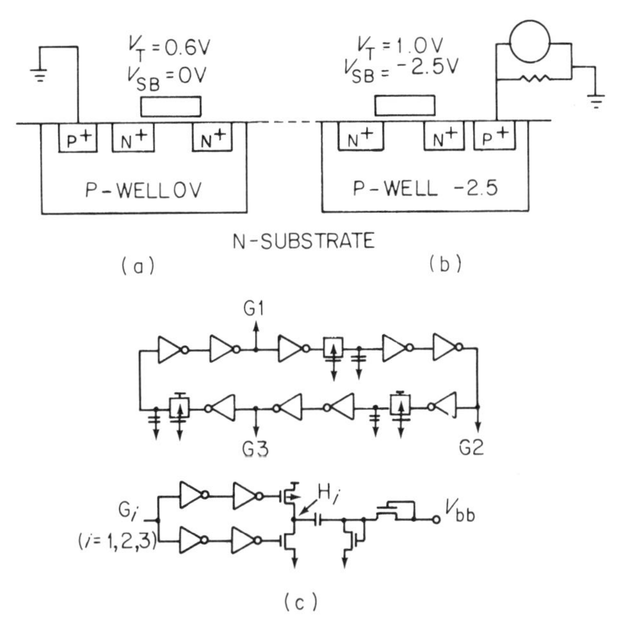

Figure 8.17
Schematic cross-section of two options for reducing substrate effects on CMOS SRAMs and a backbias generator. (a) Grounded peripheral P-well. (b) Selectively pumped P-well array. (c) Three-phase backbias generator for use in a 1 Mb SRAM. (a,b: From Wang *et al.* [45], Motorola 1987, with permission of IEEE; c: From Shimada *et al.* [55], Fujitsu 1988, with permission of IEEE.)

Back-bias generators continued to be used in some Mix-MOS SRAMs to increase transistor thresholds. Since in Mix-MOS the maximum resistor value is limited by the subthreshold currents of the transistors in the cell, high threshold transistors can be advantageous. For this reason, rather than grounding the P-well, Motorola [45] used a back gate bias supplied by a selectively pumped P-well as shown in Figure 8.17(b) to raise the threshold of the NMOS devices in the array from a nominal value of 0.6 to 1.0 V. This reduced the subthreshold current by three orders of magnitude, so that maximum load current exceeded it by at least two orders of magnitude.

It was felt that these margins were necessary to tolerate process variations in the resistor value and the subthreshold current. It was found that the selective pumping of the array p-well significantly improved cell stability, reduced the cell current, and reduced the bit-line capacitance, meanwhile allowing the peripheral devices to be optimized for speed.

A problem in the use of a back-bias generator is the current consumption in standby, and the potential for latch-up during power-up. In a subsequent paper Motorola [59] described a clamped P-well bias generator which limited the backgate bias to provide higher reliability while retaining the advantage of a selectively pumped array.

Fujitsu [54, 55] used a three-phase back-bias generator to stabilize the negative substrate bias level in their fast submicron process 1Mb SRAM. Submicron channel length transistors tended to cause substrate currents and therefore instability in the substrate bias. The fluctuation in substrate bias in turn induced a change in the threshold voltage of the transistors.

The solution was to negatively bias the substrate to control the NMOS threshold voltage without increasing the NMOS channel concentration. This resulted in a 50% increase in NMOS conductance compared to the same device in a P-well structure. The negative substrate bias also decreased the junction capacitor and improved the undershoot noise immunity. A schematic of the three-phase back-bias generator is shown in Figure 8.17(c). The purpose of the three phases was to stabilize the substrate bias in operating mode when it tended to decrease due to the substrate current.

8.8 ADDRESS TRANSITION DETECTION FOR SYNCHRONOUS INTERNAL OPERATION

8.8.1 Overview

Address Transition Detection (ATD) circuits are used to provide the initial pulse by which RAMs which are externally unclocked (asynchronous) can be operated as if internally clocked (synchronous). ATD pulses can be used to generate two classes of synchronous pulses, equalization pulses and activation pulses. Activation pulses selectively turn on particular parts of the circuitry and equalization pulses are used to reduce delay by restoring differential nodes prior to being selected.

The address transition detector generates a one-shot pulse when one or more of the inputs such as addresses or chip selects have changed. The pulse acts as an original clock for subsequent internal clocks which control the timing for various internal operations.

An early use of internal precharge circuitry in CMOS SRAMs occurred in 1979 on the RCA 16k SRAM [17]. This device used internally generated precharge signals which were transparent to the user whenever transition occurred on any address inputs. It was used, for example, during read operation both to set the bit-lines to 0.5 V_{DD} levels and to set the sense amplifiers near the toggle point to provide rapid sensing response.

8.8.2 ATD for bit-line equalization

An example from the 1987 Philips [4] 256k SRAM paper of a datapath using address transition detection for equalization is shown in Figure 8.18. At the end of the memory cycle a differential voltage existed on the bit-lines. A row address change resulted in deactivation of the word-line of the selected cell and the activation of the $\overline{\text{REQ}}$ and $\overline{\text{DEQ}}$ section equalizing signals derived from the global equalize pulse PEQ. These signals equalized the bit-lines and internal nodes of the sense amplifier and switched the global read bus driver into the high impedance state. The data remained on the global read bus since it had a static latch. After the bit-lines were equalized the new word-line became active and the $\overline{\text{REQ}}$ signal became inactive. This allowed a new differential bit-line voltage to develop.

After a short interval, the $\overline{\text{DEQ}}$ signal became inactive and the sense amplifier

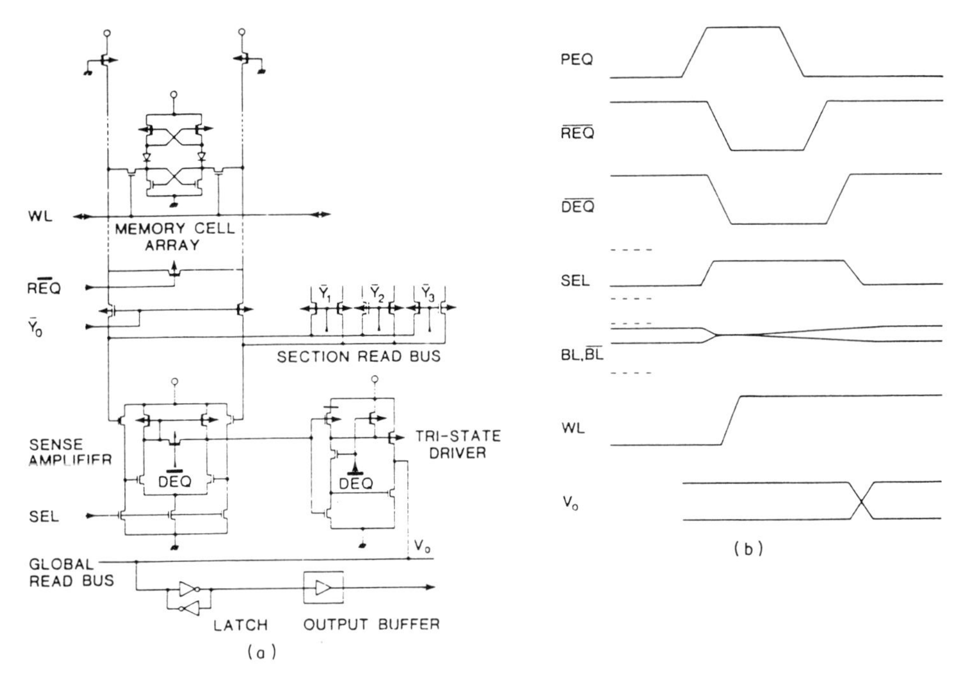

Figure 8.18
Example of a datapath using address transition detection for equalization. (a) Schematic circuit diagram of datapath. (b) Timing diagram for datapath. (From Gubbels [4], Philips 1987, with permission of IEEE.)

fed an amplified bit-line signal to the tristate driver which had been set to the low impedance state by $\overline{DEQ}$ becoming inactive. This signal was amplified in the tristate driver and put onto the global data bus where it overwrote the information stored in the latch. In the non-selected sections, the section equalizing signals were continuously active, thus isolating their read bus drivers from the global read bus.

A P-channel equalizing device was used to equalize the bit-lines after every read and write operation. Due to the high voltage level of the bit-lines, this equalizing was very efficient with a P-channel. The P-channel read pass gates were used to pass the differential bit-line voltage directly to the sense amplifier.

8.8.3 ATD pulse generators

A description of a one-shot internally generated equalizing pulse was given in 1981 by NEC [35] for a 16k SRAM. This pulse generator was used to equalize the level of bit lines, data busses, and sense amplifiers before the selection of a word-line. A symbolic circuit diagram of this address transistor detection generator is shown in Figure 8.19(a).

A similar pulse generator from Inmos [31] indicating the row address buffer with its associated ATD generator is shown in Figure 8.19(b). The row address buffer was

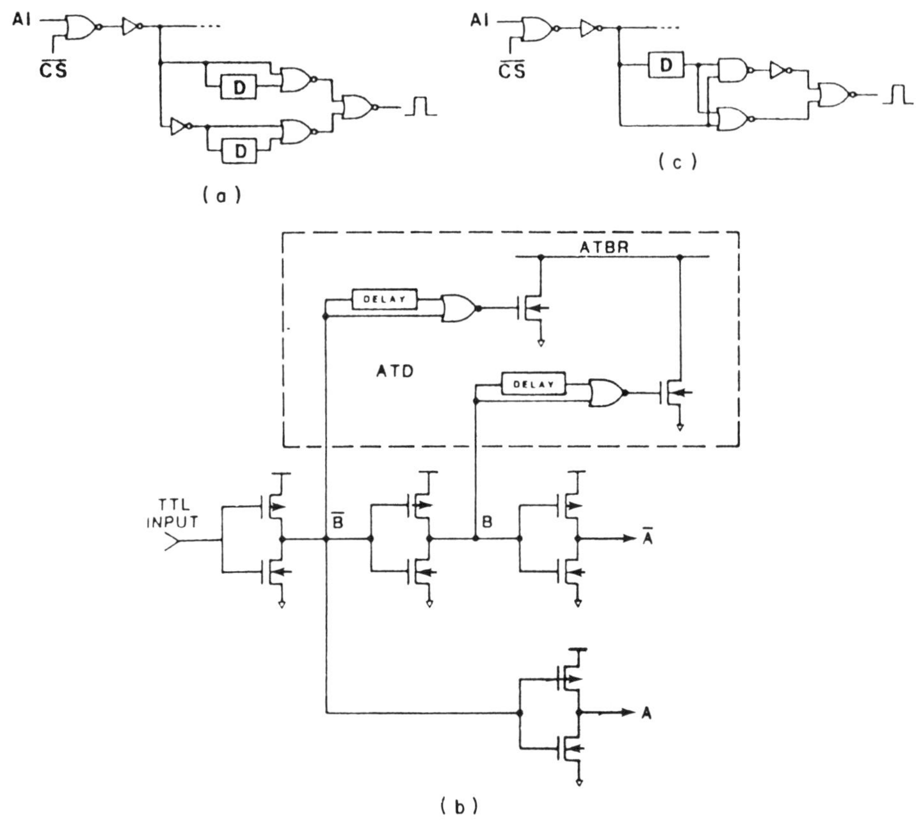

Figure 8.19
Early examples of address transition detection pulse generators. (a) Conventional for 16k SRAM. (b) Conventional with associated row address buffer. (c) Simplified for 256k SRAM. (a,c: from Kobayashi *et al.* [58], NEC 1985, with permission of IEEE; b: From Childs and Hirose [31], Inmos 1984, with permission of IEEE.)

a standard CMOS buffer with the input invertor ratioed to accept TTL levels from the input invertor. A and $\overline{A}$ were generated and routed to a group of predecoders. This was also used for a 16k SRAM in 1984. A simplified ATD generator from a later NEC [58] 256k SRAM paper is shown in Figure 8.19(c).

8.8.4 Automatic power-down using address transition detection

Since CMOS circuits dissipate power in proportion to the speed of operation, when the speed approaches zero, the power consumption drops to a low level. Only leakage current flows when the CMOS circuit is deselected and goes into standby mode.

Most SRAMs contain not only normal CMOS logic but also analog circuits for varying dc power, therefore, on most SRAMs the low power standby is controlled with the chip enable which is not a clock and does not have to be cycled. This allows the chip enable to be tied directly to system addresses and used as part of the normal decoding logic. Whenever a device is deselected, it automatically reduces its power requirements.

With multiple devices in the system, this feature adds up to significant power savings since deselected devices draw low standby power and only the active devices draw active power, thus the average power consumed by a device declines as the system size increases, asymptotically approaching the standby power level for the system.

The ac timing characteristics of the 256k SRAM in Figure 8.13 illustrated that this power down feature was obtained without any performance degradation since address access time is equal to chip enable access time.

To reduce active power still further, a widely used automatic power down technique in the late 1980s involved the use of Address Transition Detection. Clocks derived from ATD pulses turned internal circuits on and off by internal timing signals so that power down could be automatic without the need to turn off the chip enable signal.

In an early 1984 example, Toshiba [33] described a 256k CMOS SRAM in which the address transition detection circuitry was used to generate activation pulses wider than the access time effectively cutting off all the dc paths in the RAM after a read operation was complete, thus providing an automatic power down function.

The Philips 1Mb SRAM also used automatic power down controlled by an activation pulse with a pulse width slightly larger than an access time which controls the section word-line and local data path via the local section signals. This cut off all the dc current paths in the RAM after a read–write operation was complete. As a result no dc power was consumed.

8.8.5 Deselection of array sections

Much effort has gone into reducing the operating power consumption and reducing the timing delay of the higher density SRAMs. This effort consists in large part in reducing the capacitance of the increasingly longer lines on the memory device. While the focus of the following section is on reducing the power consumption, the two efforts are always linked and the trade-off between speed and power has to always be kept in mind.

A large part of the operating power consumption of a static memory is due to the column or bit-line loads when they are selected. The power can be reduced by dividing the array into blocks or columns with only the columns in a particular block selected at any one time.

Three early examples appeared in 1982. A 16k NMOS static from Mitsubishi [60] reduced power by half-word activation which cut the total bit-line current in half resulting in an active power dissipation of 275 and 22.5 mW standby power. A 64k NMOS static RAM from Intel [61] powered down unused circuit blocks during active operation to decrease power. A 64k CMOS SRAM from Toshiba [57] reduced the instantaneous precharge current by having each array independently activated during half of the clock signals while being selected by a single-column address signal.

These were followed in 1983 by a Fujitsu [62] 64k NMOS static RAM with low power consumption resulting from a power down technique for unused circuit blocks in which X and Y decoded blocks were operated selectively. This resulted in a 43% reduction in decoder current and a 50% reduction in bit-line current.

Mitsubishi [24] called their method for reducing power on high density SRAMs the 'Divided Word Line' technique in a 1983 paper on a full CMOS 64k SRAM. A schematic of the divided word line concept is shown in Figure 8.20(a). The cell array was divided into N_B blocks by the word-line segments. If the RAM had N_C columns, N_C/N_B columns were included in each block.

The divided word-line in each block was activated by switching gates which had two inputs, the horizontal row select line and the vertical block select line. Only memory cells, therefore, connected to one divided word-line within a selected block were accessed in a given cycle.

Figure 8.20(b) shows the trade-off in column power and word-line delay as a function of number of divided blocks, and Figure 8.20(c) shows the reduction in dc current distribution for the divided word-line structure over a conventional word-line structure.

This method has two major advantages. First the operating power is reduced by discharging the bit-lines. This eliminates the load of the bit-lines in unselected blocks.

Second is the reduction of the word select delay which is the sum of the row select line delay and the divided word-line delay. The capacitance of the row select line is smaller than that of a conventional word-line because it does not include the gate capacitance of access transistors in memory cells. The RC delay time in each divided word-line is small due to the short length.

Sony [41] in 1986 also used a similar divided word-line technique in which the array was divided into 16 sections with only one section activated at a time.

The 1984 Toshiba [33] 256k SRAM used a dynamic double word-line structure, similar to the Mitsubishi divided word-line technique, consisting of main word-lines and section word-lines.

In addition the hierarchical structure was also adopted for the bit sense lines with section bit sense lines leading to the section sense amplifiers which were directly across the bit-lines and main sense lines across the main sense amplifier. This double sense line structure reduced the capacitance of the main sense lines and hence reduced the sensing delay.

A subsequent Toshiba [40] part in 1986 used a 'shared word-line' structure which is a variant of the double word-line. The structure is shown in Figure 8.21 and consists for each row in the array of four second-level aluminum word-lines. Two lines are fed to the left cell plane and the other two go to the right cell plane. Only one of the four word-line dividers is activated by the word-line selector which decodes column address A_3 and A_4. The result is that only one quarter of the memory cells in each row are selected at any one time so the total cell current can be reduced by one quarter. In addition the second-level aluminum word-line structure has the effect of reducing word-line resistance and as a result word-line delay time.

The Philips [4] 256k SRAM in 1987 also used a divided word-line structure in which a global equalization pulse signal is logically combined in the section data path with the section signal to generate section equalization pulses to equalize bit-lines

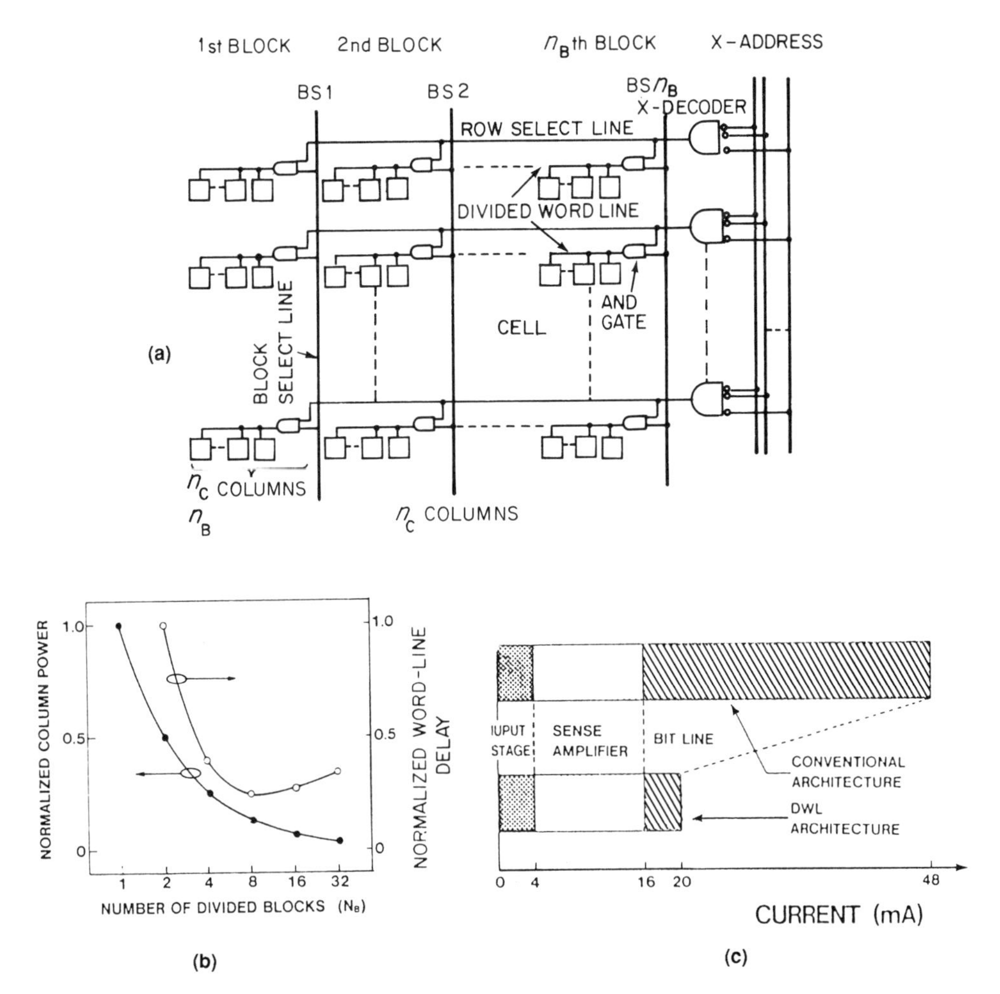

Figure 8.20
Divided word-line technique for reducing power consumption on high density SRAMs. (a) Simplified schematic concept of DWL structure. (b) Performance enhancement on 64k RAM by DWL structure. (c) dc current distribution. (From Yoshimoto *et al.* [24], Mitsubishi 1983, with permission of IEEE.)

section data read lines, sense amplifiers and control signals for the global read bus driver.

Most of the chips at the 1Mb level used some form of the segmented word-line architecture which had been developed earlier.

Sony [47] used the divided word-line architecture for the 1Mb SRAM to divide the cell array into 16 sections. Only one section, which was selected by a section address, was activated at a time resulting in 1/16 the dc current of the segmented case. This architecture also minimized the delay time due to the *RC* constant of the word line.

Mitsubishi [46] used an array divided into 16 sections for their 1Mb SRAM. Address

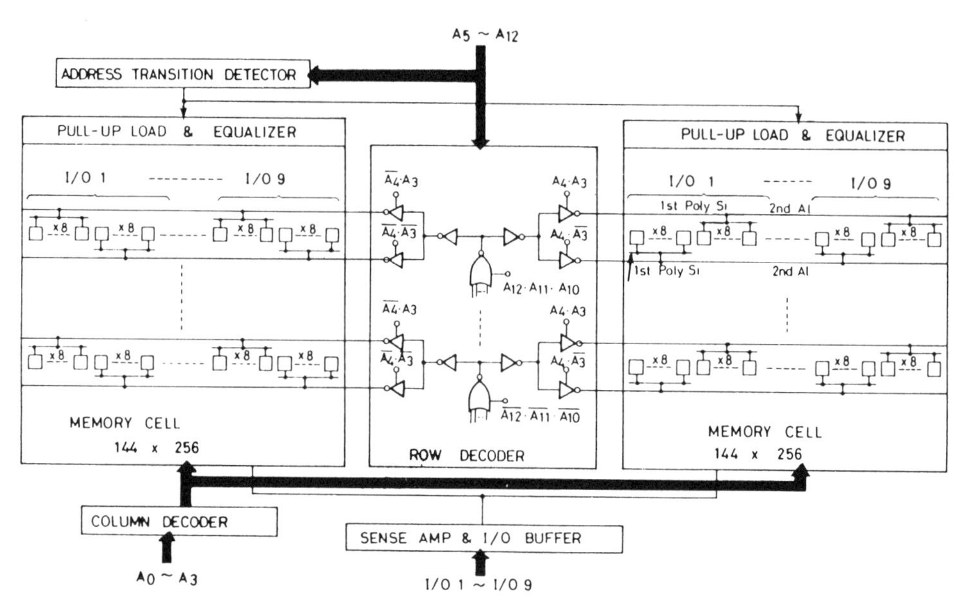

Figure 8.21
Shared word-line structure from a 72k SRAM showing row decoder and hierarchical word-line structure. (From Segawa *et al.* [40], Toshiba 1986, with permission of IEEE.)

signals were split into three groups which controlled, respectively, the row selection, column selection and block selection. A short word-line structure was combined with the tri-level word-line technique described in a previous section so that the current flowing through the bit-lines was decreased. The word-line was formed by polycide with a flow sheet resistance to reduce the word-line delay.

Mitsubishi [53] used a modification of their original divided word-line structure for their fast 1Mb SRAM to decrease bit-line capacitance and hence speed up the part. They also added a fourth signal to the three groups of address signals on the fast SRAM to control the selection between the $\times 1$ and $\times 4$ organization. In addition, each address input buffer had a local ATD pulse generator, any one of which could activate the internal clocks controlling the bit-line loads and sense amplifiers in order to speed up the read-out operation.

The Fujitsu [54, 55] 1Mb used a segmented block architecture giving them 16 blocks in the array of both of their first generation 1Mb SRAMs. The divided architecture reduced the power dissipation in the array. The delay of the word-line was reduced both by the divided architecture and in the fast 1Mb by the second aluminum word-line which connected with the polycide word-line every 32 cells.

The Philips [10] 1Mb SRAM used a divided word-line structure with the array divided into 32 sections. Only one section is active at a time. The short local word-line combined with the silicidation of the polysilicon layer kept the word-line delay at a fraction of a nanosecond. The array ratio of the chip was 66%. The addresses were predecoded in groups of three to reduce fan out capacitance on the predecoded lines and hence reduce the active power.

Toshiba [48] avoided long word lines by segmenting their chip into 32 sections using the 'dynamic double word-line' scheme used previously in their 256k SRAM.

This scheme used a main word-line which runs in second-level metal, and section select lines to enable local word-lines. They also used three-stage hierarchical decoding for column addresses which aided in the control of the sense amplifiers.

Hitachi [50] also used a double word-line technique and a divided bit-line load control which divided the chip into 18 segments.

8.8.6 Sequential activation pulses to distribute power

Both CMOS and Mix-MOS SRAMs tend to dissipate power in proportion to the speed of operation. This occurs since in read operations data transfer from cell to bit-line are accompanied by dc current dissipation through the bit-line pull-up loads.

To improve the operating power dissipation characteristic during the read operation dynamic sequential clocks can be used to selectively activate parts of the circuit. This timing can originate from an ATD pulse at the transition of an address line. When any row address input changes the circuit generates a basic clock pulse which in turn is used to generate sequential pulses as needed.

An early example was given by Toshiba [20] in 1982 for a full CMOS 64k SRAM. In this case the bit-lines were precharged by one clock pulse and the row address decoder activated by another. If the bit-line precharge and the row decoder selection were performed simultaneously it would result in an instantaneous peak current on the bit-line. To reduce this peak current, the bit-lines were precharged in sequence.

Another full CMOS 64k from NEC [25] described in 1983 used address transition detection to generate a single pulse which was then used to generate sequential internal clocks to activate the various internal circuits resulting in reduced dc current of bit-lines, sense amplifiers and I/O blocks.

In a subsequent paper in 1984 NEC [32] also described sequential, or staged, sense amplifiers. Rather than using the traditional method of having many bit-lines connect to a pair of common data bus lines which then lead to a main sense amplifier, a first-stage sense amplifier was connected directly to the bit-lines. The differential signal from the bit-line sense amplifier was then fed out to the subsequent data line sense amplifier. This was an extension of the concept of the column sense amplifier developed in 1981 by TI [34].

ATD signals were also used to equalize output data paths which both improved address access time and reduced dc power consumed in the array. The separate signals were also used to distribute the current surge over the cycle by providing timing of circuits as needed and hence reducing peak current.

A description from Motorola [26] in 1983 of a fast 16k NMOS SRAM illustrated the use of the above concepts to reduce these forms of power consumption. The schematic block diagram of this part is shown in Figure 8.22. Power is usually consumed in static RAMs in the sense amplifiers, static decoders and in the array by

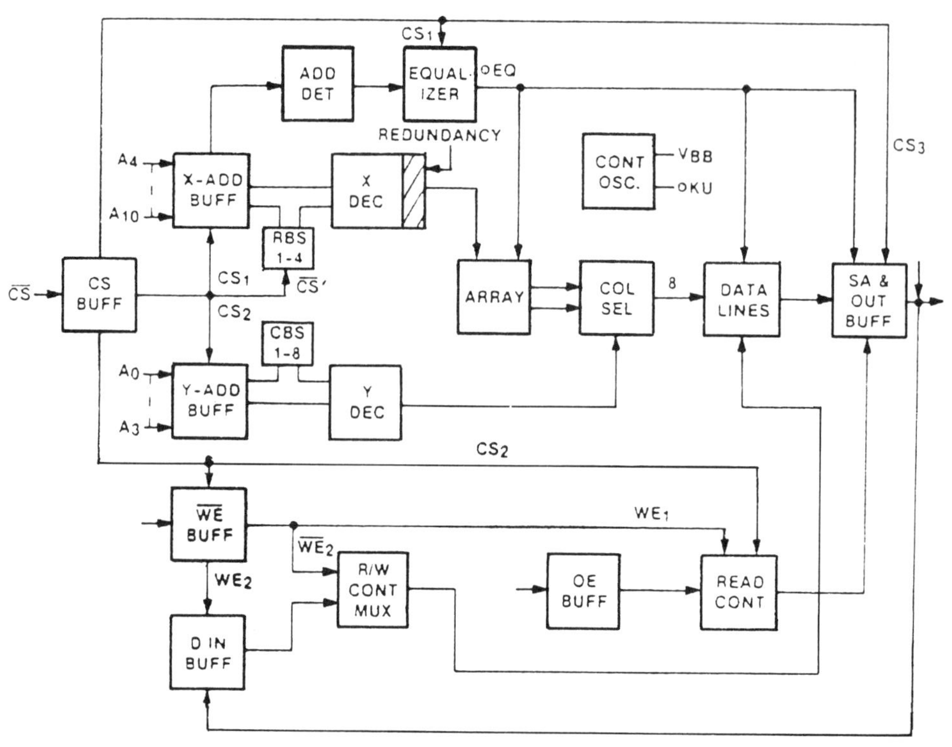

Figure 8.22
Schematic block diagram of early 16k NMOS SRAM using address transition detection for sequential activation of decoders and address buffers. (From Sood *et al.* [26], Motorola 1983, with permission of IEEE.)

the bit-line loads. In this circuit, power reduction in the main decoders and address buffers was accomplished using sequential activation in the form of one-in-four predecoders and mini-address buffers for both the row and column decoders and drivers which are shown in the figure as RBS and CBS.

This circuit illustrated tradeoffs in operating current, standby power and speed. The use of predecoders resulted in a significant saving in active power dissipation. Since they were, however, always on, they resulted in a slightly higher standby power dissipation and the addition of a predecoder path increased the access time slightly.

An example of the use of multiple parallel and sequential signals to both improve speed and optimize the distribution of power in the memory is also illustrated in the block diagram in Figure 8.22. In this case, use was made both of a sequential internal clock signal ($\overline{CS_3}$) generated from the initial external chip select ($\overline{CS}$) and of multiple parallel internal signals (CS_1, CS_2, $\overline{CS}'$) with paths of different speed to vary arrival time of these signals originating from the same external clock.

Here CS_1 activated the row address buffers and the equalization clock. This signal operated the fastest since these circuits were at the front end of the critical access time limiting path. CS_2 enabled the column address buffer, the column predecoders, and the write and output control circuits. All of these circuits could be activated later than those driven by C_1 but did not need to wait as long as would be necessary if a

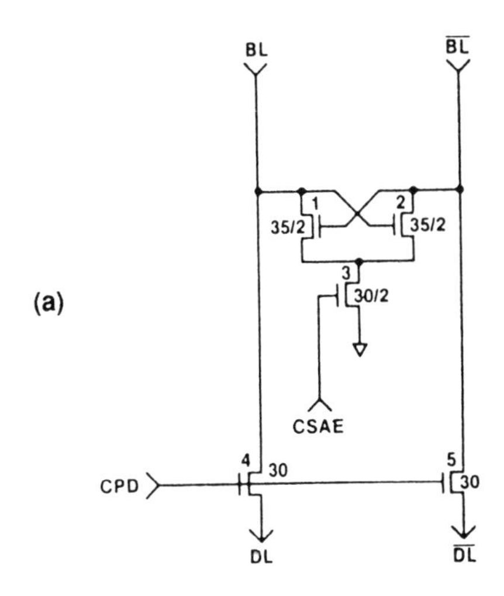

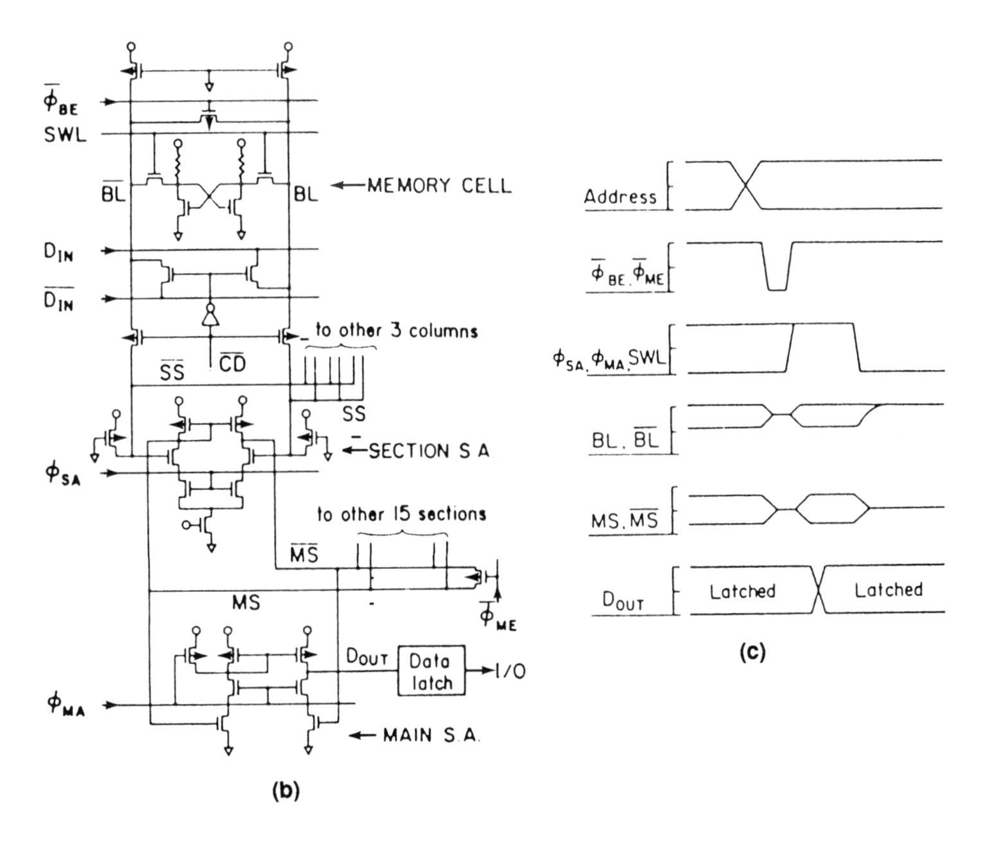

Figure 8.23

(a) Column sense amplifier to amplify the small bit-line signal from the memory cell which is then passed to the main sense amplifier (not shown). (b) Two-stage current mirror sense amplifiers and double sense line structure. (c) Timing diagram for two-stage current mirror amplifier sense structure. (a: From Sood *et al.* [36], Motorola 1985, with permission of IEEE; b,c: from Sukurai *et al.* [33], Toshiba 1984, with permission of IEEE.)

sequential pulse were used. The effect, therefore, of using the delayed parallel pulse CS_I was both to spread out the current surge thereby reducing peak current and to improve speed over the use of a sequential pulse.

The sequentially generated CS_3 pulse together with an internal write enable signal was used to power up the sense amplifier chain. A sequential signal was used since this can happen later in the operating sequence resulting in reduced initial current peak during chip select. Standby power was also reduced since the CS_3 circuitry was inactive in standby.

Reduction of the active array power was accomplished by using address transition detection for precharging the bit-lines. Once equalization is complete and a new word-line is selected the bit-lines can be effectively deselected and left floating. In the case of the Motorola part, weak sustaining diodes were left on the bit-lines to prevent the high level from being degraded by junction leakage current.

The hierarchial sense amplifier structure was illustrated in a later Motorola paper [36] in which a simple three-transistor column sense amplifier, shown in Figure 8.23(a), amplified the small bit-line signal from the memory cell which was then fed to the main data-line sense amplifiers. Toshiba [33] in 1984 also showed a hierarchial sense amplifier structure using a two-stage current mirror sense amplifier and a double sense line structure as shown in 8.23(b) along with a timing diagram (Figure 8.23(c)), showing the pulses for equalization of the bit lines and data lines and the pulses for activation of the section and main sense amplifiers.

Although the SRAMs remained externally static to the user, the internal timing had certainly become as complex as the DRAMs. For example, a fast 64k CMOS SRAM from Motorola [36] used extensive circuitry designed for internal power down. Circuits initially detected an address change (ATD) and generated a timed pulse to turn off all word-lines. Then it used a precise timing sequence to equalize all bit-lines on one side of each row decoder, synchronize the row address changes with column address changes, turn on the selected word-line, turn on the sense amplifier path to obtain valid data, turn off the bit-line loads, latched the proper data in the output devices, then shut down the sensing and output section.

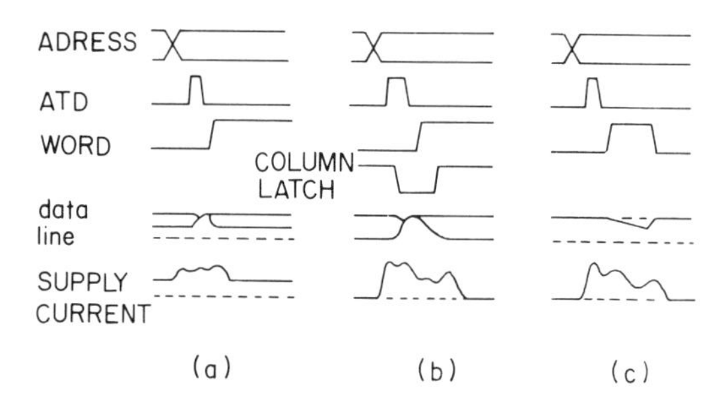

Figure 8.24
Comparison of signal waveforms for three methods of using address transition detection for power reduction and speed enhancement in SRAMs. (a) Data-line equilibration technique, (b) latched column technique, (c) pulsed word-line technique. (From Minato *et al.* [29], Hitachi 1984, with permission of IEEE.)

8.9 ATD TECHNIQUES FOR POWER REDUCTION DURING READ

Trade-offs between speed and power consumption appeared also with the use of ATD circuitry. A comparison of three ATD techniques used for power reduction during read is shown in the timing waveforms in Figures 8.24(a) to (c). These include the bit-line equalization technique, the latched column technique and the pulsed word-line technique [29].

8.9.1 Bit-line equalization technique

Bit-line equalization [56] during read used ATD generated signal waveforms to reset the bit-line before new data appeared thereby offering enhanced speed; however, it also produced high current through the memory cells.

Figure 8.24(a) shows the conventional data-line equalization technique where, during read, cells were active and drawing current while a low impedance kept the bit-line signal small.

8.9.2 Latched column technique

Another method, the latched column technique [59], achieved lower power dissipation during read by latching the signal to the column. This, however, pulled one of the bit-lines to a completely low state, and resulted in increased delay due to long bit-line recovery time.

Figure 8.24(b) shows the latched column technique where, during read, the cells were active, but bit-lines were allowed to go to ground. The cell current went to zero then bit-line equalization created large peak currents which took time.

8.9.3 Pulsed word-line technique

One solution to this problem, called the pulsed word-line technique, was investigated by Hitachi [29] in 1984 in which a pulsed word-line was used to reduce the current through the transmission gate of the cell.

Figure 8.24(c) shows the pulsed word-line technique in which cells were operated the same as in the conventional technique. A small impedance on the bit lines kept the voltage swing low. After some time, as the data were read out, the word-line was switched off and the cell current became zero. The data read out was stored in an additional latch in the data path.

The pulsed word-line technique was a new approach to the usage of ATD. The read operation itself in this case was asynchronous as in conventional SRAMs. The internal clocks were used only to activate the memory cells for a short period when they were really needed and then turned them off. This technique had a high immunity to address signal noise.

The operation was completely static since there were no control clocks. As the ATD fell all clocks were forced back to their initial states so that the bit-line was immediately precharged to prepared for subsequent reads. The minimum cycle time of the SRAM utilizing this technique was equal to the access time.

8.10 ATD TECHNIQUES FOR POWER REDUCTION DURING WRITE

The pulsed word-line technique was not applicable during the write operation since it switched off the word-lines during the memory cycle. Since there could be a change in data input late in the write cycle without a change in addresses, the chip had to remain powered-up until these data were written into the memory array, that is, during a write cycle input data changed late in the cycle must still be written into the memory cell, so the word-line had to remain activated and ac power must run in the bit-line loads. Since write in normal SRAMs occurs before the rising edge of a $\overline{\text{WE}}$ signal, a word-line should remain activated when $\overline{\text{WE}}$ is low.

Several techniques for power reduction during write were tried. These included variable impedance bit-line loads, trilevel word-lines, and data transition detection.

8.10.1 Variable impedance bit-line loads

In 1985 for a 256k SRAM, Hitachi [37] added the concept of variable impedance bit-line loads for reducing power consumption during the write operation when the pulsed word-line technique was not applicable.

Variable impedance bit-line loads as implemented by Hitachi consisted of a combination of an NMOS transistor bit-line load controlled by one clock and a PMOS transistor bit-line load controlled by a second clock as shown in Figure 8.25(a). The configuration of the clocks determined whether the load across the bit-lines at any one time was high or low impedance.

This technique suppressed the peak current of the write recovery and gave an optimized current for the read operation over a wide power supply range. This dual load was duplicated 128 times to match the 128 bit-line pairs.

8.10.2 Tri-level word-lines

Another solution for reducing power consumption during the write operation was proposed by Mitsubishi [38] also in 1985 for a 256k CMOS SRAM. This technique used a normal two-level word-line during the read cycle, but used a tri-level word-line during the write cycle.

The technique minimized bit-line swing, shortened the precharge time and depressed the transient current spike. A timing diagram showing the word-line operation during the read and write operations is shown in Figure 8.25(b).

During the read cycle the word-line went to the normal V_{CC} power supply rail and then after data sensing was complete was powered down automatically to ground. At the same time that the word-line turned off, the dc column current turned off and the bit-line loads turned on precharging the bit-lines.

During a write cycle the word-line initially was pulsed to its normal V_{CC} high level then was held at a middle level of 3.5 V which was high enough for a static write operation but saved both dc column current and lessened transient current after writing.

8.10.3 Data transition detection

Philips [4] in a 256k SRAM paper in 1987 proposed another approach to power reduction during write. This technique was called Data Transition Detection. An automatic power-up generator was used in this approach to generate the activation pulse whenever a transition occurred on either the address or the bit (data) lines. The

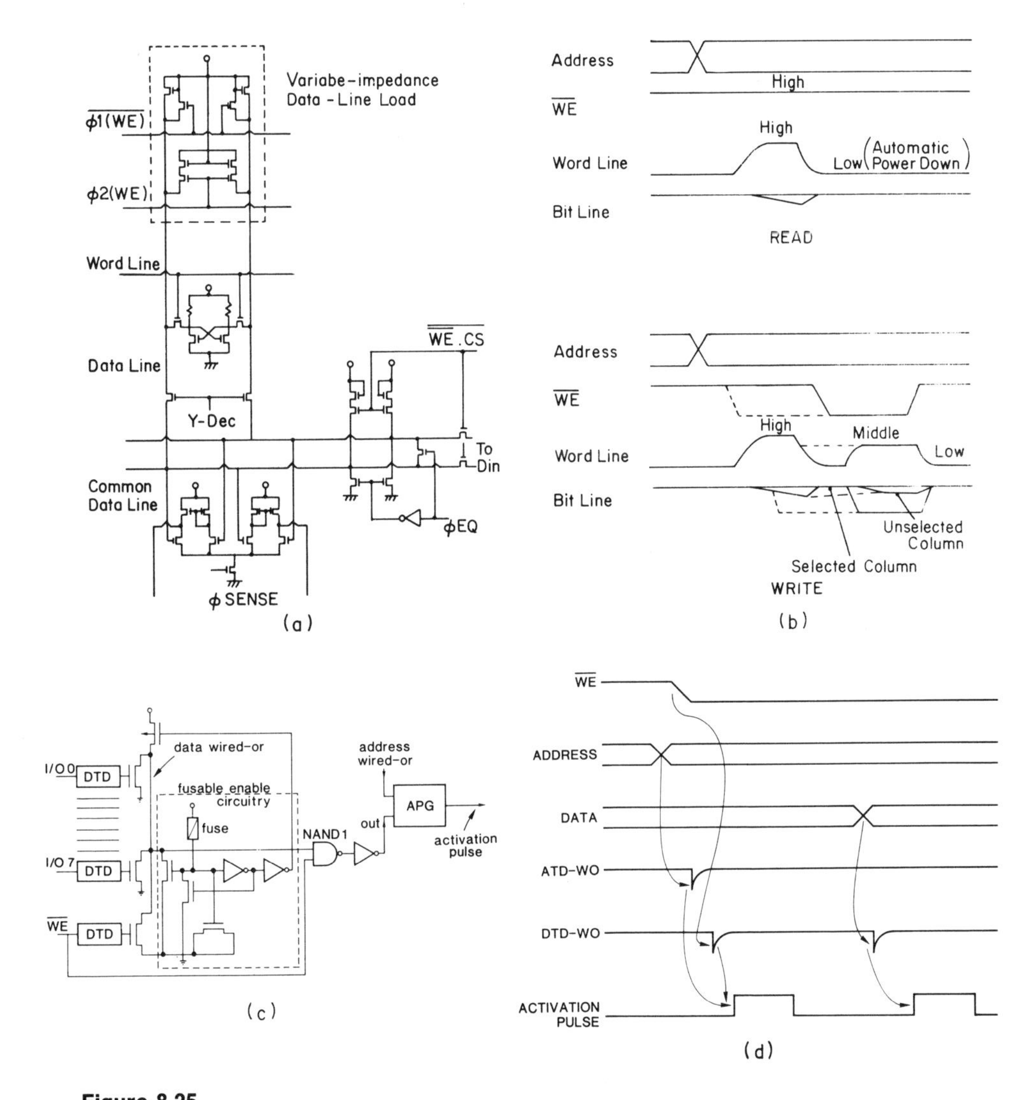

Figure 8.25
Three approaches to reducing power consumption during a write cycle. (a) Column circuit for variable impedance load technique. (b) Timing diagram for tri-level word-line technique. (c) Schematic of circuit for data transition detection technique. (d) Timing diagram for data transition technique. (a: From Yamamoto *et al.* [37], Hitachi 1987, with permission of IEEE; b: From Shinohara [38], Mitsubishi 1985, with permission of IEEE; c,d: From Gubbels [4], Philips 1987, with permission of IEEE.)

DTD circuits were placed on a separate wired-OR which is controlled by a write enable signal to avoid DTD pulses during a read cycle as shown in Figure 8.25(c). The timing diagram for the DTD circuit in Figure 8.25(d) shows the control of the activation pulse during a write cycle by both the ATD and the DTD pulses.

8.11 APPLICATIONS OF ATD ON THE 1Mb SRAMs

By the 1Mb SRAM generation address transition detection was routinely used for both equalization of various lines, for line precharge, and for activation of the various circuit elements. Both individual, sequential and summed global signals were used.

Auto-power down was accomplished by timing oscillators that were triggered by the ATD pulse, by time-out pulses, and by pulses which were wider than an access time to power down following the read cycle. For power down following a write cycle both the data detection and the variable impedance bit-line methods were used. Most of these were established techniques.

8.11.1 ATD equalization, precharge and activation on the 1Mb SRAM

Sony [47] used ATD to generate equalizing pulses for both the sense amplifier output level and the data bus level. Their simulations showed that this improved the access time by about 7 ns. Toshiba [48] also used the ATD for bit-line equalization and precharging.

In their fast part Mitsubishi [53] used equalization pulses generated from the ATD signal to equalize each of the stages of the three stage sense amplifier, along with bit-lines and I/O lines which are equalized by the same internal clock.

Hitachi [50] used an ATD generated by detecting all address and the chip select transitions and a pulsed word-line technique for sensing. In their fast SRAM Hitachi [52] used equalization for the bit-lines, common data buses, all stages of the sense amplifier and the main data buses both to attain faster data transition and to suppress incorrect data before the correct data appeared in the sense amplifiers.

Fujitsu [54] used pulses from the address transition clock for bit-line equalization and data path activation.

Philips [14] used ATD triggered by both address transitions and chip enable inputs that were summed to generate the global pulses that preconditioned the data path.

Sony [47] used an additional zero threshold NMOS transistor for providing a larger driver for fast bit-line equalizing. The additional transistor required no additional mask steps.

Fujitsu [54] also used two types of NMOS enhancement transistors with differing threshold voltages in the bit-line equalizing circuit to speed up transfers of bit-line data.

8.11.2 Auto power-down during read and write on the 1Mb SRAM

Sony [47] used an auto-power down function which utilized a ring oscillator which clocked from the ATD pulse to a set number and then powered down about 100 ns after ATD. This made this function useful up to a read cycle frequency of about 10 MHz.

Toshiba [48] used ATD for auto-power down to generate activation clocks for sense amplifiers and the cell array that had pulse width wider than an access time so that after the read operation was complete all dc paths were cut off.

Toshiba [49] in a subsequent paper in 1988 described an auto-power down for the write cycle of the 1Mb SRAM using a data transition detection technique similar to that developed by Philips for the 256k DRAM. When a write cycle began, the write enable pulse triggered the DTD pulse and generated the auto-power down timer which activated the RAM and the write operation was executed. After the power down pulse was timed out, the write amplifier first reset the input data-line and then the word-lines were shut off. An input data change generated the DTD pulse and activated the RAM again.

Philips [14] used a one-shot power-up signal to enable and disable the word-lines and sense amplifiers to reduce the active power during a long read cycle. The time-out period before auto-power down was nominally equal to two access times. In the user transparent powered down state the internal circuits were on standby but all TTL input and output buffers remained active and output data remained valid. A read bus latch maintained the valid data when the sense amp and read bus driver were disabled. To prevent stale latched data from reaching the output at the start of a chip enable access the output buffers were maintained in a high impedance state until the end of the data path initialization phase.

Hitachi [50] used ATD in a pulsed word-line technique to reduce power during the read operation and the variable impedance bit-line load technique to reduce power during the write operation. Both techniques were explained in a previous section.

Fujitsu [54, 55], in the read mode, also adopted the pulsed word-line method for auto-power down. In the write mode, they used a PMOS bit-line load that is disabled by a column select signal. The schematic of the bit-line loads is shown in Figure 8.26.

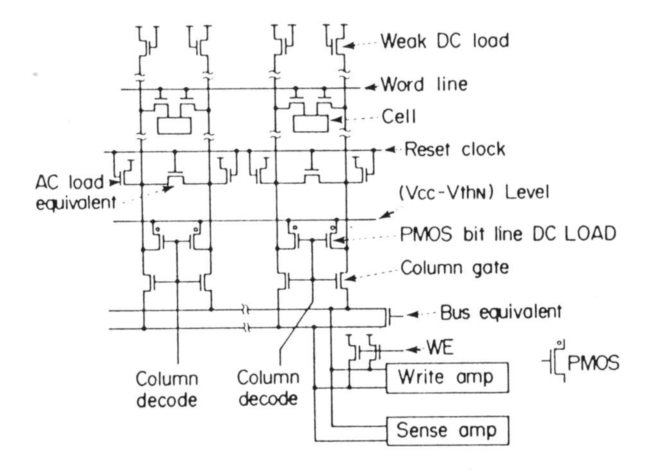

Figure 8.26
Schematic of bit-line circuitry showing PMOS bit-line dc load and weak dc load intended to reduce write bit-line current and shorten read bit-line propagation delay. (From Shimadu *et al.* [54], Fujitsu 1988, with permission of IEEE.)

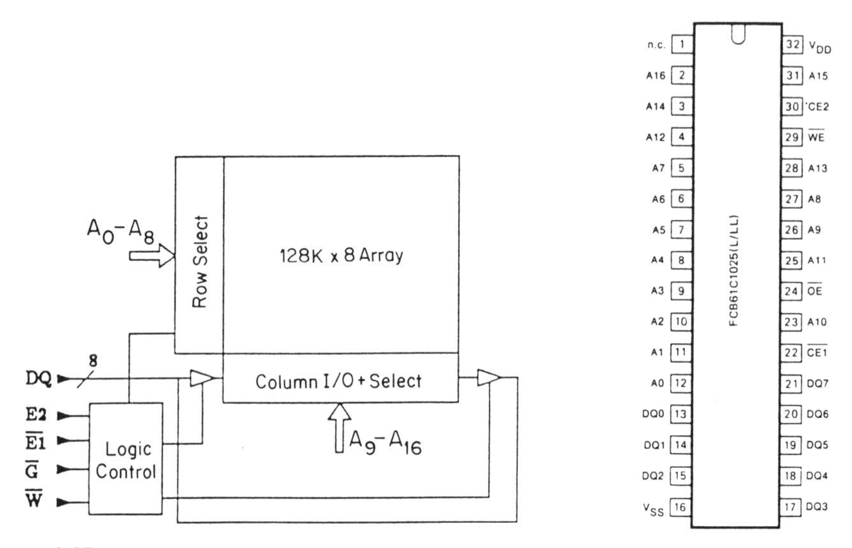

Figure 8.27
Typical block diagram and JEDEC standard pinout for 1Mb SRAM. (Reproduced with permission from Philips Memory Databook 1989 [69].)

Each bit-line had a weak conventional dc load, an ac load activated by a reset clock, and a PMOS dc load disabled by a column select signal. With the PMOS and the small NMOS load both, the potential of the bit-line stayed at the same level whether it was in the write or read mode which improved cell stability. The PMOS bit-line load also reduced the delay.

8.12 SPECIFICATION FOR A TYPICAL FIRST GENERATION 1Mb SRAM

A block diagram and pinout for a 1Mb SRAM are shown in Figure 8.27. In spite of the complex internal timing, the external block diagram remained as simple as that of the first 4k SRAMs. The part had moved into the larger 32 pin package, making it the first bytewide SRAM in three generations not in the 28 lead package. One result of the availability of additional pins was the return of the second chip enable which was given up at the 256k level. Typical ac timing characteristics are shown in Figure 8.28 along with the corresponding timing diagrams.

8.13 THE SEARCH FOR IMPROVED SPEED

As the commodity SRAMs double in density about every two years, the demands on them for speed doubles (access time halves) about every four years. Microprocessors over the 15 years of their existence have progressed from early sub 5 MHz versions to the newest 40 to 50 MHz versions. In systems using commodity SRAMs these mainstream microprocessor speed demands dominate the requirements. The effect was to move SRAM access time in this time period from the 400 ns of the early 4k SRAM to the 25 ns of the 4Mb SRAM.

Unfortunately, while scaling techniques speeded up the basic gate delay of the peripheral transistors, the accompanying increase in density added more load capacitance to the word-lines, bit-lines, and sense lines in the memory array effectively slowing the parts down.

This is illustrated in Figure 8.29 from the 1984 Toshiba [33] 256k paper which shows the percentage increase in delay component on scaled SRAMs calculated by assuming a simple scaling law with a fit at 64k bits without assuming any change in fundamental design methods to achieve the density increase.

The increase shown in word-line capacitance going from the 64k to the 256k level without an increase in number of word-line divisions was significant. Doubling the number of word-line divisions restored the balance between the components of the data path at the 256k level. The same problem then occured again at the 1Mb level.

It is clear that the word-lines constituted a major component of delay. The significant efforts on divided word-lines, bit-lines and sense lines during the mid 1980s on the 64k and 256k SRAMs not only reduced the operating power consumption but also reduced the load capacitance of these lines resulting in increased speed.

The transition from internally unclocked SRAMs to internally clocked SRAMs, made with the help of address transition detection, also contributed to the increase in speed.

A review follows of the various efforts made to increase speed in the fast SRAMs, and some of the problems associated with the increased speed such as stability, ground bounce and interface.

Three generations of 1Mb SRAMs are shown in Figure 8.30. The first generation, considered earlier in this chapter, consisted of the bytewide 1Mb SRAMs intended for microprocessor and small systems applications. The typical access time of these parts was in the 25 to 45 ns range with specified production speeds of 45 to 70 ns. This was compatible with the 15 to 20 MHz speeds of the standard microprocessors in small computers in the early 1990s.

The second generation brought the specified access time to the 25 ns range with capability for serving the higher performance 30 MHz processors used in more advanced small computers and workstations and also of serving the second-level cache applications in larger computers. The organization was predominantly $\times 1$ and $\times 4$ to void the ground bounce problems associated with higher speed and still begin to accommodate the deeper applications. The interface continued to be TTL compatible.

The third generation moved into the sub 10 ns typical speed range and was intended for high performance cache applications in workstations, mainframe computers, and super computer types of applications which needed the deep bit-wide organization. Since many of these very fast systems are ECL compatible, the ECL compatible interface predominated.

Although the minimum geometry was reduced in each generation in an effort to enhance speed, the architectural design techniques needed to design very fast SRAMs and the pitch constraints of very small geometries kept the chip sizes from decreasing accordingly.

The first SRAMs with the PMOS load transistors in the polysilicon were shown by NEC and Hitachi. These new SRAM transistors appeared in production SRAMs at the 4Mb density and will be considered further in Chapter 9.

A.C. PARAMETERS	Symbol name	−35		−45		−55	
		Min	Max	Min	Max	Min	Max
READ CYCLE							
Read cycle time (ns)	t_{RC}	35	—	45	—	55	—
Address access time (ns)	t_{AA}	—	35	—	45	—	55
Chip enable access time (ns)	t_{ACE}	—	35	—	45	—	55
Output enable access time (ns)	t_{OE}	—	20	—	25	—	30
Output enable to output Lo–Z (ns)	t_{OLZ}	5	—	5	—	5	—
Chip enable to output Lo–Z (ns)	t_{CLZ}	10	—	10	—	10	—
Output disable to output Hi–Z (ns)	t_{OHZ}	—	15	—	20	—	20
Chip disable to output Hi–Z (ns)	t_{CHZ}	—	15	—	20	—	20
Output hold time (ns)	t_{OH}	5	—	5	—	5	—
WRITE CYCLE							
Write cycle time (ns)	t_{WC}	35	—	45	—	55	—
Chip enable to end of write (ns)	t_{CW}	30	—	40	—	50	—
Address valid to end of write (ns)	t_{AW}	30	—	40	—	50	—
Address set-up time (ns)	t_{AS}	0	—	0	—	0	—
Write pulse width (ns)	t_{WP}	25	—	35	—	45	—
Write recovery time (ns)	t_{WR}	0	—	0	—	0	—
Write enable to output Hi–Z (ns)	t_{WHZ}	—	15	—	15	—	15
Data-to-write overlap (ns)	t_{DW}	20	—	25	—	25	—
Data hold from write time (ns)	t_{DH}	0	—	0	—	0	—
DQ low-Z from end of write (ns)	t_{OW}	5	—	5	—	5	—

Figure 8.28
Ac timing characteristics and corresponding timing diagrams for read and write cycles for a typical 1Mb SRAM. (Reproduced with permission from Philips Design Target Book.)

8.13.1 CMOS as opposed to resistor load cells in fast SRAMs

The full CMOS memory cell has a natural speed advantage which derives from its use of both active pull-up and active pull-down in its gates. This provides a symmetrical waveform as well as good source and sink drive for capacitive loads. The CMOS rise and fall times are short and symmetrical whereas NMOS has a short fall time, but a larger rise time because of its high resistance load. NMOS hence has slower average switching speed.

The passive resistor load also sets the standby power dissipation of the NMOS array, whereas the standby power dissipation of the CMOS array is given by leakage current only. The output current of a memory cell depends on the worst case internal high level, which is, directly after a write cycle, equal to power supply in the CMOS case and supply minus a threshold in the R-load case. The noise margins of the R-load are therefore less.

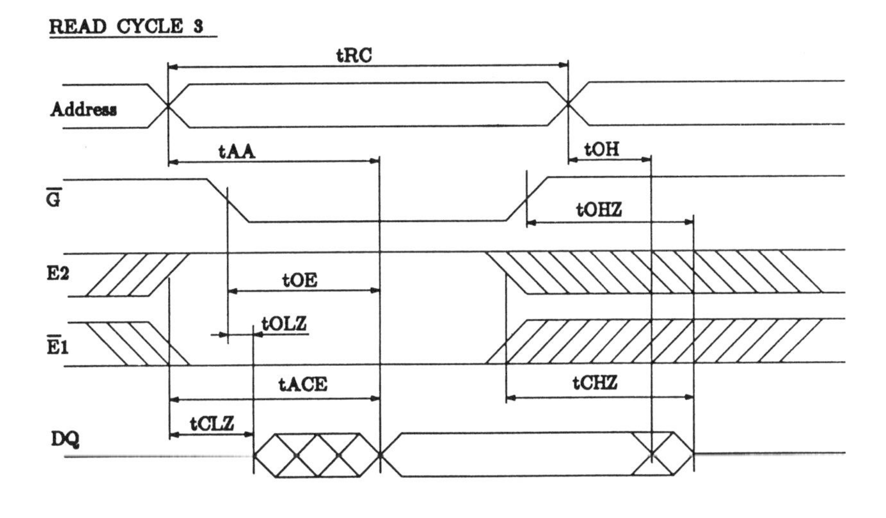

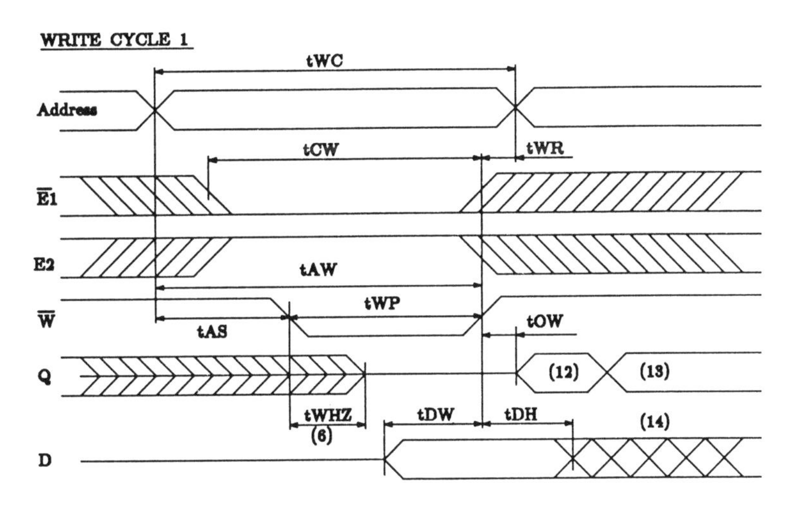

Figure 8.28 (*continued*)

8.13.2 Address transition detection for improved speed

Mostek was one of the earliest manufacturers to attempt to increase the speed of static RAMs by using dynamic interface circuitry on the chip similar to the circuitry used in the DRAMs. These chips are faster since low capacitance pre-charge methods can be employed to shorten the memory cell's *RC* time constant.

The use of address transition detection for improved access time was described in a 1983 paper from Motorola [26] for a high speed 16k SRAM. Precharge and equalization of the bit-lines was used to enhance performance.

In an internally unclocked RAM array, data from the accessed cell developed a differential voltage on the bit-lines. Typically this differential was passed onto common

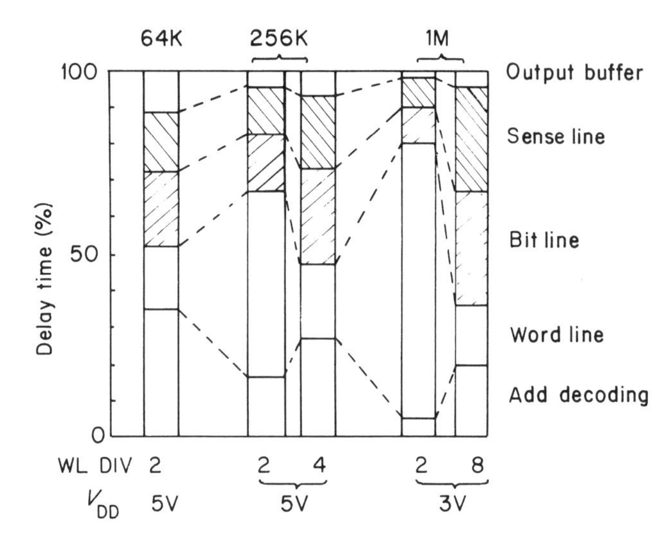

Figure 8.29
Trends of delay component in scaled CMOS SRAMs, as a function of number of word-line divisions. (From Sakurai *et al.* [33], Toshiba 1984, with permission of IEEE.)

Year	Company	Bus width	Type	I/O	Geometry (μm)	Cell size $(\mu m)^2$	Chip size (mm^2)	Speed	
								(Type.) (ns)	(Spec.) (ns)
1987	SONY	X8	4T	TTL	1.0	74	109	35	55
1987	HITACHI	X8	4T	TTL	0.8	45	79	42	70
1987	MITSUBISHI	X8	4T	TTL	0.8	44	82	34	55
1987	TOSHIBA	X8	4T	TTL	0.6	53	105	25	45
1988	PHILLIPS	X8	6T	TTL	0.7	60	93	25	45
1988	FUJITSU	X8	4T	TTL	0.8	41	94	44	70
1988	MITSUBISHI	X1/X4	4T	TTL	0.7	42	87	14	25
1988	HITACHI	X4	4T	TTL	0.8	44	94	15	25
1988	FUJITSU	X4	4T	TTL	0.7	41	90	18	25
1988	NEC	X8	6TP	TTL	0.8	41	81	35	55
1989	HITACHI	X1/X4	6TP	TTL	0.5	21	55	9	15
1989	TI	X1	6T	ECL	0.8	76	120	8	15
1989	TOSHIBA	—	—	ECL	0.8	50	107	8	15
1990	NEC	X1/X4	4T	ECL	0.8	45	113	5	15
1990	MITSUBISHI	X1	4T	ECL	0.6	39	88	7	15

* Estimated

Figure 8.30
Three generations of 1Mb SRAMs. (Soure IEEE ISSCC Proceedings 1987–1990.)

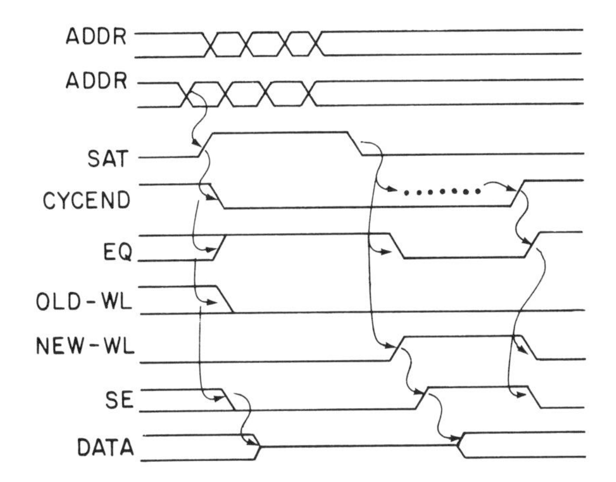

Figure 8.31
Timing control for address stability in the RAM. The old memory cycle is shut down upon detection of address transition. The new cycle does not begin until detection of address stability. (From Flannagan [43], Motorola 1986 , with permission of IEEE.)

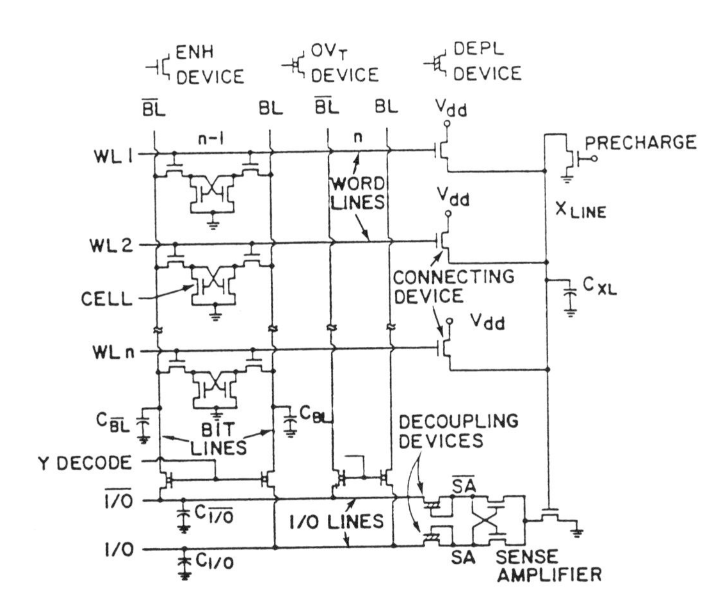

Figure 8.32
Circuit schematic of self-timed sense amplifier set from accessed word-line with possibility of rapid decoupling from the I/O and bit-line capacitance. (From Schuster [63], IBM 1984, with permission of IEEE.)

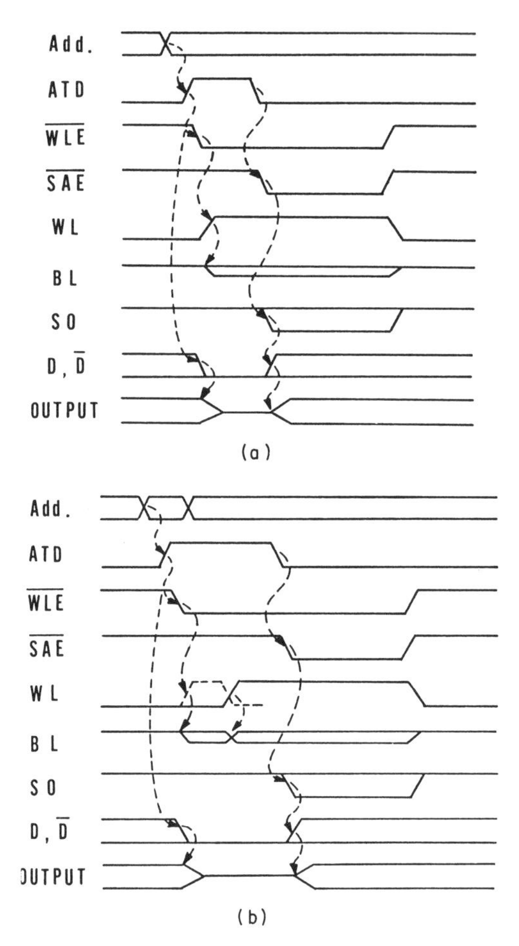

Figure 8.33
Timing diagram for a 32k × 8 SRAM showing activation from the rising edge of the ATD pulse of the word-line. Equalization of the bus lines and deactivation of the sense amplifiers for: (a) normal address input, (b) address skews. (From Okuyama [65], Matsushita 1988, with permission of IEEE.)

data-lines by column select pass devices and was then amplified to drive the output driver. When another cell, whose data were opposite that of the previous bit accessed, was addressed on a succeeding access, the data output path had to switch first to an equalized state and then to the opposite logic value. Any row change could necessitate such switching. Since the bit cell had low drive capability and the capacitance on the bit and data lines was large, the time required to switch the differential from one state to the opposite state became a large part of the overall access time.

Equalization of the bit-lines and data-lines prior to a new access could substantially reduce this time penalty. This could be done when the chip was deselected so that the chip access time could be reduced through this technique, allowing SRAMs to have equal chip select and address access times.

8.13.3 Short data hold time problems in fast SRAMs using ATD

One problem associated with ATD in fast SRAMs is that the data hold time can be too short for some systems. Data tend to be lost sooner after an address change than the system expects in a new cycle. In 1984 Motorola [28] described a solution for this case in which the data valid period was extended to the end of the equalization pulse (plus two invertor delays) rather than the beginning of this pulse.

8.13.4 Address skew in fast SRAMs using dynamic circuitry

Another problem associated with ATD in fast SRAMs was that of address stability detection and control. In the control of the RAM as shown in the timing control diagram in Figure 8.31, the old memory cycle was shut down upon detection of an address transition. The new cycle did not begin until detection of address stability.

The first edge of the summation address transition (SAT) pulse turned off the existing word-line, equalized the array and shut down the RAM. The second edge of the SAT pulse occurred after all address changes had stabilized. This second edge represented address stability detection. It was not timed from the first edge due to a lack of tracking of the first and second edges of the SAT pulse as it propagated through the circuitries. The second edge of SAT began the access cycle by turning on the new word-line, turning on the amplifiers and controlling the output buffer.

Another attempt to deal with address skew was described by IBM [63] in 1984 for a 20 ns NMOS 64k SRAM. This part used a self-timed sense amplifier which was set from the accessed word-line to eliminate the timing skews that would result if a separate timing chain were used for the sense amplifier. Fast setting of the sense amplifier was achieved by decoupling it from the capacitance of the I/O lines and the bit-lines as shown in the circuit schematic in Figure 8.32.

A 256k CMOS SRAM from Matsushita [65] used circuit techniques to reduce address skew. A double activated pulse circuit was used to improve the access time. This circuit utilized the ATD pulsed word-line technique. Timing diagrams for this circuit are shown in Figure 8.33. Figure 8.33(a) is for a normal address input and Figure 8.33(b) is with address skews. The diagrams show activation of the word-line from the rising edge of the ATD pulse, equalization of the bus lines and deactivation of the sense amplifiers.

The rising edge of the ATD pulse generated the $\overline{\text{WLE}}$ (word-line enable) pulse, which activated the word-line. This same rising edge also equalized the bus lines and inactivated the sense amplifiers. The falling edge of the ATD pulse then generated the $\overline{\text{SAE}}$ (sense amplifier enable) pulse after all address changes had stabilized. The $\overline{\text{SAE}}$ pulse activated the bus lines and the sense amplifiers. Since the word-lines had stabilized before the $\overline{\text{SAE}}$ pulse was generated, address skews were avoided.

After the data were latched to the output buffer, the $\overline{\text{WLE}}$ pulse and the $\overline{\text{SAE}}$ pulse were inactivated, the word-line was pulled down and the bus lines and the sense amplifiers were inactivated. The active power dissipation was therefore reduced during a long read cycle.

8.13.5 Fast sense amplifiers for high density SRAMs

A 17 ns 64k CMOS SRAM by Toshiba [64] in 1985 used bit-line equalization, and a three-stage sense amplifier to improve speed. The first stage of the sense amplifier was a current mirror amplifier which had a high speed data transmission rate, but a relatively small voltage amplification. This was combined with second and third stages which were Schnitt trigger latched amplifiers having large amplification ability but also large sensing delay. The combination gave high speed and sufficient amplification.

Mitsubishi [46] in their 1Mb SRAM used a two-stage current mirror sense amplifier with precharged input signals and for the fast part [53] went to a three-stage sense amplifier where the first two stages are dual input–dual output current mirror amplifiers and the third state was a normal current mirror circuit with dual inputs and a single output.

Toshiba used a hierarchial triple sense-line structure to control the sense amplifiers for their 1Mb SRAM. The first level controlled the local sense amplifiers. The outputs of every four of these section sense amplifiers in one block were connected to one set of block sense-lines and became the inputs to the second local sense amplifier. This block sense amplifier was in turn connected to the main sense-lines and input to the data latch. This triple sense-line structure, together with a sense-line equalization, optimised the sensing delay.

Hitachi [52] used a new dynamic gain controlled double-ended sense amplifier to decrease delay time in their 1Mb SRAM. Shown in Figure 8.34(a) is the conventional double-ended current mirror amplifier which had been widely used due to its fast sense speed, large voltage gain and good output voltage stability. The delay time and current of this amplifier depended on the ratio of the gate width of the NMOS transistor to that of the PMOS transistor W_n/W_p as shown in the figure.

The current in this amplifier increased rapidly as the delay time decreased as shown in Figure 8.34(b). Since more than one stage of this amplifier was normally used, the current for short delay times could be large. The new dynamic gain control double-ended amplifier is shown in Figure 8.34(c). The circuits within the dotted line are added to the conventional double-end current mirror amplifier. These additional circuits acted as a current switch to control the amplifier current by varying the effective NMOS channel widths of W_n and W_{nc}.

When input SAQ was high, transistors Q_1 and Q_2 were switched on, resulting in high current and fast sensing speed. When SAQ was low Q_1 and Q_2 were switched off resulting in just enough current to maintain the data. If SAQ was controlled with a short pulse during the sensing operation a fast sense speed and low power dissipation could be achieved simultaneously. This new sense amplifier was used in a three-stage configuration in the Hitachi fast first generation 1Mb SRAM. They also used feedback capacitances between the outputs of the first-state amplifier and the common data bus to speed up the signal transition on the common data bus.

Hitachi [66] in 1989 showed a 9 ns CMOS 1Mb SRAM in 1989 which used a three-stage cross-coupled PMOS sense amplifier along with a CMOS preamplifier to improve speed. This amplifier provided more than two times faster sensing than conventional paired current mirror amplifiers. The PMOS cross-coupled amplifier and

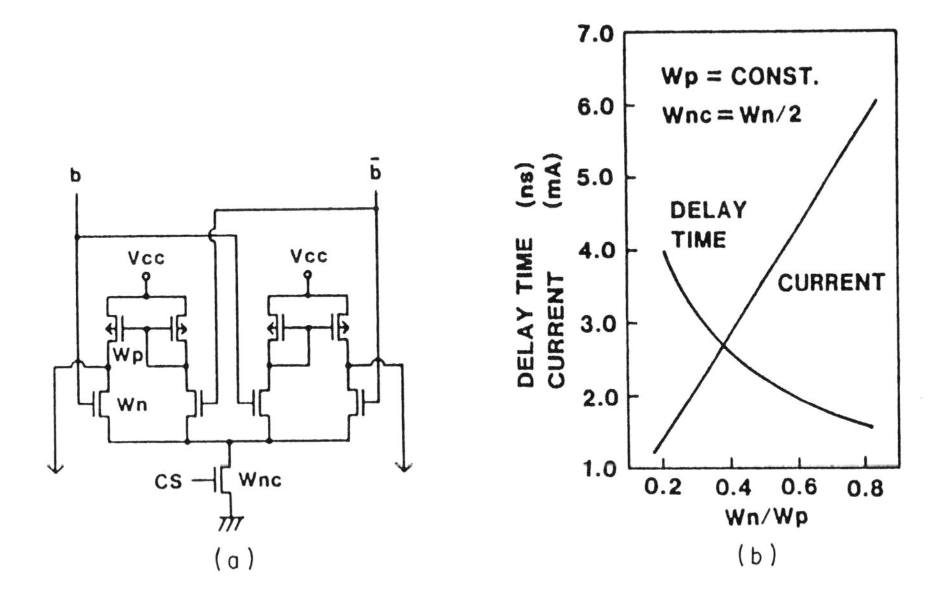

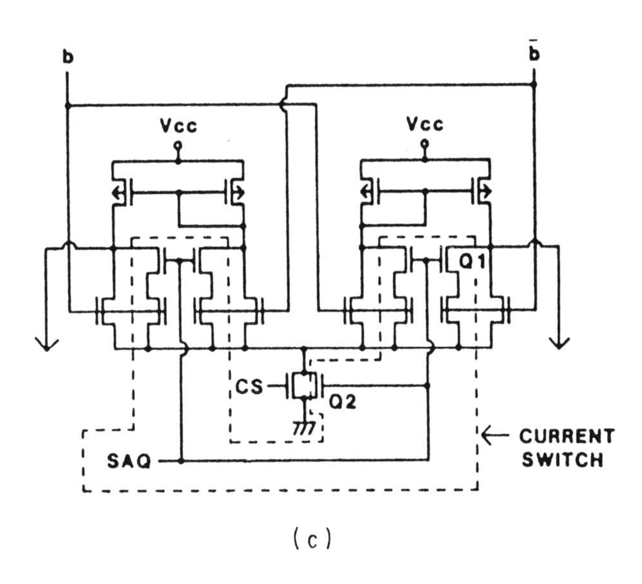

Figure 8.34
Two types of double-ended current mirror amplifiers: (a) double-end current mirror amplifier, (b) delay and current as a function of W_n/W_p for the double-end current mirror amplifier, (c) dynamic gain control double-end amplifier. (From Sasaki [52], Hitachi 1988, with permission of IEEE.)

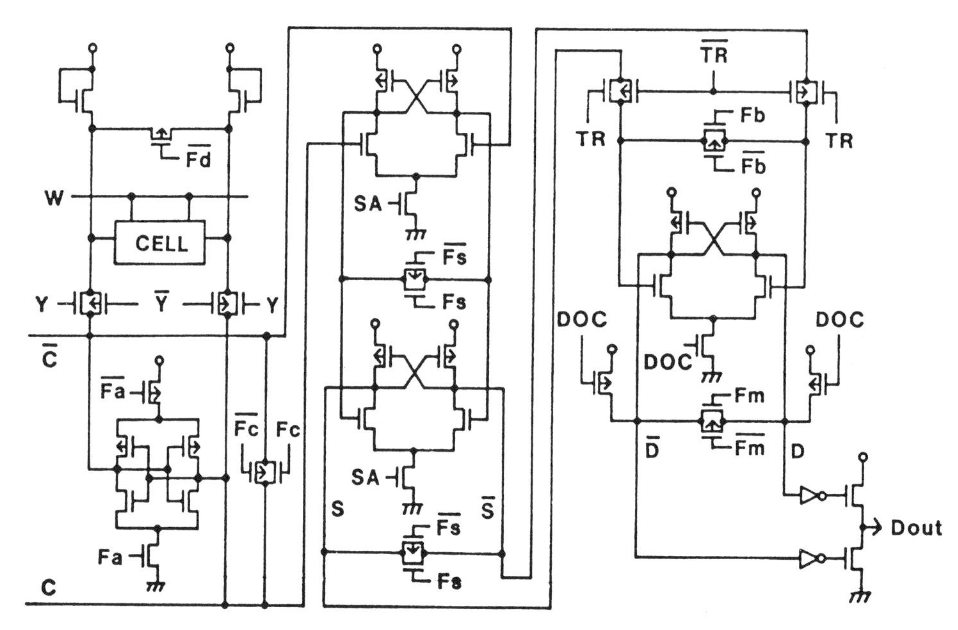

Figure 8.35
Three-stage PMOS cross-coupled sense amplifier for 1Mb SRAM. (From Sasaki [52], Hitachi 1988, with permission of IEEE.)

the current mirror amplifier from this circuit were described in Chapter 5. The complete three-stage sense amplifier is shown in Figure 8.35.

In operation the CMOS preamplifier was activated by a pulse and amplified the small signal between the common data-lines to about 0.5 V. The first PMOS cross-coupled amplifier then amplified the signal to more than 3.5 V. The second PMOS cross-coupled amplifier drove the data bus and the third amplifier amplified the signal at the end of the data bus to drive the output transistors.

Fujitsu [54, 55] in their first 1Mb SRAM used a dynamic two-stage sense amplifier clocked by two sequential pulses generated from an address transition clock. Their fast part clocked the bit-line signal through a four stage sense amplifier to achieve fast access time. This was based on the usual current mirror approach with each stage basically alike except for adjustments for the sense amp ratio and the dimensions of the transistor.

The Philips [14] 1Mb SRAM used a two-stage sense amplifier with NMOS read bus drivers to move the data from the sense amplifiers onto the read bus. The drivers were disabled during data-path initialization and only reactivated when there was valid data available. The design of the read bus drivers took into account the high fan-out due to the divided word-line architecture.

8.13.6 Short bit-lines for improved speed

The 1986 Motorola [43] 64k CMOS SRAM had, in addition to the ATD circuitry which was mentioned in a previous section also, very fast 13 ns access time. This was achieved by laying the chip out at 90° to the usual orientation resulting in short bit-lines. The reduction in bit-line capacitance allowed high speed signal development. This same technique was later used by several companies in the 1Mb SRAMs to increase speed.

A 256k 32k × 8 CMOS SRAM described by Matsushita [65] in 1988 also used short bit-lines to increase speed in addition to using a fast sense amplifier and various other techniques for speed enhancement. Mitsubishi [44] also in 1986 described a 25 ns 256k CMOS SRAM with a block architecture which used short bit-lines for increased speed.

Philips [14] and Hitachi [52] also laid their 1Mb SRAM chip out with the bit-lines and word-lines at 90° to the historical orientation to reduce the length, and resulting capacitive delay, of the bit-lines. This orientation permitted very short bit-lines to reduce the bit-line signal delay. Figure 8.36(a) shows a comparison of this new type of orientation with the traditional orientation as used in the Mitsubishi [53] 1Mb SRAM shown in Figure 8.36(b).

Another alternative was to use an architecture with blocks in which short bit-line segments run parallel to the length of the chip. This was done in a Philips [11] fast

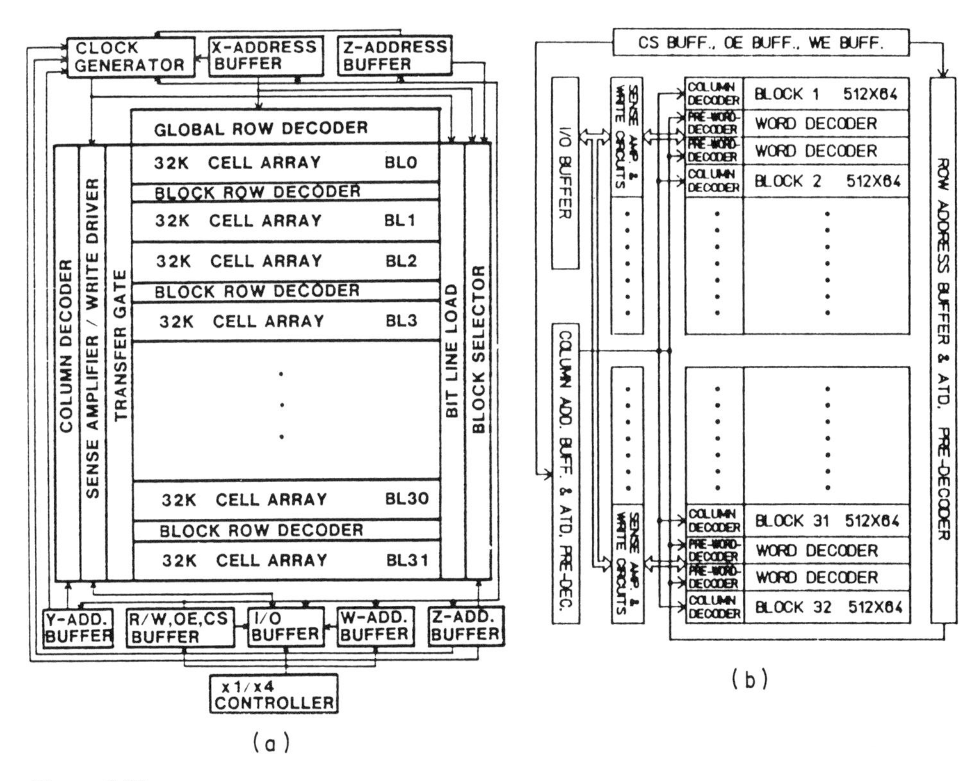

Figure 8.36
Block diagrams of 1Mb static RAMs. (a) Showing traditional orientation of word-lines and bit-lines. (b) Showing layout at 90° to traditional orientation intended to shorten the bit-lines. (a: From Kohno *et al.* [53], Mitsubishi 1988, with permission of IEEE; b: From Sasaki *et al.* [52], Hitachi 1988, with permission of IEEE.)

Figure 8.37
Photograph of a 14ns 256k × 1 SRAM structured in four blocks with short bit-lines running parallel to the chip length. (From Pfennings *et al.* [11], Philips 1988, with permission of Japan Society of Applied Physics.)

256k SRAM which is shown in the photograph in Figure 8.37. This chip was divided into four blocks which were further subdivided by use of a divided word-line technique into 16 segments.

8.13.7 Shortening the delay in the datapath

Hitachi [52] shortened the delay in the data path of their fast 1Mb by using PMOS load decoders to reduce the address decoder delay and also the decoder area. Figure 8.38(a) shows the new PMOS load decoder along with the more conventional CMOS decoder in Figure 8.38(b). The address signal line was connected only with the NMOS in the new PMOS load decoder rather than being connected with both the PMOS and the NMOS as in the CMOS decoder. This reduced the address load capacitance to about half of that in the CMOS decoder and resulted, therefore, in a shorter delay time to decode the address signals.

The trade-off was with dc power consumption which is higher in a PMOS circuit, where the PMOS transistor is always on, than in a CMOS circuit where it is off except during switching. Hitachi showed that for cycle times faster than 25 ns the ac current component dominates and the PMOS load decoder has a smaller average current as shown in Figure 8.38(c).

Mitsubishi [53] on their fast 1Mb SRAM used a modified gate-controlled data bus driver which reduced the read data bus capacitance and hence minimized the delay time on the data bus.

The Sony [47] divided word-line architecture also resulted in short data-lines and small capacitance of the input node of the sense amplifier.

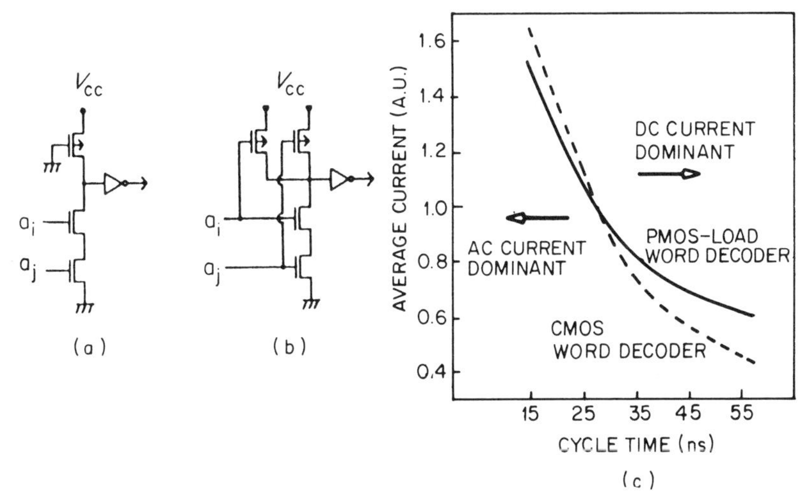

Figure 8.38
Two types of address decoders: (a) PMOS load decoder, (b) CMOS load decoder, (c) trade-offs in current and cycle time of different word decoders. (From Sasaki *et al.* [52], Hitachi 1988, with permission of IEEE.)

8.14 REDUCING SUPPLY LINE NOISE INDUCTANCE EFFECTS

Supply line noise inductance effects, as described in Chapter 5, tend to become more significant with byte wide parts than with $\times 1$ and $\times 4$ parts. This effect which depends on the inductance on the power supply lines both on chip and from chip to package is proportional to both the inductance and to the rate of change of the current with respect to time (di/dt), and causes the power supply lines to oscillate.

Since the chip has no valid output until the power supply lines have stopped oscillating, it has the effect of slowing down the potential output performance of a design. Since the more outputs that go high at one time the more current is drawn, the effect is also roughly proportional to the number of outputs on a chip. Many different design techniques have appeared in an attempt to deal with this problem.

Also with chips using address transition detection, the large rate of change in driver current which induces a large voltage drop across parasitic inductors related to the package, results in noise on the internal power buses. This noise, if large enough, can cause false address transition detection response.

One approach to this problem was taken in the Philips [4] 1987 256k SRAM in which an output buffer was used that drove the load with a constant rate of change of current with time (dI/dt). Figure 8.39(a) shows a simplified circuit schematic of this output buffer.

Another approach to this problem was taken by Motorola [45] in a 1987 256k SRAM of providing power buses for the output drivers which were separate from the rest of the circuit.

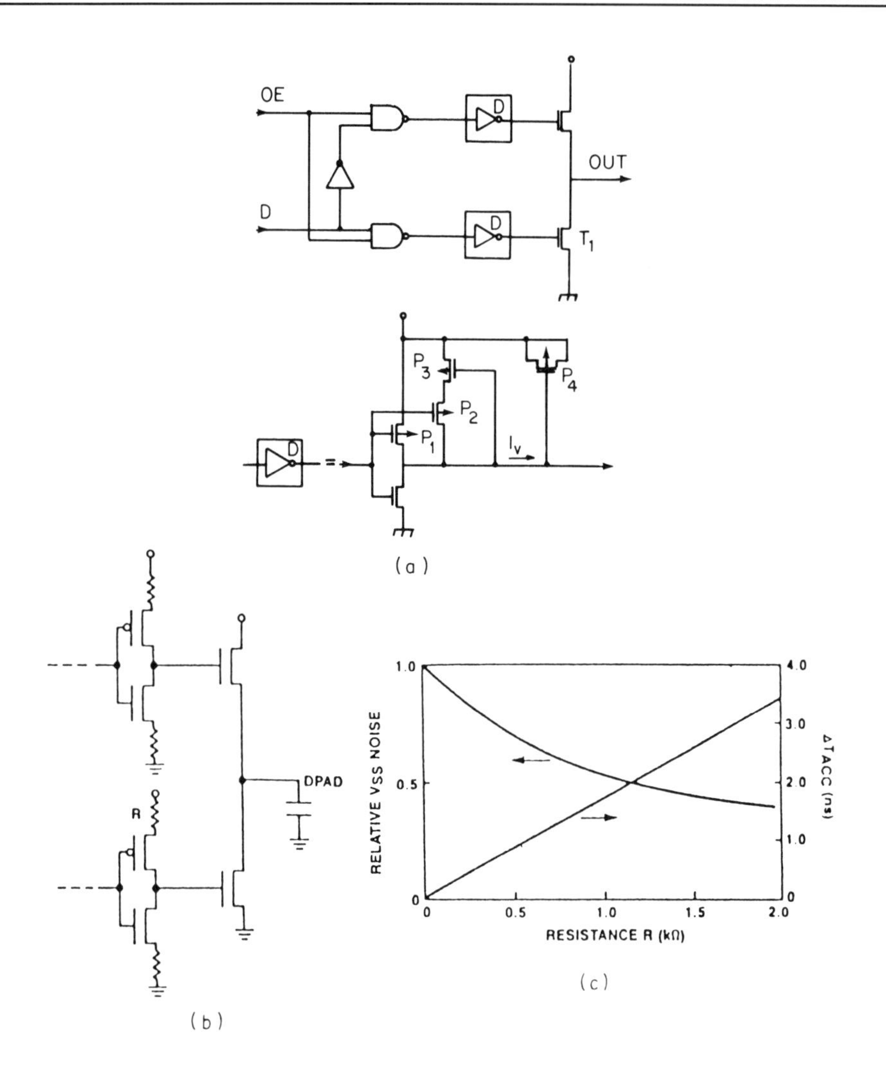
OE
D
OUT
T_1
P_3
P_2
P_1
P_4
I_v
(a)
R
DPAD
(b)
1.0
0.5
0
RELATIVE VSS NOISE
4.0
3.0
2.0
1.0
ΔTACC (ns)
0.5
1.0
1.5
2.0
RESISTANCE R (kΩ)
(c)

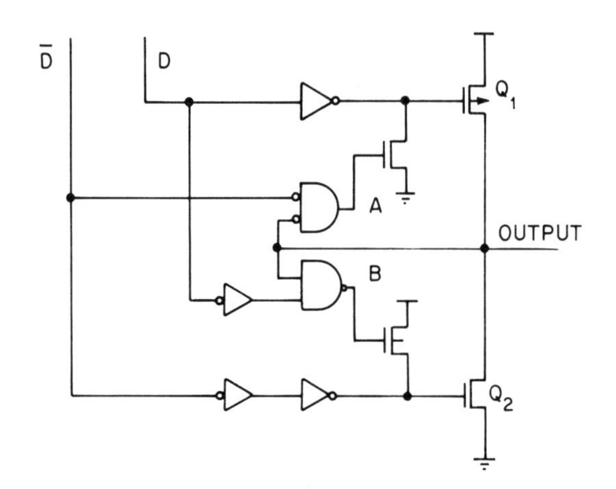
D̄
D
Q_1
A
OUTPUT
B
Q_2

Motorola [45] also used a new output buffer driver design, as shown in Figure 8.39(b), which added resistors to reduce the rate of switching of the output driver and hence reduce the inductive response of the power bus. In this case a reduction in relative noise level by 50% slowed the output down by 2 ns. Figure 8.39(c) shows a chart of power bus noise reduction and of access time penalty as a function of the output buffer resistance.

A Matshushita fast 256k SRAM [65] also used a special output buffer design, as shown in Figure 8.39(d), to reduce the peak current. In this design depending on the data that appeared on D and $\overline{D}$, either the data output level was low and node A pulled up so that the output was precharged to an intermediate level by the current through the output transistor Q_1, or the data output level was high and node B pulled down so that the data output was discharged to an intermediate level by the output transistor Q_2. The result was that the true data output always changed from the intermediate level so the transition time of the output buffer was reduced.

All of the 1Mb SRAMs struggled with how to shorten the delay of the data output without generating a large noise signal on the power and ground lines from the data output buffers. In a byte-wide RAM this noise was multiplied by eight since eight data outputs changed simultaneously.

The Philips [14] 1Mb SRAM used a new output buffer design, shown in Figure 8.40(a), with inverting circuits which were used to control the gates of the n-channel push–pull transistors N_1 and N_3 such that the output current increased or decreased linearly with time. The circuits UC and UD, shown in Figure 8.40(b) and 8.40(c), were designed so that when the output was switched from low to high, N_3 was switched off just fast enough that the current flowing through it stayed about constant during the time that the current through N_1 was increasing at its maximum rate. The same applied for N_1 when N_3 is switched on. Some crossover current flowed but it gave very little contribution to the supply line noise.

This circuit was also designed to minimize supply line noise in the case of resistive loading where N_1 could draw a considerable amount of current. Noise on the internal supply connections of the input buffers was reduced by isolating the output stages of the output buffers as much as possible from the rest of the chip by using separate bondpad connections and separate power bus lines for their supply and by doing double bonding to the package leads.

Sony [47] on the 1Mb SRAM used a dual threshold level data transfer circuit to achieve a more stable readout by avoiding double inversions of the $\overline{\text{DATA}}$ and I/O output.

Figure 8.39
Using special output buffers to reduce ground bounce effects. (a) Output buffer with constant dI/dt, (b) output buffer with added resistance, (c) graph of power bus noise reduction and access time penalty as a function of resistance for output buffer (b), (d) output buffer designed to reduce peak current by making output changes from an intermediate voltage level. (a: From Gubbels *et al.* [4], Philips 1987, with permission of IEEE; b,c: From Wang *et al.* [45], Motorola 1987, with permission of IEEE; d: From Okuyama [65], Matsushita 1988, with permission of IEEE.)

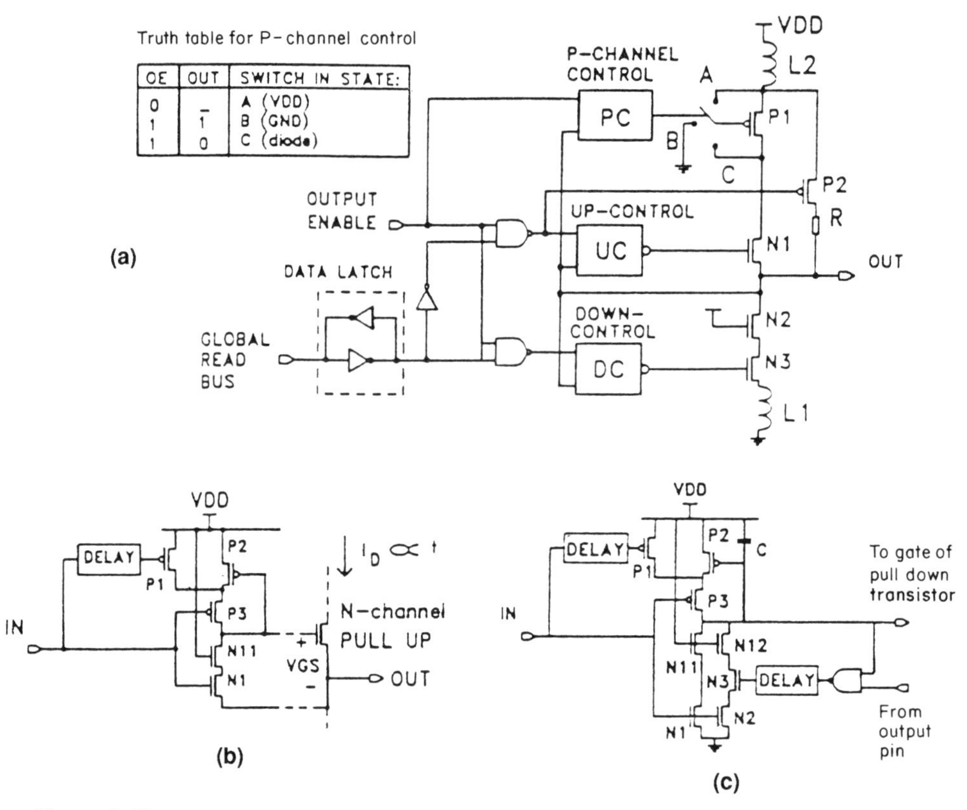

Figure 8.40
Output buffer designed to reduce ground bounce on 1Mb SRAM. (a) Schematic circuit diagram of output buffer, (b) up-control circuit (UC), (c) pull-down circuit (DC). (From Chu *et al.* [10], Philips 1988, with permission of IEEE.)

The Mitsubishi [46] 1Mb SRAM used ATD to precharge the output buffer circuitry to reduce ground bounce effects. This increased access time by 10% and reduced peak current on the power supply lines by 30% during data output buffer transition by precharging the read–write data bus and equalizing the I/O line pair.

The method used pulses generated from the ATD to both equalize the I/O lines and to set the read–write data bus to a middle level for a short time prior to transfer of the data to the data-output buffer.

BIBLIOGRAPHY

[1] *Mostek 1980 Memory Data Book and Designers Guide.*
[2] van Zanten, A. T. *et al.* (1988) *Static Memories: An Overview and Trends*, Philips.
[3] Bastiaens, J. J. J. and Gubbels, W. C. H. (1988) 'The 256-kbit SRAM: an important step on the way to submicron IC technology', *Philips Technical Review*, **44**, No.2, April, p.33.
[4] Gubbels, W. *et al.* (1987) A 40ns/100pF Low-Power Full-CMOS 256k (32k × 8) SRAM, *IEEE Journal of Solid State Circuits*, Vol. **SC-22**, No.5, October 1987, p.741.

[5] Seevinck, E. *et al.* (1987) Static-Noise Margin Analysis of MOS SRAM Cells, *IEEE Journal of Solid State Circuits*, Vol. **SC-22**, No.5, October 1987, p.748.

[6] Ohzone, T. *et al.* (1985) Ion-Implanted Thin Polycrystalline-Silicon High Value Resistors for High Density Poly Load Static RAM Applications, *IEEE Trans. Electron Devices*, **ED-32**, p.1749.

[7] Prince, B. and Burne, J. (1982), A Cost Effective HCMOS Technology, *New Electronics*, February 23.

[8] 'What is CMOS', *Engineering Design News*, June 24, 1981.

[9] *Motorola Memory Data Book*, 1982.

[10] Chu, S. *et al.* (1988) A 25ns Low-Power Full CMOS 1Mbit (128k × 8) SRAM, *IEEE Journal of Solid State Circuits*, Vol. **SC-23**, No.5, October 1988, p.1078.

[11] Pfennings, L. *et al.* (1988) A 14ns 256k × 1 CMOS SRAM with Multiple Test Modes, *VLSI Circuits Proceedings* p.42.

[12] Yaney, D. S. *et al.* (1986) Technology for the fabrication of 1Mb CMOS DRAM, *IEDM Tech. Digest, December*, p.698.

[13] Liptay, J. (1968) 'Structural aspects of the system 360 model 85, II—The cache,' *IBM System Journal*, **7**, No.1.

[14] Kawarada, K. *et al.* (1978) A Fast 7.5ns Access 1K-bit RAM for Cache Memory Systems, *IEEE Journal of Solid State Circuits*, Vol. **SC-13**, No.5, October 1978, p.656.

[15] Torimaru, Y. *et al.* (1978) DSA 4k Static Ram, *IEEE Journal of Solid State Circuits*, Vol. **SC-13**, No.5, October 1978, p.647.

[16] Glock, H. and Burker, U. (1979) An ECL 100K-Compatible 1024 × 4 bit RAM with 15ns Access Time, *IEEE Journal of Solid State Circuits*, Vol. **SC-14**, No.5, October 1979, p.850.

[17] Dingwall, A. G. F. and Stewart, R. G. (1979) 16k CMOS/SOS Asynchronous Static RAM, *IEEE Journal of Solid State Circuits*, Vol. **SC-14**, No.5, October 1979, p.867.

[18] Ohzone, T. *et al.* (1980) An 8k × 8 Bit Static MOS RAM Fabricated by n-MOS/n-Well CMOS Technology, *IEEE Journal of Solid State Circuits*, Vol. **SC-15**, No.5, October 1980, p.854.

[19] Dumbri, A. C. and Rosenzweig, W. (1980) Static RAM's with Microwatt Data Retention Capability, *IEEE Journal of Solid State Circuits*, Vol. **SC-15**, No.5, October 1980.

[20] Ochii, K. *et al.* (1982) 'An Ultralow Power 8k × 8-Bit Full CMOS RAM with a Six-Transistor Cell', *IEEE Journal of Solid State Circuits*, Vol. **SC-17**, No.5, October 1982, p.798.

[21] Minato, O. *et al.* (1982) A Hi-CMOS II 8k × 8 Bit Static RAM, *IEEE Journal of Solid State Circuits*, Vol. **SC-17**, No.5, October 1982, p.793.

[22] Uchida, Y. *et al.* (1982) A Low Power Resistive Load 64 kbit CMOS RAM, *IEEE Journal of Solid State Circuits*, Vol. **SC-17**, No.5, October 1982, p.804.

[23] Hudson, E. L. and Smith, S. L. (1982) An ECL Compatible 4K CMOS RAM, *Proceedings of the ISSCC82*, February, p.248.

[24] Yoshimoto, M. *et al.* (1983) A Divided Word-Line Structure in the Static RAM and Its Application to a 64k Full CMOS RAM, *IEEE Journal of Solid State Circuits*, Vol. **SC-18**, No.5, October 1983.

[25] Watanabe, T. *et al.* (1983) A Battery Backup 64k CMOS RAM with Double-Level Aluminum Technology, *IEEE Journal of Solid State Circuits*, Vol. **SC-18**, No.5, October 1983, p.494.

[26] Sood, L. C. *et al.* (1983) A 35ns 2k × 8 HMOS Static RAM, *IEEE Journal of Solid State Circuits*, Vol. **SC-18**, No.5, October 1983, p.498.

[27] Twaddell, W. (1981) High-speed CMOS RAMs Score Across Memory Spectrum, *EDN*, July 12, p.62.

[28] Barnes, J. J. *et al.* (1984) Circuit Techniques for a 25ns 16k × 1 SRAM Using Address-Transition Detection, *IEEE Journal of Solid State Circuits,* Vol. **SC-19**, No.4, August 1984, p.455.

[29] Minato, O. *et al.* (1984) A 20ns 64k CMOS Static RAM, *IEEE Journal of Solid State Circuits,* Vol. **SC-19**, No.6, December 1984, p.1008.

[30] Miyamoto, J. *et al.* (1984) A 28ns CMOS SRAM with Bipolar Sense Amplifiers, *ISSCC84 Proceedings,* February, 1984, p.224.

[31] Childs, L. F. and Hirose, R. T. (1984) An 18ns 4k × 4 CMOS SRAM, *IEEE Journal of Solid State Circuits,* Vol. **SC-19**, No.5, October 1984, p.545.

[32] Yamanaka, T. *et al.* (1984) A 25ns 64k Static RAM, *IEEE Journal of Solid State Circuits,* Vol. **SC-19**, No.5, October 1984, p.572.

[33] Sakurai, T. *et al.* (1984) A Low Power 46ns 256k bit CMOS Static RAM with Dynamic Double Word Line, *IEEE Journal of Solid State Circuits,* Vol. **SC-19**, No.5, October 1984.

[34] Kang, S. D. *et al.* (1981) A 30ns 16k × 1 Fully Static RAM, *ISSCC81 Proceedings,* February 18, p.18.

[35] Tsujide, T. *et al.* (1981) A 25ns 16k × 1 Static RAM, *ISSCC81 Proceedings,* February 18, p.20.

[36] Sood, L. *et al.* (1985) A Fast 8k × 8 CMOS SRAM With Internal Power Down Design Techniques, *IEEE Journal of Solid State Circuits,* Vol. **SC-20**, October 1985, p.941.

[37] Yamamoto, S. *et al.* (1985) A 256k CMOS SRAM with Variable Impedance Data-Line Loads, *IEEE Journal of Solid State Circuits,* Vol. **SC-20**, No.5, October 1985, p.924.

[38] Shinohara, H. *et al.* (1985) A 45ns 256k CMOS Static RAM with a Tri-Level Word Line, *IEEE Journal of Solid State Circuits,* Vol. **SC-20**, No.5, October 1985, p.929.

[39] Miyanaga, H. *et al.* (1986) A 0.85ns 1-kbit ECL RAM, *IEEE Journal of Solid State Circuits,* Vol. **SC-21**, No.4, August 1986, p.501.

[40] Segawa, M. *et al.* (1986) A 18ns 8KW × 9b NMOS SRAM, *ISSCC86 Proceedings,* February, p.202.

[41] Okazaki, N. *et al.* (1986) A 30ns 256k Full CMOS SRAM, *ISSCC86 Proceedings,* February, p.204.

[42] Ogiue, K. *et al.* (1986) A 13ns/500mW 64Kb ECL RAM, *ISSCC86 Proceedings,* February, p.212.

[43] Flannagan, S. *et al.* (1986) Two 13ns 64k CMOS SRAM's with Very Low Active Power and Improved Asynchronous Circuit Techniques, *IEEE Journal of Solid State Circuits,* Vol. **SC-21**, No.5, October 1986, p.692.

[44] Kayano, S. *et al.* (1986) 25ns 256k × 1/64k × 4 CMOS SRAMs, *IEEE Journal of Solid State Circuits,* Vol. **SC-21**, October 1986, p.686.

[45] Wang, K. L. *et al.* (1987) A 21ns 32k × 8 CMOS Static RAM with a Selectively Pumped p-Well Array, *IEEE Journal of Solid State Circuits,* Vol. **SC-22**, No.5, October 1987.

[46] Wada, T. *et al.* (1987) A 34-ns 1-Mbit CMOS SRAM Using Triple Polysilicon, *IEEE Journal of Solid State Circuits,* Vol. **SC-22**, No.5, October 1987, p.727.

[47] Komatsu, T. *et al.* (1987) A 35-ns 128k × 8 CMOS SRAM, *IEEE Journal of Solid State Circuits,* Vol. **SC-22**, No.5, October 1987, p.721.

[48] Matsui, M. *et al.* (1987) A 25ns 1-Mbit CMOS SRAM with Loading-Free Bit Lines, *IEEE Journal of Solid State Circuits,* Vol. **SC-22**, No.5, October 1987, p.733.

[49] Matsui, M. *et al.* (1988) Two Novel Power-Down Circuits on the 1Mb CMOS SRAM, *VLSI Circuits Symposium,* p.55.

[50] Minato, O. *et al.* (1987) A 42ns 1Mb CMOS SRAM, *ISSCC87 Proceedings,* February p.260.

[51] de Werdt, R. (1987) A 1M SRAM with Full CMOS Cells Fabricated in a 0.7μm Technology, *IEDM Proceedings,* p.532.

[52] Sasaki, K. *et al.* (1988) A 15ns 1-Mbit CMOS SRAM, *IEEE Journal of Solid State Circuits*, Vol. **23**, No.5, October 1988, p.1067.

[53] Kohno, Y. *et al.* (1988) A 14-ns 1-Mbit CMOS SRAM with Variable Bit Organization, *IEEE Journal of Solid State Circuits*, Vol. **23**, No.5, October 1988, p.1060.

[54] Shimada, H. *et a.* (1988) A 46-ns 1-Mbit CMOS SRAM, *IEEE Journal of Solid State Circuits*, Vol. **23**, No.1, February 1988, p.53.

[55] Shimada, H. *et al.* (1988) An 18-ns I-Mbit CMOS SRAM, *IEEE Journal of Solid State Circuits*, Vol. **23**, No.5, October 1988, p.1073.

[56] Hardee, K. and Sud, R. (1981) A Fault-Tolerant 30ns/375mW 16k × 1 NMOS Static RAM, *IEEE Journal of Solid State Circuits*, Vol. **SC-16**, p.435 Oct. 1981.

[57] Konishi, S. *et al.* (1982) 'A 64kb CMOS RAM', *ISSCC Dig. Tech. Papers*, February, pp.258–259.

[58] Kobayasi, Y. *et al.* (1985) A 10μW Standby Power 256k CMOS SRAM, *IEEE Journal of Solid State Circuits*, Vol. **SC-20**, No.5, October 1985, p.935.

[59] Wang, K. L. *et al.* (1988) A Low Standby Power, High Efficiency, Clamped P-Well Bias Generator for High Density Fast SRAMs, *Proceedings Japanese VLSI Conference*, p.51.

[60] Anami, K. *et al.* (1982) A 35ns 16k NMOS Static RAM. *ISSCC Proceedings*, p.250.

[61] Ebel, A. V. *et al.* (1982) A NMOS 64k Static RAM, *ISSCC Proceedings*, p.254.

[62] Tanimoto, K. *et al.* (1983) A 64k × 1b NMOS Static RAM, *ISSCC Proceedings*, p.66.

[63] Schuster, S. *et al.* (1984) A 20ns 64k NMOS RAM, *ISSCC Proceedings*, February, p.226.

[64] Ochii, K. *et al.* (1985) A 17ns 64k CMOS RAM with a Schmitt Trigger Sense Amplifier, *ISSCC Proceedings*, February, p.64.

[65] Okuyama, H. *et al.* (1988) A 7.5n 32k × 8 CMOS SRAM, *IEEE Journal of Solid State Circuits*, Vol. **23**, No.5, October 1988 p.1054.

[66] Sasaki, K. *et al.* (1989) A 9ns 1Mb CMOS SRAM, *ISSCC Proceedings*, February, p.34.

[67] Pfennings, L. *et al.* (1988) A 14ns 256k × 1 CMOS SRAM with Multiple Test Modes, *VLSI Circuits Symposium 1988*, p.11.

[68] *Motorola MOS Memory Databook* 1980.

[69] *Philips Memory Databook* 1989.

[70] *Sony Semiconductor I.C. Memory Databook* 1989.

[71] *JEDEC Standard 21C*, EIA Publication (1990).

[72] Prince, B. (1989) Life Cycles of High Technology Products, *ICCD Seminar Invited Paper*, Dec. 1989

9 FUTURE, FAST AND APPLICATION-SPECIFIC SRAMs

9.1 OVERVIEW

In this chapter the technology and architecture of the first generation of high density 4Mb SRAMs is considered along with early efforts on the 16Mb SRAM. Very fast SRAM technologies are then discussed including Bipolar, BiCMOS, CMOS ECL compatible, and GaAs. This is followed by architectural considerations for speed including: synchronous timing, wide bus configurations, center power and ground pinouts, and separate I/O structures. Finally there is an overview of various speciality SRAMs including BRAMs, dual port RAMs, FIFOs, CAMs, and cache tag RAMs.

In the latter part of the 1980s when the first generation of 1Mb SRAMs were moving into production, SRAMs did not seem like such a promising product line for the future. The 1Mb chips were almost 1 cm^2 in area and difficult to produce cost effectively. The 32 lead 600 mil DIL package was too large to be attractive in high density systems. The 4Mb pseudostatic DRAMs were about the same chip size and threatened to compete in various traditional high volume SRAM applications. The continuing requirement for 5 V external power supply led to reliability problems which limited the potential for scaling SRAMs to more cost effective dimensions.

At the cell level, the reduced noise margin of the R-load cell was becoming a problem in the submicron geometries, but a return to the traditional six-transistor cell with its better stability would have made chips which were already too big even bigger.

The fast SRAMs suffered from ground bounce problems which kept the wide buses required by the 16 and 32 bit microprocessors from being viable SRAM products.

On the applications side, however, the need for high density memories with very low power consumption for battery back-up and battery operation in the rapidly growing portable systems market was increasing. The demand was also increasing for very fast wide-bus memories in the new high performance microprocessor based portable computer systems. Second-level cache applications required high density SRAMs in the 20 to 30 ns speed range and demand for low power first-level cache in the sub-10 ns range was increasing. It was time for a generation of creativity in static memories.

The late 1980s and early 1990s saw this creativity occur. New cell concepts were developed for SRAMs carrying over many of the advanced technologies developed for DRAMs in the mid 1980s. These permitted scaling of high density SRAMs to cost effective chip dimensions. The 4 Mb SRAMs were introduced with 25 ns typical speeds and potentially cost effective chip sizes utilizing these new cell concepts and advanced technologies.

The BiCMOS technologies were developed giving high density and low power in the sub-10 ns speed range, and a low power ECL compatible CMOS technology was introduced as well as battery back-up techniques for BiCMOS SRAMs. New high speed synchronous SRAM architectures were developed to enhance systems performance. These high speed SRAMs supplemented the fast state-of-the-art processors. Static RAM speeds overall kept pace with the speed requirements of the production microprocessors as shown in Figure 9.1.

Ground bounce problems were minimized at the circuit level, and a new standard adopted by the JEDEC memory standards committee reduced the ground bounce problem at the package level.

The JEDEC 3.3 V TTL compatible and CMOS external voltage standards began to be adopted primarily in portable systems. New smaller packages were introduced for high density applications.

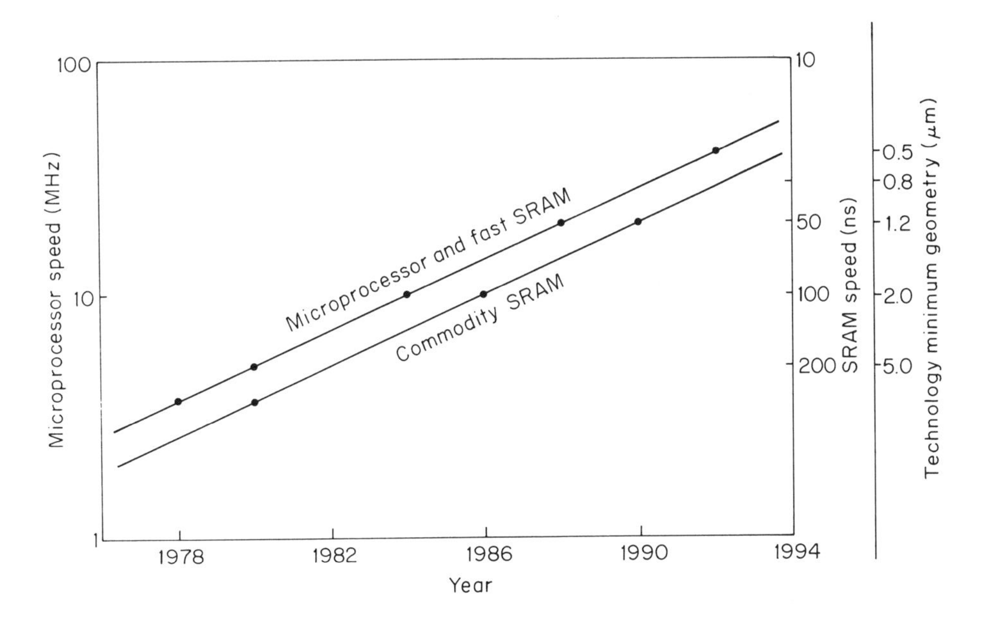

Figure 9.1
CMOS SRAM amd processor speed and minimum geometry by year. (Source ISSCC Proceedings 1976–1990.)

With the new cell concepts the 4Mb SRAM began to move toward production and 16Mb SRAM generations appeared on the horizon. These exciting developments are traced in this chapter.

9.2 4Mb, 16Mb AND FUTURE SRAMs—AN OVERVIEW

The 4Mb generation of SRAM required microamp standby power for the high density battery back-up applications such as memory cards and handheld computers. They demanded speeds in the sub 35 ns specified access time region to interface with the 40 MHz plus processors without wait states. They were also expected to meet high standards of reliability of operation and they had to be small enough to be manufacturable with reasonable yield.

The linear scaling trend which had been followed since the introduction of the R-load cell in the early 1980s was no longer possible. 1Mb SRAM chip was larger than the 70 mm^2 the historic scaling trends would have predicted as shown in Figure 9.2 which plots chip size against cell size for various generations of SRAMs. If the 0.8 μm 1Mb technology were directly scaled to 0.6 μm to make a 4Mb then the chip size of the 4Mb would have been over 200 mm^2. This is even further above the historical chip size trend of geometries possible to manufacture in 1990.

Larger chip size means a lower manufacturing yield. This in turn implies a higher cost-per-bit than the historic trend line would forecast. The historic cost-per-bit line for SRAMs is plotted in Figure 9.3 using Dataquest numbers and then projected to the 64 Mb.

This projection can be made, if we assume that the price per mm^2 for end-of-life memory silicon continues to remain fairly constant over time. The chip size can then be directly related to the cost-per-bit at end-of-life prices. This downward cost evolution effects the rate at which new applications for SRAMs are developed, and hence effects directly the growth of the SRAM market. A higher cost also encourages the development of competing products such as pseudostatics or flash memories which, if providing a better cost and performance, will also decrease the market available for the SRAMs.

Since 50% to 60% of the area of an SRAM is the memory array, the problem of reducing the size of the chip becomes one of reducing the cell size. Figure 9.4 plots cell size as a function of density for the SRAMs.

If the 1Mb SRAM cells had been scaled to the same trend as the 64k and 256k R-load cells then the cell size would have been about 35 μm^2 and, referring to Figure 9.2, the chip size would have been about 70 mm^2. This would have meant a minimum geometry of about 0.6 μm. This was prevented at the time both by scaling difficulties with the R-load cell, and hot carrier considerations for sub 0.8 μm devices with 5 V external power supply. Also the 0.8 μm process was already available for the 4 Mb DRAM and the larger chip was able to fit in the 32 lead standard 600 mil. Therefore, the larger cell size in 0.8 μm geometries was used. Since the cell size of the 1 Mb was about 45 μm^2 then to stay on the same trend line 18 μm^2 was needed for the 4 Mb and 6.5 μm^2 for a 16Mb SRAM cell.

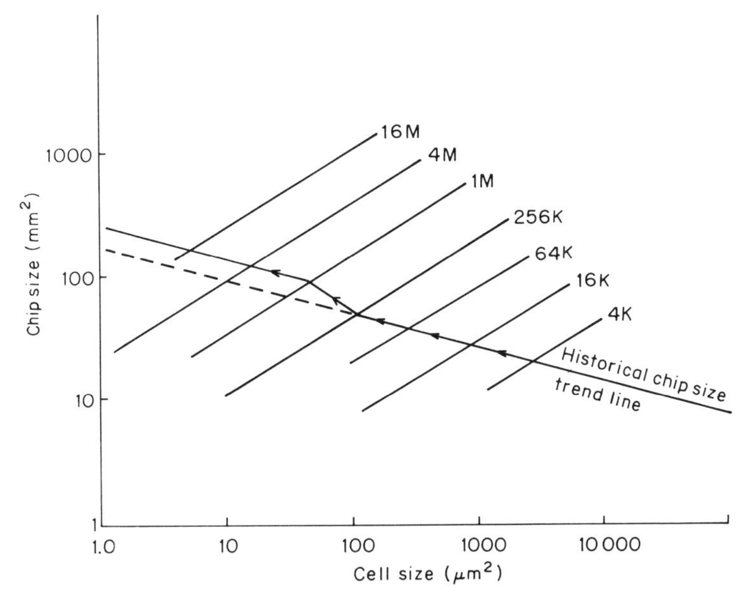

Figure 9.2
Historical chip size trend line with SRAM chip size plotted against cell size. (Source ISSCC Proceedings 1980–1990.)

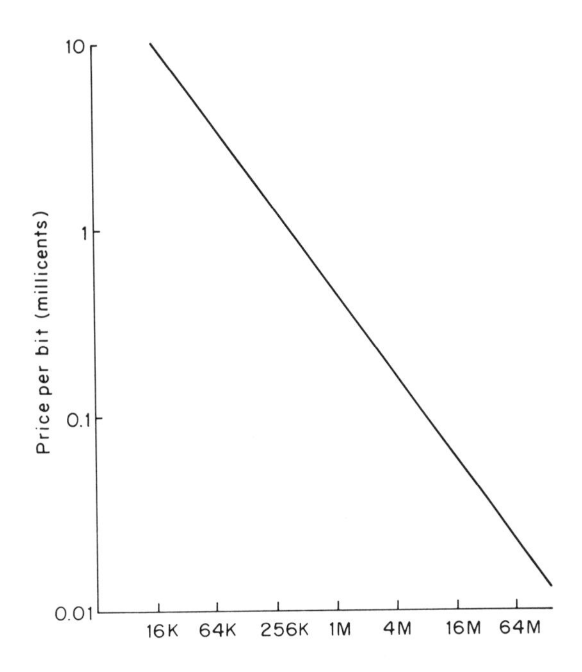

Figure 9.3
Estimated price per bit for various generations of SRAMs at early end-of-life prices. (Source Dataquest 1980–1989, ISSCC Proceedings 1978–1990.)

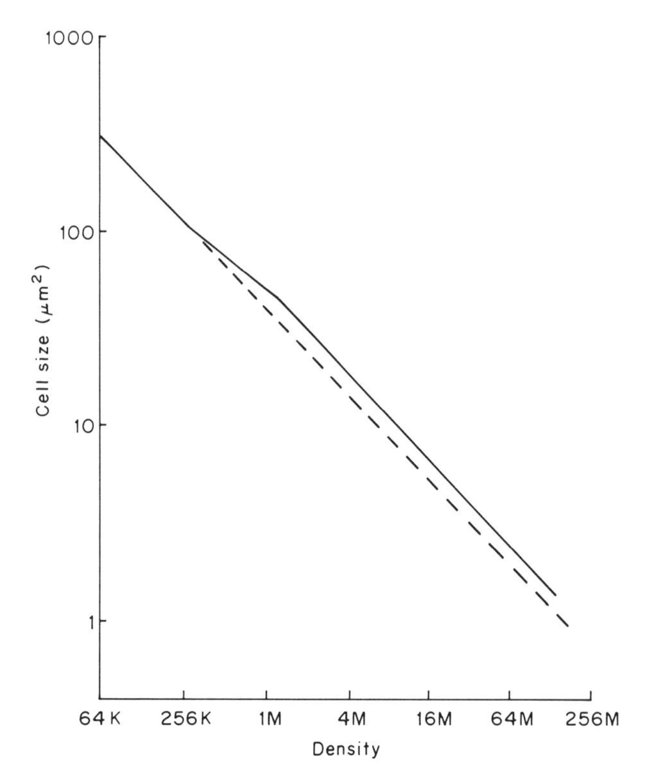

Figure 9.4
SRAM cell size plotted against density. (Source various ISSCC Proceedings.)

9.3 MULTIMEGABIT CELL TECHNOLOGY

The polysilicon load resistors which had served the SRAM commodity markets well for four generations were no longer adequate for both low power and high speed in the 4Mb generation. The low power and wide noise margins of the six-transistor cell were needed. The requirement for a small chip size demanded, however, that the memory cell be as small or smaller than the scaled poly load resistor cell would have been.

One reason for the scaling problem with the R-load cell was that the high current supply needed to retain the 'high' node level in the memory cell needed to be traded off with the low stand-by current requirement.

If the polysilicon resistors have high resistivity then the standby power will be reduced. The standby current requirement therefore sets a minimum value to the resistance. This minimum value increases with each density generation since the standby current requirement stays the same but the number of resistors increases four-fold.

On the other hand, the resistance must be small enough to provide enough current to the storage node so that the stored charge does not leak away because of the sub-threshold leakage in the storage transistors. This sets a maximum value on the resistance. Since the sub-threshold leakage of the storage transistors increases with decreasing threshold voltage and the threshold voltage drops as the geometry decreases, this maximum value of the resistance decreases with density. The gap between the minimum and maximum value of the resistance, therefore, decreases as the density of the chips increases making the process difficult to control.

Another factor in the control of the R-load cell was that the power supply of the memory array in the 0.5 to 0.6 μm technologies had to be less than 5 V for reliability reasons even though the external chip power supply was maintained at 5 V in many cases. The reduced stability and lower noise margins of the R-load cells at low voltages were discussed in Chapter 8.

The industry was looking for a cell as small as a poly load resistor cell and as stable and low power as a six-transistor cell. Clearly either innovation at the cell level or tighter process control or both were required.

9.3.1 Process developments for deep submicron transistors

Many companies developed memory cells for the 4 Mb SRAM generation and beyond. Efforts in sub 0.5 μm cells for the 4 Mb generation continued both in fine tuning the scaling of the traditional cell technology, in refinement and extension of the stacked poly-load transistor cell, and in new types of cell technologies.

Beyond that all major manufacturers continued to investigate advanced processing of deep sub-micron transistors. For example, Motorola [20] investigated a 0.4 μm process with reduced size inter-well isolation which allowed scaling of the N^+ to P^+ space to less than 2 μm, and an improved active transistor design which reduced the size of the bird's beak.

IBM [21] looked at a deposited CVD gate oxide for 0.25 μm channel length NMOS and PMOS transistors which could be deposited at low temperatures. IBM [23] also investigated the scaling properties of trench isolation used with ECL transistors. Matshushita [22] looked at forming 0.25 μm transistors with drain structures implanted at an angle to minimize the gate overlap.

9.3.2 Refinement and scaling of planar CMOS cells

Several companies including Mitsubishi, NEC and Sony, choose to continue work on optimizing the R-load SRAM cell at the 4 Mb SRAM density level.

Mitsubishi [31] in 1989 described work on the R-load SRAM cell in the 0.5 μm geometry for a 4 Mb SRAM. They used a four-level polysilicon and two-level aluminum CMOS technology with the load resistor formed in the very thin fourth level of polysilicon. This resistor was both very thin and narrow and of very high resistivity polysilicon. 0.6 μm trench isolation was used.

This cell was relatively complex due to the measures that were taken to increase the noise margins of the R-load cell under a low V_{CC} condition. Memory cell stability was increased by increasing the threshold voltage of the driver transistor, using a salicide ground line to reduce changes in V_{SS} below 0.04 V, and by decreasing the conductance of the access transistor. An extra P^+ buried barrier layer in the P-well was used because of the soft error sensitivity of the R-load cell. NEC and Sony also used R-load cells for their 4Mb SRAMs.

Philips chose the planar bulk SRAM cell with a six-transistor cell structure showing a 25.s μm^2 full CMOS cell in 1990 [79]. This cell used fully overlapping contacts in 0.5 μm technology. The bulk CMOS technology offered the low off-current, high drive capability, and cell stability typical of the full six-transistor bulk CMOS technology.

Motorola [81] meanwhile developed a fast 4Mb SRAM capability in BiCMOS using a four-transistor R-load cell in 0.5 μm technology with a three-layer polysilicon process giving a less than 20 μm^2 cell.

9.3.3 The stacked transistor CMOS cell—poly-load PMOS

Other companies worked on alternative cells for the 4Mb SRAM which would reduce the drawbacks of the R-load cell but retain its small chip size. These drawbacks include larger standby current, lower stability reflected in smaller noise margins, and lower soft error immunity. What was needed was a cell that was the size of a four-transistor R-load cell but with the advantages of the full CMOS cell.

One answer to this quest appeared in the 4Mb SRAM generation as a six-transistor cell with four NMOS transistors in the silicon substrate and the two PMOS load transistors formed in the thin film polysilicon layers above the cell much as the resistor loads were stacked in the R-load cell.

This thin film stacked PMOS load transistor cell was a natural evolution for those manufacturers who used the poly load resistor cell and wanted the stability of full CMOS in the submicron geometries. It was also a natural progression for those who had been using the six-transistor cell CMOS but wanted to reduce it in size to remain competitive. While thin film transistors had been experimented with since the early 1980s, this was their first emergence into commodity memories.

A 1Mb SRAM with a stacked transistor-in-poly cell in 0.8 μm technology was described by NEC [18] in 1988. This SRAM employed P-channel polysilicon resistors as cell load devices with offset gate–drain structure in the second-level poly with the gate below it in the first-level poly. It was stacked on N-channel driver transistors as shown in Figure 9.5(a). This vertical technology cell promised the size benefit of the R-load resistor cell and the noise margin of the six-transistor cell.

Also in 1988 Hitachi [19] showed a 25.4 μm^2 cell in 0.5 μm minimum geometry for use in 4Mb SRAMs which used PMOS load transistors in the third-level poly with the gate below it in the second-level polysilicon as shown in Figure 9.5(b). This process had three-layer poly and two-layer metal. While a cell size this small was a significant achievement for the time, a 4Mb SRAM using this cell with a 60% array efficiency would be over 170 mm^2. This was still too large to be manufacturable with

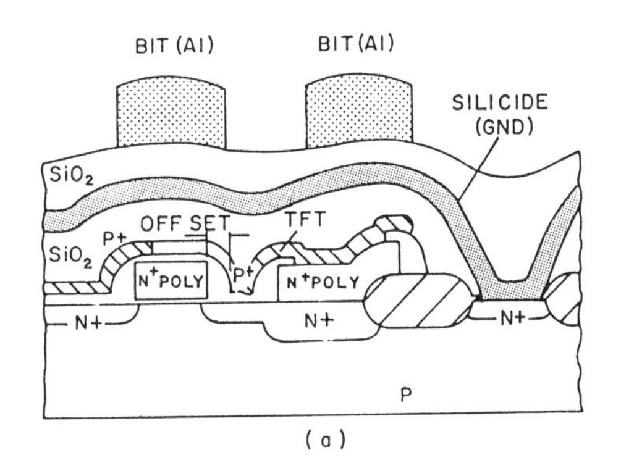

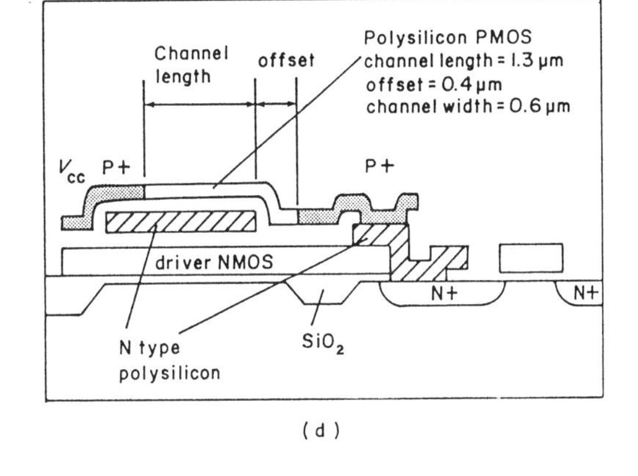

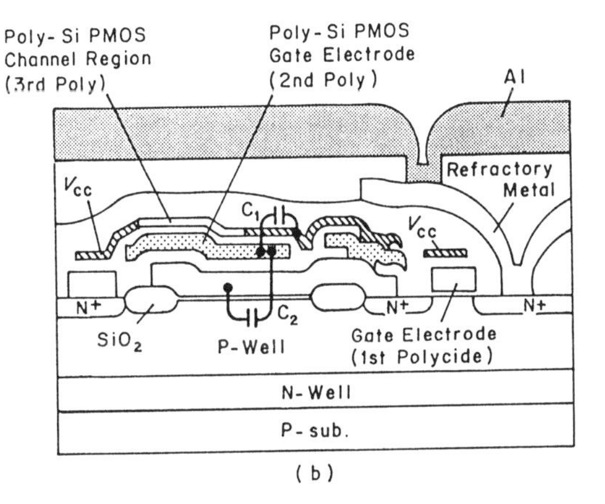

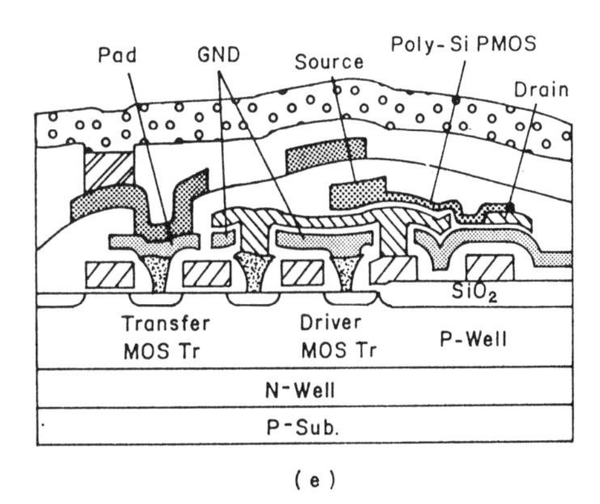

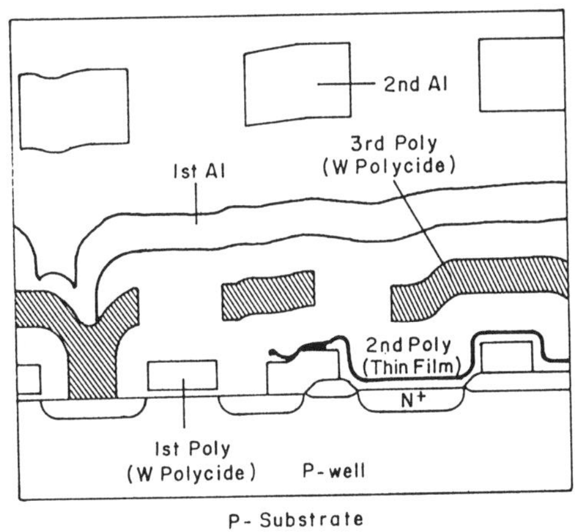

Figure 9.5
Five examples of stacked polysilicon PMOS load transistors for 4Mb and 16Mb SRAMs. (a) 4Mb SRAM cell. (From Ando *et al.* [18], NEC 1988, with permission of IEEE.) (b) 4Mb SRAM cell. (From Yananaka *et al.* [19], Hitachi 1988, with permission of IEEE.) (c) 4Mb SRAM cell. (From Hayakawa *et al.* [30], Toshiba 1990, with permission of IEEE.) (d) 4Mb SRAM cell. (From Ikeda *et al.* [78], Hitachi 1990, with permission of IEEE.) (e) 16Mb SRAM cell. (From Yamanaki, *et al.* [80], Hitachi 1990, with permission of IEEE.)

reasonable yield. The industry needed to get below 120 mm^2 for a high yielding 4 Mb SRAM manufacturability.

Early in 1989 Hitachi [26] described a very fast 9 ns 1 Mb SRAM with the new PMOS load cell also in 0.5 μm geometry and the triple poly and double metal technology, but now the cell size was 21 μm^2. Assuming the same 60% array efficiency for 4Mb a chip using this cell would be 143 mm^2. It was noted that the soft error rate of this 1Mb SRAM was reduced to less than a tenth of that of a similar chip with the R-load polysilicon load cells.

In a subsequent paper in 1990 Hitachi [27] described a 4Mb SRAM made with their third-generation polysilicon PMOS load memory cell now in a quadruple poly and double metal 0.5 μm process with the cell size now reduced to 17 μm^2. The polysilicon PMOS transistor was formed in the third and fourth poly layers. The gate electrode was formed by the third poly layer and the channel, source, and drain regions by the fourth poly layer.

With the very small cell size of this 4Mb SRAM, the chip was 122 mm^2 which is close to the trend line for manufacturable yield shown earlier in Figure 9.2. This SRAM had 18 million or so transistors on a chip just over 1 cm^2.

Later in 1990 Hitachi presented a modification of this cell [78] which is shown in Figure 9.5(d). While the cell size remained at 17 μm^2, there are changes in the process and the architecture designed to reduce the off-current thereby reducing the standby current characteristics and to increase the on-current resulting in better data retention stability and soft error immunity. These dual goals were accomplished by reducing the thickness of the channel polysilicon and gate oxide to 40 nm and by adopting a gate-to-drain offset structure. The off-set reduced the off-current by an order of magnitude at a V_{DD} of 5 V. The thin gate oxide increased the on-current. A phosphorous ion implant gave the precise control needed for the threshold voltage.

In 1990 Toshiba [30] showed a 4Mb SRAM with a thin film PMOS load transistor in 0.5 μm triple-poly and double Al CMOS technology with cell size of 20.3 μm^2. The second-layer polysilicon was used for the channel of the PMOS load transistor with the N^+ diffusion area of the driver transistors acting as a gate electrode of the PMOS load transistor as shown in Figure 9.5(c). Since the diffusion area acted both as a drain of the driver transistor and the gate electrode of the thin film PMOS poly transistor, additional gate electrode layers were not needed and therefore the layout had a planarized structure.

In 1991 Mitsubishi [89] showed another 4Mb cell, this time using a thin film polysilicon PMOS transistor to attain very low 0.4 μA standby current. The cell size was 19.5 μm^2 giving a 134 mm^2 chip. This chip was intended for use in 3 V battery operated systems so the six-transistor cell was needed for stability as well as for the low current leakage characteristics.

Hitachi [80] in late 1990 showed an early 16Mb SRAM cell with 5.89 μm^2 cell area in 0.25 μm geometry using phase-shift lithography techniques as discussed in Chapter 4. This cell was a refinement of the earlier cell with off-set polysilicon PMOS transistor. Four-layer polysilicon was used with the PMOS transistor in the fourth polysilicon layer with its gate in the third poly. A self-aligned structure was used to reduce the cell size.

9.3.4 Buried layers and silicon-on-insulator structures (SOI)

While a tightly controlled R-load cell technology or a thin film polysilicon PMOS cell appeared to be adequate for the 4Mb and 16Mb SRAM generations, the 2 μm^2 cells required in the 64Mb generation required further advances.

Many companies investigated the performance, reliability and scaling advantages of silicon-on-insulator (SOI) technology. One of the most commonly used techniques was SIMOX (Silicon IMplanted with OXide) in which oxide ions are implanted below the surface of the silicon and then annealed to form a thin layer of SiO_2 below a surface layer of single crystal silicon.

SOI transistors are attractive for deep submicron technologies because of such properties as improved subthreshold slope, reduced short channel effects, reduced electric fields, improved immunity to soft errors due to the small volume of silicon, elimination of latch-up, and total dielectric isolation [33].

Fully depleted submicron thin film SOI devices present lesser short channel effects than bulk devices and have the potential for higher transconductance. A simpler and more compact circuit design can be used since direct silicided contacts can be used between N^+ and P^+ diffusions, there are no wells so tub contacts are eliminated, and simple isolation schemes such as LOCOS can be used. A simplified cross-section of a thin film SOI CMOS invertor is shown in Figure 9.6(a).

Philips [34] in 1989 investigated the device and circuit properties of 0.5 μm CMOS on ultra-thin SOI and fabricated both a ring oscillator with 0.4 μm gate length transistors to study circuit performance and a fully functional 0.5 μm minimum geometry 2k SRAM to investigate large circuit techniques. Good performance was obtained compared to bulk CMOS as shown in the graph of delay per stage compared to power supply voltage for the ring oscillator shown in Figure 9.6(b).

A similar study by AT&T Bell Labs [36] for a ring oscillator with transistor gate length of 0.6 μm and effective channel length of 0.4 μm showed a significant improvement in delay per stage ranging from 50% at 1.25 μm to about 30% at 0.5 μm.

One of the major benefits shown by both studies was the simplified fabrication technologies possible with the thin film SOI.

Mitsubishi [36] investigated the aspect of current driveability of the thin SOI MOSFETs and showed the current driveability of the SOI technology compares satisfactorily with that of bulk silicon.

The Philips study also showed that significant investigations remained to make the SIMOX a viable device technology. Floating substrate effects and parasitic bipolar transistor action were shown to strongly influence the device properties of the thin film SOI NMOS devices. Compared to the bulk devices there was a strong reduction of the breakdown voltage and an enhanced hot carrier degradation.

9.3.5 Vertical CMOS transistors

For the 16Mb and 64Mb SRAM many companies also continued to look at new forms of vertical transistor and cell structures. As well as stacking additional

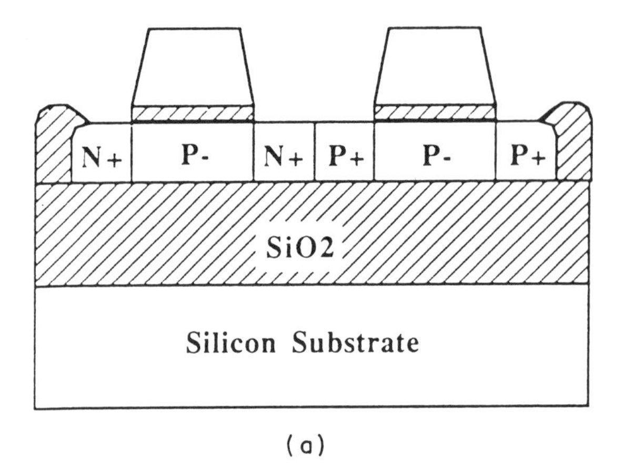

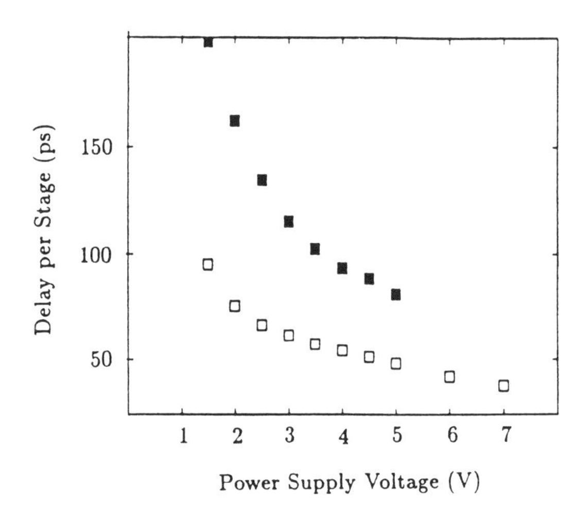

Figure 9.6
(a) Cross-section of a thin film SOI CMOS inverter. (From Colinge *et al.* [33], IMEC 1989, with permission of IEEE.) (b) Delay per stage for bulk CMOS and SIMOX for a ring oscillator. (□ SOI, ■ bulk). (From Woerlee *et al.* [34], Philips 1989, with permission of IEEE.)

interconnect layers with transistors over the surface (poly load PMOS) or burying layers under it (SIMOX), another direction is to dig vertical trenches into it or etch vertical pillars out above it.

In 1988 Toshiba [24] proposed the 'Surrounding Gate Transistor' (SGT). This is a transistor whose gate electrode surrounds a pillar silicon island and reduces the occupied area of the CMOS invertor by 50%. Conventional silicon grooving technology, as developed for DRAM trenches, is used.

A three-dimensional schematic of the SGT transistor is shown in Figure 9.7(a), typical transistor characteristics are shown in Figure 9.7(c), and the size area reduction potential for an equivalent CMOS invertor is shown in Figure 9.7(b).

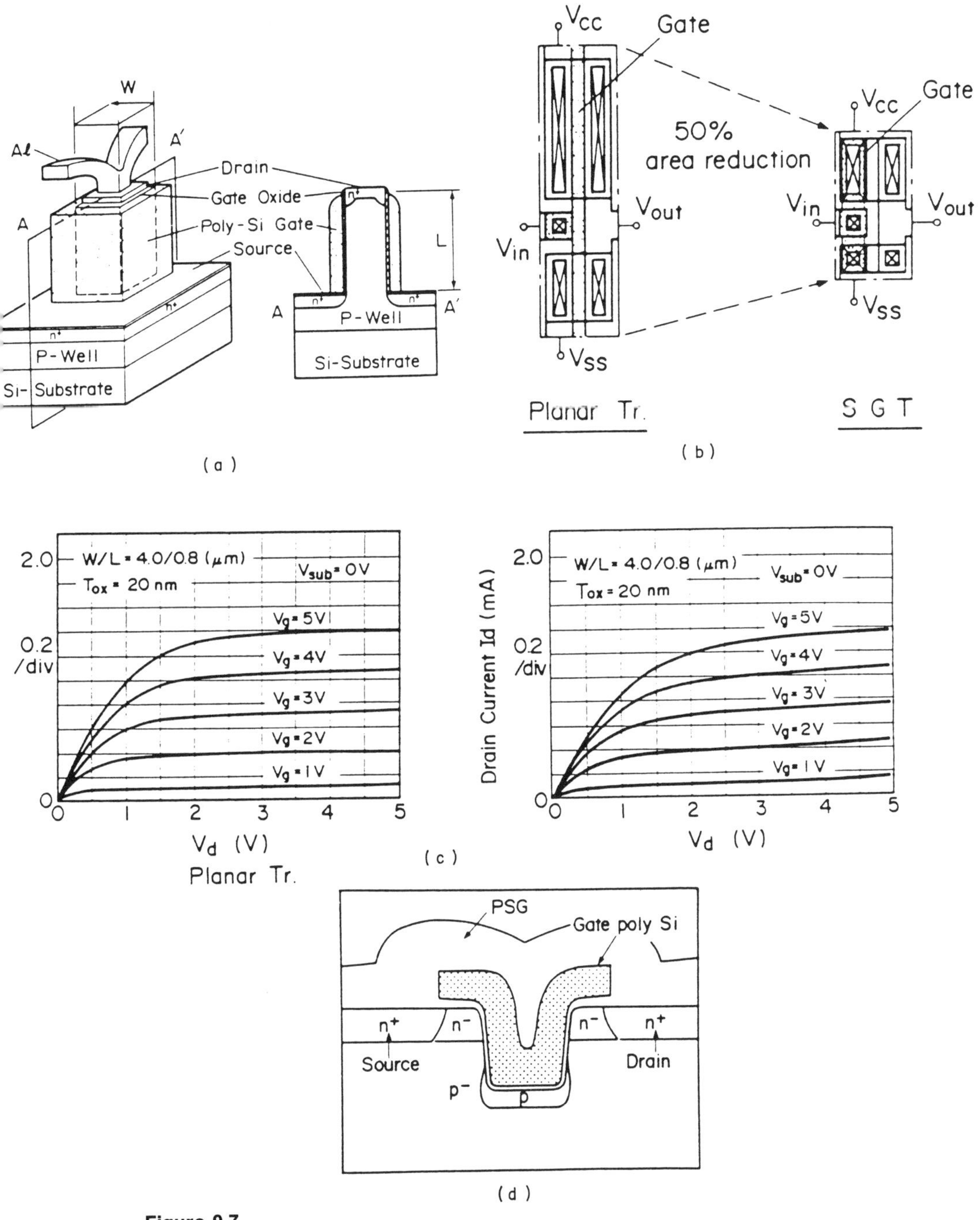

Figure 9.7
(a) Three-dimensional schematic of vertical SGT transistor. (b) Area reduction potential for SGT plotted against planar invertor. (c) Typical characteristics of SGT transistor. (d) Schematic cross-section of trench type transistor. (a,b,c: from Takato *et al.* [24], Toshiba 1988, with permission of IEEE; d: From Sunouchi *et al.* [25], Toshiba 1988, with permission of IEEE.)

In the SGT the source, gate and drain were arranged vertically using the side-walls of the pillar silicon island as the channel region. The pillar side-walls were surrounded by the gate electrode. The gate length was adjusted by the pillar height.

The SGT potentially reduces serious problems from both the short channel effects (gate length) and narrow channel effects (gate width) of planar transistors in deep submicron geometries. Short channel effects include threshold voltage lowering, increase of substrate bias effect due to impurity concentration enhancement in the channel region, and reliability degradation by hot carriers. The narrow channel effects include the decrease of current driveability and reliability degradation due to the large electric field at the edge of the LOCOS.

The results of the study showed that the characteristics of the SGT were comparable to those planar transistors. Toshiba also in 1989 showed a continuation of their work on the SGT in a DRAM memory cell which was described in Chapter 7.

In 1988 Toshiba [25] also described a trench type of transistor which they called the double LDD concave transistor for an 0.5 μm technology as shown in Figure 9.7(d). The short channel effects are improved over the planar transistor since source and drain junctions are above the channel. This transistor can operate at a 5 V supply voltage.

9.3.6 Combination vertical and stacked transistor technology

There are also possibilities in combinations of vertical and stacked transistor and cell technologies for 64Mb generations and beyond.

In 1989 Texas Instruments [29] proposed an innovative six-transistor cell with a thin film polysilicon PMOS transistor and a vertical sidewall trench NMOS transistor as shown in Figure 9.8(a). In 0.5 μm technology this gave a 23 μm^2 cell with the speed and power properties required of the 4 Mb SRAM. Since the thin film transistor could in theory be stacked over the vertical NMOS, this cell appears to have the potential for scalability to higher densities.

The vertical NMOS was made at the same time as the poly collector plug for a bipolar transistor with the buried N^+ layer acting as the grounded source. The trench formed the gate of the vertical NMOS and was patterned at the same time as the contact to the buried N^+ layer. After the trench was etched into the silicon a 120 nm gate oxide was grown on the surface of the trench followed by a doped polysilicon deposition which formed the gate of the vertical transistor and the buried N^+ contact. The polysilicon PMOS load device was formed in the third poly with an underlying layer of second-level poly acting as a bottom gate. The gate dielectric was made of LPCVD oxide with a low temperature anneal.

Another interesting 'combination' transistor structure which was investigated for the sub 0.2 μm technologies was the 'delta transistor' proposed by Hitachi in 1989. This device, shown in Figure 9.8(b), combined vertical pillar technology with fully depleted SOI transistor technology using only bulk silicon process techniques.

In this process a pillar of silicon was formed using standard trench etching techniques. Silicon nitride was deposited as the gate dielectric over this pillar and then

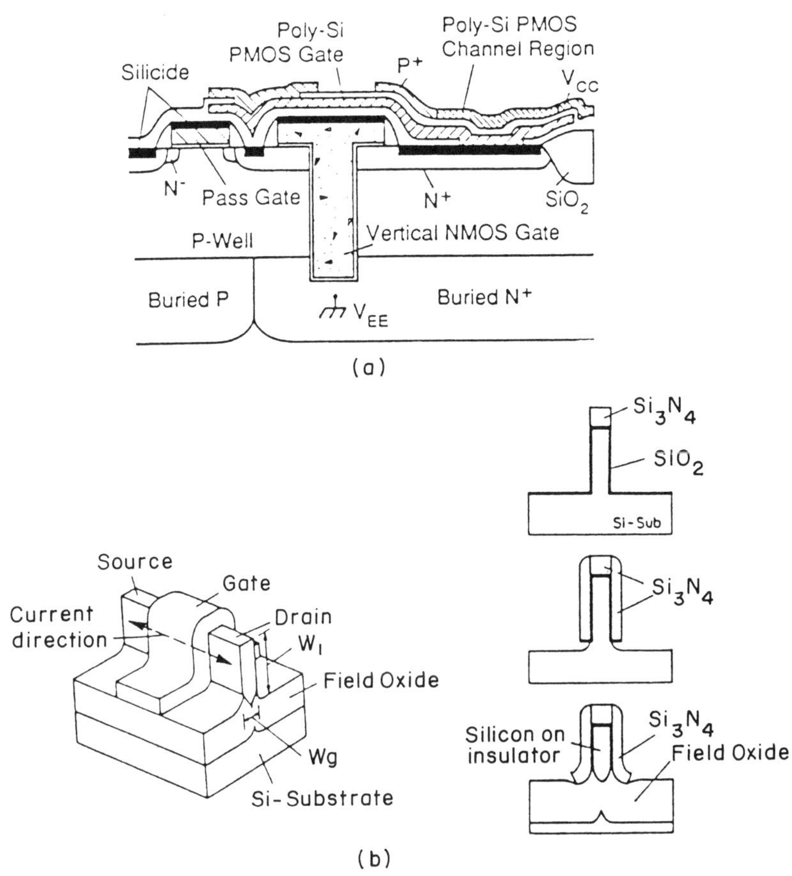

Figure 9.8
Combination vertical and stacked transistors. (a) Stacked polysilicon PMOS and trench NMOS transistor. (b) Vertical pillar transistor on field grown SOI. (a: from Tran *et al.* [28], TI 1989 , with permission of IEEE; b: from Hisamoto *et al.* [37], Hitachi 1989, with permission of IEEE.)

selective oxidation of the bulk silicon beneath the pillar isolated the silicon of the pillar and formed an 'SOI' vertical transistor above the oxide. This process eliminated the need for isolation and permitted the larger gate length and width of the vertical transistor.

Going even one step further, perhaps for the 256 Mb SRAM, one could imagine using vertical techniques to stack even more of the transistors from an SRAM. For example, in 1989 NEC [32] showed a master slice using four layers of stacked active devices. This circuit contained a PLA and a CMOS gate array stacked over a 1k CMOS SRAM using SOI laser beam recrystallization to form the SOI layers. A schematic of this chip is shown in the wafer-scale-integration section of Chapter 13.

It is not until the 1 Gb SRAM generation that derivations of currently known technology seem totally insufficient for the manufacture of SRAMs on the historical trends.

9.4 EXTERNAL POWER SUPPLY LEVEL CONSIDERATIONS

A factor which limited the extent of scaling the SRAMs to smaller geometries at the 1Mb SRAM level was the continuing systems requirement for 5 V external power supply. This meant that either the scaling was limited to about 0.8 μm minimum geometry to avoid hot carrier effects, or that smaller geometries and a lower voltage were used on the internal matrix while the external I/O circuitry remained at 5 V. Beyond the 1Mb SRAM other solutions were sought.

9.4.1 Power supply levels on the 4Mb SRAM

The problem was addressed in the 4Mb SRAM generation both by the offering of 3.3 V power supply chips, and by offering chips with low voltage memory array and 5 V periphery.

The two voltage level system had several disadvantages. It meant that the periphery, and hence the chip size, was slightly larger. It also meant that voltage down converters had to be used between the matrix and the peripheral circuitry which consumed power and required additional circuit design considerations to permit battery backup operation to be used. It also meant that two gate oxide thicknesses had to be used—one in the periphery and one in the matrix—thereby increasing the complexity of the process and making it harder to control.

While the JEDEC memory standardization body had passed a 3.3 V TTL compatible interface standard for integrated circuits in the mid 1980s [41], it was not yet widely adopted when the 16Mb DRAMs and 4Mb SRAMs were first introduced.

In 1989 the JEDEC memory standardization committee took the unusual step of publicly recommending the 3.3 V standard for the 64Mb DRAM prior to actual standardization. This, of course, also impacted industry thinking about the 16Mb SRAM which was expected to be in the same technology. It also set an environment in the industry of looking forward to the lower voltage standard which encouraged some users and suppliers to initiate conversion at the 4Mb SRAM level. Sony and Mitsubishi both showed 4Mb SRAMs which operated at 3 V only, and some of the other early 4Mb SRAMs were configured to work at 5 V or 3 V.

9.4.2 Power supply levels for the 16Mb SRAM and beyond

Background work has been done to define an optimum low voltage technology for the 16Mb SRAM and beyond.

In 1988 Texas Instruments [43] reported an 0.5 μm CMOS technology with 12 nm gate oxide thickness that gave at least a 20% speed improvement at a 3.3 V supply voltage compared to an 0.8 μm technology at 5.0 V with 20 nm gate oxide.

Toshiba [44] also investigated the optimum power supply voltage for an 0.5 μm technology and found it to be about 3.0 V. The optimum point exists since circuit speed for CMOS devices improves monotonically with power supply voltage but eventually reaches a saturation point.

While the 16Mb SRAM cell array will probably be made in a tighter technology than 0.5 μm, devices with 3.3 V external power supply voltage and lower volt internal supply can be used so that the above work is probably applicable to the 16Mb SRAM.

For the internal cell array voltage at the 16Mb densities IBM [42] in 1988 presented a paper on 0.25 μm geometry processing which focused on finding the power supply voltage which optimized the trade-offs between performance and reliability. A reduced operating voltage in the range of 2.2 to 2.5 V was found to minimize average stage delay of an experimental ring oscillator for fixed hot carrier margins.

This work is also a beginning in finding standards for the next level of circuits with perhaps internal voltages in the 2 V range and external at the new 3.3 V standard.

9.5 ARCHITECTURE FOR 4Mb SRAMs

While innovative cell development went on for the 4Mb SRAMs and beyond, interesting architectural developments also occurred.

The first 4Mb SRAM, a chip from Sony [60], appeared in 1989 and another four 4Mb SRAMs were presented at the 1990 ISSCC. All five are shown in Table 9.1. Five or six level interconnect structures were used in the technology of the 4Mb SRAM generation, with the exception of the Sony chip which continued to use only four.

With a typical (5 V, 25°C) access time of 15 to 25 ns, the specified access time at worst case voltage and temperature would be 25 to 35 ns. This was in line with the 30 to 40 MHz microprocessor speeds of the early 1990s so that these parts could work without wait states in a system.

The average size of the poly PMOS cells was slightly smaller than the R-load cells as shown in Table 9.2, and as a result the average chip size with the poly PMOS cell was also smaller than could be accounted for by the slight increase in average array efficiency. Again the Sony chip was an exception with a chip size of 130 mm^2 due to the historically high 67% array efficiency attained in this design.

The next noticeable item is that all of these 0.5 μm SRAMs used a reduced internal voltage in the array while three of the five still supported the 5 V external standard. Two of them, the Sony [60] and Mitsubishi [39] parts, went to the 3.3 V JEDEC interface standard.

Following historical trends, the microampere standby power rating of the bytewide devices indicated parts targeted at the low standby power battery operated laptop system and memory card market. The higher speed of the bitwide parts indicated an intent to service larger computers which not only could use the deeper parts, but where the data retention current mattered less.

Looking now at the architectures of these five parts, the Mitsubishi [39] 4Mb SRAM uses a further refinement of their divided word-line structure from the 1Mb SRAM which they called the Hierarchical Word Decoding architecture. Due to the increase in the number of blocks to 32, the word-line was now divided into three hierarchical levels to decrease the larger load capacitance on the global word-line as shown in Figure 9.9. A two-stage sense amplifier was also used.

Table 9.1 Characteristics of early 4Mb SRAM (Source various ISSCC Proceedings.)

Firm	Org	V_{DD} (V)	Type	Inter	Geo (μm)	TACS (ns)	Cell (μm^2)	Chip (mm^2)	Array eff(%)	I_{CC} (mA)	I_{SB} (μA)
Hitachi	X8	3/5	P-load	4P/2M	0.6	23	17.0	122	57	45(20)	0.5
Toshiba	X8	3.3/5	P-load	3P/2M	0.5	23	20.3	136	61	70(40)	1.0
Sony	X8	3.3	R-load	2P/2M	0.5	25	21.2	130	67	46(40)	5.0
Mitsubish	X1/X4	3.3	R-load	4P/2M	0.6	20	18.6	150	51	70(40)	1.5
NEC	X1/X4	4/5	R-load	3P/2M	.55	15	19.0	143	54	65(20)	N/A
Mitsubishi	N/A	3	P-load	4P/2M	0.6	—	19.5	134	60	7(10)	0.4
Fujitsu	X1/X4	3.4/5	BiCMOS	4P/2M	0.5	10	18.5	152	50	90(50	—

The Sony [60] 4Mb SRAM used the address transition detection and divided word-lines which were common in this generation of SRAM. To attain better speed Sony used a dynamic bit-line load circuit and a three stage double-ended current mirror sense amplifier which is shown in Figure 9.10.

A new noise reduction output circuit was also incorporated which effectively slows the rate of increase of the peak current when the outputs are switched resulting in a reduced level of output ringing and slightly faster time to address valid than if the circuit had not been used. The standard alternative, of course, was to delay the data valid until the ringing had stopped. Both techniques result in a slightly slower access time than if the ringing had not occurred.

The dynamic bit-line load circuit, shown in Figure 9.11(a), was developed to achieve high-speed bit-line precharge, equalization and discharge during the read operation. When an address transition occurred the bit-lines in the selected section were precharged and equalized to the V_{CC} level by the PMOS transistors Q_1, Q_2, and Q_3 during the ATR pulse as shown in the figure. Just before the ATR pulse went low, the bit-line load PMOS transistors Q_4 and Q_5 were cut off by the OFQ pulse. The result of this operation was that the bit-line load consists only of the stray capacitance rather than the usual NMOS transistor load and the stray capacitance. The memory cell could, therefore, rapidly drive the bit-line load resulting in a fast access time. A picture of the Sony 4Mb SRAM assembled in a 400 mil DIP package is shown in Figure 9.11(b).

The NEC [40] 4Mb SRAM at 15 ns typical access time was also targeted to be a faster part using an input controlled PMOS load high gain sense amplifier. A special type of precharged bit-line circuitry intended to reduce the number of precharge transistors and hence the gate capacitance load was used, thereby reducing the

Table 9.2 Array efficiency of 4Mb SRAMS.

Cell type	Average cell size (μm^2)	Average chip size (mm^2)	Average array efficiency
Poly PMOS chips	18.5	129	59%
R-load chips	19.6	141	57%

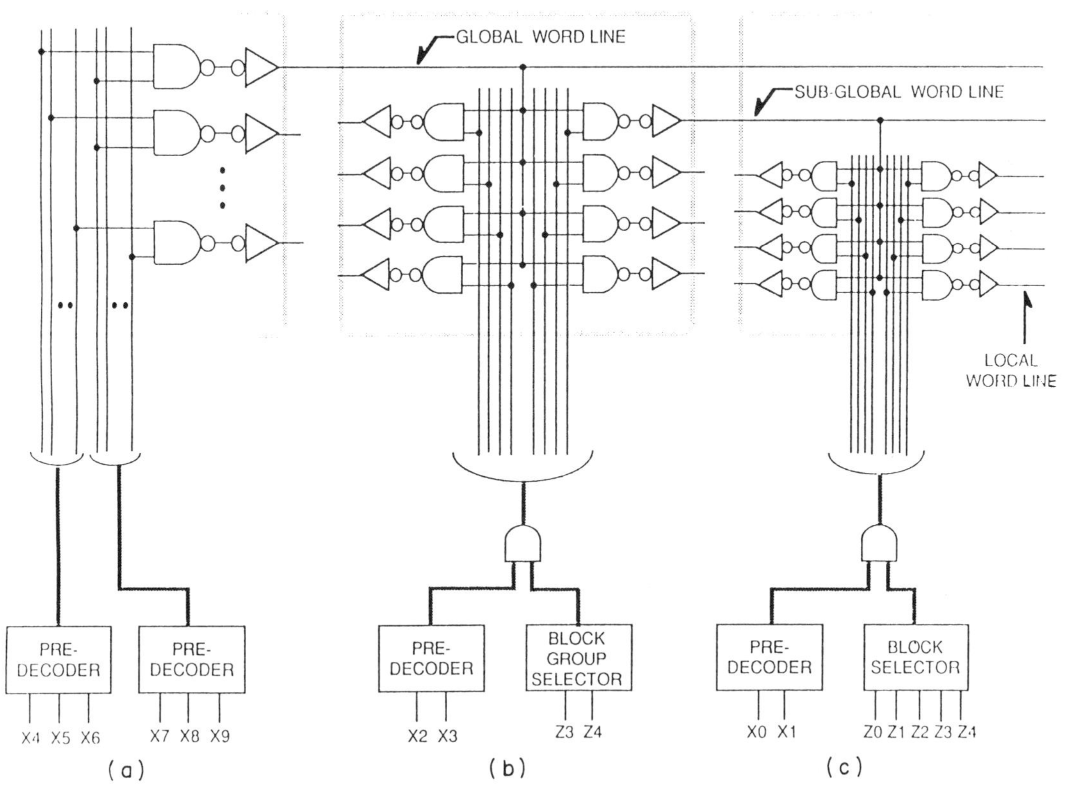

Figure 9.9
Hierarchical word decoding architecture as used in 4Mb SRAM. (a) Global row decoder, (b) sub-global decoder, (c) local row decoder. (From Hirose *et al.* [39], Mitsubishi 1990, with permission of IEEE.)

propagation delay from address change to precharge by 10%. The delay from address change to word-line activation was also shortened by the use of a new predecoder and section row decoder circuitry.

This SRAM had an electrically selectable output bit width. The bit organization selection mode was activated if V_{CC} was applied when address A_0 was low. This opened a pass gate to the $\times 4 - \times 1$ select circuit. If then either a low ($\times 4$) or a high ($\times 1$) level was applied to address pin A_1 the respective output width was elected.

The Toshiba [30] 4Mb SRAM, in order to achieve the microampere standby power, found it necessary to disable the internal voltage-down converters between the 3.3 V array and the 5 V peripheral circuitry.

The lower voltage in the array was necessary with 0.5 μm geometry channel lengths and 11 nm gate oxides to avoid impairing the reliability. In the peripheral circuitry 0.8 μm devices with 16 nm gate oxides were used which could withstand the 5 V V_{CC}.

A voltage-down converter was used consisting of a reference level generator and two differential amplifiers. One of these had fast response but higher standby power consumption and was used when operating the part. This amplifier became disabled in standby mode to save power. The other differential amplifier had a lower 50 μA

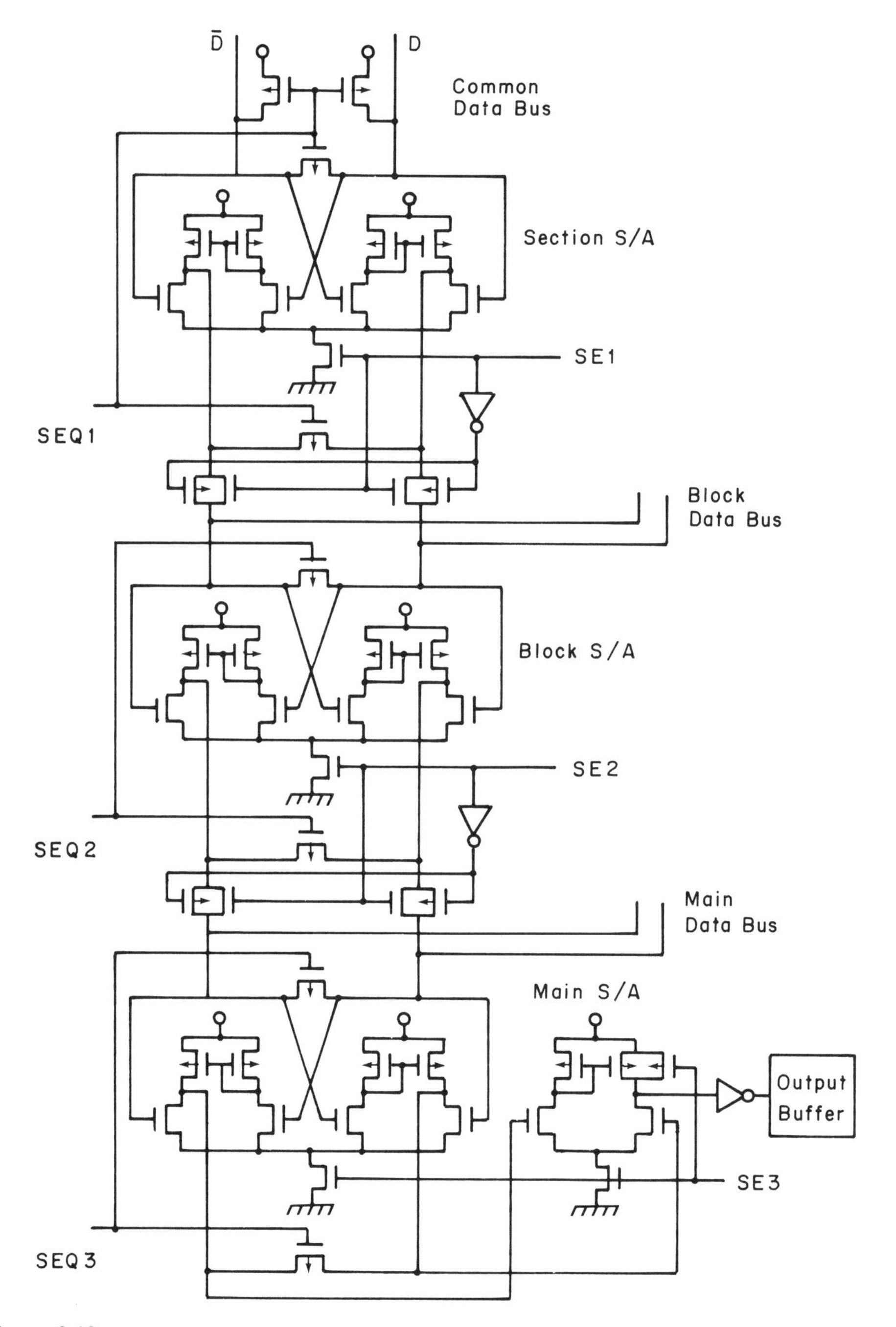

Figure 9.10
Three-stage double-ended current mirror sense amplifier circuit. (From Miyaji [60], Sony 1989, with permission of IEEE.)

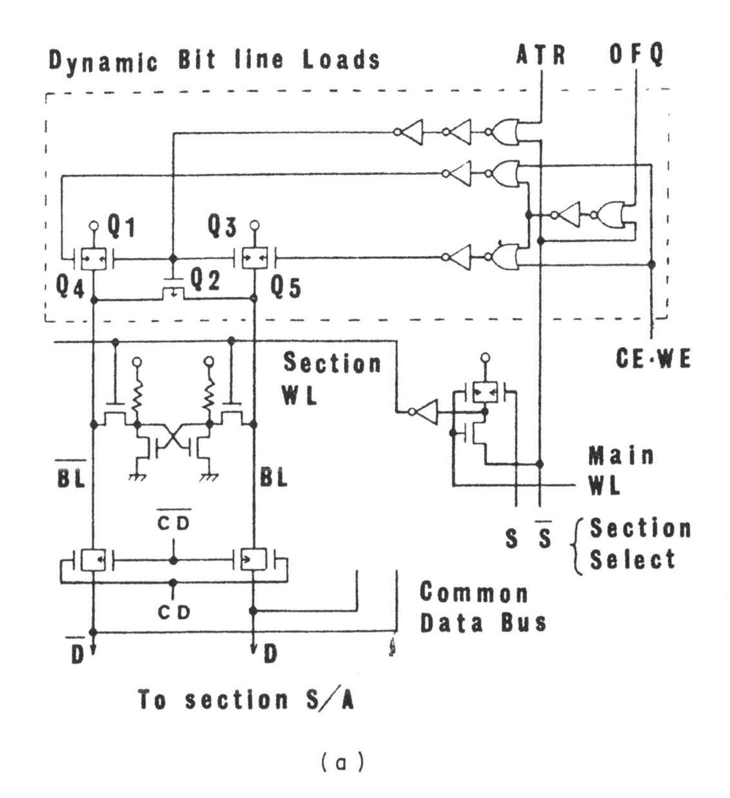

(a)

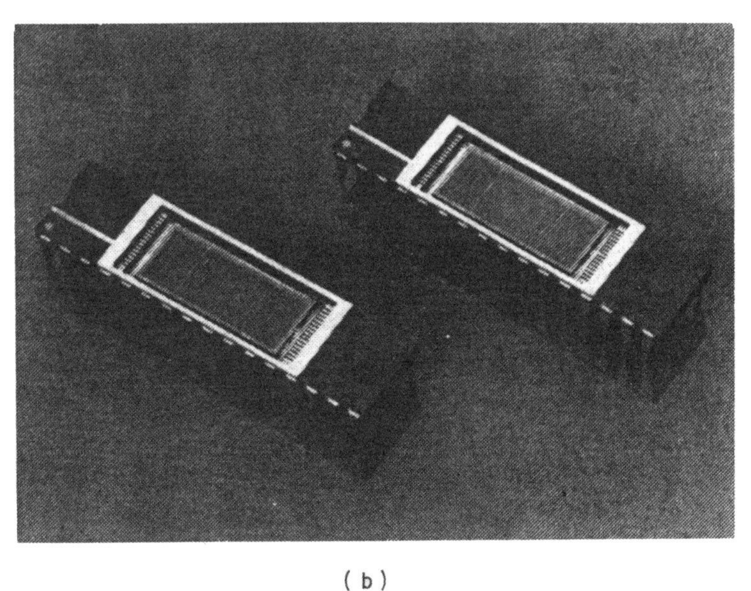

(b)

Figure 9.11
4Mb static RAM. (a) Circuit schematic showing dynamic bit-line loads, cell, and section word-line select. (b) Assembled in 400 mil DIP. (From Miyaji [60], Sony 1989, with permission of IEEE.)

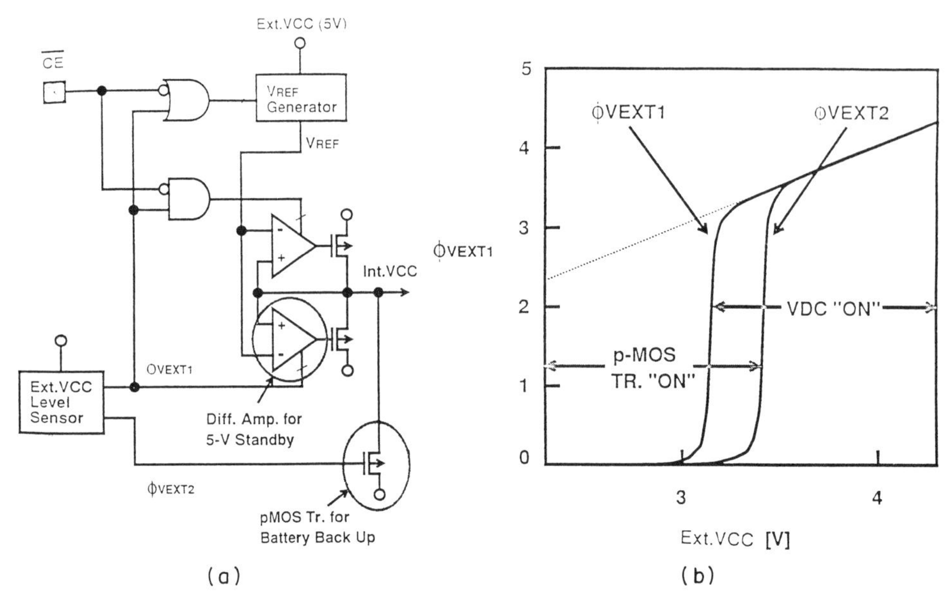

Figure 9.12

(a) Schematic diagram of voltage-down converter and the external supply level sensor. (b) Characteristics of the external supply level sensor. (From Hayakawa *et al.* [30], Toshiba 1990, with permission of IEEE.)

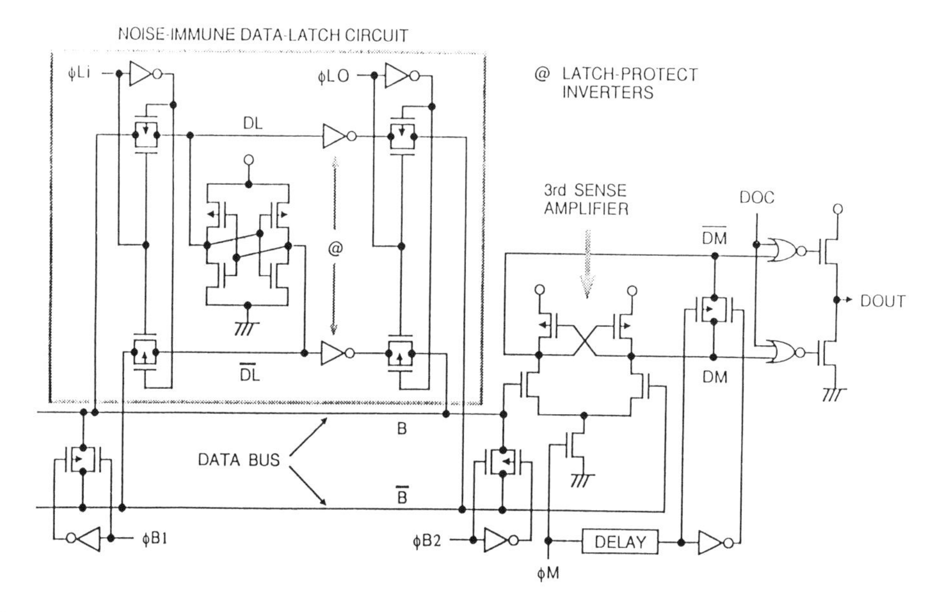

Figure 9.13

Noise-immune data latch circuit used in 4Mb SRAM. (From Sasaki [27], Hitachi 1990, with permission of IEEE.)

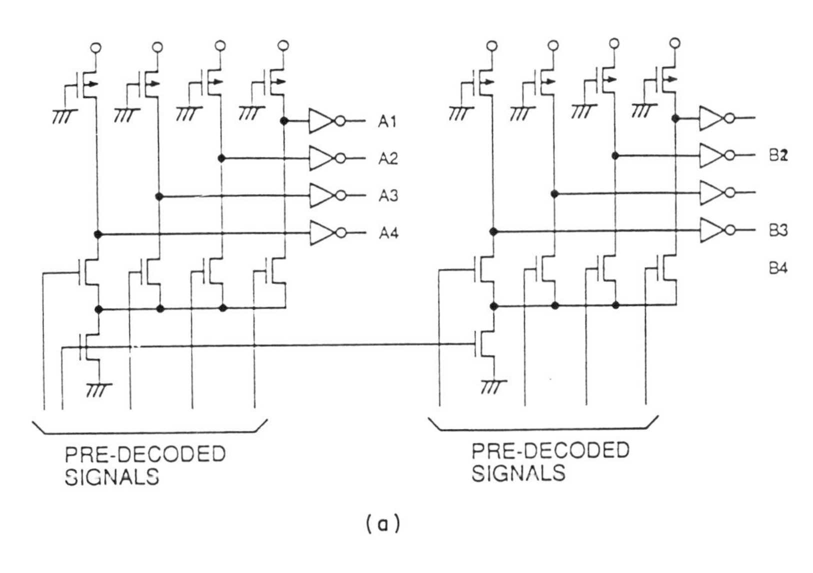

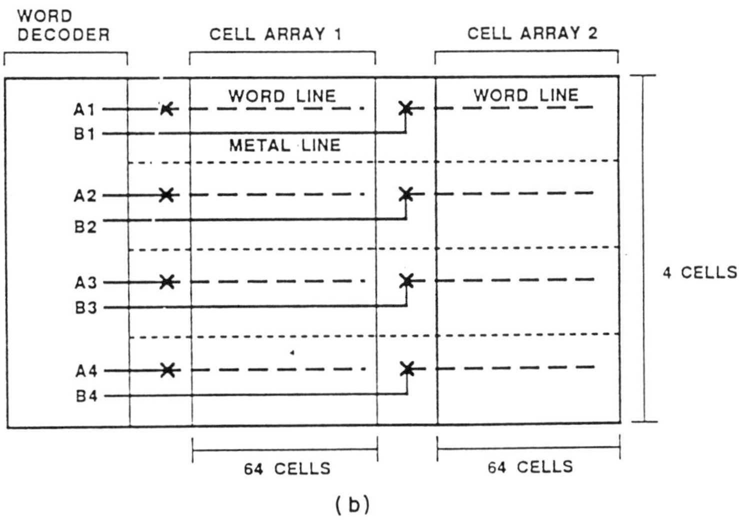

Figure 9.14
Double-array word decoder architecture. (a) Word decoder circuits for two cell arrays. (b) Layout architecture. (From Sasaki [27], Hitachi 1990, with permission of IEEE.)

current consumption and was selected during standby mode. This, however, was still not the microamp level of current needed during battery-backup data retention mode. For this purpose another approach with an external supply level sensor consisting of a circuit using a voltage reference generator was used. In this case when the external supply level became lower than 3.3 V the voltage-down converter was bypassed and the external V_{CC} was applied directly to all internal circuits. In addition, the voltage-down converter was disabled by the sensor permitting the data retention current to be in the microampere range. A schematic circuit diagram for the supply sensor circuitry is shown in Figure 9.12(a) and the characteristics of the sensor are shown in Figure

9.12(b). Architecturally a double-level hierarchical sensing circuitry was used to reduce sensing delay.

Hitachi [27] also included several interesting features on their 4Mb SRAM to reduce power consumption, minimize address noise effects on the data latch, and improve array efficiency giving a smaller chip.

Power consumption was reduced by using a pulsed signal to activate the word-lines and the sense amplifiers. The access time penalty to generate the word activation pulse signal was compensated for by using a fast cascade sense amplifier structure.

Address noise effects were minimized by a new noise immune data latch circuit shown in Figure 9.13. A data latch circuit was required in a pulsed word circuit design to store the amplified data from the memory cell for the static data output circuitry. A short address noise pulse could turn on the selected word-line and the sense amplifiers for a short period of time resulting in an unexpected signal on the data bus. If this signal turned on input gates of the data latch circuit the latch data could be destroyed. The data latch protection circuit employed latch protect invertors and an input gate control pulse circuit which passed only pulses 3 ns or longer. This prevented short pulses from destroying data in the data latch.

Array efficiency was improved by using a double array word decoder architecture in which the first metal line went across a memory cell as shown in Figure 9.14. This permitted two word decoders; A× and B× for two cell arrays to be designed together as shown in the figure, thereby reducing layout area of the word decoder.

9.6 HIGH SPEED SRAM TECHNOLOGIES

Responding to the demand for very fast sub 10 ns SRAMs for first level cache applications, a range of SRAMs in technologies other than MOS have survived including those in bipolar and GaAs.

9.6.1 Fast bipolar static RAMs

Any discussion of very fast SRAMs must begin with the bipolar SRAMs which historically developed the market for very fast cache memory.

Bipolar statics, in addition to having the ease of use common to all statics, are very fast in the below 5 ns range. They historically have been used in applications which demanded this speed, such as first-level cache memories in mainframe computer.

Although mainframe computers generally use dynamic memories for the main storage system, and CMOS SRAMs for the second-level cache, the system performance can be improved by using a small but very fast memory in a first-level cache.

The first introduction of a bipolar RAM in cache memory systems was made in 1968 in the IBM 360 model 85 [10] and achieved remarkable improvements in total system performance. Since that time, the speed, as well as the capacity of cache memory, has become one of the key characteristics of the total performance of a system wherein a memory hierarchy concept is adopted.

The usefulness of the bipolar device in this application historically made up for the high cost due to the larger chip size and the high power consumption. A cache's memory size was usually small enough that neither the cost nor the heat generated were a significant problem.

Early 1k bipolar SRAMs with sub 10 ns speeds first appeared in the later part of the 1970s. An early 1k ECL device shown by NTT [2] in 1978 had access time of 7.5 ns and power dissipation of 784 mW, which gives a power–access time rating of 5.7 pJ per bit.

The part had the separate I/O structure typical of very fast SRAMs. High speed was achieved by circuits which reduced both decoding delay time and sensing delay time. A pre-sense amplifier, which made it possible for the sense current to drive one of the pair transistors in the sense amplifier without delay, was used to obtain a fast read speed.

A 4k bit 100k ECL 15 ns SRAM by Siemens [16] followed in 1979. This part consumed 900 mW of power but had a power access time product of only 3.3 pJ per bit.

Competition for these 'slower' bipolar parts came in the form of fast 4k NMOS SRAMs such as a Cypress [6] 4k SRAM introduced to the market in 1984 which had worst case access times of 15 ns and active power consumption of 450 mW.

Bipolar SRAMs in the low densities, however, kept getting faster. Bipolar ECL SRAMs at the 4k density level reached the sub 5 ns typical access time range about 1983, with an NEC 2.3 ns typical 4k ECL RAM in 1984. 16k bipolar SRAMs reached the sub 5 ns speeds in 1986.

A 5 ns 64k density bipolar SRAM, which was discussed by Fujitsu [85] in 1987, was among the last of the full bipolar ECL SRAMs in the higher densities. Its chip size was more than three times that of the BiCMOS 64k reported the same year which had approximately the same speed.

By 1989 the BiCMOS SRAMs, which were made from combining bipolar and CMOS technologies, had reached the sub 5 ns range and by 1990 had appeared at this speed up to the 1Mb density level. The higher density, smaller chip size and lower power of the bipolar–CMOS combination parts made them much more attractive in the fast SRAM applications once the required speed was available.

There has also been a trend to take advanced BiCMOS technologies and produce lower density devices in much higher speeds than previously possible, thereby competing directly with the bipolar SRAMs.

Applications for bipolar SRAMs in the sub 5 ns range still exist as a rather limited specialty market.

9.6.2 Sub 10 ns MOS and BiCMOS static RAMs

On the leading edge of the technology in every generation of mainstream MOS SRAMs have been the smaller very fast MOS devices.

These parts are used primarily in first-level cache applications in competition with the bipolar SRAMs and tend to be upgraded to the latest technology generation as it becomes manufacturable. While most of the techniques used on the commodity SRAMs to enhance speed, as discussed in the last chapter, have been employed also with these faster parts, some are specific to them.

The history of high speed in MOS memories can well begin with one of the first of the modern submicron technology super fast SRAMs. This was an experimental NMOS device described by Bell Laboratories [14] in 1983.

This NMOS device was fabricated using 1 μm X-ray lithography techniques for minimum transistor channel lengths of 0.5 to 0.8 μm which were very aggressive in 1983. Gates were constructed of single-level titanium silicide and dry etching was used throughout. Power dissipation was kept low by having the word-line clamped to V_{SS} when the decoder was deselected. Due to the small geometries the power supply had to be reduced from the standard 5 V to 2.5 to 4.5 V. At the time JEDEC had not yet passed the 3.3 V standard.

This extremely fast 5 ns 4k $\times$ 1 static RAM rivaled the state-of-art in bipolar technology. The typical power dissipation was only 350 mW at 3.5 V typical power supply voltage which was much lower than the 1.5 W of the comparable bipolar device.

The history of higher density CMOS SRAMs was traced in the last chapter. By 1986 the speed requirements for fast SRAMs had taken the required typical speed into the sub 10 ns range. In this range the BiCMOS ECL technology which had been developing for several years became a dominant factor. This was both because the systems that required SRAMs of this speed contained ECL logic which needed ECL compatible SRAMs to interface without delays, and because the BiCMOS technology offered the additional speed needed to design high density SRAMs in this speed range.

The graph in Figure 9.15 plots typical speed by year for the high density byte-wide CMOS SRAMs, the fast TTL compatible CMOS SRAMs and the fast ECL compatible BiCMOS SRAMs. Generally the higher density commodity SRAMs tend to exhibit a balance of speed and power characteristics which reflect the performance improvement inherent in the technology. The various density levels of the very high speed CMOS and BiCMOS SRAMs are specifically designed to optimize performance often at the expense of chip size and power consumption. These have been made with both the TTL and ECL compatible I/O.

9.6.3 Sub 10 ns high density BiCMOS technology

BiCMOS SRAM technology developed into a mainstream technology in the late 1980s [55]. Previous to this time a few companies had used an occasional non-optimized bipolar transistor for higher drive capability in the output driver of CMOS SRAMs.

BiCMOS technology combines the low static power consumption and logic flexibility of CMOS with the high speed, low logic swing, and input signal sensitivity and high output drive of bipolar. Pure bipolar memory cell arrays have a higher static power dissipation than the MOS cells used in BiCMOS. Increased logic flexibility has opened many interesting possibilities for higher level integration of functions. The high drive capability can be effectively applied at large capacitive nodes such as found

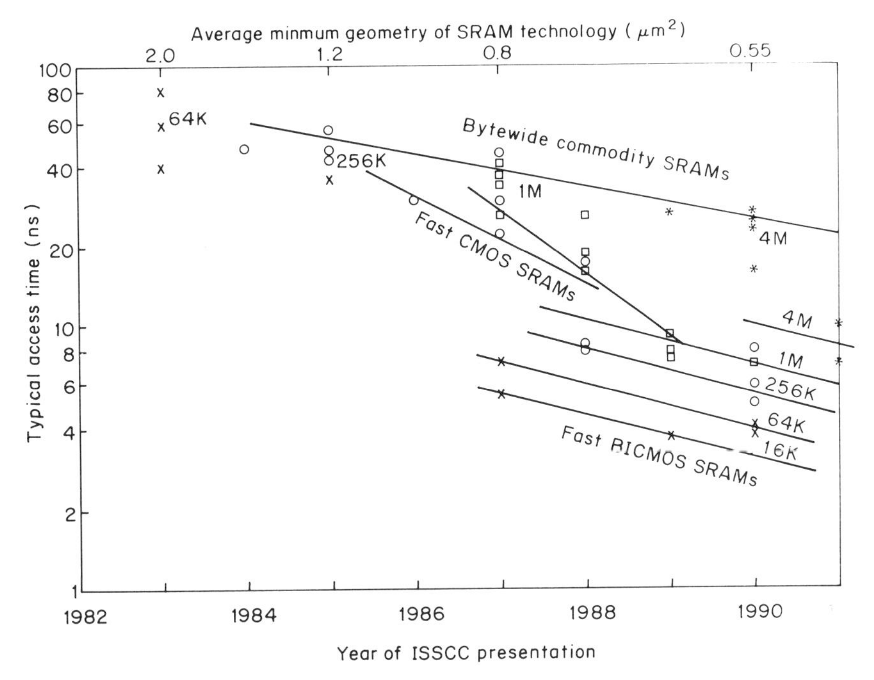

Figure 9.15
Static RAM access time plotted against year of presentation (× 64k, ○ 256k, □ 1Mb, * 4Mb). (Source various ISSCC Proceedings.)

in decoders, word-line drivers, write drivers, and output buffers. The larger transconductance of the bipolar transistors helps in the high speed sensitivity of the sense amplifiers. All of these circuit elements can benefit from the sensitive inputs and large gains provided by bipolar.

In 1984 Hitachi [8], one of the early pioneers in BICMOS described a double P-well transistor structure which permitted optimized NPN bipolar transistors to be formed in NMOS technology.

Hitachi had previously used the vertical parasitic NPN bipolar transistors normally occurring in CMOS technology as the output pull-up devices in their 4k, 16k and first generation 64k SRAMs. The complexity of the process was increased both by the formation of the thin P-well and by the use of an epitaxial layer to obtain the required high resistance N-type substrate.

A CMOS bipolar combination on the output could also make the device less speed sensitive to temperature. Since normally MOS slows down when the temperature rises and bipolar speeds up, the use on the output of both a bipolar device as a pull-up and a MOS device as a pull-down allowed the devices to compensate for each other at high temperatures.

An early example is a combination NMOS transistor pull-down and bipolar transistor pull-up from Hitachi [3] in 1982 which pulls the output node up to a high voltage of $V_{CC} - 0.5$ V for TTL compatible outputs. The bipolar transistor also provided sufficient output source current (I_{OH}) when the output voltage was 2.4 V (V_{OH}).

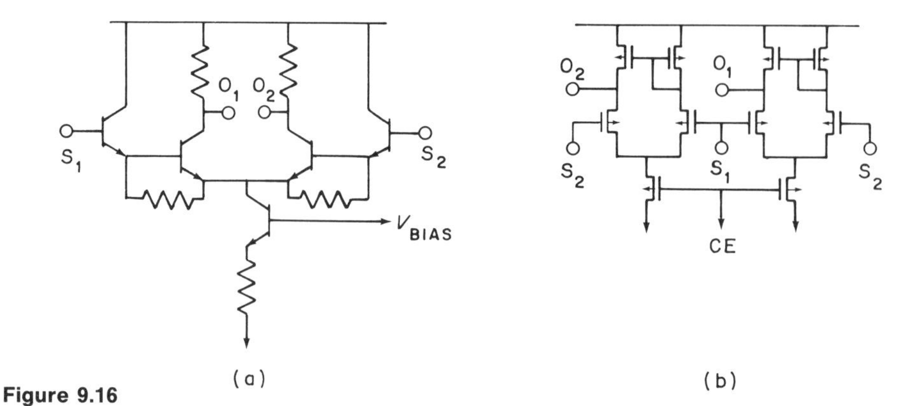

Figure 9.16
Some typical 64k SRAM sense amplifiers for fast SRAMs: (a) Bipolar differential sense amplifier, (b) CMOS differential sense amplifier. (From Miyamoto [9,52], Toshiba 1984, with permission of IEEE.)

A 1984 Toshiba [9, 52] paper explored the use of sensitive bipolar sense amplifiers as a method of reducing the bit-line swing and hence the bit-line recovery time in higher density SRAMs. They compared analytically and experimentally an emitter coupled bipolar differential amplifier and a current mirror loaded CMOS amplifier as shown in Figures 9.16(a) and (b).

Toshiba concluded that the bipolar amplifier had a capability of driving a comparatively lower impedance without sacrificing gain. This resulted in a high speed sensing capability for small input signals. A second-stage CMOS amplifier was, however, needed because a full swing output could not be obtained with the ECL amplifier. The final sense amplifier was a combination bipolar CMOS two-stage amplifier. Here the bipolar amplifier first provided rapid amplification of the small bit-line signal to an appropriate level, then the CMOS second stage amplifier converted the signal to an almost full swing. Low standby power was obtained by the use of full CMOS cell and by MOS switches which turned off the current paths of the bipolar devices in stand-by mode.

In 1986 Hitachi [11] reported on a BiCMOS 64k ECL SRAM with 30 mm^2 chip size 13 ns typical access time and 500 mW power dissipation which was about half of what would be expected for a full bipolar ECL SRAM of this density. The technology had 0.5 μm emitter base width, and conservative 2 μm N-channel gate length. It used buried twin-well double layer aluminum and double layer polysilicon with a polysilicon emitter bipolar process as shown in the cross-section in Figure 9.17.

In a 1987 paper, Hitachi [53], used a similar technology in 1.3 μm geometries and with shallower emitter and base junction depths to produce a 64k BiCMOS ECL SRAM with 7 ns access time and 350 mW power dissipation.

This circuit used the bipolar transistors throughout the memory periphery wherever their drive capability was needed [51]. For power considerations the memory cell array was pure NMOS with poly load transistors. The ECL to MOS input buffer circuit shown in Figure 9.18(a) and the circuit schematic shown in Figure 9.18(b) indicate the mix of bipolar and MOS circuitry in the decoders, word drivers, controls, sense amplifiers and output buffer. The sense amplifier and input–output buffer circuits are

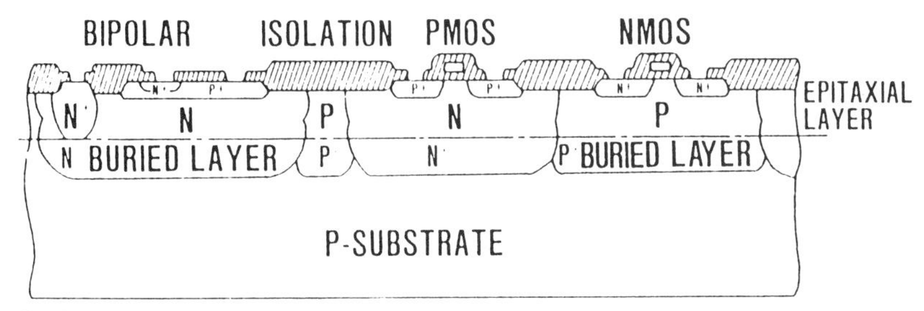

Figure 9.17
Cross-section of typical BiCMOS transistor structure showing bipolar NPN transistor, PMOS and NMOS transistors. (From Ogive [51], Hitachi 1986, with permission of IEEE.)

bipolar since they determine the speed and handle the ECL I/O signal interface. The word drivers use a bipolar–CMOS combination gate to rapidly charge up and discharge the large load capacitance.

For comparison, at the same conference in 1987 Hitachi [15] showed a pure ECL 16k SRAM with 20 mm^2 chip, 3.5 ns access time, and 2 W of power dissipation. Clearly with four times the power dissipation on a chip of one quarter the density yet $\frac{2}{3}$ the chip size of the larger chip, the only benefit of the pure bipolar part is the faster access time.

A 256k BiCMOS TTL SRAM from Intel [96] shown in 1991 had 6.2 ns access time at 5 V power supply compared to 7.8 ns for a comparable technology CMOS version. The BiCMOS SRAM maintained its speed advantage over the CMOS part at 3 V power supply through the use of BiNMOS buffers.

In 1988 Texas Instruments [57] described a 256k ECL BiCMOS SRAM with 8 ns access time and a battery back-up design technique using the full CMOS cell which gave it less than 1 μA standby current consumption. The TI BiCMOS process, shown in the cross-section in Figure 9.19(a), used an 0.8 μm technology which added just four mask steps to a baseline 0.8 μm CMOS process. This resulted in a 16 mask process with half of the mask levels patterned with 0.8 μm features.

A divided word-line structure [56] like those described in the previous chapter reduced the word-line capacitance increasing speed, while the silicided local word-lines reduced resistance resulting in increased read current. The use of these low resistance metal interconnects allowed 256 cells per bit-line pair without significantly increasing delay. A three-stage sensing scheme was used with each column having its own bipolar sense amplifier to increase sensing speed as shown in Figure 9.19(b). The circuit also used fast bipolar main sense amplifiers.

In 1989 Toshiba [61] presented an 8 ns 1Mb ECL BiCMOS SRAM in an 0.8 μm process with another new ECL to CMOS level convertor as shown in Figure 9.20(a). To save power consumption the output of the ECL input buffer was directly converted to a CMOS level without ECL predecoding. The reference level trip point was set so that the converter supplied CMOS signal levels to the output of the converter circuitry with no dc current consumption. The reference level generation did consume dc power, however. This resulted in much lower power consumption than a standard convertor which used a CMOS current mirror sense amplifier and had a dc path through the PMOS and NMOS circuitry. The outputs of the buffer were directly

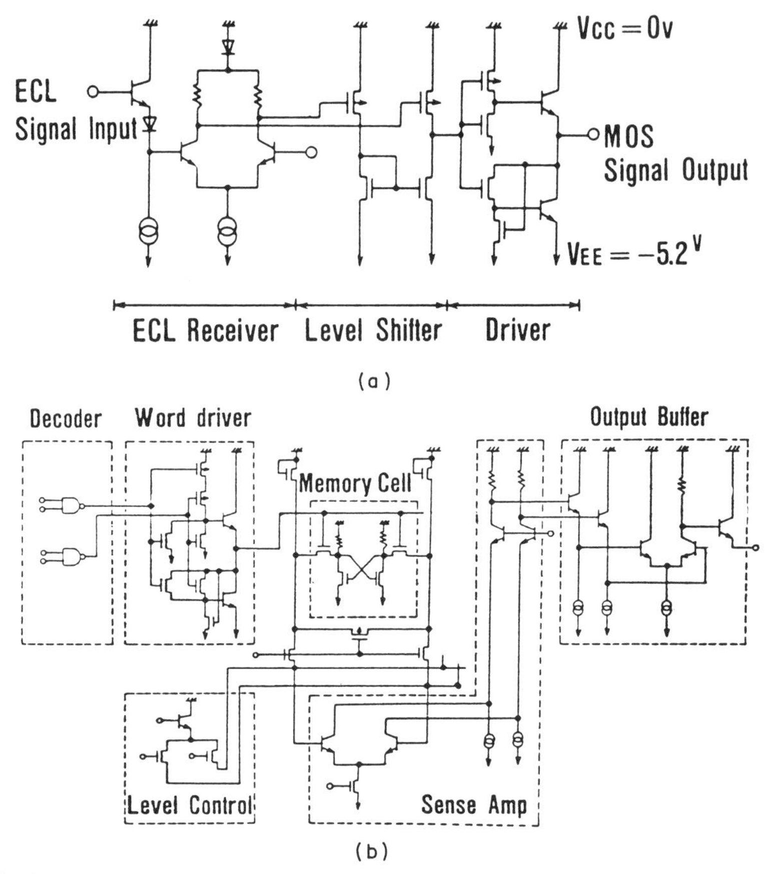

Figure 9.18
BiCMOS SRAM circuit schematics. (a) ECL to MOS input buffer, (b) BiCMOS SRAM showing mix of bipolar and MOS circuitry. (From Ogive [64], Hitachi 1986, with permission of IEEE.)

suitable for address buffer use. In 1991 Toshiba [97] showed a 1Mb TTL BiCMOS SRAM with data-line wiring delay reduction techniques.

National Semiconductor [58] also introduced a 256k BiCMOS ECL SRAM in 1988 with 12 ns typical access time and 15 ns worst case specified access time in 1.0 μm BiCMOS technology with 0.8 μm NMOS channel lengths. This chip used bipolar ECL I/O buffers and decoders with a CMOS translator to BiCMOS drivers as shown in Figure 9.20(b). An R-load MOS memory cell was used.

The read and write cycle timing diagrams of the part are shown in Figure 9.20(c) and 9.20(d) [59]. The device was timed internally so that the chip select to valid data output time was just 5 ns within the 15 ns output time giving the system designer the flexibility of a 10 ns system decode time interval.

The initial predecode of the row and column addresses was done in bipolar ECL to improve performance. The output of the translator between the ECL decoder and the BiCMOS driver achieved full CMOS levels with no dc current. It drove address line drivers. It was controlled by two reference signals one of which set the trip point of the translator while the other controlled the current of the NMOS current sources so that the latch could unlock to change states.

There was a potential problem in that the CMOS logic becomes active and stable at relatively low supply voltages compared to the ECL logic. Hence, at the interface between the ECL and CMOS, large current surges may arise during power up or power down. Since these are potentially damaging, a latch-up protection monitor circuit was included.

A special power-up circuit, shown in Figure 9.21(a), monitored the state of two reference voltages within the ECL bandgap reference generator. The V_{CS} signal was referenced to V_{EE} and controlled all ECL current sources. The V_{BB} was referenced to V_{CC} and set the trip-point of all input buffers. Until approximately a V_{BE} level existed from V'_{BB} to V'_{CS}, the power-up circuit locked out the interface to the matrix. The power-up circuit, which reached a full power-up state at about 3.7 V, disabled all rows and shut

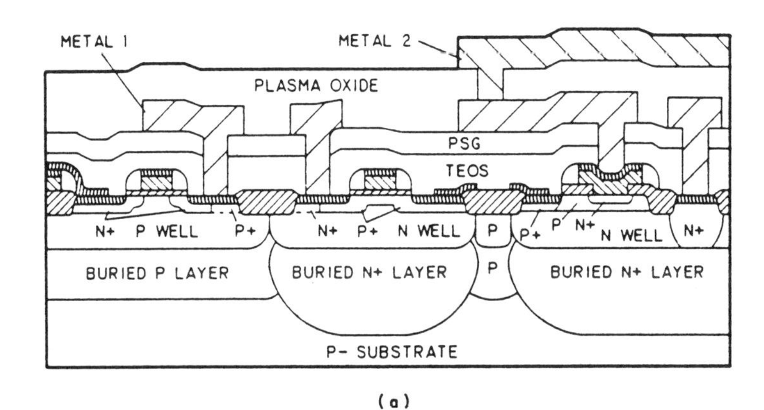

(a)

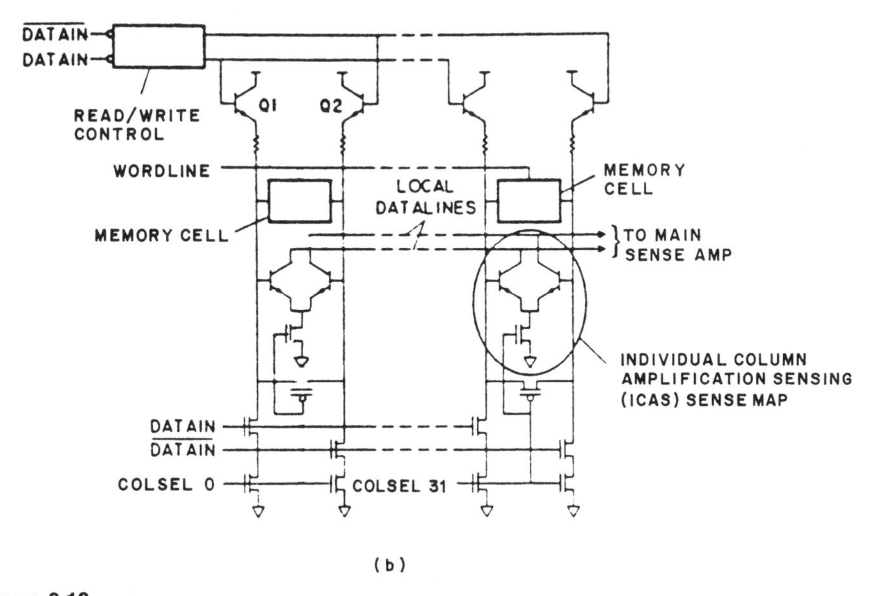

(b)

Figure 9.19
256k BiCMOS SRAM. (a) Process cross-section, (b) sensing schematic with each column having its own bipolar sense amplifier. (From Tran *et al.* [57], TI 1988, with permission of IEEE.)

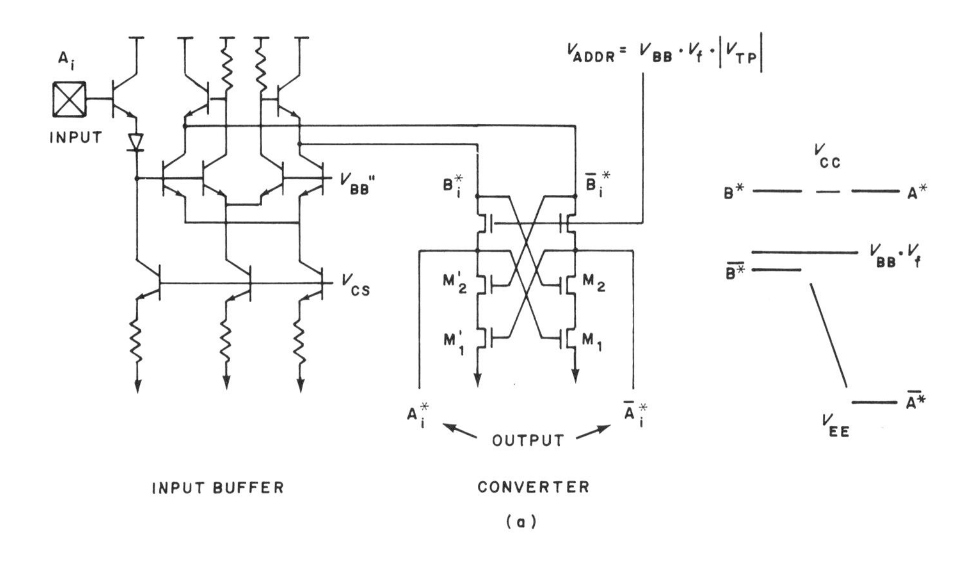

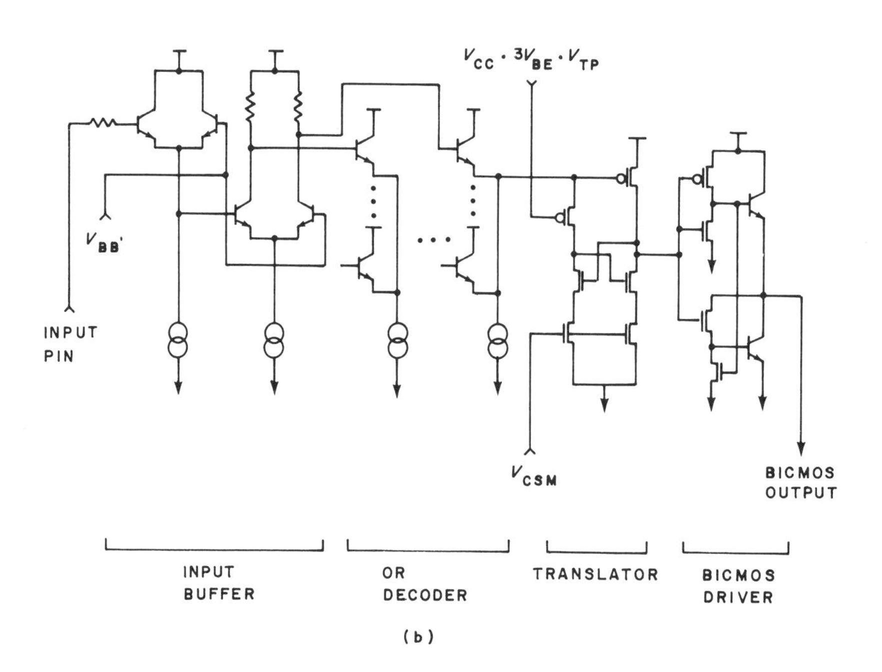

Figure 9.20
(a) 1Mb-ECL BiCMOS SRAM ECL to CMOS level converter. (b) ECL to BiCMOS translation I/O buffers and decoders with BiCMOS driver. From 256k BiCMOS SRAM timing diagrams with internal timing giving fast chip select to valid data out. (c) Read cycle showing 5 ns chip select to valid data out and 10 ns system decode time. (d) Write cycle showing 5 ns allowable for system skews. (a: from Matsui *et al.* [61], Toshiba 1989, with permission of IEEE; b: from Kertis *et al.* [58], National 1988, with permission of IEEE; c,d: from Hochstedler [59] 1988, permission of National.)

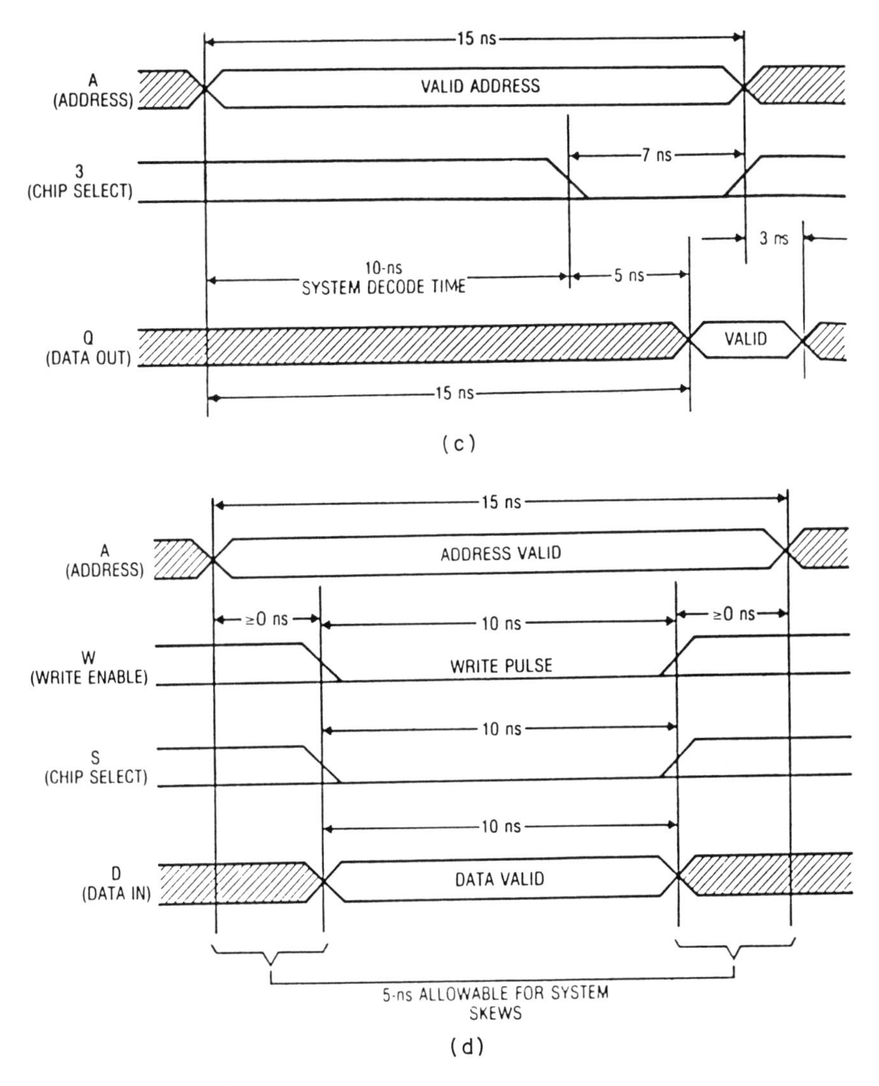

Figure 9.20 (*continued*)

down the write logic. As a result of using a power-up monitor, the SRAM would retain stored data at a 2 V supply level. The bandgap reference generator used is shown in Figure 9.21(b).

In 1990 National [76] followed with an 1Mb SRAM in an 0.8 μm BICMOS technology with 6.8 ns typical speed. This was one of the first SRAMs shown in the JEDEC standard center power and ground pin pinout for fast SRAMs. It was metal mask option compatible with both 10 K and 100 K ECL I/O and had configurable $\times 1$, $\times 2$, and $\times 4$ organization. It used a bit line structure with a CMOS self-locking load to reduce both active and standby power consumption. Independent power supply bussing was used to the ECL circuitry and to the CMOS–BiCMOS circuitry for better noise isolation. A shared BiCMOS group sense amplifier was used to save chip area

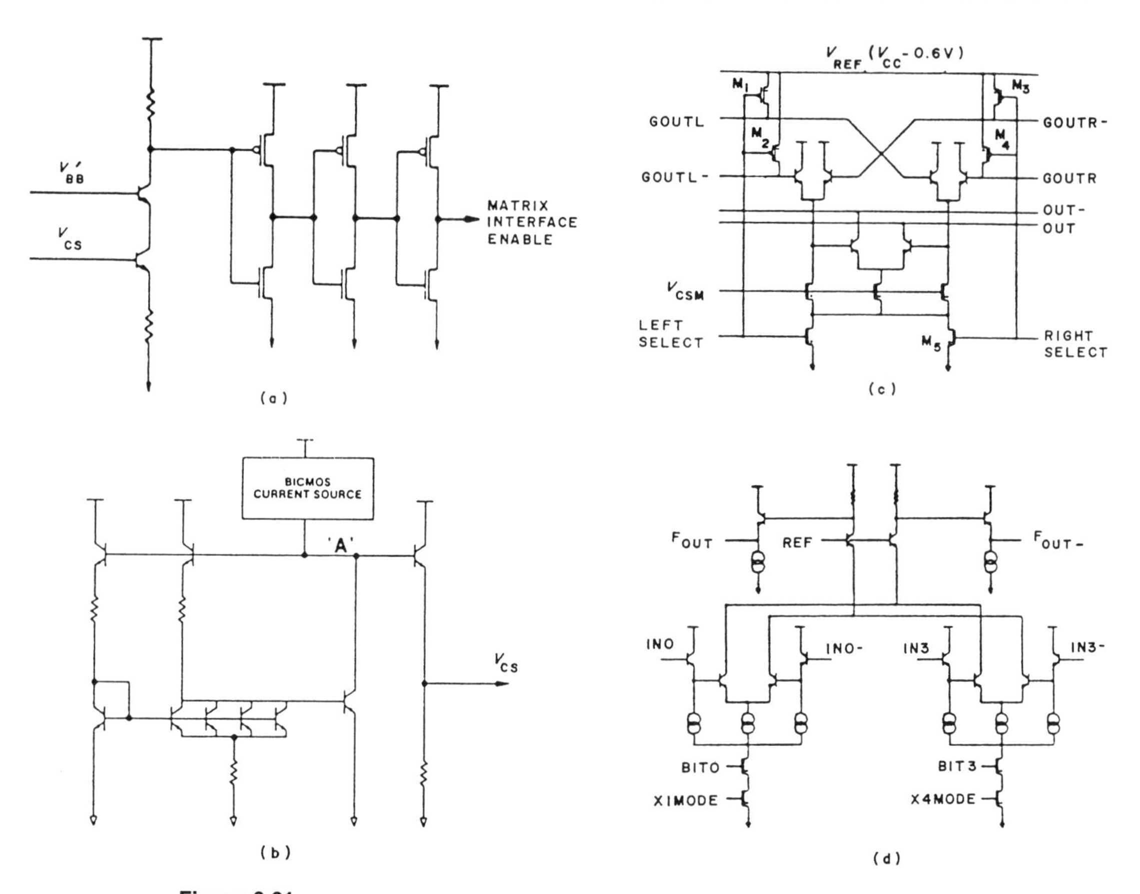

Figure 9.21
Fast 256k BiCMOS ECL level SRAM. (a) Power-up monitor circuitry. (b) Bandgap voltage reference generator. Fast 1Mb BiCMOS ECL level SRAM. (c) Shared group sense amplifier. (d) Configurable final sense amplifier. (a,b: from Kertis [58], National 1988, with permission of IEEE; c,d: from Kertis [76], National 1990, with permission of Japan Society of Applied Physics.)

as shown in Figure 9.21(c). It was staged with a configurable bipolar ECL final sense amplifier shown in Figure 9.21(d).

In 1991 Fujitsu showed two 4Mb BiCMOS SRAMs, a TTL compatible SRAM with 10 ns access time and 90 mA active current at 50 MHz [91], and an ECL compatible SRAM with 7 ns access time and 120 mA active current [93].

Also in 1991 Hitachi [94] showed a 1.5 ns 64k ECL-CMOS SRAM. This 0.5 μm part had a CMOS memory array and full ECL peripheral circuits. An ECL word driver circuit eliminated the usual level convertor and used an ECL driver to directly drive the cell array, thereby reducing the voltage swing of the word-line and increasing the speed. When used as a RAM macrocell without the output buffer delay, the speed was 1.5 ns.

IBM in 1991 [95] showed a 3.5 ns 576k BiCMOS ECL SRAM in 0.45 μm technology which also used full bipolar ECL in a BiCMOS periphery with CMOS array.

9.6.4 Power consumption limitation of ECL interface circuitry

Historically ECL components in systems had such high power consumption that they required special cooling measures to be taken. This limited their use to systems which truly required the high speed and could afford the expensive cooling measures. The liquid cooled ECL systems in the early 1980s gave way in the second half of that decade to the far less expensive air-cooled systems. Primarily ECL was used in the extremely fast upper hierarchical levels of large computer systems and is moving also into the mini-computer and workstation markets.

The memory components required in these applications historically were small ECL bipolar SRAMs which are still used in the sub 5 ns first-level cache applications. With the advent of the fast sub 10 ns high density BiCMOS SRAMs these parts were going into second-level cache in large systems and first-level cache in smaller systems. The BiCMOS SRAMs offer not only the speed but the lower power consumption of the CMOS.

9.6.5 Development of ECL I/O compatible CMOS interface circuitry

While TTL I/O compatible CMOS input and output buffers were developed in the mid 1970s, ECL compatible I/O was more difficult to make out of CMOS transistors. For this reason the majority of the industry used bipolar ECL I/O with either bipolar or CMOS arrays for those memory chips intended to interface into systems using ECL logic.

A theoretical block diagram of a CMOS ECL level compatible SRAM is shown in Figure 9.22. The challenge was to make the input converter and the output driver out of CMOS both to reduce the cost of processing the more expensive BiCMOS technology and to reduce the power consumption through the use of the lower power CMOS devices.

An early attempt to make an ECL level compatible CMOS SRAM was reported by Intel Corporation [4] in 1982 for a 4k 20 ns fast CMOS SRAM with 10k ECL compatible inputs and outputs. Delay for ECL input level to MOS level conversion was typically 4 ns and sufficient input drive was obtained to achieve full CMOS levels.

The ECL to CMOS input level converter is shown in Figure 9.23(a). While the input converter itself was pure CMOS, a bipolar ECL input switchpoint reference circuit was used for the V_{BB} reference as shown in Figure 9.23(b).

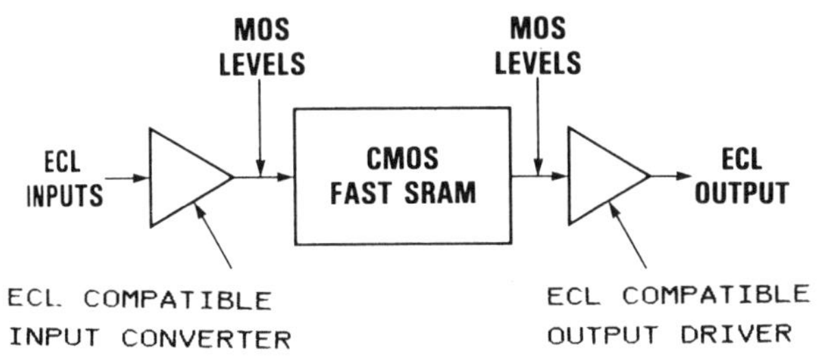

Figure 9.22
Schematic block diagram of CMOS ECL I/O level compatible SRAM.

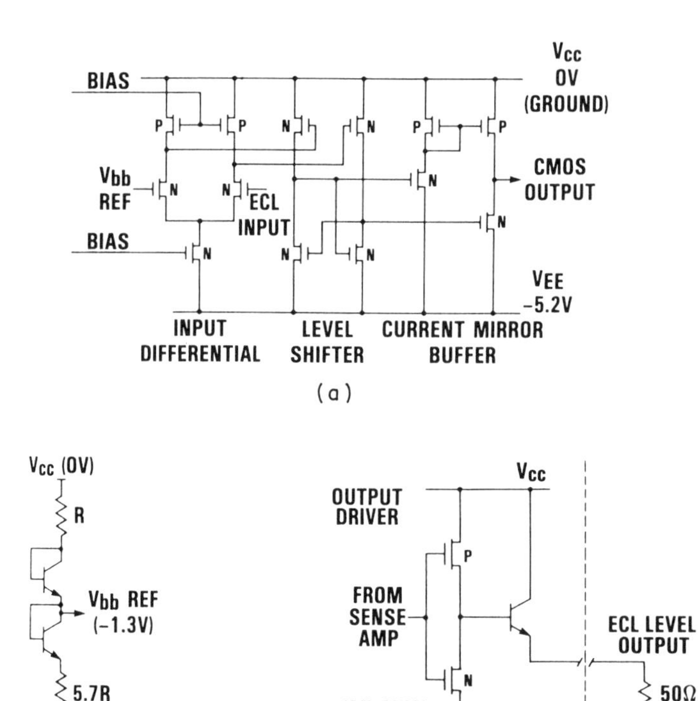

Figure 9.23
Input and output circuitry showing early attempt to make ECL compatible inputs and outputs on a CMOS SRAM. (a) ECL to CMOS input converter, (b) ECL input switchpoint reference, (c) CMOS to ECL output driver. (From Hudson and Smith [4], Intel 1982, with permission of IEEE.)

The CMOS to ECL output level convertor was mixed CMOS and bipolar as shown in Figure 9.23(c). The output driver used a large PMOS device to drive the base of the output bipolar transistor which drove a 50 Ω external load. While the output high level exactly met the 10k ECL specification, the output low voltage specification was only partially met.

A second attempt to make ECL I/O level compatible CMOS was reported by IBM in 1988 [62]. This 6.2 ns 64k CMOS RAM was made in 0.5 μm technology with nominal ECL I/O levels using only CMOS technology. Input *X*-address receivers preamplified the ECL input signal with a direct drive differential amplifier during the ECL to CMOS conversion. *Y* address, data-in, and write control inputs used a dynamic sense amplifier receiver which was slower than the *X*-address receiver. The three-state off-chip driver circuit drove a 50 Ω load and used N-channel devices consisting of an inverting device which pulled down to ground and a source follower which pulled up to an 0.86 V supply with the gate driven to 3.6 V. These levels are neither 10k or 100k ECL as commonly used in the industry.

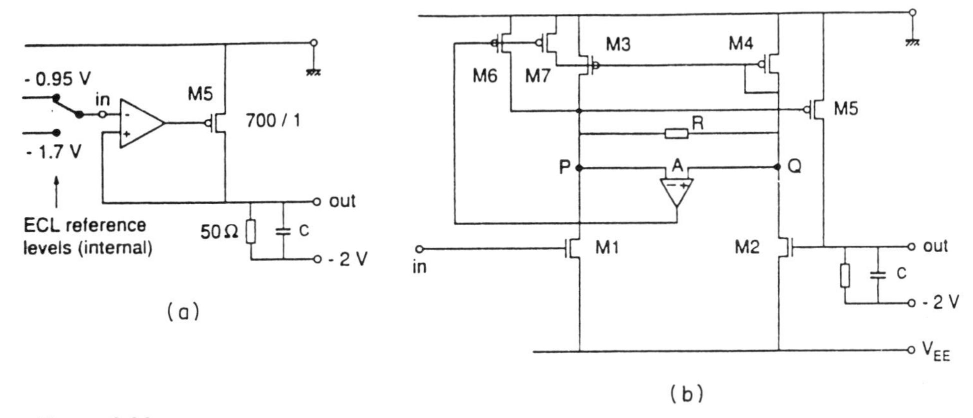

Figure 9.24
ECL compatible CMOS output buffer circuitry. (a) Feedback controlled output buffer, (b) Principle of high speed buffer with error correcting feedback loop. (From Sevinck [63], Philips 1989, with permission of Japan Society of Applied Physics.)

A true ECL level compatible CMOS technology circuit, made in an unmodified 0.7 μm CMOS SRAM process was reported by Philips [63] in 1989. Only CMOS processing was used and a 100k ECL specification was exactly met. Total delay through both the input and output circuitry was on the order of 2 ns. The technique used a negative feedback loop to automatically supply the right amount of drive independent of changes in process, temperature and supply voltage. It also used error correction to eliminate the need for external components or additional power supplies. A bandgap reference, using the parasitic bipolar transistors normally occurring in CMOS technology, was used to generate the ECL level reference sources on chip.

The basic feedback controlled ECL level output buffer circuit is shown in Figure 9.24(a). It consisted of a differential amplifier pair with M_5 as the output transistor. Figure 9.24(b) adds M_3 and M_4 as a current mirror load. An error correction circuit was added to the basic output buffer which consisted of the differential amplifier A, the PMOS transistors M_6 and M_7 and the capacitor C. The amplifier sensed the voltage difference between P and Q and drove M_6 and M_7.

The result was that the current level of transistors M_1 and M_2 could be controlled to minimize the voltage difference between P and Q. The capacitor slowed the error correction circuit down to ensure stable operation with varying capacitive loads on the output.

Another pure CMOS ECL compatible input and output buffer which could function up to 600 Mbps was described by Toshiba [99].

9.6.6 GaAs and other specialty technology super fast SRAMs

We cannot leave the topic of super fast SRAMs without glancing at those made with GaAs technology.

GaAs has a high electron mobility which makes it potentially an extremely fast technology. In 1987 alone GaAs SRAMs ranging from 1 ns for a 1k to 4k, and 5 ns for a 16k were presented.

These super fast SRAMs attained their speed in spite of suffering from a variety of technological and processing problems.

First of all normal planar processing is more difficult on a GaAs substrate because thermal oxide cannot be grown as it can on silicon. Lower quality deposited oxides must be used. This tends to limit the uniformity across the wafer and also the potential complexity of the circuit. The quality of the GaAs wafers available, and thus the threshold voltages of the transistors, vary considerably across the wafer, requiring special circuit techniques to compensate.

The cost of GaAs wafers is also higher than silicon wafers and the size is smaller. The 1k GaAs SRAM mentioned above used one inch wafers at a time when silicon processing used six to eight inch wafers (1 inch = 2.54 cm). The process was also expensive in using unusual materials such as gold which could cause contamination if used in an area where normal silicon processing was also occurring.

Another problem is that the voltages that could be supported in a sub 5 ns GaAs RAM are much lower than standard interface circuitry and special design considerations need to be taken to interface the RAM to an external ECL system environment.

A schematic cross-section of a GaAs memory cell is shown in Figure 9.25. The device was fabricated with 1.0 μm tungsten self-aligned gate field effect transistors and used double-level metal interconnect technology. The ohmic metal is AuGe–NiAu and the interconnect metals are TiAu. The dielectric insulator is phosphorus-doped CVD deposited SiN. An ion implant was used to form the channel and the source and drain regions.

The simplicity of the circuitry relative to that of a modern CMOS SRAM is shown in Figure 9.26. Figure 9.26(a) shows the data path for a 16k GaAs SRAM from Mitsubishi [69]. This SRAM used the enhancement depletion mode cell shown in the previous figure. In addition to the cell, the simple predecoder, and bit-line load circuitry are shown along with the invertors for the input control decoders.

Figure 9.26(b) shows a column sense amplifier from a Philips [70] 1k GaAs SRAM with resistive loads which were used to precharge the bit-lines. The cells in this SRAM used two supply voltages 1.5 V and −1.5 V. Figure 9.26(c) shows the bit and data-line circuitry for a 4k GaAs SRAM from Hitachi [71].

In conclusion, while very specialized systems applications requiring nanosecond speeds may continue to use GaAs RAMs, the sub 5 ns speeds of the bipolar ECL level RAMs and sub 10 ns speeds of the BiCMOS and CMOS ECL I/O compatible SRAMs will fill the majority of the sockets for the foreseeable future.

9.7 FEATURES FOR IMPROVED SPEED

When speeds in the sub 10 ns range are being considered, even fractions of a nanosecond become important. Small otherwise negligible sources of delay need to be considered and reduced. These include such factors as time to turn around a common I/O line, the delay inherent in multiplexed addressing, inductance contributions from package to chip bond wires, and internal set up times in a synchronous SRAM.

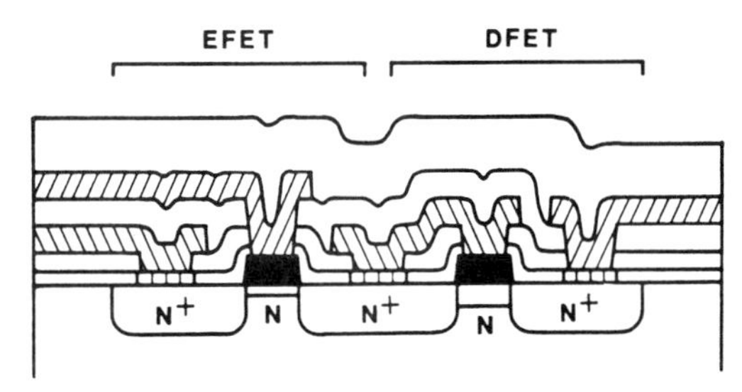

Figure 9.25
Schematic cross-section of a GaAs SRAM cell (□ insulator, ■ gate metal, ▥ ohmic metal, ▧ first metal, ▨ second metal). (From Takano [69], Mitsubishi 1987, with permission of IEEE.)

9.7.1 Separate I/O

Separate input and output lines can be used in a very fast SRAM to improve speed by eliminating the time required to turn the data around on a common system bus. A comparison of schematic block diagrams of common I/O and separate I/O parts are shown for two 10 ns access time 16k × 4 SRAMs from Cypress Semiconductor [68] in Figures 9.27(a) and (b), respectively.

The pinouts, which are also shown, illustrate that a disadvantage of separate I/O is the increase in number of pins in the package. With the modern development of the smaller lead spacing packages, this disadvantage was significantly reduced.

The separate I/O part shown was developed by Aspen Semiconductor, a subsidiary of Cypress Semiconductor.

9.7.2 Synchronous or self-timed SRAMs (STRAMs)

High performance systems require memories which can operate without skews with fast system clocks. The addition of latches on the inputs of SRAMs, i.e. synchronous SRAMs, helped meet this system need.

Basic synchronous (latched) and asynchronous (unlatched) SRAM timing techniques were described in Chapter 5. Versions with latches, registers and pipelining features are available. Latches can be used on all inputs, all outputs, or a selection of inputs and/or outputs.

An example is the Fujitsu [65] self-timed RAM (STRAM) which was available in three versions—'latch–latch' (latches on inputs and outputs), 'register–register' (registers on inputs and outputs) and 'register-latched'.

Latches, defined as level sensitive latch structures, operate on inputs and outputs as shown in the timing diagram in Figure 9.28(a). The input data (D) are controlled by the level of the clock signal (LD) input. The input data is transparent to the output (Q) when the clock level is low. When the clock level is high the latch is closed and data cannot pass through it so the data already present in the latch is maintained on

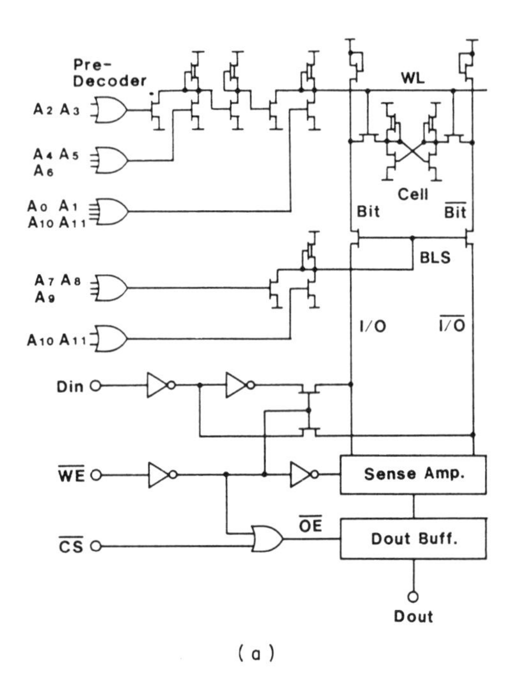

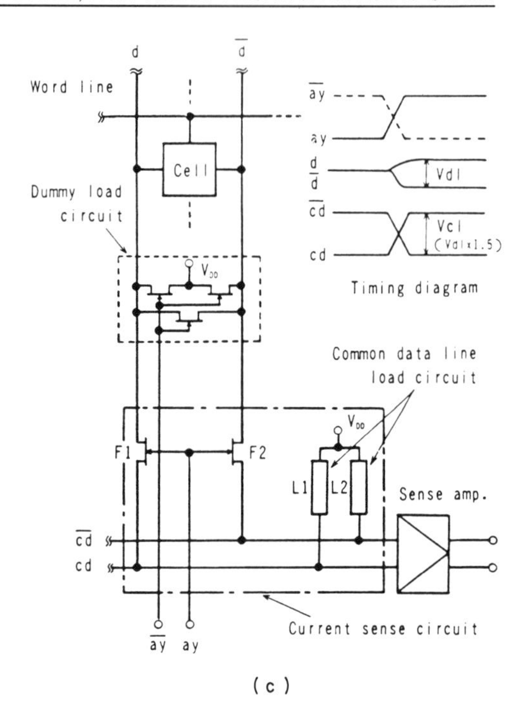

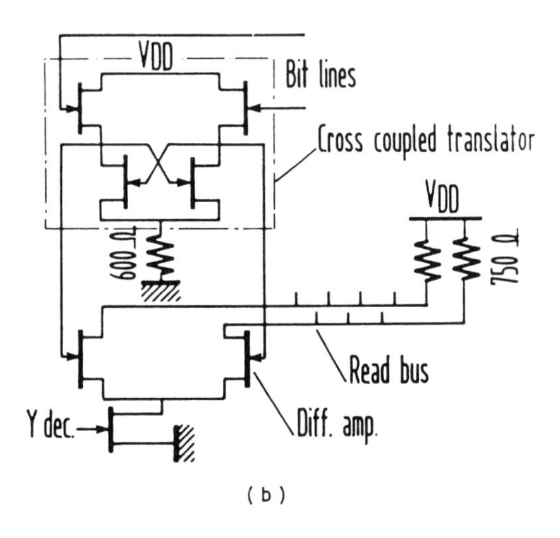

Figure 9.26
Various schematic diagrams of GaAs SRAM circuitry. (a) Data path of a 16k SRAM, (b) column sense amplifier with resistive loads, (c) bit-line and data-line circuitry. (a: from Takano [69], Mitsubishi 1987, with permission of IEEE; b: from Gabillard [70], Philips 1987, with permission of IEEE; c: from Tanaka [71], Hitachi 1987, with permission of IEEE.)

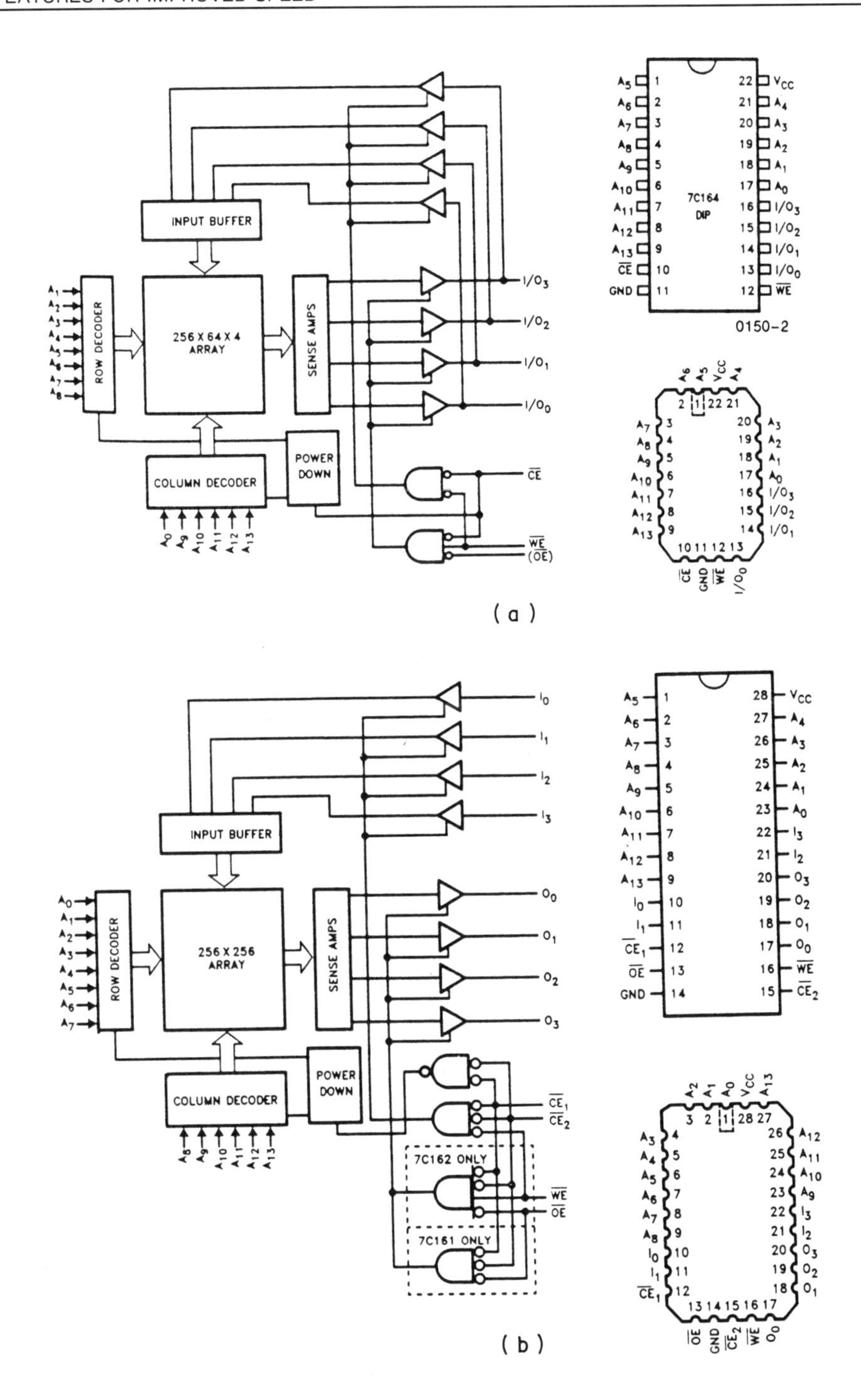

Figure 9.27
Comparison of schematic block diagrams of 16k × 4 SRAMs with (a) common I/O showing 22 lead DIP and PLCC pinouts, (b) separate I/O showing 28 lead DIP and PLCC pinouts. (Reproduced with permission of Cypress Semiconductors.)

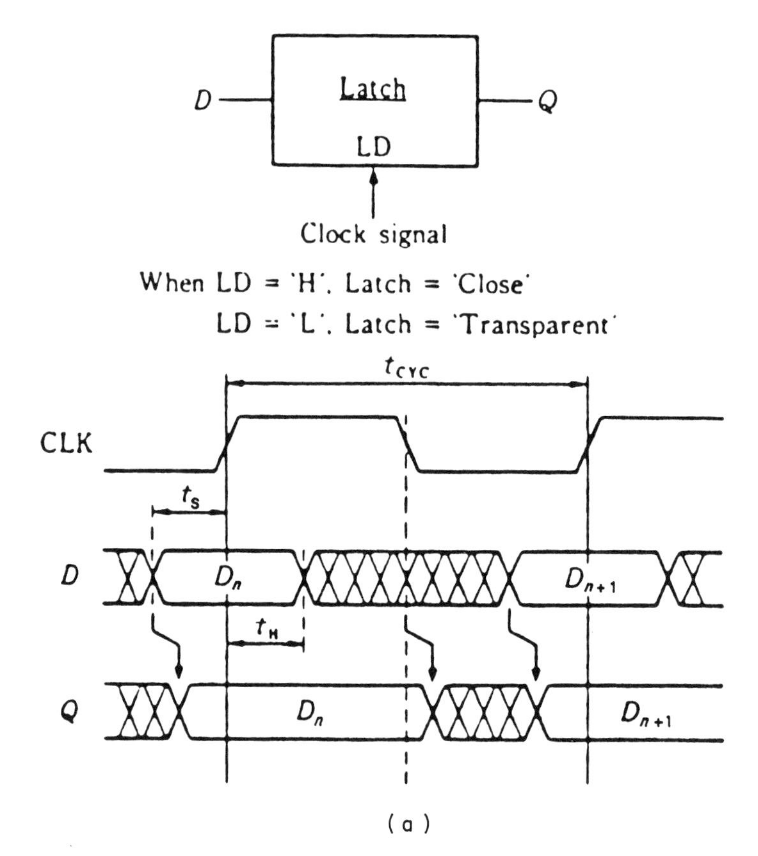

Figure 9.28
(a) Schematic block diagram of a level sensitive latch structure, and timing diagram showing latch open when the clock level is low and closed when the clock level is high. (b) Register as an edge sensitive latch structure with schematic block diagram showing register as a dual latch and timing diagram showing output stable throughout a single cycle. (From Ohno [65], Fujitsu 1988.)

the output until the clock again goes low, that is, data are latched in when the clock level is high and pass through when the clock level is low.

Registers, defined as edge-sensitive latch structures, consist of two latches in series as shown in the block diagram of Figure 9.28(b). One latch is controlled by the edge of the CLK input and the other by the edge of the $\overline{\text{CLK}}$ input. The output of the register, therefore, remains stable throughout a single clock cycle even though new data appear on the input latch as shown in Figure 9.28(b). The data from the previous cycle can be available very early in the next clock cycle giving the opportunity for a fast system read of data from the previous cycle.

At the system level, the clock can run faster with a latched (or registered) SRAM since the address data can be latched in the memory and the system address drive go on to the next address.

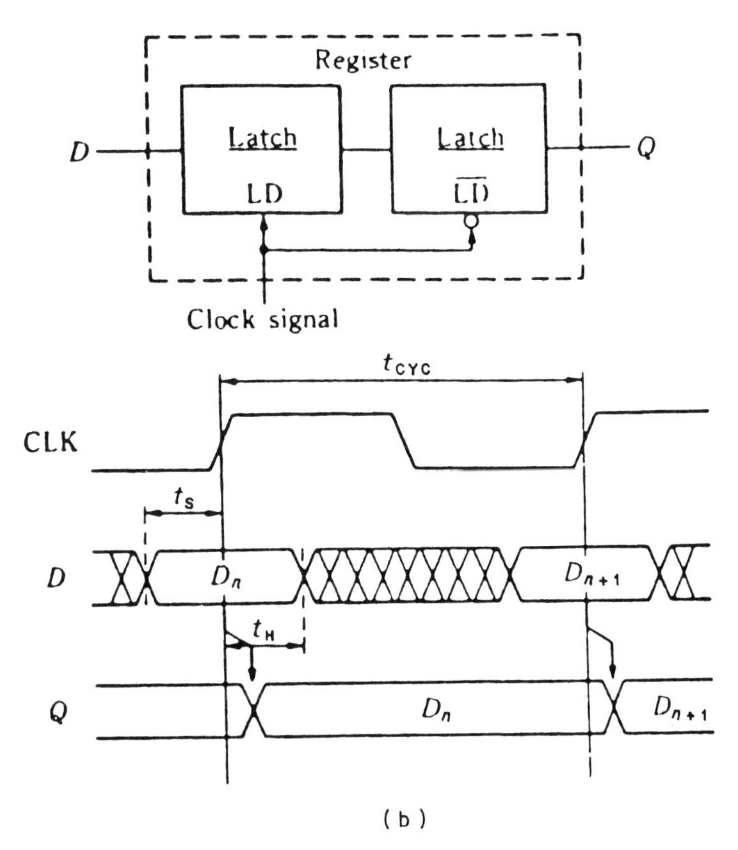

Figure 9.28 (*continued*)

Fujitsu made a 1k × 4 and 4k × 4 ECL STRAM with 7 and 10 ns address access time as shown in the diagram in Figure 9.29. Note that the power and ground pins have been moved to the center of the package to minimize the ground bounce by shortening the length of the power supply bond wires from chip to package.

Going to the chip level it is possible to use combinations of registers and latches on the inputs and on the outputs. Each additional latch in the series delays the data by half a clock cycle from input to output.

As an example registers can be used on the inputs but latches on the outputs allowing input data to be latched while giving a 'transparent output' feature. This was done on a Motorola 16k × 4 synchronous static RAM which had input registers for address data and control but latches on the outputs as shown in Figure 9.30(a). A companion part which also had registers on the outputs is shown in Figure 9.30(b). The corresponding read cycle timing diagrams for both parts are shown in Figures 9.31(a) and (b). It can be seen that the data in the register–register part are delayed half a clock cycle from those in the latch–register part.

A pipelined SRAM from AT&T Bell Laboratories [67] was presented in 1989. This part had full synchronous registering, cycle times in the 4 to 7 ns range and synchronous data access times under 2 ns. In asynchronous operation the SRAM has

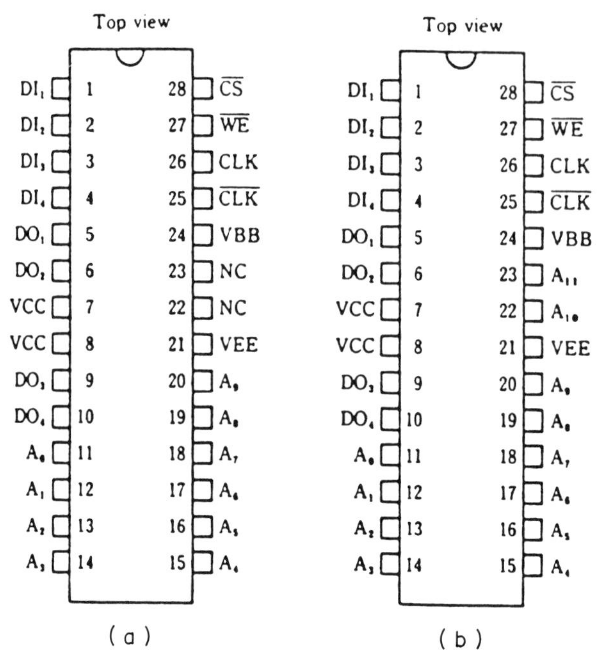

Figure 9.29
Pinouts of (a) 4k and (b) 16k STRAM from Fujitsu. (From Ohno [65], Fujitsu 1988.)

a 5 ns address access time. It had a three-stage pipeline with the memory in the front section, a byte-to-bit scanner in the second and a parity generator in the last section. The later two were intended for use in digital switching applications.

A schematic circuit diagram of the sample and hold register is shown in Figure 9.32. A complementary clock generator connected either the input or the feedback path to the load node for fast new data input on the rising clock edge. The register could be made transparent for a ripple mode operation.

National Semiconductor [86] in 1990 also showed a 2k × 9 self-timed RAM with 3.5 ns access time in BiCMOS technology. The part registered address and data on the rising clock edge and required the system to hold data valid for only 1 ns, greatly enhancing the potential for system clock speed.

9.7.3 Pinouts with multiple and center power and ground pins

In 1989 the JEDEC memory standardization committee standardized a new pinout for fast SRAMs with double power and ground pins in the center of the package. This was intended to reduce the ground bounce problem for very fast SRAMs thereby speeding up system access to the output data. The adjacent power and ground pins have a mutual inductance effect which acts to reduce the self-inductance effect causing the ground bounce when the outputs swing. The centering of the power pins on the package tends to reduce the length of the bond wires to the power supply pins thereby also reducing the self-inductance effect. Since the ground bounce effect is proportional to the number of outputs that are switching simultaneously, wide bus memories are more affected and likelier to need the next pinout. This new standard was discussed in more detail in Chapter 5.

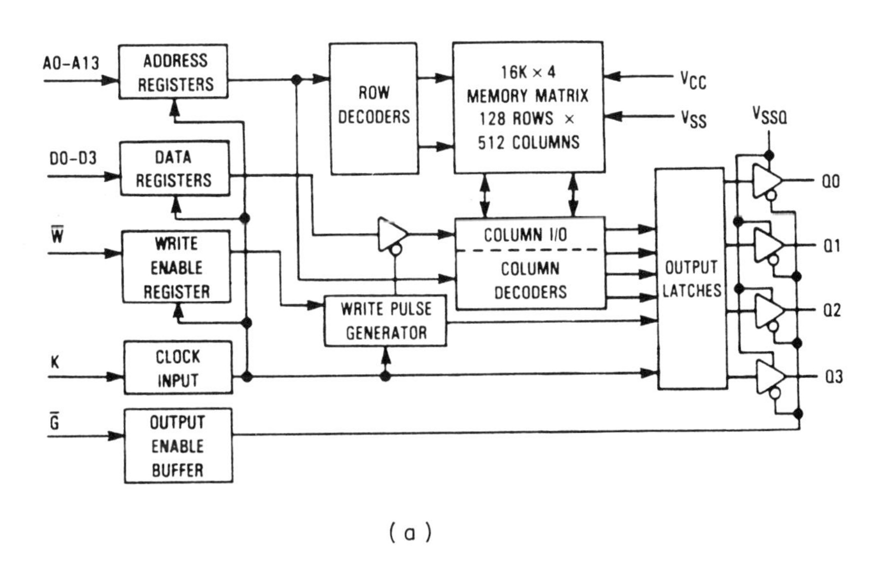

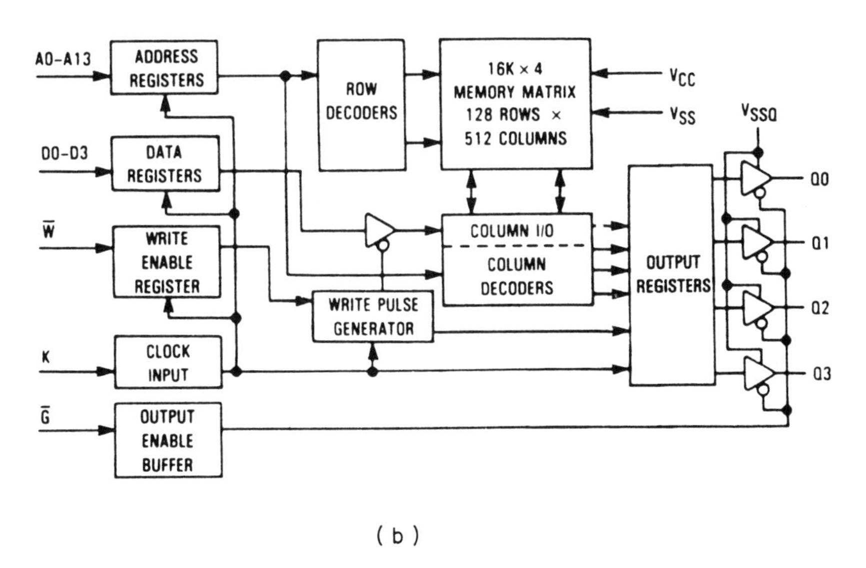

Figure 9.30
16k × 4 synchronous SRAM with (a) input registers and output latches and (b) input and output registers. (From Motorola Memory Databook [75] 1989.)

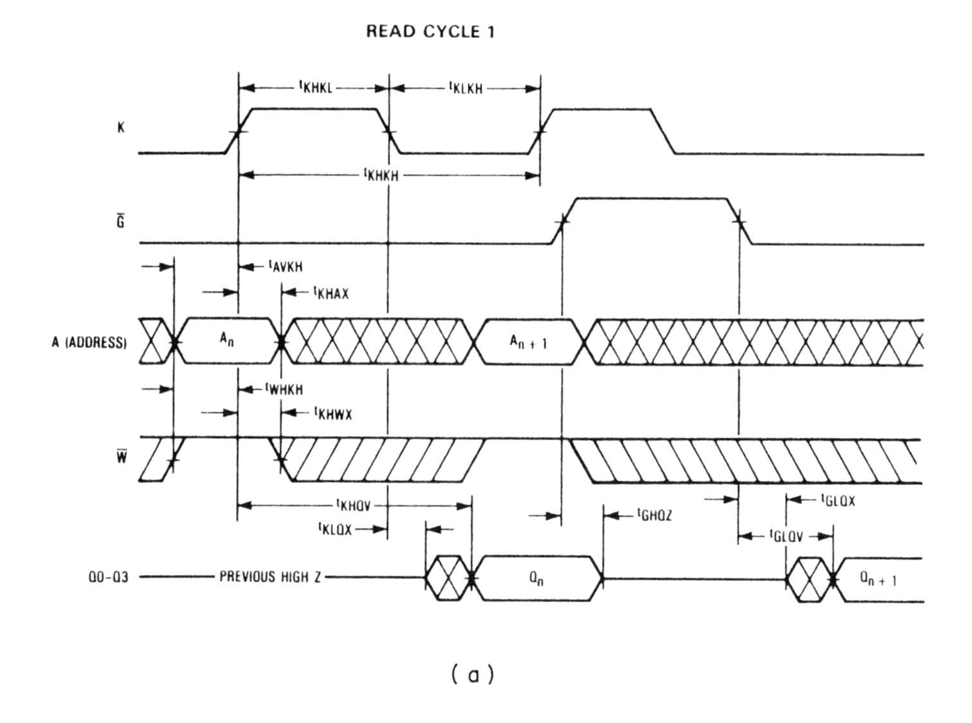

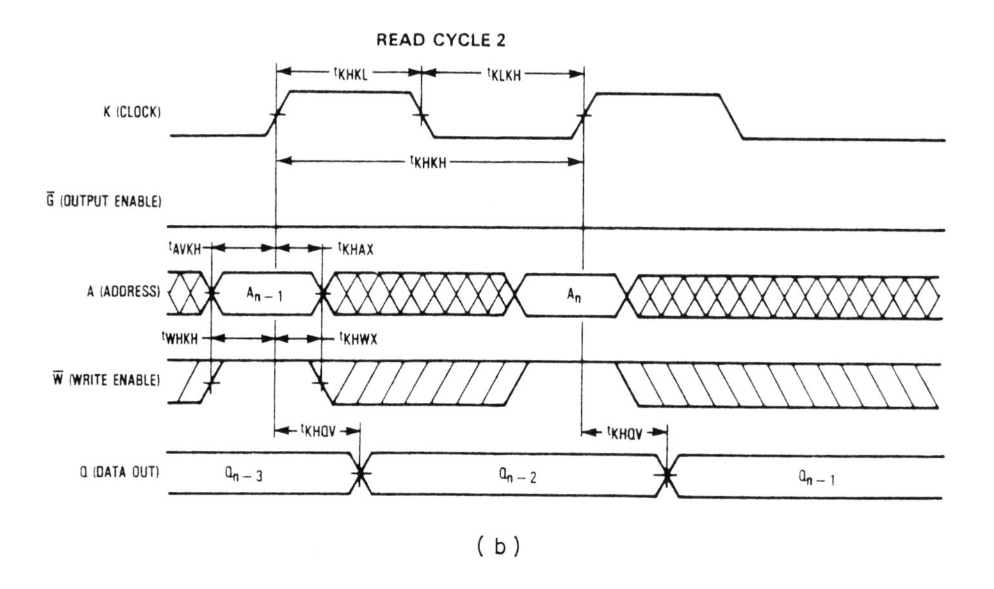

Figure 9.31
Read cycle timing diagrams for two SRAMs: (a) with input register and output latches, (b) with input and output registers. (From Motorola Memory Databook [75] 1989.)

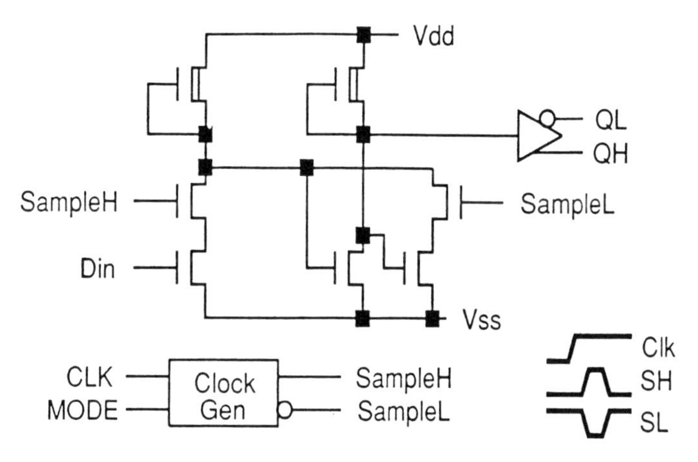

Figure 9.32
Sample and hold register. (From O'Conner [67], AT & T 1989, with permission of IEEE.)

9.7.4 A typical center-power and ground high speed SRAM

One of the first SRAMs on the market with the new JEDEC standard multiple center power and ground pinout was a fast 1Mb SRAM announced by National Semiconductor [77] in 1990. An earlier end-power and ground 256k was shown in 1988. A comparison is given in Table 9.3. of the technology and speed of the two

Table 9.3

Density	256k	1M
Year	1988	1990
Technology	1.0 μm BiCMOS	0.8 μm BiCMOS
Typical access time (ns)	12	6.8
Organization	$\times 1$	$\times 1$, $\times 2$, $\times 4$ Configurable
I/O compatibility	ECL 10k, 100k	ECL 10k, 100k (metal option)
Pinout	end power/ground	center power/ground
Number of supplies	$1 V_{EE}/1 V_{SS}$	$2 V_{EE}/2 V_{SS}$
Periphery location	end of chip	center of chip
Interconnect	2P/2M	2P/2M
Gate oxide (nm)	2000	1500
CMOS gate length (μm)	0.8	0.7
Metal pitch (μm)	2.5–2.9	1.8–1.9
Cell size (μm^2)	96.5	37.4
Cell type	R-load	R-load
Chip size (mm^2)	51.4	71.2
Array efficiency (%)	48.8	53.7%

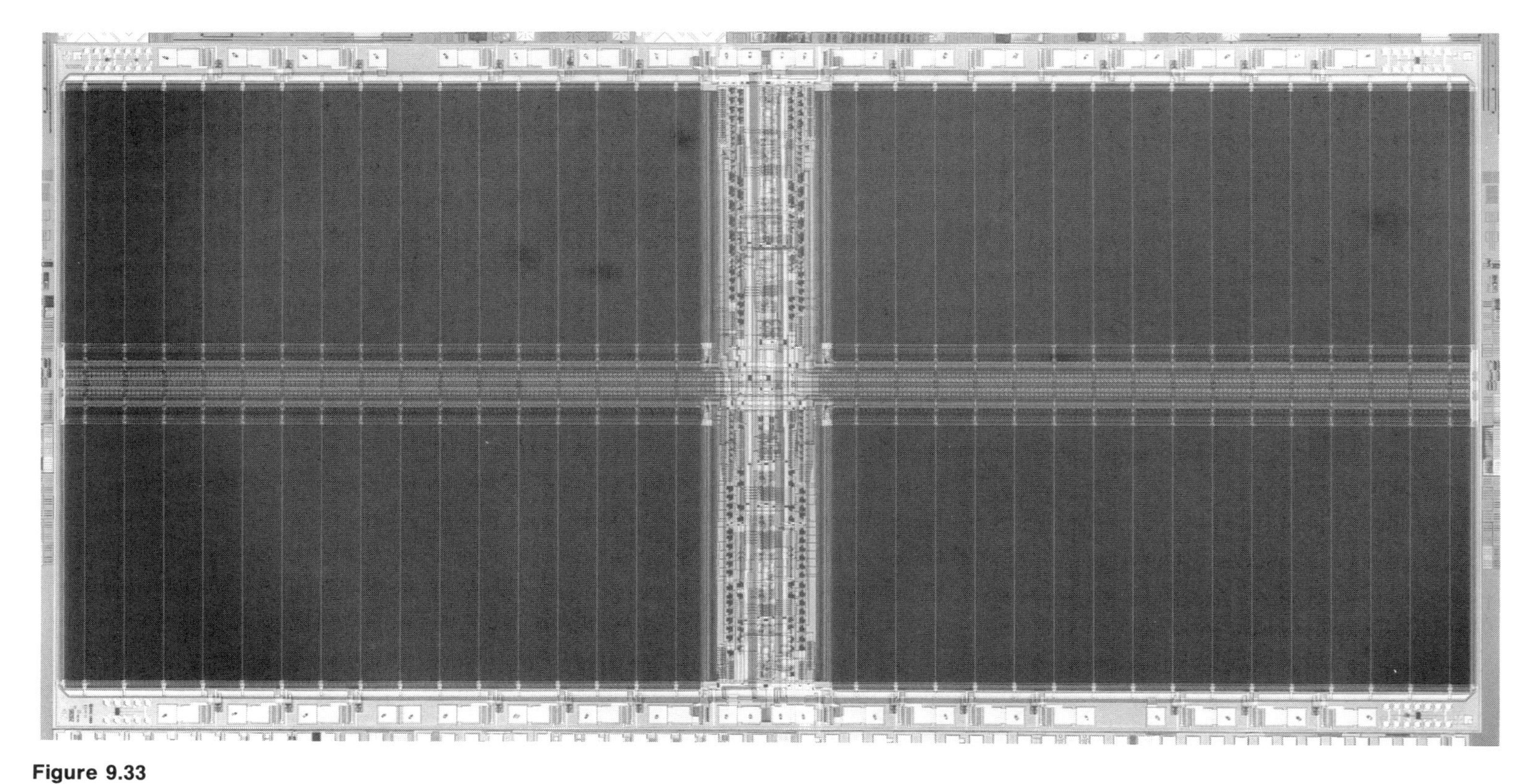

Figure 9.33
1Mb SRAM in JEDEC standard center power and ground architecture showing peripheral circuitry at center of chip and double bond pads on either side of chip center for V_{CC} and V_{SS}. (From Kertis [76] 1990, with permission of National.)

generations of BiCMOS ECL compatible SRAMs. This gives an indication of some of the architectural modifications and technological considerations made to accommodate the new center power and ground pinouts.

The die photo of this 1Mb center power and ground SRAM shown in Figure 9.33 can be compared with the photograph of a conventional end power and ground SRAM in Figure 8.38. The new architecture had the peripheral circuitry concentrated in the middle of the chip and the double bonding pads for the V_{CC} and V_{SS} can be seen at the center. The conventional architecture had the peripheral circuitry distributed in blocks and at the ends of the chip. See Chapter 5 for the pinout comparison of the two parts and for further discussion on this topic.

9.7.5 Flash write feature

A bulk write feature to enable write of the whole memory in one extended write cycle was described by Motorola [7] in 1984. With this function the entire array was written to the state of the data-in pin during one extended cycle. This feature was accessed either by an over-voltage pulse of $\overline{WE}$ or by a metal mask option separate control pin.

9.8 WIDE BUS ARCHITECTURE TRADE-OFFS

Although wider buses are in demand on fast memories, various historical concerns have kept them from being available until recently.

Historically the ground bounce problem had kept fast SRAMs $\times 1$ or at most $\times 4$. Both new circuit design concepts and the adoption of the new center power and ground pinout standards have alleviated this problem.

System granularity brings up another concern with memory organization. Granularity refers to the minimum size blocks of memory that can be used to configure a system with a given organization. A critical issue is the number of memories of a particular memory organization that must be used to implement the system configuration. For example, if a system is configured 32k $\times$ 16, then using only standard 256k SRAM organizations, one could use 16 256k $\times$ 1 SRAMs for a granularity of 256k $\times$ 16, 4 64k $\times$ 4 SRAMs for a granularity of 64k $\times$ 16 or 2 32k $\times$ 8 SRAMs for a granularity of 32k $\times$ 16. Clearly the 32k $\times$ 8 SRAMs are preferred to avoid wasted memory, all other system requirements being met.

If the cost of a 256k SRAM in a particular speed range is the same regardless of the organization, then clearly for this example the 64k $\times$ 4 configuration costs less at the chip level than the 256k $\times$ 1, and the 32k $\times$ 8 costs less than the 64k $\times$ 4.

In terms of board space in the system, 4 64k $\times$ 4 SRAMs in a standard 0.3 inch package occupies less than 25% of the board space of 16 256k $\times$ 1 SRAMs in the same 0.3 inch package.

The analogy historically did not carry on to the 32k $\times$ 8, however, since the 32k $\times$ 8 was offered in a 0.6 inch package which did not occupy significantly less board space than 4 64k × 4 SRAMs. Since the 32k × 8 also was slower due to the ground bounce problem, the ×4 configuration was used throughout the 1980s for fast wide bus systems.

For very fast systems where even the ground bounce of the ×4 configuration slowed the outputs down too much to be tolerated, or where the system organization required the greater 256k depth, the ×1 configuration continued to be used in spite of the higher cost of the multiple chips.

As very fast 32k × 8 SRAMs in 0.3 inch packages began to be offered in the latter half of the 1980s, many fast 16 and 32 bit bus systems began to use them.

Power dissipation is another consideration in selecting the granularity of the system. Since the power dissipation of an SRAM running at a particular speed is roughly the same regardless of the width of the outputs the power dissipation in this system is fairly proportional to the number of 256k SRAMs that must be run to obtain a 16 bit word of data. The two 32k × 8s consume half the power of four 64k × 4s and $\frac{1}{8}$ the power of 16 256k × 1s assuming that standby power is negligible compared to active power.

Error correction considerations provide reasoning for using the narrower bus parts. This will be considered further in the chapter on reliability and test.

Another solution to the requirement for interfaces for ever wider system buses has been the use of a serial interface such as the I^2C bus interface of Philips which implemented a byte-wide memory architecture using one serial I/O pin and one clock in addition to the necessary address pins. This simplification of the interface was at the expense of greatly reduced speed.

A 4Mb CMOS SRAM with 8 ns serial access architecture was discussed by Mitsubishi [87] in 1990. This part was intended for high speed image processing systems and had a 125 MHz serial read–write operation. The outputs of two 16 bit registers were interleaved on chip to give continuous serial access.

9.9 SPECIALITY SRAM ARCHITECTURES

The CMOS SRAM process can be completely compatible with standard logic processes, thus SRAMs are easy to add logic to for a variety of applications. A few of the more common application-specific SRAMs are considered in the following sections.

There is no clear cut distinction made here between applications specific SRAMs and SRAM arrays embedded in logic. Historically application-specific SRAMs were made in the high density, optimized memory processes with a bit of hand crafted logic added. Embedded SRAMs, on the other hand, were made from non-optimized memory cells added to logic chips using standard automated logic design techniques. Optimized SRAM processes tended to include such features as buried contacts, straps, etc which reduced the size of the memory cell.

9.9.1 Power-down protection and the battery RAM

Battery operated memory systems require not only very low operating power and battery-back-up data retention current levels from the memory, they also require logic which will protect the data in the memory if the battery voltage gets low. Critical systems using battery-back-up memory which are line operated also need circuitry to protect the data during power down if the power supply fails. Such circuitry has been commonly implemented at the system level.

In the early 1980s Mostek, later SGS-Thomson Microelectronics, added onto the memory chip the circuitry to protect against accidental writes during power down. An example is a family of byte-wide SRAMs which STM [45] called their 'zero power RAM'. These parts included a lithium battery in the SRAM package which made them, in effect, fast non-volatile SRAMs. They included a temperature compensated power fail detection circuit which monitored V_{CC} and switched from the line voltage to the back-up battery when the voltage fell below a specified level and switched back to the line voltage when the power was restored.

Figure 9.34 shows the pinout and the power-down–power-up timing diagram for the 8k × 8 version. The pinout was identical to a standard SRAM except for the optional interrupt (INT) pin. This interrupt function output provided the system with an advance warning of an impending power-fail write protect and eliminated the need for external power sensing components in applications where an orderly shutdown of the system was necessary.

In 1983 NEC [46] also showed an early 64k SRAM chip with a circuit to transfer the memory automatically to the data retention mode when a supply voltage was lowered as shown in Figure 9.35(a). This chip was internally synchronous with a one shot pulse which initiated the sequential internal clocks. In operation when the mode changed from standby to active by $\overline{CE}_1$ going low and CE_2 going high, then P_1 changed from high to low and a one shot pulse was generated which initiated the internal operation. If, however, the supply voltage was lowered to less than 3.5 V, the n-channel transistor V_{TR} was off and P_1 was high even when the chip was selected so that internal operation was inhibited. A timing chart of the autodata retention mode is shown in Figure 9.35(b).

A similar circuit concept was also shown on the Toshiba 4Mb SRAM for disabling the internal voltage-down convertors in battery-back-up mode, as was discussed in the section on 4Mb SRAM architecture.

A 256k full CMOS SRAM that operated down to one volt was shown by Fujitsu [88] in 1990. The one volt operation made this part suitable for use in hand-held computers and memory cards.

9.9.2 Dual-port RAMs

Dual-port RAM architecture can be implemented in systems which have a mismatch in speed or bandwidth between a processor and a peripheral device, or between a main memory and a set of distributed processors, and also in systems with multiple processors and servers needing to communicate with each other or along a single bus. With a standard single-port SRAM buffer, bus contention results when two processors or servers attempt to access the SRAM at the same time. One must wait until the SRAM is free. This reduces the advantage of having a high speed processor in the system. An ideal dual-port SRAM would allow simultaneous access of the same location in the memory [72].

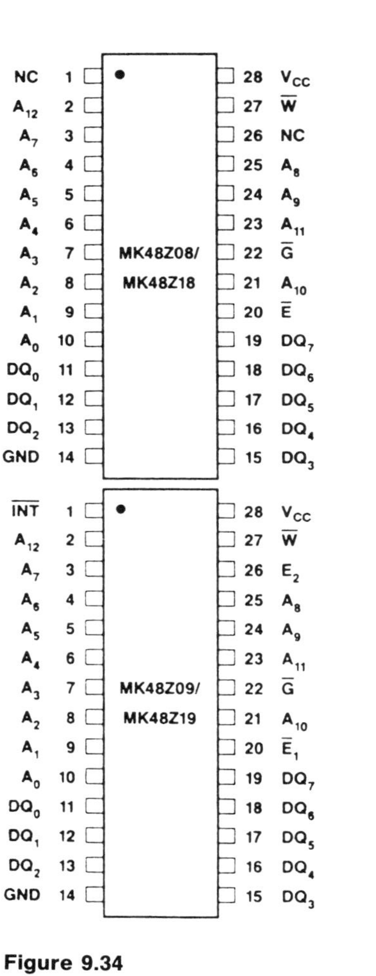

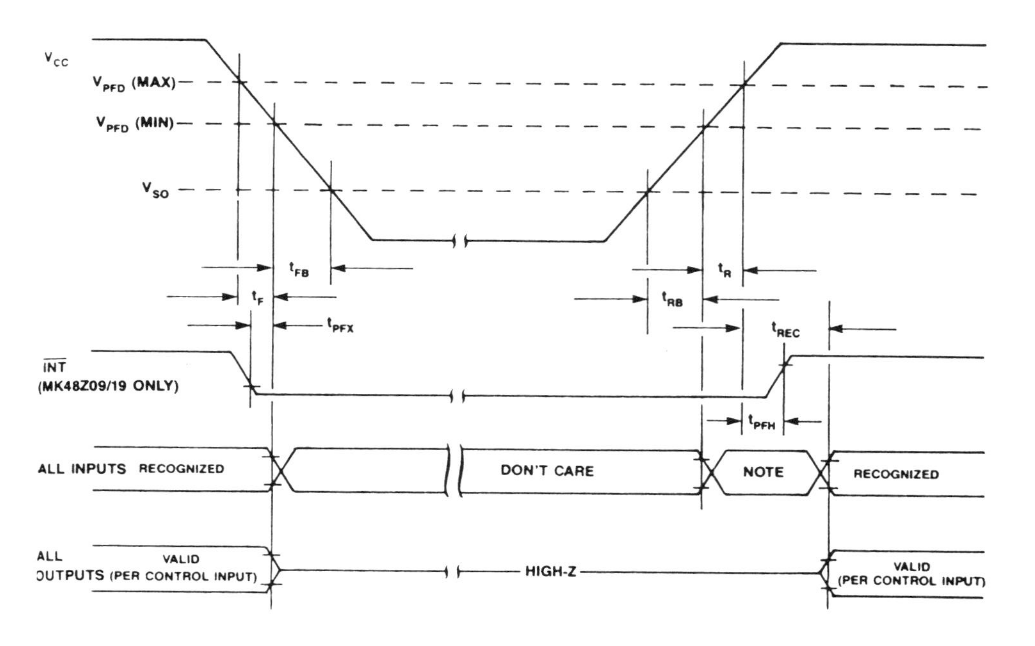

Figure 9.34
Pinout and power-down–power-up timing diagram for 8k × 8 zero power RAM. (Source SGS-Thomson Databook [88] 1988.)

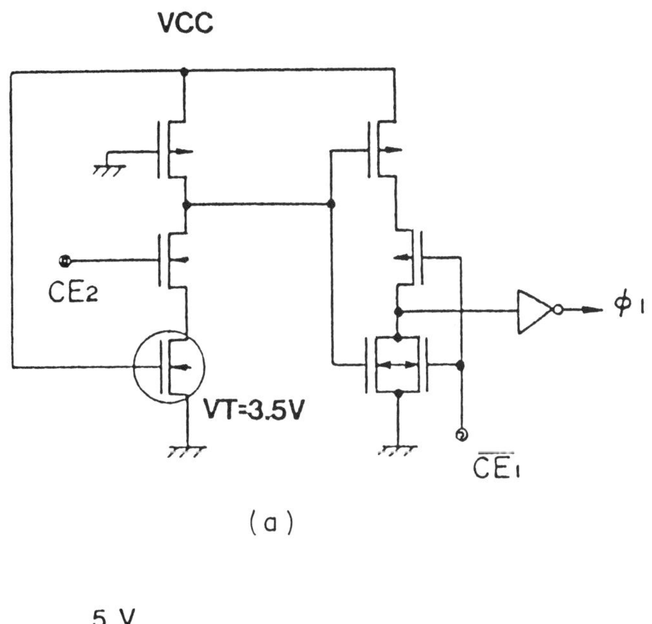

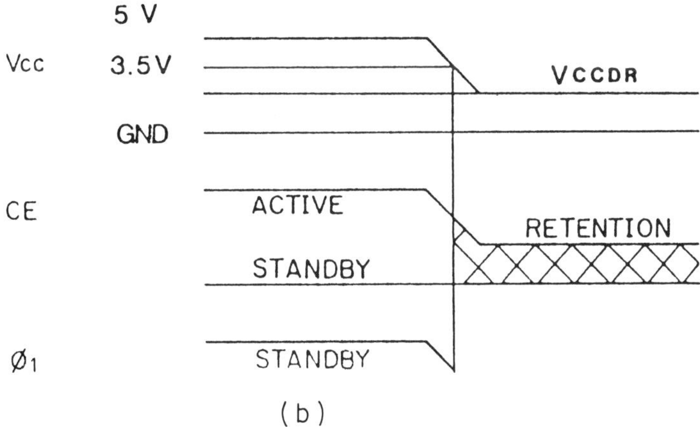

Figure 9.35
(a) Schematic circuit diagram of autodata retention circuit. (b) Timing diagram for the circuit. (From Watanabe *et al.* [46], NEC 1983, with permission of IEEE.)

Traditionally, and still for systems requiring large amounts of memory in the dual-port RAM, dual-port architecture has been implemented with discrete components. Bus contention was handled by logic configured to provide arbitration between the different access requests.

This discrete implementation, however, increases the chip count over that of a dual-port SRAM. It requires address buffers and latches, data I/O latches, an arbiter chip and standard SRAMs. When only a small memory SRAM buffer is required, the additional chips become a large proportion of the cost of the memory and the chip count can be reduced by using a single SRAM which integrates these various logic functions on chip using a dual-port approach.

For large memory buffer applications, the cost of the standard memories becomes a larger proportion of the total cost and a discrete implementation is often preferred since the cost of adding the dual-port SRAM is often higher than adding the logic and standard SRAMs.

Since bus contention is always a potential when two systems have access to the same memory array, the early dual-port RAMs used exclusive access states. This meant that when one port had access the other had to wait.

In 1983 the first SRAM with contention arbitration on the chip was described by Bell Laboratories. This small 512 × 10 NMOS RAM was claimed to replace 15 to 20 logic and single-port memory chips in a system.

One of the early commercially available dual-port SRAMs which provided contention arbitration on chip was the Synertek SY2131. This 1024 × 8 bit SRAM had 100 ns access time. Two I/O ports allowed separate and simultaneous access to common memory locations.

Internal logic handled data contention by raising busy signals when conflicts arose during attempted simultaneous accesses of the same memory address. Cross-talk

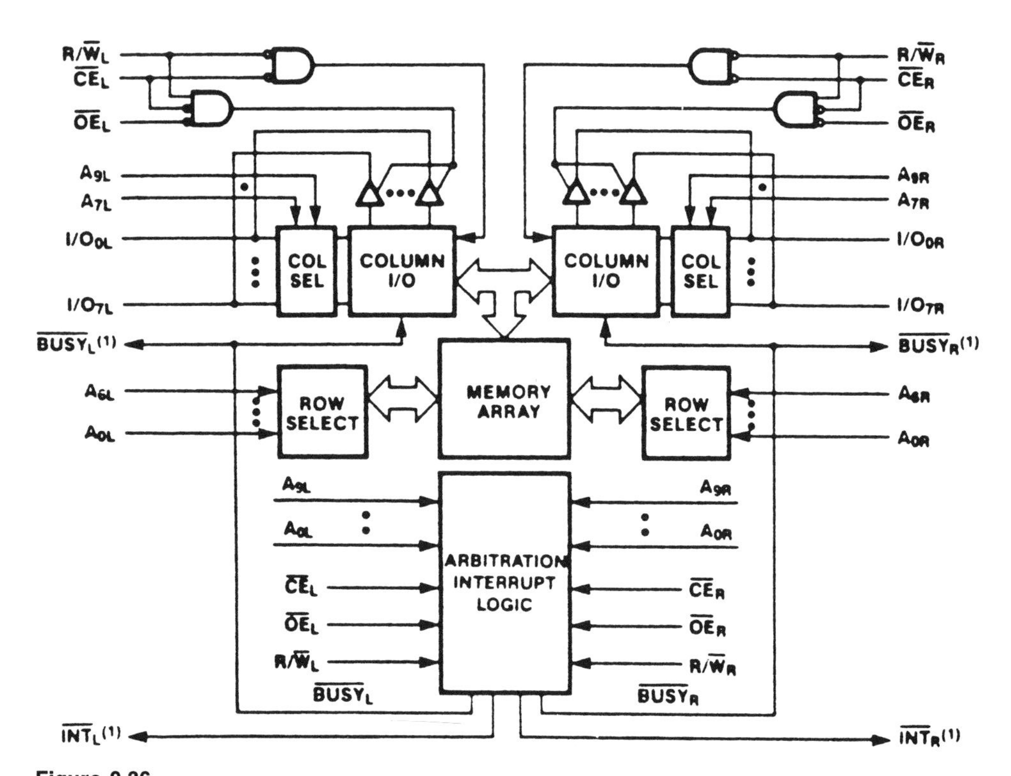

Figure 9.36
1k × 8 dual-port RAM with each port providing independent access to any location of the memory array. Three control pins and two flags are shown.(From IDT Databook [83] 1989.)

problems were also minimized because the address and data buses of the two ports were physically isolated.

Although this device provided arbitration contention on the chip, it still caused wait states for the processor if access of a single address was simultaneously requested by two processors. What was needed was a dual-port SRAM which could handle simultaneous accesses of the same memory location. This 8 bit chip also had limitations handling buses of differing widths. If a 16 bit bus device requested access, the data needed to be split and sequentially read. While Synertek no longer exists, this early 1k × 8 dual port RAM continues to be sold by other companies such as VTI who acquired the rights to it.

A later dual-port RAM designed by VHSIC, who was later acquired by VTI, had one 8 bit port and one 16 bit port and was stackable to handle wider buses. This eliminated the bus width mismatch, but still generated busy states.

Various other dual-port RAMs in various configurations mostly 16k in density or less are offered as applications specific to different systems applications.

Similar 1k × 8 dual port RAMs have been introduced by Cypress Semiconductor and Integrated Device Technology Inc. The later part is shown in Figure 9.36. The two ports on this part provided independent access to any location in the memory array. Each port had independent control pins including Write Enable, Chip Enable, and Output Enable. Two flags were provided on each port. A $\overline{\text{BUSY}}$ flag indicated that the port was trying to access the same location in memory already being accessed by the other port and an interrupt ($\overline{\text{INT}}$) flag indicated that data had been placed in a unique location by the other port.

9.9.3 FIFOs and line buffers

The serial nature of the First-In-First-Out buffer (FIFO) and the lack of address pins make it an application-specific memory in its own right. Small FIFOs are widely used in communications buffering types of applications. They are generally made from static RAM type cells if there is a requirement for the data to be maintained in the FIFO, and from dynamic RAM cells if the data simply pass through the FIFO. These dynamic cells may be one of the older three-transistor dynamic cell configurations described in Chapter 5 in order to be compatible with standard logic processing without using the capacitors found in the one-transistor DRAM cells. An example is a 910 × 8 bit FIFO made by NEC for line interleaving to reduce flicker in NTSC television sets.

Larger FIFOs, which are normally application-specific DRAMs, are used in television applications for frame storage and were already discussed under the section in Chapter 7 on frame memories.

9.9.4 Content addressable memories

Content addressable or 'associative' memories have been known since the 1960s and have been implemented with logic chips in systems. While they are rarely produced as stand-alone parts by mainstream commodity memory suppliers due to the large number of transistors in the CAM cell, they are designed and used both as embedded modules on larger VLSI chips and as stand-alone memory for specific systems applications. In the later case, it is generally the systems house itself which designed and uses the CAM. The cell structure of a CAM uses at least nine transistors as was described briefly in Chapter 5 under the section on varieties of SRAM cells.

Unlike standard memories that associate data with an address, the CAM associates an address with data. The data are presented on the inputs of the CAM which searches for a match for those data in the CAM without regard to address. When a match is found the CAM identifies the address location of the data.

Another reason that CAMs are seldom designed as standalone devices is the large number of pins associated with inputting long data words to the CAM and between CAMs in parallel. If all the CAM that is needed in a system can be contained on one standalone chip or embedded in one logic chip, then this becomes less of a problem.

Applications using CAMs include database management, network filtering, disk caching, pattern and image recognition and artificial intelligence.

An example is a 256 × 46 organized CAM offered by AMD [74] for such applications as Ethernet networks where it functioned as an address filter and performed network address look-up functions in bridges. In operation this CAM had data presented to it, called the 'comparand' and performed simultaneous compare operations on all 256 words of data. When the comparand and a word in the CAM were matched, the on-chip priority encoder generated a match word address identifying the location of the data in the CAM. If multiple matches occurred, the encoder generated the lowest matched address.

Another example of a CAM was a 256 × 64 part from Bellcore [99] which was designed for high speed look-up table applications. This part had self-test and reconfigurable features to simplify test and increase yield.

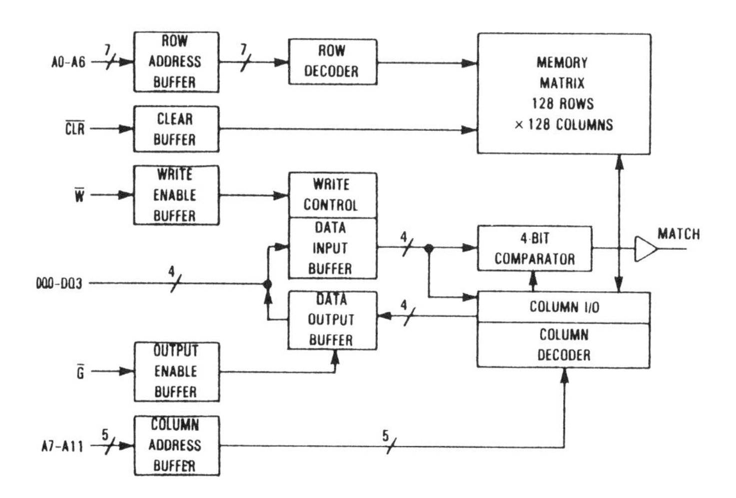

Figure 9.37
16k cache address tag comparator which integrates a 4k × 4 SRAM with an on-chip comparator. (Source Motorola MOS Memory Databook [84] 1989.)

9.9.5 Cache TAG RAM

The cache TAG RAM is memory that stores both TAG, which represents the data but in fewer bits, and the data associated with the TAG. It functions like a CAM, but is usually made with standard four- or six-transistor SRAM cells with added logic in the form of a comparats which compared the TAG of the addressed RAM location with the TAG of the data present on the inputs. When these are the same a 'match' signal is generated. The cache TAG SRAM searches for the specific contents indicated by the TAG without consideration of the location of the contents.

Several cache TAG RAMs have appeared on the market for specific applications. The block diagram of a 16k cache TAG RAM from Motorola is shown in Figure 9.37.

BIBLIOGRAPHY

[1] Dingwall, A. G. F. and Stewart, R. G. (1979) 16 k CMOS/SOS Asynchronous Static RAM, *IEEE Journal of Solid State Circuits*, Vol. **SC-14**, No.5, October 1979, p.867.

[2] Kawarada, K. *et al.* (1978) A Fast 7.5 ns Access 1K-bit RAM for Cache Memory Systems, *IEEE Journal of Solid State Circuits*, Vol. **SC-13**, No.5, October 1978, p.656.

[3] Minato, O. *et al.* (1982) A Hi-CMOSII 8k × 8 Bit Static RAM, *IEEE Journal of Solid State Circuits*, Vol. **SC-17**, No.5, October 1982, p.793.

[4] Hudson, E. L. and Smith, S. L. (1982) An ECL Compatible 4k CMOS RAM, *Proceedings of the ISSCC82*, February, p.248.

[5] Sood, L. C. *et al.* (1983) A 35 ns 2k × 8 HMOS Static RAM, *IEEE Journal of Solid State Circuits*, Vol. **SC-18**, No.5, October 1983, p.498.

[6] Twaddell, W. (1984) High-speed CMOS RAMs Score Across Memory Spectrum, *EDN*, July 12, p.62.

[7] Barnes, J. J. *et al.* (1984) Circuit Techniques for a 25ns 16k × 1 SRAM Using Address-Transition Detection, *IEEE Journal of Solid State Circuits*, Vol. **SC-19**, no.4, August 1984, p.455.

[8] Minato, O, *et al.* (1984) A 20 ns 64 k CMOS Static RAM, *IEEE Journal of Solid State Circuits*, Vol. **SC-19**, No.6, December 1984, p.1008.

[9] Miyamoto, J. *et al.* (1984) A 28ns CMOS SRAM with Bipolar Sense Amplifiers, *ISSCC84 Proceedings*, February 23, p.224.

[10] Liptay, J. (1968) Structural aspects of the system 360 model 85, II—The cache, *IBM System Journal*, **7**, No.1.

[11] Ogiue, K. *et al.* (1986) A 13 ns/500 mW 64 Kb ECL RAM, *ISSCC86 Proceedings*, February, p.212.

[12] Flannagan, S. *et al.* (1986) Two 13 ns 64 k CMOS SRAM's with Very Low Active Power and Improved Asynchronous Circuit Techniques, *IEEE Journal of Solid State Circuits*, Vol. **SC-21**, No.5, October 1986, p.692.

[13] Dual-Port Static RAMs can Remedy Contention Problems, *Computer Design*, August 1984, p.145.

[14] O'Connor, K. J. (1983) A 5 ns 4 k × 1 NMOS Static RAM, *ISSCC Proceedings*, February, p.104.

[15] Yamaguchi, K. *et al.* (1986) A 3.5 ns, 2 W, 20 mm^2 16 Kb ECL Bipolar RAM, *ISSCC Proceedings*, p.214.

[16] Glock, H. and Burker, U. (1979) An ECL 100 k-Compatible 1024 × 4 bit RAM with 15 ns Access Time, *IEEE Journal of Solid State Circuits*, **SC-14**, p.850.

[17] Shah, A. H. *et al.* (1984) A 2 μm Stacked CMOS 64 k SRAM, *Symposium on VLSI Technology*, September, p.8.

[18] Ando, M. *et al.* (1988) A 0.1 μA Standby Current Bouncing-Noise-Immune 1Mb SRAM, *Symposium on VLSI Technology*, p.49.

[19] Yananaka, T. *et al.* (1988) A 25 μm^2, New Poly-Si PMOS Load (PPL) SRAM Cell Having Excellent Soft Error Immunity, *IEDM Proceedings*, p.48.

[20] Hayden, J. *et al.* (1989) A High-Performance Sub-half Micron CMOS Technology for Fast SRAMs. *IEDM Proceedings*, p.417.

[21] Wang, L. K. *et al.* (1989) Characteristics of CMOS Devices Fabricated Using High Quality Thin PECVD Gate Oxide, *IEDM Proceedings*, p.463.

[22] Hori, T. (1989) 1/4-μm LATID (LArge-Tilt-angle Implanted Drain) Technology, *IEDM Proceedings*, p.777.

[23] Chuang, C. T. and Lu, P. F. (1989) On the Scaling Property of Trench Isolation Capacitance for Advanced High-Performance ECL Circuits, *IEDM Proceedings*, p.799.

[24] Takato, H. *et al.* (1988) High Performance CMOS Surrounding Gate Transistor (SGT) for Ultra High Density LSIs, *IEDM Proceedings*,, p.223.

[25] Sunouchi, K. *et al.* (1988) Double LDD Concave (DLC) Structure for Sub-Half Micron MOSFETs, *IEDM Proceedings*, p.227.

[26] Isomura, S. *et al.* (1989) A 36 kb/2 ns RAM with 1 kG/100 ps Logic Gate Array, *IEEE ISSCC Proceedings*, February, p.26.

[27] Sasaki, K. *et al.* (1990) A 23ns 4 Mb CMOS SRAM with 0.5 μA Standby Current *IEEE ISSCC Proceedings*, February, p.130.

[28] Tran, H. *et al.* (1989) An 89 ns BiCMOS 1 Mb ECL SRAM with a Configurable Memory Array Size, *ISSCC Proceedings*, February, p.36.

[29] Eklund, R. *et al.* (1989) A 0.5-μm BiCMOS Technology for Logic and 4 Mbit-class SRAM's, *IEDM Proceedings*, p.425.

[30] Hayakawa, S. *et al.* (1990) A 1-μA Retention 4 Mb SRAM with a Thin-Film-Transistor Load Cell, *ISSCC Proceedings*, February, p.128.

[31] Yuzuriha, K. *et al.* (1989) A New Process Technology for a 4Mbit SRAM with Polysilicon Load Resistor Cell, *VLSI Symposium Proceedings*, p.162.

[32] Kunio, T. *et al.* (1989) Three Dimensional IC's, Having Four Stacked Active Device Layers, *IEDM Proceedings*, p.837.

[33] Colinge, J. P. (1989) Thin-Film SOI Technology: The Solution to Many Submicron CMOS Problems, *IEDM Proceedings, p.817.*

[34] Woerlee, P. H. et al. (1989) Half-Micron CMOS on Ultra-Thin Silicon on Insulator, *IEDM Proceedings*, p.821.

[35] Yamaguchi, Y. *et al.* (1989) Improved Characteristics of MOSFETs on Ultra Thin SIMOX, *IEDM Proceedings*, p.825.

[36] Kamgar, A. *et al.* (1989) Ultra-High Speed CMOS Circuits in Thin SIMOX Films, *IEDM Proceedings*, p.829.

[37] Hisamoto, D. *et al.* (1989) A Fully Depleted Lean-channel Transistor (DELTA)—A Novel Vertical Ultra Thin SOI MOSFET, *IEDM Proceedings*, p.833.

[38] Rourdo, N. *et al.* (1990) Process Design for Merged Complementary BiCMOS, *IEEE IEDM Tech. Digest*, Dec. 1990, p.485.

[39] Hirose, T. *et al.* (1990) A 20 ns 4 Mb CMOS SRAM with Hierarchical Word Decoding Architecture, *ISSCC Proceedings*, February, p.132.

[40] Aizaki, S. (1990) A 15 ns 4 Mb CMOS SRAM, *ISSCC Proceedings*, February, p.126.

[41] Low Voltage TTL-Compatible Interface Standard, No.8.1, EIA JEDEC Standardization Committee.

[42] Davari, B. *et al* (1988) A High Performance 0.25 um CMOS Technology, *IEDM Proceedings*, p.56.
[43] Chapman, R. A. *et al.* (1988) 0.5 Micron CMOS for High Performance at 3.3 V, *IEDM Proceedings*, p.52.
[44] Kakuma, M. *et al.* (1988) 0.5 μm Gate 1M SRAM with High Performance at 3.3 V, 1988.
[45] *Memory Products Databook*, 1st Edn., SGS-Thomson Microelectronics, June 1988.
[46] Watanabe, T. *et al.* (1983) A Battery Back-up 64k CMOS RAM with Double Level Aluminum Technology, *IEEE Journal of Solid State Circuits*, Vol. **SC-18**, No.5, October 1983, p.498.
[47] Schuster, S. *et al.* (1984) A 20 ns 64k NMOS RAM, *ISSCC Proceedings*, February, p.226.
[48] Ochii, K. *et al.* (1985) A 17 ns 64 k CMOS RAM with a Schmitt Trigger Sense Amplifier, *ISSCC Proceedings*, February, p.64.
[49] Okuyama, H. *et al.* (1988) A 7.5 ns 32k × 8 CMOS SRAM, *IEEE Journal of Solid State Circuits*, Vol. **23**, No.5, October 1988.
[50] Sasaki, K. *et al.* (1989) A 9 ns 1 Mb CMOS SRAM, *ISSCC Proceedings*, February, p.34.
[51] Ogiue, K. *et al.* (1986) 13-ns, 500mW, 64-kbit ECL RAM Using HI-BICMOS Technology, *IEEE Journal of Solid State Circuits*, Vol. **SC-21**, No.5, October 1986, p.681.
[52] Miyamoto, J. *et al.* (1984) A High-Speed 64 k CMOS RAM with Bipolar Sense Amplifiers, *IEEE Journal of Solid State Circuits*, Vol. **SC-19**, No.5, October 1984, p.557.
[53] Miyaoka, S. *et al.* (1987) A 7-ns/350-mW 64-kbit ECL-Compatible RAM, *IEEE Journal of Solid State Circuits*, Vol. **SC-22**, No.5, October 1987, p.847.
[54] Cole, B. C. (1987) ECL's Worldwide Drive to Take over TTL sockets, *Electronics*, June 25, p.67.
[55] Cole, B. C. (1988) Is BiCMOS the Next Technology Driver? *Electronics*, February 4, p.55.
[56] Yoshimoto, M. *et al.* (1983) A 64kb Full CMOS RAM with Divided Word Line Structure, *ISSCC Proceedings*, February 23, p.58.
[57] Tran, H. V. *et al.* (1988) An 8-ns 256k ECL SRAM with CMOS Memory Array and Battery Backup Capability, *IEEE Journal of Solid State Circuits*, Vol. **23**, No.5, October 1988.
[58] Kertis, R. A. *et al.* (1988) A 12-ns ECL I–O 256k × 1-bit SRAM Using a 1-μm BiCMOS Technology, *IEEE Journal of Solid State Circuits*, Vol. **23**, No.5, October 1988, p.1048.
[59] Hochstedler, C. (1988) BiCMOS SRAM tops ECL memories, *Electronic Products*, March 1, 1988, p.32.
[60] Miyaji, F. *et al.* (1989) A 25-ns 4-Mbit CMOS SRAM with Dynamic Bit-Line Loads, *IEEE Journal of Solid State Circuits*, Vol. **24**, No.5, October 1989, p.1213.
[61] Masui, M. *et al.* (1989) An 8-ns 1-Mbit ECL BiCMOS SRAM with Double-Latch ECL-to-CMOS-Level Converters, *IEEE Journal of Solid State Circuits*, Vol. **24**, No.5, October 1989, p.1226.
[62] Chappell, T. I. *et al.* (1988) A 6.2ns 64Kb CMOS RAM with ECL Interfaces, *VLSI Circuits Proceedings*, p.19.
[63] Seevinck, E. *et al.* (1989) CMOS subnanosecond true-ECL output buffer, 1989 *Symposium of VLSI Circuits Proceedings*.
[64] Ogiue, K. *et al.* (1986) 13ns 500mW 64 k bit ECL RAM, *ISSCC Proceedings*.
[65] Ohno, C. *et al.* (1988) Self-Timed RAM—STRAM, *Fujitsu Sci. Technol.* **24**, 4 December.
[66] *Motorola Data Sheet* 1989, 16k × 18 Latched/Registered SRAM.
[67] O'Conner, K. (1989) A Prototype 2k × 8b Pipelined Static RAM, *ISSCC Proceedings*.
[68] *Cypress Semiconductor CMOS/BiCMOS Databook*, February 1, 1989.
[69] Takano, S. *et al.* (1987) A 16k GaAs SRAM, *ISSCC Proceedings*, February 26, p.140.
[70] Gabillard, B. *et al.* (1987) A GaAs 1k SRAM with 2 ns Cycle Time, *ISSCC Proceedings*, February 26, p.136.
[71] Tanaka, H. *et al.* (1987) A 4k GaAs SRAM with 1 ns Access Time, *ISSCC Proceedings*, February 26, p.138.

[72] Drumm, M. J. *et al.* (1984) Dual-port Static RAMs can Remedy Contention Problems, *Computer Design*, August, p.145.

[73] Cole, B. C. (1988) Content Addressable Memories Catch On, *Electronics*, December 1988, p.82.

[74] *AMD Databook* 1986 Am99c10, 256 × 48 Content Addressable Memory (CAM).

[75] *Motorola Memory Databook* 1989.

[76] Kertis, R. *et al.* (1990) A 6.8 ns 1 Mb ECL I–O BiCMOS Configurable SRAM, *VLSI Circuits Symposium*, June, p.39.

[78] Ikeda, S. *et al.* (1990) A Polysilicon Transistor Technology For Large Capacity SRAMs, *IEEE IEDM Tech. Digest*, Dec. 1990, p.469.

[79] Verhaar, R. D. J. *et al.* (1990) A 25 μm^2 Bulk Full CMOS SRAM Cell Technology with Fully Overlapping Contacts, *IEEE IEDM Tech. Digest*, Dec. 1990, p.473.

[80] Yamanaka, T. *et al.* (1990) A 5.9 μm^2 Super Low Power SRAM Cell Using A New Phase-shift Lithography, *IEEE IEDM Tech. Digest*, Dec. 1990, p.477.

[81] Mele, T. C. *et al.* (1990) A High Performance 0.5 μm BiCMOS triple polysilicon technology for 4Mb fast SRAMs, *IEEE IEDM Tech. Digest*, Dec. 1990, p.481.

[82] *SGS-Thomson Memory Databook* 1988.

[83] *IDT Databook* 1989.

[84] *Motorola MOS Memory Databook* 1989.

[85] Awaya, T. *et al.* (1989) A 5 ns Access Time 64Kb ECL RAM, *IEEE ISSCC Proceedings*, p.130.

[86] Wendell, D. *et al.* (1990) A 3.5 ns 2 k × 9 Self Timed SRAM, *Symposium on VLSI Circuits*, June 1990, p.49.

[87] Fujita, K., Nishimura, Y. and Anami, K. (1990) A 4-Mbit CMOS SRAM with 8 ns Serial Access Time, *Symposium on VLSI Circuits*, June 1990, p.51.

[88] Sekiyama, A. *et al.* (1990) A 1 V Operating 256kbit Full CMOS SRAM, *Symposium on VLSI Circuits*, June 1990, p.53.

[89] Murakami, S. *et al.* (1991) A 21mW 4Mb CMOS SRAM for Battery Operation, *IEEE ISSCC Proceedings*, February 1991, p.46

[90] Chappell, T. *et al.* (1991) A 2 ns Cycle, 4 ns Access 512 kb CMOS ECL SRAM, *IEEE ISSCC Proceedings*, February 1991, p.50

[91] Shimada, H. *et al.* (1991) A 10 ns 4Mb BiCMOS TTL SRAM, *IEEE ISSCC Proceedings*, February 1991, p.52

[92] Suzuki, M. *et al.* (1991) A 1.2 ns HEMT 64k SRAM, *IEEE ISSCC Proceedings*, February 1991, p.48

[93] Okajima, Y. *et al.* (1991) A 7 ns 4Mb BiCMOS SRAM with A Parallel Testing Circuit, *IEEE ISSCC Proceedings*, February 1991, p.54

[94] Nambu, H. *et al.* (1991) A 1.5 ns 64kb ECL-CMOS SRAM, *Symposium on VLSI Circuits*, June 1991, p.11.

[95] Bonges, III, H. A. *et al.* (1991) A 576k 35 ns Access BiCMOS ECL Static RAM with Array Built-in Self-test, *Symposium on VLSI Circuits.* June 1991, p.13.

[96] Young, I. *et al.* (1991) A High Performance 256k TTL SRAM using 0.8 μm Triple Diffused BiCMOS with 3 V Circuit Techniques, *Symposium on VLSI Circuits*, June 1991, p.17.

[97] Vrakama, Y. *et al.* (1991) Data-Line Wiring Delay Reduction Techniques for High-speed BiCMOS SRAMs, *Symposium on VLSI Circuits*, June 1991, p.19.

[98] Ishibe, M. *et al.* (1991) 1 Gbps Pure CMOS I/O Buffer Circuits, *Symposium on VLSI Circuits*, June 1991, p.47.

[99] McAuley, A. J. and Cotton, C. J. (1991) A Self-Testing Reconfigurable CAM, *IEEE Journal of Solid State Circuits*, **SC-26**, No. 3, March 1991, p.257.

10 MOS ROMs, PROMs AND EPLDs

10.1 OVERVIEW

Read-only memories (ROMs) are manufacturer customized parts which are basically limited to market segments and applications with a predefined and relatively long term data storage usage. There is an initial overhead cost involved in the customizing whether an explicit charge or whether amortized in the unit price. This has tended to reduce competition and lessen visibility on the mass market, leaving this memory family without the same general commercial attention as its glamorous RAM and EPROM counterparts.

One result of this reduced visibility has been less attention to the high level of standardization activity that has characterized the other MOS memories.

There has also been less attention from the technical community since ROMs can run on the same processes as other MOS memories and the architectural complexity of the ROM is also not as great as the RAMs and erasable memories. ROMs are basically just mass data storage files. The cell is as simple as the presence or absence of a single transistor. The small cell size means that the peripheral circuitry must also be streamlined to fit into the small pitch of the bit and word lines. See Figure 10.1 for a schematic circuit diagram of a 128k × 8 ROM showing the relative simplicity of the circuit elements. What technical interest exists in the ROMs involves the need to be high density, low price and fast turn-around.

While the demand for ROMs is not as great as for either DRAMs or SRAMs, it is still significant. The split of the total memory market in 1988, shown in Figure 10.2(a), indicates the mask ROM share of the $10.9 billion memory market at 7% or about $750 million.

The ROM market split by region in 1988 is shown in Figure 10.2(b). Japanese systems manufacturers used 80% of the total mask ROM market with US users at 14% and European at 2%. ROW, which is mostly Asia Pacific is about 4%. It is not surprising that the Japanese manufacturers by 1990 supplied nearly 100% of the worldwide ROM demand.

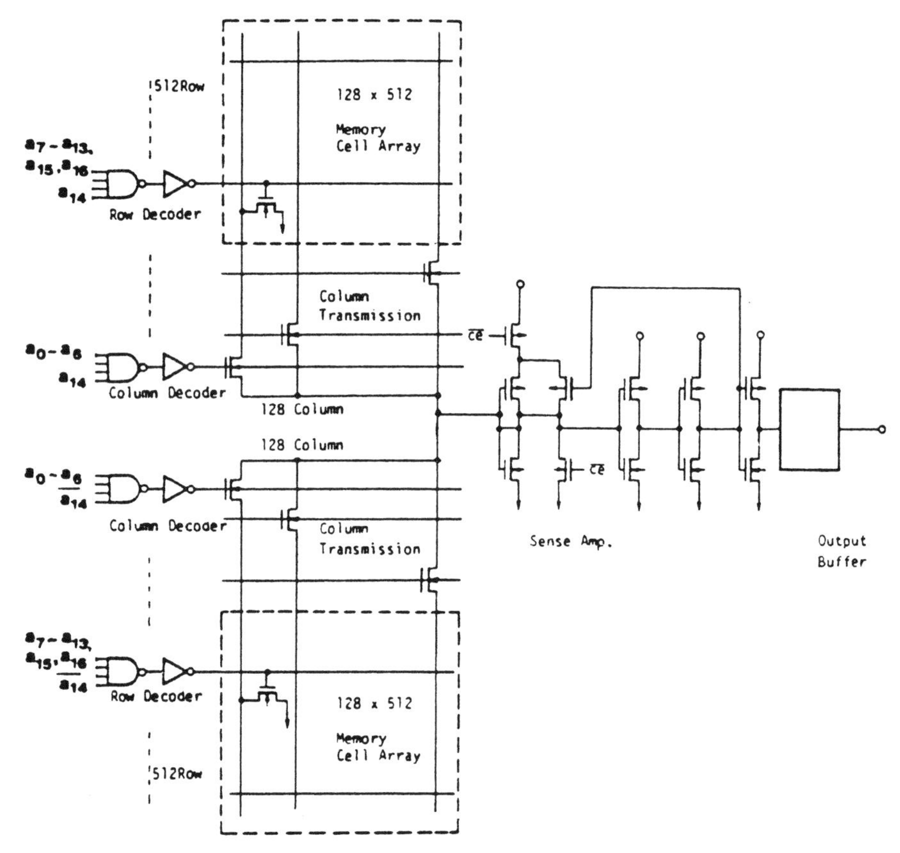

Figure 10.1
Schematic circuit diagram of a 1Mb ROM showing simple invertor row decoders, one-transistor memory cell, single-transistor chip enable, simple four-transistor sense amplifier and output circuitry. (From Masuoka *et al.* [6], Toshiba 1984, with permission of IEEE.)

10.2 HISTORY OF ROMs

During the 1970s a large number of smaller ROM types were available from a variety of manufacturers. Applications were typically in code-conversion, random logic synthesis, microprogramming, table look-up, or in other areas where specific customized data were required.

Slow P-channel metal gate parts with access times ranging from 500 ns to a couple of microseconds had, like other memory parts using the same technology, the requirement for both +5 and −12 V power supplies.

Standard devices were at the time typically organized as 1k (256 × 4 bits or 128 × 8 bits) 2k (256 × 8 bits or 512 × 4 bits) or 4k (512 × 8 bits or 1k × 4 bits). Non-standard sized devices such as 3k or 12k were also an option for braver system designers.

In addition to the above a large number of small ROMs suitable as character generators were available. This type of device was designed specifically for dot-matrix

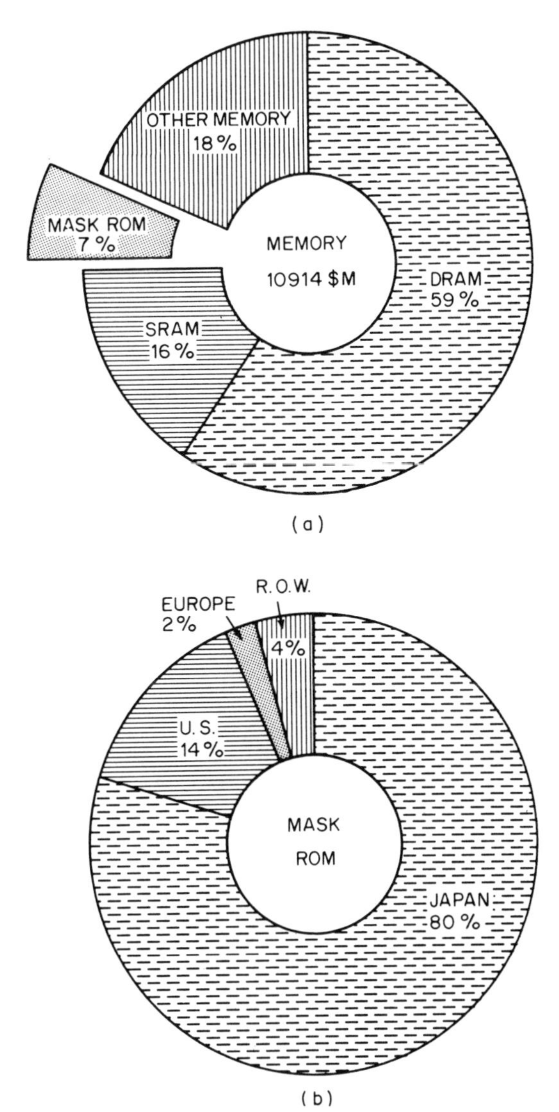

Figure 10.2
1988 Memory statistics. (a) Total memory by product type showing mask ROM at 7% ($763 million) of $10 914 million world market, (b) mask ROM by region. (With permission of Toshiba.)

character generation and to be used in CRT display systems. Each device contained a certain number of characters such as 62 or 64. The ROM size was typically in the 2k bit range.

Figure 10.3(a) shows a block diagram of an early 1k bit CMOS ROM which was housed in a 16 pin package. The memory matrix consisted of four 32 × 8 bit organized sections. The address decoder (1 of 32 and 1 of 8) would flag one bit or specific location in each of the four memory sections giving a 4 bit parallel output. Latched outputs provided a useful feature in system design.

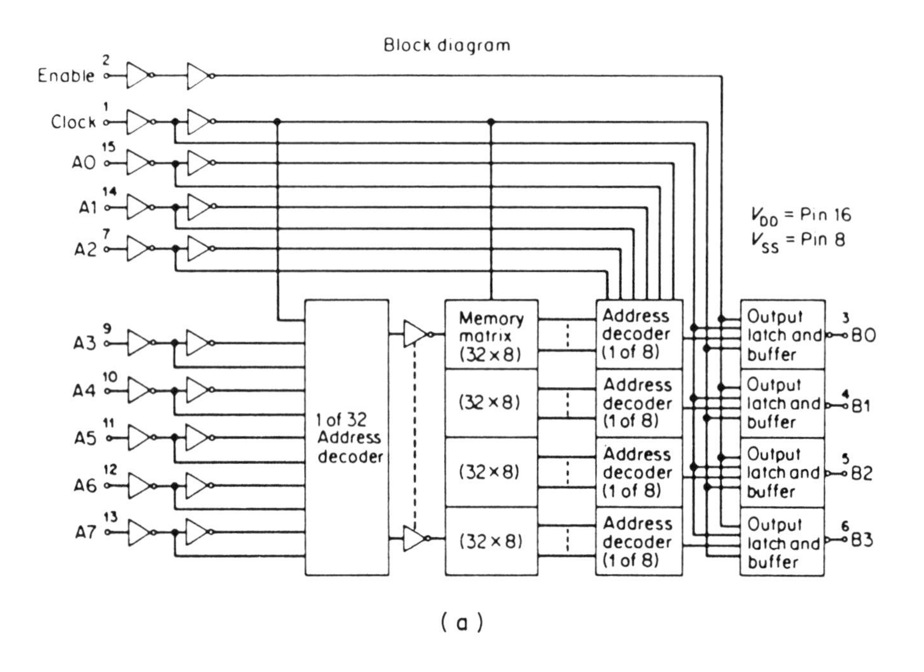

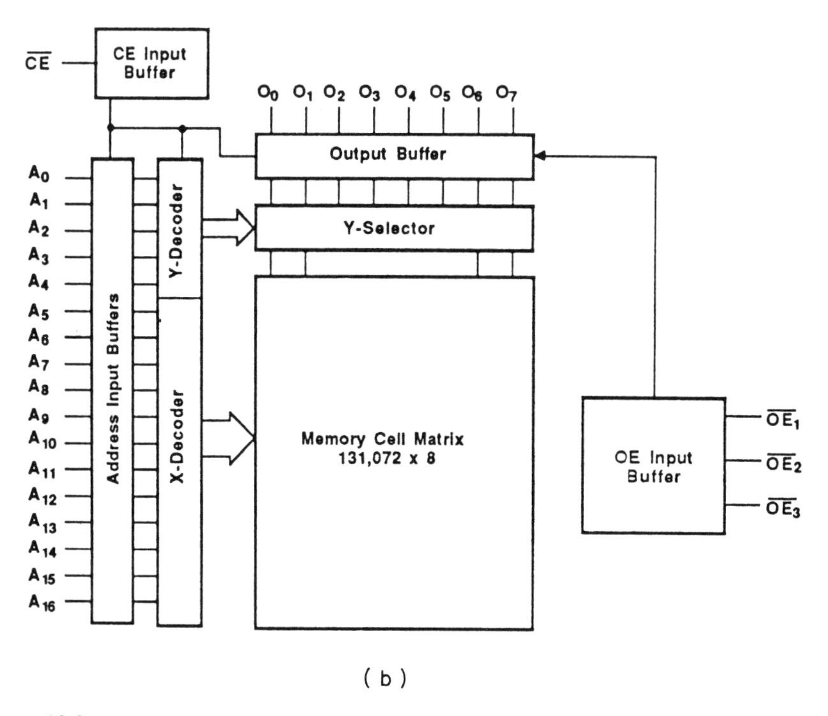

Figure 10.3
Schematic block diagrams of (a) a 1k ROM and (b) a 1Mb ROM showing similarity of architecture over time. (a: from Prince and Due-Gundersen [25] 1983; b: From NEC Databook 1989 [16].)

In the early 1980s larger 16k and 64k ROMs were developed, and in the mid 1980s 256k and 1Mb ROMs. These were used primarily as operating system storage in large computers and in the new TV games which were beginning to appear on the market. The mass volume and low cost requirement of this latter consumer usage gave a competitive nature to the ROM market which it had previously lacked. A bubble of competitive activity in the mid 1980s was followed by many suppliers leaving the ROM market.

By the late 1980s ROMs were appearing in production in densities as high as 4Mb and in technical papers up to 16Mb. Little however had changed in basic functionality or architecture from the earliest parts. ROMs provided basic mass data storage.

For comparison with the early 1k CMOS ROM shown in Figure 10.3(a), Figure 10.3(b) shows a more recent 1Mb CMOS ROM which features a chip enable clock and output latches. Little has changed over the 10 year time span and 100 fold density increase factor in the architecture of the ROM.

In the latter half of the 1980s, the ROM market increasingly became centralized in Japan where such high density memory applications as language translators, storage for speech synthesis and Kanji character generators demanded mass memory storage. The TV game market with its ROM demand also continued to be supplied by the consumer market oriented Japanese suppliers.

The end user applications in Japan in 1988 are shown in Figure 10.4. More than half were consumer oriented TV games and electronic keyboards in line with the focus of Japanese companies on the consumer market. The dataprocessing market was another 43%.

The criteria for game ROMs were high density, low power, fast turnaround and low cost. For these highly competitive consumer ROMs the requirement for performance was secondary. Turn-around was important to meet the demands of rapidly changing interests in the consumer market.

Power was also not a prime consideration except for battery operated systems such as laptop computers and hand held devices such as language translators where battery weight and lifetime made power consumption important.

Another rapidly growing application for ROMs was the portable computer market which used ROMs in large quantities for operating system storage. For these applications speed was slightly more important. High performance microprocessor and workstation software storage applications while relatively low volume did require ROMs at least fast enough to match the increasingly faster microprocessors. These applications required also the wider 16 bit databuses.

For the dataprocessing markets, printers demand a wide bus while personal computers and wordprocessors require speed and low power compromise combined with low cost.

Application areas have expanded rapidly as higher density ROM chips in smaller more cost effective packages have become available. A few of these include: speech recognition, speech synthesis, compiler and operating system storage, program storage in high performance 16 or 32 bit microprocessor or microcomputer systems, home computer applications such as games and education, look-up tables, and high density character generators for a variety of languages.

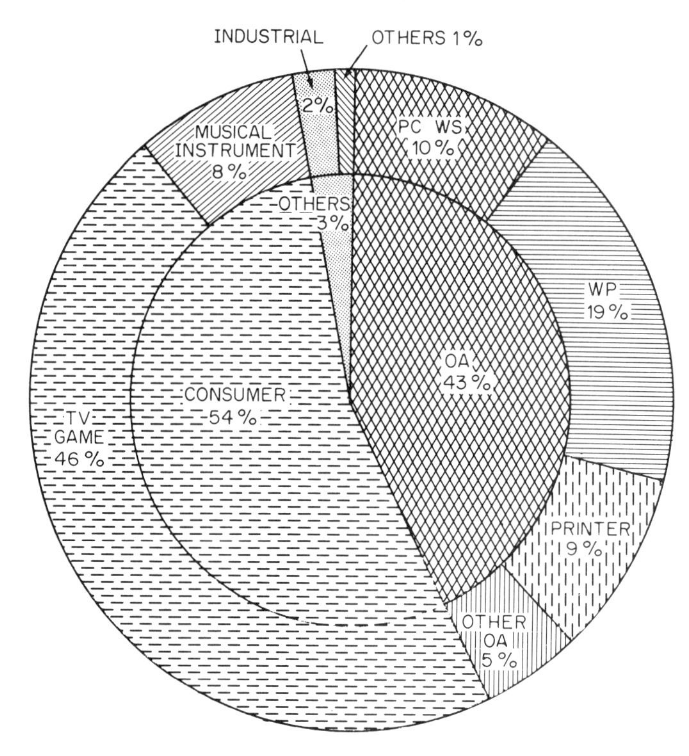

Figure 10.4
Mask ROM end use market in Japan in 1988 in units. (With permission of Toshiba.)

10.3 ROM PRODUCT TYPES

Historically ROMs could be split into two categories: customized standard parts and custom specials which were usually character generators.

10.3.1 Customized standard parts

As with RAMs and EPROMs, standardization for ROMs took place gradually. 8k, 16k, 32k, and 64k devices were introduced, all byte-wide and pin and function compatible with 24 pin or 28 pin EPROMs. This compatibility aspect is most important in the initial system design phase where all modifications can be implemented using a reprogrammable EPROM. After the system design is complete volume production of the equipment can use the less expensive EPROM compatible ROMs rather than the more costly EPROM. If a ROM manufacturer's minimum order quantity requirement is satisfied, a ROM approach can normally offer a very cost effective end-system solution.

Figure 10.5(a) shows the JEDEC standard 'Type A' ROM family ranging from 8k to 64k bits in density. All family members are industry standard and pin compatible for easy upgrading. They are organized byte-wide and each device is compatible with a similar industry standard EPROM.

Typical characteristics for a high performance 8k × 8 NMOS silicon gate ROM in the early 1980s were

- single ±10% 5 V supply
- low power dissipation such as 150 mW active and 35 mW standby
- automatic power down
- access time in the 200 to 300 ns range
- latched data inputs for simplified system design
- fully TTL compatible
- high output drive capability
- pin compatible with 8k, 16k, and 32k ROMs
- pin compatible with 64k EPROM

At the 64k density level the first standardization problem developed for ROMs and for the compatible EPROMs. The problem was that the 24 lead package ran out of pins to upgrade beyond the 64k level. Some manufacturers choose to replace one of the chip selects with the 12th address pin and remain in the smaller package, see Figure 10.5(a) type A. Others choose to keep the chip selects and move directly to the larger 28 pin package as shown in Figure 10.5(b). The 8k × 8 type A ROM was the highest density possible for the 24 lead package family. The 32k (4k × 8) type B ROM was upgradable to the 28 pin 64k (8k × 8) ROM which was the lowest density of the 28 lead package family. The double standard arose because the same chip that was used in the 28 pin package could also be sold in the 24 pin package if the type B organization was used.

From the user's point of view, it was a question of whether a particular user was trying to upgrade to the 64k level in density an older system that had 24 pin sockets, or whether a system manufacturer was designing a new system for the 64k level that would later be upgraded to the 128k and 256k density level. There clearly was a market for both ROM organizations and with this understanding JEDEC standardized both.

As for EPROMs, both 24 pin and 28 pin parts were made available. Several manufacturers introduced both 64k ROMs allowing their customer base to choose in accordance with specific requirements.

Motorola, for example offered their 64k type A ROM in two selections. One selection used the single control pin available for a chip select giving a low power

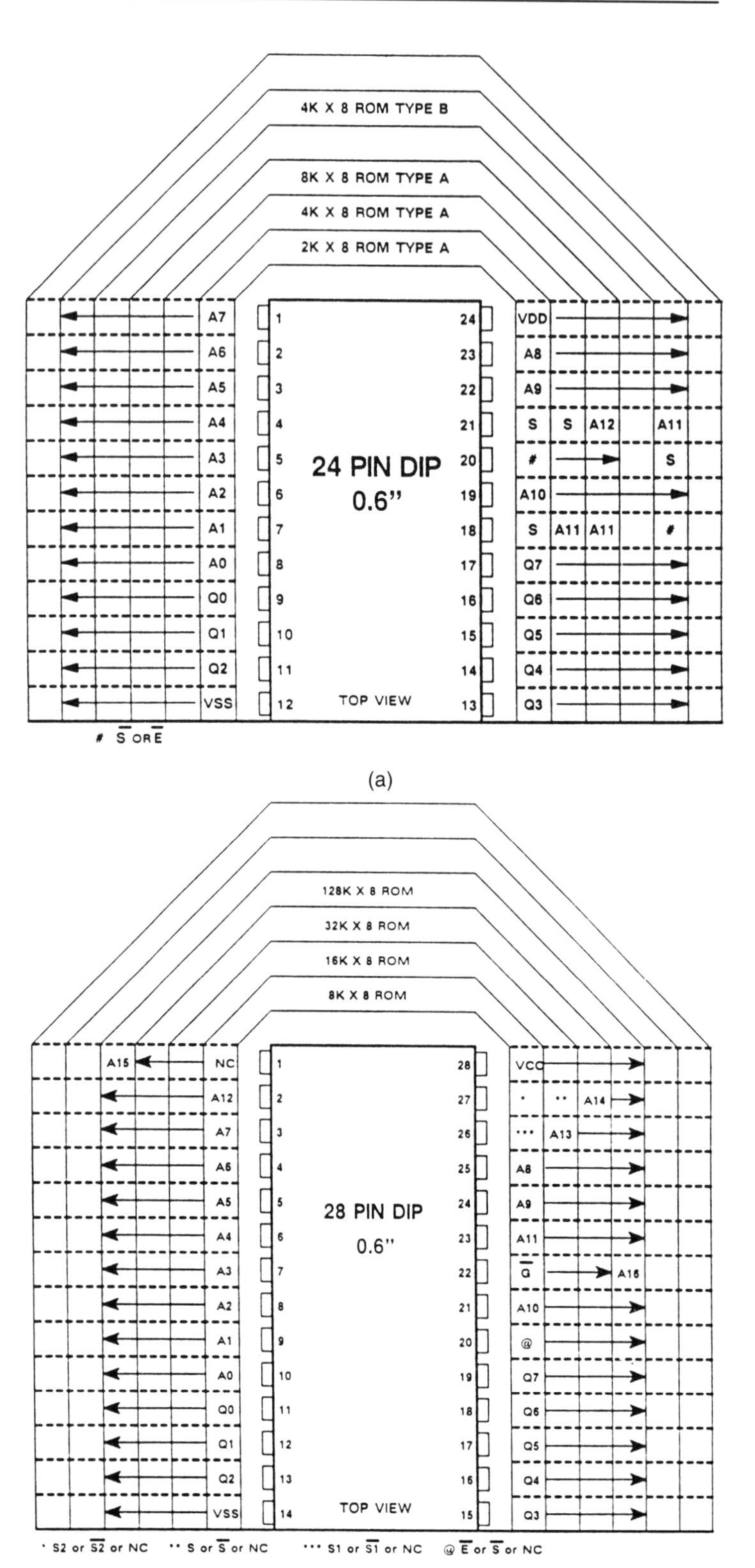
4K X 8 ROM TYPE B
8K X 8 ROM TYPE A
4K X 8 ROM TYPE A
2K X 8 ROM TYPE A
24 PIN DIP
0.6"
TOP VIEW
A7
A6
A5
A4
A3
A2
A1
A0
Q0
Q1
Q2
VSS
VDD
A8
A9
S S A12 A11
S
A10
S A11 A11 #
Q7
Q6
Q5
Q4
Q3
S̄ OR Ē

(a)

128K X 8 ROM
32K X 8 ROM
16K X 8 ROM
8K X 8 ROM
28 PIN DIP
0.6"
TOP VIEW
A15 NC
A12
A7
A6
A5
A4
A3
A2
A1
A0
Q0
Q1
Q2
VSS
VCC
* ** A14
*** A13
A8
A9
A11
Ḡ A16
A10
@
Q7
Q6
Q5
Q4
Q3
* S2 or S̄2 or NC ** S or S̄ or NC *** S1 or S̄1 or NC @ Ē or S̄ or NC

(b)

part. The other selection used it for an output enable to give an access time from output enable which was twice as fast.

Moving on to more modern parts, a similar problem arose when the ROMs outgrew the 28 lead package at the 1Mb density level. An example from NEC is shown in Figure 10.6.

The problem again is that there is only one pin left in the 28 pin package for a control pin. Figure 10.6(a) shows a 1Mb ROM which uses this pin for chip enable resulting in both a 200 ns address access time and chip enable access time. The chip enable allows the part to be powered down giving a low standby power consumption of 110 μA.

The other 1Mb ROM in the 28 pin package in Figure 10.6(b) uses the single control pin for an output enable function with fast 100 ns access from output enable. The drawback is that the chip cannot be powered down. Both of these ROMs are in accord with the JEDEC 28 pin ROM standard.

Figure 10.7 shows the double standard which then resulted for the migration of the 1Mb ROM into the 32 pin package. The 1Mb ROM in the 32 pin package in Figure 10.7(a), which is a transition part, is chip compatible with the 1Mb ROM in the 28 pin package in Figure 10.6(a) so that the same chip can be sold in both packages. It is, however, not compatible with the 1Mb ROM in 32 pin package in Figure 10.7(b) which has changed the address on pin 24 to an output enable.

While the 1Mb ROM in Figure 10.7(b) is no longer chip compatible with the lower pincount package family, it is upgradable in the system to the higher density 2Mb to 8Mb parts shown in the JEDEC standard passed for the 32 pin package shown in Figure 10.8.

The trade-offs are, on the one hand, cost effectiveness for the supplier in being able to put the same chip in two packages. On the other hand, cost effectiveness for the user at the board level in being able to upgrade the memory density of the system without redesigning the board.

10.3.2 Custom specials—the early character generators

Smaller ROMs such as an 8k bit have been commonly used as character generators for displays using western alphabets since the 1970s. These parts can be programmed with a fixed character set or ordered as customized devices. Normally this type of device will satisfy most standard display requirements.

An example from Motorola [25] is shown in Figure 10.9. This part was a preprogrammed 81 920-bit character generator containing 128 characters. A seven bit address (a_0–a_6) was used to access a specific character. Each character was defined as a

Figure 10.5
JEDEC standard memory pinouts for (a) Type A and (b) Type B ROMs in 24 and 28 pin dual-in-line packages (DIP) showing double standard at 4k × 8 and 8k × 8 density level. (From JEDEC Standard 21-C. Reproduced with permission of EIA [26].)

AC Characteristics

$T_A = -10$ to $+70\,°C$; $V_{CC} = +5.0$ V $\pm 10\%$ (Note 1)

Parameter	Symbol	Limits Min	Limits Max	Unit
Address access time	t_{ACC}		200	ns
Chip enable access time	t_{CE}		200	ns

μPD23C1000A

Pin	Signal	Pin	Signal
1	A_{15}	28	V_{CC}
2	A_{12}	27	A_{14}
3	A_7	26	A_{13}
4	A_6	25	A_8
5	A_5	24	A_9
6	A_4	23	A_{11}
7	A_3	22	A_{16}
8	A_2	21	A_{10}
9	A_1	20	$\overline{CE}$
10	A_0	19	O_7
11	O_0	18	O_6
12	O_1	17	O_5
13	O_2	16	O_4
14	GND	15	O_3

Timing Waveform

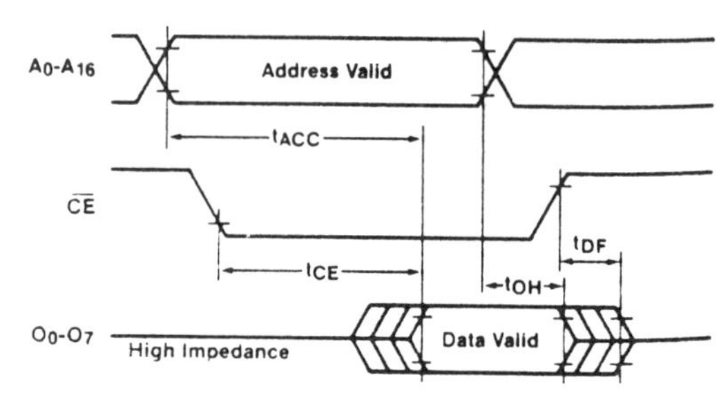

(a)

Parameter	Symbol	Limits Min	Limits Max	Unit
Address access time	t_{ACC}		200	ns
Output enable access time	t_{OE}		100	ns

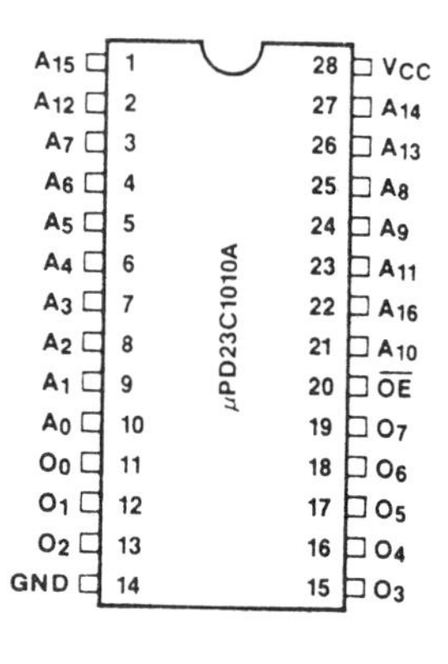

Timing Waveform

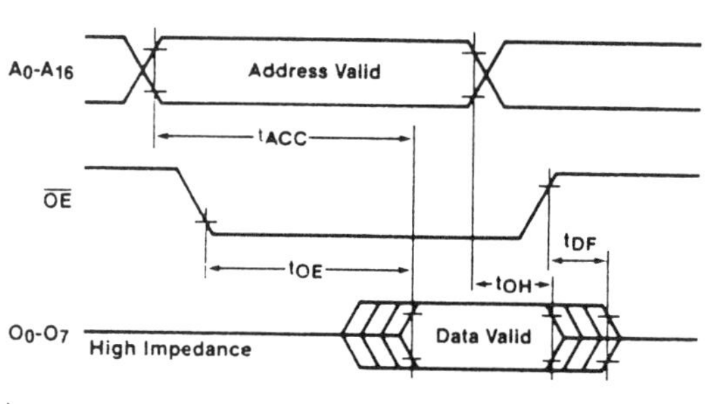

(b)

Figure 10.6
Example of JEDEC standard 1Mb ROM in 28 lead package with only one pin left for a control function. (a) Control function on pin 20 is chip enable allowing a power-down function. (b) Control function on pin 20 is output enable permitting a fast access mode [16]. (Reproduced with permission of NEC.)

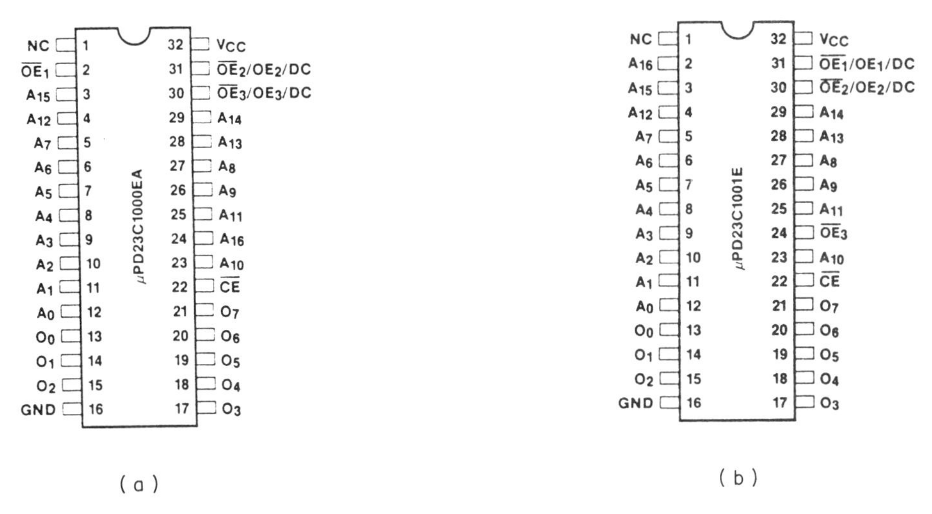

(a) (b)

Timing Waveform

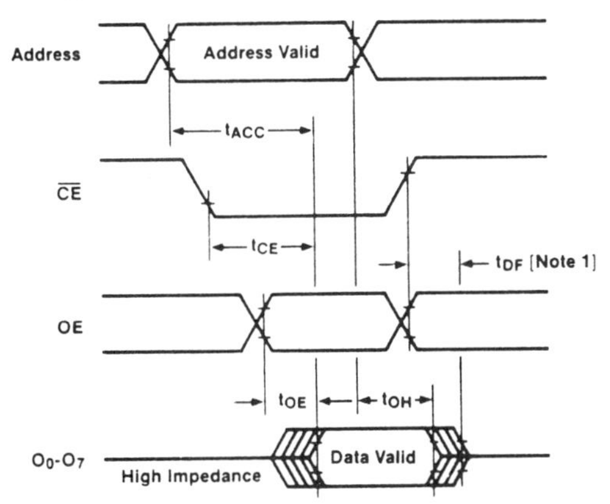

AC Characteristics

$T_A = -10$ to $+70\,°C$; $V_{CC} = +5.0\ V \pm 10\%$ (Note 1)

Parameter	Symbol	Limits Min	Limits Typ	Limits Max	Unit
Address access time	t_{ACC}			200	ns
Chip enable access time	t_{CE}			200	ns
Output enable access time	t_{OE}			100	ns
Output hold time	t_{OH}	0			ns
Output disable time	t_{DF}	0		60	ns

Note:
[1] t_{DF} is specified from OE or $\overline{CE}$, whichever occurs first.

(c)

Figure 10.7
Memory pinouts for 1 Mb ROMs in 32 pin dual-in-line packages (DIP) showing double standard which occurred in move to 32 lead package: (a) A_{16} on pin 24 making the chip (but not the package) compatible with lower density 28 pin standard. (b) Output Enable on pin 24 and Address 16 on pin 2 making chip compatible with higher density ROMs in 32 lead package (JEDEC standard). (c) Timing waveform and ac characteristics [16]. (Reproduced with permission of NEC.)

combination of logic '1's and '0's stored in a 7 × 9 dot matrix as shown in Figure 10.10(a). When a particular four-bit binary row select code was applied, a word of seven parallel bits appeared at the output. If the rows were sequentially selected nine seven-bit words defined the character and the data obtained can be used to display the character. The device also allows the user to position or move the 7 × 9 matrix vertically in a 7 × 16 array as in Figure 10.10(b) offering the capability of shifting

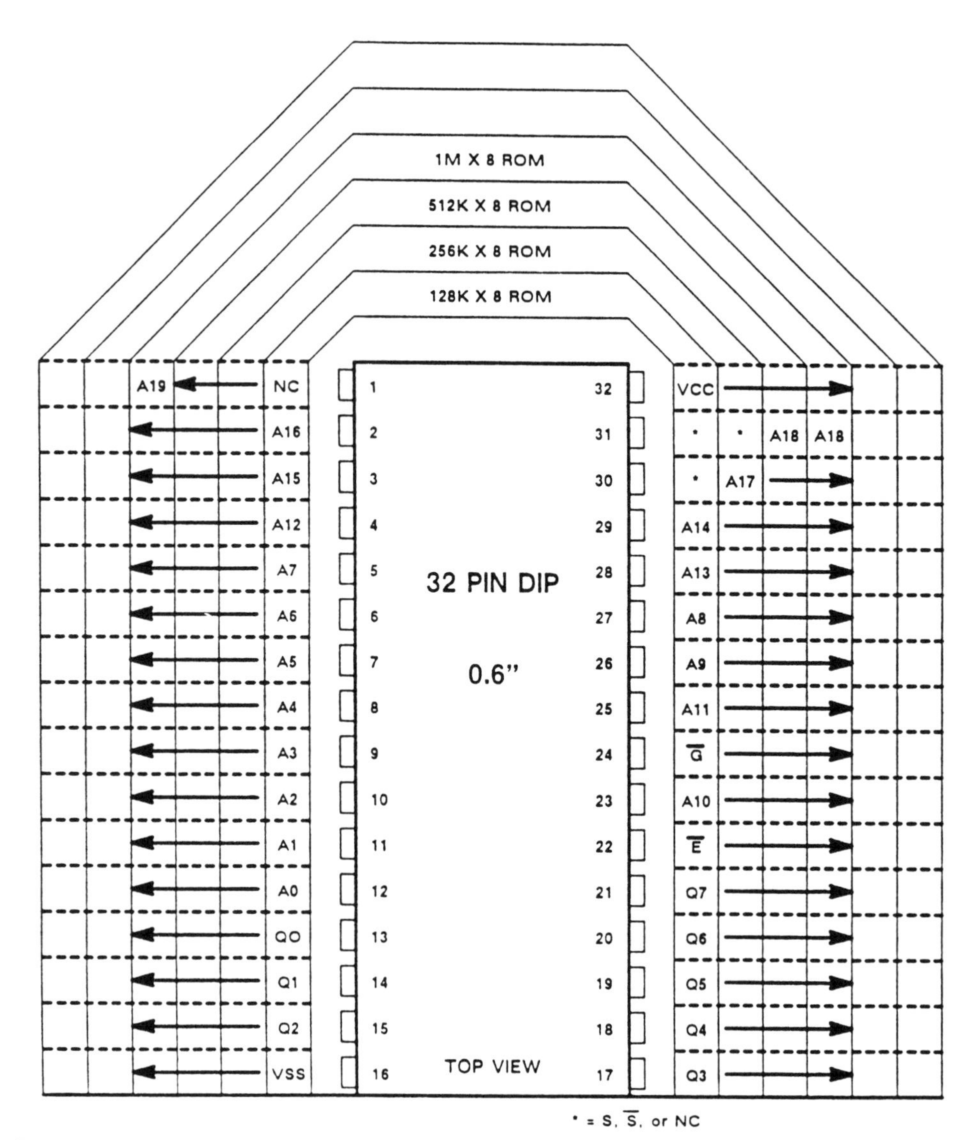

Figure 10.8
JEDEC standard for 1Mb to 8Mb byte-wide ROMs in dual-in-line package. (Source JEDEC Standard 21C. Reproduced with permission of EIA.)

certain characters that normally extend below the baseline such as 'j', 'y', 'g'. Figure 10.10(c) illustrates a device with a fixed matrix. The transition from capital to a small character is, in this case, implemented by the change in character shape only. The display effect is obviously not up to the standard of what is offered using in a shifted matrix.

The 7-bit parallel output from the character generator would normally be dealt with by the display controller which correctly formats and positions the various characters.

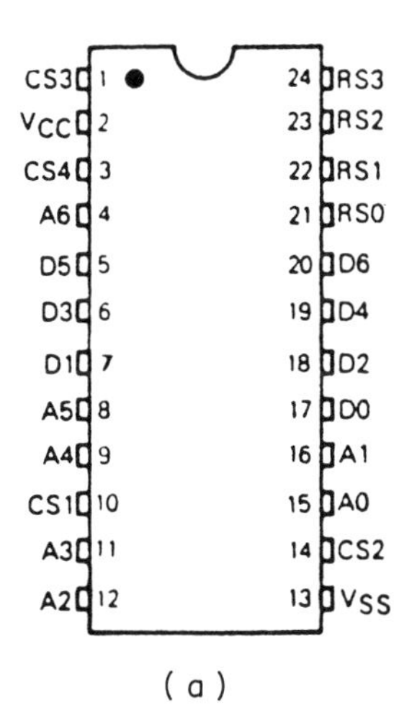

(a)

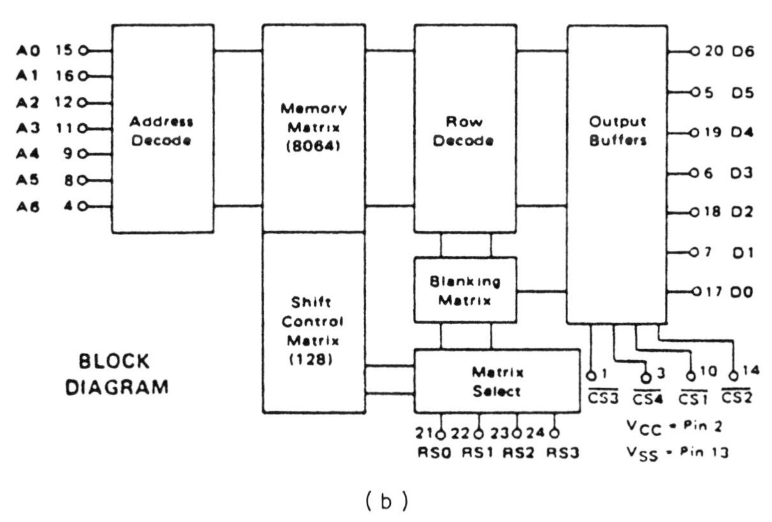

(b)

Figure 10.9
Character generator: (a) pin assignment, (b) block diagram. (From Prince and Due-Gundersen [25] 1983.)

Figure 10.10(d) shows the complete character set available for this particular Motorola chip.

10.4 UNUSUAL ROM ORGANIZATIONS AND PACKAGE SELECTIONS

The demand on the ROMs to be very high density at the same time as being very low cost has resulted in some unusual and innovative ideas. The address-data multiplexed ROM standard [26] shown in Figure 10.11(a) was an attempt in the early

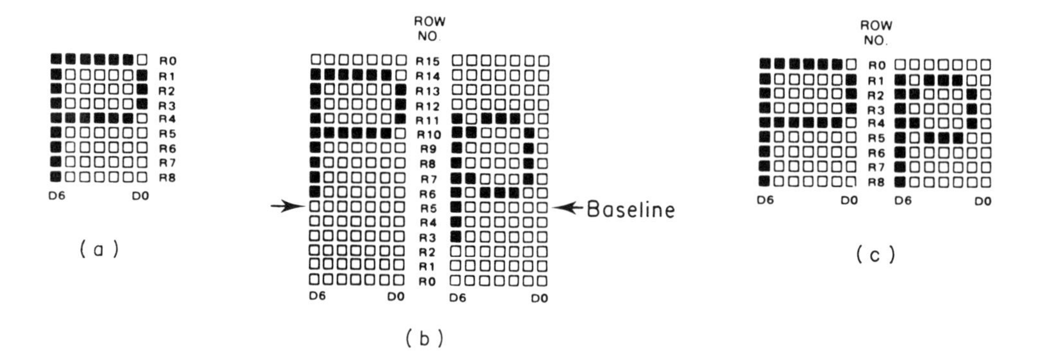

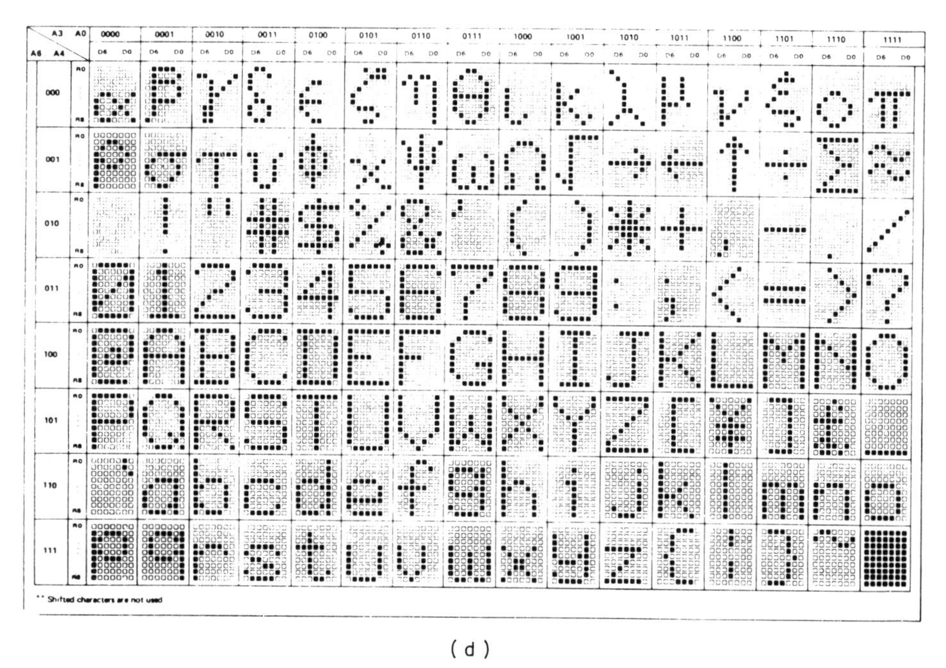

(d)

Figure 10.10
ROM character generator display. (a) 7 × 9 matrix font, (b) 7 × 9 matrix shifted in a 7 × 16 font, (c) 7 × 9 matrix non-shifted, (d) typical character display containing Greek letters. (From Prince and Due-Gundersen [25] 1983.)

1980s to have a high density word wide (× 16) 16 bit microprocessor compatible ROM which could still use the standard 28 pin ROM package.

Since some microprocessors already had address-data multiplexed buses, it seemed to offer even greater advantages. This is an example of what is known as an 'empty standard' since with the exception of a short lived introduction from Mostek this family of parts was never produced in the industry.

One reason that these unusual ROM standards did not catch on was the general acceptance of the chip carriers, such as shown in Figure 10.11(b), and small outline

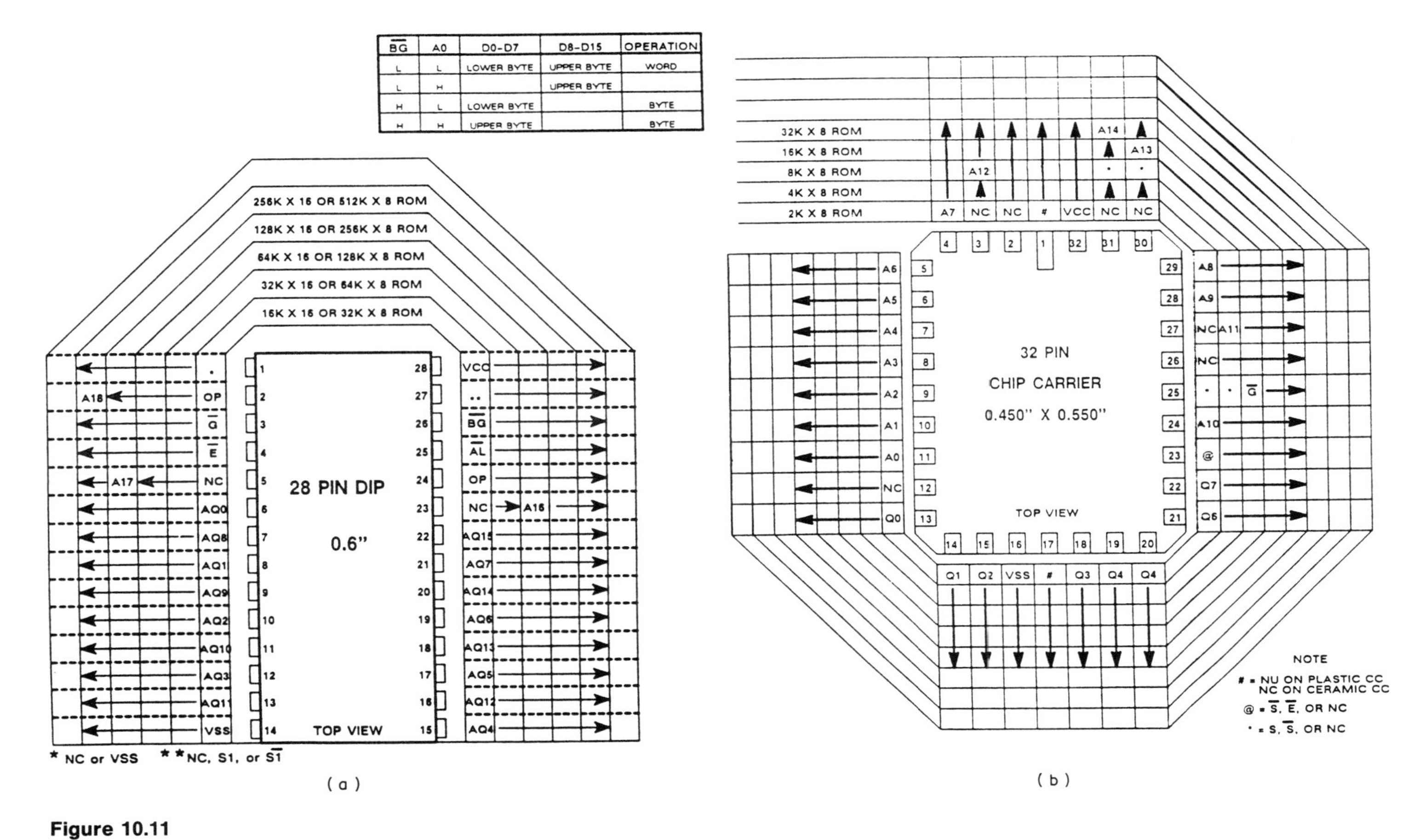

BG	A0	D0-D7	D8-D15	OPERATION
L	L	LOWER BYTE	UPPER BYTE	WORD
L	H		UPPER BYTE	
H	L	LOWER BYTE		BYTE
H	H	UPPER BYTE		BYTE

Figure 10.11
JEDEC standard for ROMs. (a) In DIP package, (b) in chip carrier package. (Source JEDEC Standard 21C, reproduced with permission of EIA.)

packages which enabled the systems manufacturers to reduce the board density and still upgrade the chip density of the ROMs without multiplexing.

Another unusual ROM organization which has been more successful is the byte-wide–word-wide switchable architectures used in multimegabit capacity ROMs primarily for Kanji character generators.

The block diagram of a 2Mb ROM from Hitachi [15] with this architecture is shown in Figure 10.12(a) along with the 40 pin package pinout in 10.12(b). The BHE pin controlled the byte-wide–word-wide function. The ROM was 16 bits wide with D_0–D_{15} active when the BHE input was high, and it was 8 bits wide with D_8–D_{14} in the high impedance state and A_1 the least significant address when the BHE input was low.

The timing diagram for the read operation in either the byte or the word mode was no different than for a standard ROM. The timing diagram for the switch from word mode to byte mode and back is shown in Figure 10.12(c). Hitachi also had the next generation 4Mb byte-wide–word-wide switchable ROM as do many of the other memory suppliers.

An example of the conversion from DIP to chip carrier package for a 4Mb word-byte ROM from Mitsubishi [17] is shown in Figure 10.13.

10.5 PRODUCTION OF CUSTOMIZED ROMs

In comparison with standard 'off the shelf' parts, the development and production of a customized device requires special attention. Since customer data and patterns are to be used, a high level of communication between vendor and manufacturers is required. Close monitoring during manufacturing is necessary due to specific requirements from each customer. For example, prototypes for pattern verification and/or special marking may be required.

10.5.1 Customer–manufacturer interface

After successfully designing and debugging the end system using, for example EPROMs, the equipment manufacturer has to transmit the correct pattern or dataset to the ROM manufacturer to initiate the production cycle.

This custom pattern can be mailed to the manufacturing source by a hard medium such as EPROM, magnetic tape, floppy disk, cards or can be transmitted using data network and communications facilities which some manufacturers have accessible from local centers near customers locations.

EPROMs as the ordering medium require the submission of multiple sets to insure integrity of the customer's pattern. After generation of the ROM data list three sets of EPROMs can be returned to the customer for verification. Magnetic tape specifications indicate the type of tape, code type, and load module data format. Floppy disk requirements specify type of floppy, number of sectors and tracks, use code and format.

Figure 10.14 shows a typical sequence of important steps involved in ROM pattern

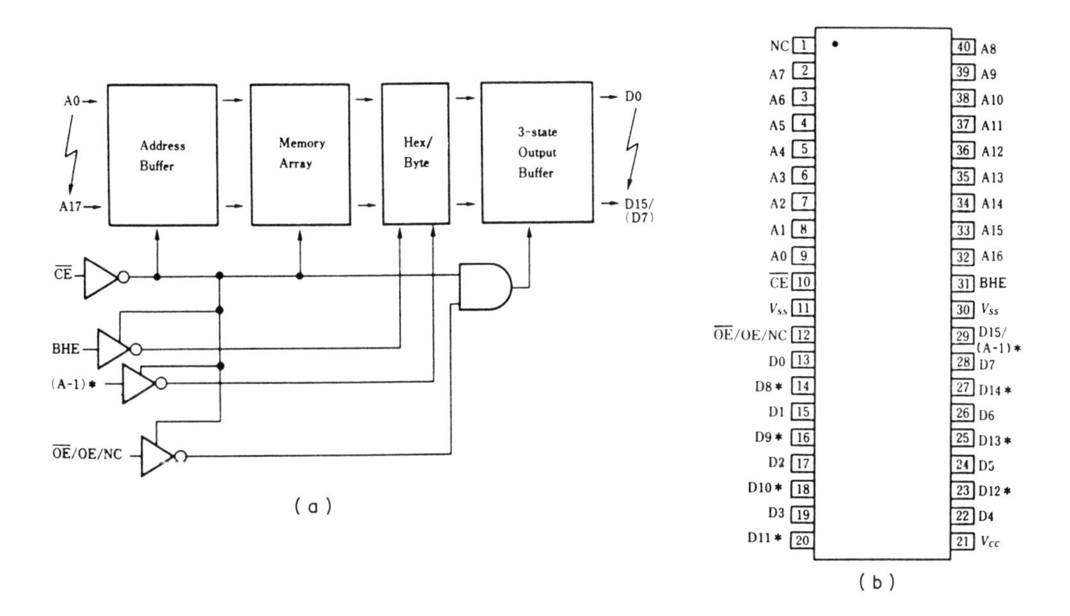

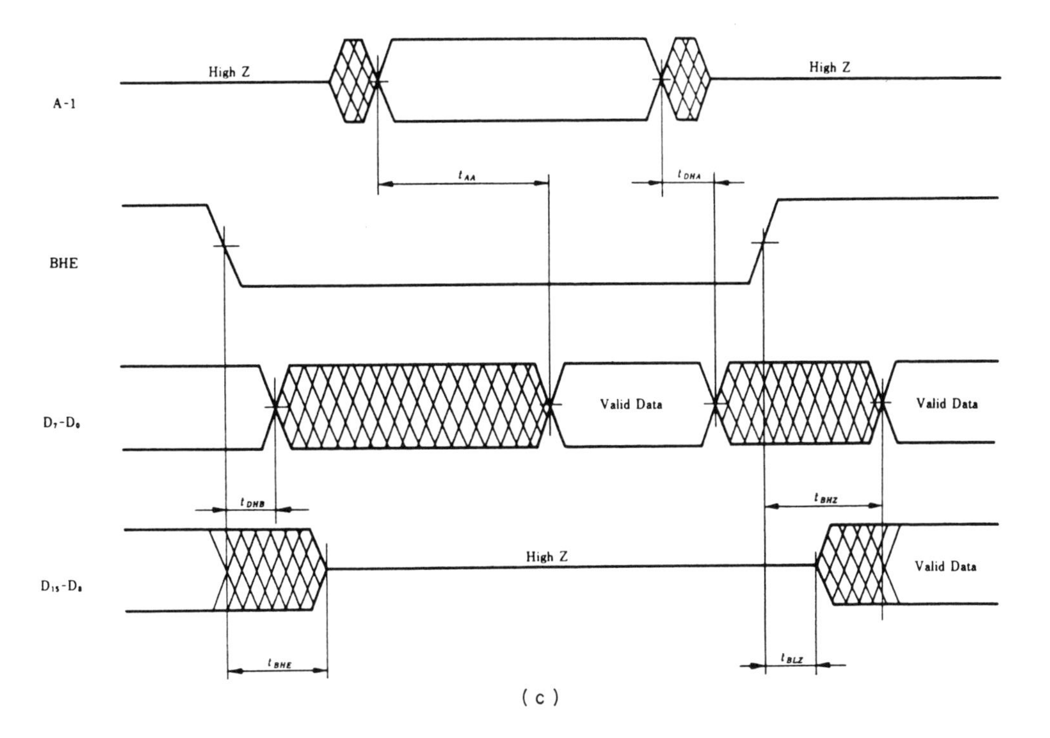

Figure 10.12
2Mb byte-wide–word-wide switchable ROM. (a) Schematic block diagram showing Hex–Byte buffer and three-state output buffer, (b) pinout, (c) timing diagram of word mode–byte mode switch [15]. (Reproduced with permission of Hitachi.)

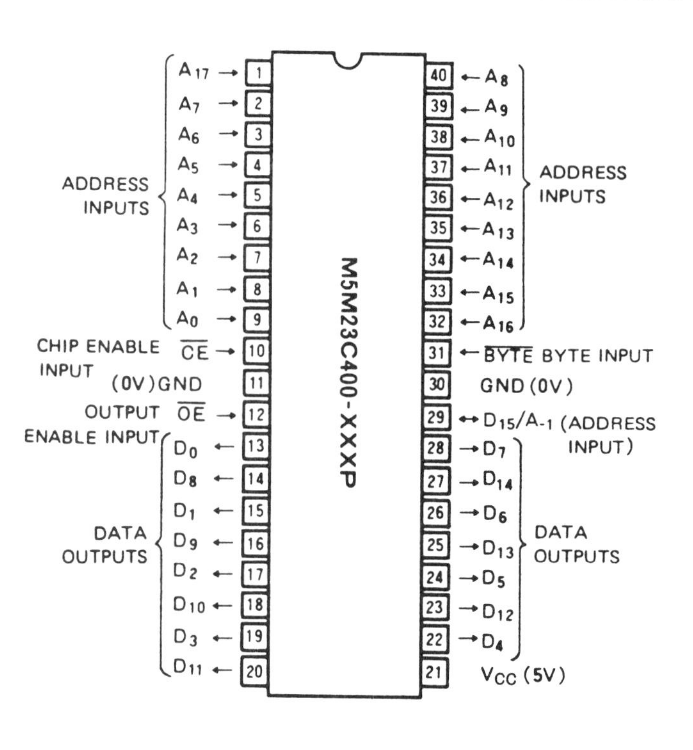
ADDRESS INPUTS
A17
A7
A6
A5
A4
A3
A2
A1
A0
CHIP ENABLE INPUT
CE
(0V) GND
OUTPUT ENABLE INPUT
OE
DATA OUTPUTS
D0
D8
D1
D9
D2
D10
D3
D11
M5M23C400-XXXP
A8
A9
A10
A11
A12
A13
A14
A15
A16
ADDRESS INPUTS
BYTE BYTE INPUT
GND (0V)
D15/A-1 (ADDRESS INPUT)
D7
D14
D6
D13
D5
D12
D4
DATA OUTPUTS
Vcc (5V)

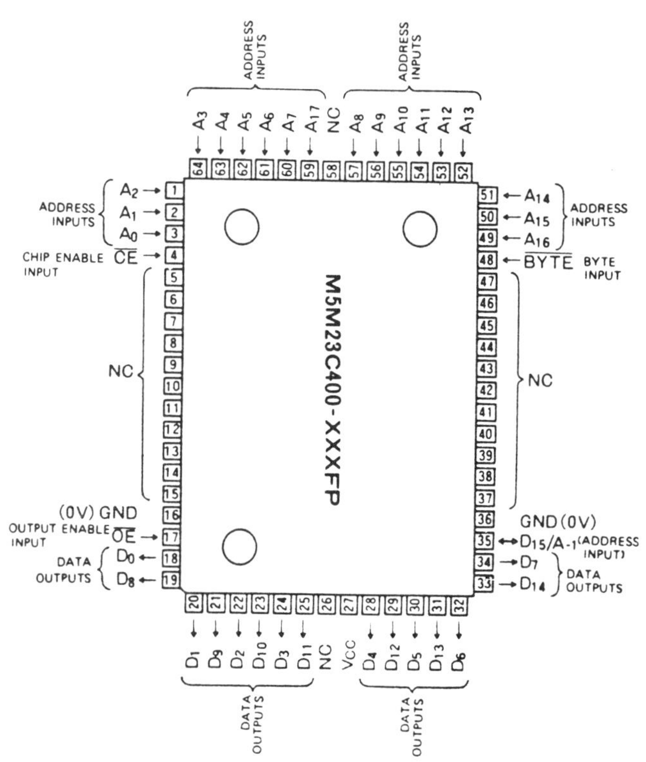
ADDRESS INPUTS
A3
A4
A5
A6
A7
A17
NC
A8
A9
A10
A11
A12
A13
A2
A1
A0
CHIP ENABLE INPUT
CE
NC
(0V) GND
OUTPUT ENABLE INPUT
OE
DATA OUTPUTS
D0
D8
M5M23C400-XXXFP
A14
A15
A16
BYTE BYTE INPUT
GND (0V)
D15/A-1 (ADDRESS INPUT)
D7
D14
D1
D9
D2
D10
D3
D11
NC
Vcc
D4
D12
D5
D13
D6

specification [17]. In this case, the data pattern input is in three sets of EPROMs from the customer which will be used in the automatic design system at the vendor to generate a ROM data list. Three sets of EPROMs with the generated pattern are then returned to the customer for verification. Following verification the ROM manufacturing process takes place.

If required, prototypes in many cases can be generated to allow verification of the final product. If this is not required (or after approval of prototypes), volume production can commence.

The time for a complete production cycle for customized ROMs will vary from manufacturer to manufacturer but is measured in weeks rather than the months or years for a full custom chip design.

A ROM requires only one customized mask layer which contains a predefined customer pattern giving permanent data storage of '1's and '0's. This customer defined mask may be applied at a very late stage of the production cycle giving a fast turnaround for the finished product. Another approach is to introduce the customer mask at an earlier point in the production process which can have various other advantages such as smaller chip size. Both approaches are looked at in the subsequent section on ROM architecture.

10.6 TYPES OF CELL PROGRAMMING

Turn around time is primarily effected by the stage in the manufacturing process at which the ROM is programmed. There are three general methods of programming: field oxide programming, threshold voltage programming and through-hole or contact programming. Each of these programming methods can be implemented at different stages of the process. A typical schematic of a ROM manufacturing process flow is shown in column 1 of Table 10.1. The stage at which each of the three methods of programming is normally implemented is shown in column 2.

10.6.1 Field oxide programming

Field oxide programming is done during the earliest part of the manufacturing process and works by differentiating the threshold voltage of the transistors by using differing thicknesses of gate oxide. In a programmed cell the gate oxide is the thickness of the field oxide so that the transistor is permanently 'off' or in a logic '0' state. In an unprogrammed cell a normal gate oxide is used so the transistor is 'on' or in a logic '1' state.

Advantages of field oxide programming are that there are large current differences between programmed and unprogrammed cells, no extra masks, and low average

Figure 10.13
Pinout conversion of a 4Mb ROM from DIP to chip carrier package. (Source Mitsubishi Semiconductor Memories 1989, reproduced with permission of Mitsubishi Electric Corp.)

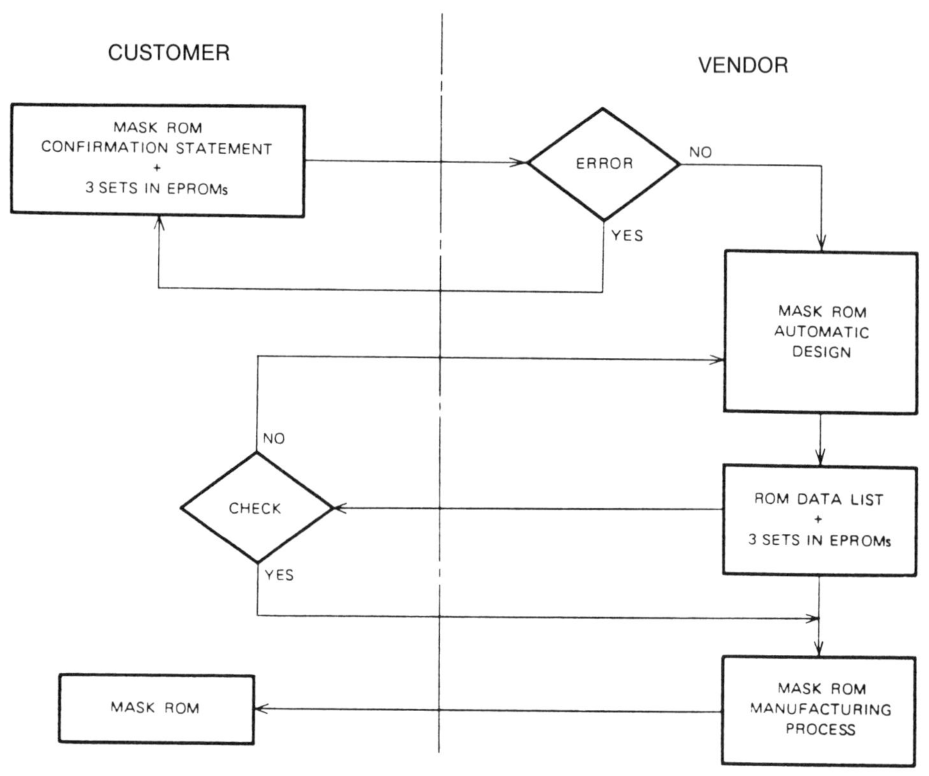

Figure 10.14
Schematic diagram showing a typical sequence of important steps involved in ROM pattern specification. (Source Mitsubishi Semiconductor Memories 1989, reproduced with permission of Mitsubishi Electric Corp.)

word-line capacitance. Since through hole contacts are not needed to the individual cells, this method gives the potential for a higher density array. A disadvantage is turn-around time after programming.

An example of a field oxide programmed cell layout and a cross section of the process is shown in Figure 10.15 for a 128k ROM shown by NTT [3] in 1979 which used the field oxide programming method. In this local oxidation technique a patterned nitride layer blocked the growth of the field oxide and defined active areas for transistor channels. In transistor regions, without the nitride mask protection, normal field oxide grows creating a non-conducting cell.

The diagram shows the two cells corresponding to a '1' (thin gate oxide) and a '0' (thick field gate oxide). A polysilicon word-line and molybdenum bit-line are used.

In this case two memory cells are connected by the same bit-line using one through hole contact so that the effective number of contacts per cell is 0.5. This results in a 62 μm^2 ROM cell in a 2 μm geometry process using the field oxide programming method. Recall that a typical DRAM cell in a 2 μm process was 85 μm^2 and a typical SRAM was 550 μm^2 and both used typically more than the two layers of interconnects used on this ROM.

Table 10.1

ROM process flow	Programming step
Field oxide growth	Field oxide programming
Gate oxide growth	
Channel implant	Implant programming
Polysilicon deposition	
Source–drain diffusion	
Passivation	
Contact hole eating	Through-hole programming
Aluminum metalization	

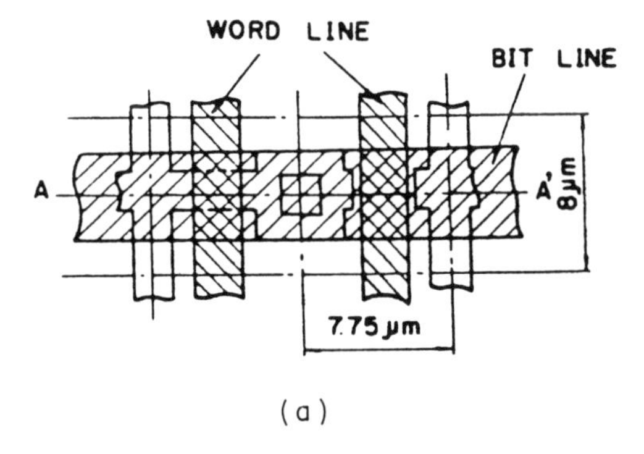

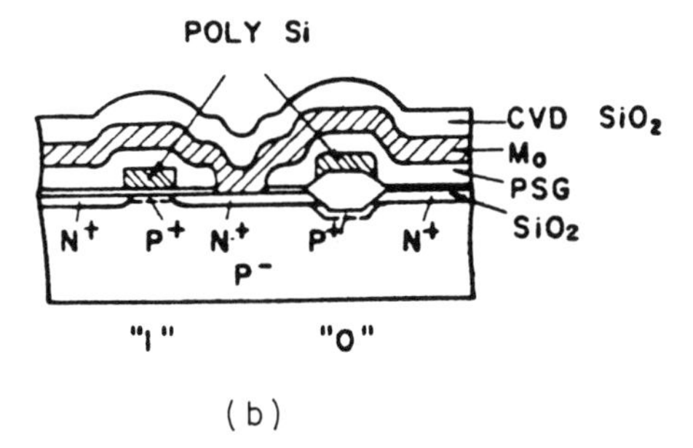

Figure 10.15
Example of field oxide threshold programmed ROM cell. (a) Memory cell layout showing polysilicon word-line and molybdenum bit-line. (b) Cross-section of two ROM cells showing, on the left, a cell with thin oxide corresponding to a logic '1' and, on the right-hand side, a cell with thick field oxide corresponding to a logic '0', (From Kiuchi *et al.* [3], NTT 1979, with permission of IEEE.)

10.6.2 Threshold voltage implant method of programming

The other method of programming that involves changing the threshold voltages is that of programming by ion implantation of some of the transistor gates to differentiate a '1' from a '0'.

This technique uses a heavy dose of boron implant in the n-channel transistors to raise their threshold voltage. The implant forces the cell permanently into an off state.

An advantage of this method is that it can be implanted through the poly gate just before contact etch. This is much later in the process than the field oxide method of programming and thus shortens the turn-around time. This method, like the field oxide method, gives a high density array by eliminating the need for individual cell contacts. This technology, therefore, offers faster turn around than the field oxide method and still retains the advantage of high density.

Disadvantages include several side effects of the heavy implant dose. One of these is possible gate oxide damage from the implant which could effect reliability. The other is a high parasitic junction capacitance between the source (drain) and channel regions of the MOS transistor resulting from the ion implantation into the channel region. This results in increased average wordline capacitance and as a result in slower speed.

If we compare this with the field oxide programming method, adoption of field oxide as a gate oxide in place of a thin gate oxide does not increase the word line capacitance and hence is faster. It also has the added advantage of making the total thin gate oxide area small potentially improving yield in high density ROMs. The penalty paid is longer turn-around time.

10.6.3 Through-hole contact programming

The third alternative is to program when the contacts to the cells are opened just before the final metal interconnect mask. This works by selectively opening the contact holes for each transistor to the drain.

This was, in fact, the historical ROM programming technique. The disadvantage of this method is that it requires a contact for every cell which increases the size of the cell array giving a less dense array thereby resulting in increased the chip cost. It gives, however, a very fast turn-around time since wafers can be programmed at this final stage of the process.

Historically a compromise was reached by having both types of ROMs available. Through-hole programmed ROMs were used to provide fast turn around for initial prototyping or for small orders. One of the other programming methods was used to provide higher density and hence lower cost for long term or high volume orders. Gradually 'one time programmable' EPROMs or EPROMs preprogrammed by the manufacturer have come to be used to supply many of the initial fast turn around orders, and also advances in these basic techniques have provided many of the benefits of both methods of programming.

For example, an advance was made in 1984 when Toshiba [6] introduced a 1Mb

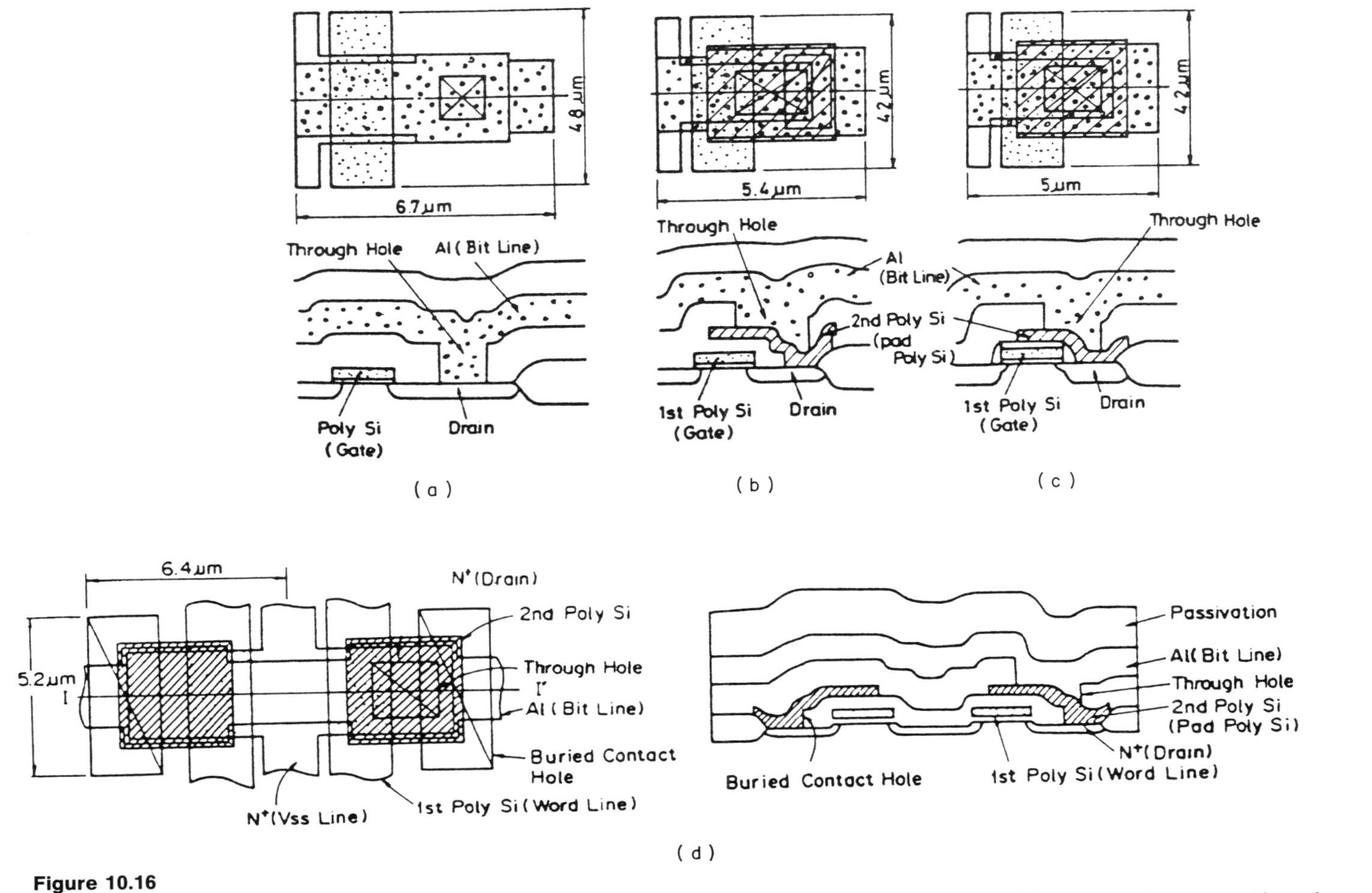

Figure 10.16
Comparison of single polysilicon ROM cell and two generations of double polysilicon ROM cells: (a) layout and cross-section of single polysilicon cell, (b) layout and cross-section of 1Mb double polysilicon ROM cell showing buried contact, (c) layout and cross-section of 2Mb double polysilicon ROM cell showing through-hole buried contact overlapping the gate region [5] (reproduced with permission of Toshiba), (d) layout and cross-section of two double polysilicon cells. The cell on the left is not programmed and the cell on the right is programmed. (From Masuoka [6], Toshiba 1984, with permission of IEEE.)

ROM with a double polysilicon ROM cell using buried contacts. This reduced the cell size significantly for a ROM using contact or through-hole programming.

Toshiba estimated that the chip area of a 1Mb ROM with through-hole programming using the same 2 μm design rules as the traditional single polysilicon cell would be 72.2 mm^2 compared to the 51.5 mm^2 for their double polysilicon cell. This was comparable to the chip size of a ROM using the early programming techniques yet still retained the turn-around time advantage of the late programming through-hole process.

A layout and cross-section of the conventional single polysilicon cell is shown in Figure 10.16(a) and new double polysilicon cell used on the 1Mb ROM in Figure 10.16(b).

In 1986 Toshiba introduced a 2Mb ROM using the double polysilicon cell but using 1.5 μm minimum design rules, where the minimum size is the contact dimension. The die size was 78 mm^2. The memory cell layout and cross section are shown in Figure 10.16(c). The cell consisted of a silicon gate LDD transistor with the second polysilicon contact pad being connected with the drain of the transistor by the buried contact using self-aligning technology. Since the second polysilicon contact pad extended over the first polysilicon electrode the space consuming contact could be placed above the first poly electrode minimizing cell area.

In Figure 10.16(d) the layout and cross section of two cells are shown, the cell on the left is not programmed, ie logic '0', while the cell on the right is programmed logic '1'. The first polysilicon forms the word-lines, the metal forms the bit-lines, and the ground line is the N^+ diffusion layer. The second polysilicon was used to connect the buried contact to the N^+ drain and the metal bit-line.

10.7 DIFFERENT ROM ARRAY STRUCTURES

Another decision which must be made on a ROM is whether to use a serial ROM cell structure which is the NAND-gate type structure which was first proposed in 1976 [3], or the more traditional parallel ROM cell structure which is the NOR gate type structure. Pattern layouts of the two types of cell structures [7] using the same technologies are shown in Figures 10.17(a) and (b). In the process shown the parallel cell size is 45.6 μm^2 for field oxide programming while the serial structure cell size is 28.9 μm^2. The cell size of the serial cell is smaller.

10.7.1 Parallel (NOR) array structure

The conventional parallel (NOR) gate ROM array [13] consists of a set of MOS transistors connected in parallel to a bit-line as shown in Figure 10.18(a).

The parallel cell structure has the advantage of high speed operation but low bit density due to the larger cell size resulting from 1 to 1.5 contact holes to every cell.

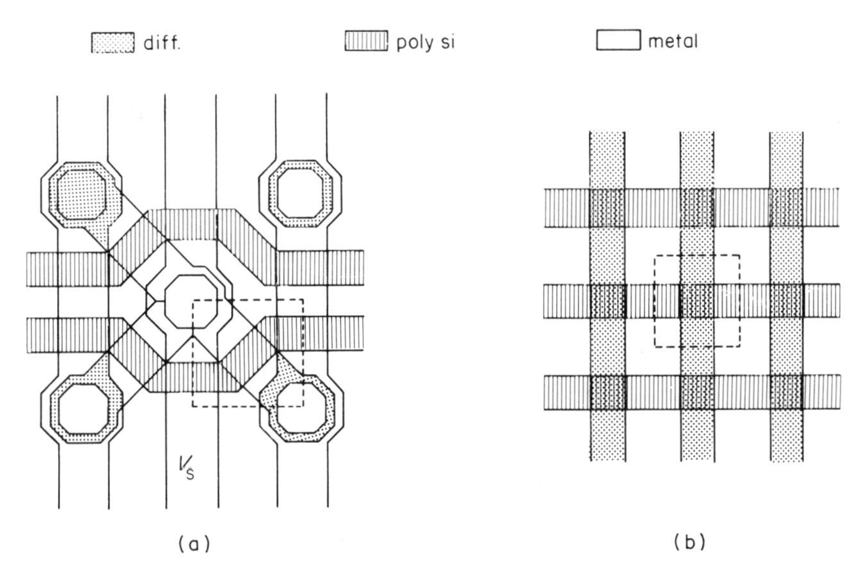

Figure 10.17
Pattern layout of two types of ROM cell structures in the same technology. (a) Parallel structure with cell size 6.75 × 6.75 μm^2. (b) Serial structure with cell size 5.25 × 5.5 μm^2. (From Cuppens and Sevat [7], Philips 1983, with permission of IEEE.)

10.7.2 Serial (NAND) array structure

A serial (NAND) array structure from a Philips 1983 paper [7] is shown in Figure 10.18(b). The packing density of the serial structure can be high since there are no contact holes required in the ROM array. It was estimated that the serial cell using threshold voltage programming resulted in a 25% to 35% smaller chip than a parallel cell with threshold or field programming. One drawback of this structure, however, is the slower speed of operation.

A trade-off between density and performance can be obtained by choosing the height of the serial memory stack. Clearly the higher the stack, the slower the ROM, but the smaller the memory area.

This trade-off is shown in Figure 10.19(a) which plots read-out delay and memory area against height of the serial MOST stacks. The read-out delay increases exponentially with increasing stack height. The memory area as the ratio of area of a 16k block to the area of 16k bits has a minimum.

The programming of the information in a serial cell structure can be done by either ion implant or field threshold technology. This implies, of course, that the turn-around time of a serial ROM is slower than that possible with a through-hole programmed parallel ROM.

One problem faced in the Philips serial ROM paper [7] was that with the very small cell size of the serial ROM cell it was not possible to get the column decoder in the pitch of the ROM cell so that a 1 out of 256 decoder was included as a preprogrammed part of the ROM. This added 16 external rows to the memory bits

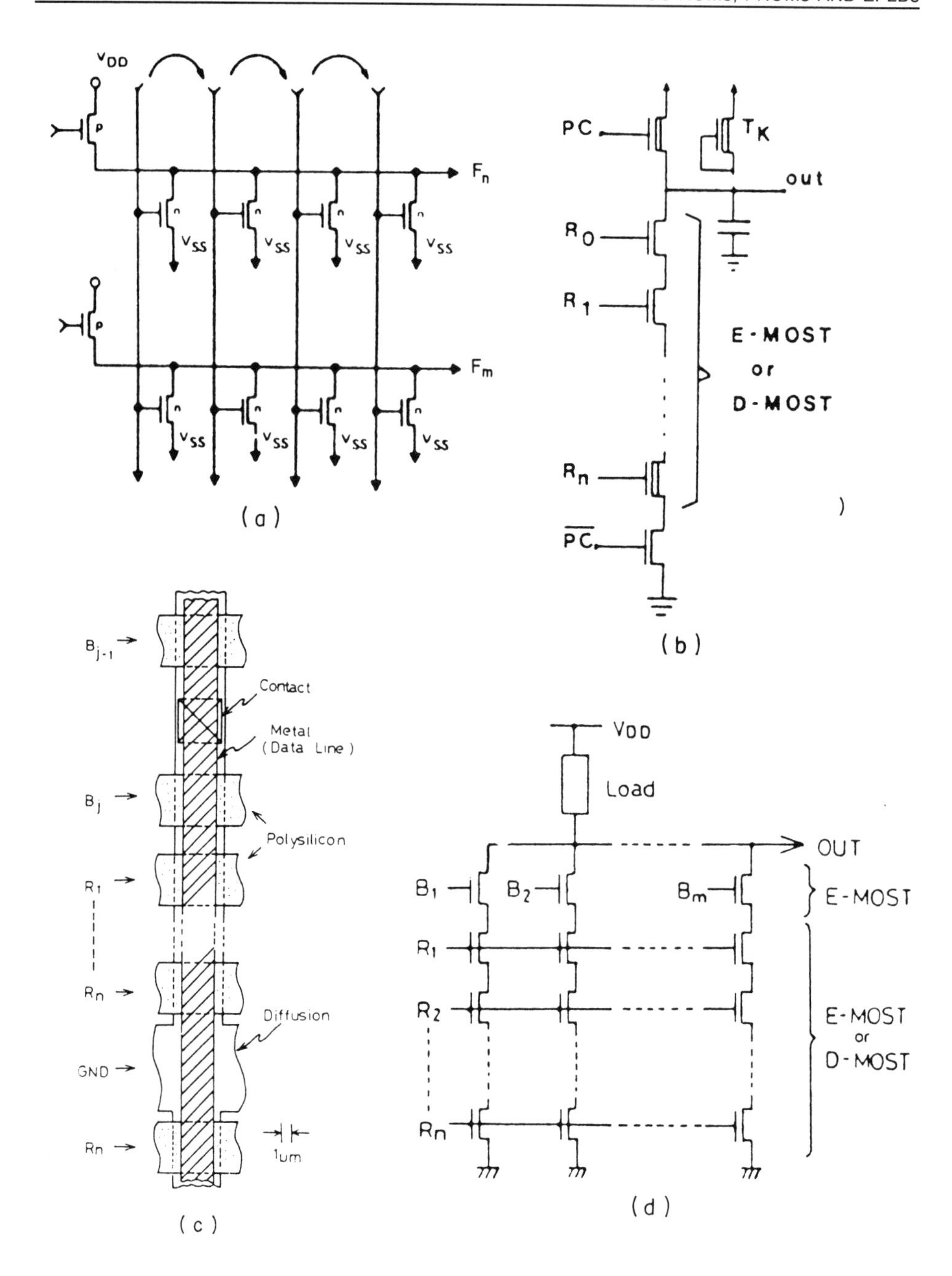

Figure 10.18
Circuit schematics of two types of ROM assay architectures. (a) Parallel (NOR). (b) Serial (NAND). (c) Layout of the serial–parallel ROM array structure showing the source–drain contact required for the parallel structure but not for the serial array structure (d) Combination serial–parallel. (a: from Haraszti [13], Hughes 1984, with permission of IEEE; b,c: from Cuppens and Sevat [7], Philips 1983, with permission of IEEE; d: from Kamuro [8], Sharp 1982, with permission of IEEE.)

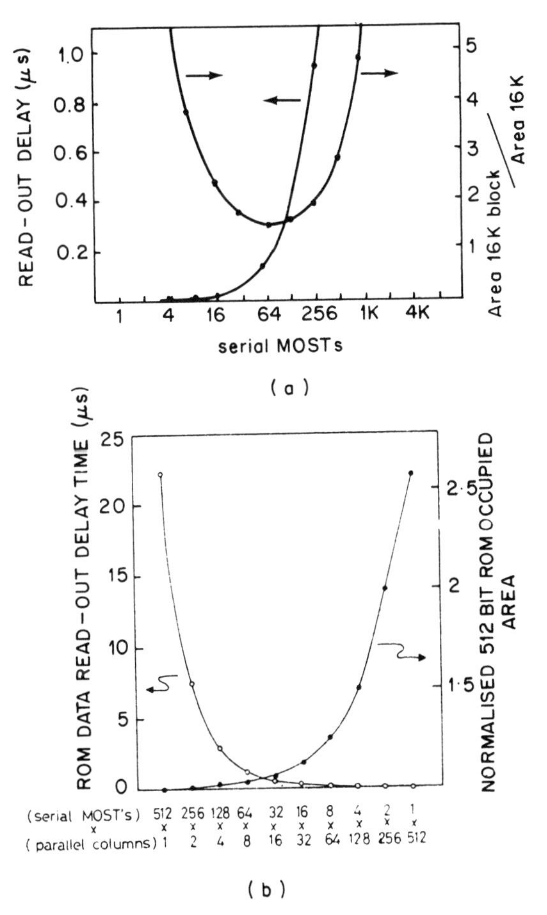

Figure 10.19
Trade-offs in various serial ROM architectures. (a) Relationship of serial MOS transistor stack height to read-out delay time and ROM area in ROM having serial array structure. (b) Relationship of serial MOS transistors and parallel columns to ROM area and read-out delay time in ROM having serial–parallel array structure. (a: from Cuppens and Sevat [7], Philips 1988, with permission of IEEE; b: from Kamuro [8], Sharp 1982, with permission of IEEE.)

giving a final NAND string stacking height of 80 which included 64 serial cells and 16 row decoder cells.

10.7.3 Combination parallel–serial array structure

In 1981 Sharp [8] proposed a combination serial–parallel ROM array structure in an attempt to trade off speed and bit density. The basic structure of the serial–parallel ROM is shown in Figure 10.18(c) and the layout of the cell structure is shown in Figure 10.18(d). The trade-offs in delay and matrix area for number of serial MOSTs plotted against parallel columns is shown in the graph in Figure 10.19(b).

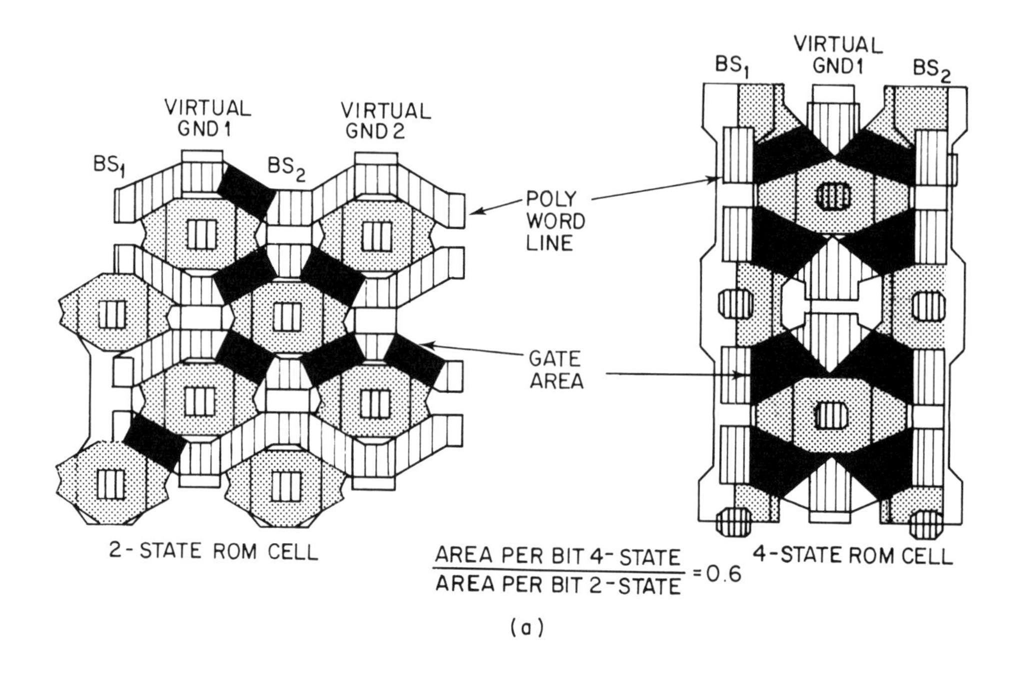

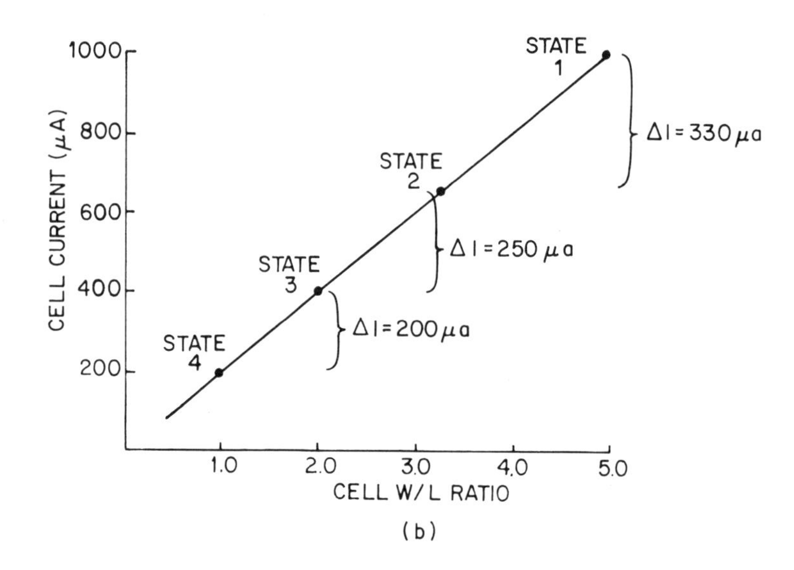

Figure 10.20
(a) Comparison of layout of four-state as opposed to two-state ROM cell. (b) Cell current variation plotted against width–length ratio for distinguishing states in four-state ROM cell. (From Donoghue *et al.* [9], Motorola 1983, with permission of Electronics.)

This 256k ROM was targeted at the new microprocessor applications requiring large amounts of fixed data such as office machines handling Chinese characters, voice synthesizers and hand-held electronic language translators. The latter requires low power and high density, and moderate speed which could not be provided by the very slow serial structured array.

10.8 MULTIPLE STATE ROM CELLS

The requirement for high bit density in ROMs led to attempts in the early 1980s to increase the number of bits per ROM cell. Several innovative cell structures were proposed.

Motorola [9] in 1983 introduced a four-state ROM cell which stored two bits per cell. The cell was constructed with continuously varying polysilicon dimensions so that the W/L ratio of the transistors varied. This was accomplished by having an active channel area which was wider at one end than at the other.

A comparison of design layouts is shown in Figure 10.20(a) between the traditional two-state ROM cell and the new four-state ROM cell. The two-state cell has constant length-to-width ratio polysilicon gates. The new four-state cell has variable length-to-width ratio polysilicon gates so each cell stores 2 bits of information.

The cell density is increased by about 40% rather than doubling since the cell is larger than a usual ROM cell.

In a standard ROM a logic '1' or logic '0' is defined by the presence or the absence of a transistor (or a transistor being on or off) giving two possible logic states per cell. The new cell allows four states per cell to be defined which gives two 'bits' per cell reducing the size of the memory array dramatically.

The four states are four detectable current levels resulting in two bits per cell. A sense amplifier comparison with a fixed reference current determines the bit current levels present. The different levels are initially generated by using a geometric programming technique that varies both the width and length of the polysilicon gates. With narrow gates the channel resistance is increased and the current is lower. If the channel length is shortened the resistance decreases and the current goes up. Figure 10.20(b) shows the cell current plotted against cell W/L ratio which differentiates between the four states.

In operation, each bit location is compared against three reference cells in three separate sense amplifiers. The width-to-length ratios of the three reference cells fall between the programmed cell dimensions; therefore, a ROM cell current level should fall between the reference cell levels to produce one of four possible states with the three sense amplifiers. These voltage states are then decoded to produce one of four possible output codes for two output bits.

G.I. [10], also in 1983, introduced another four-state ROM which reduced the matrix size by about 50% over the conventional ROM matrix. In this case the four states are encoded in the matrix by varying device thresholds using multiple ion implants. Detection of a matrix device type is determined by the length of time required for a

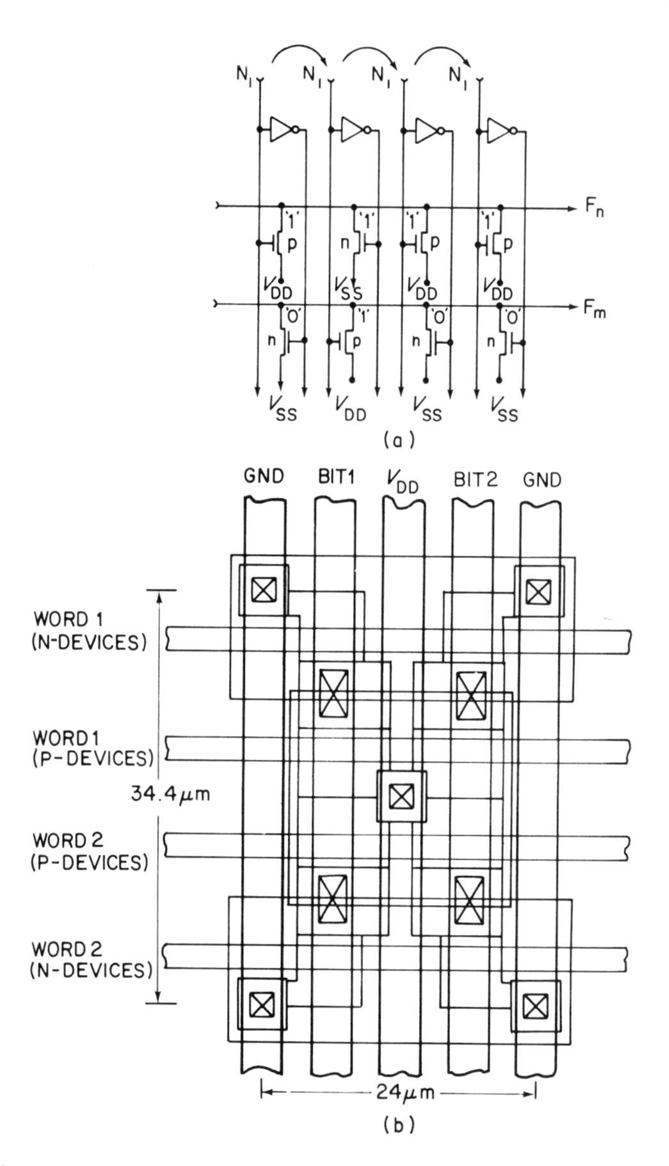

Figure 10.21
Example of ROM using complementary P-channel and N-channel transistors in the ROM cell. (a) Circuit schematic, (b) layout of four cells. (From Haraszti [13], Hughes 1984, with permission of IEEE.)

linearly ramped word-line to rise from 0 V to the point where the matrix device is turned on. This time period is measured and output as a 2 bit binary code.

A cell innovation which was developed for speed rather than for density was presented by Hughes Aircraft [13] in 1984. This cell involved the use of complementary N- and P-channel transistors in the ROM cells as shown in Figure 10.21. A logic '1' was programmed by a P-channel transistor connected to the positive supply while a logic '0' was programmed by an N-channel transistor connected to the ground. The connection of both P- and N-channel devices to a single data-line required two access lines for each individual word, one for the P-channel and one for the N-channel transistors. The complementary arrangement was capable of reversing the information on the data-line in every half clock period.

10.9 ARCHITECTURES FOR FAST ROMs

In a high density conventional ROM array the word-line signal delay would tend to be large when a single word-line is activated due to the large gate drive capacitance of many bits activated. For this reason NTT on their 128k bit ROM used a form of divided word-line structure. Rather than have 128 bits activated on a single word-line, they divided the array into eight sections with 16 bit lines activated in each word-line section. Only the section with the selected cell was activated.

Another source of delay was the use of a memory cell transistor with minimum channel width which could result in the potential of the bit-line and data bus changing very slowly. To avoid this problem a dynamic invertor with high logic threshold voltage was used as a first stage of the output buffer circuitry to increase the access time.

Toshiba [5] in their 2Mb ROM in 1986 also used dynamic circuitry to reduce the delay due to the slow charging speed of the bit-line. A precharge pulse generated by address transition detection was used to charge the bit-line. If a through hole contact existed in the selected memory cell then the data-line and the selected bit-line were discharged by the precharge pulses. If a contact did not exist in the selected memory cell then data-line and selected bit-line were maintained at the precharged level. Input and output buffers operated dynamically to reduce power dissipation. This ROM was intended for use as a Kanji character generator with 288 bits needed to store one Kanji character. Most commonly used characters can be stored in an 8 chip set, giving 1Mbyte of storage capacity.

10.10 SYNCHRONOUS ROMs FOR LOW STANDBY POWER

The Motorola ROM with the four-state cell discussed earlier was an example of a synchronous ROM. It was activated by clocking chip enable which latched the input addresses and powered up the internal peripheral circuitry. The peripheral circuitry was then powered down when the device was disabled by taking chip enable high reducing the power consumption during standby mode.

10.11 MULTIMEGABIT ROMs

10.11.1 4Mb ROM

In 1986 NTT [11] showed a 4Mb ROM intended for Kanji character generation, speech synthesizers, and language translators. This 700 mm^2 chip featured 100% redundant bits, that is, error correction by complete memory cell duplication. It ran at a very slow 7 μs access time which was adequate for the intended application.

10.11.2 16Mb ROM

Then in 1988 Sharp [12] showed a 16Mb ROM intended for use in Kanji character fonts and dictionaries in Japanese word processors. It was also intended as a game ROM for sophisticated TV games. The cell size was 3.24 μm with a 133 mm^2 chip size comparable in both cell and chip dimensions to that of a 16Mb DRAM.

This ROM used the boron ion implantation method of programming. The compact ROM cell achievable by this method is shown in Figure 10.22(a). This is a NOR gate type cell using diffused bit-lines. To reduce the bit-line series resistance the ROM array was divided into 256 banks.

The bank select architecture is shown in Figure 10.22(b). There were two kinds of bank select transistors, one for reading the cells on the even columns and the other for reading the odd columns. When one set of columns, for example the even, was selected the other, odd in this case, was deselected. The aluminum lines ran zigzag in the column direction connecting the shorted nodes in the even and odd columns alternately. This permitted the pitch of the aluminum lines to be twice that of the diffused bit-line. This reduced the possibility of metal shorts and increased the yield.

10.12 BIPOLAR PROMs

Another classification of memories with 'hard' programming that are used in custom ROM applications are fast bipolar PROMs. They differ from mask programmable ROMs only in that the customer programs them by blowing various types of fuses rather than having the vendor program them during the manufacturing process. During the late 1980s fast CMOS PROMs also were supplied into this application. These will be considered here from the programming point of view in comparison to the bipolar PROMs, and in the next chapter on EPROMs from the device and technology viewpoint.

Fast PROMs are used in applications which include microprogram control store and microprocessor program store, look-up tables, character generation, and code conversion.

The hard fuses used in the fast PROMs come in a number of varieties. There are two types in bipolar PROMs. One is the fuse blown PROM which uses the opening

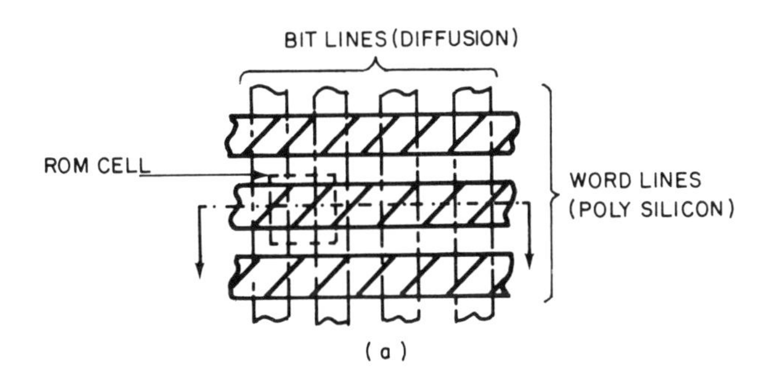

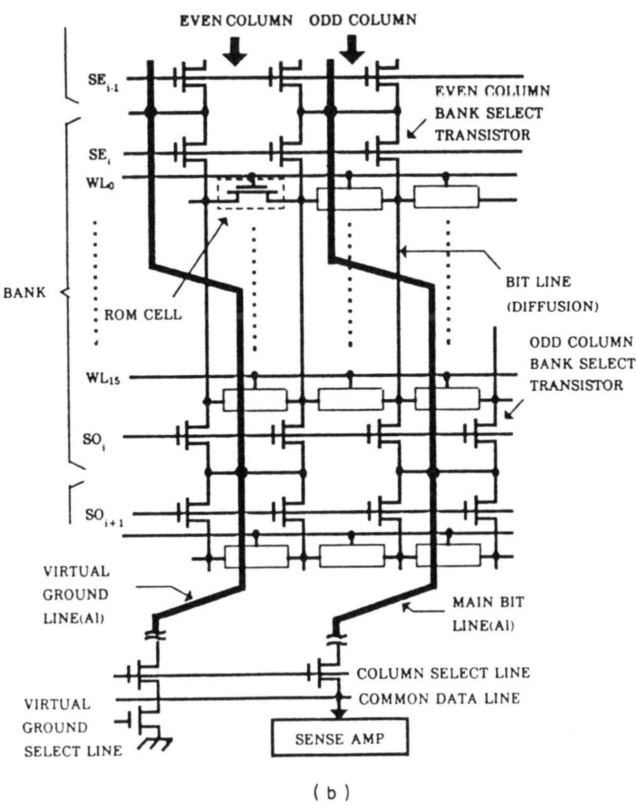

Figure 10.22
16Mb ROM. (a) Cell layout, (b) schematic diagram of the odd–even column bank select architecture. (From Okada [12], Sharp 1988, with permission of Japan Society of Applied Physics.)

of fuses by programming. The other is the junction shorting PROM which uses the shorting of P–N junction diodes by programming.

In a fuse blown PROM the fuse is deposited horizontally on the silicon surface using materials such as aluminum, nickel chromium, titanium tungsten, platinum silicide or polysilicon. An illustration of the layout for a memory cell with a lateral thin film fuse link is shown in Figure 10.23(a). The fuse shown is a horizontal NiCr link on a

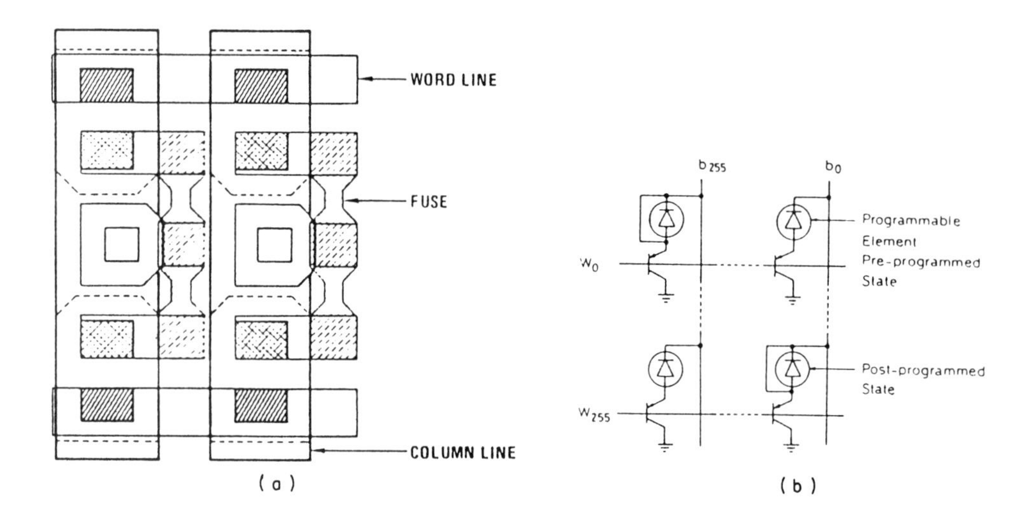

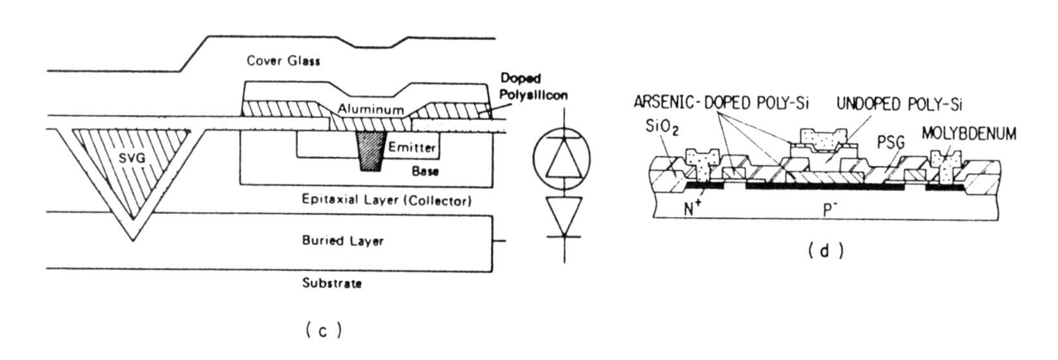

Figure 10.23
(a) Layout for a PROM cell with a lateral thin film NiCr fuse link in a junction isolated Schottky bipolar technology. (b) Shorting PROM memory array showing memory cell with PN programmable diffused junction diode and PNP transistor before and after programming. (c) Junction shorting PROM schematic cross-section with cell in its programmed state. (d) Schematic cross-section of a diffused polysilicon resistor fuse link. (a: from Fisher [20], Harris 1983, with permission of IEEE; b: from Fukushima [19], Fujitsu 1984, with permission of IEEE; c: from Fujitsu Memory Databook 1989 [24], with permission of Fujitsu; d: from Tanimoto [22], NTT 1984, with permission of IEEE.)

junction isolated Schottky bipolar technology. This fuse is from a 1982 64k bipolar PROM from Harris [20].

In a junction shorting PROM the P–N junction diode is diffused using the conventional Schottky TTL wafer fabrication process in the peripheral circuitry. Junction shorting is performed within the silicon bulk by using the stable silicon–aluminum eutectic.

The circuit diagram for a memory cell with the junction shorting programmable element before and after programming is shown in Figure 10.23(b). The memory cell consists of a programmable element which is a P–N junction diode and a vertically connected PNP transistor. The aluminum electrode to the cathode of the P–N diode is a bit-line.

In programming the selected word-line is pulled down to sink the program current while the remaining word-lines are pulled up to $PV_{BCC} = 20$ V. Reverse current pulses are applied to the selected memory cell through the single bit-line. This increases the temperature at the junction and induces the shorting of the junction using the diffused Al–Si eutectic. This fuse was used in a 1984 64k PROM from Fujitsu [18, 19]. A schematic cross-section of a junction shorting PROM in its programmed state is shown in Figure 10.23(c).

In CMOS technologies there are four different non-volatile elements available to implement programmable devices. These are blown polysilicon fuses, diffused polysilicon fuses, EPROM cells and EEPROM cells.

An example of a diffused polysilicon fuse link was shown in an NTT [22] PROM in 1982. The programmable element was a polysilicon resistor which made an irreversible resistivity transition under the effect of a low 10 V programming voltage and small 4 mA programming current.

A cross-section of the vertical polysilicon fuse link resistor is shown in Figure 10.23(d). It consisted of an undoped layer of polysilicon between two arsenic-doped layers of polysilicon. Under the programming conditions the arsenic doping was diffused across the undoped polysilicon region shorting the high resistivity undoped section of the resistor. Since the resistivity of polysilicon decreases significantly with even a small amount of impurity, the low current programming conditions were possible.

An example of a blown polysilicon fusible link using a parasitic bipolar transistor in the fuse element was shown on a 16k CMOS registered PROM from Harris [21] in 1983. The equivalent circuit and layout of the fuse element is shown in Figure 10.24(a) and Figure 10.24(b), and a schematic cross-section of the CMOS process showing the fuse element and the use of the parasitic bipolar transistor in the CMOS process is shown in Figure 10.24(c).

The fuse consisted of a thin neck between two wider tabs in a dogbone configuration. A poly fuse to N^+ buried aperture is used to connect the emitter directly to one fuse tab to lower the tab resistance which can limit the power delivered to the neck region of the fuse. There was a contact opening over the fuse to reduce the required power needed to program the element. An epitaxial layer was used in the memory to reduce the possibility of latch-up during programming.

A schematic circuit diagram of the read path and delay circuit showing a partial memory matrix with a representation of the fuse element cell array is shown in Figure 10.24(d).

PROMs having latched address registers are called registered PROMs. A block diagram of a registered PROM showing the address registers is shown in Figure 10.25(a) along with a timing diagram showing the synchronous read cycle in Figure 10.25(b).

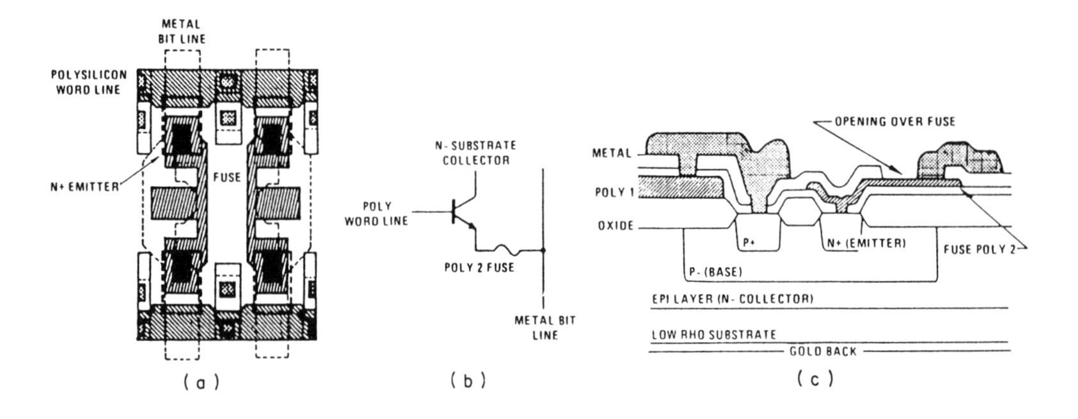

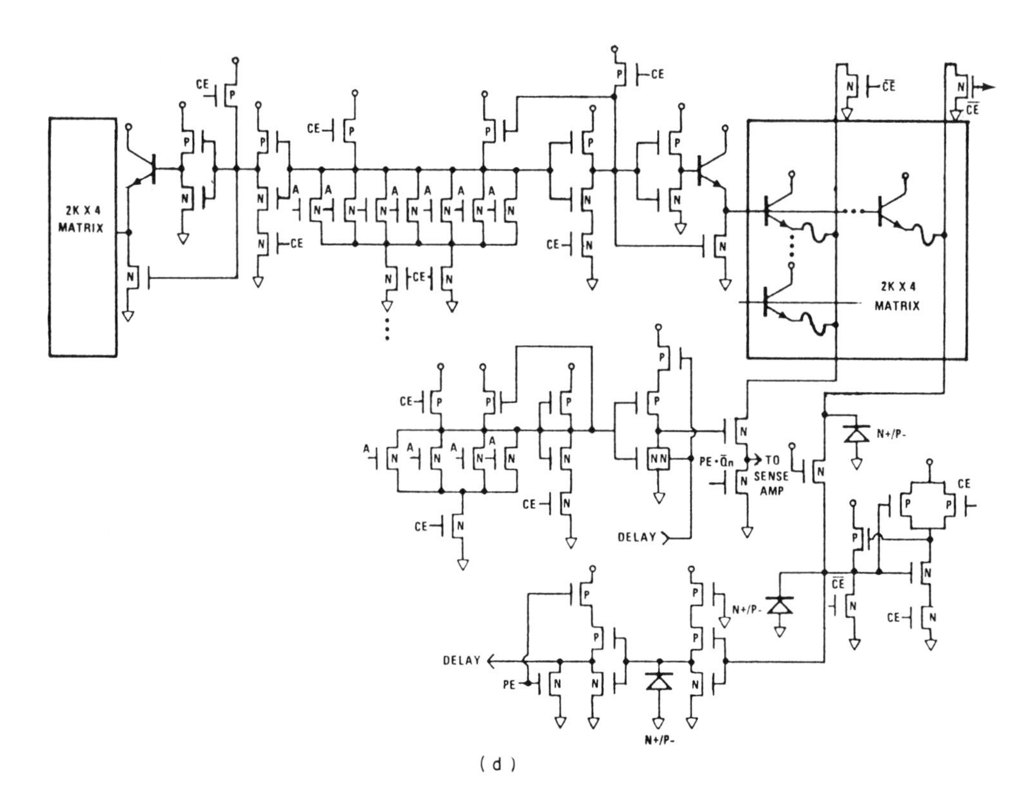

Figure 10.24
16k CMOS PROM with polysilicon fusible link memory array showing: (a) layout of the 4-bit cell, (b) single-bit equivalent circuit with fuse element and transistor, (c) schematic cross-section of the CMOS process showing the fuse element and parasitic bipolar transistor, (d) schematic circuit diagram of the read and delay path showing part of the unprogrammed memory cell array. (From Metzler [21], Harris 1983, with permission of IEEE.)

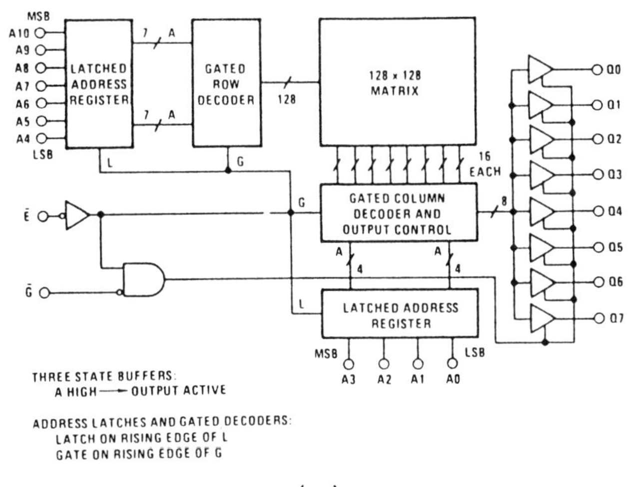

(a)

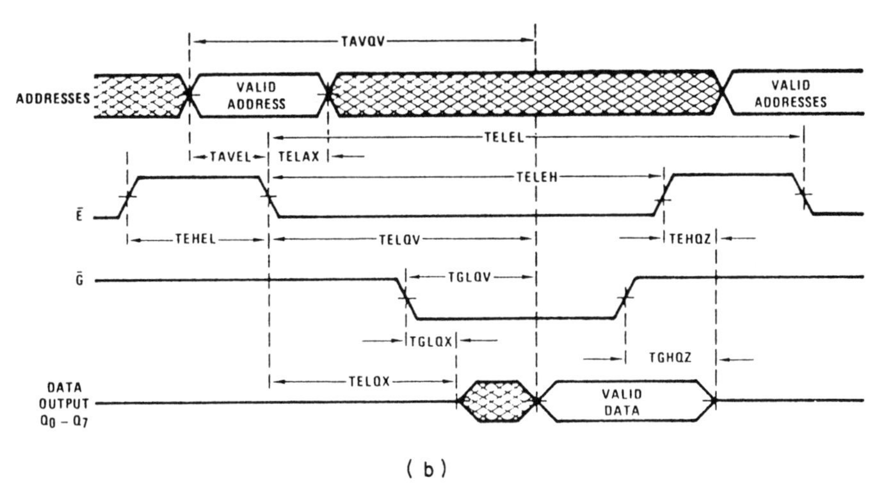

(b)

Figure 10.25
16k CMOS registered PROM showing: (a) block diagram of PROM including address registers, decoders, and 128 × 128 memory matrix and output structure, (b) synchronous read timing diagram. (From Metzler [21], Harris 1983, with permission of IEEE.)

10.13 VARIOUS SEMICUSTOM ARRAYS

As a result of the demand for higher performance through higher system integration, the use of semicustom integrated circuits is increasing. While these chips are not memories in the traditional sense of dedicated data storage arrays, many of them utilize the same type of factory systems as the ROMs.

Some semicustom circuits use memory elements such as the EPROMs and EEPROMs in erasable programmable logic devices (EPLD) arrays. An array of logic cells is like an array of memory cells in demand for density and interface structures.

10.13.1 Gate arrays

These are standard logic arrays that are processed up to the point where the interconnections are left out. The final connections of resistors, capacitors, and transistors are made by the manufacturer in accordance with customer specifications. The use of logic arrays gives a user an effective solution to designs of lower complexity levels and where the volume requirement is relatively low. They are analogous to the ROMs in user program inputs and vendor handling. The same requirements for fast turn-around time that apply for ROMs are also important for gate arrays.

10.13.2 Erasable programmable logic devices (EPLDs)

These are logic arrays which can be programmed in the field by the user and are analogous to the PROMs. Like the PROMs, EPLDs can use either bipolar or CMOS technology and similar fuse elements to the PROMs. An example of EPLD 'cells' is given by the Cypress EPLD [23] in Figure 10.26. The cells in this case are AND gates which are connected via EPROM cells to both the true and complement inputs.

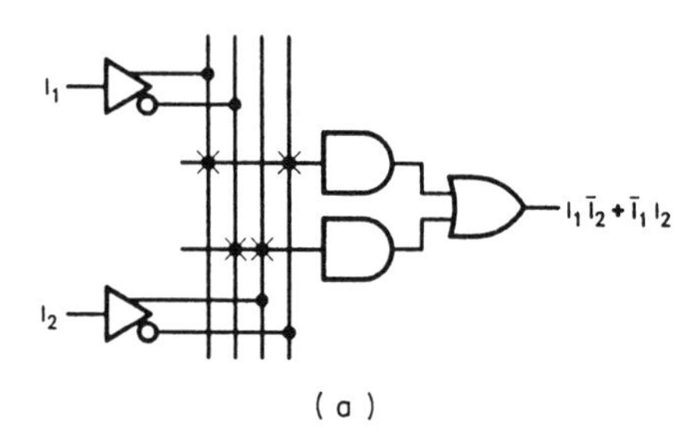

(a)

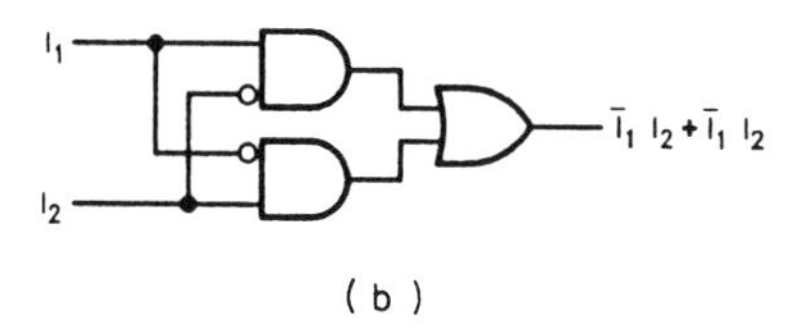

(b)

Figure 10.26
Erasable programmable logic device (EPLD) array. (a) Unprogrammed array with × representing an unprogrammed EPROM cell. (b) Array after programming. (Source Cypress Databook [23] 1989.)

When the EPROM cell is programmed the transistor is turned off and the inputs from the gate are disconnected. In Figure 10.26(a) an 'X' represents an unprogrammed EPROM cell and Figure 10.26(b) shows the gate after programming.

The design of VLSI parts with a high degree of customer involvement calls for an extensive use of computer-aided design techniques (CAD), high level software, and equipment tailor made to back these trends. Tremendous investments and developments are occurring in this area. The days when only the large traditional semiconductor houses participated in the design of integrated circuits have come to an end. Programmable logic is expected to be the fastest growing segment of the total logic market throughout the first half of the 1990s.

BIBLIOGRAPHY

[1] Lineback, R. (1982) Four State Cell Called Density Key, *Electronics*, June 30.

[2] Loesch, W. and Young, A. (1981) Multiuser IC Cell Library Buys Custom Densities At Gate Array Prices, *Electronics*, November 17.

[3] Kiuchi, K. *et al.* (1979) A 65mW 128k EB-ROM *IEEE Journal of Solid State Circuits*, Vol. **SC-14**, No.5, October 1979, p.855.

[4] Masuoka, F. *et al.* (1984) An 80ns 1 Mbit MASK ROM with a New Memory Cell, *IEEE Journal of Solid State Circuits*, Vol. **SC-19**, No.5, October 1984, p.651.

[5] Ariizumi, S. *et al.* (1986) A 70 ns 2Mb Mask ROM with a Programmed Memory Cell, *ISSCC Proceedings*, February, p.42.

[6] Masuoka, F. *et al.* (1984) An 80 ns 1Mb ROM, *ISSCC Proceedings*, February, p.146.

[7] Cuppens, R. and Sevat, L. H. M. (1983) A 256 kbit ROM with Serial ROM Cell Structure, *IEEE Journal of Solid State Circuits*, Vol. **SC-18**, No.3, June 1983, p.240.

[8] Kamuro, S. *et al.* (1982) A 256K ROM Fabricated Using n-Well CMOS Process Technology, *IEEE Journal of Solid State Circuits*, Vol. **SC-17**, No.4, August 1982, p.723.

[9] Donoghue, B. *et al.* (1983) Variable Geometry Packs 2 bits into every ROM Cell, *Electronics*, March 24, 1983, p.121.

[10] Rich, D. A. *et al.* (1984) A Four-State ROM Using Multilevel Process Technology, *IEEE Journal of Solid State Circuits*, Vol. **SC-19**, No.2, April 1984, p.174.

[11] Kohda, K. *et al.* (1986) A Giant Chip Multigate Transistor ROM Circuit Design, *IEEE Journal of Solid State Circuits*, Vol. **SC-21**, No.5, October 1986, p.713.

[12] Okada, M. *et al.* (1988) A 16Mb ROM Design Using Bank Select Architecture, *VLSI Circuits Symposium 1988*, p.85.

[13] Haraszti, T. P. (1984) Novel Circuits for High Speed ROMs, *IEEE Journal of Solid State Circuits*, Vol. **SC-19**, No.2, April 1984, p.180.

[14] *World Semiconductor Trade Statistics (WSTS)* September 1989.

[15] *Hitachi Memory Products Databook* 1987/1988.

[16] *NEC Electronics Inc. Memory Products Databook* 1989.

[17] *Mitsubishi Semiconductors Memories* 1989.

[18] Fukushima, T. *et al.* (1984) A 40 ns 64k bit Junction-Shorting PROM, *IEEE Journal of Solid State Circuits*, Vol. **SC-19**, No.2, April 1984, p.187.

[19] Fukushima, T. *et al.* (1983) A 40 ns Junction-Shorting PROM, *ISSCC Proceedings*, February, p.172.

[20] Fisher, R. M. (1982) A 64k Bipolar PROM, *ISSCC Proceedings*, February, p.114.

[21] Metzger, L. R. (1983) A 16k CMOS PROM with Polysilicon Fusible Links, *IEEE Journal of Solid State Circuits*, Vol. **SC-18**, No.5, October 1983, p.562.

[22] Tanimoto, J. *et al.* (1982) A Novel 14 V Programmable 4k bit MOS PROM Using a Poly-Si Resistor Applicable to On-Chip Programmable Devices, *IEEE Journal of Solid State Circuits*, Vol. **SC-17**, No.1, February 1982, p.62.

[23] Introduction to CMOS EPLDs *Cypress Semiconductor Databook*.

[24] *Fujitsu Memory Databook* 1989.

[25] Prince, B. and Due-Gundersen, G. (1983) *Semiconductor Memories* John Wiley & Sons, Chichester.

[26] *EIA JEDEC Standard* 21–C, 1990.

[27] *Cypress Databook* 1989.

[28] Carter, W. S. (1991) The Evolution of Programmable Logic, *VLSI Circuits Symposium*, June 1991, p.43.

11 FIELD ALTERABLE ROMs I: EPROM, OTP, AND FLASH MEMORIES

11.1 OVERVIEW OF FIELD ALTERABLE ROMs

The field alterable ROM story is a tale of the quest for the high density, fast, low cost non-volatile RAM. It was recognized early that the ideal memory would be as accessible for data changes as a RAM and retain these data indefinitely with the power off. The industry approached the problem from two directions.

One approach was to make RAMs, which can be accessed easily, low power so that they could hold data on battery backup with the power off. This led to the development of both the battery-in-the-package RAM (BRAM) as discussed in Chapter 9 and to the so-called 'Non-Volatile SRAM (NVRAM)' which combines the SRAM and EEPROM technologies and will be discussed in the next chapter. Both of these approaches suffer from high cost which has given both of them relatively minor niches of the market.

The other approach was to make ROMs, which are already big, fast, cheap and non-volatile, easier to use by making them more readily changeable in the field. This approach is traced in this chapter and the next.

This chapter traces the development of the field alterable but inconvenient and expensive UV EPROM, the low cost but still inconvenient OTP EPROM, and the new flash memory which holds out hope of being the promised ideal memory.

The next chapter traces the classical EEPROM–EAROM development from the early hopes that it would be the ideal memory, through its development as an application-specific niche part and its combination with the SRAM in the low density NV-SRAM, and then to its new role as a contender again for the ideal memory in the form of the new NV-DRAM. Various other non-volatile technologies such as the ferroelectric RAM are also covered in the next chapter.

11.1.1 Field alterable ROM applications

The field alterable ROM has a broad base of potential applications extending beyond the traditional memory market of high volume industrial memory users into every area of modern society.

ROMS and EPROMS in lower densities and speeds were used mainly for holding setup and configuration data for peripheral or bootstrap programs. As higher speeds and densities appeared they were used to hold entire operating systems, realtime kernels and canned applications programs. In addition, newer applications such as robotics, automated manufacturing, and automotive tuning systems emerged continuously.

Applications now range from the rapidly expanding microprocessor-based systems market, to the high density computer and telecommunications markets to the silicon computer file and music file markets. They include traditional applications like ROM prototyping. They include also a variety of new fields for electronics where the high endurance remote programming feature of the electrically erasable programmable ROMs (EEPROMs) and the new high density flash EPROMs are needed.

Microprocessor-based systems use a large share of field programmable memories, and in these applications small amounts of these memories are being embedded onto the microprocessor chip itself.

The field alterable ROM market can be broadly split by system flexibility between those types which must be removed from the system to be erased by ultraviolet light but which can be programmed electrically (the EPROMs), those devices which can be programmed only once but not erased (bipolar PROMs, and OTP EPROMs) and those that can be erased in the system (EEPROMs, EAROMs, flash EPROMs).

11.1.2 Field alterable ROM characteristics

The characteristics of these devices that determine their acceptance in the market are shown in Table 11.1. From the applications point of view the ROM has less flexibility, but its low cost and high density have allowed it to continue over time with a significant share of the non-volatile data storage market.

Table 11.1

	Functionality				Cost		
	Program		Field erase	System erase	Low cost package	High density	Low cost test
	1×	>1×					
ROM	no	no	no	no	yes	yes	yes
EPROM	yes	yes	yes	no	no	yes	no
EEPROM	yes	yes	yes	yes	yes	no	yes
NV-SRAM	yes	yes	yes	yes	yes	no	yes
OTP	yes	no	no	no	yes	yes	no
Flash	yes	yes	yes	yes	yes	yes	yes
NV-DRAM	yes	yes	yes	yes	yes	yes	yes

An ultraviolet (UV) EPROM is a ROM which can be electrically programmed in the system using a special 12 V programming pin, but must be erased by being removed from the system and placed under an ultraviolet light for about 20 minutes. The light reaches the EPROM through a quartz window in the ceramic package. The one-transistor cell of the UV EPROM enables it to reach high densities, but the expensive package and test limitations keep it from being low cost.

The UV EPROMs field alterability, high density, and low cost wafer manufacture have allowed it to develop a large share of the non-volatile data storage market in spite of the high package and test cost and relative inflexibility in the system.

The One Time Programmable (OTP) EPROM, a low cost plastic packaged EPROM, is functionally a PROM, or fast turn-around ROM. It can be programmed only once and cannot be erased. For this reason it is not possible to test a blank OTP after it has been packaged, a factor that has limited its market acceptance. The OTP's lack of flexibility in the applications coupled with its test problems have relegated it to a niche of this market.

The electrically erasable PROM (EEPROM) flexibility in the system, compatibility with standard CMOS logic process lines, and high reliability have given it a sizable market niche in the low densities below 256k. Its two or more transistor cell and high cost of manufacture has kept it from gaining significant market share in the higher density applications where it might have competed with the UV EPROM.

The flash is an electrically erasable and programmable ROM which shares with the UV EPROM the small one-transistor cell chip size, but is packaged in the low cost plastic package. Like the OTP it is derived from the EPROM and made in an EPROM technology with the addition of a thinner silicon dioxide layer which gives it the electrical erase property.

The flash, still in a development stage in the early 1990s, seems to offer much that this market demands: high density, low cost process and package, high system flexibility, and ease of test.

NV-SRAMs combine a six-transistor cell SRAM and a two- or four-transistor cell EEPROM into an eight- to ten-transistor cell chip which shares characteristics with EEPROMs on the above chart with the added advantage of SRAM speed and endurance and added disadvantage of low density and even higher cost of manufacture. NV-SRAMs have a niche market in peripherals requiring high speed random access with relatively small numbers of setup parameters which need to be sorted and changed only occasionally.

NV-DRAMs are, in the early 1990s, a development technology. They keep the one-transistor plus one-capacitor cell of the DRAM and add the non-volatility of an EEPROM capacitor into this combination. They hold the promise, along with the flash, of being the ideal non-volatile RAM.

Only the EEPROM, the flash and the NV-RAMs have the non-volatile RAM functionality. Only the ROMs, flash and NV-DRAMs have the required low cost and high density.

A chart of percent of dollar revenues over time for these products is shown in Figure 11.1. This chart includes both history from 1980 and a forecast to 1996 based on the current promise of the flash EPROM and the limitations of the other devices.

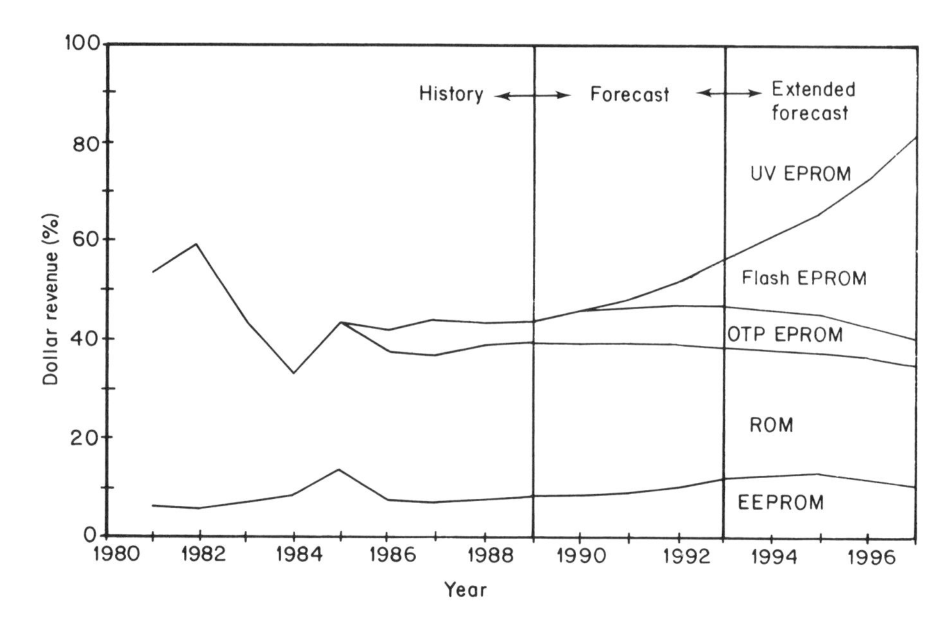

Figure 11.1
History and forecast of mask and field alterable ROM market by product in percentage of dollar revenue by year from 1981 to 1997. (Source Dataquest 1981–1990.)

Figure 11.2
Photograph of EPROM package showing quartz window. (Photograph by Mike Butterworth.) (From Prince and Due-Gundersen [98] 1983.)

11.1.3 History of field alterable ROMs

Early work on a field alterable ROM was done in 1971 at Intel by Frohman-Bentchkowsky [29], who developed the well known floating polysilicon gate avalanche injection MOS (FAMOS) process to make ultraviolet light erasable programmable ROMs (UV EPROMs).

All generations of UV EPROMs since have used this process (or the N-channel version of it, since the original EPROM was P-channel). A description of the EPROM process and cell concept was included in Chapter 4. This is also referred to as the 'hot' electron injection process.

UV EPROMs like the ROMs had a one-transistor cell which permitted them to attain high density. They were an improvement over the ROMs in that they could be programmed and erased by the user in the field; however, the necessity to erase them under ultraviolet light resulted in a number of disadvantages. To reach the chip, the ultraviolet light requires an expensive ceramic package with a quartz window in the top (see Figure 11.2 for a picture of an EPROM package). The part normally must be removed from the system to be placed under the UV light which requires sockets in the system rather than the less expensive soldering. Damage can result from the extra handling problems involved in the erase process. In addition, test time becomes significantly longer due to the lengthy erase procedure and adds to the total cost of the part. All of these factors add to the cost of the part and keep it from matching the low cost of the ROM.

Another problem is the sensitivity of the EPROM to daylight or fluorescent light which will deprogram it after a sufficiently long exposure. This problem is usually minimized by sticking paper tabs over the quartz windows after the EPROMs are programmed.

For all of these reasons the search for a low cost, high density, non-volatile memory with the flexibility of being erasable in the system continued. This search was encouraged by the trend in the systems market to remote distributed processing and automation which limited the feasibility of accessing the system manually to remove EPROMS for ultraviolet exposure when programming modification was required.

Large volume applications for EPROMs did develop, however, both as prototypes for ROMs and in high volume non-volatile storage applications as substitutes for ROMs. They are also used in some applications where the information is changed infrequently. In this latter case however it would be useful to be able to reprogram the EPROM in the system.

Work on the technology for producing electrically erasable ROMs has been ongoing for a number of years. The physical mechanism used in the majority of these devices was first developed in the 1960s by Fowler and Nordheim. Many articles exploring this technology were published between 1968 and 1969, but commercial development occurred mainly from 1977 onward.

A description of the Fowler–Nordheim 'cold electron' tunneling mechanism is included in the technology section of the next chapter. It involves a quantum mechanical effect in which the electrons pass (tunnel) through the energy barrier of a very thin dielectric such as silicon dioxide or silicon nitride.

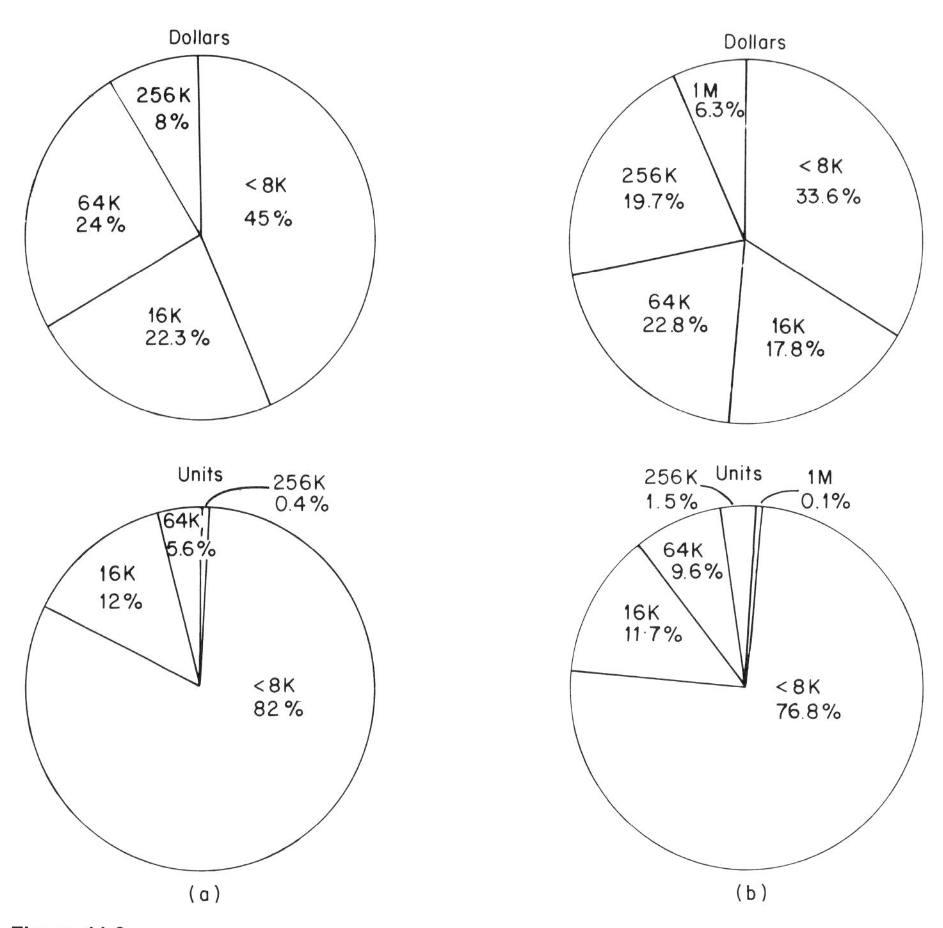

Figure 11.3
World electrically erasable ROM shipments by density in % units and dollars in (a) 1986 (total $112 million) and (b) 1990 (total $315 million). (Source Dataquest.)

Early companies to explore this technology with low density devices included General Instruments, Hughes Aircraft, Siemens, Motorola, TI, NCR, and Nitron.

The low density electrically erasable ROMs with less than 8k bits developed a sizeable market primarily in consumer applications which differentiate them as a device type in their own right as distinguished from the medium density 16k to 256k parts whose applications are primarily in the industrial and computer markets.

A comparison of the unit and dollar shipments of electrically erasable ROMs is shown in Figure 11.3 for 1986 and 1990. It is clear that the parts with less than 8k density are a major factor in this market segment with 82% of the unit market in 1986 (45% of the dollar market) and 76.8% of the unit market in 1990 (33.6% of the dollar market).

The term EAROM (Electrically Alterable ROM) was originally used for devices made in the MNOS (metal-nitride-MOS) technology which was used for many of these early low density devices. This was intended to distinguish the MNOS devices from those being made in the double polysilicon floating gate thin oxide process developed by Intel which were called EEPROMs and were offered primarily in the medium density market segment from 16k to 256k bits. In the course of time both technologies have been used in both market segments resulting in a confusion in terminology. (A more detailed description of both of these device technologies is given in the next chapter.)

We will refer to the technologies as MNOS and 'floating gate tunneling oxide' and the devices in the two market segments as EAROMs and EEPROMs.

In the aftermath of the 1981 recession, several start-up companies who specialized in the new floating gate tunneling oxide technology came to the forefront. These included Seeq and Xicor, which were spin-offs from Intel. Xicor in 1979 introduced its revolutionary 'shadow RAM' which, though a non-standard device, was the first serious attempt in the industry at a true non-volatile RAM. It combined SRAM and EEPROM technologies in a single-memory cell.

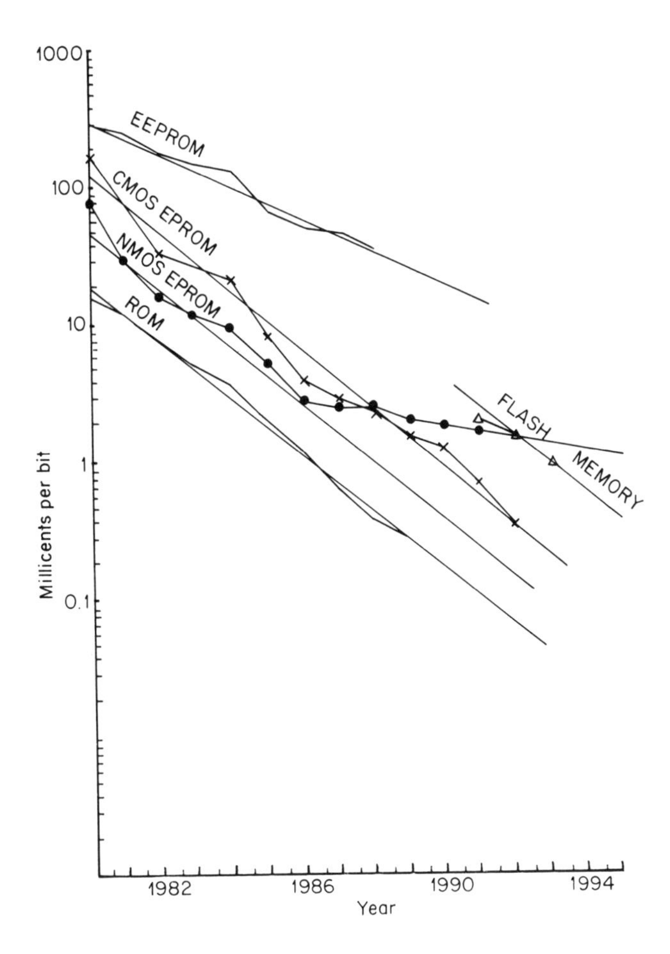

Figure 11.4
Price per bit for various field alterable ROMs. (Source Dataquest.)

The similarity of the EEPROM floating gate tunneling oxide process to the well established EPROM process and the potential for using inexpensive plastic packaging mean that the pricing of EEPROMs in volume was expected at that time to become comparable to that of EPROMs and less than that of the battery-backed up CMOS static RAMs. The attraction of a low cost, high density, field programmable replacement for the expensive EPROMs attracted many of the mainstream memory companies to this development.

By late 1982, most companies making EPROMs had either announced EEPROMs or were working on them. These included in addition to the above, Intel, NSC, Mostek, Hitachi, Mitsubishi, Panasonic, Plessey, SGS, and NEC. The EEPROM was widely expected to have the highest growth rate of any memory device with potential for replacing the established EPROM and to some extent the CMOS SRAM. A growing number of new applications outside the traditional memory market added to its attraction.

Unfortunately, due to the difficulty of manufacture and the need to have two transistors in the EEPROM cell, the parts never became competitive with the one-transistor cell EPROM. The difference in price per bit over time for the EPROM and the EEPROM is shown in the graph in Figure 11.4. The EEPROM is clearly on a different learning curve than either the EPROM or the ROM.

By the mid 1980s the EPROMs were being introduced in multimegabit densities. Meanwhile EEPROM manufacturers, basically limited to the 256k density and below, were focused on applications specific niche markets and the search was on for other alternatives for a high density low cost replacement for the EPROM.

The EPROM in plastic package came next. The plastic package reduced cost, but lacked the window through which the UV EPROM can be erased. It was originally referred to as the 'plastic EPROM' with the industry finally settling on the name 'OTP' (one-time-programmable) EPROM.

The problems inherent in final testing and doing outgoing quality inspections on a part which could only be programmed once held up the introduction of the OTP during the early 1980s. Several attempts to produce OTPs during this time period resulted in sizable recalls which gave the part a bad early reputation in the industry. There were also problems with moisture penetration of the plastic package which were later resolved with the use of oxy-nitride passivation which transmitted UV light so the parts could be erased.

The OTP EPROM finally got into production in 1985, and has retained a small niche of 5% to 7% of the overall ROM market since that time being used primarily as fast-turn-around ROMs in low cost applications. Programming in this case is normally by the manufacturer.

The major shift in the programmable ROM market in the second half of the 1980s was the shift of the UV EPROM from the NMOS to the CMOS technology as shown in Figure 11.5. UV EPROMs went from 89% NMOS in 1985 to 84% CMOS by 1990 with the shift occurring mainly at the 1Mb level as with the DRAMs. By this time the 16Mb CMOS EPROMs were being developed.

This shift is also evident in the price curves in Figure 11.4 where the NMOS EPROM trend rises above the learning curve in typical end-of-life pricing in the later part of the 1980s and the CMOS EPROM price continues to fall on its original trend line.

In the late 1980s another promising technology for a low cost electrically erasable ROM was discovered—the 'flash' EPROM memories. Like the EEPROM before them, the flash memories were derived from the UV EPROM technology. Unlike the two-transistor cell EEPROM the single-transistor flash memory cell was comparable in size to that of the UV EPROM. This meant it could be made in similar densities and at similar process costs. The flash memories used the low cost plastic package like the OTP, but being erasable could be programmed more than once and tested easily.

As was the case of the EEPROM, the forecast was for the flash type memories to have the highest growth rate of any memory device and to eventually replace the UV EPROM and a share of the CMOS SRAM market as well.

As the 1990s opened, most companies which manufactured UV EPROMs were in development on a flash memory technology although only a limited number of parts have actually appeared in production. Two versions had been introduced, those that were reprogrammable at 12 V and those reprogrammable at 5 V. The 12 V version had the smaller cell size but required an additional 12 V power supply in the system. The 5 V version had a larger cell but worked in 5 V only systems.

The designation 'flash' originated as the technical term for writing or erasing an entire memory array at one time (in bulk). Since the early flash devices were erasable only in bulk they were functionally 'flash' parts, and the name has stuck to these devices even though many of them can now be erased by section.

Another trend in the market which occurred as a result of the technology switch to CMOS was the development of fast EPROMs in the lower densities. These parts competed in speed with the power hungry bipolar PROMs and were superior in power consumption due to the low power CMOS technology.

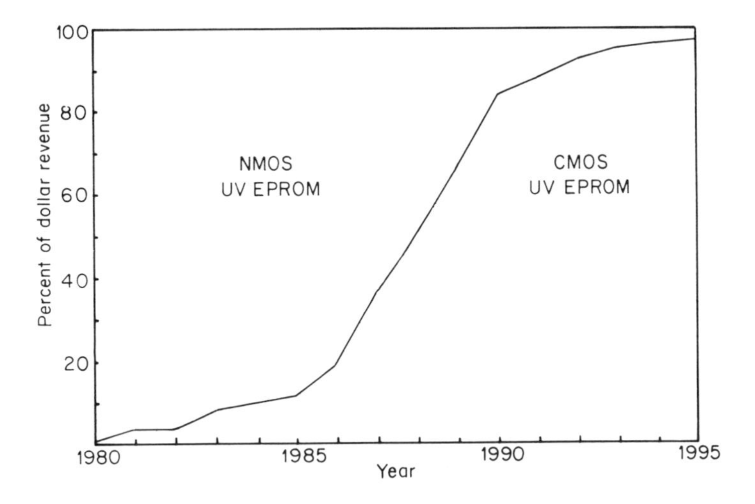

Figure 11.5
UV EPROMs percent NMOS and CMOS from 1980 to 1995 in percentage of dollar revenue. (Source Dataquest.)

11.2 EARLY NMOS UV EPROM DEVICES 1k TO 1Mb

The first UV EPROM was the Intel 1702. This 2k bit (256 × 8) device, revolutionary as it was, had slow access times with a selection from 650 ns to 1.5 μs. It required five power supplies ranging from −9 V to +48 V for the various read and program functions. It drew up to 65 mA power supply current and had no standby power function; however, in the read mode it functioned like a ROM and was, in fact, the first of a long series of true field programmable, non-volatile memories [29].

Larger and improved EPROMs followed. An 8k device, the 2708, which became an industry standard, required only four power supplies. These included a 25 V program voltage (down from the 48 V of the 1702), a +12 V power supply, a +5 V read, and a −5 V backbias supply. Access time ranged from 350 to 450 ns. While it still lacked the standby power-down of future EPROMs, it drew only 28 mA in a low power version offered by Intel. Power dissipation was rated at 1.5 W.

Two manufacturers, Motorola and TI, offered a 16k (2k × 8) device which was a direct upgrade of the 2708, called the TMS2716. This part had similar characteristics and required the same four power supplies as the 2708. The rest of the industry waited for the development of single power supply read parts before introducing devices at the 16k level.

The first of the modern EPROM devices, a 16k (2716) with single +5 V power supply for read, was introduced by Intel in 1978. The 2716 still required a high +25 V supply for programming. It became available in access times ranging from 250 to 450 ns and offered a standby power feature rated at 25 mA or below.

New design techniques permitted introduction of an on-chip backbias generator which eliminated the −5 V power supply requirement. Advances in wafer processing technology permitted use of the single +5 V power supply for read, just as it

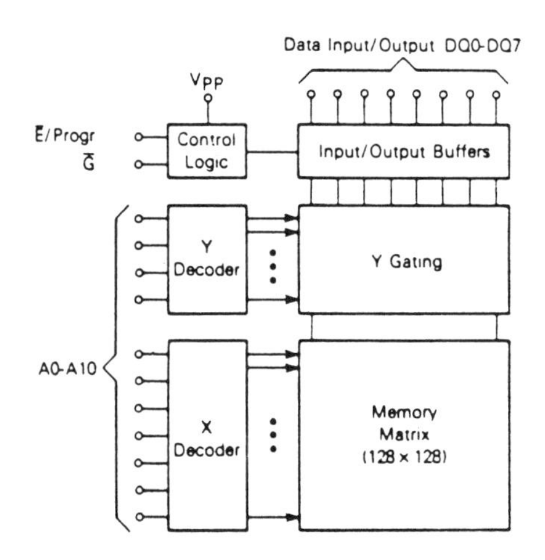

Figure 11.6
Block diagram of industry standard 2k × 8 UV EPROM (2716). (From Prince and Due-Gundersen [98] 1983.)

permitted it for the dynamic RAMs in this same time period. The 2716 was pin and function compatible with the industry standard 16k ROM and was initially used extensively for ROM prototyping.

The 2716 was fairly typical of the family of EPROMs which evolved from it. It was organized in 8 sections of 16 × 128 cells, as shown in Figure 11.6. Each section was connected to the column decode circuitry. The decoder selected one of 16 columns and routed the signal to the sense amplifier of the particular bit. Each sense amplifier was directly connected to an output amplifier. A chip select enabled and disabled the output which placed the chip in standby mode.

For programming the 25 V was applied to the V_{PP} pin and programming data were entered in 8 bit words through the data out terminals using a 2 ms programming pulse. Multiple devices could be programmed in parallel by connecting together the programming inputs and using the program inhibit mode.

32k and 64k bit EPROMs followed behind the 2716, but were made using basically the same technology and were upgrades of the same product.

At the 32k level, a standardization problem occurred with some companies adding the additional A11 address on pin 18 and others adding it on pin 21. Both retained the 24 pin package of the 2708 and 2716.

At the 64k level the lack of standardization temporarily slowed the industry acceptance of the larger EPROMs. Intel, Mostek, TI, and Motorola all offered different versions of the 64k EPROM to the market. This was the type A, type B problem discussed previously in the ROM chapter.

While most of the industry chose to go to the 28 pin package at the 64k level for the added flexibility of density upgrading to the 256k level, they did not all move in a uniform manner but perpetuated the non-uniformity from the 32k level. A few companies remained with the 24 pin package for the 64k EPROM.

The Intel 28 pin series was eventually accepted by the JEDEC memory committee and became the industry standard for 64k to 1Mb EPROMs.

At the 1982 ISSCC Intel [34] described a 23 mm^2 128k EPROM with fuse link redundancy followed a year later by a 18.4 mm^2 256k EPROM [23, 35]. This was an advance by twice the density at 80% of the size in just one year. The part compared favorably with the 18.6 mm^2 of the 256k serial ROM shown by Signetics also in 1983.

Other features of Intel's 256k EPROM included 200 ns address access, 40 nm gate oxide, and 2 μm geometries with a tight 1.0 μm channel length in the array. The part had automatic power-down to reduce standby power dissipation to 110 mW. Active power dissipation was 350 mW. The scaling of the EPROM cell allowed programming at 13 V.

Two of the last NMOS EPROM papers, before EPROM technologists converted development to CMOS, were presented in 1984. AMD [31] showed a fast 150 ns 512k EPROM which was designed to go into a plastic package as an OTP EPROM to compete with mask ROMs. The 38 mm^2 size of this chip was comparable to the size of the similar density ROMs.

NEC [32, 33] showed a 200 ns 1Mb EPROM which was 67.7 mm^2. The NEC EPROM could be configured as 64k × 16 or 128k × 8 by controlling an input signal. Operating power was 500 mW and standby power was 150 mW. Standby mode was

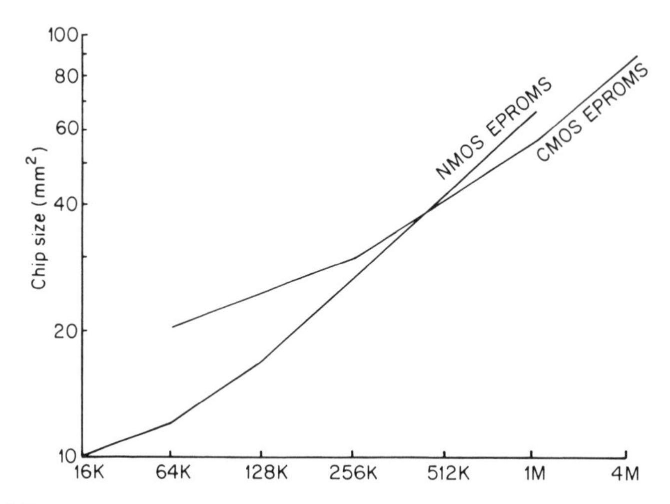

Figure 11.7
EPROM chip size plotted against density for NMOS and CMOS UV EPROMs. (Source ISSCC Proceedings 1980–1990.)

accomplished by a chip select controlled automatic power-down. Programming was at 13 V.

Table 11.2, adapted from a chart by AMD [31], indicates NMOS EPROM circuit characteristics from the 16k to the 1Mb. It is clear from this chart that the NMOS EPROMs were having both an active and a standby power problem in the higher densities.

Figure 11.7 plots chip size compared to density for the NMOS and CMOS EPROMs from the 16k to the 1Mb density level. By the 1Mb the CMOS EPROM chip size was smaller than the NMOS as well as having power advantages over it. While significant numbers of fast NMOS EPROMs continued to be shipped well into the 1990s, the major switch to CMOS occurred at the 1Mb level.

Table 11.2 Various NMOS EPROM circuit characteristics. (Adapted from Rinerson [31], AMD 1984, with permission of IEEE.)

	64k	128k	256k	512k	1Mb
Organization	8k × 8	16k × 8	32k × 8	64k × 8	128k × 8/64k × 16
Chip size (mm^2)	11.6	15.4	27.0	38.0	67.7
Cell size (μm^2)	41.0	41.0	48.8	36.6	41.0
Feature size (μm)	1.8	1.8	2.0	1.7	1.2
t_{ACC} (ns)	120	120	150	150	200
Active power (mW)	225	275	300	350	500
Standby power (mW)	60	60	60	75	150
V_{PP} (V)	21/12.5	21/12.5	12.5	12.5	12–14
t_{pp} (ave) (ms)	4	4	4	4	1

11.3 CMOS EPROMs AND THEIR OPERATION

11.3.1 Early CMOS EPROMs

While the NMOS manufacturers were moving to larger EPROMs with faster speeds, other manufacturers were developing CMOS EPROMs to help solve the problem of high power consumption. For portable systems that have a limited amount of power available, even the 150 mW worst case specified standby power of the average high density NMOS EPROM was too high. An alternative technology had to be found.

Intersil, RCA, and Harris were among the first companies to announce a CMOS UV EPROM. 16k CMOS EPROMs were introduced in 1980 by RCA and in 1981 by Intersil.

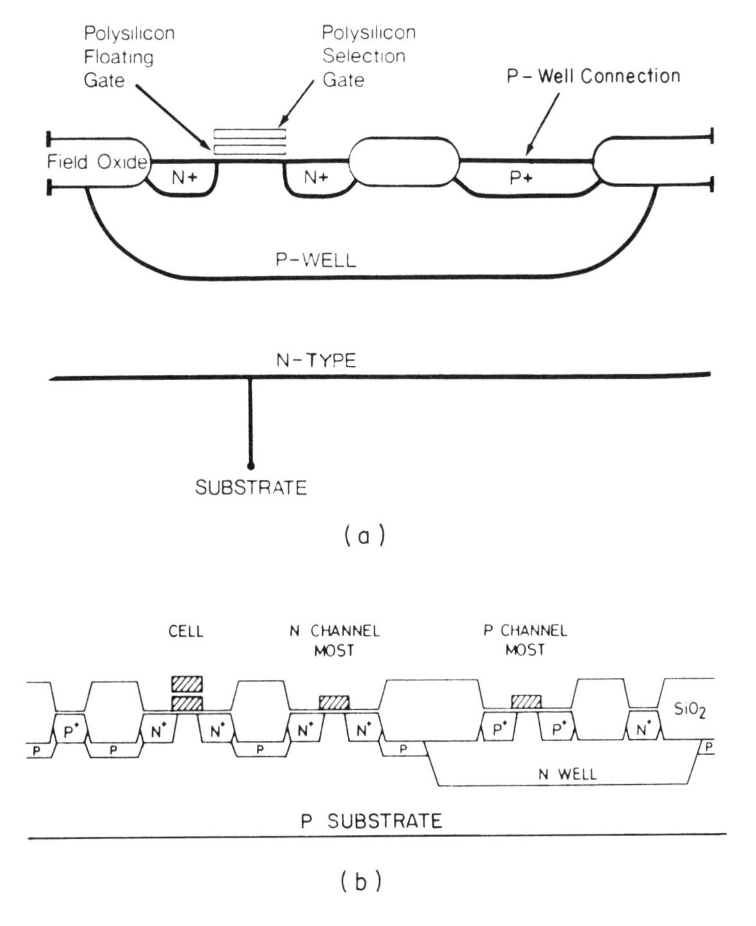

Figure 11.8
Schematic cross-sections of early CMOS EPROM cells. (a) P-well process (Cheng *et al.* [36], Intersil, with permission of IEEE). (b) N-well process (Mijasaka *et al.* [37], Fujitsu, with permission of IEEE.)

A cross-sectional view of a CMOS EPROM process in a P-well technology from the Intersil [36] part is shown in Figure 11.8(a). This part had a dual polysilicon gate single N-channel transistor EPROM cell in a P-well. All subsequent generations of CMOS EPROMs used an N-well technology. The periphery used CMOS circuitry to reduce power. This part was designed in 5 μm design rules and had an active power of 50 mW at 200 ns access time and a standby power of 5 μW. While the active power was not significantly different from what would be found on an NMOS EPROM, the standby power was an order of magnitude lower. Internally synchronous circuitry also helped reduce the standby power.

A differential sense amplifier was used and the differential inputs were precharged before sensing to increase sensitivity. Power to the sense amplifier was cut off after sensing. The X-decoder was latched during programming to prevent erroneous data entry. Since latch-up was a consideration in CMOS EPROMs during high voltage programming, this circuit clamped the programming voltage to minimize the P-well injection current which could cause latch-up.

National Semiconductor followed with 32k and 64k devices in 1982 and 1983. These parts had standby current as low as 100 μA and were offered in both synchronous and asynchronous variations. The asynchronous part was compatible with NMOS 2716s.

At the 1982 ISSCC a 150 ns CMOS 64k EPROM using N-well technology was described by Fujitsu [37]. It had an active power of 50 mW at 5 MHz and a low standby power of less than 5 μW. The 21 V programming voltage meant that guard rings were used to avoid latch-up phenomena which could be caused by carrier injection into the bulk of the cell array during programming.

A special V_{CC}–V_{PP} switching circuit was also added to eliminate forward biasing conditions on junctions and permit various ON–OFF sequences between V_{CC} and V_{PP}.

The part used N-channel transistors in the memory array, as in the standard NMOS EPROMs, with 2.5 μm CMOS channel lengths used for the peripheral circuits. A cross-section of an N-well EPROM process is shown in Figure 11.8(b). The N-well process became the standard for CMOS EPROMs due to the compatibility with the established DRAM process. The cell was 133.6 μm^2 and the chip was 20.8 mm^2.

Figure 11.7 indicates that these early CMOS EPROMs were not competitive in chip size with the established NMOS EPROM process.

In 1983 Fujitsu [38] presented the next step, a 288k CMOS EPROM with 150 ns access time which, like the 64k a year earlier, used N-well technology with CMOS peripheral circuits. The standby current was reduced into the sub-microampere range. The 288k cells provided for use as either $\times 8$ or $\times 9$ organization. The addition of the ninth bit permitted parity to be used to increase the reliability of the system. The ninth block could also be used for redundancy with polysilicon fuses provided. Four extra rows were also included for replacement.

The 34.6 mm^2 chip was made in a 1.5 μm minimum geometry process with 54 μm^2 cell size. (Minimum geometry in this case was the contact size rather than the minimum channel length as in the various RAMs.) This was only about 30% larger than the comparable NMOS EPROM. The channel lengths were reduced to 2.0 μm for the NMOST and 2.5 km for the PMOST.

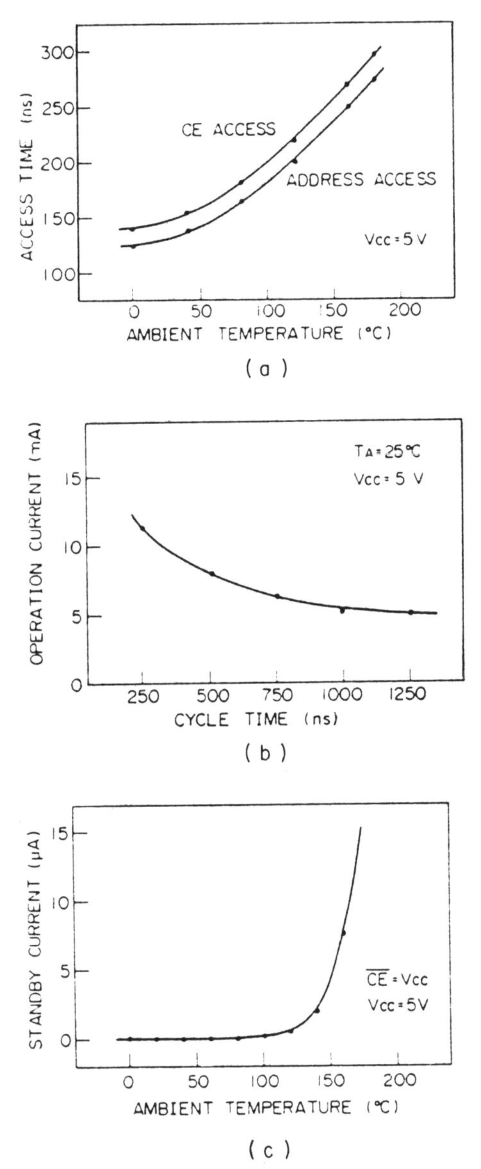

Figure 11.9
Operating characteristics as a function of ambient temperature and cycle time for a 288k CMOS EPROM. (a) Access time as a function of temperature. (b) Operation current as a function of cycle time. (c) Standby current as a function of temperature. (From Yoshiba [38], Fujitsu 1983, with permission of IEEE.)

Various operating characteristics as a function of ambient temperature and cycle time are shown in Figure 11.9. These are fairly typical for CMOS EPROMs.

Note the relatively fast sub 200 ns access times in the normal 9 to 70°C operating range, the low 10 mA operating current at 250 ns cycle time and the stable low

Table 11.3 Technological characteristics of CMOS EPROMs 16k–1Mb. (Source various ISSCC Proceedings.)

Date	Company	Density	Channel (μm)	Cell (μm^2)	Chip (mm^2)	Gateox (nm)	Interox (nm)
1981	Intersil	16k	5	—	—	65	80
1982	Fujitsu	64k	2.5	133.6	20.8	—	—
1983	Signetics	64k	—	—	18.8	65	70
1983	Fujitsu	288k	2.0	54	34.6	50	—
1984	SEEQ	256k	1.5	37.5	20.9	45	65
1984	Toshiba	256k	—	—	34.7	—	—
1985	Hitachi	256k	—	36	21.4	—	—
1985	Intel	256k	1.0	36	25.6	35	45
1988	WSI	256k	1.2	42.2	30.1	27	—
1989	Cypress	256k	0.8	39.7	25.3	24.5	30
1985	Toshiba	1Mb	1.2	28.6	60.2	28	45
1985	Hitachi	1Mb	1.0	19.3	39.4	—	—
1986	AMD	1Mb	1.5	20.25	50.9	25	30
1987	T.I.	1Mb	1.2	13.5	53.5	35	—
1987	Fujitsu	1Mb	2.0	18.9	45.4	—	—
1988	SGS–T	1Mb	1.0	18.9	45.7	25	30
1988	Hitachi	1Mb	1.2	18.5	42.7	35	—
1990	Toshiba	1Mb	0.8	30.24	77.3	—	—

Table 11.4 Electrical characteristics of CMOS EPROMs 16k to 1Mb. (Source various ISSCC Proceedings.)

Date	Company	Density	Org	V_{PP} (V)	T_{PP} (msB^{-1})	I_{AC} (MHz) (mW)	I_{SB} (μW)	t_{ACS} (25) (ns)
1981	Intersil	16k	2k × 8	20	50	(5)	5	200
1982	Fujitsu	64k	8k × 8	21	25	20(5)	5	150
1983	Fujitsu	288k	32k × 9	21	16	20(4)	<5	150
1983	Signetics	64k	8k × 8	—	—	25(5)	<1	100
1984	SEEQ	256k	32k × 8	12–16	0.5	100(5)	500	125
1984	Toshiba	256k	32k × 8	21	0.003	3(1)	0.5	100
1985	Hitachi	256k	32k × 8	12–14	0.5	12(1)	0.5	95
1985	Intel	256k	32k × 8	—	—	150(5)	500	100
1988	WSI	256k	32k × 8	—	0.8	300(18)	—	50
1989	Cypress	256k	32k × 8	—	—	—	—	23
1985	Toshiba	1Mb	64k × 16	12	0.01	50(1)	0.5	80
1985	Hitachi	1Mb	128k × 8	11–13	0.5	(7)	<5.0	140
1986	AMD	1Mb	64k × 16	11.5	0.25	250(5)	3000	150
1987	T.I.	1Mb	64k × 16	12.5	0.5	220(6)	2500	150
1987	Fujitsu	1Mb	64k × 16	—	—	75	<5.0	80
1988	SGS–T	1Mb	64k × 16	12.5	0.5	100	300	70
1988	Hitachi	1Mb	64k × 16	12.5	0.025	60(5)	—	55
1990	Toshiba	1Mb	64k × 16	—	—	85(30)	15,000	16

standby current over the 0 to 70°C ambient temperature range. These characteristics compared favorably with the typical NMOS characteristics shown earlier.

Table 11.3 shows some technical characteristics and Table 11.4 some typical electrical characteristics of various CMOS EPROMs from 1981 to 1990.

11.3.2 Data sensing in EPROMs

Data sensing in EPROMs can be done with a single-ended sensing scheme which senses the potential difference between a bit sense line and a reference line as shown in Figure 11.10(a). The bit sense line potential depends on the state of a selected EPROM cell—either erased or programmed [58, 61].

In an erased state the threshold voltage for a selected memory cell transistor is reduced to a low voltage of about 2 V. When the word-line, which is the cell transistor's gate, is selected, the cell transistor turns on. The bit-line level is determined by the current which flows through the load transistor M_1 and the selected memory cell transistor.

If the cell is programmed, then its threshold voltage is increased and the cell is OFF even if its word-line is selected. As the threshold voltage for a cell transistor becomes higher, the selected bit-line potential becomes higher. If the bit-line is charged up to $V_{bias}-V_{thres}$, then the pass transistor between the sense line and the bit-line is off.

The reference line potential is determined by the current flowing through load transistor M_2, and a reference cell transistor which has the same structure as the memory cell except that is in permanently in the erased state. The conductance of M_2 is set larger than that of M_1.

In an erased state, the memory cell and the reference cell conductance values are the same so the sense line potential is lower than the reference line. When a programmed cell is selected, the sense line becomes higher than the reference as shown in Figure 11.10(b).

The bit-line can be discharged to ground level through an erased cell transistor on the bit-line when one of the common gate cells is selected. When the discharged bit-line is selected, the sense line is instantly pulled down by the high capacitance bit line as shown in Figure 11.10(c). Then the sense line is charged up through the load transistor by the bit-line charging.

As shown in Figure 11.10(c), the zero read access time is worst when the bit-line must be charged from ground. If a programmed cell is selected, the data zero is sensed correctly after the sense level passes the reference level (which is not affected by the address change). Access time for data zero depends then on the bit-line charging time.

The reference level is, however affected by the cell read current which in turn affects the zero read speed. The reference line level is set by the current that flows through the load transistor and reference cell transistor. Large cell current causes a low reference level and small cell current causes a high reference level. Number 1 in the diagram in Figure 11.10(c) shows the case where the reference line level is low due to large cell current and 2 shows the case where the reference line is high due to small cell current. When the reference line level is low, data zero will be sensed

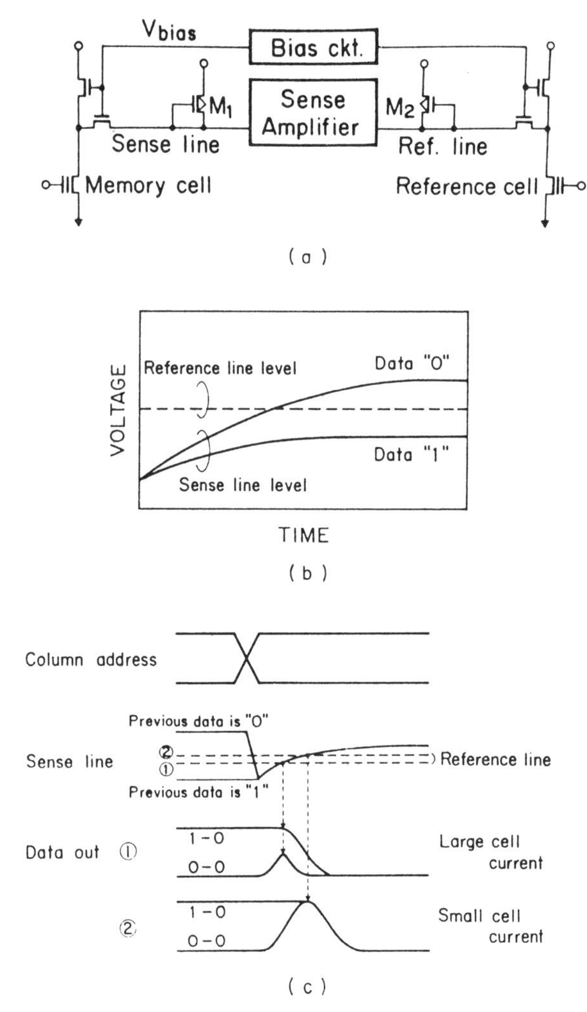

Figure 11.10
Data sensing in EPROMs. (a) Single-ended data sensing scheme. (b) Time dependent change of reference and sense line levels. (c) Waveform comparison between data from zero to zero and from one to zero for large and small cell current. (From Imamiya *et al.* [61], Toshiba 1990, with permission of IEEE.)

rather quickly so the 0–0 waveform will turn lower before it reaches the 1–0 waveform. As the reference level is set higher, data zero sensing time becomes longer and the turning point for the 0–0 waveform comes near the 1–0 waveform. So if the cell read current is large, 0–0 is faster than 1–0. If it is small 0–0 can be as slow as 1–0.

11.3.3 Address transition detection in CMOS EPROMs

A fast 100 ns 64k CMOS EPROM was shown in 1983 by Signetics [39] using a double-layer polysilicon N-channel structure with a 18.8 mm^2 chip size. The standby power dissipation was less than 1 μW which made this also a very low standby power part. The part was internally synchronous and used clock signals generated by address transition detection to reduce power consumption.

Several of the considerations inherent in using ATD for EPROMs were addressed. One was that the ATD circuit must ensure that the shortest clock pulse created is adequate to trigger subsequent circuits. A second is the need to insure a valid sensing period after the final address transition occurs. If incoming address transitions are not coincident the ATD circuit needs to extend the clock pulse to insure a valid sensing period after the final address transition occurs. A third consideration is glitch sensitivity. The circuitry needs to screen out address pulses shorter than a specified minimum width to ensure that they do not initiate a read cycle.

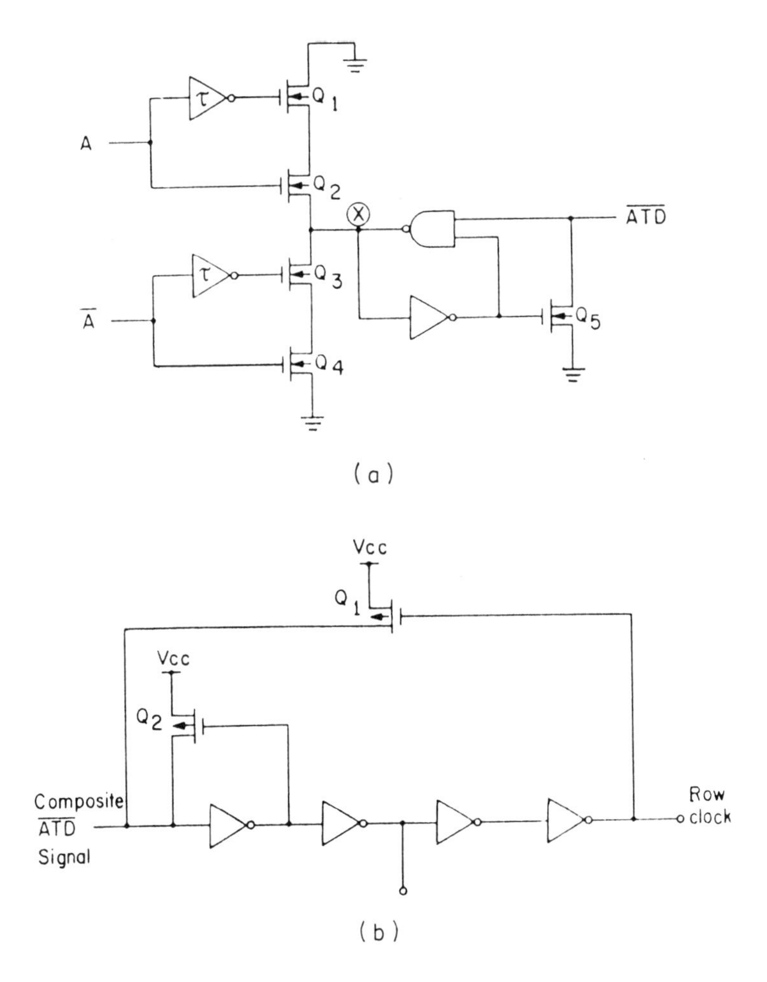

Figure 11.11
Schematic ATD circuitry from a typical 64k CMOS EPROM. (a) Address transition detection circuitry. (b) Clock generation circuitry. (From Knecht [39], Philips 1983, with permission of IEEE.)

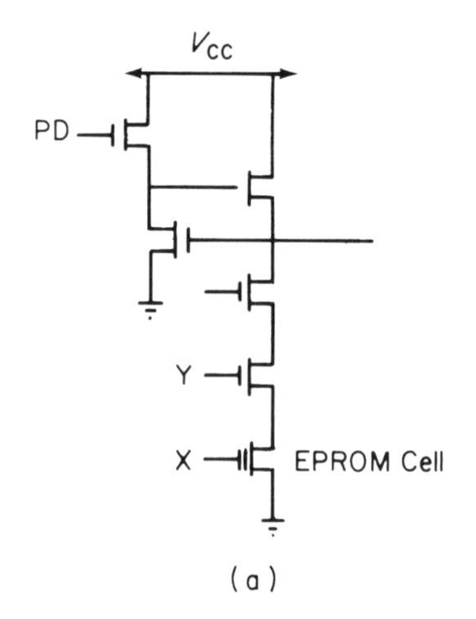

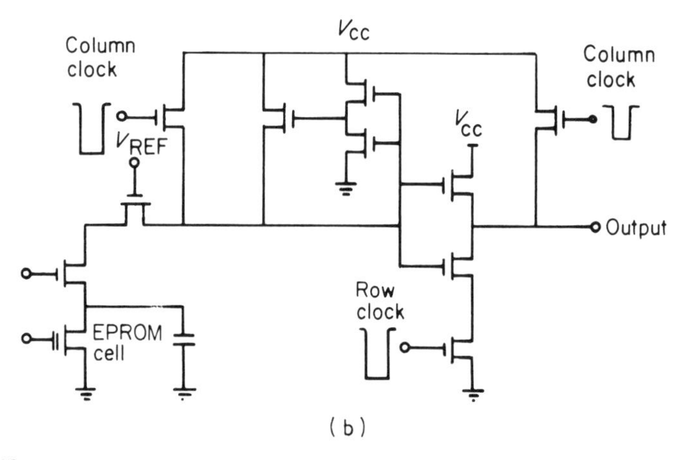

Figure 11.12
Schematic sense amplifier circuitry for a typical 64k CMOS EPROM. (a) Static sense amplifier. (b) Clocked sense amplifier. (From Knecht [39], Philips 1983, with permission of IEEE.)

The address transition detection circuitry used in the Signetics EPROM is shown in Figure 11.11(a). The minimum address pulse width that would cause $\overline{ATD}$ to be pulled low could be varied by altering the W/L ratio of the transistors Q_2 and Q_4 in the ATD circuitry. In this case they were set so that the minimum valid address pulse width was designed to be 5 ns.

The clock generation circuit, shown in Figure 11.11(b), is designed to guard against address skew. When $\overline{ATD}$ was pulled low by one of the address buffers Q_2 was turned off and the column clock went low followed by the row clock. After the row clock has fallen, Q_1 was turned on and pulled $\overline{ATD}$ high. Since Q_1 was weaker than Q_5 in the ATD circuitry, any new address transition would keep $\overline{ATD}$ low and prolong the clock cycle. For this reason non-coincident address transitions did not cause the circuit to malfunction.

A clocked row decoder circuit used in this chip had zero static current consumption. Power was dissipated in the selected decoder only during the brief time after an address transition when the row clock was low. A conventional static sense amplifier circuit which drew dc current during sensing is shown in Figure 11.12(a) compared to the clocked sense amplifier circuit used in this 64k EPROM which is shown in Figure 11.12(b). The use of the dynamic sense amplifier greatly reduced the power consumption.

This was also one of the first of the EPROMs to incorporate special test features into the circuit to reduce test time. Another CMOS EPROM which featured on-chip test circuits was a 256k from Toshiba [40] shown in 1985. Typically testing is very time consuming for EPROMs constituting a major component of production cost. These ease-of-test features are considered at more length in the chapter on memory testing.

11.4 HIGH SPEED CMOS EPROMs

11.4.1 ATD for speed

An example of using address transition detection for improving access time was shown in a Toshiba [44] 80 ns 1Mb CMOS EPROM in 1985. In this chip a precharge pulse was used to minimize the bit-line and sense line delays. The precharge pulse was initiated from the ATD circuit and terminated by a delayed pulse through a dummy word-line. Only the selected bit-line was precharged to minimize peak current during the precharge cycle. The sense line was also equalized to a reference level. A type of divided word-line structure was used dividing the array into four blocks to reduce the word-line delay. A two-stage sense amplifier was used with a high sensitivity preamplifier which was sensitive to small voltage differentials. This minimized the required bit-line swing.

The ATD was used also to provide an automatic power-down function with the chip being automatically disabled by an internal pulse 360 ns after chip activation.

Another chip which used ATD to improve access time was an Intel 256k EPROM introduced in 1985. This part was particularly notable for its 'unerasable EPROM (UPROM)' cells which will be covered in the chapter on yield and redundancy.

Two other 1Mb EPROMs used ATD to improve speed, an 80 ns part from Fujitsu [48] and a 70 ns part from STM [49]. The STM chip used four techniques for enhanced speed. The memory array was divided into four quadrants to reduce the bit-line capacitance. The row decoder was interleaved. This helped improve performance since normally the very small pitches between cells in EPROMs require very simple row decoders that are not optimized for speed. By interleaving the row decoders SGS-T had the space of two cell pitches in which to fit an optimized row decoder. Address transition detection from column address was used for the bit-line precharge and equalization.

A fast 1Mb 55 ns part was shown by Hitachi [50]. This 64k × 16 chip used tungsten polycide to reduce word-line delay. It was shown that a single sense amplifier, in this case of a 16 bit wide EPROM, was faster than the delay through several staged sense amplifiers. Ground bounce was reduced by a special output buffer arrangement resulting in less delay in the output buffers and hence faster read operation.

11.4.2 Double-layer metal strapping for speed

A fast 23 ns 256k EPROM from Cypress [51] in 1989 used a double-layer metal technology to strap word lines in the array and to bus signals in the periphery. Bit-line length was reduced to 256 cells. In addition, speed was enhanced by the use of differential sensing, address transition detection and a ground switched decoding scheme. The address transition detection circuits were divided between the top and bottom of the chip to reduce the parasitic capacitance. These signals merged to form a composite signal that triggered the equalization signal. The outputs were latched during address transition to improve noise immunity.

Another very fast part, a 16 ns 1Mb EPROM from Toshiba [72] shown in 1990, also used double-layer metal technology to strap word-lines in the array to enhance speed. The cell array was divided into eight sections to reduce bit-line capacitance. The word-line delay was reduced by a double word-line structure similar to that used in the static RAMs. The row main decoder used a feedback circuit to guarantee high speed without reducing voltage. The section word-line was selected using a NOR gate whose inputs were a second aluminum main word-line and one of the section selection lines. The section word-lines in non-selected sections were set low to save power. The section word-line was made of MoSi to reduce delay. A three-stage differential sense amplifier scheme was used. Complementary nodes in the sense amplifier were equalized by an ATD pulse to speed sensing.

11.4.3 New transistor structures for speed

One approach to enhancing read access time and programming time was a high read current split gate cell developed by Waferscale Integration [52] in 1987. The cross-section of this cell is shown in Figure 11.13(a) and the equivalent circuit in Figure 11.13(b). The cell was comprised of a MOS transistor in series with a floating gate transistor merged into a single composite device which was essentially a 1.5-transistor cell.

The cell had a high read current of over 150 μA which provided access times as low as 35 ns and a fast programming rate of less than 1 ms. The programming mechanism of the split gate EPROM cell was channel hot-electron injection. The fast programming rate was due to the short channel length of the floating gate transistor.

A two-transistor cell high speed 16k CMOS EPROM was introduced to the market by TI [70] in 1987. The two transistors were used for differential sensing. Because of the differential cell the sense amplifier could detect a programmed or unprogrammed state with a smaller voltage swing which implied a faster access time.

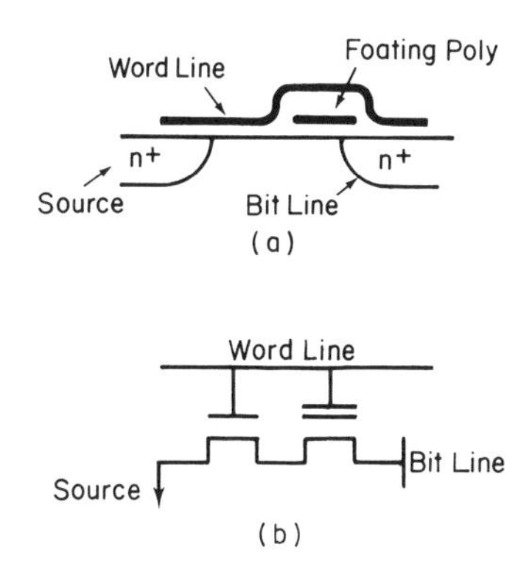

Figure 11.13
Schematic cross-section (a) and equivalent circuit diagram (b) of a high read current split gate cell with the select transistor integrated in series with the floating gate transistor. (From Ali *et al.* [52], Waferscale Integration 1988, with permission of IEEE.)

The differential sensing scheme also meant that the outputs of the unprogrammed EPROM were in an indeterminate state which can be sensed without programming by using a 'blank check' mode which will be described in the section of Chapter 14 on EPROM programming. This eased the test difficulties of one-time-programmable EPROMs. The drawback was that both the logic '0' and the logic '1' states need to be programmed.

Another two-transistor cell for high speed EPROMs was developed by Cypress [70]. This cell had separate read and program transistors with the floating gates connected. It used a single-ended sensing scheme. The sense amplifier used a reference voltage on one input and the read transistor on the other. This single-ended sensing had the effect of causing an erased device to contain all '0'.

For lower density high speed EPROMs, Cypress [68, 69] used a four-transistor differential memory cell. This cell was optimized for high read current and fast programmability. This was accomplished by separating the read and program transistors as shown in Figure 11.14(a). The program transistor had a separate implant to maximize the generation and collection of hot electrons while the read transistor implant dose was chosen to provide a large read current. The problem of injecting unwanted charge onto the floating gate during the read operation was eliminated since the read and program transistors were separate and the program transistor drain could be grounded during the read operation.

The two sets of read and program transistors were used for differential sensing with a three-stage sense amplifier which is shown in Figure 11.14(b). The read and program paths were separated to optimize speed. The *X* and *Y* decoding paths were predecoded to optimize the power delay product.

These fast EPROMs could be used in 'zero wait state' systems with high performance microprocessors. An example of this requirement for various processor clock frequencies [70] is shown in Table 11.5.

These lower density high speed EPROMs competed with bipolar PROMs in applications such as control stored in bit-slice based designs and in statemachine

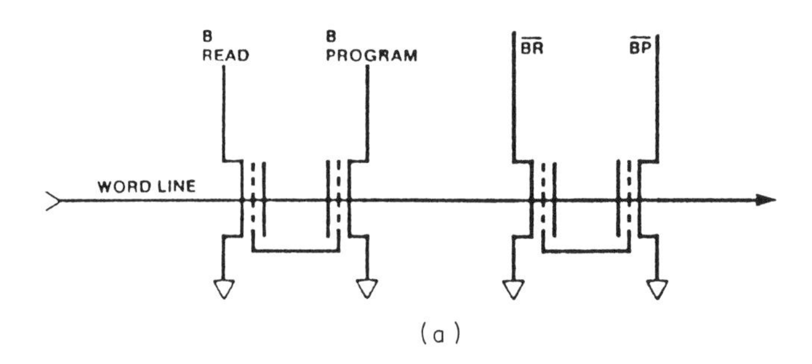

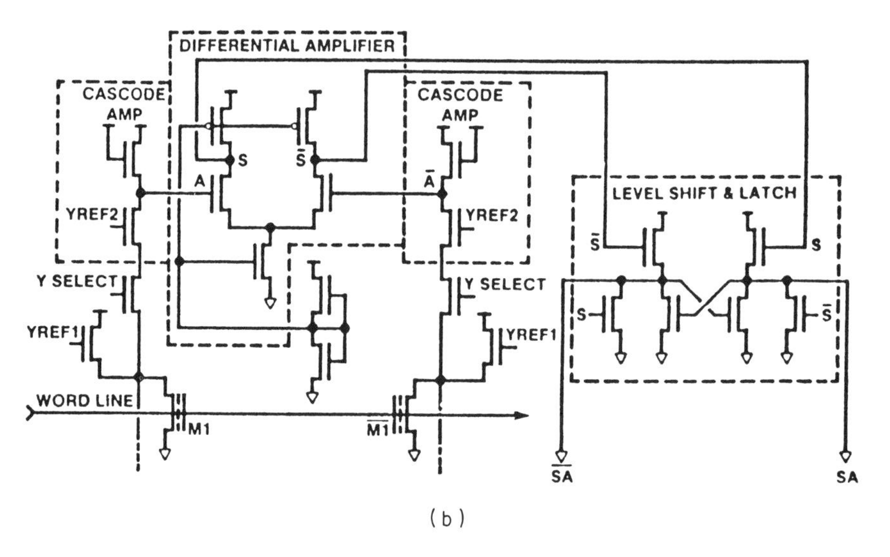

Figure 11.14
(a) Circuit diagram showing four-transistor cell for fast low density 16k EPROM with read and program transistors implanted separately. (b) Circuit schematic showing differential sensing scheme with four-transistor cell. (From Pathak *et al.* [68], Cypress 1985, with permission of IEEE.)

Table 11.5

Microprocessor	Clock frequency (MHz)	EPROM access time (Max. in ns)
80386	16	70
68020	20	70
32020	20	75
56000	20	55
320C25	40	

implementations replacing random logic. These parts were discussed from the point of view of custom programmability in Chapter 10.

While the access times of the CMOS and bipolar parts were the same, the power dissipation of the CMOS EPROMs was significantly less. A comparison of the dc electrical characteristics of a 64k CMOS OTP EPROM and a 64k TTL bipolar PROM, with similar 45 and 55 ns access times, are shown in Figure 11.15. These fast CMOS EPROMs had the pinout configurations of the bipolar PROMs which were different from those of the commodity EPROMs of the same densities.

11.5 PROGRAMMING TIME AND TEST MODES IN CMOS EPROMs

A 64k EPROM shown by Signetics in 1983 included several special test modes. A gang programming mode allowed two or four bytes to be programmed in parallel in order to reduce the total device programming time. In this mode each data input buffer was connected to four columns allowing 32 EPROM cells to be programmed simultaneously. The parallel programming mode was initiated by applying a high voltage greater than 12 V to two address pins.

This chip also included a full array stress mode which allowed all word-lines to be held simultaneously at V_{PP}. This subjected all EPROM cells to V_{PP} stress at the same time. The burn-in time was thus reduced by a factor of 256 while maintaining a constant accumulative voltage stress time for each cell. This mode was initiated by applying a greater than 12 V input to CE.

A test for individual cell threshold was also included which allowed the threshold of an individual cell to be measured while the peripheral circuits continued to operate with their standard supply voltages.

A Toshiba [43] 256k CMOS EPROM shown in 1984 reduced the programming time to 0.003 ms/byte, which was several orders of magnitude less than had previously been available. This was accomplished with a new cell structure that they called the 'Diffusion Self-Aligned' cell (DSA). In this cell, which is shown in Figure 11.16(a), the high ion concentration region was localized in the vicinity of the drain while the rest of the channel region was lightly doped like the peripheral transistors. This cell could both program faster and draw more current than a conventional cell.

A graph which shows programming time characteristics for the Toshiba cell compared to those of a conventional EPROM cell is given in Figure 11.16(b). An 8k byte simultaneous programming circuit was also included to speed up programming. Special on-chip test circuits were also included on this part and on a second Toshiba 256k in 1985. These test circuits are described in the test chapter.

Programming speed was improved at the 1Mb level by a Hitachi [45] part that had fast page mode programming. In page mode programming the EPROM accepted four bytes of input data sequentially and programmed them all at one time as shown in the timing diagram in Figure 11.17(a). The four bytes of input data were stored in 32 latches. Gates of switching transistors on the high voltage paths to the memory bit lines were controlled by latch outputs when the program enable signal was applied as shown in the schematic circuit diagram of the programming circuit in Figure 11.17(b).

Symbol	Parameter	Test conditions	Limits Min.	Typ.	Max.	Unit
Input Current						
I_{IH}	High	$V_{IN} = V_{CC}$			10	μA
I_{IL}	Low	$V_{IN} = 0.45$ V			10	μA
Output Current						
I_{LO}	Leakage	$V_{OUT} = 0$ to V_{CC}			± 10	μA
I_{OS}	Output short-circuit current	$V_{OUT} = 0$ V, $\overline{G} = V_{IL}$	-15		-70	mA
Supply Current						
I_{CC}	V_{CC} operating current	$\overline{G} = V_{IH}$, $O_{1-8} = 0$ mA, $f = 20$ MHz			110	mA
Capacitance						
		$f = 1$ MHz, $T_A = 25°C$ $V_{CC} = 5.0$ V				
C_{IN}	Input	$V_{IN} = 0$ V			6	pF
C_{OUT}	Output	$V_{OUT} = 5.0$ V			12	pF

(a) Dc electrical characteristics: $0°C \leqslant T_A \leqslant +70°C$, $4.5\text{ V} \leqslant V_{CC} \leqslant 5.5$ V.

Symbol	Parameter	Test conditions	Limits Min.	Typ.	Max.	Unit
Input current						
I_{IL}	Low	$V_{IN} = 0.45$ V			-250	μA
I_{IH}	High	$V_{IN} = 5.25$ V			40	μA
Output current						
I_{OZ}	Hi-Z State	$\overline{CE}_1 = $ High, $V_{OUT} = 0.5$ V			-40	μA
		$\overline{CE}_1 = $ High, $V_{OUT} = 5.25$ V			40	
I_{OS}	Short circuit	$\overline{CE}_1 = $ Low, $V_{OUT} = 0$ V	-15		-70	mA
Supply current						
I_{CC}		$V_{CC} = 5.25$ V		130	175	mA
Capacitance						
		$\overline{CE}_1 = $ High $V_{CC} = 5.0$ V				
C_{IN}	Input	$V_{IN} = 2.0$ V		5		pF
C_{OUT}	Output	$V_{OUT} = 2.0$ V		8		pF

(b) Dc electrical characteristics: $0°C \leqslant T_A \leqslant +75°C$, $4.75\text{ V} \leqslant V_{CC} \leqslant 5.25$ V.

In conventional programming mode the latches are bypassed and the input data controls the switching transistors.

Other parts that used a fast programming mode similar to the page mode programming described above included a 64k × 16 AMD [46] chip shown in 1986. In this case two 16 bit words were stored in data-in latches and simultaneously programmed into two locations in the memory. TI [47] also used programming of 32 bits of data in parallel in their 1Mb shown in 1987 to allow the array to be programmed in less than 15 seconds. The program timing cycle of the TI part will be discussed in the test chapter.

The STM [49] 1Mb mentioned in the previous section used an offset current approach to speed up programming. This method entailed a reference cell current being applied to a P-channel load while the array current is applied to a mirror load. The reference cell is equal to a virgin cell and the load size is the same. The current imbalance necessary to sense a virgin cell was provided by adding an offset current on the matrix side. The voltage difference on the output nodes was sensed by a differential amplifier.

This offset current method can be compared to the load ratio approach which is frequently used in which the imbalance to sense a virgin cell is obtained by connecting a number of loads on the reference side while the mirror load on the matrix side is unique. This method divides the reference cell current by the load ratio. Schematic circuit diagrams of the two approaches are shown in Figures 11.18(a) and (b).

The erased and written cell currents and the virtual reference current are plotted in Figures 11.19(a) and (b). A matrix cell with current below the virtual reference current will be sensed as a '0' when one with higher current will be sensed as a '1'. In the load ratio approach, a cell written with a limited threshold shift at a given gate voltage will be sensed incorrectly as a '1' since the reference and write currents cross. In the offset current approach the three characteristics run parallel making the sensing theoretically independent of the gate voltage and of V_{CC}. This means that a limited threshold shift is sufficient for sensing which speeds up the required programming time.

Several other parts with fast utilizing fast programming techniques are discussed in the next section on latch-up.

11.6 LATCH-UP PROTECTION TECHNIQUES IN CMOS EPROMs

Latch-up needs to be guarded against in CMOS EPROMs due to the high voltages incurred during programming. Several of the standard techniques for reducing the possibility of latch-up have already been mentioned. These include the use of an epitaxial substrate or the use of guard rings around the PMOS transistors.

SEEQ [42], in 1984, announced one of the first CMOS EPROMs with 0.5 ms/byte programming time, a factor of 10 improvement over the previous 5 ms/byte.

Figure 11.15
Dc electrical characteristics of fast PROMs. (a) 64k CMOS EPROM. (b) 64k bipolar PROM. (From Philips Memory Databook 1989.)

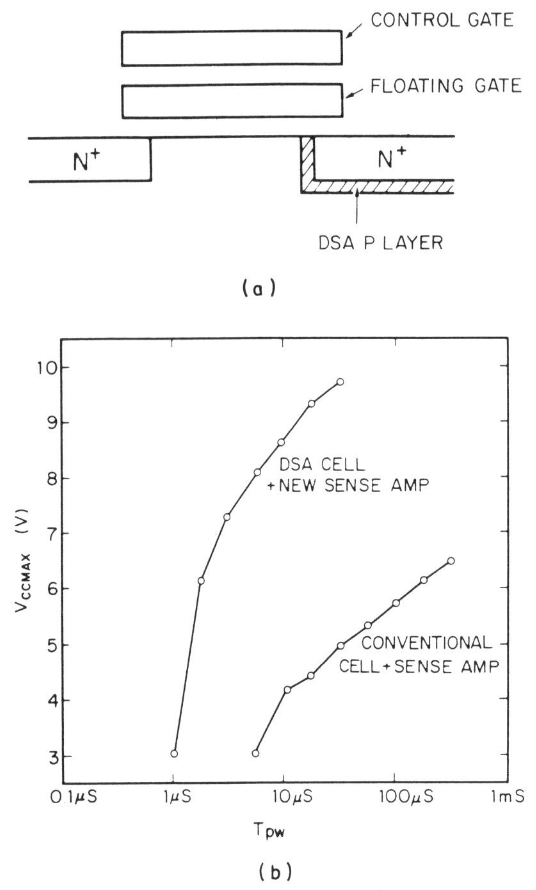

Figure 11.16
256k CMOS EPROM with new cell for rapid programming. (a) Circuit schematic of rapid programming cell using a high ion concentration drain structure. (b) Programming characteristics compared with a conventional EPROM cell. (From Tanaka *et al.* [43], Toshiba 1989, with permission of IEEE.)

However, as a result, they had to take careful precautions against latch-up during programming. All of the traditional protections against latch-up were implemented including using an epitaxial layer, and employing guard rings around the array and in the input and output circuitry.

Hitachi [41] in 1985 developed a 256k EPROM which operated with the EPROM transistor biased in the snap-back region. This shortened the programming time by a factor of 10 to 0.5 ms/byte, but resulted in large substrate currents which could trigger latch-up. Their method of reducing the potential for latch-up was simply to get the PMOS transistor in the high voltage programming circuit out of the substrate.

To accomplish this they used a polysilicon PMOS transistor as has already been described in the chapter on future generation SRAMs. A cross-section of the polysilicon PMOS transistor and the NMOS memory cell is shown in Figure 11.20(a).

A schematic circuit diagram of the programming circuit using the polysilicon PMOS transistor is shown in Figure 11.20(b). In the programming mode $\overline{PGM}$ and $\overline{DIN}$ went low and the polysilicon PMOS was activated. The selected data switching and Y-select transistors were pulled up to an internally generated high voltage and the word-line was pulled up to V_{PP}.

11.7 HIGH DENSITY EPROM ORGANIZATIONS AND STANDARDS

Since EPROMs have traditionally been used in microprocessor and microcontroller based systems, they have had the bytewide organization. As 16 bit and 32 bit microprocessors became more common there was a requirement for wider EPROMs

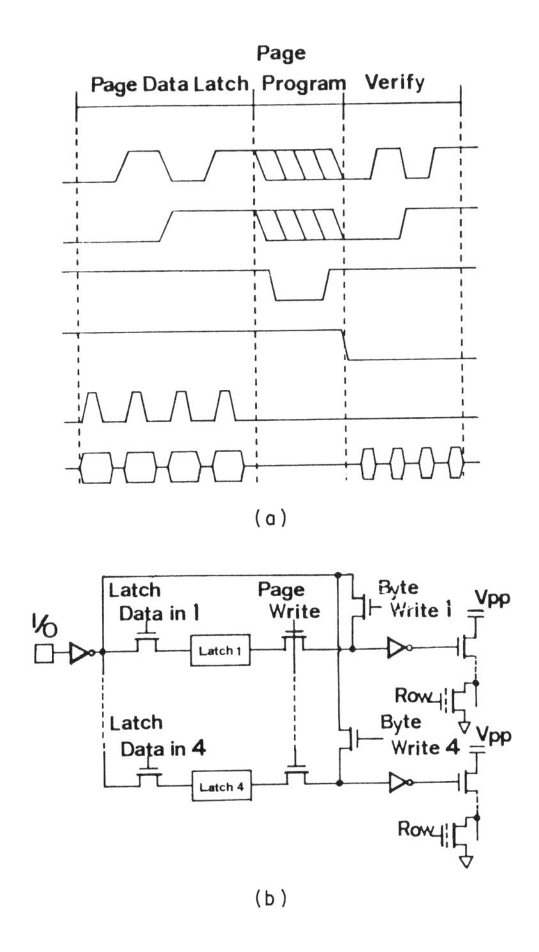

Figure 11.17
1Mb EPROM with fast page mode programming. (a) Timing diagram of fast page mode showing four byte data latching and programming. (b) Program schematic circuit showing data latches controlling memory bit-lines. (From Hagiwara and Matsuo [45], Hitachi 1985, with permission of IEEE.)

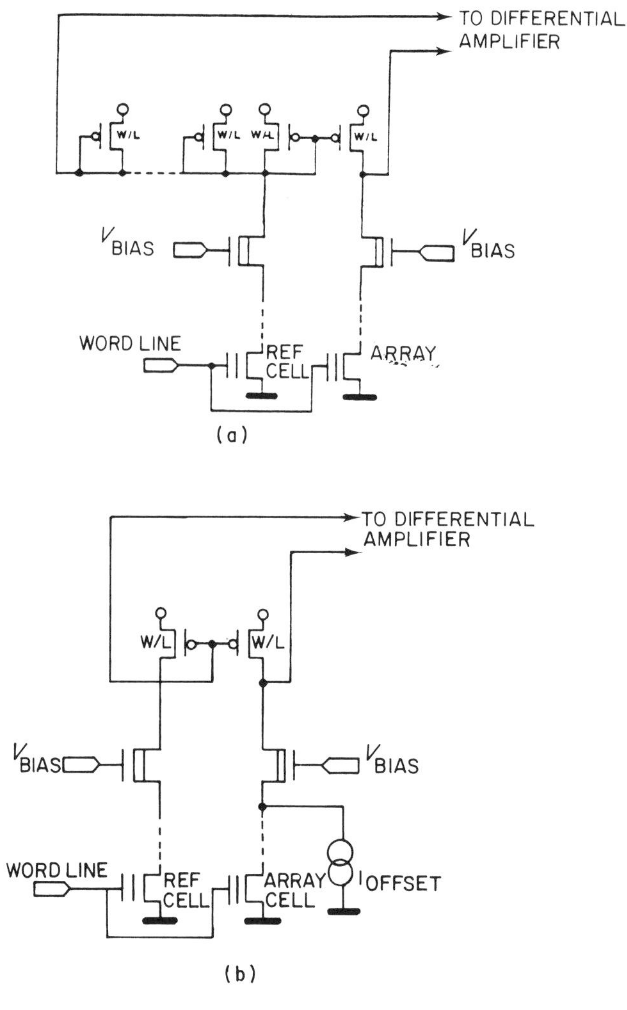

Figure 11.18
Two different approaches to EPROM sensing. (a) Load ratio sensing. (b) Offset current sensing. (From Gastaldi *et al.* [49], STM 1988, with permission of IEEE.)

also; however, the larger packages that went with an increase in number of outputs made the chips more expensive. There were considerations of trying to reduce the number of leads by, for example, multiplexing the address and data pins. These were not widely produced.

11.7.1 Byte-wide EPROM organizations and standards

Pinout standards from the JEDEC Standard 21C for byte-wide EPROMs in DIP are shown in Figure 11.21. Figure 11.21(a) is the low density for 8k byte to 64k byte

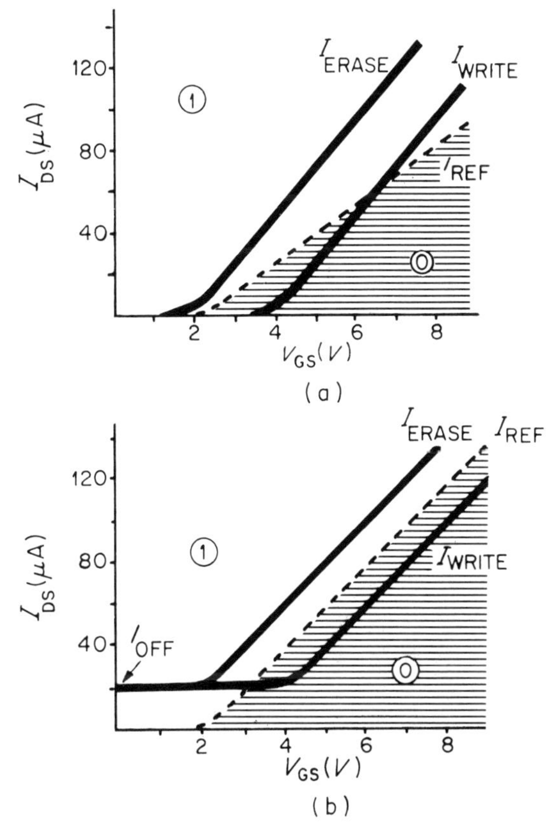

Figure 11.19
I_{DS} plotted against V_{GS} characteristics for EPROMs using different sensing schemes using (a) load ratio sensing and (b) offset current sensing. (From Gastaldi *et al.* [49], STM 1988, with permission of IEEE.)

density, and Figure 11.21(b) is the higher density standard for 128k bytes to 1M bytes. While the two do not overlap note that the location of the V_{PP} pin has changed. The corresponding JEDEC standard for the SOJ package is shown in Figure 11.22(a). This runs from 32k bytes to 512k bytes and follows the high density placement standard for V_{PP}. A dual standard pinout with the DIP package therefore existed at the 32k and 64k byte levels. Many manufacturers, wanting to supply the same chip in both packages, followed the SOJ pinout also in DIP. This resulted in a *de facto* double standard for the two generations of EPROMs.

11.7.2 Word-wide EPROM organizations and standards

One of the first of the word-wide, 16 bit wide, EPROMs was the 1985 Toshiba [44] 1Mb described above in the section on EPROM speed. This part had an optional $\times 8$ and $\times 16$ organization. When operated in the byte mode the part used lower byte

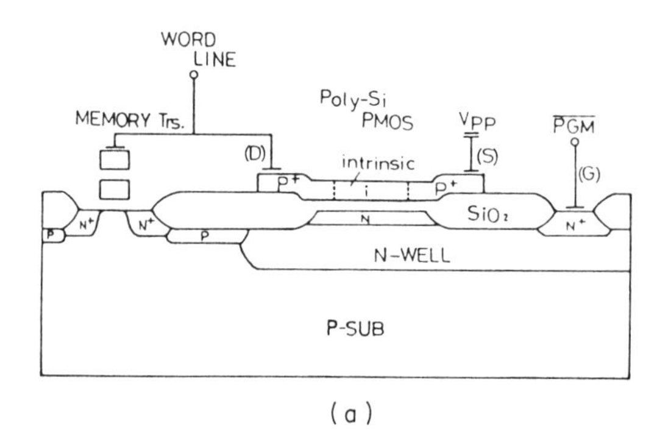

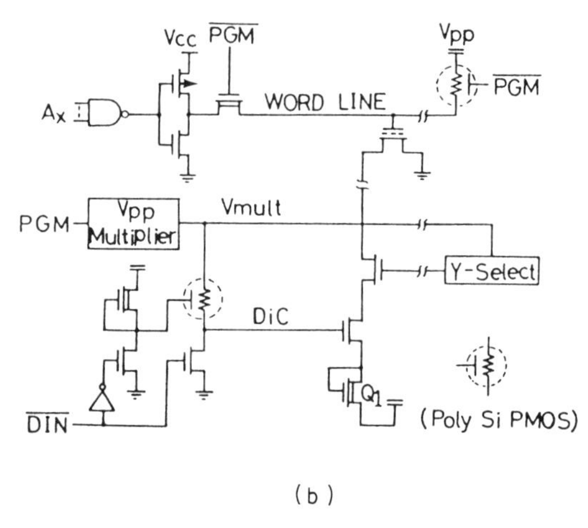

Figure 11.20
256k EPROM using polysilicon PMOST to reduce potential for latch-up with high voltage programming in the snapback region of the storage transistor. (a) Cross-section of polysilicon PMOS and memory cell. (b) High voltage programming circuits. (From Yoshizaki *et al.* [41], Hitachi 1985, with permission of IEEE.)

(LB) and upper byte ($\overline{UB}$) pins. The word-wide organization became common at the 1Mb EPROM density for ease of use with 16 bit processors.

The 1986 AMD 1Mb tried a change in the placement of the lower V_{SS} power pins to the center of the package to minimize inductive ground bounce effects. This concept was standardized in the JEDEC Standard 21C for word-wide EPROMs as shown in Figure 11.22b.

A pinout for a byte-wide page select EPROM also appears in the JEDEC standard 21C along with functional truth tables for page selection.

An address-data multiplexed word-wide 1Mb EPROM was also sourced in the small 28 pin package. One such chip from Fujitsu [48] follows the JEDEC standard pinout discussed in the ROM chapter. The pin assignment for this part is shown in Figure 11.23(a) showing the electrically changeable control signal $\overline{BHE}$ and the ADD pin for word or byte transfer. A schematic block diagram is shown in Figure 11.23(b) and the functional and output truth tables are shown in Figure 11.23(c).

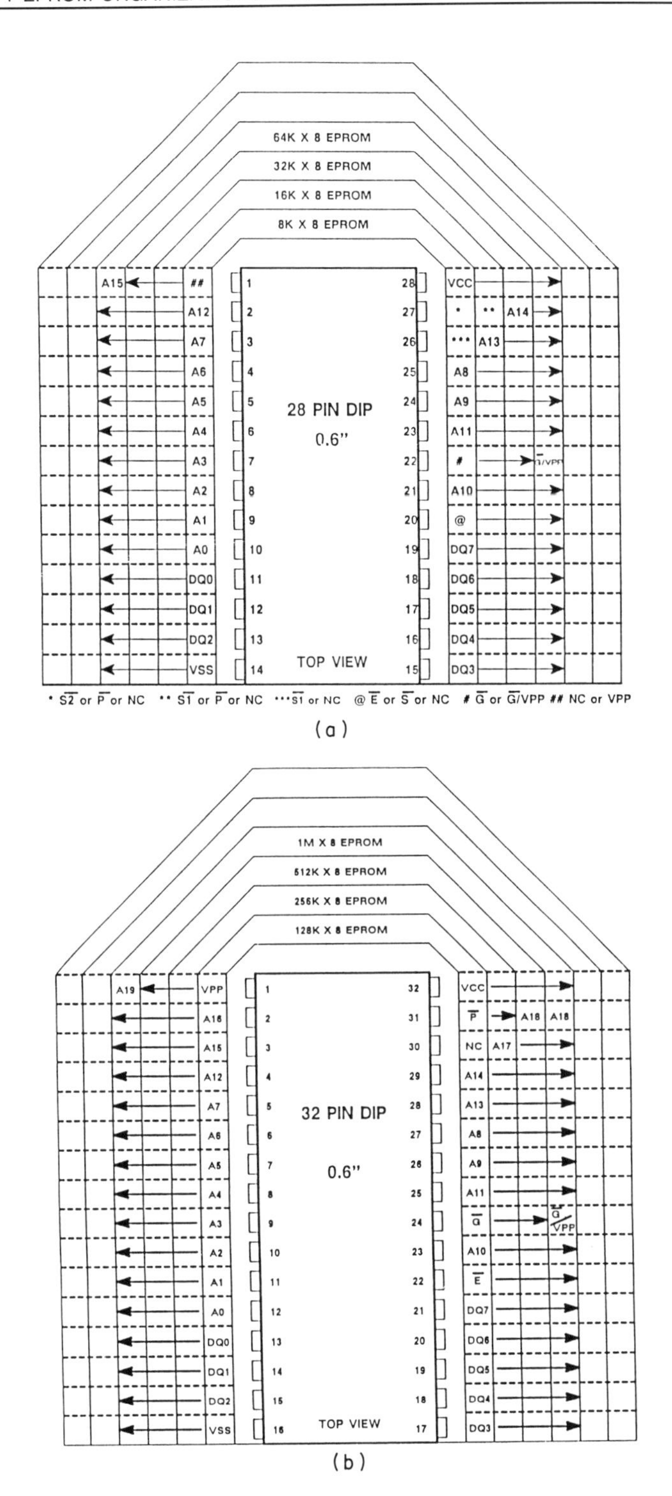

Figure 11.21
Byte-wide EPROM standards in dual-in-line-package (DIP). (a) 8k bytes to 64k bytes, (b) 128k bytes to 1M bytes. (Source JEDEC Standard 21-C, with permission of EIA.)

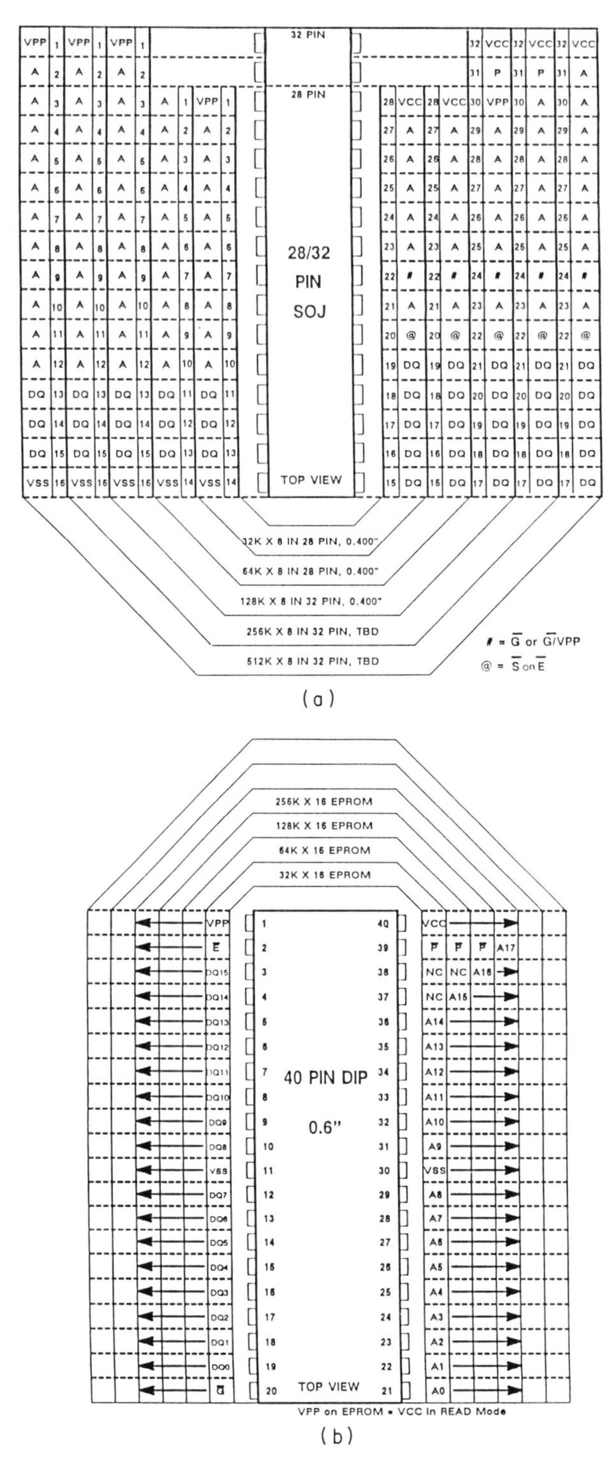

Figure 11.22
(a) Byte-wide EPROM standards in small outline 'J' leaded package (SOJ) 32k bytes to 512k bytes. (b) Word-wide (16 bits) standard with center ground pins in DIP. (Source JEDEC Standard 21-C, with permission of EIA.)

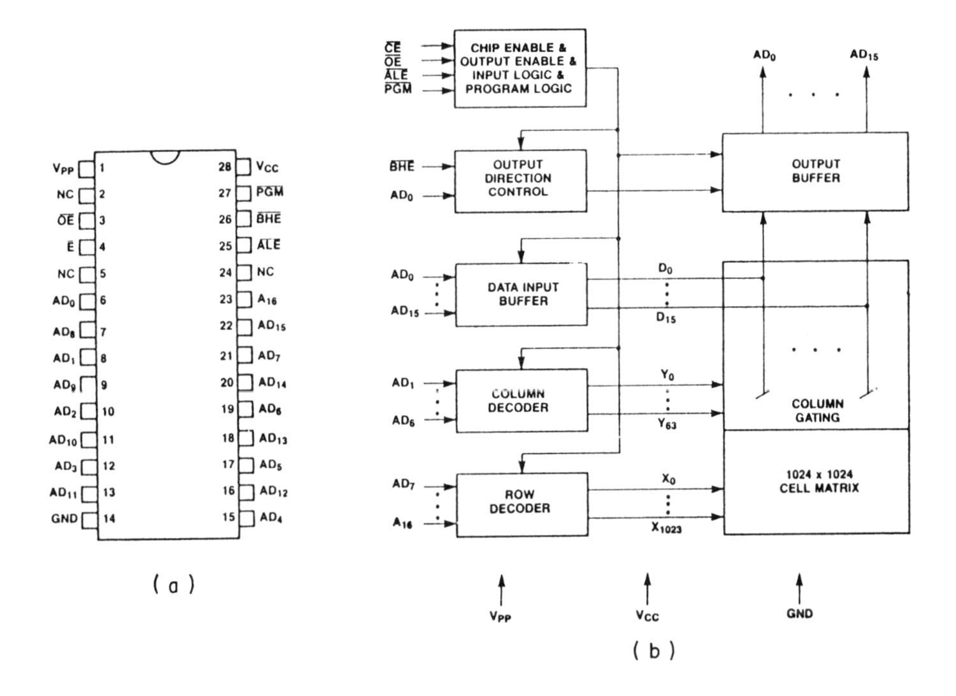

FUNCTIONAL TRUTH TABLE

	AD_0	A_1–A_{15}	A_{16}	$\overline{ALE}$	$\overline{BHE}$	$\overline{E}$	$\overline{OE}$	$\overline{PGM}$	V_{CC}	V_{PP}
STANDBY	HIGH-Z	HIGH-Z	X	X	X	V_{IH}	X	X	5V	5V
READ ADDRESS LATCH	•	A_{IN}	A_{IN}	↓	•	V_{IL}	V_{IH}	V_{IH}	5V	5V
READ	OUT*	OUT*	X	V_{IL}	X	V_{IL}	V_{IL}	V_{IH}	5V	5V
OUTPUT DISABLE	HIGH-Z	HIGH-Z	X	V_{IL}	X	V_{IL}	V_{IH} / X	X / V_{IL}	5V	5V
PROGRAM ADDRESS LATCH	V_{IL}	A_{IN}	A_{IN}	↓	V_{IL}	V_{IL}	V_{IH}	V_{IH}	6V	12.5V
PROGRAM	IN	IN	X	V_{IL}	X	V_{IL}	V_{IH}	V_{IL}	6V	12.5V
PROGRAM VERIFY	OUT	OUT	X	V_{IL}	X	V_{IL}	V_{IL}	V_{IH}	6V	12.5V
PROGRAM INHIBIT	HIGH-Z	HIGH-Z	X	X	X	V_{IH}	X	X	6V	12.5V

X: Either V_{IL} or V_{IH}
A_{IN}: Address Input
IN: Data Input
OUT: Data Output

*OUTPUT TRUTH TABLE

AD_0	$\overline{BHE}$	AD_0–AD_7	AD_8–AD_{15}
V_{IL}	V_{IL}	D_0–D_7	D_8–D_{15}
V_{IL}	V_{IH}	D_0–D_7	HIGH-Z
V_{IH}	V_{IL}	HIGH-Z	D_8–D_{15}
V_{IH}	V_{IH}	HIGH-Z	HIGH-Z

(c)

Figure 11.23
Address-data multiplexed word-wide 1Mb EPROM. (a) JEDEC standard pinout. (b) Schematic block diagram. (c) Functional and output truth tables. (From Yoshida *et al.* [48], Fujitsu 1987, with permission of IEEE.)

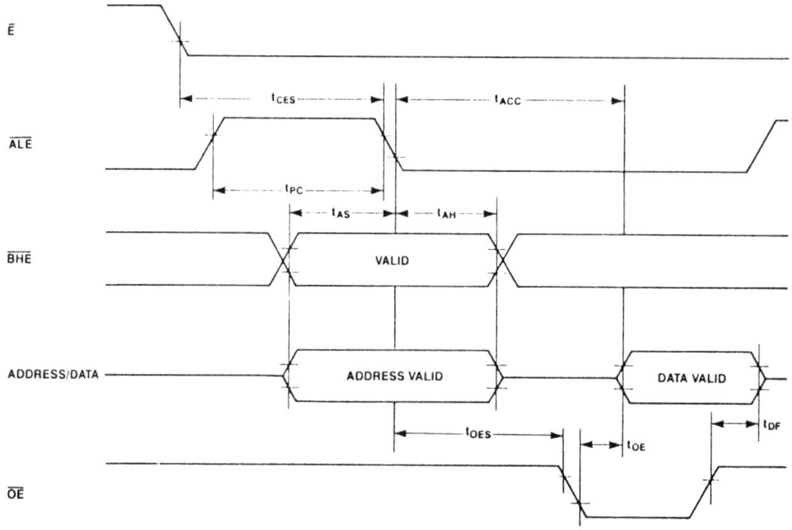

$\bar{E}$

t_{CES} t_{ACC}

$\overline{ALE}$

t_{PC} t_{AS} t_{AH}

$\overline{BHE}$ VALID

ADDRESS/DATA ADDRESS VALID DATA VALID

t_{OES} t_{OE} t_{DF}

$\overline{OE}$

(a)

t_{CES}

$\bar{E}$ V_{IH} V_{IL}

t_{PC} t_{ACC} t_{ALH}

$\overline{ALE}$ V_{IH} V_{IL}

t_{AS} t_{AH}

$\overline{BHE}$ V_{IH} V_{IL} LOW HIGH LOW HIGH

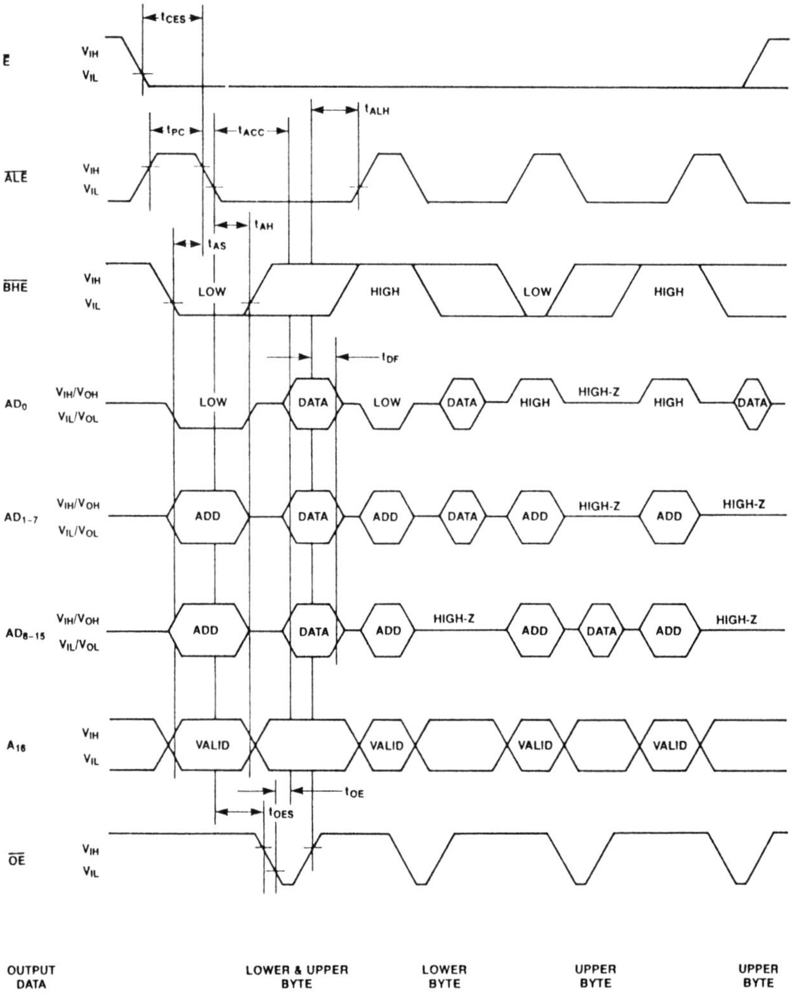

t_{DF}

AD_0 V_{IH}/V_{OH} V_{IL}/V_{OL} LOW DATA LOW DATA HIGH HIGH-Z HIGH DATA

AD_{1-7} V_{IH}/V_{OH} V_{IL}/V_{OL} ADD DATA ADD DATA ADD HIGH-Z ADD HIGH-Z

AD_{8-15} V_{IH}/V_{OH} V_{IL}/V_{OL} ADD DATA ADD HIGH-Z ADD DATA ADD HIGH-Z

A_{16} V_{IH} V_{IL} VALID VALID VALID VALID

t_{OE} t_{OES}

$\overline{OE}$ V_{IH} V_{IL}

OUTPUT DATA — LOWER & UPPER BYTE — LOWER BYTE — UPPER BYTE — UPPER BYTE

(b)

The read cycle timing diagrams are shown in Figure 11.24. Figure 11.24(a) shows the read cycle for the word-wide organization and Figure 11.24(b) shows the read cycle for the byte-wide organization showing transfer from upper to lower byte. This part also used the UPROM cell to program the redundant circuit elements.

11.8 A TYPICAL 1Mb CMOS EPROM IN TWO ORGANIZATIONS

The 1Mb CMOS EPROM was offered by most manufacturers in both the 128k × 8 and the 64k × 16 organizations. The parts used 5 V read and had access times ranging from 100 to 300 ns. Programming voltage (V_{PP}) was typically 12.5 V. Standard pinouts for the 128k × 8 and 64k × 16 parts as shown above were used throughout the industry.

A typical second generation 1Mb EPROM from STM operated from a single +5 V power supply, had 120 ns read access time, 50 mA maximum active current during read at 120 ns and 1 mA maximum standby current. Programming voltage (V_{PP}) was 12.5 V.

For use with 10 MHz processors the part had separate chip enable and output enable control as shown in the pinouts for the byte-wide and word-wide parts in Figure 11.25(a). The addition of output enable was intended to eliminate bus contention in multiple bus microprocessor systems. A schematic block diagram is shown in Figure 11.25(b) and an operating mode truth table in Figure 11.25(c) including the usual read standby and program modes. An electronic signature mode was also included for vendor identification.

Dc operating characteristics and ac timing conditions for the read operation are shown in Figures 11.26(a) and (b). The fast selection has 120 ns address and chip enable access time and 60 ns output enable access time permitting the part to be used in high performance processor systems. A read mode timing diagram is shown in Figure 11.26(c).

In the read mode both chip enable and output enable must be active. Chip enable is the power control and should be used for device selection. Output enable is the output control and should be used to control the data to the output pins independent of whether the part is selected or not.

Various programming modes for byte-wide and word-wide EPROMs are described in detail in Chapter 13 in the section on EPROM test.

Figure 11.24
Address-data multiplexed word-wide 1Mb EPROM. (a) Read cycle for the word-wide organization. (b) Read cycle for the byte-wide organization showing transfer from upper to lower byte. (From Yoshida *et al.* [48], Fujitsu 1987, with permission of IEEE.)

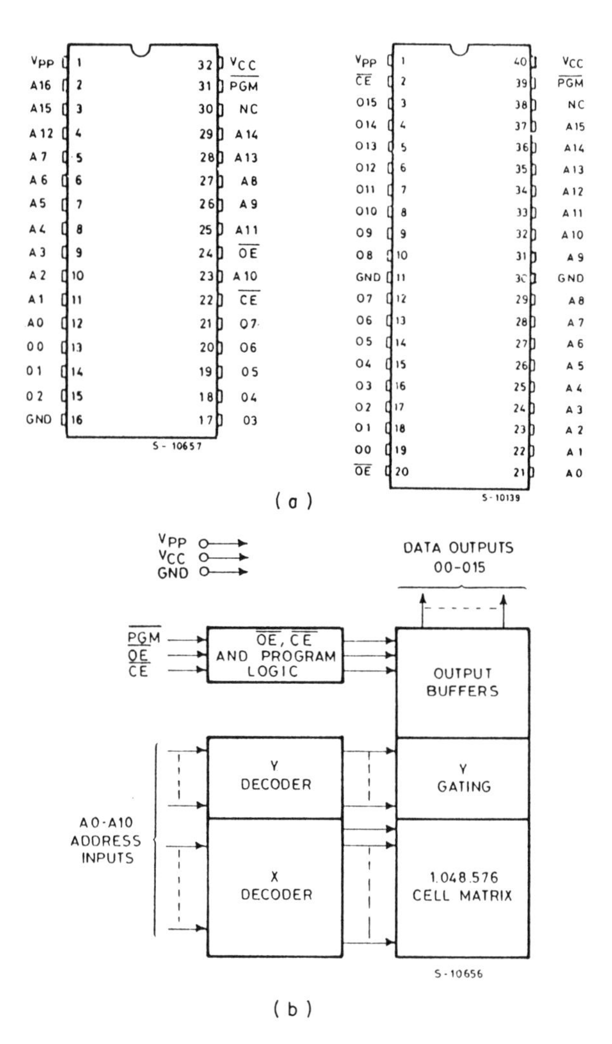

OPERATING MODES

MODE \ PINS	$\overline{CE}$ (2)	$\overline{OE}$ (20)	A9 (31)	$\overline{PGM}$ (39)	V_{PP} (1)	OUTPUTS
READ	L	L	X	H	V_{CC}	D_{OUT}
OUTPUT DISABLE	L	H	X	H	V_{CC}	HIGH Z
STANDBY	H	X	X	X	V_{CC}	HIGH Z
PROGRAM	L	X	X	L	V_{PP}	D_{IN}
PROGRAM VERIFY	L	L	X	H	V_{PP}	D_{OUT}
PROGRAM INHIBIT	H	X	X	X	V_{PP}	HIGH Z
ELECTRONIC SIGNATURE	L	L	V_H	H	V_{CC}	CODE

NOTE: X = DON'T CARE; V_H = 12V ±0.5V; H = HIGH; L = LOW

(c)

Figure 11.25
Typical 1Mb EPROM pinouts. (a) Byte-wide organization and word-wide organization, (b) schematic block diagram, (c) operating mode truth table. (Source STM Memory Databook.)

Selection Code	−12XF1/−15XF1 −20XF1/−25XF1	−12F1/−15F1 −20F1/−25F1	−15XF6/−20XF6 −25XF6
Operating Temperature Range	0 to 70°C	0 to 70°C	−40 to 85°C
V_{CC} Power Supply (1,2)	5V ±5%	5V ±10%	5V ±5%
V_{PP} Voltage (2)	$V_{PP} = V_{CC}$	$V_{PP} = V_{CC}$	$V_{PP} = V_{CC}$

Symbol	Parameter	Test Conditions	Values			Unit
			Min.	Typ. (2)	Max.	
I_{LI}	Input Load Current	$V_{IN} = 5.5V$			10	μA
I_{LO}	Output Leakage Current	$V_{OUT} = 5.5V$			10	μA
I_{CC1}	V_{CC} Current Standby	$\overline{CE} = V_{IH}$			1	mA
I_{CC2}	V_{CC} Current Active	$\overline{CE} = \overline{OE} = V_{IL}$ @f = 8MHz		20	50	mA
V_{IL}	Input Low Voltage		−0.1		+0.8	V
V_{IH}	Input High Voltage		2.0		$V_{CC} + 0.5$	V
V_{OL}	Output Low Voltage	$I_{OL} = 2.1$ mA			0.45	V
V_{OH}	Output High Voltage	$I_{OH} = -400$ μA	2.4			V

(a)

Symbol	Parameter	$V_{CC} \pm 10\%$	27C1024-12/		27C1024-15/		27C1024-20/		27C1024-25/		Unit
		$V_{CC} \pm 5\%$	27C1024-12X		27C1024-15X		27C1024-20X		27C1024-25X		
		Test Condition	Min	Max	Min	Max	Min	Max	Min	Max	
t_{ACC}	Address to Output Delay	$\overline{CE} = \overline{OE} = V_{IL}$		120		150		200		250	ns
t_{CE}	$\overline{CE}$ to Output Delay	$\overline{OE} = V_{IL}$		120		150		200		250	ns
t_{OE}	$\overline{OE}$ to Output Delay	$\overline{CE} = V_{IL}$		60		60		75		75	ns
$t_{DF(3)}$	$\overline{OE}$High to Output Float	$\overline{CE} = V_{IL}$	0	60	0	60	0	90	0	90	ns
t_{OH}	Output Hold from Address $\overline{CE}$ or $\overline{OE}$ Whichever Occurred First	$\overline{CE} = \overline{OE} = V_{IL}$	0		0		0		0		ns

(b)

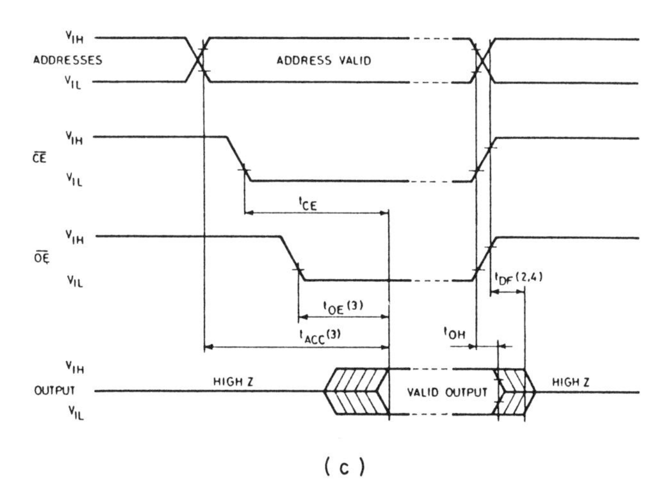

(c)

Figure 11.26
Typical 1Mb EPROM read operation. (a) Maximum and dc operating characteristics, (b) ac timing conditions, (c) ac waveforms. (Source STM Memory Databook.)

11.9 ELECTRONIC SIGNATURE MODE

Electronic signature, developed by SEEQ, is a method by which vendor identification can be indicated electrically on the chip. Signature mode provides access to a binary code identifying the manufacturer and device type.

An example of the use of signature mode on a 1Mb EPROM is given in Table 11.6 which shows the binary manufacturer code and device code for TI's 1Mb EPROM. The hex identifiers are in accordance with ANSI/IEEE Standard 91-1984 and IEC Publication 617–12. The mode is activated when A_9 (pin 26) is forced to 12.0 V. The two identifier bytes are accessed by toggling A_0. All addresses must be held low.

Table 11.6 Signature mode†.

	Pins S									
Identifier†	A_0	Q_8	Q_7	Q_6	Q_5	Q_4	Q_3	Q_2	Q_1	HEX
Manufacturer code	V_{IL}	1	0	0	1	0	1	1	1	97
Device code	V_{IH}	1	0	0	0	0	1	1	0	86

† $\overline{E} = \overline{G} = V_{IL}$, $A_9 = V_H$,A_1–$A_8 = V_{IL}$, A_{10}–$A_{15} = V_{IL}$, $V_{PP} = V_{CC}$, $\overline{PGM} = V_{IH}$ or V_{IL}.

11.10 MULTIMEGABIT AND FUTURE EPROMs

The demands for high density non-volatile, programmable memory continue to increase. The high performance microprocessors require fast read and write, and prefer the 'word-wide' organization which was established as an industry organization at the 1Mb level. New equipment for office automation, medical and communications equipment continue to develop including fast language translators, optical character recognition systems, on-line frame storage for medical and communication systems. EPROMs are used both in ROM prototyping and as ROM substitute devices in volume applications.

Test cost which has always been a significant factor in EPROM cost has become more so since it depends strongly on the programming time of the individual bit and is proportional to the number of bits in the memory. Test of the OTP EPROM is very difficult since the device cannot be enabled once it is in the UV impenetrable plastic package. Thus self-test techniques had to be developed and will be covered in the chapter on memory test.

Read performance is limited by the word- and bit-line delays and which increase as the number of cells increases and demand innovative architectures.

Manufacturing yield is impacted by the reliance of the EPROMs on oxide integrity which thins continually with scaling to smaller cell sizes. This led to a requirement for redundancy which will be covered in the chapter on manufacturing yield.

Table 11.7 Technological characteristics of early CMOS EPROMS 4Mb–16Mb.

Date	Company	Density	Channel (μm)	Cell (μm^2)	Chip (mm^2)	Gateox (nm)	Interox (nm)
1987	Toshiba	4Mb	0.9	9.0	87.4	20	30
1988	Intel	4Mb	1.0	11.9	83.4	25	28
1989	Toshiba	4Mb	0.9	9.0	86.0	20	30
1990	Hitachi	4Mb	0.8	7.8	75.3		
1990	NEC	16Mb	0.9	3.6	121.4	20	20

Table 11.8 Electrical characteristics of early CMOS EPROMS 4Mb–16Mb. (Source 1987–1990 ISSCC Proceedings.)

Date	Company	Density	Org	V_{PP} (V)	T_{PP} (s)	I_{AC} (MHz) (mW)	I_{SB} (μW)	t_{ACS} (25) (ns)
1987	Toshiba	4Mb	512k × 8	12.5		5(1)	0.05	120
1988	Intel	4Mb	256k × 16	12.5	3	100(1)	50	90
1989	Toshiba	4Mb	512k × 8	12.5	1			68
1990	Hitachi	4Mb	256k × 16	12.5	1			55
1990	NEC	16Mb	2Mb × 8/1Mb × 16	12.5	10	18(8.3)	5.0	85

Technological and electrical characteristics of 4Mb to 16Mb EPROMs are shown in Tables 11.7 and 11.8.

11.10.1 Innovative cell developments for 4Mb EPROMs

Both new process technologies and innovative cell structures were considered in the course of developing the tiny 6 to 7 μm^2 cells required for cost effective 4Mb EPROMs.

One of the first 4Mb EPROMs, shown in 1987 by Toshiba [55, 57], used the 9.0 μm^2 cell shown in Figure 11.27(a). Relatively tight 0.9 μm design rules, obtained with advanced stepper lithography, were used along with RIE. Programming was by channel hot election injection to the floating gate.

This cell used some of the most advanced manufacturing techniques available on the conventional cell technology. It used LOCOS to isolate individual bits, and a half contact per cell to connect drain diffusions to the metal bit lines. Even with a self aligned contact and self aligned floating gate, there were still alignment tolerances involved which tended to limit the reduction in cell size.

Hitachi [59] managed, using scaling to 0.8 μm minimum geometries with similar conventional but highly sophisticated processing techniques, to create an even smaller 7.8 μm^2 cell.

The process was becoming very expensive and sophisticated. It was time for some innovation in the basic cell structure and technology.

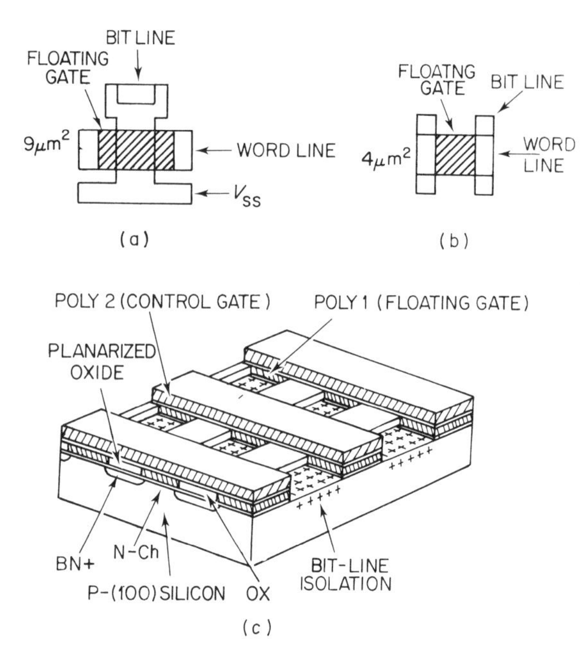

Figure 11.27
Early 4Mb EPROM cell structures. (a) 9.0 μm^2 conventional cell in 1.0 μm design rules. (b) 4.0 mm^2 cross-point cell in 1.0 μm design rules. (c) Three-dimensional diagram of EPROM array using cell shown in (b). (From Mitchell *et al.* [53], TI 1987, with permission of IEEE.)

Two innovative cell structures for 4Mb EPROMs were shown by Texas Instruments [53] and Toshiba [54] the same year. The TI cell used clever process techniques to make a true cross-point EPROM cell. The Toshiba cell used clever NAND gate cell architecture. The TI [53] cell, shown in Figure 11.27(b), did not require use of LOCOS isolation or contacts in the array, and was fully self aligned. Only masking levels, bit-line and word-line, were needed for fabrication of the cell. The cell was about half the size of the conventional cell using the same 1.0 μm design rules.

A three-dimensional schematic diagram of an EPROM array using this cell is shown in Figure 11.27(c). The stacking of the poly 1 interspersed with oxide and the poly 2 in the direction of the poly 2 bit-lines was achieved by a planarization process which leveled the poly 1 and oxide before deposition of the poly 2. A schematic representation of this fabrication process is shown in Figure 11.28.

The Toshiba [54] cell changed from the normal NOR structure used in EPROMs to a NAND structure such as described previously in the ROM chapter. A comparison between the Toshiba NAND cell and the conventional EPROM cell is shown in Figures 11.29(a) and (b). An equivalent circuit of the NAND structure cell with 4 bits is shown in Figure 11.29(c). The new cell is only about 70% of the size of the conventional cell in the same 1.0 μm design rules.

The NAND structure cell can be programmed by hot electron injection to the floating gate and be erased by either UV irradiation or by electric field emission of

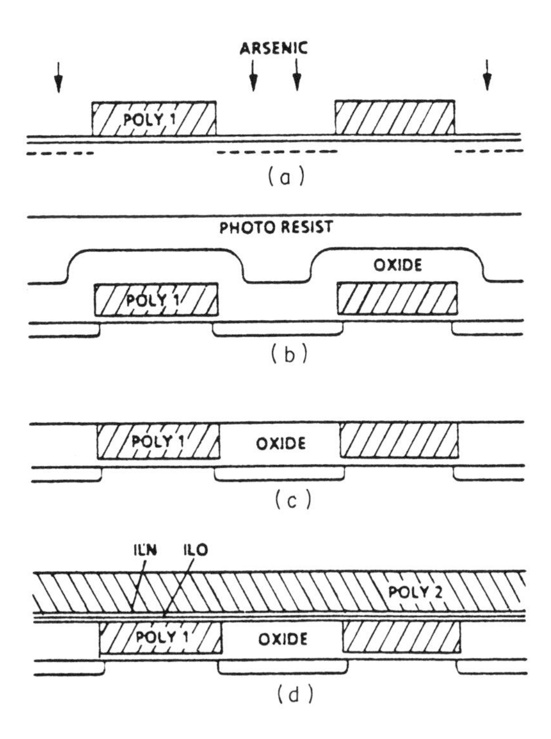

Figure 11.28
Schematic diagram of fabrication process for cross-point cell showing: (a) bit-line formation, (b) photoresist and oxide prior to planarization, (c) post planar etch, (d) interlevel oxide, interlevel nitride and second polysilicon deposition. (From Mitchell *et al.* [53], TI 1987, with permission of IEEE.)

electrons from the floating gate—making it also one of the new flash EPROM cells which will be discussed in a later section.

It is possible to write to only the selected bit in the NAND cell as illustrated in Figure 11.30(a). In program mode, 9 V is applied to the bit line, and 10 V is applied to the word-line of the selected bit and 20 V is applied to the other three word-lines of unselected bits in the NAND cell structure. The bit-line voltage is mostly supplied between the drain and source of the selected bit which operates in saturation mode so that hot electrons are generated in the channel. Hot electrons are not generated for the three unselected bits. A graph of cell threshold against programming time, as illustrated in Figure 11.30(b), shows the programming only of cell 4 by the technique described above without corresponding programming of cells 1, 2, and 3 which are in series with cell 4.

Since Toshiba chose to introduce this new cell as a flash EEPROM which is erased by field emission rather than as an EPROM which is erased by UV light, we will delay consideration of the 4Mb device using this cell until the section on flash EEPROMs.

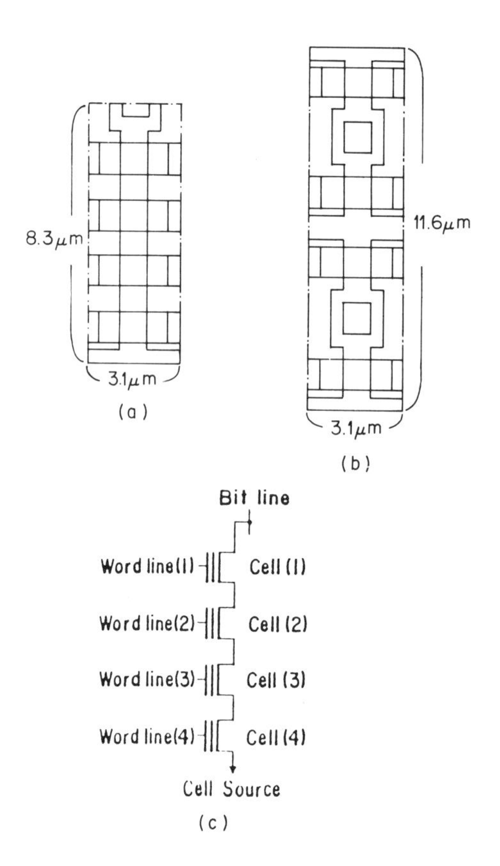

Figure 11.29
Cell structures for the 4Mb EPROM. (a) NAND cell structure, (b) conventional NOR cell structure, (c) equivalent circuit diagram of the 4-bit NAND cell. (From Masuoka *et al.* [54], Toshiba 1987, with permission of IEEE.)

11.10.2 An early 4Mb EPROM

The key requirements for the multimegabit EPROMs were fast programming time, fast access time, low power dissipation, and competitive cost. The latter factor translated into a chip size for the 4Mb EPROM of around 85 mm^2. This in turn required a cell size around 8 μm^2 as can be seen in the charts in Figure 11.31. The graph in Figure 11.31(a) plots chip size against cell size while Figure 11.31(b) shows cell size trends by density for CMOS EPROMs.

Low cost also required, for the ROM prototyping application, that the EPROM be sold in the low cost plastic packaging of the one-time-programmable (OTP) EPROM and be testable while in this low cost package. The testability issue of the EPROM and the OTP EPROM along with efforts at on-chip self-test will be covered in Chapter 13 on memory test.

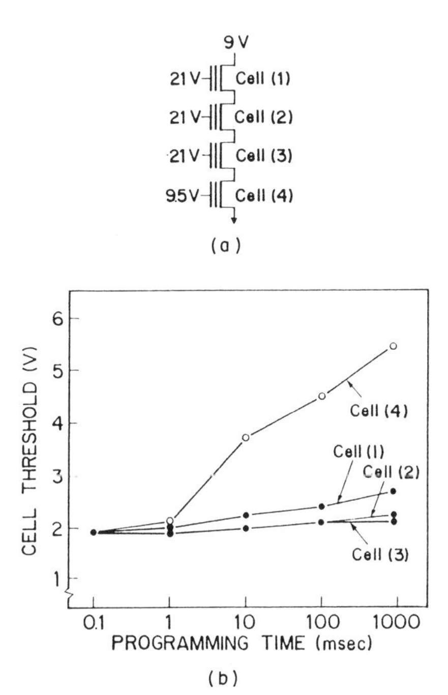

Figure 11.30
Programming a bit in the 4Mb EPROM NAND cell. (a) NAND cell with voltages applied for selectively programming cell 4. (b) Cell threshold voltage as a function of programming time showing cell 4 selectively programmed. All cells were initially erased by UV light. (From Masuoka *et al.* [54], Toshiba 1987, with permission of IEEE.)

11.10.3 Speed issues on the 4Mb EPROM

The early 120 ns Toshiba [55, 57] 4Mb addressed primarily the speed issue. It used sensitive differential current mirror amplifiers shown in Figure 11.32(a) and bit-line swing was restricted below 0.2 V to improve access time. The sense amplifier was composed of a bias circuit, a p-channel load transistor, a reference circuit, and an n-channel current mirror difference amplifier.

Soft writing during the read operation was avoided by limiting the bit-line voltage. This unintentional writing could happen as a result of the scaling of the cell transistor gate length to 0.9 μm if the drain voltage of the cell is not scaled also. To accomplish this scaling the bias voltage was set to 2.5 V during reading. This in turn clamped the bit-line level to below 1.2 V which was low enough to avoid soft writing and also restricted the bit-line swing.

The restricted bit-line swing was then amplified to the level required for the sense line swing using the W_2/Z p-channel load transistor. The W/L ratio of this transistor needed to be larger than that of the sense line W_1/Z in order to set the reference voltage to the intermediate level of the sense line. An N-channel current mirror was then used since it has a high sensitivity for both programmed and unprogrammed states and a large output swing.

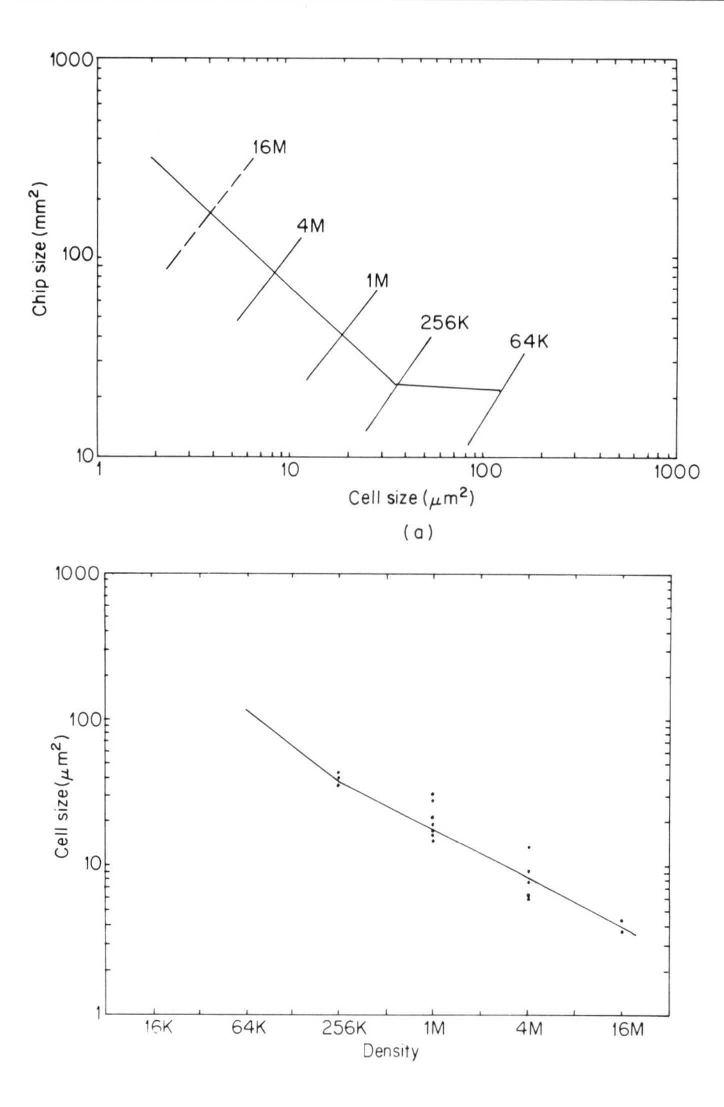

Figure 11.31
(a) Chip size plotted against cell size for CMOS EPROMs. (b) Cell size plotted against density for CMOS EPROMs. (Source various ISSCC Proceedings.)

Programming characteristics for the 4Mb in 0.9 μm technology with 10.5 V V_{PP} are compared to those of the 1Mb in 1.2 μm technology and 12.5 V V_{PP} in Figure 11.32(b). For a given power supply voltage the programming time of the 4Mb was almost a factor of 10 faster.

A 55 ns page-program Hitachi part introduced in 1990 reduced access time by both process and architectural techniques. This part decreased word-line resistance both by using a polycide metal structure and by dividing the memory array into four blocks

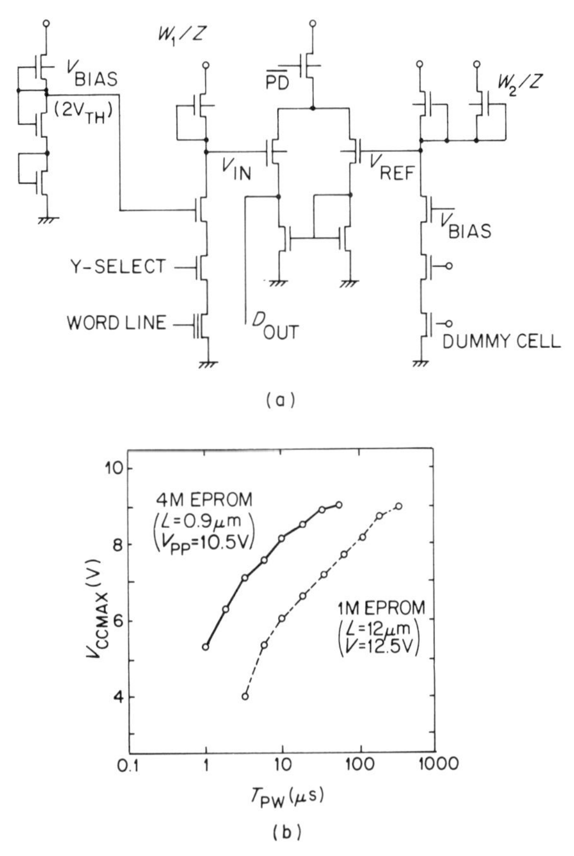

Figure 11.32
Architectures for speed in a 4Mb EPROM. (a) Differential current mirror amplifier using dummy reference cells, (b) Graph of maximum supply voltage against programming time showing improved program speed of 4 Mb with 10.5 V V_{PP} compared to 1Mb with 12.5 V V_{PP}. (From Atsumi *et al.* [55], Toshiba 1987, with permission of IEEE.)

with one X-decoder for each pair of planes. This resulted in a 25% improvement in access time but cost a 10% enlargement in die size.

It also used a variable sensitivity sense amplifier to be able to monitor the memory transistor in both the high voltage but slow 'verify' mode and in the high sensitivity but low voltage read mode. To extend the monitor range a load transistor which operated only in the verify mode was added and the verify load current was made smaller than the read load current. The result was that a high programmed V_t can be verified without sacrificing sensitivity in read mode.

To reduce programming time, this chip used an 'overvoltage' applied to the OE pin to enter page program mode and latch the input data as shown in the timing diagram in page program timing chart in Figure 11.33(a). The page program control circuit is shown in Figure 11.33(b). There were four latches in each I/O block to hold input data with two additional latches for redundancy.

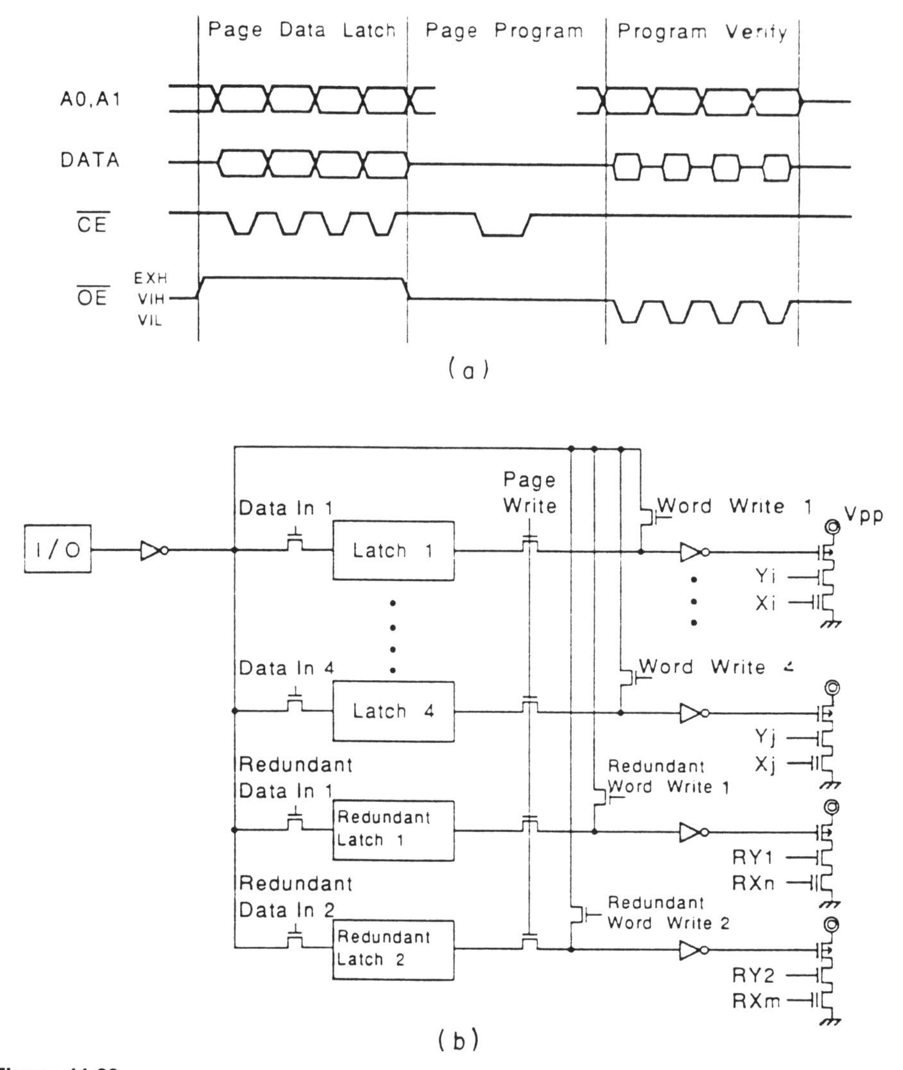

Figure 11.33
4Mb EPROM with page mode programming. (a) Schematic timing diagram showing overvoltage applied to $\overline{OE}$ pin to enter programming mode and latch the input data. (b) Page program schematic circuit diagram showing four latches plus two redundant latches in one I/O block. (From Nakamura *et al.* [59], Hitachi 1990, with permission of IEEE.)

11.10.4 Noise immunity problems in fast 4Mb CMOS EPROMs

The speed of this early word-wide 4Mb EPROM was slowed down by the problem of internal noise generation during operation. A second 4Mb EPROM with a faster 68 ns access time but using basically the same technology was shown by Toshiba [58, 61] in 1989. Speed was further enhanced by going to 0.8 μm channel lengths and molybdenum polycide word-lines. This part added noise suppression circuit techni-

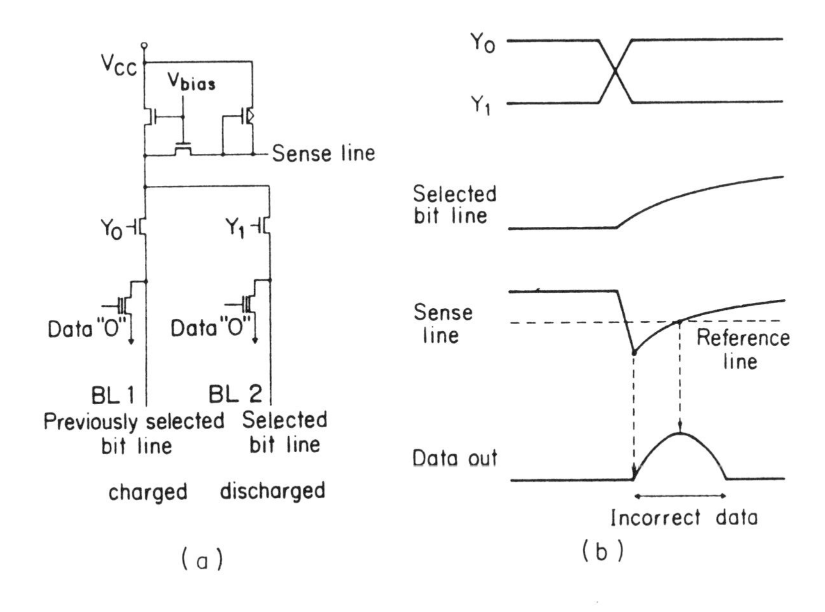

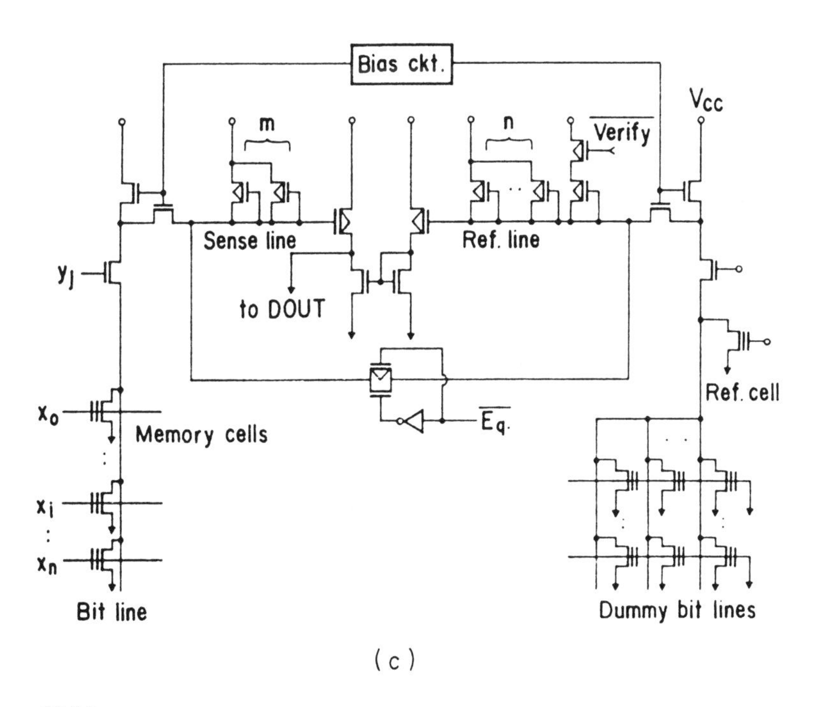

Figure 11.34
Example of source and one solution of output noise in EPROM due to temporary changes in the sense line level when a new bit-line is selected. (a) Illustration of sense line and bit-line structure. (b) Waveforms of selected bit-line, sense line and data output showing incorrect data temporarily appearing on the data-out line. (c) Balanced sense amplifier to reduce noise by insuring that the sense line and reference line oscillate in phase at the same frequency. (From Imamiya *et al.* [61], Toshiba 1990, with permission of IEEE.)

ques including divided bit-lines, a symmetric sense amplifier configuration and a chip-enable transition detector to improve stable data sensing and provide high speed access time.

Output noise reduces the TTL input noise margin and can cause erroneous addresses to be read. It also can effect the data sensing circuitry, particularly in EPROMs with a single-ended sensing scheme, as described in Chapter 5, and cause erroneous data to be read.

Large current flows in the data-out buffers due to temporarily incorrect data are one source of internal noise in an EPROM. Figure 11.34(a) shows a sense line and bit-line structure. Figure 11.34(b) shows the potential changes by the Y_0 and Y_1 column select signals along with the waveforms of the selected bit-line, sense line and data out on the selected bit-line. When a programmed cell in bit-line BL_1 is selected, the bit-line is charged to $V_{bias}-V_{thres}$. The sense line potential is raised higher than the reference line level and the data out are low level.

Next, the column address changes and a programmed cell on the discharged bit-line BL_2 is selected causing the sense line level to drop instantly pulled down by the low bit-line level. As the bit-line gradually charges, the sense line comes up so that even if the cell's data are '0', data out changes temporarily from '0' to '1', then back to '0' as the sense-line level exceeds the reference level. A similar waveform occurs when the $\overline{CE}$ signal activates the chip and data '0' is read out.

Another source of internal noise in an EPROM is the phase mismatch in single-ended sense circuitry which is due to the sense line and reference line not having the same structure. Many memory cell transistors are attached to the sense line but not to the reference line so the sense line has a significantly higher capacitance than the reference line. This means that when a voltage bump occurs the two lines oscillate at different frequencies and are hence out of phase; thus the sense line level surpasses the reference line level periodically causing incorrect data to be read out.

These noise sources were addressed in the 1989 Toshiba 4Mb EPROM by several methods. One was using a divided bit-line technique which decreased the bit-line charging time and permitted the cell to have a sufficient current read margin with a smaller cell read current. Another was to use a balanced sense amplifier with dummy bit-lines, as shown in Figure 11.34(c), so that the sense line and reference line oscillated in phase at the same frequency.

A 90 ns 4Mb EPROM from Intel [56] also took design measures against output noise sources which delayed chip speed. This part included a balanced sense amplifier and an output buffer circuit that ensured a mutually exclusive condition of the N-channel drives. In addition, the outputs were switched in succession rather than all at once to reduce the effective inductive dI/dT.

11.11 THE EARLY 16Mb CMOS EPROM CELL DEVELOPMENT AND DEVICES

Several companies described possible cells in the 0.6 μm geometry which appeared to be required for the sub 4 μm^2 cells required for the 16Mb EPROM. These included STM, Toshiba and NEC. An early 16Mb EPROM was shown by NEC in 1990.

STM [63] in 1989 presented a contactless cross-point cell with overlapping floating gate which was self aligned to the field oxide. The traditional 'T' cell is shown compared to the new cross-point cell in Figures 11.35(a) and (b).

The new cell forms a virtual ground array which is shown compared to a standard common ground array in Figures 11.35(c) and (d). This cell, like the TI cross-point cell mentioned earlier, used a planarization step after the stacked gate definition. This cell, however, maintained the use of field oxide.

In a separate paper STM [62] described the 16Mb EPROM process which featured less than 0.2 μm 'bird's beak' isolation, LDD N-channel and P-channel transistors, 20 nm interpoly dielectrics, and cold processing involving rapid thermal annealing. This produced a cell of the standard T-shaped type that was less than 4.5 μm^2 and presumably an even smaller cell of the new cross-point type. The soft write endurance for various geometry processes is shown in Figure 11.36.

Toshiba [64] also presented a virtual ground array cell for the 16Mb EPROM in 1988. This cell used an asymmetrical lightly doped source cell structure which is shown in Figure 11.37(a) along with a circuit schematic for the equivalent virtual ground array shown in Figure 11.37(b). They found that the asymmetrical cell had a higher immunity to the soft-write and write-disturb problems that appear to be

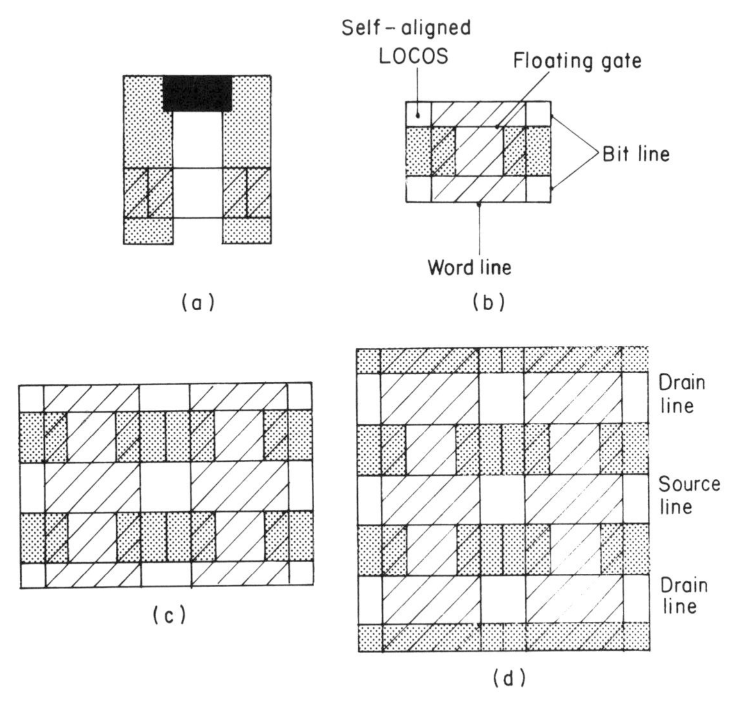

Figure 11.35
Layouts of cell structures for 16Mb EPROMs. (a) Traditional 'T' cell structure. (b) New cross-point cell structure. (c) Four of the cross-point cells in a 'virtual' ground array. (d) Four 'T' cells in a common ground array. (From Bellezza *et al.* [63], STM 1989, with permission of IEEE.)

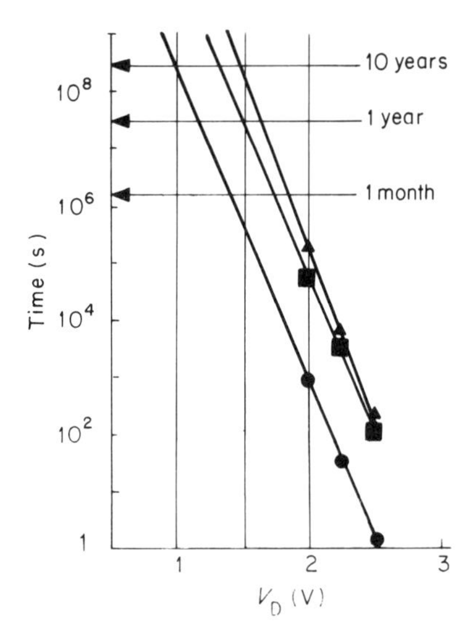

Figure 11.36
Soft write endurance for various geometries of 'T' type cell in a 16Mb process technology with ΔV_τ of 200 mV and $V_{GS} = 5.5$ V. (● $L_{drawn} = 0.4\ \mu$m, ■ $L_{drawn} = 0.5\ \mu$m, ▲ $L_{drawn} = 0.6\ \mu$m). (From Bergemont *et al.* [62], STM 1989, with permission of IEEE.)

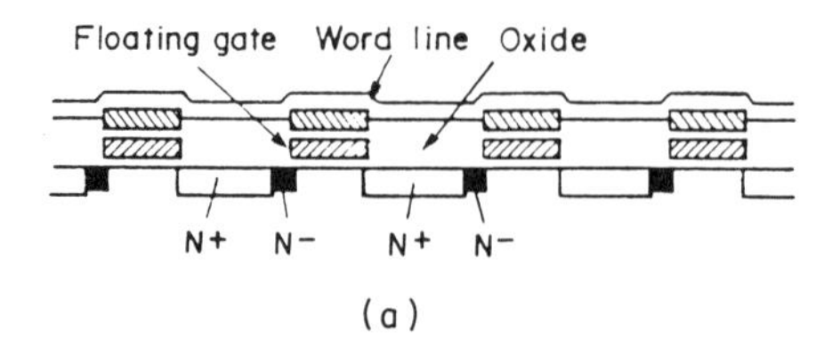

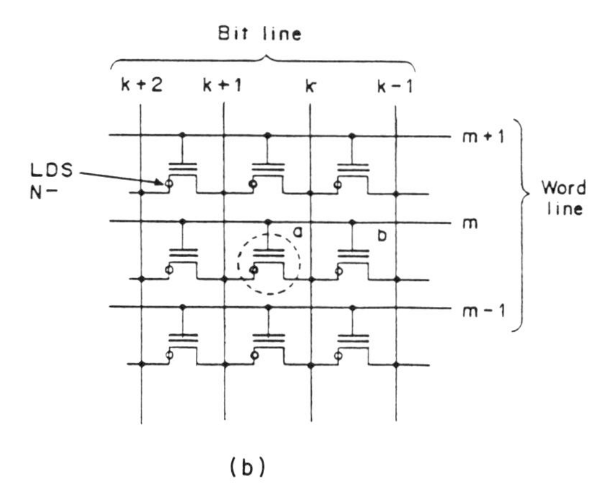

Figure 11.37
16Mb EPROM asymetric lightly doped source cell structure. (a) Schematic cross-section. (b) Circuit schematic for the virtual ground array formed by these cells. (From Yoshikawa *et al.* [64], Toshiba 1988, with permission of IEEE.)

associated with virtual ground designs. In 1990 Toshiba [99] showed an asymmetrical cell for the 16Mb CMOS EPROM in 0.6 μm technology. This cell, which is shown in Figure 11.37(c), was a modification of the earlier diffusion self-aligned cell used on the 4Mb EPROM. The multiple diffusions were accomplished with large tilt angle ion implantation of boron. This cell was also usable as a flash EPROM cell with minor modifications.

Another virtual ground EPROM architecture was described by Wafer Scale Integration [105] in 1991 for a 1Mb EPROM. This part improved speed by reducing the bit-line capacitance by using a segmented diffused bit-line scheme connected to a global metal bit-line. Address transition detection and a programmable signal development timer were also used.

Toshiba [65] in addition in 1989 investigated and proposed a new set of scaling rules for high density EPROMs as well as making modules of a scaled 3.85 μm^2 EPROM cell.

The scaling rules used, as prior constraints, the two required cell characteristics of reliability and performance. Independent scaling factors were introduced using 'k' for lateral dimensions and 'h' for vertical dimensions. An attempt was then made to empirically find a relation between k and h that satisfied both the reliability and performance constraints simultaneously. The results are shown in Figure 11.38.

Key issues which were considered included drain stress, gate stress, soft-write, cell read current, programming and punch-through.

Drain stress is a form of program disturb involving deprogramming of unselected cells on the same bit (drain) line as the selected cell. This was avoided by requiring that the electric field be maintained as in earlier generations of EPROMs. Since the field across the thin oxide (E_{ox1}) depends on the ratio of 'delta V_t/T_{ox1}', this requirement is that 'delta V' be scaled as $1/h$.

Gate stress is another form of program disturb involving deprogramming of an unselected cell on the same word line. Leakage through the interpoly dielectric is the main cause here and requires maintaining the electric field across the interpoly oxide. This translated to V_{PP} and V_t being scaled as $1/h$.

Soft write, which was discussed earlier, is a form of read disturb in which unselected cells are unintentionally written during the read operation. The reliability requirement of 10 years lifetime here required that V_{dr} be scaled for read disturb resistance.

One other interesting result of this study was that the programming voltage is optimized at V_{PP} below 5 V which could mean that the 16Mb EPROM will have internally generated V_{PP}.

Another process involved a 3.6 μm^2 16Mb EPROM cell which NEC [66, 67] showed in 1989 and used in 1990 in an early 16Mb EPROM. The technology included self-aligned trench isolation to reduce the spacing between the 3.6 μm^2 cells. Oxide–nitride–oxide (ONO) was used for the interpoly dielectric. A selectively deposited tungsten plug was used for the bit-line contact. While the gate length was 0.9 μm, the contacts used an 0.6 μm technology. A three-dimensional cross-section of the cell is shown in Figure 11.39(a) along with the top cell layout shown in Figure 11.39(b).

The 16Mb EPROM had a 121 mm^2 chip size which is right on the trend line shown

Issues	Expressions	Scaling	This Model	Constant Voltage	Constant Field
Lateral dimensions	L, W, X_j	$1/k$	$1/k$	$1/k$	$1/k$
Vertical dimensions	t_{ox1}, t_{ox2}	$1/h$	$1/\sqrt{k}$	$1/k$	$1/k$
Program-disturb 1 (Drain stress)	$\text{Box 1} \sim \frac{C_2}{C_T} * \frac{\Delta V_t}{t_{ox1}} + \frac{(C_1 + C_2)}{C_T} * \frac{V_{dp}}{t_{ox1}}$	$\sim \left(\frac{2}{3} \Delta V_{dp}\right) * h$	1	k	1
Program-disturb 2 (Gate stress)	$\text{Box 2} \sim \frac{C_2}{C_T} * \frac{\Delta V_t}{t_{ox2}} + \frac{C_1}{C_T} * \frac{V_{pp}}{t_{ox2}}$	$\sim \left(\frac{2\Delta V_t + V_{pp}}{3}\right) * h$	1	k	1
Read-disturb (Soft-write)	$I_{gate}/W \sim \exp(-C/E_{m,r})$, $E_{m,r} \sim \frac{(V_{dr} - V_{dsat})}{3\sqrt{t_{ox1} * X_j}}$	$(V_{dr} - V_{dsat}) * (kh)^{1/3}$	1	$k\sqrt{k}$	$1/\sqrt{k}$
	$\ln(T) \sim \ln(LW/t_{ox2}) + 1/E_{m,r}$	$\sim 1/E_{m,r}$	1	$1/k\sqrt{k}$	$\sqrt{k}$
Cell read current (Cell Vt)	$I_{cell} \sim \frac{W}{L} * \frac{C_2}{C_T} (V_{cc} - V_{tcell}) \frac{V_{dr}}{t_{ox1}}$	$V_{dr} * h$	1	k	$1/k$
	$V_{tcell} \sim t_{ox1} * \sqrt{N}$	$\sqrt{N}/h$	1	1	$1/k$
Program	$\Delta V_t \geq AV_{cc} + B(V_{cc} - V_{tcell})$	ΔV_t	$1/\sqrt{k}$	1	$1/k$
	$\sim \int_L^{t_{ox1}} * \frac{I_{gate}}{W} dt$	$\sim (\ \ )^{1/E_{m,p}}$	$(\ \ )$	$(\ \ )^{k\sqrt{k}}$	$(\ \ )^{1/\sqrt{k}}$
	$I_{gate}/W \sim \exp(-C/E_{m,p})$, $E_{m,p} \sim \frac{(V_{dp} - V_{dsat})}{3\sqrt{t_{ox1} * X_j}}$	$(V_{dp} - V_{dsat}) * (kh)^{1/3}$	1	$k\sqrt{k}$	$1/\sqrt{k}$
Punchthrough (Bulk)	$V_{pt} - N * L^2$	N/k^2	$1/k$	1	$1/k$
(Drain turn-on)	$I_{dl} \sim \frac{W}{L} * \frac{1}{N} \sqrt{N} \exp\left(\frac{q}{kt} V_{FG} + \frac{V_{dp}}{D}\right)$,	$\sim (\ \ )^{V_{dp}}$	$(\ \ )^{1/\sqrt{k}}$	$(\ \ )$	$(\ \ )^{1/k}$
	$V_{FG} \sim (Xj/L * V_{dp}$	V_{dp}	$1/\sqrt{k}$	1	$1/k$

Notation: C_T: Total capacitance of floating gate.
C_1: Floating gate-sub capacitance.
C_2: Control gate–floating gatecapacitance.
V_{dsat}: Drain saturation voltage.
N: Impurity conc. of channel.
I_{gaye}: gate current.
V_{FG}: Floating gate voltage.
$E_{m,r}$: Maximum channel field in reading.
$E_{m,p}$: Maximum channel field in programming.
V_{tcell}: Cell threshold voltage.
D_{Vt}: Minimum threshold voltage shift after programming.
E_{ox1}: Cell gate oxide electric field.
E_{ox2}: Interpoly ONO electric field.
T: Soft-write lifetime.
$1/B$: Sense amp. gain.
I_{dl}: Drain turn-on leakage current.
A, C, D: constant.

previously. It could be configured either 1Mb × 16 or 2Mb × 8 by controlling an input signal as shown in the timing diagram in Figure 11.40. The part was compatible with the 16Mb ROM.

The part had a typical access time of 85 ns. Typical active current was 18 mA and typical standby current was 1 μA. The word-lines were pumped to a level above V_{PP} for programming by a charge pump. Total programming time for the chip was less than 10 s.

For redundancy the device used both EPROM cells and UPROM cells. The EPROM cells were used to select the redundant cells at wafer sort before the chip was packaged. The UPROM cells were programmed at final test. This allowed the chip to be repaired both at wafer sort and at final test.

In 1991 Toshiba [104] showed a 16Mb CMOS EPROM with 62 ns access time. The part used address transition detection for sense line equalization and a data-out latching scheme. Noise immunity was improved by the use of a bit-line backbias generator. V_{PP} was 12.5 V.

11.12 APPLICATION-SPECIFIC AND EMBEDDED EPROMs

Small EPROM arrays have been included as embedded memories on microprocessor chips for several years. They are also becoming widely available as macro-cells in standard cell libraries.

The recent trend to lower power supply voltages for some applications has led to increasing interest in low voltage integrated circuits using the 3.3 V low voltage TTL compatible standard and the 2.8 V standard for battery operated systems. As the market requires processors and cell libraries to work at the lower voltage standard, low voltage embedded EPROMs will also be needed. Applications include both the rapidly growing battery operated portable computer and games market, and the more traditional telephony applications which require integrated circuits in the 2.0 to 5.5 V range.

An example of development in this area is given by a Toshiba [73] paper on a low voltage EPROM for embedded ASIC applications. In this paper Toshiba demonstrated that their asymmetrical drain cell, which was discussed earlier, significantly improved soft write error and hence improved data retention time in the low voltage range. At 3.3 V the read retention time was shown to be 1000 times longer than that of the conventional symmetrical drain-source EPROM cell.

11.13 ONE-TIME-PROGRAMMABLE (OTP) EPROMs

The one-time-programmable EPROM was developed in the attempt to reduce the cost of the EPROM so it could be cost competitive in the fast turn-around ROM market.

Figure 11.38
Proposed scaling rules for high density EPROMs. (From Yoshikawa *et al.* [65], Toshiba 1989, with permission of IEEE.)

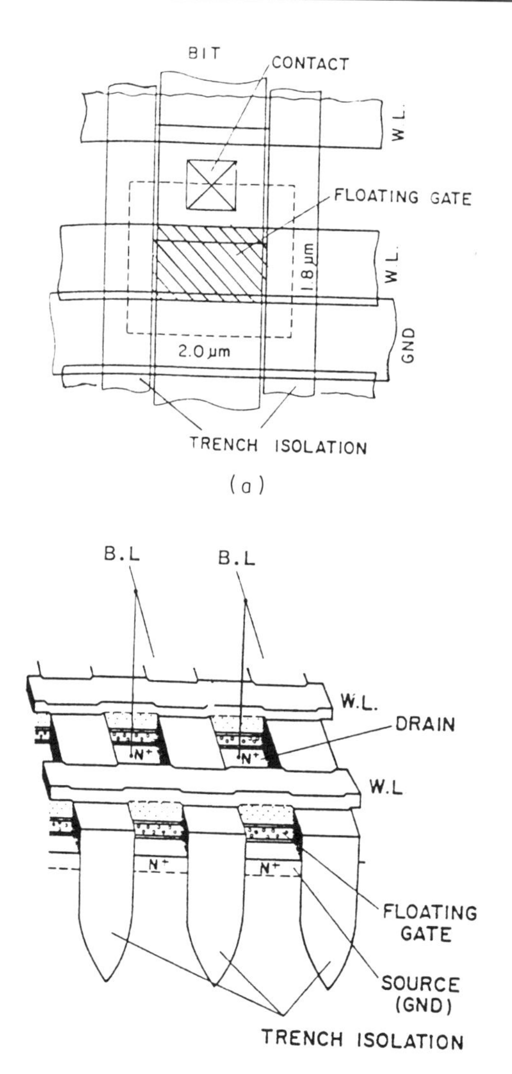

Figure 11.39
16Mb trench isolation EPROM cell with ONO interpoly dielectric. (a) Three-dimensional cross-section. (b) Layout. (From Higuchi [67], NEC 1990, with permission of IEEE.)

The high cost of the EPROM results from package and test costs. The quartz windowed ceramic package in which it must be packaged to permit UV light to reach the chip for erasure is more expensive than the standard molded plastic package. The high test costs result from the handling difficulties and time consumed in removing a chip from the system and erasing it by UV light during test.

The OTP EPROM solves the first problem by simply putting the EPROM chip in a plastic package. Unfortunately this makes the test problem worse since now the part cannot be erased at all after assembly. This means it is difficult to check the

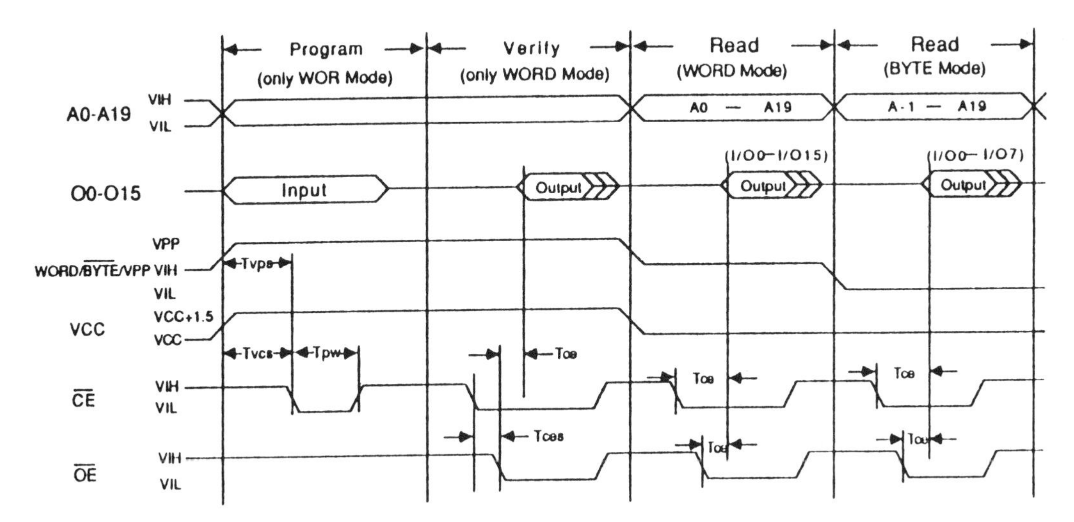

Figure 11.40
Timing diagram of 16Mb EPROM showing word and byte mode. (From Higuchi [67], NEC 1990, with permission of IEEE.)

programmability of the cells before shipment, or to do a final erase in case testing has partially programmed some of the cells.

Some manufacturers solved this problem by using OTPs only as fast turn-around ROMs and doing the programming themselves to the customer's data pattern. The parts can then be tested like ordinary ROMs before shipment. An additional cost saving can be made by using a less expensive surface coating on the EPROM which is not transparent to UV light since it will not be exposed to UV light.

Another partial solution is to include on the chip additional rows of cells. These are used at final test after assembly to check the operation of the peripheral circuitry and also the programmability of cells. While this method does not actually check the cells that will be used in the part, it has proved a fairly acceptable method of outgoing test for a low cost part.

The fast EPROMs with the two- or four-transistor cell differential sensing techniques allow the working cells to be checked without actually reading them. A special test mode called 'blank check' testing is used which is described in the chapter on memory test. In the case of these parts, however, the low cost of the plastic package is partially offset by the higher cost of the two- or four-transistor cell chip.

The OTP is not a perfect solution to the quest for a low cost plastic packaged testable EPROM. Over the last 15 years other potential candidates for a low cost EPROM challenger to the ROM have appeared.

EEPROMs were widely expected to fill this gap in their early days, but the multi-transistor cell technology has proved too expensive and they have migrated into the diverse realm of applications specific memories.

The one-transistor cell 'flash' memory which is now considered a potential low cost candidate to fill ROM and EPROM sockets as well as for high density EEPROM applications will be discussed in the following section.

11.14 ELECTRICALLY FLASH REPROGRAMMABLE ROMs

Toward the end of the 1980s the industry discovered how to erase the EPROM electrically and a new generation of memories targeted at the low cost, high density EPROM and ROM market evolved. These innovative next generation EPROMs generally were distinguished by three characteristics. They were derived from one-transistor EPROM cells and circuits and made in advanced EPROM technology. They were electrically reprogrammable a few times ranging up to many thousand times. They were therefore able to be packaged in the low cost plastic package and be used in the next generation of high density EPROM and ROM sockets.

The name 'flash' was given to these parts as a result of the electrical erasability which initially was only possible as a bulk phenomena. The term 'flash' historically had been used to describe a mode of programming or erasing an entire memory array at one time. Although designers were quickly able to make flash memories that were section erasable, the name stuck and came to designate rather those electrically erasable PROMs which were made with one-transistor cells like an EPROM rather than two or more transistor cells like the low density conventional EEPROM.

Figure 11.41 indicates a forecast for the ROM market in 1994 with the flash memories broken out by density. While these parts are expected to exceed the conventional EEPROM in sales, they are not forecast by then to have seriously eroded the EPROM or mask ROM markets.

11.14.1 A short history of flash memories

The earliest flash memory was shown by Toshiba [78, 98]. This part had a cell with the split gate so common in later flash memories and will be described in the following section on split gate parts.

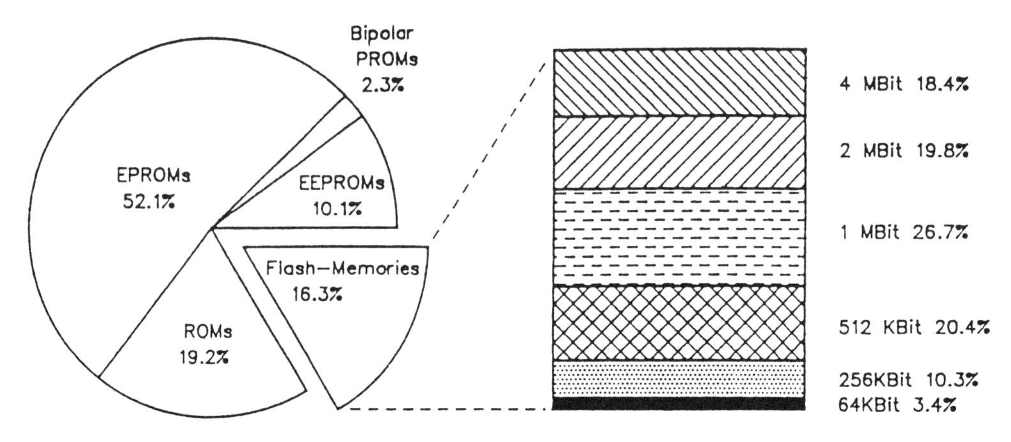

Figure 11.41
Market projection for 1994 for split of ROM market by product and of flash market by density. (Source Dataquest.)

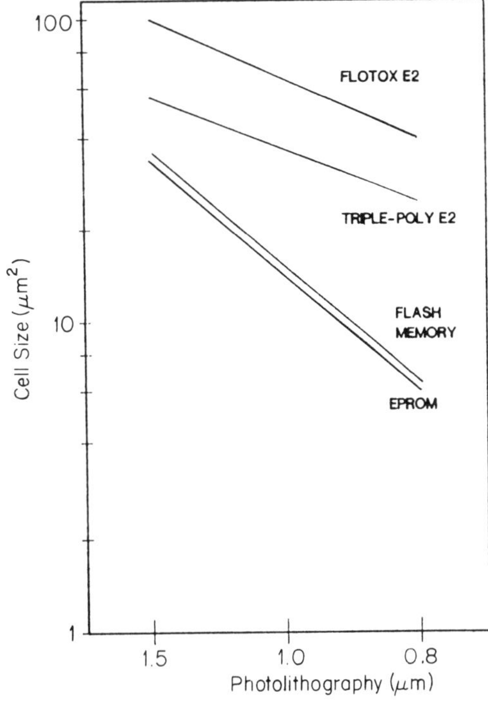

Figure 11.42
Comparison of cell size to minimum geometry for various electrically erasable PROM technologies: conventional EEPROM, triple poly EEPROM, flash memory and EPROM showing cell size of flash memory nearly the same as that of the conventional EPROM and a factor of 10 smaller than the two-transistor conventional EEPROM. (From Kynett *et al.* [84], Intel 1988, with permission of IEEE.)

Many of the early flash memories were basically modified one-transistor cell EPROMs that could be erased electrically in bulk. They required an external high voltage supply for programming like a conventional EPROM. The electrical erasure was by either hot electron or cold electron injection from the floating gate. The erase–write cycle endurance, while comparable to that of the EPROM, was several orders of magnitude less than that of the EEPROM.

These early parts were targeted at embedded control applications where the system central processing unit provided occasional code updates over the lifetime of the part.

11.14.2 Single-transistor flash memories as an EPROM compatible technology

An early 256k EPROM derivative flash memory discussed by Intel [76] in 1988 used 12 V programming and offered 100 cycle write–erase endurance. The 32k × 8 organized part had a 192 mm^2 chip and a 36 μm^2 one transistor cell in a 1.5 μm EPROM based technology. A comparison chart of cell size with lithography is shown in Figure 11.42. The single-transistor flash cell is shown to be only slightly larger than the EPROM cell.

The Intel flash memory cell structure was similar to the EPROM floating gate cell structure with the exception of the use of a high quality of tunnel oxide under the floating poly gate. It was programmed by hot electron injection like an EPROM. In the flash erase, all cells were erased simultaneously by cold electron tunneling from the floating gate to the source by applying 12 V to the source junctions and grounding the select gates. The parts were made in the EPROM technology with the exception of the tunnel oxide.

In 1989 [85] Intel showed a scaled 1.0 μm version of the technology with a 15.2 μm^2 cell in a 153 mm^2 1Mb chip. The endurance by this time was in the range of 1 million write–erase cycles. The part had a 90 ns typical read access time and a 900 ms array erase time and a 10 μs/byte programming rate. A 12 V power supply for programming and erase continued to be used. An internal erase verify reference circuit was used in conjunction with a specified erase algorithm to insure that all bytes in the array were erased to a 3.2 V maximum threshold voltage.

In 1990 [97] Intel showed a new single-transistor cell using contactless array technology which allowed a 4.48 μm^2 cell for a 16Mb flash EPROM to be manufactured in an 0.8 μm technology. A buried N^+ bit line was used rather than contacts. Programming was by channel hot electron injection at the drain edge with an 11.4 V control gate. Erase was by Fowler–Nordheim tunneling with a high voltage at the source and the gate at ground.

Hitachi [90] in 1987 showed a flash cell consisting of a single-transistor double poly stacked gate, also without a select transistor. The cell size was 9.3 μm^2 in an 0.8 μm technology adequate for a 4Mb EPROM compatible chip size. It was programmed by hot electron injection at the drain edge similar to the EPROM and erased by Fowler–Nordheim tunneling from the floating gate to the source.

Asymmetric source and drain regions were used to enable fast program–erase operation as shown in Figure 11.43. Erase was performed by tunneling electrons from the floating gate to the N^+ source region with the control gate grounded. A high erase voltage could be applied to the source because of the high P^+ concentration in this

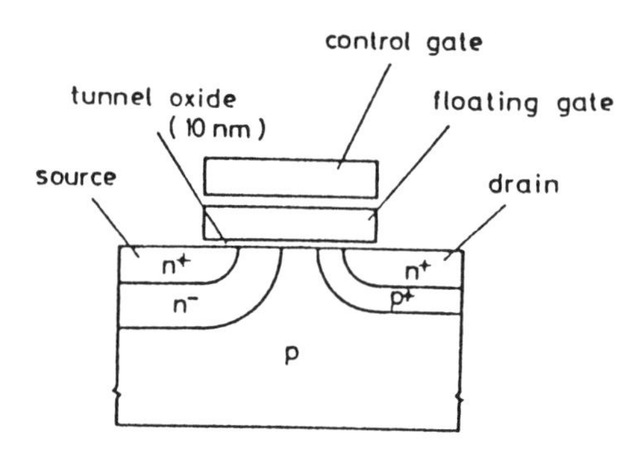

Figure 11.43
Schematic cross-section of flash memory cell with asymmetric source and drain structure. (From Kume *et al.* [90], Hitachi 1987, with permission of IEEE.)

region. The high N^+ concentration in the drain region suppressed punch-through current during erase.

Endurance characteristics were checked for a single cell out to 10^4 write–erase cycles.

11.14.3 Flash memories with split gate single-transistor cells

The split gate cell was used in many of the early flash memories. This cell merged the select gate with the storage gate over the channel. This permitted a single merged transistor cell rather than the two-transistor cell used commonly on the conventional EEPROM. The select gate over part of the channel region prevented the programmed cell from turning on as a result of positive charge on the floating gate which could result from over-erasing. This measure had not been necessary on the conventional EPROM since the UV erase is self-limiting and over-erasure not likely.

The early 256k flash memory from Toshiba [77, 78] was shown first in 1984. This one-transistor cell part had a 64 μm^2 cell in 2 μm geometries and a 32.9 mm^2 chip size. A comparison chart of this flash part with the UV-EPROM, OTP EPROM and conventional EEPROM is shown in Figure 11.44. The advantages of the flash were plastic package, fast erase and program time, cell and chip area comparable to the EPROM. The disadvantage in this case was a more complicated cell technology than the EPROM.

The cell [78] used a triple polysilicon one-transistor floating gate cell with layout and two perpendicular cross-sections as shown in Figure 11.45. Figure 11.45(a) shows the layout and (b) shows the separate first polysilicon erase gate and the stacked structure floating gate which has part on the channel and part on the erase gate. To minimize cell area, two cells share one erase gate.

Figure 11.45(c) shows the split cell concept which merged the enhancement mode select transistor with the storage transistor over the channel to prevent the cell from going into depletion mode as a result of overerasing.

Compatibility with the UV EPROM pinout was maintained by using the A_{10} address pin also as a 12 V erase detect indicator. Since a flash (bulk) erase was used, addressing was not needed during erase. The erase circuitry is shown in Figure 11.46(a). The erase mode occurred when 12 V was applied to the A_{10} pin and the V_{PP} pin was raised above 12 V (21 V erase was used). At this time all of the word select lines became 0 V and a high voltage was applied to all the erase gates. Programming was done at 21 V to maintain compatibility with the UV EPROMs.

Programming time was 200 μs/byte and erase time was less than 100 ms per chip. Endurance was indicated to be at least 200 write–erase cycles which was comparable to that of a UV EPROM but several magnitudes lower than the conventional EEPROM.

The cell was programmed by hot carrier injection like a UV EPROM. Flash erasing was done by field emission of hot electrons from the floating gate to the erase gate.

To achieve a 90 ns read access time a differential sense amplifier with a negative feedback bias circuit was used as shown in Figure 11.46(b). The reference voltage was

	Current device			New device
	UV-EPROM	One time PROM	Current EEPROM	Flash EEPROM
Package	Ceramic with window ×	Plastic ○	Plastic ○	Plastic ○
Erase time	20 min ×	No erase ×	1 ms ○	100 ms ○
Program time (ms)	<1 ○	<1 ○	<1 ○	0.2 ○
Cell area (μm^2) (2 μm ground rule)	64 ○	64 ○	270 ×	64 ○
Chip area 256K (mm^2) (2 μm rule)	32.9 ○	32.9 ○	$\simeq$98 ×	32.9 ○
Reliability	Screening ○	No screening ×	Screening ○	Screening ○
Eraser	UV LIGHT ×	No need ○	Electrically ○	Electrically ○
Structure	Double POLY-Si	Double POLY-Si	Double POLY-Si	Triple POLY-Si

Figure 11.44
Comparison of flash memory with UV-EPROM, and conventional EEPROM. (From Masuoka *et al.* [77], Toshiba 1987, with permission of IEEE.)

designed to be half of the voltage swing of the memory section to provide rapid detection.

Another split gate flash memory was shown by SEEQ [79] also in 1987. This 128k part had a 15.6 mm^2 chip size and required a 21 V power supply for programming and erasing. It used hot electron injection for programming and cold electron tunneling from the floating gate to the drain for erase.

The 43 μm^2 cell was 20% larger than an EPROM cell in the same 2.5 μm NMOS technology due to the split cell arrangement merging the select transistor and the floating gate of the storage transistor over the channel as shown in the layout and cell cross-section in Figures 11.47(a) and (b).

Erasure took about one second with 19 V applied to the drain. Programming time deteriorated after an electrical erase due to positive charge collection on the floating gate. To reduce this effect shorter programming pulses were investigated and the results are shown in Figure 11.48.

This part had a single-cell endurance of several thousand write–erase cycles which was still significantly less than that of the conventional EEPROM. The wearout

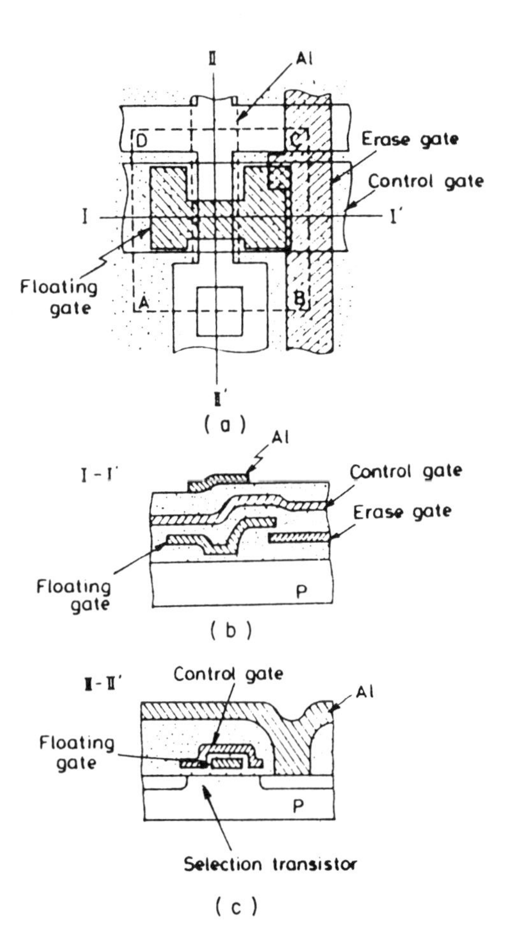

Figure 11.45
Cell structure of 256k flash memory with triple level polysilicon. (a) Layout. (b,c) Two perpendicular schematic cross-sectional views. (From Masuoka *et al.* [78], Toshiba 1984, with permission of IEEE.)

mechanism was believed to be electron trapping of hot electrons during programming in the gate oxide near the drain.

A comparison of cell size plotted against year of introduction for the four-transistor cell EEPROM, the two-cell transistor EEPROM, the one-transistor cell UV EPROM and this early flash is shown in Figure 11.49.

A year later in 1988 SEEQ introduced a 512k flash to the market which also used the split-level cell architecture. The external high voltage power supply requirement was down from 21 to 12 V which was still above the 5 V only EEPROMs available at the time. The 64k $\times$ 8 part was in a 1.5 μm CMOS process with a 20 μm^2 one-transistor cell.

The part was synchronous with input latches on address, data and control ports added for ease of use with host microprocessors.

A block erase mode was added so 512 byte sections could be erased separately. Block erase time was 7.5 s and the entire chip could be erased in 1 minute.

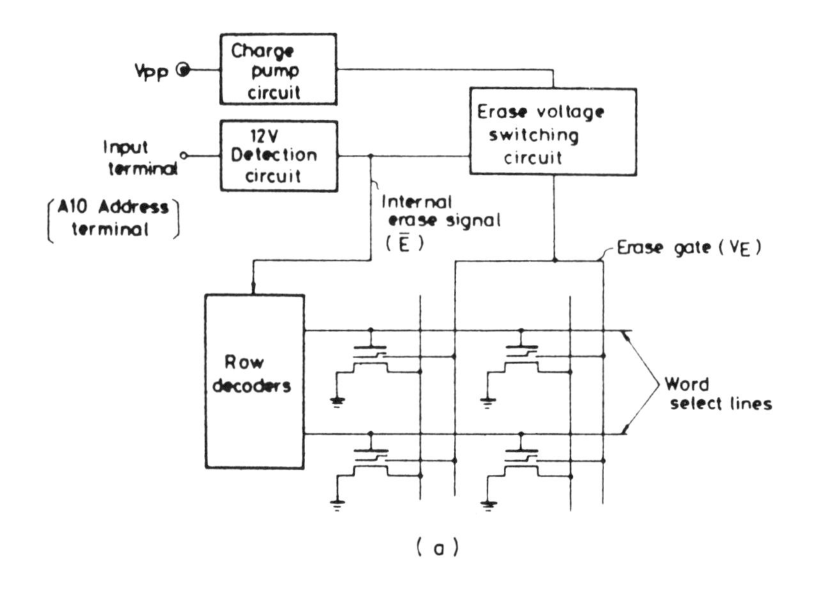

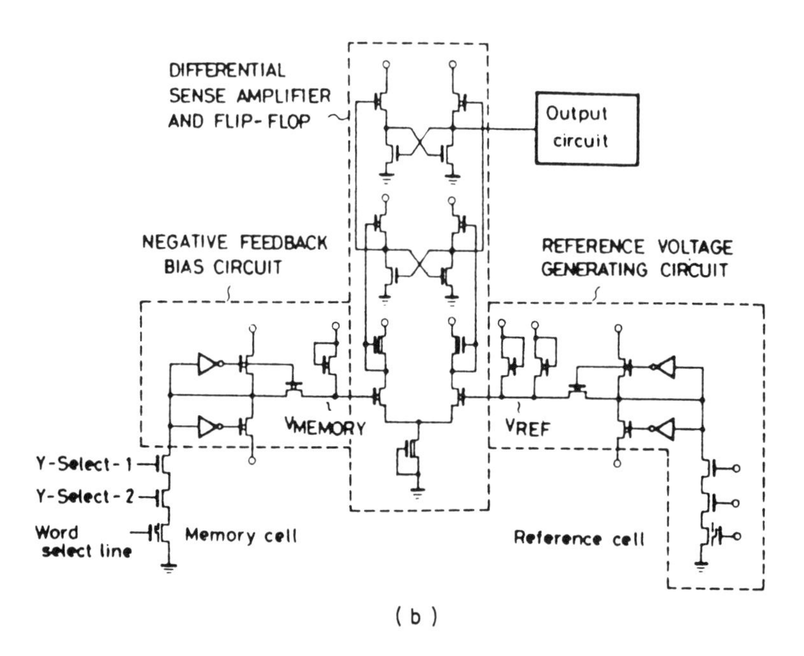

Figure 11.46

256k flash memory circuit architecture including, (a) erase system showing detection circuit for external 12 V power supply on pin A_{10}. (b) differential sense amplifier with negative feedback bias circuit and reference voltage generator circuit. (From Masuoka *et al.* [77], Toshiba 1984, with permission of IEEE.)

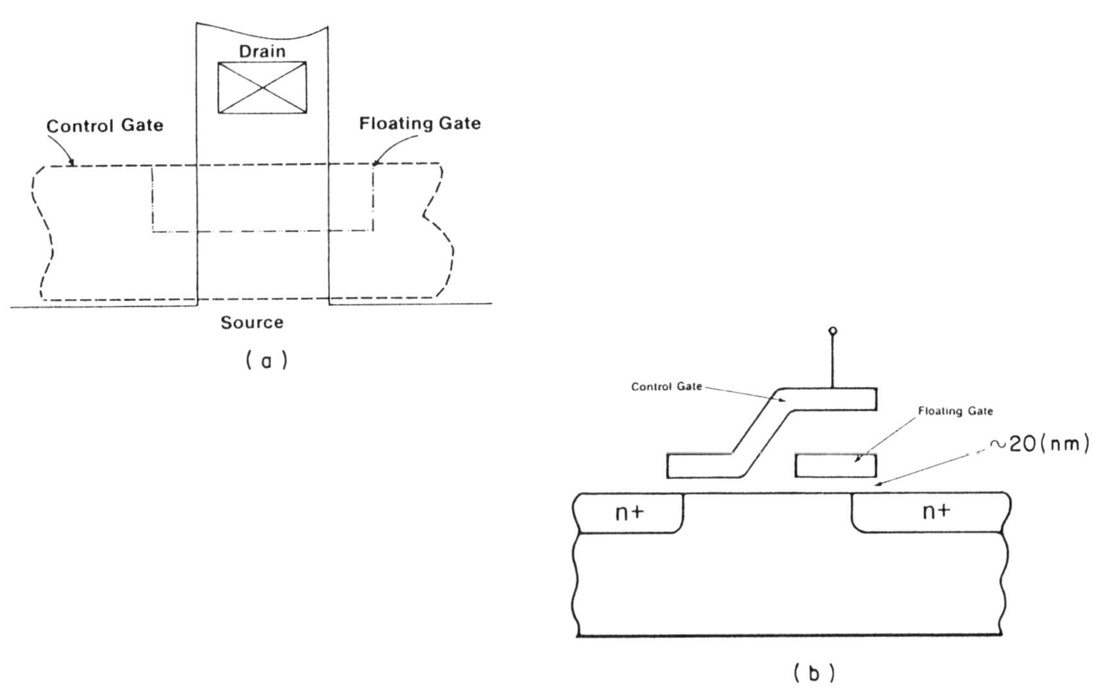

Figure 11.47
128k double polysilicon flash memory cell. (a) Layout, (b) schematic cross-section. (From Samachisa *et al.* [79], SEEQ 1987, with permission of IEEE.)

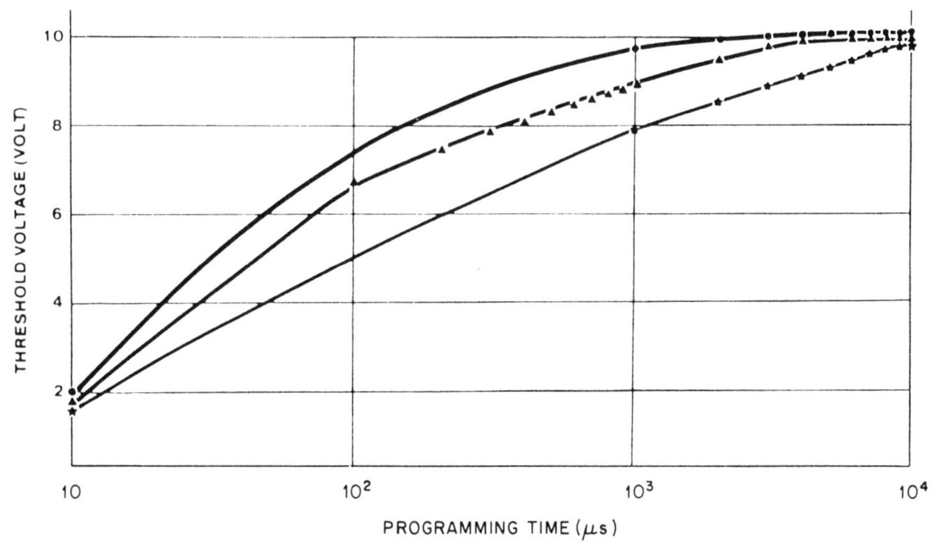

Figure 11.48
Programming characteristics after UV erase and flash erase with various pulse widths, (● after UV erase 1 ms pulse width, ★ after flash erase, 1 ms pulse width, ▲ after flash erase, 100 μs pulse width). (From Samachisa *et al.* [79], SEEQ 1987, with permission of IEEE.)

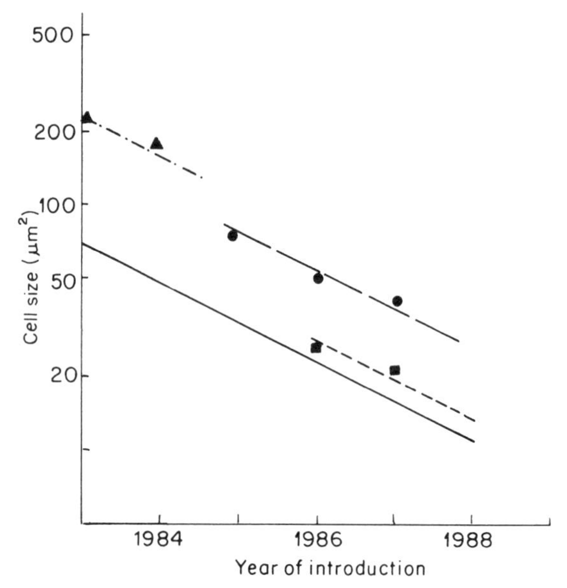

Figure 11.49
Comparison of scaling trends of UV-EPROM (———), two-transistor cell EEPROM (●), four-transistor cell EEPROM (▲) and flash memory (■) showing flash about 20% larger than EPROM and about 50% smaller than the two-transistor EEPROM. (From Samachisa *et al.* [79], SEEQ 1987, with permission of IEEE.)

Programming at 1 ms/byte continued to be by hot electron injection like the EPROM and erase by cold electron tunneling like the EEPROM.

Endurance was specified as a minimum of 100 write–erase cycles but a special selection was offered at 1000 cycles. Access time was specified at 200 ns and power dissipation was 60 mA active and 100 μA standby at CMOS input levels.

11.14.4 A typical 1Mb flash memory

In 1989 SEEQ [82] presented their 1Mb flash memory now in a 1.2 μm process with 24.6 μm^2 cell size. Timing and features were similar to the 512k device. Special circuitry was added to regulate the timing of the internal operations to minimize voltage stress on the cell, provide the required waveforms for programming and erase and prevent unintentional erasing of the memory. Power-up protection circuitry protected against unintended write or erase.

A block diagram in Figure 11.50(a) shows the address and input control latches for writing and erasing along with the timer. A pinout is shown in 11.50(b) with the high voltage V_{PP} pin and the program ($\overline{PGM}$) control pin indicated. A mode select table is in 11.50(c) and a read cycle timing diagram in 11.50(d) showing the synchronous EPROM–ROM compatible read.

A flow chart for the byte-write algorithm and a byte-write timing diagram showing programming of successive bytes are shown in Figures 11.51(a) and (b).

A flowchart for the algorithm for sector erase and the sector erase timing diagram are shown in Figures 11.52(a) and (b) [81].

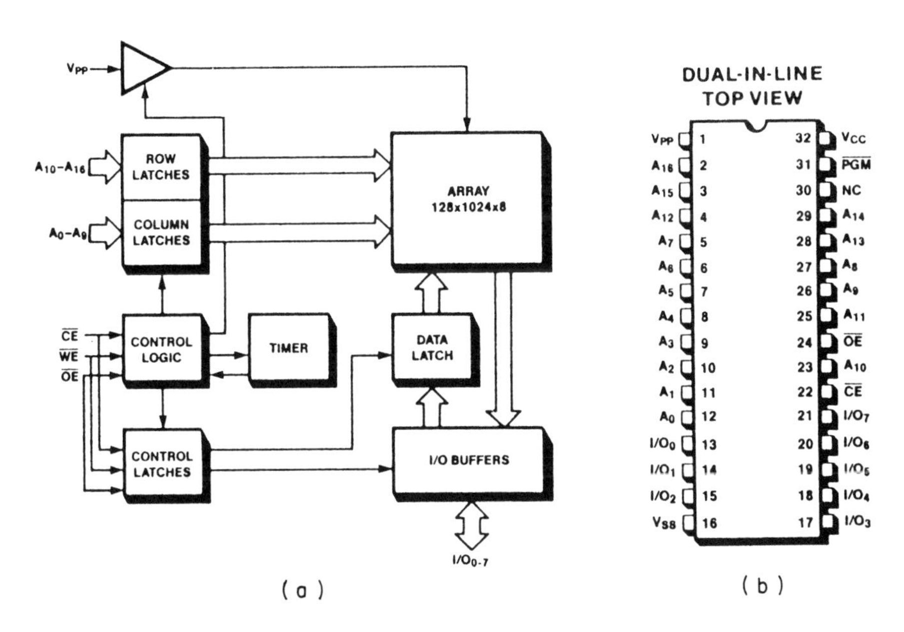

MODE	$\overline{CE}$	$\overline{OE}$	$\overline{PGM}$	V_{PP}	A_{10-16}	A_{0-9}	I/O_{0-7}
Read	V_{IL}	V_{IL}	V_{IH}	X	Address	Address	D_{OUT}
Standby	V_{IH}	X	X	X	X	X	HI-Z
Byte write	V_{IL}	V_{IH}	V_{IL}	V_P	Address	Address	D_{IN}
Chip erase select	V_{IL}	V_{IH}	V_{IL}	TTL	X	X	X
Chip erase	V_{IL}	V_{IH}	V_{IL}	V_P	X	X	'FF'
Sector erase	V_{IL}	V_{IH}	V_{IL}	V_P	Address	X	'FF'

(c)

ADDRESS, I/O_{0-7}, $\overline{CE}$, $\overline{OE}$, $\overline{PGM}$; t_{RC}, t_{AA}, t_{CE}, t_{DF}, t_{OE}, t_{OH}

(d)

Figure 11.50
1Mb flash memory. (a) Block diagram showing input latches for writing and erasing. (b) EPROM compatible pinout showing high voltage V_{PP} and Program ($\overline{PGM}$) pin. (c) Mode selection table defining pin functions. (d) Read cycle timing diagram showing synchronous EPROM–ROM compatible Read. (Source SEEQ Databook [81], 1989.)

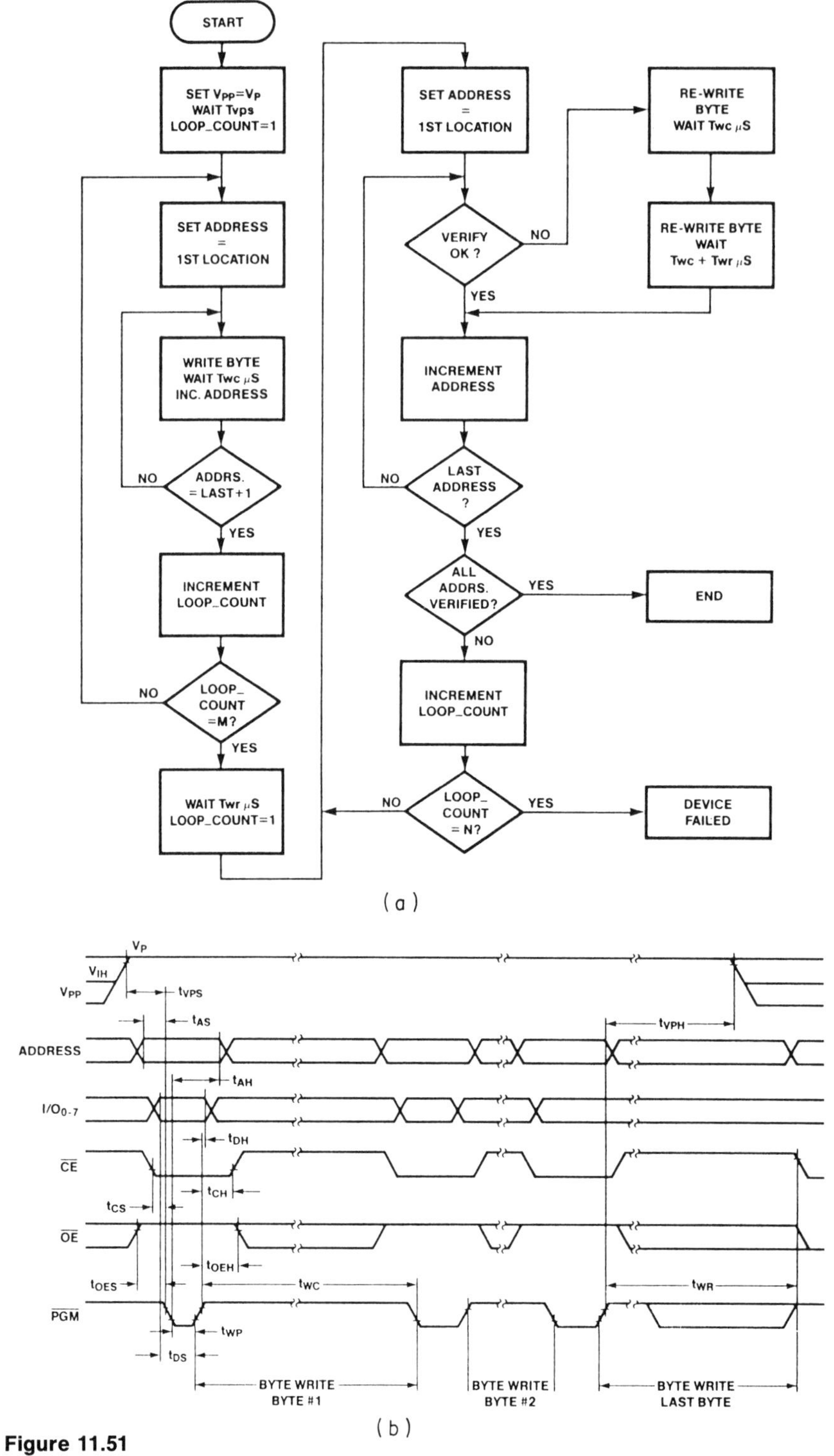

Figure 11.51
1Mb flash memory. (a) Typical flowchart for a byte-write algorithm, (b) byte-write timing diagram. (From SEEQ Databook [81], 1989.)

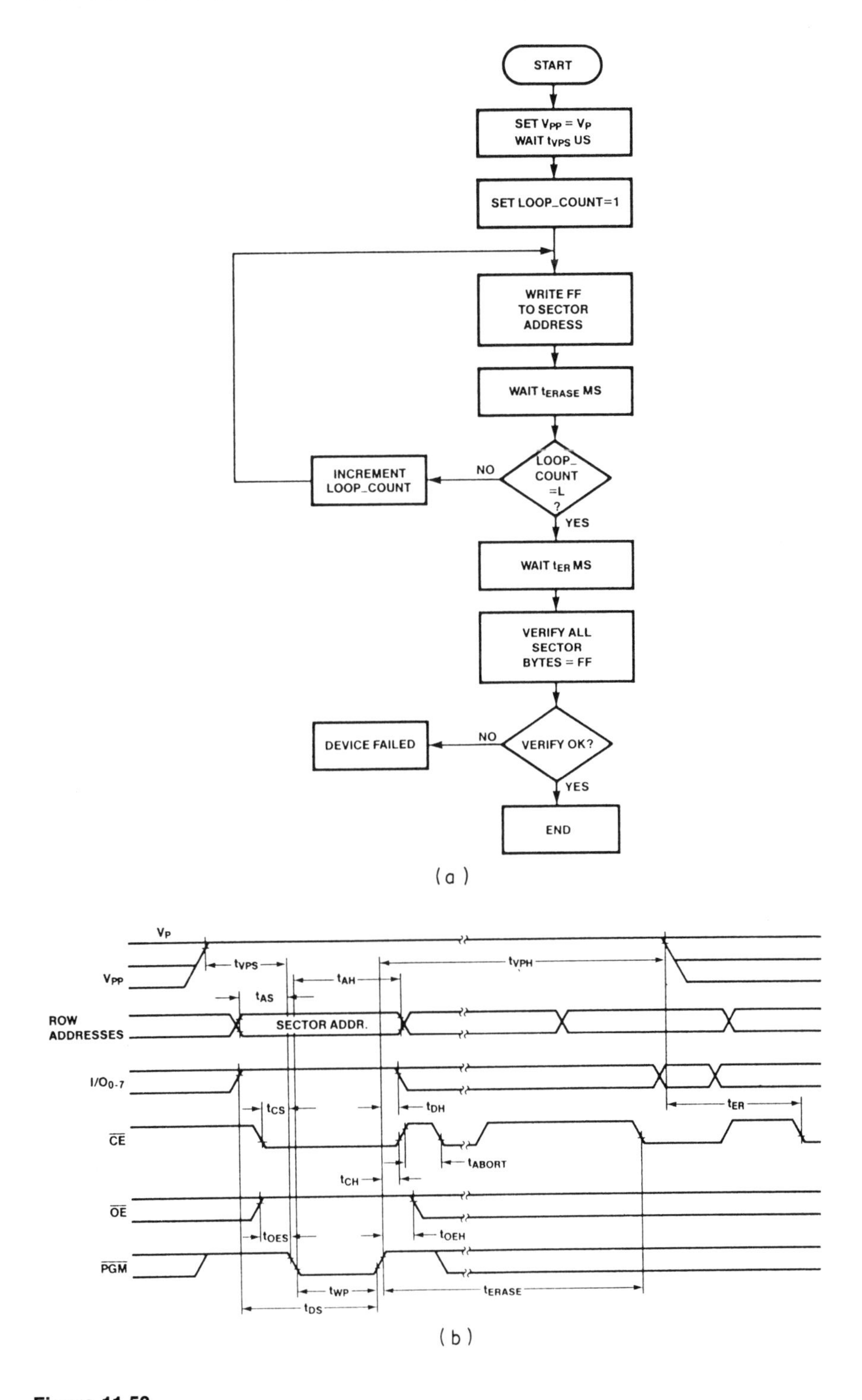

Figure 11.52
1Mb flash memory. (a) Typical flowchart for a sector erase algorithm, (b) sector erase timing diagram. (From SEEQ Databook [81], 1989.)

11.14.5 5 V only flash memories

An early 5 V only flash memory which still programmed and erased by hot electron injection was shown by Waferscale Integration in 1988. Erase was through poly-poly oxide at greater than 27 V. Charge pumping was used on chip to provide the required power supplies. A split-level cell arrangement was used.

A cell size of 18 μm^2 was obtained with 1.2 μm minimum geometries by using a staggered virtual ground array as shown in the cross-section in Figure 11.53(a) and the layout in Figure 11.53(b) which shows that one erase line was shared between two word-lines and runs across the whole array. A schematic circuit diagram is shown in Figure 11.53(c). Interpoly oxide erase was used with the shared second poly erase gate.

Two 5 V flash memories from TI used Fowler–Nordheim tunneling for erase and program. These included an embeddable 256k flash EPROM test module presented in 1988 [87] and a 256k flash EPROM shown in 1989 [88].

All operating voltages were generated internally from the 5 V supply with the use of charge-pump circuitry. These included -18 V applied to the gate for programming along with 5 V applied to the source, -11 V applied to the gate for erase and $+3$ V applied to the gate and 1 V applied to the drain for read of the selected word-line.

Fowler–Nordheim tunneling through a 10 nm tunneling oxide over the drain region was used for both program and erase as shown in Figure 11.54(a). The single-transistor split cell arrangement was used with select gate and floating gate over the channel.

The small 40 μm^2 cell size in a 1.5 μm technology was permitted by the use of a contactless cross-point array with buried N^+ source and drain diffusions for the bit-lines. This cross-point array differs from the one described from TI in an earlier section in that field oxide isolation was used.

An example of operation of a single cell in the array is shown in Figure 11.54(b). The bulk erase operation was performed by applying a negative high voltage pulse to the word-lines and 5 V to the bit-lines simultaneously.

Protection against accidental program or erase was given by a required sequence of three commands for programming and six commands for erase which were loaded in an internal register and had to be detected before the respective operations were enabled. Protection against inadvertent changes during voltage variations was also included.

Endurance of at least 10^5 write erase cycles was shown which is comparable to the EEPROM devices which also use Fowler–Nordheim tunnels for both write and erase.

In 1991 TI [10] discussed a 4Mb flash memory which offered sector erase as well as bulk erase. This 5 V part also used the Fowler–Nordheim tunneling for program and erase used in the earlier device. The sector erase was implemented using a segmented architecture which allowed erasure of 16k bytes at a time.

In 1989 Mitsubishi [92] also presented a 5 V only one-transistor 256k electrically erasable memory using the traditional floating gate construction with thin tunneling oxide over the drain region. The cross-section of the cell is shown in Figure 11.55(a). This cell construction permitted the part to be programmed and erased by Fowler–Nordheim tunneling. Designed in a 1.5 μm technology the cell size was 56 μm^2 and the chip size was 39.1 mm^2. The select gate was integrated with the storage gate and

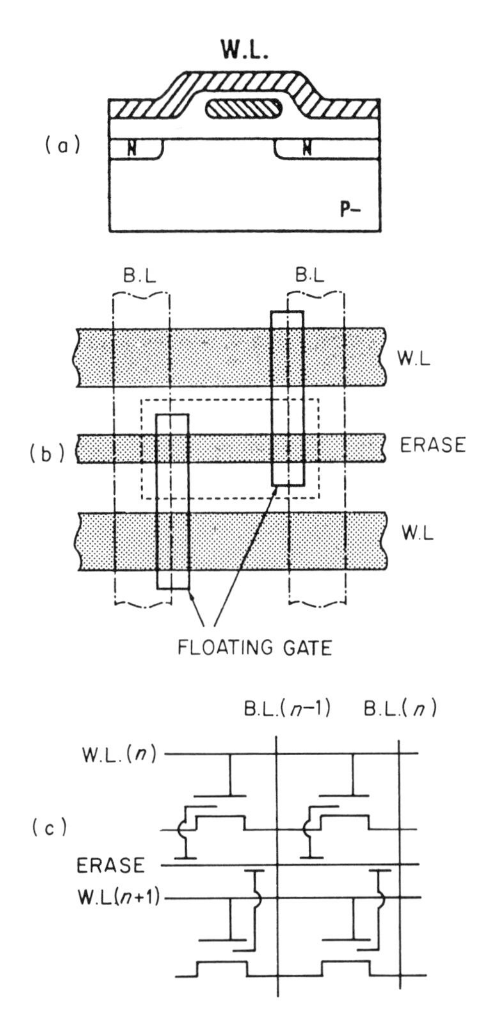

Figure 11.53
Flash memory cell with split level control gate and shared erase gate. (a) Schematic cross section showing split level gate with merged select transistor. (b) Layout showing one erase line shared between two word-lines. (c) Schematic circuit diagram of array structure showing shared erase line. (From Kazcrounian *et al.* [91], Waferscale Integration 1988, with permission of IEEE.)

was over the channel at the source to avoid current leakage through programmed depletion memory transistors.

An erase architecture similar to that of the conventional EEPROM was used giving a page mode erase capability. The programming and erase scheme of the device is shown in Figure 11.55(b).

In 1990 Mitsubishi added a new sensing architecture for multimegabit flash memories which included a divided memory array to halve the bit-line capacitance and dummy reference cells. The sense amplifiers were shared by two pairs of bit-lines so they were placed in the pitch of two bit-lines. Address transition detection was used for internal timing.

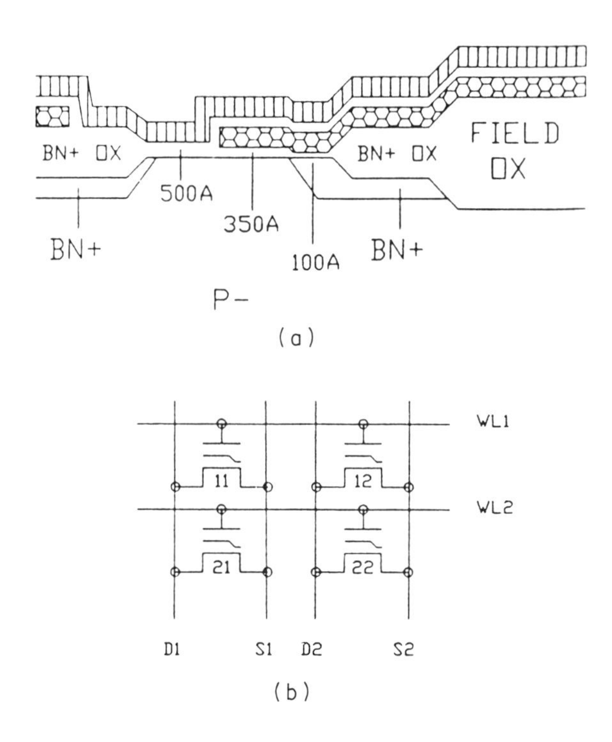

	WL_1	WL_2	S_1	D_1	S_2	D_2
Erase	−11 V	−11 V	5 V	FLOAT	5 V	FLOAT
Program	7 V	18 V	0 V	FLOAT	7 V	FLOAT
Read	0	3	0	1 V	0	1 V

(From D'Arrigo *et al.* [88], TI 1989, with permission of IEEE.)

Figure 11.54
256k only CMOS flash memory cell in contactless array with Fowler–Nordheim program and erase. (a) Schematic cross-section of memory cell showing integrated select transistor over source side of channel and tunneling oxide region over drain. (b) Operating condition for programming a single cell in the contactless array.

A sense timing control circuit compensates for cell current fluctuations caused by process deviations and repeated programming cycles. A program inhibit scheme is used to prevent false programming of unselected cells.

11.14.6 A stacked three-layer polysilicon flash memory cell

In 1989 Mitsubishi [74] described a one-transistor cell 1Mb electrically erasable PROM with a 91.4 mm^2 chip size. The 30.4 μm^2 cell used a 1.0 μm triple polysilicon double metal stacked cell as shown in the schematic cross-section in Figure 11.56(a) [74, 75].

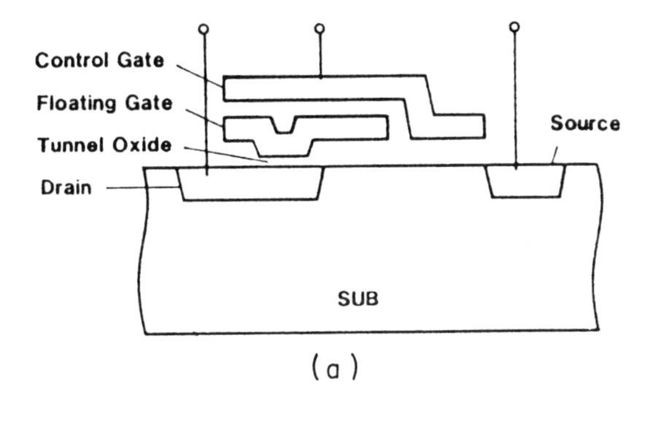

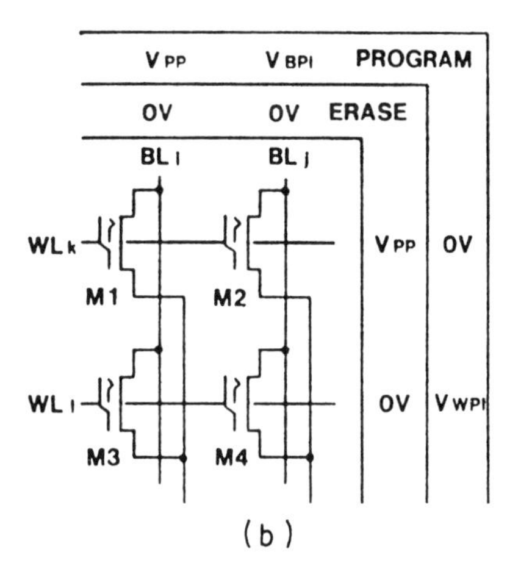

Figure 11.55
256k only one-transistor flash memory using Fowler–Nordheim program and erase. (a) Cross-section of merged select transistor cell showing tunneling oxide over the drain region. (b) Set up for programming and erasing the cell. (From Nakayama *et al.* [92], Mitsubishi 1989, with permission of IEEE.)

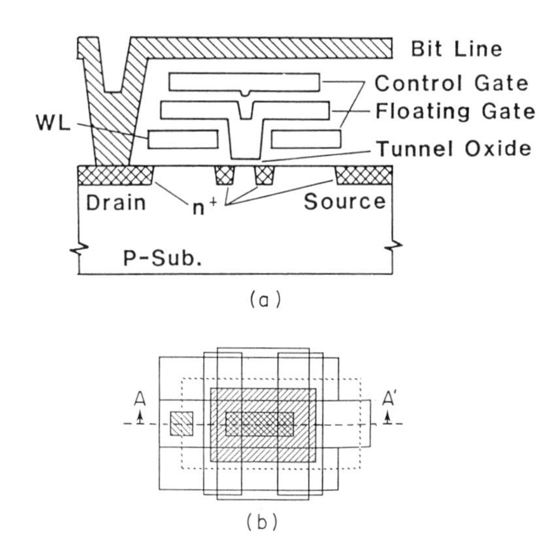

Figure 11.56
1Mb flash memory cell using triple level polysilicon. (a) Schematic cross-section of cell showing floating gate sandwiched between two sections of control gate. (b) Layout. (a: From Terada *et al.* [74], Mitsubishi 1989, with permission of IEEE; b: From Arima *et al.* [35], Mitsubishi 1988, with permission of IEEE.)

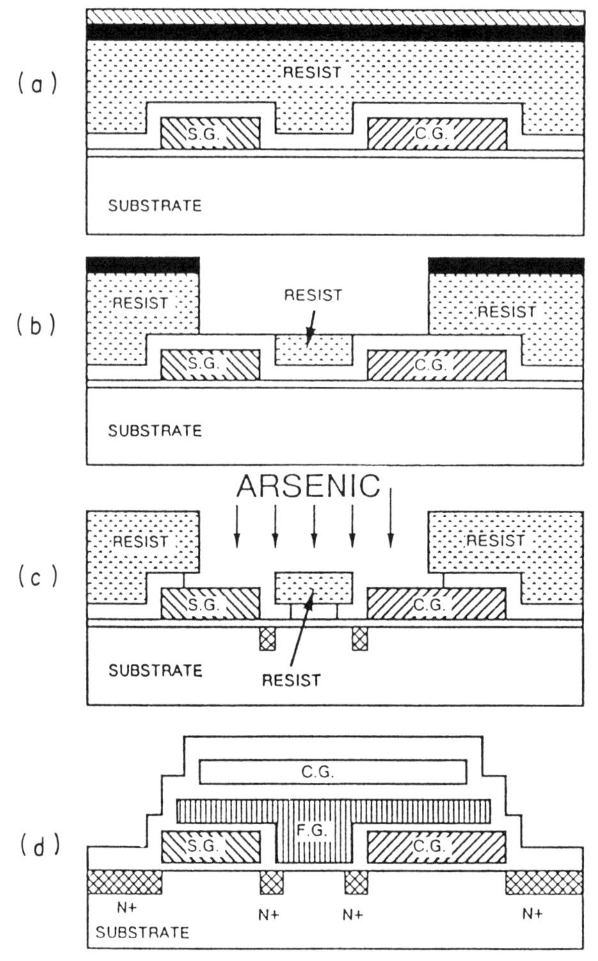

Figure 11.57
Schematic diagram of process steps in manufacture of the 1Mb triple level polysilicon flash memory cell shown in Figure 11.56. (From Arima *et al.* [75], Mitsubishi 1988, with permission of IEEE.)

The first polysilicon layer was used as both the select gate and the lower control gate. The second polysilicon was the floating gate and the third poly layer was the upper control gate. A layout is shown in Figure 11.56(b).

The floating gate was sandwiched between two layers of control gate. The capacitance between floating gate and control gate was effectively increased since the capacitance of the floating gate with the upper and lower controls gates was additive. The capacitance, however, between the floating gate and the substrate was decreased by the shielding effect of the lower control gate and select gate, therefore a low capacitance ratio is achieved resulting in a high drive current.

A schematic representation of the cell fabrication process is shown in Figures 11.57(a) to (d). A differential sensing scheme was used in which the memory array was divided in two halves and the selected cell was compared to the unselected array as a reference level.

Mitsubishi in 1991 [103] showed a 16Mb flash EPROM. The part attained a 60 ns access time both by dividing the chip into 64 blocks and by using a differential sensing scheme with dummy cells. It also featured simultaneous block level erase and verify operations.

11.14.7 A contactless NAND-structured flash memory cell

The NAND cell structure used by Toshiba, which was described in an earlier section as applicable to either EPROM or flash memory, was used in a 4Mb 5 V only electrically erasable programmable ROM. Using 1.0 μm design rules the average cell area per bit was 12.9 μm^2. Block erase, successive programming and random reading were achieved with a new NAND-cell control circuit. Typical erase time was 1.0 ms and page program time was 4.0 ms or 1.0 μs/bit. Typical read access time was 1.6 μs. Chip size was 163.7 mm^2.

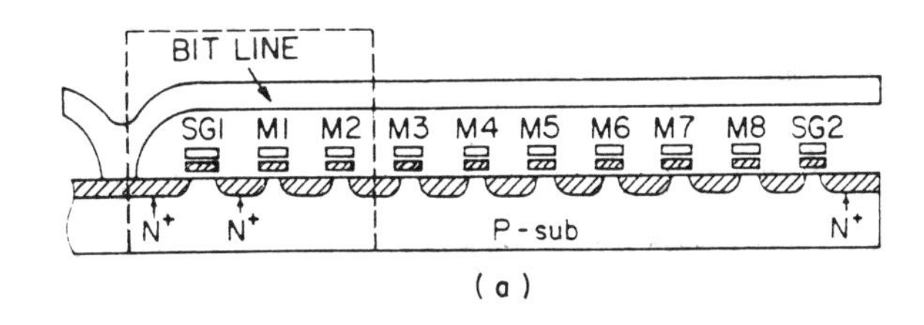

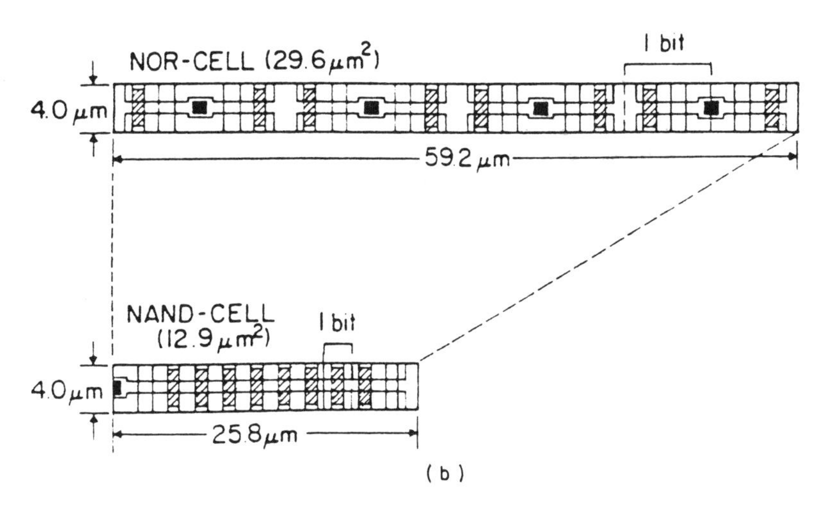

Figure 11.58
4Mb flash electrically erasable PROM with NAND cell structure. (a) Schematic cross-section of cell. (b) Comparison of conventional NOR cell with NAND cell in same technology. (From Momodomi *et al.* [60], Toshiba 1989, with permission of IEEE.)

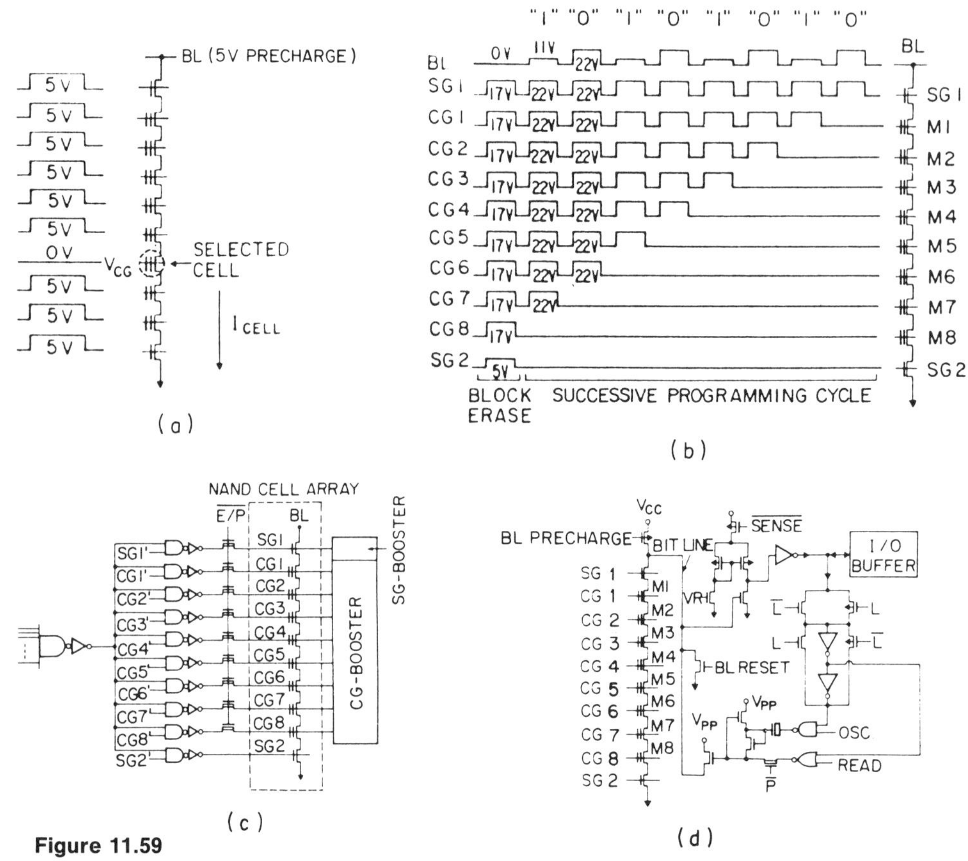

Figure 11.59
4Mb flash electrically erasable PROM with NAND cell structure. (a) Control voltages for read operation, (b) control voltages for erase–program operation, (c) ROW decoder, (d) sense amplifier and bit-line control circuitry. (From Momodomi *et al.* [60], Toshiba 1989, with permission of IEEE.)

A cross-sectional view of the NAND structured cell and a comparison of the conventional NOR cell with the NAND cell are shown in Figures 11.58(a) and (b).

Control voltages for the read and the block erase and eight successive programming operations are shown in Figures 11.59(a) and (b). A row decoder for the NAND cell and a sense amplifier and bit-line control circuitry are shown in Figures 11.59(c) and (d). A tight programmed threshold voltage distribution was controlled by a new program verify algorithm [108].

In 1990 Toshiba [101] showed a 2.3 μm^2 NAND flash EPROM cell in 0.6 μm technology for a 16Mb device. Sixteen transistors were used in parallel which reduced the size 20% from the previous structures with 8 transistors in parallel.

11.14.8 3 V flash memories

In 1991 Mitsubishi [106] showed low voltage circuitry for a 64Mb flash memory which used boosted word-lines and negative gate biased erasing to achieve fast 3 V operation and 5 V erase.

BIBLIOGRAPHY

[1] Miyasaka, K. *et al.* (1982) A 150 ns CMOS 64K EPROM using n-well Technology, *IEEE International Solid State Circuits Conference.*

[2] Spaw, W. *et al.* (1982) A 128K EPROM With Redundancy, *IEEE International Solid State Circuits Conference.*

[3] Bursky, D. (1980) UV EPROMS and EEPROMS Crash Speed and Density Limits, *Electronic Design,* November 22.

[4] Designers Reference, *Electronic Design,* July 19, 1980.

[5] Schneider, B. *et al.* (1980) UV-EPROMS, *Elektronikcentralen,* November.

[6] Woods, M. H. (1980) An EPROMS Integrity Starts With Its Cell Structure, *Electronics,* August 14, 1980.

[7] Stewart, R. B. and Plus, D. (1982) A 40 ns CMOS EEPROM, *IEEE International Solid State Circuits Conference.*

[8] Yaron, G. *et al.* (1982) A 16K EEPROM, *IEEE International Solid State Circuits Conference.*

[9] Gupta, A. *et al.* (1982) A 5 Volt Only 16K EEPROM Utilizes Oxynitride Dielectrics and EEPROM Redundancy, *IEEE International Solid State Circuits Conference.*

[10] Kuo, C. *et al.* (1982) A Sub 100 ns 32K EEPROM, *IEEE International Solid State Circuits Conference.*

[11] Prince, B. (1982) A Fast 32K EEPROM, *New Electronics,* August 17.

[12] Yaron, G. *et al.* (1982) 16K EEPROM With New Array Architecture, *Electronic Engineering,* June.

[13] Vchiumi, K. (1981) 16K EEPROM Keeps MNOS In The Running, *Electronics,* February 24.

[14] 5 V only EEPROM—Springboard For Autoprogrammable Systems, *Electronics,* February 10.

[15] Johnson, W. S. *et al.* (1980) 16K EEPROM Relies On Tunnelling For Byte-Erasable Program Storage, *Electronics,* 28 February.

[16] Prince, B. (1981) Serial 1/0 Reduces EEPROM Package, *Electronic Product Design,* December.

[17] Kalet, M. *et al.* (1982) N-Channel Process Increases Speed And Density Of MNOS EEPROMS, *Electronics,* 5 May.

[18] Duthie, I. (1982) Novel Applications of Non-Volatile Memories, *Semiconductor International.*

[19] *EEPROM Family Applications Handbook* Intel, April 1981.

[20] Ahrens, T. *et al.* (1982) Simplify Alterable Memories With A Versatile 32K Bit EEPROM, *Engineering Design News,* 1 September.

[21] Ford, D. (1982) New Fast 16K and 32K EEPROMS Offer System Ease-Of-Use Features, *Electro 1982.*

[22] Capece, R. P. and Posa, J. G. (1979) Static RAM Tunnels For Non-volatility, *Electronics,* 13 September.

[23] Van Buskirk, M. A. *et al.* (1983) EPROMS Graduate To 256K Density With Scaled N-Channel Process, *Electronics,* February 24.

[24] Dham, V. K. *et al.* (1983) A 5 V-Only E^2PROM Using 1.5 μ Lithography, *ISSCC Proceedings.*

[25] Lancaster, A. *et al.* (1983) A 5 V-Only EEPROM With Internal Program/ Erase Control, *ISSCC Proceedings.*

[26] Donaldson, D. D. *et al.* (1983) +5 V-Only 32K EEPROM. *ISSCC Proceedings.*

[27] Becker, N. J. *et al.* (1983) A 5 V-Only 4K Nonvolatile Static RAM. *ISSCC Proceedings.*

[28] Yoshida, M. *et al.* (1983) A 150 ns 288K CMOS EPROM With Redundancy. *ISSCC Proceedings.*

[29] Frohman-Benchkowsky, D. (1971) A fully decoded 2048-bit electrically programmable FAMOS read-only memory, *IEEE Journal of Solid State Circuits,* Vol. **SC-6**, Oct. 1971, p.301.

[30] Special Report on Semiconductor Memories, *Computer Design*, August 1984, p.114.
[31] Rinerson, D. *et al.* (1984) 512K EPROMs, *ISSCC Proceedings*, February p.136.
[32] Okumura, K. *et al.* (1984) 1Mb EPROM, *ISSCC Proceedings*, February, p.140.
[33] Kanauchi, S. *et al.* (1984) A High Performance 1 Mbit EPROM, *IEEE Journal of Solid State Circuits*, **SC-19**, No. 5, October 1984, p.646.
[34] Spaw, W. *et al.* (1982) A 128K EPROM with Redundancy, *ISCC Proceedings* February, p.112.
[35] Van Buskirk, M. A. *et al.* (1983) A 200 ns 256k HMOSII EPROM, *ISSCC Proceedings*, February, p.162.
[36] Cheng, P. Y. *et al.* (1981) A 16k CMOS EPROM, *ISSCC Proceedings*, February, p.160.
[37] Miyasaka, K. *et al.* (1982) A 150 ns CMOS 64k EPROM using N-Well Technology, *ISSCC Proceedings*, p.182.
[38] Yoshida, M. *et al.* (1983) A 288K CMOS EPROM with Redundancy, *IEEE Journal of Solid State Circuits*, Vol. **SC-18**, No.5, October 1983, p.544.
[39] Knecht, M. W. *et al.* (1983) A High-Speed Ultra-Low Power 64K CMOS EPROM with On-Chip Test Functions, *IEEE Journal of Solid State Circuits*, Vol. **SC-18**, No.5, October 1983, p.554.
[40] Tanaka, S. *et al.* (1984) A Programmable 256k CMOS EPROM with On-Chip Test Circuits, *ISSCC Proceedings*, February, p.148.
[41] Yoshizaki, K. *et al.* (1985) A 95 ns 256k CMOs EPROM, *ISSCC Proceedings*, February, p.166.
[42] Ip, W. *et al.* (1984) 256Kb CMOS EPROM, *ISSCC Proceedings*, February, p.138.
[43] Tanaka, S. *et al.* (1984) A Programmable 256K CMOS EPROM with On-Chip Test Circuits, *ISSCC Proceedings*, February, p.148.
[44] Saito, S. *et al.* (1985) A Programmable 80 ns 1Mb CMOS EPROM, *ISSCC Proceedings*, p.176.
[45] Hagiwara, T. and Matsuo, H. (1985) Page Mode Programming 1Mb CMOS EPROM, *ISSCC Proceedings*, February, p.174.
[46] Venkatesh, B. *et al.* (1986) A CMOS 1Mb EPROM, *ISSCC Proceedings*, February, p.40.
[47] Coffman, T. *et al.* (1987) A 1Mb CMOS EPROM with a 13.5 μm^2 Cell, *ISSCC Proceedings*, February, p.72.
[48] Yoshida, M. *et al.* (1987) An 80 ns Address-Data Multiplex 1Mb CMOS EPROM, *ISSCC Proceedings*, February p.70.
[49] Gastaldi, R. *et al.* (1988) A 1-Mbit CMOS EPROM with Enhanced Verification, *IEEE Journal of Solid State Circuits*, Vol. **23**, No.5, October 1988, p.1150.
[50] Fukuda, M. *et al.* (1988) A 55 ns 64k × 16b CMOS EPROM, *ISSCC Proceedings*, February, p.122.
[51] Hoff, D. *et al.* (1989) A 23 ns EPROM with Double-Layer Metal and Address Transition Detection, *ISSCC Proceedings*, February, p.130.
[52] Ali, S. B. *et al.* (1988) A 50-ns 256k CMOS Split-Gate EPROM, *IEEE Journal of Solid State Circuits*, Vol. **23**, No.1, February 1988, p.79.
[53] Mitchell, A. T. *et al.* (1987) A New Self-Aligned Planar Array Cell for Ultra High Density EPROMs, *IEDM Proceedings*, p.548.
[54] Masuoka, F. *et al.* (1987) New Ultra High Density EPROM and Flash EEPROM with NAND Structure Cell, *IEDM Proceedings*, p.552.
[55] Atsumi, S. *et al.* (1987) A 120 ns 4Mb CMOS EPROM, *ISSCC Proceedings*, February, p.74.
[56] Canepa, G. *et al.* (1988) A 90 ns 4Mb CMOS EPROM, *ISSCC Proceedings*, February, p.120.
[57] Ohtsuka, N. *et al.* (1987) A 4-Mbit CMOS EPROM, *IEEE Journal of Solid State Circuits*, Vol. **SC-22**, No.5, October 1987, p.669.
[58] Imamiya, K. *et al.* (1989) A 68 ns 4Mbit CMOS EPROM with high noise immunity design, *Symposium on VLSI Circuits Digest, 1989*, p.37.

[59] Nakamura, Y. *et al.* (1990) A 55 ns 4Mb EPROM with 1-Second Programming Time, *ISSCC Proceedings*, February, p.62.
[60] Momodomi, M. *et al.* (1989) An Experimental 4-Mbit CMOS EEPROM with a NAND-Structured Cell, *IEEE Journal of Solid State Circuits*, Vol. **24**, No.5, October 1989, p.1238.
[61] Imamiya, K. *et al.* (1990) A 68 ns 4Mb CMOS EPROM with High Noise Immunity Design, *IEEE Journal of Solid State Circuits*, Vol. **25**, No.1, February 1990, p.72.
[62] Bergemont, A. *et al.* (1989) A High Performance CMOS Process for Submicron 16Mb EPROM, *IEDM Proceedings*, p.591.
[63] Bellezza, O. *et al.* (1989) A New Self-Aligned Field Oxide Cell for Multimegabit EPROMs, *IEDM Proceedings*, p.579.
[64] Yoshikawa, K. *et al.* (1988) An Asymmetrical Lightly-Doped Source (ALDS) Cell for Virtual Ground High Density EPROMS, *IEDM Proceedings*, p.88.
[65] Yoshikawa, K. *et al.* (1989) 0.6 μm EPROM Cell Design Based on a New Scaling Scenario, *IEDM Proceedings*, p.587.
[66] Hisamune, Y. S. *et al.* (1989) A 3.6 μm^2 Memory Cell Structure for 16MB EPROMs. *IEDM Proceedings*, p.583.
[67] Higuchi, M. (1990) An 85 ns 16Mb CMOS EPROM with Alterable Word Organization, *ISSCC Proceedings*, February, p.56.
[68] Pathak, S. *et al.* (1985) A 25 ns 16K CMOS PROM using a 4-Transistor Cell, *ISSCC Proceedings*, February, p.162.
[69] Pathak, S. *et al.* (1985) A 25 ns 16K CMOS PROM Using a Four-Transistor Cell and Differential Design Techniques, *IEEE Journal of Solid State Circuits*, Vol. **SC-20**, No. 5, October 1985, p.964.
[70] Wright, M. (1987) High-Speed EPROMs, *EDN*, September 17, p.133.
[71] Markowitz, M. C. (1989) Nonvolatile Memories, *EDN*, September 1, p.94.
[72] Atsumi, S. *et al.* (1990) A 16 ns 1Mb CMOS EPROM, *ISSCC Proceedings*, February, p.58.
[73] Maruyama, T. *et al.* (1988) Wide Operating Voltage Range and Low Power Consumption EPROM Structure for Consumer Oriented ASIC Applications, *Proceedings IEEE 1988 Consumer Integrated Circuits Conference*, p.4.1.2.
[74] Terada, Y. *et al.* (1989) 120 ns 128k $\times$ 8b/64k $\times$ 16b CMOS EEPROMs, *ISSCC Proceedings*, February, p.136.
[75] Arima, H. *et al.* (1988) A Novel Process Technology and Cell Structure for Mega Bit EEPROM, *IEDM Proceedings*, December, p.420.
[76] Cole, B. C. (1988) Intel's First Flash EEPROMs Make In-Circuit Reprogramming Easy, *Electronics*, April 14, p.103.
[77] Masuoka, F. *et al.* (1987) A 256k bit Flash EEPROM Using Triple-Polysilicon Technology, *IEEE Journal of Solid State Circuits*, Vol. **SC-22**, No. 4, August 1987, p.548.
[78] Masuoka, F. *et al.* (1984) A new flash EEPROM cell using triple polysilicon technology, *IEDM Proceedings*, p.464.
[79] Samachisa, G. *et al.* (1987) A 128k Flash EEPROM Using Double-Polysilicon Technology, *IEEE Journal of Solid State Circuits*, Vol. **SC-22**, No. 5, October 1987, p.676.
[80] Lineback, J. R. (1988) SEEQ's 512k bit Flash EEPROMs Support In-System Programming on 12 V Supply, *Electronics*, March 17, p.149.
[81] *SEEQ Data Book* 1989.
[82] Cernea, R. A. *et al.* (1989) A 1Mb Flash EEPROM, *IEEE ISSCC Proceedings*, February, p.138.
[83] Cole, B. C. (1986) The Exploding Role of Nonvolatile Memory, *Electronics*, August 21, p.47.
[84] Kynett, V. N. *et al.* (1988) An In-System Reprogrammable 32k $\times$ 8 CMOS Flash Memory, *IEEE Journal of Solid State Circuits*, Vol. **23**, No. 5, October 1988, p.1157.
[85] Kynett, V. N. *et al.* (1989) A 90 ns One-million Erase/Program Cycle 1M bit Flash Memory, *IEEE Journal of Solid State Circuits*, Vol. **24**, No. 5, October 1989, p.1259.

[86] Cole, B. C. (1987) The Changing Face of Non-volatile Memories, *Electronics*, July 9, p.61.
[87] Gill, M. *et al.* (1988) A 5-Volt Contactless Array 256k bit Flash EEPROM Technology, *IEDM Proceedings*, p.428.
[88] D'Arrigo, S. *et al.* (1989) A 5 V-Only 256k Bit CMOs Flash EEPROM, *IEEE ISSCC*, February, p.132.
[89] Esquivel, J. *et al.* (1986) High Density Contactless, Self Aligned EPROM Cell Array Technology, *IEDM Proceedings*, p.592.
[90] Kume, H. *et al.* (1987) A Flash-Erase EEPROM Cell with an Asymmetrical Source and Drain Structure, *IEDM Proceedings*, p.560.
[91] Kazerounian, R. *et al.* (1988) A 5 Volt High Density Poly-Poly Erase Flash EPROM Cell, *IEDM Proceedings*, p.436.
[92] Nakayama, T. *et al.* (1989) A 5 V Only One Transistor 256k EEPROM with Page Mode Erase, *IEEE Journal of Solid State Circuits*, Vol. **24**, No.4, August 1989, p.911.
[93] Kobayashi, K. *et al.* (1990) A High-Speed Parallel Sensing Architecture for Multi-Megabit Flash EEPROMs, *IEEE Journal of Solid State Circuits*, Vol. **25**, No.1, February 1990, p.79.
[94] Masuoka, F. *et al.* (1987) New Ultra High Density EPROM and Flash EEPROM with NAND Structure Cell, *IEDM Proceedings*, p.552.
[95] Kobayashi, K. *et al.* (1989) A Self-Timed Dynamic Sensing Scheme for 5 V Only Multi-Mb Flash EEPROMs, *VLSI Symposium*, p.39.
[96] Marktubersicht: EPROMs, EEPROMs, Flash Memories (1989) *Markt & Technik*, Nr. 35 vom 25, August 1989, p.53.
[97] Woo, B. J. *et al.* (1990) A Novel Memory Cell Using Flash Array Contactless EPROM (FCE) Technology, *IEEE IEDM Tech. Digest*, Dec. 1990, p.91.
[98] Prince, B. and Due-Gundersen, G. (1983) *Semiconductor Memories* John Wiley, Chichester.
[99] Ohshima, Y. *et al.* (1990) Process and Device Technology for 16Mbit EPROMs with Large Tilt Angle Implanted P-Pocket Cell', *IEEE IEDM Tech. Digest*, Dec. 1990, p.95.
[100] Bez, R. *et al.* (1990) A Novel Method for the Experimental Determination of the Coupling rations in submicron EPROM and Flash EEPROM cells, *IEEE IEDM Tech. Digest*, Dec. 1990, p.99.
[101] Shirota, R. *et al.* (1990) A 2.3 μm^2 Memory Cell Structure for 16Mb NAND EEPROMs, *IEEE IEDM Tech. Digest*, Dec. 1990, p.103.
[102] Aritome, S. *et al.* (1990) A Reliable Bi-Polarity Write/Erase Technology in Flash EEPROMs, *IEEE IEDM Tech. Digest*, Dec. 1990, p.111.
[103] Nakayama, T. *et al.* (1991) A 60 ns 16Mb Flash EEPROM with Program and Erase Sequence Controller, *IEEE ISSCC Proceedings*, February 1991, p.260
[104] Ohtsuka, N. *et al.* (1991) A 62 ns 16Mb CMOS EPROM with Address Transition Detection Technique, *IEEE ISSCC Proceedings*, February 1991, p.262.
[105] Kammerer, W. *et al.* (1991) A New Virtual Ground Array Architecture for Very High Speed, High Density EPROMs, *Symposium on VLSI Circuits*, June 1991, p.83.
[106] Miyawaki, Y. *et al.* (1991) A New Erasing and Row Decoding Scheme for Low Supply Voltage Operation of 16Mb and 64Mb Flash EEPROMs, *Symposium on VLSI Circuits*, June 1991, p.85.
[107] McConnell, M. *et al.* (1991) An Experimental 4Mb Flash EEPROM with Sector Erase, *IEEE Journal of Solid State Circuits*, **SC-26**, No. 4, April 1991, p.484.
[108] Momodomi, M. *et al.* (1991) A 4Mb NAND EEPROM with Tight Programmed V_C Distribution, *IEEE Journal of Solid State Circuits*, **SC-26**, No. 4, April 1991, p.492.

12 FIELD ALTERABLE ROMs II: EEPROM, EAROM, NV-RAM

12.1 OVERVIEW OF ELECTRICALLY ERASABLE PROMs

The electrically erasable PROM market has divided for historical reasons into four fairly distinct product segments. Different types of devices have been developed for specific applications requirements in each of these segments. These include the low density devices which we will call by their historical name of EAROMs to distinguish them from the medium density EEPROMs, the embedded EEPROM–EAROMs and the non-volatile SRAMs. These divisions are reflected in the market numbers and the trends and are illustrated in Figure 12.1. In addition two new electrically erasable devices have been developed recently, the flash memory which was discussed in the last chapter, and the non-volatile DRAM which is discussed in this chapter.

The low density below 8k EAROMs have been used predominantly in such applications as consumer radio tuners, automotive engine controllers, price storage in point of sale terminals and postage meters, line driven telephony systems, and calibration and set up parameters for industrial automation systems. The companies supplying and using parts in this specialized applications market have tended to be the consumer or industrial oriented companies such as Mitsubishi, G.I., Hitachi, NCR, and Philips.

The devices supplied for this low density market are increasingly being embedded into microcontrollers as the technology for embedding of reasonable sized arrays of electrically erasable cells and the ability to generate the high voltage needed for program and erase on the chip have been developed. The spread of this market between stand-alone and embedded device types, both of which are growing markets, tends to make this usage much larger than shown in the market figures for the stand-alone parts only.

Medium density EEPROMs have been required by microprocessor driven applications such as distributed systems or changeable program store. These parts have, as a result, been developed with a focus on the high endurance and high speed requirements of this market as well as the need for applications specific features for

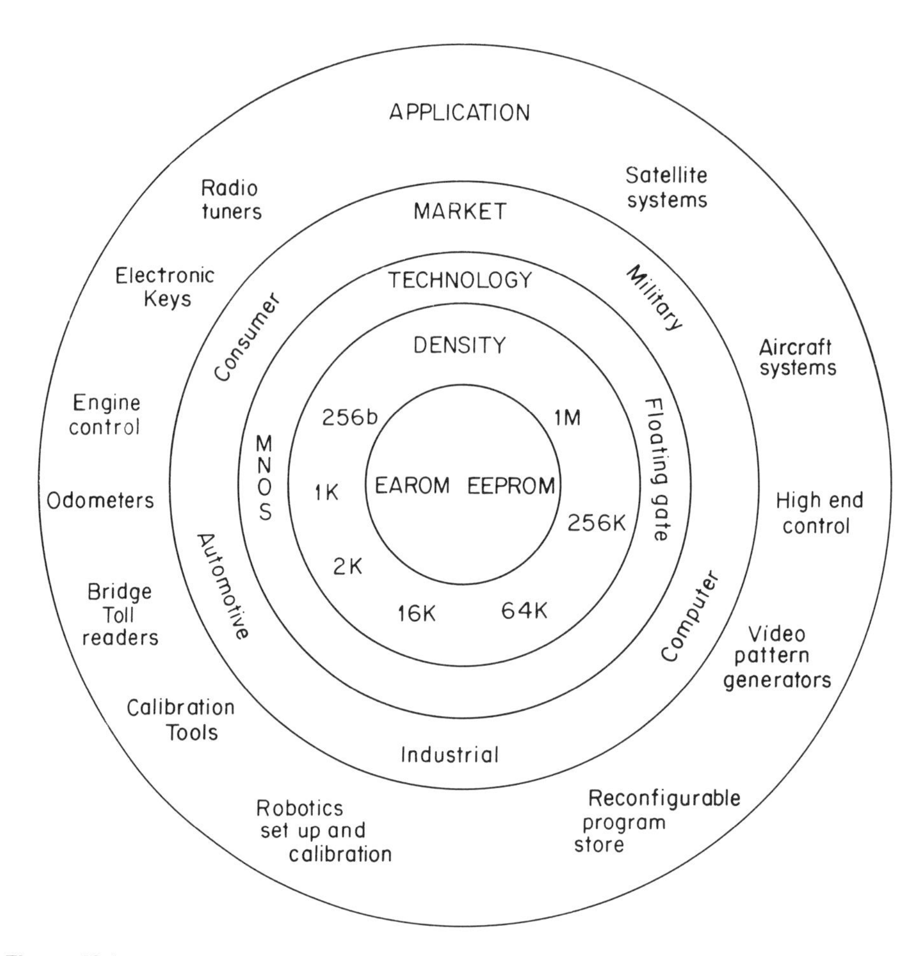

Figure 12.1
The market for electrically erasable reprogrammable ROMs.

making the parts easier to use in microprocessor controlled systems. Suppliers have tended to be the microprocessor oriented suppliers such as Intel, and two Intel spin-offs—SEEQ and Xicor. Other applications include adaptive robotics, programmable video pattern generators, programmable dataloggers, and high speed process controllers.

Military requirements for remote reprogramming in aircraft and satellites have also driven this market segment with a focus on extended temperature and on high reliability. The requirement of this application for radiation hardness has also encouraged the development of such new technologies as the ferroelectric memories.

NV-SRAMs have been developed for those applications requiring a small amount of memory with SRAM read and write speeds and non-volatile back-up in case of power failure. Since they have large chip sizes and hence are relatively expensive, they are used predominantly in applications requiring only a small amount of non-volatile memory.

Finally, the original market foreseen for the EEPROMs, that of a low cost, high density replacement for the UV EPROMs and ROMs remains and is the target of such new technology developments as the Flash memories. These are predominantly being developed by the historic UV EPROM and ROM suppliers such as Intel, Toshiba, TI, and Mitsubishi, and the EEPROM suppliers hoping to hold their historical market share in mid-density EEPROMs such as SEEQ and Waferscale.

The new NV-DRAMs are high density and offer non-volatile backup. It remains to be seen if they become a significant device.

While much of the technology used in these parts is similar, the devices that have developed for these widely differing market segments are quite applications specific and will be considered separately in the following sections.

12.2 ELECTRICALLY ERASABLE PROM TECHNOLOGY AND TRENDS

The four basic technologies used to manufacture electrically reprogrammable ROMs all utilize to some extent Fowler–Nordheim tunneling which is cold electron tunneling through the energy barrier at a silicon–silicon dioxide interface and into the oxide conduction band.

Cold electron tunneling requires an oxide thickness of 10 nm or less depending on the particular technology. This is as opposed to the avalanche or channel injection mechanism of hot electrons over the energy barrier as found in the standard UV EPROM process which can occur with oxides as thick as 100 nm. These mechanisms were discussed in more detail in Chapter 5.

The thicker oxide of the UV EPROM process permits programming by hot electron injection but allows no similar deprogramming mechanisms. This necessitates the use of ultraviolet light to discharge the gate.

Fowler–Nordheim tunneling, however, is a reversible process dependent only on the ability of the industry to process thin oxide layers of high integrity.

12.3 THE MNOS PROCESS AND TRENDS

The earliest electrically reprogrammable ROM process in the early 1970s utilized a metal–nitride–oxide silicon combination (MNOS) for the gate region of a P-channel storage cell producing devices called EAROMs (electrically alterable ROMs). Figure 12.2 shows a schematic cross-section of an early MNOS EAROM cell with programming (a) and deprogramming (b) conditions indicated. The thin (5 nm) silicon dioxide layer allowed charge to tunnel through when a voltage in the +25 to −30 V range was applied to the gate. This charge was trapped in the silicon dioxide to silicon nitride interface and remained trapped there since both the materials are high quality insulators.

With the excess negative charge trapped at the interface, the cell was in a deprogrammed low threshold state. The application of a negative read signal to the gate turned the gate on more readily than if the trapped electrons were not present.

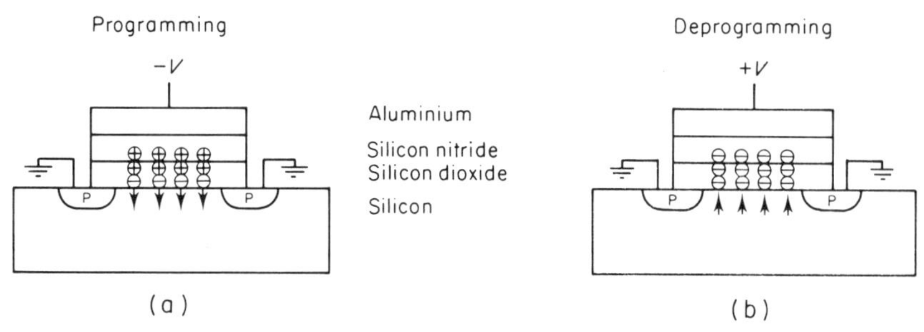

Figure 12.2
Schematic cross-section of an early P-channel MNOS EAROM cell showing (a) programming with negative voltage applied to the gate and the source and drain grounded, and (b) deprogramming with a positive voltage applied to the gate and the source and drain grounded. (From Prince and Due-Gundersen [56] 1983.)

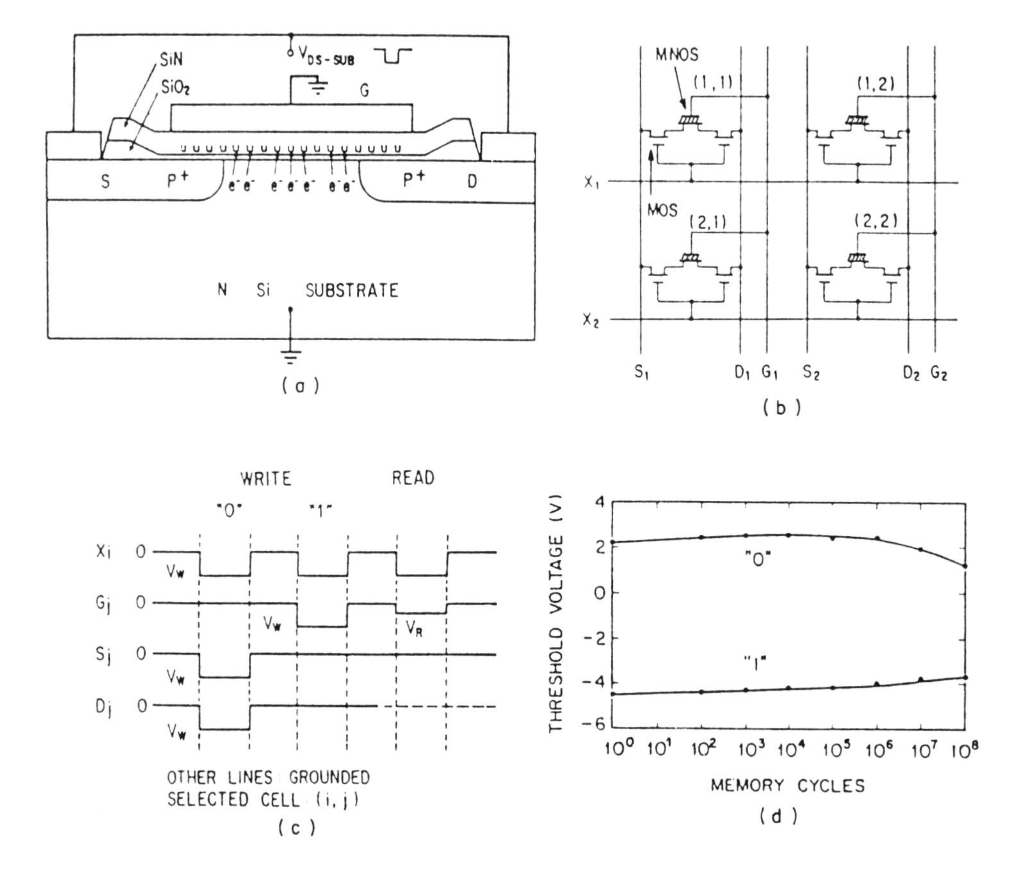

Figure 12.3
Early MNOS EAROM using only positive voltage to write and erase cell. (a) Schematic cross-section of storage transistor showing bias for writing. (b) Three-transistor cell consisting of storage transistor and two select transistors in a four-cell memory array. (c) Write and read timing diagram for selected cell (*i*, *j*) in the memory array shown in (b). (d) Write–erase cycle endurance showing change in threshold voltage of programmed and erased cell. (From Uchida *et al.* [12], Toshiba 1975, with permission of IEEE.)

Conversely, a high negative voltage (−25 to −30 V) applied to the gate drove electrons from (holes to) the interface region resulting in a programmed high threshold state. An applied negative read signal would be opposed by the effective positive gate bias and not turn the cell on.

These first EAROMs tended to be low performance and suffered from the severe density limitations inherent in PMOS processing.

While earliest MNOS devices used both positive and negative voltages to write and erase, an MNOS EAROM shown by Toshiba [12] in 1975 utilized only voltages of the same polarity. This PMOS storage cell used a combination of avalanche tunneling for erasing (logic '0') and direct tunneling for writing (logic '1').

For writing a '0' (erasing) a negative pulse of −35 V was applied to the source and drain electrodes while the substrate and gate were grounded as shown in the biased schematic cross-section of the MNOS transistor in Figure 12.3(a). The potential at the central portion of the channel became almost the same as that of the drain and source so that avalanche tunneling of electrons occurred from the silicon to the nitride through the thin oxide layer and the electrons were trapped in the nitride. This made the threshold of the MNOS transistor shift in the positive direction so it conducted more easily. This was the logic '0' state.

Programming, which is not shown in the figure, consisted of writing a '1' by forcing the emission of electrons from the traps in the nitride by applying a negative pulse of −35 V to the gate electrode while the source and drain were grounded.

To enable individual bits to be written without affecting unselected bits, a three transistor cell structure was used as shown in Figure 12.3(b). The timing diagram for writing, erasing and reading is shown in Figure 12.3(c).

The thickness of the nitride was 43 nm and the oxide was 1.8 nm. The read voltage was 12 V, the read access time was a slow 600 ns and the write cycle time for a single bit was 10 to 100 μs. The write–erase cycle endurance of a single cell is shown in Figure 12.3(d) as about 10^8 write–erase cycles, although the total device was guaranteed at 10^4 write–erase cycles. The chip was relatively large due both to the PMOS periphery and the three transistor per bit memory array.

This density limitation could be reduced by using one-transistor cell structures. In this case, the same transistor both stored information and was interrogated by applying a voltage directly to its gate. Each such read operation diminished the level of stored charge until eventually it was depleted (typically after 10^9 read cycles).

As a result, the MNOS memories using PMOS one-transistor cells suffered from slow read speeds, very high voltage power supplies, nonstandard pin configurations and read cycle limitations. They were manufactured nevertheless because of the smaller chip size and resulting lower cost.

In the late 1970s, several manufacturers including General Instruments, NCR, Inmos, and Hitachi, developed N-channel processes using the MNOS technology which improved performance and density and were an order of magnitude more immune to read disturbance due to a lower read voltage (4 V versus 9 V) which took off less stored charge. Even this degree of data loss was, however, found unnecessary as the reduced density technology permitted use of a two-transistor cell which could be read nearly indefinitely without degradation of the stored information by interrogating only the select transistor and not the storage transistor.

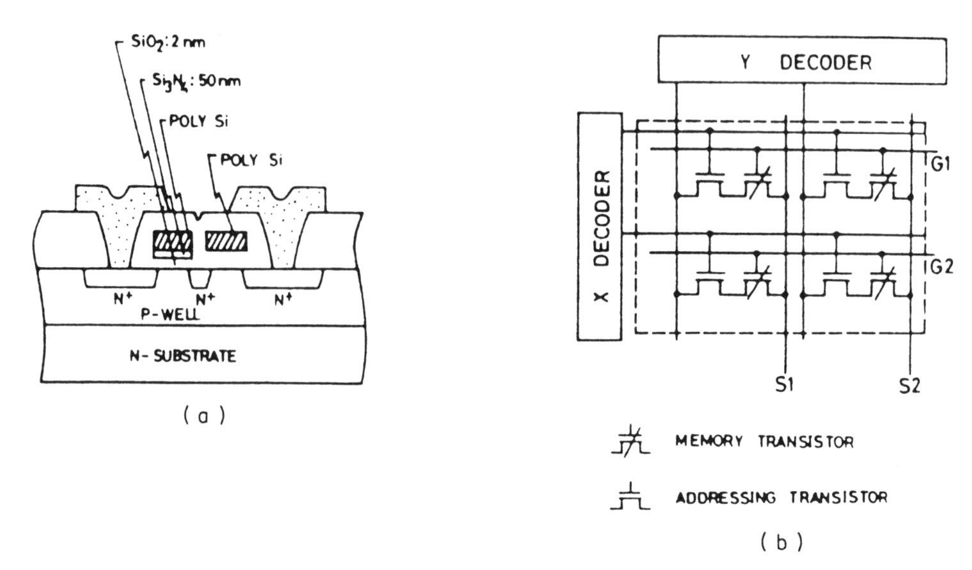

Figure 12.4
16k two-transistor cell EAROM in silicon gate MNOS technology. (a) Schematic cross-section. (b) Circuit diagram of memory array showing select and storage transistors. (From Hagiwara *et al.* [13], Hitachi 1980, with permission of IEEE.)

A Hitachi [13] 16k device using N-channel silicon gate MNOS cell technology was shown in 1980. It used a two-transistor cell consisting of an MNOS storage transistor and a NMOS read transistor as shown in the schematic cross-section in Figure 12.4(a). This cell configuration reduced data loss caused by direct reading of the MNOS cell in one-transistor structures since the gate of the MNOS was grounded in the read mode. It also significantly improved the read access time over that of the earlier three transistor structure due to reduction of the load capacitance and increase in the read current. A schematic circuit diagram of the memory indicating the array, X and Y decoders and select and ground lines is shown in Figure 12.4(b).

A +25 V program–erase voltage and +5 V read voltage were externally applied. A +20 V source was generated internally. 10 year data retention and endurance of about 10^5 cycles were demonstrated.

The high program–erase voltage required measures to prevent MOS breakdown. Guard ring structures were used to prevent latch-up. High voltage DMOS transistors, which were stable to over 35 V, were used for the load devices. The wells of the DMOS transistors were separated from other wells to eliminate substrate effects.

The read access time was 140 ns, programming time was 1 ms per byte, and erase time was 100 ms. Power dissipation was 210 mW typical.

In 1983 a +5 V only MNOS EAROM was presented by Inmos [20]. This part used a cell with double polysilicon gates as shown in the layout and cross-section in Figure 12.5. The first polysilicon was used as an isolation to allow the source diffusion to be common to all memory cells in the array. This allowed a small 167 μm^2 cell area with 3 μm design rules.

The first poly also extended over the channel region to be used as the select gate. The storage part of the transistor consisted of a thin oxide layer followed by a nitride layer topped by a second polysilicon storage gate. The array was in one P-well and the peripheral circuitry in another. All high voltages required were generated on chip.

Bulk and row erase were offered with a row latch register available to store the data from a row before erasure. 64 byte page program was offered.

NCR also in 1983 announced a 32k part which had +5 V only operation calling their version of the MNOS technology SNOS (silicon–nitride–oxide–silicon). An on-chip pump generated the necessary program and erase voltages from the +5 V external supply.

This part had a page write feature which enabled data to be loaded and latched either 1 byte at a time or in multiple bytes up to maximum of a page of 16 bytes for writing in one 10 ms cycle. The entire part could be written in less than 3 s. Previous devices with 10 ms per byte write cycles required about 40 s to program. Erasure was possible by 16 byte page or in bulk for the entire chip.

It was also one of the first chips with on-chip margining circuitry. This allows data retention time and number of write–erase cycles remaining to be characterized by measuring written voltage levels in the device cells.

Margining circuitry can be used at any point in the device's life to measure the remaining data retention and endurance capacity of the part. Two external power supplies, +22 and −5 V were required to use this special feature.

Data retention over the military temperature range was guaranteed at 10 years and endurance at 10^4 write–erase cycles. A four micron geometry process was used giving

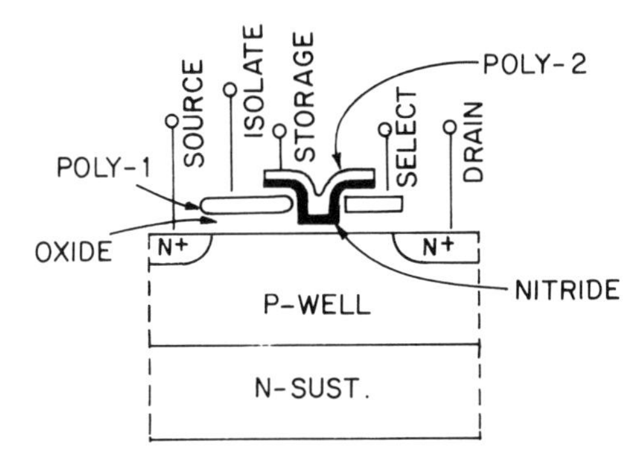

Figure 12.5
Layout and schematic cross-section of a 5 V only EAROM using double polysilicon and nitride–oxide dielectric as the tunneling gate oxide. (From Lancaster *et al.* [20], Inmos 1983, with permission of IEEE.)

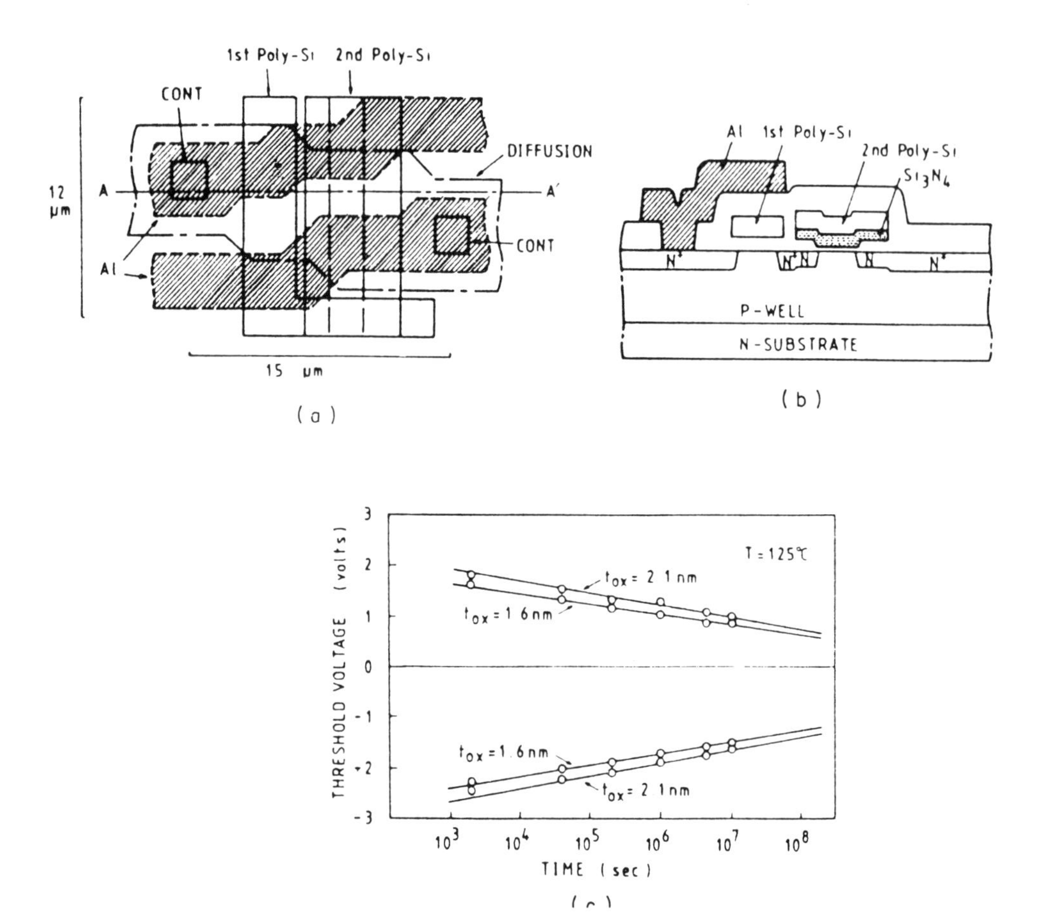

Figure 12.6
64k MNOS electrically erasable PROM. (a) Layout pattern. (b) Schematic cross-section showing select and MNOS storage transistors. (c) Retention characteristics for high voltage MNOS devices with 16 nm and 21 nm thin oxide films ($T = 125°C$). (From Yatsuda *et al.* [18], Hitachi 1985, with permission of IEEE.)

a 26 mm^2 chip. Access time was 300 ns with address latches to free the microprocessor during the long write cycle. Active power dissipation was 350 mW active and 80 mW standby.

There was considerable unresolved controversy in the industry over the relative reliability of the MNOS devices which tended to be specified at 10^4 write–erase cycles and the floating gate devices, to be discussed next, which in the same time frame were specified at 10^6 write–erase cycles. Some clarification was needed over the conditions under which this particular specification was given.

In 1985 a byte erasable 5 V only 64k MNOS electrically erasable PROM was presented by Hitachi [18] in a 2 μm minimum geometry technology. The cell size was 180 μm^2 and the chip size 31.5 mm^2. The 16 V programming voltage was generated on chip. Access time was 150 ns typical, power dissipation was 55 mA typical, write–erase times were less than 1 ms per byte. The endurance was specified at less than 1% failure after 10^4 write–erase cycles.

A double polysilicon process was used in a two-transistor cell consisting of the usual MNOS transistor and a select transistor as shown in the cross-section and layout in Figure 12.6. This structure was similar to Hitachi's single polysilicon two-transistor cell shown earlier.

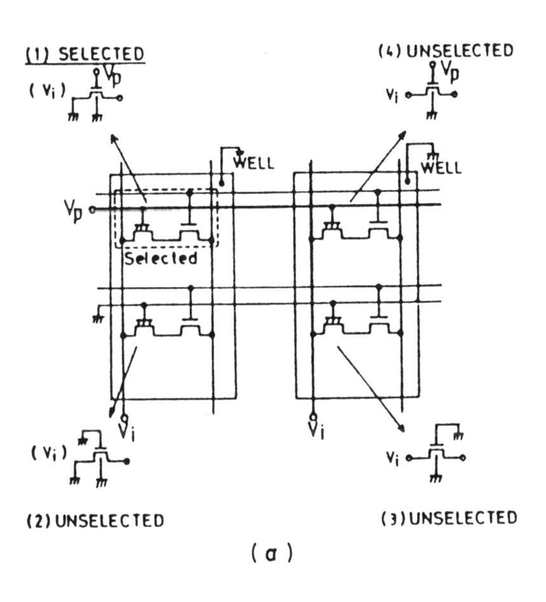

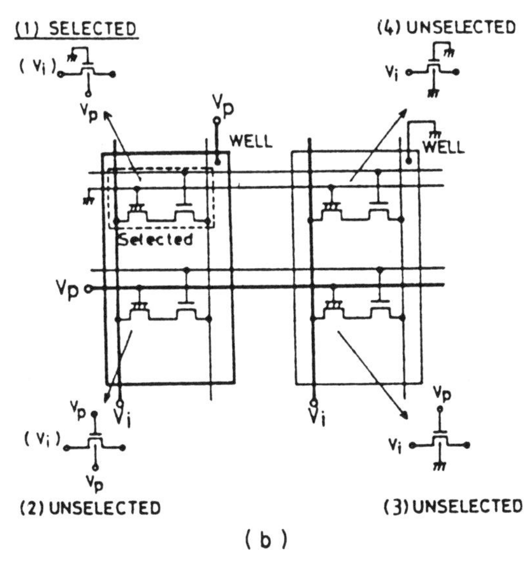

Figure 12.7
64k electrically byte erasable PROM bias conditions for (a) byte writing and (b) byte erasing. (From Yatsuda *et al.* [18], Hitachi 1985, with permission of IEEE.)

The programming voltage was reduced by decreasing the nitride thickness to 50 nm. Erase time was shortened by decreasing the thin oxide to 2.1 nm and maintaining a 25 V erase voltage. Programming time, being dependent on cold electron tunneling, remained the same as in the earlier unscaled 16k devices. The measured retention characteristics of MNOS devices with 1.6 nm and 2.1 nm thin oxides are shown in the graph in Figure 12.6(c).

The requirement that the substrate of the MNOS transistor be biased made byte write and byte erase difficult with this technology. One solution would be to put each cell in a separate well. This, however, would have increased the size of the chip significantly.

Hitachi used a compromise method in their 64k. Byte write and byte erase were accomplished by placing each of 32 8 × 256 bit memory cell arrays in 32 separate wells. Eight bits are on one word-line corresponding to one byte with eight related sense amplifiers and I/O buffers for this byte. Byte write is then accomplished as shown in Figure 12.7(a) and byte erase as shown in Figure 12.7(b). Care was needed in sequencing the application of the high voltage to the various parts of the structure to avoid inadvertent writing or erasing of unselected cells.

In addition to the difficulty in performing byte erase, the MNOS technology also had the disadvantage of requiring the silicon nitride material in addition to the standard MOS materials. Silicon nitride was historically a difficult material to work with, although modern plasma nitride deposition and reactive ion etching have made use of silicon nitride more feasible. Today it is found in both MNOS and floating gate type devices.

The electrically erasable PROMs in the MNOS technology have also been put into CMOS to reduce the power dissipation and permit the ease of logic design for the many applications specific features such as data-polling and ready–busy signal which have appeared on more modern EEPROMs. A comparison of the dc operating conditions of two NMOS and CMOS 64k MNOS electrically erasable PROMs from Hitachi's *1988 Memory Databook* shows a standby power consumption with chip enable high of 200 mW for the NMOS chip and 5 mW for the CMOS chip.

While both chips are erasable by byte or by block, the CMOS part has input, address and data latches. It offers the extra features of data polling, and Ready–Busy signals. These will be discussed in a later section. In addition, data protection circuitry for power up and power down cancelled noise with width less than 20 ns in program mode.

12.4 THE FLOATING GATE THIN OXIDE PROCESS AND TRENDS

12.4.1 Process, operation and reliability

Many of the mainstream NMOS memory suppliers chose a thin oxide floating gate process to make an electrically erasable PROM. This process, which was developed by Intel [16], had the advantage both of using the traditional MOS materials and of being a derivative of the well understood floating gate UV EPROM process. Intel called their version of the process FLOTOX (FLOating gate Tunneling OXide). It was also called FETMOS, SIMOS, etc. and differing slightly from company to company.

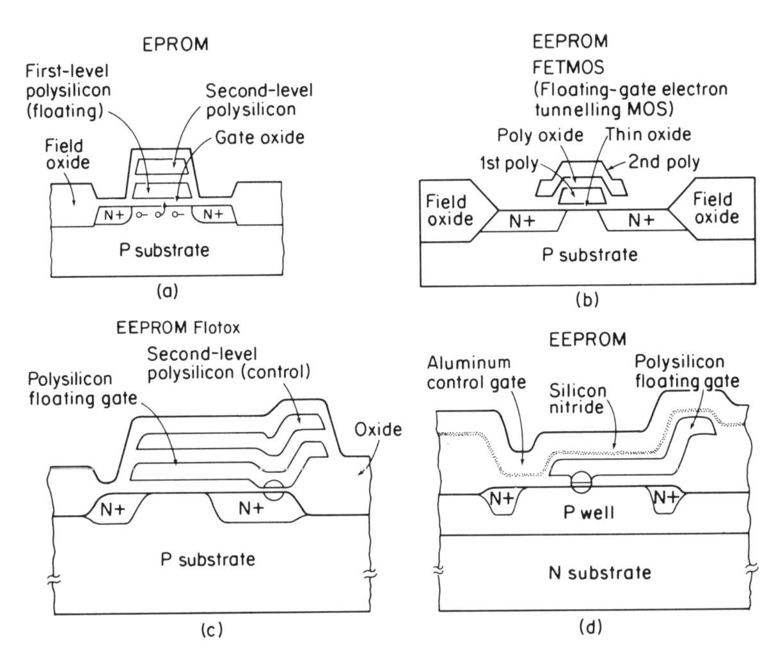

Figure 12.8
Variations of the thin oxide floating gate process. (a) Original floating gate EPROM process; (b) tunneling with thin oxide over the entire gate region (reproduced by permission of Motorola Inc.); (c) tunneling with thin oxide over the N^+ drain region only (reproduced by permission of Intel International); (d) tunneling with thin oxide over a small area of the channel (Reproduced by permission of Hughes Aircraft). (From Prince and Due-Gundersen [56] 1983.)

The basic storage cell consists of an access transistor and a double polysilicon storage cell with a floating polysilicon gate isolated in silicon dioxide capacitively coupled to a second polysilicon control gate which is stacked above it. Several early floating gate EEPROM cells are shown in Figure 12.8(b) to (d). The process was used initially with NMOS periphery and later with CMOS.

The storage transistor is programmed by Fowler–Nordheim tunneling of electrons through a thin oxide layer between the gate and the drain of the transistor. Electrons tunneling from the floating gate to the drain leave the floating gate relatively more positively charged. This shifts the threshold voltage in the negative direction so that in the READ mode the transistor will be 'on'. The programmed state corresponds to a logical '0' state in the cell.

In the erase mode the control gate is at the high voltage while the drain is grounded. Electrons tunnel to the floating gate and the threshold voltage shifts in the positive direction so that in the read mode the transistor will be 'off'. The erased state corresponds to a logical '1' stored in the cell.

The thin tunneling oxide, generally about 10 nm, is usually isolated to a small area over the drain region as illustrated in Figure 12.8(c) for the Intel Flotox process. It can also be a small area over the channel as shown in Figure 12.8(d) for the Hughes process but, in this case, the ability to program individual bytes is foregone for bulk erase of the entire chip. Another early process, the Motorola FETMOS process, had oxide which extended under the entire gate region as shown in Figure 12.8(b). This had the advantage of compatibility with the existing UV EPROM process but the disadvantage of low yields due to the larger area of thin oxide.

A more complete description of the operation of this cell was given in Chapter 5.

12.4.2 Early floating gate EEPROM cells and devices

The SIMOS cell developed by Siemens [11] in 1977 for the floating gate thin oxide process was a single-transistor stacked cell as shown in Figure 12.9 consisting of an NMOST with a short effective channel length and a split polysilicon control gate over the floating gate. This cell had the advantage over the contemporary MNOS devices that it could be programmed, read, and erased with positive voltages only. It was similar to the flash memory cells described in the last chapter.

The floating gate extended over part of the drain region and part of the channel. The other part of the channel is covered by the control gate. Programming was achieved by applying a high voltage to the drain and the control gate and grounding the source. Hot electrons generated in the channel were injected into the floating gate

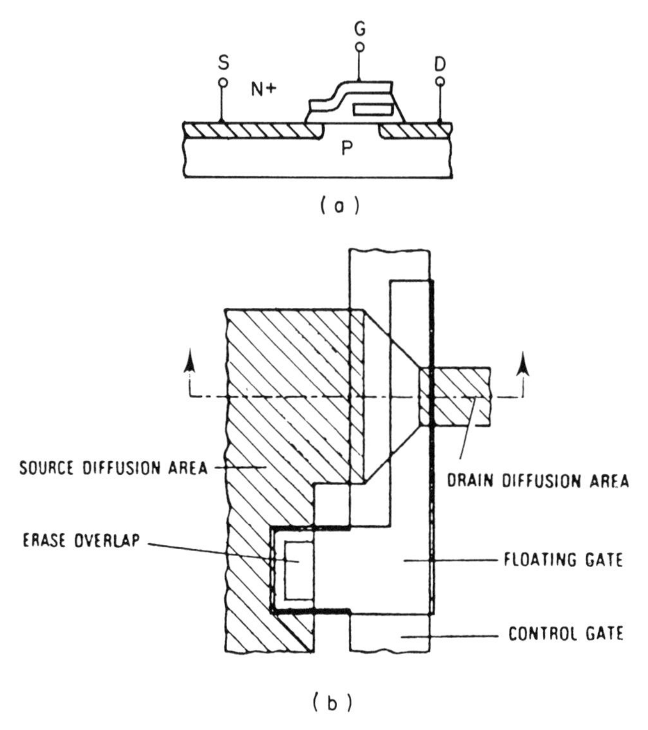

Figure 12.9
Floating split gate cell for 8k EEPROM. (a) Schematic cross-section. (b) Cell layout. (From Muller *et al.* [11], Siemens 1977, with permission of IEEE.)

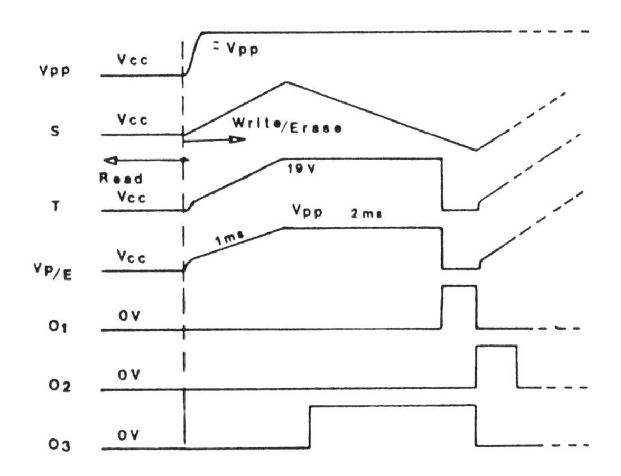

Figure 12.10
Timing diagram for V_{PP} ramp generator on early 16k EEPROM. (From Gee *et al.* [15], Intel 1982, with permission of IEEE.)

resulting in an upward shift in threshold voltage. Erasure was by cold electron emission of electrons from the floating gate when the high voltage was applied to the source, the drain was left floating and the control gate was grounded.

This cell had the drawback that the erase was not self limiting and could result in a net positive charge on the floating gate which could turn the transistor on. This was avoided by the control gate being directly over part of the drain so that even a positively charged floating gate could not turn the transistor on. It was also necessary to make the gate oxide thin enough that the applied voltage could be low enough to keep hot electron injection from occurring during erase.

Voltages applied during programming included $V_{DD} = +12$ V, $V_{BB} = -5$ V, and $V_{PP} = +26$ V. Read voltages were $V_{DD} = +12$ V, $V_{BB} = -5$ V, and $V_{PP} = +5$ V. The classical FLOTOX cell from Intel [16] followed in 1980 as described above.

Intel [15] in 1982 on a 16k EEPROM used their same thin oxide cell but added various on-chip control features. A V_{PP} ramp generator, produced a waveform, shown in Figure 12.10, which was at V_{CC} during the Read operation and ramped to V_{PP} for the program operation. After a 2 ms program pulse it ramped down for 30 to 50 ns while an on-chip data comparison cycle took place to see if the cell was yet programmed. Depending on the outcome of the test it would either ramp back up for another program pulse or remain at V_{CC}.

A timing control generator was included to sequence and latch the various inputs and control functions properly for a write or erase cycle. At the end of the cycle a Ready Signal register, which was triggered by the data comparator, signaled a return to read state and reset the latches accordingly.

A program line generator was included to generate the various voltage levels needed for operation of the chip. These included V_{PP} for erase, 0 V for write, V_g for normal read, $V_g + \Delta V$ for margin '1' and $V_g - \Delta V$ for margin '0'.

A ready–busy register, which is discussed further in a later section, compared the

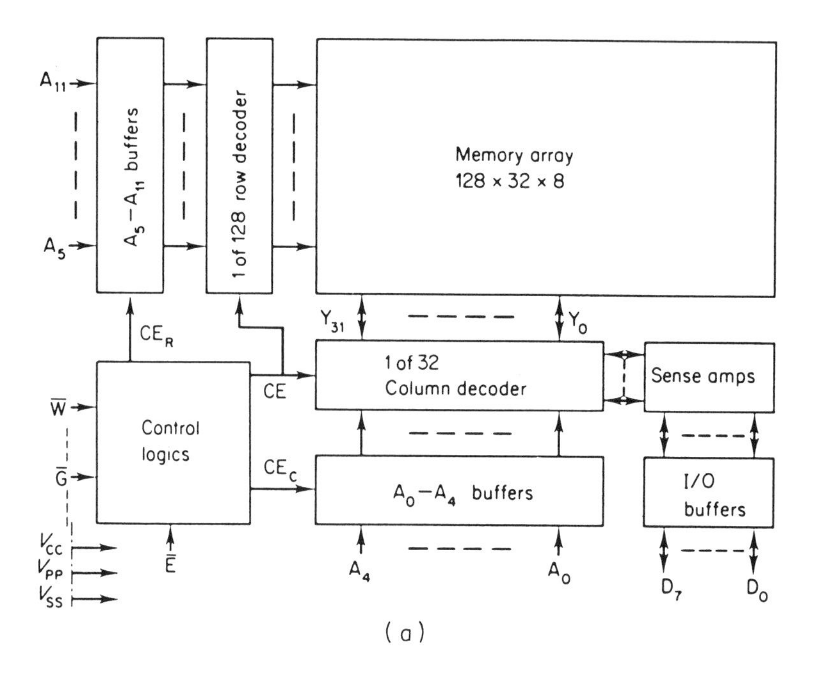

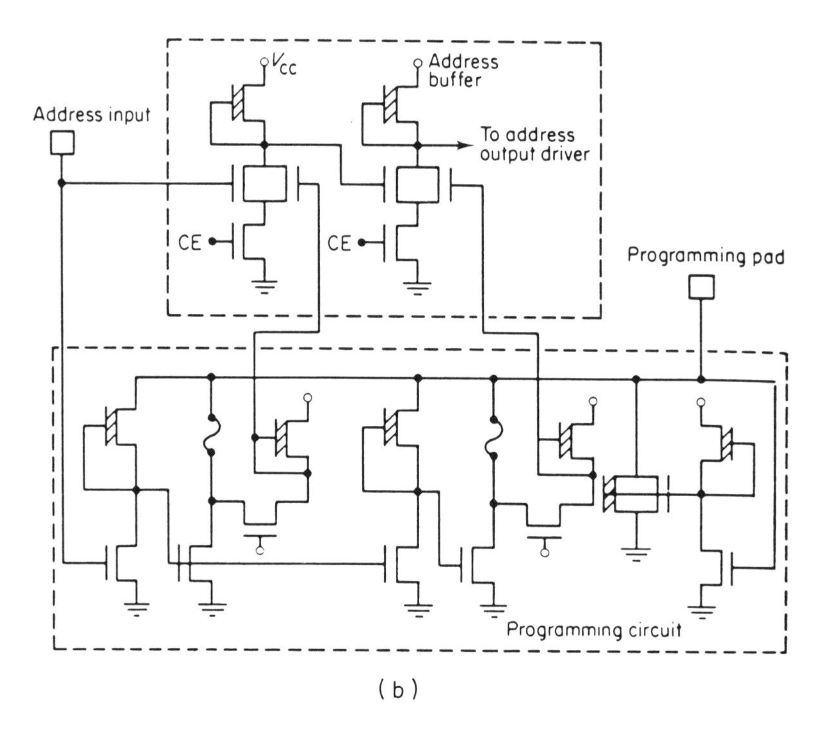

Figure 12.11
32k EEPROM. (a) Block diagram. (b) Circuit which permitted programming of 16k and 8k partials at test. (From Kuo *et al.* [17], Motorola 1982, with permission of IEEE.)

data in the cell with the data intended to be written and gave a 'not ready' signal if the two differed.

In 1983 Intel [23] gave the 16k EEPROM with the same floating gate cell 5 V only operation. The pulse shaping and ramp generator was retained and a charge pump both generated the on-chip high voltage and with a feedback loop handled the worst case loading during chip erase at high temperatures.

This Intel part had a 15 mm^2 chip, +5 V supply, access time of 150 ns, active power dissipation of 450 mW and standby power of 175 mW. Write and chip erase times were 10 ms. Endurance was 10^4 write–erase cycles and data retention over 10 years.

In 1982 a 32k EEPROM from Motorola [17] was presented which used a variation of the floating gate process, which they called FETMOS (floating gate electrically erasable tunneling MOS) with thin oxide under the entire gate region as shown previously in Figure 12.8(b). The part continued the use of an on-chip voltage shaping circuit, and the latched inputs.

Features added on this chip were a page erase mode as well as the usual byte and bulk erase options. This was accomplished by the use of internally generated row and column address controls which permitted selection of a byte, a row consisting of 32 bytes, a column consisting of 128 bytes, or bulk erase of all 4096 bytes on the chip. The block diagram is shown in Figure 12.11(a). It also added a feature called transparent partial programming which permitted 16k and 8k 'partials' to be selected at test by blowing polysilicon fuses to disconnect the good part from the defective circuitry, thereby increasing the effective yield of the chip. The circuit schematic for this feature is shown in Figure 12.11(b).

Margin test circuits were included to detect marginally programmed or erased cells at test so that chips with potential early cell failures could be eliminated thereby improving the overall reliability of the part. These margin test circuits worked by checking the voltage at which an individual cell programmed for '0' switched the sense amplifier on (margin '0' test) and the voltage at which a cell programmed for '1' switched the sense amplifier off (margin '1' test).

A 16k EEPROM from National [24] shown in 1983 combined all of the voltage and control features developed to that point in one EEPROM. This 5 V only chip used a two-transistor floating gate cell with a self-timed program cycle with automatic erase before write, address and data latches, and a 'ready' signal. A voltage multiplier circuit generated the high voltage on chip, and a self timed pulse generator produced and shaped the high voltage pulses required for programming and erase.

12.4.3 Early cell trends in floating gate nitride barrier technology

Several companies also tried nitride barrier floating gate cells. These combined the features of the double polysilicon floating gate technology with nitride–oxide combinations for the dielectric.

Fujitsu [19] reported in 1982 a 2k EAROM with a two-transistor cell. The double polysilicon storage transistor had 10 nm oxidized thermal nitride film forming a low energy barrier gate oxide under the floating polysilicon storage gate as shown in the

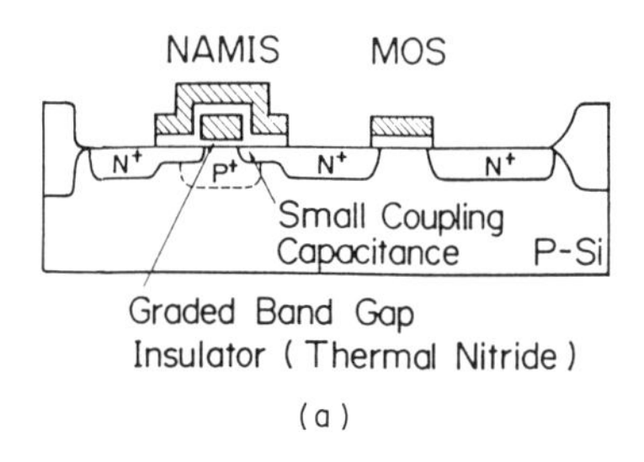

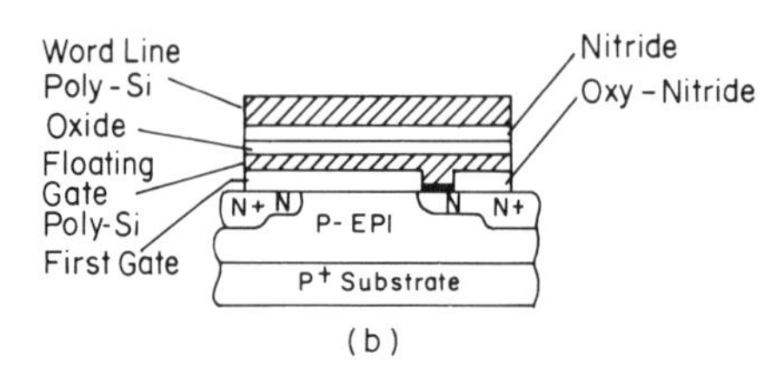

Figure 12.12
(a) Schematic cross-section of a nitride barrier electrically erasable ROM storage transistor and NMOS select transistor for a 2k EAROM. (b) Storage cell of 5 V only 16k EEPROM using oxynitride as the thin dielectric. (a: From Hijiya *et al.* [19], Fijitsu 1982, with permission of IEEE; b: From Mohrotra *et al.* [54], SEEQ 1984, with permission of IEEE.)

diagram in Figure 12.12(a). Programming and erase were by hot carrier injection to the floating gate from the drain. While historically avalanche injection EAROMs need negative gate bias to inject hot holes for writing, this device obtains a hole injection field under zero gate voltage by inhibiting the capacitive coupling of the floating gate to the drain by processing techniques allowing hot hole injection with a grounded gate and a drain voltage of 10 V. Erase was done by injecting electrons to the floating gates.

The advantage of this part was the low programming voltage required of 12 V compared to the conventional MNOS approaches which used 25 V and the tunneling oxide approaches which used between 17 and 21 V. Standard write erase endurance of 10^6 cycles and data retention of 10 years was obtained.

The current 64k EEPROM from the Fujitsu *1987 Memory Databook* has a specification very much like that of the Hitachi 64k shown earlier with data polling, ready busy signal and automatic erase before write.

Another 1982 16k EEPROM from SEEQ [21] used the low voltage properties of the oxide–nitride film combination to produce a 5 V only EEPROM. The high voltages needed were generated on chip by charge-pumping techniques. A floating double polysilicon gate structure was combined with a thin 'oxynitride' tunnel dielectric over the drain region. In 1984 on their 64k EEPROM SEEQ added also an oxide–nitride interpoly dielectric to improve programming efficiency. This cell with both nitride layers is shown in Figure 12.12(b). The memory cell had a four-transistor structure designed to eliminate read–disturb problems and reduce programming and erasing current to near zero. The part was both byte and bulk erasable. A feature was added on this part, which permitted the tracing of each device back to its manufacturing source. Redundancy was implemented by EPROM fuses.

Oxynitride dielectric was used again by SEEQ [37] on their 256k EEPROM which appeared in 1987 in 1.25 μm process with a 50 nm gate and interpoly oxide, and a 54 μm^2 cell. The intrinsic endurance of the part had been extended to 10^8 write–erase cycles before closure of the threshold window. The layout and cross-section of the two-transistor cell are shown in Figure 12.13.

12.5 THICK OXIDE PROCESS AND EARLY CELL TRENDS

A third process, which is a variation of the previous processes, has also been used to manufacture electrically erasable ROMs. This development came from an early undesirable side-effect of MOS processing. It was found that the conduction through

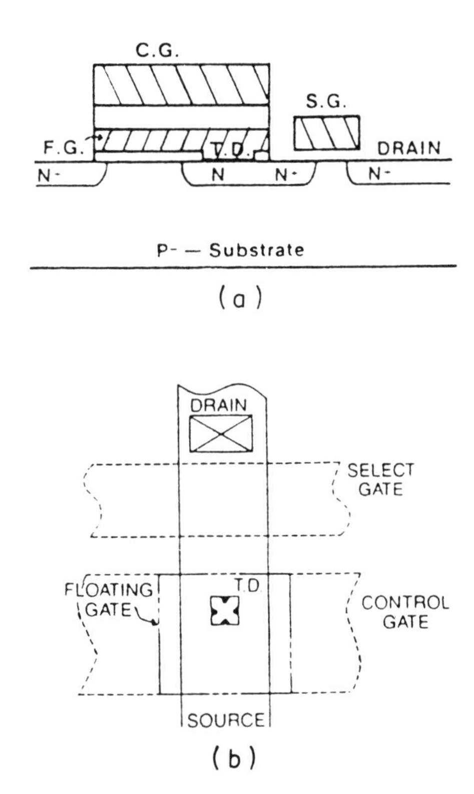

Figure 12.13
256k two-transistor cell EEPROM using oxynitride. (a) Cell schematic cross-section showing thin dielectric over the drain region. (b) Cell layout. (From Cloaca *et al.* [37], SEEQ 1987, with permission of IEEE.)

insulating silicon dioxide grown thermally over deposited polysilicon is markedly higher than that measured when the oxide is grown thermally over single-crystal silicon. A brief review of the early development of this process follows.

12.5.1 Early process development

This effect was used as early as 1979 by RCA [10] for the electrical erasure of a small electrically erasable ROM. The erase was by cold electron tunneling through a 400 nm field oxide which consisted of a thin layer of thermally grown oxide overlaid with a thicker layer of deposited oxide. The non-uniformity of the field oxide combined with the surface roughness of the polysilicon surface increased the localized field strength near the surface of the floating gate so that Fowler–Nordheim tunneling started at lower voltages than would have been expected for such a thick oxide.

IBM then discovered that electron tunneling was encouraged by sharp needle-like protrusions extending from the surface of the polysilicon into the thermal oxide. These asperities, as they were called, concentrate fields at their tips and supported enhanced localized conduction which is as much as an order of magnitude greater than on a asperity-free silicon surface.

The asperities occur because oxidation progresses faster along some crystal directions than others. Since crystal orientation is random in deposited polysilicon, there are points on the surface where oxide growth is enforced.

The temperature of the oxide growth controls the size and shape of the asperities which makes this technology process dependent and to some extent difficult to control.

While the barrier height of the silicon–silicon dioxide interface is 3.2 eV, it has been shown that the barrier height of an interface with low temperature grown ($<1000^{\circ}C$) oxide can be as low as 0.4 eV due to the presence of the localized asperities. This permits Fowler–Nordheim tunneling to occur in the thicker oxide.

This phenomenon is used by Xicor to enable them to use thicker silicon dioxide in the EEPROM transistor and still discharge the floating gate. Figure 12.14 shows a cross-section of the Xicor EEPROM cell showing the 'textured' surfaces containing the asperities. The polysilicon electrodes are separated by oxide layers greater than 100 nm thick. Without the asperities 100 V would need to be applied to produce appreciable tunnel current. The voltage required for tunneling with asperities is low enough to generate easily on-chip.

Figure 12.14(a) shows the programming by electron injection from poly 1 to poly 2. Figure 12.14(b) shows the erase operation by asperity enhanced tunneling from poly 2 to poly 3 through the thick oxide. One of the first 5 V only program–erase EEPROMs used asperities.

12.5.2 Early cell trends in asperity assisted Fowler–Nordheim process

The CMOS–SOS cell developed by RCA [10] in 1979 used two transistors for select and store and an asperity erase process to achieve block erase.

TI [14], also in 1979, discussed a 16k +5 V read EEPROM with a split floating

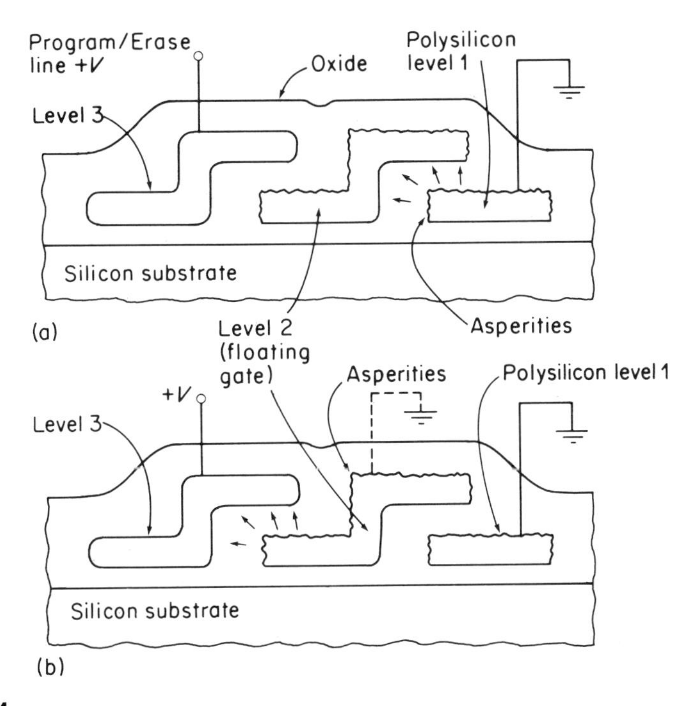

Figure 12.14
Cross-section of an EEPROM cell using asperities to enhance the tunneling. (a) Charge tunnels to higher potential of the floating gate to store a bit. (b) When the gate potential approaches ground, charge tunnels to the third level erasing the bit. (From Prince and Due-Gundersen [56] 1983.)

gate structure as shown in Figure 12.15. Programming occurred from the channel to the floating gate with the control gate at voltages from 30 to 35 V and the drain and source at ground. The injection process occurs within a very small region near the drain junction.

The erase method used the asperity assisted Fowler–Nordheim tunneling through thick oxide from the floating gate to the control gate. To form the interlevel oxide with asperities, oxidation took place at temperatures slightly less than 1000°C. This erase method is not self-limiting as in the case of the EPROMs but the section of the control gate directly over the channel insured that even if the floating gate became positively charged, the transistor would not turn on. TI measured a barrier height at the interface with the asperities at 0.4 eV.

A good analytical treatment of the various processes involved is given in this TI paper.

12.6 ELECTRICALLY ERASABLE ROM DEVICES

EEPROM and EAROM devices fall into two general overlapping categories: small, low performance special purpose devices which we will call EAROMs are generally below 16k density and are used mainly for consumer and automotive, and industrial

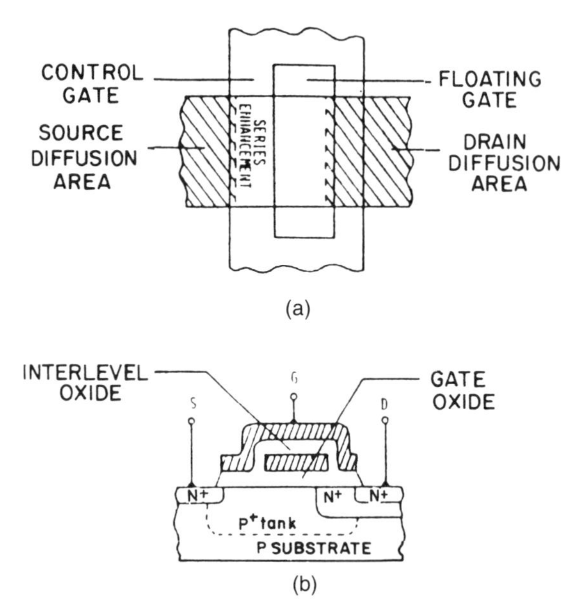

Figure 12.15
16k 5 V EEPROM using a split floating gate structure and asperity assisted electron tunneling for erase. (a) Layout. (b) Schematic cross-section of cell. (From Guterman *et al.* [14], TI 1979, with permission of IEEE.)

tuning and control store applications. Higher density EEPROM devices in EPROM–ROM compatible pinouts to 256k density are intended mainly for specific applications in high performance or high reliability microprocessor based systems.

12.6.1 Low density EAROM applications and characteristics

Many of the low density electrically erasable ROMs are used in such non-traditional applications as TV tuners and automotive odometers. An example of a typical TV tuner circuit using a small 16 × 16 bit EAROM, is shown in Figure 12.16. The system offered digital tuning of a TV received between 80 MHz and 1000 MHz using a phase lock loop.

The tuner's local oscillator output was first divided by 64 in an external prescaler and then fed to a 14 bit programmable divider in the UAA 2000A, where it was further divided by a number determining the frequency which was controlled by a MC6805P2 microcomputer. The divider output was compared with a crystal derived reference frequency and the result used as the control for the local oscillator varicap. The UAA 2000A also supplied band switching information to the tuner for up to four bands, which it also received from the MCU (Microcomputer).

In order to selectively recall certain programs, it was therefore necessary to store 16 bits of information, 14 bits for the frequency corresponding to that program and 2 bits for the band. The EAROM, which was organized in 16 words of 16 bits with serial

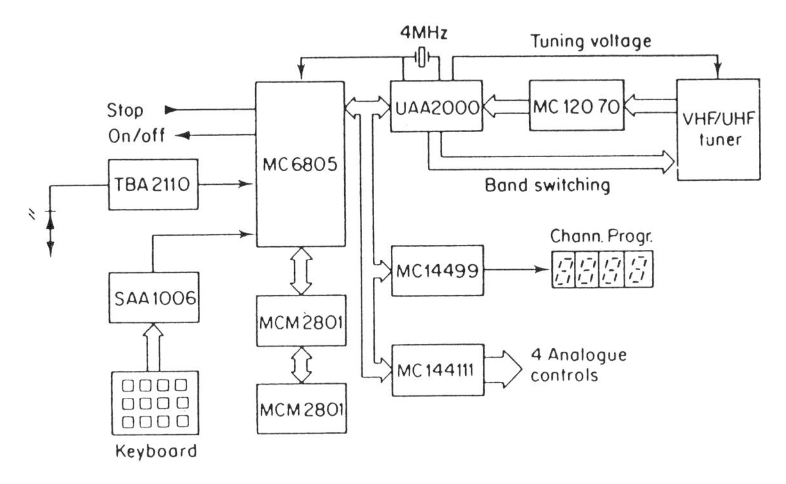

Figure 12.16
Typical TV tuning systems using a 256 bit EAROM, the 2801, and the 6805 microprocessor with phase locked loop. (From Prince and Due-Gundersen [56] 1983.)

output, was ideally suited to allow this, without any need for battery back-up when the receiver was switched off.

Another interesting application was an electronic 'dog tag' used by the US military which contained 2k of EEPROM. Civilian uses included access keys to secure facilities and computer terminals. It also has been used as a debit card for vending and copy machines.

Several companies have made small EAROMs for such applications including Motorola, G.I., N.C.R., and National. Both the MNOS and thin oxide floating gate tunneling processes are used.

Another example of a low density electrically erasable ROM application is shown in the dashboard system in Figure 12.17. Here an EAROM–MCU combination, together with appropriate sensor, interface and display, are used in a trip computer to display total vehicle or trip distance in miles and kilometers, road speed, engine speed, etc.

12.6.2 Medium density EEPROM characteristics and applications

The medium density 16k to 256k electrically erasable ROMs shared with the small EAROMs the advantages of in-system reprogramming possibilities. These include cost, plastic packaging, immunity from accidental deprogramming from daylight or fluorescent lighting, ability to be used with automated insertion equipment which was difficult with the fragile ceramic quartz window EPROM package, and single erase–reprogram in or out of the system. EEPROMs used for EPROM replacement can be bulk erasable—that is, capable only of erasing the entire memory at one time.

EEPROM suppliers were, however, becoming clear by the 64k density level that the parts were not going to be competitive with the EPROM technology which was

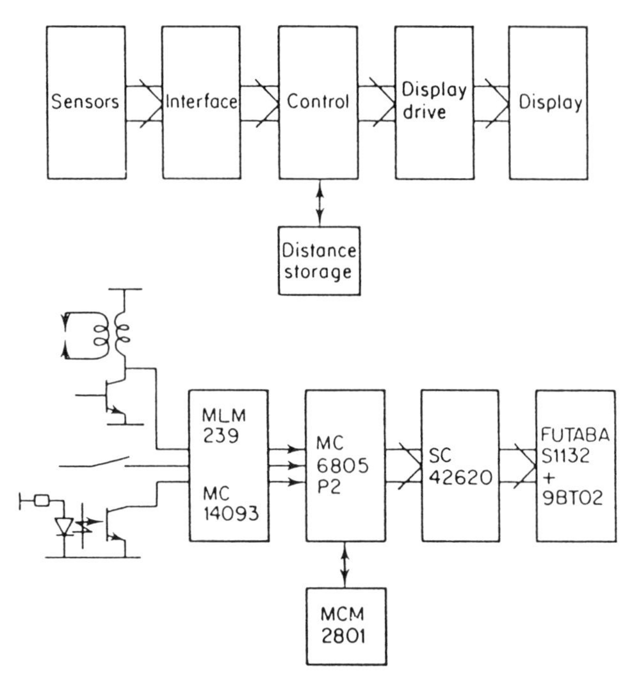

Figure 12.17
A 256 bit EAROM; used for odometer distance storage in an automotive application. Shown is a general dashboard block diagram and dashboard system construction. (From Prince and Due-Gundersen [56] 1983.)

then approaching the 1Mb density. This was not only because of the larger size of the two transistor cell but also because of the significant amount of control logic necessary on the EEPROM chip.

This realization led the EEPROM suppliers to concentrate their efforts on applications specific markets such as high performance microprocessor systems, low power portable instrumentation, and extended temperature applications. The embedded EEPROM on a processor chip was also in demand for the controller type of applications.

An early microprocessor system applications is shown in Figure 12.18, using the Motorola 32k EEPROM discussed earlier along with a parallel interface adapter (PIA) and a bus-compatible timer with the EEPROM. The PIA controls the memory operating modes and the time clocks the EEPROMs erase and program cycles. Enable and level decode circuits translate TTL-compatible control words to this EEPROM's early three-level logic format.

12.6.3 Change from NMOS to CMOS technology

At the 64k level of density the EEPROMs began to change from the NMOS to the CMOS technology with SEEQ, Exel, and Lattice all showing CMOS parts in 1984

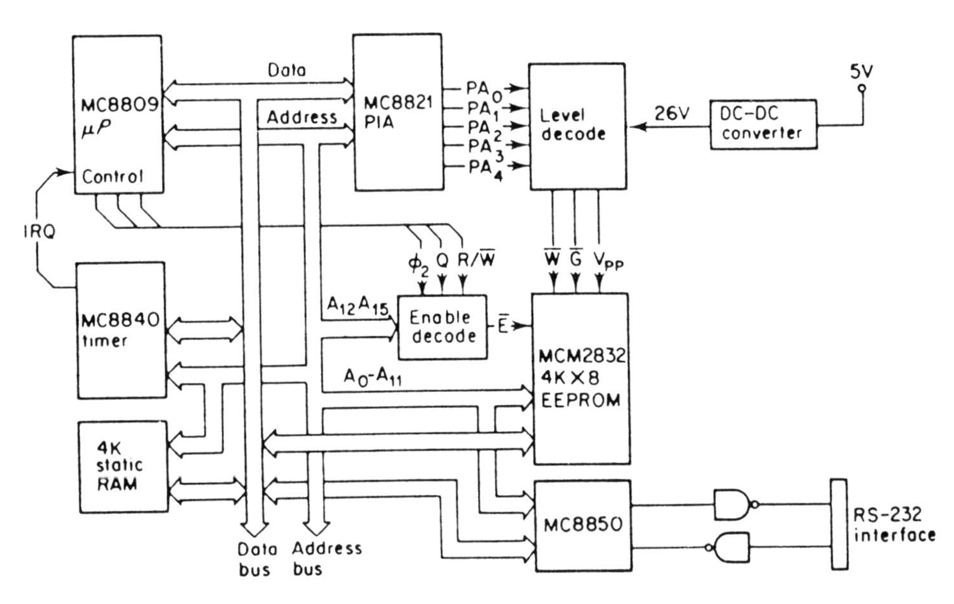

Figure 12.18
A typical microprocessor system using the 32k EEPROM from Motorola. (From Prince and Due-Gundersen [56] 1983.)

and 1985. The move to CMOS took the operating power dissipation into the 50 mA range and standby power dissipation into the microamp range. The benefit of the CMOS was not only in the lower power but also in the ease and compactness of design of the logic needed in the periphery of an EEPROM for control functions.

12.6.4 High speed features on EEPROMs

A fast 55 ns read access time was produced in the Exel [26] 32k CMOS EEPROM in 1984 by several techniques taken from the static RAM technology. Differential, or double ended, sensing of two bits per cell was used with each complementary bit-line including a standard two-transistor EEPROM cell, that is, the cell included four transistors, two selects and two storage transistors. Address transition detection was used to initialize the sense amplifier and to equalize the bit-lines. N-channel pass transistors are used to block the high voltage from the decoders and sensitive sense amplifiers.

The part was also scaled from 3 to 1.5 μm design rules. A scaled layout comparison of the 16k EEPROM cell at 3 μm and the 32k EEPROM cell with 1.5 μm design rules is shown in Figure 12.19(a). A cell schematic comparing the EEPROM cell to an SRAM cell is given in Figure 12.19(b).

Another fast CMOS EEPROM, a 16k 30 ns part from SEEQ [41], used a six-transistor differential cell which was also fault tolerant. Two-stage differential sense amplifiers and address transition detection circuitry were used, along with a regulated on-chip substrate bias generator, to improve speed and latch-up immunity. High speed

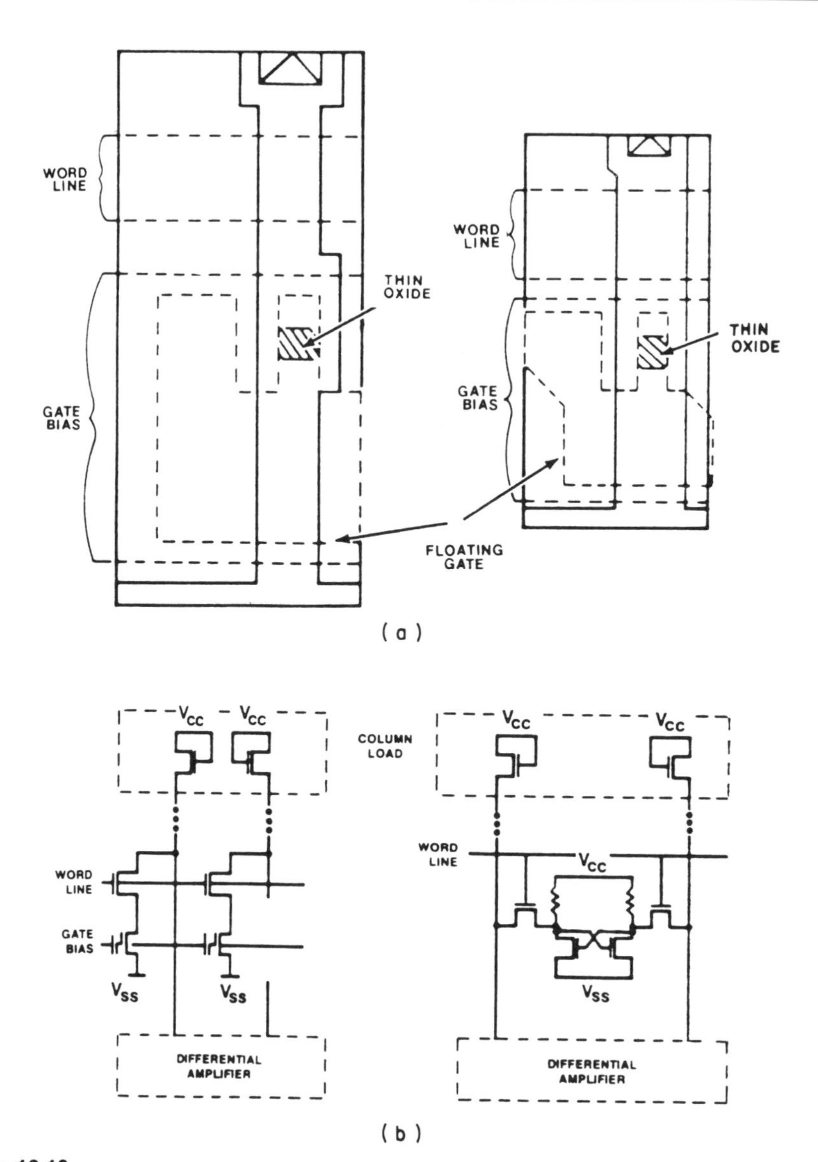

Figure 12.19
16k EEPROM. (a) Cell size comparison of 16k EEPROM in 3 μm design rules, and a scaled EEPROM cell in 1.5 μm design rules. (b) Schematic cell diagram of EEPROM compared to a SRAM cell. (From Zeman [26], Exel 1984, with permission of IEEE.)

data and address latching were also used to make this circuit look like an SRAM. The high voltage required for erase and write was generated on chip using an eight-stage voltage multiplier. The fault tolerant part of this chip will be discussed in the following section on endurance enhancement where an illustration of the cell is also included.

In contrast a 35 ns 64k EEPROM from Lattice [27] in 1985 used single-ended sensing combined with fine tuned address transition detection and static bootstrapping techniques, all borrowed from the SRAM development of the day, to achieve high performance. Substrate bias was also used to improve performance of N-channel transistors, to reduce junction capacitance and eliminate potential latch-up due to voltage undershoot.

The theory of the single-ended sensing technique's advantage over double ended sensing for high speed included the reduction in cell capacitance of two transistors instead of four. A fast bootstrapped NOR-type decoder was used to speed up word-line selection. Both enhancement and depletion load devices also added to the optimised speed of the access path.

The read speed achieved on this part allowed it to compete in the high speed bipolar PROM market. The part had a metal mask option allowing it to be configured either as a SRAM compatible or a bipolar PROM compatible part.

The lattice part also used 32 byte page mode programming for fast write speed. A byte latch was used to store address selection to avoid clearing and rewriting unselected bytes in the page.

12.6.5 Standards for system required features on EEPROMs

Due to the application-specific markets of the EEPROM types of devices, standardization was difficult to obtain. It was never totally achieved at the 16k and 64k levels. At the 256k it was finally achieved. Required standard features for the 256k EEPROM from the JEDEC Standard 21C are

- operate with a primary power supply of 5.0 V nominal
- operate in conformance with the standard truth table
- have read and write timing cycles which are consistent with the standard timing diagrams
- contain data and address latches for write cycles
- operate with self timed write cycles
- operate with input levels between 0 and 5 V

In addition certain optional features are outlined which if implemented must operate in certain defined ways to be in compliance with the standard. These include: page write mode, minimum page size of 16 bytes in page write mode, $\overline{\text{DATA}}$ polling, software data protect option, hardware mass erase, and software mass erase. The reader is referred to the current edition of the Standard 21 for further details.

These and other special features for EEPROMs are described further in the following section.

Ready–Busy Signal. The long, 10 ms per byte, typical write time of the early EEPROMs, meant that some method had to be originated to flag the microprocessor that the EEPROM was still busy writing.

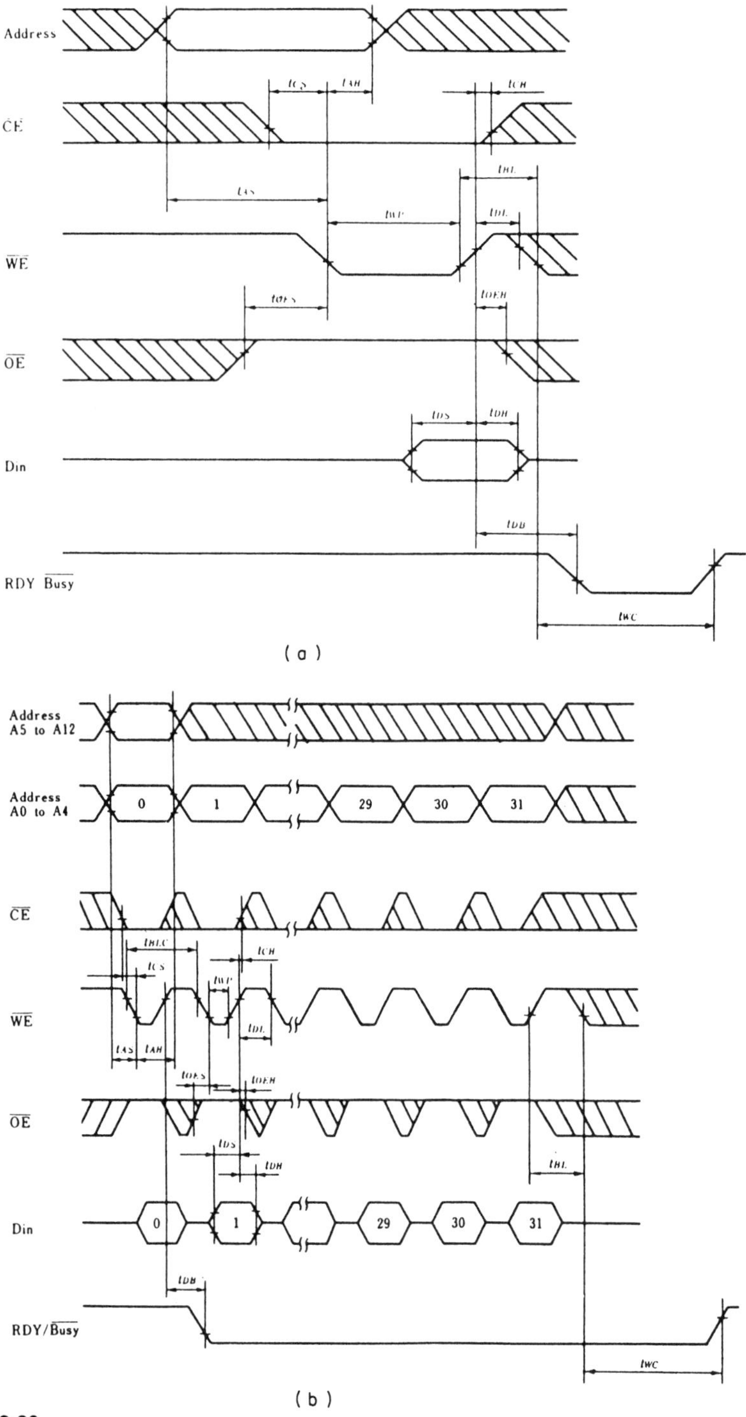

Figure 12.20
Two timing diagrams showing use of ready–busy signal. (a) Byte erase and byte write for $\overline{\text{WE}}$ controlled write cycle. (b) Page erase and page write for $\overline{\text{WE}}$ controlled write cycle. (Source Hitachi 1988 Databook.)

A ready–busy signal on pin 1 was tried on some 16k and 64k. This signal sent an interrupt to the microprocessor when the write cycle was complete. Figure 12.20 shows use of the ready–busy signal during a byte erase–write cycle and a page erase–write cycle for a 16k EEPROM.

The JEDEC standard 256k EEPROM from the Standard 21C does not have a ready–busy signal.

Data polling. This was a software method, originated by Xicor, of flagging the microprocessor that the memory was still busy writing. It was originated at the 64k level to keep the pin 1 free for upward migration to the 256k. Data polling allows the processor to look at the data being written to see if a write is still in progress.

During a write cycle an attempted read of the last byte written results in the complement data of that byte appearing at the output. This has three advantages. It frees the processor to perform other tasks during the write cycle, it allows the system to optimize the actual time consumed by write operations, and it saves one pin on the device package. In a system with large numbers of EEPROMs, the time savings may be large.

Page mode. Another feature of the 64k and larger EEPROMs was page mode which was originated to reduce the time required to write the entire memory. In a 64k with 10 ms per byte write time, the time to write the entire memory is over 80 s. Page mode, rather than latching single bytes as in the lower density EEPROMs, buffers and latches groups of bytes and writes each group at one time.

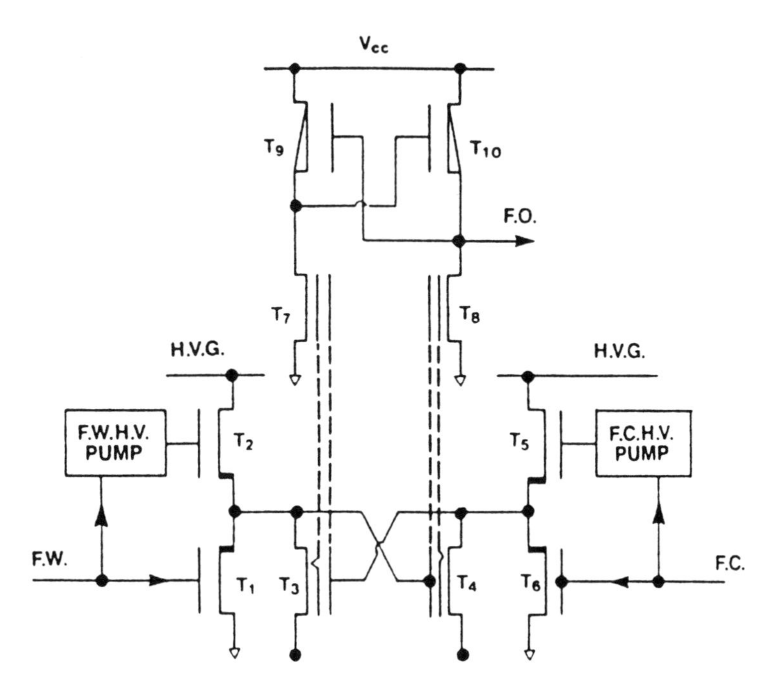

Figure 12.21
Test programmable EEPROM fuse used to select 10 or 5 ms page write time in a 256k EEPROM. (From Ciaoca *et al.* [37], SEEQ 1987, with permission of IEEE.)

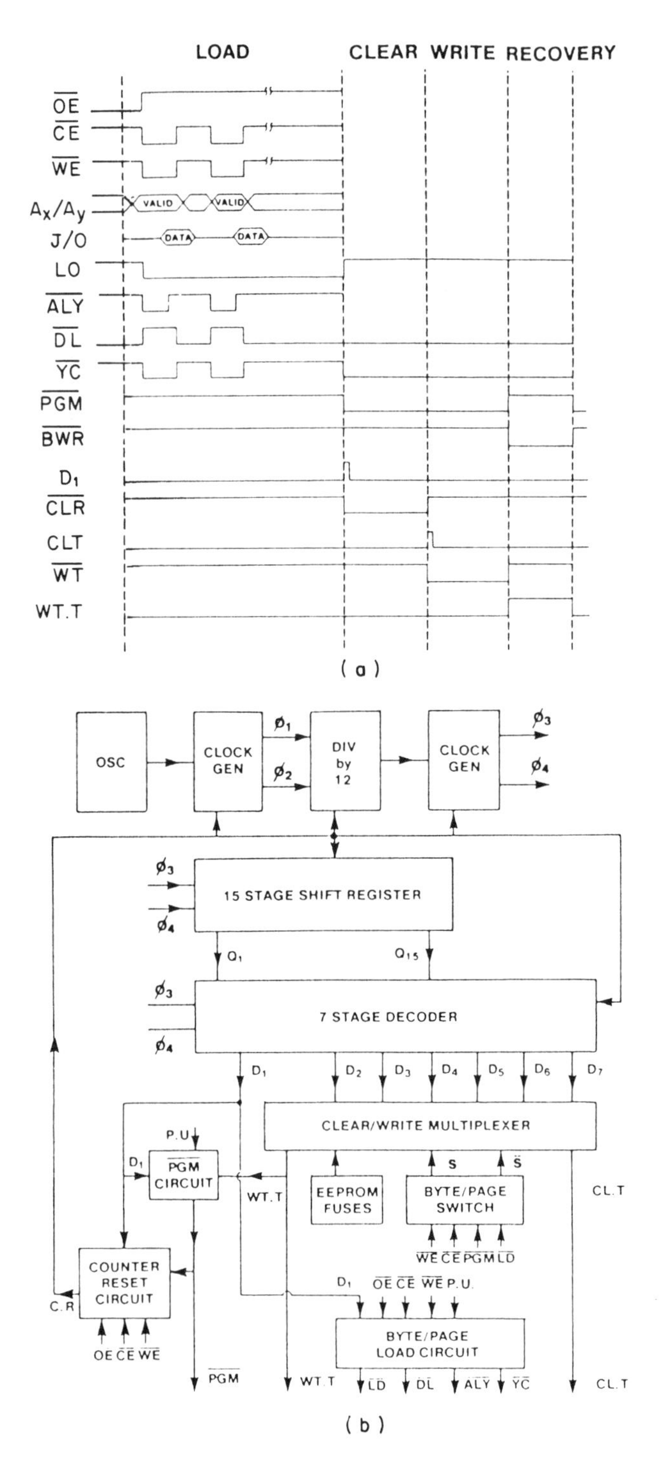

Figure 12.22

(a) Page mode programming signals. Page loading signals are controlled by $\overline{OE}$, $\overline{CE}$, and $\overline{WE}$. The signals needed during clear, write and recovery are controlled by the timer. (b) Timer block diagram. (From Ciaoca *et al.* [37], SEEQ 1987, with permission of IEEE.)

Using page mode, Xicor, for example, on their 64k EEPROM specified a 5 ms write cycle time, with 16 bytes latching and writing at a time, the time to completely write 64K was 2.5 s. The speed advantage is clear, the disadvantage of page mode is the unnecessary writing of locations not requiring change. With endurance typically at 10 000 write–erase cycles, the use of page mode could seriously reduce the expected endurance of an EEPROM.

The standard which developed at the 64k level for page mode operation is a 32 byte page and a 10 ms page write time. This resulted in a 32 μs effective byte write time.

On the SEEQ [37] 1987 256k EEPROM an option of the 64k page mode timing or a 64 byte page with 5 ms write time was offered. This gave an 80 μs byte write time. An EEPROM fuse which was programmable at test was used to select the 10 or 5 ms page write time as shown in Figure 12.21.

The 64 byte page reduced writing time and simplified system implementation. A retriggerable page mode eliminated the fixed page load time constraint so that a designer could load as many bytes as wanted up to a maximum of 64 as fast as 350 ns per byte or as slow as 300 μs per byte. Data was loaded into the 64 bit register and then into the main memory.

When the first byte was loaded in internal 300 μs timer started counting. This timer was reset if another byte was entered before it timed out. When it timed out the internal write began. The same page location could be written into several times with only the last byte loaded being written into the EEPROM. Only the bytes that were loaded were transferred to the EEPROM with bytes not loaded remaining unchanged. This increased the effective endurance of the storage array. The main page mode programming signals were included in the timing diagram in Figure 12.22(a).

A schematic diagram of the timer block from this 256k EEPROM is shown in Figure 12.22(b) which indicates EEPROM fuses, byte–page switch, byte–page load circuits along with the decoder and shift registers.

Status word. An auxiliary software function to page mode was offered by Exel in its 'status word'. Using the status word feature, a set of on-chip registers could be read or written by the microprocessor when the status word pin is low thus providing ready–busy status information to the processor when it reads a register during SW active. The processor can also write commands into the registers to set the EEPROM for page mode, fast write mode, or for chip erase mode.

Write protect. To protect an EEPROM against inadvertent writes during power-up and power-down a combination of voltage threshold sensors, noise filters and times-outs are employed. It is necessary to disable the write until memory reaches V_{CC} and for a short time thereafter to allow V_{CC} to stabilize. A time-out was also employed by most designs before the write is enabled, of typically 100 ms on initial power-up to allow V_{CC} to stabilize.

12.6.6 A typical 256k EEPROM with required and optional features

Microchip technology [38] in 1988 described a 5 V CMOS 256k EEPROM which implemented all of the JEDEC standard required and optional extended standards

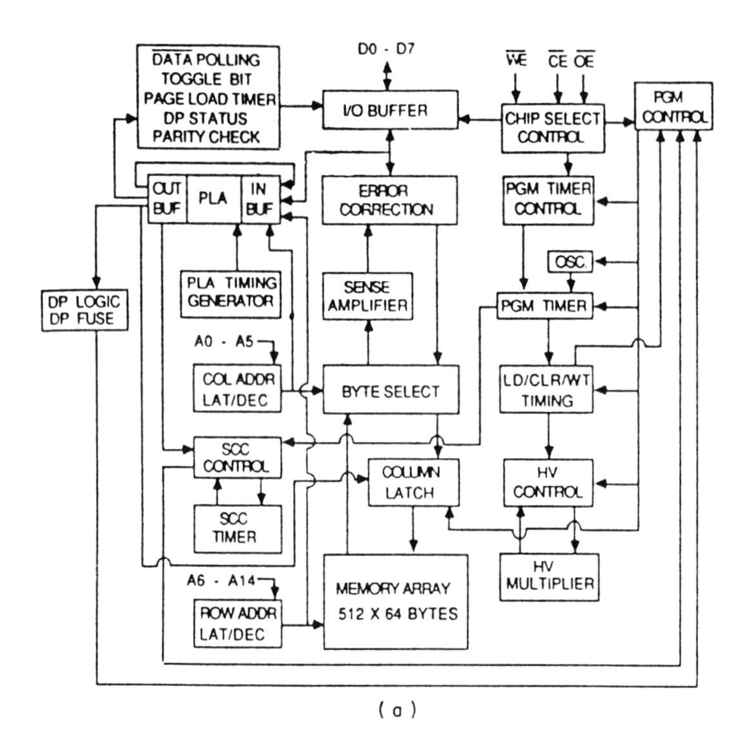

(a)

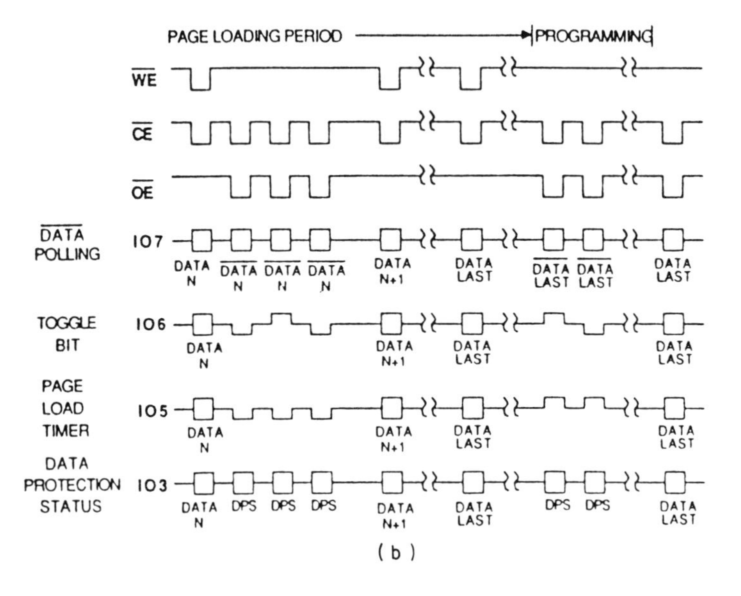

(b)

Figure 12.23

5 V CMOS 256k EEPROM with required and optional extended standards. (a) Block diagram. (b) Timing diagram for status bits. (From Ting *et al.* [38], Microchip Technology, with permission of IEEE.)

including: software data protection, chip clear, parity check, toggle bit, page-load timer, and data protection status bit. This part had 50 ns typical read access time, 1 ms page programming time and used a double poly, double metal 1.25 μm process. The functional block diagram is shown in Figure 12.23(a).

In addition to the special functions which were described in the last section, software algorithms permitted loading of instruction codes in the first three or six bytes. Legal code sequences were defined and once recognized will instruct the device to perform the designated functions according to the instruction. Instructions included were software data protection, software chip clear and the parity check mode.

The software data protection mode was set with an EEPROM switch. Once the mode had been set no further programming could be done unless the same code instruction was issued at the beginning of each loading period. A six code instruction could be used to reset the data protection mode.

The software chip clear replaced the conventional 12 V hardware chip clear and permitted this operation with a 5 V power supply.

Parity check mode sets I–O 4 so it is dedicated to signaling whether there was an error bit in the addressed byte. This worked by storing an original set of parity bits in the array. When a new set of parity bits was generated and compared with the original stored set, any single bit errors would result in a mismatch which would be indicated on I–O 4.

Four status bits were used with the timing diagram shown in Figure 12.23(b). These were designed to enhance the communication of the EEPROM with the system. $\overline{\text{Data}}$ polling and toggle bit were used to detect the completion of the programming cycle. The page load time status bit indicated the end of the loading period. The data protection status bit indicated whether the data protection mode was being used.

Another 256k electrically erasable PROM which included the required and optional standard features was a part from G.I. This part used a floating gate cell with composite interpoly dielectric less than 30 nm thick. A two-step intermetal oxynitride deposition was used to reduce hillock stress and give conformal coverage. The 50 ns read access time was attributed to the use of double metal, a fast sense amplifier and the high current cell design.

12.6.7 A typical 256k EEPROM timing diagram

Typical timing diagrams for a 256k EEPROM are shown in Figure 12.24. These include read cycle timing in Figure 12.24(a). $\overline{\text{CE}}$ controlled write cycle in Figure 12.24(b) and page mode write cycle in Figure 12.24(c).

The page mode write feature in this case allows the entire memory to be written in 1 s. Page write allows 2 to 64 bytes of data to be consecutively written prior to beginning the internal programming cycle. The destination address for a page write operation must reside on the same page, that is A_6 to A_{14} must not change. Page mode can be entered during any write operation.

This Xicor [39] 256k EEPROM also provides a type of hardware write protection to protect against inadvertent writes. Noise protection is offered since a $\overline{\text{WE}}$ pulse

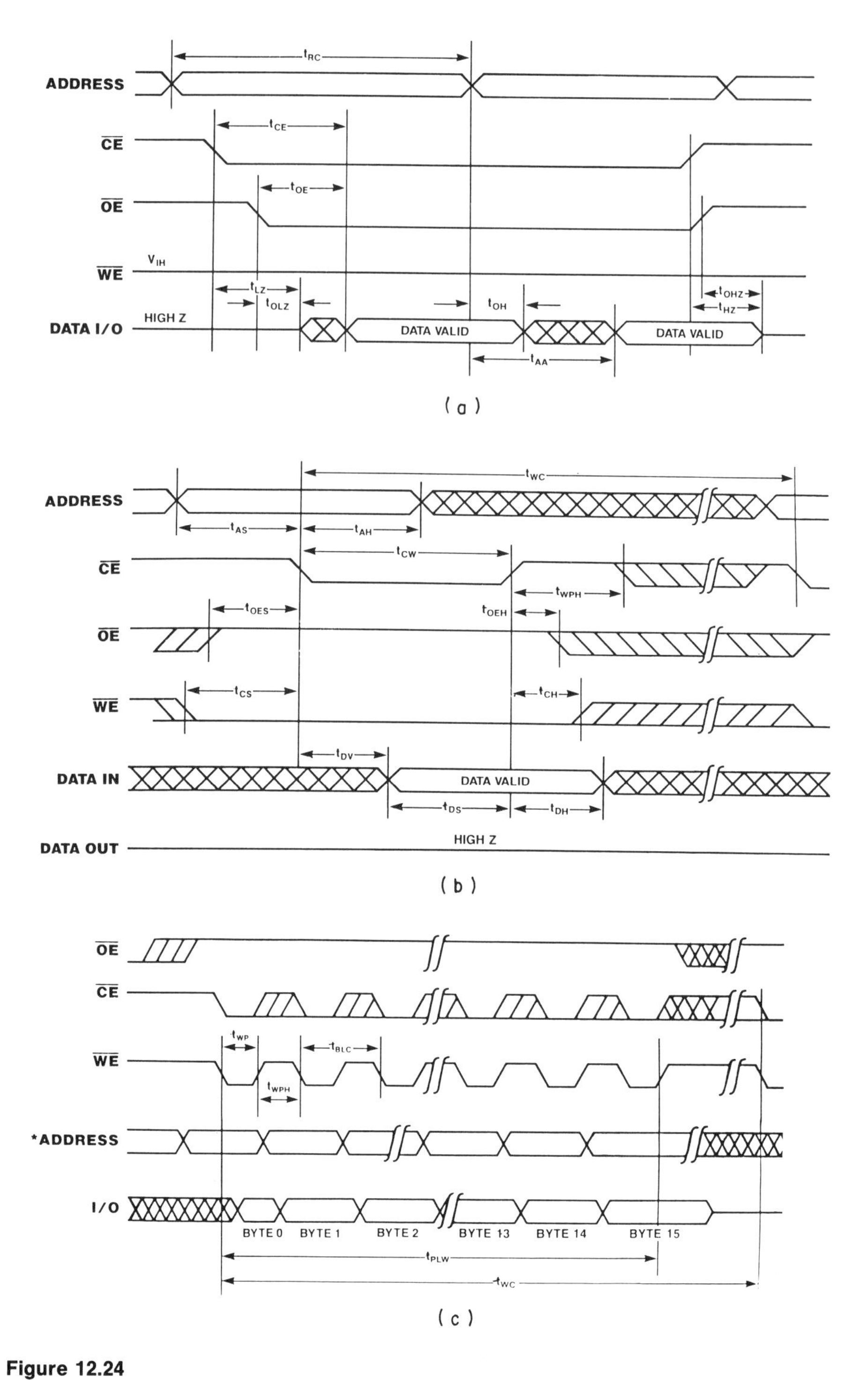

Figure 12.24
Typical operational timing diagrams for a 256k EEPROM. (a) Read cycle. (b) $\overline{CE}$ write cycle. (c) Page mode write cycle. (Source Xicor Databook.)

of less than 20 ns will not initiate a write cycle. A V_{CC} sense function inhibits all functions when V_{CC} is less than 3 V. A write inhibit function implemented by holding either $\overline{OE}$ low, $\overline{WE}$ high, or $\overline{CE}$ high will prevent inadvertent write cycle during power on and power off to maintain data integrity during this time.

Another chip which offered an innovative write protection was a 4k EEPROM from Catalyst [40] Semiconductor. This EEPROM, in addition to other intelligent features, permitted a software controlled splitting of the array into two blocks. A memory pointer indicated an address above which values are read only while addresses below and including the pointer require the security code to be accessed.

12.7 YIELD AND ENDURANCE ENHANCEMENT SPECIAL FEATURES ON EEPROMs

Error correction codes such as Hamming codes which added four parity bits per byte were used on the EEPROMs as were full error correction techniques such as the SEEQ 'Q-Cell'.

Redundancy for yield improvement was also used extensively on EEPROMs. While general error correction and redundancy techniques will be covered in the chapter on yield and redundancy, since endurance enhancement was considered a special feature of an applications specific nature in the EEPROMs, the specific details will be considered here.

12.7.1 The SEEQ Q-cell

With the introduction of the Q-cell on the 256k EEPROM, SEEQ extended the endurance of the EEPROM out to a million cycles. The Q-cell was a four-transistor cell which consisted of a pair of standard two-transistor EEPROM cells which were read out through separate sense amplifiers as shown in Figure 12.25. The sense amplifier outputs in turn were combined into a NOR gate.

The effect was that the correct information would be read out as long as at least one of the two cells was not defective. Since single-bit defects due to thin oxide failures are the commonest failure mechanism on the EEPROM the probability of two adjacent cells wearing out is extremely low. So that the number of actual defective storage transistors in the array could be determined at test, modes of test were built in which allowed reading of all of the storage cells.

The disadvantage of the Q-cell method of enhancing endurance was that the number of transistors in the array was doubled so that the chip size was considerably enlarged. On the average assuming a 50% array efficiency one would expect that doubling the size of the array and maintaining the size of the periphery would add about 50% to the size of the chip or that the extra array area would be about 33% of the chip area.

This technique could be used to reduce the required external testing of the part by testing the outputs of the Q-cell and then specifying the endurance at the intrinsic

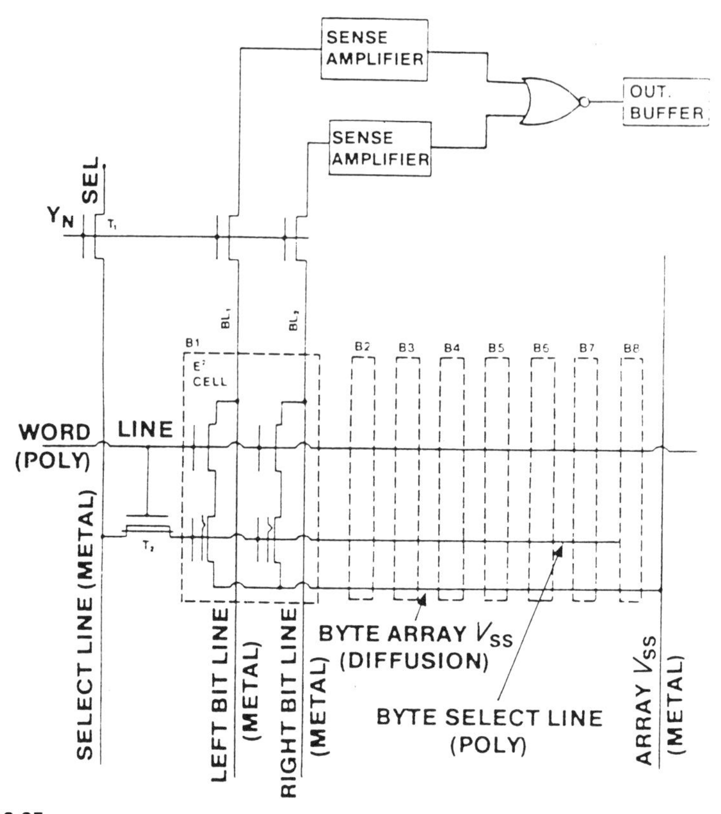

Figure 12.25
256k EEPROM using four-transistor Q-cell. (From Cioaca *et al.* [37], SEEQ 1987, with permission of IEEE.)

endurance of the array. This had the effect of increasing the yield which tended to decrease the cost of the part and balance the effect of the increased chip size.

The other option was to test that all bits in the array were functional and then specify increased endurance since every bit has a back-up. This gave a higher cost chip with an added feature to increase the selling price to balance the higher cost.

Another application of the Q-cell concept was in a fast 16k EEPROM from SEEQ [41], which was mentioned in an earlier section. The six-transistor differential cell, which was used for speed on this chip, is shown in Figure 12.26. Two floating gate storage cells and a select transistor were in series in each branch of the differential cell, if one of the two devices failed, then the branch current flow would be controlled by the other device.

Other features which were included to enhance reliability include protection against accidental write by filtering out noise in the $\overline{WE}$ input and using a temperature compensated reference voltage generator to prevent accidental writing at supply voltages below 4 V. Special test modes were activated by high voltage input levels. These included block write modes to reduce the time for endurance screening, and margin stress mode for reliability screening.

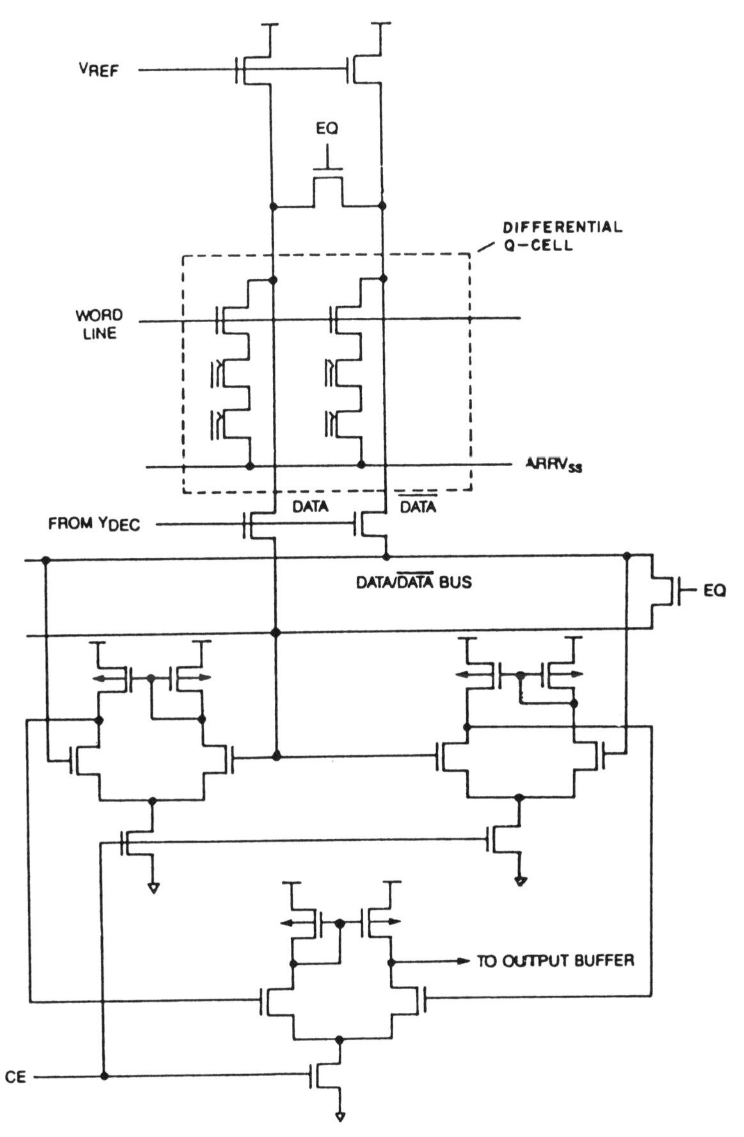

Figure 12.26
Sensing path of fast EEPROM with six-transistor differential Q-cell. (From Vancu *et al.* [41], SEEQ 1988, with permission of IEEE.)

12.7.2 A 10 million cycle endurance EEPROM design

An improved endurance EEPROM cell was designed by Toshiba [54] in 1989 by doing a computer generated simulation of the electric field in the channel and source regions of the EEPROM. Injection of hot holes into the gate oxide, which can happen due to lateral field components in an EEPROM, generate electric traps, Toshiba reduced this injection level by designing to reduce the lateral field component in the

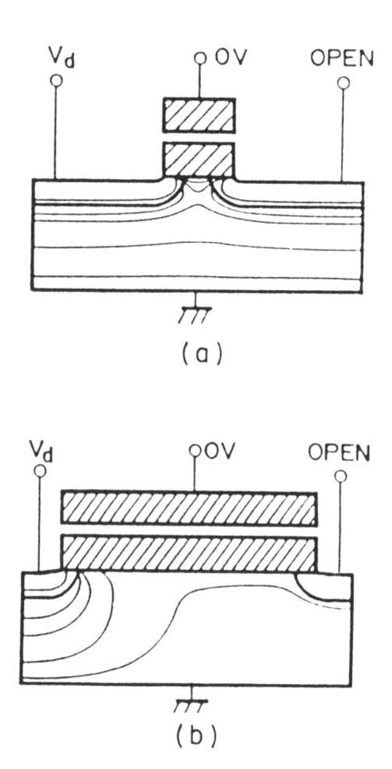

Figure 12.27
Cross-section of EEPROM cells showing electric field during erase of (a) new design developed by computer simulation whose source N^+ region is located within the depletion region of the surface channel area during erase operation and (b) conventional device. (From Endoh *et al.* [54], Toshiba 1989, with permission of IEEE.)

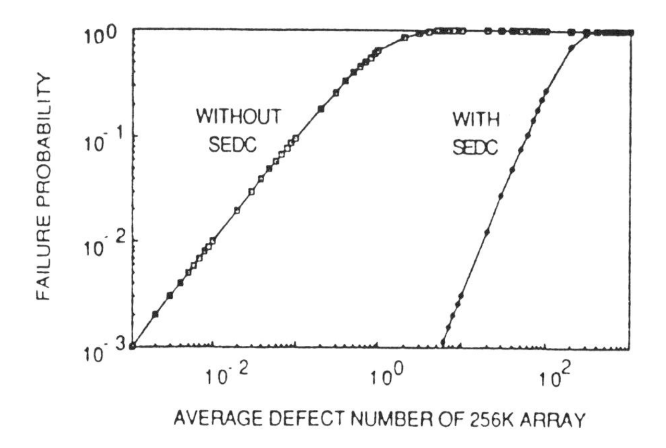

Figure 12.28
Failure probability plotted against average number of defects in a 256k EEPROM array using Hamming code type of error correction (with SEDC) and without error correction. (From Ting *et al.* [38], Microchip Technology 1988, with permission of IEEE.)

depletion region of the surface channel during erase operations. The field during erase of the new device is shown in Figure 12.27(a), and of the conventional floating gate EEPROM in Figure 12.27(b). The new cell was reported to have 10 million cycle endurance.

12.7.3 Hamming codes to improve endurance

A Hamming code was used in the 1988 Microchip Technology 256k EEPROM. Figure 12.28 illustrates the mathematical model used. Each parity bit was generated as an Exclusive-OR of five input data. Each check bit was an Exclusive-OR of a parity output and five data sense outputs. The parity bit generator and the check bit generator shared the same Exclusive-OR circuits and a multiplexer was used to input the proper inputs to the X-OR gates in accordance with the read–write status. This error correction circuit occupied 23% of the chip area.

Another 256k CMOS EEPROM with error correction circuitry using a modified Hamming code was shown by Samsung Semiconductor [43] in 1988. The ECC scheme used four parity cells per byte. The parity code generator, core array data bits, and core array parity bits could be tested in separate modes.

Margining circuits for checking individual cell threshold voltage were included. A fault tolerant EEPROM fuse circuit was used which is shown in Figure 12.29. The fuse can tolerate a single cell failure out of its four cells. The required and optional feature set were included.

12.7.4 Extended temperature

Military applications for EEPROMs such as data updates in systems used for remote processing in aircraft and satellites led to a demand for extended temperature parts.

A temperature tolerant 64k EEPROM from AMD [28] was described in 1985. The part used special circuit techniques to achieve extended temperature operation. These

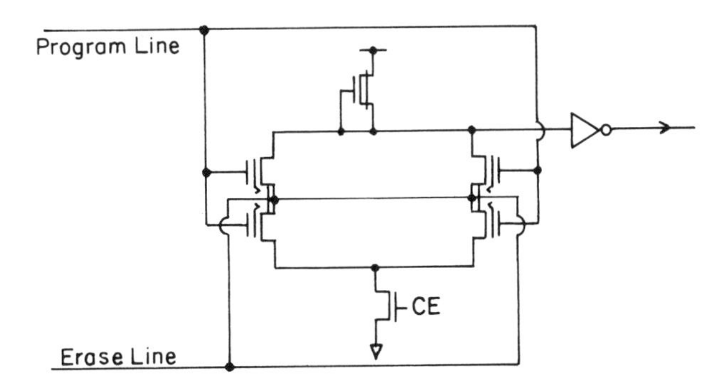

Figure 12.29
Fault tolerant EEPROM fuse. (From Do *et al.* [43], Samsung 1988, with permission of Japan Society of Applied Physics.)

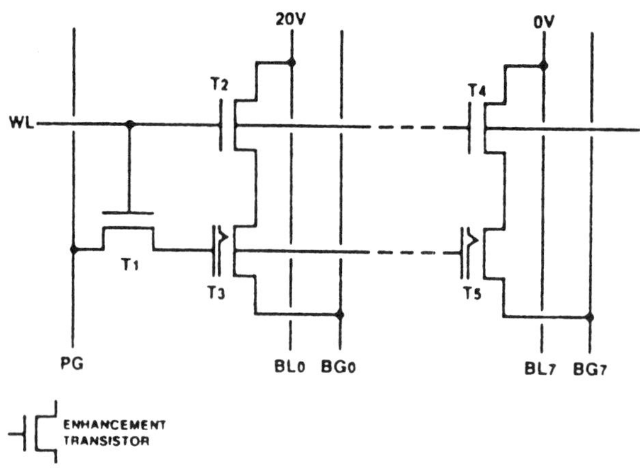

Figure 12.30
Extended temperature tolerent 64k EEPROM showing separate parallel ground and bit-lines in each cell. (From Bill *et al.* [28], AMD 1985, with permission of IEEE.)

included a feedback controlled substrate bias generator to reduce field transistor leakages and a stable voltage reference to accurately control the programming voltage over the full military temperature range. The write timing pulse was trimmed by EEPROM fuses.

The high voltage decoder pumps tended to limit the operating temperature range of 5 V only EEPROMs. These were used during the write cycle to charge up the selected word and bit-lines. This chip, therefore, used a pump on every word and bit-line in the array. Each individual pump could not under worst case temperature and voltage conditions source more than a few nanoamperes of current. This current sourcing limitation led to the choice of a cell design with separate ground lines running parallel to the bit-lines which avoided punchthrough between adjacent cells as shown in Figure 12.30.

12.7.5 Low operating voltage

Low voltage applications for reprogrammable non-volatile memory in portable applications include such systems as handheld computers, smart cards, pagers, beepers, language translators and many telecommunications devices. Historically a low voltage EEPROM has been difficult to manufacture. In 1987 Catalyst [55] announced a 3 V 64k EEPROM with low voltage access time of 120 ns and 5 V access time of 60 ns. Active current is a low 7.5 mA at 8 MHz. Various design techniques included a low voltage tolerant differential sense amplifier and a dual clocked high voltage switching circuit which was insensitive to voltage fluctuations. The part used five types of transistors including PMOS, and NMOS enhancement mode in the 3 V array and enhancement, depletion and non-ion implant NMOS in the charge pumping circuitry that converted the 3 V supply to 18 V on chip.

There were in addition the floating gate storage transistors. The chip size was 34 mm^2.

12.8 EEPROM COMBINATIONS AND EMBEDDED EEPROMs

Examples of how EEPROM technology is combined with other circuitry are the devices with large amounts of ROM with a small amount of EEPROM on chip. Both standalone memory combinations, such as a Motorola device which combines 115k ROM and 18k EEPROM, and the more common ROM and EEPROM embedded on microcontroller, such as the Philips 80C51 and Motorola 68HC805C4, are examples. This combination is used due to the much larger size of the EEPROM cell than the ROM cell which makes it more expensive. If the EEPROM were cost competitive with the ROM the need for such combinations would vanish.

A 2k EEPROM cell, which was specifically designed for embedding in a standard logic process, was discussed by Philips [32] in 1985. The requirement for a memory cell to be embedded in logic is that the logic process is not changed and only 5 V power supply is required. In this case the logic process was single polysilicon and single level metal which meant that the usual double polysilicon cell approach could not be used for the EEPROM. The solution was a single polysilicon cell using a large tunnel-oxide MOS capacitor which coupled the control gate, which was an N^+ diffusion, with the storage cell. This is shown in the layout and cross section shown in Figure 12.31(a). The N^+ diffusion is shared between two cells. The schematic circuit diagram of a conventional double polysilicon EEPROM cell and the single poly EEPROM cell with the thin oxide coupling capacitor is shown in Figure 12.31(b).

The program voltage was reduced from the usual 20 V to 13 V by decreasing the thickness of the injector oxide and by increasing the capacitive coupling between the control and floating gate. This increased the size of the cell over that of the stacked gate cell.

The coupling capacitor between floating gate and control gate was made much larger than that between floating gate and drain so that only a small fraction of the program voltage was applied between the floating gate and the drain. This meant that tunneling through this capacitor was negligible and this cell operated the same as the usual double polysilicon gate cell.

Only one extra mask step for definition of the thin oxide region was needed.

The endurance was also similar to the double polysilicon device. The data retention, however, was smaller due to the large proportion of oxide area in this embedded cell. Yield tended to be increased by the simpler single polysilicon process and decreased by the large thin oxide area.

Special circuit design techniques were also used to eliminate parasitic leakage paths that can occur at voltages below the 13 V used for programming. The program voltage limiting effects could be punchthrough, snap-back in short channel MOSTs and the turn-on of the parasitic field MOS transistors.

Diffusion to diffusion punchthrough was reduced by increasing the minimum diffusion spacings. Snapback only occurs in transistors that conduct a current, so it was eliminated by biasing the NMOS transistors during programming so that no currents flowed. Turn-on of parasitic field transistors was suppressed by applying a small voltage to the source of the transistors during the programming cycle. This acted as an artificial substrate bias effectively increasing their threshold voltage.

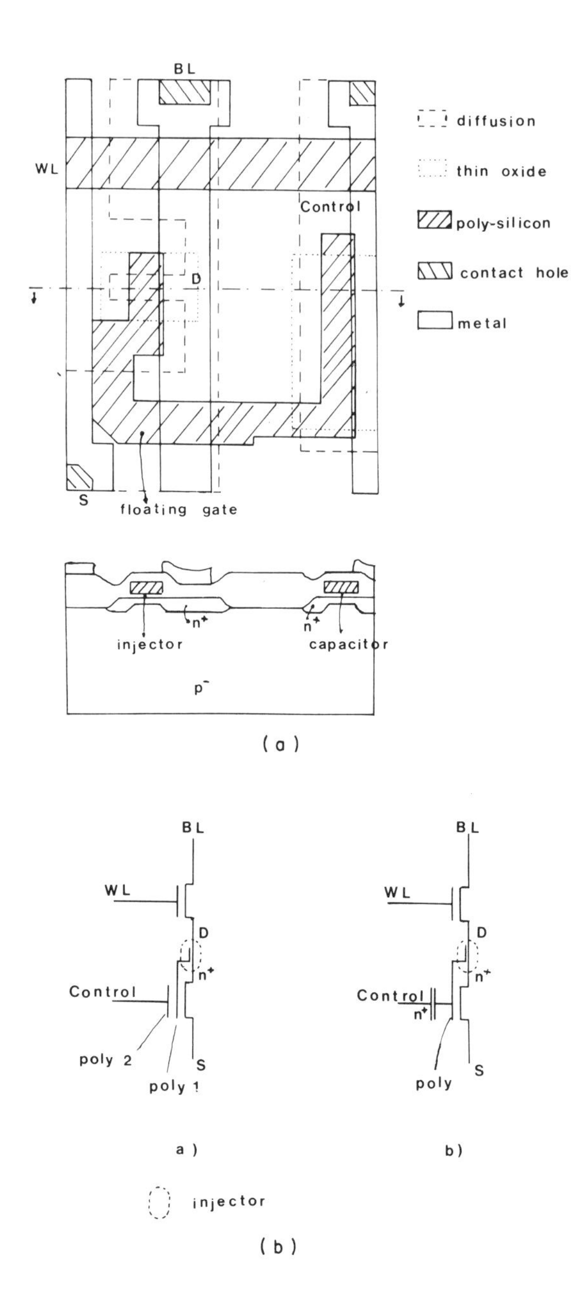

Figure 12.31
2k single polysilicon EEPROM cell designed for embedding in a standard CMOS logic process. (a) Top view and cross-section of cell. (b) Schematic circuit diagram comparing a double polysilicon EEPROM cell and the single poly EEPROM cell. (From Cuppens [32], Philips 1985, with permission of IEEE.)

Another Philips [42] embedded EEPROM cell was shown in 1989. This cell continued the use of the single polysilicon cell but added a titanium silicide control gate with a single oxynitride dielectric layer. Both a 70 μm^2 two transistor EEPROM and a 30 μm^2 single-transistor flash EEPROM cell were processed together. Program and erase characteristics of the two cells along with endurance characteristics are shown in Figure 12.32(a) to (e). The EEPROM cell gives an acceptable threshold voltage window with 14 V programming for 2 ms. The flash cell can be programmed by a 13 V and 5 ms pulse at the drain and erased with a 15 V 100 ms pulse on the control gate with a 3.5 V threshold window.

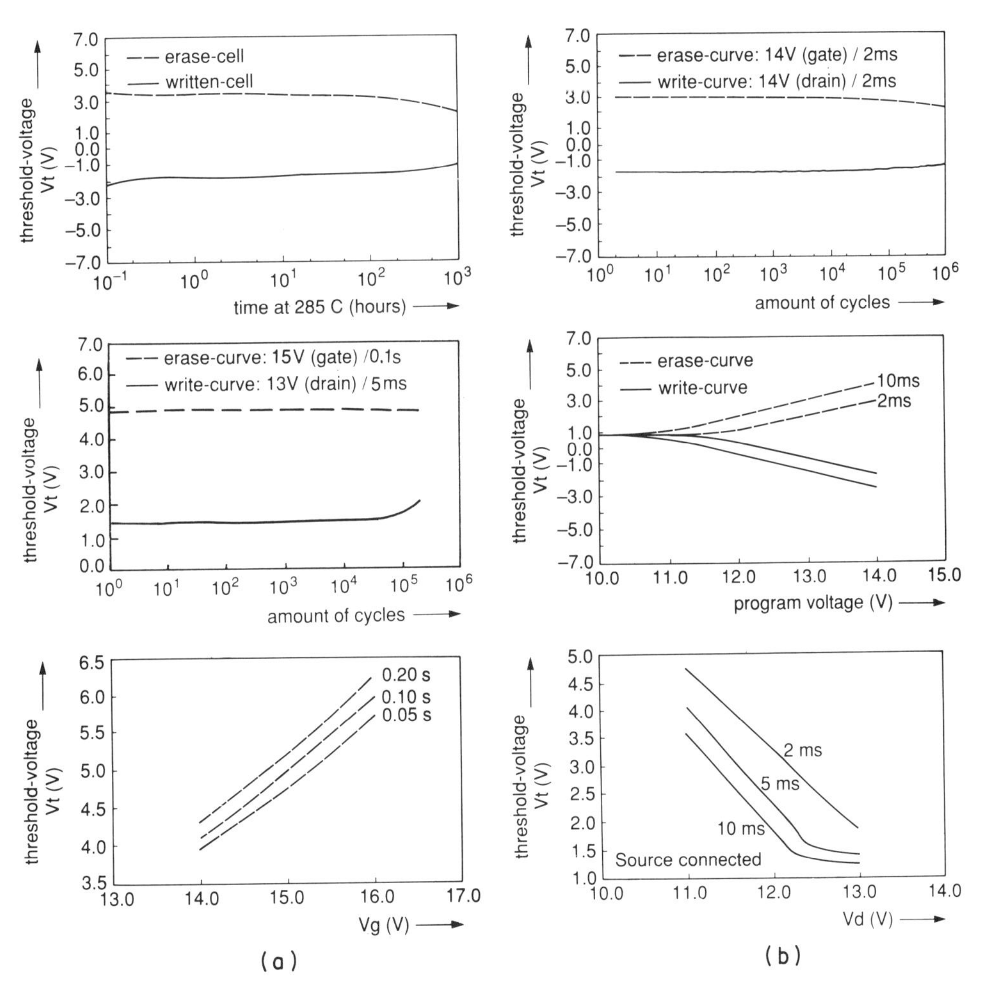

Figure 12.32
Characteristics of two electrically erasable single polysilicon embedded cells using titanium silicide control gate and oxynitride dielectric layer. (a) Two-transistor EEPROM. (b) Single-transistor flash memory. (From Vollebregt *et al.* [42], Philips 1989, with permission of IEEE.)

An example of a larger EEPROM embedded in logic is given by the 32k EEPROM module that Motorola embedded in one of their 6805 family of microcomputers. The part included an on-board charge pump for generating the 18 V programming voltage from a single 3 V to 6 V source. An application for this part was described in Chapter 3.

Another company designing EEPROMs as standard cells for use in ASIC chips for such systems as workstations is Sierra Semiconductor [36]. The cells include both EEPROM arrays and a family of high voltage interface cells which permit generation of both 5 V and 18 V programming voltages on chip. They have also worked on a set of EEPROM standard cells intended for portable applications which permit operation down to 3 V.

A 512k flash EEPROM cell was shown by Motorola [57] in 1991. It was intended for embedding in a 32 bit microcontroller. A split gate cell structure was used to eliminate the overerase problems of the single-transistor flash cell. Programming was through hot electron injection on the drain side and erase was through cold electron tunneling on the source side to the source diffusion by means of an extension of the floating gate. A charge pump was used to generate the 14 V erase voltage.

12.9 SOME GENERAL THOUGHTS ON EEPROMs

The EEPROMs have become a significant niche product in the memory industry with well defined applications for which they are particularly suited. The slow write speed and endurance trade-offs with cost have kept this product from fulfilling its original promise of entering the mainstream of the memory market.

12.10 NON-VOLATILE SRAMs

12.10.1 The NV-SRAM or shadow RAM

The user who needs a non-volatile memory with faster write-cycle times or more endurance than the EEPROMs can provide, might be able to use a non-volatile SRAM. These come in two forms: the SRAM with a battery in the package (BRAM) which was discussed in Chapter 9, and the non-volatile SRAM which is a combination of SRAM and EEPROM cells on a chip. NV-SRAMs are useful for retrieving or saving critical CPU default values during system power-up or power-down. Other potential applications include office video terminals, point of sale terminals, industrial controllers and robotics.

The first non-volatile SRAM, the 'shadow RAM' was introduced in 1979 by Xicor, then a start-up company. While not widely second sourced by major memory suppliers, the shadow RAM heralded the new age of non-volatile memories to come. It contained a 256 bit static RAM overlaid bit-for-bit with a 256 bit EEPROM. A functional block diagram of a later 64k 5 V only NVRAM from Xicor [39] is shown in Figure 12.33.

Shadow RAMs combine a normal static RAM array with an EEPROM array. When the chip is powered up, the contents of the EEPROM are written into the RAM,

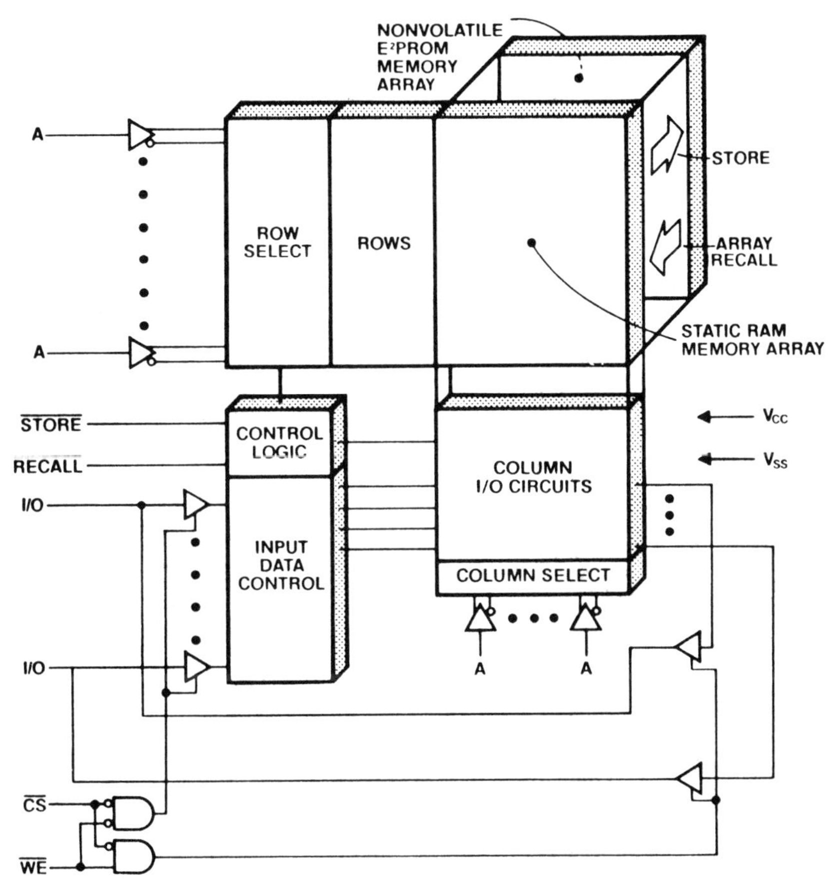

Figure 12.33
64k × non-volatile SRAM functional diagram showing static RAM memory array and EEPROM background array. (Source Xicor Databook 1985.)

providing a default configuration. Parameters can be changed while the system is on, and the EEPROM can be reprogrammed to change the default setting.

Data can be stored in the EEPROM and simultaneously independent data can be accessed in the RAM. Data could be transferred between the RAM and EEPROM on the Xicor part using only a 5 V power supply due to the inclusion of a charge pump on the chip. The unique use by Xicor of asperities on the polysilicon as a tunnel enhancement mechanism was discussed in a previous section.

In 1983 two 4k shadow RAMs were introduced by Intel [45, 47] who described a non-volatile static RAM with 4k of memory. The cell was a nine-transistor structure consisting of a standard six-transistor depletion load static RAM cell, an EEPROM storage transistor, and two additional transistors used as gating devices as shown in Figure 12.34(a).

An on-chip high voltage power supply gave 5 V only operation. In two timed 10 ms steps the EEPROM array could be erased and the current contents of the static RAM stored. The complex control functions required for proper operation were integrated on chip. A single control pin differentiated between standard and static

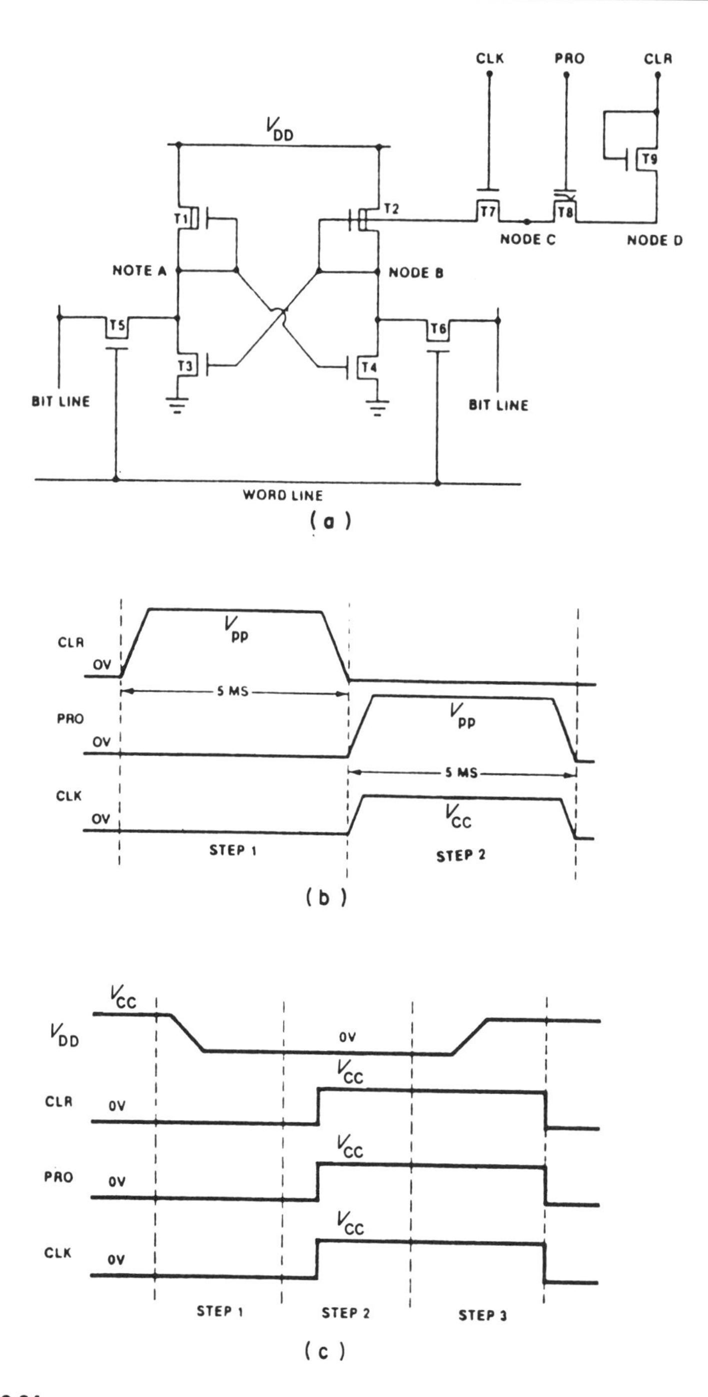

Figure 12.34
(a) NV-SRAM with a nine-transistor cell structure consisting of six-transistor depletion load SRAM cell, an EEPROM storage transistor and two select transistors. (b,c) Control timing waveforms for STORE and RECALL functions. (From Lee *et al.* [45], Intel 1983, with permission of IEEE.)

RAM operation with 200 ns access time and the store–recall cycles. The STORE waveforms and RECALL waveforms are shown in Figures 12.34(b) and (c).

NCR [44] in 1982 showed a 1k × 8 three power supply NV-SRAM using the SNOS process which was described earlier in the section of MNOS processes. An endurance of 10^4 cycle was shown with minimum 1 year data retention.

A circuit diagram of a cross-section of the SNOS NVRAM cell is shown in Figure 12.35(a) and the R-load SRAM and the EEPROM parts of the ten-transistor cell are shown in Figure 12.35(b).

NCR [48] in 1983 introduced a 5 V only shadow RAM also in a 4k single power supply version. This chip had 200 ns typical access time and was made in 4 μm minimum geometries. The chip was 28 mm^2 in area.

The NV-SRAM has been limited to very low densities by the large cell size resulting from combining an SRAM cell with an EEPROM cell—two devices which already

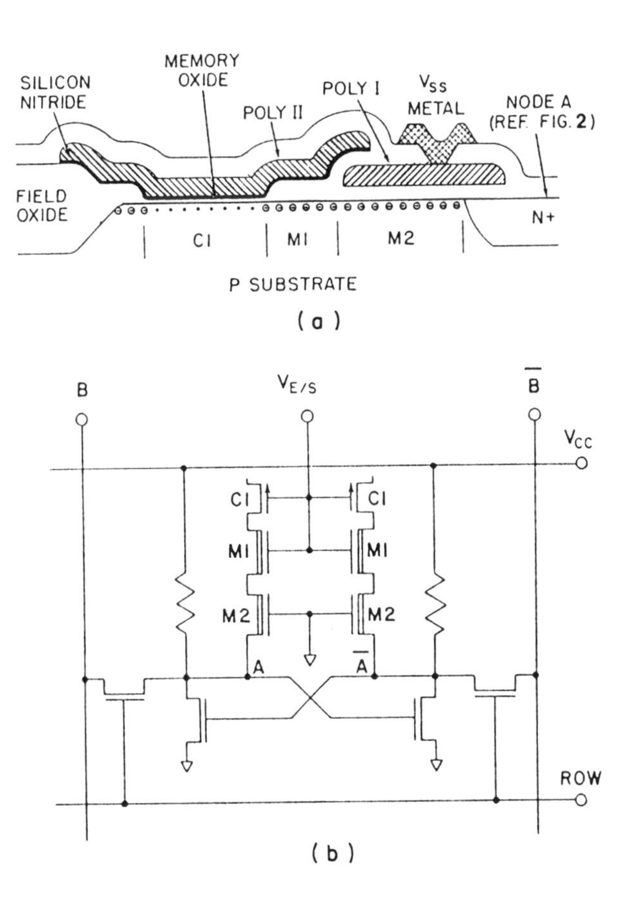

Figure 12.35
NV-SRAM in MNOS (SNOS) technology. (a) Cross-section of non-volatile memory element. (b) Ten-transistor NV-SRAM cell consisting of R-load SRAM and six-transistor non-volatile element. (From Donaldson *et al.* [44], NCR 1982, with permission of IEEE.)

individually suffered from large cell sizes relative to the DRAM and EPROM. The NV-SRAM has become, therefore, another of the many memory niche products which has applications but will never enter the mainstream of the technology.

The search for a high density, fast, low cost, low power, non-volatile RAM has continued in some rather novel directions.

12.10.2 The ferroelectric non-volatile RAM

An unusual non-volatile RAM concept was presented at the ISSCC in 1988 by Ramtron [49] Corporation and also discussed by Krysalis (now National) [50], also in 1988. This ferroelectric non-volatile RAM is formed of a standard RAM cell over which is deposited a small 'battery' consisting of a ferroelectric material sandwiched between two metal electrodes.

The ferroelectric is a dielectric with two useful properties: one is having a dielectric constant about a factor of 100 larger than silicon dioxide or silicon nitride and the other is having two stable states between which the material can switch under the effect of an applied voltage. The first property means that capacitors made of PZT, a commonly used ferroelectric for non-volatile SRAMs, can store 100 times the charge of a silicon nitride capacitor of the same thickness or be a factor of 100 smaller.

The second property means that a ferroelectric capacitor can be used as a digital memory storage device. A typical hysteresis *I–V* switching loop for PZT is shown in Figure 12.36. The stable states are at the top and bottom of the loop.

The Ramtron non-volatile SRAM used the ferroelectric capacitors as non-volatile

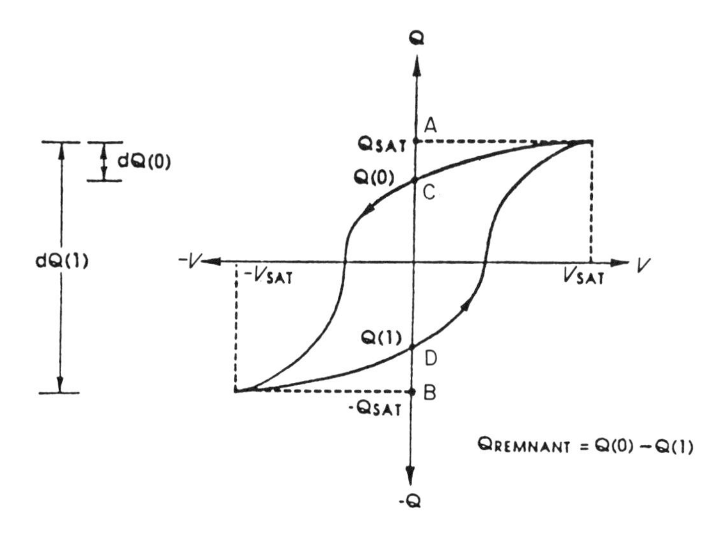

Figure 12.36
Schematic of typical hysteresis curve of ferroelectric capacitor. (From Evans and Womack [50], National Semiconductor [1988], with permission of IEEE.)

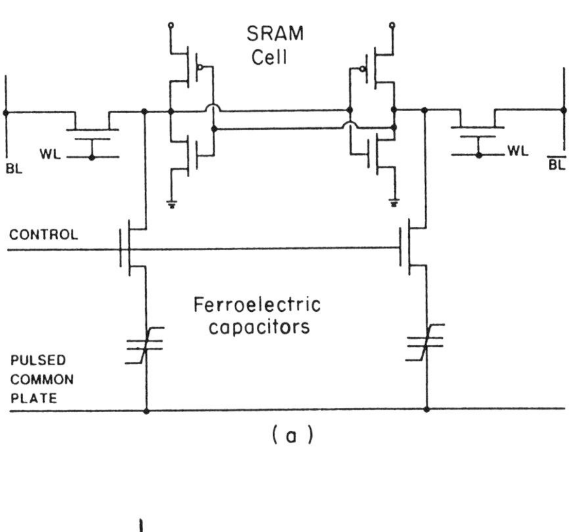

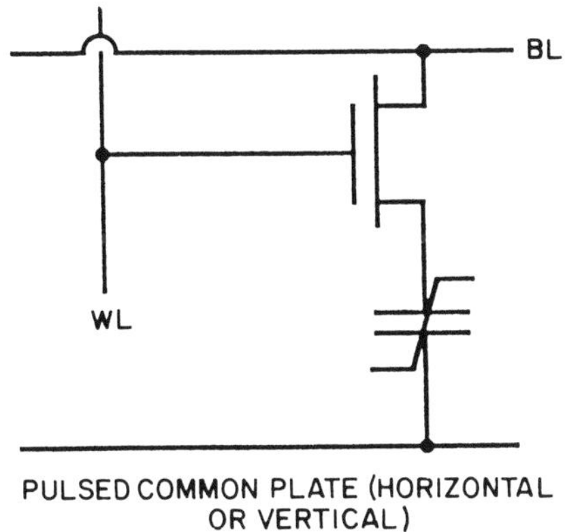

Figure 12.37
Memory cells using the ferroelectric capacitor. (a) Six-transistor SRAM plus ferroelectric cell consisting of select transistors and two ferroelectric capacitors. (b) Two-element NV-DRAM cell consisting of select transistor plus ferroelectric capacitor as used in a four-element double-ended differential sensing scheme. (From Eaton *et al.* [49], Ramtron 1988, with permission of IEEE.)

storage devices in conjunction with an SRAM in the memory cell as shown in Figure 12.37(a). A DRAM type one-transistor cell using a pulsed common plate ferroelectric capacitor is shown in 12.37(b).

A non-volatile memory that used the ferroelectric capacitor as a storage cell in a non-volatile DRAM type of configuration was described by Krysalis [50] in 1988. The memory cell of this 512 bit part consisted of a pass transistor and a ferroelectric capacitor. The part used in a double-ended differential sensing scheme.

Various research aspects of ferroelectric materials are being pursued by several companies. Questions of data retention and endurance are still in the early stages of being investigated.

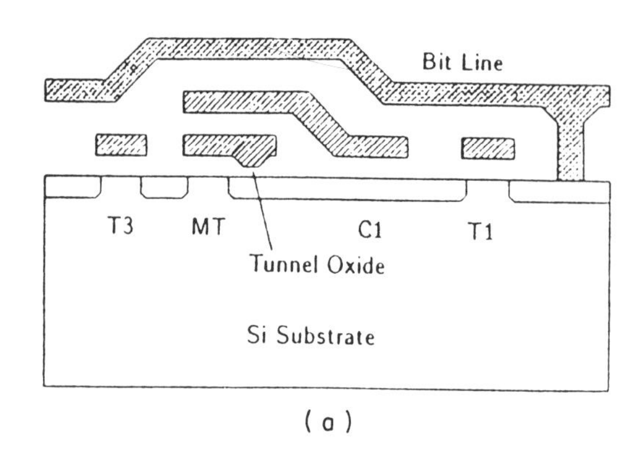

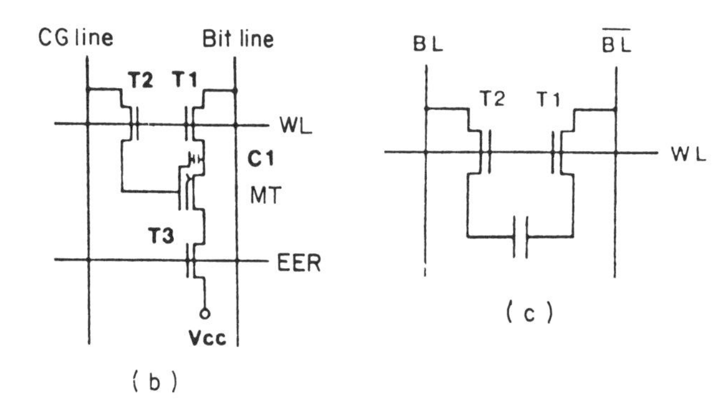

Figure 12.38
Non-volatile DRAM. (a) Schematic cross-section of NV-DRAM cell. (b) Equivalent circuit of NV-DRAM cell. (c) Equivalent circuit of NV-DRAM cell in DRAM mode. (From Terada *et al.* [51], Mitsubishi 1988, with permission of IEEE.)

12.11 THE NON-VOLATILE DRAM

Another promising direction has been recent research on non-volatile DRAMs which combine the single transistor of the DRAM with the electron injection properties of the electrically erasable capacitor found in the EEPROM or EAROM.

In 1988 Mitsubishi described a cell which combined a floating gate EEPROM with a DRAM capacitor. A cross-section of the cell is shown in Figure 12.38(a). An equivalent circuit of the NVRAM cell is shown in Figure 12.38(b). The electrodes of the DRAM capacitor are the control gate and the drain of the floating gate storage cell. Transistors T_1 and T_2 are the DRAM transistors. T_3 is the mode selection transistor which selects EEPROM or DRAM mode.

An equivalent circuit of the NVRAM cell in the DRAM mode is shown in Figure 12.38(c). The two transistors T_1 and T_2 can eliminate the dummy cell in the DRAM mode.

A circuit diagram of the NVRAM is shown in Figure 12.39(a). The same sense amplifiers are used in the DRAM and in the EEPROM mode. To operate in the DRAM mode, clock EER is kept low and T_3 is turned off. This makes the equivalent

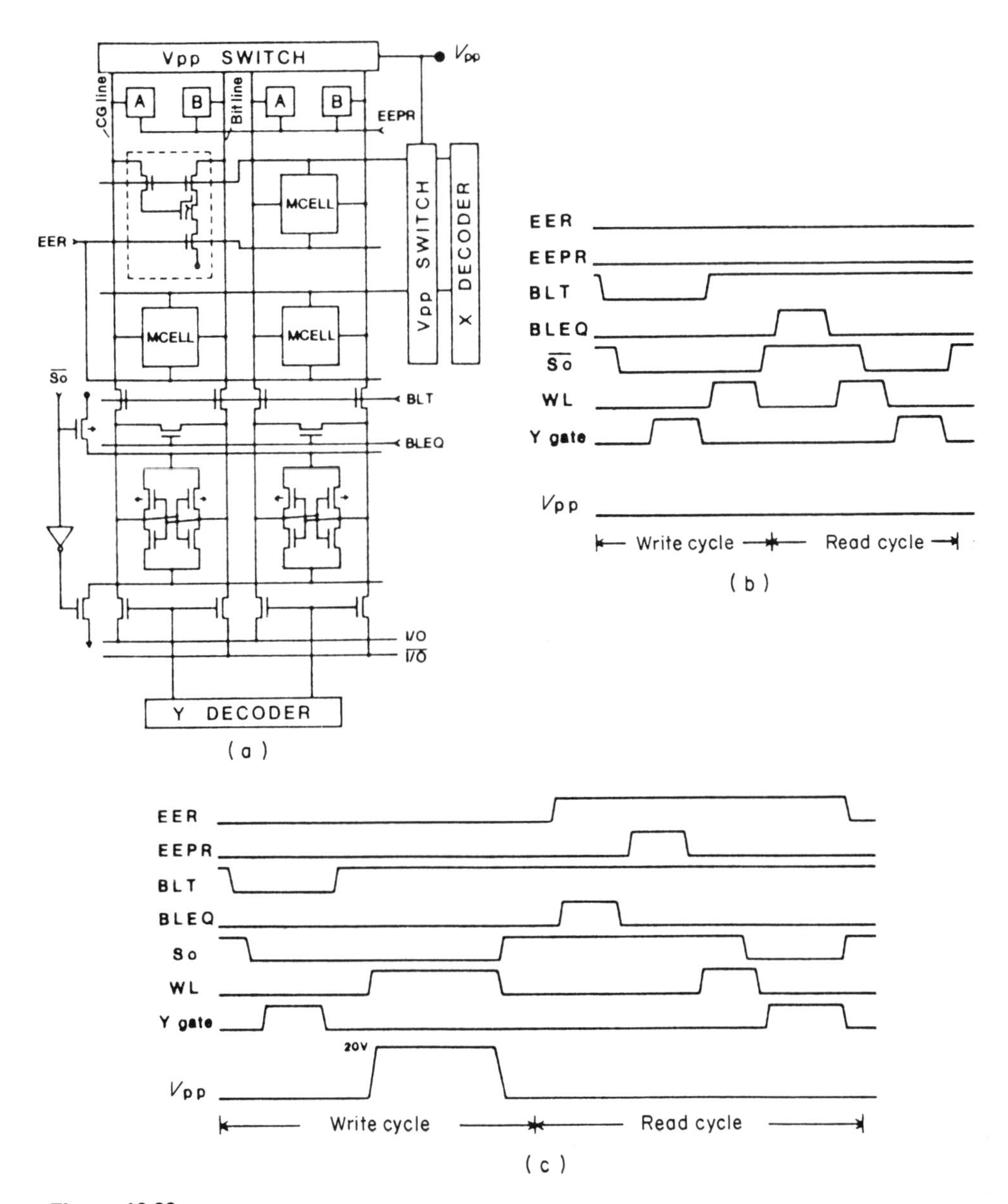

Figure 12.39
Non-volatile DRAM. (a) Circuit diagram of NV-DRAM, (b) DRAM mode write and read cycle. (c) EEPROM mode write and read cycle. (From Terada *et al.* [51], Mitsubishi 1988, with permission of IEEE.)

circuit of the cell the two transistors and capacitor shown before in Figure 12.38(c). The capacitance value of the DRAM capacitor depends on the stored charge on the floating gate affecting the apparent value of the dummy cell.

In the EEPROM mode the sense amplifier acts as a data latch and in the write cycle the input data are latched in the sense amplifier. The V_{PP} is then applied through

A and B. Depending on the latched data either the bit-line or the control gate line is pumped up to the V_{PP} level, thereby programming the cell.

Data are transferred between the DRAM and the EEPROM through the sense amplifier. Data stored on the DRAM are sensed and latched in the sense amplifier which then acts as an input for the EEPROM. Clock timing diagrams for the DRAM and EEPROM mode are shown in Figures 12.39(b) and (c).

In 1988 Sharp [52] also presented work on a non-volatile DRAM cell which combined a DRAM cell with a floating gate EEPROM cell. The cell structure is shown in Figure 12.40(a). The cell consists of two transistors (T_1 and T_2), a floating gate storage transistor (MT) and a DRAM type storage capacitor (C). T_1 is the word-line select transistor and T_2 is the mode select transistor. The gate of T_2 is grounded during data transfer from the 'DRAM' to the 'EEPROM'.

Figure 12.40(b) shows a cross-section of the cell with a DRAM data state of '0' and Figure 12.40(c) a DRAM data state of '1'. Operating bias tables for the 'store'

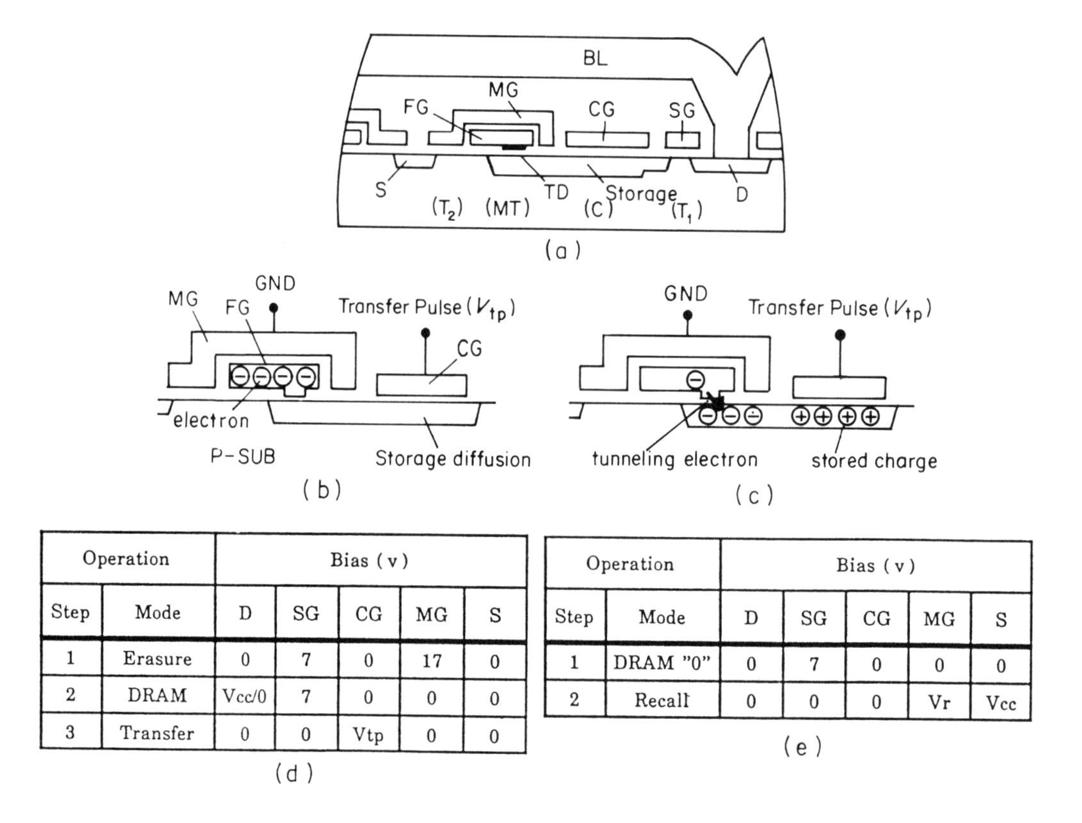

Operation		Bias (v)				
Step	Mode	D	SG	CG	MG	S
1	Erasure	0	7	0	17	0
2	DRAM	Vcc/0	7	0	0	0
3	Transfer	0	0	Vtp	0	0

(d)

Operation		Bias (v)				
Step	Mode	D	SG	CG	MG	S
1	DRAM "0"	0	7	0	0	0
2	Recall	0	0	0	Vr	Vcc

(e)

Figure 12.40
Non-volatile DRAM. (a) Schematic cross-section showing cell structure. (b) Schematic cross-section showing DRAM data state of '0'. (c) Schematic cross-section showing DRAM data state of '1'. (d) DRAM to EEPROM store operation. (e) EEPROM to DRAM recall operation. (From Yamauchi *et al.* [52], Sharp 1988, with permission of IEEE.)

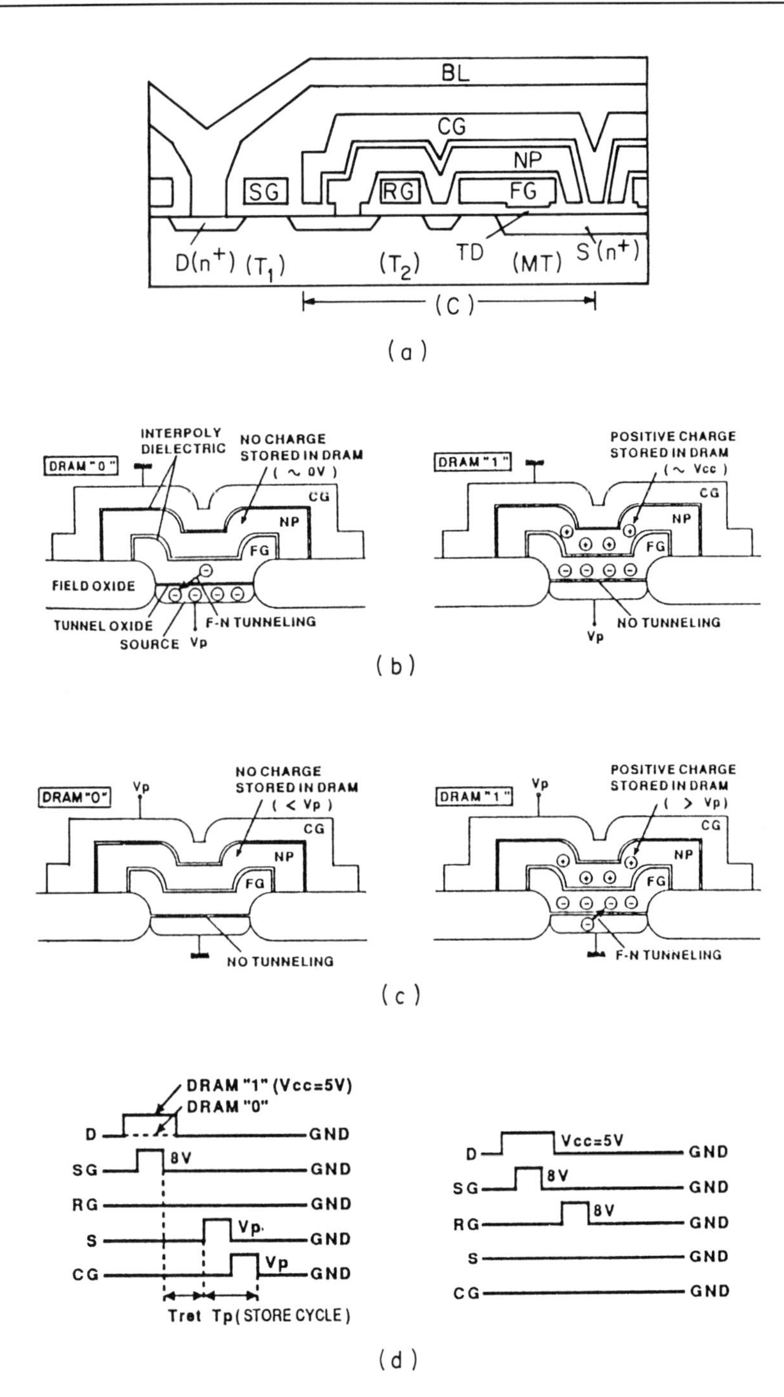

Figure 12.41
Second generation non-volatile DRAM cell in 0.8 μm technology. (a) Schematic cross-section. (b) Erase in the first half of the store cycle. (c) Program in the second half of the store cycle. (d) Timing diagrams for store and recall operations. (From Yamauchi *et al.* [53], Sharp 1989, with permission of IEEE.)

operation from DRAM to EEPROM and the 'recall' operation from EEPROM to DRAM are shown in Figures 12.40(d) and (e). This cell is made in 1.2 μm technology and is 78 μm^2 in cell size. This is about 30% of the size of an SRAM cell in this same technology and only slightly larger than a 256k DRAM cell. It is potentially a significant step in cost reduction for a non- volatile RAM. Questions of an operational and reliability nature, such as potential for read–disturb problems remained to be answered.

In 1989 Sharp [53] took the next step with this concept with another non-volatile DRAM cell. This cell was made in 0.8 μm technology and has a cell area of 35 μm^2 which could make it competitive at the 1Mb DRAM density level.

A schematic cross-section of the cell is given in Figure 12.41(a). Cross-sections showing the 'erase' and 'Program' mechanisms are given in Figures 12.41(b) and (c) along with timing diagrams for the store and recall operations.

This cell differed from the first Sharp cell in that it stacked the control gate and added a 'recall' gate. The potential for read–disturb were minimized in this part. During the first part of the store cycle, if the DRAM is in the '0' state then the EEPROM is erased by applying a high voltage pulse to the source and grounding the control gate. If the DRAM is in the '1' state the voltage applied between the source and the DRAM storage node NP is reduced due to the effect of the positive charge stored in the DRAM. This inhibits the EEPROM from being erased.

During the second part of the store cycle, if the DRAM is in the '1' state, then the EEPROM is programmed by applying a high voltage pulse to the control gate and grounding the source. If the DRAM is in the '0' state the EEPROM is inhibited from being programmed due to its low capacitive coupling ratio. Therefore whether the EEPROM is programmed or erased depends on whether the DRAM is '1' or '0'. The transistors T_1 and T_2 are turned off during the store operation.

On chip high voltage generation is possible due to the low current drain of the cell during the store operation.

Endurance of a single EEPROM cell was measured at less than 10% closure of the threshold voltage window after 10^6 cycles. The endurance is improved in that the EEPROM is not programmed if it is in the same state as the DRAM at the initiation of the store operation.

BIBLIOGRAPHY

[10] Stewart, R. G. (1979) CMOS/SOS EAROM Memory Arrays, *IEEE Journal of Solid State Circuits*, Vol. **SC-14**, No.5, October 1979, p.860.

[11] Muller, R. G. *et al.* (1977) An 8192-bit Electrically Alterable ROM Employing a One-Transistor Cell with Floating Gate, *IEEE Journal of Solid State Circuits*, Vol. **SC-12**, No.5, October 1977, p.507.

[12] Uchida, Y. *et al.* (1975) A 1024-Bit MNOS RAM Using Avalanche-Tunnel Injection, *IEEE Journal of Solid State Circuits*, Vol. **SC-10**, No.5, October 1975, p.288.

[13] Hagiwara, T. *et al.* (1980) A 16k bit Electrically Erasable PROM Using N-Channel Si-Gate MNOS Technology, *IEEE Journal of Solid State Circuits*, Vol. **SC-15**, No.3, June 1980, p.346.

[14] Guterman, D. C. *et al.* (1979) An Electrically Alterable Nonvolatile Memory Cell Using a Floating-Gate Structure, *IEEE Journal of Solid State Circuits,* Vol. **SC-14**, No.2, April 1979, p.498.

[15] Gee, L. *et al.* (1982) An Enhanced 16k EEPROM, *IEEE Journal of Solid State Circuits,* Vol. **SC-17**, No.5, October 1982, p.828.

[16] Johnson, W. S. *et al.* (1980) A 16 kbit Electrically Erasable Nonvolatile Memory, *ISSCC Proceedings,* February, p.152.

[17] Kuo, C. *et al.* (1982) An 80 ns 32k EEPROM Using the FETMOS Cell, *IEEE Journal of Solid State Circuits,* Vol. **SC-17**, No.5, October 1982, p.821.

[18] Yatsuda, Y. *et al.* (1985) Hi-MNOS II Technology for a 64k bit Byte-Erasable 5 V Only EEPROM, *IEEE Journal of Solid State Circuits,* Vol. **SC-20**, No.1, February 1985, p.144.

[19] Hijiya, S. *et al.* (1982) A Nitride-Barrier Avalanche-Injection EAROM, *ISSCC Proceedings,* February, p.116.

[20] Lancaster, A. *et al.* (1983) A 5 V-Only EEPROM with Internal Program/Erase Control, *ISSCC Proceedings,* February, p.164.

[21] Gupta, A. *et al.* (1982) A 5 V Only 16k EEPROM utilizing Oxynitride Dielectrics and EPROM Redundancy, *ISSCC Proceedings,* February, p.184.

[22] Lucero, E. M. *et al.* (1983) A 16k bit Smart 5 V Only EEPROM with Redundancy, *IEEE Journal of Solid State Circuits,* Vol. **SC-18**, No.5, October 1983, p.539.

[23] Dham, V. K. *et al.* (1983) A 5 V-Only EEPROM using 1.5 μm Lithography, *ISSCC Proceedings,* February, p.166.

[24] Witters, J. S. *et al.* (1987) Programming Mode Dependent Degradation of Tunnel Oxide Floating Gate Devices, *IEDM Proceedings,* p.554.

[25] Harris, M. A. (1983) Sturdy Memory Device Key to Army Contract, *Electronics,* May 19, p.49.

[26] Zeman, R. *et al.* (1984) A 55 ns CMOS EEPROM, *ISSCC Proceedings,* February, p.144.

[27] Jolly, R. D. *et al.* (1985) A 35 ns 6k EEPROM, *IEEE Journal of Solid State Circuits,* Vol. **SC-20**, No.5, October 1985, p.971.

[28] Bill, C. S. *et al.* (1985) A Temperature and Process Tolerant 64k EEPROM, *IEEE Journal of Solid State Circuits,* Vol. **SC-20**, No.5, October 1985, p.979.

[29] Cormier, D. (1985) Erasable Programmable Solid-State Memories, *EDN,* November 14, p.145.

[30] *JEDEC Standard 21-C,* Configuration for Solid State Memories, Release 1, January 1990.

[31] Special Report on Semiconductor Memories, *Computer Design,* August 1984, p.105.

[32] Cuppens, R. (1985) An EEPROM for Microprocessors and Custom Logic, *IEEE Journal of Solid State Circuits,* Vol. **SC-20**, No.2, April 1985, p.603.

[33] Wilson, D. (1985) EEPROM Technologies Diversify and Find New Applications, *Digital Design,* July, p.71.

[34] Silvey, J. (1985) With EEPROM on a Single-Chip μC, Changing Programs is a Snap, *Electronics Products,* October 15, p.35.

[35] Shamshirian, M. and Jenq, C. (1986) 256k EEPROM Opens New Doors for Designers, *Digital Design,* March, p.57.

[36] Cole, B. C. (1987) Sierra's EEPROM Cell Move into Designs, *Electronics,* April 16, p.75.

[37] Cioaca, D. *et al.* (1987) A Million-Cycle CMOS 256k EEPROM, *IEEE Journal of Solid State Circuits,* Vol. **SC-22**, No.5, October 1987, p.684.

[38] Ting, T. J. *et al.* (1988) A 50 ns CMOS 256k EEPROM, *IEEE Journal of Solid State Circuits,* Vol. **SC-23**, No.5, October 1988, p.1164.

[39] *Xicor Databook* 1985.

[40] Bursky, D. (1988) Protect your EEPROM Data and Gain More Flexibility, *Electronic Design,* May 25, p.43.

[41] Vancu, R. *et al.* (1988) A 30 ns Fault Tolerant 16k CMOS EEPROM, *ISSCC Proceedings*, February, p.128.
[42] Vollebregt, F. *et al.* (1989) A New EEPROM Technology with a TiSi2 Control Gate, *IEDM Proceedings*, December, p.607.
[43] Do, J. Y. *et al.* (1988) A 256k CMOS EEPROM with Enhanced Reliability and Testability, Symposium on VLSI Circuits, p.83.
[44] Donaldson, D. D. *et al.* (1982) SNOS 1k × 8 Static Nonvolatile RAM, *IEEE Journal of Solid State Circuits*, Vol. **SC-17**, No.5, October 1982, p.847.
[45] Lee, D. J. *et al.* (1983) Control Logic and Cell Design for a 4k NVRAM, *IEEE Journal of Solid State Circuits*, Vol. **SC-18**, No.5, October 1983, p.525.
[46] *Fujitsu MOS Memory Databook* 1987.
[47] Becker, N. J. *et al.* (1983) A 5 V-Only 4K Nonvolatile Static RAM, *ISSCC Proceedings*, February, p.170.
[48] Iversen, W. R. (1983) 4-K Shadow RAMs Use Single 5-V Supply, *Electronics*, January 27, p.128.
[49] Eaton, S. S. *et al.* (1988) A Ferroelectric Nonvolatile Memory, *ISSCC Proceedings*, February, p.130.
[50] Evans, J. T. and Womack, R. (1988) An Experimental 512-bit Nonvolatile Memory with Ferroelectric Storage Cell, *IEEE Journal of Solid State Circuits*, Vol. **23**, No.5, October 1988, p.1171.
[51] Terada, Y. (1988) A New Architecture for the NVRAM—An EEPROM Backed-Up Dynamic RAM, *IEEE Journal of Solid State Circuits*, Vol. **23**, No.1, February 1988, p.86.
[52] Yamauchi, Y. *et al.* (1988) A Novel NVRAM Cell Technology for High Density Applications, *IEDM Proceedings*, December, p.417.
[53] Yamauchi, Y. *et al.* (1989) A Versatile Stacked Storage Capacitor on Flotox Cell for Megabit NVRAM Applications, *IEEE IEDM Proceedings*, December, p.595.
[54] Endoh, T. *et al.* (1989) New Design Technology for EEPROM Memory Cells with 10 Million Write/Erase Cycling Endurance, *IEDM Proceedings*, December, p.599.
[55] Cole, B. C. (1987) The Changing Face of Nonvolatile Memories, *Electronics*, July 9, p.67.
[56] Prince, B. and Due-Gundersen, G. (1983) *Semiconductor Memories*, John Wiley & Sons, Chichester.
[57] Kuo, C. *et al.* (1991) A 512kb Flash EEPROM for a 32 bit Microcontroller, *Symposium on VLSI Circuits*, June 1991, p.87.
[58] Hayashikoshi, M. *et al.* (1991) A Dual-mode Sensing Scheme of Capacitor Coupled EEPROM Cell for Super High Endurance, *Symposium on VLSI Circuits*, June 1991, p.89.

13 PACKAGING—SINGLE, MODULE AND WAFER SCALE INTEGRATION

13.1 OVERVIEW

Although the history, development, and technology of packaging is common to most semiconductor components, it has been and will continue to be an important factor in the field of memories where the mix of volume shipments, reliability, performance, package size, and cost are key ingredients.

In a variety of applications the selection, testing, and use of a specific package may in fact be of the same importance and complexity as the semiconductor component itself.

A satisfactory package must mechanically be sufficiently strong to withstand external stress caused during testing, handling, and end-application. Electrically the package must not exhibit parasitic effects and the overall capacitance must be kept to a minimum. Additional requirements are the insulation of the die from the external environment (such as extreme temperatures, electrical field, and humidity) and constraints with respect to thermal conductivity and thermal expansion.

A sequence of milestones have influenced the development of semiconductor memory packaging going from the original side brazed ceramic dual-in-line packages used for the first memory devices to the thin very small-outline packages in development today. Beyond these are the modules and bare chip packaging which are today's closest approach to wafer scale integration.

One factor driving the trend to denser packaging techniques for memories has been the rapid growth of portable computer applications including such diverse systems as laptop computers, portable language translators and personal time management systems.

Another factor driving the development of advanced packaging has been the need to reduce the length of interconnects in fast systems to reduce ground bounce effects which tend to slow the system down. While memories have gained in speed as they have gone to smaller dimensions, the interconnects between chip and package and between packages on the board have not kept pace.

The smaller minipackages have reduced the chip-to-package interconnects. Package-to-package, or chip-to-chip, interconnects have been reduced by both the multilayer printed circuit boards and the shorter leads of the miniature packages.

Packaging techniques which mount chips directly on the substrate have reduced the chip-to-chip interconnects even further. Finally vertical packaging methods which stack high density substrates provide another level of density for large and fast systems.

13.2 SINGLE-CHIP THROUGH-HOLE PACKAGES

The historical packaging technique not only for memories but for all semiconductors uses wire leads which extend from the electrical connections on the chip inside the package through to the outside of the package. These are soldered into holes in printed circuit boards for assembly into systems and are therefore called 'through-hole' packaging.

13.2.1 Dual-in-line package (DIL or DIP)

The most common of the single chip 'through-hole' packages have been the DIP. The initial DIP package was introduced during the late 1960s for TTL logic. The original 14 pin package was soon expanded and DIP use, popularity, and application areas grew rapidly. DIP packages are available in the entire range of semiconductor components from simple gates up to complex VLSI memory and microprocessor chips. The packages historically have had 0.1 inch (100 mil) spacing between the leads although recently DIP packages with smaller 75 mil spacing between the leads have been experimented with.

Standard DIP packages used for memories are 16 pin, 18 pin, 22 pin, 24 pin, 28 pin, 32 pin, and 40 pin DIP. The body widths range from the original 600 mil to 400 mil and 300 mil. There have also been rare offerings in 900 mil width—usually for module packaging. A 'shrink' DIP with 75 mil spacing has also been offered.

13.2.2 DIP hermetic

The ceramic DIP was the original memory package. Originally two different types of hermetic package were produced—CERDIP (ceramic DIP) and solder sealed (side-brazed).

Both of these are more expensive than plastic but historically offered higher reliability. Ceramic packaging is frequently used in military applications, in tele-

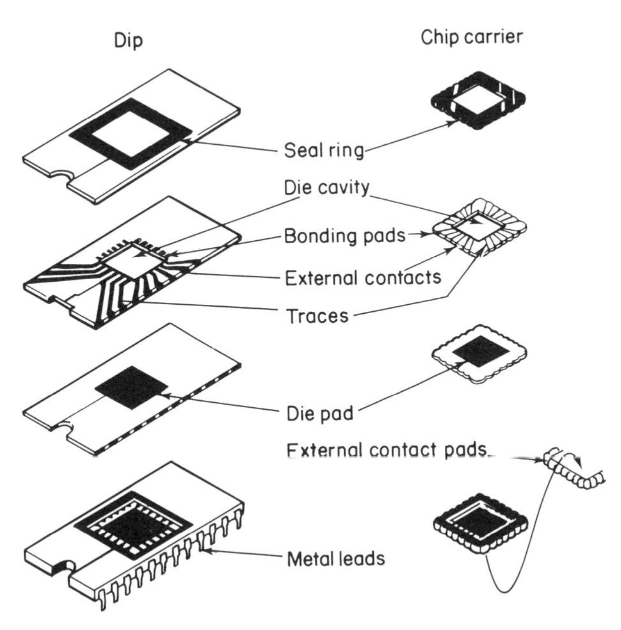

Figure 13.1
Schematic assembly flow for the ceramic side brazed package and the ceramic leadless chip carrier. (From Prince and Due-Gundersen [26] 1983.)

communications products, and generally where the environmental operating conditions of the end product are expected to be extreme.

The side-brazed ceramic packages are the most expensive but have the highest reliability. Three layers of alumina are used to form the package body with a cavity for the die. The lead frame is brazed to the sides of the package (giving the name side-brazed). A metal or ceramic lid is used to hermetically seal the cavity containing the die. Figure 13.1 illustrates the ceramic side-brazed assembly flow along that of the ceramic leadless chip carrier which will be discussed in a later section.

Although seldom used in production any more, these packages are still frequently used for prototyping during the development stage of a new memory product since it is easy to remove the lid to access the chip for visual inspection or testing such as accelerated radiation testing. The lid can then be taped back on to protect the chip during further tests. The package is, of course, no longer hermetic at that point.

The CERDIP (ceramic dual-in-line package) is composed of two layers of black alumina with the seal formed between these by a glass frit layer. The lead frame is embedded in the glass and both the lid and the base alumina have recesses to provide a cavity for the die. The leads of a CERDIP package are a part of the frame embedded in the glass seal. Figures 13.2(a) and (b) illustrate the CERDIP package materials specification and process flow.

The CERDIP is less expensive than the side-braze package but can suffer from cracking of the glass layer holding the two halves of the package together. It is also

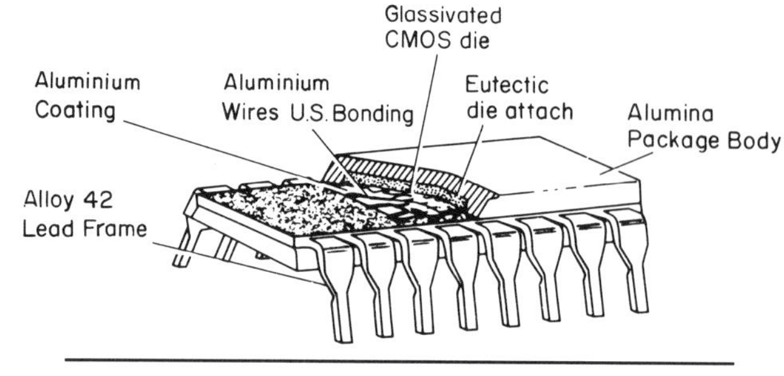

Package description		
CERDIP	—	Frit seal ceramic dual inline package
Die	...	Glassivated
Leadframe	...	Alloy 42
Die bond	...	Eutectic
Wire	...	Aluminum
Wire bond	...	Ultrasonic
Package body	...	Alumina
Lead finish	...	Tin plate or solder dip
Side braze	—	Laminated ceramic package
Die	...	Glassivated
Die bond	...	Eutectic
Wire	...	Aluminum
Wire bond	...	Ultrasonic
Seal	...	Solder with gold plated Kovar lid
Package body	...	Laminated alumina
Metalization	...	Gold-plated tungsten
Leads	...	Gold- or tin-plated Kovar

(a)

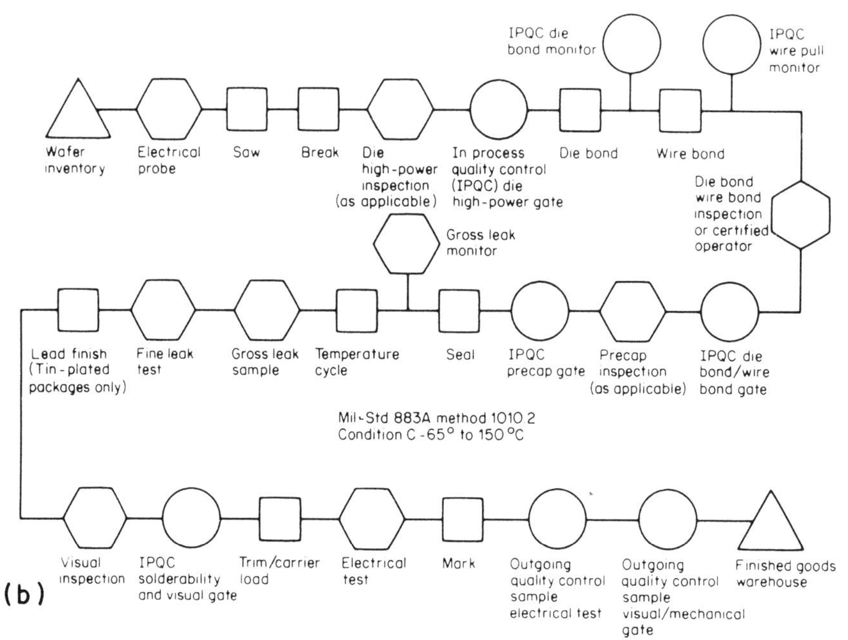

(b)

Figure 13.2
(a) Package material specification for the CERDIP package. (b) CERDIP package process flow. (From Prince and Due-Gundersen [26] 1983.)

more expensive than the plastic DIP and is therefore used only when the application absolutely requires a hermetic package.

The exception is the EPROM, which requires a quartz window in the package and is usually assembled in either a side-braze or a CERDIP package with such a window. A photograph of several packages including the quartz windowed CERDIP EPROM package and the 28 lead 600 mil plastic DIP package is shown in Figure 13.3(a).

13.2.3 DIP plastic (PDIP)

By far the most common memory package up to the present time is the PDIP or plastic DIP. Plastic packages are normally transfer molded using a plastic compound and a metal lead frame with semiconductor chips attached to the lead frame as shown in the photograph in Figure 13.3(b). The assembly process flow for a plastic package is shown in Figure 13.4(a) and the plastic package material specification is shown in Figure 13.4(b). This packaging technique is relatively simple and cheap resulting in high volumes and low cost. The plastic is also light and has a high resistance to physical damage.

In contrast to the ceramic packages, however, the PDIP package is not hermetic. This is much less of a problem presently since the passivation on the modern memory chip itself makes it practically hermetic. Also various techniques such as coating the chip with a hermetic material, such as silicone gel, prior to molding the package have been developed. In the past contamination, thermal performance, wire bond integrity, and moisture resistance (corrosion) have been significant concerns with the plastic packages.

Most of the early plastic problems have been eliminated today and extensive lifetesting has shown that the plastic package has a reliability that is suitable for almost any application. Plastic is, therefore, the highest volume package for the memory mass market where high volume and low cost considerations are key factors.

13.2.4 Zigzag-in-line package (ZIP)

For through-hole assemblies requiring high density layout on the printed circuit board, the ZIP package is ideal. The plastic package has the same body shape as a PDIP but has leads only along one edge. Two ZIP packages are shown in the assortment of plastic memory packages in Figure 13.5.

These leads are staggered in a zigzag pattern which gives the package its name. Since a ZIP stands on its side, the width, or footprint on the printed circuit board, is only 100 mil as compared to the 300 mil width of the 'slimline DIP'. This means that the memory density of a board can be more than doubled. The height is, however, over 400 mil compared to slightly over 200 mil for a DIP. The height dimension, which is the spacing between boards, is usually not so critical, however.

(a)

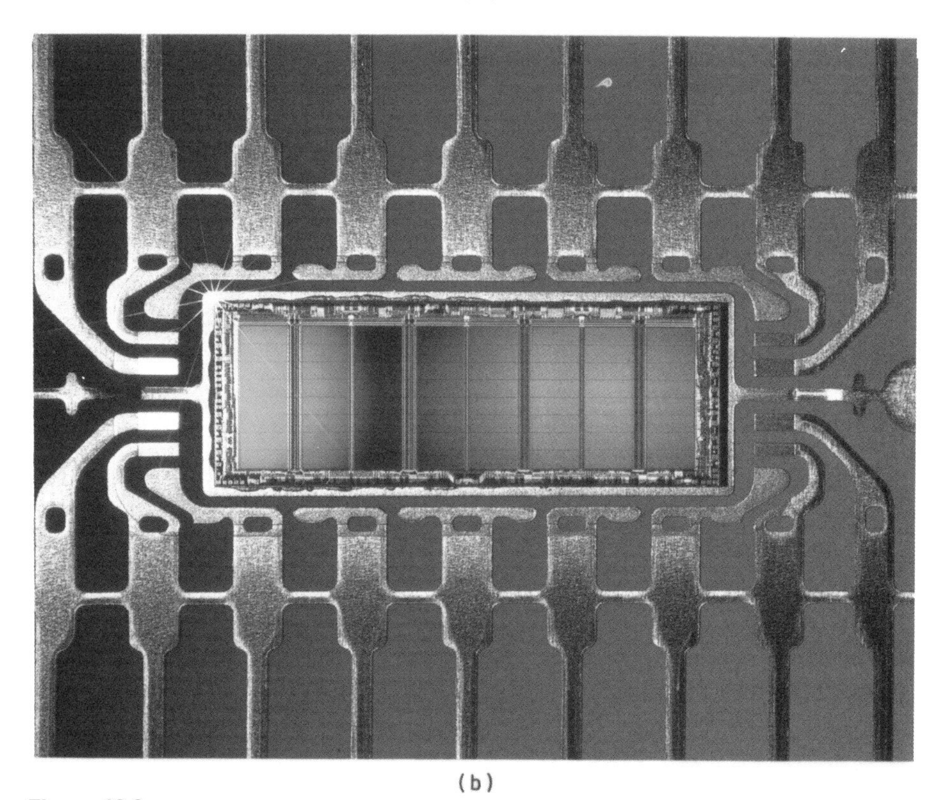

(b)

Figure 13.3
(a) Various packaged memory chips showing EPROM quartz window package, DRAMs in SOJ mounted on a SIMM carrier, standard PDIP, and PLCC. A silicon wafer with memory chips is in the background. (Reproduced with permission of Texas Instruments.) (b) Example of 4Mb DRAM chip on a PDIP lead frame prior to the plastic molding process. (Reproduced with permission of Siemens.)

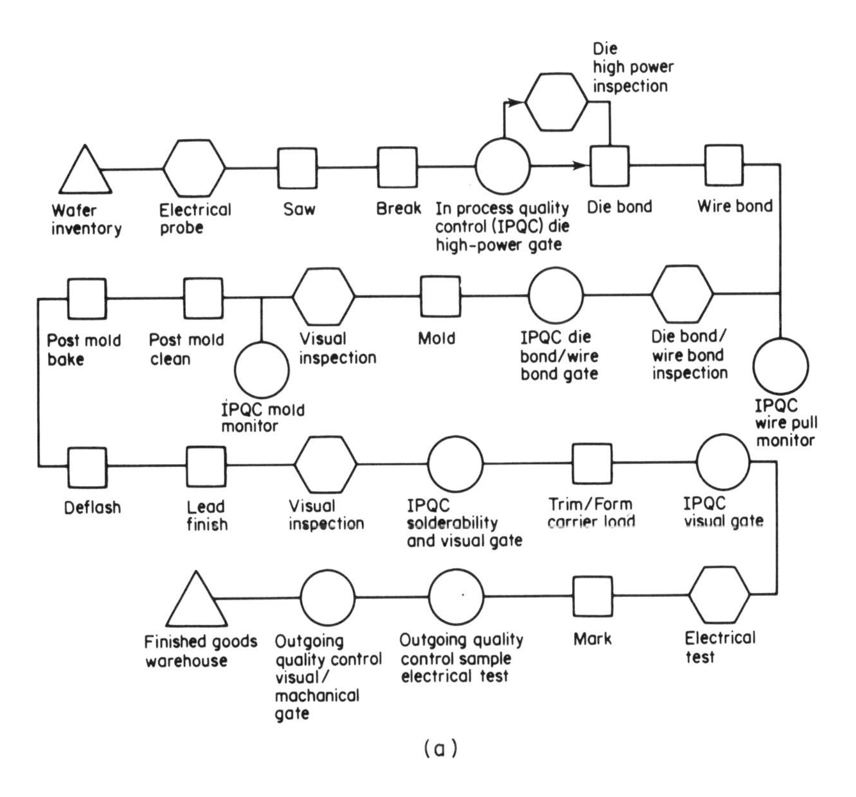

(a)

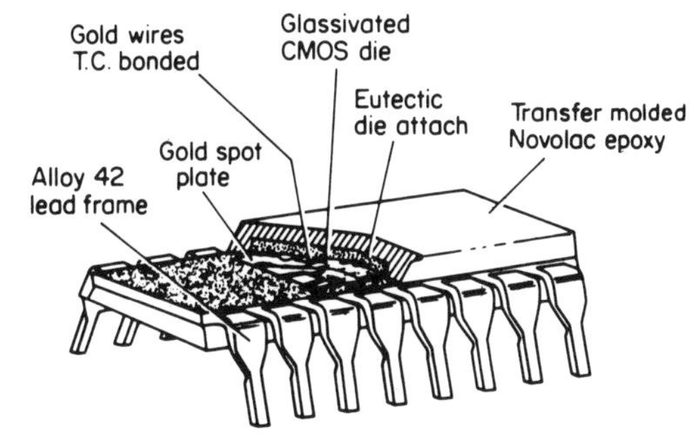

Package description		
Dia	...	Glassivated
Lead frame	...	Alloy 42
Wire	...	Alloy 42
Wire	...	Gold
Wire bond	...	Thermocompression
Molding compound	...	Epoxy Novolac
Mold method	...	Transfer
Lead finish	...	Tin plate or solder dip

(b)

Figure 13.4
(a) Plastic package process assembly flow. (b) Plastic package material specification. (From Prince and Due-Gundersen [26] 1983.)

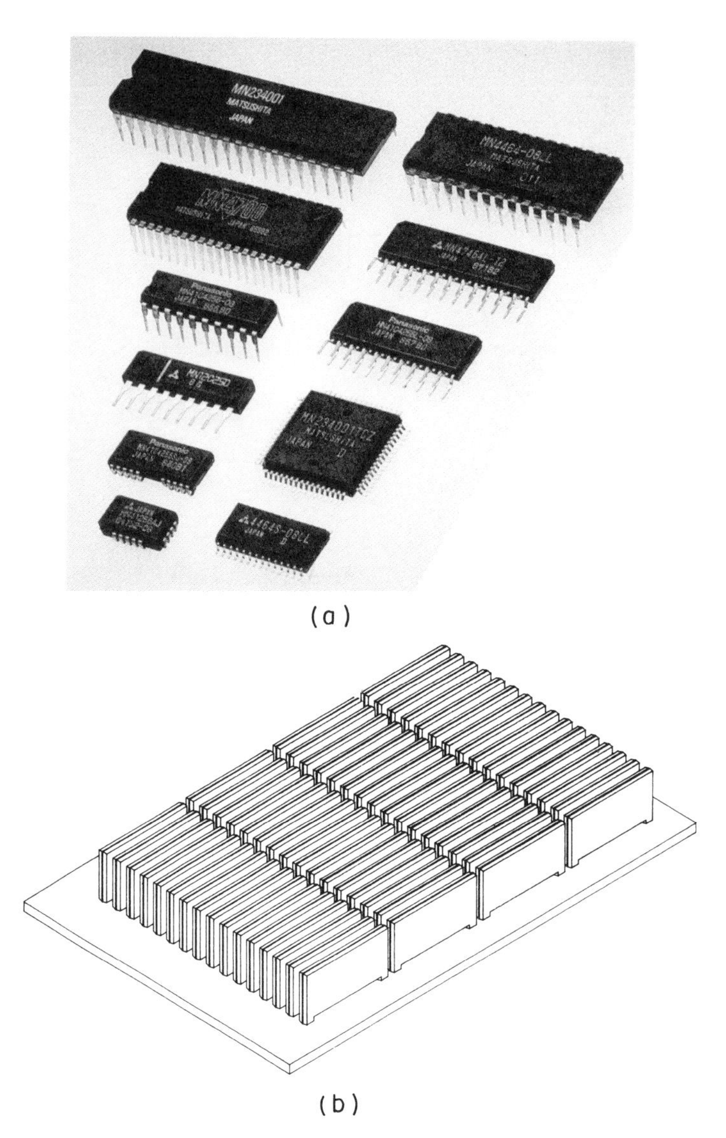

Figure 13.5
(a) Examples of plastic packages used for memories including PDIP, ZIP, SIP, Flat Pack, SOJ, PLCC and SOG. (Reproduced with permission of Matsushita.) (b) Example of high density surface mount assembly of DRAMs in VPAK packaging. (Reproduced with permission of Texas Instruments.)

13.2.5 Single-in-line package (SIP)

There is also a SIP package which is like the ZIP only the leads are in a straight line. A SIP package is also shown in the picture in Figure 13.5(a). This package is much less common than the ZIP as a single chip memory package. It is more common for memories as a multi-chip package and will be discussed more extensively in the section under multi-chip modules.

13.3 SURFACE MOUNT PACKAGES

Surface mount packages are much smaller than their standard DIP through-hole counterparts and are therefore able to increase the density of printed circuit board assemblies by a factor of two or more. A DIP has a 100 mil lead spacing while surface mount components range from the 50 mils of the SO packages to the 20 and 25 mil spacings of the miniature SO packages. In addition the surface mount packages can be mounted on both sides of the printed circuit board whereas the through-hole packages can only be mounted on one side.

In systems like computers that are memory intensive this can result in significant size reduction of the system. Portable consumer systems also benefit from the size reduction possible using surface mount component packaging.

A comparison is shown in Figure 13.6 of component density for two through-hole components and for the SOJ package in a single-sided and double-sided board configuration. While the ZIP gives a denser board layout than the single-sided SOJ board, the double-sided SOJ board is significantly denser. The ZIP board and the double-sided SOJ board are higher with the ZIP almost twice the height of the DIP and over 20% higher than the double-sided SOJ. The SOJ will be discussed in more detail in this section.

Electrical performance is enhanced by the shorter lead lengths and since the assembled boards are smaller, multilayer printed circuit boards can be used without significantly increasing board cost.

An example is shown in Figure 13.5(b) of a memory module with 16 rows by 4 columns of 4Mb DRAMs in a vertical surface mount package, the Texas Instruments VPAK, which provides even greater density than the SOJ. The VPAK is discussed in more detail in a later section.

Surface mount technology has come into use more rapidly in Japan and Europe with their traditional emphasis on consumer systems. US usage is encouraged by the personal computer, communications and automotive electronics segments, all of whom are major users of the smaller surface mount packages.

It has been estimated that usage of surface mount by personal computer makers in 1988 was over 70% and new systems coming into production in that time frame used over 80% surface mount. Figure 13.7(a) shows the estimated percent of ICs consumed in surface mount packages in the major regions of the world for 1987 and forecast for 1992. The emphasis on consumer systems with their high density requirements has encouraged Japanese suppliers to convert to surface mount much earlier than the North American or European suppliers. By 1992 the split by region is expected to be more balanced.

One drawback of surface mount packaging that has slowed the emergence of this market is the high initial investment required in surface mount assembly equipment.

Component	X–Y comp. pitch (in)	Number comp./sq. in.	Board thickness W/comp. (in)
DIP	0.4 × 1	2.5	0.245
ZIP	0.2 × 1.1	4.5	0.412
SOJ: SS†	0.4 × 0.7	3.6	0.202
SOJ; DS†	0.4 × 0.7	7	0.342

DS SMT offers the highest component density
SM boards allow better spacing and cooling in systems (based on 0.5 in. centers between boards)
† SS-single sided, DS-double sided

Figure 13.6
Comparison of component density on the board for two through-hole components and for the SOJ package in a single-sided and double-sided board configuration. (Reproduced with permission of Texas Instruments.)

An automated placement machine capable of fine pitch assembly costs $500 000 or more. CAD tools for placement and routing of interconnects on the boards also need to be purchased, as well as special handlers for test equipment.

Figure 13.7(b) shows the estimated consumption of various surface mount and through-hold IC packages in North America in 1987 and estimated for 1992 in millions of units. Surface mount usage is expected to grow from 8% to 36% of the total IC packaging usage by 1992.

13.4 SMALL OUTLINE (SO) SURFACE MOUNT PACKAGES

Miniature SO surface mount packages which historically have had a high market penetration in discrete, linear and logic circuits which are used in consumer products. Their use with memories has historically been rather limited due to the relatively large size of the memory chips and due to the small number of memories which historically have been used in consumer systems. There have also been concerns about reliability.

The hermiticity considerations for the plastic package already discussed could be more severe with the small outline packages because there is even less package thickness protecting the chip. Since MOS memory technology is a surface process, it tends to be susceptible to moisture. This concern, as in the case of the PDIP, has become less as the hermeticity of the chip itself and of various chip and wafer coatings have been improved.

In 1988 GE Solid State conducted extensive reliability tests on PDIPs and SOPs and concluded in their published results that there was no difference found in device reliability between parts in the two packages. A second concern has been that a memory chip is relatively large and brittle and may be prone to cracking in a thin shell-like package. Another consideration that delayed the conversion of the market to surface mount technology was the expense of the automated assembly equipment required to mount packages with 25 or 50 mil lead spacings accurately. Finally, test

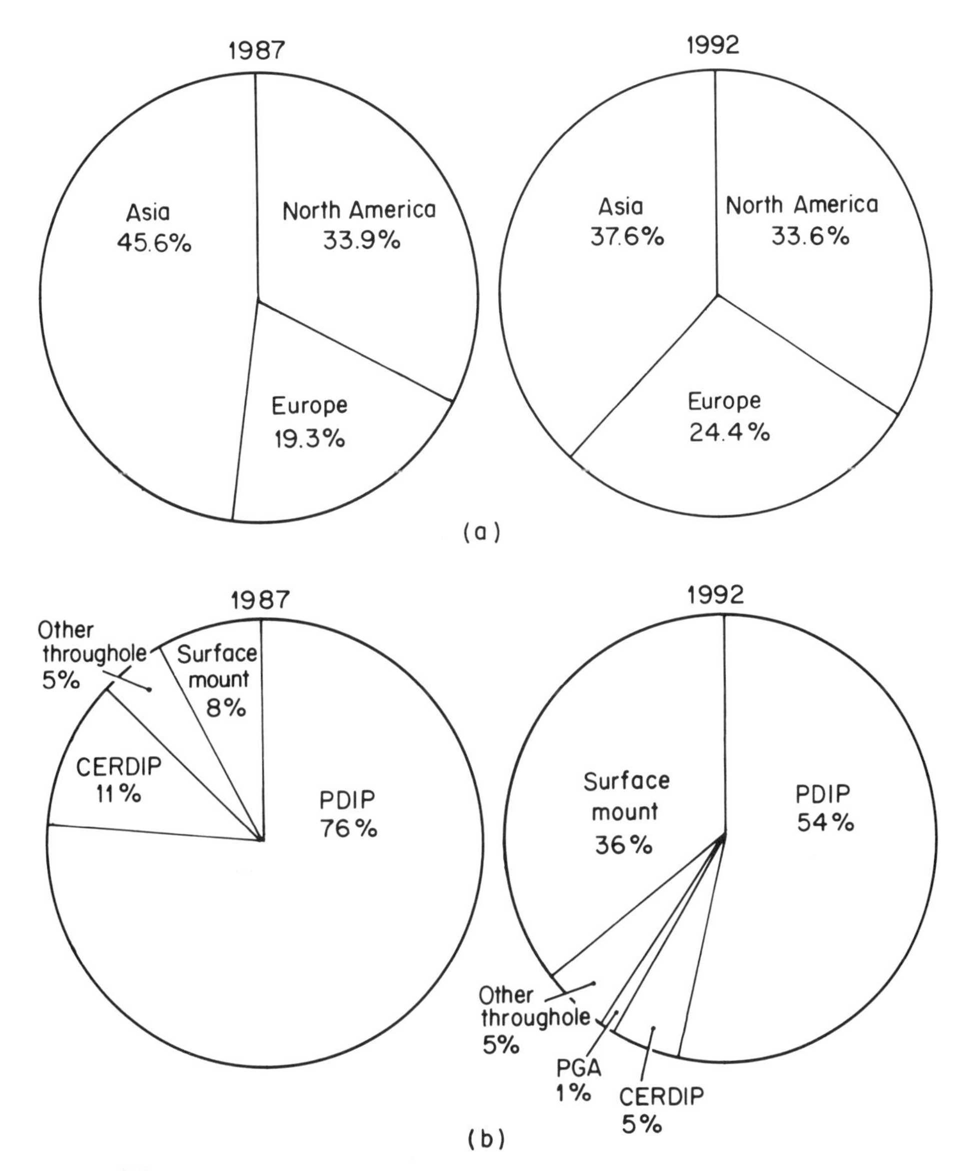

Figure 13.7
(a) Estimated percent of integrated circuits consumed in surface mount packages by region of the world. (b) Estimated consumption of various surface mount and through-hole integrated circuit packages in North America in 1987 and 1992 in millions of units. (Source Dataquest.)

and handling equipment needed to be developed and purchased for the various surface mount packages.

All of these barriers have been overcome and presently an increasing percentage of all memories shipped worldwide are in surface mount packages. The benefit of board space saved is great and the need for high density packaging in such growing market applications as laptop computers and other portable consumer systems is significant.

Other factors favoring the conversion to surface mount devices for high volume commodity type systems include the difficulty of using automated insertion tools for large DIPs. Fast systems have converted due to the capacitive delays inherent in the longer internal bond wires of the larger DIP package and the longer system interconnections on a larger printed circuit board.

13.4.1 SOP–SOIC

The small outline package which has the leads spread outward in a 'gull wing' style was developed originally by Philips and is generally referred to simply as the SOP (small outline package) or the SOIC (small outline I.C.). The SOP has 50 mil spacings between the leads compared to the 100 mil spacings of the DIP package. It is also thinner than the DIP. It is widely used in consumer applications because of its small size. Memories packaged in the SOP tend to retain the same pinouts as a comparable lead DIP since the same chips are used in both packages.

A picture of a 28 lead SOP is shown in Figure 13.5 in the lower right-hand corner. It can be compared to the size of the 28 lead DIP package shown in the upper right-hand corner of the same figure. The size advantage is obvious. This is a standard high volume package for 64k and 256k commodity SRAMs.

One of the benefits of the 'gull wing' lead configuration is that after mounting the leads can be visually inspected for good attachment to the printed circuit board.

13.4.2 SOJ

Another small outline package, which was developed more recently during the 1980s especially for memories, is called the small outline 'J-bend lead' (SOJ). This package also has leads on two sides with 50 mil spacings between the leads which are curved back under the package in the shape of a 'J' as shown in the package outline specification in Figure 13.8 and the photograph in Figure 13.5(a). Notice that this DRAM package specification has a gap in the leads, that is, it is a 26 lead package with the center six leads missing and is hence referred to as a 20/26 lead package.

The 20/26 lead SOJ package is also shown in the photograph in Figure 13.5 where it can be seen to be comparable in package area to the SOP. The SOJ is, however, slightly thicker than the SOP and as a result is believed to be more stable with a large chip like a multi-megabit DRAM. It was derived from the PDIP.

The SOJ has been widely used in the computer industry for surface mounting of high density memory boards, and is the surface mount package which has been standardized by the EIA JEDEC memory standardization committee. The SOJ package without missing leads has also been standardized and used for SRAMs. This package has been standardized up to 46 leads. Above that lead count the PLCC package which will be discussed later is preferred.

Three advantages cited for the SOJ are that during automated mounting the

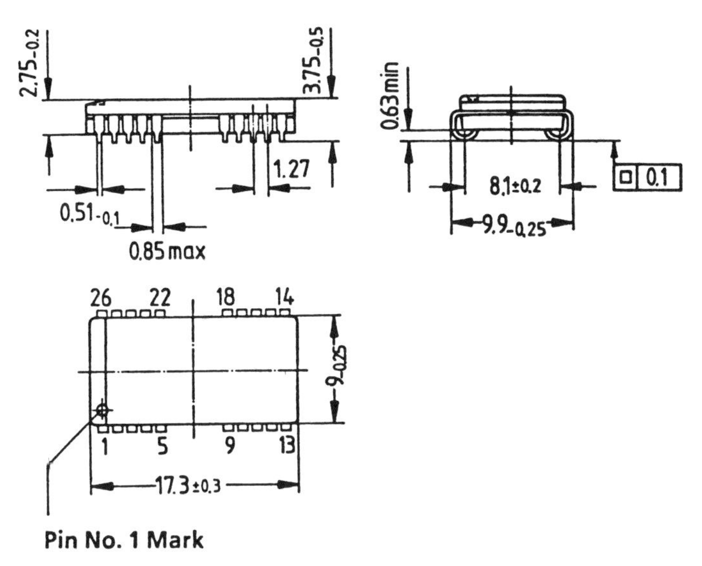

Figure 13.8
Package outline for an SOJ DRAM package showing the missing center package leads and the curvature of the J-leads under the package (dimensions in mm). (Source Siemens 4Mb DRAM Data Sheet 1989.)

flexibility of the leads improves contact, the footprint of the package on the printed circuit board is smaller than that of the SOP which is an advantage in high density memory systems such as used in large computers and the thicker package is more stable with a large chip.

A disadvantage frequently given by those who prefer the SOP is that it is difficult to observe the quality of attachment of the 'J' leads and the flexibility of the leads to match with the printed circuit board under thermal stress tends to be less. Both of these concerns are more relevant to consumer type systems where the SOP memory package is predominantly used.

13.5 MINIATURE SMALL OUTLINE PACKAGES

In recent development are even smaller SO packages which have less than 50 mil spacings between the leads ranging down as low as 20 mil and also thinner SO packages. The packing density on the printed circuit board of these tiny parts is very high leading to significant speed and density improvement at the systems level. There are, however, board level problems remaining to be solved before these packages will come into general usage. These include lead control, solder control, placement accuracy, and inspectability.

13.5.1 Thin small outline package (TSOP)

The TSOP II is just a thinner version of the SOP. It was developed originally for such applications as memory smart cards where density is required and package height is restricted. It retains the standard dual-in-line pinout. Versions with lead spacings of 25 mil rather than 50 mil have also been developed.

A TSOP I miniature package is infrequently used which has the leads on the short side.

13.5.2 Vertical surface mount package (VPAK)

This package, which was shown in Figure 13.5(b), is a miniature SOP that has leads only on one side. These leads are bent in a half arc and soldered to the surface of the printed circuit board. The package is stabilized by a post at each end.

13.6 CHIP CARRIER SURFACE MOUNT PACKAGES

Chip carriers come in an assortment of sizes, shapes, lead types, and packaging materials. They are all rectangular surface mount packages with a four sided pinout. The packages can be ceramic or plastic, leaded or leadless, and with gull wing or J-bend leads. They are used for memories predominantly for high pincount chips such as used for high density or wide bus applications.

Size reduction is the most obvious advantage in chip carriers. 40 and 50 mil lead pitch packages are standard. 20 and 25 mil versions are not uncommon, offering additional benefits in board density reduction.

Electrical performance is already improved by shorter and more uniform conductor length to the chip bonding pads. This results in lower resistance, less capacitance, and reduced inductance—all well known parasitic parameters which have traditionally

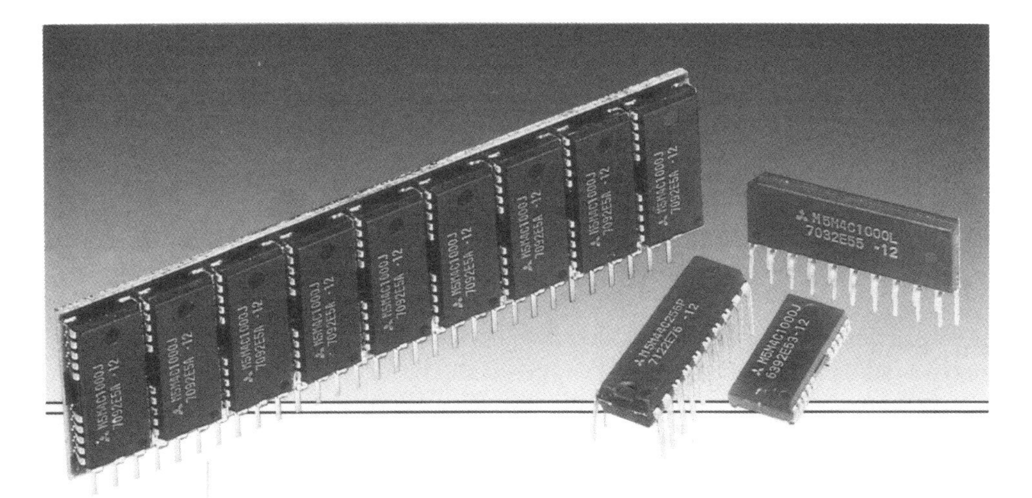

Figure 13.9
Examples of plastic packages used for memories including SIMM, PDIP, SOJ, and ZIP. Units mounted on the SIMM are the SOJ. (Reproduced with permission of Mitsubishi Electric Corp.)

limited fast switching and high device operating speed. Device operation in the 4 GHz range is possible and represents a significant improvement on the 500 MHz limit on DIP. Figure 13.9 illustrates a size and electrical performance comparison between the two packaging types.

The three most common types of chip carrier—the ceramic leadless, the plastic leaded and the flatpack—are discussed below.

13.6.1 Flatpack

Flatpacks are high pincount rectangular chip carrier packages with gull wing leads on four sides of the package. They are well established in ceramic in the military market and in plastic versions in the consumer market. Offering 50 mil and smaller spacing between the leads, a flatpack requires only 30% of the space needed by the corresponding standard DIP. A 64 lead memory flatpack is shown in the photograph in Figure 13.5.

While flatpacks historically have not been used for the relatively low pincount packages used for memories, as high density and wide bus memories require packages in the 50 to 100 lead range, their usage is increasing.

13.6.2 Ceramic leadless chip carrier (LCC)

The original chip carrier was a ceramic, leadless package which offered the usual hermetic package reliability advantages giving a package well suited for high performance and military applications. The assembly flow for the LCC package was shown in Figure 13.1.

While an LCC can be mounted in a socket, this reduces the density advantage. More common is soldering the LCC directly onto the substrate. A consideration in this case is that the coefficient of expansion of the package and the substrate match or bad connections or open circuits could result. This package had also on occasion been referred to as the RCC (rectangular leadless chip carrier).

13.6.3 Plastic leaded chip carrier (PLCC)

Plastic leaded chip carriers started to penetrate the memory market in the early 1980s. Their general acceptance was delayed due to the development and widespread acceptance of the SO packages shortly thereafter for the high volume commodity SRAMs and DRAMs. Today they are becoming more widely used both for the very high lead count packages such as the word-wide (16 bit wide) organizations of SRAMs and EPROMs, and for mounting on modules—a use that will be discussed in a following section. The PLCC package has 'J-bend' leads as was shown in Figure 13.3.

Plastic packages are much cheaper and lighter than ceramic versions and are more robust with respect to handling. They can be attached using sockets, using mother board (which in turn is mounted on the circuit board) and by direct attachment.

The same hermeticity considerations addressed under the small outline packages apply for the plastic chip carriers.

13.7 MULTICHIP MODULES

Multichip modules have become a very common high density through-hold-memory packaging technique. This has been encouraged by both the demand for higher density memories than can be produced on a single chip, and the need to have these memories with very short interconnections to minimize delay in high speed systems.

A multichip module can be single sided or double sided. While it can be flat on the substrate in a dual-in-line configuration, there is a significant density advantage to the single-in-line configuration in which the module stands up on the printed circuit board.

The major drawback of higher density three-dimensional packaging is the power consumption. The conversion of all of the major memory technologies to CMOS in the second half of the 1980s has reduced, but not eliminated, this problem.

13.7.1 Single-in-line memory modules (SIMMs) for DRAMs

A significant percentage of all DRAMs shipped today are in the form of standard multichip module SIMMs (single-in-line memory modules). A photograph of a DRAM SIMM with nine SOJ carriers mounted on it is shown in Figure 13.9. Single chip DRAMs in DIP, ZIP, and SOJ are shown for comparison.

The JEDEC committee has been active since the late 1980s standardizing these configurations. The number format used to describe a SIMM is designated by JEDEC as shown in the following excerpt from the JEDEC Standard 21C.

4.1.2 Number Format

The description number designation shall consist of 8 fields with the form nnSccbbDttlpp where :
nn = the number of longitudinal positions on the module: 4, 5, 8, 9.
S = the number of sides on the module stated as "single or double" : S, D
cc = the capacity of the memory chip stated in terms of the log(2) of the capacity (i.e.-the number of address bits needed for the chip): 16, 18, 20, 22
bb = the number of data bits in the interface: 1, 2, 4, 5, 8, 9, 10, 17
D = the data interface configuration, common, separate, or mixed: C, S, M
tt = total number of words stated as log(2) of the capacity: 16--26
I = mechanical interface: P = pins, E = edge card connector
pp = number of pins or pads

4.1.3 Number Example

A module with the following attributes:
9 chips long
Double sided
18 bit data interface
Separate I/O for parity bits, common I/O for other bits
1M X 1 memory chips
Pin interface
Architectural number: 9D2018M20P40

An example of a standard DRAM SIMM package is shown in the pinout in Figures 13.10(a) along with the functional block diagram. Figure 13.10(a) is a 2Mb which used 256k DRAMs in the PLCC package mounted on a SIP substrate with decoupling capacitors. Figure 13.10(b) is a 9Mb DRAM module using nine 1Mb DRAMs in SOJ. The ninth bit is for parity.

An advantage of the SIMM, in addition to high density and small board footprint, is that it is a through-hole part and can be used without the investment on the part of the system manufacturer in the costly surface mount assembly equipment. The cost of the surface mounted chips on the SIMM board falls back on the memory manufacturer. Initially, however, contract assembly companies have provided this service for a number of semiconductor manufacturers thereby providing the scale of volume needed to amortize the expensive assembly equipment.

13.7.2 Modules for high density SRAMs

Next generation high density SRAMs can be made available from current generation standard parts by means of modules. Two examples of 4Mb SRAMs made from 256k SRAMs along with block diagrams are shown in Figures 13.11(a) and (b). Figure 13.11(a) shows a 45 ns commodity 512k × 8 SRAM SIMM made from 16 32k × 8 SRAMs assembled in plastic chip carriers. Figure 13.11(b) shows a 25 ns fast 4Mb SRAM DIP module made from 16 fast 256k × 1 SRAMs. In this case the module substrate is ceramic and the 256ks are assembled in ceramic leadless chip carriers so that the entire assembly is hermetic and can be used in high reliability applications.

Various companies have made these SRAM modules including Harris, Electronic Designs, Micron Technology and Cypress.

13.7.3 Application-specific memories using SIMMs

Another advantage of the multichip module is the potential for offering multiple organizations and configurations of high density memories without the need to specifically design each of the chips. This can be done by building these out of smaller standard memories. This gives the opportunity to offer relatively fast turn-around on application-specific memories without the significant investment involved in designing a new single high density chip.

An example of an application-specific SRAM SIMM is shown in Figure 13.12. This 15 ns cache tag SRAM is organized 1k × 32 and is constructed from 8 resettable 1k × 4 SRAMs in SOJ package.

Clearly a number of types of applications can be served. For example a ceramic LCC can be mounted on a SIMM and give a high density hermetic package for high reliability applications.

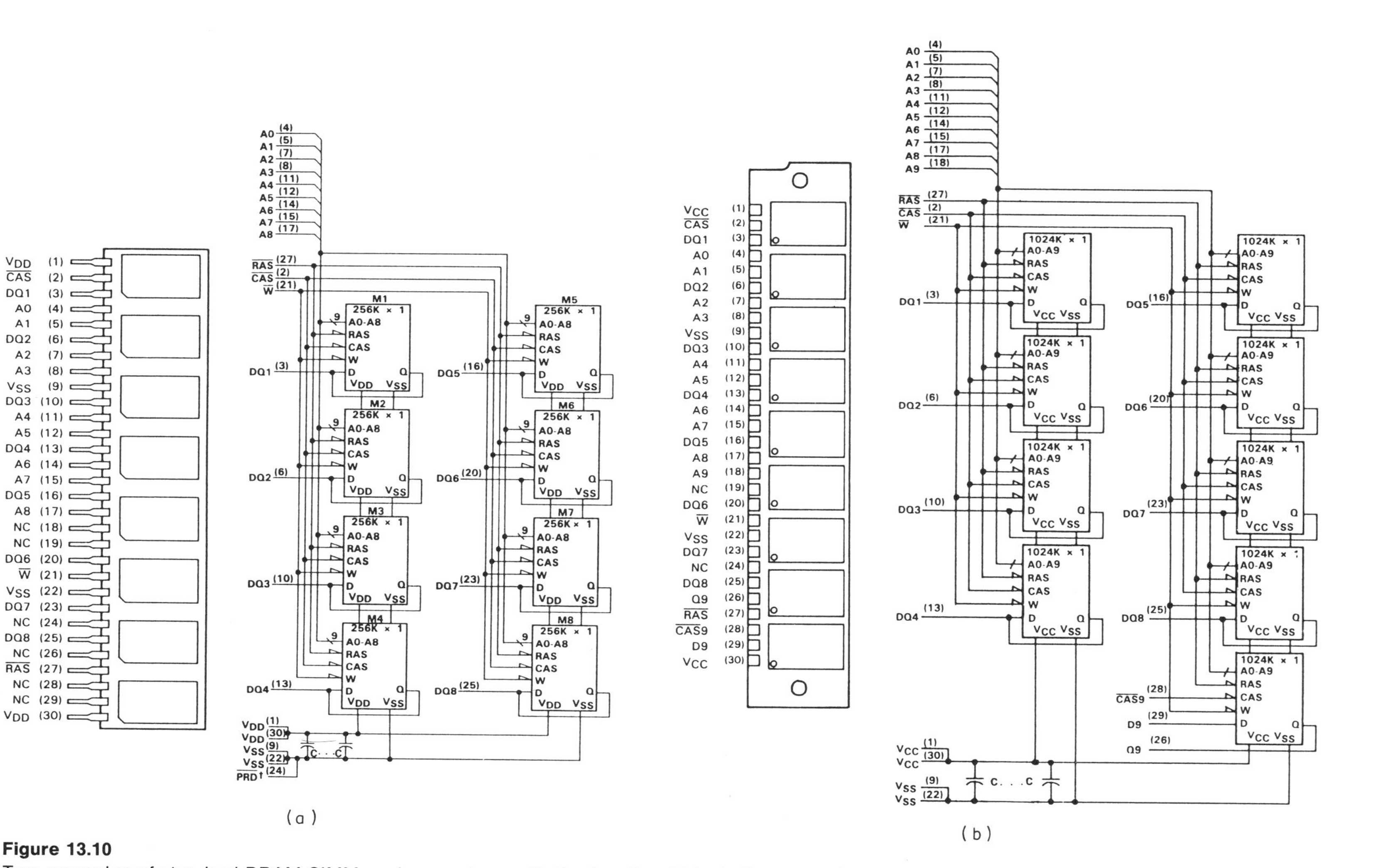

Figure 13.10

Two examples of standard DRAM SIMM packages along with the functional block diagrams of the parts. (a) 2Mb DRAM module using 256k DRAMs in the PLCC package mounted on a SIP substrate with decoupling capacitors. (b) 9Mb DRAM module using nine 1Mb DRAMs in SOJ. The ninth bit enables a parity check to be done. (Source Texas Instruments Databook. Reproduced with permission of Texas Instruments.)

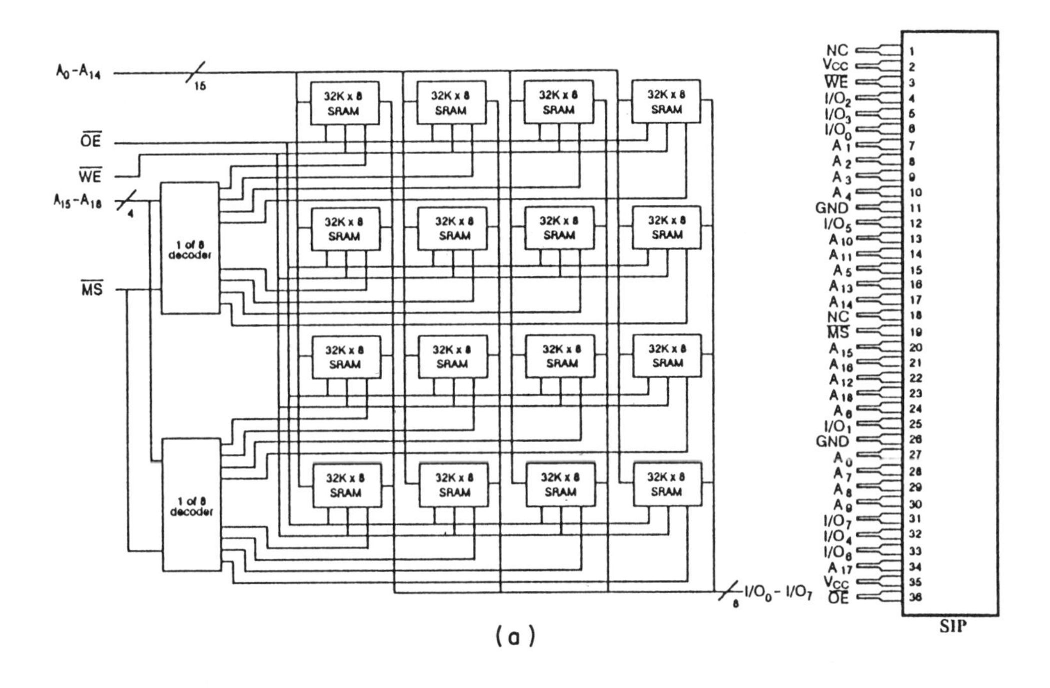

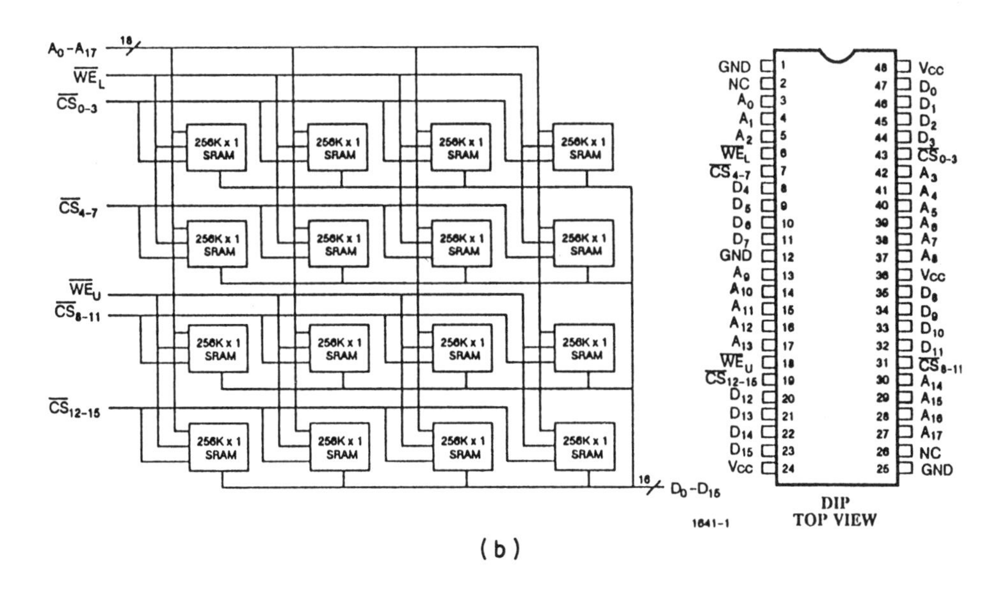

Figure 13.11

Examples of 4Mb SRAMs made from 256k SRAMs along with block diagrams. (a) 512k $\times$ 8 SRAM SIMM made from 16 32k $\times$ 8 SRAMs in PLCC. (b) 4Mb SRAM DIP module made from 16 fast 256k $\times$ 1 SRAMs in ceramic LCC. The entire module is hermetic. (Reproduced with permission of Cypress Semiconductor.)

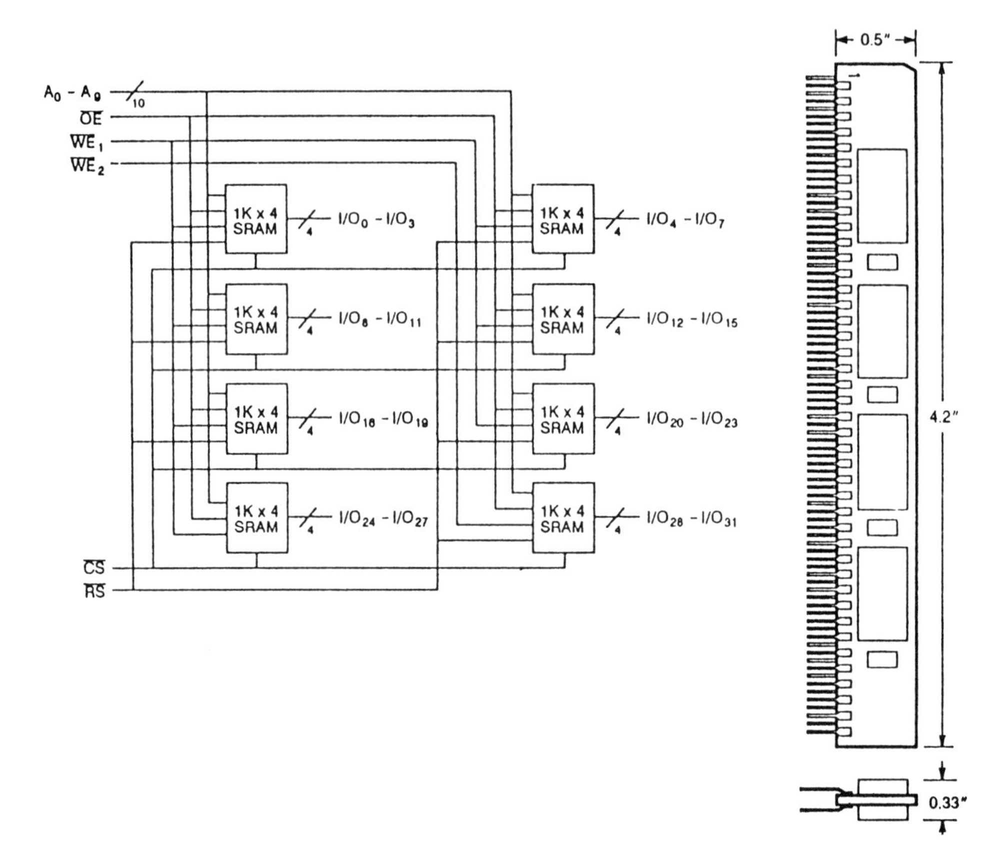

Figure 13.12
Illustration of an application-specific SRAM dual SIMM which is a 15 ns cache tag SRAM organized 1k × 32 constructed from eight resettable 1k × 4 SRAMs in SOJ package mounted on two sides of the substrate. (Reproduced with permission of Cypress Semiconductor.)

13.7.4 Direct and reverse image packages

A packaging technique introduced by Mitsubishi for use on double-sided SIMMs is the reverse image package which can be used in high density assemblies to simplify interconnections on a two-sided board. An example is shown in Figure 13.13.

13.7.5 Memory intensive modules

The next step in SIMMs is clearly to add additional rows of memories to the single row on the SIMM making a memory array. An example of a TI 'Memory Intensive' module is shown in Figure 13.14.

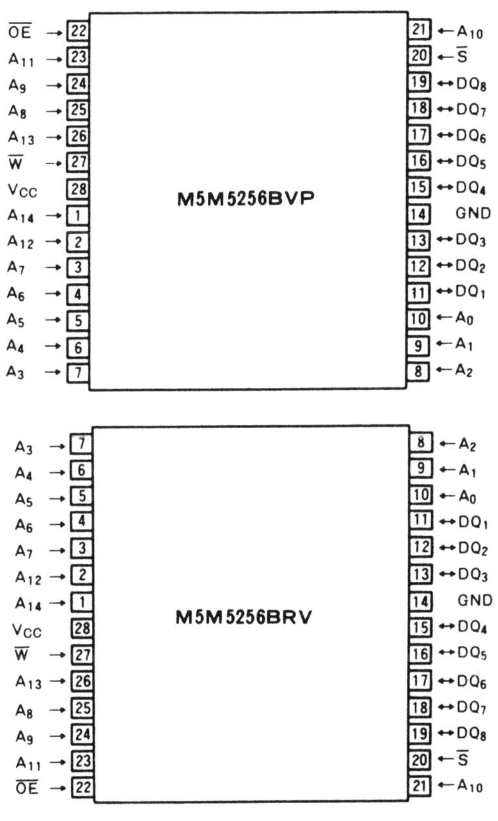

Figure 13.13
Illustration of the two sides of a reverse image package used in high density assemblies to simplify interconnections on a two-sided printed circuit board. (Source Mitsubishi Databook. Rreproduced with permission of Mitsubishi Electric Corp.)

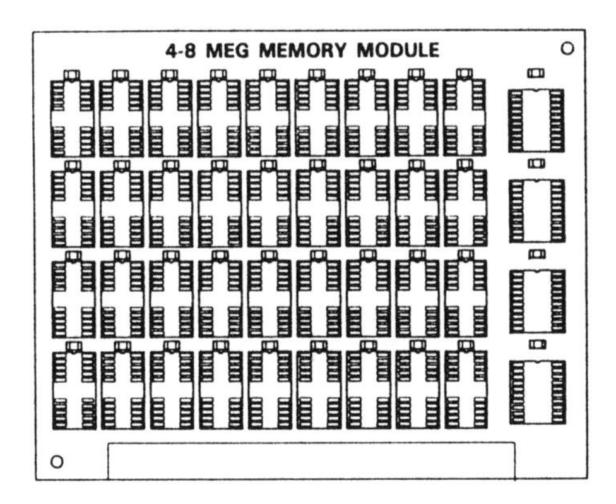

Figure 13.14
Memory intensive module which adds additional rows of surface mount packages to the single row contained in a standard SIMM. (DRAM TMS4C1024DJ, capacitor 0.10–0.22 μf, component pitch $X = 0.425$, $Y = 0.8000$, area = 18.80 in^2. (Reproduced with permission of Texas Instruments.)

13.8 BARE CHIP MODULES

The next potential step for getting the memories closer together is to remove the package from the individual chips altogether. In this approach many bare chips are mounted in one larger package or on a printed circuit board with a connector to enable it to interface electrically with the system. A third approach is to build up a multilayer module package out of conductive and dielectric thin film layers in which the bare memory chips are embedded.

The use of bare chip assemblies has various advantages and disadvantages. Advantages include increasing the packaging density of the system, and reducing the system delays due to interconnects still further.

There are several disadvantages which are yet to be fully overcome. Speed testing at the wafer or chip level prior to assembly is difficult due to the inductance of the wafer probe leads. Protective coatings are needed for the memory chips to avoid moisture related problems.

Power dissipation in a fast application can also be quite high due to the density of memory chips on the module and attention needs to be paid to efficient heat removal at all points in the module. Delays due to transmission line effects increase in multilayer thin films because of the high series resistance of thin conductive lines and the high capacitance of these lines to ground due to thin dielectrics.

13.8.1 Conventional hybrids

Bare-chip-on-substrate hybrids using wire bond assembly are expensive but have been around for a long time. They have primarily been used in systems where density and speed are absolute requirements and cost is no object. The expense is due not only to attaching the individual wire from each bond pad to the substrate, but the difficulties inherent in chip handling and testing.

Applications in aircraft and space instrumentation have used hybrid technology extensively. For more down-to-earth applications other methods are being explored. Figure 13.15 shows a high density hybrid module assembly.

One historical attempt to use hybrids for consumer applications was the old 'chip-on-board' modules for game ROMs in the early 1980s. These had sufficient manufacturing problems that the technology did not become widespread at the time. The collapse of the game ROM market in the mid 1980s probably also contributed to this early attempt not really taking hold.

13.8.2 Flip-chip on substrate

One attempt to gain the speed and density advantages of hybrids while reducing cost by using a type of 'gang-bonding' are the bumped flip-chip-on-substrate modules.

Figure 13.15
Example of large scale hybrid assembly with chips mounted face up and wire bonded to substrate. (Source IEEE Spectrum, with permission of IEEE.)

The flip-chip-on-substrate technique has been extensively explored in recent years by such major semiconductor memory users as AT&T and IBM. The technique, however, dates back at least as far as 1968 flip-chip modules made by Philco-Ford [25]. This technique involves plating a metal 'bump', normally gold or lead solder, onto the bonding pad of the memory chip.

A substrate is formed of interconnect layers with reverse bonding pads matching those on the memory chips. This thin film interconnect layer is deposited on a base substrate which can be either ceramic or silicon. Silicon has the advantage of matching

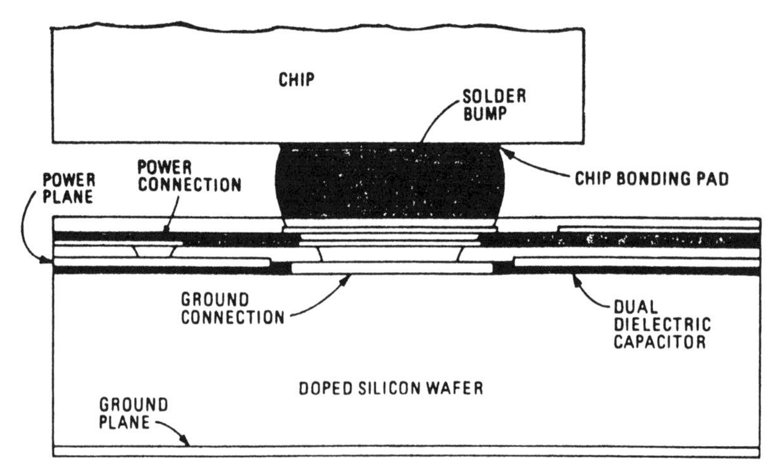

Figure 13.16
Flip-chip-on-silicon schematic cross-section of solder bump attach between silicon substrate and active chip showing the interconnect layers including ground and power planes and a dielectric capacitor. (From Shande [11], AT & T 1987, with permission of Electronics.)

the coefficient of thermal expansion of the silicon memory chip, but ceramic substrates with similar coefficients of expansion are available and can be made in sizes not necessarily available in silicon wafers.

Memory chips are then soldered up-side-down to the substrate as shown in Figure 13.16. Since the interconnect layers can be made with standard semiconductor processes, they can be of similar dimensions to those used on the device.

The number of chips per unit area is almost as high using this technique as for a single wafer. The number of working chips per unit area is, however, normally much higher since good chips are preselected rather than having to route around defective chips.

Two examples are shown in Figure 13.17 of experimental high density memory modules composed of standard SRAMs which are gold bumped and flipped onto silicon substrates containing interconnects. Figure 13.17(a) is an 8k × 40 module from Philips composed of five standard 8k × 8 SRAM chips. One chip is left off to show the interconnects and bonding pads on the substrate. Figure 13.17(b) is a fast 1Mb SRAM module made of 4 256k × 1 SRAMs as described by Philips in 1989. Again one of the chips has been left off to show the substrate [28].

Figure 13.18 shows the 8k × 40 SRAM module from Figure 13.17(a) being sample tested [27] by a wafer probe.

It is essential that an effective pretest is done since the yield for five chips on a substrate is statistically significantly less than the yield for one chip. Assume that the average yield for one chip to be bonded to the substrate is 99%, then the average expected yield for five chips bonded to that substrate is 95%. A reasonable solution is to have a rework procedure available. This is the densest assembly technique currently known of functional memory chips in two dimensions. The extension to three

(a)

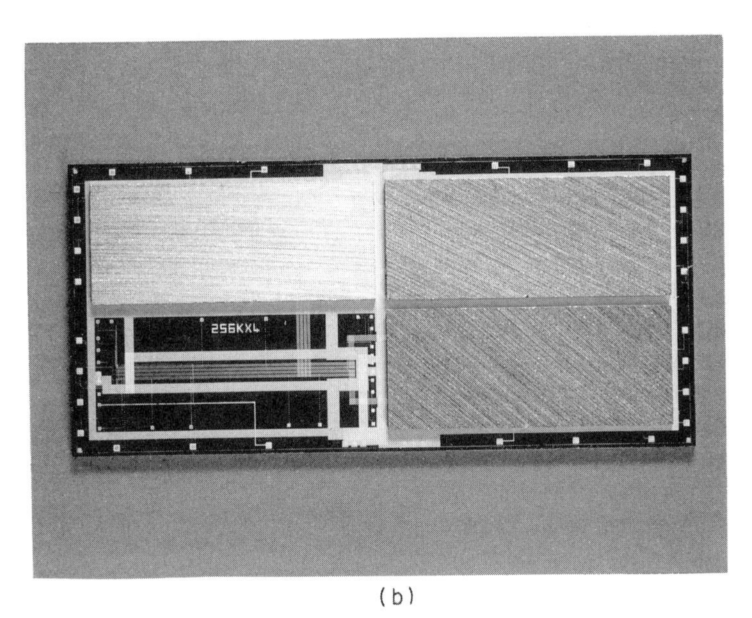

(b)

Figure 13.17
High density memory modules using flip-chip-on-silicon technology. (a) 8k × 8 SRAMs mounted in 8k × 40 configuration with one chip missing to show substrate. (b) 256k × 4 SRAM consisting of four 256k chips with one missing to show substrate. (Reproduced with permission of Philips Research.)

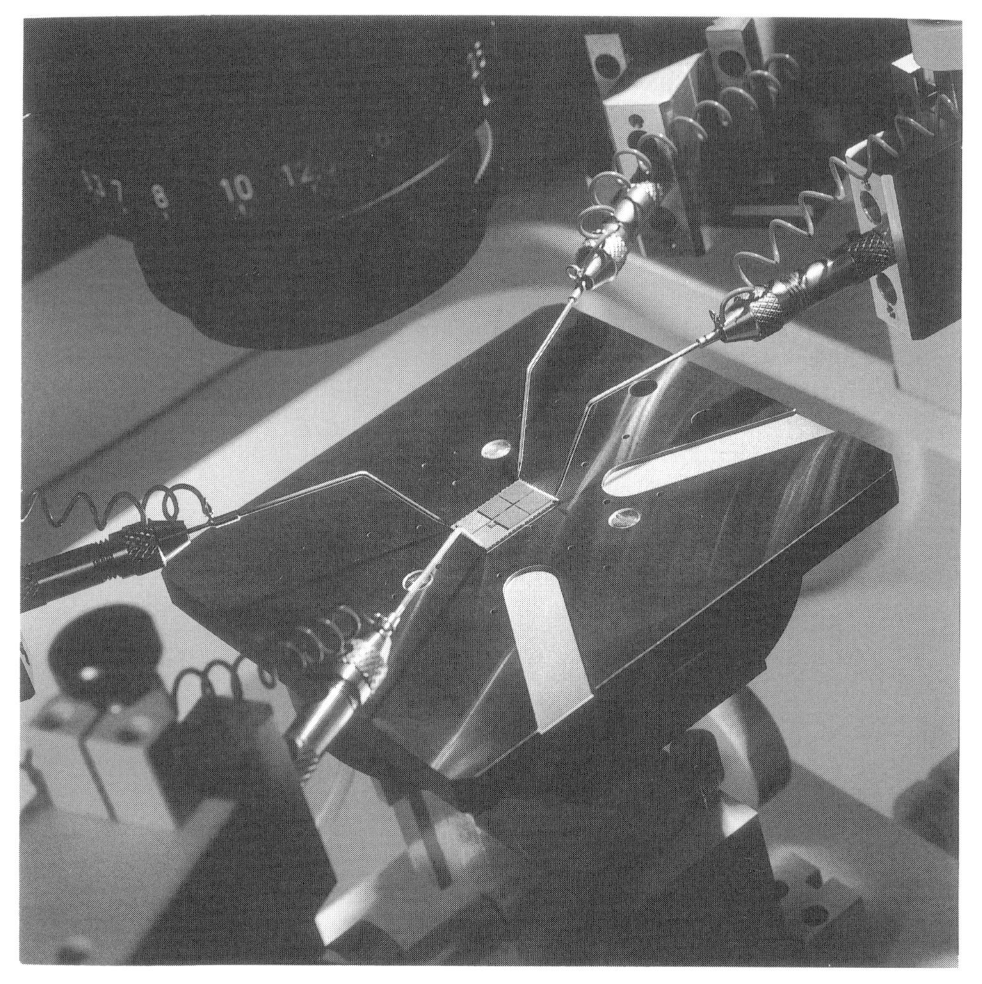

Figure 13.18
Sample testing of 8k × 40 SRAM multichip module. (Reproduced with permission of Philips Research.)

dimensions by using vertically stacked substrates has also been demonstrated by Hughes Aircraft and NEC.

The assembly time can be significantly reduced by this method since the traditional automatic wirebonding is a serial, time consuming process connecting each bonding pad on the die to the package one at a time. The bonding time is, therefore, proportional to the number of pins for the historical method, but constant for any die for the gang-bonded method.

The interconnect delay of such an assembly is significantly reduced. The transmission line effects between interconnects can be reduced by such techniques as using a metal ground plane such as the copper ground plane used by AT&T.

This technique is already well known in consumer chip assemblies such as the flip-chip-on-glass assemblies of LCD drivers for television monitors and in TAB assemblies. Volume commercial production in the memory area has not yet occurred probably due to the greater complexity of the memory chips.

13.8.3 Tape-automated bonding (TAB)

Tape-automated-bonding (TAB) is another bare chip packaging method in which the chip is mounted on tape rather than on a solid substrate. The leadframe for the chip is printed on the tape, and the tape is wound on reels for automated processing. TAB techniques allow simultaneous connection of the bonding pads to a specially prepared film giving a constant bonding time. The throughput is high and is not a function of the number of connections. The bonding pad pitch (pad to pad separation) can be as small as 2 mil compared to 25 mil for the TSOP package. Also the performance is improved due to the short interconnects. TAB uses a flexible metal foil to connect each bond pad on the die to the lead frame or substrate. The foil is usually made of copper. The foil is connected to the chip with bumps which are either plated or thermally compressed onto the die or plated onto the tape. The bumps can be either gold or solder. To join tape and die the gold is usually thermal compression bonded and the solder is reflowed. A chip assembled on a TAB leadframe is shown in Figure 13.19.

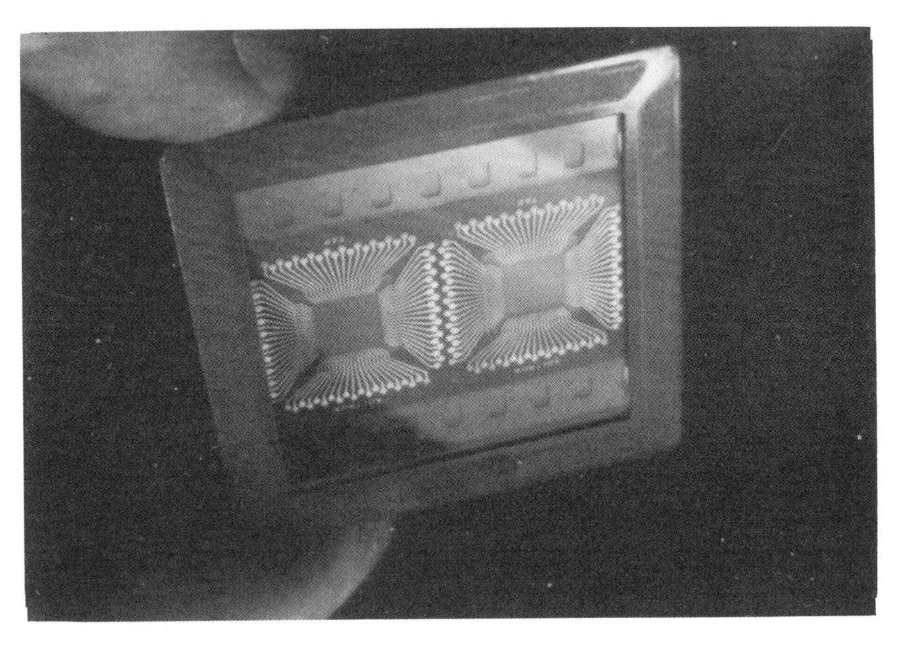

Figure 13.19
Tape automated bonding lead frame with chip attached (photograph by Mike Butterworth). (From Prince and Due-Gundersen [26].)

A TAB leadframe can either be molded into a quad flat package or it can be attached directly to a board. One of the advantages of TAB over flip-chip assembly is that an assembled TAB chip can be tested prior to assembly easier than a bare chip can be tested.

13.8.4 A multilayer module for high performance computers

An example of the multilayer packaging concept is packaging technology developed by DEC for high performance processors. The module is built up from layers of copper interconnections and polyamide insulation. The copper is deposited onto the polyamide using sputtering and plating techniques and semiconductor type of photolithography. Line connections between layers are made by masking the polyamide and etching a hole with a plasma etcher. The holes are filled with copper at the time of the next copper deposition. Sites for the semiconductors are cut out by lasers and TAB bonding is used to mount the chips in the module. A 4 inch by 4 inch module can hold as much logic as four 15 inch by 12 inch circuit boards. A module has nine layers for power and signal distribution.

13.8.5 High density memory cards

Another high density application of bare chip assembly technology is the high density memory card which is used as a RAM card in laptop PCs. These cards are thin and require the high density of very close chip placement. Various assembly techniques used in these cards are bare-chip-on-substrate, TAB-on-board (TOB), and the new TSOP surface mount packages. A picture of a memory card is shown in Figure 13.20 along with a view of the flatpack surface mount packages assembled inside the memory card.

13.9 WAFER SCALE INTEGRATION

There is basically not a lot of difference between putting the memory chip onto a layer of interconnects and putting a layer of interconnects onto a memory chip. While the latter is frequently referred to as 'wafer scale integration' (WSI), the difference in most ways is in perception and not in kind.

One difference, however, that is significant between the two approaches is that the WSI approach can probably not reach the density and performance levels of the modular approach. This is because of the need to wire in redundant circuit elements to replace defective parts on the single wafer, and also the need for longer interconnects to bypass the defective parts of the chip. With the modular approach all chips, in theory, are tested good prior to assembly. The result is that interconnects do not have to bypass defective devices and are therefore on the average shorter.

Figure 13.20
High density 1Mb memory card package with a view of the inside of the package. (Reproduced with permission of Texas Instruments.)

13.9.1 Programmed interconnections–discretionary wiring

Large scale integration on a monolithic silicon slice by means of programmed interconnection involves identifying the functioning elements on a wafer, deactivating those that are defective, and integrating the remaining segments onto a working system by interconnect wiring. This can be done with either an additional layer of metal or by fusible links. The defective elements are permanently isolated from the system by the hard wiring of the interconnects. The programmable logic arrays, discussed in Chapter 10, are examples of large scale integration using fusable links.

An early wafer scale integration effort using discretionary wiring was made at TI in 1970 using the pioneering efforts of Jack Kilby who, in addition to inventing the integrated circuit, also developed the concept of discretionary wiring for interconnection of systems directly on a wafer from small clusters of gates that tested good.

Once the good cells were located a computer progam mapped the desired logic function and created routing to interconnect them. These techniques were used in 1970 to produce a 32k memory using wafer scale integration by discretionary wiring.

An example of this technique is a memory array incorporating programmed interconnect redundancy. As a defective element is found spare elements are brought in.

Another example of this technology was the 128k × 8 SRAM of Inova Semiconductor which was formed from a basic chip containing 40 32k arrays in two rows of 20 blocks each. These arrays were tested after first metal to select chips having no more than eight blocks defective. These arrays were deselected at this stage by blowing fuses. The final metal interconnect layer was then applied. Only the good 128k × 8 SRAM formed from the selected blocks was present at this point.

While this technique used a much greater amount of silicon than would be used by a single good 128k × 8 SRAM chip, at the time Inova introduced the part there were no single chip 1Mb SRAMs yet on the market. The early life pricing was high enough to justify the larger chip at the time. The process used a standard 1.2 μm technology compared to the 0.7 to 0.9 μm geometries used for the standard 1Mb SRAMs.

For comparison the silicon used was also larger than is occupied by a modular flip-chip-on silicon 1Mb SRAM such as that shown previously in Figure 13.17(b). This chip is also made up of 256k chips but only the space of four working chips and the relatively small distance between them is occupied on the silicon. The area ratio of 1Mb SRAMs using programmed interconnect, a flip-chip-on-substrate multiple wafer integration, and monolithic die is in this case roughly 3:2:1.

13.9.2 Multichip module with single substrate

An alternative approach to the late interconnect deposition technique which eliminates the presence of defective but unwired chips is the chip-in-silicon technique. In this case a single silicon substrate has holes etched in it into which the functional active chips are placed. The surface is planarized and the interconnects between these chips are then deposited on the surface and etched using standard wafer lithographic techniques. An example is shown in Figure 13.21.

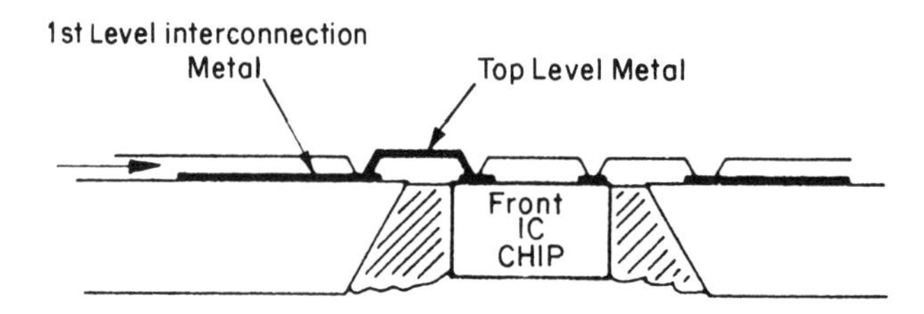

Figure 13.21
Large scale integration using chip-in-wafer processing with integrated circuit chips inserted in holes in a wafer which is mounted on an optically flat surface. After planazation double metal interconnects are deposited and etched. (From Johnson [13], Auburn University 1986, with permission of IEEE.)

(a)

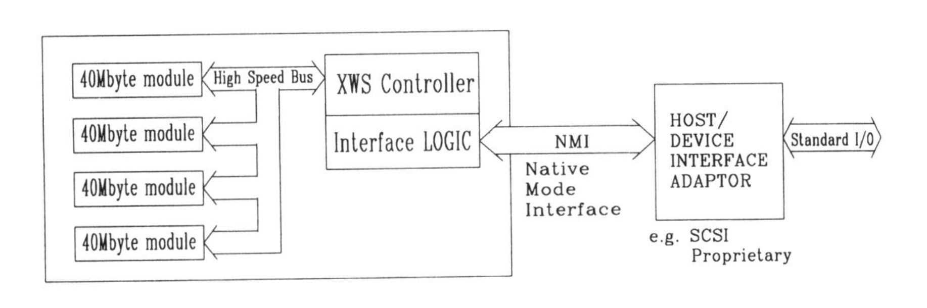

(b)

Figure 13.22
(a) Photograph of a 1Mb DRAM wafer using the spiral algorithm method of wafer-scale integration. The wafer is mounted on a printed circuit card which is mounted in a hard drive dimensioned assembly. (b) The system configuration of this 160Mb assembly which contains eight of these wafers is also shown. (Reproduced with permission of Anamartic.)

13.9.3 The spiral algorithm technique

One of the first examples of real production memory wafer scale integration was the serial wafer integration developed by Anamartic [17] in the UK. This technique interconnects relatively standard DRAM chips on a single wafer. Each DRAM chip is connected with four neighboring chips through logic elements that are added to each of the DRAMs. Interconnection is done in software by programming the logic elements to connect each DRAM to two of its neighbors. The software algorithm defines a serial path through the working chips which spirals inward toward the center of the wafer. Partially working arrays as well as fully functional chips are usable which increases the effective 'yield' of the wafer. The path through the wafer is then inscribed in a non-volatile medium such as an EEPROM which is shipped with the wafer, see Figure 13.22 for a photograph of a wafer using the spiral algorithm technique made by Anamartic in conjunction with Fujitsu using 1Mb DRAMs. The 202 1Mb DRAMs on each wafer used by Anamartic contain a maximum potential data storage of 25.25M bytes of memory. 20M bytes is obtained from each wafer by using both fully and partially functional chips resulting in a yield of about 80%.

The wafer and EEPROM are mounted flat on a PC board with an end connector which can be assembled with other such units in a hard disk format. The final 160M byte assembly consists of eight 6 inch wafers each of which contains around 200 1Mb DRAMs together with control logic and interconnections. Since the wafer is software programmable, it can be reconfigured and the EEPROM holding the configuration data changed.

This method of single wafer chip assembly produces a serial memory which, at 220 ns typical access time, is relatively slow compared to RAM semiconductor memories. For such applications as replacement of 20 ms hard disks, however, it is a high performance approach.

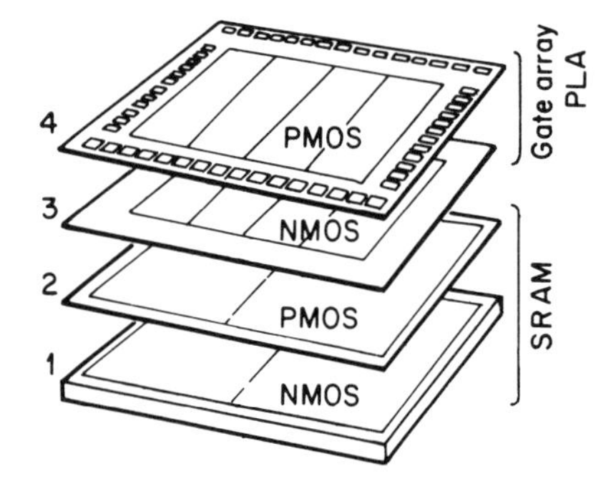

Figure 13.23
Schematic diagram of stacked polysilicon transistors forming a gate array stacked on a SRAM. (Reproduced with permission of NEC.)

13.9.4 Vertical wafer scale integration

Another development which is further out is the potential of stacking layers of integrated circuits in multiple layers of interconnects on a single chip. An example of this technique was given by NEC in a 1989 paper in a circuit which used three layers of polysilicon transistors arranged as a gate array stacked above a SRAM as shown in Figure 13.23.

BIBLIOGRAPHY

[1] *Hitachi Memory Databook* 1988.
[2] *T.I. Memory Databook.*
[3] *Cypress Databook.*
[4] Johnson, R. R. (1990) Multichip Modules: Next-Generation Packages, *IEEE Spectrum*, March, p.34.
[5] Iversen, W. R. (1988) Surface-Mount Technology catches on with U.S. Equipment Makers—Finally, *Electronics*, December, p.116.
[6] Bindra, A. (1988) SMTs in U.S. Ready to Take Off, *Electronic Engineering Times*, November 7, p.T4.
[7] Laybhen, N. and Walker, J. (1988) Fine Pitch puts SMT on TAB, *Electronic Engineering Times*, November 7, p.T11.
[8] Curran, L. (1989) Packaging Breakthrough Shrinks CPU delays, *Electronic Design International*, November, p.24.
[9] Lyman, J. (1987) PC Board will Go the Way of the Vacuum Tube, *Electronics*, April 2, p.90.
[10] Clifford, L. (1988) Stacking up the Memory Power, *Electronics Weekly*, November 1, p.18.
[11] Shandle, J. (1987) Trying to Keep Up With Past-Moving Chips, *Electronics*, October 15, p.133.
[12] Neugebauer, C. A. *et al.* (1988) High Performance Interconnections between VLSI Chips, *Solid State Technology*, June, p.93.
[13] Johnson, R. W. (1986) Silicon Hybrid Wafer-Scale Package Technology, *IEEE Journal of Solid State Circuits*, Vol. **SC-21**, No.5, October 1986, p.845.
[14] Cole, B. C. (1987) Inova Brings Wafer-Scale Integration to Market, *Electronics*, January 8, p.91.
[15] Bently, L. and Jesshope, C. R. (1989) The Implementation of a Two-Dimensional Redundancy Scheme in a Wafer Scale High Speed Disk Memory, *Wafer Scale Integration.*
[16] Wolff, H. (1988) The Megabit Static RAM is here, and a U.S. Startup Jumps in Front, *Electronics*, June, p.29.
[17] Baba, F. *et al.* (1989) High Capacity Whole Wafer Memory, *ISSCC Proceedings*, February, p.240.
[18] Raffel, J. I. *et al.* (1985) A Wafer Scale Digital Integrator Using Restructurable VLSI, *IEEE Journal of Solid State Circuits*, Vol. **SC-20**, No.1, February 1985 p.399.
[19] Ferris-Probho, A. V. (1985) Defect Size Variations and their Effect on the Critical Area of VLSI Devices, *IEEE Journal of Solid State Circuits*, Vol. **SC-20**, No.4, August 1985.
[20] Maly, W. (1985) Modeling of Lithography Related Yield Losses for CAD of VLSI Circuits, *IEEE Trans. Computer Aided Design*, **CAD-4**.
[21] Cohen, C. L. (1984) Full Wafer Memory of Color Displays has 1.5Mb Capacity, *Electronics*, January 26, p.77.

[22] McDonald, J. F. (1984) The Trials of Wafer-Scale Integration, *IEEE Spectrum*, October 1984, p.34.
[23] Ueoka, Y. (1984) A Defect-Tolerant Design for Full Wafer Memory LSI, *IEEE Journal of Solid State Circuits*, Vol. **SC-19**, No.3, June 1984, p.319.
[24] Clifford, L. 91988) Stacking up the Memory Power, *Electronics Weekly*, November.
[25] Thornton, G. G. (1968) LSI Subsystems Assembled by the Silicon Wafer Chip Technique, *ISSCC Proceedings*, p.42.
[26] Prince, B. and Due-Gundersen, G. (1983) *Semiconductor Memories*, John Wiley & Sons, Chichester.
[27] 1988 Annual Report of N.V. Philips Gloeilampen Fabrieken.
[28] Salters, R. (1989) Presentation at Fast SRAM Panel Discussion, *IEEE ISSCC*, February 1987.

14 MEMORY ELECTRICAL AND RELIABILITY TESTING

14.1 OVERVIEW OF MEMORY TESTING

During the last decade the ever increasing complexity of semiconductor components has been matched by similar developments in test and reliability philosophies and practices. A large capital investment is necessary to establish a high performance test area which balances high quality, high yields, high throughput, and provides meaningful feedback to the wafer processing area. This is a critical parameter in the cost of semiconductor manufacturing.

The historical tradeoff between throughput and quality no longer exists since throughput must be kept high and users expect competitively high quality and low cost too. Overall quality assurance practices are demonstrating that these two demands are mutually complementary.

Development of high density memories introduced a new dimension in testing complexity. In the early days of logic circuitry, device performance was basically judged on parametric measurements such as voltage and current levels. With the increase in device complexity it was, however, necessary to introduce functional testing (such as timing specifications) as well as parametric testing. This was followed by the development of algorithm functional testing which is currently widely used in memory manufacturing.

This type of testing involves the use of patterns of an algorithmic nature. A 100% test for all possible operating combinations of a high density memory would be financially impossible, due to the extremely long test times required. Algorithmic testing will effectively perform test combinations which are sufficient to weed out failures and still minimize test time.

A sophisticated test area which includes equipment, software, and skilled engineering expertise is a sizable investment. This overhead is required, however, for any manufacturer wanting to become, or remain, a leading memory supplier.

Historically memory test was performed for a variety of reasons. Memory manufacturers tested to guarantee performance within predefined specifications and screen out those components not meeting the specification. Today, although screening remains, the philosophy of screening is being replaced with a philosophy of testing to a statistical process control distribution and rejecting those parts which fall outside the normal distribution. The idea is that if a process is stable then stable test results must follow. The basis of this reasoning is that any abnormality in the test results implies an abnormality in the process and should be eliminated.

A high volume memory user historically did incoming inspection on the components to monitor quality and performance before product assembly. This is also rapidly becoming a thing of the past as more users expect the parts to be delivered just-in-time (JIT) for use and with effective zero defect rate.

An end equipment manufacturer may screen ready assembled boards of systems to his specifications. This seldom however involves a specialized tester.

The first part of this chapter is concerned primarily with test activities ongoing within the component manufacturer such as electrical characterization, probe testing to sort wafers electrically, and final test. The second part of the chapter discusses reliability concepts and methods.

14.1.1 Characterization

Characterization of device parameters takes place before a new device type is introduced, or when a well established product has gone through a design, process or package modification.

A characterization includes extensive testing of the electrical characteristics of a representative sample. A characterization program will provide data on dc and ac parameters and absolute measurements of voltage margins within which the device will operate. The characterization is performed over a wide temperature range with a large number of patterns and gives a complete picture of device operating characteristics. This detailed component knowledge is also important at a later stage when the compilation of a short but efficient production final test program is required.

When the part is in production it is of vital importance to monitor the status by characterization exercises on a regular basis.

14.1.2 Probe

Historically probe, or wafer sort, involved only functional and gross parametric testing of each die on a wafer before assembly in order to filter out gross rejects and give rapid feedback to the wafer processing area on serious problems. 'Gross reject' testing is becoming a thing of the past for several reasons. Assembly and package costs, which used to be a small fraction of the total cost of a memory component, are becoming a much more significant percentage of the total cost.

Also the feedback required in the wafer manufacturing area has increased in complexity along with the increase in complexity of the memory process.

With the reduction in cost of computer power, statistical analysis has become a major part of the work of the memory product and test engineer. Small trend variations caused by slight process shifts can be quickly analyzed and many gross failure modes experienced in the early days of MOS memories can be avoided completely.

The result has been that more emphasis is being placed on probe test and more effective tests are being devised.

The probe test will normally include screens for dc parameters and a few patterns for ac parameters. Accurate ac testing is difficult at probe since the long cables on the test machine and the long needles of the probe card add a significant amount of inductance into the circuit. Apart from the rejection of failures before assembly the probe test will provide information related to speed distribution. These early warning data are important for planning and scheduling of the finished product and give valuable feedback to wafer processing areas. More recently stress testing at probe has been used to weed out marginal devices and reduce the level of early reliability failures after assembly and final test.

14.1.3 Final test

This is the verification that the ready assembled product will meet its specification before shipping to the customer and will continue to do so over some period of time. The final test can basically be split into various sections including screens for continuity, dc parameters, and ac functional performance. The latter is sometimes split into a quick 'easy functional' test to screen gross rejects and a longer more extensive functional test involving many complex test patterns. All specified data sheet ac and dc characteristics are normally checked.

(a) Continuity testing will confirm electrical connections between pins and bonding pads. No power is applied to the device. Historically discontinuity was the most common failure mode. Modern assembly practices have made this less true.

(b) dc parametric testing verifies the performance of dc electrical characteristics such as leakage currents, input–output levels, active and standby power supply currents, etc.

(c) ac parametric tests are functional tests which expose the component to a 'real' operating environment, checking for number of working bits, presence of on-chip logic on application-specific memories, access times, refresh time, cycle times, etc.

A final test program will normally consist of the following four sequential subsections.

- easy functional
- continuity
- dc parametric
- functional–ac parametric

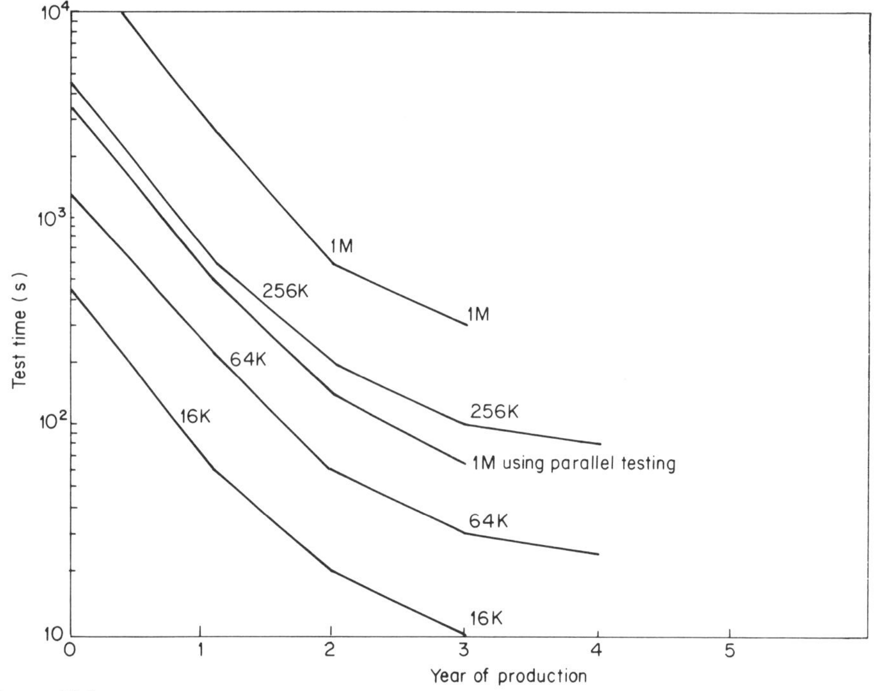

Figure 14.1
Final test time for various DRAM densities by year of production (assuming similar test flow).

Highly sophisticated test algorithms have been developed with the aim of reducing test time while maintaining the required quality.

Figure 14.1 illustrates test timing improvements for various densities of DRAMs from the point of device introduction followed by an engineering experience curve until an optimum and stable situation is reached in about the third year of production.

The average test time required for the different types of MOS memory products varies considerably. DRAMs and SRAMs of similar densities have similar test times. Products such as EPROMs take a factor of 10 increase in test time. Since test time impacts the throughput of the product line and is allocated in overhead on expensive test equipment it adds considerably to the cost of production. Figures 14.2(a) and (b) show memory test equipment being used.

14.2 FAILURE MODES AND TEST PATTERNS

The rapid growth in circuit complexity has greatly increased the difficulty and cost of testing memories.

Semiconductor memory testing can be basically reduced to four different groups:

Cell test Every cell must be capable of storing a logic '0' and a logic '1' for a given minimum amount of time and, if the memory is writable, must be capable of being changed from one state to the other.

(a)

(b)

Figure 14.2
Examples of memory test. (a) Volume production final electrical test showing memory test machines and handlers. (b) Sample electrical test at outgoing inspection. ([60] 1990, with permission of NMB Semiconductor.)

Data in–out test Sense lines and data in–out lines must be capable of recovering from read–write operations. Sense amplifiers have to operate within the small, specified voltage or charge levels.

Address decoder test Every cell must be correctly and uniquely addressed by the decode logic.

Disturb testing The accessing of one part of the memory array must not affect any other part.

Special test Functions specified for different memory types must be checked and verified operational. This will include the testing of control lines such as chip enable, output enable, chip select, or special logic circuits.

The design of a sequence of test patterns to check the necessary failure condition for a memory device calls for not only thorough component knowledge, but also an in-depth understanding of various MOS failure mechanisms.

It is important to test for the maximum number of failure modes for the test pattern chosen, while at the same time avoiding combinations of standard patterns which result in double testing, or in checking for unnecessary conditions.

A few common memory failure modes highlighting the above failure groups are listed.

- Cell: 'stuck-at' failures, open or short circuits, leakage, adjacent cell disturbance.
- Access times: minimum and maximum to specification.
- Address decoder: open or short circuits, noise, high sensitivity.
- Sense amplifier: recovery time.
- Refresh times: minimum.
- Clocks.
- Write: recovery time.

'Stuck-at' cell failures are one of the commonest forms of failure. In this mode the single cell bit is simply 'stuck' at '1' or '0'. Complex tests are not necessary to determine this type of failure.

Additional failure modes specifically related to different memory families also have to be screened for and will add to the above list.

A variety of standard test patterns are commonly used for screening out most known failures. Figure 14.3(a) illustrates some test pattern types with corresponding test time constants and failure descriptions.

The test time required is calculated by substituting N with memory size (or number of bits) followed by multiplication by the cycle time used. It can be seen that most tests fall within three different categories: $2N$, $2N^{3/2}$, or N^2. Test time is mostly dependent on memory size, cycle time, and test pattern performed.

Figure 14.3(b) illustrates test times as a function of memory size for $2N$, $2N^{3/2}$, and $2N^2$ type patterns.

A typical test pattern sequence might include the following.

1. All '1' or all '0'.
2. Checkerboard

Pattern	Constant	Failure
March	$10N$	Cell access
Walking pattern	$2(N^2+N)$	Multiple address sense amplifier recovery
Galloping pattern	$2(N^2+N)$	Multiple address sense amplifier recovery
Sliding diagonal	$2(3N^{3/2}+5N)$	Cell testing, diagonal
Galloping column	$2(3N^{3/2}+6N)$	Cell testing, columns

(a)

Pattern	1k	4k	16k	64k	256k
$2N$	0.41	1.64	6.55	26.2	76.8
$2N^{3/2}$	13.57	104.8	839	6710.9	53,687
$2N^2$	439	6710.9	107 347	1717 987	27 487 792

(b)

Figure 14.3
(a) Typical test patterns with corresponding test time constants and failure descriptions 'N' is the device bit density. (b) Illustrative test times as a function of density. (Reproduced with permission of John Wiley.)

3. Stripe
4. Marching
5. Galloping
6. Sliding diagonal
7. Waling
8. Ping-Pong.

Numbers 1 to 4 are called 'N' patterns. These can check one sequence of N bits of memory by at most using the given pattern several times. Numbers 5 to 7 are called N^2 patterns. These need several times of N^2 patterns to check one sequence of N bits of memory. N^2 patterns have a long test time for high density memories. For example, a 64k RAM takes about 30 minutes to test with a galloping pattern and a 1Mb RAM takes over 6 hours.

The first three patterns in the sequence shown can check the array but are not sufficient to check the decoder circuits. The marching pattern is the simplest pattern which will check out the function of the memory. A description follows as an example of a typical marching diagonal test pattern.

A marching pattern is a pattern in which '1's march into all '0's. The procedure is as follows.

1. Clear all bits to '0'.
2. Read '0' from '0'th address and check that read data are '0'.
3. Write '1' on 0th address.
4. Read '0' from 1st address, and check read data are '0'.

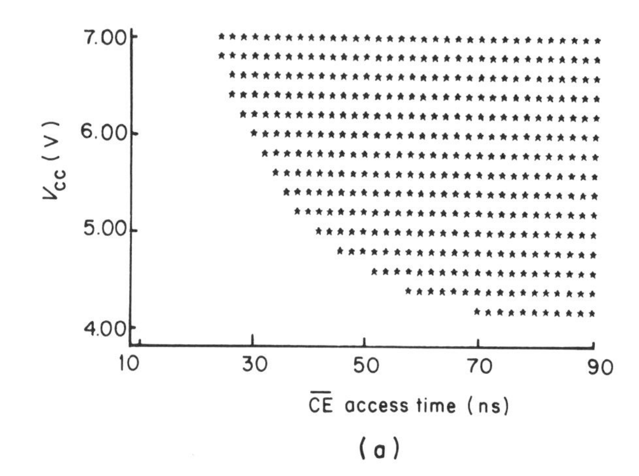

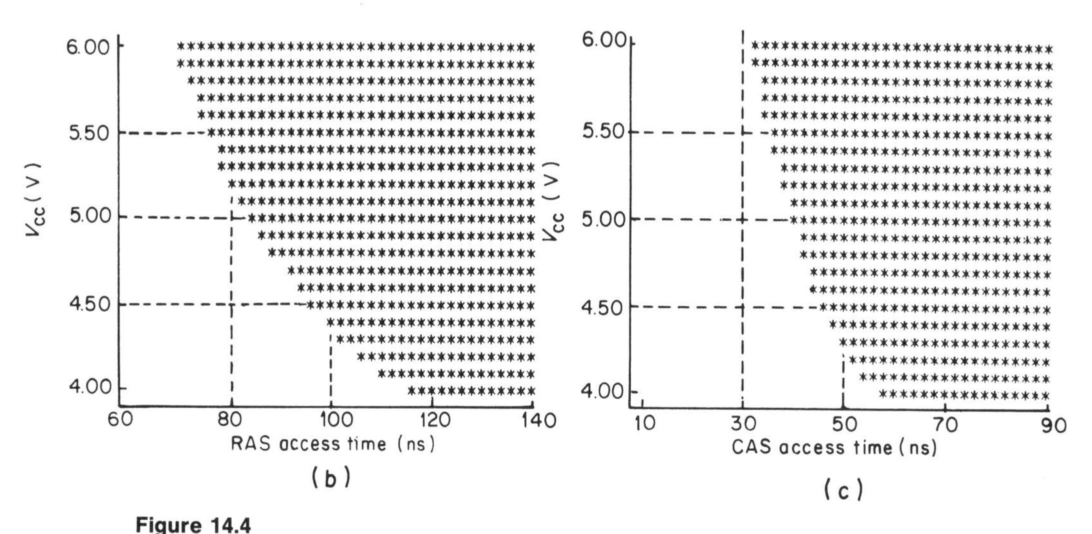

Figure 14.4
Various Shmoo plots showing functional device parameter regions for DRAMs. (a) V_{CC} plotted against $\overline{CE}$ (From Benevit *et al.* [37], AT & T 1982, with permission of IEEE), (b) V_{CC} plotted against $\overline{RAS}$, (c) V_{CC} plotted against $\overline{CAS}$. (From Kantz [45], Siemens 1984, with permission of IEEE.)

5. Write '1' on 1st address.
6. Repeat for N addresses until all are '1's.
7. Repeat 2 to 6 reading '1's and writing '0's until all data are '0'.

The commonest kind of 'stuck-at' failures can be found by this type of test. Test information can be presented as follows.

- Shmoo plots show the interaction between device parameters by plotting the range of functional parameter sets. Figure 14.4(a) shows the relation between V_{CC} and $\overline{CE}$ access time on a 256k DRAM. The plot shows a 40 ns chip enable access time at 5 V for the device under test. Figure 14.4(b) shows V_{CC} as a function of $\overline{RAS}$ access time and Figure 14.4(c) shows V_{CC} as a function of $\overline{CAS}$ access time.
  Asterisks denote the regions where the part is functional. Shmoos are generally used for device characterization, intersystem correlation, test program development, process and device correlation, and failure analysis.

- Accumulative Shmoo plots are derived from superimposing several Shmoos from different components.

- Wafer mapping is shown in Figure 14.5(a). This is used for probe yield analysis, wafer fabrication defects analysis, mask defect analysis, alignment problems, and system correlation. This is an accumulative representation of several wafers where the total failures for each die location are given as a percentage.

- Bit mapping: this gives fail–pass information for every bit or cell location of the memory under test. The results can be displayed on a color graphics terminal or printed. Real time bit mapping is fast and very effective during failure analysis, process evaluation, probe yield enhancement, pattern sensitivity detection, data retention (EPROMs) testing, pattern verification (ROMs) and device characterization. Figure 14.5(b) for example, shows a bit map of a 16k RAM as a 178 by 178 matrix. The vertical dotted line illustrates a number of faulty bits.

14.3 TESTING DRAMs AND SRAMs

The various memory types of families such as ROMs, EPROMs, or RAMs will have different testing requirements due to differences in device characteristics, failure mechanisms, and operating modes. We will discuss special test considerations for dynamic and static RAMs, EPROMs, EEPROMs, and embedded memories.

14.3.1 Fault coverage considerations

The major problem in RAM testing is obtaining good fault coverage. As RAM density increases and advanced technologies and circuit techniques bring with them more complex failure modes, testing time increases rapidly.

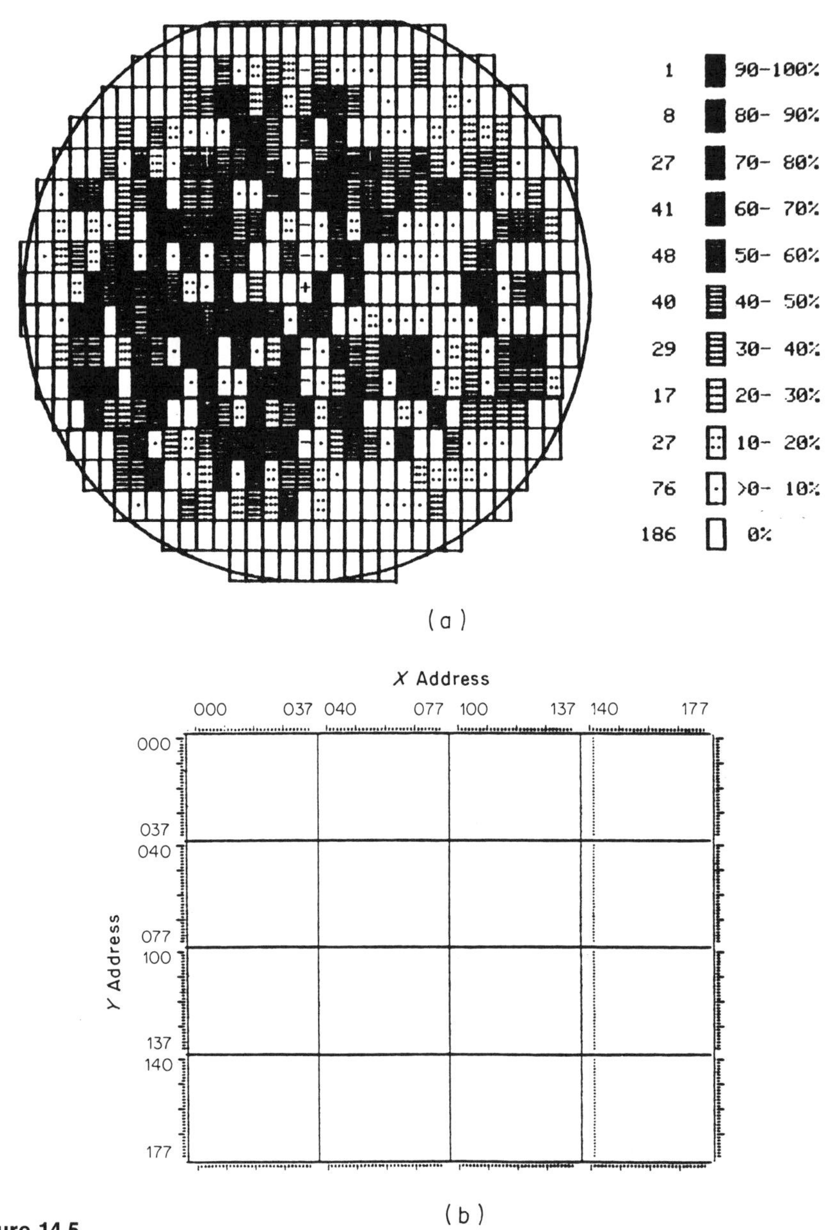

Figure 14.5
(a) Wafer mapping of 16k DRAM. (b) Bit map of 16k DRAM. (From Prince and Due-Gundersen [64] 1983.)

For example a simple sliding diagonal test requires several hours to perform on a single 1Mb RAM chip. The more complex galloping pattern (GALPAT) test requires a testing time proportional to N^2 for an N-bit RAM. If testing a 1k RAM takes a few seconds, then testing a 1Mb RAM may require several days. The extent of fault coverage in these tests is also not easy to define.

Table 14.1 Normalized test times for dynamic RAMs.

Density	Normalized test time
65k	1.0
256k	3.1
1M	10.4

Normalized test times for various generations of DRAM are shown in Table 14.1 taken from a paper by TI [50] where the test time of the 64k DRAM is taken as '1' and it is assumed that the same test flow is used for all. If 8 bit parallel testing is used for the 1Mb DRAM then the test time is reduced from a factor of 10.4 to 2.0. Parallel testing is described more fully in the later section on DRAM test modes.

14.3.2 Failure modes

Most failure modes related to RAMs are well known with corresponding test patterns for effective testing. Typical RAM failure modes include

- 'stuck at' faults
- pattern sensitivity
- multiple writing
- refresh sensitivity (DRAMs)
- open–short circuits
- leakage current faults
- sense amplifier recovery
- access time
- voltage bump (DRAMs)

A combination of well known $N^{3/2}$ and N type test patterns will generally provide the solution to effective screening for most of these failure modes. When the characteristics of a specific device are known it is often possible to delete parts of the initial test program and obtain a substantial reduction of test times.

Mechanisms, which are frequently involved in these failure modes, are reviewed below.

Gate oxide defects can cause stuck-at faults such as stuck-at 0, stuck-at word line, bit-word line crosstalk, transmission line effects, and row decoder failures.

The major sources of pattern sensitive failure modes in DRAMs are neighborhood interference faults, sense amplifier recovery problems and bit-line imbalance faults.

Neighborhood interference faults are frequently due to leakage current mechanisms which can depend on the pattern of stored data in the neighboring cells. When the leakage current flowing between one cell and another is sufficient to destroy the contents of the cell, a neighborhood interference fault is said to occur. The worst case is when all surrounding cells have the opposite state to the tested cell.

Sense amplifier recovery problems can be caused by parasitic capacitance and resistance, which can also cause slow sense amplifiers, and transmission line effects. Sense amplifier recovery faults occur when, after repeated writing of the same cell, the data read out from that cell are independent of the contents of the cell.

Bit-line imbalance faults are caused by the difference in the total leakage associated with the cells connected to the two bit-lines involved.

14.3.3 Voltage bump test for DRAMs

Voltage bumping, or fluctuations of the power supply voltage, can cause erroneous data to be read out of the RAM. The voltage bump problem demands some special testing and the voltage bump test is a particularly rigorous test for a DRAM.

A positive V-bump is defined as V_{CC} during the write operation being lower than V_{CC} during the read operation. This fluctuation of V_{CC} lowers the read out voltage from the memory cell and may cause a read error.

If a V_{CC} level cell plate is used, the positive V-bump raises the stored level in the cell. Since the dummy cell level is kept at V_{SS} in spite of the bump, a low level in the cell may read erroneously as a high level.

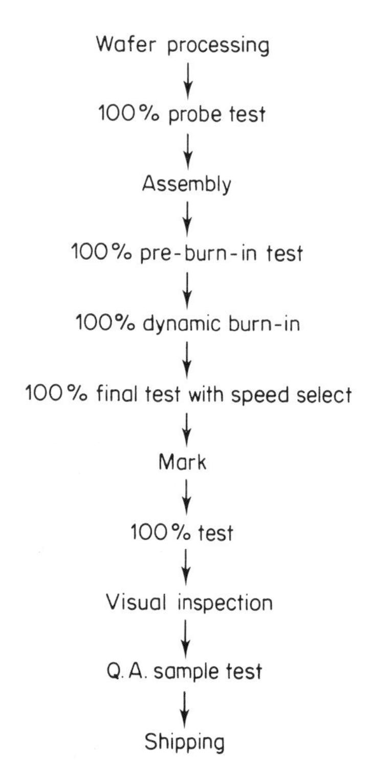

Figure 14.6
Typical dynamic RAM production flow charts. (From Prince and Due-Gundersen [64] 1983.)

If a V_{SS} level cell plate is used, the positive V-bump causes no change of the stored level or of the dummy level. In a read operation the lowered voltage on the word line due to the bump may cause a high level to be misread as a low.

It was this reasoning which led in part to the development in the early 1980s of the half V_{CC} biased cell plate which minimized the effects of a voltage bump [25].

Figure 14.6 shows a typical dynamic RAM production flow chart. Points or tests related to this table not previously discussed are the following.

(1) Pre-burn-in test which is purely introduced to filter out gross failures and save burn-in capacity

(2) 100% test after marking which is performed to check for both damage during handling and that the memory is correctly marked.

14.3.4 Addressing considerations when testing a RAM

To properly test a random access memory a detailed description of the internal topology and address decoding of the device is required in order to run complex disturb patterns and to optimize testing procedures.

Before a particular memory cell location can be accessed, it is necessary to know the internal addresses corresponding to the actual physical memory cell locations. These are frequently different from the addresses applied to the external address pins of the device. An address decoder scrambler provides the conversion from the external address to the internal address. An example of an address decoder scrambling is given in Figure 14.7 for the Siemens 1Mb DRAM [6]. A transformation must be made in accordance with the logic diagram from internal rows and columns to external addresses.

A bit map then relates an external address location to an internal physical location on the chip. This information is required for testing the device since it is necessary to test for interactions between physically adjacent cells.

14.3.5 Datapolarity

It is also necessary to know the internal data polarity before a test program can be written. For example, if balanced sense amplifiers are used on a DRAM, then the data on one side of the sense amplifiers must be stored at an inverted polarity from that on the other side. This inversion is transparent to the user, but must be known, for example, to write all bits to the same internal state.

14.3.6 Dual-port (video) DRAM

The dual-port video DRAM is a dynamic RAM with both a random port for interfacing with the CPU and a serial port for interfacing with the video display. A

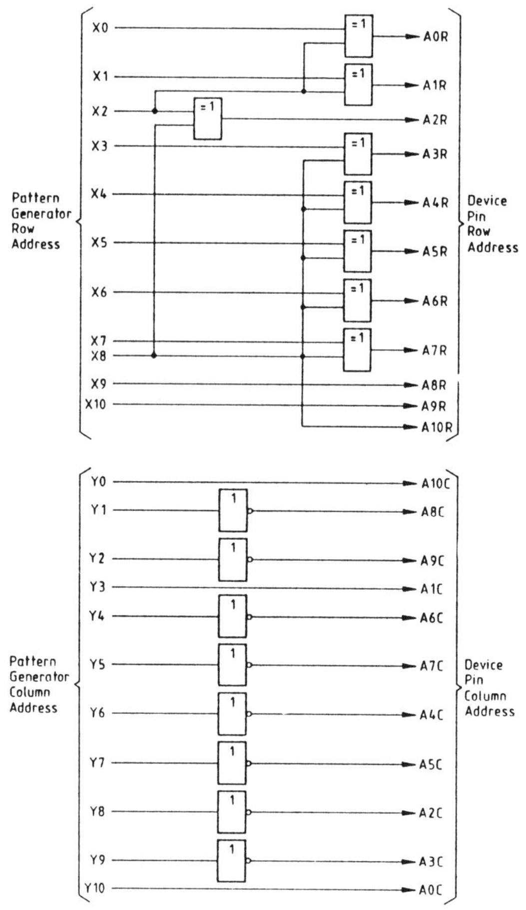

Figure 14.7
Address decoder scrambler for converting from internal rows and columns to external device row and column address pins for a typical 4Mb DRAM. (Source Siemens 4Mb DRAM Datasheet 1989.)

typical sequence is that the random port receives video display data from the CPU and stores it in the RAM array. The serial port receives data from the RAM array at a high speed and sequentially outputs it.

Testing this device differs from standard DRAMs in several ways. The dual-port RAM has to be tested by matching the tester cycle time to the faster serial port cycle time. Special test patterns must be used to test the serial port. Since the two ports operate asynchronously they must be tested asynchronously. The 'write-per-bit' function, which is unique to the video DRAM must be tested. This is a masking function in which selected bits in an addressed word are masked from being written.

Test methods have been developed to handle all of these problems. See, for example, the 1985 paper on dual-port RAM testing by Fujieda and Arai [17].

14.3.7 DRAM test modes

The EIA JEDEC Standardization body has specified special test and operational modes for address multiplexed DRAMs in their Standard 21C [44] for solid state memories. The standard defines the logic interface required to enter, control and exit from the special modes. If these special modes are used at all then it is required that the basic test mode must be implemented.

The special modes are initiated by the Write Enable and CAS before RAS clock sequence shown in the example in Figure 14.8. When this clock sequence is generated, the state of the eight low order row address bits define the mode to be selected. The mode is latched and remains in effect until the device is returned to its normal operational state by the application of any normal refresh cycle.

The special test mode performs a type of address space compression in which data that are written into the memory will be written into multiple locations. When the data are recovered they include the same set of data bits on the parallel data bus. The states of the bits are compared and if they are all equal the 'Q' pin will equal '1'. If any internal data bits are not equal then 'Q' is zero. This is called the '1/0/ = ' test algorithm.

Using this sequence the test address can be reduced by a factor of from 4 to 16

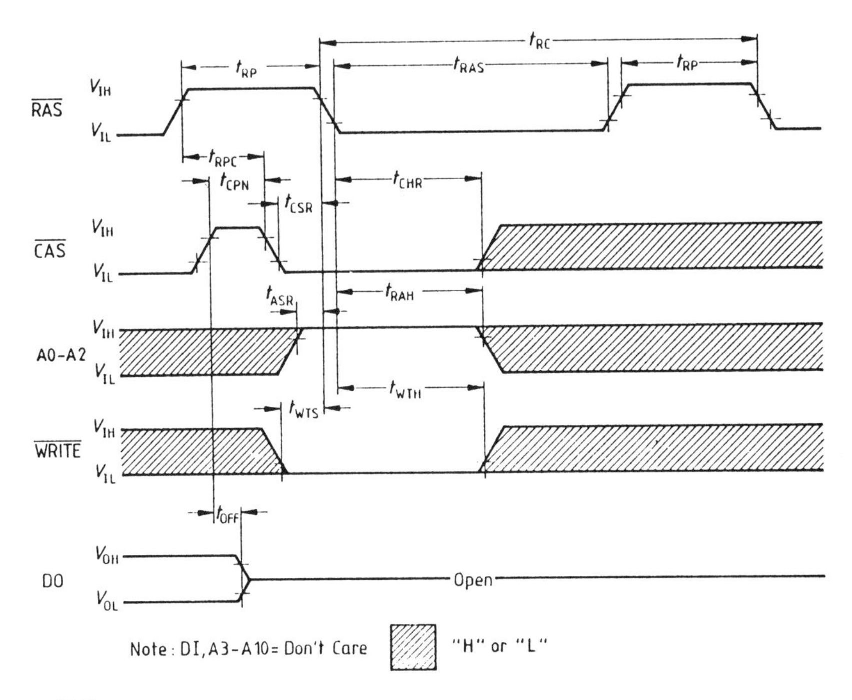

Figure 14.8
Standard test mode entry cycle for a 4Mb DRAM. (Source Siemens 4Mb DRAM Datasheet 1989.)

depending on the device architecture. This results in a test time reduction of three times over that required for a full $\times 1$ organized part.

In another example of special test modes, post redundancy repair modes were used by Siemens [45] in 1984 on their 256k DRAM. These included a 'signature' mode (developed originally by SEEQ) and a 'roll call' mode. 'Signature' test indicates whether redundancy has been used and the roll call test indicates which elements were repaired.

The 'signature' test was performed by a short RAS only cycle in which the data-out of unrepaired devices remains in the normal high impedance state and repaired devices show either a '1' or a '0' on the output.

The 'roll call' allows repaired rows and columns to be detected by their actual address and position in the cell array. A complete detection by role call involved 512 RAS-only cycles, and 512 early-write cycles. The change in address descrambling involved in substituting redundant rows and columns can also be known so that complete post repair diagnostics can be performed.

The redundancy test circuitry is shown in Figure 14.9(a). Test conditions for signature and role call mode are shown in the timing diagram in Figure 14.9(b). The sequence and test times for a complete descrambled redundancy test are shown in Figure 14.9(c).

In 1985 Mitsubishi [18] used a multibit test mode on a 1Mb DRAM which reduced test time to $\frac{1}{4}$. In this test mode the nibble decode devices were used to write four

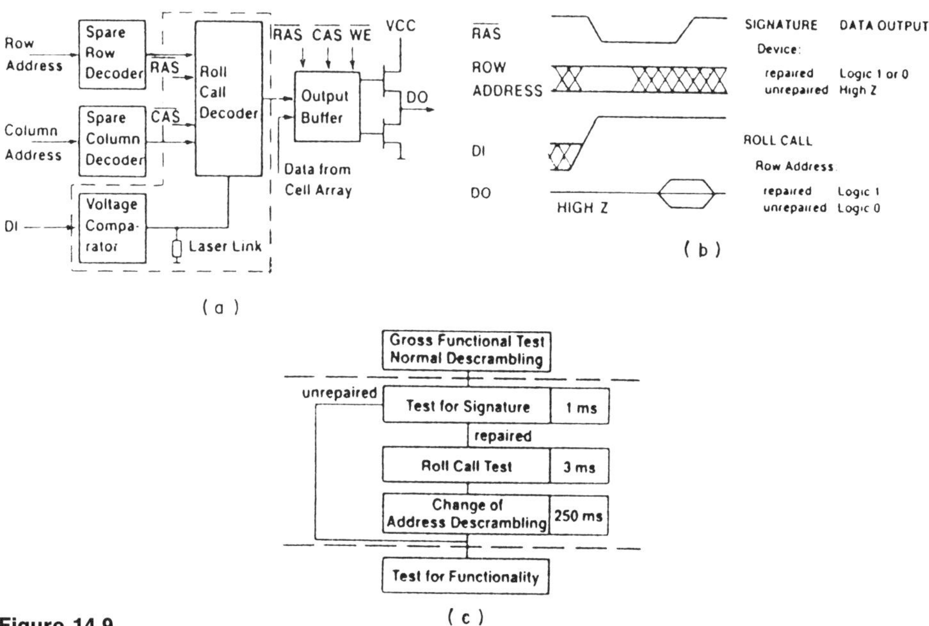

Figure 14.9
Special test mode circuitry on a typical 256k DRAM. (a) Redundancy test circuitry. (b) Timing diagrams showing test conditions for signature and role call mode. (c) Sequence and test times for a complete descrambled redundancy test. (From Rantz *et al.* [45], Siemens 1981, with permission of IEEE.)

different bits at a time and then in the read cycle the logical AND of these 4 bits is read out and compared with what was written in. What appears on the D_{out} pin depends on whether the data are the same as what were written in. This allows a 1Mb DRAM to be tested as if it were a 256k DRAM.

Test times and initial characterization for DRAMs can be shortened significantly by using parallel test modes. TI [57] in 1986 used a 16 bit parallel test mode in their 4Mb DRAM in which four arrays of sense amplifiers are activated simultaneously with each of these arrays addressing four bits at a time giving 16 bits accessible concurrently for parallel test. 16 bits are written in and then read out. If the 16 bits read out do not match with those read in, then the complement of the expected data appears on the output pin. Mitsubishi [51] on a 4Mb DRAM in 1987 used a similar 8 bit parallel test mode.

In another TI [50] paper in 1986, various design for test considerations were described for a 1Mb DRAM. The starting point was an analysis of tests that take large amounts of test time to screen a small percentage of failing product. The attempt was made to include these tests on the chip and access them by means of a test mode.

It was necessary to have a means of entering test mode that would not interfere with the functioning of the part in normal system operation. An overvoltage condition on $\overline{RAS}$ was chosen as the method of entering test mode since this ensured that the part was in a precharged and inactive state and did not add time consuming logic to the normal data access path. The individual modes were selected by the address configuration when the overvoltage was applied. The list of test functions available on this chip included the following.

0. eight bit parallel read and write
1. static refresh disturb
2. field leakage (static)
3. field leakage (dynamic)
4. external sense amplifier timing
5. redundancy roll call
6. sense amplifier margin
7. reset to memory mode
8. external oscillator to VBB pump
9. overvoltage detector test.

The JEDEC JC 42 Committee has now standardized a DRAM test mode [44].

14.3.8 Self-testing of DRAMs

A self-test generator for DRAMs was proposed in 1985 by You and Hayes [49]. After extensive fault analysis of the typical DRAM, the test sequence required to give coverage for the expected faults was created and an embedded RAM test generator designed. The structure of the conventional DRAM also needed modification to

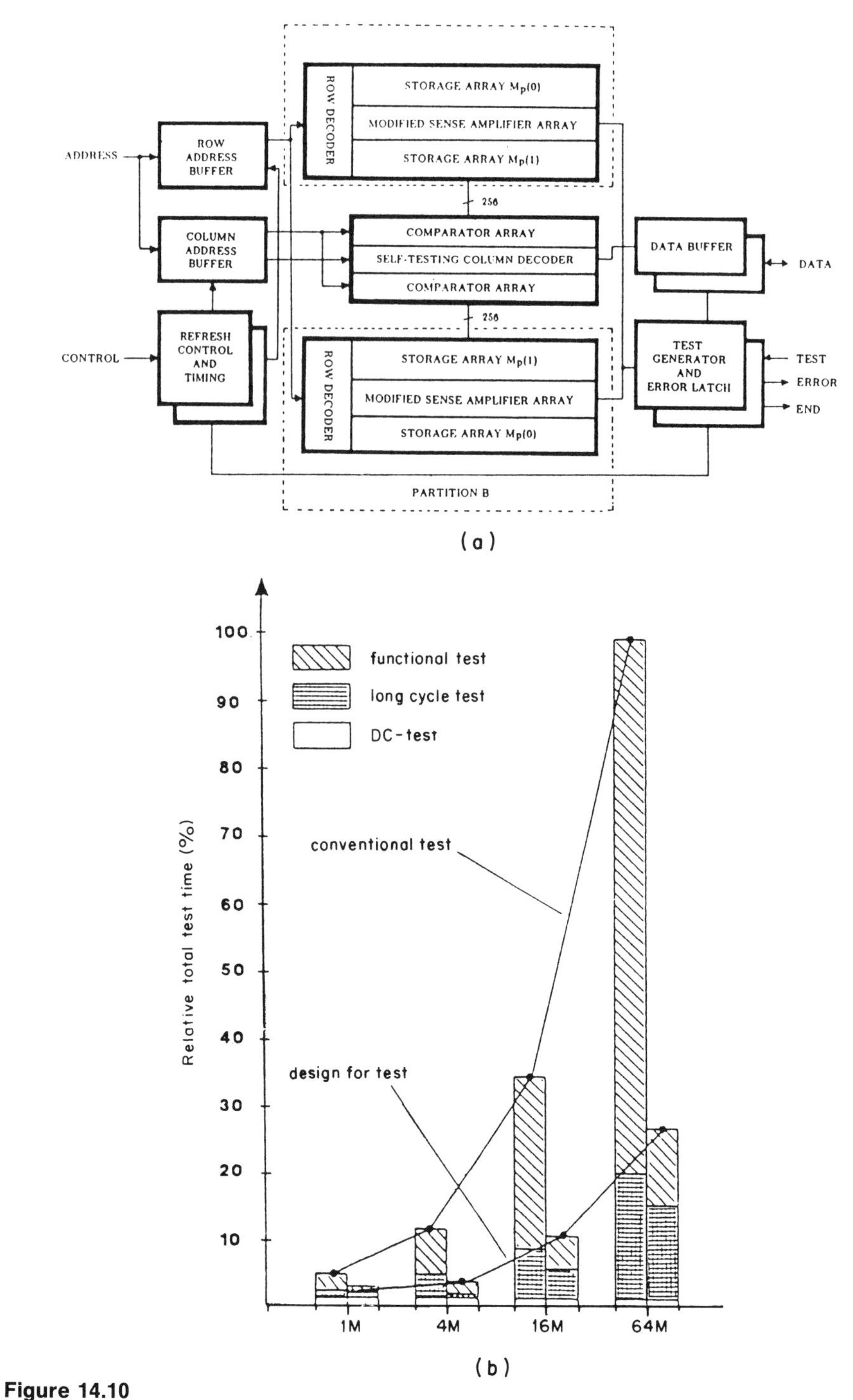

Figure 14.10
(a) Block diagram of a self testing DRAM. (b) The savings in total test time possible for different DRAM generations assuming eight bit parallel testing in the 'design for test' curve. (From You *et al.* [49], Siemens 1985, with permission of IEEE.)

implement a special shift operation required to generate neighborhood data patterns. On-chip logic was added to generate test sequences. A block diagram of the self-testing DRAM is shown in Figure 14.10(a). Test time for a 1Mb DRAM using the on-chip pattern generator was expected to be less than 1 second. Overhead in chip area for the test circuitry was expected to be 12% for a 4k DRAM and about 5% for a 1Mb DRAM. The generator used here could be extended to use in SRAMs.

Such on-chip generators are pattern generators which only check for functionality and do not check ac or dc specification limits.

A chip with both a 16 bit parallel test mode structure and a built-in self-test generator was a 4Mb DRAM shown by Toshiba [52] in 1987. The built-in test generator enables a simultaneous and automatic test of all the memory devices on a memory board by writing a test pattern provided by the on chip data generator, comparing the readout data with the expected pattern, and providing an external error flag if an error is detected. This function reduces memory board testing time. The test pattern in the self test generator checks the array functionally only and not

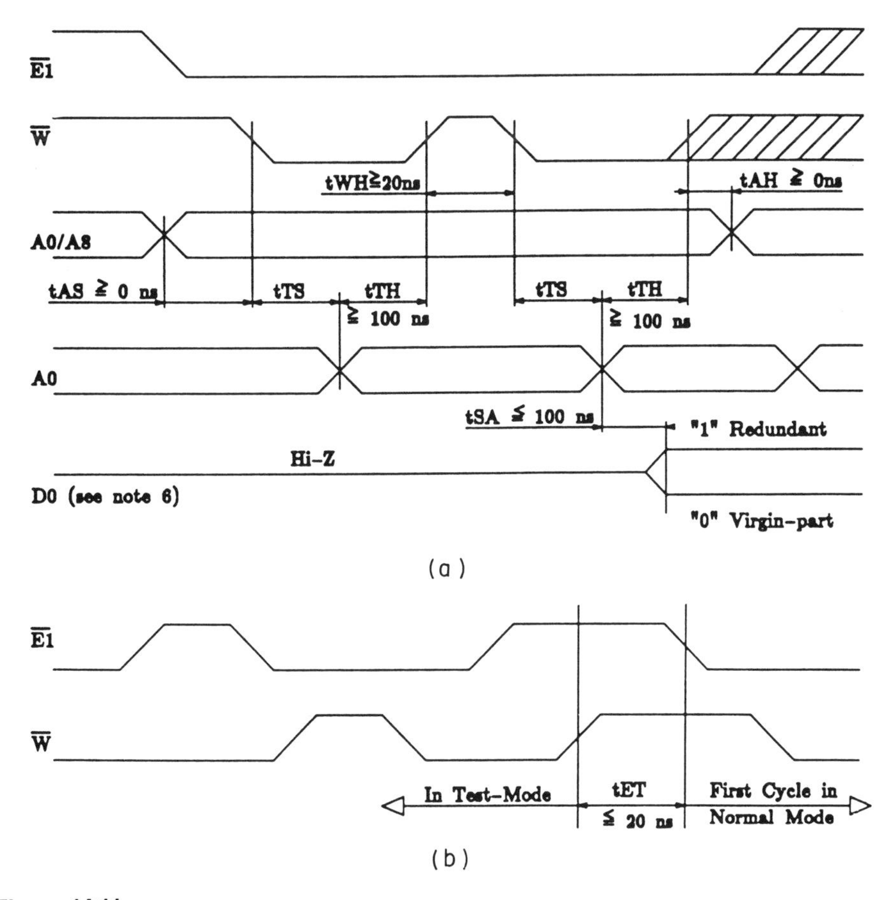

Figure 14.11
A static RAM 'test mode'. (a) Timing diagram for entering 'test mode', (b) Logic configuration for leaving 'test mode'. (Reproduced with permission of Philips Research.)

the decoders or sense amplifiers. Like all test generators it does not test ac or dc specified limits.

Figure 14.10(b) indicates the savings in total test time possible in one case for different DRAM generations assuming eight bit parallel testing in the design for test curve.

14.3.9 SRAM test modes

A test mode for SRAMs was described by Philips in 1989. This mode used a chip enable ($\overline{CE}$) low and write enable ($\overline{WE}$) low logic entry with specific tests designated by the configuration of the address pins. The logic entry mode permitted standard 5 V power supplies to be used as opposed to the standard DRAM test mode which required overvoltage supplies to enter test mode. To ensure that the system did not randomly power-up in test mode the $\overline{WE}$ pin was clocked twice before the test mode was entered as shown in Figure 14.11(a). To leave test mode $\overline{WE}$ and $\overline{CE}$ needed only to be high at the same time as shown in Figure 14.11(b). In the example shown, the result which appeared on the output pin in test mode was an indication of whether the redundancy on the chip had been used. Other test modes such as a bypass for the voltage regulator for stress testing were also implemented.

14.4 EPROMs

The EPROM technology differs from the standard MOS structure used in DRAMs due to the use of the floating gate structure and the requirement for UV erase which introduced new and complex failure mechanisms specific to EPROMs. These need to be considered during component testing.

14.4.1 Common problems and failure modes in EPROMs

Some of the most common problems are briefly described below.

- Underprogramming: this occurs when the gate charge is in an unspecified region and may introduce an unwanted data bit.

- Light sensitivity or spurious erasure: light of lower frequencies than UV light may affect the content of an EPROM.

- Electron trapping: a potential build up of charge in the presence of imperfect oxide may cause problems during sensing and programming.

- Charge loss: this could cause reliability problems related to data retention.

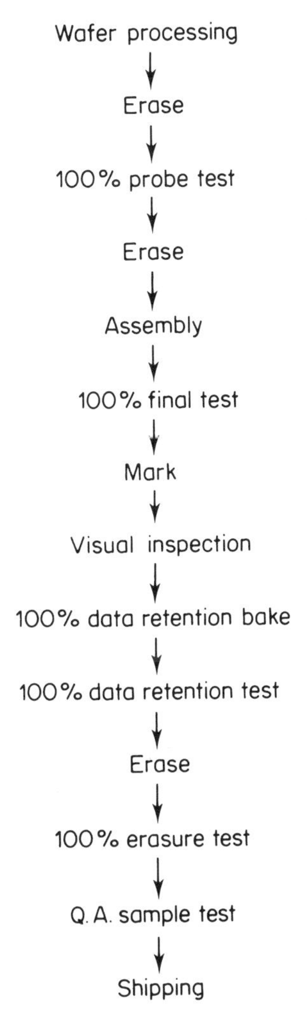

Figure 14.12
Typical EPROM test flow. (From Prince and Due-Gundersen [64] 1983.)

Figure 14.12 shows a typical production test flow for EPROMs. It should be noted that data retention bake, data-retention tests, and erasure tests have been added to those tests which were shown in Figure 14.4 which illustrated a similar test for dynamic RAMs. These screens are all related to the specific EPROM failure modes indicated above.

Erasure of the complete memory are performed after processing, probe, and data retention test. This establishes a fixed reference level during testing and guarantees that parts to be shipped are 'clear' and ready for customer programming.

Special test patterns have been developed for EPROMs since the normal read and write test patterns took too long due to the long UV erase cycle of the EPROM [18].

14.4.2 Programming of EPROMs

EPROMs are programmed by the injection of high energy electrons into the floating storage gate [6]. This injection is accomplished by raising the voltage on the select gate and drain to attract the high energy electrons from the substrate through the thin gate oxide region and onto the dielectrically isolated storage gate where they are trapped. The trapped charge causes the cell threshold voltage to shift upward and keeps the cell turned off.

Since a programmed cell is off, it is defined to be at logic 'zero'. The programming state can be determined by sensing the threshold voltage of the cell. A schematic diagram of a floating gate n-channel EPROM memory cell showing the select gate and floating gate is shown in Figure 14.13(a). The threshold voltages in the programmed and unprogrammed state are shown in Figure 14.13(b) which plots the current through the cell transistor as a function of select gate voltage.

Unfortunately if there is any ion surface charge contamination the threshold of the cell will tend to drift after extended operation due to movements of the ions under the effects of the electric fields during operation.

14.4.3 UV erase

If an EPROM is exposed to ordinary light sources such as sunlight or fluorescent light over a period of time it will tend to deprogram since electrons in the conduction and valence energy bands of the programmed storage gate will tend to absorb photons and photo-emit from the storage gate into the surrounding oxide. There they are swept away into the substrate or select gate depending on the orientation of the internal electric field. Frequently an opaque label is placed on the window of the EPROM to block out the light.

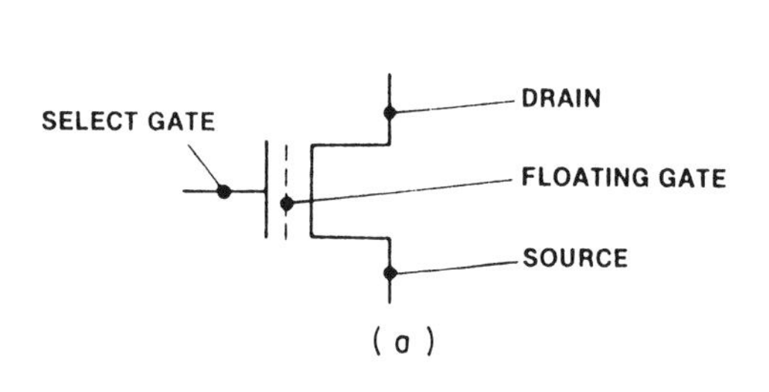

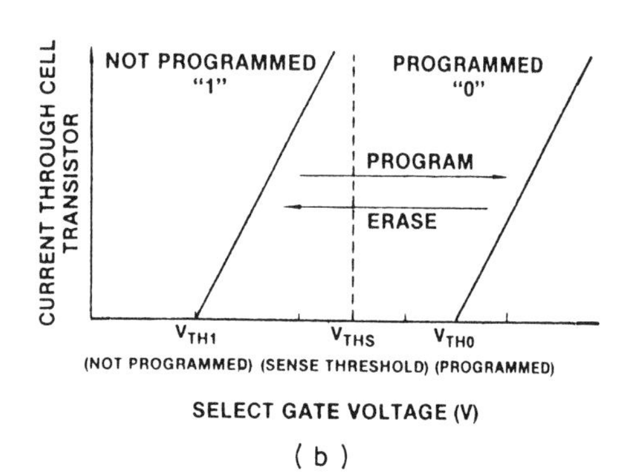

Figure 14.13
(a) Schematic diagram of a floating gate n-channel EPROM memory cell. (b) Threshold voltages in the programmed and unprogrammed state. (Reproduced with permission of Fujitsu.)

The entire array must be erased before reprogramming is attempted. A light dosage of 15 W s cm^{-2} is required to entirely erase an EPROM array. A commonly used source of high energy photons is a mercury vapour ultraviolet lamp which emits UV light with a wavelength of 253.7 nm. This is equivalent to 4.9 eV which if placed one inch away from the array will erase it in about 15 to 20 minutes. Light sources with lower intensity will require correspondingly longer times to accomplish complete erase.

14.4.4 Test time for an EPROM

The time required to do full electrical test on an EPROM is substantial, since the nominal programming time per byte of an EPROM is long. Early 64k EPROMs took as long as 50 ms per byte to program. This translates to about 400 ms per device. The 1Mb EPROMs brought this down by a factor of 100 to 0.5 ms per byte or 64 ms per device.

Since a normal test program needs a sequence of tests to check for the various types of failure modes, each device must be reprogrammed several times in the course of a single test sequence. Each erasure of an EPROM by UV light takes from 10 to 20 minutes and must be done between programming cycles. The result is that the entire test time for an EPROM is significant. It is thus a major factor in overall manufacturing costs.

A parallel programming mode is usually designed into an EPROM which allows several bytes to be programmed in parallel in order to reduce the time required to fully program the device. These are described in more detail in a later section.

14.4.5 Failure mechanisms for an EPROM

Failure mechanisms common in EPROMs include ionic contamination, polarization, microcracks, silicon defects, oxide defects, and oxide breakdown.

Ionic contamination, which is associated with a 1.0 to 1.4 eV activation energy, appears to start on an edge bit. It then disperses throughout the array affecting more bits. Ionic impurities which have penetrated beneath the protective passivation layers diffuse along the storage gate connections. These impurities compensate for the electrons in the gates resulting in a net loss in charge and therefore a drop in threshold voltage.

Oxide defects, which are associated with an 0.3 eV activation energy appear to affect random bits resulting in either a positive or negative threshold shift. This failure is frequently associated with manufacturing defects and appears during the infant mortality stage of the device life.

Silicon defects are associated with an 0.3 eV activation energy and are similar in occurrence to the oxide defects but are characterized by only negative threshold shifts.

Thermochemical reactions primarily during high temperature storage or data retention bakes and are characterized by a loss of charge. This results from thermal emission from the storage gate.

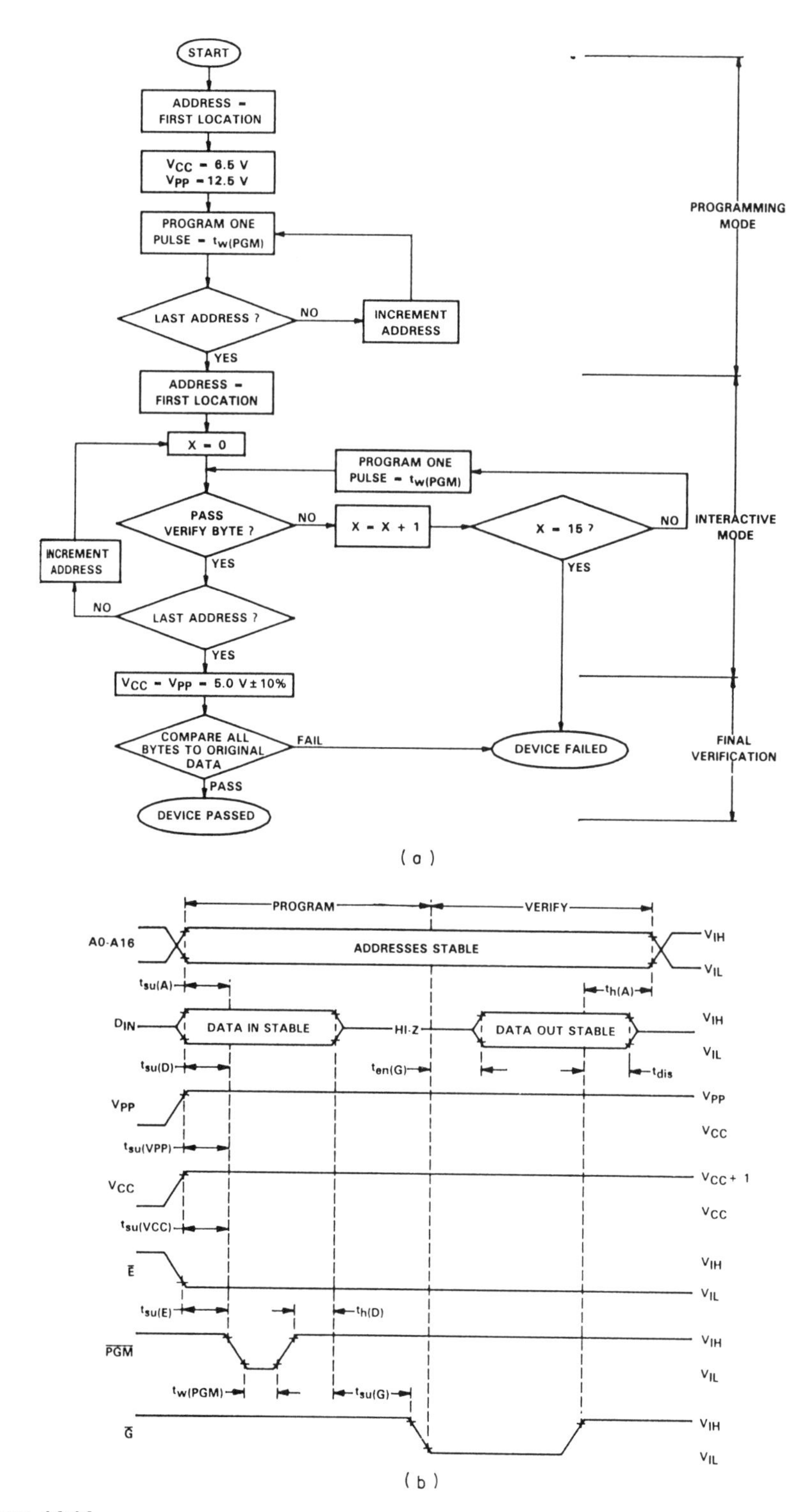

Figure 14.14
Byte program algorithm. (a) Flow chart. (b) Timing diagram showing program and verify modes. (Source Texas Instruments MOS Memory Databook 1989.)

Intrinsic wear-out is associated with activation energies of 1.4 eV or higher and is associated with electrons that during programming become trapped in the gate oxide layer without reaching the floating gate. These tend to be cumulative over repeated programming eventually causing the wear-out phenomena to occur.

14.5 STANDARD EPROM PROGRAMMING METHODS

14.5.1 Single bit programming

Single bit programming was the standard method of programming in the lower density devices in which an EPROM bit could be programmed by a 50 ms pulse technique. The programming mode is entered when an over-voltage, specified by the manufacturer, is applied to the V_{PP} pin. Chip select ($\overline{E}$) remains at V_{IL} throughout programming. The Program pin $\overline{P}$ is initially at V_{IL}. The address to be programmed is applied to the proper address pins. Bit patterns are placed on the outputs. TTL voltage levels are maintained. When both the address and data are stable, a 50 ms TTL low level pulse is applied to the $\overline{P}$ input to program the programming. This procedure can be done either manually address by address, or automatically on a test machine.

14.5.2 Byte programming algorithm

To shorten test time on higher density byte-wide EPROMs most manufacturers also made available an 8 bit fast programming algorithm using a 1 ms pulse which significantly improved test time and has become standard in the industry at the 256k and 1Mb densities and above. Figure 14.14(a) shows a flow chart for the byte program algorithm.

This algorithm requires $V_{CC} = 6$ V and $V_{PP} = 12$ V during byte programming. A sequence of 1 ms pulses is applied. A byte is verified after each pulse. In addition when the algorithm cycle has been completed, all bytes must be read at $V_{CC} = V_{PP} = +5$ V. The timing diagram for this algorithm is shown in Figure 14.14(b).

14.5.3 16 bit programming algorithm

At the 1Mb density level 16 bit wide (word-wide) EPROMs were used and a 16 bit programming mode analogous to the 8 bit programming mode on the byte-wide devices was used. In this mode one 16 bit word at a time is latched into the EPROM and programmed.

14.5.4 32 bit page programming algorithm

At the 1Mb and 4Mb EPROM level a 32 bit page programming algorithm has come into standard usage. For the 128k × 8 organized type, the device is programmed in

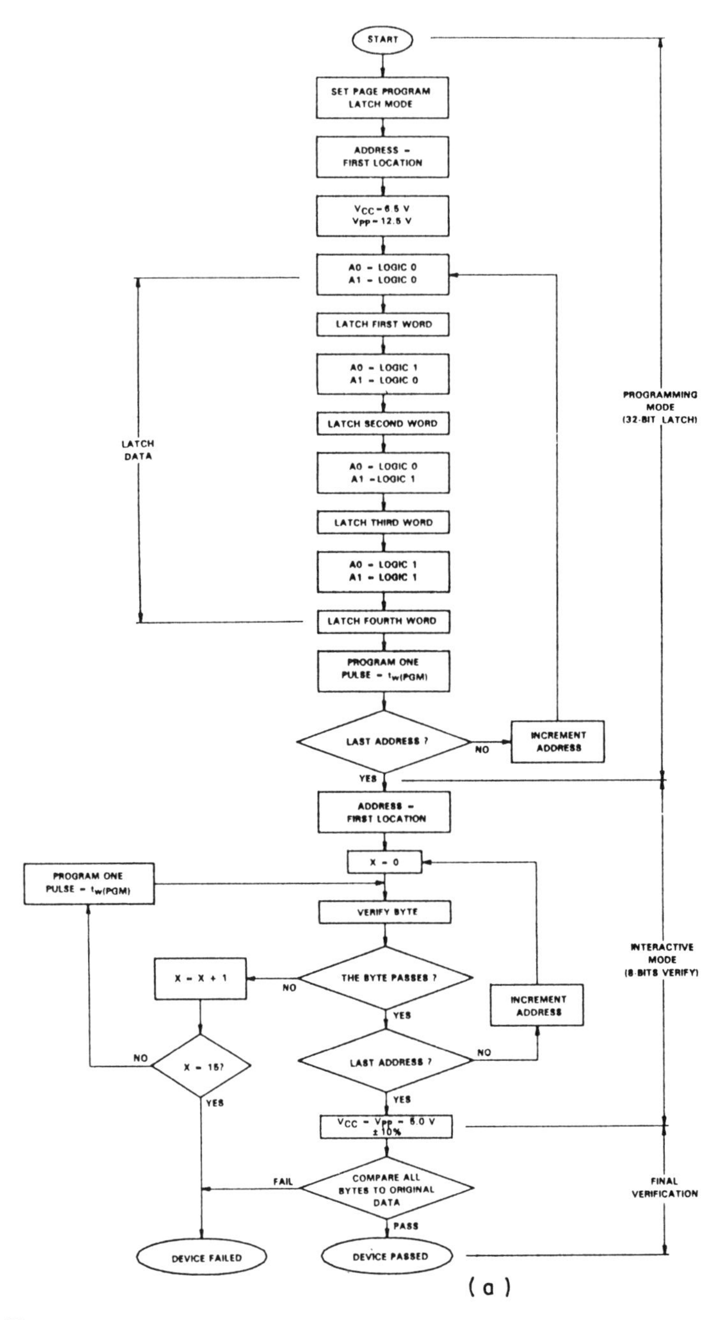

Figure 14.15

128k × 8 EPROM 32 bit programming algorithm. (a) Flow chart of typical programming sequence for this algorithm. (b) A timing diagram for the progam cycle timing. (Source Texas Instruments MOS Memory Databook [63] 1989.)

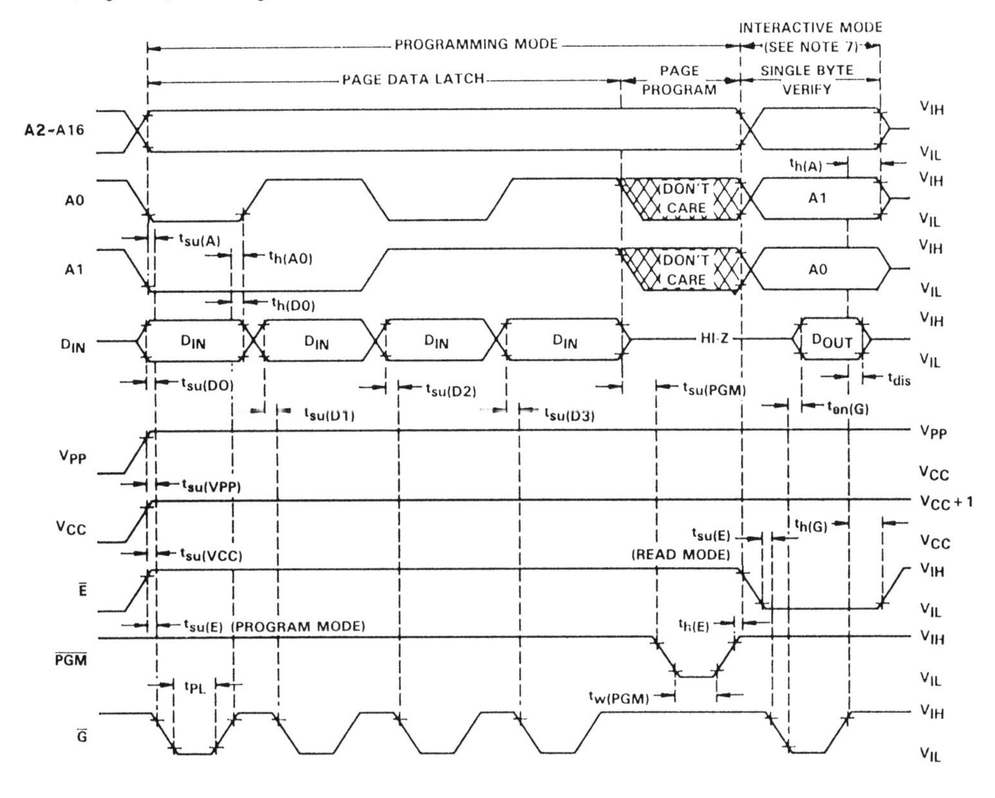

NOTE 7: 8-bit programming is applied in Interactive Mode when needed.

(b)

Figure 14.15 (*continued*)

32 bit blocks by latching four successive bytes. A typical programming sequence for this algorithm is shown in the flow chart in Figure 14.15(a).

The initial setup is $V_{PP} = 12.5$ V and $V_{CC} = 6.5$ V. $\overline{E} = \overline{G} = \overline{PGM} = V_{IH}$. The first location is defined as any address when $A_0 = A_1 =$ logic 0 and the first-order data are the corresponding eight bits of parallel data. Subsequently the next three bytes of data are defined when $A_0 =$ logic 1 and $A_1 =$ logic 0 and so on.

Starting at the first location the four bytes of data are presented in sequence to the data pins. $\overline{G}$ is then pulsed low and the byte is latched on the rising edge of $\overline{G}$. The 32 bit parallel programming takes place when $\overline{PGM}$ is pulsed low with a pulse duration of T_w. The sequence is then repeated up to a maximum of 15 times. When the part is fully programmed all bytes are verified with $V_{CC} = V_{PP} = 5.0$ V.

A timing diagram for the 128k × 8 EPROM for the program cycle timing for 32 bit program is shown in Figure 14.15(b). The programming mode in which the page data is latched and the page programmed is shown followed by the interactive mode in which single bytes are verified.

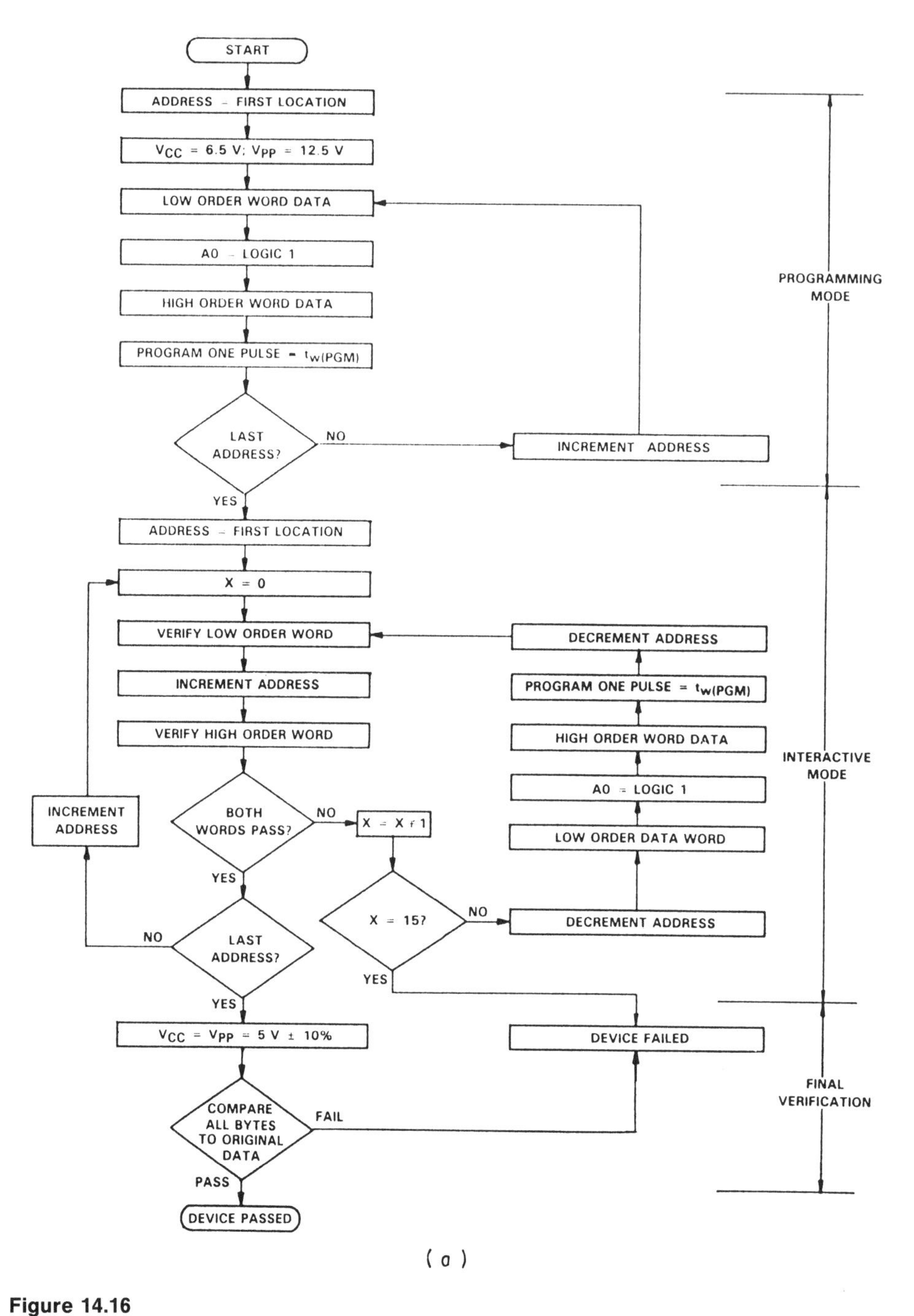

Figure 14.16
32 bit programming for a 64k × 16. (a) Flow chart showing low and high order word selection. (b) Timing diagram. (Source Texas Instruments MOS Memory Databook [63] 1989.)

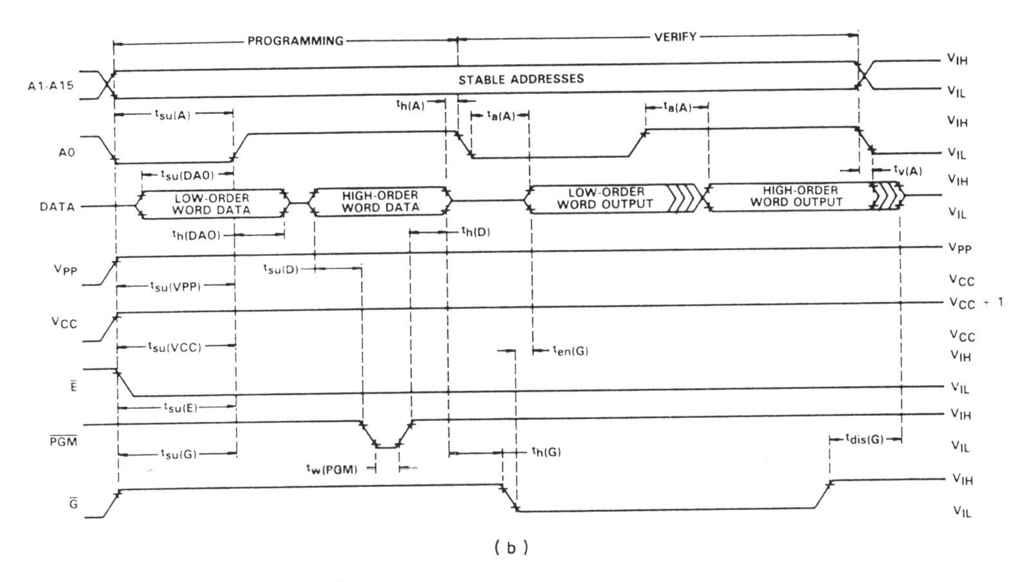

(b)

Figure 14.16 (*continued*)

A similar method can be used for 32 bit programming of a 1Mb EPROM which is organized 64k × 16 as shown in the test flow chart in Figure 14.16(a) and the timing diagram in Figure 14.16(b). In this case two 16 bit words are addressed. The low order word data are addressed when A_0 = logic 0, and the high order word data when A_0 = logic 1. These two 16 bit words are latched into the data pins sequentially and when the data is stable the 32 bit parallel programming takes place by pulsing $\overline{PGM}$ low. Every location is programmed once before going into the interactive mode.

The interactive mode in this case consists of a sequence of verification and programming in which two 16 bit words corresponding to low and high order word data are verified and reprogrammed if not correct. As before a final verification at 5 V is performed to check all bits.

14.5.5 EPROM on-chip test modes

EPROMs with special test modes on chip have also been designed. An example is a 1Mb CMOS EPROM from TI which included the following.

1. Word-line stress to test the interpoly dielectric integrity.
2. Bit-line stress to test the first gate oxide dielectric.
3. BVDSS measurement on addressed array columns.
4. Column shorts to test for source–drain shorts and leakage.

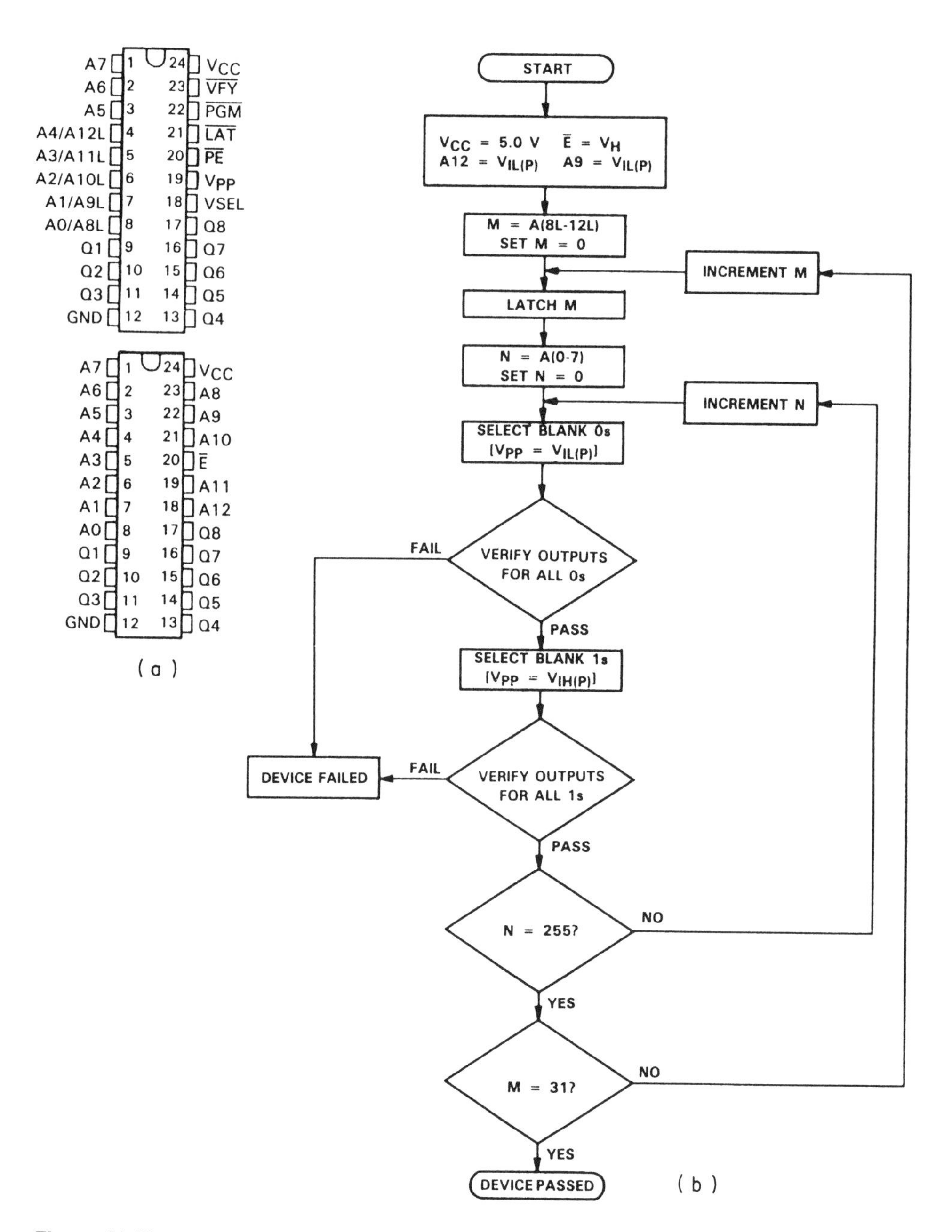

Figure 14.17

Blank check mode for checking erasure of differential cells in a 64k OTP EPROM. (a) Comparison of read pinout and blank check–program pinouts. (b) Blank check mode flow chart. (Source Texas Instruments MOS Memory Databook [63] 1989.)

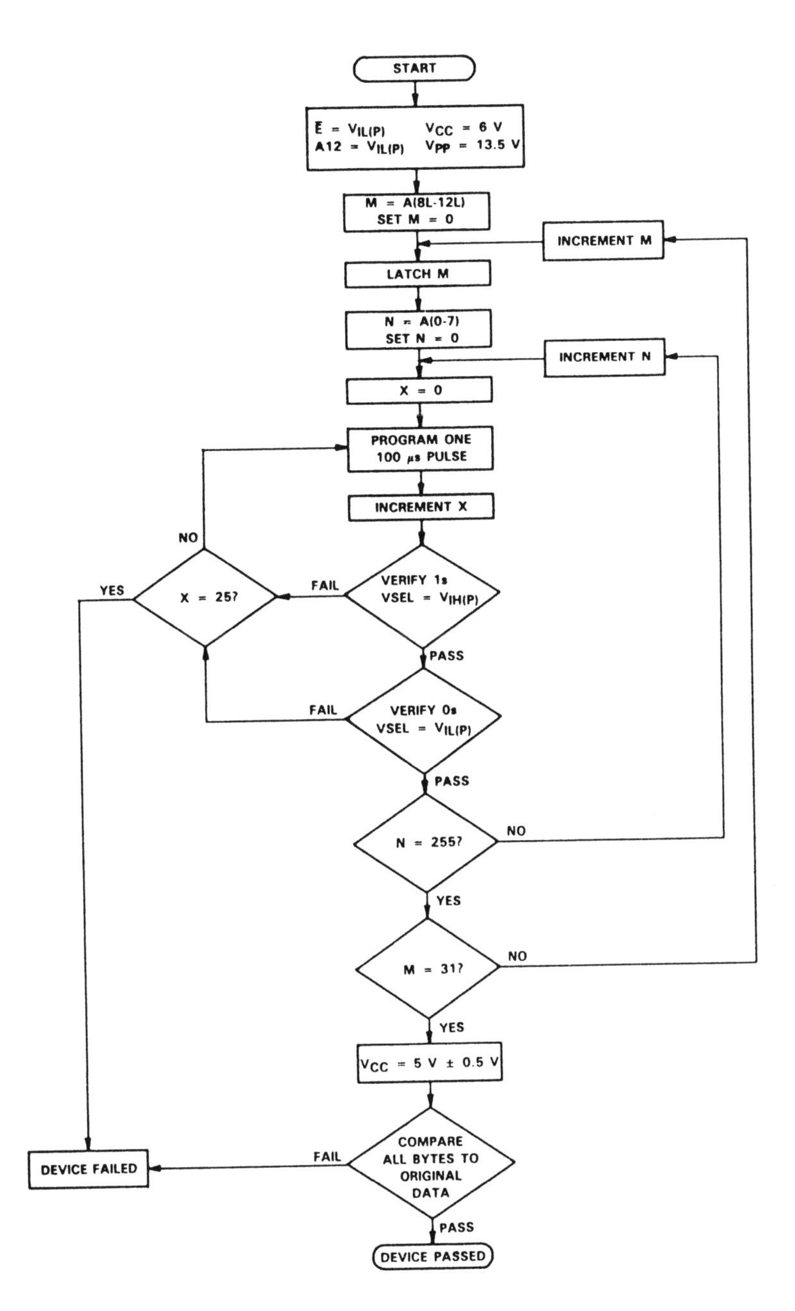

Figure 14.18
Flowchart for programming '1's and '0's in the differential cells of a fast OTP 64k EPROM. (Source Texas Instruments MOS Memory Databook [63] 1989.)

5. Program VT to measure the program threshold on addressed bits.

6. Programming mode to program 64 bits simultaneously to reduce test time.

7. A PROM row which is an extra row that is included for OTP testing.

Another example is a Toshiba [56] 256k CMOS EPROM shown in 1984 which included on-chip test circuits which were activated by setting address or control pins to a voltage greater than V_{CC}. These circuits included the following.

1. A data retention test circuit to raise all word-lines to V_{CC} and stress the control gates of all memory cells at the same time. This reduced the test time of bias temperature testing by 1/512 on this chip.

2. Stress test circuits which raised all word-lines to the programming voltage. They also could raise all bit-lines with all word-lines remaining at ground level. This reduced the test time for gate oxide leakage current in unselected cells during programming.

3. A threshold voltage shift detector circuit which provided the exact threshold voltage shift during the test routine (added in 1985 on a second 256k EPROM [57]).

14.6 OTP TEST CONSIDERATIONS

Testing one-time-programmable EPROMs is complicated by the fact that a UV EPROM in a plastic windowless package can not be erased after programming. This precludes direct testing of assembled parts by programming the cells.

Several methods are in use to overcome this limitation. The simplest is for a memory supplier to use OTPs as fast turn-around ROMs and program them to the customer's specification. After they are programmed they can, of course, be tested before shipment.

A second is the technique mentioned on the TI part in the last section—to design an extra row on the chip that can be used for final testing after assembly. This technique can check on ac and dc parameters and check out peripheral circuitry to some extent. The limitation is clearly that the memory array is not tested. If it is assumed that defects in the array were screened out at wafer probe then this method can provide some degree of certainty.

For one-time-programmable fast EPROMs that use the two-transistor or four-transistor cell differential sensing structure it is easier to check that all cells are in the unprogrammed state since one of the transistors in the cell must be programmed for data to be read. If both are programmed or both are erased then an undefined condition appears on the outputs.

The differential cell EPROM differs also in that both the '1's and the '0's must be programmed.

An illustration of erasure check and programming is given in Figures 14.17 and 14.18 for a 64k fast OTP EPROM from TI. The altered pinout and flowchart for the 'blank check' mode of checking erasure of the cells is shown in Figure 14.17. Since an unprogrammed device has ambiguous state in all address locations, prior to programming the blank check mode can be used to verify that both sides of the differential cell are erased.

The pinout and flow chart for programming mode is shown in Figure 14.18. In the programming mode A_{12} becomes V_{SEL}, A_{11} becomes V_{PP}, chip enable ($\overline{E}$) becomes program enable ($\overline{PE}$), A_{10} becomes latch input ($\overline{LAT}$), A_9 becomes program input ($\overline{PGM}$) and A_8 becomes verify ($\overline{VFY}$) input.

14.7 STATISTICAL TEST ANALYSIS

Although testing of the memory devices to ensure performance to the electrical specification is critical, no test is 100% perfect. Real world test equipment has a finite probability of error, just as handling of devices in process can cause defects even after final test. For this reason a sample check of outgoing quality levels is usually done prior to shipment to provide actual statistical outgoing quality measurements.

14.7.1 Average outgoing quality (AOQ)

This refers to the number of devices per million that are outside specification limits at the time of shipment of the product. It is calculated as follows. AOQ is measured in units of ppm (parts per million)

$$\text{AOQ} = \text{Process Average} \times \text{Lot Acceptance Rate}$$

where

$$\text{Process average} = \text{Fraction Estimated Reject Devices} \times \text{Lot size/Total number of devices}$$

$$\text{Fraction Estimated Reject Devices} = \text{Defects in Sample–Sample size}$$

$$\text{Total number of devices} = \text{sum of all the units in each lot.}$$

$$\text{Lot Acceptance Rate} = 1 - \frac{\text{Number of lots rejected}}{\text{Number of lots tested.}}$$

14.7.2 Standard deviation from the mean (sigma)

Any manufacturing process experiences some variation between the ideal and the actual from time to time and unit to unit. It is usually assumed that these variations follow a Gaussian (or Poisson) distribution. Variation of the process is measured in standard deviations from the mean.

One measure of the performance of a product is determined by the amount of margin that exists between the process that the product was designed for and the actual process in which it is made. Within any real world manufacturing process some variation occurs.

The variation of the process is usually measured in 'standard deviation' (sigma) from the mean. Normal variation, defined as process width, is $\pm 3\sigma$ about the mean. This means that 97.3% of all parts (or steps in making a part) are within the expected distribution (or alternately that 2700 parts per million will fall outside the normal distribution variation).

In the past when integrated circuit process were much simpler it was sufficient for designs to be targeted to perform at least $\pm 3\sigma$ from the mean. See Figure 14.19(a)

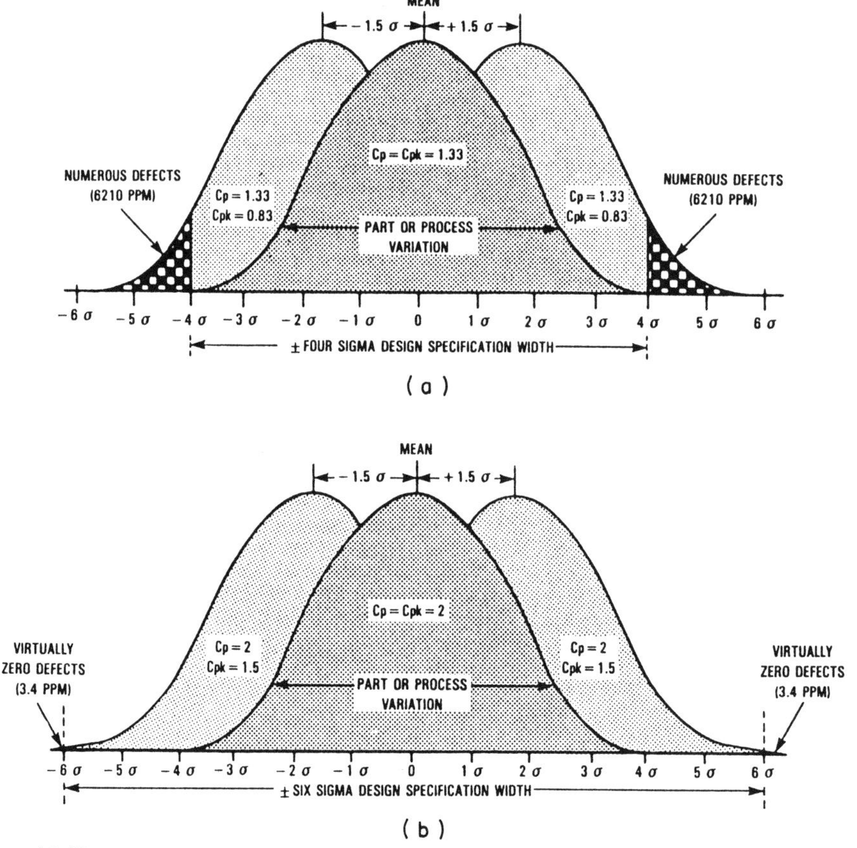

Figure 14.19
(a) Poisson distribution showing process shift of $\pm 1.5\sigma$ together with a design specification tolerance width of $\pm 4\sigma$. (b) The same process variance as (a) together with a design tolerance of $\pm$ 6σ resulting in virtually zero defects. (Reproduced with permission of Motorola.)

for an example of a process tolerance of 3σ ($\pm 1.5\sigma$) together with a design variation tolerance of $\pm 4\sigma$ resulting in significant probability of defects.

An illustrative example follows from the Motorola *1989 Memory Databook* for a complex process. Assume we are building a product containing 1200 parts (or process steps). We can expect 3.24 defects per unit (1200×0.0027) on the average. This would result in a cumulative yield of less than 4%. Clearly in the modern high density memory area we need to have better than the traditional $\pm 4\sigma$ tolerance in memory design.

Figure 14.19(b) shows the same process variance together with a design tolerance of $\pm 6\sigma$ resulting in virtually zero defects.

14.8 RELIABILITY AND QUALITY

The requirement for memory manufacturers to ship highly reliable product has resulted in an overall philosophy of reliability and quality assurance at most memory manufacturers. Care is taken to design reliability in during the product development stage.

Incoming inspection of all parts and materials is done normally along with an increasing trend to working closely with equipment vendors and material suppliers to assure uniform and timely supply.

In-process controls assure uniform processes resulting in uniform quality and reliability in volume production with quality built into all steps in the manufacturing process. Meanwhile sample inspection is done regularly at frequent process steps to insure that no variation in the procedure has occurred. Finally feedback from stress testing and electrical testing is fed back into the manufacturing process on a regular basis.

Failure analysis is also done to determine the underlying cause of failures so that process control monitors can prevent that failure mechanism from reoccurring.

Close communication and relationships between manufacturers and users have been established. The need for correlating the results of outgoing quality results with incoming inspection criteria is the first step along this direction. When a user's incoming quality levels are used as a reference for manufacturer's outgoing quality test, a lot of work can be reduced on both sides. Reliability and quality feedback from the user to the memory manufacturer is an important link.

The concept of doing it right the first time has been generally accepted in modern VLSI memory production. Rework and low yields are not acceptable in this competitive situation.

Design, layout, and processing rules are being modified to cope with the new concepts. Wider margins and 'safer' tolerances have been implemented, as have closer reliability and quality monitoring of the complete production process.

Industry response to the quality challenge has been such that quality has become a way of life in the semiconductor business. A complete quality assurance system is viewed as incorporating the entire process and product development cycle and quality and reliability control of mass production.

The latter includes quality control of materials, control of the manufacturing environment, control of manufacturing equipment and measuring devices, process quality control, product quality control including product inspections, reliability assurance, on-time-delivery, and customer service.

14.9 RELIABILITY CONSIDERATIONS

In the highly competitive market surrounding memories and semiconductors in general every parameter which can be affected or improved by the manufacturers has to be fully optimized. This includes price, delivery performance, product range, and last, but definitely not least, reliability and quality.

As complexity levels of VLSI components increase and the reliability performance of high technology and equipment is prioritized, attention is focused on the performance of each individual semiconductor component.

Good memory reliability is an obvious criterion considering the high volumes normally required in most applications.

14.9.1 General reliability

Most manufacturers are committed to shipping highly reliable products with reliability results well within specified levels. To ensure these standards, on-going reliability testing is normally performed both in order to monitor device performance and to accumulate statistical data base information.

Samples from finished goods are subjected to a series of tests such as life tests and others. The result of these tests provide the basis for production decisions and the generation of product reliability of this system; however, the test methods and conditions by which the data are obtained, and the proper interpretation of these data, will be essential in projecting the overall reliability of a system with a high degree of confidence.

The failure rate for semiconductor devices normally follows a predictable curve called the bathtub curve which is shown in Figure 14.20. This particular curve can be split into three different regions for reliability analysis.

1. *Infant mortality* A region with 'Early Life Failures' showing an initial high value. The failures are normally related to manufacturing defects.

2. *Normal life* This region shows a relatively stable performance representing the useful time period of device life. A low level of random failures will usually occur in this region.

3. *Wear out* After the normal life period the failure rate will increase rapidly due to wear out.

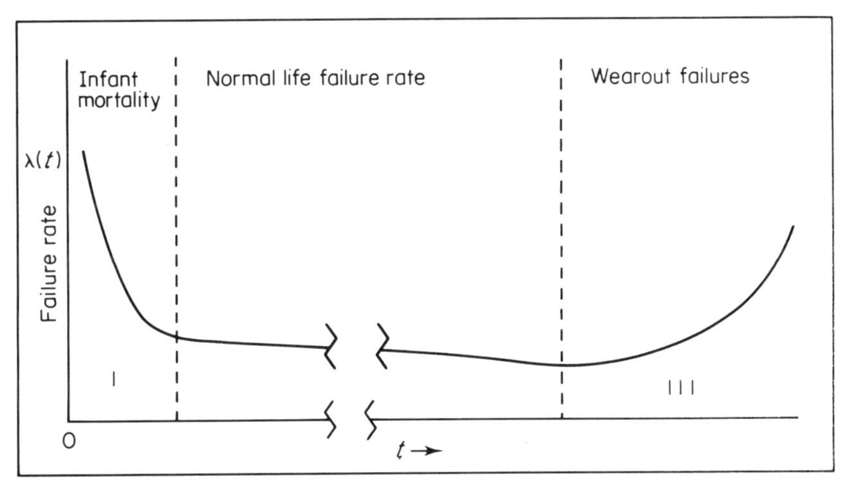

Figure 14.20
Empirical bathtub curve used to represent typical time of failure of memory devices including the early infant mortality stage and the final wear-out stage. (From Prince and Due-Gundersen [64] 1983.)

As an indication in operation the infant mortality period would typically be a few weeks while the stable or normal life period could be up to 25 years or so. Historically most memory vendors have subjected their parts to a short period of accelerated stress testing prior to shipment in an effort to eliminate the infant mortality stage. There is a serious effort also with many vendors to eliminate the infant mortality phenomena at final test by some form of stress testing at wafer probe. Most semiconductor parts will have been replaced with new designs and new technologies before the wear out stage is reached.

In general manufacturers have programs for MOS reliability in evaluation of new products, design changes, and process changes. Evaluation of new packages, vendors, and packaging techniques is equally important. Even after the introduction of a new product or package, on-going monitoring of reliability and quality normally continues throughout the production lifecycle.

14.9.2 Reliability tests and test conditions

A number of stress tests have been developed to accelerate the effects of various failure mechanisms. By subjecting a device to extreme operating conditions for a shorter time, one can with some specified confidence level predict life device performance under more normal conditions. Some of the most common tests are listed below.

(a) *Dynamic operating test* The device is operated at high temperature and/or a power supply voltage above 5 V. The objective is to accelerate the effects of potential defects, such as contamination, mechanical stress, circuit sensitivity, and normal MOS failure mechanisms by measuring the resistance to thermal and electrical stress during operation. Typical test conditions for an accelerated operating test might be $T_a = 25°C$, $V_{CC} = 0.5$ V.

(b) *High temperature bias* This test measures the effect of thermal stress under static bias conditions. Typical test conditions might be $T_a = 150°C$ and $V_{CC} = 7.0$ V. The ambient temperature can be higher in this case without damaging the part since the device is not generating additional heat from operation.

(c) *Temperature cycle* The device is subjected to changing temperatures (such as −65 to 25°C to 150°C, air to air) in the time span of a few minutes. The net result is to accelerate the effects of thermal expansion mismatch as they apply particularly to wire bonds and pattern sensitive die stress.

(d) *Thermal shock* This is a more severe version of temperature cycling. Liquid is used rather than air as described above.

(e) *Mass spectrometer* This technique is used to determine moisture content of the package cavity and distribution of other gases. It is also called Residual Gas Analysis (RGA) and is only applicable to cavities of ceramic packages.

(f) *Constant acceleration* The device is subjected to, say, 30 kg, so that internal connections are effectively tested.

(g) *Static damage* By applying increasing charges to the inputs one can evaluate the effectiveness of input protection circuits.

(h) *Temperature humidity bias test (THB)* The object of this test is to determine the performance of the device under extreme environmental conditions such as 85% relative humidity and 85°C with an applied bias. This type of test is industry standard for plastic packages and is used to accelerate electrolytic moisture induced failure mechanisms, such as corrosion.

(i) *Autoclave (or pressure cooker test)* Applies high pressure in addition to high humidity and high temperature but without bias to accelerate simple chemical corrosion failure mechanisms. Test conditions are, for example, 100% humidity, two atmospheres of pressure, and a temperature of 121°C. Biased pressure cooker test (HAST) is also used as an accelerated version of the THB test.

Other tests related to specific product groups such as data retention test for EPROMs and soft error testing for dynamic RAMs and static RAMs are also performed.

14.9.3 Failure rate calculations

As previously mentioned, the accelerated lifetest will give good indications of the long life performance of a device type. By extreme operating conditions during a short time period one is, however, only simulating long term 'standard environment' performance. Since the value of lifetest data is dependent on many variables and the interpretation and manipulation of the involved parameters can produce a wide range of results, it is imperative that one has an adequate understanding of how they contribute to the reported failure rate.

The most important variables include

- temperature
- voltage
- biasing technique
- failure criteria
- acceleration factor
- confidence limit

Several mathematical models are used in calculating the failure rate obtained by lifetesting.

The statistical failure rate is determined by the chi square distribution

$$\lambda \leqslant \frac{\chi^2(\alpha; 2r + 2)}{2t}$$

where λ is the failure rate, χ the chi square function, $\alpha = 1 -$ (confidence limit/100), r the number of failures and t the number of equivalent device hours.

The term $2r + 2$ is normally called the degrees of freedom. The application of confidence levels is a statement of how 'confident' the population in general is (e.g. 60%). χ in the chi square function can be selected from a chi square distribution table with the appropriate degrees of freedom and confidence levels. The number of equivalent device hours, t, means the transformation of actual device hours (at say 125°C) used, which is the normal maximum specified temperature for a standard memory, e.g. 151 000 hours at 125°C is equivalent to 9.2 million hours at 70°C. Equivalent device hours can be calculated by $t = F_a(t)$ where F_a is the acceleration factor.

F_a equates test temperature with operating temperature and is derived for the Arrhenius relationship

$$F_a = \exp[(\phi/K)(1/T_o - 1/T_\tau)]$$

where ϕ is the activation energy. This is normally a value in the 0.8 to 1.0 range and is related to the specific failures expected. K is 8.62×10^{-5} (eV/K^{-1}), T_o the operating temperature in K and T_t the test temperature in K.

T_0 and T_t are junction temperatures rather than ambient temperatures. The transformation from ambient to junction values can be performed by the following relationship

$$T_j = T_a + P_d\theta_{ja}$$

where T_j is the junction temperature in °C, T_a the ambient temperature in °C, P_d the average power dissipation in Watts and θ_{ja} the thermal resistance minus junction to ambient in $°C\,W^{-1}$.

Figure 14.21(a) shows an Arrhenius plot of junction temperatures against acceleration factor for different activation energies, while Figure 14.21(b) is a plot of failure rate for the junction temperatures derived from one activation energy curve using the failure rate formula shown above.

One caution is that acceleration factor calculations must be done separately for each applicable failure mechanism and the results then summed rather than attempting to develop an average or equivalent acceleration energy.

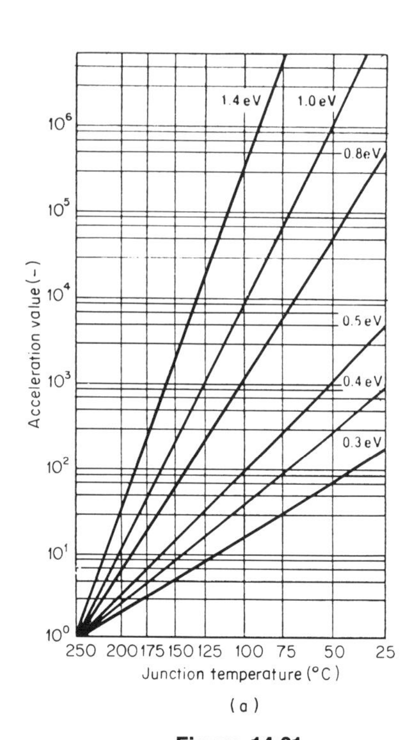

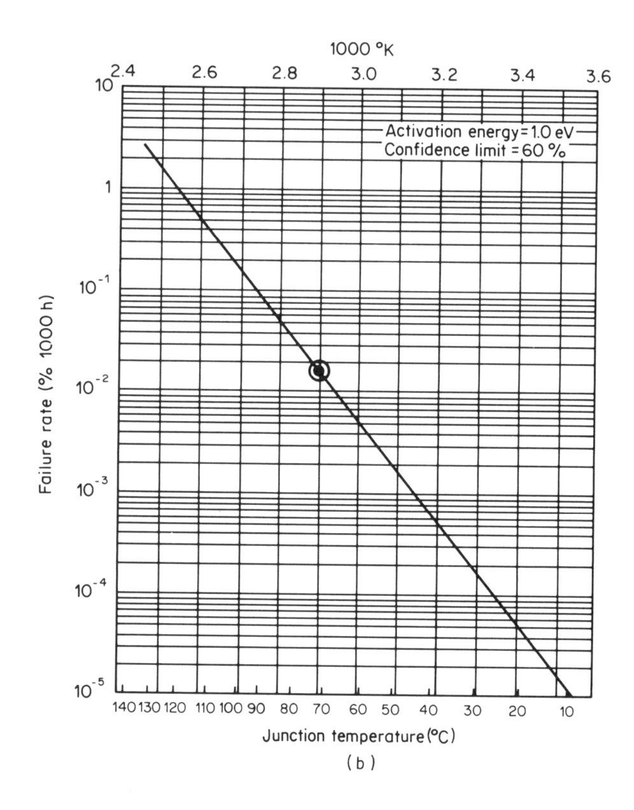

Figure 14.21
(a) Arrhenius plot of junction temperatures against acceleration factor for different activation energies (with permission of Fujitsu). (b) Plot of failure rate for the junction temperatures derived from one activation energy curve using the failure rate formula. (From Prince and Due-Gundersen [64] 1983.)

14.9.4 Mean time between failures

The inverse of the failure rate is called the mean time between failures (MTBF) and is a normal measure of the statistical failure rate at a particular temperature where

$$\text{MTBF} = 1/\text{failure rate} = 2T/\chi^2$$

14.10 GENERAL RELIABILITY FAILURE MECHANISMS

14.10.1 Hot electron effects in MOS transistors

The term 'hot electron effects' refers to the phenomenon of device degradation due to channel hot electron injection which induces substrate currents at the drain high field region where channel electrons are heated by the high localized channel electric field.

Reliability degradation due to hot electron effects are common to all MOS memory devices. This includes overloading of the on-chip substrate bias generator, threshold variation due to changing substrate potential, drain–source breakdown, DRAM refresh time degradation and device degradation. All of these mechanisms are directly related to the amount of substrate current generated in the circuit.

Substrate current results from the hole generation by the channel hot electron through impact ionization. Excessive substrate current can overload the on-chip substrate bias generators. Potential variations in the substrate produced by the flow of substrate current can cause threshold variations and in extreme cases avalanche breakdown (or snap-back) of the MOSFET. They can also cause latch up in CMOS circuits.

The injected electrons may also be trapped in the oxide causing threshold drift or generating interface traps effecting subthreshold characteristics.

Substrate electrons may be collected by nearby nodes as excess leakage currents which can cause DRAM refresh time degradation and can discharge other charge storage nodes.

Hot electrons also indirectly effect the transistor current–voltage (*IV*) characteristics by adding to the gate controlled surface current at the drain causing the *IV* curve to rise. Avalanche breakdown can result.

Substrate current generation from the cell load device is dependent on the technology of the device. PMOS FETs generate negligible substrate current. Both depletion and enhancement mode NMOS transistors generate hot electrons when the device is under high drain source bias.

Short channel devices generate significantly more substrate current than long channel devices due to the high field resulting from power supply voltages which remain at 5 V while the rest of the device is being scaled to smaller geometries.

Process and design techniques to reduce hot electron effects have been mentioned throughout the product chapters and include lightly doped drain MOSFETs, elimination of substrate bias generators and isolation of DRAM capacitors from the substrate.

14.10.2 Electromigration of metal interconnects

Electromigration is a phenomenon in which metal atoms are moved by a large current in the metal interconnect. When ionized atoms collide with the current of scattering electrons an 'electron' wind is produced. This wind moves the metal atoms in the opposite direction from the current flow. The result is the generation of voids at the negative electrode and hillocks and whiskers at the positive one. The voids generated increase wiring resistance and cause excessive currents to flow in some areas increasing the probability of the metal trace breaking. The whiskers may cause shorts with other metal lines. This problem is particularly severe when the metal passes over steps which tend to accentuate the electric field and also tend to have reduced metal thickness due to deposition related problems.

14.10.3 Soft errors

Dynamic RAMs and, to a lesser extent, static RAMs are affected by soft errors. These errors are called 'soft' since they result only in loss of data but do not entail actual damage to the device. A cell which has undergone a soft error or 'event' can be reprogrammed satisfactorily. These events can be generated by noise from capacitive coupling of lines in the device, system noise, or by impact ionization from alpha particles. Internal line noise was considered in the SRAM chapters. Alpha particle related soft errors will be considered here.

Alpha particles are helium nuclei which originate primarily from the decay of radioactive elements in the package itself. Alpha particles penetrate the die surface, creating a cloud of electron–hole pairs along the track of the alpha particle. About 10^6 pairs are created along a particle range of 25 μm. In DRAMs storage regions such as potential wells can collect sufficient numbers of the generated minority carriers to change the stored state of the cell. Diffusions and sense amplifiers are also sensitive. See Figure 14.22 for an illustration of charge collection at a memory cell due to an alpha hit.

In static RAMs alpha particle hits can cause the high node of the cross-coupled flipflop to be discharged causing the cell state to change. Changes to either '1' or '0' state are possible. This problem is particularly severe for the resistor load SRAM cells as was discussed in Chapter 8.

An alpha particle hit results in a 'soft' error since there is no physical damage to the cell. Only the contents have been changed and new information can be written to the cell and retained. Failure rates for soft errors are measured as 'Failures in Time' (FITs). (Fit: 1000 fits = 1000 failures in 10^9 hours.)

To help protect against alpha induced soft errors, many manufacturers coat the surface of the RAM chip with a coating which absorbs alpha radiation and lowers the soft error rates. This coating can be applied either to the chip during the assembly process when it is known as 'chip coat' or 'die coat', or it can be applied to the wafer during the manufacturing process when it is called a 'wafer coat'.

An alternative or additional method of improving soft error rate in DRAMs is to increase the capacitance of the memory cell. A large capacitor requires more charge

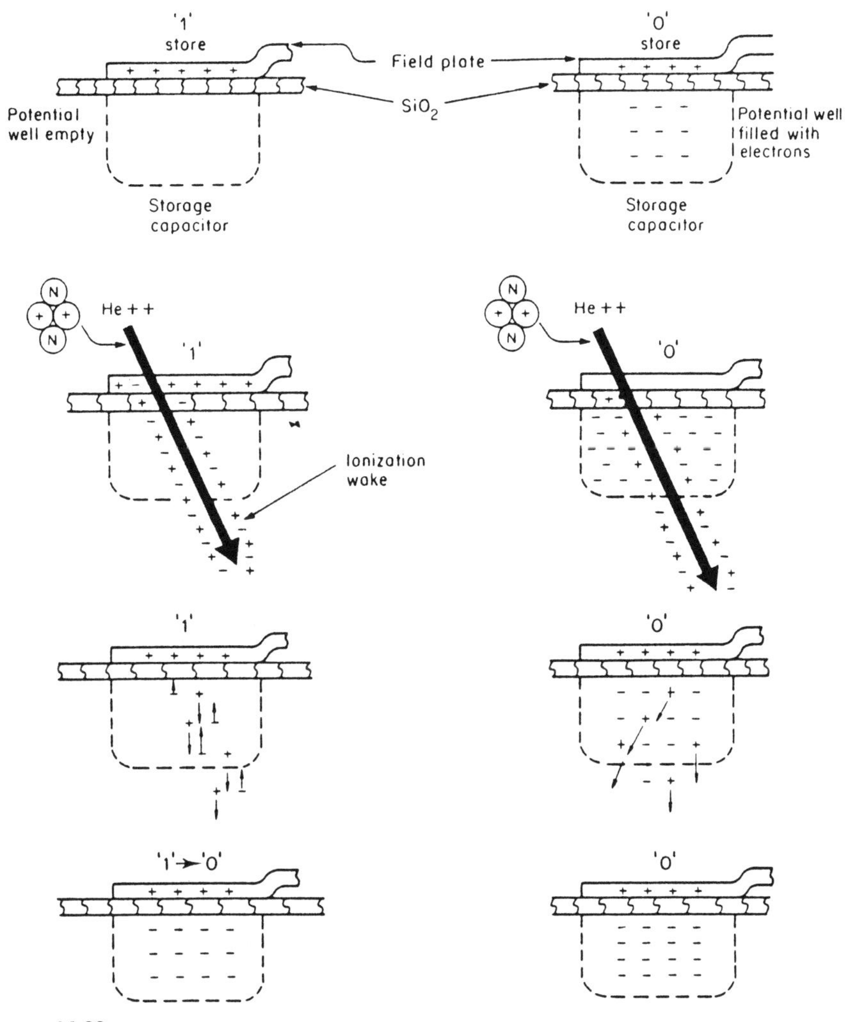

Figure 14.22
Illustration of charge collection at a memory cell due to an alpha hit. (From Prince and Due-Gundersen [64] 1983.)

to switch the logic state. An alpha particle has a known energy range and thus the cell can be designed with sufficient capacitance to enable the charge generated from an alpha hit to be absorbed without reaching the critical level.

A capacitor of 50 fF charged to 5 V holds 250 fC of charge. This corresponds to about 1.5 million electrons. This is about the minimum signal that can withstand an alpha particle hit.

Capacitance can be maintained either by scaling down the gate oxide thickness as the DRAM cell is scaled laterally, or by using a gate material with a higher dielectric constant than silicon dioxide such as silicon nitride. Materials which hold promise of even higher dielectric constant are being investigated such as the new ferroelectric materials.

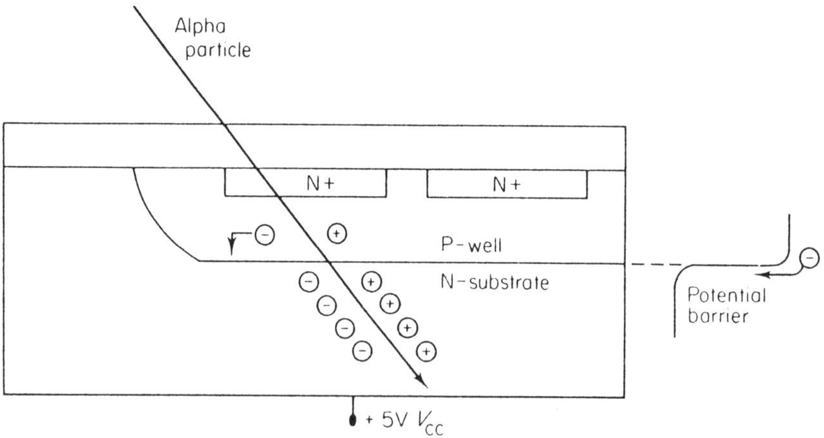

Figure 14.23
CMOS P-well structure showing the well forming a barrier preventing electron–hole pairs formed in the substrate by alpha radiation hits from reaching the memory cell. (Reproduced with permission of John Wiley.)

CMOS is naturally immune to alpha particle sensitivity due to the shallow well. Any electron–hole pairs formed by alpha particles striking the chip will be primarily formed in the substrate below the well. The well forms an electrical barrier to the electrons (or holes in an n-well technology) and prevents them from reaching the gate as illustrated in Figure 14.23. Any carriers generated in the shallow well recombine quickly or get lost in the flow of majority carriers.

Testing for alpha particle sensitivity can be done using an alpha particle radiation source such as ^{241}Am placed on a packaged DRAM or SRAM which has the cap removed so that the source is directly over the chip. If a large area source is used so the entire package opening is covered, then the chip will be exposed to alpha particles at all angles of incidence. The error rate can then be measured by running suitable test patterns with a read–modify–write cycle. The measured error rate can then be scaled by a factor which relates the alpha particle flux of the source to that typical of package environment emissions.

An example of the results of such a test on a DRAM measured in soft error rate against cycle time is shown in Figure 14.24(a) for two power supply voltages. The drop in error rate with increasing cycle time is characteristic of errors resulting from alpha particle hits on bit lines and sense amplifiers. For 4.5 V and long cycle times the curve levels. In this region we are seeing errors in the individual cells also.

Figure 14.24(b) shows the variation of soft error rate with cycle time for an NMOS 64k DRAM, a resistor load NMOS 16k SRAM and a resistor load 64k CMOS SRAM. The variation of the DRAM SER with cycle time is again shown. The 16k SRAM as shows a pronounced dependence of SER on cycle time. This has been explained as a result of the long time constant for recharging the stored '1' high voltage on the cell via the large cell pull-up resistor.

One would expect that the 64k SRAM would be even more soft error sensitive than the 16k SRAM since it also has load resistors and in addition has lower cell

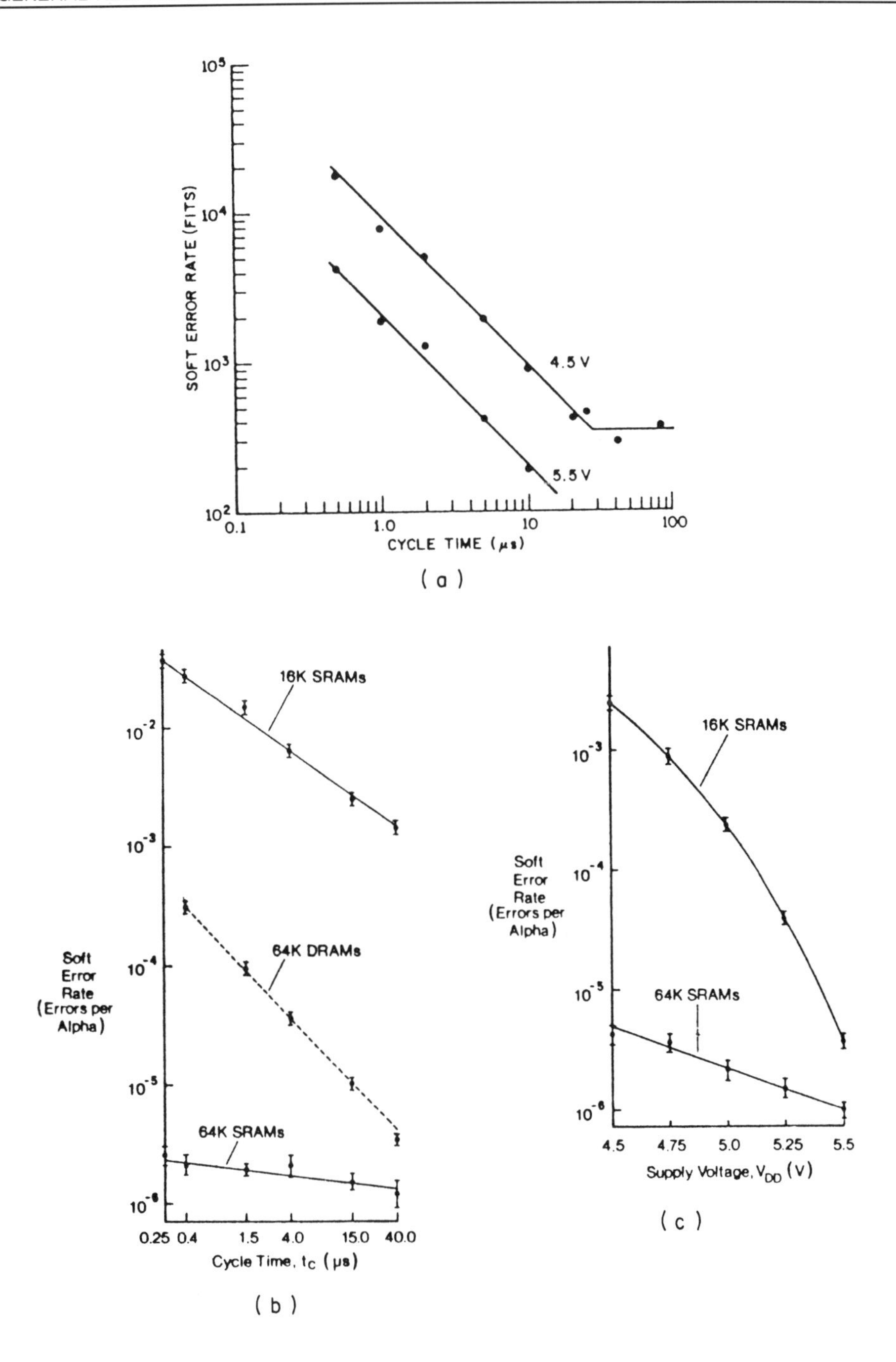

Figure 14.24
Examples of dependence of soft error rate on various parameters. (a) Soft error rate plotted against cycle time for two different voltages, (b) soft error rate plotted against cycle time for a 16k NMOS SRAM, a 64k NMOS DRAM and a 64k CMOS SRAM, (c) soft error rate plotted against supply voltage for a 16k NMOS SRAM and a 64k CMOS SRAM. (From Carter and Wilkins [38], with permission of IEEE.)

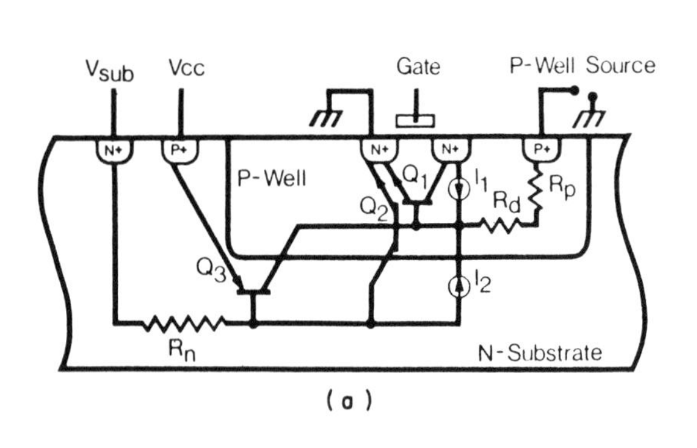

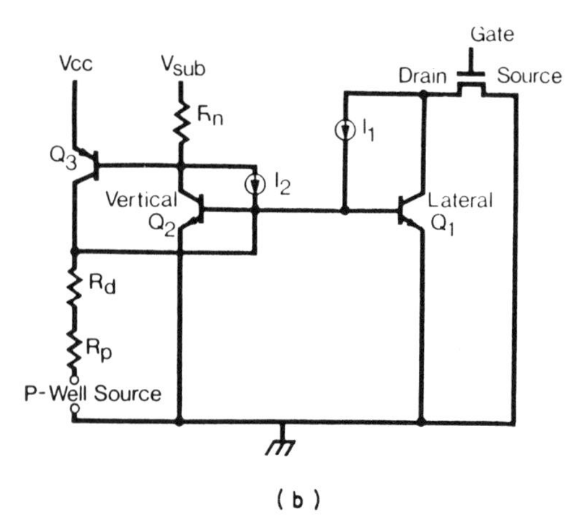

Figure 14.25
Illustration of latch-up phenomena in a CMOS device showing parasitic thyristor action. (a) CMOS structure showing parasitic circuit, (b) equivalent circuit model. (From Cheng *et al.* [62], G.E. 1981, with permission of IEEE.)

capacitance due to the effects of being scaled to smaller geometries. The lower sensitivity actually shown of the higher density SRAM is attributed to the use of the CMOS process. The cell is in a P-well with a potential barrier being formed at the interface between the well and the substrate which helps prevent any alpha induced electrons produced below the well from reaching the sensitive device nodes. Analytical discussions and spice modeling for soft error rates in RAMs are available [38, 39].

Figure 14.24(c) shows the variation of soft error rate with supply voltage for a 16k NMOS SRAM and a 64k CMOS SRAM. For both technologies the soft error rate increases at lower power supply voltages.

Mitsubishi [41] in 1988 showed that for higher density SRAMs the soft error rate could be reduced by increasing the threshold voltage of the driver and access transistors, and also by setting the selected word line level lower than V_{CC}.

14.10.4 Latch-up susceptibility

Historically CMOS was used on the SRAMs before being used on the DRAMs and other memories. One of the main disadvantages of CMOS SRAMs historically has been the tendency to exhibit a phenomena called 'latch-up' which is inherent in the CMOS process and can destroy the part.

The already larger chip size of the CMOS SRAMs has been increased even more to provide extra well spacings, well contacts, and guardbands to prevent latch-up.

Latch-up results from a parasitic bipolar action which causes the device to draw large currents and destroy itself. It occurs in the thyristor type PNPN layers as shown in Figure 14.25. When this thyristor latches, it becomes self-maintaining and the power supply must be disconnected to stop it.

There are various process modifications which can decrease the tendency to latch-up conditions. The impurity doping in the well can be increased in order to decrease minority carrier lifetime in the parasitic base regions. This leads, however, to higher capacitance and slower speed. Guard rings can be used on the wells but they increase the size of the chip as they are space consuming. Twin-well CMOS technology also provides a further improvement in latch-up susceptibility.

Historically many of these methods have been combined together with careful power supply sequencing to reduce the susceptibility to latch-up.

14.10.5 Electrostatic discharge (ESD)

While MOS circuits are optimized internally for voltages around 5 V, they can encounter much higher voltages on the inputs. MOS is particularly sensitive to high voltage discharges due to the liability of rupture of the thin gate oxide.

High electrostatic voltages can originate from the human body, which can generate charges exceeding 2000 V, or from poorly grounded equipment such as test machines. They can also originate from noisy environments such as are encountered in automotive applications or consumer applications such as personal computers. For this reason input protection circuitry designed to withstand voltages on the order of 2000 V is common in CMOS circuits.

Input protection circuitry must withstand both automated and human handling. For this reason, two models are available, the machine model and the human body model. The machine model provides a short duration, high voltage pulse as might be encountered if a pin of the memory touched a grounded conductor. The human body model provides a relatively long duration, high current pulse such as experienced during human handling. Typical design targets for these models differ significantly with the machine model being typically less than 500 V and the human body model being designed as high as 4000 V.

14.11 NOISE SOURCES IN DYNAMIC RAMs

Various noise sources found by small signal detection are listed in Figure 14.26(a). The three basic causes of noise in a DRAM [59] are imbalance of the dynamic sense amplifiers, memory array noise and peripheral circuits, such as timing pulse generators, which act as a noise source.

14.11.1 Sense amplifier noise

Sense amplifier noise occurs due to imbalance of bit-line pair capacitances, imbalance of the storage transistors in the memory cell flipflop, and imbalance in the bit-line precharge voltage.

	NOISE SOURCES	Ref. FIG.1
CELL NOISE	CELL PLATE V_{CC} BUMP	①
SENSE AMP. NOISE	UNBALANCE OF BIT LINE PAIR CAPACITANCES	②
	OF FLIP-FLOP PAIR MOSFETS	③
	OF FLIP-FLOP PRE-CHARGE VOLTAGE	④
ARRAY NOISE	BIT-WORD FEED-BACK CAPACITANCE	⑤
	VARIATION OF CELL PLATE VOLTAGE	⑥
	OF SUBSTRATE VOLTAGE	⑦
	OF DUMMY CELL GROUND VOLTAGE	⑧
	OF WORD LATCH GROUND VOLTAGE	⑨
EXTERNAL NOISE	CHARGE INJECTION BY CHARGE PUMPING	⑩
	BY INPUT UNDERSHOOT	⑪
	BY α-PARTICLE RADIATION	⑫
	ON-CHIP SUBSTRATE BIAS STABILITY	⑬

(a)

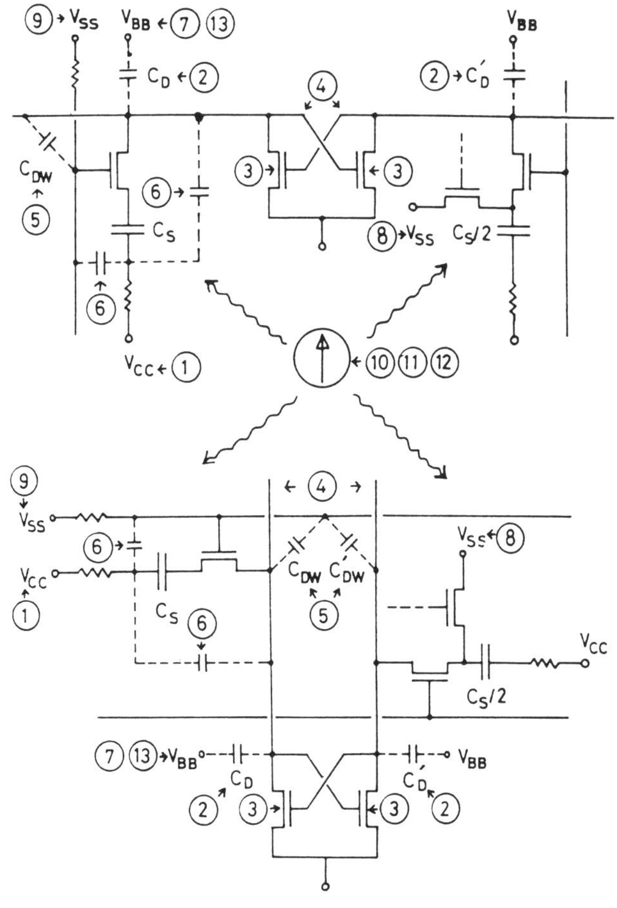

(b)

Figure 14.26
(a) Table of noise sources of NMOS DRAMs. (b) Circuit schematics of noise sources of open and folded bit memory cell arrays in the DRAM indicating sources of noise. (From Masuda *et al.* [59], Hitachi 1980, with permission of IEEE.)

14.11.2 Array noise

Memory array noise is generated during sensing in NMOS DRAMs in open bit cell arrays due to a ringing effect during word-line selection when many bit-lines swing from V_{CC} to 0 V simultaneously as for example in performing a checkerboard pattern test. A general array noise effect occurs due to the capacitive coupling of all the other bit-lines to the selected bit-line pair producing a negative feedback effect on the selected bit-line pair.

14.11.3 External noise

External causes of noise include charge injection into the substrate from charge pumping, for example, from substrate bias generators. Alpha particles from the memory packaging material are another source of external noise.

14.11.4 A bit-line model of DRAM noise

A memory array model developed by Masuda of Hitachi [59] of open and folded bit cells is shown in Figure 14.26(b). The open bit-line array gives differential mode noise response while the folded bit-line array gives a common mode noise response which is less due to common mode rejection.

The folded bit sense line architecture was used for DRAMs to improve the noise immunity and provide larger layout pitches for the sense amplifiers. The important noise source is the coupling noise between bit-line and word-line. A return to open bit sense lines occurs only when device densities and architectures are high enough that coupling between adjacent bit-lines becomes more of a problem than non-common mode noise on a single bit-line.

Experimental results have shown the folded bit cell to reduce array noise by as much as 90% of that of the open bit-line if sense amplifier noise is assumed the same.

The open bit sense line architecture also has the advantage of providing for a smaller geometry memory cell over the folded bit-line architecture.

Analytic expressions for the noise in open and closed bit-line structures were developed by Masuda [59].

14.12 RELIABILITY OF EPROMs AND EEPROMs

The introduction of the floating gate technology gave birth to a new product range, new applications, and new markets. At the same time, this technology introduced new failure mechanisms which all had to be considered from a reliability point of view. Hot or cold electron programming–erase with high voltage energies, coupled in the case of EPROMs with the use of UV light, was new ground where unexpected problems could easily occur.

Underprogramming, overprogramming, electron trapping, spurious erasure, polar-

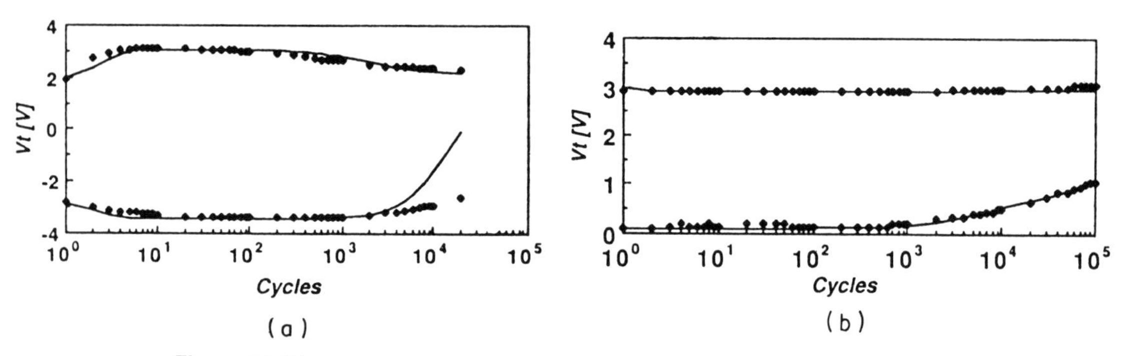

Figure 14.27
Analysis of erase–write endurance degradation characteristics. (a) Uniform writing, (b) non-uniform writing. (From Witters *et al.* [58], IMEC 1987, with permission of IEEE.)

ization, oxide hopping, and others are common failure mechanisms. Effective failure screens which attempted to weed out early life problems were successfully introduced by trial and error and experience. EPROMs are now considered to be both a good and reliable product.

For electrically erasable PROMs special reliability attention had to be given to the endurance which refers to the number of read, and write–erase cycles which are specified. A typical endurance curve showing high and low thresholds of the storage transistor as a function of number of write–erase cycles is shown in Figure 14.27.

A study [58] of the degradation of tunnel oxide in electrically erasable ROMs by the IMEC Institute in Belgium indicated reasons for the shape of the degradation curves shown. The externally measured threshold voltage of a floating gate transistor is determined by the amount of charge stored on the floating gate and by the intrinsic threshold voltage when there is no charge on the floating gate. IMEC found that the degradation curve is formed by changes in both of these factors.

During the erase operation of an EEPROM a positive voltage is applied to the control gate causing electrons to tunnel through the gate oxide to the floating gate. If a negative charge located in the thin oxide this will reduce the number of electrons stored on the gate after the erase which will tend to decrease the decrease threshold voltage. The electronics located in the thin oxide will, however, tend to increase the intrinsic threshold voltage. The two effects nearly balance so that there is no net effect on the erase operation from electrons trapped in the thin oxide.

During the write operation a positive voltage is, however, applied to the drain causing the electrons to tunnel through the gate oxide toward the drain. The charge trapped in the thin oxide will tend again to diminish this flow resulting in a more negative potential on the gate and higher threshold voltage. In addition the electrons trapped in the thin oxide also cause a higher intrinsic threshold voltage so that the two effects are cumulative in increasing the threshold voltage after writing. In the case of uniform distribution of flow over the channel region only the write operation would show a degradation as shown in Figure 14.27(a).

IMEC showed that for a non-uniform flow distribution with flow concentrated in the drain region the degradation curves look like Figure 14.27(b) which is what the actual experimental curves show.

The mechanism is that a hole current is generated during non-uniform write which leads to trapping of holes in the thin oxide so that a low voltage tail shows up at both low and high threshold voltages. After many cycles electron trapping at the drain side caused by the higher voltage in this region induces a window closing. The positive charge in the channel causes the closure effect to start first for the high threshold voltage.

Two type of lifetesting are also necessary. The first is similar to standard lifetesting as previously discussed and the second is used for data retention testing. The latter is high temperature (250°C) bake to study data retention capability, checking that data stored does not alter. Figure 14.28 shows a typical reliability test plan and illustrates the significance of write–erase and data retention testing.

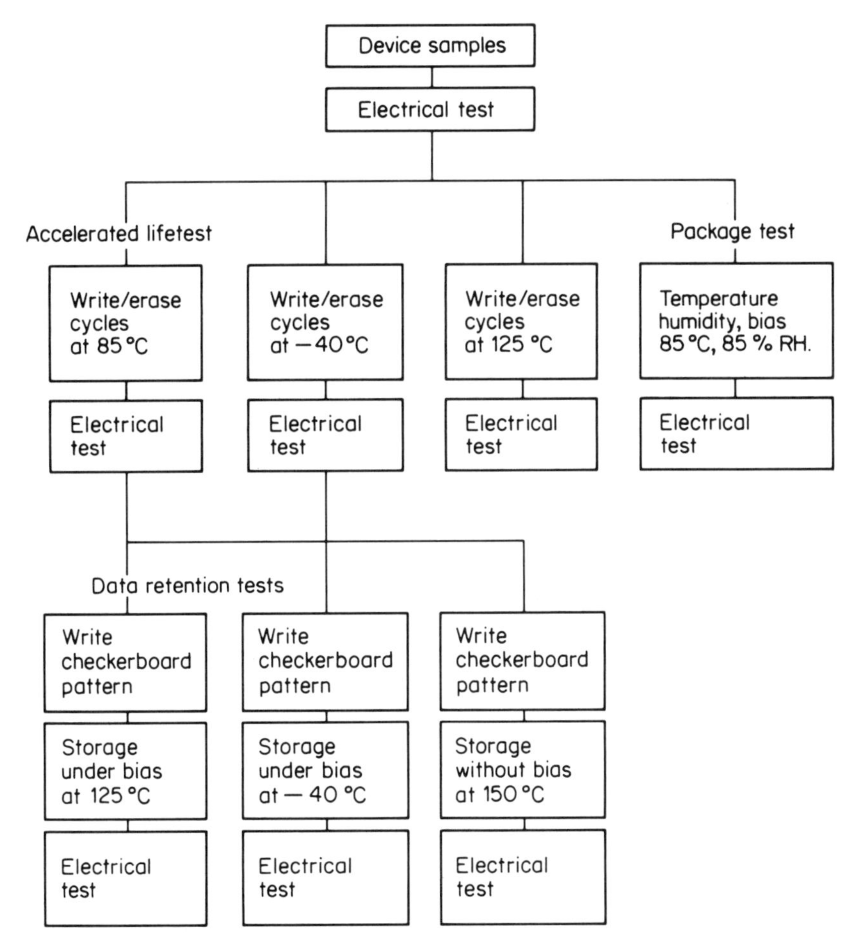

Figure 14.28
Sample reliability test flow for an EPROM.

14.13 FAILURE ANALYSIS

In addition to process controls to prevent defects from occurring and testing to be sure defects have not occurred, it is also necessary to analyze failures to prevent them from reoccurring.

Failure analysis includes both electrical and visual analysis to determine the failure mode and understand the failure mechanism so that corrective action can be taken.

A typical failure analysis flow chart [60] is shown in Figure 14.29. Following electrical test and a visual microscopic examination of the package, a search is made for records of similar cases so that the failure mode can be categorized and potential failure mechanisms determined.

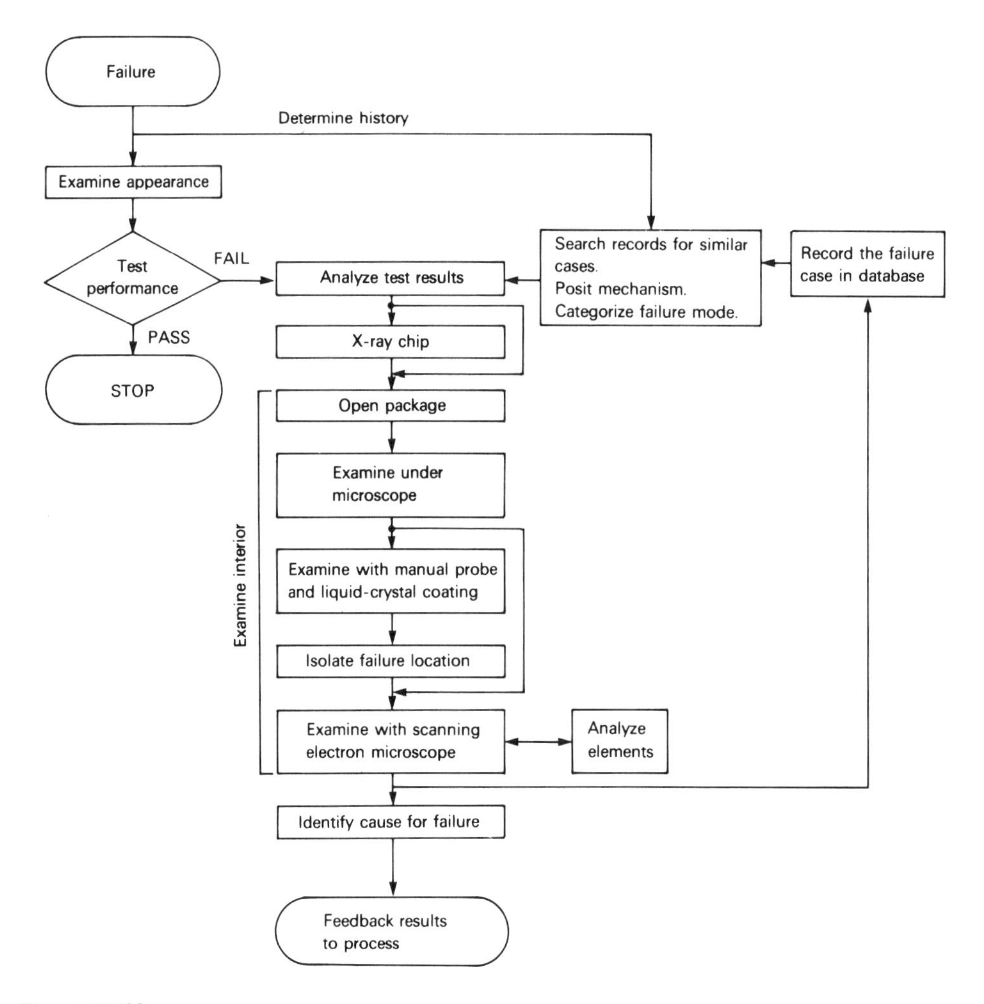

Figure 14.29
Typical failure analysis flow chart. (From [60], with permission of NMB Semiconductor 1990.)

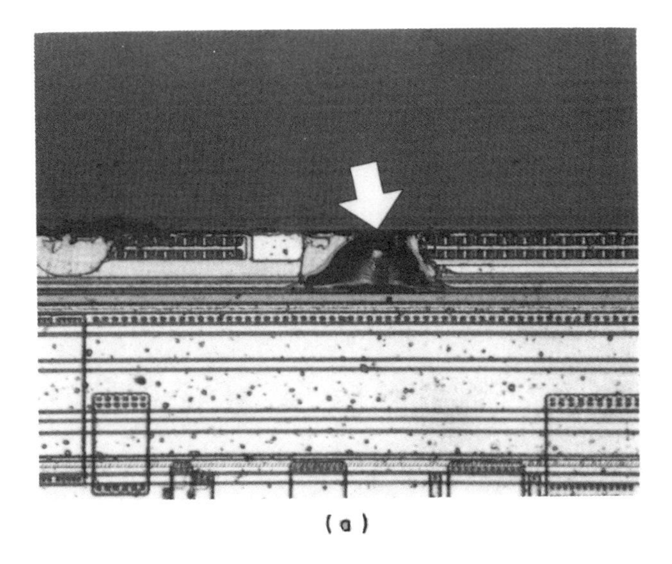

(a)

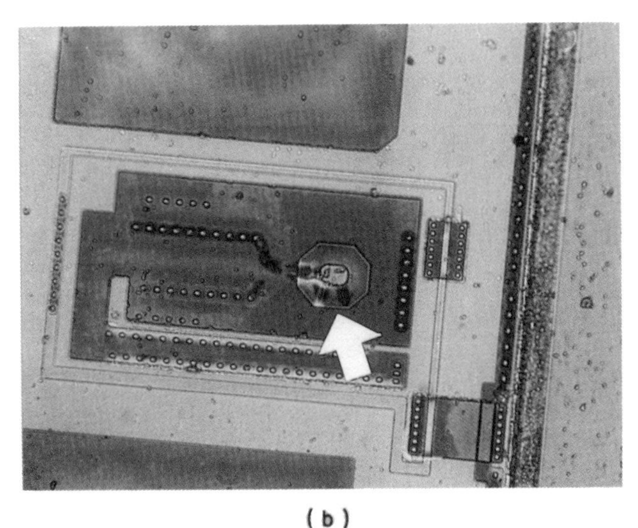

(b)

Figure 14.30
Two typical causes of chip failure. (a) Die crack causing a leakage failure mode. The mechanism is stress due to sawing the chips apart. (b) Damage to input protection diode. The mechanism is damage due to electrostatic discharge. (From [60], with permission of NMB Semiconductor.)

The mechanism expected determines the test performed. In this case the package is X-rayed to check for breaks in leads or bond wires. The package is then opened and the bare chip is examined under the microscope for visible defects such as the die cracks shown in Figure 14.30(a) or the damage to the input diode due to electro-static discharge as shown in Figure 14.30(b).

A liquid crystal coating may now be applied and an electrical probe test of a particular suspected location on the chip be done. The liquid crystal will show hot

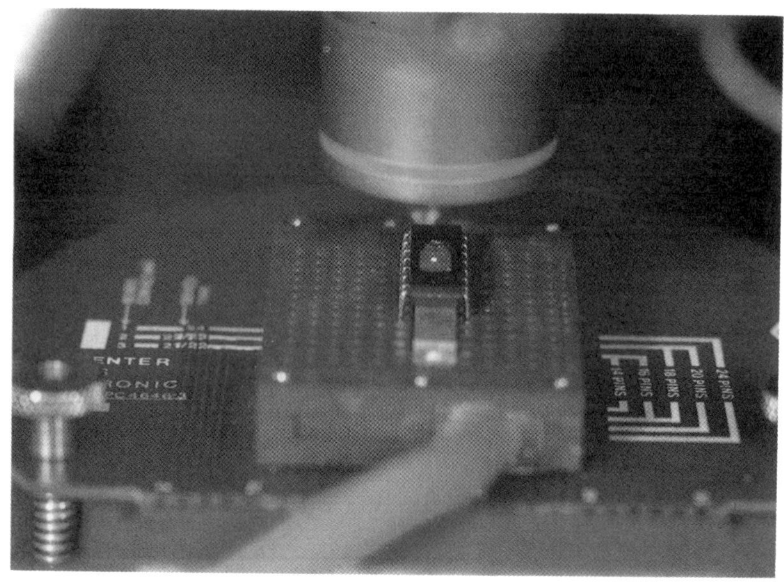

(a)

(b)

Figure 14.31
Two advanced tools for failure analysis of memory chips. (a) Liquid crystal failure analysis system. An encapsulated device is set on the table for analysis using a laser beam. (b) Scanning electron microscope. A powerful tool for failure analysis providing up to 100 000 magnification. (From [60], with permission of NMB Semiconductor.)

spots such as the shorts due to electrostatic discharge visibly. The photograph in Figure 14.31(a) shows an encapsulated device with the laser beam from a liquid crystal failure analysis system focused on it. An examination with a scanning electron microscope such as shown in Figure 14.31(b), may then be done to provide further evidence for final identification of the cause of failure. These results are then fed back to the manufacturing area so that corrective action can be taken.

14.14 PACKAGE RELIABILITY

As larger and more sophisticated memory chips are housed in ever smaller packages, package reliability becomes an ever more important part of the overall chip reliability. The package stresses on the chip can affect mechanical integrity and electrical characteristics. Frequently a coating is put on the chips before assembly to relieve some of these stresses. A poor leadframe design can lead to moisture entering the package and corroding the leads. Thermal stresses occur which change with time and with environmental conditions and must be compensated for by good package design. Finally the trend to bare chip assembly requires that the chip itself be nearly hermetic.

14.15 MEMORY TEST AND BURN-IN EQUIPMENT

Development of high technology semiconductor components has caused a similar development of highly sophisticated test equipment. The current range goes from inexpensive volume testers in the $200 000 to $500 000 range up to high performance engineering testers costing one million dollars or more.

The selection criteria include data rates, word length, device size, pin count, timing resolution, and software requirements. A memory tester will normally fall in one of three different categories.

- LSI tester
- dedicated memory tester
- general purpose tester

High volumes are often achieved by dedicated test equipment used for a limited number of product lines. A wide product range and low volume mix is the criterion for choosing a general purpose tester.

BIBLIOGRAPHY

[1] Jensen, E. and Schneider, B. (1979) Characterization of Random Access Memories, *Electronikcentralen* November 1979.

[2] Hansen, C., Fregil, F. and Schneider, B. (1980) UV-EPROMs, *Electronikcentralen*, November 1979.

[3] *NMOS Capability Manual Motorola*. East Kilbride, Scotland.
[4] Rosenfield, P. (1979) Memory testing characterization timing and patterns, *Electronics and Power*, January 1979.
[5] *Motorola Memory Databook* 1989.
[6] *Fujitsu Memories Databook* 1986–1987.
[7] *NEC Memory Products Databook* 1989.
[8] *Micron Technology MOS Databook* 1988.
[9] *Sharp Memory Databook* 1988.
[10] *64K Dynamic RAM Qualification Manual* Motorola Inc.
[11] *64K Dynamic RAM Reliability Design Manual* Dick Brunner, Motorola Inc.
[12] *CMOS Standard Logic Reliability* Motorola Inc. 1980.
[13] *Reliability Report 8139* Motorola Inc.
[14] *MOS Capability Manual* Motorola Semiconductor Limited, East Kilbride, Scotland.
[15] *Reliability Reports* Motorola Semiconductor Limited, East Kilbride, Scotland.
[16] *Hitachi MOS Memory Databook* 1987/1988.
[17] Fujieda, T. and Arai, N. (1985) Considerations of the Testing of RAMs with Dual Ports, *IEEE International Test Conference Proceedings*, p.456.
[18] Kondo, R. (1979) Test Patterns for EPROMs, *IEEE Journal of Solid State Circuits*. Vol. **SC-14**, No.4, August 1979, 730.
[19] Hamming, R. (1950) Error detecting and correcting codes, *Bell System Technology Journal*, **29**, 147–160.
[20] Davis, H. L. (1985) A 70 ns Word Wide 1-Mbit ROM with On-Chip Error-Correction Circuits, *IEEE Journal of Solid State Circuits*, Vol. **SC-20**, No.5, October 1985.
[21] Vancu, R. *et al.* (1990) A 35 ns 256k CMOS EEPROM with Error Correcting Circuitry, *ISSCC Proceedings*, February, p.64.
[22] Asakura, M. *et al.* (1989) An Experimental 1Mb Cache DRAM with ECC, *Symposium on VLSI Circuits*, p.43.
[23] Arimoto, K. *et al.* (1989) A Speed Enhancement DRAM Array Architecture with Embedded ECC, *Symposium on VLSI Circuits*, p.111.
[24] Smayling, M. C. and Maekawa, M. (1983) 256k Dynamic RAM is more than just an upgrade, *Electronics*, August 25, p.135.
[25] Fujishima, K. *et al.* (1983) A 256k Dynamic RAM with Page-Nibble Mode, *IEEE Journal of Solid State Circuits*, Vol. **SC-18**, No.5, October 1983, p.470.
[26] Price, J. E. (1970) A new look at yield of integrated circuits, *Proceedings IEEE*, **58**, p.1290.
[27] Fujishima, K. *et al.* (1983) A 256k Dynamic RAM with Page-Nibble Mode, *IEEE Journal of Solid State Circuits*, Vol. **SC-18**, No.5, October 1983, p.470.
[28] Mano, T. *et al.* (1982) A Redundancy Circuit for a Fault-Tolerant 256k MOS RAM, *IEEE Journal of Solid State Circuits*, Vol. **SC-17**, No.4, August 1982, p.726.
[29] Smith, R. T. *et al.* (1981) Laser Programmable Redundancy and Yield Improvement in a 64k DRAM, *IEEE Journal of Solid State Circuits*, Vol. **SC-16**, October 1981, p.506.
[30] Spaw, W. *et al.* (1982) A 128k EPROM with Redundancy, *IEEE ISSCC Proceedings*, February, p.112.
[31] Kokkonen, K. *et al.* (1981) Redundancy Techniques for Fast Static RAMs, *ISSCC Proceedings*, February, p.80.
[32] Naruka, Y. *et al.* (1989) A 16Mb Mask ROM with Programmable Redundancy, *ISSCC Proceedings*, February 16, p.128.
[33] Gaw, H. *et al.* (1985) A 100 ns 256k CMOS EPROM, *ISSCC Proceedings*, February, p.164.
[34] Yamada, J. *et al.* (1984) A Submicron 1 Mbit Dynamic RAM with a 4-bit-at-a-Time Built-in ECC Circuit, *IEEE Journal of Solid State Circuits*, Vol. **SC-19**, No.5, October 1984, p.627.

[35] Osman, F. I. (1982) Error-Correction Technique for Random-Access Memories, *IEEE Journal of Solid State Circuits*, Vol. **SC-17**, No.5, October 1982, p.877.

[36] Edwards, L. (1981) Low Cost Alternative to Hamming Codes corrects Memory Errors, *Computer Design*, July, p.132.

[37] Benevit, C. A. *et al.* (1982) A 256k Dynamic Random Access Memory, *IEEE Journal of Solid State Circuits*, Vol. **SC-17**, No.5, October 1982, p.857.

[38] Carter, P. M. and Wilkins, B. R. (1989) Influences of Soft Error Rates in Static RAMs, *IEEE Journal of Solid State Circuits*, Vol. **SC-22**, No.3, June 1987, p.430.

[39] Chappell, B. *et al.* (1985) Stability and SER Analysis of Static RAM Cells, *IEEE Journal of Solid State Circuits*, Vol. **SC-20**, No.1, February 1985, p.383.

[40] Noorlag, D. J. W. (1980) The Effect of Alpha-Particle-Induced Soft Errors on Memory Systems with Error Correction, *IEEE Journal of Solid State Circuits*, Vol. **SC-15**, No.3, June 1980, p.319.

[41] Murakami, S. *et al.* (1988) Improvement of Soft Error Rate in MOS SRAMs, VLSI Circuits Symposium, 1988, p.67.

[42] Khan, A. (1983) Fast RAM Corrects Errors on Chip, *Electronics*, September 8, p.126.

[43] Tanaka, S. *et al.* (1984) A Programmable 256k CMOS EPROM with On-Chip Test Circuits, *ISSCC Proceedings*, February, p.148.

[44] *EIA JEDEC Standard 21-C* Configurations for Solid State Memories, 1990.

[45] Kantz, D. (1984) A 256k DRAM with Descrambled Redundancy Test Capability, *IEEE Journal of Solid State Circuits*, Vol. **SC-19**, No.5, October 1984, p.596.

[46] Atsumi, S. *et al.* (1985) Fast Programmable 256k Read Only Memory with On-Chip Test Circuits, *IEEE Journal of Solid State Circuits*, Vol. **SC-20**, No.1, February 1984, p.422.

[47] Shah, A. H. (1986) A 4-Mbit DRAM with Trench-Transistor Cell, *IEEE Journal of Solid State Circuits*, Vol. **SC-21**, No.5, October 1986, p.618.

[48] Kumanoya, M. *et al.* (1985) A Reliable 1-Mbit DRAM with a Multi-Bit-Test Mode, *IEEE Journal of Solid State Circuits*, Vol. **SC-20**, No.5, October 1985, p.909.

[49] You, Y. (1985) A Self-Testing Dynamic RAM Chip, *IEEE Journal of Solid State Circuits*, Vol. **SC-20**, No.1, February 1985, p.428.

[50] McAdams, H. *et al.* (1986) A 1-Mbit CMOS Dynamic RAM With Design-For Test Functions, *IEEE Journal of Solid State Circuits*, Vol. **SC-21**, No.5, October 1986, p.635.

[51] Mashiko, K. *et al.* (1987) A 4-Mbit DRAM with Folded-Bit-Line Adaptive Sidewall-Isolated Capacitor (FASIC) Cell, *IEEE Journal of Solid State Circuits*, Vol. **SC-22**, No.5, October 1987, p.643.

[52] Ohsawa, T. *et al.* (1987) A 60 ns 4-Mbit CMOS DRAM with Built-In Self-Test Function, *IEEE Journal of Solid State Circuits*, Vol. **SC-22**, No.5, October 1987, p.663.

[53] Minato, O. *et al.* (1982) A Hi-CMOSII 8k × 8 Bit Static RAM, *IEEE Journal of Solid State Circuits*, Vol. **SC-17**, No.5, October 1982, p.793.

[54] Higuchi, M. *et al.* (1990) An 85 ns 16Mb CMOS EPROM with Alterable Word Organization, *ISSCC Proceedings*, February, p.56.

[55] Anami, K. *et al.* (1983) Design Considerations of a Static Memory Cell, *IEEE Journal of Solid State Circuits*, Vol. **SC-18**, No.4, August 1983, p.414.

[56] Tanaka, S. *et al.* (1984) A Programmable 256k CMOS EPROM with On-Chip Test Circuits, *ISSCC Proceedings*, February, p.148.

[57] Venkatesh, B. *et al.* (1986) A CMOS 1Mb EPROM, *ISSCC Proceedings*, February, p.40.

[58] Witters, J. S. *et al.* (1987) Programming Mode Dependent Degradation of Tunnel Oxide Floating Gate Devices, *IEDM Proceedings*, December, p.544.

[59] Masuda, H. *et al.* (1980) A 5 V-Only 64k Dynamic RAM Based on High S/N Design, *IEEE Journal of Solid State Circuits*, Vol. **SC-15**, October 1980, p.846.

[60] Quality and Reliability Assurance for NMBS CMOS Memory IC's, NMB Semiconductor, 1990.
[61] Lyman, J. (1988) TI Finds A New Way to Predict Package Reliability, *Electronics*, March 31, p.87.
[62] Cheng, P. Y. *et al.* (1981) *IEEE ISSCC Proceedings*, p.160.
[63] *Texas Instruments MOS Memory Databook* 1989.
[64] Prince, B. and Due-Gundersen, G. (1983) *Semiconductor Memories*, John Wiley & Sons, Chichester.

15 YIELD, COST AND THE MODERN FACTORY

15.1 OVERVIEW

With memory wafer facilities costing up to $1 billion for a 64Mb DRAM factory, the yield on the processed silicon wafer becomes one of the major determinants of cost in the memory industry. For this reason a lot of effort has gone into both predicting yield and improving it both in the design process and in the wafer manufacturing process. Effort has also gone into reducing the basic cost of a memory factory without reducing the yield of the wafers produced in it. These efforts are traced in this chapter.

15.2 YIELD THEORY

Chip yield can be estimated by Price's [5] (Seed's) equation*

$$\text{Yield} = 1/(1 + AD_0)$$

This says that yield is inversely proportional to the area of the chip (A) and the defect density of the manufacturing line (D_0).

Poisson's equation can also be used to estimate chip yield

$$\text{Yield} = \exp(-AD_0)$$

For Poisson's equation the total yield using redundancy is [6]

$$Y = \sum_{k=0}^{i} (A_M D_0)^k (1/k!) \exp(-A_M D_0) \exp(-A_s k D_0)$$

---

* While this equation is also referred to as Seed's [14] Equation (1976), it is more frequently referenced in memory literature as Price's equation (1980).

where A is the total chip area, i the number of spare elements (redundant word–bit lines), A_m the repairable area, D_0 the defect density of the fabrication area and A_s the spare element area.

Another model which adds real life boundary conditions is a two-parameter clustering model of yield which is given by the equation

$$Y = (1 + AD/c)^{-c}$$

where Y is the yield as a percentage of good die, A the die area, D the average defect density and c the clustering coefficient.

This is an extension of Price's equation and takes into account the fact that defects in many processes seem to occur in clusters rather than randomly distributed on the chip as point defects.

A further extension of this 'cluster' model done by Philips [13] assumes that not every defect causes a fault, but that every defect type has a given probability P_j of causing a fault in region A_j. The chip is assumed to be divided into regions with different probability of a fault causing a defect so that the formula for yield for that type of defect summed over the different regions of the chip becomes

$$Y = \left(1 + \sum_j A_j D P_j / c\right)^{-c}$$

If we now designate defect type 'i' as having defect density D_i, and cluster coefficient c_i and probability in region j of P_{ij}, then the formula summed over the different types of faults becomes

$$Y = \prod_i \left(1 + \sum_j A_j D_i P_{ij} / c_i\right)^{-c_i}$$

The product is used since it is assumed that the different defect types are independent and that the chip will only work if none of them cause a fault.

This model is applied to the special case of a defect occurring which causes a break in an interconnect and to the general case of the wafer yield as a function of the shrink factor which the interested reader might want to look up.

15.3 YIELD ENHANCEMENT

After finding methods for determining the yield of a memory chip, the next step is to find methods to improve it. This is typically done with either error correction which is electrical or 'soft' correcting, or it is done with replacement of defective circuit elements with redundant elements included on the chip which is 'hard' correcting.

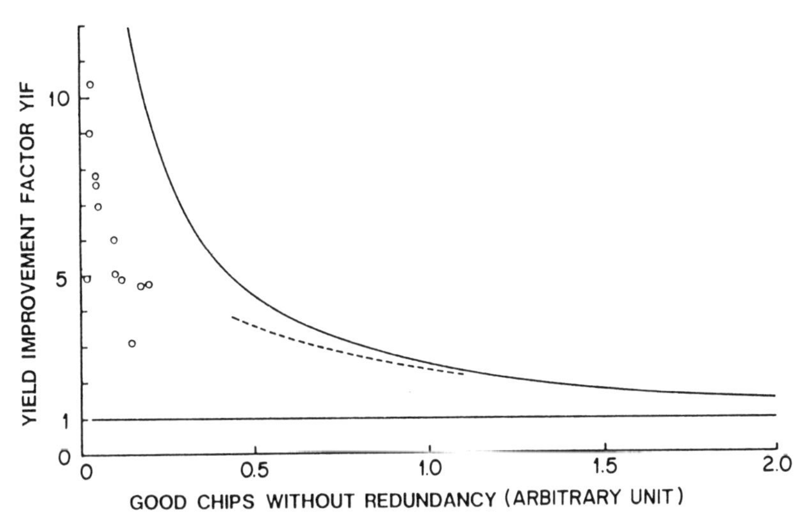

Figure 15.1
Estimated yield improvement factor for: (i) Price's equation $1/(1 + AD_0)$ (---), (ii) Poisson's equation $\exp(-AD_0)$ (——), (iii) actual data for a 256k DRAM with redundancy (○) (A is the die size and D_0 the defect density). (From Fujishima *et al.* [6], Mitsubishi 1983, with permission of IEEE.)

15.3.1 Redundancy

The use of redundant elements on a memory chip to replace defective elements can result in lower capital manufacturing costs and earlier introduction of new products on existing wafer fab lines or in new process technologies.

An example of the theoretical yield improvement factor compared to the number of good chips without redundancy for one case of a 256k DRAM [6] for Price's and for Poisson's equations is shown in Figure 15.1. The actual measured values are plotted as circles.

The theoretical yield improvement factor is defined as total good chips after applying redundancy divided by the number of good chips without redundancy. This is an important consideration since the number of good chips without redundancy increases from the early production stage when it is quite low to the later production stages as volume production is achieved and experience is gained in the manufacturing area. The conclusion from the figure is that the use of redundancy is much more important in the early stages of production than later when the basic yield of the process is high.

An interesting example of the impact of redundancy on the shrinking line widths of higher density DRAMs was given by IBM [18] in 1978. Let us assume that the defect density is proportional to the inverse square of the line width used in a given technology wafer fab. Then if a 4k DRAM has 4 μm line width and a 64k DRAM has 1 μm line width, and the total number of defects in the 4k DRAM is 1, then the number of defects expected on the 64k is 16.

We refer now to Figures 15.2(a) and (b) which plot percentage of effective array

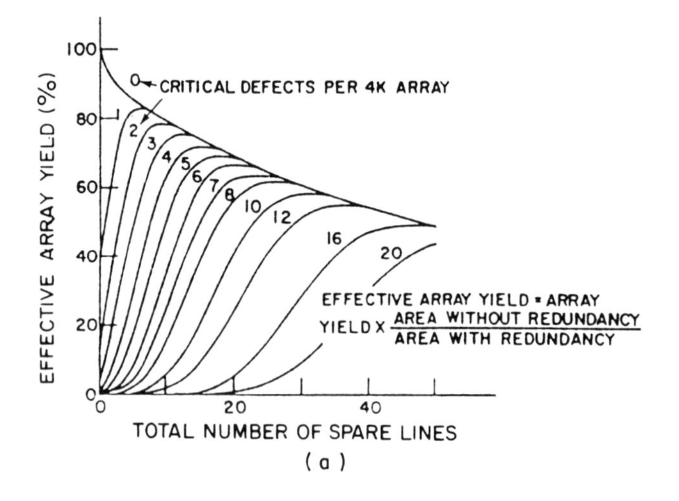

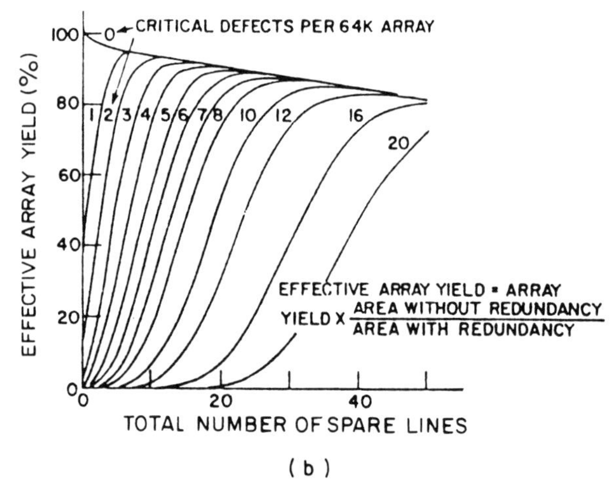

Figure 15.2
Effective array yield improvement with multiple word–bit-line redundancy plotted against total number of spare lines as a function of number of critical defects on the chip for (a) 4k DRAM array and (b) 64k DRAM array. (From Schuster [18], IBM 1978, with permission of IEEE.)

yield against number of spare redundant lines available as a function of number of critical defects. These were plotted from a Poisson type of formula. The array yield with no redundancy expected for the 4k with 1 critical defect is 37%, and for the 64k with 16 critical defects is effectively 0%. With redundancy, however, the effective array yield can be increased to over 80% in both cases.

The number of redundant lines which needed to be added for the 4k was 3 and for the 64K was 24. Of course as the defect density of a wafer area is improved, the intrinsic yield of the higher density memory increases and less redundancy is needed to give the needed yield improvement. Since redundant elements occupy silicon area and impair speed and power, it improves chip performance and reduces chip cost to be able to decrease their number.

We can carry the above reasoning on to another interesting conclusion. Assume as above that a 64k in 1 μm line width has 16 defects and a 4k in 4 μm line width has 1. If we take a 4k segment ($\frac{1}{16}$) out of the 64k, we would expect on the average that it would have 1 defect. That is, a 4k in 4 μm line width has the same expected defect density as a 4k in 1 μm line width. This leads to the conclusion that the yield penalty from random defects for a chip does not change when its area is shrunk by decreasing the minimum design rules. (Of course this assumes that a representative percentage of peripheral overhead and array comes with the 4k segment of the 64k chosen.)

From this one could conclude that yield can be modeled as a function of number of circuits on the chip independent of area if the defect density of the manufacturing area is constant.

IBM [20] confirmed this experimentally in 1986 for mask ROMs by showing measured relative yield data as a function of chip capacity for several factories with known defect densities. These data are shown in Figure 15.3. The data from factory 2 were taken during four different time periods. The points which fall off the curve were taken during periods of low volume. The chips involved were designed in different locations and used different design rules.

This is a strong argument for shrinking memory chips to improve revenue. If the yield in a given wafer fab is given by the number of bits per chip regardless of the design rules then if the chips are shrunk, more good chips per wafer will result. The revenue per wafer will therefore be higher. A mathematical treatment of this result is given in another IBM paper [21] and an extension to include yield variations in a 1986 University of California paper [22]. IBM then used these results and the cluster model given in the first section to determine the cluster parameter 'c' for various factories.

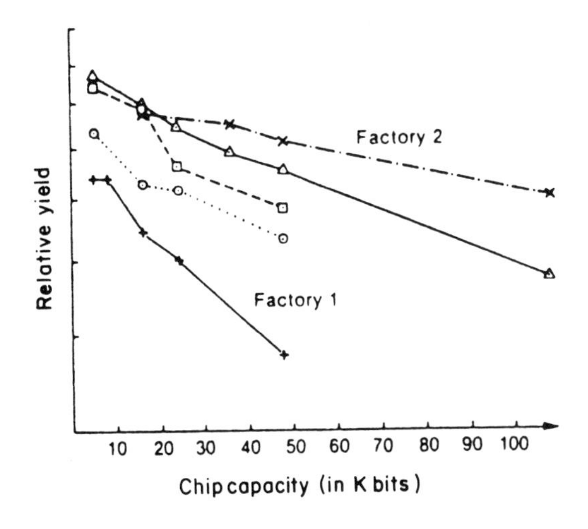

Figure 15.3
Plot of manufacturing data for yield against chip capacity for factory 1 in one time period and factory 2 in four different time periods. (From Stapper [20], IBM 1986, with permission of IEEE.)

Another conclusion from this study is that the yield for higher capacity chips is lower regardless of chip area. This points to the need for redundancy in higher capacity chips.

To this point we have assumed that all defects in the array are correctable. Toshiba [19] in 1982 considered the case where not all of the defects in the array are repairable. A repairability factor P_r was introduced so using Poisson's equation the yield with redundancy becomes

$$Y_r = Y_0 \sum_{i=0}^{n} \frac{(P_r D)_i}{i!}$$

where n is the number of spare cells, D the average number of critical defects per chip and $Y_0 = \exp(-D)$. For $P_r = 1$ we would expect to see the set of curves shown in the last example. Referring to Figure 15.4, these are the curves drawn with the dotted lines. The significant yield improvement with redundancy is again evident.

Toshiba now took the results from actual failure analysis of attempted repairs and extracted a repairability factor of 0.5. The yield with redundancy calculated with this repairability factor is shown in the solid lines in this figure. This shows that there is a limit to the yield improvement as a function of the number of spare cells.

In a slightly more complex example, Figure 15.5(a) shows the theoretical dependence of yield on number of redundant lines when Poisson's equation for yield was applied to a 256k DRAM by NTT [11] in 1982 with the boundary condition that the area of the chip was split into three sections: non-replaceable areas, replaceable areas, and redundant areas. Five to ten redundant lines seem to be optimum in this case with the improvement being greater in the higher technology low defect density fabrication areas.

It can be seen from the figure that the yield with and without redundancy in lower technology (higher defect density) wafer fabs is significantly less than in higher technology lines. For example an unrepaired yield of 10% on a $D = 7\ \text{cm}^{-2}$ line goes to a yield of 70% with redundancy. For lines of lower technology the unrepaired

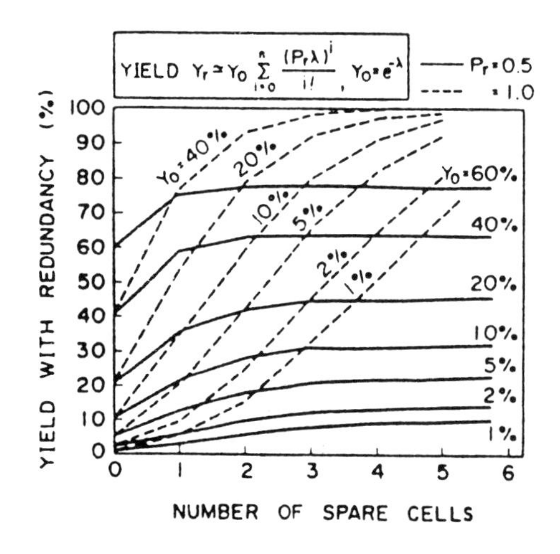

Figure 15.4
Calculated yield improvement with redundancy plotted against number of spare cells for two values of the repairability factor P_r. Dotted lines are $P_r = 1.0$ and solid lines are $P_r = 0.5$. Repairability factor is the probability that a potentially repairable chip can be repaired. (From Uchida *et al.* [19], Toshiba 1982, with permission of IEEE.)

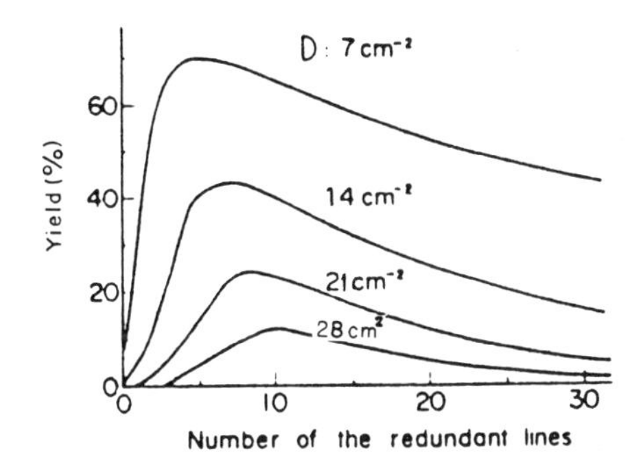

Figure 15.5
Yield as a function of number of redundant lines for different wafer fab defect densities when the chip is partitioned into repairable and non-repairable areas. (From Mano *et al.* [11], NTT 1982, with permission of IEEE.)

yield of this chip would be expected to be zero without redundancy but finite with redundancy. Use of redundant elements on a chip would therefore permit it to yield on an older less expensive wafer fab line than would be required without redundancy.

15.3.2 Redundancy and speed considerations

The implementation of redundant lines on a memory array can impair the speed if these lines are a significant distance away from the lines they are replacing. The attempt is usually made, therefore, to locate the redundant elements in blocks of the array near the locations they will replace.

15.3.3 Redundancy and power considerations

One of the considerations in implementing redundancy is to limit the dc current leakage in the defective element which was replaced.

One solution proposed by Toshiba for a 64k full CMOS SRAM [7] was to both replace a defective row pair and associated V_{CC} line with a spare row pair and

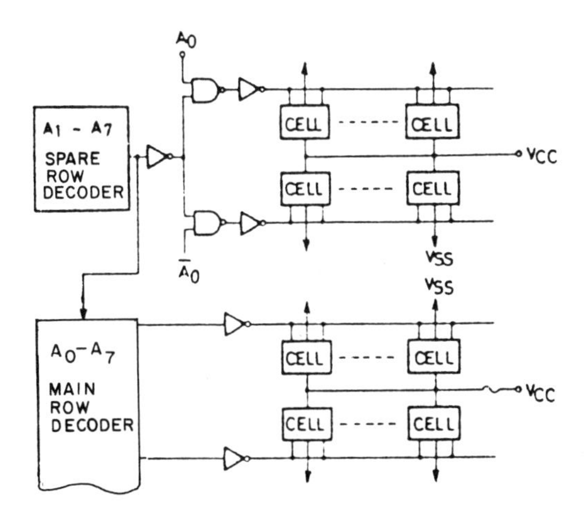

Figure 15.6
Redundancy for low standby current CMOS SRAM using row pairs and the included V_{CC} line for replacement elements. (From Ochii [7], Toshiba 1982, with permission of IEEE.)

associated V_{CC} line and also isolate the defective row electrically by cutting the corresponding V_{CC} line. The main and spare row pairs along with the fuse on the V_{CC} line is shown in Figure 15.6. Standby power dissipation of this part was 50 nW.

Another partial solution is a flag circuit which indicates whether the chip has had the redundancy used or not. If it was not used then the chip is expected to function without degradation of speed and power. This technique was used by NTT [11] in 1982 and by others.

Redundant elements which are typically included on most high density DRAMs, SRAMs and many EPROMs and EEPROMs can be used to replace sections failing during test. These can include redundant rows, columns, decoders, and even sense amplifiers.

15.3.4 Methods of replacing redundant elements

Several methods have been developed for effecting replacement of redundant elements. These include laser blow fuses, laser annealed diffusions, electrically blown links, EEPROM, EPROM or UPROM cells, metal fuses, etc. These divide into those replacement methods which are changeable in the field (EEPROM) and those which are not (fuses, and PROMs). If the only purpose of having the redundancy is to improve factory yield, then either type is adequate. If there is an advantage in periodically correcting hard errors occurring in the field, for example life failures in a large system, then the reprogrammable type might be preferred.

Laser blown polysilicon fuses, which were developed by Bell Laboratories [23], were among the earliest redundant elements. They have been widely used in DRAMs and SRAMs since they occupy a small amount of area and require no extra logic on the chip for programming them. The lack of extra logic on chip also means that access time loss in the extra logic circuitry is avoided. They have several disadvantages including the opening of the surface passivation of the wafer by the laser, and the extra laser process, which is not a standard MOS silicon process, in the clean room. The reliability of laser blown fuses was considered in a different Bell Laboratories report [24].

The hardware required for laser programming includes a computer controlled memory tester, a wafer prober, a laser and a laser positioning system. The memory tester must be configured to record errors during test for subsequent analysis [25].

In 1984 Siemens [27] used a $TaSi_2$ fuse which was laser blown on their 256k DRAM.

Another electrically programmable redundancy element which has been used on a 256k DRAM by NTT [11] is a diffusion fuse link consisting of two polysilicon P–N diodes with a lateral N^+PN^+ structure protected by a molybdenum layer as shown in the layout and schematic cross-section in Figure 15.7(a). The element was programmed by applying a voltage over 11 V and a current over 7 mA. This has the effect of connecting the N^+ regions and reducing the resistivity from more than $10^9\ \Omega$ to less than $3 \times 10^3\ \Omega$. The programming takes about 100 μs by applying a few pulses of 10 μs width. It has the benefit of being compatible with standard MOS processing.

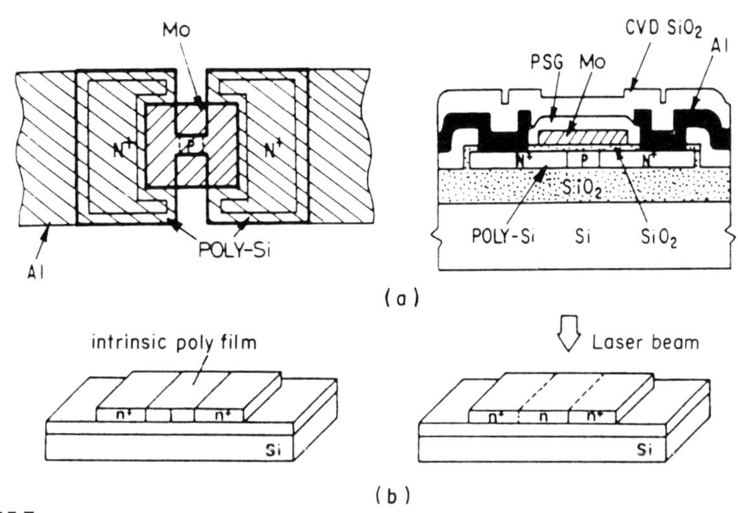

Figure 15.7
Two examples of diffusion fuse links used to implement redundancy in SRAMs. (a) Electrically programmable fuse line layout and cross-section, (b) laser programmable fuse line (cross-section of unprogrammed and programmed link. (a: From Mano *et al.* [11], NTT 1982, with permission of IEEE; b: From Minato [17], Hitachi 1982, with permission of IEEE.)

A similar diffusion fuse which is laser programmable has been used on a 64k SRAM by Hitachi [17]. The N^+NN^+ structure of this polysilicon film device is shown in Figure 15.7(b). To avoid damage to the chip surface from a high intensity pulse of the laser, multiple pulses were used to obtain a larger window than a single low intensity pulse.

An option which has also been used for SRAMs [26] is an electrically blown polysilicon fuse. The address of a faulty element is programmed into the spare element by blowing these fuses during wafer probe.

Electrically blown programmable polysilicon fuses were used on a 16Mb ROM by Toshiba [8] in 1989. In this case the spare element not only had to be inserted in the circuit but also had to be programmed since every cell has unique programming on a ROM.

Another option which has been used on EPROMs is a polysilicon fuse associated with a content addressable memory (CAM) circuit element such as was used on a 128k EPROM by Intel [9] in 1982. The fuse resistance is changed from about 200 Ω to an open circuit by applying a high voltage pulse to the fuse. A fuse protection circuit was used to protect the fuse from the high voltage applied to the chip during programming. CAMs are used to program the redundant row decoders to respond to the addresses of any row. This technique has the benefit that implementing the redundancy can be done either at electrical wafer sort or after packaging at final test.

EPROM redundancy has also been effected by Intel [10] on a 256k EPROM which used UPROM (un-erasable ROM) cells as fuses to link redundant circuit elements. These are EPROM cells that have a metal shield over them to prevent them from being erased by ultraviolet light as shown in Figure 15.8(a).

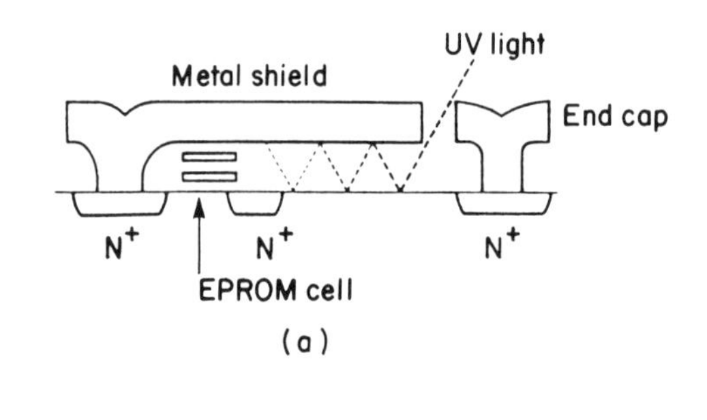

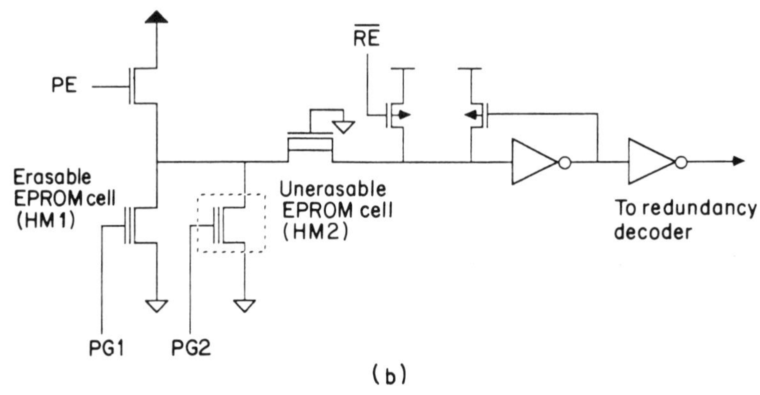

Figure 15.8
(a) Schematic cross-section of UPROM cell showing basic EPROM storage transistor with metal shield to prevent UV light from erasing the programmed cell. (b) Redundancy address circuit using UPROM cell. (a: From Gaw *et al.* [10], Intel 1985, with permission of IEEE; b: From Higuchi *et al.* [12], NEC 1990, with permission of IEEE.)

NEC [12] on a 16Mb EPROM in 1990 also used the UPROM but in conjunction with an EPROM on the redundant elements. For in process test before and after assembly the EPROM was programmed. Then at final test the UPROM can be programmed. This technique permits complete testing at each stage of the manufacturing process. Figure 15.8(b) shows the parallel EPROM and UPROM cells on the redundancy address circuit.

Redundancy on a EEPROM has been included by Intel [3] in 1983 with the use of an EEPROM cell and a CAM. In this case each redundant row has one CAM associated for each row select address. In addition it has a CAM which can be programmed to select that row for use at the address encoded in the CAMs. During testing the number of defective rows and their addresses are remembered. If this is equal or less than the number of redundant rows available, then the redundant rows are tested and, if good, are programmed in. Lattice [4], in 1985, also described a fast EEPROM using EEPROM programmable fuse elements.

15.4 ERROR CORRECTION

Redundancy is used primarily for correcting 'hard' errors in the memory circuits to improve the production yield. With error correction it is possible to detect and correct

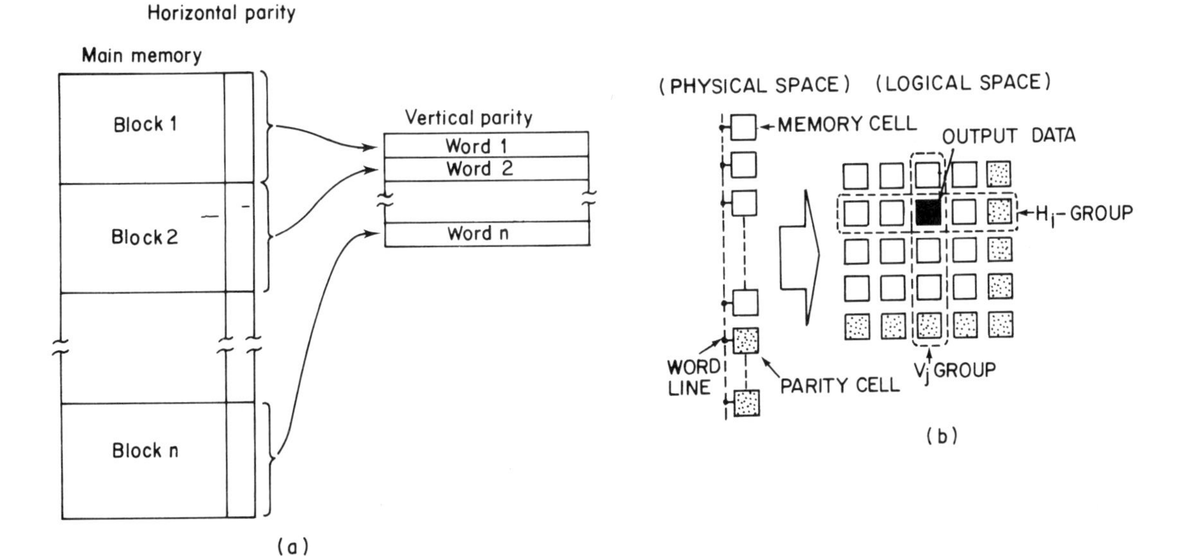

Figure 15.9
Basic bidirectional parity code technique showing: (a) memory divided into blocks of fixed size with a horizontal parity bit added to each byte and a vertical parity bit stored for each block. (b) memory cell with parity cells shown in physical space and horizontal–vertical (*H–V*) groups in logical space. (a: From Osman [29], Unisys 1987, with permission of IEEE; b: From Yamada *et al.* [28], NTT 1987, with permission of IEEE.)

both soft and hard errors occurring either in manufacturing or in the field and so improve the reliability of the memory device in the system as well as the yield. The two methods can be used effectively in combination to give improvement in long term in-system yield.

15.4.1 Error correction theory

Error correction involves two steps. One is detecting the error and the other is correcting it. The simplest method of error correction is a simple complete duplication of every bit in the array. The drawback is, of course, the increase in the chip size. It is possible to use various coding techniques which take less area on the chip.

Hamming codes [1], which were first developed by Hamming in 1950, are frequently used for this purpose. With a Hamming code, error correction can be performed on many different bit field sizes. This entails using additional parity bits in the array. To determine the number of parity bits needed to correct a given number of data bits, the following inequality must be satisfied

$$2k \geqslant m + k + 1$$

where 'm' is the number of data bits to be corrected and 'k' is the number of parity bits needed for correction.

There is a trade-off in that small bit fields allow fast error detection but require a large number of extra parity bits in the memory matrix. Large bit fields permit use of a smaller amount of extra matrix for the parity cells but error detection is slower.

With single error correction, a memory word with one bad bit can be corrected to recover the original information, whereas two bad bits can be detected, but not corrected.

Another type of error correction code which has been used on memories is the 'horizontal–vertical' error correction technique was developed by Edwards [30] in 1981 and proposed for DRAMs by Burroughs [29]. This method is also called bi-directional error correction.

In this method the memory is divided into blocks of a fixed size as shown in Figure 15.9(a). To each byte, a horizontal parity bit is added. A vertical parity is also generated and stored. During normal operation only the horizontal parity is checked which improves the time delay due to the error correction.

If an error is detected when the horizontal parity bit is checked then the vertical parity bit is checked. The error is then found at the intersection of the two parities. For wide bus memories the overhead involved in using bi-directional parity is less than that of the Hamming codes.

The basic concept is shown generalized in Figure 15.9(b). Memory cells and parity cells associated with each word line are arranged into a two-dimensional logical space which forms horizontal and vertical groups. Parity information for each data group is written in the appropriate parity cell. Error correction is done by checking the parity of the associated pair of horizontal and vertical parity bits. This technique allows simple ECC operation with smaller error correction circuitry than does Hamming code.

In order to compose a bidirectional parity code $k + m + 1$ parity cells are added to $k \times m$ memory cells on each word-line. The same number of bit-lines are also added to read these parity cell data. A pair of horizontal and vertical data selectors are arranged close to the cell array to select a pair of H and V groups including the output data.

The error detection circuit consists of only a '$k + 1$' bit, a horizontal parity checker, a '$m + 1$' bit vertical parity checker, and a two input AND gate. Two kinds of parity checking are carried out simultaneously and then the two input AND gate operates as a decoder which detects a bit error from the results of the parity checking and supplies the error correcting signal. Using this signal the data is corrected through the XOR gate. The improvement in access time and chip area overhead as a function of data bit length of the bidirectional (H–V) error correction code are shown in Figures 15.10(a) and (b). The figure also shows the additional improvement of a variation of the bidirectional error correction code called selector line merged ECC which reassigns the H–V logical space diagonally along word-lines to permit short selector lines for both the horizontal and vertical selectors.

Error correction is useful in systems not only to correct soft errors due to noise or alpha particle hits, but also for hard error correction. A soft error can be corrected by reprogramming the bit. A hard error is a 'stuck at' failure. Since only one error in a word can be corrected, a system failure with error correction present occurs only if two errors occur in the same word. This occurrence can be lessened by clearing the soft errors more frequently and by regularly replacing parts with hard errors.

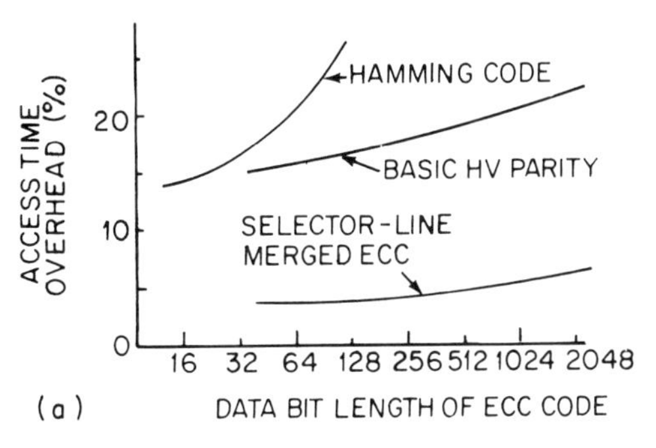

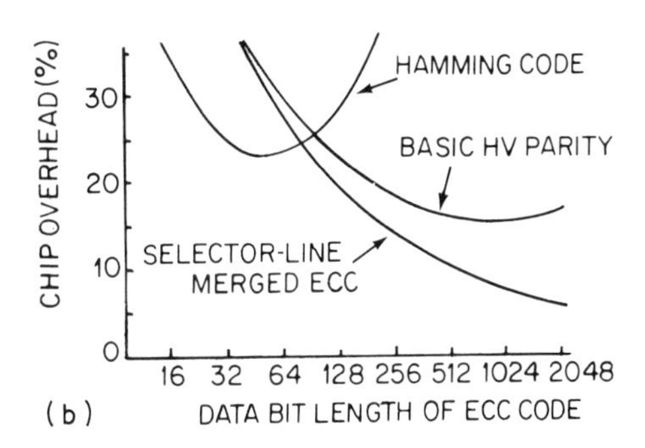

Figure 15.10
Comparison, for various error correction codes, of (a) access time and (b) chip area overhead as a function of data bit length of the ECC code. (From Yamada *et al.* [28], NTT 1987, with permission of IEEE.)

15.4.2 Error correction on ROMs

While redundancy can be used on a RAM for yield enhancement, other methods are normally used for ROMs since every row and column of a ROM is uniquely programmed and can be different from every other row.

If redundancy is used, then the entire ROM array must be duplicated thereby doubling the size of the array. This was done on the NTT 4Mb ROM discussed in Chapter 10.

A method of having essentially the same yield improvement while occupying less area on the chip is to use an error correction technique. The most common error correction technique involves the use of Hamming code as described above.

In an example from a 1985 1Mb ROM paper by Mostek [2], error correction is performed on 64 data bits using seven parity bits for a total data field of 71 bits. This requires only about 11% extra matrix area.

In this case a single error in the 71 bit field can be corrected, but more than one error in any given 71 bit field cannot be corrected. The 71 bits were therefore spaced throughout the array in such a way as to maximize the error correction coverage.

The drawback of this method is that the 71 bit field requires seven 33-input EXCLUSIVE-OR, (XOR) gates for error correction which can be very slow. To improve speed Mostek used two-tier staged XOR gates as shown in the diagram in Figure 15.11(a). The 33 input gate is broken up into four first tier eight input gates whose outputs are combined with one parity bit in a second tier five input gate as shown in the diagram. This reduces the maximum number of series XOR gates to eight. The circuit schematic of the eight input XOR gate is shown in 15.11(b). In this gate it takes about 10 ns to detect and correct a bad bit.

An example of a fast 35 ns 256k EEPROM with error correcting circuitry was given by SEEQ Technology [21] at the 1990 ISSCC. In this case error correction with four parity bits per byte was used with an associated access time penalty of about 3 ns to enhance the endurance to 100k cycles. For yield improvement redundancy was also used with four electrically fusible redundant rows and two byte-wide redundant columns being used.

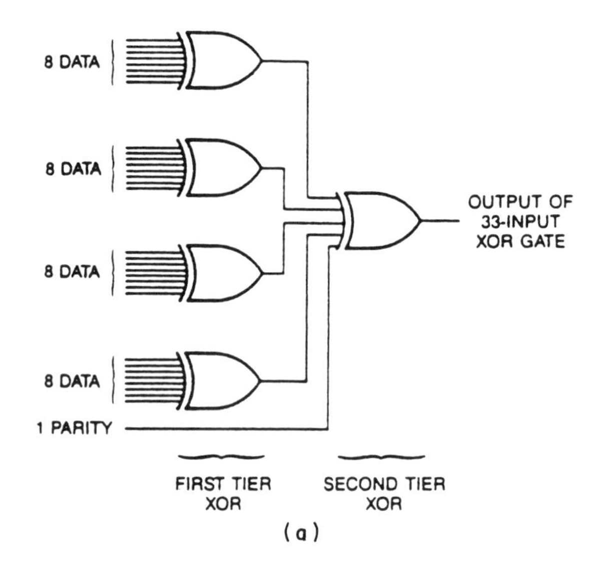

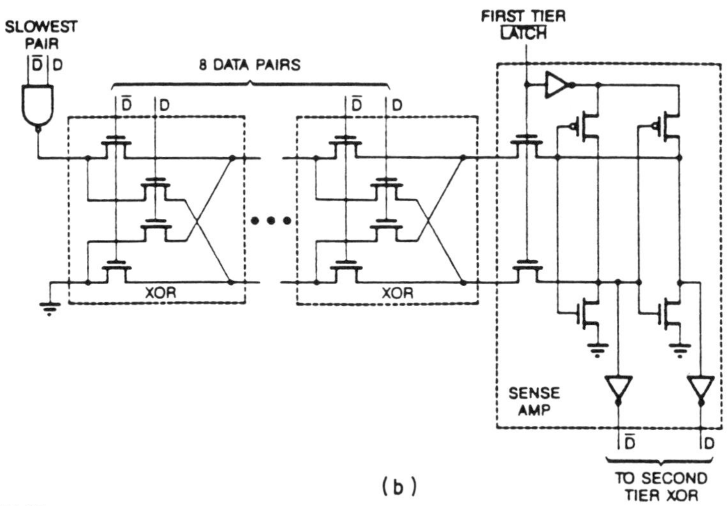

Figure 15.11
Circuitry for improving memory speed with Hamming code type of error correction. (a) Two tier XOR gates reducing 33 input XOR gate to four 8 input XOR gates. (b) Schematic circuit diagram of one of the eight input XOR gates. (From Davis [2], STM (Mostek) 1985, with permission of IEEE.)

15.4.3 Error correction on DRAMs

Error correction has become more commonly used in the high density multimegabit DRAMs for correction of soft errors in addition to the redundancy circuitry used to improve yield.

An example of an application-specific 1Mb cache DRAM with error correction was given by Mitsubishi [22] in 1989. This chip had double error detection and single error correction using 8 parity bits for a block of 32 data bits. While only a single error could be corrected with this number of parity bits, a flag for an uncorrectable error was set if a double error was detected.

Another application of error detection and correction from Mitsubishi [32] the same year was a 16Mb DRAM with background error correction in parallel with dual stage amplification which minimized the access time delay due to the error correction.

A data field of 136 bits consisting of 8 parity bits and 128 data bits was used to construct a Hamming matrix. The matrix required 65 transmission type XOR gates to generate the signals to assign an error position. For speed reasons, a double tier XOR scheme was used with eight XOR gates in the first tier and 9 gates in the second tier. A circuit diagram of the memory cell array is shown in Figure 15.12(a).

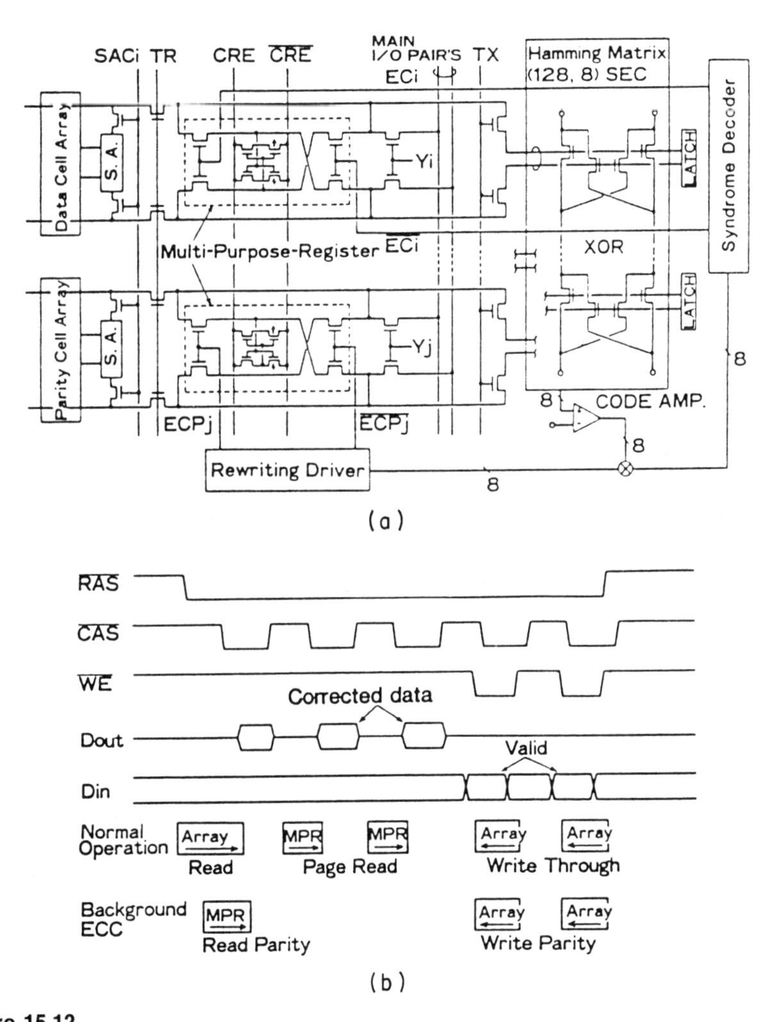

Figure 15.12
16Mb DRAM with background error correction. (a) Circuit diagram of a memory cell array using background Hamming code in parallel with dual stage amplification. (b) Timing diagram of fast page mode showing background error correction. (From Arimoto *et al.* [32], Mitsubishi 1988, with permission of IEEE.)

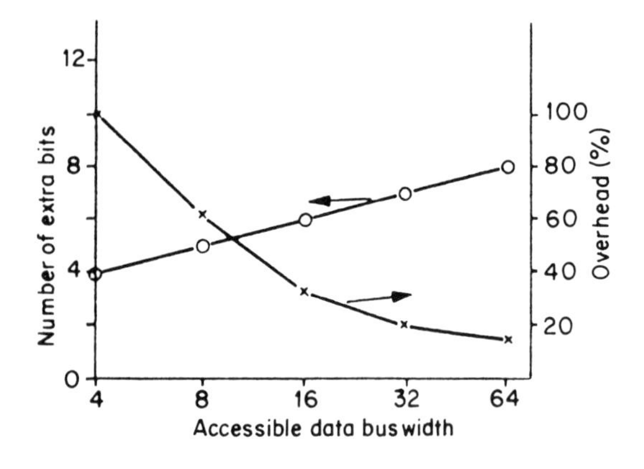

Figure 15.13
Memory overhead for implementing a Hamming code showing number of extra bits and percent overhead as a function of accessible data bus width. (From Osman [29], Unisys 1982, with permission of IEEE.)

In this part the data bits on the selected word-line were checked and corrected in the read and refresh cycles, while the parity bits were corrected or rewritten in the write cycles. The timing diagram of the fast page mode for this DRAM in Figure 15.12(b) shows the background error correction operation. In the read cycle the data after the second page mode cycle are corrected with no degradation of access time. In write cycle there is a small access time penalty for write parity.

All of the above circuits used the well known Hamming code techniques. The drawback of Hamming codes is that they are slow and consume substantial area. A chart of the memory overhead for implementing a Hamming code is shown in Figure 15.13. This graph plots number of extra bits and percent of overhead against data bus width for which one error is corrected and two detected. While overhead drops as the bus width increases, 20% overhead for a 32 bit data bus width is still significant.

Several companies have used the bidirectional or 'horizontal–vertical' (*H–V*) error correction code on high density wide bus DRAMs. This approach both offers an overhead reduction and a speed improvement over Hamming codes.

NTT [28] developed a '4-bit-at-a-time' bidirectional parity code with a self checking function in a 1984 1Mb DRAM paper which used the horizontal–vertical parity concept. This code took 98k parity cells and a built-in ECC circuit which together occupied about 12% of the chip area and took 20 ns for ECC operation. This compares favorably with the 100% bit overhead required using the Hamming code technique for 4 bit bus width as was shown in Figure 15.13.

A concurrent error correction using the bidirectional parity code for DRAMs with multiple bit data outputs was described by Matsushita [31] on their 4Mb DRAM in 1987. This error correction circuit could concurrently check 16 bit errors and correct 1 error in 20 ns.

Rather than checking the vertical direction parities separately as in the conventional bidirectional parity code, the concurrent ECC checks all vertical directional parities at the same time. This means that the concurrent method eliminates the need for vertical directional selectors since all vertical cell data are directly transferred to the vertical parity check circuits. Also it needs only one error correcting group for multiple data output. Fewer parity cells and no vertical directional selectors are needed means that a smaller chip can be realized for DRAMs with multiple bit data outputs.

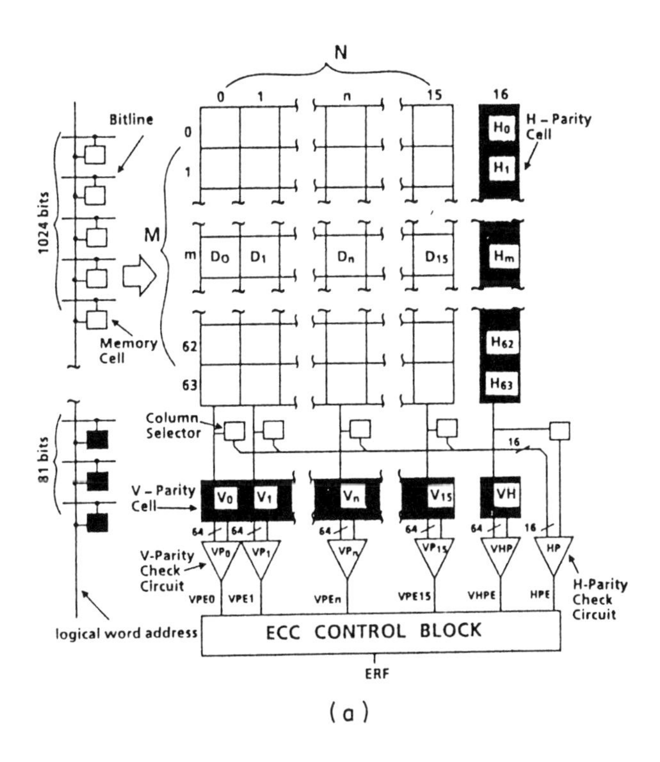

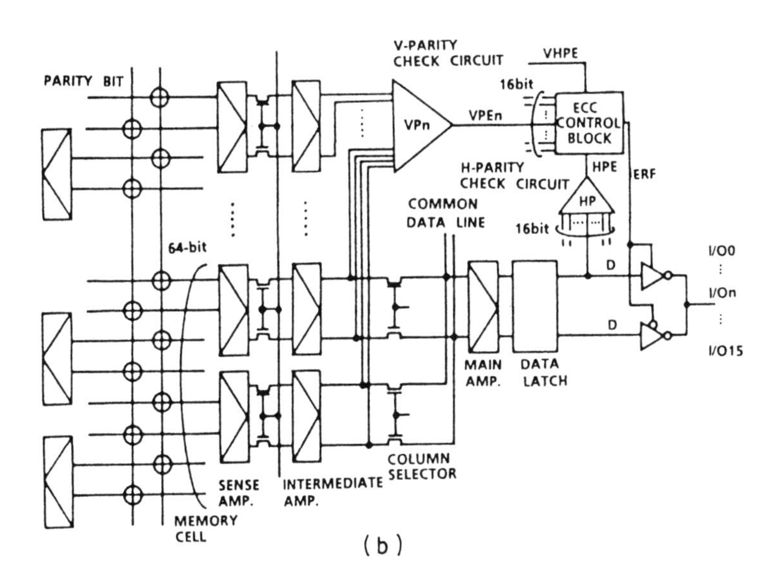

Figure 15.14
Bidirectional error correction method for multiple bit data outputs in which all vertical parities are checked concurrently. (a) Schematic of logical vertical–horizontal space. (b) Schematic circuit. (From Yamada *et al.* [33], Matsushita 1988, with permission of IEEE.)

The ECC circuit has 16 vertical parity check circuits, one vertical–horizontal parity check circuit and one horizontal parity check circuit. The parity code is applied to 1024 memory cells and 81 parity cells which consist of 16 vertical and 64 horizontal parity cells and 1 self checking vertical–horizontal parity cell. A schematic of the error correction concept and the circuit configuration corresponding to the fundamental 1024 cells and 16 bit outputs is shown in Figures 15.14(a) and (b).

In all 331 476 parity cells are included on the chip. The error correction circuitry adds only 7.5% to the area of the chip and hence is feasible to use on multimegabit DRAMs.

Most on-chip error correction schemes have a means of being disabled so that the memory can be fully tested for hard errors and feedback given to the wafer fabrication area.

15.5 COMBINATION REDUNDANCY AND ERROR CORRECTION

As memory chips continue to increase in density and grow in area, neither error correction nor redundancy alone may be sufficient to give acceptable yields.

A combination of error correction and redundancy was used by IBM [7] in their 16Mb DRAM. A single error correct and double detect Hamming code was used for the error correction. Word level replacement redundancy was used.

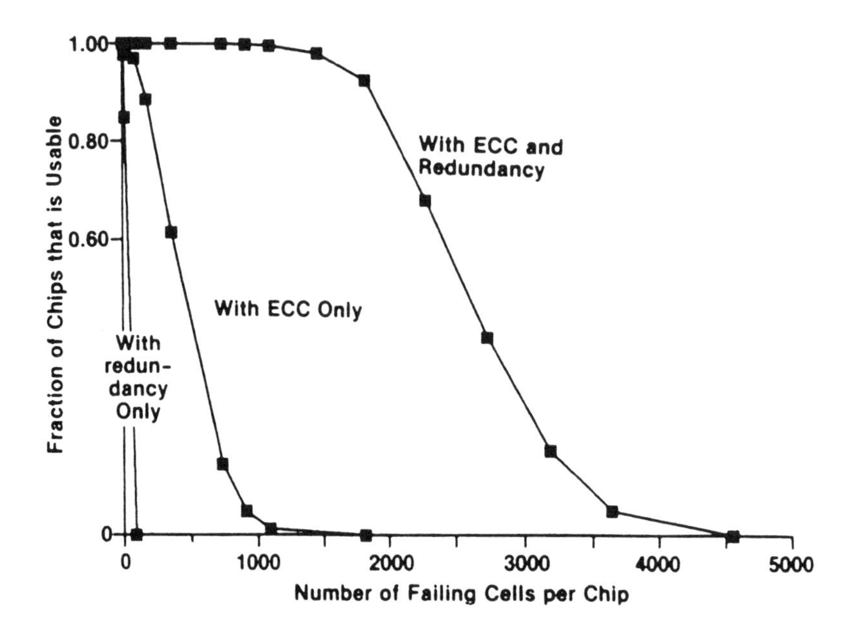

Figure 15.15
Comparison of fraction of usable 16Mb DRAM chips as a function of failing cells per chip for redundancy only, error correction only and with error correction and redundancy combined. (From Kalter *et al.* [16], IBM 1990, with permission of IEEE.)

The error correction was provided to improve the single bit failure type of mode and the redundancy for fixing bit-line and word-line failures. Figure 15.15 shows the significantly improved yield obtained with the use of both ECC and redundancy.

15.6 YIELD IMPROVEMENT OVER TIME IN A TYPICAL MEMORY WAFER FABRICATION

The cost of a typical memory product decreases over time as 'learning' takes place on a wafer manufacturing line. This learning consists of several parts.

Fault reduction in the wafer fabrication is one of the areas where learning occurs. It results in improved yields through learning the process steps as the product moves into volume production and through ongoing engineering efforts targeted at identifying and analyzing faults which occur and eliminating the source.

Figure 15.16 shows an example from IBM [20] of the reduction in average number of faults occurring over time in similar memory products in two factories.

Faults due to defects on the processed wafers can result from a number of causes. Particulate count in the air is a major cause and can result in particles settling on the wafer during the lithography operation where they can result in unexposed resist spots and later holes or hillocks following the subsequent etch operation. Another source of faults is line registration error or displacement of artwork edges caused by such process phenomena as under and over etching, lateral diffusions and lithography misalignment.

Ongoing process engineering identification and analysis of fault is required to give the type of fault reduction shown in the two IBM factories. Fault reduction leads to yield improvement. Figures 15.17(a) and (b) show the forecast and actual yield increases of a 256k DRAM and a 1Mb DRAM over a three year period from initial production in two IBM factories.

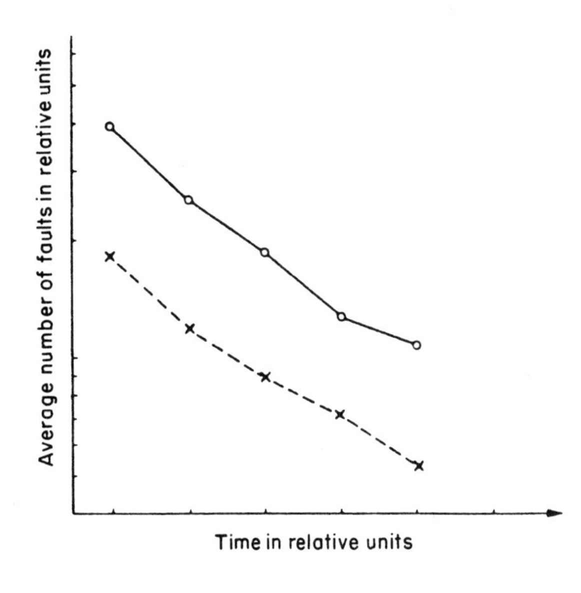

Figure 15.16
Wafer fab fault reduction over time showing 'learning' effect. (From Stapper [20], IBM 1986, with permission of IEEE.)

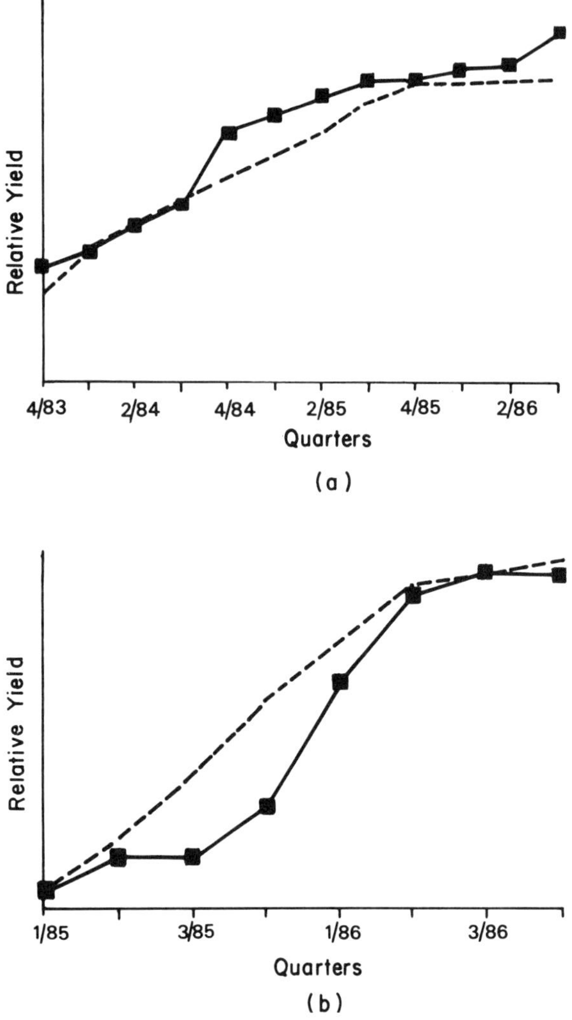

Figure 15.17
Yields in first three years of production of high density DRAMs. (a) 256k DRAM, (b) 1Mb DRAM (--- planned yield, -■-■- yield). (From Stapper [20], IBM 1989, with permission of IEEE.)

If the chip size was also reduced during this three year time frame resulting in more parts out for the same yield, then further cost reduction would result.

15.7 A SIMPLE COST ANALYSIS OF A WAFER PROCESSING LINE

There are many components of cost which make up the learning curve over time in a typical memory wafer fab.

A simple cost analysis model of a memory wafer processing line over a 5 year period is shown in Table 15.1. This is part of a larger cost model constructed by

Table 15.1 Typical 1Mb DRAM cost model.

Year of Production	1	2	3	4	5
Chip area (mm^2)	62.90	54.20	54.20	48.35	42.50
Gross die per wafer	220	250	250	290	330
Price per wafer ($)	60	60	56	54	54
Fab utilization (%)	28	76	94	100	100
Wafer Fab Yield (%)	85	90	94	95	95
Cost per unit ($)	22.23	3.79	2.29	1.57	1.29
E-sort die yield (%)	22	48	62	72	75
Cost per unit ($)	3.30	0.57	0.41	0.29	0.24
Assembly yield	91	93	94	95	96
Cost per Unit ($)	1.27	0.32	0.30	0.29	0.28
Final test yield	80	87	90	92	94
Cost per Unit	5.90	1.43	0.98	0.69	0.55
G & A of manufacturing	1.94	0.21	0.13	0.09	0.08
Total yield	13.2	34.3	48.8	59.2	64.3
Total cost per unit from manufacturing	34.66	6.34	4.13	2.94	2.46
R & D cost per unit	4.16	0.76	0.50	0.35	0.29
Other overhead cost/u					
Total yield (%)	13.2	34.3	48.8	59.2	64.3
Total units per wafer	29	86	122	172	212
Total cost per unit	46.09	8.33	5.41	3.85	3.21

Albert Maringer of Siemens as representative of a typical 1Mb DRAM line. This model should be understood as representative of the industry and not specifically as the actual cost of any particular production line.

Chip area is shown decreasing over time as learning in the manufacturing process makes shrinks feasible. This results in the increase in gross die per wafer. Yield increase through learning occurs in each operation. The resulting units out per wafer from 29 to 212 give a seven-fold increase over the five year time period.

The initial low fab utilization is a contributing factor to initial high costs since overhead is spread over a smaller number of wafers. The wafer fab contributes more than 40% to the total unit cost.

The source of the cost reduction shown in the bottom line is due to several factors. The reduction in die area permits the gross die per wafer to increase and hence the good die out to increase. The factory utilization increases as volume production occurs permitting the overhead costs to be spread over more units. The wafer fab yield which is number of wafers output as a function of number of wafers input increases with experience with the process. Good chip yield per wafer at wafer sort increases as a result of fault reduction and design modifications.

Table 15.2 Costs of operation (million of $).

Year of Production	1	2	3	4	5
Wafer fab					
Capital cost	25.12	32.57	31.27	29.97	28.67
Maintenance	0.63	0.63	0.63	1.25	3.40
Direct labour	4.06	5.51	6.00	6.18	6.37
Indirect labour	2.69	3.30	3.82	3.93	4.05
Total material	7.17	19.01	23.44	26.37	25.30
Other					
Total cost	45.22	61.64	65.78	67.33	68.66
Wafer sort:					
Capital cost	3.65	5.21	6.60	7.08	7.03
Total cost	6.70	9.31	11.62	12.23	12.60
Final test					
Capital cost	6.36	15.05	17.71	18.98	18.18
Total cost	12.00	23.29	28.06	29.70	29.38

Table 15.3 Capital investment (millions of $).

Year	−1	0	1	2	3	4	5
Wafer fab							
Building	5.0	18.0	2.0	0.0	0.0	0.0	0.0
Clean room	0.0	25.0	5.0	0.0	0.0	0.0	15.0
Infrastructure	0.0	7.0	3.0	0.0	0.0	0.0	5.0
Equipment	0.0	60.0	25.0	0.0	0.0	0.0	25.0
Total	5.0	110.0	35.0	0.0	0.0	0.0	45.0
Wafter sort							
Building	0.0	5.0	0.0	0.0	0.0	0.0	0.0
Clean room	0.0	5.0	0.5	0.0	0.0	0.0	3.0
Infrastructure	0.0	1.0	0.5	0.0	0.0	0.0	1.0
Equipment	0.0	6.7	5.8	6.4	3.0	1.0	4.0
Total	0.0	17.7	6.8	6.4	3.0	1.0	8.0
Final test							
Building	0.0	8.0	2.0	0.0	0.0	0.0	0.0
Test room	0.0	10.0	5.0	0.0	0.0	0.0	0.0
Infrastructure	0.0	5.0	2.0	0.0	0.0	0.0	1.0
Equipment	0.0	7.5	28.0	13.0	8.0	0.0	6.0
Total	0.0	30.5	37.0	13.0	8.0	0.0	7.0
Total investment	5.0	158.2	78.8	19.4	11.0	2.0	60.0

All of these factors combine to give a total yield for the product involved through wafer sort. Yields for assembly, final test and mark are added to give a bottom line yield for the manufacturing process which goes from 13.2% to 64.3% in five years.

The manufacturing cost per unit goes from $5.90 to $0.55 in the five years as a result of the yield improvement. The total cost per unit goes from $46.09 to $3.21 as a result of adding in various overhead and R&D expenditures.

Costs which are calculated in each operation include amortized capital cost, maintenance, direct labour, indirect labour and material. These are shown in more detail in Table 15.2.

Direct labour is a large part of total labour indicating the relatively labour intensive nature of the semiconductor operation. As factory automation increases, indirect labour should increase as a percentage of the total.

Materials costs are about 30% of the wafer fab costs. Table 15.3 shows the distribution of materials costs for year four which was the first year with full capacity utilization. These include wafers (53%), chemicals (9.3%), processed gases (8.9%), distilled water (1.2%), supplies (9.7%), electricity (7.9%), water–sewage (3.1) and compressed air for vacuums (6.6). The cost of the silicon wafer is by far the most significant factor.

Capital costs, which are shown amortized over the life of the product, are more than 50% of the total cost. Table 15.3 shows the dollar investment in the year it was made.

Note that the clean rooms needed to reduce defect density in the conventional wafer fab are a significant part of the total cost.

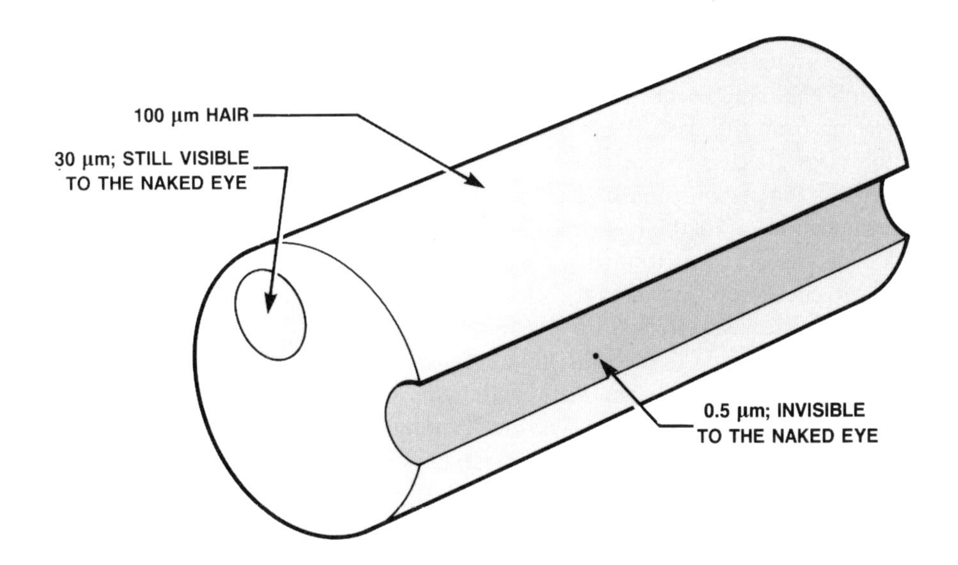

Figure 15.18
Relative side of submicron features. (Reproduced with permission of Philips Research.)

15.8 TRENDS IN THE MODERN MEMORY FACTORY

Clean room costs are a larger part of the capital investment in the wafer fab than is the building itself. The clean room is required to reduce air particles which can cause defects in devices with the minimum line widths, in this example, which range from 1.2 to 0.8 μm. This means that air particles of submicron dimensions can cause defects. Figure 15.18 shows the relative dimensions of a human hair, and a submicron particle which is invisible to the naked eye.

The level of cleanliness of wafer fabs is usually indicated as 'class' of clean room. The class of a clean room is given in terms of number of particles permitted per cubic meter of air. Figure 15.19 shows a schematic cross-section of a typical building housing a Class 10 clean room with Class 1 wafer handling areas and a class 1000 service tunnel through which the clean room is reached for maintenance. Notice that the actual manufacturing area, which is the class 10 clean room is a very small percentage of the total facility required to insure the cleanliness of such a room.

The services, other than air, required for a class 1 factory which can produce 5000 wafers per week are shown in Figure 5.19.

15.8.1 The traditional 'clean room factory'

Figure 15.20 shows the advances in clean room air handling techniques from a typical class 10 000 factory in 1980 to a typical class 10 factory in 1990.

In the class 10 000 factory shown in Figure 15.20(a) there is a central air processing system and central air circulation through the various production tunnels in the clean room. While the air class in both the clean room and the service tunnels is class 10 000, the air class in the wafer handling area is class 100.

Clean oversuits without air breathing apparatus are worn by personnel to cover clothes and hair. Equipment is fully in the clean room with access to the clean room for servicing from the class 10 000 auxiliary tunnel.

In the class 10 clean room shown in Figure 15.19(b), separate air processing is provided for the personnel area (class 10), the wafer handling area (class 1) and the service tunnel (class 1000). The clean tunnels are partitioned away from each other. The floor is raised and perforated so that air flow is from ceiling to floor so that no upward air circulation occurs in the clean room area. There is a full ceiling HEPA filter system with air return through the raised access floor. The air flow is at 90 ft min^{-1}.

The personnel wear full 'space suits' with respirator to isolate them from the clean area. The equipment is in the service tunnel mounted flush with the clean room to give access for processing. All equipment repair is done in the service tunnel.

Other requirements than air handling include ultra pure services such as gases and DI water, strict environmental control, electromagnetic control measures such as planar electrical grounds and vibration controls.

Figure 15.21 shows the minimum feature size, wafer diameter and clean room class for SRAM and DRAM manufacturing from 1970 to 1994.

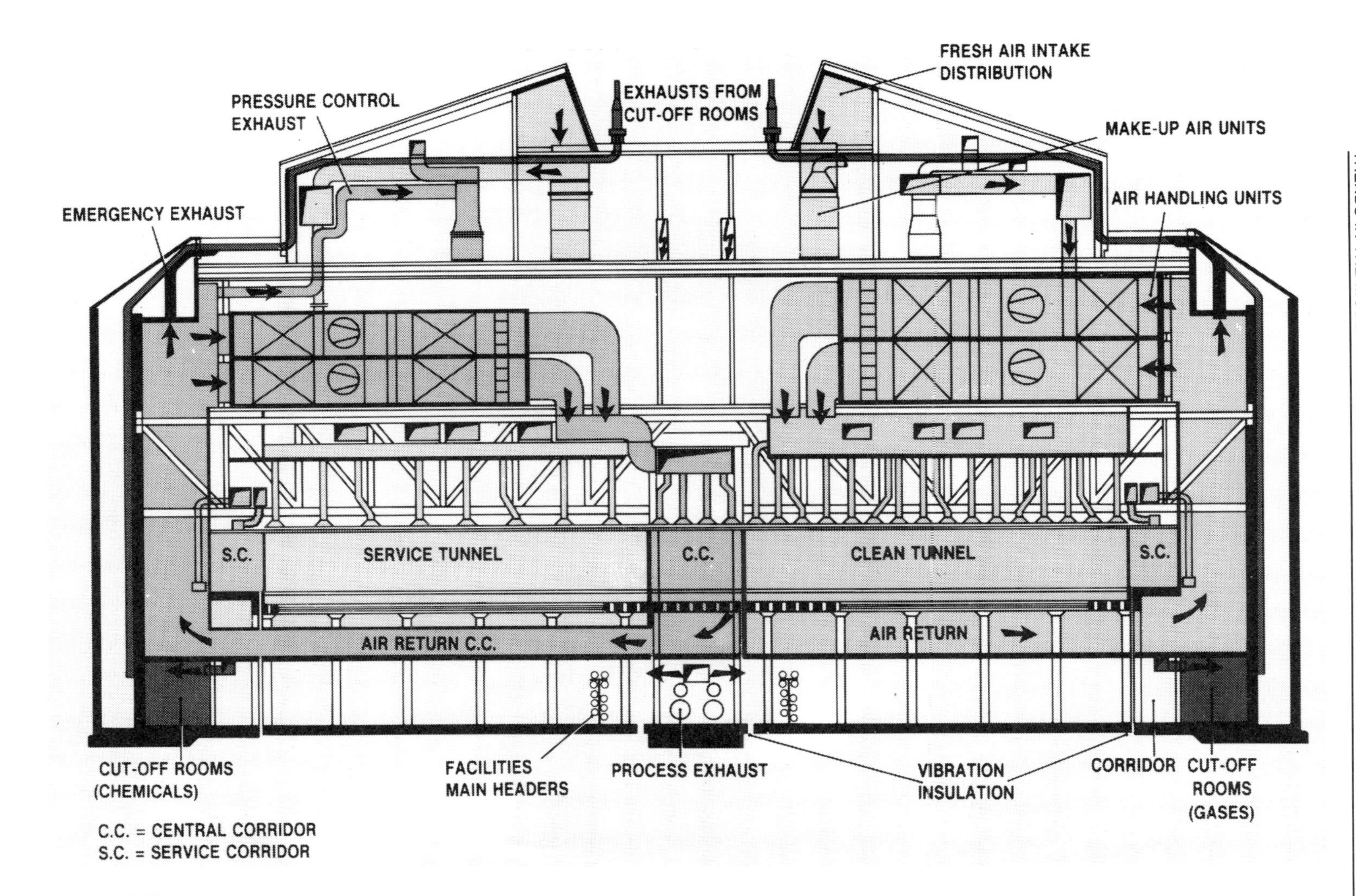

Figure 15.19
Air handling system for submicron wafer fab with a class 10 clean room. (Reproduced with permission of Philips Research.)

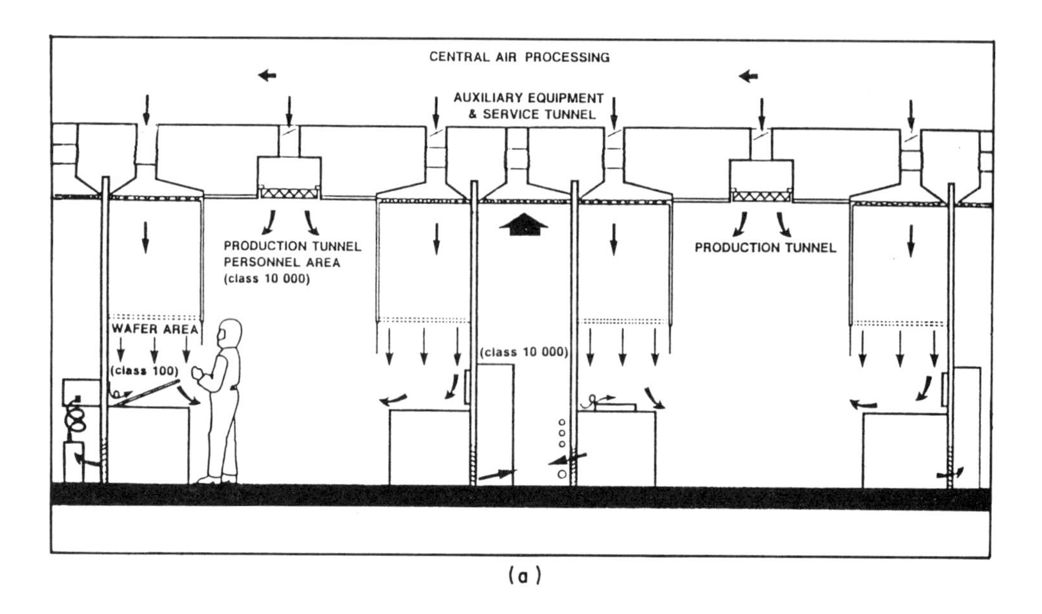

(a)

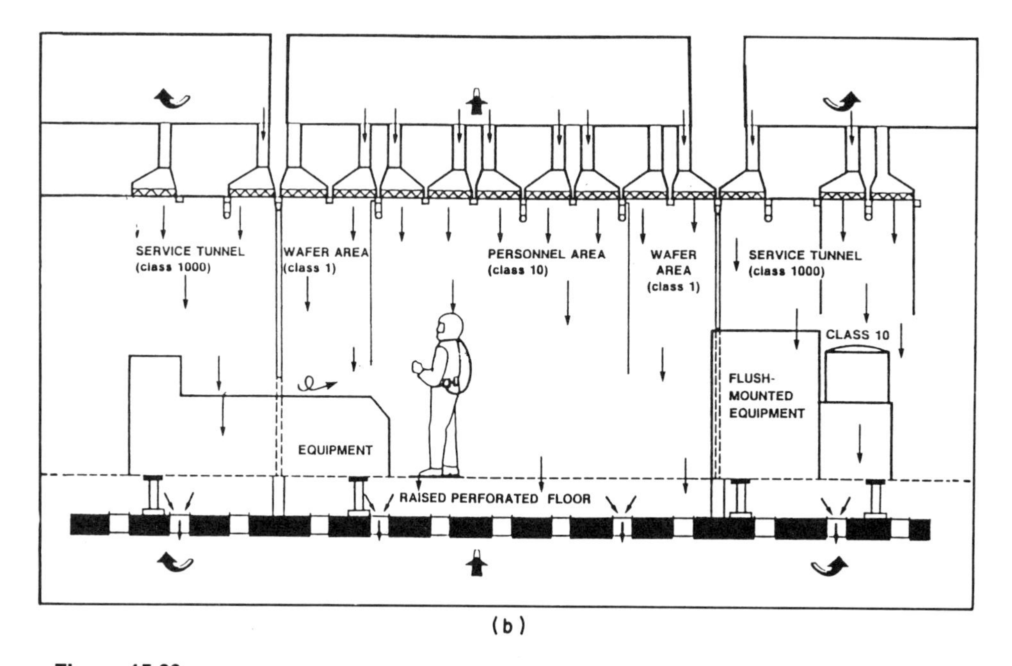

(b)

Figure 15.20
State-of-the-art clean room air handling in: (a) 1980 tunnel concept, (b) 1990 100% down-flow. (Reproduced with permission of Philips Research.)

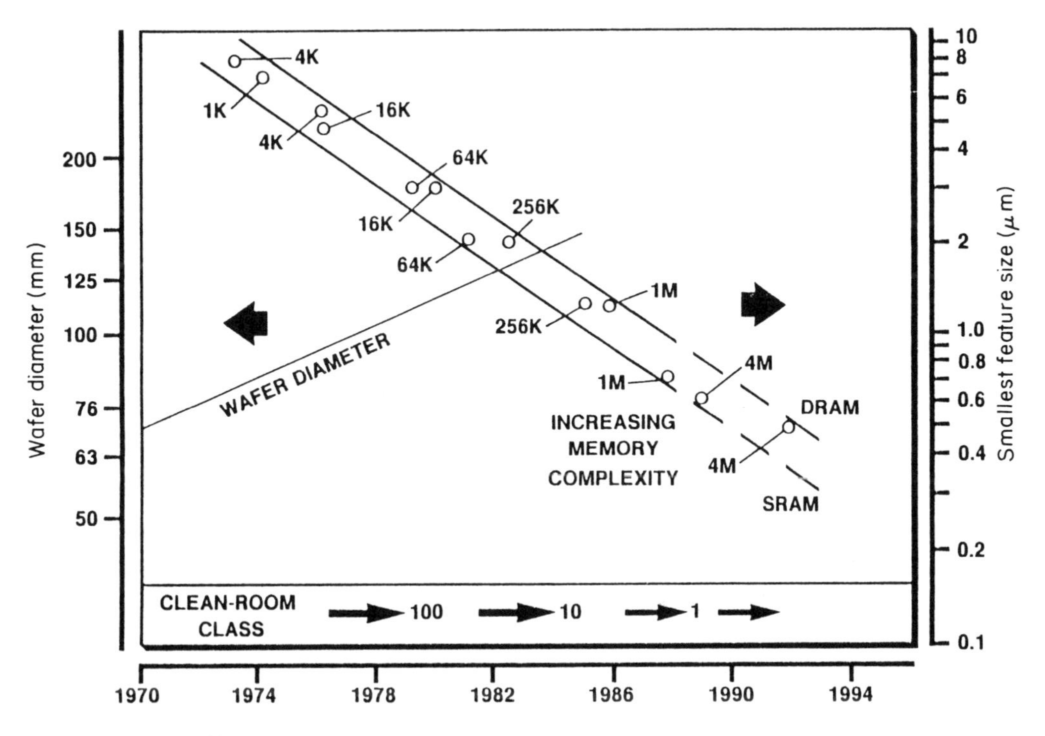

Figure 15.21
Wafer diameter and smallest feature size by year for various densities of SRAM and DRAM. (Reproduced with permission of Philips Research.)

In light of these extreme requirements for processing of memories with submicron features, the $700 million investment required for a 0.5 μm 16Mb DRAM factory becomes understandable. The next question has to be: is it necessary? Such extreme measures seem to beg for an alternate solution. One such solution would be total automation of the wafer fab, a goal which to date has eluded the semiconductor industry, although extensive automation is found in such factories. Another solution is to attempt to isolate the wafers from the clean room and have only 'clean' machines to worry about. The 'SMIF' method is an attempt in this direction.

15.8.2 A new concept in clean processing

The SMIF (Standard Mechanical InterFace) system of Asyst Technologies is an alternative technology for achieving better than class 10 environment for the wafers while relaxing the clean room requirements outside of the system.

The components of this system include a clean wafer enclosure, a mechanical arm, and the equipment enclosures which together serve to isolate the wafers and equipment from the external environment. The schematic of the system is shown in Figure 15.22 which shows two cassettes of wafers in storage being transferred from inside the processing equipment enclosure.

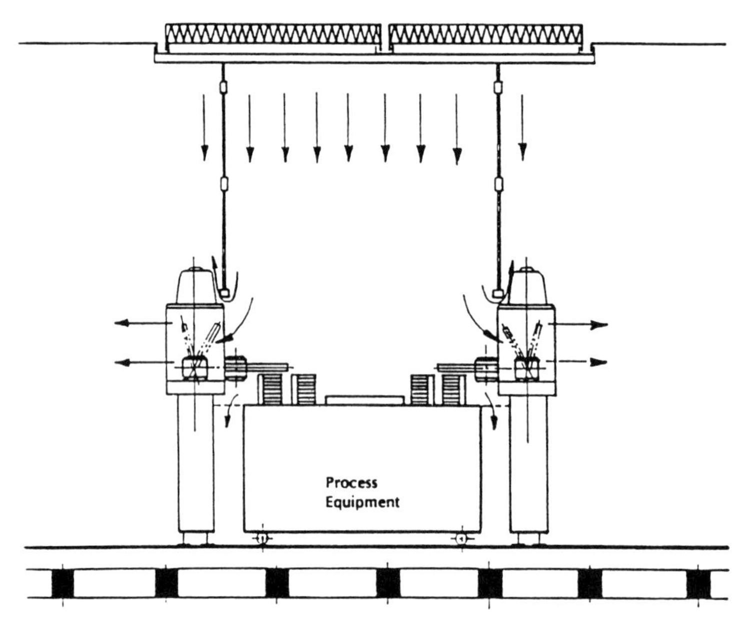

Figure 15.22
The Asyst SMIF automated wafer transfer process showing wafer cassettes being transferred from the processing equipment enclosure by mechanical arms. (Reproduced with permission of Asyst Technologies.)

A small sealed container is used for storing and transporting wafer cassettes. These containers have a specially designed port which is used to transfer cassettes to and from the process equipment. The pod has a disposable inner shield and a wafer lock mechanism which secures the wafers in the cassette during storage and transport. The inner shield totally surrounds the wafer cassette to give improved isolation from the environment.

The mechanical arm is a robotic mechanism for automated transfer of cassettes to and from the equipment enclosures. The equipment enclosures isolate the process equipment from the external environment. Clean air to class 10 or better requirements is supplied inside the equipment enclosure from dedicated air filter units.

The results of an experiment by Asyst Technologies [39] in which wafers were transferred in different environments with and without the mechanical interface are shown in Figure 15.23. These results indicated that automated handling under even relaxed class 20 000 external conditions is better than the conventional human cassette handling in a class 10 environment with transport in a box.

The results improve also with cleanliness of the clean room environment, so that in the deep submicron geometries, it may be necessary to have both the external clean room environment and a contained transport method.

One of the benefits of SMIF is that it can extend the life of current clean room facilities for a number of years thereby reducing capital investment. Another benefit

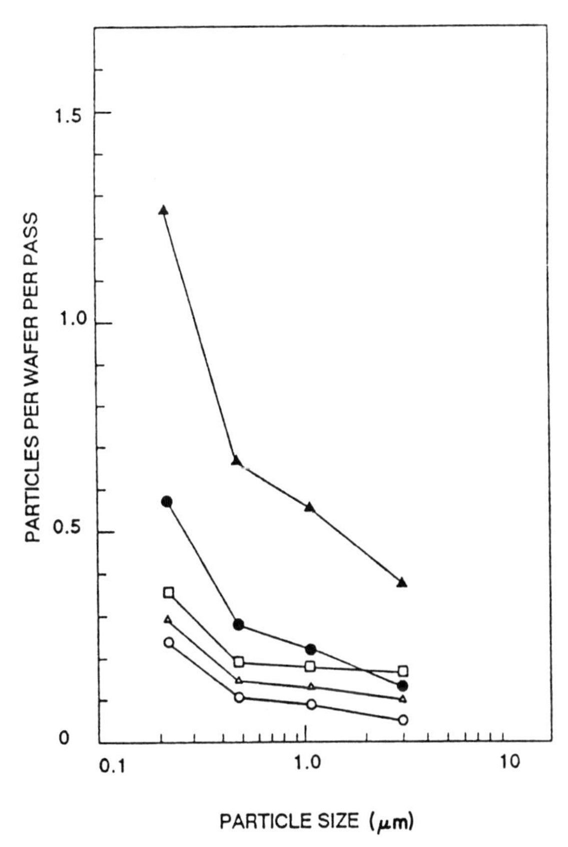

Figure 15.23
Experimental comparison of particles per wafer pass plotted against particle size for SMIF and human handling in various class environments.

	CLEANLINESS CLASS	CASSETTE HANDLING	CASSETTE TRANSPORT
○	10	SMIF	SMIF-POD
△	1000	SMIF	SMIF-POD
□	200000	SMIF	SMIF-POD
●	10	HUMAN	BOX
▲	10	HUMAN	OPEN

(From Ortiz [39], Asyst Technologies.)

is that the capital investment in new clean room facilities can be reduced. A human benefit is that the potential exposure of human operators to toxic chemicals can be reduced by isolating the air handling for the mechanical enclosures from the general air circulation system.

A final benefit of the Asyst type of system is that it is itself a robotic system and can be integrated into an overall automated environment with minimum difficulties.

BIBLIOGRAPHY

[1] Hamming, R. (1950) Error Detecting and Correcting Codes, *Bell System Technology Journal*, **29**, p.147.

[2] Davis, H. L. (1985) A 70-ns Word-Wide 1–Mbit ROM with On-Chip Error-Correction Circuits, *IEEE Journal of Solid State Circuits*, Vol. **SC-20**, No.5, October 1985, p.958.

[3] Gongwer, G. S. and Gudger, K. H. (1983) A 16k EEPROM Using EE Element Redundancy, *IEEE Journal of Solid State Circuits*, Vol. **SC-18**, No.5, October 1983, p.550.

[4] Jolly, R. D. *et al.* (1985) A 35-ns 64k EEPROM, *IEEE Journal of Solid State Circuits*, Vol. **SC-20**, No.5, October 1985, p.971.

[5] Price, J. E. (1970) A New Look at Yield of Integrated Circuits, *Proceedings IEEE*, **58**, 1290.

[6] Fujishima, K. *et al.* (1983) A 256k Dynamic RAM with Page-Nibble Mode, *IEEE Journal of Solid State Circuits*, Vol. **SC-18**, No.5, October 1983, p.470.

[7] Ochii, K. *et al.* (1982) An Ultralow Power 8k × 8-Bit Full CMOS RAM with a Six-Transistor Cell, *IEEE Journal of Solid State Circuits*, Vol. **SC-17**, No.5, October 1982, p.798.

[8] Naruke, Y. *et al.* (1989) A 16Mb Mask ROM with Programmable Redundancy, *ISSCC Proceedings*, February 1989, p.128.

[9] Spaw, W. *et al.* (1982) A 128k EPROM with Redundancy, *1982 ISSCC Proceedings*, February, p.112.

[10] Gaw, H. *et al.* (1985) A 100 ns 256k CMOS EPROM, *ISSCC Proceedings*, February, p.164.

[11] Mano, T. *et al.* (1982) A Redundancy Circuit for a Fault-Tolerant 256k MOS RAM, *IEEE Journal of Solid State Circuits*, Vol. **SC-17**, No.4, August 1982, p.726.

[12] Higuchi, M. *et al.* (1990) An 85 ns 16Mb CMOS EPROM with Alterable Word Organization, *ISSCC Proceedings*, February, p.56.

[13] Yoshida, M. *et al.* (1983) A 150 ns 288k CMOS EPROM with Redundancy, ISSCC Proceedings, February, p.174.

[14] Seeds, R. B. (1967) Yield, Economic and Logistical Models for Complex Digital Arrays, *IEEE International Convention Rec.*, Part 6, 1967, p.60.

[15] Kooperberg, C. (1988) Circuit Layout and Yield. *IEEE Journal of Solid State Circuits*, **23**, 887.

[16] Kalter, H. *et al.* (1990) A 50 ns 16Mb DRAM with a 10 ns Data Rate, *ISSCC Proceedings*, February, p.232.

[17] Minato, O. (1982) A Hi-CMOSII 8k × 8 Bit Static RAM, *IEEE Journal of Solid State Circuits*, Vol. **SC-17**, No.5, October 1982, p.793.

[18] Schuster, S. E. (1978) Multiple Word/Bit Line Redundancy for Semiconductor Memories, *IEEE Journal of Solid State Circuits*, Vol. **SC-13**, No.5, October 1978, p.698.

[19] Uchida, Y. *et al.* (1982) A Low Power Resistive Load 64k bit CMOS RAM, *IEEE Journal of Solid State Circuits*, Vol. **SC-17**, No.5, October 1982, p.804.

[20] Stapper, C. H. (1986) The Defect-Sensitivity Effect of Memory Chips, *IEEE Journal of Solid State Circuits*, Vol. **SC-21**, No.1, February 1986, p.193.

[21] Stapper, C. H. (1984) Modeling of defects in integrated circuit photolithographic patterns, *IBM J. Res. Develop.*, Vol. **28**, July 1984, p.461.

[22] Flack, V. F. (1986) Estimating Variation in IC Yield Estimates, *IEEE Journal of Solid State Circuits*, Vol. **SC-21**, No.2, April 1986, p.362.

[23] Cenker, R. P. *et al.* (1979) A Fault-Tolerant 64k Dynamic RAM, *ISSCC Proceedings*, February, p.150.

[24] Rand, M. J. (1979) Reliability of LSI Memory Circuits Exposed to Laser Cutting, *IEEE International Reliability Physics Symposium, 17th Annual Proceedings*, p.220.

[25] Bindels, J. F. M. *et al.* (1981) Cost-Effective Yield Improvements in Fault-Tolerant VLSI Memory, *ISSCC Proceedings*, February, p.82.

[26] Kokkonen, K. *et al.* (1981) Redundancy Techniques for Fast Static RAMs, *ISSCC Proceedings*, February, p.80.

[27] Kantz, D. (1984) A 256k DRAM with Descrambled Redundancy Test Capability, *IEEE Journal of Solid State Circuits*, Vol. **SC-19**, No.5, October 1984, p.596.

[28] Yamada, J. *et al.* (1984) A Submicron 1 Mbit Dynamic RAM with a 4-bit-at-a-Time Built-in ECC Circuit, *IEEE Journal of Solid State Circuits*, Vol. **SC-19**, No.5, October 1984, p.627.

[29] Osman, F. I. (1982) Error-Correction Technique for Random-Access Memories, *IEEE Journal of Solid State Circuits*, Vol. **SC-17**, No.5, October 1982, p.877.

[30] Edwards, L. (1981) Low Cost Alternative to Hamming Codes corrects Memory Errors, *Computer Design*, July, p.132.

[31] Kotani, H. *et al.* (1987) 4M bit DRAM Design Including 16-bit-Concurrent ECC, *VLSI Symposium Proceedings*, p.87.

[32] Arimoto, K. *et al.* (1988) A Speed Enhancement DRAM Array Architecture with Embedded ECC, *VLSI Symposium Proceedings*, 1988, p.111.

[33] Yamada, T. *et al.* (1988) A 4-Mbit DRAM with 16-bit Concurrent ECC, *IEEE Journal of Solid State Circuits*, Vol. **23**, No.1, February 1988, p.20.

[34] Khan, A. (1983) Fast RAM Corrects Errors on Chip, *Electronics*, September 8, p.126.

[35] Briner, D. and Yeaman, M. D. (1988) Cost/Benefit Analysis of two SMIF Alternatives Compared to a Conventional Class 10 Cleanroom, *9th ICCCS Proceedings*, Institute of Environmental Sciences, 1988, p.137.

[36] Parikh, M. *et al.* (1988) Contamination Control Performance of SMIF Isolation Techniques in Installed Production Facilities, *9th ICCCS Proceedings*, Institute of Environmental Sciences, 1988, p.503.

[37] Kopp, R. J. (1989) What's Ahead in Wafer Processing and Materials, *Semiconductor International*, January, p.54.

[38] Will there be Cleanrooms in the Future?, A Perspective from Dataquest, *Solid State Technology*, December, p.27.

[39] Ortiz, G. (1977) Exceeding Class 10 via an Integrated SMIF System, *ICCCS Proceedings*, Institute of Environmental Sciences, 1977, p.338.

[40] Gupta, A. and Lathrop, J. W. (1972) Yield Analysis of Large Integrated-Circuit Chips, *IEEE Journal of Solid State Circuits*, Vol. **SC-7**, October 1972, p.389.

[41] Maly, W. (1985) Modeling of Lithography Related Yield Losses for CAD of VLSI Circuits, *IEEE Transactions on Computer-Aided Design*, **CAD-4**, p.166.

[42] Stapper, C. H. (1989) *Proc. IEEE Int. Conference on Microelectronic Test Structures*, Vol. **2**, No. 1, March 1989.

16 MEMORY TRENDS IN THE FUTURE

16.1 FUTURE APPLICATIONS

The memory of the future will be driven by the new applications of the future. The advent of ISDN on the communications side and HDTV on the consumer side will use large quantities of memories. Smart cards with multimegabytes of memory will be the floppy disk replacement in laptop and handheld computer systems. Solid state stereo systems will permit new dimensions in sound.

The home will be automated. The smart house will use large amounts of non-volatile and RAM memory in its distributed control system.

The factory of the future will require large quantities of memory for computer integrated manufacturing (CIM) systems and will move to intelligent automation manned by artificially intelligent robotics. CIM will integrate factories away from repetitive labor. Manufacturability will be emphasized. The trend will be to more human manufacturing environments.

By the year 2000 silicon disks and cards will have a large share of the archival computer market. Computer interfaces will become human friendly with speech processing, vision processing, character recognition, and communications interfaces, all of which require large amounts of memory. Scratch pad computers and pocket computers will require both high density and low power memories to run off lightweight batteries.

Automotive systems will feature intelligent navigational systems and human comfort features using artificially intelligent memory dominated computer control systems.

Communications will progress rapidly with standards issues becoming a thing of the past. Wireless worldwide cellular phone systems, together with electronic simultaneous language translators and sophisticated vision systems, made possible by high memory densities coupled with processors, will move the problem of communications to content rather than method. Communications will then make the office in the home or in Hawaii a reality, allowing every person to live and work where they want.

All of these new applications will be made possible by the inclusion of massive amounts of memory in increasingly specialized architectures.

For the applications of the next century new memory technologies will be developed that will replace the current silicon based technology. By 2000 the memory technology of today will be totally changed by the new breakthroughs which will occur between now and then.

16.2 MARKETING TRENDS

What will the decade of the 1990's hold for memories? Thinking about marketing trends first, the globalization of this industry will continue and become more truly international. Global companies are emerging. To preserve innovation they will sponsor smaller companies in fast-moving and advanced technical concepts that might get lost in the structure of a larger company.

The distinction between corporations by national origin will cease to be as important as before as each of the major memory companies supply local markets from the major regions of the world. New memory companies will continue to emerge in developing technology areas such as Korea and Taiwan with the aid of sympathetic governments.

The stable factors in memories will continue. These include, economically, the stable rate of increase of usage of memory bits and the stable rate of decline of the cost per bit which results from it. From the cost standpoint, the memory industry will continue to follow the learning curve. The technology which allows these trends to continue will be available though not necessarily foreseeable.

The slope of the learning curve may be somewhat slower, constrained by the need to control the rapid decline of memory prices on submicron products and preserve the profit margins necessary to fund the huge capital investments needed for deep submicron manufacturing capability and development of the next generation technology.

Stability from the past volatility of the market will be provided by more careful control of inventories and the new computer based production control and marketing environments. It will also be enhanced by the closer relationships between vendors and customers which will level out some of the ups and down historically found in this market. Competition, however, will certainly not diminish.

16.3 MEMORY SYSTEMS

From the system standpoint, power supply voltages will evolve through a transition system stage in the early 1990's during which some systems will be 5 V and some 3.3 V, although few will be mixed. The 3.3 V will prevail for a limited period of time and other lower voltage standards will follow.

The trend toward systems integrated on the chip or module will continue. Microprocessors and gate arrays will begin to be embedded in cost effective memory chips, reducing cost and enhancing system performance by reducing the interface and bandwidth problems. Perhaps the ability to integrate systems on a chip will lead to new structures and partitioning of the memory. Memory testing will be provided by logic on the chip.

The memory will stop being looked on as an auxiliary to the microprocessor, and embedded microprocessors will start being looked on as aids to optimized memories. The day of the standard average memory with tradeoffs which make it generally good for all systems and optimized for none will draw to a close. The day of memories optimized and, in fact, integrated into the specific system will prevail. The increased bandwidth will lead to significant improvements in system performance.

16.4 MEMORY PRODUCTS

16.4.1 DRAMs

From the product point of view, the rate of increase in density from generation to generation will continue with the DRAMs reaching the 1G bit sampling stage by the year 2000 and the SRAMs reaching the 256Mb SRAM by that time.

The cost and density advantage of the one-transistor cell will allow the DRAMs to continue as the cost effective workhorse of the industry, successively cannibalizing other memory markets as they have in the past.

DRAMs will move to lower voltages driven by technology, by the handheld systems market and by the high density system power dissipation limitations. They will continue to reach higher densities, targeting the silicon file market. Perhaps they will also become serial for this market or files will become random.

DRAMs will add pseudostatic modes as standard features which will permit them to gradually replace the high volume part of the commodity SRAM market. Very low power DRAMs will even compete well against low power SRAMs.

Fast access modes on DRAMs will target keeping pace with the rapidly increasing microprocessor speeds. These, and varieties of cache DRAMs, will compete with fast SRAMs to some extent.

DRAMs will go on chips for high density embedded arrays. VLSI chips will begin to use the high density memory technologies as microcomputer chips become predominantly memory on the chip.

In materials, DRAM capacitors will benefit from high dielectric constant materials, such as ferroelectrics, being able to retain capacitance without the large size usually needed.

16.4.2 SRAMs

SRAMs will also evolve, becoming either very fast or very low power. Both varieties will move to lower power supply ranges.

The very fast SRAMs will have low swing interfaces such as ECL, although the technology will be both the traditional bipolar and the lower power CMOS.

The full CMOS six-transistor cell SRAM will continue to be the best option for very low power sub-μA standby power dissipation. New cells of the stacked variety will be refined to reduce the chip size.

SRAMs will increasingly be integrated with other logic functions to produce ASIC and intelligent logic products.

16.4.3 Non-volatiles

EPROMs will continue increasing in density and will gradually be phased out as the flash versions become of equal density and approach the EPROM in cost. The flash EPROMs will take a huge slice of the silicon file market. The density trends of EPROMs will be continued by one-transistor flash memories.

The specialty markets of the medium range EEPROMs, especially the very high reliability markets of the asperity aided EEPROMs, will continue as application-specific product segments. Increasingly EEPROM cells will be embedded in processors and other VLSI chips to provide on-chip configurable non-volatile memory storage.

The merging of technologies for use in each of the product families will continue. The flexibility of being able to combine different technologies such as bipolar and CMOS will become assumed naturally by the designer of the future. Analog will become a more important part of the memory designer's portfolio of options.

16.5 PACKAGING

As the linear scaling trends slow, packaging will pick up the slack. Packaging will move to high density miniature packages, smartcard formats, multichip modules and perhaps even three-dimensional wafer scale integration.

Packaging will go toward systems on a chip. Multichip modules will make wafer scale integration at last a reality. Good chips on a flipped chip substrate will be integrated in close proximity, leading to very high speed systems. Other three-dimensional chip integration methods will be explored. The chip itself will become hermetic.

The memory chips will stop increasing in size. The increase in size with each generation was permitted in the past by the increase in the size of the package, which grew to permit the lead spacings needed by the higher density generations of memories. With the development of finer lead pitch packages, the chip size will no longer be determined by the size of the allowed package and the competitive trends to smaller chip size and reduced board spacing will predominate, giving us smaller chips.

Perhaps the ability to integrate systems on a chip will lead to new structures and partitioning of memory. This trend will also be encouraged by embedded memories. Embedded memories begin to be designed by memory designers in optimized memory technologies.

16.6 MEMORY PROCESS TECHNOLOGY

Memory technologies will continue to move in the direction of increased numbers of vertical layers and deeper trenches as the ability to scale linearly on silicon slows. Multiple layers of polysilicon and metal will gave way to multiple materials. Trenches for various purposes will become a well understood process technology.

Lithography will move from the excimer laser to perhaps X-ray and then to ion beam technologies by the time the gigabit DRAM appears around the year 2000.

SOI will become more important as layers of transistors replace the single-layer transistors of today and stacked logic and memories come into being. SRAM arrays with transistors and the periphery stacked will be developed. Horizontal and vertical devices will be used.

Finally, before the end of the decade a totally new technology media for storing logic 1's and 0's will be developed, probably at the atomic level. This will take us into the next decade, gradually replacing the old semiconductor technologies of the past 20 to 30 years over a 10 to 20 year period.

16.7 MEMORY MANUFACTURING TECHNOLOGY

In wafer processing the current trend to modular manufacturing equipment will continue so that a single piece of equipment will become the complete site for wafer processing of the future. Perhaps the final technology will be to go from the clean room to the clean machine to the clean chip. By the year 2000 we may know how to process a memory wafer so that the clean room is no longer needed. Perhaps we will go from processing on the surface of the wafer to processing in the bulk of the wafer in three-dimensional chips that will be completely contained within the wafer bulk.

Labor costs will be less relevant as automated memory factories come into increased use so that manufacturing will move to areas with a high percentage of technically trained people who can provide support for the hardware and software requirements of automated development and manufacturing. Capital requirements will continue to increase. Data storage in the year 2000 will use technologies known today. But the technology and manufacturing in development at that time will be a total revolution from what we know today.

INDEX